The GeoDiscoveries website is a rich electronic medium for exploring the core concepts of physical geography and the tools geographers use to answer geographic questions. Move beyond the textbook with interactive exercises, animations, videos, self-assessments, and a Virtual Globe. GeoDiscoveries offers unique present/interact/assess activities that will help you visualize spatial relationships and think critically about your interactions with the Earth's atmosphere and landscapes. Guided tours of the internet challenge you to gather real-time information and use geographer's tools to explore the science and systems of the human environment. Enjoy your adventure!

www.wiley.com/college/strahler

Third Edition

PHYSICAL GEOGRAPHY

SCIENCE AND SYSTEMS OF THE HUMAN ENVIRONMENT

Third Edition

PHYSICAL GEOGRAPHY

SCIENCE AND SYSTEMS OF THE HUMAN ENVIRONMENT

Alan Strahler

Boston University

Arthur Strahler

John Wiley & Sons, Inc.

Acquisitions Editor	Jerry Correa
Associate Editor	Denise Powell
Developmental Editor	Suzanne Thibodeau/Denise Powell
Marketing Manager	Clay Stone
Production Editor	Sandra Dumas
Senior Designer	Karin Kincheloe
Senior Illustration Editor	Sandra Rigby
Electronic Art Enhancement	Precision Graphics
Senior Photo Editor	Jennifer MacMillan
Photo Researcher	Teresa Romito
Production Management Services	Hermitage Publishing Services
Cover Photo	Yann Arthus-Bertrand/Altitude

This book was typeset in 10/12 Times New Roman by Hermitage Publishing Services and printed and bound by Von Hoffmann Press, Inc. The cover was printed by Von Hoffmann Press, Inc.

The paper in this book was manufactured by a mill whose forest management programs include sustained yield harvesting of its timberlands. Sustained yield harvesting principles ensure that the number of trees cut each year does not exceed the amount of new growth.

This book is printed on acid-free paper. 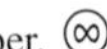

Library of Congress Cataloging in Publication Data:
Strahler, Alan, H.
Physical Geography: Science and Systems of the Human Enviroment, Third Edition / Alan Strahler, Arthur Strahler.

ISBN 0-471-48053-3
WIE ISBN 0-471-65764-6

Printed in the United States of America.

10 9 8 7 6 5 4 3 2 1

PREFACE

With great pleasure, we present the Third Edition of *Physical Geography: Science and Systems of the Human Environment.* Our third edition builds on the considerable strengths of the first edition, but also adds much that is new:

- ***Expanded coverage of global change,*** including four new *Eye on Global Change* features centered on predictions of global change expected as the human impact on climate increases.
- ***Four new* Focus on Remote Sensing *features***, including two with global themes of mapping global productivity and global land cover.
- **Earth from Above *chapter and part opener photos***, selected from the acclaimed collection of aerial landscape photographs taken by Yann Arthus-Bertrand and published in his 462-page, full-color folio book, *Earth from Above* (Abrahms, 1999, 2002). These stunning views, acquired from light aircraft at low altitudes, introduce students to the beauty and diversity of the Earth's varied landscapes.
- ***New* A Closer Look *feature*** that positions longer essays between chapters. These interchapter features include *Geographer's Tools, Eye on the Environment, Eye on Global Change*, and *Focus on Remote Sensing* boxes.
- ***Expanded "Showcase" art pieces***, now appearing in seven more chapters. These two-page layouts group photos together to communicate concepts visually in a direct and striking fashion.
- ***Additional* Eye on the Landscape *features***, using many *Earth from Above* chapter openers as well as other new photos. In this feature, an inset image marks locations of interest on landscape photos and queries students about what geographers might observe or infer from them, helping students to learn how to view the landscape in an informed manner.
- ***A new location for our multimedia learning supplement,* GeoDiscoveries,** which is now accessible on our web site. Material now available includes the animations, interactivities, film and video clips, quizzes, and other activities previously available on our *GeoDiscoveries* CD-ROM, as well as new and updated ones. Special materials for instructors are available on a second tier of the web site. More detail is shown on the following pages.
- ***A new location for our* Geographers at Work *features***, which are now referenced by text lead-ins, but appear in full on our book companion web site. These short essays, written by young physical geographers from American and Canadian geography departments, focus on their research interests. They provide interest and motivation for further study in geography by showing the range of research in physical geography and its relevance to human and environmental needs.

Yann Arthus-Bertrand

The images of Yann Arthus-Bertrand, photographer and author of *Earth from Above*, have received world-wide acclaim for their unique vision of the Earth and its inhabitants. His probing lens captures forests, grasslands, tundra, deserts, mountains, islands, coastlines, rivers, lakes, and human settlements from cities to desert oases in extraordinary compositions of light and form. We are indeed fortunate to be able to use his photos to illustrate the third edition of *Physical Geography*. As you can see from the thumbnail gallery opposite, they are of striking beauty and diversity.

Population growth and technological progress have shifted the natural equilibrium of our planet in many ways that now seem to threaten the future habitability of Earth. It is this emerging crisis of habitability now facing humanity that inspired the mission of photographer Arthus-Bertrand to create his portrait gallery of the world, highlighting both its natural beauty and its precarious condition. Yan Arthus-Bertrand's partner in producing *Earth from Above*, Anne Jankeliowitch, provided the statement below for inclusion here in our preface.

TOWARD A SUSTAINABLE DEVELOPMENT

Since 1950, economic growth has been considerable, and world production of goods and services has multiplied by a factor of 7. During this same period, while the world's population has only doubled, the volume of fish caught and meat produced has multiplied by 5. So has the energy demand. Oil consumption has multiplied by 7, and carbon dioxide emissions, the main cause of the greenhouse effect and global warming, by 4. Since 1900, fresh water consumption has multiplied by 6, chiefly to provide for agriculture.

And yet, 20 percent of the world's population does not have drinkable water, 40 percent lacks access to improved sanitation, 40 percent is without electricity, 826 million people are underfed, and half of humanity lives on less than $2 a day.

In other words, a fifth of the world's population lives in industrialized countries, consuming and producing in excess and generating massive pollution. The remaining four-fifths live in developing countries and, for the most part, in poverty. To provide for their needs, they make heavy demands upon the Earth's natural resources, causing a constant degradation of our planet's ecosystem and limited supplies of fresh water, ocean water, forests, air, arable land, and open spaces.

This is not all. By 2050, the Earth will have close to 3 billion additional inhabitants. These people will live, for the most part, in developing countries. As these countries develop, their economic growth will jockey for position with that of industrialized nations—within the limits of ecosystem Earth.

The Earth's situation is not irreversible, but changes need to be made as soon as possible. We have the chance to turn toward a sustainable development, one that allows us to improve the living conditions of the world's citizens and to satisfy the needs of generations to come. This development would be based on an economic growth respectful both of man and the natural resources of our unique planet.

Such development requires improving production methods and changing our consumption habits. With the active participation of all the world's citizens, each and every person can contribute to the future of the Earth and mankind, starting right now.

Anne Jankeliowitch
Earth from Above

Overview and Application of Our Text

Although many instructors reading this preface will already be familiar with the goals and learning objectives of *Physical Geography: Science and Systems of the Human Environment, Third Edition,* we would like to restate them here using words largely taken from our first edition.

In writing *Physical Geography: Science and Systems of the Human Environment,* our objective was to provide a view of physical geography as a key discipline in understanding the Earth's diverse environments and how they are modified by global change. This goal required a new and fresh approach to the subject matter, one that emphasized not only the basic science of physical geography, but also the interrelations among the diverse processes that act at the Earth's surface and thus condition the human environment. In response, we chose to rely on the concept of systems—more particularly, flow systems of energy and matter—as a unifying theme. In this way we provide a paradigm for understanding the underlying scientific principles and common elements of the many processes that constitute the science of physical geography.

As we developed the book chapter by chapter, it became clear that with systems as an ongoing theme, it would be easy to integrate some simple quantitative materials into the text. Accordingly, we have added a sampling of quantitative concepts as a new feature. These concepts are selected to explore quantitative relationships and why they are useful, rather than to cover all, or even the most important, quantitative relationships in physical geography. In this way, we hope to open the door for students in a population that is traditionally less oriented toward mathematics to realize how quantitative methods can contribute to the understanding of scientific principles.

Global change is also a recurring theme in our text. Geographers, perhaps more than scientists from other fields, are particularly aware of human impact on the Earth and its environment. Global change is, of course, a two-way street. Humans change the Earth in many ways—by clearing and cultivating its lands, harvesting its wildland resources, damming its rivers, and polluting its air. But the Earth changes the situations of its human inhabitants in return when volcanic eruptions bury human settlements, tornadoes and hurricanes level towns and villages, and floods erode agricultural lands. An understanding of global change requires not just a familiarity with the latest climatic change projections, but an understanding of the many processes that shape the human physical environment, both natural and human-induced. Our text focuses on these processes directly, providing both the breadth and depth necessary to appreciate how humans are changing, and are changed by, the Earth.

Course Goals

Physical Geography: Science and Systems of the Human Environment is designed for a one- or two-semester course in introductory physical geography. Its overall objective is to introduce the science of physical geography and its related themes of systems and environment. A secondary goal is to expose the student to quantitative tools used by physical geographers to explore and model the phenomena they observe. It is targeted at the student with some science preparation at the secondary school level, and can be used for courses suited to both majors and nonmajors.

In structure and organization, our text follows the standard sequence of topics that has evolved over the years, beginning with atmosphere-surface interactions and concluding with ecosystem geography. This pattern is eminently teachable and should fit existing lecture structures quite easily. The sequence of 24 chapters is broken into five major parts: *Introduction; Weather and Climate Systems; Systems and Cycles of the Solid Earth; Systems of Landform Evolution;* and *Systems and Cycles of Soils and the Biosphere.*

Although the sequence of topics may be traditional to the instructor, students need continuing orientation and explanation of the structure of the material they are learning. Throughout the text, particular emphasis is placed on establishing this context at all levels. Within chapters, major topics are introduced by context-providing opening statements. Each chapter begins and ends with paragraphs that place the chapter in the perspective of the book as a whole. Each part opener includes text that explains the structure of the chapter sequence to come and relates it to parts past. Our experience as classroom instructors has shown the importance of providing this context to help students master the course as a whole, not just as a sum of individual parts.

The Learning Environment

The learning environment in *Physical Geography: Science and Systems of the Human Environment* includes a number of features expressly designed to further its goals.

Systems Treatment

At the heart of our presentation of physical geography is a systems treatment. Systems concepts are introduced throughout the text as unifying concepts that aid in understanding key principles and ideas, not as separate topics that merely add an additional level of complexity to the book. Systems treatment begins in Part 1, with Chapter 2 providing basic ideas about flow systems and cycles of energy and matter. It continues through the remainder of the text, in contextual and transitional paragraphs, where it is applied to relevant topics as they occur. Systems concepts are highlighted in *Focus on Systems* boxes. These features provide specific applications and examples of systems appropriate to the material at hand. And, systems are called out specifically in the text of part openers, thus providing the orientation and perspective on the next set of chapters to come.

Environmental Features

Although physical geography is largely concerned with systems and processes that shape the human environment, there are specific topics that can be targeted as primarily environmental. These are presented in *Eye on the Environment* features, which are set off from the text in a distinctive boxed treatment. Specific review questions focusing on these features appear at the end of the chapter.

Quantitative Features

Since physical geography is a quantitative science, it is appropriate to introduce students to simple quantitative concepts and techniques as a way of enhancing the understanding of particular science principles. However, the tradition in introductory physical geography has been to provide a largely descriptive treatment at the beginning level. We have broken with that tradition, but have been careful to separate quantitative material from the main text in a series of *Working It Out* boxes that are independent of the text. In this way, the instructor has the freedom to select any or all of these, depending on the readiness of the students and the quantitative topics to be emphasized.

The majority of the *Working It Out* boxes are accessible with simple algebraic skills—being able to substitute values in a computational formula, for example. A few require natural and common logarithms and exponentiation. All have associated problems that are placed in the closing matter of the chapter in which they appear. Problems are solved in a *Problem Answers* section in the back of the book. In developing the boxed features, we tried to focus on the application of each quantitative concept in some useful way, rather than simply presenting a formula to be evaluated. The problems that accompany each feature are designed to extend this focus on the quantitative approach as an aid to learning rather than as a test of arithmetic skills.

Canadian Features

Our previous physical geography texts have been used widely in Canada, and in writing this book we have paid special attention to Canadian needs. In particular, we have expanded our coverage of periglacial processes and landforms in Chapter 15. We have also revised and expanded our treatment of permafrost. Wherever possible, we have extended U. S. maps into southern Canada, providing coverage of most of Canada's populated areas, or have used North American maps showing the full extent of both countries. We have also increased our use of Canadian photos and examples in the text. And, we provide a six-page supplement on the Canadian system of soil classification as an appendix.

Terms

Vocabulary is an essential part of any science. To help organize the terms that we have included, we have set off the *Key Terms*—those dozen or more terms that are most important in each chapter—in **boldface,** and listed them at the close of each chapter. Terms of lesser importance are set in *italics.* All terms, whether bold or italic, are defined in the *Glossary* at the close of the book.

End of Chapter Materials

Each chapter concludes with a number of aids to facilitate student learning. The *Chapter Summary* is worded like a scientific abstract, succinctly covering all the major concepts of the chapter. The list of *Key Terms* indicates the most important concepts to study to understand the chapter material. The *Review Questions* are designed as oral or written exercises that require description or explanation of important ideas and concepts. *Visualizing Exercises* utilize sketching or graphing as a way of motivating students to visualize key concepts. Questions are provided for text, *Focus on Systems* boxes, and *Eye on the Environment* boxes. Also provided are *Essay Questions* that require more synthesis or the reorganization of knowledge in a new context. Last of all are problems for *Working It Out* boxes, which invite the student to apply quantitative skills in a manner that emphasizes learning from quantitative principles rather than computational skills.

Text and Illustrations

Today, science must be taught in an open, accessible manner to reach students, and written material needs to communicate directly and simply. Moreover, today's students are accustomed to visual learning and to an interactive methodology. The text and illustrations of *Physical Geography: Science and Systems of the Human Environment* draw on the heritage of *Introducing Physical Geography* (Wiley, 1994), which devised entirely new treatments of topics in physical geography that are accessible and inviting, yet provide clear descriptions of scientific principles. In our text, each sentence is worded to ensure that it communicates its meaning clearly and simply. Every map and line drawing is carefully designed and styled to make sure that its message is clear, obvious, and direct. Every photo is selected not only to provide a fine illustration of the relevant science concept, but also to demonstrate it effectively and strikingly. In summary, we have worked very hard to produce an illustrated text of which we are very proud.

GEODISCOVERIES

The GeoDiscoveries media program offers a rich electronic medium for exploring the core concepts of physical geography. This robust assembly of media resources allows students and professors to move beyond the textbook with interactive exercises, animations, videos, self-assessments and an interactive globe.

Building off the success of our present/interact/assess framework, our GeoDiscoveries program has been integrated into our on-line book companion web sites. Web sites tailored to the needs of both students and instructors include the following resources:

Resources specific to the study of physical geography have been created and developed by S. Mary P. Benbow of the University of Manitoba.

For Students:

Animated Globe: This 3D interactive globe allows you to overlay different layers of geographic information, including plate tectonics, topography, political boundaries, and population to help visualize the spatial relationships between these features and elements.

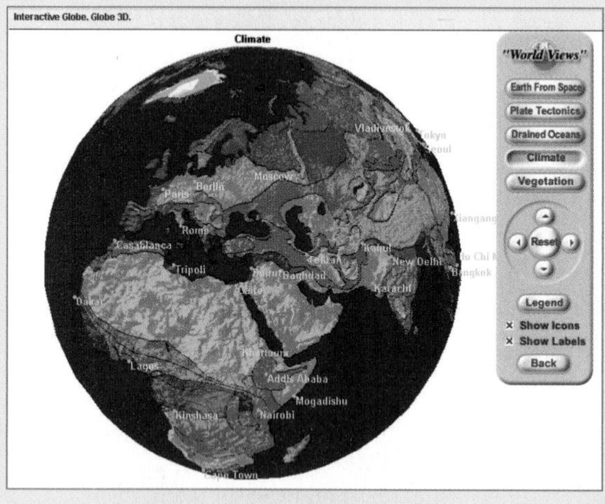

Animations: Key diagrams and drawings from our rich signature art program have been animated to provide a virtual experience of difficult concepts. These animations have proven crucial to the understanding of this content for visual learners.

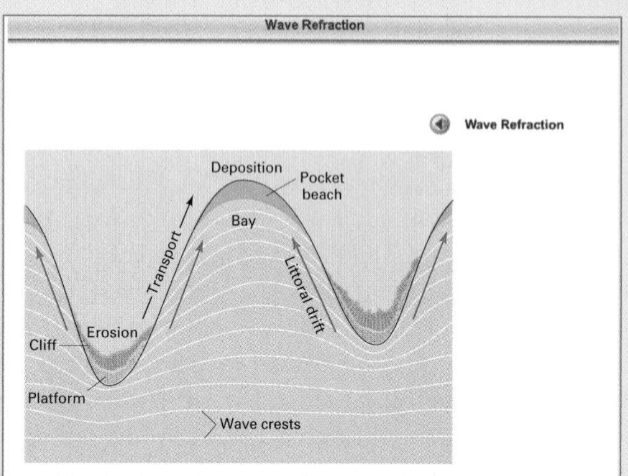

Videos: Brief video clips provide real-world examples of geographic features, and put these examples into context with the concepts covered in the text.

Simulations: Computer-based models of geographic processes allow you to manipulate data and variables to explore and interact with virtual environments.

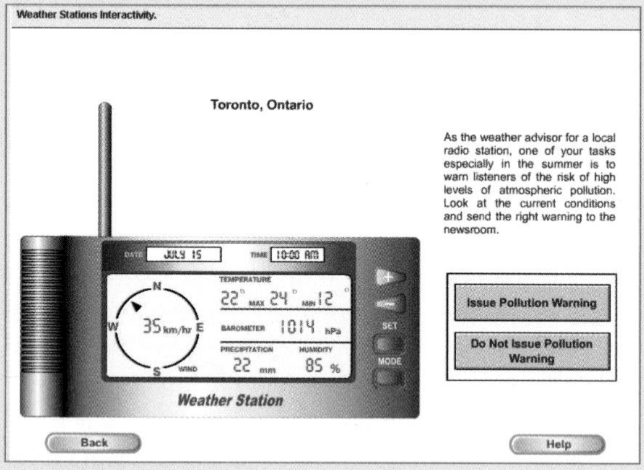

Interactive Exercises: Learning activities and games build off our presentation material. They give students an opportunity to test their understanding of key concepts and explore additional visual resources.

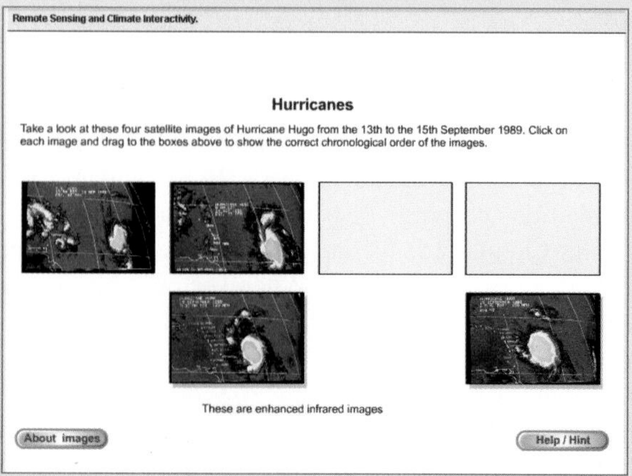

For Instructors:
Our animations, videos, simulations, and interactive exercises provide ideal in-class presentation tools to help engage students in the classroom. All these resources can be easily integrated into instructors' customized PowerPoint presentations.

In addition to our multi-media content, we offer an on-line image gallery, containing both line art and comprehensive photographic materials. These resources can be easily uploaded into PowerPoint presentations, course management systems, and on-line tutorials and web sites.

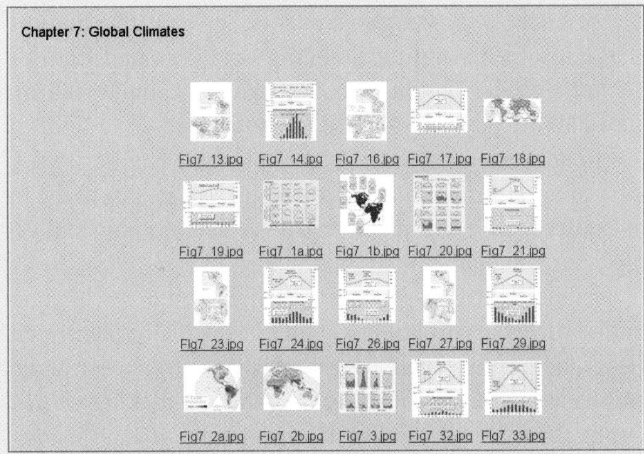

We also offer prepared PowerPoint presentations centered on key concepts and hot topics in the field of physical geography. These lecture materials can be customized to suit the instructor's individual needs, or can be used as is to add context and relevancy to any lecture.

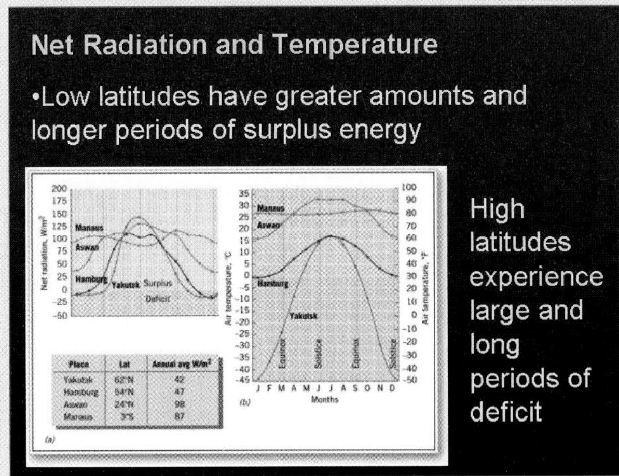

GeoDiscoveries also offers a wealth of assessment and assignment tools that can be used for tutorials as well as quizzes purposes. These question banks are closely tied to our multi-media materials to help gauge understanding of the resources. Easy integration of this content into course management systems and homework assignments gives instructors the opportunity to integrate multimedia with their syllabuses and with more traditional reading and writing assignments.

Book Companion Site

In addition to our GeoDiscoveries multimedia content, our student companion site offers a wealth of study and practice materials, including:

Student On-line Resources:

- *Self-Quizzes*—chapter based multiple choice and fill-in-the-blank questions
- *Annotated Weblinks*—useful weblinks selected to enhance chapter topics and content
- *Chapter Summaries*—brief overviews of chapter content
- *Flashcards*—drill and practice application to help study key terms
- *Learning Objectives*—list of key concepts covered in each chapter
- *Learning Styles Survey*—a survey developed to help determine your own particular way of learning, and study skill suggestions to help each kind of learner
- *Geographers at Work*—weblinks to geographers' web sites to give you an opportunity to explore research being done in the field today
- *Virtual Field Trips*—web sites devoted to the exploration and virtual experience of landscapes and environments around the world.

Instructor Resources:

- *PowerPoint Lecture Slides*—chapter-oriented slides including lecture notes and text art
- *Computerized Test Bank*—including multiple choice and fill-in-the-blank test items
- *Teaching Tips*—for enhancing the classroom experience
- *On-line Essay and In-class Activities*—Materials for in-class projects, with corresponding essay-based homework assignments
- *Instructor's Manual*—including lecture notes, chapter objectives, and guides to additional resources

Acknowledgments

The preparation of *Physical Geography: Science and Systems of the Human Environment* was greatly aided by many reviewers who read and evaluated the various parts of the manuscript. They include

Peter D. Blanken, University of Colorado at Boulder
Gail L. Chmura, McGill University
C. Mark Cowell, Indiana State University
Richard W. Dixon, Southwest Texas State University
Claude M. Epstein, Richard Stockton College of New Jersey
Donald A. Friend, Minnesota State University, Mankato
Douglas Gamble, Mississippi State University
Douglas Goodin, Kansas State University
Robert B. Howard, California State University, Northridge
Dorleen Jenson, Salt Lake City Community College
Jeffrey A. Lee, Texas Tech University
Elliot McIntire, California State University, Northridge
Katrina A. Moser, University of Utah
Darren Sjogren, University of Calgary
Susan C. Slowey, Blinn College
Robert Wingate, University of Wisconsin, LaCrosse

Reviewers for the Third Edition:
Christopher Justice, University of Maryland
Kurt Kipfmueller, University of Arizona
Roger Brown, University of Maryland
Ronald Janke, Valparaiso University

Special thanks also go to the contributors to our *Geographers at Work* series for the time and effort they spent in introducing their work to beginning students. They include

Dar A. Roberts, University of California, Santa Barbara
James A. Voogt, University of Western Ontario
Andrew C. Comrie, University of Arizona
Katherine Klink, University of Minnesota
Lisa Naughton, University of Wisconsis
Kirsty Duncan, University of Windsor, Ontario
Joe Meert, Indiana State University
Kathy Young, York University, Toronto
Molly Marie Pohl, San Diego State University
Kurt M. Cuffey, University of California, Berkeley
Christina Tague, San Diego State University

It is with particular pleasure that we thank the staff at Wiley for their careful work, encouragement, and sense of humor in the preparation and production of *Physical Geography: Science and Systems of the Human Environment,* Third Edition. They include our editor Jerry Correa, developmental editors Suzanne Thibodeau and Denise Powell, photo researcher Teresa Romito, senior photo editor Jennifer MacMillan, designer Karin Kincheloe, senior illustration editor Sandra Rigby, illustration studio Precision Graphics, marketing manager Clay Stone, new media project manager Tom Kulesa, editing and production services, Hermitage Publishing Services, and our production editor Sandra Dumas.

Alan Strahler
Boston, Massachusetts

About the Authors

It is with great sadness that we mark the passing of Arthur Strahler in December, 2002. Through five decades of bringing science to geography students, he held the highest of standards in his writing, making accuracy and clarity his primary goals. Throughout, he strived to make the many subfields of physical geography both accessible and exciting to successive generations of learners. His heritage persists not only in our books written for Wiley, but in the table of contents of every major textbook used in physical geography today.

Arthur Strahler will also be remembered for his contributions to geomorphology, which are still frequently cited. He pioneered the quantitative approach to geomorphology, emphasizing the direct study of landform-making processes rather than the descriptive interpretation of landforms. He was among the first researchers to apply statistical techniques to analysis of landforms, and his work on stream characteristics laid the foundation for most of modern fluvial geomorphology. He later applied general systems theory to geomorphology, bringing a new paradigm to the field. His research achievements were honored by his election to the rank of Fellow in both the Association of American Geographers and the Geological Society of America.

He was truly a giant of geographers.

Alan Strahler received his B.A. degree in 1964 and his Ph.D degree in 1969 from The Johns Hopkins University, Department of Geography and Environmental Engineering. He has held academic positions at the University of Virginia, the University of California at Santa Barbara, and Hunter College of the City University of New York, and is now Professor of Geography at Boston University. With Arthur Strahler, he is a coauthor of seven textbook titles with nine revised editions on physical geography and environmental science. He has published over 250 articles in the refereed scientific literature, largely on the theory of remote sensing of vegetation, and has also contributed to the fields of plant geography, forest ecology, and quantitative methods. His work has been supported by over $6 million in grant and contract funds, primarily from NASA. In 1993, he was awarded the Association of American Geographers/Remote Sensing Specialty Group Medal for Outstanding Contributions to Remote Sensing. In 2000, he received the honorary degree Doctorem Scientiarum Honoris Causa (D.S.H.C) from the Université Catholique de Louvain, Belgium, for his academic accomplishments in teaching and research. In 2004, he was honored by election to the rank of Fellow in the American Association for the Advancement of Science.

Arthur Strahler (1918–2002) received his B.A. degree in 1938 from the College of Wooster, Ohio, and his Ph.D. degree in geology from Columbia University in 1944. He was appointed to the Columbia University faculty in 1941, serving as Professor of Geomorphology from 1958 to 1967 and as Chairman of the Department of Geology from 1959 to 1962. He was elected as a Fellow of both the Geological Society of America and the Association of American Geographers for his pioneering contributions to quantitative and dynamic geomorphology, contained in over 30 major papers in leading scientific journals. He was the author or coauthor with Alan Strahler of 16 textbook titles with 15 revised editions in physical geography, environmental science, the Earth sciences, and geology.

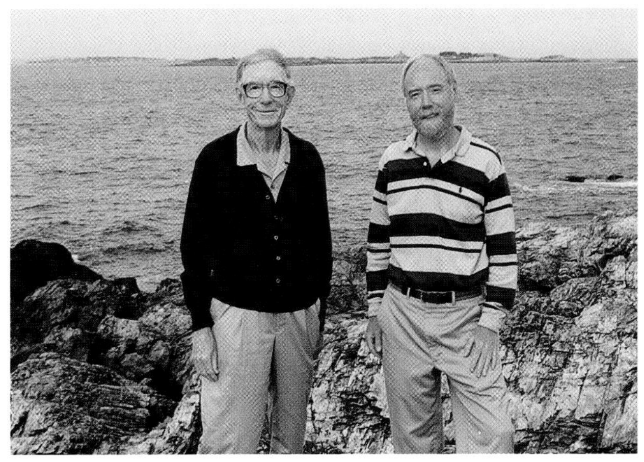

BRIEF CONTENTS

CONTENTS

Chapters in Part 1

Part 1 | Introduction

Traditional village, north of Antananarivo, Madagascar The circular plan of this village, with radial paths leading outward, is found early in the development of many civilizations. It also echoes the vision of a human society of peoples living in unity and harmony with their environment.

Geography is a rich and ancient discipline of study. It is concerned with both the places of the Earth and the physical and human processes that differentiate those places and make them unique. If you haven't studied geography before, you may think of geography as a collection of names of cities, rivers, mountains, and the like. However, these are just geographic facts. They are no more geography than history is a collection of dates. Just as history captures the flow in time of human events, ideas, and aspirations, geography captures the spatial connections among human activities as they occur on the Earth's physical landscape. And, in contrast to the humanities of history, literature, and art, geography is a science—both social and physical—that searches for common principles and processes explaining human and natural phenomena. ■ The ability to model and predict human and natural spatial phenomena makes today's geography a vital discipline as never before. Human impact on our environment increases inexorably as our population grows and as ever-higher standards of living make increasing demands on the Earth's natural resources. If we are to live in harmony with our environment, hard choices lie ahead, and understanding geography can help us make them, individually and collectively. ■ Part 1, our introduction to geography, includes two chapters. The first provides an orientation to geography as a discipline and explains the role of physical geography in understanding global change. The second chapter introduces some big ideas that carry through all of physical geography. Among these is the concept of flow systems of energy and matter, which is a theme that will appear in many chapters. With these topics under your belt, you will be set to embark on your main journey through our text—learning about the systems and processes of physical geography.

1 INTRODUCING PHYSICAL GEOGRAPHY

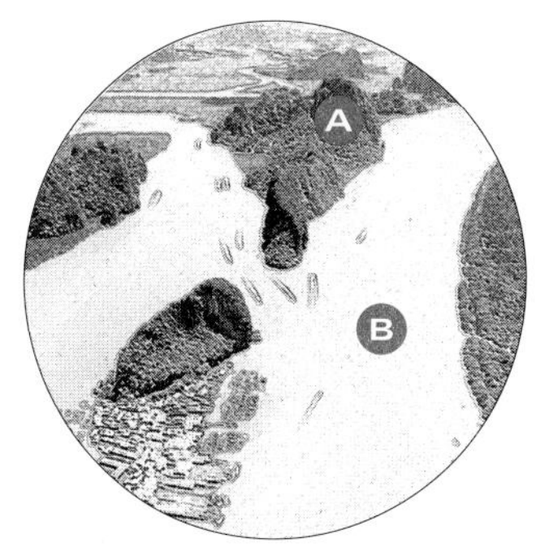

EYE ON THE LANDSCAPE

Phang Nga Bay, Thailand. The fishing village of Koh Pannyi floats on bamboo poles in a spectacular landscape of limestone peaks.

What else would the geographer see?...Answers at the end of the chapter.

You're invited! To a learning journey devoted to physical geography—the science of life environments from local to global scale. Your learning experience will take you from the top layers of the Earth's atmosphere to the rocks underlying the ocean basins to the forests of the farthest continents. It will focus primarily on the land surface, helping you understand and explain many common natural phenomena that you experience every day—ranging from the formation of thunderstorms to the building of sand dunes on a beach. When your journey is complete, you will come away with both a new appreciation of the landscape around you and a better understanding of the global environmental problems faced by the human race as its population expands into the twenty-first century.

This introductory chapter and the next will provide you with some key concepts and ideas of physical geography that you will use along your journey. Some of the most important ideas concern how we can view natural phenomena from the perspective of systems (Chapter 2). But before we begin that discussion, you may be curious about what geography is and what geographers study.

INTRODUCING GEOGRAPHY

What is geography?[1] **Geography** is the study of the evolving character and organization of the Earth's surface. It is

[1] For descriptions and perspectives of geography, see *Rediscovering Geography: New Relevance for Science and Society* (Washington, DC: National Academy Press, 1997), 234 pp., and *Geography in America,* G. L. Gaile and Cort J. Wilmott, eds. (Columbia, OH: Merrill, 1989), 840 pp., which were used in preparing this introduction.

1.1 *Eye on the Landscape* ***Vancouver, British Columbia*** *This cosmopolitan city enjoys a spectacular setting on the Strait of Georgia, flanked by the Pacific and Vancouver Island Ranges.* **What else would the geographer see?...Answers at the end of the chapter.**

about how, why, and where human and natural activities occur and how these activities are interconnected.

To get a better understanding of geography, think of it as having two sides. One side, which we can term **regional geography,** is concerned with how the Earth's surface is differentiated into unique **places.** Take Vancouver, British Columbia, for example (Figure 1.1). What makes Vancouver unique? Is it its spectacular setting where the Pacific Ranges meet the Pacific Ocean? the marine west-coast climate that provides its mild and rainy winters and blue summer skies? its position as a seaport gateway to Asia? its English roots combined with dashes of French, Asian, and even Russian culture? In fact, all of these attributes contribute to making Vancouver the unique place that it is.

Although places are unique, the physical, economic, and social processes that form them are not. Thus, geographers are concerned with discovering, understanding, and modeling the processes that differentiate the Earth's surface into places. This is the other side of geography, which we can term **systematic geography.** Why are pineapples cheap in Hawaii and expensive in Toronto? oranges cheap in Florida and expensive in North Dakota? steak cheap in Kansas City and expensive in Boston? These are examples of a simple principle of economic geography—that prices include transportation costs and that when goods travel a longer distance, they are usually more expensive. Discovering such principles and extending them to model and predict spatial phenomena is the domain of systematic geography. To summarize, geographers study both the "vertical" integration of characteristics that define a place and the "horizontal" connections between places.

Realms of Geography

As a field of study, systematic geography is often divided into two broad realms—**human geography,** which deals with social, economic, and behavioral processes that differentiate places, and **physical geography,** which examines the natural processes occurring at the Earth's surface that provide the physical setting for human activities. Figure 1.2 is a diagram showing the principal fields of physical and human geography. Reading downward from the left, we see five fields of physical geography, from climatology to biogeography, which are illustrated in Figure 1.3. These are the main focus of this text.

Climatology is the science that describes and explains the variability in space and time of the heat and moisture states of the Earth's surface, especially its land surfaces. Since heat and moisture states are part of what we call

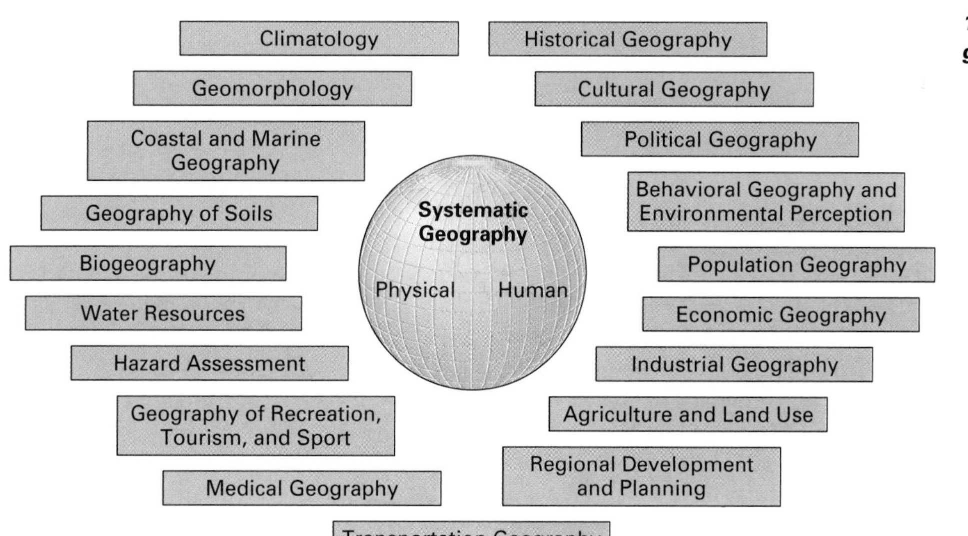

1.2 Fields of systematic geography

weather, we can think of climate as a description of average weather and its variation at places around the world. Chapters 3–11 will familiarize you with the essentials of climatology, including the processes that control the weather we experience daily. Climatology is also concerned with climate change, both past and future. One of the most rapidly expanding and challenging areas of climatology is global climate modeling, which we touch on in several chapters. This field attempts to predict how human activities, such as converting land from forest to agriculture or releasing CO_2 from fossil fuel burning, will change global climate.

Geomorphology is the science of Earth surface processes and landforms. The Earth's surface is constantly being altered under the combined influence of human and natural factors. The work of gravity in the collapse and movement of Earth materials, as well as the work of flowing water, blowing wind, breaking waves, and moving ice, acts to remove and transport soil and rock and to sculpt a surface that is constantly being renewed though volcanic and tectonic activity. The middle chapters of our book (Chapters 12–20) describe these geomorphic processes as well as the basic geologic processes that provide the raw material. Modern geomorphology also focuses on modeling landform-shaping processes to predict both short-term, rapid changes, such as landslides, floods, or coastal storm erosion, and long-term, slower changes, such as soil erosion in agricultural areas or as a result of strip mining.

Coastal and marine geography combines the study of geomorphic processes that shape shores and coastlines with their application to coastal development and marine resource utilization. Chapter 19 describes these processes and provides some perspectives on problems of human occupation of the coastal zone.

Geography of soils includes the study of the distribution of soil types and properties and the processes of soil formation. It is related to both geomorphic processes of rock breakup and weathering, and to biological processes of growth, activity, and decay of organisms living in the soil (Chapter 21). Since both geomorphic and biologic processes are influenced by the surface temperature and availability of moisture, broad-scale soil patterns are often related to climate.

Biogeography is the study of the distributions of organisms at varying spatial and temporal scales, as well as the processes that produce these distribution patterns. Local distributions of plants and animals typically depend on the suitability of the habitat that supports them. In this application, biogeography is closely aligned with *ecology,* which is the study of the relationship between organisms and environment (Chapter 22). Over broader scales and time periods, the migration, evolution, and extinction of plants and animals are key processes that determine their spatial distribution patterns (Chapter 23). Thus, biogeographers often seek to reconstruct past patterns of plant and animal communities from fossil evidence of various kinds. *Biodiversity*—the assessment of biological diversity from the perspective of maintaining the diversity of life and life-forms on Earth—is a biogeographic topic of increasing importance as human impact on the environment continues. The present global-scale distribution of life-forms as the great biomes of the Earth provides a basic context for biodiversity (Chapter 24).

In addition to these five main fields of physical geography, two others are strongly involved with applications of physical geography—water resources and hazards assessment. **Water resources** is a broad field that couples both basic study of the location, distribution, and movement of water, for example, in river systems or as groundwater, with the utilization and quality of water for human use. This field involves many aspects of human geography, including regional development and planning, political geography, and agriculture and land use. We touch on water resources briefly in this book by discussing water wells, dams, and water quality in Chapters 16 and 17.

Hazards assessment is another field that blends physical and human geography. What are the risks of living next to a river, and how do inhabitants perceive those risks? What is the role of government in protecting citizens from floods or assisting them in recovery from flood damages?

1.3 Fields of Physical Geography

a...***Climatology*** *Climate is largely dependent on atmosphere-surface interaction.*

b...***Geomorphology*** *Geomorphology is the study of landform-making processes.*

c...***Eye on the Landscape*** ***Coastal and marine geography*** *Coastal and marine geography examines coastal processes, marine resources, and their human interface.* **What else would the geographer see?...Answers at the end of the chapter.**

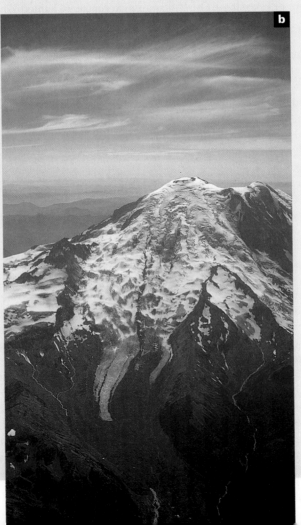

d...*Eye on the Landscape*

Biogeography *Biogeography examines the distribution patterns of plants and animals and relates them to environment, migration, evolution, and extinction.* **What else would the geographer see?...Answers at the end of the chapter.**

e...***Geography of soils*** *Soils are influenced by their parent material, climate, biota, and time.*

Answering questions such as these requires not only knowledge of how physical systems work, but also how humans perceive and interact with their physical environment as both individuals and as societies. In this text, we develop an understanding of the physical processes of floods, earthquakes, landslides, and other disaster-causing natural events as a background for appreciating hazards to humans and their activities.

The many remaining fields of human geography are also shown in Figure 1.2. Although they are not covered in this book, most have linkages with physical geography. For example, climatic and biogeographic factors may determine the spread of disease-carrying mosquitoes (medical geography). Or mountain barriers may isolate populations and increase the cost of transporting goods from one place to another (cultural geography, transportation geography). Or unique landforms and landscapes may be the objects of tourism (geography of recreation, tourism, and sport). Nearly all human activities take place in a physical environment that varies in space and time, so the physical processes that we examine in this text provide a background useful for further learning in any of geography's fields.

Tools in Geography

Because geographers deal with spatial phenomena, they use unique tools to represent spatial information (Figure 1.4). The map is a common example. A **map** is a paper representation of space showing point, line, or area data—that is, locations, connections, and regions. It typically displays some characteristics or features of the Earth's surface that are positioned on the map in much the same way that they occur on the surface. The map's scale links the true distance between places with the distance on the map. The art and science of mapmaking is called **cartography.** Chapter 3 provides some basic information about maps as well as an interchapter feature, *A Closer Look: Geographer's Tools 3.5 • Focus on Maps,* that adds depth on this important subject.

Maps, like books, are very useful methods for storing information, but they have limitations. In the past two decades, striking advances in data collection, storage, analysis, and display have led to the development of **geographic information systems (GISs).** These spatial databases rely on computers for analysis, manipulation, and display of spatial data. We will return to an examination of GISs in Chapter 3 with our special supplement, *Geographer's Tools 3.3 • Geographic Information Systems.*

Another important geographic technique for acquiring spatial information is **remote sensing,** in which aircraft or spacecraft provide images of the Earth's surface. Depending on the scale of the remotely sensed image, the information obtained can range from fine local detail—such as the arrangement of cars in a parking lot—to a global-scale picture—for example, the "greenness" of vegetation for an entire continent. As you read this textbook, you will see many examples of remote sensing, especially images from orbiting satellites, as well as boxed features highlighting applications of remote sensing in physical geography. We will return to remote sensing in our interchapter feature, *A Closer Look: Geographer's Tools 4.7 • Remote Sensing for Physical Geography,* which follows Chapter 4.

Tools in geography also include **mathematical modeling** and **statistics.** Using math and computers to model geographic processes is a powerful approach to understanding both natural and human phenomena. Statistics provides methods that can be used to manipulate geographic data so that we can ask and answer questions about differences, trends, and patterns. Because these tools rely heavily on specialized knowledge, we only touch on them lightly in this book. Our **Working It Out** boxes, found in most chapters, provide simple examples of how mathematical models and statistical techniques can be applied to physical geography.

Systems in Physical Geography

Geography emphasizes the interconnection of processes that occur on the Earth's land surface. An important perspective on physical geography that emphasizes these interconnections for natural processes is the **systems approach.** As we will see in Chapter 2, this approach directs our attention to how and where matter and energy flow in natural systems. For example, using the systems approach, we see the flow of water in a river channel as matter in motion, powered by gravity. But we also see as part of the system the rain that falls on the uplands, providing water above sea level so that gravity can do its work. In turn, the rainfall occurs because solar energy has evaporated water from the oceans and moist land surfaces, releasing it to the atmosphere. So stream flow is really powered by the Sun. Throughout this book we highlight our use of a systems approach with boxed features, entitled **Focus on Systems,** that apply systems thinking to topics in physical geography.

Understanding Physical Geography

From the start of your study of physical geography, you will find that it differs from other sciences in that it focuses on the world around you, from the changes in the weather to the landforms you travel over every day. When you finish your study, you should be able to see the landscape in a new way. To help you preview what you will learn about our environment, we provide a new feature—**Eye on the Landscape**—which takes our chapter-opener photos and identifies prominent features that a physical geographer would recognize and understand.

Another way to better understand physical geography is to examine research topics and problems that physical geographers pursue. To accomplish this, our web site includes a number of **Geographers at Work** features that introduce you to young scientists working within physical geography and highlight their work. These features not only describe the specific interests of these researchers, but also place their work in the context of global and environmental change issues. Watch for the short descriptions of each geographer at work at appropriate places in the text.

Physical Geography, Environment, and Global Change

Physical geography is concerned with the natural world around us—in short, with the human environment. Because natural processes are constantly active, the Earth's environments are constantly changing (Figure 1.5). Sometimes the changes are slow and subtle, as when crustal plates move over geologic time to create continents and ocean basins. At other times, the changes are rapid, as when hurricane winds flatten vast areas of forests or even tracts of houses and homes.

Environmental change is now produced not only by the natural processes that have acted on our planet for millions of years but also by human activity. The human race has populated our planet so thoroughly that few places remain free of some form of human impact. Global change, then, involves not only natural processes, but also human processes that interact with them. Physical geography is the key to understanding this interaction.

Environment and global change are sufficiently important that we have set off these topics from ordinary text in this book by placing them in special sections identified with **Eye on the Environment** and **Eye on Global Change.** As you read, watch for these sections as your key to the application of physical geography to environmental and global change topics.

What are some of the important topics in global change that lie within physical geography? Let's examine a few.

Global Climate Change

Are human activities changing global climate? It seems that almost every year we hear that it has been the hottest year, or one of the hottest years, on record. But climate is notoriously variable. Could such a string of hot years be part of the normal variation? This is the key question facing scientists studying global climate change. Over the past decade, most scientists have come to the opinion that human activity has, indeed, begun to change our climate. How has this happened?

The answer lies in the greenhouse effect. As human activities continue to release gases that block heat radiation from leaving the Earth, the greenhouse effect intensifies. The most prominent of these gases is CO_2, which is released by fossil fuel burning. Others include methane (CH_4), nitrous oxide (NO), and the chlorofluorocarbons that until recently served as coolants in refrigeration and air conditioning systems and as aerosol spray propellants. Taken together, these gases are acting to raise the Earth's surface temperature, with consequences including dislocation of agricultural areas, rise in sea level, and increased frequency of extreme weather events, such as severe storms or record droughts.

Climate change is a recurring theme throughout this book, ranging from the urban heat island effect that tends to raise city temperatures (Chapter 5) to the El Niño phenomenon that alters global atmospheric and ocean circulation (Chapter 7), to the effect of clouds on global warming (Chapter 8), and to rising sea level due to the expansion of sea water with increasing temperature (Chapter 19).

The Carbon Cycle

One way to reduce human impact on the greenhouse effect is to slow the release of CO_2 from fossil fuel burning. In fact, the Kyoto Protocol of 1997, an international agreement first proposed at the 1992 Earth Summit in Rio de Janeiro, commits the world's industrialized nations to limit current CO_2 emissions to levels at or below those of 1990. But since modern civilization depends on the energy of fossil fuels to carry out almost every task, reducing fossil fuel consumption to stabilize the increasing concentration of CO_2 in the atmosphere is not easy. However, some natural processes reduce atmospheric CO_2. Plants withdraw CO_2 from the atmosphere by taking it up in photosynthesis to construct plant tissues, such as cell walls and wood. In addition, CO_2 is soluble in sea water. These two important pathways, by which carbon flows from the atmosphere to lands and oceans, are part of the **carbon cycle.** So, biogeographers and ecologists are now focusing in detail on the global carbon cycle in order to better understand the pathways and magnitudes of carbon flow. They hope that this understanding will suggest alternative actions that can reduce the rate of CO_2 buildup without penalizing economic growth.

As nations work to determine how much CO_2 they release and to develop plans to reduce that amount, scientists are focusing on the role of CO_2 uptake by natural vegetation. An important question now under study concerns the balance between forest growth and organic matter decay in the northern hemisphere forest zone. At present, there are large areas of young forests that are actively growing and taking up significant amounts of CO_2. In an opposite effect, this uptake is offset by the decay of organic matter in soils, which releases CO_2. Because the decay process depends on temperature, global warming is thought to be increasing the rate of decay and thus CO_2 release. Scientists have not yet determined the balance between these two pathways but are intensively studying these flows.

The processes of carbon fixation and release are covered in detail in Chapter 22. Chapter 23 also describes how changes in temperature and CO_2 concentration affect plant growth and decay, and documents the process of succession by which plant communities change through time.

Biodiversity

Among scientists, environmentalists, and the public, there is a growing awareness that the diversity in the plant and animal forms harbored by our planet—the Earth's **biodiversity**—is an immensely valuable resource that will be cherished by future generations. One important reason for preserving as many natural species as possible is that, over time, species have evolved natural biochemical defense mechanisms against diseases and predators. These defense mechanisms involve bioactive compounds that can sometimes be very useful, ranging from natural pesticides that increase crop yields to medicines that fight human cancer.

1.4 Tools of Physical Geography

a...Cartography *A portion of the U.S. Geological Survey 1:24,000 topographic map of Green Bay, Wisconsin. Using symbols, the map shows creeks and rivers, a bay, swampy regions, urban developed land, streets, roads, and highways.*

b...Geographic information systems (GIS) *Computer programs that store and manipulate geographic data are essential to modern applications of geography.*

c...Remote sensing *Remote sensing includes observing the Earth from the perspective of an aircraft or spacecraft. Wildfires on the Greek island of Peloponnesos, in July 2000, are an example.*

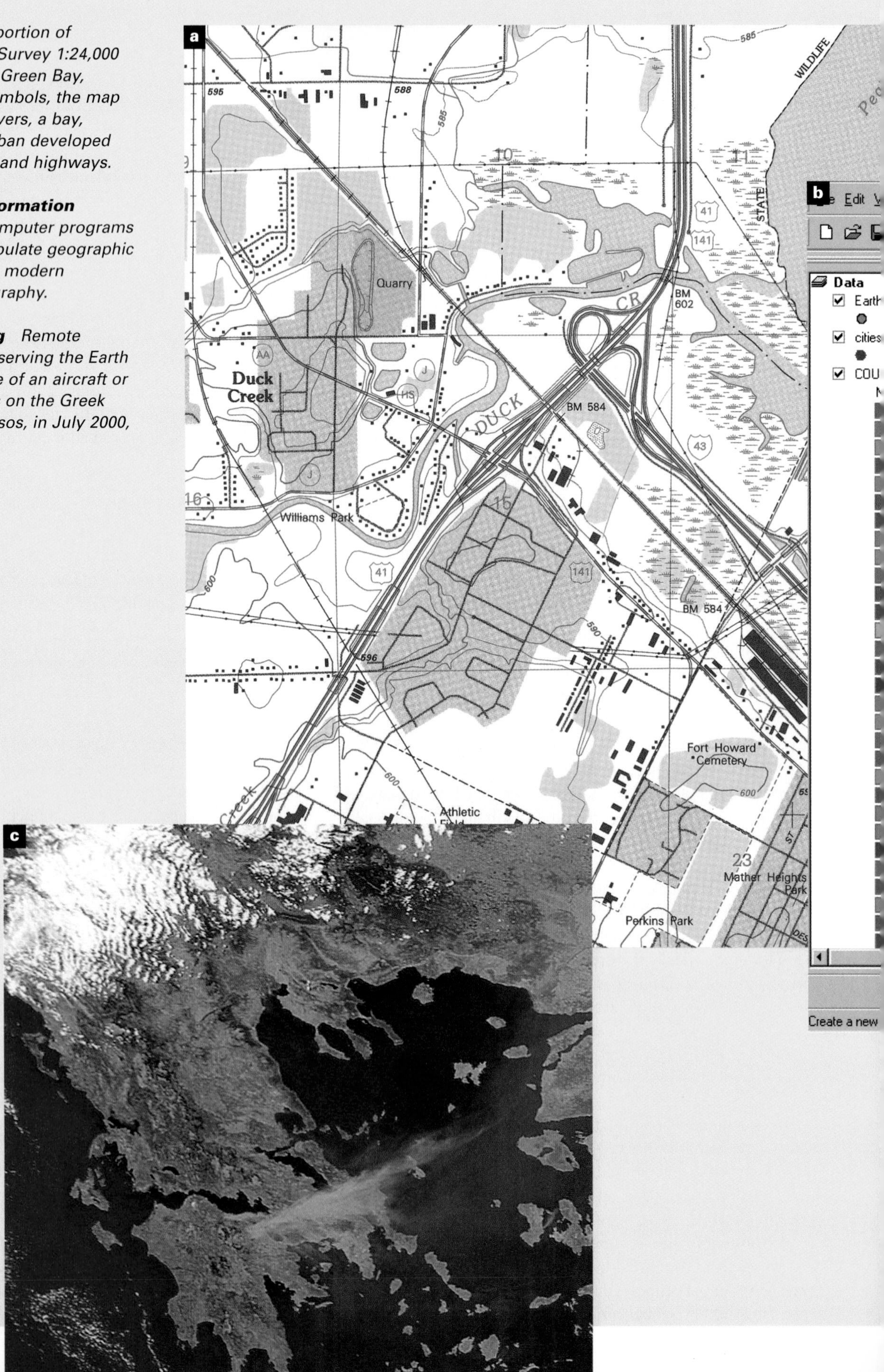

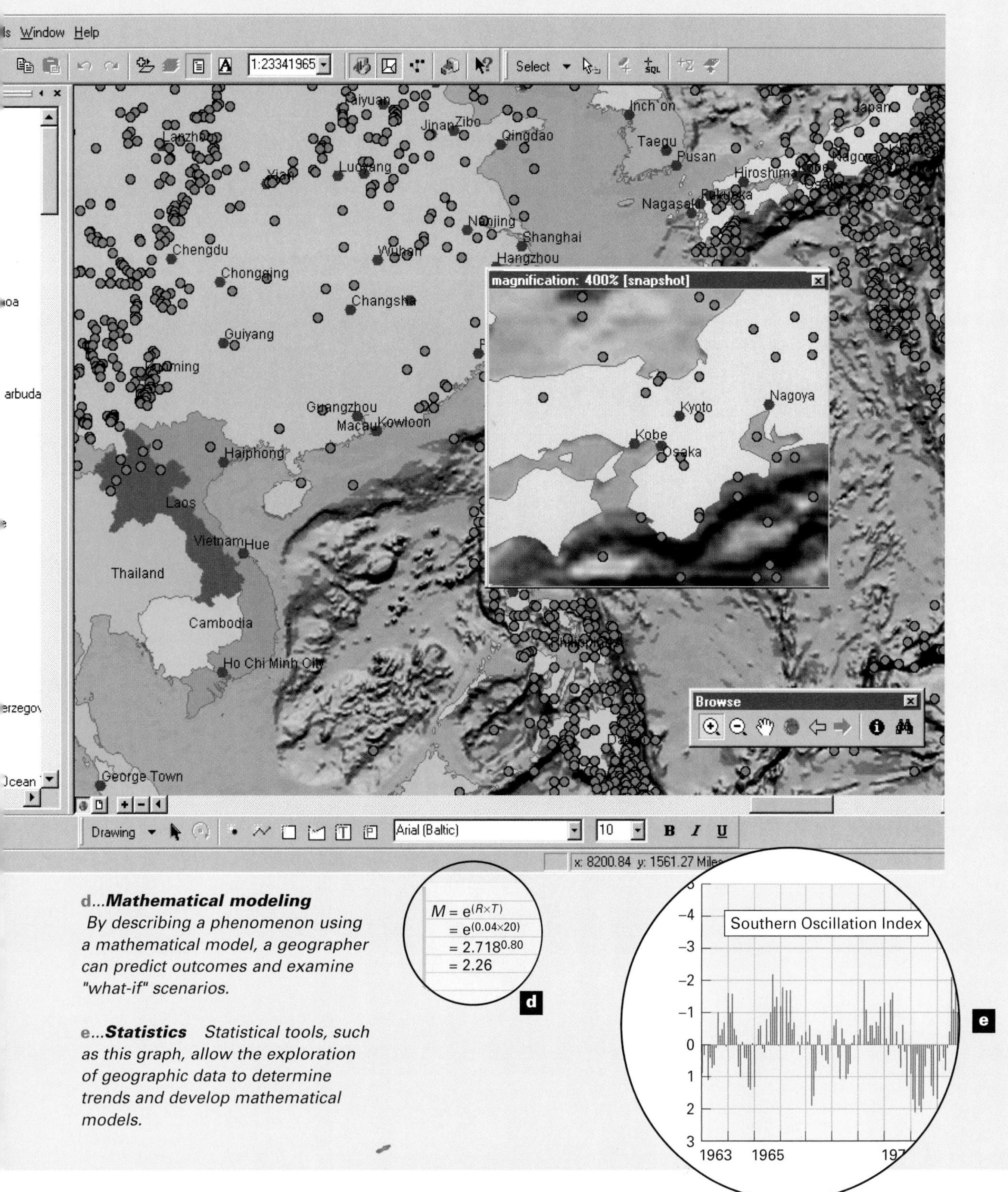

d...***Mathematical modeling*** *By describing a phenomenon using a mathematical model, a geographer can predict outcomes and examine "what-if" scenarios.*

$$M = e^{(R \times T)} = e^{(0.04 \times 20)} = 2.718^{0.80} = 2.26$$

d

e...***Statistics*** *Statistical tools, such as this graph, allow the exploration of geographic data to determine trends and develop mathematical models.*

e

1.5 Dimensions of Global Change

a...Global climate change *Is the Earth's climate changing? Nearly all global change scientists have concluded that human activities have resulted in climate warming and that weather patterns, shown here in this satellite image of clouds and weather systems over the Pacific ocean, are changing.*

b...Carbon cycle *Clearcutting of timber, shown here near the Grand Tetons, Wyoming, removes carbon from the landscape,*

c...Biodiversity *Reduction in the area and degradation of the quality of natural habitats is reducing biodiversity. The banks of this stream in the rainforest of Costa Rica are lined with several species of palms.*

d...Pollution *Human activity can create pollution of air and water, causing change in natural habitats as well as impacts on human health. The discharge from this pulp mill near Port Alice, British Columbia, is largely water vapor, but pulp mill pollutants often include harmful sulfur oxides.*

e...Extreme events *Hurricanes, severe storms, droughts, and floods may be becoming more frequent as global climate warms. A tornado flattened this neighborhood in Camilla, Georgia, in February 2000.*

Geographers at Work

The Changing Face of the Earth: Mapping the Land Surface Using Remote Sensing

by Dar A. Roberts *University of California, Santa Barbara*

One of the most important dimensions of global change is change in land use and land cover. Human alteration of the landscape not only changes the prior natural ecosystem, but also modifies other aspects of the environment. Deforestation of the equatorial Amazon rainforest, for example, not only affects the global carbon cycle by reducing carbon stocks on the land, but may also alter the climatic regime by reducing transpiration. This, in turn, reduces the moisture in the air, which can produce a drier regional climate less suited to forest cover.

Dar Roberts has an avid interest in natural history in addition to Geography. This photograph, taken in Lencõis, Brazil, shows Dr. Roberts holding a 2.1-meter Southern Indigo snake.

Dar Roberts studies how, where, and when land covers change and investigates how the changes impact both the carbon cycle and global climate. His work has taken him from dangling from a crane in the treetops of the conifer forests of the Pacific Northwest to climbing towers rising over the Brazilian rainforest. Read more about Dar's work on our web site.

Another important reason for maintaining biodiversity is that complex ecosystems with many species tend to be more stable and to respond better to environmental change. If human activities inadvertently reduce biodiversity significantly, there is more risk of unexpected and unintended human effects on natural environments.

Biogeographers focus on both the processes that create and maintain biodiversity and the existing biodiversity of the Earth's many natural habitats. These topics are treated in Chapters 23 and 24.

Pollution

As we all know, unchecked human activity can degrade environmental quality. In addition to CO_2 release, fuel burning can release gases that are hazardous to health, especially when they react to form such toxic compounds as ozone and nitric acid in photochemical smog. Water pollution from fertilizer runoff, toxic wastes of industrial production, and acid mine drainage can severely degrade water quality. Such degradation impacts not only the ecosystems of streams and rivers, but also the human populations that depend on rivers and streams as sources of water supply. Ground water reservoirs can also be polluted or turn salty in coastal zones when drawn down excessively.

Environmental pollution, its causes, its effects, and the technologies used to reduce pollution, form a subject that is broad in its own right. As a text in physical geography that emphasizes the natural processes of the Earth's land surface, we touch on air and water pollution in several chapters. Our interchapter feature *A Closer Look: Eye on the Environment 6.2 • Air Pollution* provides an overview of this topic, while we cover surface water pollution, irrigation effects, and ground water contamination in Chapter 16.

Extreme Events

Catastrophic events—floods, fires, hurricanes, earthquakes, and the like—can have great and long-lasting impacts on both human and natural systems. Are human activities increasing the frequency of these **extreme events**? As our planet warms in response to changes in the greenhouse effect, global climate modelers predict that weather extremes will become more severe and more frequent. Droughts and consequent wildfires and crop failures will happen more often, as will spells of rain and flood runoff. In the last decade, we have seen numerous examples of extreme weather events, from Hurricane Andrew in 1992—the most costly storm in U.S. history—to the southeast drought of 2000, which devastated crops in large parts of the southeastern United States. Is human activity responsible for the increased occurrence of these extreme events? The answer is not yet clear, but significant evidence points in that direction.

Other extreme events, such as earthquakes, volcanic eruptions, and seismic sea waves (wrongly called tidal waves), are produced by forces deep within the Earth that are not affected by human activity. But as the human population continues to expand and comes to rely increasingly on a technological infrastructure ranging from skyscrapers to the Internet, we are becoming more sensitive to damage and disruption of these systems by extreme events.

This text describes many types of extreme events and their causes. In Chapters 6 and 8, we discuss thunderstorms, tornadoes, cyclonic storms, and hurricanes. Droughts in the

African Sahel and the American Midwest are presented in Chapters 10 and 11. Earthquakes, volcanic eruptions, and seismic sea waves are covered in Chapter 14. Floods are described in Chapter 16, including the great Mississippi flood of 1993.

A LOOK AHEAD The past few pages have presented an introduction to geography, to physical geography, and to some of the tools and approaches that physical geographers use in studying the landscape. We have also introduced some of the key environmental and global change topics that will appear in our text.

Our next chapter focuses on some grand ideas that cut across all of physical geography. These include the great *spheres,* or Earth realms, that physical geographers study; the *scales,* from local to global, that characterize Earth processes; the flow *systems* that power the natural processes of physical geography; and the *cycles* of processes that repeat in time and space. As you read the chapters that follow, these ideas will recur again and again. By focusing on spheres, scales, systems, and cycles, we provide an overall structure for the detailed knowledge of physical geography that you will acquire.

Humans are now the dominant species on the planet. Nearly every part of the Earth has felt human impact in some way. As the human population continues to grow and rely more heavily on natural resources, our impact on natural systems will continue to increase. Each of us is charged with the responsibility to treat the Earth well and respect its finite nature. Understanding the processes that shape our habitat as they are described by physical geography will help you carry out this mission.

Chapter Summary

- **Geography** is the study of the evolving character and organization of the Earth's surface. **Regional geography** is concerned with how the Earth's surface is differentiated into unique **places,** while **systematic geography** is concerned with the processes that differentiate places in time and space.
- Within systematic geography, **human geography** deals with social, economic, and behavioral processes that differentiate places, and **physical geography** examines the natural processes occurring at the Earth's surface that provide the physical setting for human activities. **Climatology** is the science that describes and explains the variability in space and time of the heat and moisture states of the Earth's surface, especially its land surfaces. **Geomorphology** is the science of Earth surface processes and landforms. **Coastal and marine geography** combines the study of geomorphic processes that shape shores and coastlines with their application to coastal development and marine resource utilization. **Geography of soils** includes the study of the distribution of soil types and properties and the processes of soil formation. **Biogeography** is the study of the distributions of organisms at varying spatial and temporal scales, as well as the processes that produce these distribution patterns. **Water resources** and **hazards assessment** are applied fields that blend both physical and human geography.
- Important tools for studying the fields of physical geography include **maps, geographic information systems, remote sensing, mathematical modeling,** and **statistics.** A **systems approach** to physical geography helps in understanding the interconnections in natural processes.
- Physical geography is concerned with the natural world around us—the human environment. Natural and human processes are constantly changing that environment. **Global climate change** occurs in response to human impacts on the greenhouse effect. Global pathways of carbon flow within the **carbon cycle** can influence the greenhouse effect and are the subject of intense research interest. Maintaining global **biodiversity** is important both for maintaining the stability of ecosystems and guarding a potential resource of bioactive compounds for human benefit. Unchecked human activity can degrade environmental quality and create **environmental pollution. Extreme events** take ever-higher tolls on life and property as populations expand. Extreme weather—storms and droughts, for example—may be more frequent with global warming caused by human activity.

Key Terms

geography
regional geography
place
systematic geography
human geography
physical geography
climatology
geomorphology
coastal and marine geography
geography of soils
biogeography
water resources
hazards assessment
map
cartography
geographic information system (GIS)
remote sensing
mathematical modeling
statistics
systems approach
climate change
carbon cycle
biodiversity
environmental pollution
extreme events

Review Questions

1. What is geography? regional geography? systematic geography?
2. How does human geography differ from physical geography?
3. Identify and define three important fields of science within physical geography.
4. Identify and describe three tools that geographers use to acquire and display spatial data.
5. Explain how a systems approach shows that stream flow is actually powered by the Sun.
6. Identify and describe two interacting components of global change.
7. How is global climate change influenced by human activity?
8. Why are current research efforts focused on the carbon cycle?
9. Why is loss of biodiversity a concern of biogeographers and ecologists?
10. How does human activity degrade environmental quality? Provide a few examples.
11. How do extreme events affect human activity? Is human activity influencing the size or reoccurrence rate of extreme events?

Essay Question

1. Go to the extended table of contents for this book. With a pencil, check off every topic listed that you know something about. Now think about where you learned about these topics. Which of the following sources has provided you with the most information about physical geography? television, books and magazines, schoolwork, courses in college or university, learning by experience, other? How do you think the information sources by which you have learned about physical geography have affected what you have learned?

Eye on the Landscape

A

B

Chapter Opener Phang Nga Bay, Thailand (A). These distinctive, steep-sided peaks are formed as rainwater dissolves the underlying limestone over long periods of time in a warm, humid climate. We'll cover solution weathering in Chapter 15. **(B)** The green color indicates high concentrations of phytoplankton and algae. These primary producers lie at the bottom of the marine food chain and ultimately support the fish population of the bay. Primary production and the food chain are covered in Chapter 22.

B

A

1.1 Vancouver, British Columbia With its deep embayments and steep topography, this is a coastline of submergence **(A)**. Much of the terrain was carved into steep slopes and wide valleys by glacial ice descending from nearby peaks during the Ice Age, when sea level was as much as 100 m (about 300 ft) lower. With the melting of ice sheets, sea level rose, drowning the landscape. Also note the snowcapped peaks in the far distance **(B)**. Rugged landscapes like this often indicate recent plate tectonic activity in which immense crustal plates collide to push up mountain chains like the Cascade Range. Chapters 19 and 20 describe coastlines and glacial erosion features. Plate tectonics is covered in Chapter 13.

B

A

A

1.3c Coastal and marine geography (A) This rock platform, shown at low tide, is a feature formed by wave action and is found on many rocky shorelines. At times of high water and surf, waves attack the cliffs, releasing rock fragments that are carried back and forth by the waves, thus cutting the flat bench. Chapter 19 provides more information on wave action. At **(B)**, horizontal lines revealed in the rock faces of these cliffed promontories mark individual layers (beds) of sedimentary rock. These beds were probably formed when silt and clay settled from turbid waters in a shallow arm of the ocean or in an inland sea. After millions of years of burial and consolidation into rock, the beds were uplifted and are now being eroded by waves. Sedimentary rock formation is covered in Chapter 12, and uplift (tectonic processes) is covered in Chapters 13 and 14.

A

I-3d Biogeography These patches of fog at **(A)** illustrate that fog is simply cloud touching the ground. They are probably patches of morning fog, formed when the moist air of the rainforest is chilled below the dew point by radiation cooling on a clear night. As the air warms, they will evaporate. Chapter 6 covers clouds, fog, and radiation cooling.

2 SPHERES, SCALES, SYSTEMS, AND CYCLES

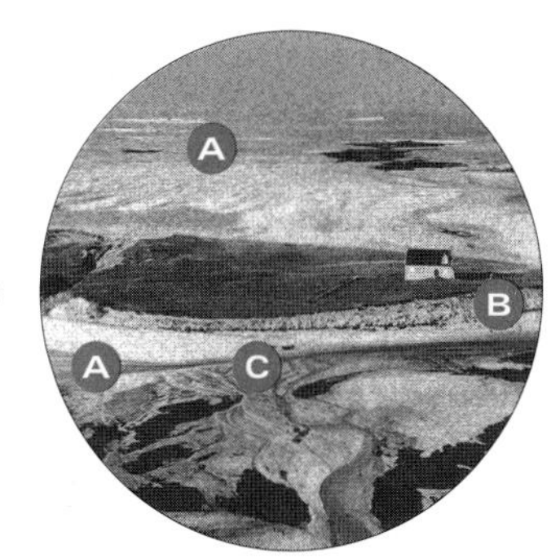

EYE ON THE LANDSCAPE
An isolated outpost on the English Channel coast of Brittany, Finistère, France. The four great realms of lithosphere, hydrosphere, atmosphere, and biosphere are all visible in this dramatic image. **What else would the geographer see?... Answers at the end of the chapter.**

Before turning to the principles and concepts of physical geography, it's useful to take a look at the big picture and examine some ideas that arch over all of physical geography—that is, spheres, scales, systems, and cycles. The first of these ideas is that of the four great physical realms, or *spheres* of Earth—atmosphere, lithosphere, hydrosphere, and biosphere. These realms are distinctive parts of our planet with unique components and properties. They converge at the life layer—the interface between the four realms located at the Earth's surface.

Another important idea is that of *scales* in physical geography, which range from global to local. When we look at the Earth at different scales, we often see different processes operating. For example, at the global scale, we can focus on how solar energy is received by the Earth and atmosphere, redistributed by reflection and absorption, and ultimately returned to space. At the local scale, we can see how erosion alters a hill slope or how weeds invade and conquer a fallow field.

But the biggest idea in this overview is that of systems—*flow systems* of energy and matter. By looking for flow systems in studying the processes of physical geography, we are naturally directed to thinking about how things move, change, and interact. Using a systems approach, we find it easier to explore questions like "What would happen if human activity enhances the greenhouse effect?" Or "How does building a dam on a river affect local climate?"

The last big idea is that of *cycles*—regular changes in energy and matter flows in a system that recur in a fixed period of time. The seasonal changes in the landscape provide an example of a cycle that is driven by the increase and decrease in the Sun's energy flow to Earth that occurs through the year at a particular latitude. On a longer time

Focus on Systems | 2.1

The Value of Systems Thinking

Scientists spend a lot of time on systems thinking. They observe systems as a whole, as well as the pathways, interconnections, power sources, and other components of systems. They formulate theories about how systems are organized, and they design experiments to test their understanding of systems. They use math and computers to model the behavior of systems. In fact, there is a body of knowledge called *systems theory* that attempts to explain how systems work generally and how system characteristics can be predicted from components and interconnections.

Systems thinking is very important for the scientists who study natural processes—geographers, geologists, ecologists, oceanographers, atmospheric scientists, and others. Because they study Earth and life phenomena, their ability to conduct experiments is limited. For example, it is not possible to reduce the Sun's energy output by, say, 10 percent in order to see how climate would be affected. The ecosystem of even a small pond or bog is too complex to be reproduced in the laboratory for any meaningful experiments on the ecosystem as a whole. Thus, scientists who study natural systems must do their work largely by treating the Earth as their laboratory—that is, by observing the Earth and its processes at different times and places and on different scales. By using a systems approach to guide these observations, natural scientists can understand the components and connections within the systems they study and can come to understand and predict the behavior of these systems as wholes.

Whatever the application, thinking critically about how things work and how they are related to one another is a skill that everyone relies on. One objective of this book is to help you build those skills and apply them to the principles and processes of natural systems.

scale, the Earth enters and departs from ice ages in a cycle driven by regular changes in the Earth's orbit around the Sun.

As you read this chapter, keep in mind that systems thinking will probably be new to you and may perhaps take a bit more effort to master than some of the other concepts in physical geography. But systems thinking is a very important tool for scientists, and it will ultimately give you a lot of insight into the processes that shape the Earth's varied landscapes. *Focus on Systems 2.1 • The Value of Systems Thinking* provides more perspectives on the value of systems thinking.

THE FOUR GREAT EARTH REALMS

The natural systems that we will encounter in the study of physical geography operate within the four great realms, or spheres, of the Earth. These are the atmosphere, the lithosphere, the hydrosphere, and the biosphere (Figure 2.1).

The **atmosphere** is a gaseous layer that surrounds the Earth. It receives heat and moisture from the surface and redistributes them, returning some heat and all the moisture to the surface. The atmosphere also supplies vital elements—carbon, hydrogen, oxygen, and nitrogen—that are needed to sustain life-forms.

The outermost solid layer of the Earth, or **lithosphere,** provides the platform for most Earthly life-forms. The solid rock of the lithosphere bears a shallow layer of soil in which nutrient elements become available to organisms. The surface of the lithosphere is sculpted into landforms. These features—such as mountains, hills, and plains—provide varied habitats for plants, animals, and humans.

The liquid realm of the Earth is the **hydrosphere,** which is principally the mass of water in the world's oceans. It also includes solid ice in mountain and continental glaciers, which, like liquid ocean and fresh water, is subject to flow under the influence of gravity. Within the atmosphere, water

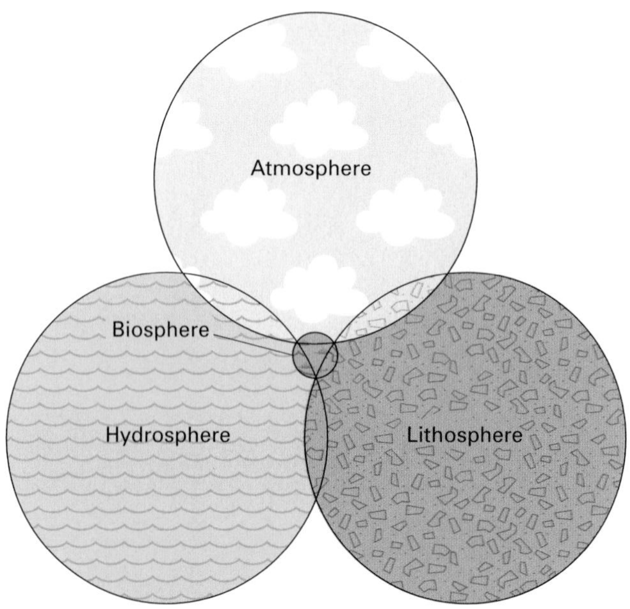

2.1 The Earth realms *The Earth realms, shown as intersecting circles.*

*2.2 **The life layer** As this sketch shows, the life layer is the layer of the Earth's surface that supports nearly all of the Earth's life. This includes the land and ocean surfaces and the atmosphere in contact with them.*

occurs as gaseous vapor, liquid droplets, and solid ice crystals. In the lithosphere, water is found in the uppermost layers in soils and in ground water reservoirs.

The **biosphere** encompasses all living organisms of the Earth. Life-forms on Earth utilize the gases of the atmosphere, the water of the hydrosphere, and the nutrients of the lithosphere, and so the biosphere is dependent on all three of the other great realms. Figure 2.1 diagrams this relationship.

Most of the biosphere is contained in the shallow surface zone called the **life layer.** It includes the surface of the lands and the upper 100 meters or so (about 300 ft) of the ocean (Figure 2.2). On land, the life layer is the zone of interactions among the biosphere, lithosphere, and atmosphere. The hydrosphere is represented on land by rain, snow, still water in ponds and lakes, and running water in rivers. In the ocean, the life layer is the zone of interactions among the hydrosphere, biosphere, and atmosphere. The lithosphere is represented by nutrients dissolved in the upper layer of sea water.

Throughout our exploration of physical geography, we will often refer to the life layer and the four realms that interact within it. We will find that the themes and subjects of physical geography concern systems and cycles that involve the atmosphere, hydrosphere, lithosphere, or biosphere—and usually two or three of these realms simultaneously.

Scales in Physical Geography

The processes of the four great realms and the life layer operate at various **scales** (Figure 2.3). On a *global scale,* our planet is a nearly spherical body that turns endlessly within the radiance of our Sun. Because the Sun is the energy source that powers most of the phenomena that occur within the life layer, Earth–Sun relationships are very important. In studying these relationships, we need to consider the planet and its global energy balance system as a whole and view the Sun and Earth from a vantage point far from the Earth itself.

The Sun's energy is not absorbed evenly by the Earth's land and water surface. Unequal solar heating produces currents of air and water, just as a candle flame creates a current of hot, rising air. These currents constitute the global atmospheric and oceanic circulation system. To study this system, we need to move to the *continental scale,* where we can distinguish continents and oceans and track winds and ocean currents.

Moving still closer to the Earth to the *regional scale,* we see the cloud patterns of weather systems and their regular movements over time. These movements, along with solar control of surface temperature, form the basis for the climates of the world.

Regional climate strongly influences the nature of the Earth's cover of plants and the soil layers beneath them. But factors at the *local scale* are important in determining the exact patterns of vegetation and soils. Is the location on a mountain slope or in a flat valley? Is it being farmed or used as grazing land? This information is too specific to use at the global, continental, or regional scales, but it is important when we consider the nature of the vegetation cover and soils within a specific region.

At the very finest scale, we see *individual-scale* landscape features such as a grassy sand dune on a beach or a moss-covered bank on the side of a river. These individual landforms and their associated plant and animal communities are produced by unique activities of wind or water, and develop distinctive biological communities and soil properties.

Time scales vary widely, too. It may take millions of years for Earth forces to produce a chain of mountains like the Himalayas. But a fracturing of the Earth's crust and the resulting earthquake may last only a minute or two.

The point here is that natural processes act over a wide range of scales in time and space. Some are best understood by considering the Earth as a whole. The global circulation of the atmosphere is a good example. Still others are concerned with very local phenomena, such as sand grains being blown along a beach.

If you look at the table of contents for this book, you will note that our study of physical geography will generally take us from the coarse scale to the fine—from global phenomena to local phenomena. For example, in Part 2, Weather and Climate Systems, we begin with Earth–Sun relationships and global atmospheric circulation, then move to precipitation and weather systems, and finally consider individual climates within specific environments. Similarly, in Part 3, we consider the structure of the Earth as a whole, discuss the formation of continents, ocean basins, and other crustal features, and then describe specific landforms of volcanic and tectonic activity. There are exceptions to this overall plan, but understanding the focus of this organization is helpful in learning about the systems and processes of physical geography.

Systems in Physical Geography

The processes that interact within the four realms to shape the life layer and differentiate global environments are varied and complex. As mentioned in the beginning of this chapter, a helpful way to understand the relationships among these processes is to study them as **systems.** "System" is a common English word that we use in everyday speech. It

2.3 Scales in Physical Geography

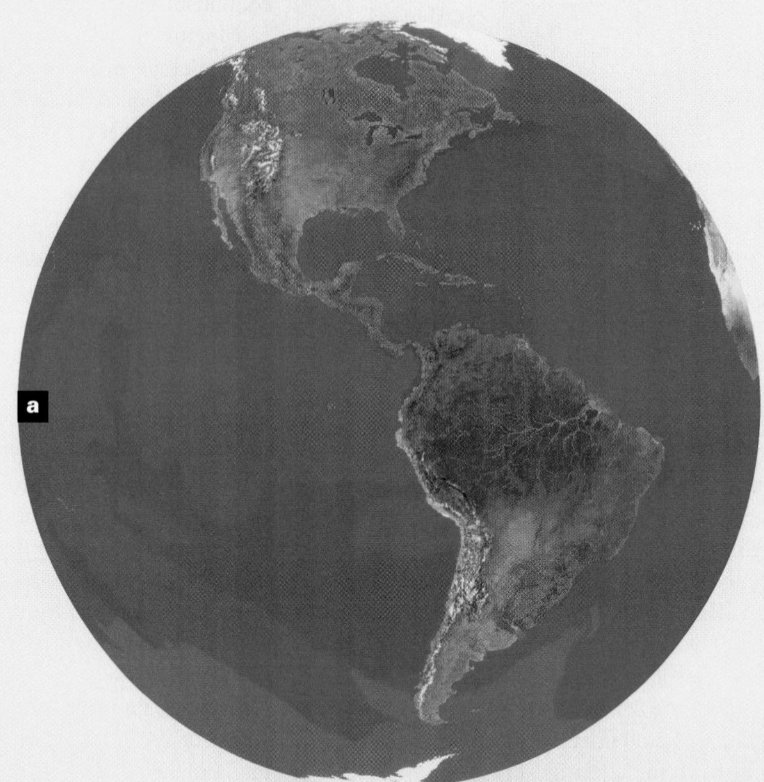

a...***Global scale*** *A model of the Earth's topography as it would be seen from a viewpoint in space above the Earth's equator.*

b...***Continental scale*** *A topographic model of the conterminous United States, with clouds added elsewhere for realism.*

c...***Regional scale*** *Topography of the western United States and southwestern Canada.*

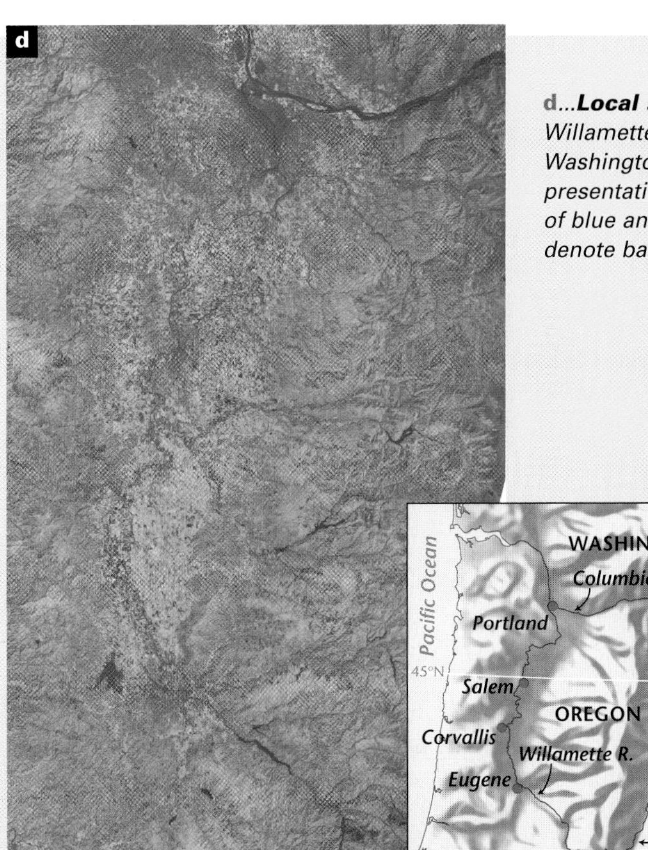

d...***Local scale*** *A Landsat image of the Willamette River Valley, Oregon and Washington. In this false-color presentation, vegetation appears in tones of blue and green, while reddish tones denote bare ground and urbanized areas.*

e...***Eye on the Landscape***
Individual-scale landscape
The Willamette River, near Newburg, Oregon, seen from the air. **What else would the geographer see?...Answers at the end of the chapter.**

typically means a set or collection of things that are somehow related or organized. An example is the solar system—a collection of planets that revolve around the Sun. Another is the geographic grid, which we will encounter in Chapter 3 as a system of meridians and parallels. In the text, we will use the word *system* in this way quite often. Sometimes it refers to a scheme for naming things. For example, we will introduce a climate system in Chapter 9 and a soil classification system in Chapter 21.

When we study the processes of physical geography as systems, however, we refer to a specific type of system—a **flow system** in which matter, energy, or both, move through time from one location to another. Understanding flow systems in physical geography is important because it helps to explain how things are connected—that is, how all the processes that affect climate, for example, are related and influence each other. So as you read the text remember that when we use the term *flow system,* we have this special meaning of *system* in mind.

To explain flow systems more clearly, let's start with a familiar example—a superhighway system of interconnected freeways and expressways. In this human-engineered system, individual roadways serve as pathways or tracks for the flow of vehicles (Figure 2.4). The roadways are connected by interchanges that permit vehicles to move from one roadway to another. The vehicles on the roads and interchanges—cars, trucks, and motorcycles—are part of the system, too. They are tangible objects—pieces of matter—that are in motion in the superhighway system. Thus, we can distinguish this system as a *matter flow system,* a system in which matter is in motion. In some systems of physical geography, it will be energy that is in motion. For example, we may be concerned with a system that describes the flow of solar energy from the Sun to the Earth and its atmosphere. In that case, we will be studying an *energy flow system.*

Flow systems have a structure in which **pathways** of flow are connected. For the superhighway system, the individual roadways are the pathways. However, pathways of a natural system are not always quite so obvious. For example, we may regard the reflection of solar radiation from the top of a bright, white cloud as a pathway by which a flow of solar energy is turned away from the Earth and back toward space.

We refer to the pattern of the pathways and their interconnections as the **structure** of the system. It is also convenient to use the term *components* to refer to the parts of a system, such as the pathways, their connections, and the types of matter and/or energy that flow within the system.

Many flow systems have **inputs** and **outputs.** In the highway system example, the flow of vehicles fed in by on-ramps is the input, and the flow of vehicles leaving the system by exit ramps is the output.

Each flow system needs some sort of **power source.** For the highway system, it is easy to identify the power source—the motor fuels burned by cars, trucks, and motorcycles on the highways.

2.4 A superhighway system as a flow system *A superhighway system is a simple example of a flow system that has many of the properties of the flow systems of energy and matter that we will examine in this text.*

The flow systems we will study in this book are *natural flow systems* rather than artificial, human-engineered systems such as the superhighway system. Natural systems are powered largely or completely by natural power sources. These sources include the flow of energy from the Sun to the Earth, the power stored in the inertia of the Earth's rotation, and the outward flow of heat from the Earth's interior that ultimately produces movements of the Earth's crust.

What types of natural flow systems will we encounter in physical geography? An example is a river system, a matter flow system of water in a set of connected stream channels (Figure 2.5). In this system, the stream channels—pathways—are connected in a structure—the channel network—that organizes the flow of water from high lands to low lakes or oceans. Another is the food chain of an ecosystem, in which energy in the form of food flows among plant and animal components. Yet a third example is the global energy balance system. In this energy flow system, the heat of the Sun is distributed around the Earth by currents of warm water and moist air. Thus, the flow in this system is one of heat energy.

To summarize, a flow system consists of a structure of connected pathways within which matter or energy flows—as light, heat, air, water, ozone, pollutants, or sand grains, for example. Flow systems can have input and output streams in which flows enter and exit from the system. All flow systems require power sources to run them.

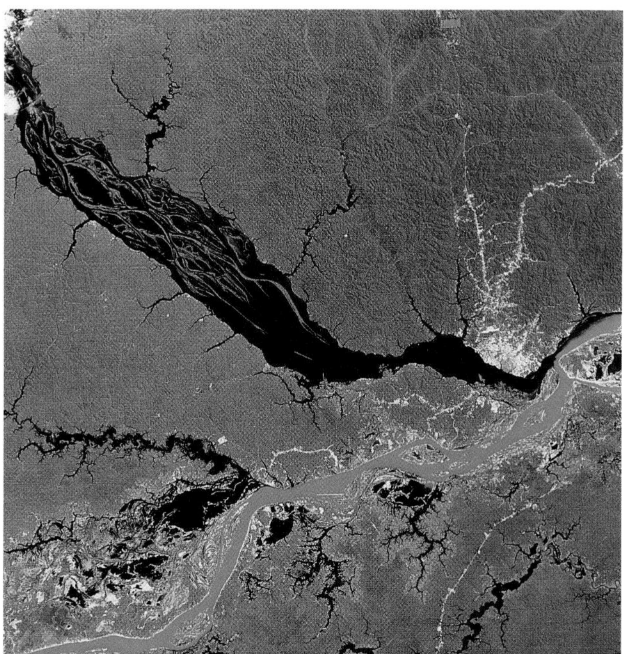

2.5 River channels as flow system pathways *A river and its tributaries provide an example of a natural flow system. This satellite image of the Manaus region, Brazil, shows the confluence of the Rio Negro (dark water) and Amazon (blue-green). Dense vegetation appears bright red. The city of Manaus is the blue-gray region on the northern bank of the Rio Negro near the junction.*

Open and Closed Flow Systems

The superhighway system is an example of an **open flow system,** one in which there are inputs and outputs of energy and matter. In this system, cars enter and leave the system freely, powered by fuel brought in to service stations from outside the system. Among natural systems, a river system is also an open flow system. Precipitation provides an input, and water is output to the ocean (or lake) where the river terminates. Power for the flow of water is provided by the climate system, which condenses water above the landscape as rain or snow, and by gravity, which pulls the water downhill.

There is also a type of matter flow system in which there are no input or output flows of matter. Instead, the flowing materials in the system move endlessly in a series of interconnected paths or loops. Because there is no input or output, we refer to this system as a **closed flow system.** This type of system is also known as a **cycle,** or more fully, a **material cycle.**

The closed flow system is rather like an automobile race around a track—the Indy 500, for example. The vehicles move around the track, making loop after loop, powered by the flow of motor fuel in their tanks (Figure 2.6). During the race, the cars remain on the track (except for brief stops in the pit), so there is no input or output flow of cars as there is in the superhighway system. Most natural closed flow systems are not as simple as this example, of course. They have more complicated structures consisting of many looping pathways that are interconnected in many places. Note also that a closed matter flow system requires a flow of energy to sustain itself—in this case, the refueling that occurs in pit stops. So although a material cycle may be closed, energy flow systems will always be open.

What closed flow systems do we encounter in physical geography? One is the hydrologic cycle, which describes global flows of water. The loops in this system are flow paths of water in solid, liquid, and gaseous forms—for example, water moving as solid ice in glaciers, water as a liquid in rivers, streams, or ocean currents, and as a gas as water vapor in flows of moist air. We will develop the hydrologic cycle as a flow system in more detail in Chapters 6 and 16. Another example of a closed flow system is the global carbon cycle (Chapter 22). Here, the loops describe how the element carbon travels in different chemical forms between carbon-bearing rocks, ocean waters, the atmosphere, and the bodies of plants and animals.

Whether a matter flow system is open or closed depends partly on where we draw the boundary around the system. As an example, consider a single river network as a simple system, represented in part (*a*) of Figure 2.7. Water enters the network when it falls on the land as precipitation and runs off into the river system. We regard this water as an input that crosses the system boundary and enters the system. Water exits from the network at the river mouth, where it enters the ocean. This is a system output, crossing another system boundary located here. Thus, the river system is an open matter flow system in which water enters and leaves the system as input and output flows.

2.6 A road race as a closed material flow system *A road race on a closed circuit, such as the Monte Carlo Grand Prix shown here, demonstrates the concept of a closed system.*

2.7 A river system as a flow system *(a) Diagram of a river system as an open matter flow system. (b) When the boundary is moved to enclose the Earth and atmosphere, the system becomes a closed, global flow system—the hydrologic cycle.*

(a)

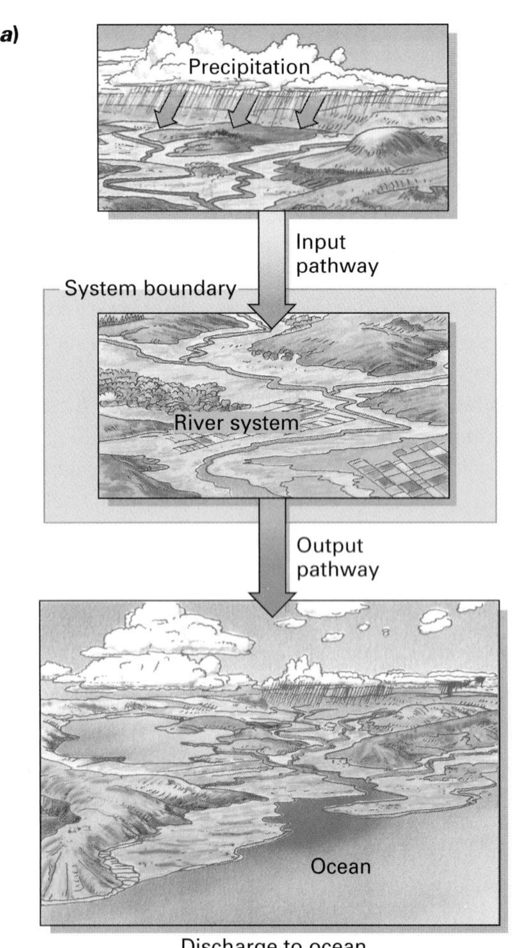

(b)

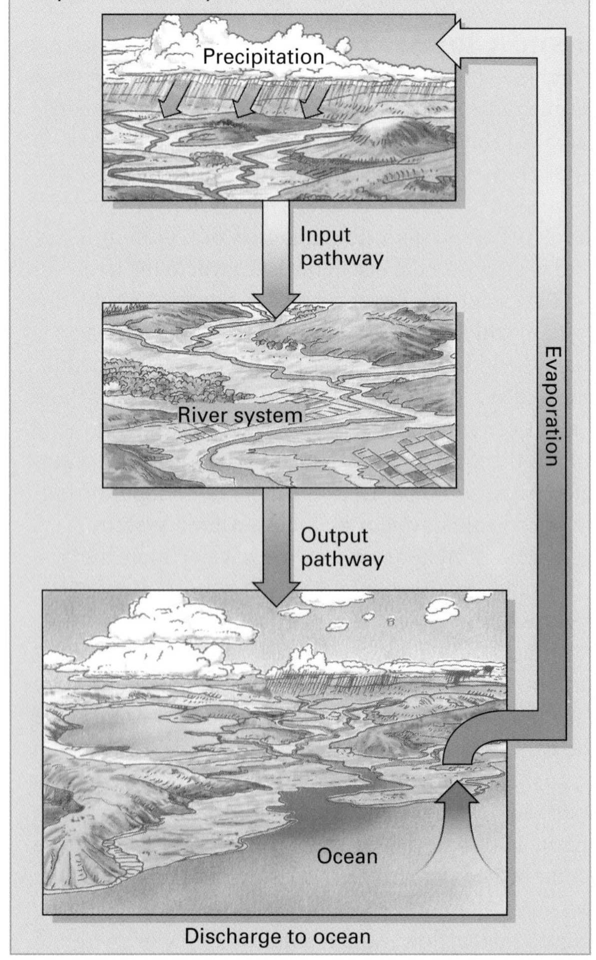

Let's now redraw the system boundary to include the whole Earth and its atmosphere. This situation is shown symbolically in part (*b*) of the figure. For this case, a new pathway must be added—the return flow of water from the oceans to the atmosphere by evaporation. There is no input or output because water does not leave the Earth or enter from space. Thus, the system is closed. The river system becomes the global hydrologic cycle, a closed matter flow system, which is described above.

Indeed, any global matter flow system must be closed, since only a minuscule amount of matter flows from Earth to space (wandering gas molecules at the edge of the atmosphere) or from space to the Earth (meteors, meteorites, and solar particles). Thus, the global carbon, nitrogen, and oxygen cycles (described further in Chapter 22) are all closed matter flow systems.

What about energy flow systems? They are always open. Why? Because, as we shall see in Chapter 4, all objects that are warmer than the depths of space emit radiant energy, and some fraction of that energy ultimately leaves the Earth. Thus, there is always an output energy flow, even when the system boundary is drawn around the whole Earth and atmosphere. All objects also absorb some portion of the radiant energy they receive (such as solar energy), so there is always an energy input.

In short, matter flow systems may be open (have inputs and outputs) or closed (lack inputs and outputs). Closed systems are referred to as cycles. Global matter flow systems, such as the carbon cycle, are always closed. Energy flow systems are always open.

Feedback and Equilibrium in Flow Systems

In this section we touch on two other important concepts regarding flow systems. One is **feedback** in a flow system, which occurs when the flow in one pathway acts either to reduce or to increase the flow in another pathway. Feedback is probably most familiar to us as the squealing noise made by a public address system when the volume is turned up too high. In this case, a voice is picked up by a microphone, and the sound is amplified and transmitted by loudspeakers. If the sound from the speakers is too great, it reaches the microphone and becomes reamplified. The process instantly repeats itself, making the sound even louder and producing an unpleasant squeal. This is a *positive feedback* because it reinforces the flow of sound energy in the pathway between the microphone and the speakers.

An example of *negative feedback* is provided by a thermostat that controls a home heating system. When the heating system is on, the room warms. Eventually, the warming trips a temperature-sensitive switch that turns the heating system off. Thus, the heat flow from the furnace provides a negative feedback that reduces its amount.

The second concept is **equilibrium.** By this term, we mean a steady state in which the flow rates in the various pathways of a system remain about the same. A lake within a closed basin in an arid climate—like the Great Salt Lake, Utah—is a simple example of an equilibrium system (Figure 2.8, part *a*). This type of lake has no stream outlets and would dry up completely if not fed by streams that arise in nearby high mountains. Water enters the lake from rivers and streams that feed it, while water leaves the lake by evaporation. Note that the amount of evaporation depends mainly on the surface area of the lake—the larger the surface area, the more evaporation occurs. Of course, local climate is a factor, too, in that air temperature and humidity will also affect the rate of evaporation from the lake's surface.

Suppose that the climate becomes a bit wetter and the input of rivers that feed the lake increases (part *b*). The water level then rises, and the area of the lake expands. Because of the greater area, evaporation is greater. Eventually, the level rises to the point where the increased evaporation rate equals the increased inflow rate. That is, the level reaches an equilibrium. If the climate changes again and the input is reduced, the lake's level will fall, surface area will decrease, and evaporation will decrease. Eventually, the lake will move to a new, and lower, equilibrium level. You will also recognize the coupling between input, surface area, and evaporation as a negative feedback. Systems that come to an equilibrium are normally stabilized by negative feedback loops or pathways.

An example of feedback and equilibrium occurs in the global climate system. This system has numerous pathways of energy flow that include several important negative feedbacks. Thus, the global climate system tends to an equilibrium in which global surface temperature fluctuates around a mean. Climate researchers have recently agreed that human activity has modified the global climate system, and they predict that global temperatures will soon be rising in response.

The effect of clouds on the global climate system points out the role of negative and positive feedback on an equilibrium system. What is the effect of clouds on surface temperature? Low clouds are large white bodies that reflect sunlight back to space much more efficiently than dark land or ocean surfaces below (Figure 2.9). They provide an energy flow pathway in which a portion of the solar energy flow is turned backward and redirected toward space. This pathway tends to cool the surface, and so it acts as a negative feedback. High clouds are different, however. They tend to absorb the outgoing flow of heat from the Earth to space and redirect it earthward. Thus, high clouds provide a positive feedback that warms the surface.

Now consider a small increase in the global surface temperature. If the Earth is warmer, then more water will evaporate from the oceans and moist land surfaces, and so more clouds will form. Will this increase in cloud cover cause surface cooling or warming? The best calculations now suggest that if surface temperatures increase, more high clouds than low clouds will form. Thus, the effect will be a positive feedback that will tend to make the surface even warmer. The effect of clouds on climate is complex, and we will return to this topic in Chapter 6.

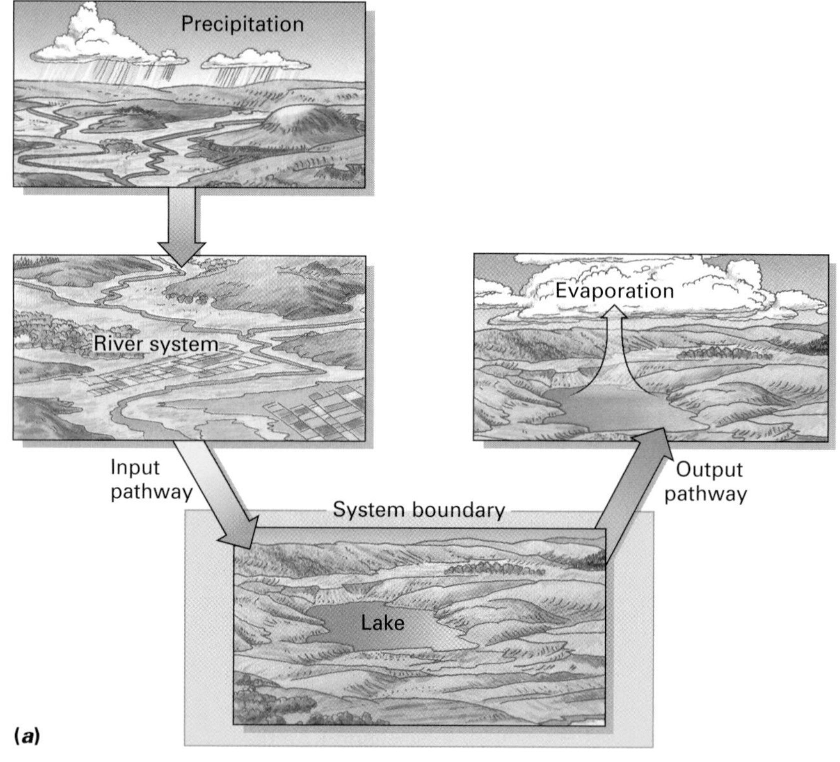

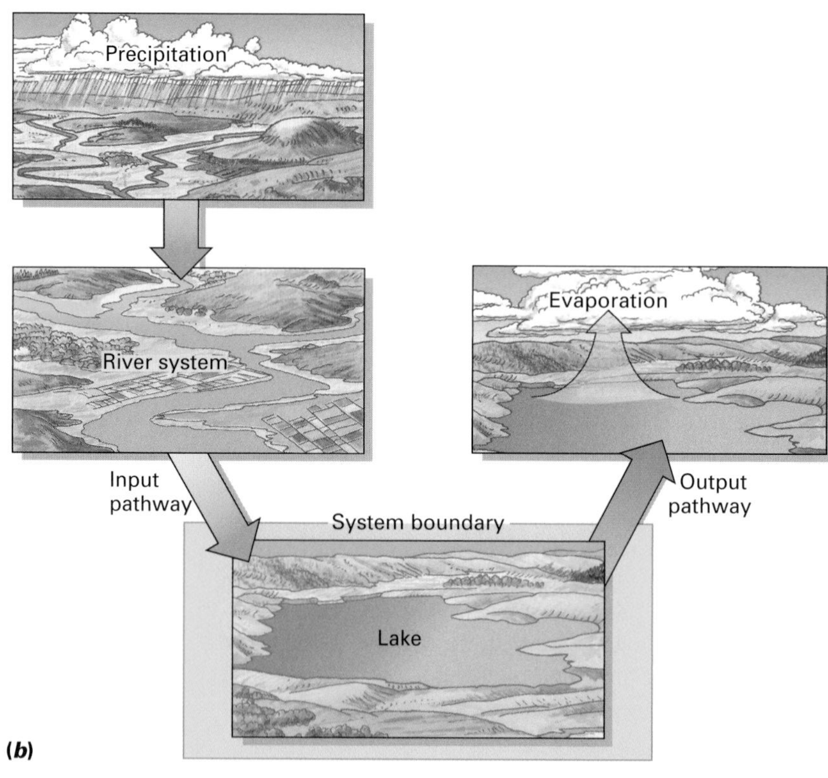

2.8 Systems equilibrium demonstrated by a lake in a closed basin *(a) A lake in a closed basin as a matter flow system in equilibrium. In (b), precipitation increases, the lake level rises, and because the surface area of the lake increases, evaporation also increases to balance the greater input.*

2.9 Eye on the Landscape* *Cloud cover reflecting sunlight *Low white clouds, such as these over the Hawaiian islands of Molokai, Lanai, and Maui, reflect much more of the Sun's energy back to space than the darker surfaces of land and water below them.* **What else would the geographer see?...Answers at the end of the chapter.**

Time Cycles

Any system, whether open or closed, can undergo a change in the rates of flow of energy or matter within its pathways. Flow rates may grow faster and faster or may slow down. These changes in activity can be reversed at intervals of time—that is, a rate can alternately speed up and slow down—in what we call a **time cycle.**

In many natural systems, there is a rhythm of increasing and decreasing flow. The annual revolution of the Earth around the Sun generates a time cycle of energy flow in many natural systems. We speak of this cycle as the rhythm of the seasons. The rotation of the Earth on its axis sets up the night-and-day cycle of darkness and light. The moon, in its monthly orbit around the Earth sets up its own time cycle. We see the lunar cycle in the range of tides, with higher high tides and lower low tides ("spring tides") occurring both at full moon and at new moon.

The astronomical time cycles of Earth rotation and solar revolution will appear at several places in our early chapters. Other time cycles with durations of tens to hundreds of thousands of years describe the alternate growth and shrinkage of the great ice sheets. Still others, with durations of millions of years, describe cycles of the solid Earth in which supercontinents form, break apart, and reform anew.

A LOOK AHEAD The past few pages have presented a rather heavy dose of some fairly abstract ideas. But as you will see in further chapters of this book, examining the phenomena of physical geography as natural flow systems of energy and matter will enhance your understanding of the processes that shape the world around us, from the daily changes in weather to the generation of earthquakes in the Earth's crust.

With the next chapter, we begin Part 2 of this book, which focuses on the processes of the atmosphere, leading ultimately to weather and climate. But first we need to examine some basic concepts about the Earth as a rotating planet. That takes us to Chapter 3.

Chapter Summary

- Spheres, scales, systems, and cycles are four overarching themes that appear in physical geography. The four great Earth realms are **atmosphere**, **hydrosphere**, **lithosphere**, and **biosphere**. The **life layer** is the shallow surface layer where lands and oceans meet the atmosphere and where most forms of life are found. The systems of interaction between the realms can be examined at different **scales**, including *global*, *continental*, *regional*, *local*, and *individual.*
- A helpful way to understand the relationships among the processes that occur in the life layer is to study them as **flow systems**. We can distinguish matter flow systems and energy flow systems in which we are concerned with flows of matter and energy, respectively. Flow systems are composed of **pathways** of energy and/or matter flow that are interconnected in a **structure**. All flow systems have a **power source**.
- **Open flow systems** have **inputs** and **outputs**, while **closed flow systems** do not. Closed matter flow systems are also called **cycles**. In a cycle, materials move in an endless series of interconnected pathways or loops. The hydrologic cycle and the carbon cycle are examples of cycles encountered in

physical geography. Although *matter flow systems* may be open or closed, *energy flow systems* are always open.

- **Feedback** in a flow system occurs when the flow in one pathway acts either to reduce or to increase the flow in another pathway. *Positive feedback* enhances or increases the flow within a pathway, while negative feedback reduces it. A system with *negative feedback* loops or pathways tends to be self-stabilizing and move to an **equilibrium**—a steady state in which flow rates in system pathways remain about the same. The role of clouds in the global climate system provides an example of feedback and equilibrium.
- Systems may undergo periodic, repeating changes in flow rates that constitute time cycles. Important **time cycles** in physical geography range in length from hours to millions of years.

Key Terms

atmosphere
lithosphere
hydrosphere
biosphere
life layer
scale
system
flow system
pathway
structure (of a system)
input
output
power source
open flow system
closed flow system
cycle
material cycle
feedback
equilibrium
time cycle

Review Questions

1. Identify four overarching ideas in physical geography and briefly explain each.
2. Name and describe each of the four great physical realms of Earth. What is the life layer?
3. Provide two examples of processes or systems that operate at each of the following scales: global, continental, regional, local, and individual.
4. What is a flow system? Provide a simple example. Distinguish two types of flow systems.
5. Identify the key components of a flow system.
6. What distinguishes an open flow system from a closed flow system?
7. What is a cycle (material cycle)? Identify one or more cycles that are studied in physical geography.
8. Describe the concepts of feedback and equilibrium as applied to systems. Provide an example drawn from a natural system.
9. What is a time cycle as applied to a system? Give an example of a time cycle evident in natural systems.

Focus on Systems 2.1 The Value of Systems Thinking

1. What is systems thinking?
2. Why is systems thinking important in studying natural processes?

Essay Question

1. Select a flow system with which you are familiar. Identify its components and describe any negative or positive feedback pathways or loops it may contain. Does the system tend to an equilibrium? Possible examples: the system of electrical wiring or plumbing in a house or apartment; a bodily system, such as the respiratory system; a small ecosystem, such as an aquarium. Imagine some changes to the structure of the system, and describe how they might affect it.

Eye on the Landscape

Chapter Opener **Coast of Brittany** The tide is out, shown by the boats lying high and dry **(A)**. You can also notice this from the waves breaking on exposed sand beds, far from the shore **(A)**. The large expanse of tidal sediment indicates a large tide range—in fact, it is about 8 m (26 ft) here. The sandy bluff **(B)** is being cut away by the erosion of the adjacent tidal channel, threatening the dwelling on the vegetated dune above. See Chapter 19 for more on tides, as well as bluff and coastal erosion. The pattern of tidal channels at **(C)** is called a braided channel network, and is characteristic of rivers carrying heavy sediment loads. We provide more detail on stream channels in Chapter 17.

2.3e **Individual-scale landscape** Did you ever notice that the best farmlands are next to rivers? That's because larger rivers meander back and forth in their floodplains over long periods of time, leaving flat terraces **(A)** of fertile soil that are ideal for cultivation. We describe this process further in Chapter 17. Another feature of most rivers is a bank covered with vegetation **(B)**. Often these are species that can survive occasional immersion and mechanical damage by floods. See Chapter 23 for more on how species are adjusted to their environment.

2.9 **Cloud cover reflecting sunlight** These puffy, white cumulus clouds **(A)** indicate warm, moist air, lifted above its condensation level as it passes over the island's topography. Orographic precipitation is a subject of Chapter 6.

Chapters in Part 2

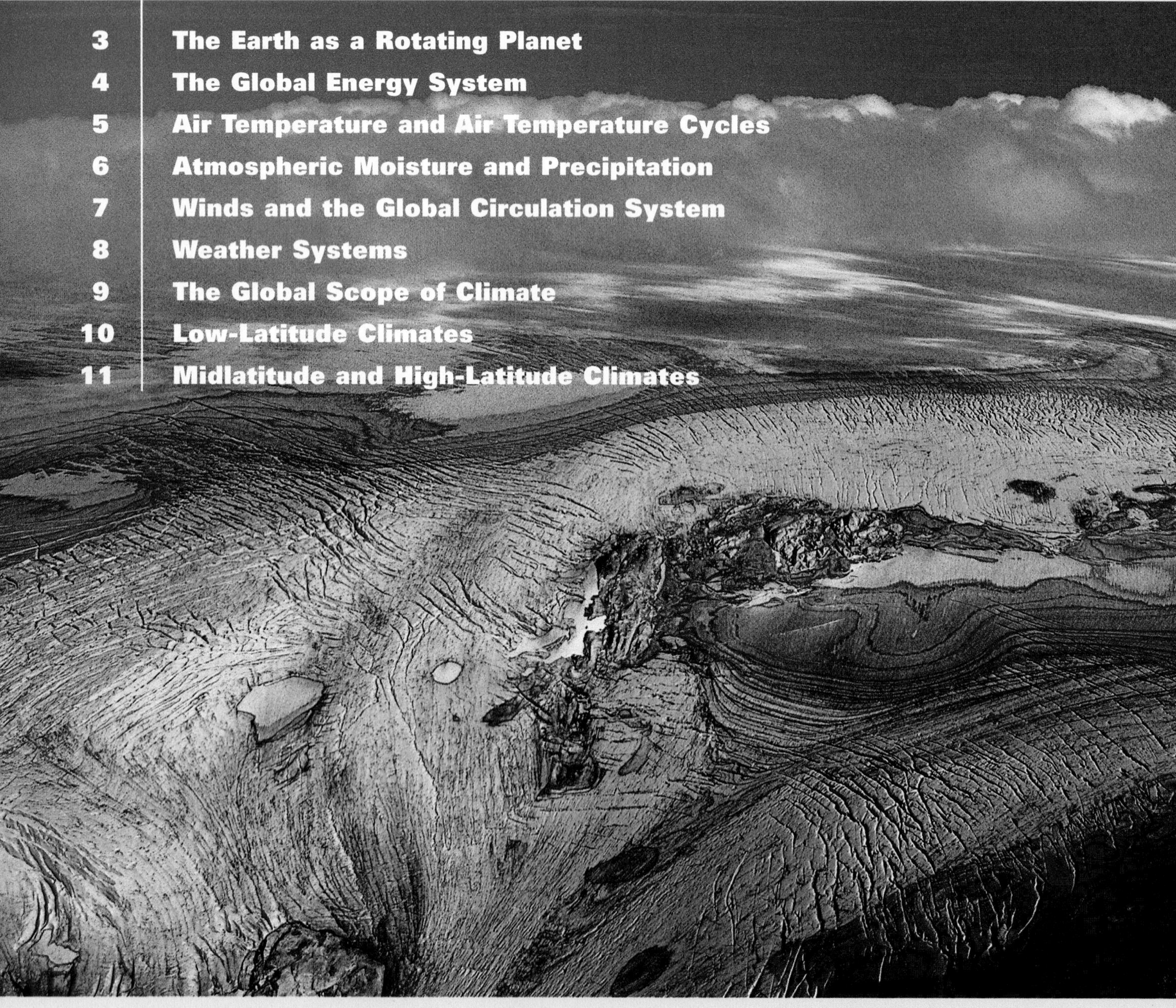

Part 2 | Weather and Climate Systems

Folgefonni Glacier on the high plateaus of Sorfjorden, Norway In the cold but wet climate of Norway, thousands of years of snows have created this plateau glacier. Cracked and fissured, the ice flows around rock masses that color the ice with sediment and rock debris.

The flow of energy from the Sun to the Earth powers a vast and complex system of energy and matter flows within the atmosphere, oceans, and at the land surface. In this part, we explore how these flows are linked to weather and to climate. In Chapter 3 we examine how the Earth's rotation on its axis and its revolution around the Sun induce the daily and seasonal rhythms in energy flow that mark the passage of time on the human scale. The pathways of energy flow are complex, but their main features can be easily understood using simple physical principles. We present these principles and the broad picture of global energy flows in Chapter 4. ■ Flows of energy into and out of the layer of air close to the surface control a characteristic of the weather that is very important in our daily lives—air temperature—which we treat in Chapter 5. An important pathway for solar energy flow lies in the evaporation of water from the ocean or from moist land surfaces and its later release in the form of precipitation, the subject of Chapter 6. ■ Another feature of Earth-Sun relationships is that some parts of the Earth are heated more intensely that others, such as the equatorial regions. This unequal heating, coupled with the Earth's eastward rotation, produces the pattern of global wind circulation that is discussed in Chapter 7. The global pattern of wind circulation, in turn, when coupled with the global positions of oceans and continents, produces organized weather systems in which warm and cool air masses are in contact. This contact causes fronts and storm systems, which are the subjects of Chapter 8. ■ The last three chapters of Part 2 are devoted to climate—the average cycle of weather experienced at a location through the year. These average weather cycles are grouped in to a set of distinctive climates that cover the Earth's land surface. Chapter 9 provides the basic principles for climate classification, while Chapters 10 and 11 cover specific climates from the equator to the poles. ■ As you study the chapters of Part 2, keep in mind that nearly all the phenomena we describe are connected in some way to the grand, global flow system of energy and matter at the Earth's surface that is powered by the Sun.

3 THE EARTH AS A ROTATING PLANET

EYE ON THE LANDSCAPE

An illuminated greenhouse in southwestern Finland. Illuminating the greenhouse extends the day length, thus shortening the long polar night for the plants inside. **What else would the geographer see?...Answers at the end of the chapter.**

Since the dawn of civilization, humans have observed the movements of the Sun in the sky. However, only since the time of Copernicus—the sixteenth century—have we known that the daily and seasonal motions of the Sun are a consequence of the motion of the Earth rather than the Sun. From day to day, the Earth's rotation on its axis produces the alternation of the light and darkness that we experience in 24 hours. From month to month, the revolution of the Earth around the Sun produces slow changes in the length of daylight that create the rhythm of the seasons. These two great cycles continue, day after day, year after year, regulating the processes of Earthly life. We begin our systematic study of physical geography by devoting a chapter to the motions of the Earth and Sun and the implications of these motions for Earth location, timekeeping, and the seasons. GEODISCOVERIES

THE SHAPE OF THE EARTH

The Earth's shape is very close to that of a sphere, as we all learn early in school (Figure 3.1). Pictures taken from space by astronauts and by orbiting satellites also show us that the Earth is a round body. We learn this fact so early in life that it seems quite unremarkable.

Many of our ancestors, however, were not aware of our planet's spherical shape. To sailors of the Mediterranean Sea in ancient times, the shape and breadth of the Earth's oceans and lands were unknown. On their ships and out of sight of land, the sea surface looked perfectly flat and bounded by a circular horizon. Given this view, many sailors concluded that the Earth had the form of a flat disk and that their ships would fall off if they traveled to its edge. Of course, had the telescope been invented, they

3.1 Our Spherical Earth

a...***Photo of Earth curvature*** *This astronaut photo shows the Earth's curved horizon from low-Earth orbit.*

b...***Eye on the Landscape*** ***Space view of Earth*** *From a geostationary weather satellite, the Earth is a disk obscured by swirls of moving clouds.* **What else would the geographer see?...Answers at the end of the chapter.**

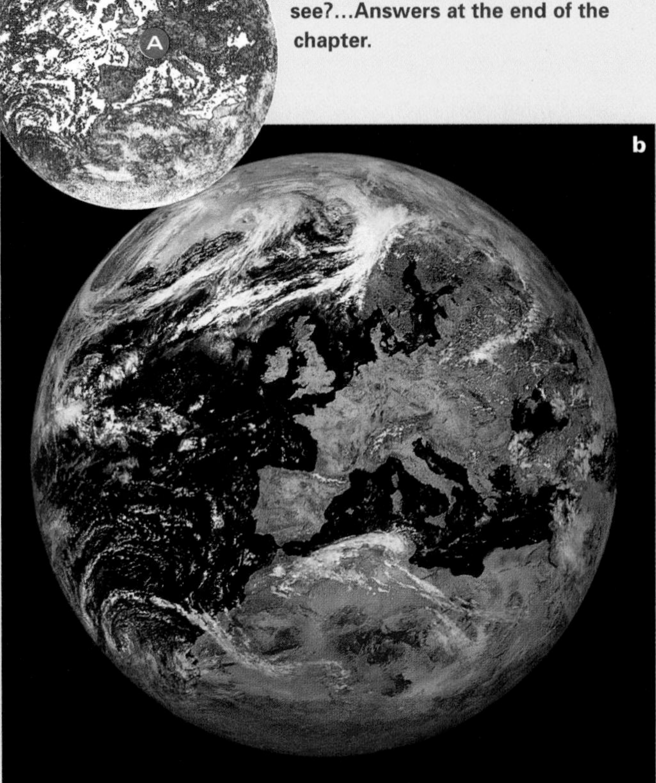

c...***Distant Ship*** *Seen through a telescope, the decks of a distant ship seem to be under water.*

d...***Sunset photo*** *In this dramatic sunset photo, the far distant clouds are still directly illuminated by the Sun's last red rays. For the clouds directly overhead, however, the Sun has left the sky.*

could have seen a distant ship with its decks seemingly below the water line—a demonstration of the curvature of the Earth's surface.

The curvature of the Earth is also evident as you view a sunset with clouds in the sky (Figure 3.1*d*). Although the Sun is below the horizon from your viewpoint and no longer illuminates the land around you, it is above the horizon at the altitude of the clouds and still bathes them in its red and pinkish rays. As the Sun progressively washes the clouds overhead in red light, the red band of illumination slowly moves farther and farther toward your horizon. Soon only the far distant clouds can still "see" the Sun. This movement of solar illumination across the clouds is easily explained by a rotating spherical Earth.

Actually, the Earth is not perfectly spherical. Because the outward force of the Earth's rotation causes the Earth to bulge slightly at the equator and flatten at the poles, the Earth assumes the shape of an *oblate ellipsoid.* ("Oblate" means flattened.) Thus, the Earth's equatorial diameter, about 12,756 km (7926 mi), is very slightly larger than the polar diameter, about 12,714 km (7900 mi). The difference is very small—about three-tenths of 1 percent.

Scientists still study the Earth's shape and attempt to measure it as precisely as possible. Satellite navigation systems need this important information for aircraft, ocean vessels, and ground vehicles seeking to determine their exact location (see *Remote Sensing 3.2 • The Global Positioning System,* later in this chapter).

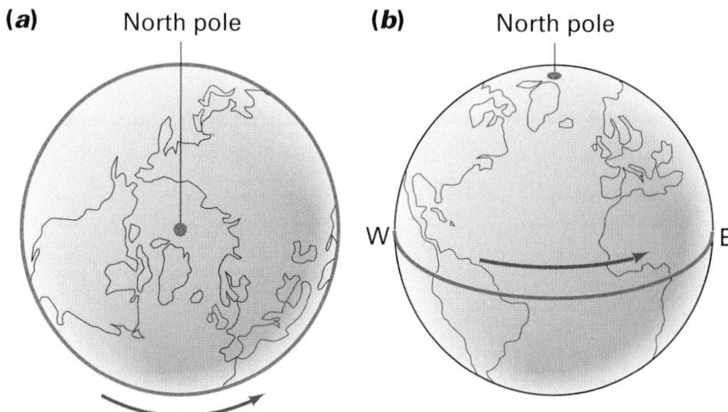

3.2 Direction of Earth rotation *The direction of rotation of the Earth can be thought of as (a) counterclockwise at the north pole, or (b) from left to right (eastward) at the equator.*

Earth Rotation

Another fact about our planet that we learn early in life is that it spins slowly, making a full turn with respect to the Sun every day. We use the term **rotation** to describe this motion. One complete rotation with respect to the Sun defines the *solar day.* By convention, the solar day is divided into exactly 24 hours.

The Earth rotates on its *axis,* an imaginary straight line through its center. The two points where the axis of rotation intersects the Earth's surface are defined as the poles. To distinguish between the two poles, one is called the *north pole* and the other, the *south pole.*

Direction of Rotation

To determine the direction of Earth rotation, you can use one of the following guidelines (Figure 3.2):

- Imagine yourself looking down on the north pole of the Earth. From this position, the Earth is turning in a counterclockwise direction (Figure 3.2*a*).
- Imagine yourself off in space, viewing the planet much as you would view a globe in a library, with the north pole on top. The Earth is rotating from left to right or in an eastward direction (Figure 3.2*b*).

The Earth's rotation is important for three reasons. First, the axis of rotation serves as a reference in setting up the geographic grid of latitude and longitude, which we will discuss later in the chapter. Second, it provides the day as a convenient measure of the passage of time, with the day in turn divided into hours, minutes, and seconds. Third, it has important effects on the physical and life processes on Earth.

Environmental Effects of Earth Rotation

The first—and perhaps most obvious—effect of the Earth's rotation is that it imposes a daily, or *diurnal,* rhythm in daylight, air temperature, air humidity, and air motion.

All surface life responds to this diurnal rhythm. Green plants receive and store solar energy during the day and consume some of it at night. Among animals, some are active during the day, others at night. The daily cycle of incoming solar energy and the corresponding cycle of fluctuating air temperature are topics for analysis in Chapters 4 and 5.

A second environmental effect is that the flow paths of both air and water are turned consistently in a sideward direction because of the Earth's rotation. Flows in the northern hemisphere are turned toward the right and in the southern hemisphere toward the left. This phenomenon is called the *Coriolis effect.* It is of great importance in studying the Earth's systems of winds and ocean currents and is discussed in Chapter 7.

A third physical effect of the Earth's rotation is the movement of the tides. The Moon exerts a gravitational attraction on the Earth, while at the same time the Earth turns with respect to the Moon. These forces induce a rhythmic rise and fall of the ocean surface known as the *tide.* The tide in turn causes water currents to flow in alternating directions in the shallow salty waters of the coastal zone. This ebb and flow of tidal currents is a life-giving pulse for many plants and animals that live in coastal saltwater environments. The tidal cycle is a clock regulating many daily human activities in the coastal zone as well. When we examine the tide and its currents further in Chapter 19, we will see that the Sun also has an influence on the tides.

THE GEOGRAPHIC GRID

The geographic grid provides a system for locating places on the Earth's surface. Because the Earth's surface is curved, and not flat, we cannot divide it into a rectangular grid, like a sheet of graph paper. Instead, we divide it using imaginary circles set on the surface that are perpendicular to the axis of rotation in one direction and parallel to the axis of rotation in the other direction.

Parallels and Meridians

Imagine a point on the Earth's surface. As the Earth rotates, the point traces out a path in space, following an *arc*—that is, a curved line that forms a portion of a circle. With the completion of one rotation, the arc forms a full circle. This circle is known as a parallel of latitude, or a **parallel** (Figure 3.3*a*). Imagine cutting the globe much as you might slice an onion to produce onion rings—that is, perpendicular to the onion's main axis. Each cut creates a circle or parallel crosswise through the globe. The Earth's longest parallel of latitude lies midway between the two poles and is designated the **equator.** The equator is a fundamental reference line for measuring the position of points on the globe.

Imagine now slicing the Earth with a plane that passes through the axis of rotation instead of across it. This is the way you might cut up a lemon to produce wedges. Each cut outlines a circle on the globe passing through both poles. Half of this circular outline, connecting one pole to the other, is known as a meridian of longitude, or, more simply, a **meridian** (Figure 3.3*b*).

Meridians and parallels define geographic directions. Meridians are north-south lines, so if you walk north or south you are following a meridian. Parallels are east-west lines, and so if you walk east or west you are following a parallel. Any number of parallels and meridians are possible. Every point on the globe is associated with a unique combination of one parallel and one meridian. The position of the point is defined by their intersection. The total system of parallels and meridians forms a network of intersecting circles called the **geographic grid.**

Looking more closely at meridians and parallels, we see that they are made up of two types of circles—great and small (Figure 3.4). A **great circle** is created when a plane passing through the center of the Earth intersects the Earth's surface. A great circle always has the Earth's center as its own center, and it bisects the globe into two equal halves. A **small circle** is created when a plane passing through the Earth, but not through the Earth's center, intersects the Earth's surface. By looking at Figure 3.3, you can easily see that meridians are actually halves of great circles, while all parallels, except the equator, are small circles. Great circles can be aligned in any direction on the globe. This means that if we choose two points on the globe, we can always find a great circle that passes through both points. As we will see shortly in our discussion of map projections, the portion of the great circle between the two points is the shortest distance between them.

3.3 Parallels and meridians *(a) Parallels of latitude divide the globe crosswise into rings. (b) Meridians of longitude divide the globe from pole to pole.*

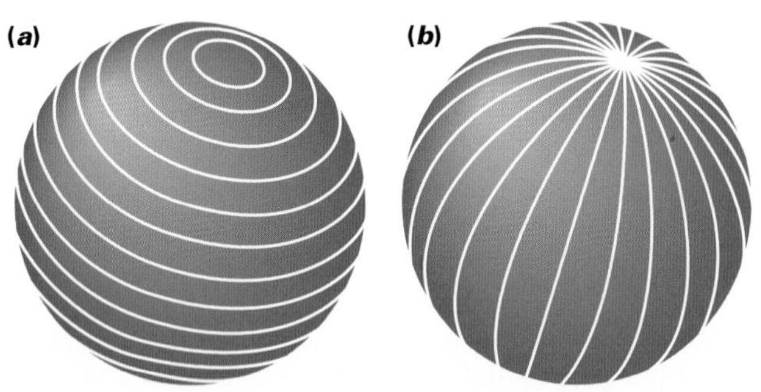

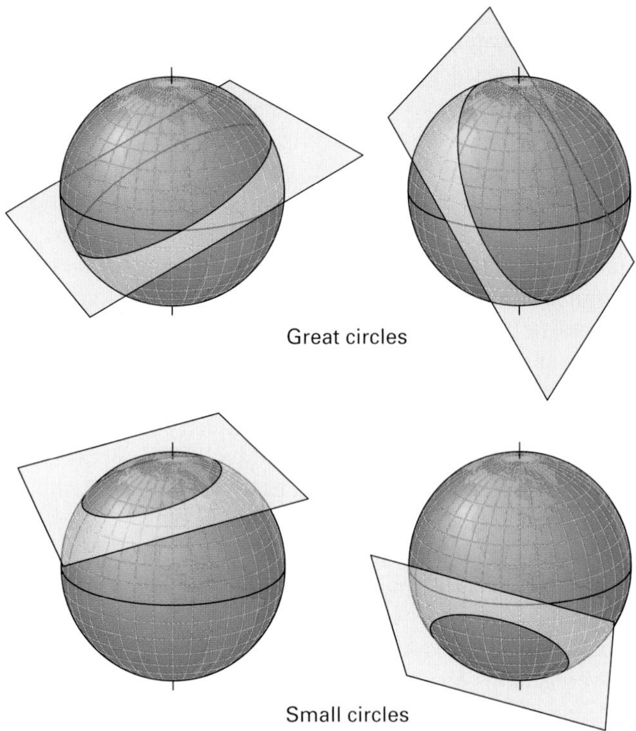

3.4 Great and small circles *A great circle is created when a plane passes through the Earth, intersecting the Earth's center. Small circles are created when a plane passes through the Earth but does not intersect the center point.*

Latitude and Longitude

To label parallels and meridians, we use a special system—latitude and longitude. Parallels are identified by latitude and meridians by longitude. **Latitude** is an indicator of how far north or south of the equator a parallel is situated. The latitude of a parallel is measured by the angle between a point on the parallel, the center of the Earth, and a point on the equator intersected by a meridian passing through the point of interest on the parallel (Figure 3.5*a*). Note that the equator divides the globe into two equal portions, or *hemispheres.* All parallels north of the equator—that is, in the *northern hemi-*

sphere—are designated as having north latitude, and all points south of the equator—in the *southern hemisphere*—are designated as having south latitude (N or S).

Longitude is a measure of the position of a meridian eastward or westward from a reference meridian, called the *prime meridian.* As Figure 3.5*b* shows, longitude is the angle, measured in degrees, between a plane passing through the meridian and a plane passing through the prime meridian. The prime meridian passes through the location of the old Royal Observatory at Greenwich, near London, England (Figure 3.6). For this reason it is also referred to as the Greenwich meridian. It has the value long. 0°. The longitude of a meridian on the globe is measured eastward or westward from the prime meridian, depending on which direction gives the smaller angle. Longitude thus ranges from 0° to 180°, east or west (E or W).

3.5 Latitude and longitude angles *(a) The latitude of a parallel is the angle between a point on the parallel (P) and a point on the equator at the same meridian (Q) as measured from the Earth's center. (b) The longitude of a meridian is the angle between a point on that meridian at the equator (P) and a point on the prime meridian at the equator (Q) as measured at the Earth's center.*

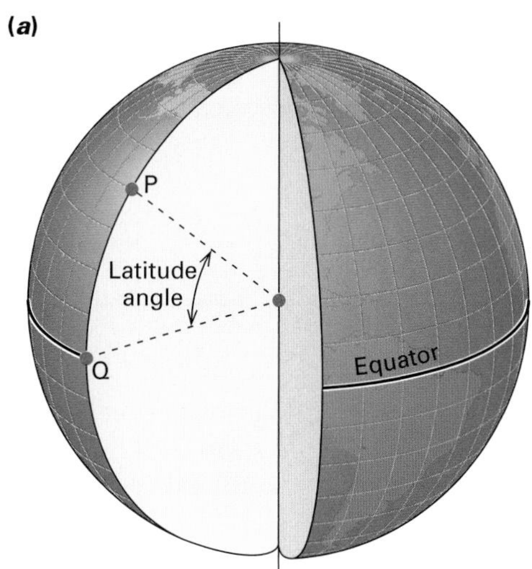

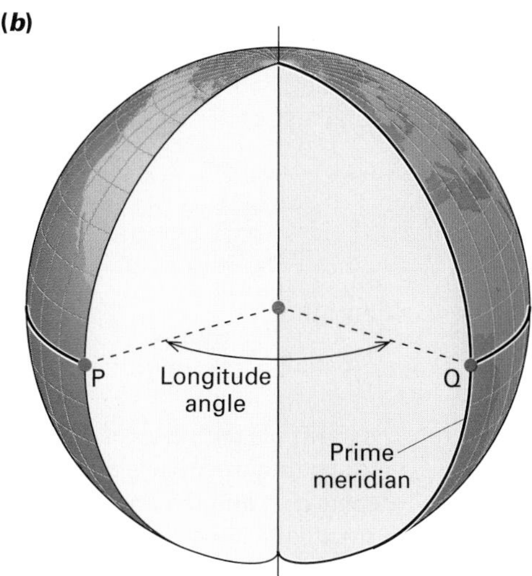

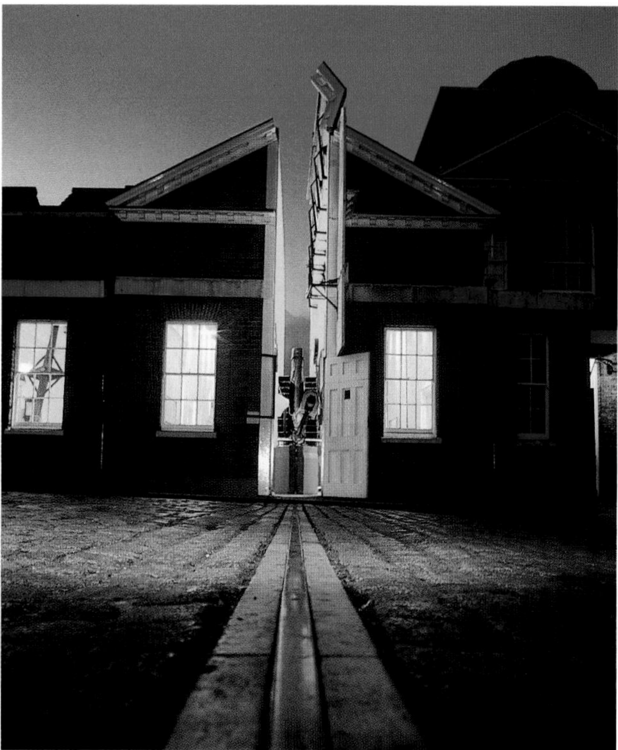

3.6 The prime meridian *This photograph, taken at dusk at the old Royal Observatory at Greenwich, England, shows the prime meridian, which has been marked as a stripe on the forecourt paving.*

Figure 3.7 shows how the location of a point on the Earth's surface is determined by its latitude and longitude. The point *P* lies on the parallel of latitude at 50 degrees north, which we can abbreviate as lat. 50° N. It also lies on the meridian of longitude at 60 degrees west, which we can abbreviate as long. 60° W. When both the latitude and longitude of a point are known, it can be accurately and precisely located on the geographic grid.

When latitude or longitude angles are measured other than in full-degree increments, *minutes* and *seconds* can be used. A minute is 1/60 of a degree, and a second is 1/60 of a minute, or 1/3600 of a degree. Thus, the latitude 41°, 27 minutes (′), and 41 seconds (″) north (lat. 41° 27′ 41″ N) means 41° north plus 27/60 of a degree plus 41/3600 of a degree. This cumbersome system has now largely been replaced by decimal notation. In this example, the latitude 41° 27′ 41″ N translates to 41.4614° N.

Degrees of latitude and longitude can also be used as distance measures. A degree of latitude, which measures distance in a north-south direction, is equal to about 111 km (69 mi). The distance associated with a degree of longitude, however, will vary over the globe because meridians converge toward the poles. *Working It Out 3.1 • Distances from Latitude and Longitude* provides more information on this topic and shows how to convert change in latitude and longitude into distance.

The latitude and longitude of any point can now be determined quickly and accurately with the help of the *Global Positioning System (GPS)*—a system of satellites that con-

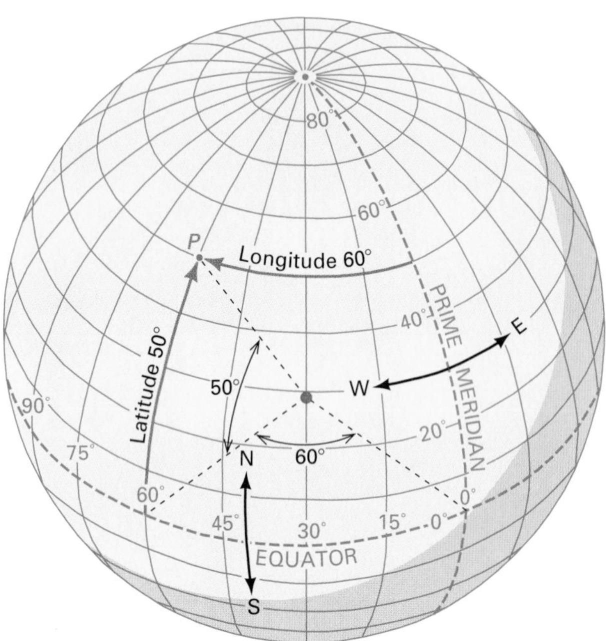

3.7 Latitude and longitude of a point *The point P lies on the parallel of latitude at 50° north (50° from the equator) and on the meridian at 60° west (60° from the prime meridian). Its location is therefore lat. 50° N, long. 60° W. The red dot denotes the Earth's center.*

stantly sends radio signals to Earth with information that allows a GPS receiver to calculate its position on the Earth's surface. *Focus on Remote Sensing 3.2 • The Global Positioning System* provides more information about GPS and how it works.

MAP PROJECTIONS

With a working understanding of the geographic grid, we can consider how to display the locations of continents, rivers, cities, islands, and other geographic features on maps. This will take us briefly into the realm of *cartography,* the art and science of making maps. The discussion in the main text will focus on a few simple types of maps that are used widely and found in our chapters. We also provide an interchapter feature, *A Closer Look: Geographer's Tools 3.5 • Focus on Maps* with more information about how maps are made and how they display information.

As we observed earlier, the Earth's shape is nearly spherical, and so the Earth's surface is curved. However, maps are flat. It is impossible to copy a curved surface onto a flat surface without cutting, stretching, or otherwise distorting the curved surface in some way. So, making a map means devising an orderly way of changing the globe's geographic grid of curved parallels and meridians into a grid that lies flat. We refer to a system for changing the geographic grid to a flat grid as a **map projection.** GEODISCOVERIES

Working It Out | 3.1
Distances from Latitude and Longitude

Statements of latitude and longitude do not describe distances in kilometers or miles directly. However, for latitude, you can estimate conversions from degrees into kilometers quite easily. One degree of latitude is approximately equivalent to 111 km (69 mi) of surface distance in the north-south direction. This value can be rounded off to 110 km (or 70 mi) for multiplying in your head. For example, if you live at lat. 40° N (on the 40th parallel north), you are located about 40 × 110 = 4400 km (40 × 70 = 2800 mi) north of the equator.

East-west distances cannot be converted so easily from degrees of longitude into kilometers or miles because the meridians converge toward the poles (Figure 3.3*b*). Only at the equator is a degree of longitude equivalent to 111 km (69 mi). At lat. 60° N or S, meridians are twice as close as at the equator, so a degree of longitude is reduced to half its equatorial length, or about 56 km (35 mi). As the pole is approached, the length goes to zero.

We can use a simple formula to determine the length of a degree of longitude:

$$L_{LONG} = \cos(lat) \times L_{LAT}$$

where L_{LONG} is the length of a degree of longitude; cos is the trigonometric cosine function, evaluated in degrees; *lat* is the latitude of the location at which the length is to be calculated, in degrees; and L_{LAT} is the length of a degree of latitude, that is, 111 km or 69 mi.

For example, a degree of longitude at 30° lat. has length

$$\begin{aligned} L_{LONG} &= \cos(lat) \times L_{LAT} \\ &= \cos(30°) \times 111 \text{ km} \\ &= (0.866) \times 111 \text{ km} \\ &= 96.1 \text{ km} \end{aligned}$$

Here's another: suppose we want to find the distance between two cities located on the 35th parallel that are separated by 6° of longitude. First, we determine the length of a degree of longitude at 35° lat. That is,

$$\begin{aligned} L_{LONG} &= \cos(lat) \times L_{LAT} \\ &= \cos(35°) \times 111 \text{ km} \\ &= (0.819) \times 111 \text{ km} \\ &= 90.9 \text{ km} \end{aligned}$$

Then,

$$6° \text{ long} \times \frac{90.9 \text{ km}}{1° \text{ long}} = 546 \text{ km}$$

These simple conversions help demonstrate the nature of the geographic grid and the convergence of meridians toward the poles.

Focus on Remote Sensing | 3.2

The Global Positioning System

The latitude and longitude coordinates of a point on the Earth's surface describe its position exactly. But how are those coordinates determined? For the last few hundred years, we have known how to use the position of the stars in the sky coupled with an accurate clock to determine the latitude and longitude of any point. Linked with advances in mapping and surveying, these techniques became highly accurate, but they were impractical for precisely determining desired locations in a short period of time.

Thanks to new technology originally developed by the U.S. Naval Observatory for military applications, there is now in place a *global positioning system (GPS)* that can provide location information to an accuracy of about 20 meters within a minute or two. The system uses 24 satellites that orbit the Earth every 12 hours, continuously broadcasting their position and a highly accurate time signal.

GEODISCOVERIES

To determine location, a receiver listens simultaneously to signals from four or more satellites. The receiver compares the time readings transmitted by each satellite with the receiver's own clock to determine how long it took for each signal to reach the receiver. Since the radio signal travels at a known rate of speed, the receiver can convert the travel time into the distance between the receiver and the satellite. Coupling the distance to each satellite with the position of the satellite in its orbit at the time of the transmission, the receiver calculates its position on the ground to within about 20 m (66 ft) horizontally and 30 m (98 ft) vertically.

The accuracy of the location is affected by several types of errors. First are small perturbations in the orbits of the satellites, caused by such unpredictable events as solar particle showers. These cause errors in the information about satellite position. Another source of error is small variations in the atomic clock that each satellite carries. A larger source of error, however, is the effect of the atmosphere on the radio waves of the satellite signal as they pass from the satellite to the receiver. The layer of charged particles at the outer edge of the atmosphere (ionosphere) and water vapor in the lowest atmospheric layer (troposphere) act to slow the radio waves. Since the conditions in these layers can change within a matter of minutes, the speed of the radio waves varies in an unpredictable way. Another transmission problem is that the radio waves may bounce off local obstructions and then reach the receiver, causing two slightly different signals to arrive at the receiver at the same time. This "multipath error" creates noise that confuses the receiver.

There is a way, however, to determine location within about 1 m (3.3 ft) horizontally and 2 m (6.6 ft) vertically. The method uses two GPS units, one at a base station and one that is mobile and used to determine the desired locations. The base station unit is placed at a position that is known with very high accuracy. By comparing its exact position with that calculated from each satellite signal, it determines the small deviations from orbit of each satellite, any small variations in each satellite's clock, and the exact speed of that satellite's radio signal through the atmosphere at that moment. It then broadcasts that information to the GPS field unit, where it is used to calculate the position more accurately. Because this method compares two sets of signals, it is known as *differential GPS.*

Differential GPS is now in wide use for coastal navigation, where a few meters in position can make the difference between a shipping channel and a shoal. It is also required for the new generation of aircraft landing systems that will allow much safer instrument landings with equipment that is much lower in cost than existing systems.

As GPS technology has developed, costs have fallen exponentially. It is now possible to buy a small, hand-held GPS receiver for less than $100. Worldwide sales of GPS products and services reached the $15 billion mark in 2003 and continue to rise rapidly. Besides plotting your progress on a computer-generated map as you drive your car or sail your boat, GPS technology can even keep track of your children at a theme park. With the coupling of cellular telephones and GPS, you may never have the chance to claim you're at a client meeting when the readout shows your employer that you're at the movies!

A GPS satellite as it might look in orbit high above the Earth.

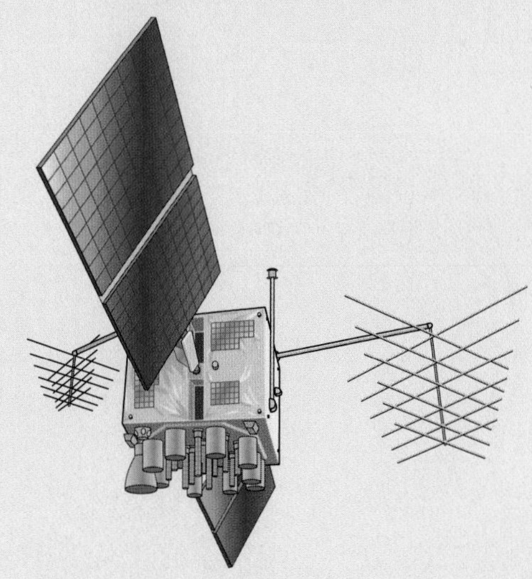

Associated with every map is a *scale fraction*—a ratio that relates distance on the map to distance on the Earth's surface. For example, a scale fraction of 1:50,000 means that one unit of map distance equals 50,000 units of distance on the Earth.

Because a curved surface cannot be projected onto a flat surface without some distortion, the scale fraction of a map holds only for one point or a single line on the map. Away from that point or line, the scale fraction will be different. However, this variation in scale is a problem only for maps that show large regions, such as continents or hemispheres.

We will concentrate on the three most useful map projections. The first is the polar projection, which pictures the globe from top or bottom and is essential for scientific uses like weather maps of the polar regions. Second is the Mercator projection, a navigator's map invented in 1569 by Gerhardus Mercator. It is a classic that has never gone out of style. Third is the Goode projection, named for its designer, Dr. J. Paul Goode. It has special qualities not found in the other two projections.

Polar Projection

The *polar projection* (Figure 3.8) can be centered on either the north or the south pole. Meridians are straight lines radiating outward from the pole, and parallels are nested circles centered on the pole. Spacing of the parallels increases outward from the center. The map is usually cut off to show only one hemisphere so that the equator forms the outer edge of the map. Because the intersections of the parallels with the meridians always form true right angles, this projection shows the true shapes of all small areas. That is, the shape of a small island would always be shown correctly, no matter where it appeared on the map. However, because the scale fraction increases in an outward direction, the island would look larger toward the edge of the map than near the center.

3.8 A polar projection *The map is centered on the north pole. All meridians are straight lines radiating from the center point, and all parallels are concentric circles. The scale fraction increases in an outward direction, making shapes toward the edges of the map appear larger.*

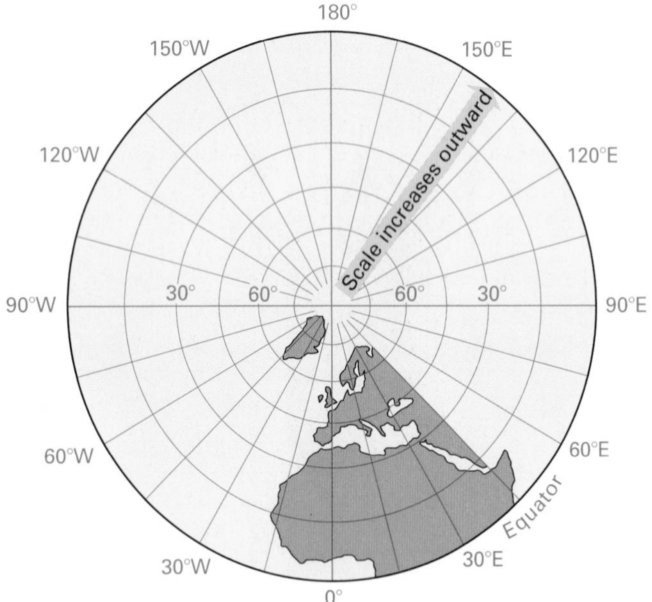

Mercator Projection

The *Mercator projection* (Figure 3.9) is a rectangular grid of meridians as straight vertical lines and parallels as straight horizontal lines. Meridians are evenly spaced, but the spacing between parallels increases at higher latitude so that the spacing at 60° is double that at the equator. Closer to the poles, the spacing increases even more, and the map must be cut off at some arbitrary parallel, such as 80° N. This change of scale enlarges features when they near the pole, as can easily be seen in Figure 3.9. There Greenland appears larger than Australia and is nearly the size of Africa! In fact, Greenland is very much smaller, as you can see on a globe.

The Mercator projection has several special properties. One is that a straight line drawn anywhere on the map is a line of constant compass direction. A navigator can therefore simply draw a line between any two points on the map and measure the *bearing,* or direction angle of the line, with respect to a nearby meridian on the map. Since the meridian is a true north-south line, the angle will give the compass bearing to be followed. Once aimed in that compass direction, a ship or an airplane can be held to the same compass bearing to reach the final point or destination.

This line will not necessarily follow the shortest actual distance between two points. As we noted earlier, the shortest path between two points on the globe is always a portion of a great circle. On a Mercator projection, a great circle line curves (except on the equator) and can falsely seem to represent a much longer distance than a compass line.

Because the Mercator projection shows the true compass direction of any straight line on the map, it is used to show many types of straight-line features. Among these features are flow lines of winds and ocean currents, directions of crustal features (such as chains of volcanoes), and lines of equal values, such as lines of equal air temperature or equal air pressure. This explains why the Mercator projection is chosen for maps of temperatures, winds, and pressures.

Goode Projection

The *Goode projection* (Figure 3.10) uses two sets of mathematical curves (sine curves and ellipses) to form its meridians. Between the 40th parallels, sine curves are used, and beyond the 40th parallel, toward the poles, ellipses are used. Since the ellipses converge to meet at the pole, the entire globe can be shown. The straight, horizontal parallels make it easy to scan across the map at any given level to compare regions most likely to be similar in climate.

The Goode projection has one very important property—it indicates the true sizes of areas of the Earth's surface. That is, if we drew a small circle on a sheet of clear plastic and

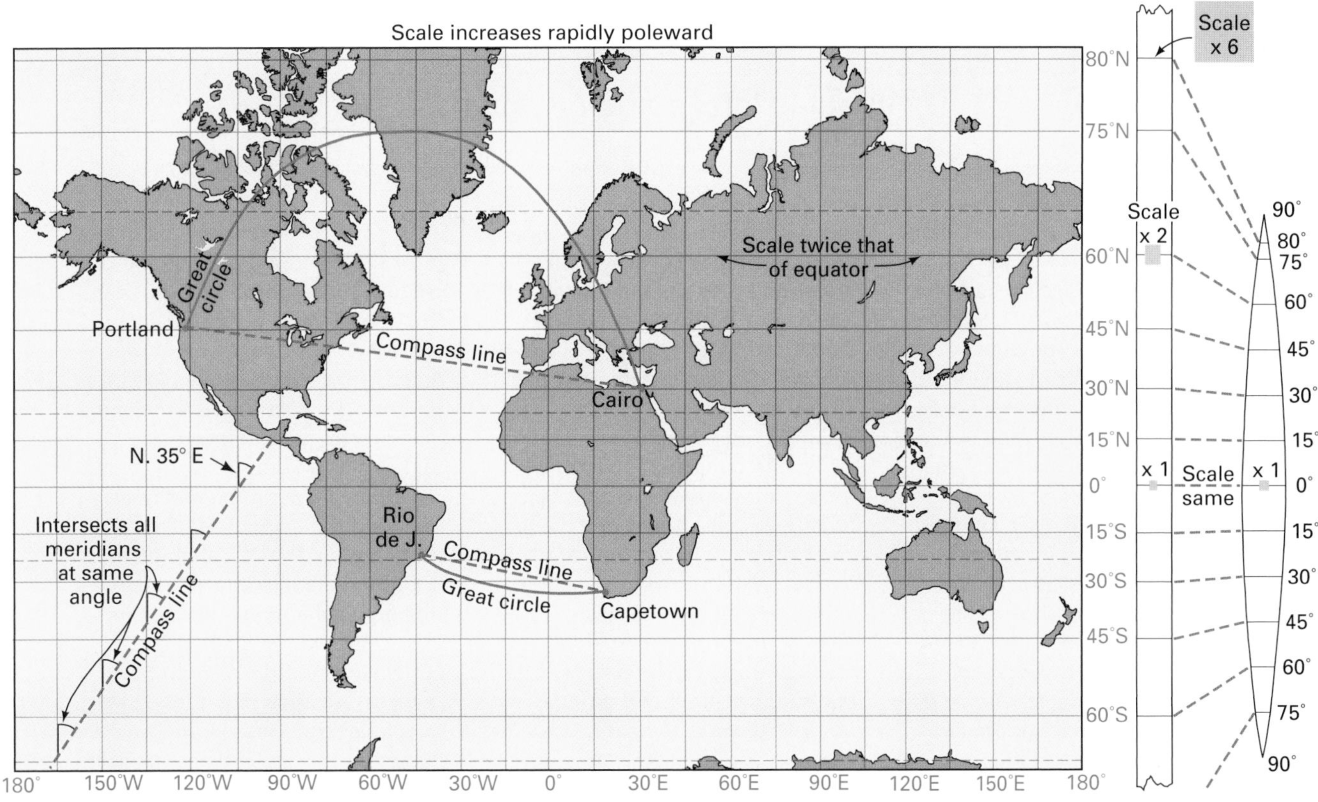

3.9 The Mercator projection *The compass line connecting two locations, such as Portland and Cairo, shows the compass bearing of a course directly connecting them. However, the shortest distance between them lies on a great circle, which is a longer, curving line on this map projection. The diagram at the right shows how rapidly the map scale increases at higher latitudes. At lat. 60°, the scale is double the equatorial scale. At lat. 80°, the scale is six times greater than at the equator.*

3.10 The Goode projection *The meridians in this projection follow sine curves between lat. 40° N and lat. 40° S, and ellipses between lat. 40° and the poles. Although the shapes of continents are distorted, their relative surface areas are preserved. (Copyright © by the University of Chicago. Used by permission of the Committee on Geographical Studies, University of Chicago.)*

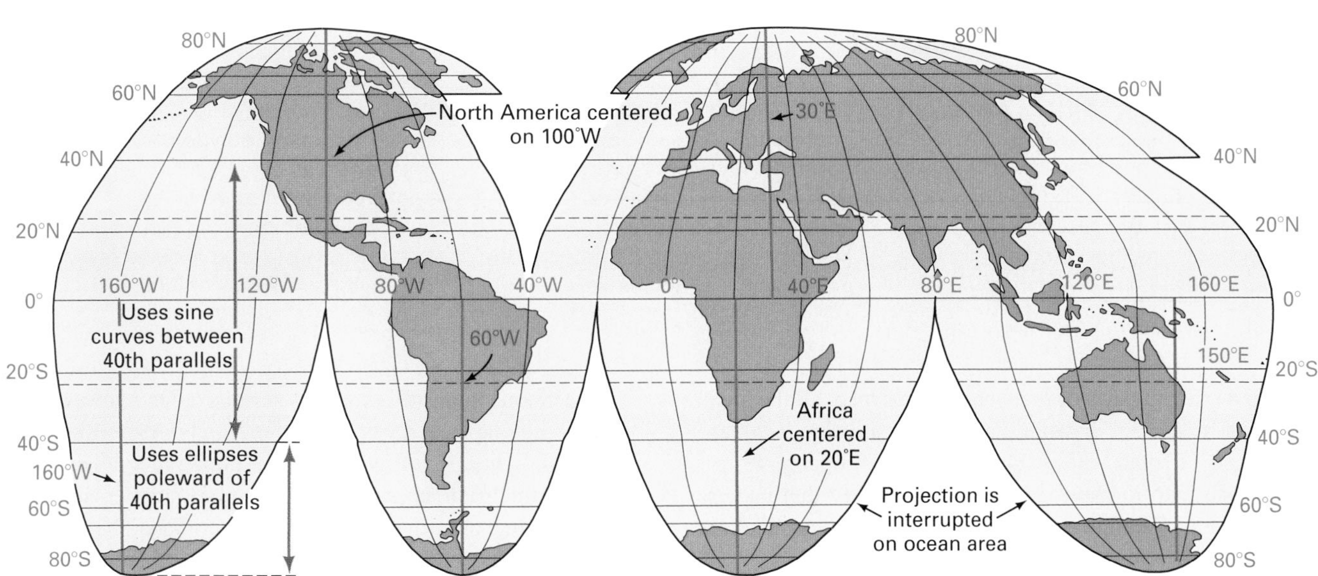

Geographers Tools | 3.3

Geographic Information Systems

A primary activity of geographers is the acquisition, processing, and display of *spatial data*—pieces of information that are in some way associated with a specific location or area of the Earth's surface. Recent advances in computing capability have enabled geographers to develop a powerful new tool to acquire, analyze, and display large volumes of spatial data—the *geographic information system (GIS).* Geographic information systems have allowed geographers, geologists, geophysicists, ecologists, planners, landscape architects, and others to develop applications of spatial data processing ranging from planning land subdivisions on the fringes of suburbia to monitoring the deforestation of the Amazon Basin. This special supplement provides a brief introduction to some of the key concepts of geographic information systems.

GEODISCOVERIES

What Is a GIS?

What is a geographic information system? In essence, it is a system for acquiring, processing, storing, querying, creating, analyzing, and displaying spatial data. A simple example of a GIS is a map overlay system. Imagine that a planner is deciding how to divide a tract of land into building lots. Appropriate inputs would include a topographic map, giving contour lines of equal elevation; a vegetation map, showing the type of existing vegetation cover; a map of existing roads and trails; a map of streams and watercourses; a map of wetlands areas; a map of utility corridors crossing the area (power transmission lines and gas pipelines); and so forth. From these, the planner would locate the positions of roads and bridges; identify steep slopes, utility corridors, wetlands, and regions near streams as conservation areas; and then lay out individual lots having a minimum area of land suitable for building, access to roads and utilities, and other necessary attributes.

To do this, the planner needs to be able to overlay the various maps. Normally this would be done by tracing them onto clear layers, then stacking the layers as needed while drawing the subdivision plans on the top layer. However, some problems may be encountered in this process. One common one is that the maps may not be drawn to the same scale or compiled on the same map projection. Another is that sometimes some maps need further work, such as, for example, to identify areas of steep slopes given a map of topographic contours. Modern, computer-based geographic information systems are designed to solve these problems and allow ready manipulation of spatial data for diverse applications.

Spatial Objects in Geographic Information Systems

Geographic information systems are designed to manipulate spatial objects. A *spatial object* is a geographic area to which some information is attached. This information may be as simple as a place name or as complicated as a large data table with many types of information. A spatial object will normally have a boundary, which is described by a *polygon* that outlines the object. Some spatial objects are illustrated in the figure on top of the facing page.

Spatial objects also include points and lines. A *point* can be thought of as a special type of spatial object with no area. A *line* is also a spatial object with no area, but it has two points associated with it, one for each end of the line. These special points are often referred to as *nodes.* Normally a line is straight, but it can also be defined as a smooth curve having a certain shape. If the two nodes marking the ends of the line are differentiated as starting and ending, then the line has a direction. If the line has a direction, then its two sides can be distinguished. This allows information to be attached to each side—for example, labels for land on one side and water on the other. Lines connect to other lines when they share a common node. A series of connected lines that form a closed chain is a polygon.

By defining spatial objects in this way, computer-based geographic information systems allow easy manipulation of the objects and permit many different types of operations to compare objects and generate new objects. As an example, suppose we have a GIS data layer composed of conservation land in a region represented as polygons and another layer containing the location of preexisting water wells as points within the region, as shown in the figure on the facing page. It is very simple to use the GIS to identify the wells that are on conservation land. Or the conservation polygons containing wells may also be identified and even output as a new data layer. By comparing the conservation layer with a road network layer portrayed as a series of lines, we

moved it over all parts of the Goode world map, all the areas enclosed by the circle would have the same value in square kilometers or square miles. Because of this property, we use the Goode map to show geographical features that occupy surface areas. Examples of useful Goode projections include maps of the world's climates, soils, and vegetation.

The Goode map suffers from a serious defect, however. It distorts the shapes of areas, particularly in high latitudes and at the far right and left edges. To minimize this defect, Dr. Goode split his map apart into separate, smaller sectors, each centered on a different vertical meridian. These were then assembled at the equator. This type of split map is

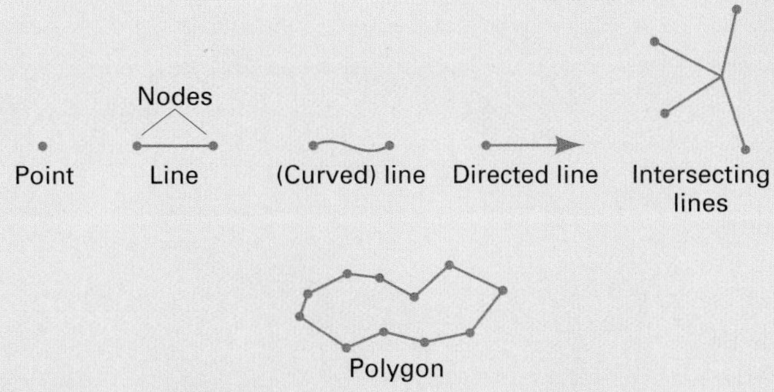

Spatial objects *Spatial objects in a GIS can include points, lines of various types, intersecting lines, and polygons.*

could identify the conservation polygons containing roads. We could also compare the conservation layer to a layer of polygons showing vegetation type, and tabulate the amount of conservation land in forest, grassland, brush, and so forth. We could even calculate distance zones around a spatial object, for example, to create a map of buffer zones that are located within, say, 100 meters of conservation land. Many other possible manipulations exist.

Key Elements of a GIS

A geographic information system consists of five elements: data acquisition, preprocessing, data management, data manipulation and analysis, and product generation. Each is a component or process needed to ensure the functioning of the system as a whole. In the *data acquisition* process, data are gathered together for the particular application. These may include maps, air photos, tabular data, and other forms as well. In *preprocessing,* the assembled spatial data are converted to forms that can be ingested by the GIS to produce data layers of spatial objects and their associated information. The *data management* component creates, stores, retrieves, and modifies data layers and spatial objects. It is essential to proper functioning of all parts of the GIS. The *manipulation and analysis* component is the real workhorse of the GIS. Utilizing this component, the user asks and answers questions about spatial data and creates new data layers of derived information. The last component of the GIS, *product generation,* produces output products in the form of maps, graphics, tabulations, statistical reports, and the like that are the end products desired by the users. Taken together, these components provide a system that can serve many geographic applications at many scales.

Many new and exciting areas of geographic research are associated with geographic information systems, ranging from development of new ways to manipulate spatial data to the modeling of spatial processes using a GIS. An especially interesting area is understanding how outputs are affected by errors and uncertainty in spatial data inputs, and how to communicate this information effectively to users.

Geographic information systems is a new, emerging field of geographic research and application that relies heavily on computers and computer technology. Given the rate at which computers become ever more powerful as technology improves, we can expect great strides in this field in future years.

Data layers in a GIS *A GIS allows easy overlay of spatial data layers for such queries as "Identify all wells on conservation land."*

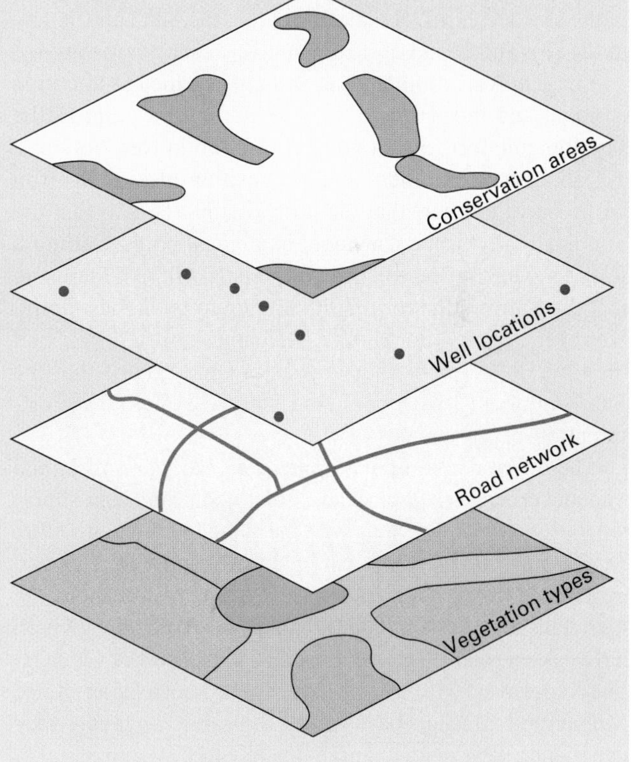

called an *interrupted projection.* Although the interrupted projection greatly reduces shape distortion, it separates parts of the Earth's surface that actually lie close together, particularly in the high latitudes.

Maps are in wide use today for many applications as a simple and efficient way of compiling and storing spatial information. However, in the past two decades, maps are being supplemented by more powerful computer-based methods for acquiring, storing, processing, analyzing, and outputting spatial data. These are contained within *Geographic Information Systems (GISs).* Our special supplement, *Geographer's Tools 3.3 • Geographic Information*

Systems, presents some basic concepts of geographic information systems and how they work. GEODISCOVERIES

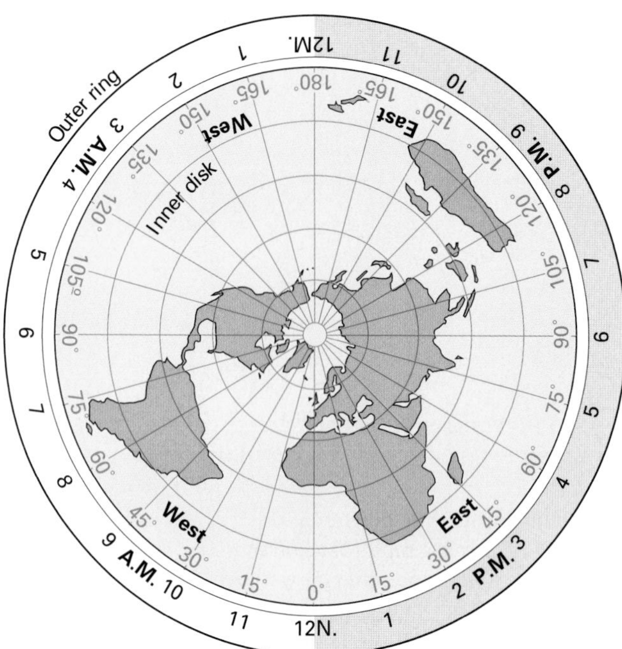

3.11 The relation of longitude to time *The outer ring shows the time at locations identified by longitude meridians on the central map. The diagram is set to show noon conditions in Greenwich, England—that is, on the prime meridian. Clock time is earlier to the west of Greenwich and later to the east.*

GLOBAL TIME

Maps and map projections are a practical application of the Earth's geographic grid. Another practical application, which involves both the grid and the Earth's rotation, is global time.

Our planet requires 24 hours for a full rotation with respect to the Sun. Put another way, humans long ago decided to divide the solar day into 24 units, called hours, and devised clocks to keep track of hours in groups of 12. Yet, different regions set their clocks differently. For example, when it is 10:03 A.M. in New York, it is 9:03 A.M. in Chicago, 8:03 A.M. in Denver, and 7:03 A.M. in Los Angeles. Note that these times differ by exactly one hour. How did this system come about? How does it work?

Our global time system is oriented to the Sun. Think for a moment about how the Sun appears to move across the sky. In the morning, the Sun is low on the eastern horizon, and as the day progresses, it rises higher until at *solar noon* it reaches its highest point in the sky. If you check your watch at that moment, it will read a time somewhere near 12 o'clock (12:00 noon). After solar noon, the Sun's elevation in the sky decreases. By late afternoon, the Sun appears low in the sky, and at sunset it rests on the western horizon.

Imagine for a moment that you are in Chicago, the time is noon, and the Sun is at or near its highest point in the sky. Imagine further that you call a friend in New York and ask about the time there and the position of the Sun. You will receive a report that the time is 1:00 P.M. and that the Sun is already past solar noon, its highest point. Calling a friend in Vancouver, you hear that it is 10:00 A.M. there and that the Sun is still working its way up to its highest point. However, a friend in Mobile, Alabama, will tell you that the time in Mobile is the same as in Chicago and that the Sun is at about solar noon. How can we explain these different observations?

The difference in time between Chicago, New York, and Vancouver makes sense because solar noon can occur simultaneously only at locations with the same longitude. In other words, only one meridian can be directly under the Sun and experience solar noon at a given moment. Locations on meridians to the east of Chicago, like New York, already will have passed solar noon, and locations to the west of Chicago, like Vancouver, will not yet have reached solar noon. Since Mobile and Chicago have nearly the same longitude, they experience solar noon at approximately the same time.

Figure 3.11 indicates how time varies with longitude. In this figure, the inner disk shows a polar projection of the world, centered on the north pole. Meridians are straight lines (radii) ranging out from the pole. The outer ring indicates the time in hours. The figure shows the moment in time when the prime meridian is directly under the Sun—that is, the 0° meridian is directly on the 12:00 noon mark. This means that, at this instant, the Sun is at the highest point of its path in the sky in Greenwich, England. The alignment of meridians with hour numbers tells us the time in other locations around the globe. For example, the time in New York, which lies roughly on the 75° W meridian, is about 7:00 A.M. In Los Angeles, which lies roughly on the 120° W meridian, the time is about 4:00 A.M.

Notice that 15° of longitude equates to an hour of time. Since the Earth turns 360° in a 24-hour day, the rotation rate is 360° ÷ 24 = 15° per hour.

Standard Time

We've just seen that locations with different longitudes experience solar noon at different times. But consider what would happen if each town or city set its clocks to read 12:00 at its own local solar noon. All cities and towns on different meridians would have different local time systems. In these days of instantaneous global communication, chaos would soon result.

The use of standard time simplifies the global timekeeping problem. In the **standard time system,** the globe is divided into 24 **time zones.** All inhabitants within a zone keep time according to a *standard meridian* that passes through their zone. Since the standard meridians are usually 15 degrees apart, the difference in time between adjacent zones is normally one hour. In some geographic regions, however, the difference is only one-half hour.

Seven time zones cover the United States and its Caribbean possessions. Six zones cover Canada. Their names and standard meridians of longitude are as follows:

U.S. Zones	Meridian	Canadian Zones
	$52^1/_2°$	Newfoundland
Atlantic	60°	Atlantic
Eastern	75°	Eastern
Central	90°	Central
Mountain	105°	Mountain
Pacific	120°	Pacific-Yukon
Alaska-Bering	135°	
Hawaii	150°	

If carried out strictly, the standard time system would consist of belts exactly 15°, extending to meridians $7^1/_2°$ east and west of each standard meridian. However, this system could be inconvenient since the boundary meridians could divide a state, county, or city into two different time zones. As a result, time zone boundaries are often routed to follow agreed-upon natural or political boundaries.

Figure 3.12 presents a map of time zones for the contiguous United States and southern Canada. From this map, you can see that most time zone boundaries are conveniently located along an already existing and widely recognized line. For example, the eastern time–central time boundary line follows Lake Michigan down its center, and the mountain time–Pacific time boundary follows a ridge-crest line also used by the Idaho–Montana state boundary.

World Time Zones

Figure 3.13 shows the 24 principal standard time zones of the world. In the figure, 15° meridians are dashed lines, while the $7^1/_2°$ meridians, which form many of the boundaries between zones, are bold lines. The figure also shows the time of day in each zone when it is noon at the Greenwich meridian. The country spanning the greatest number of time zones from east to west is Russia, with 11 zones, but these are grouped into eight standard time zones. China spans five time zones but runs on a single national time using the standard meridian of Beijing.

A few countries keep time by a meridian that is midway between standard meridians, so that their clocks depart from those of their neighbors by 30 or 90 minutes. India and Iran are examples. The Canadian province of Newfoundland and the interior Australian states of South Australia and Northern Territory are examples of regions within countries that keep time by $7^1/_2°$ meridians.

World time zones are numbered to indicate the number of hours difference between time in a zone and time in Greenwich. A number of +7, for example, indicates that adding seven hours to local time will give Greenwich time, while a –3 indicates that subtracting three hours from local time will give Greenwich time. Given the zone numbers of two dif-

3.12 Time zones of the conterminous United States and southern Canada *The name, standard meridian, and number code are shown for each time zone. Note that time zone boundaries often follow preexisting natural or political boundaries.*

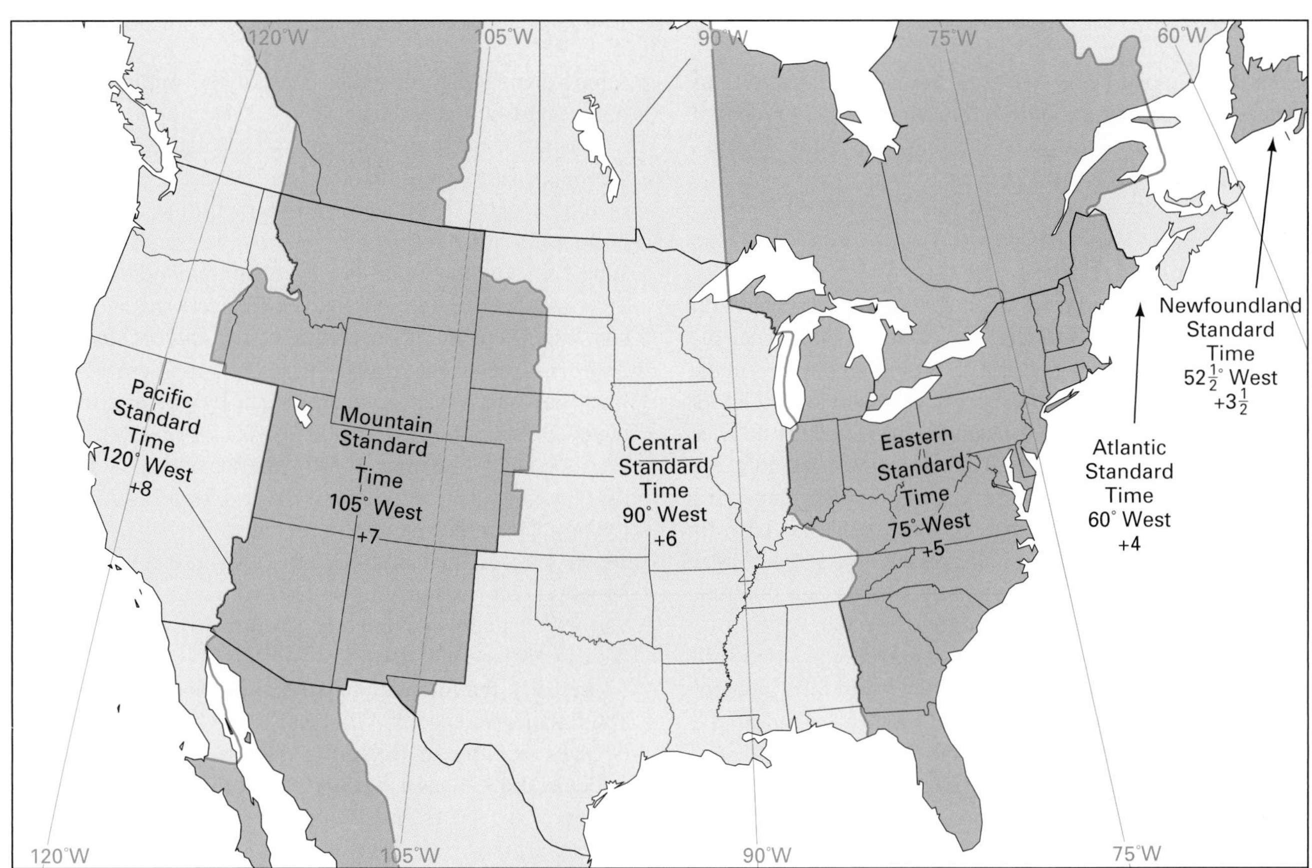

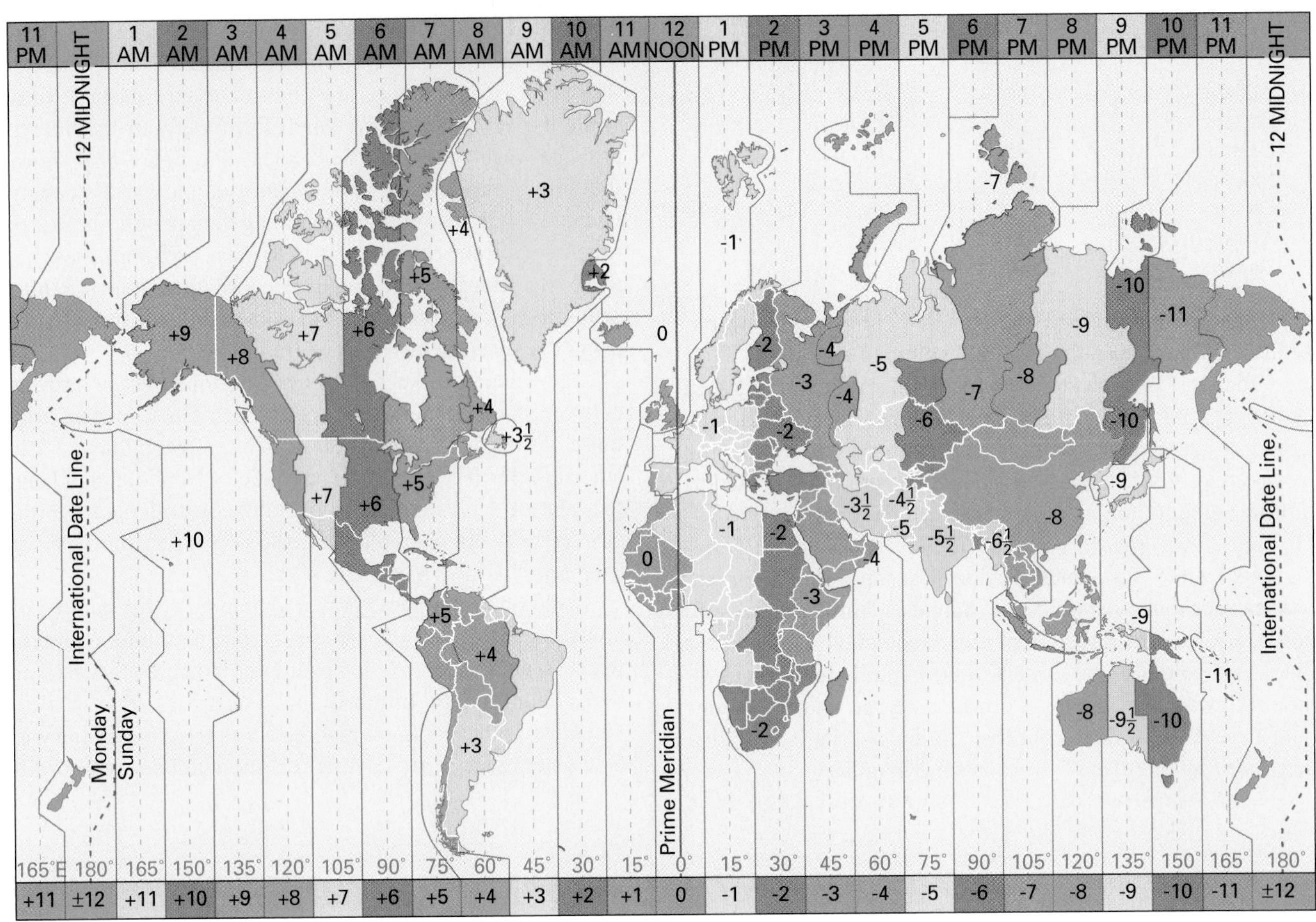

3.13 Time zones of the world *Dashed lines represent 15° meridians, and bold lines represent 7½° meridians. Alternate zones appear in color. (U.S. Navy Oceanographic Office.)*

ferent time zones, it is easy to convert the time in one of them to the time in the other. *Working It Out 3.4 • Global Timekeeping* shows how to do this and provides some practical examples for the world traveler.

International Date Line

When we take a world map or globe with 15° meridians and count them in an eastward direction, starting with the Greenwich meridian as 0, we find that the 180th meridian is number 12 and that the time at this meridian is therefore 12 hours later than Greenwich time. Counting in a similar manner westward from the Greenwich meridian, we find that the 180th meridian is again number 12 but that the time is 12 hours earlier than Greenwich time. How can the same meridian be both 12 hours ahead of Greenwich time and 12 hours behind? This paradox is explained by the fact that different days are observed on either side of this meridian.

Imagine that you are on the 180th meridian on June 26th and it is exactly midnight. Let's stop the world for a moment and examine the situation. On the 180th meridian at the exact instant of midnight, the same 24-hour calendar day covers the entire globe. Stepping east will place you in the very early morning of June 26, while stepping west will place you very late in the evening of June 26. You are in the same calendar day on both sides of the meridian but 24 hours apart in time.

Doing the same experiment an hour later, at 1:00 A.M., stepping east you will find that you are in the early morning of June 26. But if you step west you will find that midnight of June 26 has passed, and it is now the early morning of June 27. So on the west side of the 180th meridian, it is also 1:00 A.M. but one day later than on the east side. For this reason, the 180th meridian serves as the *International Date Line.* This means that if you travel westward across the date line, you must advance your calendar by one day. If traveling eastward, you set your calendar back by a day.

Air travelers between North America and Asia cross the date line. For example, flying westward from Los Angeles to Sydney, Australia, you may depart on a Tuesday evening and arrive on a Thursday morning after a flight that lasts only 14 hours. On an eastward flight from Tokyo to San Francisco, you may actually arrive the day before you take off, taking the date change into account!

Actually, the International Date Line does not follow the 180th meridian exactly. Like many time zone boundaries, it deviates from the meridian for practical reasons. As shown in Figure 3.13, it has a zigzag offset between Asia and North America, as well as an eastward offset in the South Pacific to keep clear of New Zealand and several island groups.

Working It Out 3.4
Global Timekeeping

Your eight-hour flight to Rome leaves New York at 8:30 P.M. What time will it be in Rome when you arrive? To help you answer this and other practical problems involving timekeeping, time zones are numbered away from the time zone of the Greenwich meridian—negatively in an eastward direction and positively in a westward direction. These numbers are shown on the map in Figure 3.13. The numbers tell you how many hours to change your clock to get Greenwich time.

For example, New York is in zone +5, so you must add five hours to New York time to get the time in Greenwich. Rome is in zone −1, so in Rome you subtract an hour to get Greenwich time. Thus, when it is 8:30 P.M. in New York, it must be 1:30 A.M. in Greenwich. And if it is 1:30 A.M. in Greenwich, it must be 2:30 A.M. in Rome, since Romans have to subtract an hour to get Greenwich time. With the knowledge that you are departing New York at 2:30 A.M. Rome time, you can see that after an eight-hour flight, you will arrive in Rome at 10:30 A.M. the next morning.

It is easy to come up with a simple formula for global timekeeping:

$$D = Z_{HOME} - Z_{AWAY}$$

Here D is the difference—positive for ahead, negative for behind—that you add to change the time; Z_{HOME} is the time zone whose time you already know; and Z_{AWAY} is the time zone whose time you want to know. So, for New York and Rome, $D = +5 - (-1) = +6$, meaning you must add six hours to New York time to get the time in Rome.

Of course, this simple formula works only if both locations are on standard time or are on daylight saving time. If they differ, then you need to take the difference into account. For example, if it's daylight time in Rome but not in the United States, then they've "sprung ahead" and there will be seven hours difference.

This formula works when you cross the date line as well. Consider a 10-hour flight from Tokyo (−9) to Los Angeles (+8). The flight leaves at 4:00 P.M. Tokyo time on Tuesday, December 12. From the formula, $D = (-9) - (+8) = -17$, so you subtract 17 hours from Tokyo time, and the time in Los Angeles when you depart from Tokyo is 11:00 P.M. on Monday, December 11. Add 10 hours for your flight time, and you will arrive in Los Angeles at 9:00 A.M. on Tuesday—well before you left Tokyo!

Daylight Saving Time

Especially in urban areas, many human activities begin well after sunrise and continue long after sunset. Therefore, we adjust our clocks during the part of the year that has a longer daylight period to correspond more closely with the modern pace of society. This adjusted time system, called *daylight saving time,* is obtained by setting all clocks ahead by one hour. The effect of the time change is to transfer the early morning daylight period, theoretically wasted while schools, offices, and factories are closed, to the early evening, when most people are awake and busy. Daylight saving time also yields a considerable savings in power used for electric lights. In the United States and Canada, daylight saving time comes into effect on the first Sunday in April and is discontinued on the last Sunday of October. Although daylight saving time is in wide use, it is not observed at all locations.

Precise Timekeeping

Many scientific and technological applications require precise timekeeping. Today, a worldwide system of master atomic clocks measures time to better than one part in 1,000,000,000,000. However, our Earth is a much less precise timekeeper, demonstrating small changes in the angular velocity of its rotation on its axis and variations in the time it takes to complete one circuit around the Sun. As a result, constant adjustments to the timekeeping system are necessary. The legal time standard recognized by all nations is Coordinated Universal Time, which is administered by the Bureau International de l'Heure, located near Paris.

THE EARTH'S REVOLUTION AROUND THE SUN

So far, we have discussed the importance of the Earth's rotation on its axis. Another important motion of the Earth is its **revolution,** or its movement in orbit around the Sun. GEODISCOVERIES

The Earth completes a revolution around the Sun in 365.242 days—about one-fourth day more than the calendar year of 365 days. Every four years, the extra one-fourth days add up to about one whole day. By inserting a 29th day in February in leap years, we largely correct the calendar for this effect. Further minor corrections are necessary to perfect the system.

The Earth's orbit around the Sun is shaped like an ellipse, or oval. This means that the distance between the Earth and Sun varies somewhat through the year. The Earth is nearest to the Sun at *perihelion,* which occurs on or about January 3. It is farthest away from the Sun at *aphelion,* on or about July 4. However, the distance between Sun and Earth varies only by about 3 percent during one revolution because the elliptical orbit is shaped very much like a circle. For most purposes we can regard the orbit as circular.

3.14 Revolution of the Earth and Moon *Viewed from a point over the Earth's north pole, the Earth both rotates and revolves in a counterclockwise direction. From this viewpoint, the Moon also rotates counterclockwise.*

In which direction does the Earth revolve? Imagine yourself in space, looking down on the north pole of the Earth. From this viewpoint, the Earth travels counterclockwise around the Sun (Figure 3.14). This is the same direction as the Earth's rotation.

The Moon rotates on its axis and revolves about the Sun in the same direction as well. However, the Moon's rate of rotation is such that one side of the Moon is always directed toward the Earth. Thus, astronomers had never seen the far side of the Moon until a Soviet spacecraft passing the Moon transmitted photos back to Earth in 1959. The phases of the Moon are determined by the position of the Moon in its orbit around the Earth, which in turn determines how much of the sunlit Moon is seen from the Earth. In Figure 3.14, the Moon is behind the Earth and we see the Moon as full. In Figure 3.15, the Moon is nearly full. From the way that the Sun illuminates the Moon as a sphere, it is easy to see that the Sun is to the right.

Tilt of the Earth's Axis

The seasons we experience on Earth are related to the orientation of the Earth's axis of rotation as it revolves around the Sun. We usually describe this situation by stating that the Earth's axis is tilted with respect to the *plane of the ecliptic*—the plane containing the Earth's orbit around the Sun. Figure 3.16 shows the plane of the ecliptic as it intersects the Earth, and Figure 3.17 shows the full Earth orbit traced on the plane of the ecliptic.

Now consider the angle of the axis of the Earth's rotation as shown in Figure 3.16. Note that the axis is tilted at an

3.15 Eye on the Landscape ***Midnight in June, Lake Clark National Park, Alaska*** *Although it is midnight, the Sun is only just below the horizon, bathing the scene in soft twilight. The way the spherical Moon is lit by the Sun also shows that the Sun is located below the horizon and to the right.* **What else would the geographer see? ... Answers at the end of the chapter.**

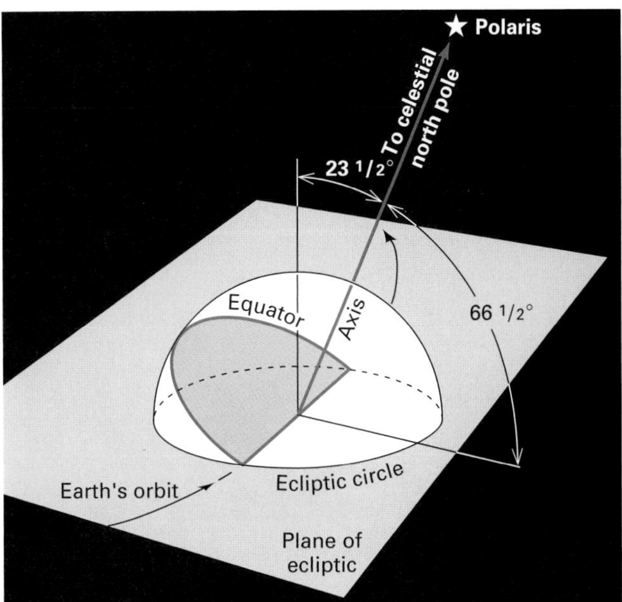

3.16 The tilt of the Earth's axis of rotation with respect to its orbital plane *As the Earth moves in its orbit on the plane of the ecliptic around the Sun, its rotational axis remains pointed toward Polaris, the north star, and makes an angle of 66$^1/_2$° with the ecliptic plane.*

angle of 23$^1/_2$° away from a right angle to the plane of the ecliptic. That is, the angle between the axis and the plane of the ecliptic is 66$^1/_2$°, not 90°. In addition, the direction toward which the axis points is fixed in space—it aims toward Polaris, the north star. The direction of the axis does not change as the Earth revolves around the Sun. Let's investigate this phenomenon in more detail.

Solstice and Equinox

Figure 3.17 diagrams the Earth as it revolves in its orbit through the four seasons. Because the direction of the Earth's axis of rotation is fixed, the north pole is tilted away from the Sun during one part of the year and is tilted toward the Sun during the other part. Consider first the event on December 22, which is pictured on the far right. On this day, the Earth is positioned so that the north polar end of its axis leans at the maximum angle away from the Sun, 23$^1/_2$°. This event is called the **winter solstice.** (While it is winter in the northern hemisphere, it is summer in the southern hemisphere, so you can use the term *December solstice* to avoid any confusion.) At this time, the southern hemisphere is tilted toward the Sun and enjoys strong solar heating. GEODISCOVERIES

Six months later, on June 21, the Earth is on the opposite side of its orbit in an equivalent position. At this event, known as the ***summer solstice*** *(June solstice),* the north polar end of the axis is tilted at 23$^1/_2$° toward the Sun. Thus, the north pole

3.17 The four seasons *The four seasons occur because the Earth's tilted axis keeps a constant orientation in space as the Earth revolves about the Sun. This tips the northern hemisphere toward the Sun for the summer solstice and away from the Sun for the winter solstice. Both hemispheres are illuminated equally at the spring equinox and the fall equinox.*

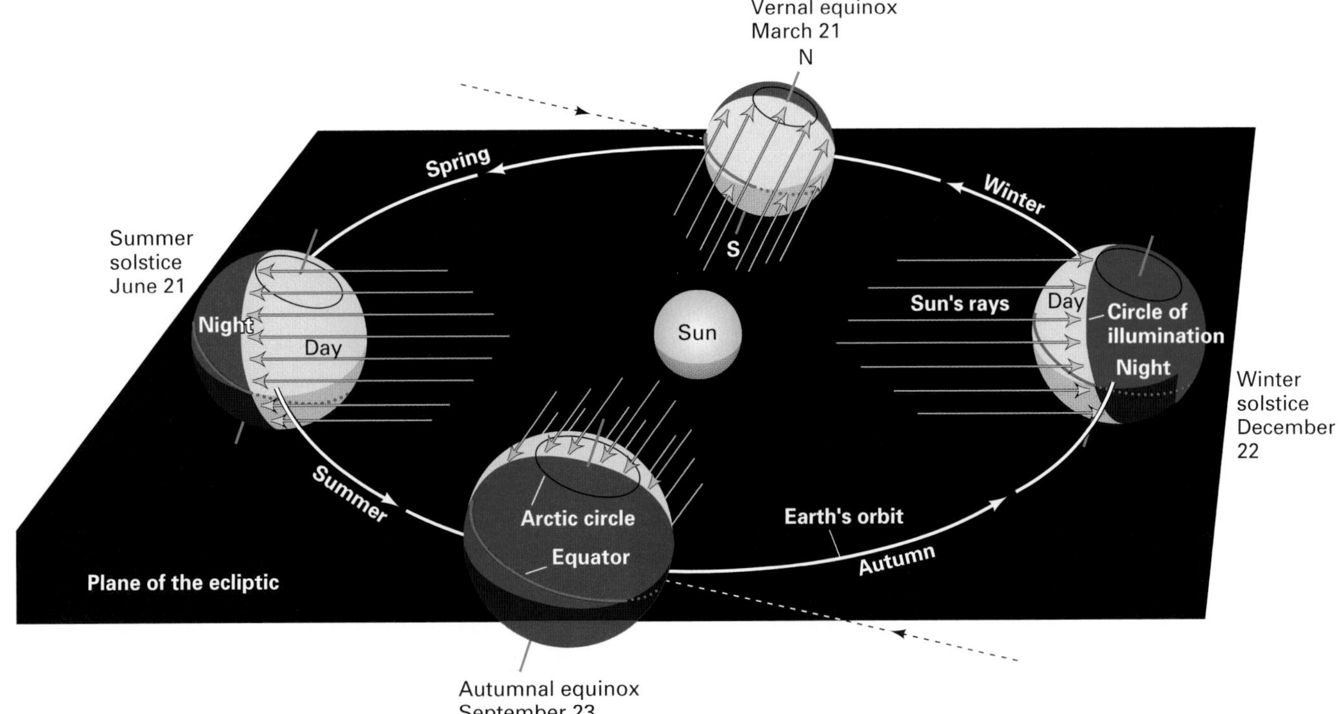

and northern hemisphere are tilted toward the Sun, while the south pole and southern hemisphere are tilted away.

Midway between the solstice dates, the equinoxes occur. At an **equinox,** the Earth's axial tilt is neither toward nor away from the Sun. The axis makes a right angle with a line drawn to the Sun, and neither the north nor south pole is tilted toward the Sun. The *vernal equinox* occurs on March 21, and the *autumnal equinox* occurs on September 23. Conditions are identical on the two equinoxes as far as Earth–Sun relationships are concerned. We should also note that the date of any solstice or equinox in a particular year may vary by a day or so, since the revolution period is not exactly 365 days. Let's look at equinoxes and solstices in more detail.

Equinox Conditions

The conditions at an equinox, shown in Figure 3.18, form the simplest case. The figure illustrates two important concepts of global illumination that we use for describing equinoxes and solstices. The first concept is the *circle of illumination.* Note that the Earth is always divided into two hemispheres with respect to the Sun's rays. One hemisphere (day) is lit by the Sun, and the other (night) lies in the darkness of the Earth's shadow. The circle of illumination is the circle that separates the day hemisphere from the night hemisphere. The second concept is the *subsolar point,* the single point on the Earth's surface where the Sun is directly overhead at a particular moment.

At equinox, the circle of illumination passes through the north and south poles, as we see in Figure 3.18. The Sun's rays graze the surface at either pole, and the surface there receives little or no solar energy. The subsolar point falls on the equator. Here, the angle between the Sun's rays and the Earth's surface is 90°, and the solar illumination is received in full force. At an intermediate latitude, such as 40° N, the rays of the Sun at noon strike the surface at a lesser angle. This *noon angle*—the elevation of the Sun above the horizon at noon—can be easily determined. Some simple geometry shows that the noon angle is equal to 90° minus the latitude, or 50° in this example, for equinox conditions.

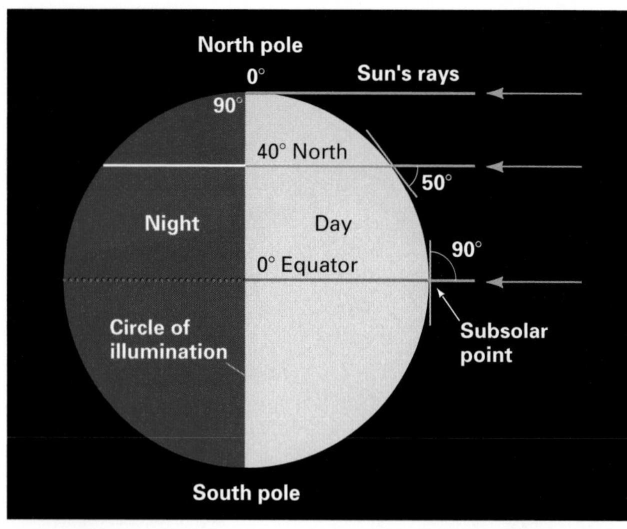

3.18 Equinox conditions *At this time, the Earth's axis of rotation is exactly at right angles to the direction of solar illumination. The subsolar point lies on the equator. At both poles, the Sun is seen at the horizon. Note that the viewpoint for this diagram is away from the plane of the ecliptic, as shown in Figure 3.17, so that both poles may be seen.*

Imagine yourself at a point on the Earth, say, at a latitude of 40° N. Visualize the Earth rotating from left to right, so that you turn with the globe, completing a full circuit in 24 hours. At the equinox, you spend 12 hours in darkness and 12 hours in sunlight. This is because the circle of illumination passes through the poles, dividing every parallel exactly in two. Thus, one important feature of the equinox is that day and night are of equal length everywhere on the globe.

Solstice Conditions

Now examine the solstice conditions shown in Figure 3.19. Summer solstice is on the left. Consider yourself back at a point on the lat. 40° N parallel. The circle of illumination

3.19 Solstice conditions *At the solstice, the north end of the Earth's axis of rotation is fully tilted either toward or away from the Sun. Because of the tilt, polar regions experience either 24-hour day or 24-hour night. The subsolar point lies on one of the tropics, at lat. $23^1/_2$° N or S.*

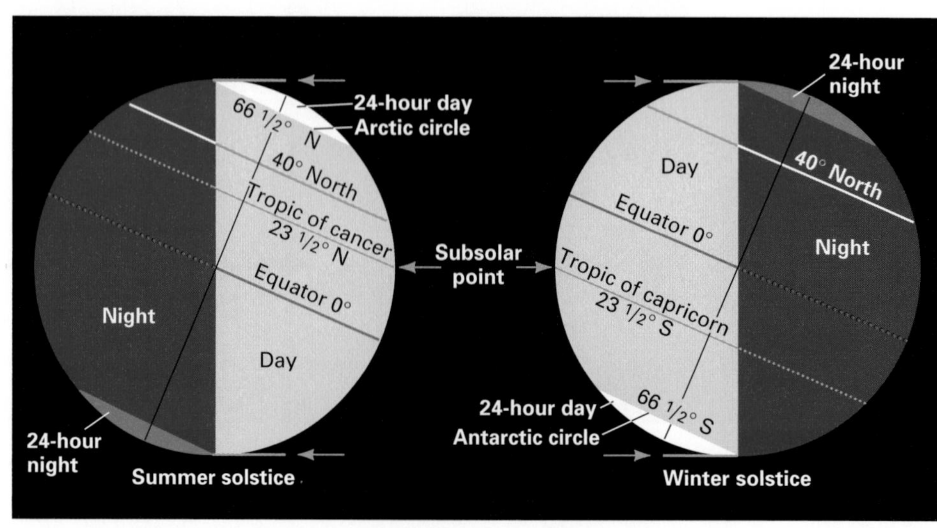

does not divide your parallel in equal halves because of the tilt of the northern hemisphere toward the Sun. Instead, the larger part is in daylight. For you, the day is now considerably longer (about 15 hours) than the night (about 9 hours).

The farther north you go, the more the effect increases. In fact, the entire area of the globe north of lat. 66$^1/_2$° is on the daylight side of the circle of illumination. This parallel is known as the **arctic circle.** Even though the Earth rotates through a full cycle during a 24-hour period, the area north of the arctic circle remains in continuous daylight. We also see that the subsolar point is at a latitude of 23$^1/_2$° N. This parallel is known as the **tropic of cancer.** Because the Sun is directly over the tropic of cancer at this solstice, solar energy is most intense here.

At the winter solstice, conditions are exactly reversed from those of the summer solstice. If you imagine yourself back at lat. 40° N, you find that the night is now about 15 hours long while daylight lasts about 9 hours. All the area south of lat. 66$^1/_2$° S lies under the Sun's rays, inundated with 24 hours of daylight. This parallel is known as the **antarctic circle.** The subsolar point has shifted to a point on the parallel at lat. 23$^1/_2$° S, known as the **tropic of capricorn.**

The solstices and equinoxes represent conditions at only four times of the year. Between these times, the latitude of the subsolar point travels northward and southward in an annual cycle between the tropics of cancer and capricorn. We refer to the latitude of the subsolar point as the Sun's *declination.* Figure 3.20 plots the Sun's declination through the year. As the year progresses, the declination varies between 23$^1/_2$° S lat. at the December solstice and 23$^1/_2$° N lat. at the June solstice.

As the seasonal cycle progresses in polar regions, areas of 24-hour daylight or 24-hour night shrink and then grow. At other latitudes, the length of daylight changes slightly from one day to the next, except at the equator. In this way, the Earth experiences the rhythm of the seasons as it continues its revolution around the Sun.

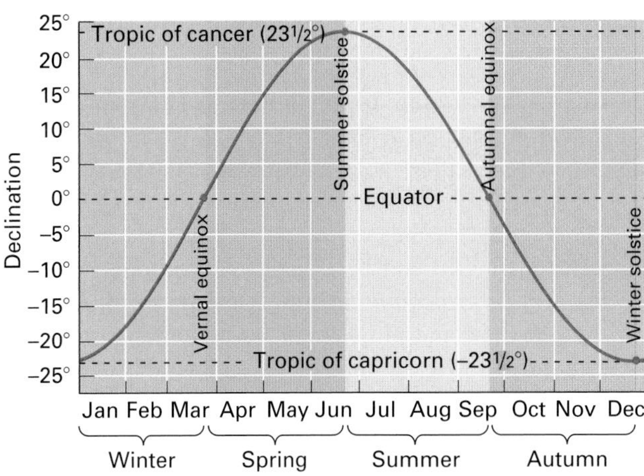

3.20 The Sun's declination through the year *The latitude of the subsolar point marks the Sun's declination, which changes slowly through the year from −23$^1/_2$° to +23$^1/_2$° to −23$^1/_2$°. Solstices and seasons are labeled for the northern hemisphere.*

A Look Ahead

The flow of solar energy to the Earth powers most of the natural processes that we experience every day, from changes in the weather to the work of streams in carving the landscape. Our next chapter examines in detail solar energy and its interaction with the Earth's atmosphere and surface. The solar radiation intercepted by the Earth constitutes the biggest and most important energy flow system for us, the land dwellers of the Earth. What have we learned about the Sun–Earth energy flow system thus far? Here are two simple, important facts we can glean from this chapter:

- *Half of the Earth is always receiving solar energy.* So, there is a constant flow of light and heat from Sun to Earth.
- *However, the solar energy flow is not received uniformly over the surface.* This is because of the Earth's spherical shape, its constant rotation on its axis, the tilt of its axis, and its constant revolution around the Sun.

As you will see in Chapter 4, the amount of incoming solar energy at a location depends on the length of the daylight portion of the day and on the height of the Sun's path in the sky during the daylight period. Both of these factors depend on the latitude and season of the location. So, your understanding of the Earth's rotation and revolution will come in quite handy in the chapter to come.

Chapter Summary

- The **rotation** of the Earth on its axis and the revolution of the Earth in its orbit about the Sun are fundamental topics in physical geography. The Earth is nearly spherical—it is slightly flattened at the poles into the shape of an *oblate ellipsoid.* It rotates on its *axis* once in 24 hours. The intersection of the axis of rotation with the Earth's surface marks the *north* and *south poles*. The direction of rotation is counterclockwise when viewed from above the north pole.
- The Earth's rotation provides the first great rhythm of our planet—the daily alternation of sunlight and darkness. The tides, and a sideward turning of ocean and air currents, are further effects of the Earth's rotation.
- The Earth's axis of rotation provides a reference for the system of location on the Earth's surface—the **geographic grid**, which consists of **meridians** and **parallels**. This system is indexed by our system of **latitude** and **lon-**

gitude, which uses the **equator** and the *prime meridian* as references to locate any point on Earth.

- We require a **map projection** to display the Earth's curved surface on a flat map. The *polar projection* is centered on either pole and pictures the globe as we might view it from the top or bottom. The *Mercator projection* converts the curved geographic grid into a flat, rectangular grid and best displays directional features. The *Goode projection* distorts the shapes of continents and coastlines but preserves the areas of land masses in their correct proportion.
- We monitor the Earth's rotation by daily timekeeping. Each hour the Earth rotates by 15°. In the **standard time system**, we keep time according to a nearby *standard meridian*. Since standard meridians are normally 15° apart, clocks around the globe usually differ by whole hours. At the *International Date Line*, the calendar day changes—advancing for westward travel, dropping back a day for eastward travel. *Daylight saving time* advances the clock by one hour.
- The seasons are the second great Earthly rhythm. They arise from the **revolution** of the Earth in its orbit around the Sun, combined with the fact that the Earth's rotational axis is tilted with respect to its orbital plane. The **solstices** and **equinoxes** mark the cycle of this revolution. At the **summer** (June) **solstice**, the northern hemisphere is tilted toward the Sun. At the **winter** (December) **solstice**, the southern hemisphere is tilted toward the Sun. At the equinoxes, day and night are of equal length.

Key Terms

rotation
parallel
equator
meridian
geographic grid
great circle
small circle
latitude
longitude
map projection
standard time system
time zone
revolution
winter solstice
summer solstice
equinox
arctic circle
tropic of cancer
antarctic circle
tropic of capricorn

Review Questions

1. What is the approximate shape of the Earth? How do you know? What is the Earth's true shape?
2. What is meant by Earth rotation? Describe three environmental effects of the Earth's rotation.
3. Describe the geographic grid, including parallels and meridians.
4. How do latitude and longitude determine position on the globe? In what units are they measured?
5. Name three types of map projections and describe each briefly. Give reasons why you might choose different map projections to display different types of geographical information.
6. Explain the global timekeeping system. Define and use the terms *standard time, standard meridian,* and *time zone* in your answer.
7. What is meant by the "tilt of the Earth's axis"? How is the tilt responsible for the seasons?

Geographer's Tools 3.3 • Geographic Information Systems

1. What is a geographic information system?
2. Identify and describe three types of spatial objects.
3. What are the key elements of a GIS?

A Closer Look:
Geographer's Tools 3.5 • Focus on Maps

1. Explain three types of map "projections" as they might occur by projecting a wire globe onto a flat sheet of paper.
2. What is the scale fraction of a map or globe? Can the scale of a flat map be uniform everywhere on the map? Do large-scale maps show large areas or small areas?
3. What types of symbols are found on maps, and what types of information do they carry?
4. How are numerical data represented on maps? Identify three types of isopleths. What is a choropleth map?

Visualizing Exercises

1. Sketch a diagram of the Earth at an equinox. Show the north and south poles, the equator, and the circle of illumination. Indicate the direction of the Sun's incoming rays and shade the night portion of the globe.
2. Sketch a diagram of the Earth at the summer (June) solstice, showing the same features. Also include the tropics of cancer and capricorn, and the arctic and antarctic circles.

Essay Question

1. Suppose that the Earth's axis were tilted at 40° to the plane of the ecliptic instead of $23^1/_2$°. What would be the global effects of this change? How would the seasons change at your location?

Problems

Working It Out 3.1 • Distances from Latitude and Longitude

1. Chicago and Mobile are located on about the same meridian, but they are about 11° lat. apart. What is the approximate distance between Chicago and Mobile?
2. Ottawa and Portland, Oregon, are both located very close to the 45th parallel, but their longitudes are 76° W and 124° W, respectively. What is the approximate distance between the two cities, measured along the parallel? (In using the formula, be sure that your calculator is set to use degrees when finding the cosine.)
3. A map of a region close to the equator shows an area of 1° of latitude by 1° of longitude. About how many square kilometers are covered by the map? How many square miles? How do these areas compare with those of a 1° by 1° map near Winnipeg (50° N lat.)?

Working It Out 3.4 • Global Timekeeping

1. A flight from Chicago (+6) to Beijing (–8) takes 16 hours and 30 minutes (since it has a three-hour layover in Tokyo). You leave Chicago at 3:30 P.M. on Saturday, June 14. When are you scheduled to arrive, Beijing time?
2. A one-stop flight from Anchorage to Washington leaves at 1:59 A.M. and arrives at 3:51 P.M. The flight includes a 1-hour-and-25-minute layover in Denver. How long will you be in the air (or taxiing), if the flight holds to the schedule? (Use Figure 3.13 to obtain time zone numbers.)

Eye on the Landscape

Chapter Opener Illuminated greenhouse This Finnish winter landscape shows trees of both deciduous **(A)** and evergreen **(B)** habit. Pine (evergreen) and birch (deciduous) trees are abundant in boreal forests, often growing in nearly pure stands. Their lumber is a primary export of Finland. Boreal forests are discussed in Chapter 24.

3.1b Space view of Earth Unusually clear summer weather provides a good look at Europe **(A)** in this image acquired by the Meteosat geostationary meteorological satellite. It's easy to recognize Great Britain and Ireland, as well as the Italian Peninsula, by their distinctive shapes. Further east, Greece, Turkey, and the Black Sea are visible. To the north of Germany lie Denmark and Sweden, enclosing the western end of the Baltic Sea. The Pyrenees, at the top of the Iberian Peninsula, and the Alps, above Italy, are white with either clouds or snow. They delineate the edges of crustal Earth plates that are slowly colliding to push up mountain chains (see plate tectonics, covered in Chapter 13). The comma-shaped cloud patterns **(B)** mark a progression of cyclones with trailing frontal boundaries as they move eastward in the midlatitudes. We'll see more cloud images in Chapters 7 and 8.

3.15 Lake Clark National Park, Alaska These mountain landforms **(A)** show the effects of glacial ice and frost action during the most recent Ice Age. The patches of snow mark the sites where small glaciers once formed, carving shallow basins in the bedrock of the peaks (see mountain glaciers, Chapter 20). At the base of the mountains lie sloping fields of angular rock fragments, called talus, that are broken loose by frost action and carried downslope by gravity (covered in Chapter 15, Weathering and Mass Wasting). The gravel in the foreground **(B)** has a particularly fresh look and was probably deposited recently by running water, perhaps during the spring flood season. Chapter 17 provides more information on how water in streams and rivers moves sediment such as this.

A CLOSER LOOK

Geographer's Tools 3.5 | Focus on Maps

Maps play an essential role in the study of physical geography because much of the information content of geography is stored and displayed on maps. Map literacy—the ability to read and understand what a map shows—is a basic requirement for day-to-day functioning in our society. Maps appear in almost every issue of a newspaper and in nearly every TV newscast. Most people routinely use highway maps and street maps. The purpose of this special supplement is to provide additional information on the subject of cartography, the science of maps and their construction.

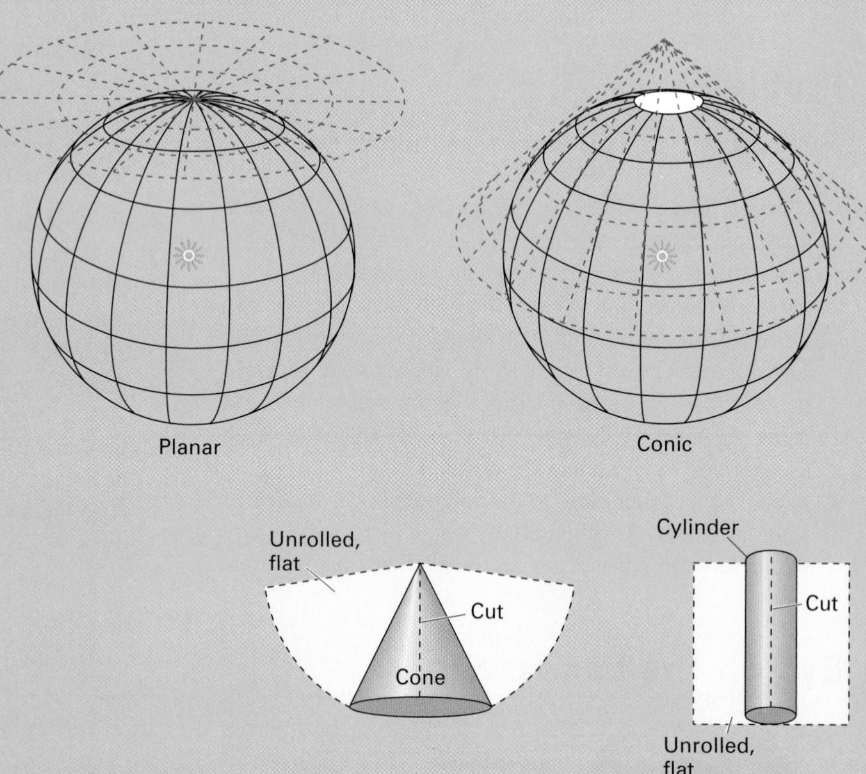

S3.1 Simple ways to generate map projections *Rays from a central light source cast shadows of the spherical geographic grid on target screens. The conical and cylindrical screens can be unrolled to become flat maps. (A. N. Strahler.)*

More about Map Projections

Recall from the chapter text that a *map projection* is an orderly system of parallels and meridians used as a base on which to draw a map on a flat surface. As noted earlier, a projection is needed because the Earth's surface is not flat but, rather, curved in a shape that is very close to the surface of a sphere. All map projections misstate the shape of the Earth in some way. It's simply impossible to transform a spherical surface to a flat (planar) surface without violating the true surface as a result of cutting, stretching, or otherwise distorting the information that lies on the sphere. GEODISCOVERIES

Perhaps the simplest of all map projections is a grid of perfect squares. In this simple map, horizontal lines are parallels and vertical lines are meridians. They are equally spaced in degrees. This grid is often used in modern computer-generated world maps displaying data that consist of a single number for each square, representing, for example, 1, 5, or 10 degrees of latitude and longitude. A grid of this kind can show the true spacing (approximately) of the parallels, but it fails to show how the meridians converge toward the two poles. This grid fails dismally in high latitudes, and the map usually has to be terminated at about 70° to 80° north and south.

Early attempts to find satisfactory map projections made use of a simple concept. Imagine the spherical Earth grid as a cage of wires (a kind of bird cage). A tiny light source is placed at the center of the sphere, and the image of the wire grid is cast upon a surface outside the sphere. This situation is like a reading lamp with a lampshade. Basically, three kinds of "lampshades" can be used, as shown in Figure S3.1.

First is a flat paper disk balanced on the north pole. The shadow of the wire grid on this plane surface will appear as a combination of concentric circles (parallels) and radial straight lines (meridians). Here we have a polar-centered projection. Second is a cone of paper resting point-up on the wire grid. The cone can be slit down the side, unrolled, and laid flat to produce a map that is some part of a full circle. This is called a conic projection. Parallels are arcs of circles, and meridians are radiating straight lines. Third, a cylinder of paper can be wrapped around the wire sphere so as to be touching all around the equator. When slit down the side along a meridian, the cylinder can be unrolled to produce a cylindrical projection, which is a true rectangular grid.

Take note that none of these three projection methods can show the entire Earth grid, no matter how large a sheet of paper is used to receive the image. Obviously, if the entire Earth grid is to be shown, some quite different system must be devised. Many such alternative solutions have been proposed. In our main chapter text, we described three types of projections used throughout the book—

the *polar projection;* the *Mercator projection,* which is a cylindrical projection; and the *Goode projection,* which uses a special mathematical principle.

Scales of Globes and Maps

All globes and maps depict the Earth's features in much smaller size than the true features they represent. Globes are intended in principle to be perfect scale models of the Earth itself, differing from the Earth only in size. The *scale* of a globe is the ratio between the size of the globe and the size of the Earth, where "size" is some measure of length or distance (but not of area or volume).

Take, for example, a globe 20 cm (about 8 in.) in diameter, representing the Earth, with a diameter of about 13,000 km. The scale of the globe is the ratio between 20 cm and 13,000 km. Dividing 13,000 by 20, we see that one centimeter on the globe represents 650 kilometers on the Earth. This relationship holds true for distances between any two points on the globe.

Scale is more usefully stated as a simple fraction, termed the *scale fraction.* It can be obtained by reducing both Earth and globe distances to the same unit of measure, which in this case is centimeters. (There are 100,000 centimeters in one kilometer.) The advantage of the scale fraction is that it is entirely free of any specified units of measure, such as the foot, mile, meter, or kilometer. It is usually written as a fraction with a numerator of one using either a colon or with the numerator above the denominator. For the example shown above, the scale fraction is obtained by reducing 20/1300000000 to 1/65000000 or 1:65,000,000.

Being a true-scale model of the Earth, a globe has a constant scale everywhere on its surface, but this is not true of a map projection drawn on a flat surface. In flattening the curved surface of the sphere to conform to a plane surface, all map projections stretch the Earth's surface in a nonuniform manner, so that the map scale changes from place to place. So we can't say about any world map: "Everywhere on this map the scale is 1:65,000,000." It is, however, possible to select a meridian or parallel—the equator, for example—for which a scale fraction can be given, relating the map to the globe it represents. In Figure 3.9, the scale of the Mercator projection along its equator is about 1:325,000,000.

Small-Scale and Large-Scale Maps

When geographers refer to small-scale and large-scale maps, they refer to the value of the scale fraction. For example, a global map at a scale of 1:65,000,000 has a scale fraction value of 0.00000001534, which is obtained by dividing 1 by 65,000,000. A hiker's topographic map might have a scale of 1:25,000, for a scale value of 0.000040. Since the global scale value is smaller, it is a small-scale map, while the hiker's map is a large-scale map. Note that this contrasts with many persons' use of the terms *large-scale* and *small-scale.* When we refer in conversation to a large-scale phenomenon or effect, we typically refer to something that takes place over a large area and that is usually best presented on a small-scale map.

Maps of large scale show only small sections of the Earth's surface. Because they "zoom in," they are capable of carrying the enormous amount of geographic information that is available and that must be shown in a convenient and effective manner. Most large-scale maps carry a *graphic scale,* which is a line marked off into units representing kilometers or miles. Figure S3.2 shows a portion of a large-scale map on which sample graphic scales in miles, feet, and kilometers are superimposed. Graphic scales

S3.2 Graphic scales on a topographic map *A portion of a modern, large-scale topographic map for which three graphic scales have been provided. (U.S. Geological Survey.)*

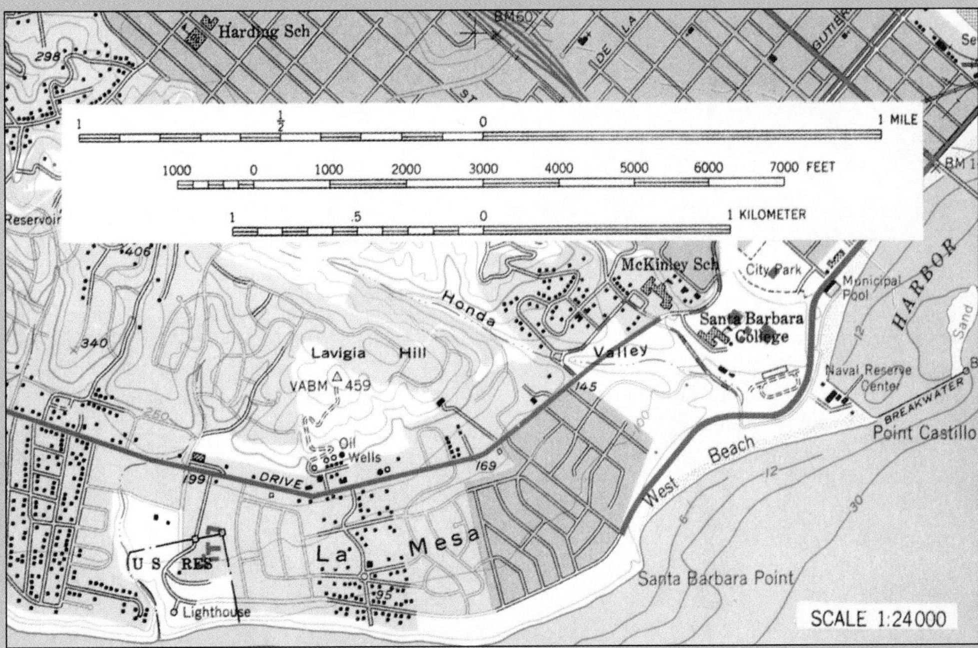

make it easy to measure ground distances.

For practical reasons, maps are printed on sheets of paper usually less than a meter (3 ft) wide, as in the case of the ordinary highway map or navigation chart. Bound books of maps—atlases, that is—consist of pages usually no larger than 30 by 40 cm (about 12 by 16 in.), whereas maps found in textbooks and scientific journals are even smaller.

Informational Content of Maps

The information conveyed by a map projection grid system is limited to one category only: absolute location of points on the Earth's surface. To be more useful, maps also carry other types of information. Figure S3.2 is a portion of a large-scale *multipurpose map.* Map sheets published by national governments, such as this one, are usually multipurpose maps. Using a great variety of symbols, patterns, and colors, these maps carry a high information content. A larger example of a multipurpose map is reproduced on the last pages of this book. It is a portion of a U.S. Geological Survey topographic quadrangle map for San Rafael, California.

In contrast to the multipurpose map is the *thematic map,* which shows only one type of information, or *theme.* We use many thematic maps in this text. Some examples include Figure 6.23, mapping the frequency of severe hailstorms in the United States; Figure 7.17, atmospheric surface pressures; Figure 9.5, mean annual precipitation of the world; and Figure 9.9, world climates.

Map Symbols

Symbols on maps associate information with points, lines, and areas. To express points as symbols, they can be renamed "dots." Broadly defined, a dot can be any small device to show point location. It might be a closed circle, open circle, letter, numeral, or a little picture of the object it represents (see "church with tower" in Figure S3.3). A line can vary in width and can be single or double. The line can also consist of a string of dots or dashes. A specific area of surface can be referred to as a "patch." The patch can be shown simply by a line marking its edge, or it can be depicted by a distinctive pattern or a solid color. Patterns are highly varied. Some consist of tiny dots and others of parallel, intersecting, or wavy lines.

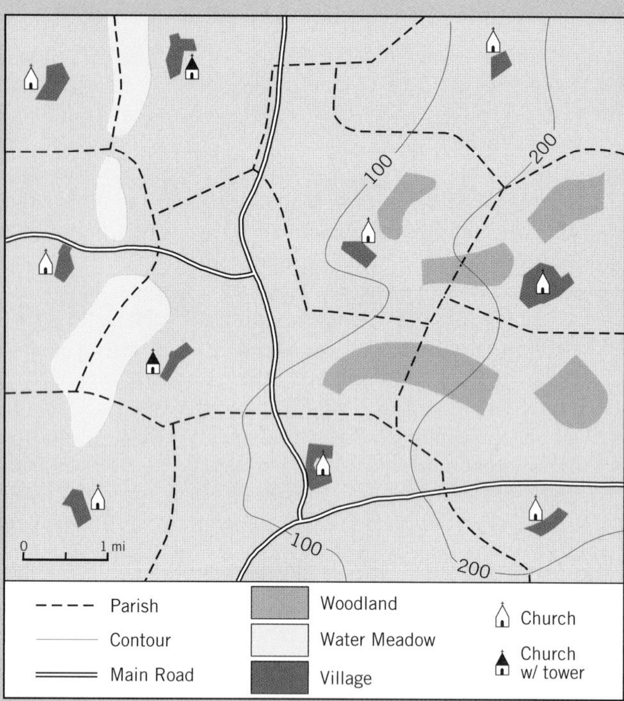

S3.3 Multipurpose map *A multipurpose map of an imaginary area with 10 villages illustrating the use of dots, lines, and patches. (After J. P. Cole and C.A.M. King, Quantitative Geography, copyright © John Wiley & Sons, London. Used by permission.)*

A map consisting of dots, lines, and patches can carry a great deal of information, as Figure S3.3 shows. In this case, the map uses two kinds of dot symbols (both symbolic of churches), three kinds of line symbols, and three kinds of patch symbols. Altogether, eight types of information are offered. Line symbols freely cross patches, and dots can appear within patches. Two different kinds of patches can overlap. For more examples of symbols, consult the display of topographic map symbols facing the map of San Rafael on the last pages of this book.

The relation of map symbols to map scale is of prime importance in cartography. Maps of very large scale, along with architectural and engineering plans, can show objects to their true outline form. As map scale is decreased, representation becomes more and more generalized. In physical geography, an excellent example is the depiction of a river, such as the lower Mississippi. Figure S3.4 shows the river channel at three scales, starting with a detailed plan, progressing to a double-line symbol that generalizes the channel form, and ending with a single-line symbol. As generalization develops, the details of the river banks and channel bends are simplified as well. The level of depiction of fine details is described by the term *resolution.* Maps of large scale have much greater resolving power than maps of small scale.

Presenting Numerical Data on Thematic Maps

How is numerical data shown on maps? In physical geography, much of the information collected about particular areas is in the form of numbers. The numbers might represent readings taken from a scientific instrument at various places throughout the study area. A simple example is the collection of weather data, such as air temperature, air pressure, wind speed, and amount of rainfall. Another example is the set of measurements of the elevation of a land surface, given in meters above sea level. Another category

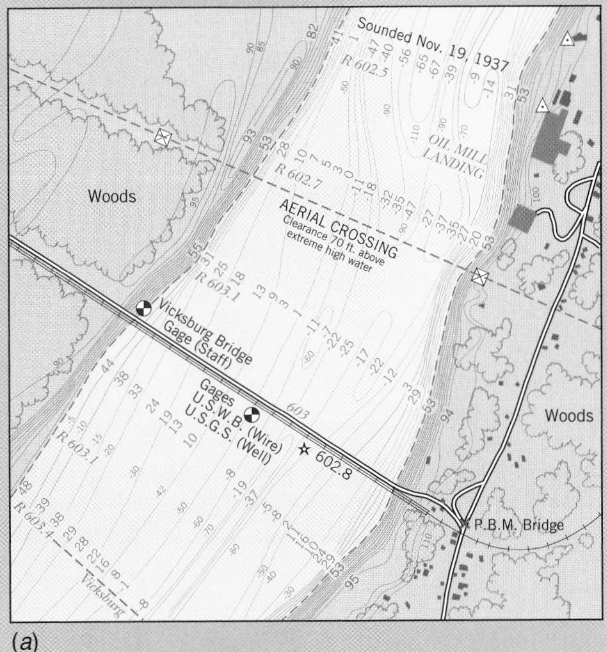

(a)

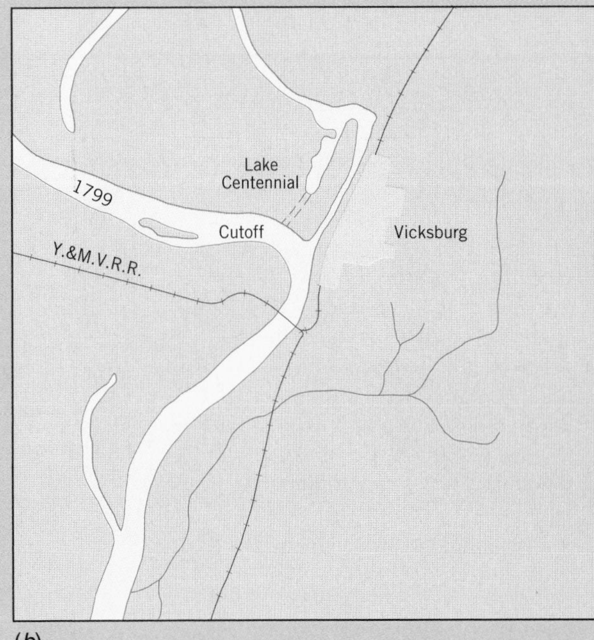

(b)

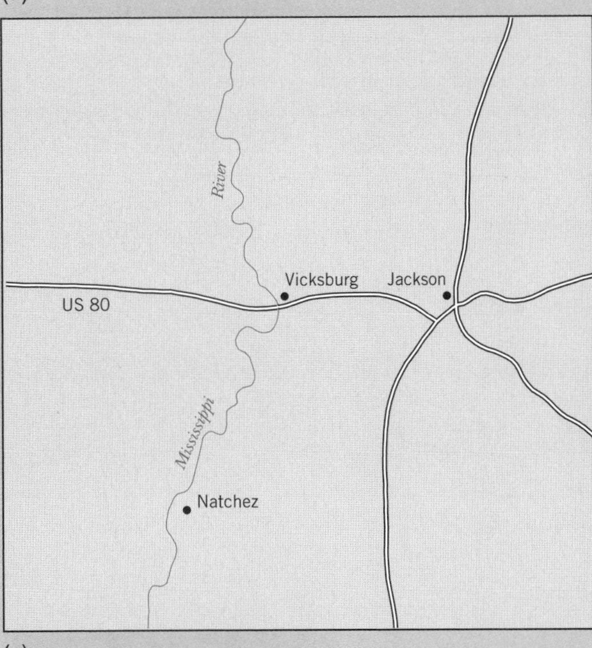

(c)

S3.4 Map scale and information content *Maps of the Mississippi River on three scales. (a) 1:20,000. Channel contours give depth below mean water level. (b) 1:250,000. Waterline only shown to depict channel. (c) 1:3,000,000. Channel shown as a solid line symbol. (Maps slightly enlarged for reproduction.) (Modified from U.S. Army Corps of Engineers.)*

of information consists merely of the presence or absence of a quantity or attribute. In such cases, we can simply place a dot to mean "present," so that when entries are completed, we have before us a field of scattered dots (Figure S3.5).

In some scientific programs, measurements are taken uniformly, for example, at the centers of grid squares laid over a map. For many classes of data, however, the locations of the observation points are predetermined by a fixed and nonuniform set of observing stations. For example, data of weather and climate are collected at stations typically located at airports. Whatever the sampling method used, we end up with an array of numbers and dots indicating their location on the field of the base map.

Although the numbers and locations may be accurate, it may be difficult to see the spatial pattern present in the data being displayed. For this reason, cartographers often simplify arrays of point values into isopleth maps. An *isopleth* is a line of equal value (from the Greek *isos,* "equal," and *plethos,* "fullness" or "quantity"). Figure 5.19 shows how an isopleth map is constructed for temperature data. In this case, the isopleth is an *isotherm,* or line of constant temperature. In drawing an isopleth, the line is routed among the points in a way that best indicates a uniform value, given the observations at hand.

Isopleth maps are important in various branches of physical geography. Table S3.1 gives a partial list of isopleths of various kinds used in the Earth sciences, together with their special names and the kinds of information they display. Examples are cited from our text. A special kind of isopleth, the *topographic contour* (or isohypse), is shown on the maps in Figures S3.2, S3.4*(a)*, and in the portion of the San Rafael topographic map at the close of the book. Topographic contours show the configuration of land surface features, such as hills, valleys, and basins.

In contrast to the isopleth map is the *choropleth* map, which identifies information in categories. Our global maps of soils (Figure 21.13) and vegetation (Figure 24.3) are examples of thematic choropleth maps.

Cartography is a rich and varied field of geography with a long history of conveying geographic information accurately and efficiently. If you are interested in maps and map-making, you might want to investigate cartography further.

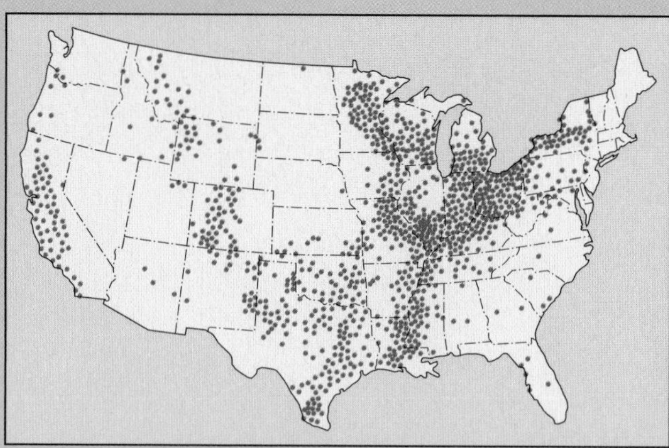

S3.5 Dot map *A dot map showing the distribution of soils of the order Alfisols in the United States. (From P. Gersmehl, Annals of the Assoc. of Amer. Geographers, vol. 67. Copyright © Association of American Geographers. Used by permission.)*

Table S3.1 | **Examples of Isopleths**

Name of Isopleth	Greek Root	Property Described	Examples in Figures
Isobar	*barros,* weight	Barometric pressure	7.17
Isotherm	*therme,* heat	Temperature of air, water, or soil	5.19
Isotach	*tachos,* swift	Fluid velocity	7.26
Isohyet	*hyetos,* rain	Precipitation	6.17
Isohypse (topographic contour)	*hypso,* height	Elevation	20.9, 20.10

4 THE GLOBAL ENERGY SYSTEM

EYE ON THE LANDSCAPE

Dromedary caravans near Nouakchott, Mauritania. The Earth's bright deserts, like the Sahara shown here, reflect back to space almost half of the solar energy they receive. **What else would the geographer see?...Answers at the end of the chapter.**

Lounging on an exotic tropical beach, you might be particularly aware of the Sun. Basking in its glow, you might be impressed by its ability to light your world and warm your body across millions of miles of black space. Actually, the Sun's power is even more impressive, considering that most natural phenomena that take place at the Earth's surface are directly or indirectly solar-powered. From the downhill flow of a river to the movement of a sand dune to the growth of a forest, solar radiation drives nearly all of the natural processes that shape the world around us. As we will see here and in many other chapters of this book, it is the power source for wind, waves, weather, rivers, and ocean currents. This chapter explains how solar radiation is intercepted by our planet, flows through the Earth's atmosphere, and interacts with the Earth's land and ocean surfaces.

A primary topic of this chapter is the *energy balance* of the Earth. The energy balance refers to the balance between the flows of energy reaching the Earth, which includes land and ocean surfaces and the atmosphere, and flows of energy leaving it. The Earth's energy balance controls the seasonal and daily changes in the Earth's surface temperature. Differences in energy flow rates from place to place also drive currents of air and ocean water. All these, in turn, produce the changing weather and rich diversity of climates we experience on the Earth's surface.

Human activities now dominate many regions of the Earth, and we have irreversibly modified our planet by changing much of its surface cover and adding carbon dioxide to its atmosphere. Have we shifted the balance of energy flows? Is our Earth absorbing more solar energy and becoming warmer? Is it absorbing less and becoming cooler? Before we can understand human impact on the Earth–atmosphere system, we must examine the global energy balance in detail.

Focus on Systems 4.1

Forms of Energy

Cosmologists—those scientists who study the universe—tell us that everything that exists is composed of either matter or energy. Energy is what causes matter to change through time within the universe. That goes, of course, for our planet and its inhabitants. On Earth, energy drives the life cycle of every individual organism as it comes into existence, grows, reproduces, and dies. Every action of an organism through that life span requires the intake and expenditure of energy. Whatever you learn here about energy and its relationship to matter will apply not only to our physical geography topics but to every personal life experience and every event you observe going on in the environment around you.

The term *energy* is a comparative newcomer to physics. It was first used in 1807, but the idea is much older, going back to Sir Isaac Newton's laws of motion, published in 1687. Leibnitz, a great philosopher and mathematician, proposed in 1695 the existence of a "vital force." This force, acting over distance (force times distance), is equal to one-half of mass times velocity squared ($^1/_2\, mv^2$). Today we call the vital force *kinetic energy*—the energy of mass (or "matter") in motion.

Physicists define *energy* as the ability to do work. So what is work? Work, again according to physicists, is equal to the change in kinetic energy in a body. This obviously circular definition is of no real help to us in understanding the nature of energy. We already know that energy is somehow tied in with matter in motion through space and time, but what *is* energy?

Whatever energy is, it does seem to "move" or "travel" with (or within) forms of matter. For example, we understand that the gasoline engine in our car transfers the energy of fuel combustion into forward motion of the entire structure of the car, including the engine, by means of the transmission and wheels. This entire moving object is said to possess kinetic energy. Kinetic energy is one of the forms of energy that comes under the general heading of *mechanical energy.* Should this speeding automobile crash head-on into a massive power pole, it will demonstrate its ability to do work upon its own body, upon its passengers, and upon the pole as well. The energy released in the collision will increase in direct proportion to the mass of the car, and it will also increase with the square of the auto's speed, since, as Liebnitz suggested, "vital force," or kinetic energy, is equal to $^1/_2 mv^2$.

Now the crushed car lies silent and still. What happened to all that energy of motion? Energy cannot just be destroyed and disappear. If you had been on the spot immediately after the crash you would have found the car body to be extremely hot, even though it did not catch fire. Kinetic energy of motion has turned into *heat energy.* In this form of energy, the atoms and molecules within a solid (or liquid, or gas) are in extremely fast motion invisible to us. In the metal of the car, the individual molecules are "vibrating" in place. The hotter the metal, the faster they vibrate. So here we find another form of kinetic energy. The proper name for it is "kinetic energy of atoms in motion," but it is also called *sensible heat* because it can be sensed by touch as well as being measured by a thermometer. This form of energy is extremely important in physical geography, and we will encounter it over and

Electromagnetic Radiation

Our study of the global energy system begins with the subject of radiation—that is, **electromagnetic radiation.** This form of energy is emitted by all objects. Light and radiant heat are two familiar examples of electromagnetic radiation. Light is radiation that is visible to our eyes. Heat radiation, though not visible, is easily felt when you hold your hand near a warm object, such as an oven or electric stove burner. Electromagnetic radiation is one of several fundamental forms of energy. Other forms that are important in physical geography are discussed in *Focus on Systems 4.1 • Forms of Energy.*

Think of electromagnetic radiation as a collection, or spectrum, of waves of a wide range of wavelengths traveling quickly away from the surface of an object. *Wavelength* describes the distance separating one wave crest from the next wave crest (Figure 4.1). The unit we will use to measure wavelength is the *micrometer* (formerly called the micron). A micrometer is one millionth of a meter, or 10^6 m. This is such a small unit that the tip of your little finger is about 15,000 micrometers wide. In this text, we use the abbreviation μm for the micrometer. The first letter is the Greek letter μ, or mu. It is used in metric units to denote micro-, meaning one-millionth (10^{-6}). Radiant energy can exist at any wavelength. Heat and light are identical forms of electromagnetic radiation except for their wavelengths.

Figure 4.2 shows how electromagnetic waves differ in wavelength throughout their entire range, or *spectrum.* At the short wavelength end of the spectrum are *gamma rays* and *X rays.* Their wavelengths are normally expressed in *nanometers.* A nanometer is one one-thousandth of a micrometer, or 10^{-9} m, and is abbreviated nm. Gamma and X rays have high energies and can be hazardous to health. Gamma and X-radiation grade into *ultraviolet* radiation, which begins at about 10 nm and extends to 400 nm or over in studying temperature changes in air, water, and soil.

If you had stood close to the collapsed auto just after the impact, you might have felt a glow of heat on the exposed skin of your face and hands. This is yet another form of energy: *radiant energy.* It is described and explained in detail in this chapter because it is the form of energy we receive from the Sun and it "runs the whole show" on our planet. We know radiant energy best in the form of sunlight, which travels in straight lines through space and easily penetrates transparent substances such as air, clear water, and even clear solids such as glass.

Let's move to another form of energy, one that depends on the attractive force of gravity. Imagine dragging or rolling a heavy object—a boulder, for example—to the top of a hill. Feeling the spirit of Galileo atop the Leaning Tower of Pisa, as he dropped objects to the ground, you nudge the boulder off a cliff at the top of the hill and watch it drop to the ground below. It lands with a satisfying thud.

Now, analyze the energy changes so far. In sweating the boulder to the top of the hill (an expenditure of energy released from your body), you overcame the ever-present downward pull of gravity. In so doing, you endowed the boulder with a certain amount of *stored energy.* This quantity of stored energy belongs to the energy class known as *potential energy.* With each meter of vertical distance the boulder falls, it converts a certain fixed quantity of that potential energy into kinetic energy. (We can disregard the small energy loss by friction with the surrounding air.) Upon impact, the kinetic energy is converted to heat energy, which quickly dissipates.

An important point is that potential energy, present because of gravity, must always be evaluated in terms of a given reference level, or *base level.* The standard base level is the average level of the ocean, called "mean sea level." In nature, any kind of matter on the land or in the atmosphere can be said to hold potential energy with respect to sea level. Examples are raindrops and hailstones formed in a storm cloud and the mineral particles of ash or cinder spewed from an erupting volcano. As these substances fall toward the Earth, their potential energy is converted to kinetic energy. Potential energy can also be converted to kinetic energy as substances descend along sloping and winding pathways. Flow of water in a river and of ice in a glacier are two examples.

Finally, in this introduction to energy found in the natural realms of our life environment, we must recognize *chemical energy,* which is absorbed or released by matter when chemical reactions take place. These reactions involve the coming together of atoms to form simple molecules, the recombining of simple molecules into new and more complex compounds, and the reverse changes back into simpler forms. Most of the time, chemical energy is in the stored form, held within the molecule. An important example is the way in which green plants absorb radiant energy of the Sun's rays to produce those complex organic molecules we refer to as carbohydrates.

Chemical energy is also stored for long periods of geologic time in the hydrocarbon compounds we use as fuels. The gasoline that powered our automobile to its tragic end represents stored chemical energy. As the gasoline burns, its molecules combine with oxygen of the air, releasing heat energy as well as many simpler molecules, such as carbon dioxide and water.

At this point, the realm of energy in all its forms may seem complicated and hard to grasp. However, we will return to the topic of energy many times in this book, and as you work with concepts of energy, they will become clearer.

0.4 μm. Like gamma and X rays, ultraviolet radiation can be damaging to living tissues.

The **visible light** portion of the spectrum starts at about 0.4 μm with violet. Colors then grade through blue, green, yellow, orange, and red, reaching the end of the visible spectrum at about 0.7 μm. Next is *near-infrared* radiation, with wavelengths from 0.7 to 1.2 μm. This radiation is very similar to visible light in that most of it comes from the Sun. However, human eyes are not sensitive to radiation beyond about 0.7 μm, so we do not see near-infrared light. Healthy plant leaves and green tissues reflect near-infrared light strongly, a fact that is often used in remote sensing with near-infrared light.

Short-wave infrared radiation lies in the range of 1.2 to 3.0 μm. Like near-infrared light, it comes mostly from the Sun. In remote sensing, this radiation can distinguish clouds from snow. It is also useful for differentiating rock types, which may have different "colors" in this portion of the spectrum. From 3 μm to about 6 μm is *middle-infrared* radiation. It comes from the Sun as well but is also emitted by fires burning at the Earth's surface, such as forest fires or gas well flames.

At wavelengths between 6 and about 300 nm, we have *thermal infrared* radiation. This radiation includes the heat radiation given off by bodies at temperatures normally found at the Earth's surface. As we'll see shortly, most heat radiation lies in the range of 8 to 12 μm. Beyond infrared wavelengths lie the domains of microwaves, radar, and communications transmissions, such as radio and television.

Thermal infrared radiation is no different in principle from visible light radiation. Although photographic film is not sensitive to thermal radiation, it is possible to acquire an image of thermal radiation using a special sensor.

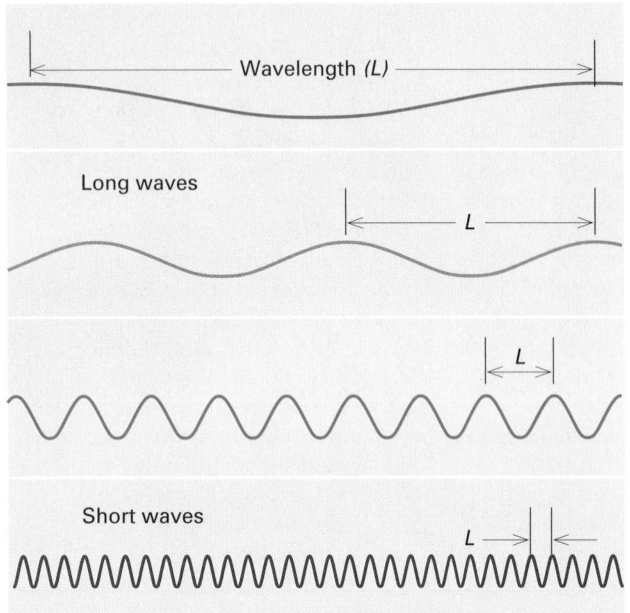

4.1 Wavelength of electromagnetic radiation *Electromagnetic radiation can be described as a collection of energy waves with different wavelengths. Wavelength L is the crest-to-crest distance between successive wave crests.*

Figure 4.3 shows a thermal infrared image of a suburban scene obtained at night. Here the image brightness is displayed using color, with red tones indicating the warmest temperatures and black tones the coldest. Windows appear red because they are warm and radiate more intensely. House walls are intermediate in temperature and appear blue. Roads and driveways are cool, as are the trees, shown in purple tones. Ground and sky are coldest (black). GEODISCOVERIES

Radiation and Temperature

There are two important physical principles that concern the emission of electromagnetic radiation. The first is that an inverse relationship exists between the range of wavelengths of the radiation that an object emits and the temperature of the object. For example, the Sun, a very hot object, emits radiation with short wavelengths. In contrast, the Earth, a much cooler object, emits radiation with longer wavelengths.

The second principle is that hot objects radiate more energy than cooler objects—much more. In fact, the flow of radiant energy from the surface of an object is directly related to the absolute temperature of the surface raised to the fourth power. (Absolute temperature is temperature measured on the Kelvin scale, with zero being the absence of all heat.) Thus, if the absolute temperature of an object is doubled, the flow of radiant energy from its surface will be 16 times larger. Because of this relationship, a small increase in temperature can mean a large increase in the rate at which radiation is given off by an object or surface. For example, water at room temperature emits about one-third more energy than when it is at the freezing point. These two principles are described in more depth in *Working It Out 4.2 • Radiation Laws*. Understanding them will help you to master the global energy balance.

Solar Radiation

Our Sun is a ball of constantly churning gases that are heated by continuous nuclear reactions. It is about average in size compared to other stars, and it has a surface temper-

4.2 The electromagnetic spectrum *Electromagnetic radiation can exist at any wavelength. By convention, names are assigned to specific wavelength regions as shown in the figure.*

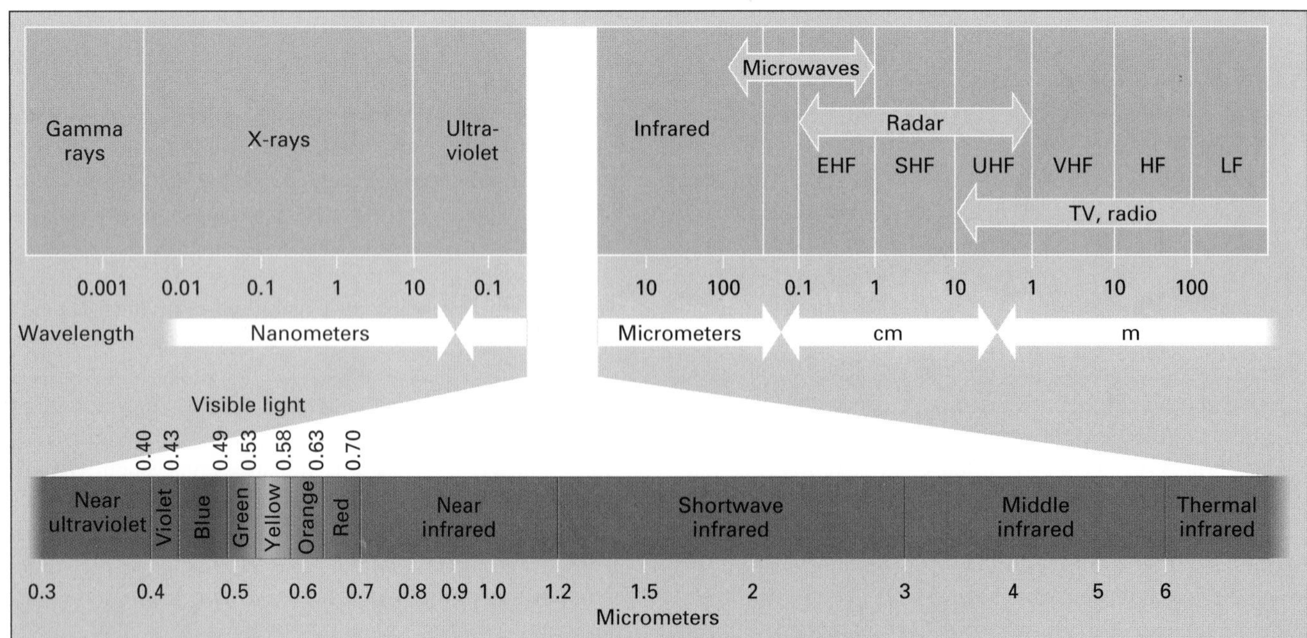

4.3 A thermal infrared image *This suburban scene was imaged at night in the thermal infrared spectral region. Black and blue tones show lower temperatures, while yellow and red tones show higher temperatures. Ground and sky are coldest, while the windows of the heated homes are warmest.*

ature of about 6000°C (about 11,000°F). Like all objects, it emits energy in the form of electromagnetic radiation. The energy travels outward in straight lines, or rays, from the Sun at a speed of about 300,000 km (about 186,000 mi) per second. At that rate, it takes the energy about $8^1/_3$ minutes to travel the 150 million km (93 million mi) from the Sun to the Earth.

As solar radiation travels through space, none of it is lost. However, the rays spread apart as they move away from the Sun. This means that a planet farther from the Sun, like

Working It Out 4.2

Radiation Laws

Physicists use two simple laws, expressed as formulas, to describe the principles that hotter objects radiate more energy and that the energy is of shorter wavelengths. First is the Stefan-Boltzmann Law. It describes the relation of flow of energy emitted by a surface to its temperature:

$$M = \sigma T^4$$

Here, M is the energy flow from the surface, in W/m^2; σ (the Greek letter sigma) is the Stefan-Boltzmann constant, 5.67×10^{-8} W/m^2/K^4; and T is the temperature of the surface, in °K.

The watt, W, is a measure of energy flow that is explained within the chapter text. The symbol K denotes absolute temperature, which is measured in kelvins. On this scale, 0 indicates the absence of heat energy. Absolute temperature is related to temperature in °C by the formula $K = C + 273$, where K is absolute temperature and C is Celsius temperature.

The second law applied to radiant energy is Wien's Law, which identifies the wavelength at which radiation emitted by a surface will be greatest, given the temperature:

$$\lambda_{MAX} = \frac{b}{T}$$

where λ_{MAX} is the wavelength at which radiation is at a maximum, in μm; b is a constant equal to 2898 μm K; and T is the temperature of the surface in K. At wavelengths to either side of $\lambda_{MAX,}$ the radiant energy flow rate will be lower.

Both of these laws apply only to a perfectly radiating surface, known as a *blackbody.* However, most natural surfaces are good radiators and do not differ too much from a blackbody.

As an example of the use of these formulas, let's calculate the flow of energy emitted by a surface and then determine the wavelength at which radiation from the surface is greatest. We will use a temperature of 450 K (177°C, 351°F) for the surface, which is approximately that of a home oven while baking. Using the Stefan-Boltzmann equation, substituting 450 K for T yields

$$\begin{aligned} M &= \sigma T^4 = (5.67 \times 10^{-8}\ \text{W/m}^2\text{K}^4) \times (450\ \text{K})^4 \\ &= 5.67 \times 10^{-8}\ \text{W/m}^2\text{K}^4 \times 4.10 \times 10^{10}\ \text{K}^4 \\ &= 2.32 \times 10^3\ \text{W/m}^2 \\ &= 2320\ \text{W/m}^2 \end{aligned}$$

Applying Wien's Law, we have

$$\lambda_{max} = \frac{b}{T} = \frac{2898\ \mu\text{m K}}{450\ \text{K}} = 6.44\ \mu\text{m}$$

In comparison, a similar calculation for a surface at room temperature (293 K, 273°C, 68°F) shows an emission rate of 418 W/m^2 and a wavelength of maximum emission of 9.89 μm. Thus, the oven surface emits more than five times as much energy as the room temperature surface, and the wavelength of maximum emission is somewhat shorter.

Mars, receives less radiation than one located nearer to the Sun, like Venus. The Earth intercepts only about one-half of one-billionth of the Sun's total energy output.

The Sun's interior is the source generating solar energy. Here, hydrogen is converted to helium at very high temperatures and pressures. In this process of nuclear fusion, a vast quantity of energy is generated and finds its way to the Sun's surface. Because the rate of production of energy is nearly constant, the output of solar radiation is also nearly constant. So, given the average distance of the Earth from the Sun, the amount of solar energy received on a small, fixed area of surface held at right angles to the Sun's rays is almost constant. This rate of incoming energy, known as the *solar constant,* is measured beyond the outer limits of the Earth's atmosphere, before energy has been lost in passing through the atmosphere. The solar constant has a value of about 1370 watts per square meter (W/m^2).

The *watt* (W), which describes a rate of energy flow, is a familiar measure of power. You've probably seen it applied to light bulbs or to home appliances, such as stereo amplifiers or microwave ovens. When we use the watt to measure the intensity of a flow of radiant energy, we must specify the unit of surface area that receives or emits the energy. This cross section is assigned the area of 1 square meter (1 m^2). Thus, the measure of intensity of received (or emitted) radiation is given as watts per square meter (W/m^2). Because there are no common equivalents for this energy flow rate in the English system, we will use only metric units.

Characteristics of Solar Energy

Let's look in more detail at the Sun's output as it is received by the Earth, which is shown in Figure 4.4. Energy intensity is shown on the graph on the vertical scale. Note that it is a logarithmic scale—that is, each whole unit marks an intensity 10 times greater than the one below. Wavelength is shown on the horizontal axis, also on a logarithmic scale.

The left side of Figure 4.4 shows how the Sun's incoming electromagnetic radiation varies with wavelength. The uppermost line shows how a "perfect" Sun would supply solar energy at the top of the atmosphere. By "perfect," we mean a Sun radiating as a *blackbody,* which follows physical theory

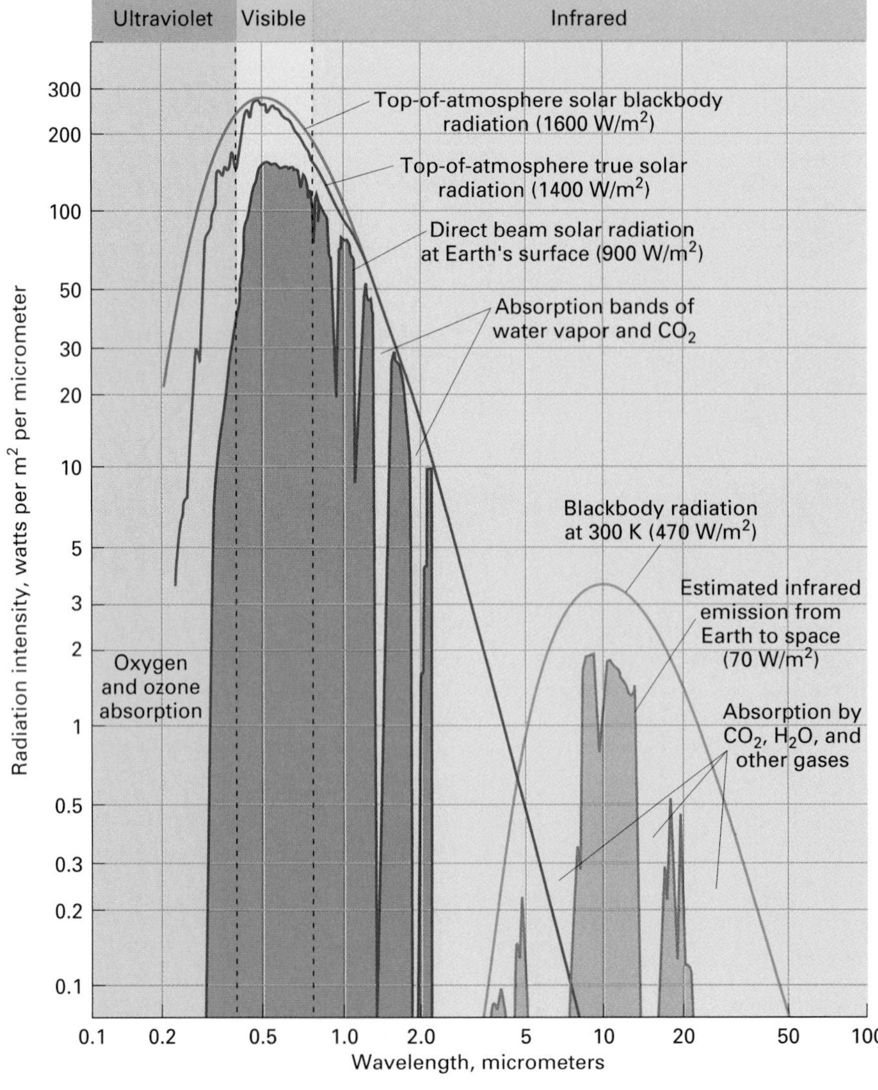

4.4 Spectra of solar and Earth radiation *This figure plots both shortwave radiation, which comes from the Sun (left side), and longwave radiation, which is emitted by the Earth's surface and atmosphere (right side). (After W. D. Sellers, Physical Climatology, University of Chicago Press. Used by permission.)*

exactly. (See *Working It Out 4.2 • Radiation Laws* for more details on blackbody radiation.) The solid line shows the actual output of the Sun as measured at the top of the atmosphere. It is quite close to the "perfect" Sun, except for ultraviolet wavelengths, where the real Sun emits less energy. Note that the Sun's output peaks in the visible part of the spectrum. Thus, our human vision system is adjusted to the wavelengths where solar light energy is most abundant.

The line showing solar radiation reaching the Earth's surface is quite different from the others, however. As solar radiation passes through the atmosphere, it is affected by **absorption** and scattering. Absorption occurs when molecules and particles in the atmosphere intercept and absorb radiation at particular wavelengths. This absorption is shown by the steep "valleys" in the graph—for example, at about 1.3 μm and 1.9 μm. At these wavelengths, molecules of water vapor and carbon dioxide absorb solar radiation very strongly and keep nearly all of it from reaching the Earth's surface. Note also that oxygen and ozone absorb almost all of the ultraviolet radiation at wavelengths shorter than about 0.3 μm. Atmospheric absorption is important because it warms the atmosphere directly and constitutes one of the flows of energy in the global energy balance that we will discuss toward the end of this chapter.

Solar radiation passing through the atmosphere can also be scattered. By **scattering,** we refer to a turning aside of radiation by a molecule or particle so that the direction of a scattered ray is changed. Scattered rays may go upward toward space or downward toward the surface. The line on Figure 4.4 is for direct beam energy only, so it does not include any scattered radiation that makes its way to the surface.

From the wavelengths shown on the horizontal axis, we see that solar energy received at the surface ranges from about 0.3 μm to 3 μm. This is known as **shortwave radiation.** These wavelengths are "short" as compared to the longer wavelengths of energy that are emitted by the Earth and atmosphere, which we discuss next.

Longwave Radiation from the Earth

Recall that both the range of wavelengths and the intensity of radiation emitted by an object depend on the object's temperature. Since the Earth's surface and atmosphere are much colder than the Sun's surface, we can deduce that our planet radiates less energy than the Sun and that the energy emitted has longer wavelengths.

The right side of Figure 4.4, which shows outgoing radiation, confirms these deductions. The upper line shows the radiation of a blackbody at a temperature of about 300 K (23°C, 73°F), which is not a bad approximation for the Earth as a whole. At this temperature, radiation ranges from about 3 to 30 μm and peaks at about 10 μm in the thermal infrared region. We refer to the thermal infrared radiation emitted by the Earth as **longwave radiation.**

Beneath the blackbody curve is an irregular line, shaded underneath, that shows upwelling energy emitted by the Earth and atmosphere as measured at the top of the atmosphere. Comparing this line with the blackbody curve shows that much of the Earth's surface radiation is absorbed by the atmosphere, especially between about 6 and 8 μm and between 14 and 17 μm, where absorption is nearly complete. Above 21 μm, absorption is also essentially complete. Absorption of Earth radiation by the atmosphere is an important part of the greenhouse effect, which we will discuss shortly. Water and carbon dioxide are the primary absorbers.

The atmospheric absorption regions of the spectrum leave three regions in which outgoing energy flow is significant—4 to 6 μm, 8 to 14 μm, and 17 to 21 μm. We can call these *windows* through which longwave radiation leaves the Earth and flows to space.

An examination of Figure 4.4 shows that significant amounts of radiant energy, both shortwave by scattering and longwave by radiation leave the Earth's surface and pass upward through the atmosphere. This energy, whose wavelength ranges from visible to thermal infrared, is therefore available for aircraft and satellite instruments to measure and compose into images of the Earth's surface through the process of remote sensing. Our interchapter feature, *A Closer Look: Geographer's Tools 4.8 • Remote Sensing for Physical Geography* provides some basic concepts of remote sensing that will prove useful in understanding both our *Focus on Remote Sensing* features and the uses of remotely sensed images in the main text.

The Global Radiation Balance

As we have seen, the Earth constantly absorbs solar shortwave radiation and emits longwave radiation. Figure 4.5 presents a diagram of this energy flow process, which we refer to as the Earth's **global radiation balance.**

The Sun provides a nearly constant flow of shortwave radiation toward Earth that is received at the top of the atmosphere. Part of this radiation is scattered away from the Earth and heads back into space without absorption. Clouds and dust particles in the atmosphere contribute to this scattering. Land and ocean surfaces also reflect some shortwave radiation back to space. The shortwave energy from the Sun that is not scattered or reflected is absorbed by either atmosphere, land, or ocean. Once absorbed, solar energy raises the temperature of the atmosphere as well as the surfaces of the oceans and lands.

The atmosphere, land, and ocean also emit energy in the form of longwave radiation. This radiation ultimately leaves the planet, headed for outer space. The longwave radiation outflow tends to lower the temperature of the atmosphere, ocean, and land, and thus cool the planet. In the long run, these flows balance—incoming energy absorbed and outgoing radiation emitted are equal. Since the temperature of a surface is determined by the amount of energy it absorbs and emits, the Earth's overall temperature tends to remain constant.

Using the radiation laws presented in *Working It Out 4.2,* we see that it is not difficult to calculate the magnitude of the energy flows involved in the global radiation balance. *Work-*

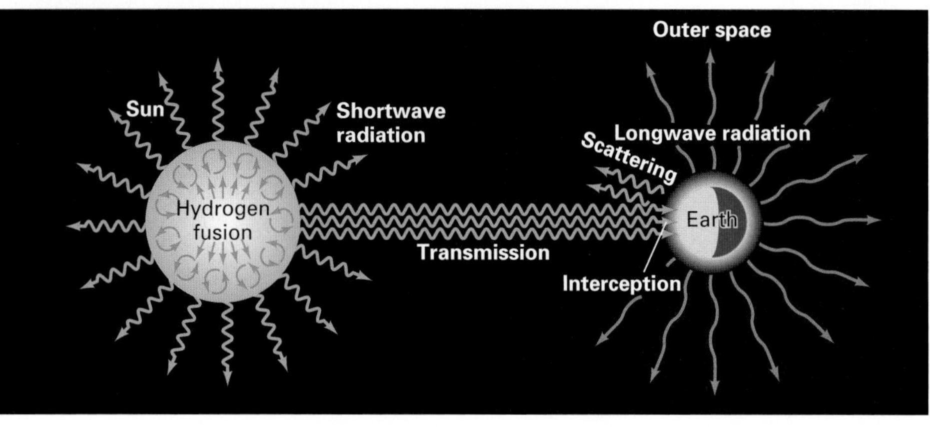

4.5 The global radiation balance *Shortwave radiation from the Sun is transmitted through space, where it is intercepted by the Earth. The absorbed radiation is then ultimately emitted as longwave radiation to outer space. (From A. N. Strahler, "The Life Layer," Journal of Geography, vol. 69, p. 72. Copyright 1970 © by The Journal of Geography. Used by permission.)*

ing It Out 4.3 • Calculating the Global Radiation Balance shows how this can be done.

INSOLATION OVER THE GLOBE

The flow of solar radiation to the Earth as a whole remains constant, but the amount received varies from place to place and time to time. We turn now to this variation and its causes. In this discussion, we will refer to **insolation** (**in**coming **sol**ar radi**ation**)—the flow of solar energy intercepted by an exposed surface assuming a uniformly spherical Earth with no atmosphere. Insolation is a flow rate and has units of watts per square meter (W/m^2). Daily insolation refers to the average of this flow rate over a 24-hour day. Annual insolation, discussed in a later section, is the average flow rate over the entire year. Insolation is important because it measures the amount of solar power available to heat the land surface and atmosphere.

Insolation depends on the angle of the Sun above the horizon. Figure 4.6 shows that insolation is greatest when the Sun is directly overhead and the Sun's rays are vertical (position *A* in the figure). When the Sun is lower in the sky, the same amount of solar energy spreads over a greater area of ground surface, so insolation is lower (position *B* in the figure).

Daily insolation at a location depends on two factors: (1) the angle at which the Sun's rays strike the Earth, and (2) the length of time of exposure to the rays. Recall from Chapter 3 that both of these factors are controlled by the latitude of the location and the time of year. For example, in midlatitude locations in summer, days are long and the Sun rises to a position high in the sky, thus heating the surface more intensely. Let's investigate these variations in more detail.

Refer back to Chapter 3 and Figure 3.18, showing equinox conditions. Only at the subsolar point does the Earth's spherical surface present itself at a right angle to the Sun's rays. To a viewer on the Earth at the subsolar point, the Sun is directly overhead. But as we move away from the subsolar point toward either pole, the Earth's curved surface becomes turned at an angle with respect to the Sun's rays. To the Earthbound viewer at a new latitude, the Sun appears to be nearer the horizon. When the circle of illumination is reached, the Sun's rays are parallel with the surface. That is, the Sun is viewed just at the horizon. Thus, the angle of the Sun in the sky at a particular location depends on the latitude of the location, the time of day, and the time of year. GEODISCOVERIES

Insolation and the Path of the Sun in the Sky

How does the angle of the Sun vary during the day? The angle depends on the Sun's path in the sky. When the Sun is high above the horizon—near noon, for example—the Sun's angle is greater, and so insolation will be greater. Figure 4.7 shows the Sun's daily path in the sky at various latitudes and how this path changes from season to season. Let's look at part *b* first, since it is likely to be the most familiar. This diagram is drawn for latitude 40° N and is typical of conditions found in midlatitudes in the northern hemisphere—for example, at New York or Denver. The diagram shows a small area of the Earth's surface bounded by

4.6 Insolation and Sun angle *The angle of the Sun's rays determines the intensity of insolation on the ground. The energy of vertical rays A is concentrated in square a by c, but the same energy in the slanting rays B is spread over a larger rectangle, b by c.*

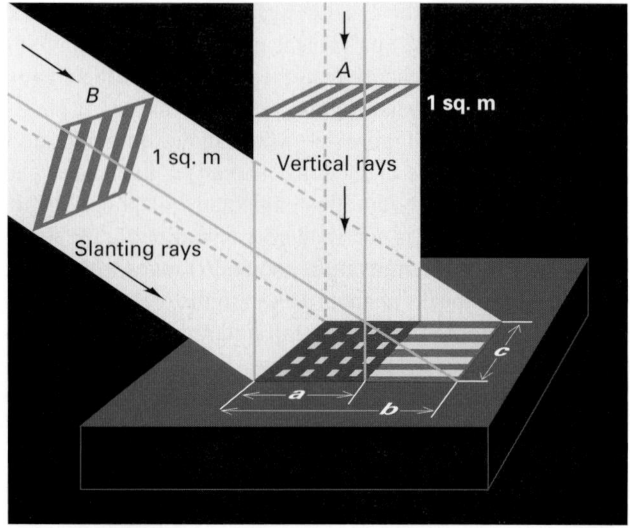

Working It Out | 4.3

Calculating the Global Radiation Balance

The Earth's global radiation balance, as diagrammed in Figure 4.5, is an example of an energy flow system. With a few simple facts and formulas, it is easy to estimate the magnitude of flow in the two principal pathways of this system—incoming shortwave radiation and outgoing longwave radiation.

First, what is the input flow of solar radiation to the Earth? To determine that, we need to know the rate of flow and the area that receives the flow. Recall that the rate is described by the solar constant, which has a value of about 1370 W/m^2 or 1.37 kW/m^2, where kW is the kilowatt, equal to 10^3 W. Now, the Earth's radius is about 6400 km. That means that the Earth presents to the parallel rays of the Sun a disk of radius 6400 km (see figure at right). The disk has an area $\pi r^2 = 3.14 \times (6.40 \times 10^3 \text{ km})^2 = 1.29 \times 10^8 \text{ km}^2$. Since 1 km^2 = 10^6 m^2, this is equal to 1.29×10^{14} m^2.

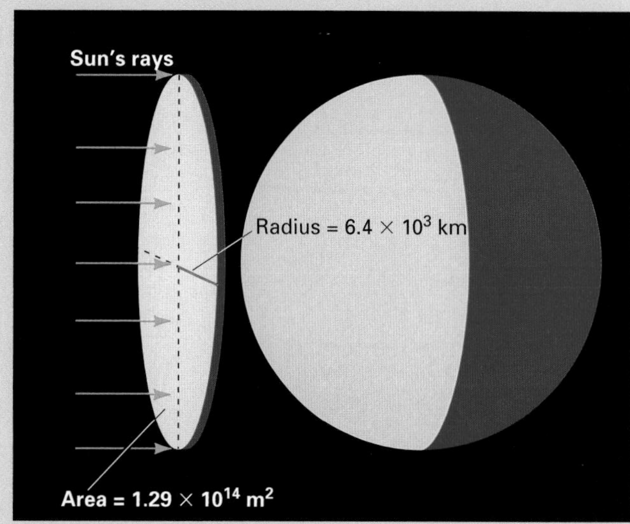

Interception of solar radiation *The Earth intercepts the parallel rays of the Sun in an area equal to a disk of Earth radius.*

Each square meter of the disk intercepts solar energy flow at the rate of the solar constant, so the energy flow from the Sun to the Earth is about 1.37 kW/m^2 × 1.29×10^{14} m^2 = 1.77×10^{14} kW. In other words,

$$\text{Flow per Unit Area} \times \text{Area of Earth Presented to Sun} = \text{Sun-Earth Energy Flow}$$

$$1.37 \text{ kW/m}^2 \times 1.29 \times 10^{14} \text{ m}^2 = 1.77 \times 10^{14} \text{ kW}$$

What about the outflows? These take the form of (1) reflected shortwave radiation and (2) emitted longwave radiation. The Earth as a planet reflects about one-third of the solar radiation that it receives, or about 1.77×10^{14} kW ÷ 3 = 0.589×10^{14} kW = 5.89×10^{13} kW. The remaining two-thirds, 1.17×10^{14} kW, is absorbed by the Earth and atmosphere and ultimately emitted as longwave radiation.

A simple diagram of this radiant energy flow system is shown in the second figure. Energy flows are shown by arrows. The Earth and its atmosphere receives an inflow of solar shortwave radiation. Within the Earth and atmosphere, some of the energy flow is transformed from shortwave radiation to longwave radiation. Both shortwave and longwave radiation flow outward to space.

Radiant energy flow system for the Sun and Earth–atmosphere system *The input is the flow of shortwave radiation from the Sun. Outputs are flows of reflected shortwave radiation and emitted longwave radiation.*

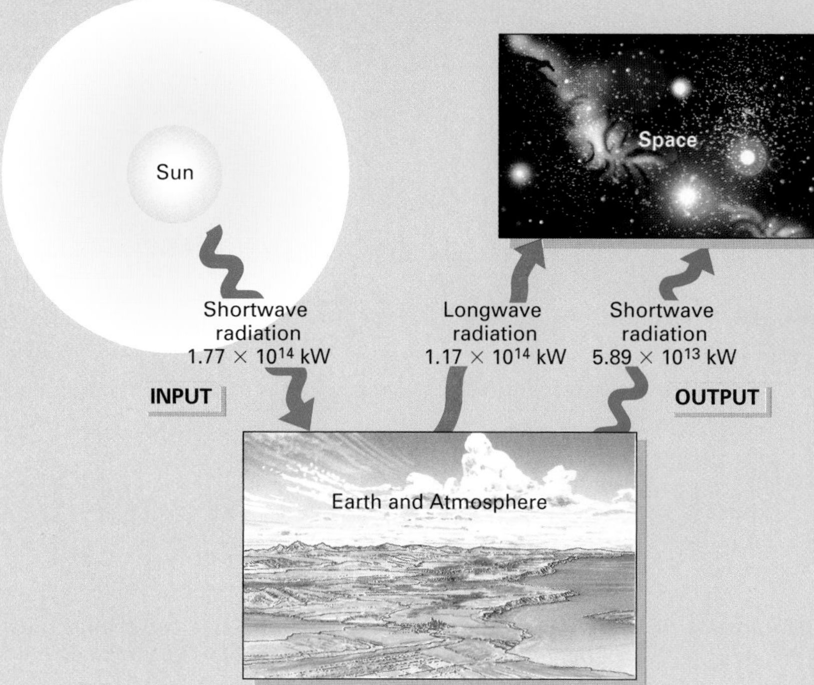

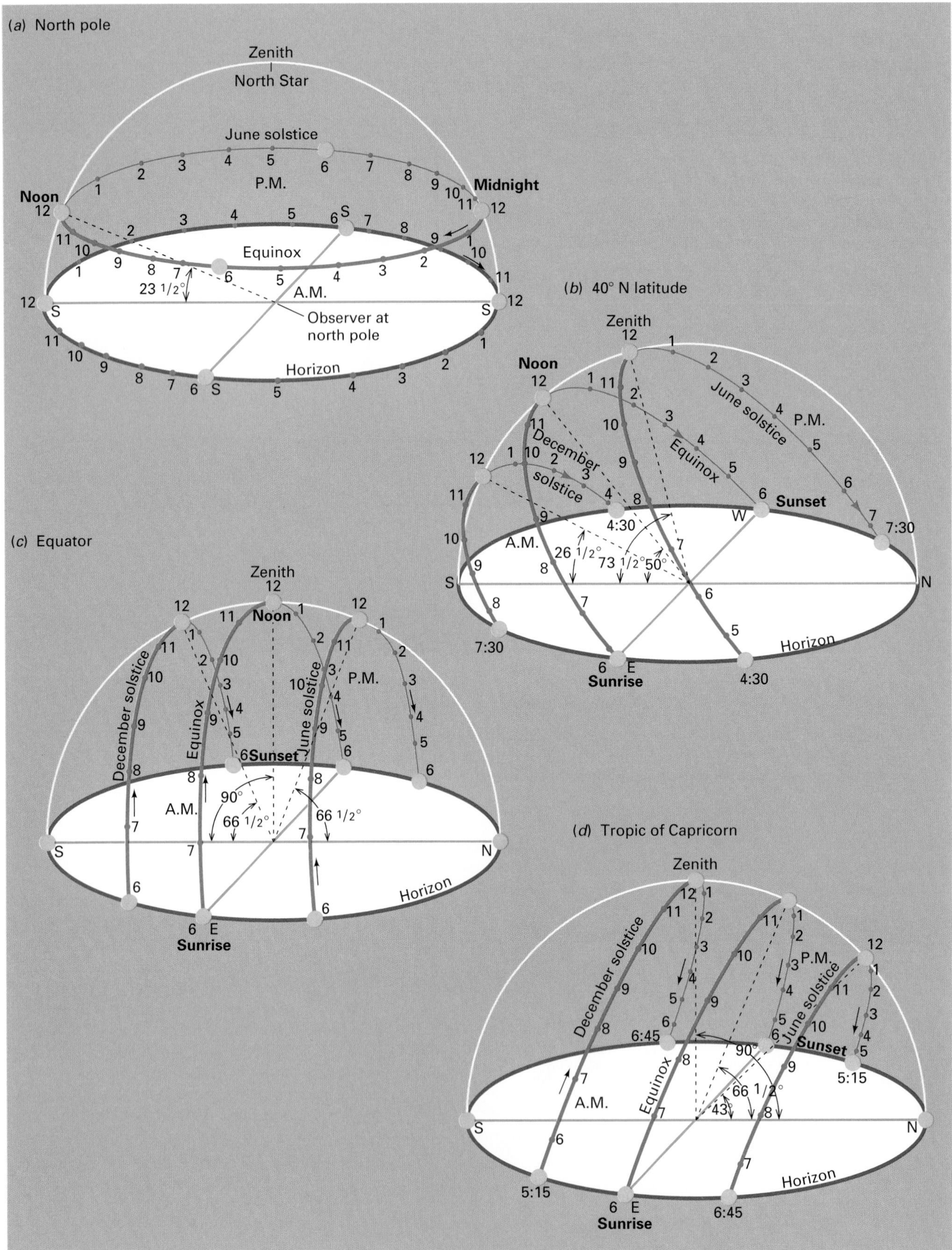

***4.7 The path of the Sun in the sky at different latitudes and seasons** The Sun's path in the sky can change greatly in position and height above horizon through the seasons. (a) North pole. (b) 40° N latitude. (c) Equator. (d) Tropic of capricorn. (Parts a, c, d, Copyright © A. N. Strahler. Used by permission.)*

a circular horizon. This is the way things appear to an observer standing on a wide plain. The Earth's surface appears flat, and the Sun seems to travel inside a vast dome in the sky. GEODISCOVERIES

At equinox, the Sun rises directly to the east and sets directly to the west. The noon Sun is positioned at an angle of 50° above the horizon in the southern sky. The Sun is above the horizon for exactly 12 hours, as shown by the hour numbers on the Sun's path. At the June solstice, the Sun's path rises much higher in the sky—at noon it will be $73^1/_2$° above the horizon. The Sun is above the horizon for about 15 hours, and it rises and sets at points on the horizon that are well to the north of east and west. Clearly, daily insolation will be greater at the June solstice than at the equinox since the Sun is in the sky longer and reaches a higher angle at noon. At the December solstice, the Sun's path is low in the sky, reaching only $26^1/_2$° above the horizon, and the Sun is visible for only about 9 hours. Sunrise and sunset are to the south of east and west. Thus, daily insolation reaching the surface at the December solstice will be less than at the equinox and much less than at the June solstice.

Figure 4.8 shows daily insolation values for a selection of latitudes in the northern hemisphere. Following the curve for 40° N, we see that daily average insolation ranges from about 160 W/m^2 at the December solstice to about 460 W/m^2 at the June solstice. Equinox values are around 350 W/m^2. Thus, the actual values confirm our expectation, based on our analysis of how high the Sun rises during the day and how long it stays above the horizon.

Part *a* of Figure 4.7 depicts conditions at the north pole. At this latitude, the Sun seems to trace a daily circle in the heavens that is parallel to the horizon. This motion is actually part of a long, flat spiral, as shown in Figure 4.9. At the June solstice, the Sun is elevated at $23^1/_2$ degrees above the horizon. As time proceeds to the September equinox, the Sun's elevation slowly falls to the horizon. Between the September and March equinoxes, the Sun is below the horizon. At the March equinox, it emerges above the horizon to work its way slowly up to the June solstice position during the next three months.

From this analysis of the Sun's motion, we see that daily insolation is greatest at the June solstice, when the Sun is at its highest position in the sky (Figure 4.7). Insolation then falls to zero at the September equinox, remains at zero between September and March equinoxes, and then

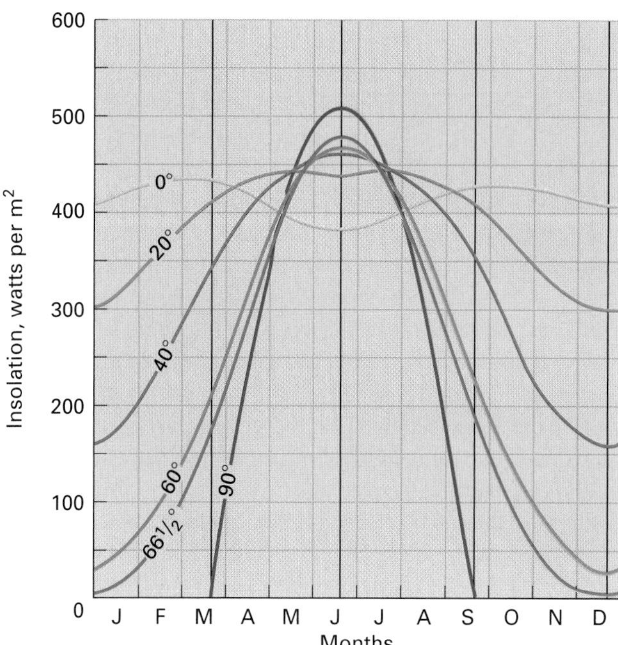

4.8 Daily insolation through the year at various latitudes (northern hemisphere) *Black lines mark the equinoxes and solstices. Latitudes between the equator (0°) and tropic of cancer ($23^1/_2$° N) show two maximum values; others show only one. Poleward of the arctic circle ($66^1/_2$° N), insolation is zero for at least some period of the year. (Copyright © A. N. Strahler. Used by permission.)*

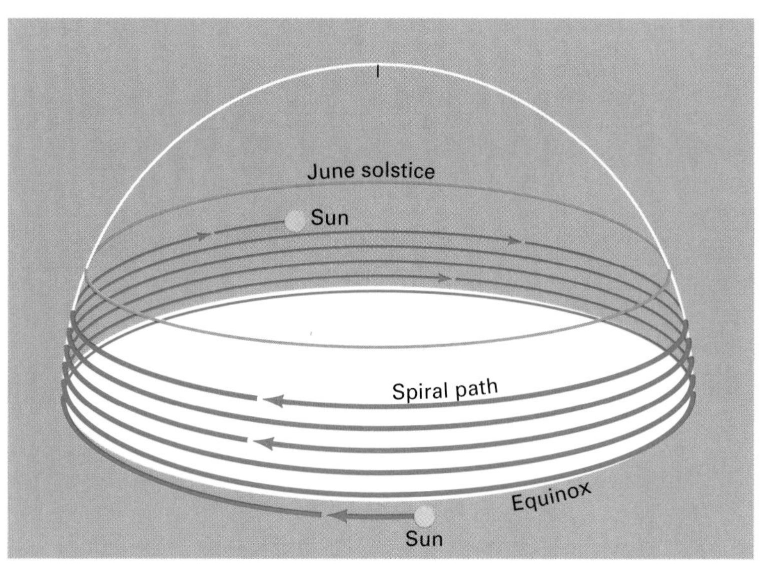

4.9 Sun's path in the sky at the north pole *At the poles, the Sun's daily path traces part of a long, fine spiral beginning on the horizon at the equinox, proceeding to $23^1/_2$° above the horizon at the solstice, and then descending to the horizon again at the following equinox. (For the south pole, the direction of the spiral is reversed.) (Copyright © A. N. Strahler. Used by permission.)*

increases as the Sun rises above the horizon to reach its position at the June solstice again. Note from Figure 4.8 that summer solstice daily insolation, over 500 W/m^2, is higher at the pole than at any other location at any other time of the year. This shows the importance of day length in determining daily insolation.

At the equator (part *c*), the Sun at the September equinox rises exactly east at 6:00 A.M., moves to the zenith, or position directly overhead, at noon, and then moves to set due west at 6 P.M., 12 hours later. At the December solstice, the Sun also rises at 6 A.M., but from a point south of east. At noon, it is not directly overhead but at an elevation of 66$^1/_2$° above the horizon. It sets at 6:00 P.M. at a point south of west. Comparing the solstice and equinox, we see that the Sun spends the same amount of time above the horizon (12 hours) but is not as high above the horizon on the solstice. Daily insolation is therefore smaller at the solstice.

As December moves to March, the Sun returns to the equinox position and then passes on to the June solstice position. Here the situation is the same as the December solstice but reversed in that the Sun rises to the north of east and sets to the north of west. Again, daily insolation is less at solstice than at equinox.

Note that for the equator, there are two maximums of insolation—one at each equinox—and two minimums—one at each solstice. This means that for locations near the equator, two hot seasons may be experienced. Figure 4.8 clearly shows the two peaks in daily insolation, each centered around an equinox. (Note that the two peaks are not quite equal and are not exactly at the equinoxes. These small effects occur because the Earth's orbit is not exactly circular.) There are actually two maximum daily insolation values at all latitudes between the tropic of cancer (23$^1/_2$° N) and the tropic of capricorn (23$^1/_2$° S). However, as either tropic is approached in a poleward direction, the two maximum periods get closer and closer in time, and then merge into a single maximum. The latitude 20° line in Figure 4.8 shows this clearly, with two maximums about two months apart.

The tropic of capricorn (part *d*), at 23$^1/_2$° S. lat., provides an example from the southern hemisphere with one maximum in daily insolation. At the equinox, the Sun rises directly east, moves to a noon position 66$^1/_2$° above the horizon to the north, and then sets directly west. On the December solstice, the Sun rises to the south of east at about 5:15 A.M. and moves to the zenith point, where it is directly overhead at noon. During the afternoon, it moves toward its sunset point, at a position to the south of west at about 6:45 P.M. Day length is about 13$^1/_2$ hours. At the June solstice, the Sun only rises to 43° above the horizon, and day length is only about 10$^1/_2$ hours. So the summer (December) solstice is the time of maximum daily insolation at the tropic of capricorn.

Daily Insolation Through the Year

Returning to Figure 4.8, what conclusions can we draw about daily insolation as it varies with latitude and season?

- Latitudes between the tropics of cancer and capricorn and arctic and antarctic circles show a wavelike pattern of greater daily insolation at the summer solstice and lower daily insolation at the winter solstice.
- Poleward of the arctic and antarctic circles, the Sun is below the horizon for at least part of the year, and daily insolation drops to zero during that period.
- Daily insolation is greatest of all latitudes at the pole at the summer solstice.
- There are two maximums and two minimums in daily insolation at the equator, occurring at the equinoxes and solstices, respectively.
- Poleward of the equator but between the two tropics, there are also two maximum and minimum daily insolation values, but as the tropic is approached, the two maximum periods get closer and closer in time, and then merge into a single maximum at the tropic.

These differences in daily insolation are important because daily insolation measures the flow of solar power available to heat the Earth's surface. Thus, it is the most important factor in determining air temperatures, as we will see in later chapters. The change in daily insolation with the seasons at a location is therefore a major determinant of climate.

Annual Insolation by Latitude

What is the effect of latitude on annual insolation—the rate of insolation averaged over an entire year? Figure 4.10 shows two curves of annual insolation by latitude—one for the actual case of the Earth's axis tilted at 23$^1/_2$° and the other for an Earth with an untilted axis. Let's look first at the

4.10 *Annual insolation from equator to pole for the Earth* *The red line shows the actual curve. The blue line shows what the annual insolation would be if the axis were perpendicular to the plane of the ecliptic.*

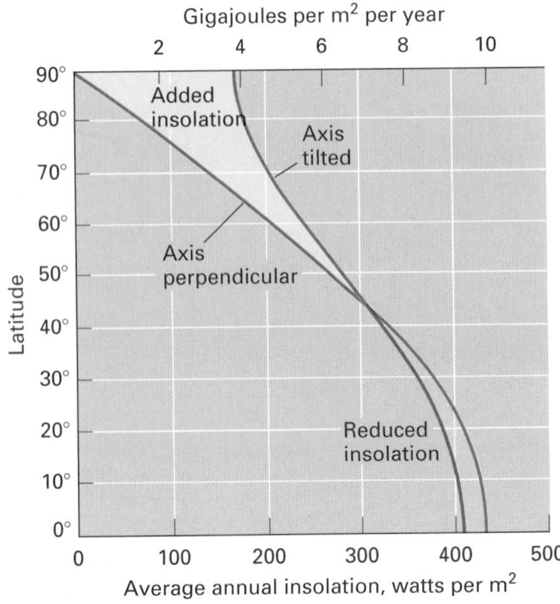

real case of a tilted axis. Annual insolation varies smoothly from the equator to the pole. As we might expect, annual insolation is greater at lower latitudes. But the high latitudes still receive a considerable flow of solar energy—the annual insolation value at the pole is about 40 percent of the value at the equator.

The case of the untilted axis—that is, for which the Earth's axis is perpendicular to the plane of the ecliptic—is quite different. In this case, there are no seasons. The Sun traces its equinox path in the sky every day of the year at every location. Under these circumstances, annual insolation changes from a high value at the equator, where the Sun is directly overhead at noon every day, to zero at the poles, where the Sun is always just on the horizon. The curve of annual insolation thus shows a progression from a value near 440 W/m^2 at the equator to zero at 90° latitude.

The comparison of the two cases shows that the tilting of the Earth's axis redistributes a very significant portion of the Earth's insolation from the equatorial regions toward the poles. Although the pole is in the dark for six months of the year, it still receives nearly half the amount of solar radiation received at the equator. Without a tilted axis, our planet would be quite a different place!

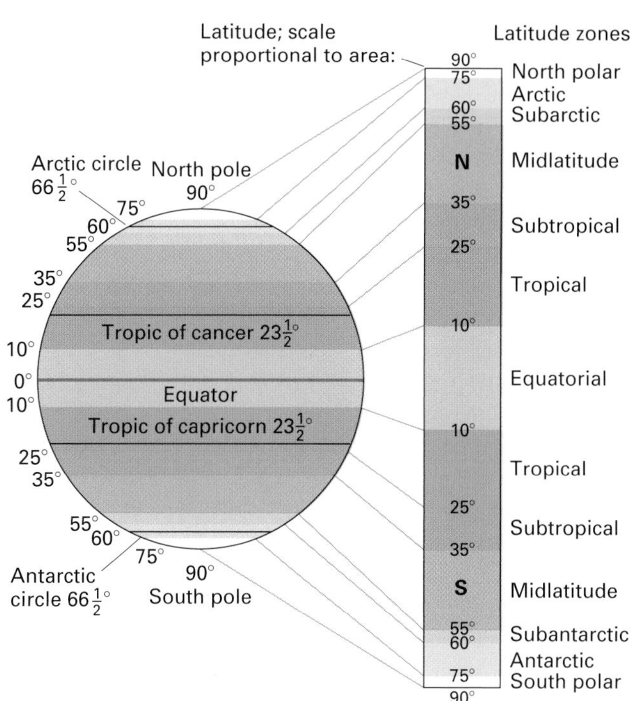

4.11 World latitude zones *A geographer's system of latitude zones, based on the seasonal patterns of daily insolation observed over the globe.*

World Latitude Zones

The seasonal pattern of daily insolation can be used as a basis for dividing the globe into broad latitude zones (Figure 4.11). The zone limits shown in the figure and specified below should not be taken as absolute and binding, however. Rather, this system of names is a convenient way to identify general world geographic belts throughout this book.

The *equatorial zone* encompasses the equator and covers the latitude belt roughly 10° north to 10° south. Here the Sun provides intense insolation throughout most of the year, and days and nights are of roughly equal length. Spanning the tropics of cancer and capricorn are the *tropical zones,* ranging from latitudes 10° to 25° north and south. A marked seasonal cycle exists in these zones, combined with high annual insolation.

Moving toward the poles from each of the tropical zones are transitional regions called the *subtropical zones.* For convenience, we assign these zones the latitude belts 25° to 35° north and south. At times we may extend the label "subtropical" a few degrees farther poleward or equatorward of these parallels. These zones have a strong seasonal cycle combined with an annual insolation nearly as large as the tropical zones.

The *midlatitude zones* are next, lying between 35° and 55° north and south latitude. In these belts, the Sun's height in the sky shifts through a wide range annually. Differences in day length from winter to summer are also large. Thus, seasonal contrasts in insolation are quite strong. In turn, these regions can experience a large range in annual surface temperature.

Bordering the midlatitude zones on the poleward side are the *subarctic* and *subantarctic zone,* 55° to 60° north and south latitudes. Astride the arctic and antarctic circles from latitudes 60° to 75° N and S lie the *arctic* and *antarctic zones.* These zones have an extremely large yearly variation in day lengths, yielding enormous contrasts in insolation from solstice to solstice.

The *polar zones,* north and south, are circular areas between about 75° latitude and the poles. Here the polar regime of a six-month day and six-month night is predominant. These zones experience the greatest seasonal contrasts of insolation of any area on Earth.

Composition of the Atmosphere

A main goal of this chapter is to explain the flows of energy within the atmosphere and between the atmosphere and the Earth's surface. To understand this topic, you need to know some basic facts about the atmosphere.

The Earth's atmosphere consists of *air*—a mixture of various gases surrounding the Earth to a height of many kilometers. This envelope of air is held to the Earth by gravitational attraction. Almost all the atmosphere (97 percent) lies within 30 km (19 mi) of the Earth's surface. The upper limit of the atmosphere is at a height of approximately 10,000 km (about 6000 mi) above the Earth's surface, a distance approaching the diameter of the Earth itself.

From the Earth's surface upward to an altitude of about 80 km (50 mi), the chemical composition of air is highly uniform in terms of the proportions of its gases. Pure, dry air consists largely of nitrogen, about 78 percent by volume, and oxygen, about 21 percent (Figure 4.12). Other gases account for the remaining 1 percent.

Nitrogen in the atmosphere exists as a molecule consisting of two nitrogen atoms (N_2). Nitrogen gas does not enter easily into chemical union with other substances and can be thought of mainly as a neutral substance. Very small amounts of nitrogen are extracted by soil bacteria and are made available for use by plants. Otherwise, it is largely a "filler," adding inert bulk to the atmosphere.

In contrast to nitrogen, *oxygen gas* (O_2) is highly active chemically. It combines readily with other elements in the process of oxidation. Combustion of fuels represents a rapid form of oxidation, whereas certain forms of rock decay (weathering) represent very slow forms of oxidation. Living tissues require oxygen to convert foods into energy.

These two main component gases of the lower atmosphere are perfectly diffused so as to give pure, dry air a definite set of physical properties, as though it were a single gas. The remaining 1 percent of dry air is mostly argon, an inactive gas of little importance in natural processes. In addition, there is a very small amount of *carbon dioxide* (CO_2), amounting to about 0.035 percent. As we saw in Figure 4.4, CO_2 absorbs incoming shortwave radiation and outgoing longwave radiation in specific wavelength regions. Longwave absorption is particularly important for the greenhouse effect, in which the lower atmosphere is warmed by this absorption and reradiates a portion of its heat back to the surface, thus making surface temperatures warmer. (We'll return to the greenhouse effect in more detail later in this chapter.) Carbon dioxide is also used by green plants in the photosynthesis process. During photosynthesis, CO_2 is converted into chemical compounds that build up the plant's tissues, organs, and supporting structures.

Another important component of the atmosphere is *water vapor,* the gaseous form of water. Individual water molecules in the form of vapor are mixed freely throughout the atmosphere, just like the other atmospheric component gases. Unlike the other component gases, however, water vapor can vary highly in concentration. Usually, water vapor makes up less than 1 percent of the atmosphere. Under very warm, moist conditions, as much as 2 percent of the air can be water vapor. Since water vapor, like carbon dioxide, is a good absorber of heat radiation, it also plays a major role in warming the lower atmosphere.

The atmosphere also contains dust and tiny floating particles that absorb and scatter radiation. These are discussed more fully in Chapter 6.

4.12 Component gases of the lower atmosphere *Values show percentage by volume for dry air. Nitrogen and oxygen form 99 percent of our air, with other gases, principally argon and carbon dioxide, accounting for the final 1 percent.*

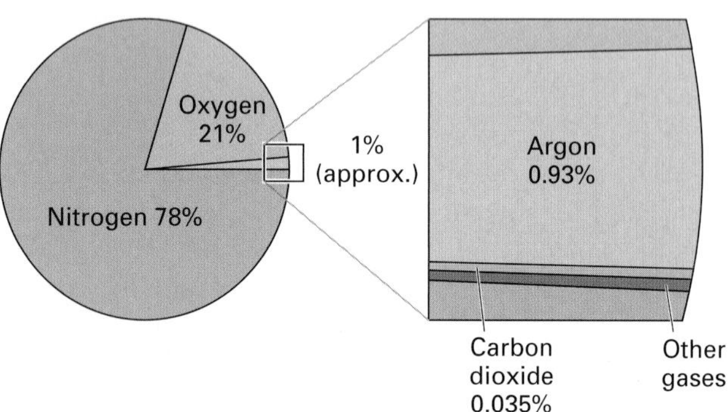

Ozone in the Upper Atmosphere

Another small, but important, constituent of the atmosphere is **ozone,** a form of oxygen in which three oxygen atoms are bonded together (O_3). Ozone is found mostly in the upper part of the atmosphere, in a layer termed the *stratosphere.* This layer of the atmosphere lies about 14 to 50 km (9 to 31 mi) above the surface, and is described in more detail in the next chapter. Ozone in the stratosphere absorbs ultraviolet radiation from the Sun as this radiation passes through the atmosphere, thus shielding the life layer from its harmful effects.

Ozone is most concentrated in a layer that begins at an altitude of about 15 km (about 9 mi) and extends to about 55 km (about 34 mi) above the Earth. It is produced by gaseous chemical reactions in the stratosphere. The details of these chemical reactions are complicated, but in summary they are quite simple. An oxygen molecule (O_2) absorbs ultraviolet light energy and splits into two oxygen atoms ($O + O$). A free oxygen atom (O) then combines with an O_2 molecule to form ozone, O_3.

Once formed, ozone can also be destroyed. Like O_2, ozone absorbs ultraviolet radiation, and it splits to form $O_2 + O$. Furthermore, two oxygen atoms ($O + O$) can join to form O_2. The net effect is that ozone (O_3), molecular oxygen (O_2), and atomic oxygen (O) are constantly formed, destroyed, and reformed in the ozone layer, absorbing ultraviolet radiation with each transformation.

The absorption of ultraviolet radiation by the ozone layer protects the Earth's surface from this damaging form of radiation. If the concentration of ozone is reduced, fewer transformations among O, O_2, and O_3 occur, and so ultraviolet absorption is reduced. If solar ultraviolet radiation were to reach the Earth's surface at full intensity, bacteria exposed on the Earth's surface would be destroyed, and unprotected animal tissues would be severely damaged. The presence of the ozone layer is thus essential to maintaining a viable environment for life on this planet.

Certain forms of air pollution are currently reducing ozone concentrations substantially in some portions of the atmosphere at certain times of the year. This threat to the ozone layer is described in *Eye on Global Change 4.4 • The Ozone Layer—Shield to Life.*

Ozone is also found in the lowest layers of the atmosphere, where it is an air pollutant. It occurs in photochemical smog, and is very damaging to lungs and other living tissues. *Eye on the Environment 6.6 • Air Pollution* provides more information about tropospheric ozone and smog.

Sensible Heat and Latent Heat Transfer

Thus far, we have discussed energy in the form of shortwave and longwave radiation. However, energy can be transported in other ways. Two of them—sensible heat and latent heat transfer—are very important in discussing the global energy balance.

The most familiar form of heat storage and transport is known as **sensible heat**—the quantity of heat held by an object that can be sensed by touching or feeling. This kind of heat is measured by a thermometer. When the temperature of an object or a gas increases, its store of sensible heat increases. When two objects of unlike temperature contact each other, heat energy moves by *conduction* from the warmer to the cooler. This type of heat flow is referred to as **sensible heat transfer.** Sensible heat transfer can also occur by *convection.* In this process, a fluid is heated by a warm surface, expands, and rises, creating an upward flow. This flow moves heat away from the surface and into the fluid. We will return to atmospheric convection in Chapter 6.

Unlike sensible heat, **latent heat**—or hidden heat—cannot be measured by a thermometer. It is heat that is taken up and stored in the form of molecular motion when a substance changes state from a solid to a liquid, from a liquid to a gas, or from a solid directly to a gas. Water is an important example. For liquid water to make the transition to water vapor, energy is required. As the water evaporates from liquid to gas, it draws up heat from the surroundings. This is the latent heat. It is carried in the form of the fast random motion of the free water vapor molecules as they enter the gaseous state during the process of evaporation. For example, consider the cooling you feel in a breeze when your skin is wet. As the water evaporates from liquid to gas, it draws up the heat required to make the change of state and carries it away from your skin, keeping you cool. (*Working It Out 6.1 • Energy and Latent Heat,* in Chapter 6, provides more information about the amount of heat required or released in changes of state of water.) GEODISCOVERIES

Although the latent heat of water vapor in air cannot be measured by a thermometer, energy is stored there just the same. When water vapor reverts to a liquid or solid, the latent heat is released, making the surroundings warmer. In the Earth–atmosphere system, **latent heat transfer** occurs when water evaporates from a moist land surface or open water surface. This process transfers heat from the surface to the atmosphere. On a global scale, latent heat transfer by movement of moist air provides a very important mechanism for transporting large amounts of energy from one region of the Earth to another.

The Global Energy System

The flow of energy from the Sun to the Earth and then back out into space is a complex system involving not only radiant energy flow, but also energy storage and transport. Keep in mind that solar energy is the ultimate power source for the Earth's surface processes, ranging from wind systems to the cycle of evaporation and precipitation. So when we trace the energy flows between the Sun, surface, and atmosphere, we are really studying how these processes are powered. GEODISCOVERIES

Solar Energy Losses in the Atmosphere

Let's examine the flow of insolation through the atmosphere on its way to the surface. This is shown in Figure 4.13, which gives typical values for losses of incoming solar energy. As the shortwave radiation penetrates the atmosphere, its energy is absorbed or diverted in various ways. The thin outer layers of the atmosphere almost completely absorb gamma and X rays. Much of the ultraviolet radiation is absorbed as well, particularly by ozone.

As radiation moves through deeper and denser atmospheric layers, gas molecules cause some radiation to be scattered. As we noted earlier, scattered radiation is radiation that is turned aside in all directions. It is unchanged except for the direction of its motion. Dust and other particles in the air can also cause scattering. Scattered radiation moving in all directions through the atmosphere is known as *diffuse radiation.* Because scattering occurs in all directions, some energy is sent back into outer space, and some flows down to the Earth's surface. When the scattering turns radiation back to space, we call this process *diffuse reflection.* Under clear sky

4.13 Fate of incoming solar radiation *Losses of incoming solar energy are much lower with clear skies (left) than with cloud cover (right). (Copyright © A. N. Strahler. Used by permission.)*

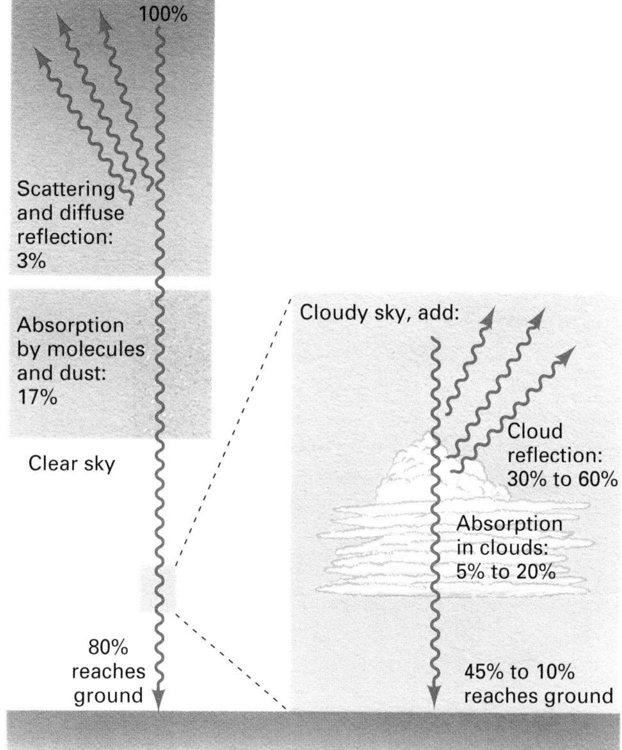

Eye on Global Change | 4.4

The Ozone Layer—Shield to Life

The release of *chlorofluorocarbons,* or *CFCs,* from the Earth's surface into the atmosphere poses a serious threat to the ozone layer. CFCs are synthetic industrial chemical compounds containing chlorine, fluorine, and carbon atoms. Although household uses of CFCs in aerosol sprays were banned in most parts of the world in the 1970s, CFCs are still in wide use as cooling fluids in existing refrigeration systems. When these appliances using CFCs are disposed of, their CFCs can be released to the air.

Molecules of CFCs in the atmosphere are very stable close to the Earth. They move upward by diffusion without chemical change until they eventually reach the ozone layer. There, they absorb ultraviolet radiation and are decomposed to chlorine oxide (ClO) molecules. The chlorine oxide molecules in turn attack molecules of ozone, converting them in large numbers into ordinary oxygen molecules by a chain reaction. In this way, the ozone concentration within the ozone layer is reduced, and so there are fewer ozone molecules to absorb ultraviolet energy in the stratosphere.

Other gaseous molecules can act to reduce the concentrations of ozone in the stratosphere as well. Nitrogen oxides, bromine oxides, and hydrogen oxides are examples. As with chlorine oxides, the concentrations of these molecules in the stratosphere are influenced by human activity.

Not all threats to the ozone layer can be attributed to human activity. Aerosols inserted into the stratosphere by volcanic activity also can act to reduce ozone concentrations. The June 1991 eruption of Mount Pinatubo, in the Philippines, reduced global ozone in the stratosphere by 4 percent during the following year, with reductions over midlatitudes of up to 9 percent.

Studies of the ozone layer based on satellite data during the 1980s showed a substantial decline in total global ozone. By late in the decade, scientists had agreed that the decrease in the ozone layer was escalating far faster than had been predicted.

Compounding the problem of global ozone decrease was the discovery in the mid-1980s of a "hole" in the ozone layer over the continent of Antarctica. Here, seasonal thinning of the ozone layer occurs during the early spring of the southern hemisphere, and ozone reaches a minimum during the month of October.

The thinning occurs after the formation of a polar vortex in the stratosphere during the winter period. This vast whirlpool of wind traps the air it contains and keeps this air out of the Sun during the months of the long polar night. The air in the vortex eventually becomes very cold, and clouds of crystalline water (ice) or other water-containing compounds form within it. The crystals are important because they provide a surface on which chemical reactions can take place. One of these reactions allows chlorine to be converted from a stable form to chlorine oxide. As we noted with CFCs, ClO can rapidly destroy ozone in the presence of sunlight. As the southern hemisphere spring approaches, the polar vortex slowly becomes illuminated by the Sun, and the chlorine oxide reacts with ozone. Ozone concentrations are reduced, and the ozone hole is formed.

The figure on the facing page presents a satellite-derived map of stratospheric ozone over the south pole on September 6, 2000, clearly revealing the ozone hole. In this south polar projection, transparent

conditions, this diffuse reflection sends about 3 percent of the incoming solar radiation back to space.

What about absorption? As we saw earlier, molecules and particles can absorb radiation as it passes through the atmosphere. Carbon dioxide and water are the two primary absorbers, but because the water vapor content of air can vary greatly, absorption can vary from one global environment to another. Absorption results in a rise in air temperature. In this way, incoming solar radiation causes some direct heating of the lower atmosphere. Absorption accounts for about 15 percent of the incoming solar radiation. Thus, when skies are clear, diffuse reflection and absorption combined total about 17 percent, leaving as much as 80 percent of the solar radiation to reach the ground.

The presence of clouds can greatly increase the amount of incoming solar radiation reflected back to space. The bright, white surfaces of thick, low clouds are extremely good reflectors of shortwave radiation. Cloud reflection can account for a direct turning back to space of 30 to 60 percent of incoming radiation (Figure 4.13). Clouds also absorb radiation, perhaps as much as 5 to 20 percent. Under conditions of a heavy cloud layer, as little as 10 percent of incoming solar radiation may actually reach the ground.

The amount of global cloud cover, including cloud type and altitude, is an important factor in determining our global climate. In Chapter 8, we will discuss this topic further.

Albedo

The proportion of shortwave radiant energy scattered upward by a surface is called its **albedo.** For example, a surface that reflects 40 percent of the shortwave radiation it

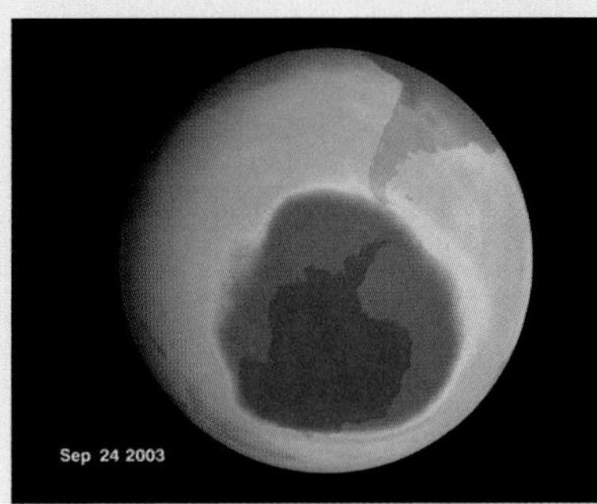

Map of ozone concentration for the south polar region on September 24, 2003 *Ozone concentration is measured in Dobson units. The lowest values over central Antarctica are about 90 units (blue). Prior to 1979, values were in the range of 250 to 300 units (green). Data were acquired by the Total Ozone Mapping Spectrometer (TOMS) on NASA's Earth Probe TOMS-EP Satellite. (Courtesy NASA.)*

color overlies lighter oceans and darker continents to depict ozone concentration. Blue tones are lowest, while red-brown tones are highest. The largest ozone hole through 2003 was recorded on September 10, 2000, and covered 28.3 million km^2 (11 million mi^2). The year 2001 saw an ozone hole almost as large. In 2002, due to unusual stratospheric weather, the ozone hole was weaker and split into two parts. The year 2003 saw the second-largest ozone hole on record. In typical years, the antarctic ozone hole forms in mid-September, grows to a maximum extent around the first of October, then slowly shrinks and ultimately disappears in early December.

In the northern hemisphere, conditions for the formation of an ozone hole are not as favorable. An arctic polar vortex forms as well but is much weaker than the antarctic vortex and is less stable. As a result, an early-spring ozone hole is not usually observed in the Arctic. However, in 1993, 1996, 1997, and 1999, notable arctic ozone holes occurred, and atmospheric computer models have projected more such events in the period 2010–2019.

Although the most dramatic reductions in stratospheric ozone levels have been associated with arctic and antarctic ozone holes, reductions have been noted elsewhere as well—including the northern midlatitudes. A number of studies during the 1990s showed reductions of 6 to 8 percent in stratospheric ozone concentrations occurring over North America.

As the global ozone layer thins, the rate of incoming ultraviolet solar radiation reaching the Earth's surface should increase. Recent studies of satellite data have documented this effect. Since 1978, the average annual exposure to ultraviolet radiation increased by 6.8 percent per decade along the 55° N parallel. At 55° S, the increase was 9.8 percent per decade. Over most of North America, the increase was about 4 percent per decade.

How might declining ozone concentrations affect the Earth? An increase in the incidence of skin cancer in humans from enhanced ultraviolet radiation is one of the predicted effects. Other possible effects include reduction of crop yields and damage to some forms of aquatic life.

Responding to the global threat of ozone depletion and its anticipated impact on the biosphere, 20 nations signed the Vienna Convention for the Protection of the Ozone Layer in 1985. In 1987, they agreed in the Montreal Protocol to specific reductions in CFCs, with developed nations phasing out CFC production by 2000. Halting the use of CFCs in developing nations, scheduled for 2010, is the next step. In 1999, developed nations pledged one billion dollars to help developing nations switch to safer alternatives.

In operation for more than 15 years, the international agreements have had an effect. In 2003, scientists using three NASA satellite instruments and three international ground stations detected a notable slowing in the rate of ozone loss in the stratosphere since 1997. While not a reversal of ozone loss, the trend is encouraging. Current predictions show the ozone layer restored by the middle of the century.

receives has an albedo of 0.40. Albedo is an important property of a surface because it measures how much incident solar energy will be absorbed. A surface with a high albedo, such as snow or ice (0.45 to 0.85), reflects much or most of the solar radiation and absorbs only a smaller amount. A surface with a low albedo, such as black pavement (0.03), absorbs nearly all the incoming solar energy. Since the energy absorbed by a surface can warm the air immediately above it by conduction and convection, surface temperatures will be warmer over a low albedo surface (pavement) than over a high albedo surface (snow or ice).

The albedo of a water surface varies with the angle of incoming radiation. It is very low (0.02) for nearly vertical rays on calm water. However, when the Sun shines on a water surface at a low angle, much of the radiation is directly reflected as Sun glint, producing a higher albedo. For fields, forests, and bare ground, albedos are of intermediate value, ranging from as low as 0.08 to as high as 0.3 percent.

Certain orbiting satellites are equipped with instruments to measure the energy levels of shortwave and longwave radiation emerging at the top of the atmosphere. Both incoming radiation from the Sun and outgoing radiation from the atmosphere and Earth's surfaces below are measured. Data from these satellites have been used to estimate the Earth's average albedo. This albedo includes reflection by the Earth's atmosphere as well as all its surfaces, so it includes diffuse reflection by clouds, dust, and atmospheric molecules. The albedo values obtained in this way vary between 0.29 and 0.34. This means that the Earth–atmosphere system directly returns to space slightly less than one-third of the solar radiation it receives. *Focus on Remote Sensing 4.5 • CERES—Clouds and the Earth's Radiant Energy System* describes how

Focus on Remote Sensing | 4.5

CERES—Clouds and the Earth's Radiant Energy System

The Earth's global radiation balance is the primary determinant of long-term surface temperature, which is of great importance to life on Earth. Yet this balance can also be affected by human activities, such as converting forests to pasture lands or releasing greenhouse gases into the atmosphere. Thus, it is important to monitor the Earth's radiation budget over time as accurately as possible.

For nearly 20 years, the Earth's global radiation balance has been the subject of NASA missions that have launched radiometers—radiation-measuring devices—into orbit around the Earth. These devices scan the Earth and measure the amount of shortwave and longwave radiation leaving the Earth at the top of the atmosphere. An ongoing NASA experiment entitled CERES—Clouds and the Earth's Radiant Energy System—is placing a new generation of these instruments in space to monitor the Earth's global radiation balance well into the twenty-first century. GEODISCOVERIES

The CERES instruments scan the Earth from horizon to horizon, measuring outgoing energy flows of reflected solar radiation (0.3 to 5.0 μm), outgoing longwave radiation (8 to 12 μm), and outgoing total radiation (0.3 to 100 μm). From these observations, a team of scientists at NASA's Langley Research Center, led by Dr. Bruce Barkstrom, produces daily global maps of the Earth's upwelling radiation fields. The two images presented in the figure show global reflected solar energy and emitted longwave energy averaged over the month of March 2000, as obtained from the CERES instrument aboard NASA's Terra satellite platform.

The top image shows average shortwave flux ("flux" means "flow"), ranging from 0 to 210 W/m^2. The largest flows occur over regions of thick clouds near the equator, where the bright, white clouds reflect much of the solar radiation back to space. The Amazon Basin of South America is an example. In the midlatitudes, persistent cloudiness during this month also shows up as light tones. Tropical deserts, the Sahara for example, are also bright. Snow and ice surfaces in polar regions are quite reflective, but in March the amount of radiation received in polar regions is much less than at the equator or in the midlatitudes. As a result, they don't appear as bright in this image. Oceans, especially where skies are clear, absorb solar radiation and thus show low shortwave fluxes.

Longwave flux is shown in the bottom image on a scale from 100 to 320 W/m^2. Cloudy equatorial regions have low values, showing the blanketing effect of thick clouds that trap longwave radiation beneath them. Warm tropical oceans in regions of clear sky emit the most longwave flux. Poleward, surface and atmospheric temperatures drop,

satellite instruments map and monitor the reflection of solar shortwave radiation and the emission of longwave radiation by the Earth–atmosphere system.

Counterradiation and the Greenhouse Effect

The amount of shortwave energy absorbed by a surface is an important determinant of a surface's temperature. But the surface is also warmed significantly by longwave radiation emitted by the atmosphere and absorbed by the ground. Let's look at this in more detail. Figure 4.14 is a diagram showing energy flows between the surface, atmosphere, and space. On the left is the flow of shortwave radiation from the Sun to the surface that we saw in more detail in Figure 4.13. Some of this radiation is reflected back to space (as measured by the albedo of the surface), but much is absorbed, warming the surface.

Looking now at longwave radiation, we see that the surface emits radiation upward as flows labeled A and B. Some of this radiation escapes directly to space (A), but the remainder is absorbed by the atmosphere (B). What about longwave radiation emitted by the atmosphere? Although the atmos-

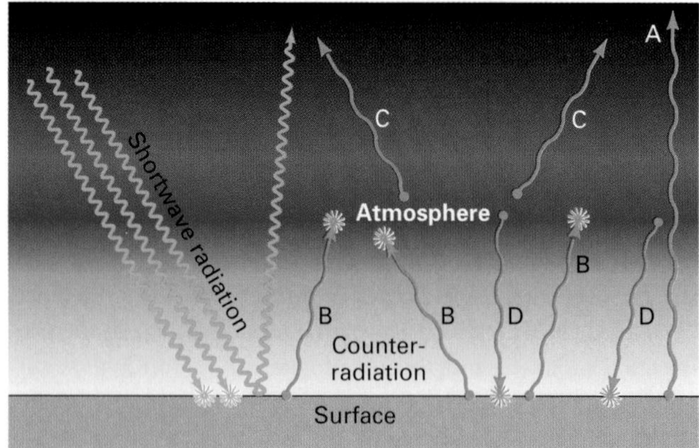

4.14 Counterradiation and the greenhouse effect *Shortwave radiation (left) passes through the atmosphere and is absorbed at the surface, warming the surface. The surface emits longwave radiation. Some of this flow passes directly to space (A), but most is absorbed by the atmosphere (B). In turn, the atmosphere radiates longwave energy back to the surface as counterradiation (D) and also to space (C). The return of outbound longwave radiation (B) by counterradiation (D) constitutes the greenhouse effect.*

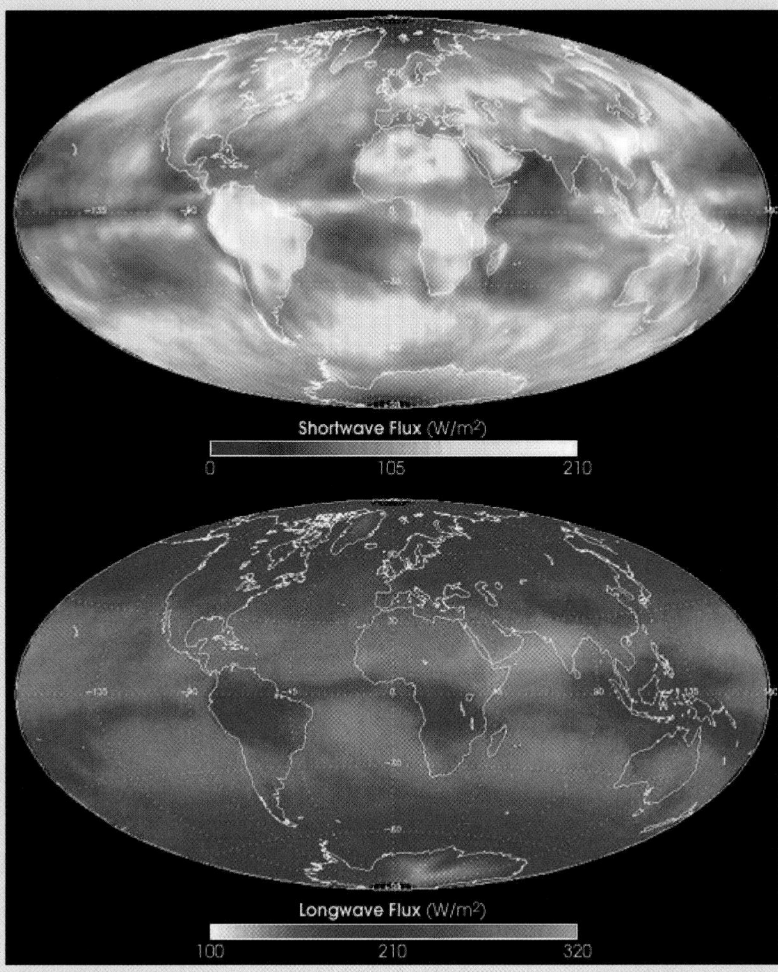

Global shortwave and longwave energy fluxes from CERES *These images show average shortwave and longwave energy flows from Earth for March 2000, as measured by the CERES instrument on NASA's Terra satellite platform. (Courtesy NASA.)*

so longwave energy emission also drops significantly.

As you can see from these images, clouds are very important determiners of the global radiation balance. A primary goal of the CERES experiment is to learn more about the Earth's cloud cover, which changes from minute to minute and hour to hour. This knowledge can be used to improve global climate models that predict the impact of human and natural change on the Earth's climate.

The most important contribution of CERES, however, is continuous and careful monitoring of the Earth's radiant energy flows. In this way, small, long-term changes, induced by human or natural change processes, can be detected in spite of large variations in energy flows from place to place and time to time caused by clouds.

phere is colder than the surface, it emits longwave radiation as well. This radiation is shown by flows C and D. It is emitted in all directions, and so some radiates upward to space (C) while the remainder radiates downward toward the surface (D). Since this downward flow is the opposite direction to longwave radiation leaving the surface, we call this flow **counterradiation.** It acts to replace some of the heat emitted by the surface, and so it makes the surface warmer.

The amount of counterradiation depends strongly on the presence of carbon dioxide and water vapor in the atmosphere. Recall from our discussion of longwave radiation earlier in this chapter that much of the longwave radiation emitted upward from the Earth's surface is absorbed by these two atmospheric constituents. This absorbed energy raises the temperature of the atmosphere, which causes it to emit more counterradiation. Thus, the lower atmosphere, with its longwave-absorbing gases, acts like a blanket that traps heat underneath it. Because liquid water is also a strong absorber of longwave radiation, cloud layers, which are composed of tiny water droplets, are even more important than carbon dioxide and water vapor in producing a blanketing effect.

This mechanism, in which the atmosphere traps longwave radiation and returns it to the surface, is termed the **greenhouse effect.** It is similar to the principle used in greenhouses and in homes with large windows to entrap solar heat. In the greenhouse, window glass permits entry of shortwave energy but absorbs and blocks the exit of longwave energy. This phenomenon is even better demonstrated by the intense heating of air in a closed car left parked in the Sun.

Global Energy Budgets of the Atmosphere and Surface

Thus far we have described a number of important pathways by which incoming solar radiation is scattered, absorbed, and reradiated as longwave radiation. It is an important principle of physics that, except for nuclear reactions, energy is neither created nor destroyed. It is therefore possible to follow the initial stream of solar energy and account for its diversion into different system pathways and its conversion into different energy forms. A full accounting of all the important energy flows among the Sun, atmosphere, surface, and space forms the global energy budget of the Earth's

Focus on Systems | 4.6

Energy and Matter Budgets

An important tool in the understanding of systems is a *budget*—an accounting of the energy and matter flows that enter, move within, and leave a system. It relies on the principle that matter and energy are conserved. So when energy or matter flow into a system at a certain rate, we should be able to account for the entire flow in pathways within the system and in flows leaving the system.

As a familiar example of a flow system, consider a checking account. Money flows into the account as you make deposits and flows out as you make withdrawals or write checks. Given a monthly salary of $2000, a budget might record $500 spent for rent, $300 for food, and so forth.

Suppose that energy or matter flows into a system at a fixed rate but flows out of a system at a lesser rate. This can happen if energy or matter is flowing into storage inside the system. For a checking account, if you deposit more than you withdraw, the account balance grows. Your budget will show this as inputs exceed outflows. Conversely, if outflows exceed inflows, energy or matter must flow out of storage to make up the difference. If you withdraw more than you deposit, the account balance will shrink. The budget will show that your stock of money is decreasing.

Recall from Chapter 2 that a flow system in equilibrium is one in which the flows in its various pathways tend toward a steady value. Unless you are accumulating money in your account, or spending it down over a period of time, your checking account is probably like this. Each month, inflow and output are about the same, and so the amount of money in the account stays about the same. At equilibrium, a budget will show all the flows entering a system to be balanced by flows leaving the system.

Energy and matter budgets are complicated by the fact that energy and matter can change forms inside a system. For example, radiant energy becomes sensible heat when it is absorbed by matter, as when the Sun warms the Earth's surface. When the surface cools, energy is transformed from sensible heat to radiant energy. Matter changes form when it changes state (that is, changes among solid, liquid, or gas) and when it enters into chemical reactions. As you may know from studying chemistry, the products of a chemical reaction may be very different—in state, color, or density, for example—from the reactants. However, the atoms of the reactants are fully conserved in the products, and no matter is lost. Because energy and matter change form, we must account in our budget for flow in all forms.

One of the most important reasons for constructing a budget for a flow system is that it directs us toward finding out what happens to all the matter and energy entering the system. In this way, new pathways of flow and new conversions of matter and energy to other forms are uncovered, leading to a better understanding of the workings of the entire flow system.

atmosphere and surface. Like a household budget, it must balance, at least over the long term. *Focus on Systems 4.6 • Energy and Matter Budgets* discusses energy and matter budgets more fully. GEODISCOVERIES

Incoming Shortwave Radiation

The global energy budget for the Earth's surface and atmosphere is diagrammed in Figure 4.15. We begin with a discussion of solar shortwave radiation, shown in the left part of the figure. Here we will trace only the downward flow of solar radiation and its fate. Reflection by molecules and dust, clouds, and the surface (including the oceans) totals 31 percentage units. Expressed as a proportion (0.31), this is the albedo of the Earth and atmosphere as a combined system. The right side of this diagram shows values for absorption in the atmosphere. The combined losses through absorption by molecules, dust, and clouds average 20 percentage units. With 31 percentage units of the incoming solar energy flow reflected and 20 percentage units absorbed in the atmosphere, 49 units are left. These units are absorbed by the Earth's land and water surfaces. We have now accounted for all of the incoming flow of insolation.

Surface Energy Flows

Let's now turn to energy flows to and from the surface. We've noted the flow of 49 units of shortwave radiation absorbed by the surface, but what about longwave radiation? The right part of Figure 4.15 shows the components of outgoing longwave radiation for the Earth's surface and the atmosphere. The large arrow on the left shows total longwave radiation leaving the Earth's land and ocean surface. The long, thin arrow indicates that 12 percentage units are lost to space, while 102 units are absorbed by the atmosphere. (These flows are equivalent to flows A and B in Figure 4.14.) The loss is therefore equivalent to 114 percentage units.

This means that the surface emits 114 percent of the total incoming solar radiation! How can this be correct? To answer this question, note that the surface also receives a flow of 95 units of longwave counterradiation from the atmosphere. This will be in addition to the 49 units of shortwave radiation that are absorbed (left figure). So the surface receives 95 (longwave) + 49 (shortwave) = 144 units in all. Thus, a loss of 114 units by longwave radiation out of a gain of 144 units of longwave and shortwave radiation is certainly possible.

On the far right of the figure are two smaller arrows. These show the flow of energy away from the surface as

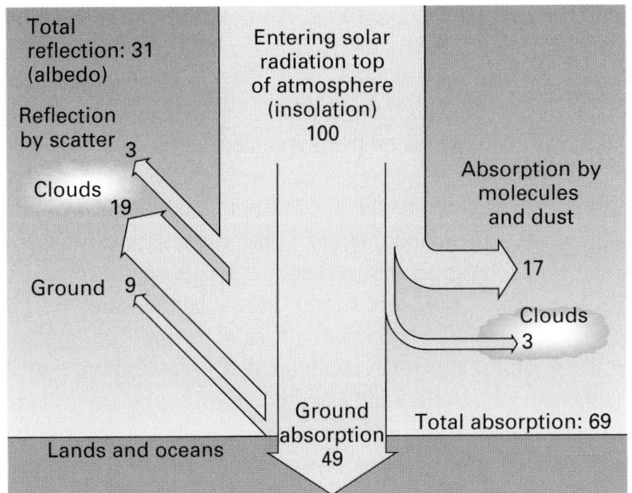

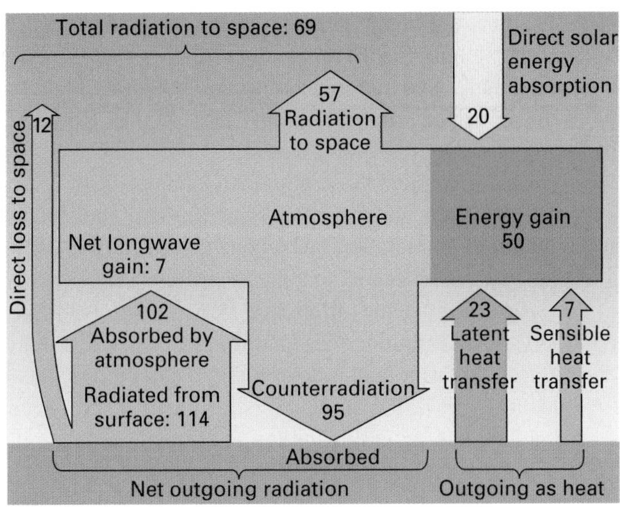

4.15 Diagram of the global energy balance *Values are percentage units based on total insolation as 100. The left figure shows the fate of incoming solar radiation. The right figure shows longwave energy flows occurring between the surface and atmosphere and space. Also shown are the transfers of latent heat, sensible heat, and direct solar absorption that balance the budget for Earth and atmosphere. Data of Kiehl and Trenberth, 1997.*

latent heat (23 units) and sensible heat (7 units). Evaporation of water from moist soil or ocean transfers latent heat from the surface to the atmosphere, as explained earlier. Sensible heat transfer occurs when heat is directly conducted from the surface to the adjacent air layer. When air is warmed by direct contact with a surface and rises, it takes its heat along with it by convection. In this process, heat flows from the surface to the atmosphere.

These last two heat flows are not part of the radiation balance, since they are not in the form of radiation. However, they are a very important part of the total energy budget of the surface, which includes all forms of energy. Taken together, these two flows account for 30 units leaving the surface. With their contribution, the surface energy balance is complete. Total gains are 95 (longwave) + 49 (shortwave) = 144. Total losses are 114 (longwave) + 23 (latent heat) + 7 (sensible heat) = 144.

Energy Flows to and from the Atmosphere

What about the atmosphere? Let's first look at energy flowing into the atmosphere. Like the Earth's surface, the atmosphere gains energy by absorption of shortwave radiation, amounting to 20 units. The atmosphere also gains energy from latent heat transfer (23 units) and sensible heat transfer (7 units). In longwave radiation, the atmosphere absorbs 102 units emitted by the surface. (These are the flows labeled B in Figure 4.14.) These units total to 152.

Looking at energy leaving the atmosphere, we see a loss of 57 units of longwave radiation to space (flows labeled C in Figure 4.14) and a loss of 95 units in counterradiation to the surface (flows labeled D in Figure 4.14). Since these total 152 units, we see that the atmosphere's energy budget is balanced.

This analysis helps us to understand the vital role of the atmosphere in trapping heat though the greenhouse effect. Without the 95 units of counterradiation of the atmosphere, the surface would have only 49 units of absorbed shortwave radiation to emit. The temperature needed to radiate so few units is well below freezing. So without the greenhouse effect, our planet would be a cold, forbidding place.

Climate and Global Change

Our analysis also helps us to understand how global change might affect the Earth's climate. Suppose that clearing forests for agriculture and turning agricultural lands into urban and suburban areas decreases surface albedo. In that case, more energy would be absorbed by the ground, raising its temperature. That, in turn, would increase the flow of surface longwave radiation to the atmosphere, which would be absorbed and then boost counterradiation. So the effect would probably be to amplify the warming through the greenhouse effect. What if air pollution causes more low, thick clouds to form? Since low clouds increase shortwave reflection back to space, the effect will be to cool the surface and atmosphere. What if increasing condensation trails from jet aircraft cause more high, thin clouds? High, thin clouds absorb more longwave energy than they reflect shortwave energy, so they should make the atmosphere warmer. That should boost counterradiation, increasing the greenhouse effect. The energy flow linkages between the Sun, surface, atmosphere, and space are critical components of our climate system, and understanding them will help you better understand global climate change. GEODISCOVERIES

Net Radiation, Latitude, and the Energy Balance

Solar energy is intercepted by our planet, and because some of it is absorbed, the heat level of our planet tends to rise. At the same time, our planet radiates energy into outer space, a process that tends to reduce its level of heat energy. Over time, as we have seen, these incoming and outgoing radiation flows must balance for the Earth as a whole. However, incoming and outgoing flows do not have to balance at any given surface location. At night, for example, there is no incoming radiation, yet the Earth's surface and atmosphere still emit outgoing radiation. In any one place and time, then, more radiant energy can be lost than gained, or vice versa.

Net radiation is the difference between all incoming radiation and all outgoing radiation. In places where radiant energy flows in faster than it flows out, net radiation is a positive quantity, providing an energy surplus. In other places, where radiant energy is flowing out faster than it is flowing in, net radiation is a negative quantity, yielding an energy deficit. Our analysis of the Earth's radiation balance has already shown that for the entire Earth and atmosphere as a unit, net radiation is zero on an annual basis.

In Figure 4.8, we saw that solar energy input varies strongly with latitude. What is the effect of this variation on net radiation? To answer this question, we will look at the net radiation profile spanning the entire latitude range from 90° N to 90° S. In this analysis we will use yearly averages for each latitude, so that the effect of seasons is concealed.

The lower part of Figure 4.16 presents a global profile of net radiation from pole to pole. Between about 40° N lat. and 40° S lat. there is a net radiant energy gain labeled "energy surplus." In other words, incoming solar radiation exceeds outgoing longwave radiation throughout the year. Poleward of 40° N and 40° S, the net radiation is negative and is labeled "energy deficit"—meaning that outgoing longwave radiation exceeds incoming shortwave radiation.

If you carefully examine the lower part of Figure 4.16, you will find that the area on the graph labeled "surplus" is equal in size to the combined areas labeled "deficit." The matching of these areas confirms the fact that net radiation for the Earth as a whole is zero and that global incoming shortwave radiation exactly balances global outgoing longwave radiation.

The pattern of energy surplus at low latitudes and energy deficit at high latitudes creates a flow of energy from low latitudes to high (upper part of Figure 4.16). This energy flow is in the form of sensible and latent heat in poleward movements of warm ocean water and warm, moist air that are part of the global circulation patterns of the ocean and the atmosphere. Heat is exchanged when oceanic and atmospheric currents move warm water and warm, moist air poleward and cooler water and cooler, drier air equatorward. GEODISCOVERIES

We'll return to these flows in later chapters, but keep in mind that this **poleward heat transfer,** driven by the imbalance in net radiation between low and high latitudes, is thus the power source for ocean currents and broad-scale atmospheric circulation patterns. Without this circulation, low latitudes would heat up and high latitudes would cool down until a radiative balance was achieved, leaving our planet with much more extreme temperature contrasts. Our Earth would be a far different place than it is now. Figure 4.17 illustrates some of the ways that natural processes and human uses are driven by solar power.

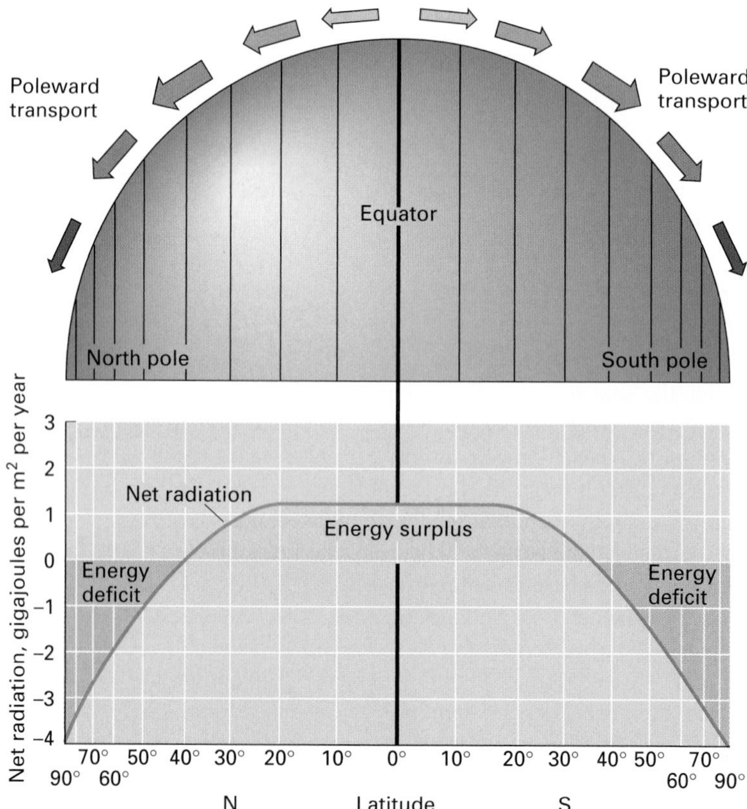

4.16 Annual surface net radiation from pole to pole *Where net radiation is positive, incoming solar radiation exceeds outgoing longwave radiation. There is an energy surplus, and energy moves poleward as latent heat and sensible heat. Where net radiation is negative, there is an energy deficit. Latent and sensible heat are lost in the form of outgoing longwave radiation.*

a...***Solar-powered call box***
This emergency telephone is powered by the solar cell atop its pole.

b...***Eye on the Landscape***
Wave erosion *Ocean waves, powered by the Sun through the Earth's wind system, have eroded the eastern bluffs of Cape Cod by hundreds of meters since the end of the Ice Age.* **What else would the geographer see?...Answers at the end of the chapter.**

c...***Eye on the Landscape*** ***Tropical cyclone*** *Solar power also indirectly powers severe storms like Typhoon Odessa, shown here in a space photo.* **What else would the geographer see?...Answers at the end of the chapter.**

d...***Water power*** *The hydrologic cycle, powered by solar evaporation of water over oceans, generates runoff from rainfall that erodes and deposits sediment.*

Eye on the Environment | 4.7

Solar Power

As we saw in *Working It Out 4.3 • Calculating the Global Radiation Balance,* our Earth intercepts solar energy at the rate of 1.77×10^{14} kW, or about $1^1/_2$ quadrillion megawatt-hours per year. (A megawatt, mW, is one million—10^6—watts.) This quantity of energy is about 28,000 times as much as all the energy presently being consumed each year by human society. Thus, an enormous source of energy is near at hand waiting to be used. In addition to its abundance, solar energy has other advantages. Unlike fossil fuel consumption, solar power generation does not release carbon dioxide. As we will see in Chapter 5, CO_2 concentrations have been steadily increasing from fossil fuel combustion for the last century, and it is predicted that they will cause climatic warming through an enhancement of the greenhouse effect. Also, utilization of solar power does not involve the release of pollutant gases and airborne particles as does the burning of fossil fuel.

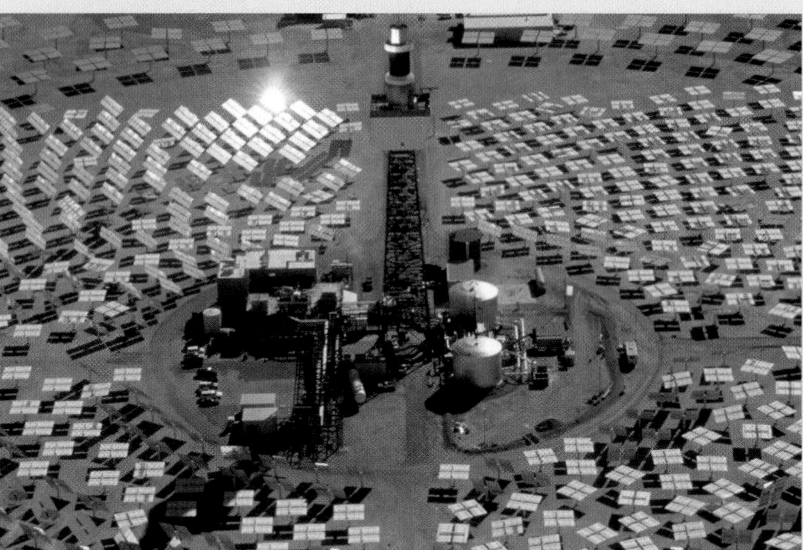

Solar II *This solar power plant, built as a demonstration project in 1994–1995, generated 100 megawatts of electric power when in operation. The mirrors, called heliostats, were moved by computer-controlled motors to focus the Sun's ray on a reservoir of molten nitrate salt located atop the central tower. Heat from the salt was exchanged to create steam that was used to generate the power.*

The simplest form of solar heating involves the use of large glass panes to admit sunlight into a room—the greenhouse principle. An overhanging roof admits solar rays during the winter, when the Sun's path is low in the sky, but blocks them in summer. Practical solar heating of interior building space and hot-water systems makes use of solar collectors. Panels of solar collectors are usually placed on the roof of the building. The flat-plate collector consists of a network of aluminum or copper tubes carrying circulating water. The tubes, painted black, absorb solar energy and conduct it to the water. Water is heated to a temperature of about 65°C (about 150°F) and can be used to transfer heat to a hot-water supply, as well as to a conventional space-heating system.

A few solar energy power plants also operate by direct absorption of solar energy using reflecting mirror systems. Many movable mirrors, called heliostats, reflect solar rays to the top of a central tower where a steam boiler and electric generator are located. The black metal of the boiler readily absorbs the heat. A heliostat power plant constructed in 1981 near Barstow, California, named Solar I, provided steam for five years, delivering 10 megawatts of electric power. In 1994 and 1995, the plant was modified by adding more heliostats and converting it to use molten nitrate salt to store the solar heat instead of steam. Its capacity was also increased to 100 megawatts. The power plant is now known as Solar II (see photo).

Another way to produce electricity from solar radiation is to generate it directly with solar cells. These cells use special materials, such as crystalline silicon, that absorb solar radiation and convert it into electricity. Using present technology, arrays of solar cells would need to occupy large areas of ground surface to produce significant outputs of energy. Storage of the electricity is another problem, since the solar cells will not generate power at night. Storage as hydrogen fuel is an attractive possibility for large-scale systems, while batteries can be used to store power in small systems, such as those of homes, farms, and ranches. One small-system use is to drive electric water pumps needed for irrigation of fields in the sunny dry season of tropical countries, for example, on the wheat farms of Pakistan and northern India.

Note that the Sun also powers the natural motion of wind and water and that these moving fluids can be harnessed for power generation as well. We will touch on wind, water, and wave power in later chapters.

Solar radiation to Earth is a truly huge energy flow, dwarfing human energy production by fossil fuel burning or other means of power generation. In the long run, it seems likely that we will have to tap more of the flow of the Sun's energy to reduce the buildup of CO_2 in the atmosphere that is caused by fossil fuel burning and threatens to warm global climate. *Eye on the Environment 4.7 • Solar Power* provides more information on how solar power is harvested.

A LOOK AHEAD Although our analysis of the Earth's energy balance is not yet complete, it is clear that the balance is a sensitive one involving a number of factors that determine how energy is transmitted and absorbed. Have our industrial activities already altered the components of the planetary radiation balance? An increase in carbon dioxide will increase the absorption of longwave radiation by the atmosphere, enhancing the greenhouse effect. Will this change cause a steady and permanent rise in average atmospheric temperature?

An increase in atmospheric particles at upper levels of the atmosphere will increase the scattering of incoming shortwave radiation and thus reduce the shortwave energy available to warm the surface. On the other hand, increased dust content at low levels will act to absorb more longwave radiation and so raise the surface temperature. Has either change occurred?

Human habitation, through cultivation and urbanization, has profoundly altered the Earth's land surfaces. Have these changes, which affect surface albedo and the transfer of latent and sensible heat to the atmosphere, modified the global energy balance? Answers to these questions require further study of the processes of heating and cooling of the Earth's atmosphere, lands, and oceans. Our next chapter concerns air temperature and how and why it varies daily and annually depending on the surface energy balance.

Chapter Summary

- **Electromagnetic radiation** is a form of energy emitted by all objects. The wavelength of the radiation determines its characteristics. The hotter an object, the shorter the wavelengths of the radiation and the greater the amount of radiation that it emits.
- Radiation emitted by the Sun includes *ultraviolet, visible, near-infrared,* and *shortwave infrared radiation. Thermal infrared radiation,* which is emitted by Earth surfaces, is familiar as heat. The atmosphere **absorbs** and **scatters** radiation in certain wavelength regions. Radiation flows are measured in watts per square meter.
- The Earth continuously absorbs solar **shortwave radiation** and emits **longwave radiation**. In the long run, the gain and loss of radiant energy remain in balance, and the Earth's average temperature remains constant.
- **Insolation**, the rate of solar radiation flow available at a location at a given time, is greater when the Sun is higher in the sky. *Daily insolation* is also greater when the period of daylight is longer. Between the tropics and poles, the Sun rises higher in the sky and stays longer in the sky at the summer solstice than at the equinox than at the winter solstice. Near the equator, daily insolation is greater at the equinoxes than at the solstices. Annual insolation is greatest at the equator and least at the poles. However, the poles still receive 40 percent of the radiation received at the equator.
- The pattern of annual insolation with latitude leads to a natural naming convention for latitude zones: *equatorial, tropical, subtropical, midlatitude, subarctic (subantarctic), arctic (antarctic), and polar.*
- The Earth's atmosphere is dominated by *nitrogen* and *oxygen gases. Carbon dioxide* and *water vapor*, though lesser constituents by volume, are very important because they absorb longwave radiation and enhance the greenhouse effect.
- **Ozone** (O_3) is a tiny but important constituent of the atmosphere that is concentrated in a layer of the upper atmosphere. It is formed from oxygen (O_2) by chemical reactions that absorb ultraviolet radiation, thus sheltering the organisms of the Earth's surface from the damaging effects of ultraviolet rays.
- **Latent heat** and **sensible heat** are additional forms of energy. Latent heat is taken up or released when a change of state occurs. Sensible heat is contained within a substance. It can be transferred to another substance by *conduction* or *convection*.
- The global energy system includes a number of important pathways by which insolation is transferred and transformed. Part of the solar radiation passing through the atmosphere is absorbed or scattered by molecules, dust, and larger particles. Some of the scattered radiation returns to space as *diffuse reflection*. The land surfaces, ocean surfaces, and clouds also reflect some solar radiation back to space.
- The proportion of radiation that a surface absorbs is termed its **albedo**. The albedo of the Earth and atmosphere as a whole planet is about 30 percent.
- The atmosphere absorbs longwave energy emitted by the Earth's surface, causing the atmosphere to **counterradiate** some of that longwave radiation back to Earth,

thereby creating the **greenhouse effect**. Because of this heat trapping, the Earth's surface temperature is considerably warmer than we might expect for an Earth without an atmosphere.

- **Net radiation** describes the balance between incoming and outgoing radiation. At latitudes lower than 40 degrees, annual net radiation is positive, while it is negative at higher latitudes. This imbalance creates **poleward heat transport** of latent and sensible heat in the motions of warm water and warm, moist air, and thus provides the power that drives ocean currents and broad-scale atmospheric circulation patterns.

Key Terms

electromagnetic radiation
visible light
absorption
scattering
shortwave radiation
longwave radiation
global radiation balance
insolation
ozone
sensible heat
sensible heat transfer
latent heat
latent heat transfer
albedo
counterradiation
greenhouse effect
net radiation
poleward heat transport

Review Questions

1. What is electromagnetic radiation? How is it characterized? Identify the major regions of the electromagnetic spectrum.
2. How does the temperature of an object influence the nature and amount of electromagnetic radiation that it emits?
3. What is the solar constant? What is its value? What are the units with which it is measured?
4. How does solar radiation received at the top of the atmosphere differ from solar radiation received at the Earth's surface? What are the roles of absorption and scattering?
5. Compare the terms *shortwave radiation* and *longwave radiation*. What are their sources?
6. How does the atmosphere affect the flow of longwave energy from the Earth's surface to space?
7. What is the Earth's global energy balance, and how are shortwave and longwave radiation involved?
8. How does the Sun's path in the sky influence daily insolation at a location? Compare summer solstice and equinox paths of the Sun in the sky for 40° N lat. and the equator.
9. What influence does latitude have on the annual cycle of daily insolation? on annual insolation?
10. Identify the three largest components of the dry air. Why are carbon dioxide and water vapor important atmospheric constituents?
11. How does the ozone layer protect the life layer?
12. Describe latent heat transfer and sensible heat transfer.
13. What is the fate of incoming solar radiation? Discuss scattering and absorption, including the role of clouds.
14. Define albedo and give two examples.
15. Describe the counterradiation process and how it relates to the greenhouse effect.
16. Discuss the energy balance of the Earth's surface. Identify the types and sources of energy flows that the surface receives and do the same for energy flows that it loses.
17. Discuss the energy balance of the atmosphere. Identify the types and sources of energy flows that the atmosphere receives and do the same for energy flows that it loses.
18. What is net radiation? How does it vary with latitude?
19. What is the role of poleward heat transport in balancing the net radiation budget by latitude?

Focus on Systems 4.1 • Forms of Energy

1. What is energy?
2. Distinguish among these forms of energy: kinetic, potential, mechanical, radiant, chemical.

Eye on Global Change 4.4 • The Ozone Layer—Shield to Life

1. What are CFCs, and how do they impact the ozone layer?
2. When and where have ozone reductions been reported? Have corresponding reductions in ultraviolet radiation been noted?
3. What is the outlook for the future health of the ozone layer?

Focus on Systems 4.6 • Energy and Matter Budgets

1. What important physical principle permits the construction of budgets of energy and matter flow?
2. Using the example of a checking account, explain how monitoring the flows of money with a budget can show whether or not a flow system is in equilibrium.
3. Why is budgeting an important tool for studying flow systems?

Eye on the Environment 4.7 • Solar Power

1. Why is solar power in principle more desirable than power produced by burning fossil fuels?
2. Describe two different ways of harnessing solar power.

A Closer Look:
Geographer's Tools 4.8 • Remote Sensing for Physical Geography

1. What is remote sensing? What is a remote sensor?
2. Compare the reflectance spectra of water, vegetation, and a typical soil. How do they differ in visible wavelengths? in infrared wavelengths?
3. What color is vegetation on color infrared film? Why?
3. What is emissivity, and how does it affect the amount of energy radiated by an object?
4. Is radar an example of an active or a passive remote sensing system? Why?
5. What is a digital image? What advantage does a digital image have over a photographic image?
6. Describe two ways of acquiring a digital image.
7. How does a Sun-synchronous orbit differ from a geostationary orbit? What are the advantages of each type?

Visualizing Exercises

1. Place yourself in Figure 4.7*b*. Imagine that you are standing in the center of the figure where the N–S and E–W lines intersect. Turn so that you face south. Using your arm to point at the Sun's position, trace the path of the Sun in the sky at the equinox. It will rise exactly to your left, swing upward to about a 50° angle, and then descend to the horizon exactly at your right. Repeat for the summer and winter solstices, using the figure as a guide. Then try it for the north pole, equator, and tropic of capricorn.
2. Sketch the world latitude zones on a circle representing the globe and give their approximate latitude ranges.
3. Sketch a simple diagram of the Sun above a layer of atmosphere above the Earth's surface, somewhat like Figure 4.14. Using Figure 4.15 as a guide, draw arrows indicating flows of energy among Sun, atmosphere, and surface. Label each arrow using terms from Figure 4.15.

Essay Questions

1. Suppose the Earth's axis of rotation was perpendicular to the orbital plane instead of tilted at $23^1/_2$° away from perpendicular. How would global insolation be affected? How would insolation vary with latitude? How would the path of the Sun in the sky change with the seasons?
2. Imagine that you are following a beam of either (a) shortwave solar radiation entering the Earth's atmosphere heading toward the surface, or (b) a beam of longwave radiation emitted from the surface heading toward space. How will the atmosphere influence the beam?

Problems

Working It Out 4.2 • Radiation Laws

1. Assuming that the Sun behaves as a blackbody, what is the flow rate of energy leaving its surface, if the surface is at a temperature of 5950 K (5677°C, 10,251°F)? At what wavelength will the emitted radiation be greatest?
2. In 1995, the average surface temperature of the Earth was reported as 15.4°C (59.7°F). Assuming that the surface radiates perfectly, what is the flow rate of energy emitted by the surface? At what wavelength will the radiation be greatest?
3. Using the answers to problems 1 and 2, determine the ratio of the flow rate of energy emitted by the Sun's surface to the flow rate of energy emitted by the Earth's surface.

Working It Out 4.3 • Calculating the Global Radiation Balance

1. Venus has a radius of about 6050 km. Because it is closer to the Sun than the Earth, the solar radiation flow it receives is 1.92 times stronger than that received by the Earth. The atmosphere of Venus consists of dense clouds of carbon dioxide vapor, which reflect about 65 percent of the incoming solar radiation. Diagram the energy flow system for Venus and calculate the rates of inflow and outflow of shortwave and longwave radiation.

Eye on the Landscape

Chapter Opener Dromedary caravans near Nouakchott, Mauritania Having just learned about the path of the Sun in the sky in Chapter 3, what can we infer from the elongated shadows of the dromedaries **(A)**? Clearly, the Sun is very low on the horizon, so the time of day is either near sunrise or sunset. If sunset, the Sun should be in the west, so the camera would be looking southeastward. If sunrise, then northwestward. Also, the sharpness of the shadows suggests that the air is very clear, lacking moisture and dust. While the Sahara is very dry, it can be quite dusty at certain times of year. (Atmospheric dust is covered in Chapter 5.)

4.17b Wave erosion Note the low vegetation near these ocean bluffs **(A)**. In severe storms, heavy salt spray is driven across this area, killing young salt-intolerant plants and leaving only the salt-tolerant ones. The environment is also very dry, since the sandy soils of these bluffs don't hold much rainwater. As a result, a distinctive community of plants is found here, including such shrubby species as bayberry, beach plum, and wild rose. Environmental effects on plants are covered in Chapter 23.

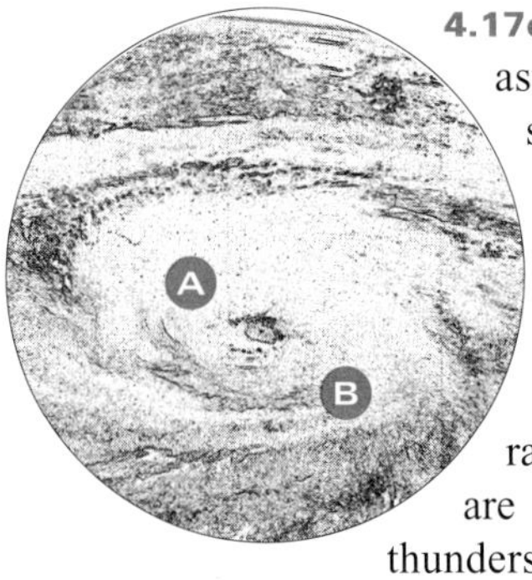

4.17c Tropical cyclone This astronaut photo shows the structure of the hurricane, or typhoon, very nicely. You can easily see the central eye **(A)**, where air descends rapidly and ground wind speeds are light and variable. The radiating arms of the storm **(B)** are formed by bands of severe thunderstorms and rain spiraling inward toward the center of the storm. Hurricane structure and development are topics discussed in Chapter 8, Weather Systems.

A CLOSER LOOK

Geographer's Tools 4.8 Remote Sensing for Physical Geography

Remote sensing refers to gathering information from great distances and over broad areas, usually through instruments mounted on aircraft or orbiting space vehicles. These instruments, or *remote sensors,* measure electromagnetic radiation coming from the Earth's surface and atmosphere as received at the aircraft or spacecraft platform. The data acquired by remote sensors is typically displayed as images—photographs or similar depictions on a computer screen or color printer—but is often processed further to provide other types of outputs, such as maps of albedo, vegetation condition or extent, or land-cover class.

All substances, whether naturally occurring or synthetic, are capable of reflecting, transmitting, absorbing, and emitting electromagnetic radiation. For remote sensing, however, we are only concerned with energy that is reflected or emitted by an object and that reaches the remote sensor. For remote sensing of reflected energy, the Sun is the source of radiation in many applications. As we saw earlier in this chapter, solar radiation reaching the Earth's surface is concentrated in wavelengths from about 0.3 to 2.1 μm in visible, near-infrared, and shortwave infrared wave bands (Figures 4.2, 4.4). Remote sensors are commonly constructed to measure radiation in all or part of this range. For remote sensing of emitted energy, the object or substance itself is the source of the radiation, which is related largely to its temperature.

Colors and Spectral Signatures
Most objects or substances at the Earth's surface possess color to the human eye. This means that they reflect radiation differently in different parts of the visible spectrum. Figure S4.1 shows typical reflectance spectra for water, vegetation, and soil in the solar wavelength range as they might be viewed through the atmosphere. (Note that water vapor in the atmosphere absorbs radiation strongly at wavelengths from about 1.2 to 1.4 μm and 1.75 to 1.9 μm, so it is not possible for a remote sensor to "see" the surface at those wavelengths.)

Let's examine the three spectral reflectance curves. Water surfaces are always dark but are slightly more reflective in the blue and green regions of the visible spectrum. Thus, water appears blue or blue-green to our eyes. Beyond the visible region, water absorbs nearly all shortwave radiation it receives and so looks black in images acquired in the near-infrared and shortwave infrared regions.

Vegetation appears dark green to the human eye, which means that it reflects more energy in the green portion of the visible spectrum while reflecting somewhat less in the blue and red portions. But vegetation also reflects very strongly in near-infrared wavelengths, which the human eye cannot see. Because of this property, vegetation will be bright in near-infrared images. This distinctive behavior of vegetation—appearing dark in visible bands and bright in the near-infrared—is the basis for much of vegetation remote sensing, as we will see in many examples of remotely sensed images throughout this book.

The soil spectrum shows a steady increase of reflectance across the visible and near-infrared spectral regions. Looking at the

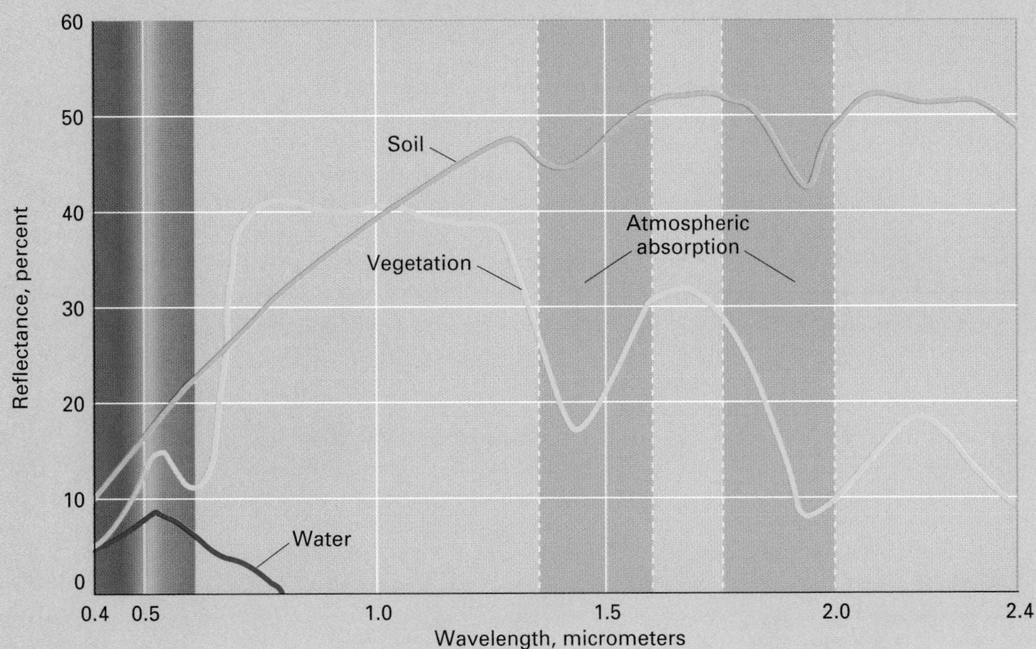

S4.1 Reflectance spectra of vegetation, soil, and water

visible part of the spectrum, we see that soil is brighter overall than vegetation and is somewhat more reflective in the orange and red portions. Thus, it appears brown. (Note that this is just a "typical" spectrum—soil color can actually range from black to bright yellow or red.)

We refer to the pattern of relative brightness within spectral bands as the *spectral signature* of an object or substance. Spectral signatures can be used to recognize objects or surfaces in remotely sensed images in much the same way that we recognize objects by their colors. In computer processing of remotely sensed images, spectral signatures can be used to make classification maps, showing, for example, water, vegetation, and soil.

Aerial Photography

Aerial photography is the oldest form of remote sensing. Air photos have been in wide use since before World War II. Commonly, the field of one photograph overlaps the next along the plane's flight path, so that the photographs can be viewed stereoscopically for a three-dimensional effect. Because of its high resolution (degree of sharpness) and low cost, aerial photography is a widespread application of remote sensing.

Aerial photography often makes use of color infrared film. This special film is sensitive to near-infrared wavelengths in addition to visible wavelengths. Red color in the film is produced as a response to near-infrared light, green color is produced by red light, and blue color by green light. Because healthy, growing vegetation reflects much more strongly in the near-infrared than in the red or green regions of the spectrum, vegetation has a characteristic red appearance. Figure S4.2 shows a color infrared photo image of an agricultural region in California. The distinctive red color identifies crop fields. The image also shows areas of abnormally high groundwater and high soil salinity, both of which are agricultural problems in this region. Other examples of color infrared photos and images will be found in later chapters of this book.

Photography has been extended to greater distances through the use of cameras on orbiting space vehicles. An example is the large-format camera, which was designed to produce very large, very detailed transparencies of the Earth's surface suitable for precise topographic mapping. An excellent example, acquired in a mission aboard the Space Shuttle, is shown in Figure S4.3. Astronauts have also made a number of striking photos of the Earth from space using hand-held cameras.

Thermal Infrared Sensing

Recall from Figure 4.2 that radiation leaving the Earth's surface is concentrated in the thermal infrared spectral region, ranging from about 8 to 12 μm. In *Working It Out 4.2 • Radiation Laws,* we noted that the amount of radiation emitted by the surface of an object or substance is proportional to the fourth power of its absolute temperature. This means that warm objects emit more thermal radiation than cold ones, so warmer objects appear brighter in thermal infrared images. We have already noted this principle in our discussion of Figure 4.3 in the main chapter text.

Besides absolute temperature, the intensity of infrared emission depends on the *emissivity* of an object or substance. Whereas the blackbody is an ideal perfect radiator of energy, all substances are imperfect radiators and emit less energy. Emissivity is the ratio of the actual energy emitted by an object or substance to that of a blackbody at the same temperature. For most natural Earth surfaces, emissivity is comparatively high—between 0.85 and 0.99. Differences in emissivity affect thermal images. For example, two different surfaces might be at the same temperature, but the one with the higher emissivity will look brighter because it emits more energy.

Some substances, such as crystalline minerals, show different emissivities at different thermal infrared wavelengths. In a way, this is like having a particular color, or spectral signature, in the thermal infrared spectral region. In Chapter 12 we will see examples of how some rock types can be distinguished and mapped using thermal infrared images from several wavelengths.

Radar

There are two classes of remote sensor systems: passive and active. *Passive systems* acquire images without providing a source of wave energy. The most familiar passive system is the camera, which uses film that is sensitive to solar energy reflected from the scene. *Active systems* use a beam of wave energy as a source, sending the beam toward an object or surface. Part of the energy is reflected back to the source, where it is recorded by a detector. A simple analogy would be the use of a spotlight on a dark night to illuminate a target, which then reflects light back to the eye.

Radar is an example of an active sensing system. Radar systems in remote sensing use the *microwave* portion of the electromagnetic spectrum, so named because the waves have a short wavelength compared to other types of radio waves (Figure 4.2). (However, these wavelengths are much, much longer than those of solar shortwave radiation.) Radar systems emit short pulses of microwave radiation and then "listen" for a returning microwave echo. By analyzing the time and strength of the return pulse, an image is created showing the surface as it is illuminated by the radar beam.

An advantage of radar systems is that they use wavelengths that are not significantly absorbed by liquid water. This means that radar systems can penetrate clouds to provide images of the Earth's surface in any weather. At some short wavelengths, however, microwaves are scattered by water

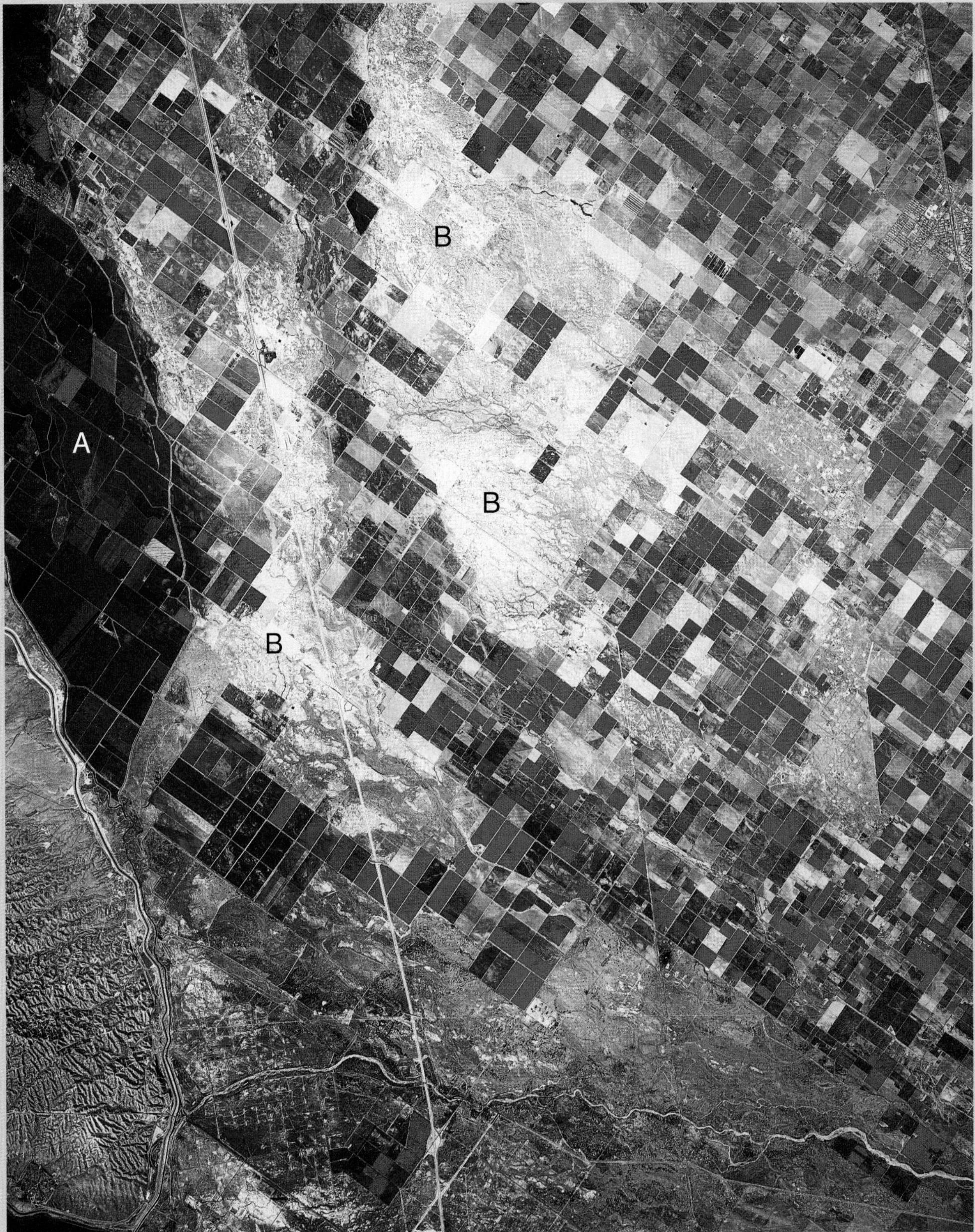

S4.2 High-altitude color infrared photograph *This color infrared photo of an area near Bakersfield, California, in the southern San Joaquin Valley, was taken by a NASA aircraft flying at approximately 18 km (60,000 ft). Photos such as this can be used to study problems associated with agriculture. Problems affecting crop yields arise in abnormally high ground water areas (A), which appear dark, and areas of high soil salinity (B), which appear light. The various red tones are associated with different types of crops. (Courtesy NASA, compiled and annotated by John E. Estes and Leslie W. Senger.)*

S4.3 Color infrared photograph from low Earth orbit *A special large-format camera carried on a flight of the NASA Space Shuttle acquired this color infrared photograph. The area shown, about 125 km (77 mi) in width, includes the Sulaiman Range in western Pakistan (upper left). The Indus and Sutlej rivers, fed by snowmelt from the distant Hindu Kush and Himalaya Ranges, cross the scene flowing from northeast to southwest. A mottled pattern of green fields (red) interspersed with barren patches of saline soil (white) covers much of the lower fourth of the area. (Courtesy Geonex/Chicago Aerial Survey, Inc.)*

droplets and can produce a return signal sensed by the radar apparatus. This effect is the basis for weather radars, which can detect rain and hail and are used in local weather forecasting.

Figure S4.4 shows a radar image of the folded Appalachian Mountains in south central Pennsylvania. It is produced by an airborne radar instrument that sends pulses of radio waves downward and sideward as the airplane flies forward. Surfaces oriented most nearly at right angles to the slanting radar beam will return the strongest echo and therefore appear lightest in tone. In contrast, those surfaces facing away from the beam will appear darkest. The effect is to produce an image resembling a three-dimensional model of the landscape illuminated at a strong angle. The image shows long mountain ridges running from upper right to lower left and casting strong radar shadows to emphasize their three-dimensional form. The ridges curve and turn sharply, revealing the geologic structure of the region. Between the ridges are valleys of agricultural land, which is distinguished by its rougher texture in the image. In the upper left is a forested plateau that has a smoother appearance.

S4.4 Side-looking radar image from south-central Pennsylvania *The image shows a portion of the folded Appalachians with zigzag ridges and intervening valleys. The area shown is about 40 km (25 mi) wide. Compare this with the corresponding photo in* Focus on Remote Sensing 18.1 • Landsat Views Rock Structures, *which is a Landsat image of this region. (SAR image courtesy of Intera Technologies Corporation, Calgary, Alberta, Canada.)*

Digital Imaging

Modern remote sensing relies heavily on computer processing to extract and enhance information from remotely sensed data. This requires that the data be in the form of a *digital image* (Figure S4.5). In this format, the picture consists of a very large number of grid cells arranged in rows and columns to cover the scene. Each grid cell records a brightness value and is referred to as a *pixel,* a term that arises as a contraction of "picture element." Normally, low values code for dark (low reflectance), and high values code for light (high reflectance). To create an image that is viewable, as, for example, on a computer screen, the brightness values are fed to a special computer chip that generates a corresponding television signal fed to the monitor. Or the image is processed and sent to a computer printer to produce a hard copy print or transparency.

The great advantage of digital images over images on photographic film is that they can be processed by computer, for example, to increase contrast or sharpen edges (Figure S4.6). *Image processing* refers to the manipulation of digital images to extract, enhance, and display the information that they contain. In remote sensing, image processing is a very broad field that includes many methods and techniques for processing remotely sensed data.

Digital Imaging Systems

Many remotely sensed digital images are acquired by scanning systems, which may be mounted in aircraft or on orbiting space vehicles. *Scanning* is the process of receiving information instantaneously from only a very small portion of the area being imaged (Figure S4.7). The scanning instrument senses a very small field of view that runs rapidly across the ground scene. Light from the field of view is focused on a detector that responds very quickly to small changes in light intensity. Electronic circuits read out the detector at very short time intervals and

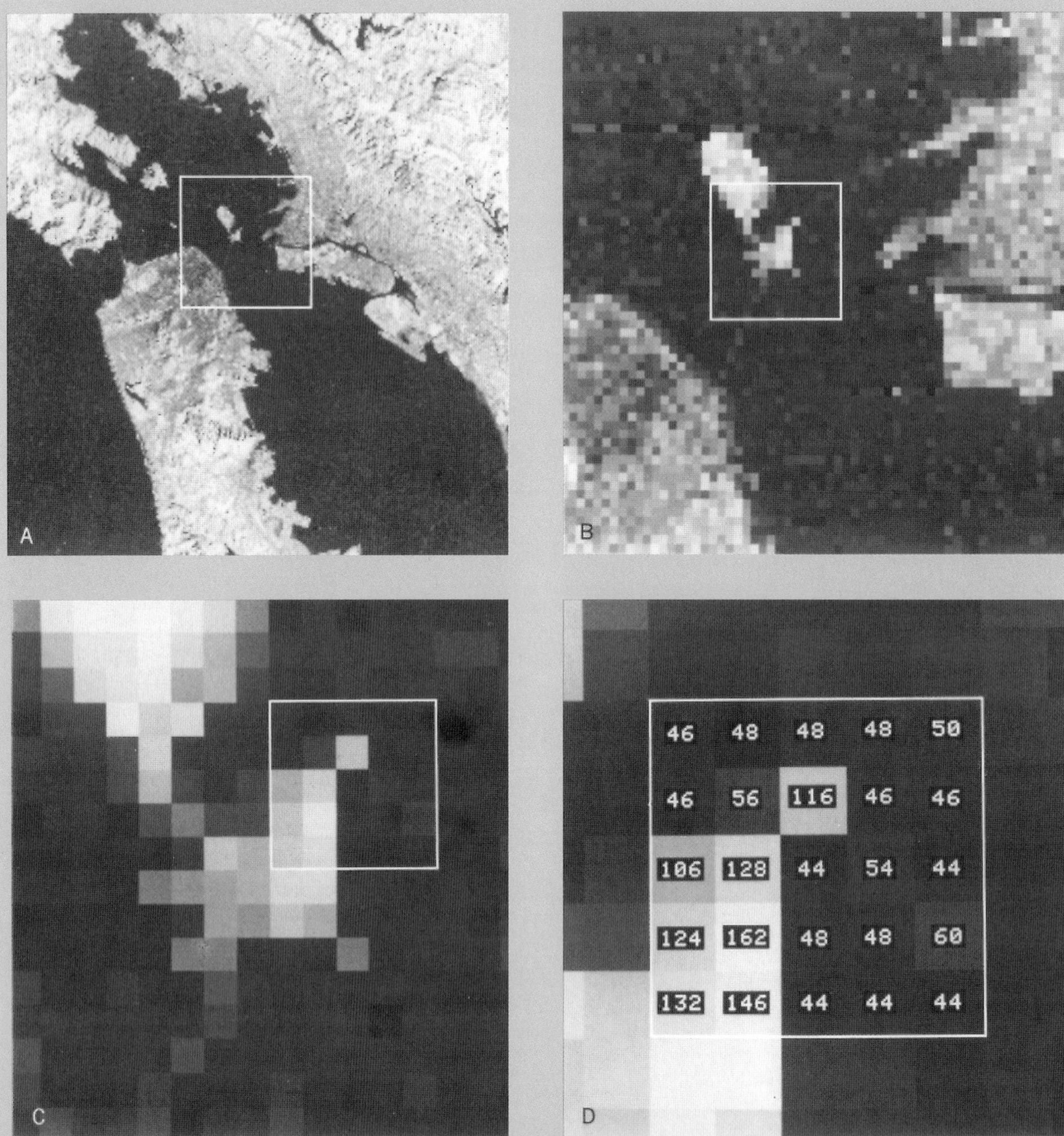

S4.5 Zooming in on a digital image *These four views of the San Francisco Bay Bridge and Yerba Buena Island acquired by Landsat in the near-infrared band show how a digital image is composed of individual pixels with different brightness values. (A–C) Progressively smaller subimages zooming on the east end of the island. (D) Actual brightness values for a small array of 25 pixels, scaled to range from 0 (darkest) to 255 (lightest). (A. H. Strahler.)*

record the intensities. Later, the computer reconstructs a digital image of the ground scene from the measurements acquired by the scanning system.

Most scanning systems in common use are *multispectral scanners.* These devices have multiple detectors and measure brightness in several wavelength regions simultaneously. An example is the Thematic Mapper instrument used aboard the Landsat series of Earth-observing satellites. This instrument simultaneously collects reflectance data in seven spectral bands. Six wavebands sample the visible, near-infrared and shortwave infrared regions, while a seventh records thermal infrared emissions. Figure S4.8 shows a

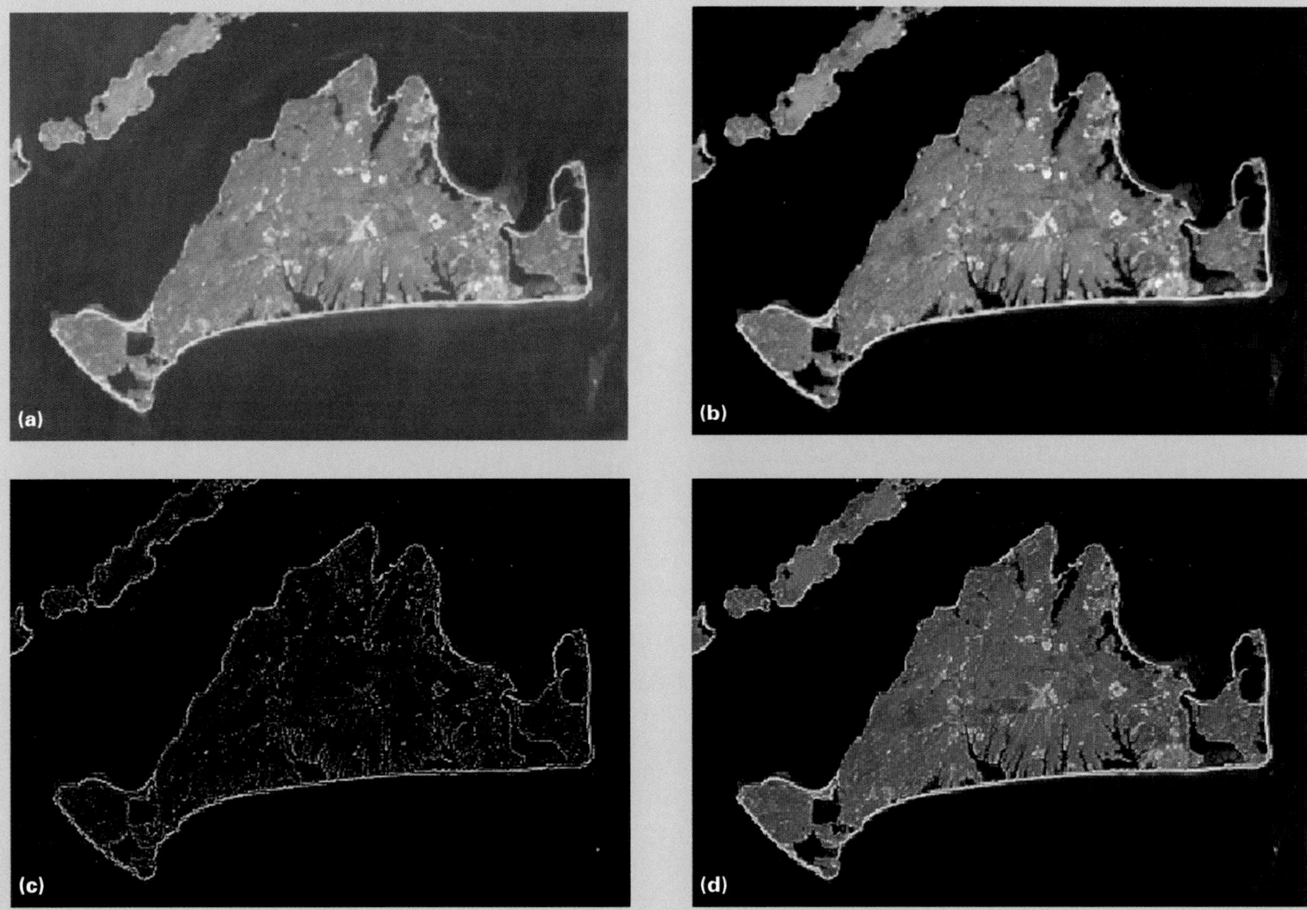

S4.6 Image processing *These four panels show an image of the island of Martha's Vineyard, Massachusetts, acquired by the Landsat Enhanced Thematic Mapper instrument on August 26, 2000. As originally acquired (a), the image lacks contrast. In (b), the color scales have been adjusted to show a wider range of colors. In (c), an edge enhancement computation shows edges within the image as bright pixels. When the edge image is added to to contrast-enhanced image, the result is an image that appears clearer and sharper than the orginal (d).*

color composite of the Boston region acquired by the Landsat Thematic Mapper on September 27, 1991. The image uses red, near-infrared, and shortwave infrared wavebands to show vegetation in green, beaches and bare soils in pink, and urban surfaces in shades of blue.

Scanning provides images using a small number of detectors that are read out very quickly. An alternative to scanning is *direct digital imaging* using large numbers of detectors arranged in a two-dimensional array (Figure S4.9). This technology is in common use in digital cameras. Instead of a layer of light-sensitive film, a digital camera has an array of millions of tiny detectors arranged in rows and columns that individually measure the amount of light they receive during an exposure. Electronic circuitry reads out the measurement made by each detector, composing the entire image within a second or two. The image is stored on a memory chip for later downloading to a computer. Although photographic film in a conventional camera can record more fine detail than a digital camera in a typical application, the digital camera measures brightness more accurately and outputs a digital image that can be read directly and processed by computer. Note that it is possible to scan a photograph or transparency to produce a digital image as well. This allows the photographic image to be processed in ways that could never be attempted in a photographic darkroom.

Orbiting Earth Satellites

With the development of orbiting Earth satellites carrying remote sensing systems, remote sensing has expanded into a major branch of geographic research. Because orbiting satellites can image and monitor large geographic areas or even the entire Earth, global and regional studies have become

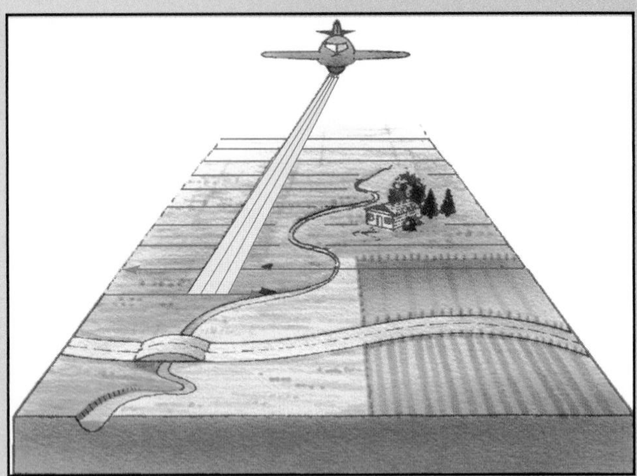

S4.7 Multispectral scanning from aircraft *As the aircraft flies forward, the scanner sweeps from side to side. The result is a digital image covering the overflight area. (A. H. Strahler.)*

S4.9 An area array of detectors *The center part of this computer chip is covered by an array of tiny light detectors arranged in rows and columns.*

possible that could not have been carried out in any other way.

How does a satellite sensor image the entire Earth? One way is to use a polar orbit. As the satellite circles the Earth, passing over each pole, the Earth rotates underneath it, allowing all of the Earth to be imaged after repeated passes. However, a polar orbit remains fixed in space with respect to the stars, not to the Sun. So as the Earth revolves around the Sun in its yearly cycle, a satellite in true polar orbit passes overhead about four minutes later each solar day. This means that illumination conditions slowly change from day to day and quite a lot from month to month or season to season.

S4.8 Landsat image of Boston *An image of Boston, acquired by Landsat Thematic Mapper on September 27, 1991. In this false-color composite, the red color is from the shortwave infrared band (1.55 to 1.75 μm), the green from the near-infrared (0.79 to 0.91 μm), and the blue from the red band (0.60 to 0.72 μm). (A. H. Strahler.)*

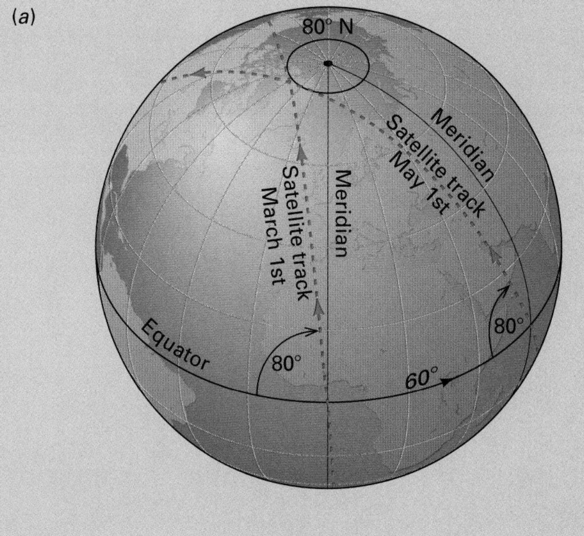

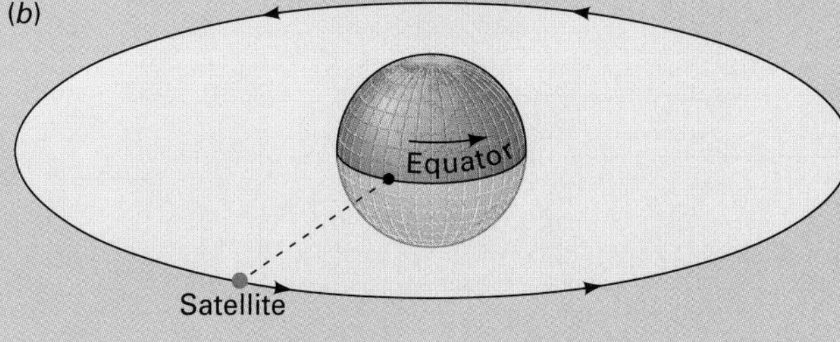

S4.10 Satellite orbits *(a) Earth track of a Sun-synchronous orbit. With the Earth track inclined at about 80° to the equator, the orbit slowly swings eastward at about 30° longitude per month, maintaining its relative position with respect to the Sun. Between March 1 and May 1 (shown) the orbit moves about 60°. (© A. N. Strahler.) (b) Motion of a geostationary satellite. Because the satellite revolves around the Earth above the equator at the same rate as the Earth's rotation, it appears fixed in the sky above a single point on the equator.*

A solution to this problem is to use a *Sun-synchronous orbit* (Figure S4.10*a*). This type of near-polar orbit is oriented at a small angle to the polar axis and so crosses the equator at an angle less than 90 degrees. Recall from Chapter 3 that the Earth is not a perfect sphere but is slightly thicker at the equator than at the poles. Thus, the Earth's gravity is slightly greater at the equator. When the orbit crosses the equator at an angle, the difference in gravity acts to push the orbit very slightly eastward. This keeps the satellite orbit in time with the Sun instead of the stars, so that satellite overpasses continue to occur at the same time of day. Typical Sun-synchronous orbits take 90 to 100 minutes to circle the Earth and are located at heights of about 700 to 800 km (430 to 500 mi) above the Earth's surface.

Another orbit used in remote sensing is the *geostationary orbit* (Figure S4.10*b*). Instead of orbiting around the poles, a satellite in geostationary orbit constantly revolves above the equator. The orbit height, about 35,800 km (22,200 mi), is set so that the satellite makes one revolution in exactly 24 hours in the same direction that the Earth turns. Thus, the satellite always remains above the same point on the equator. From its high vantage point, the geostationary orbiter provides a view of nearly half of the Earth below.

Geostationary orbits are ideal for observing weather, and the weather satellite images readily available on television and the Internet are obtained from geostationary remote sensors. Geostationary orbits are also ideal for communications satellites. Since a geostationary orbiter remains at a fixed position in the sky for an Earthbound observer, a high-gain antenna can be pointed at the satellite and fixed in place permanently, providing high-quality, continuous communications.

Remote sensing is an exciting, expanding field within physical geography and the geosciences in general. This special supplement has provided some basic information about remote sensing that will come in handy as you read our special boxed features, *Focus on Remote Sensing,* and as you view the various examples of remotely sensed images throughout the text.

5 Air Temperature and Air Temperature Cycles

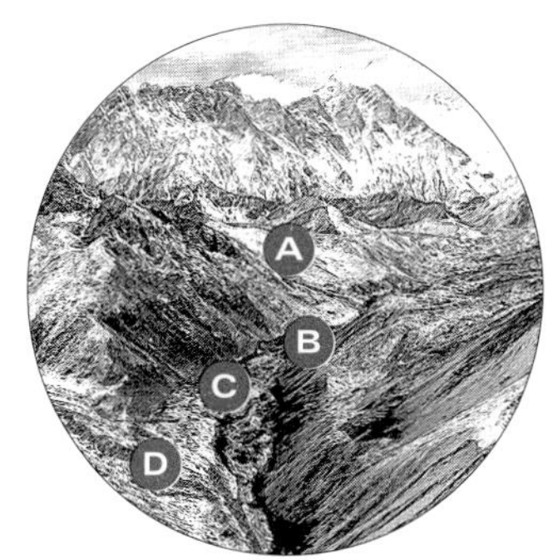

Eye on the Landscape
Mount Everest, Himalayas, Nepal. Mountains are cooler than surrounding plains, and atop Everest, the highest peak on Earth, the temperature is always bitterly cold.
What else would the geographer see?...Answers at the end of the chapter.

This chapter focuses on **air temperature**—that is, the temperature of the air as observed at 1.2 m (4 ft) above the ground surface. Air temperature conditions many aspects of human life, from the clothing we wear to the fuel costs we pay. Air temperature and air temperature cycles also act to select the plants and animals that make up the biological landscape of a region. Even geological weathering and soil-forming processes are dependent on temperature over the long term. And air temperature, along with precipitation, is a key determiner of climate, which we will explore in more depth in Chapters 9–11.

Is air temperature changing? You may have heard that the Earth is becoming increasingly warmer, and that as a result, sea level is rising, climate boundaries are shifting, and severe weather is becoming more frequent. Global warming and its effects are a complicated story that we have already touched upon at several points and will return to again in future chapters. Toward the end of this chapter, we will examine the effect of increasing CO_2 and other gases on global warming and place the present record warmth in the perspective of the Earth's natural time cycles. But first you will need to understand how and why air temperature changes from day to day, month to month, and year to year.

What factors influence air temperatures? Five factors are important.

1. **Insolation.** The two motions of the Earth—its daily rotation on its axis and its annual revolution in its solar orbit—create the daily and annual cycles of insolation that we described in Chapter 4. In turn, these cycles produce the cycles of air temperature that distinguish day and night, winter and summer.

2. **Latitude.** As we noted in Chapter 4, daily and annual cycles of insolation vary systematically with latitude, causing air temperatures and air temperature cycles to vary as well. As shown in Figure 4.10, yearly insolation decreases toward the poles, so less energy is available to heat the air. Thus, temperatures generally fall as we move poleward. Temperatures also become more variable over the year as latitude increases. Recall from Figure 4.8 that as much, or more, solar energy may be received in a summer day at high latitudes than is received at the equator, while little or no solar energy is received at high latitudes during the winter. Because of this annual variation, high latitudes experience a much greater range in air temperatures through the year.
3. **Surface type.** Urban surfaces of asphalt, roofing shingles, and rubber are dry compared to the moist soil surfaces of rural areas and forests. They heat more rapidly because solar energy cannot be taken up in evaporation of water. They also absorb a greater portion of the Sun's energy than vegetation-covered surfaces. Because of these factors, urban air temperatures are generally higher than rural temperatures. The same is true for areas of barren or rocky soil surfaces, such as those of deserts.
4. **Coastal vs. interior location.** Locations near the ocean experience a narrower range of air temperatures than locations in continental interiors. The reason is that water heats and cools more slowly than land, so air temperatures over water bodies tend to be less extreme than temperatures over the land surface. And since winds can easily cause air to flow from water to land, a coastal location will more often feel the influence of the adjacent water.
5. **Elevation.** At high elevation, there is less atmosphere above the surface. This means that the greenhouse effect provides a less effective insulating blanket. Average temperatures are cooler because more surface heat is lost to space. This allows snow to accumulate and remain longer on high peaks, for example. The reduced greenhouse effect also results in greater daily temperature variation.

Understanding air temperatures will be easy if you keep these five factors in mind: insolation, latitude, surface type, location, and elevation.

Focus on Systems 5.1

The Surface Energy Balance Equation

Using the principles of budgeting, it is easy to devise a simple equation for the energy balance of a surface as an energy flow system. For this equation, we consider the surface as a thin layer positioned between the atmosphere and the soil. It acts as an energy transfer layer that changes one type of energy flow to another. The surface is too thin to hold any heat itself. There are five flows of energy to and from the surface. They are related by the surface energy balance equation:

$$R_{SHORT} + R_{LONG} + H_{LATENT} + H_{SENSIBLE} + H_{SOIL} = 0$$

Here, R_{SHORT} is the flow of shortwave radiation absorbed by the surface; R_{LONG} is the net flow of longwave radiation between the surface and the atmosphere; H_{LATENT} is the latent heat flow between the surface and the atmosphere; $H_{SENSIBLE}$ is the sensible heat flow between the surface and the atmosphere; and H_{SOIL} is the sensible heat flow between the surface and the soil. In using the equation, we consider outgoing flows to be positive and incoming flows to be negative. From the principle of conservation of energy, incoming energy flow must balance outgoing energy flow, so the terms of the equation must add up to zero.

The equation includes two radiation terms. R_{SHORT} is the flow of shortwave radiation to the surface from the Sun. Note that this flow includes only radiation that is absorbed by the surface. Shortwave radiation that is reflected directly back toward the atmosphere is not included. R_{SHORT} will be negative during the day, since by our convention flows coming into the surface are negative. At night, this flow will drop to zero. R_{LONG} is the net longwave radiation. Because it is a net term, it represents the balance between outgoing longwave radiation emitted by the surface and incoming longwave radiation from the atmosphere above. Because the surface is usually warmer than the atmosphere, even at night, the flow will normally be away from the surface, which is taken as positive. As we saw in Chapter 4, the sum of R_{SHORT} and R_{LONG} is the net radiation flow.

Three energy flows remain. H_{LATENT} is the latent heat flow, which arises when water changes state from a solid or liquid at the surface and diffuses into the atmosphere as water vapor. During the day, it will be positive as soil water evaporates. At night, condensation or deposition may occur, yielding dew or frost. If so, latent heat will be released at the surface, providing a heat flow to the surface (negative).

$H_{SENSIBLE}$ is the sensible heat flux to the atmosphere, which arises when the surface conducts heat to the air at the surface–atmosphere boundary, and that heat is carried upward by convection. It will be positive when the air is warmed by the surface, which is the normal condition

SURFACE TEMPERATURE

Temperature is a familiar concept. It is a measure of the level of sensible heat of matter, whether it is gaseous (air), liquid (water), or solid (rock or dry soil). We know from experience that when a substance receives a flow of energy, its temperature rises. Similarly, if a substance loses energy, its temperature falls. This energy flow moves in and out of the substance at its **surface**—for example, the very thin surface layer of soil that actually absorbs solar shortwave radiation and radiates longwave radiation out to space.

The temperature of a surface is determined by the balance among the various energy flows that move across it. As we saw in Chapter 4, **net radiation**—the balance between incoming shortwave radiation and outgoing longwave radiation—produces a radiant energy flow that can heat or cool a surface. Recall that during the day, incoming solar radiation normally exceeds outgoing longwave radiation, so the net radiation balance is positive and the surface warms. Energy flows through the surface into the cooler soil below. At night, net radiation is negative, and the soil loses energy as the surface temperature falls and the surface radiates longwave energy to space.

Energy may also move to or from a surface in other ways. **Conduction** describes the flow of sensible heat from a warmer substance to a colder one through direct contact. When heat flows into the soil from its warm surface during the day, it flows by conduction. At night, heat is conducted back to the colder soil surface. **Latent heat transfer** is also important. When water evaporates at a surface, it removes the heat stored in the change of state from liquid to vapor, thus cooling the surface. When water condenses at a surface, latent heat is released, warming the surface. Another form of energy transfer is **convection,** in which heat is distributed in a fluid by mixing. If the surface is in contact with a fluid, such as a soil surface with air above, upward and downward flowing currents can act to warm or cool the surface. (We'll return to convection in more detail in Chapters 6 and 7.)

Since energy is neither created nor destroyed—just transformed—the energy flows occurring at a surface must be in balance. The *surface energy balance equation* describes how net radiation, latent heat, and sensible heat flows as conduction and convection all balance for a surface. This equation is discussed more fully in *Focus on Systems 5.1 • The Surface Energy Balance Equation.* It is important

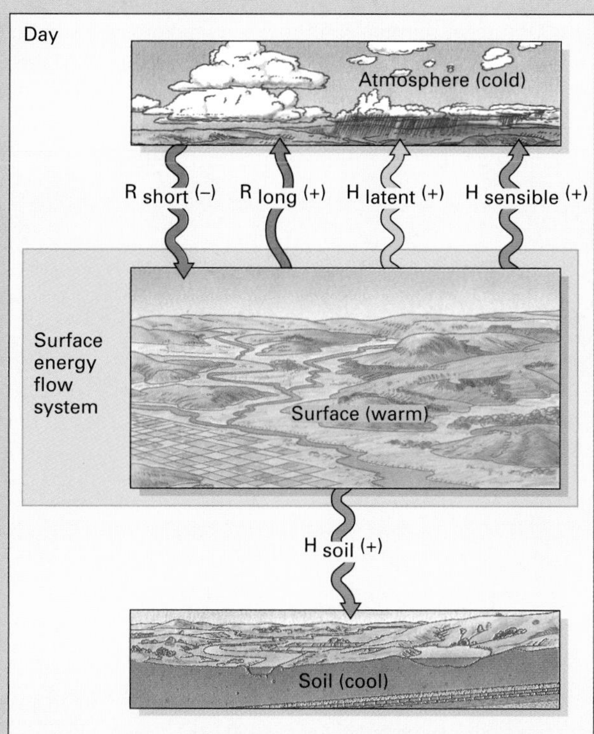

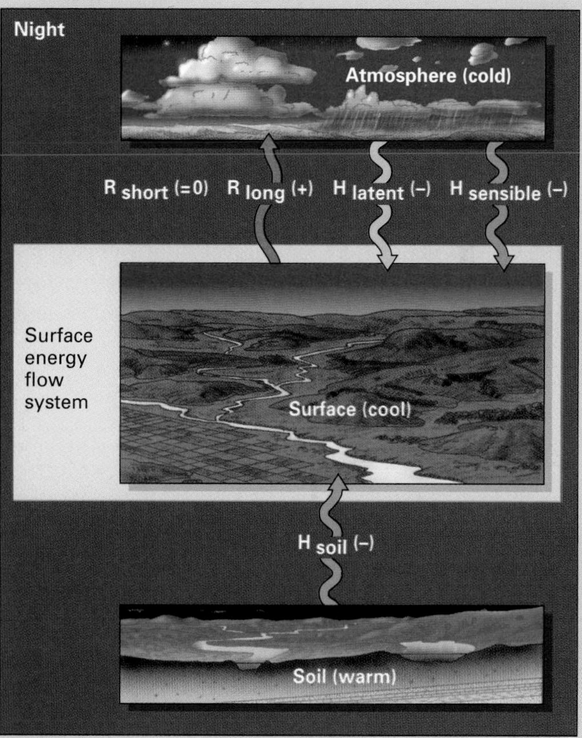

Diagrams of the surface energy balance equation for typical day and night conditions

during the day. At night, the surface can become colder than the air it contacts, so the heat flow may be from the air to the surface (negative).

The last flow, H_{SOIL}, is the flow of heat by conduction from the surface to the soil below. This flow will normally be away from the surface (positive) during the day, as heat is conducted from the warm surface into the soil. At night, however, the surface will normally cool enough that heat will be conducted upward. This will then be a negative flow.

because it explains the temperature of the surface. In its latent heat transfer term, the surface energy balance equation also describes how water evaporates from the soil. This helps us understand how soil moisture changes with time, a subject that is of great interest in agriculture.

AIR TEMPERATURE

Thus far, we've discussed the temperature of surfaces. However, in this chapter we are concerned with air temperature, which is measured a short distance above the ground surface. As we will see shortly, air temperature can be different from surface temperature. Walking across a parking lot in sandals on a clear summer day, you will certainly notice that the pavement is quite a lot hotter than the air you feel on the upper part of your body. But because conduction and convection transfer sensible heat to and from the surface to the air, air temperatures will tend to follow surface temperatures. So we can safely conclude that the temperature changes and trends experienced at the ground surface will be reflected in the air temperatures that are observed and recorded a short distance above the surface.

Measurement of Air Temperature

Air temperature is a piece of weather information that we encounter daily. In the United States, temperature is still widely measured and reported using the Fahrenheit scale. The freezing point of water on the Fahrenheit scale is 32°F, and the boiling point is 212°F. This represents a range of 180°F. In this book, we use the Celsius temperature scale, which is the international standard. On the Celsius scale, the freezing point of water is 0°C and the boiling point is 100°C. Thus, 100 Celsius degrees are the equivalent of 180 Fahrenheit degrees (1°C = 1.8°F; 1°F = 0.56°C). Conversion formulas between these two scales are given in Figure 5.1 and in *Working It Out 5.2 • Temperature Conversion.*

You are probably familiar with the *thermometer* as an instrument for measuring temperature. In this device, a liquid inside a glass tube expands and contracts with temperature, and the position of the liquid surface indicates the temperature. Since air temperature can vary with height, it is measured at a standard level—1.2 m (4 ft) above the ground. A *thermometer shelter* (Figure 5.2*a*) is a louvered box that holds thermometers or other weather instruments at proper height while sheltering them from the direct rays of the Sun. Air circulates freely through the louvers, ensuring that temperatures inside the shelter are the same as the outside air.

Liquid-filled thermometers are being replaced by newer instruments for the routine measuring of temperatures. Perhaps you have used a digital fever thermometer, which reads out your body temperature directly as a number on a display. It uses a device called a *thermistor,* which changes its electrical resistance with temperature. By measuring this resistance, the temperature may be obtained electronically. Many weather stations are now equipped with temperature measurement systems that use thermistors (Figure 5.2*b*).

Although some weather stations report temperatures hourly, most stations only report the highest and lowest temperatures recorded during a 24-hour period. These are the most important values in observing long-term trends in temperature.

Temperature measurements are reported to governmental agencies charged with weather forecasting, such as the U.S. Weather Service or the Meteorological Service of Canada. These agencies typically make available daily, monthly, and yearly temperature statistics for each station using the daily maximum, minimum, and mean temperature. Note that the mean daily temperature is defined as the average of the maximum and minimum daily values. These statistics, along with others such as daily precipitation, are used to describe the climate of the station and its surrounding area.

THE DAILY CYCLE OF AIR TEMPERATURE

Because the Earth rotates on its axis, incoming solar energy at a location can vary widely throughout the 24-hour period, while outgoing longwave energy remains more constant. During the day, net radiation is positive, and the surface

5.1 Celsius and Fahrenheit temperature scales compared *At sea level, the freezing point of water is at Celsius temperature (C) 0°, while it is 32° on the Fahrenheit (F) scale. Boiling occurs at 100°C, or 212°F.*

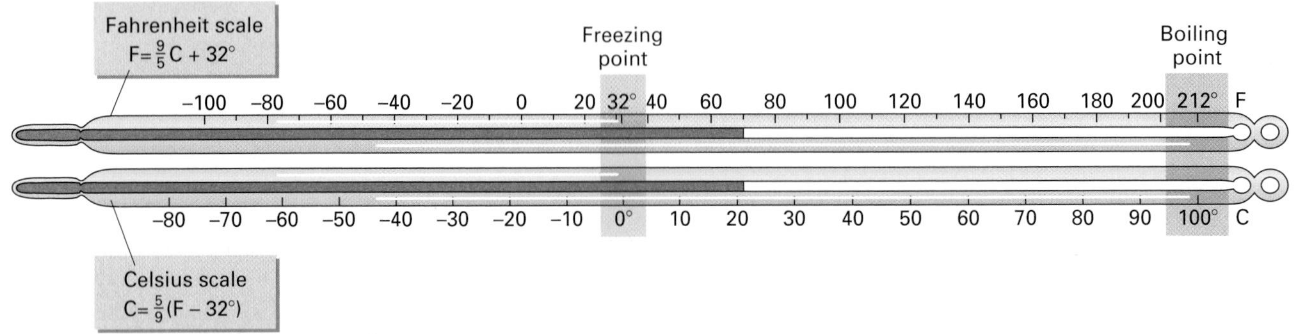

Working It Out | 5.2
Temperature Conversion

Converting from Fahrenheit to Celsius and the reverse can be done simply and quickly using the following formulas:

$$C = \frac{5}{9}(F - 32)$$

$$F = \frac{9}{5}C + 32$$

Here C is the temperature in degrees Celsius, and F is the temperature in degrees Fahrenheit. The fractions $\frac{5}{9}$ and $\frac{9}{5}$ arise from the fact that 100 Celsius degrees covers the span of 180 degrees Fahrenheit. These values are in the ratio of 5 to 9 and vice versa.

As an example, let's convert 50°F to Celsius and 50°C to Fahrenheit. For the first conversion, we have

$$C = \frac{5}{9}(F - 32) = \frac{5}{9}(50 - 32) = \frac{5}{9}(18)$$

$$= 10°C$$

and for the second,

$$F = \frac{9}{5}C + 32 = \frac{9}{5}(50) + 32 = 90 + 32$$

$$= 122°F$$

gains heat. At night, net radiation is negative, and the surface loses heat by radiating it to the sky and space. Since the air next to the surface is warmed or cooled as well, air temperatures follow the same cycle. This results in the daily cycle of rising and falling air temperatures.

Daily Insolation and Net Radiation

Let's look in more detail at how insolation, net radiation, and air temperature are linked in this daily cycle. The three graphs in Figure 5.3 show average curves of daily insolation, net radiation, and air temperature that we might expect for a typical observing station at lat. 40° to 45° N in the interior of North America. The time scale is set so that 12:00 noon occurs when the Sun is at its highest elevation in the sky.

Graph (*a*) shows daily insolation. At the equinox (middle curve), insolation begins at about sunrise (6 A.M.), rises to a peak value at noon, and declines to zero at sunset (6 P.M.). At the June solstice, insolation begins about two hours earlier (4 A.M.) and ends about two hours later (8 P.M.). The June peak is much greater than at equinox, and the total insolation for the day is also much greater. At the December solstice,

5.2 Weather recording instruments *la) An instrument shelter housing a pair of thermometers. The shelter is constructed with louvered sides for ventilation and painted white to reflect solar radiation. (b) A thermister-based maximum-minimum recording system (left) and a recording rain gauge (right).*

(a)

(b)

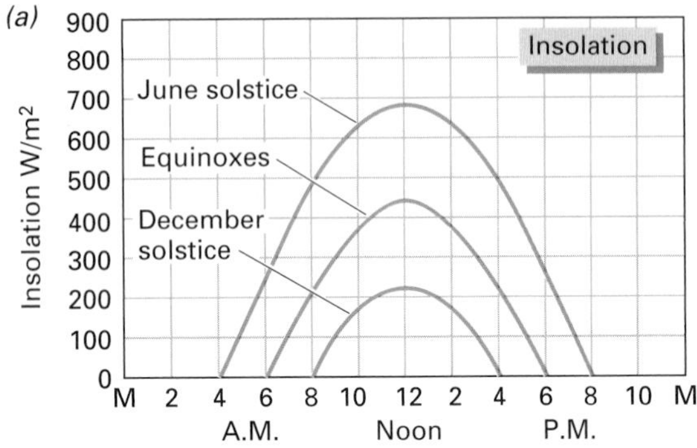

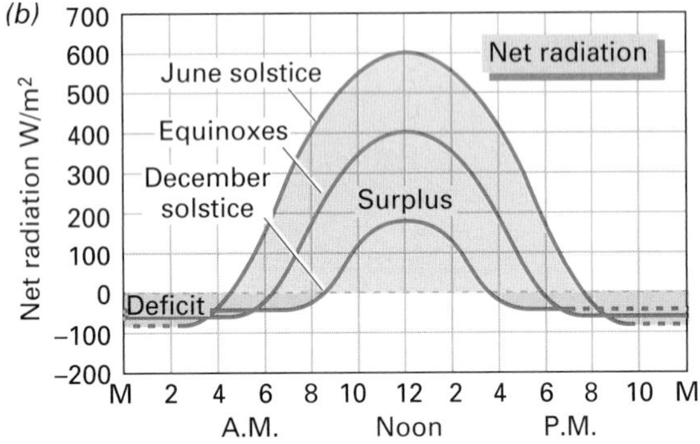

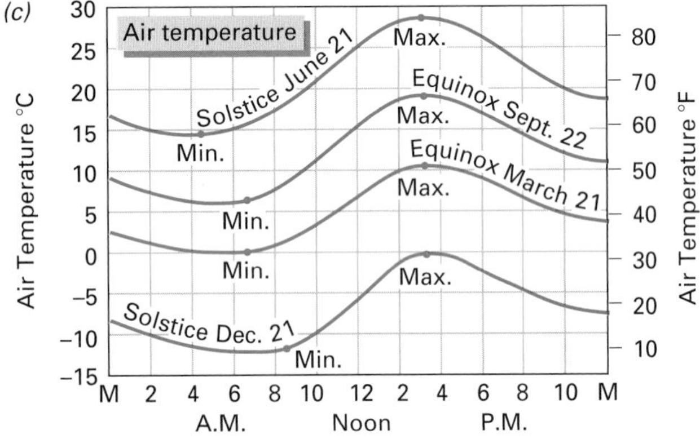

5.3 Daily cycles of insolation, net radiation, and air temperature *These three graphs show idealized daily cycles for a midlatitude station at a continental interior location. Insolation (a) is a strong determiner of net radiation (b). Air temperatures (c) respond by generally increasing while net radiation is positive and decreasing when it is negative.*

insolation begins about two hours later than the equinox curve (8 A.M.) and ends about two hours earlier (4 P.M.). Both the peak intensity and daily total insolation are greatly reduced in the winter solstice season.

Graph (*b*) shows net radiation for the surface. Recall that when net radiation is positive, the surface gains heat, and when negative, it loses heat. The curves for the solstices and equinox generally resemble those of insolation. Net radiation begins the 24-hour day at midnight as a negative value—a deficit. The deficit continues into the early morning hours. Net radiation shows a positive value—a surplus—shortly after sunrise and rises sharply to a peak at noon. In the afternoon, net radiation decreases as insolation decreases. A value of zero is reached shortly before sunset. With no incoming insolation, net radiation then becomes negative, showing a deficit, where it remains until morning.

Although the three net radiation curves show the same general daily pattern—negative to positive to negative—they differ greatly in magnitude. For the June solstice, the positive values are quite large, and the surplus period is much larger than the deficit period. This means that for the day as a whole, net radiation is positive. At the December solstice, the surplus period is short and the surplus is small. The total deficit, which extends nearly 18 hours, outweighs the surplus. This means that the net radiation for the entire day is negative. As we will see, this pattern of positive daily net radiation in the summer and negative daily net radiation in the winter drives the annual cycle of temperatures.

Daily Temperature

Graph (*c*) shows the typical, or average, air temperature cycle for a 24-hour day. The minimum daily temperature usually occurs about a half hour after sunrise. Since net radiation has been negative during the night, heat has flowed from the ground surface, and the ground has cooled the surface air layer to its lowest temperature. As net radiation becomes positive, the surface warms quickly and transfers heat to the air above. Air temperature rises sharply in the morning hours and continues to rise long after the noon peak of net radiation.

We should expect the air temperature to rise as long as net radiation is positive. However, another process begins in the early afternoon on a sunny day. Large convection currents develop and mix the air within several hundred meters of the surface, carrying heated air aloft and bringing cooler air downward toward the surface. Therefore, the temperature peak usually occurs in the midafternoon. The peak is shown in the figure at about 3 P.M., but it typically occurs between 2 and 4 P.M., depending on local conditions. By sunset, air temperature is falling rapidly. It continues to fall, but at a decreasing rate, throughout the night.

Note also that the general level of the temperature curves varies with the seasons. In the summer, the daily curve is high, showing warm temperatures. In winter, the curve is low, showing cold temperatures. In between are the equinoxes, with their intermediate temperatures. Since the temperatures lag behind the seasonal changes in net radiation, the September equinox shows temperatures that are considerably warmer than the March equinox. Even though net radiation is the same for the two, the curves are not the same because each reflects earlier seasonal conditions.

Temperatures Close to the Ground

We noted at the beginning of this chapter that air temperatures, as measured at 1.2 m (4.0 ft) above the surface, generally reflect the same trends as those of the ground surface, but that ground surface temperatures are likely to be more extreme. Let's examine how soil, surface, and air tempera-

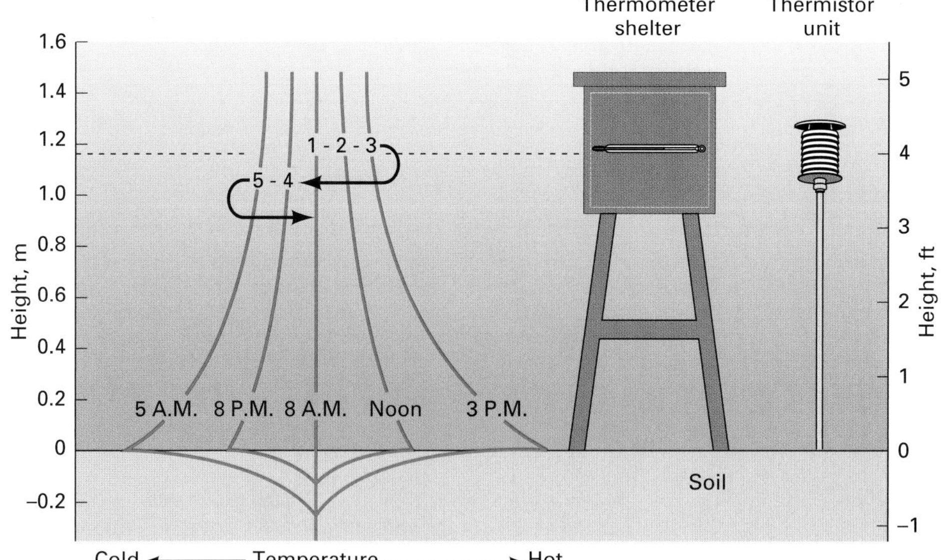

5.4 Daily temperature profiles close to the ground *This simplified diagram shows the temperature profile close to the ground surface at five times of day. The soil surface becomes very hot by midafternoon but cools greatly at night.*

tures within a few meters of the ground might change through the day for a bare, dry soil surface.

Figure 5.4 presents a series of generalized temperature profiles from about 30 cm (12 in.) below the surface to 1.5 m (4.9 ft) above it at five times of day. At 8 A.M. (curve 1), the temperature of air and soil is uniform, producing a vertical line on the graph. By noon (curve 2), the surface is considerably warmer than the air at standard height, and the soil below the surface has been warmed as well. By 3 P.M. (curve 3), the soil surface is very much warmer than the air at standard height. By 8 P.M. (curve 4), the surface is cooler than the air, and by 5 A.M. (curve 5), it is much colder.

This analysis shows that daily temperature variation is greatest at the surface, while air temperature at standard height is less variable. Note also that within the soil, the daily cycle weakens with depth to a point where the temperature does not change during the 24-hour cycle.

Environmental Contrasts: Urban and Rural Temperatures

Human activity has altered much of the Earth's land surface. Cities are an obvious example. To build a city, vegetation is removed, and soils are covered with pavement or structures. Using the principles of the surface energy balance we have learned, we can understand the effects of these changes on the temperatures of urban areas as compared to rural regions.

Figure 5.5 illustrates some of the differences between urban and rural surfaces. In rural areas, the land surface is normally covered with a layer of vegetation. A natural activity of plants is **transpiration.** In this process, water is taken up by plant roots and moved to the leaves, where it evaporates. This cools leaf surfaces, which in turn tend to keep nearby air cooler. Water also evaporates directly from the soil, again cooling the surface and keeping air temperatures down.

The cooling effect of vegetation is even greater in a forest. Not only are transpiring leaves abundant, but also solar radiation is intercepted by a layer of leaves stretching from the tops of the trees to the forest floor. Thus, solar warming is not concentrated intensely at the ground surface, but rather serves to warm the whole forest layer.

Another feature of the rural environment is that the soil surfaces tend to be moist. During rainstorms, precipitation seeps into the soil. When sunlight reaches the soil surface, the water evaporates, thus removing heat and keeping the soil cool. We refer to the combined effects of transpiration and evaporation as **evapotranspiration.**

In contrast are the typical surfaces of the city. Rain runs easily off the impervious surfaces of roofs, sidewalks, and streets. The rainwater is channeled into storm sewer systems, where it flows directly into rivers, lakes, or oceans, instead of soaking into the soil beneath the city surfaces. Because the impervious city surfaces are dry, the full energy of insolation warms the surfaces, with evaporation taking place only where street trees, lawns, and bare soil patches occur.

City surfaces are also darker and more absorbent than rural surfaces. In fact, asphalt paving and roofing absorb about twice as much solar energy as vegetation. Heat absorption is also enhanced by the many vertical surfaces in cities, which reflect radiation from one surface to another. Since some radiation is absorbed with each reflection, the network of vertical surfaces tends to trap solar energy as well as radiant heat more effectively than a single horizontal surface. Another factor is the ability of concrete, stone, and asphalt to conduct and hold heat better than soil, even when the soil is dry. Therefore, during the day, more heat can be stored by the city's building materials. At night, the heat is conducted back to the surface, keeping nighttime temperatures warmer.

Yet another important factor in warming the city is fuel consumption and waste heat. In summer, city temperatures are raised through the use of air conditioning. This equipment pumps heat out of buildings, releasing the heat to the air. The power used to run the air conditioning systems is also released as heat.

5.5 Urban and Rural Surface Temperature Contrasts

a...*Eye on the Landscape* ***Urban surfaces*** *Urban surfaces are composed of asphalt, concrete, building stone, and similar materials. Sewers drain away rainwater, keeping urban surfaces dry. Because evapotranspiration is limited, surface and air temperatures are hotter.* **What else would the geographer see?...Answers at the end of the chapter.**

b...*Eye on the Landscape* ***Rural surfaces*** *Rural surfaces are composed of moist soil, covered largely by vegetation. Evapotranspiration keeps these surfaces cooler.* **What else would the geographer see?...Answers at the end of the chapter.**

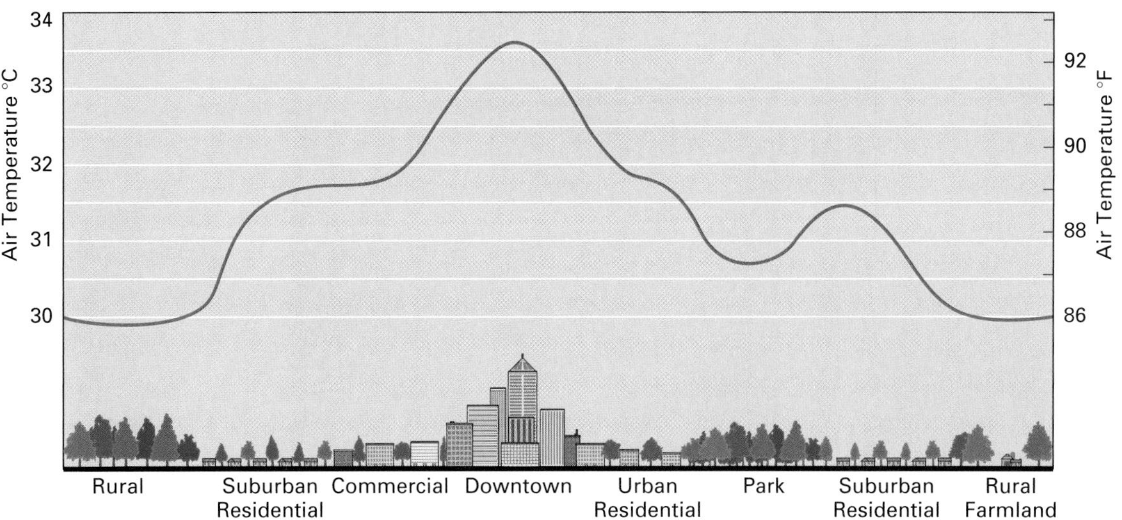

5.6 Urban heat island profile *This diagram shows how air temperatures might vary on a summer afternoon across an urban area. Downtown and commercial areas are warmest, while rural farmland is coolest. (After EPA.)*

The Urban Heat Island

As a result of these effects, air temperatures in the central region of a city are typically several degrees warmer than those of the surrounding suburbs and countryside. Figure 5.6 shows a sketch of a temperature profile across an urban area in the late afternoon. Temperatures are hottest in the downtown area, followed by commercial, urban-residential, and suburban-residential areas. Parks and rural areas are coolest. The elevated temperatures identify an urban **heat island.** The heat island persists through the night because of

Geographers at Work | How Cities Alter Weather and Climate

by James A. Voogt ***University of Western Ontario***

Are you an urban dweller? If so, you probably know that the climate of your city is different from the climate of the suburbs and rural areas nearby. For one thing, it's hotter. And air quality is generally lower. But did you know that city air is also typically drier?

Urban climatology is a new and expanding field of geography that focuses on how human activities affect urban climate. Urban climatologists are concerned with a range of problems, including atmospheric processes in an urban environment; urban climate and health; improvement of urban climate by adding parks and green spaces; urban climate and urban planning; and use of remote sensing to observe urban areas.

Urban climatology is the specialty of James Voogt, shown at work in the photo at right. He focuses on the key issue of measuring urban temperatures accurately using radiometers—devices that measure surface temperature by absorbing longwave radiation coming from surfaces, such as streets, plazas, and lawns. Through the day and night, different types of surfaces will have different temperatures, and a radiometer looking down on a city from atop a tower will see a mixture of different surfaces and temperatures that varies with the look angle. By sorting out these effects, James can study how urban surface temperatures relate to urban air temperature measured by thermometers; how radiometers on aircraft and orbiting satellites sense the urban heat island; and how to plan urban areas to consider the impact of surface temperatures on the local climates that affect residents. Read more about James's work on our web site.

James Voogt in an extendable tower preparing instruments for an urban climate study in Tokyo, Japan.

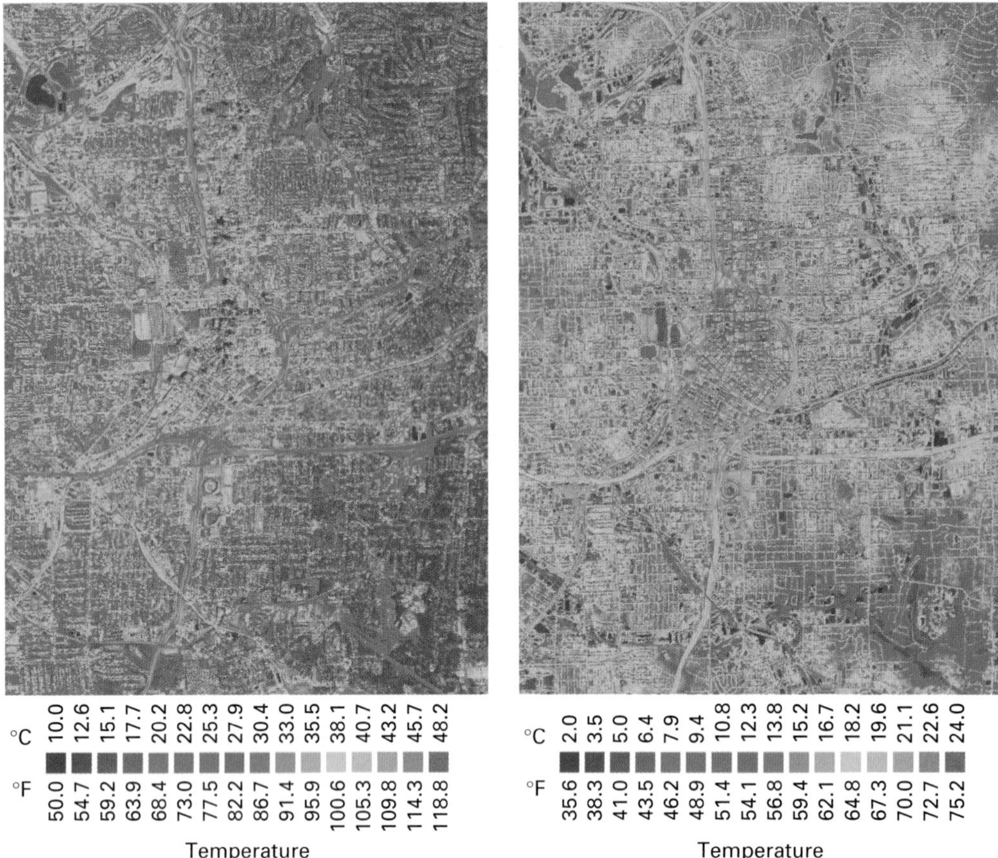

***5.7 Thermal infrared images of downtown Atlanta** These thermal infrared images, acquired in May 1997, over downtown Atlanta, Georgia, show the urban heat island. On the left is a daytime image, while a nighttime image is on the right. (Courtesy NASA/EPA. Provided by Dr. Dale Quattrochi, Marshall Space Flight Center.)*

the availability of a large quantity of heat stored in the ground during the daytime hours.

Figure 5.7 shows two remotely sensed images of the Atlanta central business district that demonstrate the heat island effect. The images were acquired by an aircraft with a thermal infrared scanning system that images longwave energy radiated by the surface. During the day, the main city area, in tones of red and yellow, is clearly warmer than the suburban area, which is to the right in tones of blue and green. At night, the contrast is even greater. The street pattern of asphalt pavement is shown very clearly as a red grid, with many of the downtown squares also filled with red.

The urban heat island effect has important economic consequences. Higher temperatures demand more air conditioning and more electric power in the summer. The fossil fuel burned to generate this power contributes CO_2 and air pollutants to the air. The increased temperatures are also more conducive to smog formation, which is unhealthy and damaging to materials. To reduce these effects and the magnitude of the heat island, cities are now taking steps to plant more vegetation and to encourage the use of reflective surfaces, such as concrete or bright roofing materials, where they can reflect solar energy back to space.

Note that the heat island effect does not necessarily apply to cities in desert climates. In the desert, the evapotranspiration of the irrigated vegetation of the city may actually keep the city cooler than the surrounding barren region.

Temperature Structure of the Atmosphere

So far, we have discussed temperatures in the layer of the atmosphere that surrounds us—up to about 2 m (6 ft) above the ground surface. How different are air temperatures at increasing heights through the atmosphere? In general, temperatures are lower. Why is air cooler at higher altitudes? Recall from Chapter 4 that most of the incoming solar radiation penetrates the atmosphere and is absorbed at the surface. Thus, the atmosphere is largely warmed from below. In general, the farther the air is away from the Earth's surface, the cooler the air will be.

The decrease in measured air temperature with increasing altitude is called the **lapse rate.** This rate measures the drop in temperature in degrees C per 1000 m (or degrees F per 1000 ft). Figure 5.8 shows how temperature varies with alti-

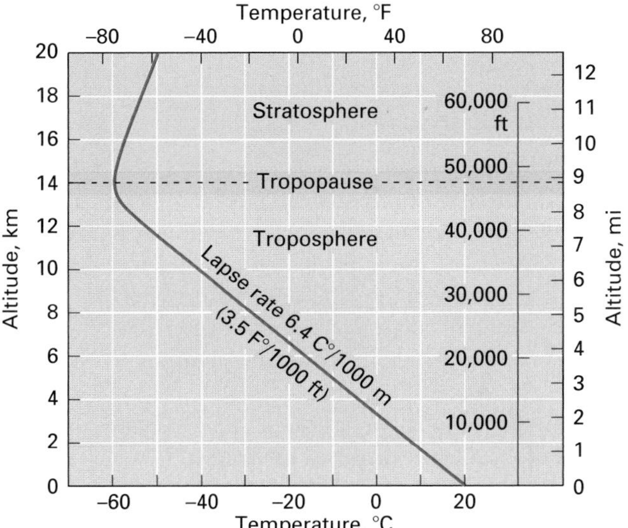

5.8 A typical atmospheric temperature curve for a summer day in the midlatitudes *Temperature decreases with altitude in the troposphere. The rate of temperature decrease with elevation, or lapse rate, is shown at the average value of 6.4°C/1000 m (3.5°F/1000 ft), which is known as the environmental temperature lapse rate. At the tropopause, the decreasing trend stops. In the stratosphere above, temperature is constant or increases slightly with altitude.*

tude for a typical summer day in the midlatitudes. Altitude is plotted on the vertical axis and temperature on the horizontal axis. The curve shows average values—if we sent up a balloon that radioed the temperature of the air back to us every minute or two, and sent up many balloons over a long period of time, we would obtain an average profile of temperature very much like that shown in the figure. Temperature drops with altitude at an average rate of 6.4°C/1000 m (3.5°F/1000 ft). This average value is known as the **environmental temperature lapse rate.** For example, when the air temperature near the surface is a pleasant 21°C (70°F), the air at an altitude of 12 km (40,000 ft) will be a bone-chilling −55°C (−67°F). Keep in mind that the environmental temperature lapse rate is an average value and that on any given day the observed lapse rate might be quite different.

Figure 5.8 shows another important feature of the atmosphere. For the first 12 km (7 mi) or so, temperature falls with increasing elevation. However, at 12 to 15 km (7 to 9 mi) in height, the temperature stops decreasing. In fact, above that height, temperature slowly increases with elevation. This feature has led atmospheric scientists to distinguish two different layers in the lower atmosphere—the troposphere and the stratosphere.

Troposphere

The **troposphere** is the lowest atmospheric layer, in which temperature decreases with increasing elevation. It is thickest in the equatorial and tropical regions, where it ranges from sea level to about 16 km (10 mi), and it thins toward the poles, where it is about 6 km (4 mi) thick. Since almost all human activity occurs in this layer, it is of primary importance to us. Everyday weather phenomena, such as clouds or storms, occur mainly in the troposphere.

One important feature of the troposphere is that it contains significant amounts of water vapor. When the water vapor content is high, vapor can condense into water droplets, forming low clouds and fog, or the vapor can be deposited as ice crystals, forming high clouds. When condensation or deposition is rapid, rain, snow, hail, or sleet—collectively termed *precipitation*—may be produced and fall to Earth. Regions where water vapor content is high throughout the year will therefore have moist climates. In desert regions, where water vapor is present only in small amounts, precipitation is infrequent. In Chapter 4, we also described the important role of water vapor in absorbing and reradiating heat emitted by the Earth's surface. In this way, water vapor helps to create the greenhouse effect, which warms the Earth to habitable temperatures.

The troposphere contains countless tiny particles that are so small and light that the slightest movements of the air keep them aloft. These are called *aerosols.* They are swept into the air from dry desert plains, lakebeds, and beaches, or they are released by exploding volcanoes. Oceans are also a source of aerosols. Strong winds blowing over the ocean lift droplets of spray into the air. These droplets of spray lose most of their moisture by evaporation, leaving tiny particles of watery salt as residues that are carried high into the air. Forest fires and brushfires are another important source of aerosols, contributing particles of soot as smoke. Meteors contribute dust particles as they vaporize from the heat of friction upon entering the upper layers of air. Industrial processes that incompletely burn coal or fuel oil release aerosols to the air as well.

The most important function of aerosols is that certain types serve as nuclei, or centers, around which water vapor condenses to form tiny droplets. When these droplets grow large and occur in high concentration, they are visible to the eye as clouds or fog. As we noted in Chapter 4, aerosol particles also scatter sunlight, thus brightening the whole sky while reducing slightly the intensity of the solar beam.

The height at which the troposphere gives way to the stratosphere above is known as the *tropopause.* Here, temperatures stop decreasing with altitude and start to increase. The altitude of the tropopause varies somewhat with season. This means that the troposphere is not uniformly thick at any location.

Stratosphere and Upper Layers

Above the tropopause lies the **stratosphere,** in which the air becomes slightly warmer as altitude increases. The stratosphere extends to a height of roughly 50 km (about 30 mi) above the Earth's surface. It is the home of strong, persistent winds that blow from west to east. There is little mixing of air between the troposphere and stratosphere, and so the stratosphere normally holds very little water vapor or dust.

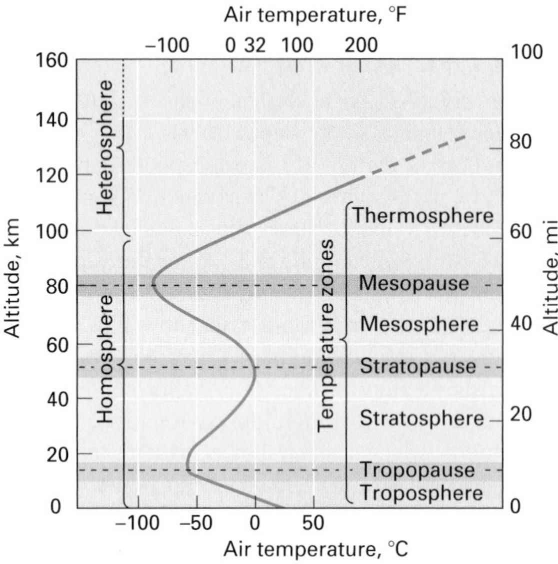

5.9 Temperature structure of the upper atmosphere *Above the troposphere and stratosphere are the mesosphere and thermosphere. The homosphere, in which air's chemical components are well mixed, ranges from the surface to nearly 100 km altitude.*

One important feature of the stratosphere is that it contains the ozone layer. As we saw in *Eye on Global Change 4.4 • The Ozone Layer—Shield to Life,* the ozone layer absorbs solar ultraviolet radiation and thus shields Earthly life from this intense, harmful form of energy. In fact, the warming of the stratosphere with altitude is caused largely by the absorption of solar energy by ozone molecules.

Above the stratosphere are two other layers—the mesosphere and thermosphere. They are shown in Figure 5.9. In the *mesosphere,* temperature falls with elevation. This layer begins at the *stratopause,* the altitude at which the stratospheric temperature ceases to increase with altitude. It ends at the *mesopause,* the level at which temperature ceases to fall with altitude. Above the mesosphere is the *thermosphere,* a layer of increasing temperature. However, at the altitude of the thermosphere, the density of air is very thin and the air holds little heat.

From the viewpoint of gas composition, the atmosphere is uniform for about the first 100 km of altitude, which includes the troposphere, stratosphere, mesosphere, and lower portion of the thermosphere. This region is referred to as the *homosphere.* Above 100 km, gas molecules tend to become increasingly sorted into layers by molecular weight and electric charge. This region is referred to as the *heterosphere.*

High-Mountain Environments

Did you ever walk or drive to the top of a high mountain? Mount Whitney, in California, perhaps (Figure 5.10)? or New Hampshire's Mount Washington? or British Columbia's Mount Garibaldi? If so, you have probably made the simple observation that temperatures generally drop as you go up in elevation. We've already noted this fact in the discussion of

5.10 ***Eye on the Landscape*** ***Clear air at high elevation*** *Mount Whitney (elevation 4418 m, 14,495 ft) and surrounding peaks of the Southern Sierra Nevada provide a stunning panorama. The excellent visibility and dark blue sky are products of the thin air at this elevation.* **What else would the geographer see? ... Answers at the end of the chapter.**

the environmental temperature lapse rate. But what else might you notice as you climb? You might notice that you sunburn a bit faster. Perhaps you also find yourself getting out of breath more quickly. And if you camp out, you may notice that the nighttime temperature gets lower than you might expect, even given that temperatures are generally cooler.

These effects are related to the fact that at high elevations, there is significantly less air above you. With fewer air molecules and aerosol particles to scatter and absorb the Sun's light, the Sun's rays will be stronger. There is also less carbon dioxide and water vapor, and so the greenhouse effect is reduced. With the reduced warming effect, temperatures will tend to drop lower at night. At high elevations, the air pressure is lower because there is a smaller mass of air above you. This means that there are fewer gas molecules in a unit volume of air. You feel short of breath on a high mountain simply because the oxygen pressure in your lungs is lower than you are used to at lower elevations.

What effect does elevation have on the daily temperature cycle? Figure 5.11 shows temperature graphs for five stations ascending the Andes Mountain range in Peru during the same 15 days in July. The 15-day mean for each station is shown as a horizontal line. Mean temperatures clearly decrease with elevation, from 16°C (61°F) at sea level to –1°C (30°F) at 4380 m (14,370 ft). The daily range also increases with elevation, except for Cuzco. This large city does not experience nighttime temperatures as low as you might expect because of its urban heat island.

Temperature Inversion

So far, air temperatures seem to decrease with height above the surface. Is this always true? Think about what happens on a clear, calm night. Under clear conditions, the ground surface radiates longwave energy to the sky, net radiation becomes negative, and the surface cools. This means that the air near the surface will also be cooled, as we saw in curves 4 and 5 on Figure 5.4 for the air within a few meters of the ground. However, if the surface continues to stay cold, a layer of cooler air can accumulate underneath a layer of warmer air.

This situation is illustrated in a graph of temperature with height (Figure 5.12). The straight, slanting line of the normal environmental lapse rate bends to the left in a "J" hook. In this example, the air temperature at the surface, point A, has dropped to –1°C (30°F). This value is the same as that at point B, some 750 m (about 2500 ft) aloft. As we move up from ground level, temperatures become warmer up to about 300 m (about 1000 ft). Here the curve reverses itself, and normal lapse rate takes over. The lower portion of the lapse rate curve is called a low-level **temperature inversion.** Here, the normal cooling trend is reversed and temperatures increase with height.

In the case shown in Figure 5.12, the temperature of the lowermost air has fallen below the freezing point, 0°C (32°F). For sensitive plants during the growing season, this temperature condition is called a killing frost (even though actual frost may not form). Perhaps you have seen news reports about killing frosts damaging fruit trees or crops in Florida or California. Growers commonly use several methods to break up an inversion. Most commonly, large fans are used to mix the cool air at the surface with the warmer air above (Figure 5.13). Where air pollution is not a problem, oil-burning heaters are sometimes used to warm the surface air layer and create air circulation.

Low-level temperature inversions often occur over snow-covered surfaces in winter. Inversions of this type

5.11 The effect of elevation on air temperature cycles *These graphs show daily maximum and minimum air temperatures for mountain stations in Peru, lat. 15° S. All data cover the same 15-day observation period in July. As elevation increases, the mean daily temperature decreases and the temperature range increases.*

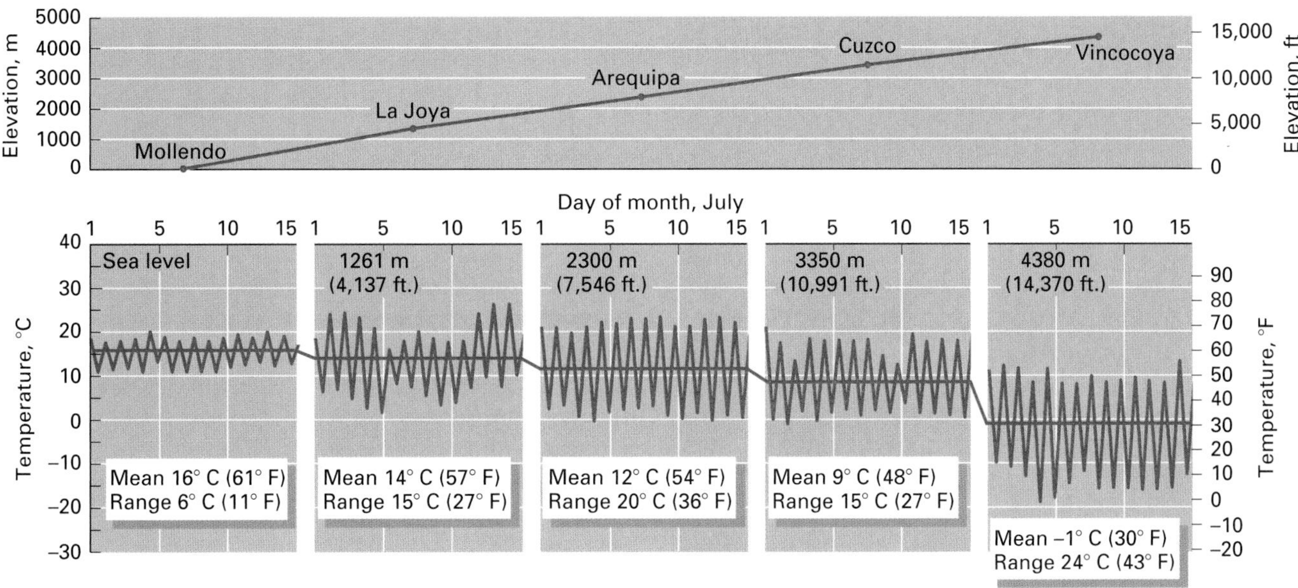

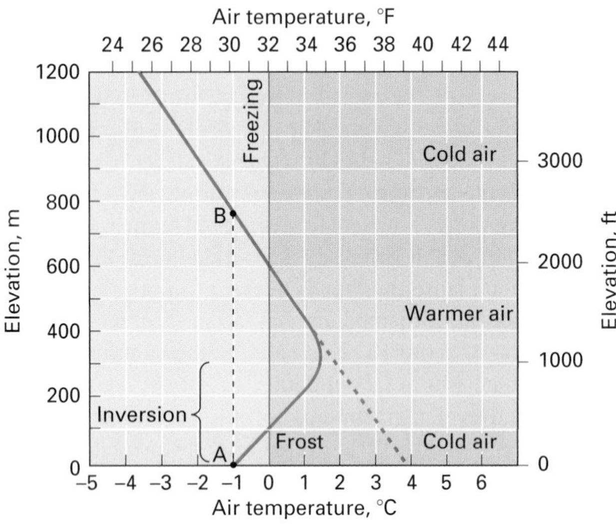

5.12 A low-level temperature inversion with frost *While air temperature normally decreases with altitude (dashed line), in an inversion temperature increases with altitude. In this example, the surface temperature is at –1°C (30°F), and temperature increases with altitude (solid line) for several hundred meters (1000 ft or so) above the ground. Temperature then resumes a normal, decreasing trend with altitude.*

are very intense and can extend thousands of meters into the air. They build up over many long nights in arctic and polar regions, where the solar heat of the short winter day cannot completely compensate for nighttime cooling. Inversions can also result when a warm air layer overlies a colder one. This type of inversion is often found along the west coasts of major continents, and we will discuss it in more detail in Chapter 6. We will also see in that chapter that temperature inversions suppress vertical mixing of the air and so can act to seal air pollution within a thin surface layer.

5.13 Frost-buster fan *This large fan is used to mix the air at low levels in a navel orange orchard to protect blossoms and fruit from freezing.*

THE ANNUAL CYCLE OF AIR TEMPERATURE

As the Earth revolves around the Sun, the tilt of the Earth's axis causes an annual cycle of variation in insolation. This cycle produces an annual cycle of net radiation, which, in turn, causes an annual cycle to occur in mean monthly air temperatures. Although the annual cycle of net radiation is the most important factor in determining the annual temperature cycle, we will see that location—maritime or continental—also has an important influence. That is, places located well inland and far from oceans generally experience a stronger temperature contrast from winter to summer. Let's begin our study of the annual temperature cycle by looking at the relationship between net radiation and temperature at four stations, ranging from the equator almost to the arctic circle.

Net Radiation and Temperature

Graph (*a*) of Figure 5.14 shows the yearly cycle of the net radiation rate for the four stations. The average value of net radiation flow rate for the month is plotted, in units of W/m^2. Graph (*b*) shows mean monthly air temperatures for these same stations. We will compare the net radiation graph with the air temperature graph for each station, beginning with Manaus, a city on the Amazon River in Brazil.

At Manaus, located nearly on the equator, the average net radiation rate is strongly positive in every month. However, there are two minor peaks. These coincide approximately with the equinoxes, when the Sun is nearly straight overhead. A look at the temperature graph of Manaus shows uniform air temperatures, averaging about 27°C (81°F) for the year. The annual temperature range, or difference between the highest and lowest mean monthly temperature, is only 1.7°C (3°F). In other words, near the equator the temperature is similar each month. There are no temperature seasons.

We go next to Aswan, Egypt, a very dry desert location on the Nile River at lat. 24° N. The positive net radiation rate curve shows that a large radiation surplus exists for every month. Furthermore, the net radiation rate curve has a much stronger annual cycle, with values for June and July that are almost double those of December and January. The temperature graph shows a corresponding annual cycle, with an annual range of about 17°C (31°F). June, July, and August are terribly hot, averaging over 32°C (90°F).

Moving farther north, we come to Hamburg, Germany, lat. 54° N. The net radiation rate cycle here is also strongly developed. The rate is positive for nine months, providing a radiation surplus. When the rate becomes negative for three

(*a*) **Net radiation**

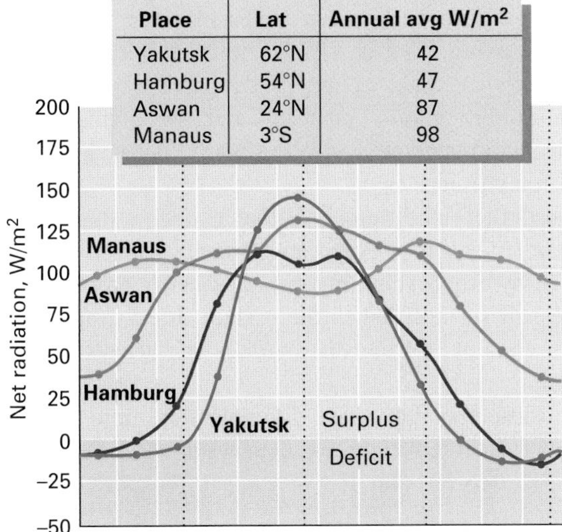

Place	Lat	Annual avg W/m²
Yakutsk	62°N	42
Hamburg	54°N	47
Aswan	24°N	87
Manaus	3°S	98

(*b*) **Monthly mean air temperature**

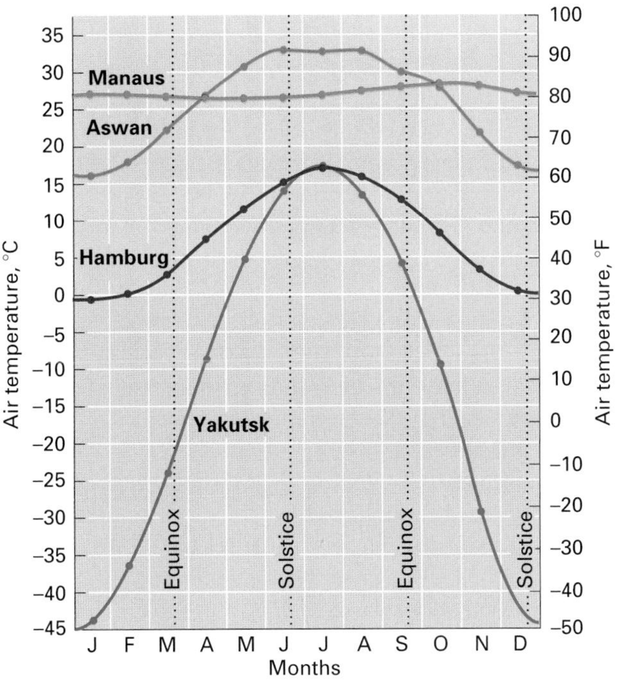

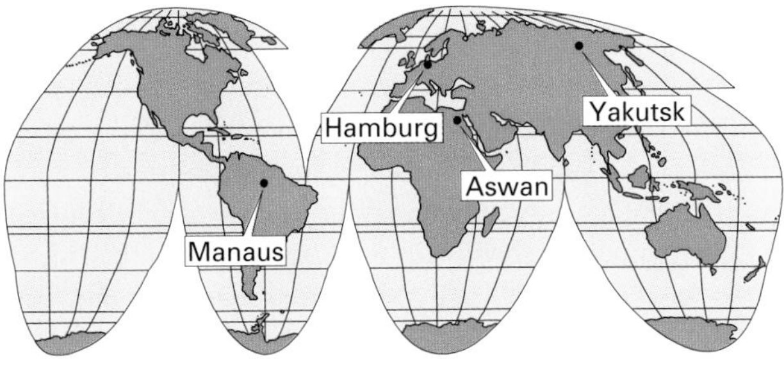

5.14 Net radiation and temperature cycles *Net radiation and temperature for Aswan (Egypt), Hamburg (Germany), and Yakutsk (Siberia) all show strong annual cycles with summer maximums and winter minimums. In contrast is Manaus (Brazil), very near the equator, which shows two net radiation peaks and a nearly uniform temperature. (Data courtesy of David H. Miller.)*

winter months, a deficit is produced. The temperature cycle reflects the reduced total insolation at this latitude. Summer months reach a maximum of just over 16°C (61°F), while winter months reach a minimum of just about freezing (0°C or 32°F). The annual range is about 17°C (31°F), the same as at Aswan.

Finally, we travel to Yakutsk, Siberia, lat. 62° N. During the long, dark winters, the net radiation rate is negative, and there is a radiation deficit that lasts about six months. During this time, air temperatures drop to extremely low levels. For three of the winter months, monthly mean temperatures are between –35 and –45°C (about –30 and –50°F). This is actually one of the coldest places on Earth. In summer, when daylight lasts most of a 24-hour day, the net radiation rate rises to a strong peak. In fact, this peak value is higher than those of the other three stations. As a result, air temperatures show a phenomenal rise beginning in the spring to summer-month values of over 13°C (55°F). Because of Yakutsk's high latitude and continental interior location, its annual temperature range is enormous—over 60°C (108°F).

Land and Water Contrasts

Have you ever visited San Francisco? If so, you probably noticed that this magnificent city has quite a unique climate. Fog is frequent, and cool, damp weather prevails for most of the year. The cool climate is due to its location—on the tip of a peninsula, with the Pacific Ocean on one side and San Francisco Bay on the other. Ocean and bay water temperatures are quite cool, since a southward-flowing current sweeps cold water from Alaska down along the northern California coast. Winds from the west move cool, moist ocean air, as well as clouds and fog, across the peninsula, keeping summer air temperatures low and winter temperatures above freezing. Figure 5.15 shows a typical record of

5.15 Maritime and continental temperatures *A recording thermometer made these continuous records of the rise and fall of air temperature for a week in summer at San Francisco, California, and at Yuma, Arizona. At San Francisco, on the Pacific Ocean, the daily air temperature cycle is very weak. At Yuma, a station in the desert, the daily cycle is strongly developed.*

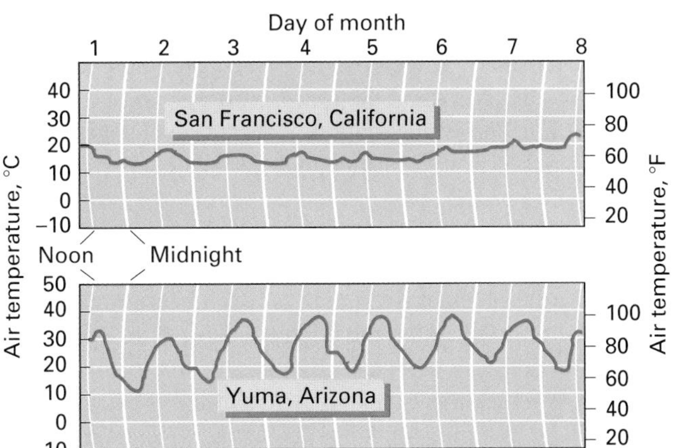

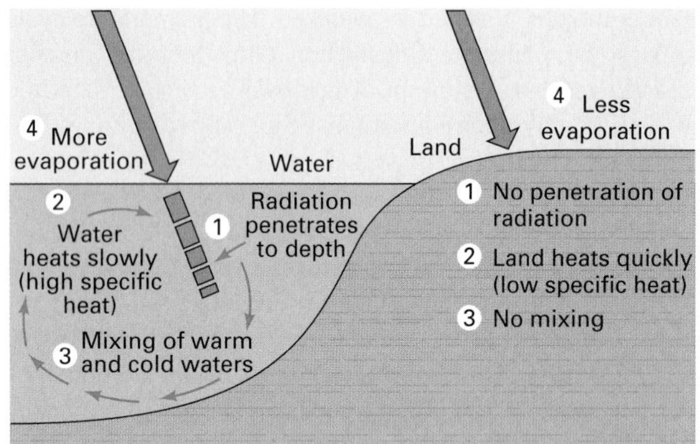

5.16 Land–water contrasts *These four differences illustrate why a land surface heats more rapidly and more intensely than the surface of a deep water body. As a result, locations near the ocean have more uniform air temperatures—cooler in summer and warmer in winter.*

temperatures for San Francisco for a week in the summer. Temperatures hover around 13°C (55°F) and change only a little from day to night.

In contrast is a location far from the water, like Yuma, Arizona, also shown in Figure 5.15. Located in the Sonoran desert, air temperatures here are much warmer on the average—about 28°C (82°F). Clearly, no ocean cooling is felt in Yuma! The daily range is also much greater, nearly 20°C (36°F). Hot desert days become cool desert nights, with the clear, dry air allowing the ground to lose heat rapidly.

Why do these differences occur? The important principle is this: the surface layer of any extensive, deep body of water heats more slowly and cools more slowly than the surface layer of a large body of land when both are subjected to the same intensity of insolation. Because of this principle, daily and annual air temperature cycles will be quite different at coastal locations than at interior locations.

Four important thermal differences between land and water surfaces account for the land–water contrast (see Figure 5.16). The most important difference is that much of the downwelling solar radiation penetrates water, distributing the absorbed heat throughout a substantial water layer. In contrast, solar radiation does not penetrate soil or rock, so its heating effect is concentrated at the surface. The radiation therefore warms a thick water surface layer only slightly, while a thin land surface layer is warmed more intensely.

A second thermal factor is that water is slower to heat than dry soil or rock. The *specific heat* of a substance describes how the temperature of a substance changes with a given input of heat. Consider an experiment using a small volume of water and the same volume of rock. Let's suppose we warm each volume by the same amount—perhaps one degree. If we measure the amount of heat required, we will find that it takes about five times as much heat to raise the temperature of the water one degree as it does to raise the temperature of the rock one degree. That is, the specific heat of water is about five times greater than that of rock. The same will be true for cooling—after losing the same amount of heat, the temperature of water falls less than the temperature of rock.

A third difference between land and water surfaces is related to mixing. That is, a warm water surface layer can mix with cooler water below, producing a more uniform temperature throughout. For open water, the mixing is produced by wind-generated waves. Clearly, no such mixing occurs on land surfaces.

A fourth thermal difference is that an open water surface can be cooled easily by evaporation. Land surfaces can also be cooled by evaporation, but only if water is present at or near the soil surface. When the surface dries, evaporation stops. In contrast, a free water surface can always provide evaporation.

With these four important differences in mind, we can easily see that air temperatures above water will be less variable than those over land. But just what effect does the contrast in thermal characteristics have on air temperature?

DAILY TEMPERATURE CYCLE Let's examine the daily cycle of air temperature first. Two sets of daily air temperature curves are shown in Figure 5.17—El Paso, Texas, and North Head, Washington. Four months are presented—January, April, July, and October.

The El Paso curves show the temperature environment of an interior desert in midlatitudes. Because soil moisture content is low and vegetation is sparse, evaporation and transpiration are not important cooling effects. Cloud cover is generally sparse. Under these circumstances, the ground surface heats intensely during the day and cools rapidly at night. Air temperatures show an average daily range of 11 to 14°C (20 to 25°F). This type of variation represents an interior temperature environment—typical of a station located in the interior of a continent, far from the ocean's influence.

North Head is located on the Washington coast. Here, prevailing westerly winds sweep cool, moist air off the adjacent Pacific Ocean. The average daily range at North Head is a mere 3°C (5°F) or less. Persistent fogs and cloud cover also contribute to the minimal daily range. Note further that the annual range is much restricted, especially when compared to El Paso. North Head exemplifies a coastal temperature environment, typical of a station located in the path of oceanic air.

ANNUAL TEMPERATURE CYCLE Let's now turn to the effects of land–water surface contrasts on the temperature cycle for the year. We have already noted for El Paso and North Head that the temperature cycle for the four months plotted shows a greater range for the interior station than for the coastal one. Let's look in more detail at the annual cycle for another pair of stations—Winnipeg, Manitoba, located in the interior of the North American continent, and the Scilly Islands, off the southwestern tip of England, which are surrounded by the waters of the Atlantic Ocean. This time, the two stations chosen are at the same latitude, 50° N. As a result, they have the same insolation cycle and receive the same potential amount of solar energy for surface warming. Figure 5.18

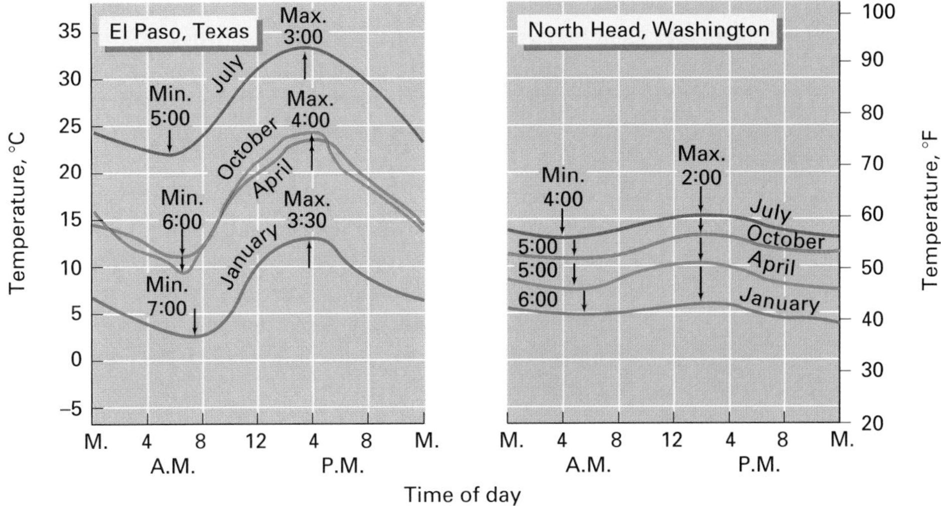

5.17 Continental and maritime daily air temperature cycles *The average cycle of air temperature throughout the day for four different months shows the effect of continental and maritime location. Daily and seasonal ranges are great at El Paso, a station in the continental interior, but only weakly developed at North Head, Washington, a station on the Pacific coast. The seasonal effect on overall temperatures is stronger at El Paso.*

shows their annual cycles of temperature as well as the insolation curve common to both.

The temperature graphs for the Scilly Islands and Winnipeg confirm the effects we have already noted for North Head and El Paso—that the annual range in temperature is much larger for the interior station (39°C, 70°F) than for the coastal station (8°C, 14°F). Note that the nearby ocean waters keep the air temperature at the Scilly Islands well above freezing in the winter, while January temperatures at Winnipeg fall to near –20°C (–4°F).

Another important effect of land–water contrasts concerns the timing of maximum and minimum temperatures. Insolation reaches a maximum at summer solstice, but it is still strong for a long period afterward. This means that heat energy continues to flow into the ground well after the solstice. Therefore, the hottest month of the year for interior regions is July, the month following the solstice. Similarly, the coldest month of the year for large land areas is January, the month after the winter solstice. This is because the ground continues to lose heat even after insolation begins to increase.

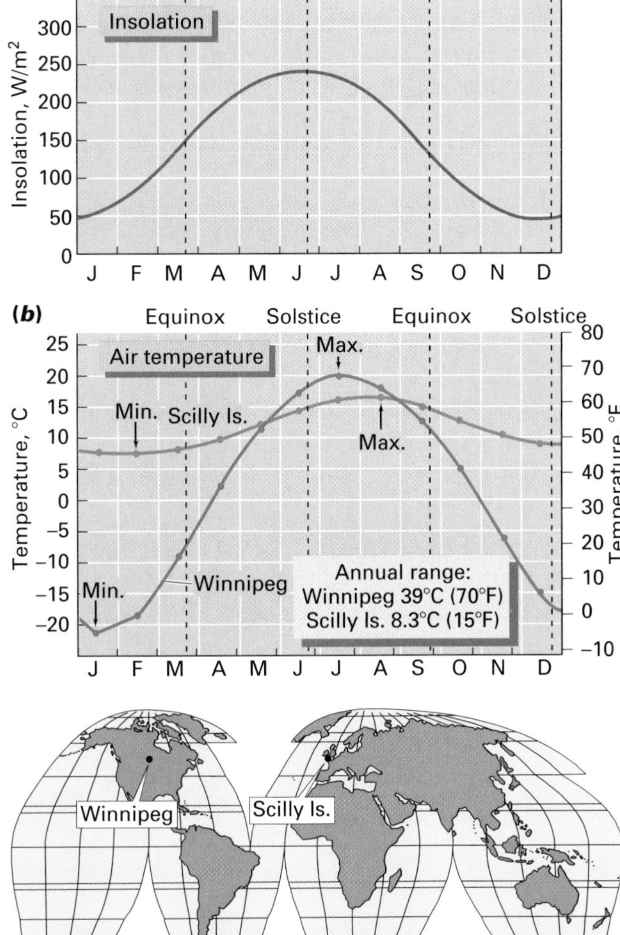

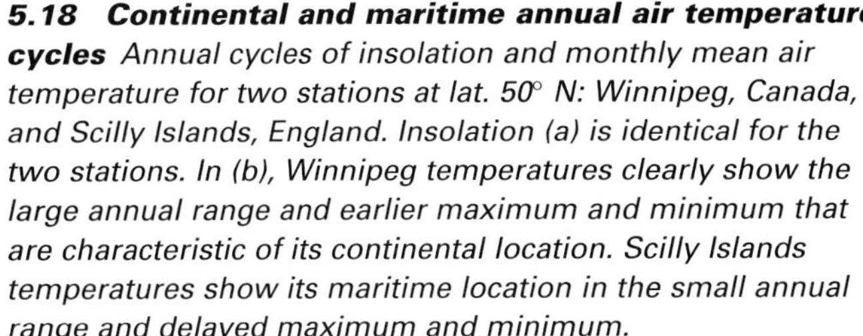

5.18 Continental and maritime annual air temperature cycles *Annual cycles of insolation and monthly mean air temperature for two stations at lat. 50° N: Winnipeg, Canada, and Scilly Islands, England. Insolation (a) is identical for the two stations. In (b), Winnipeg temperatures clearly show the large annual range and earlier maximum and minimum that are characteristic of its continental location. Scilly Islands temperatures show its maritime location in the small annual range and delayed maximum and minimum.*

Over the oceans and at coastal locations, maximum and minimum air temperatures are reached a month later than on land—in August and February, respectively. Because water bodies heat or cool more slowly than land areas, the air temperature changes more slowly. This effect is clearly shown in the Scilly Islands graph, where February is slightly colder than January.

World Patterns of Air Temperature

We have learned some important principles about air temperatures in this chapter. Some of these principles are local, such as the effect of urban and rural surfaces on air temperatures. Others are on a continental scale, such as the effect of maritime or continental location. Still others are global, such as the effect of latitude or elevation on temperature. We now turn to world temperature patterns, tying our discussion closely to these principles. First, however, we will need a quick explanation of air temperature maps and their meaning.

The distribution of air temperatures is often shown on a map by **isotherms**—lines drawn to connect locations having the same temperature. Figure 5.19 shows a map on which the observed air temperatures have been recorded and placed at their proper location on the map. These may be single readings, such as a daily maximum or minimum, or they may be averages of many years of records for a particular day or month of a year, depending on the purposes of the map. The isotherms are constructed by drawing smooth lines through and among the points in a way that best indicates a uniform temperature, given the observations at hand. Usually, isotherms representing 5- or 10-degree differences are chosen, but they can be drawn at any convenient temperature interval.

5.19 ***Isotherms*** *Isotherms are used to make temperature maps. Each line connects points having the same temperature. Where temperature changes along one direction, a temperature gradient exists. Where isotherms close in a tight circle, a center exists. This example shows a center of low temperature.*

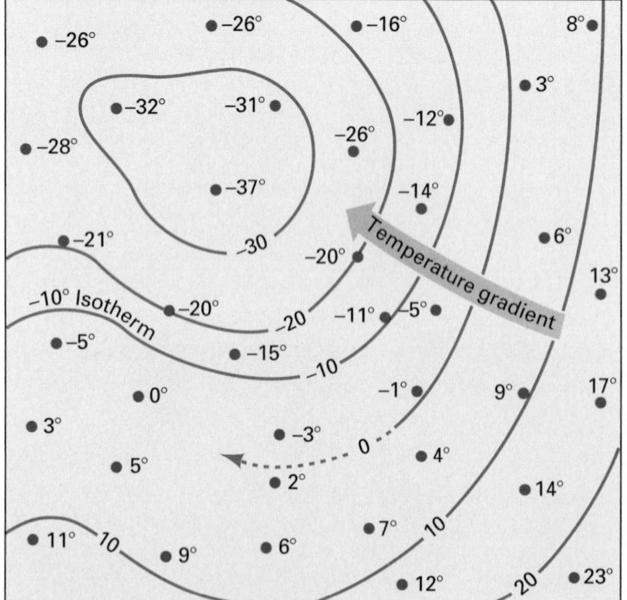

Isothermal maps are valuable because they show the important features of the temperature pattern clearly. Centers of high or low temperatures are visible, as are *temperature gradients*—directions along which temperature changes. Centers and gradients give the broad pattern of temperature, as shown in the following discussion of world patterns.

Factors Controlling Air Temperature Patterns

World patterns of isotherms are largely explained by three factors that we have already discussed. The first of these factors is latitude. As latitude increases, average annual insolation decreases, and so temperatures decrease as well, making the poles colder than the equator. The effect of latitude on seasonal variation is also important. For example, more solar energy is received at the poles at the summer solstice than at the equator. We must therefore consider the time of year as well as the latitude in understanding world temperature patterns.

The second factor is that of coastal-interior contrasts. As we've noted, coastal stations that receive marine air from prevailing winds have more uniform temperatures—cooler in summer and warmer in winter. Interior stations, on the other hand, show a much larger annual variation in temperature. Ocean currents, discussed further in Chapter 7, can also have an effect. By keeping coastal waters somewhat warmer or cooler than expected, temperatures at maritime stations will be influenced in a similar manner.

Elevation is the third important factor. At higher elevations, temperatures will be cooler. Therefore, we expect world temperature maps to show the presence of mountain ranges, which will be cooler than surrounding regions.

World Air Temperature Patterns for January and July

With these factors in mind, let's look at some world temperature maps in more detail. Figure 5.20 consists of maps of world temperatures for two months—January and July. The Mercator projections show temperature trends from the equator to the midlatitude zones, and the polar projections give the best picture for high latitudes. From the maps, we can make six important points about the temperature patterns and the factors that produce them. Be sure to follow along by examining the maps carefully.

1. **Temperatures decrease from the equator to the poles.** Annual insolation decreases from the equator to the poles, thus causing temperatures to decrease. This temperature gradient is most clearly seen in the polar

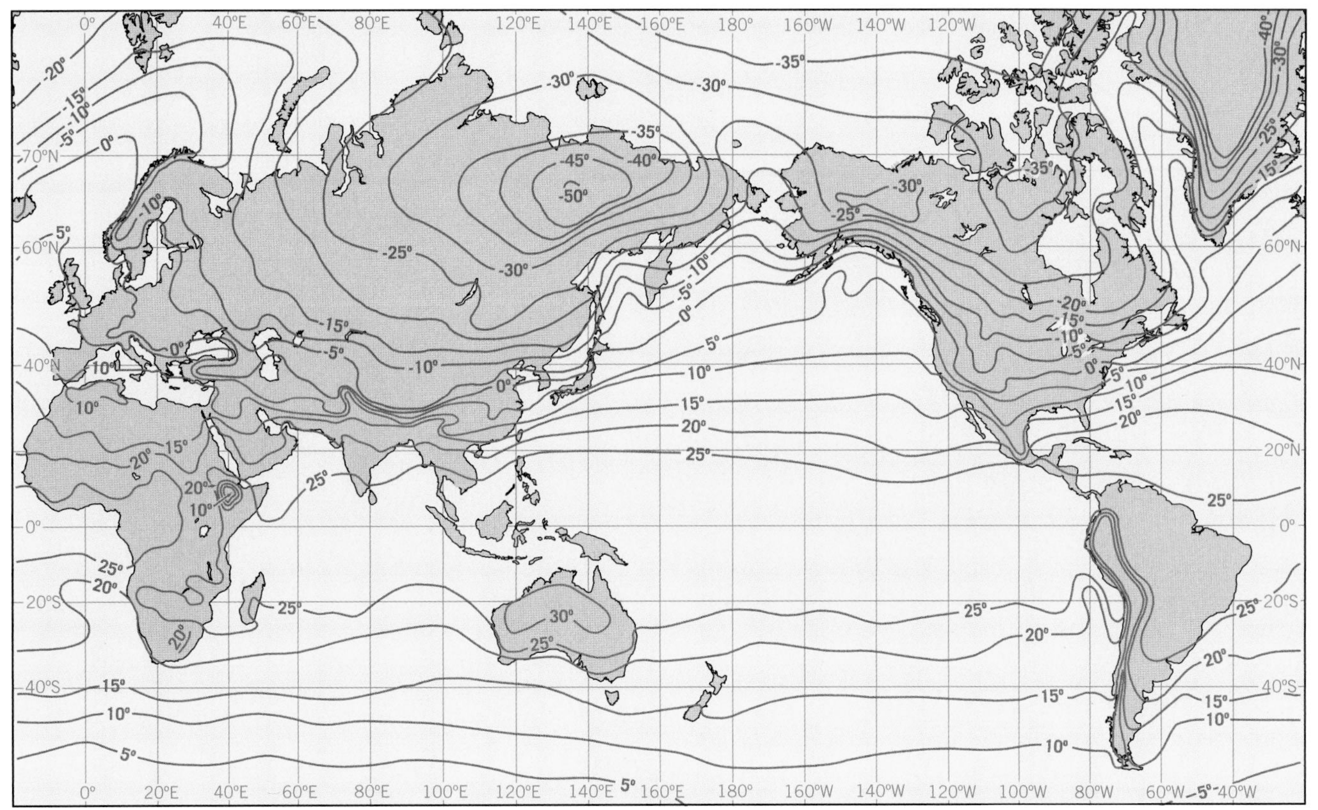

JANUARY

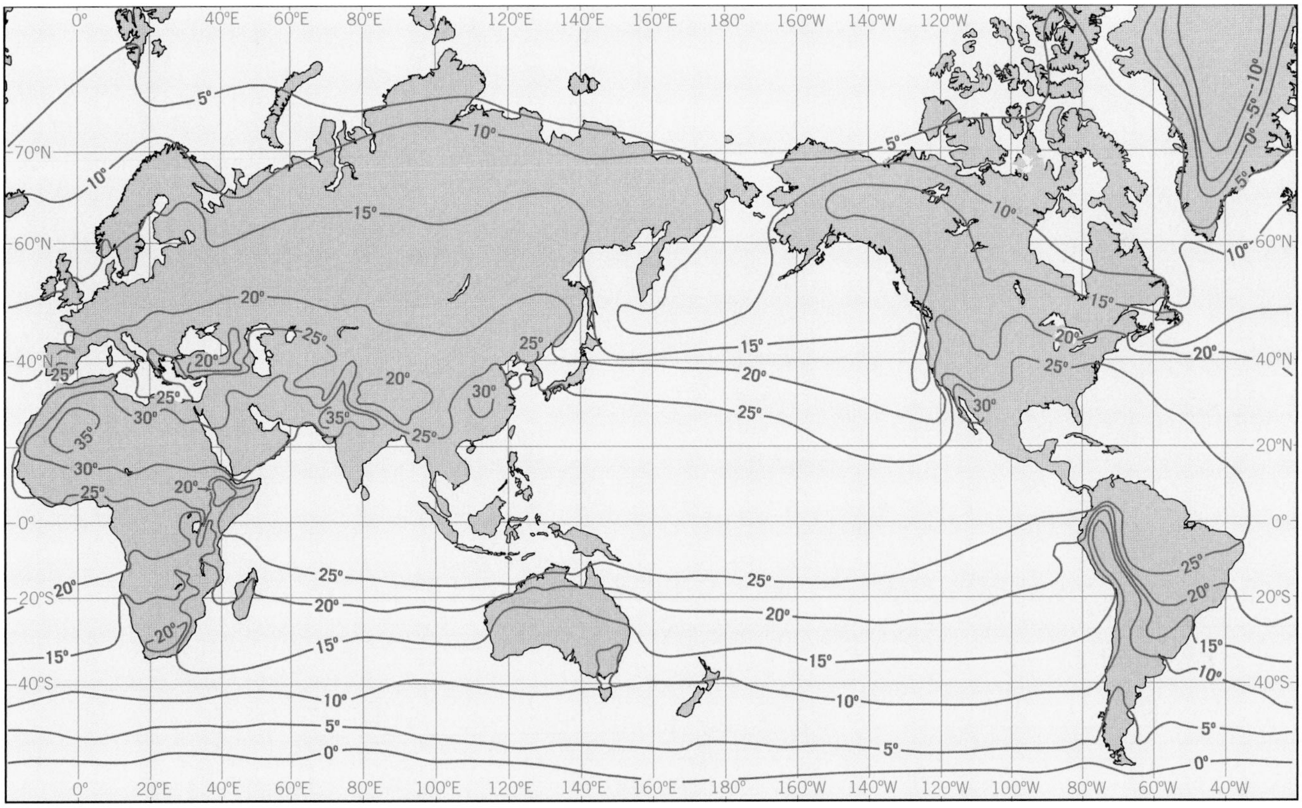

JULY

5.20 Mean monthly air temperatures (°C) for January and July, Mercator and polar projections *(Compiled by John E. Oliver.) (Figure continues)*

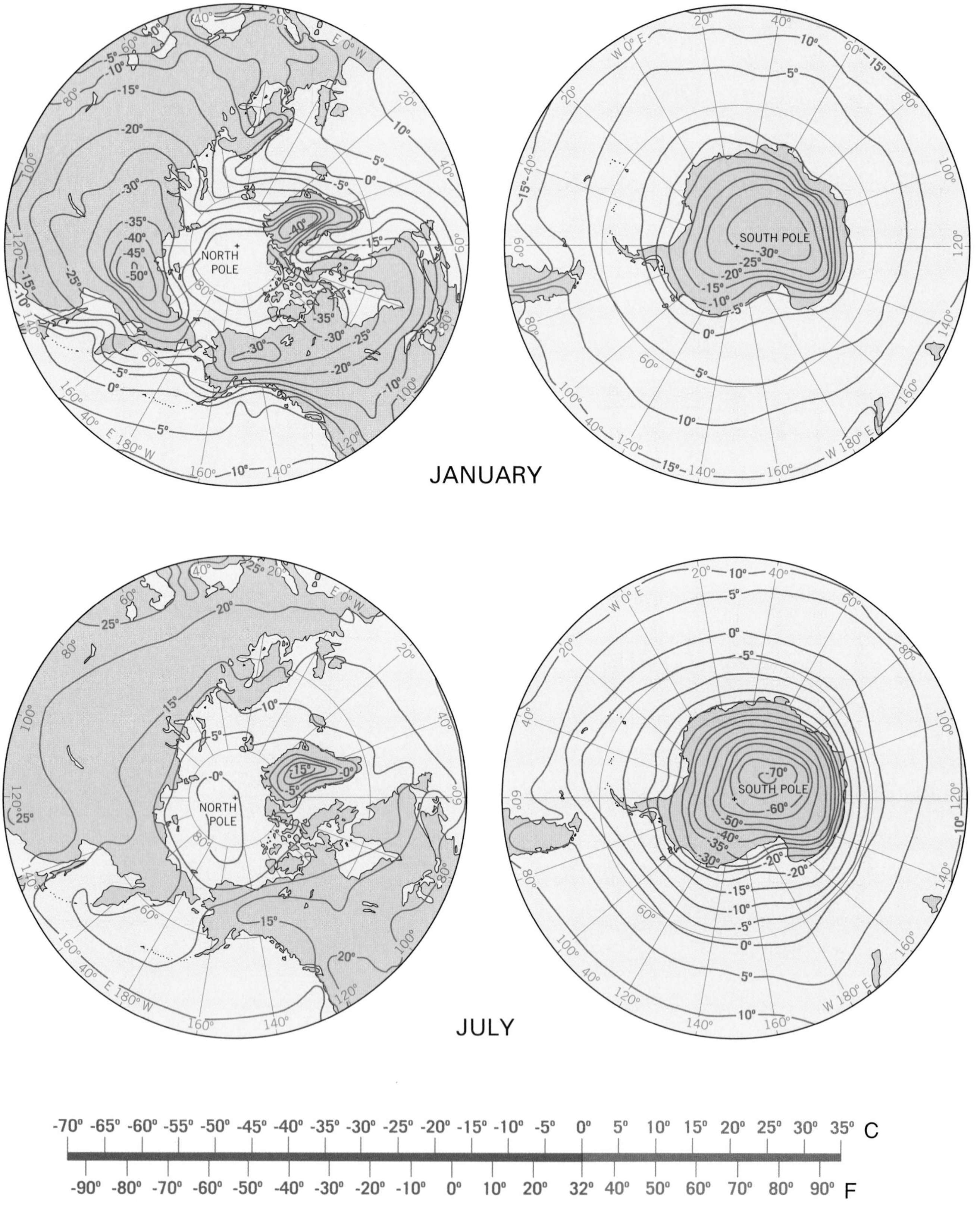

JANUARY

JULY

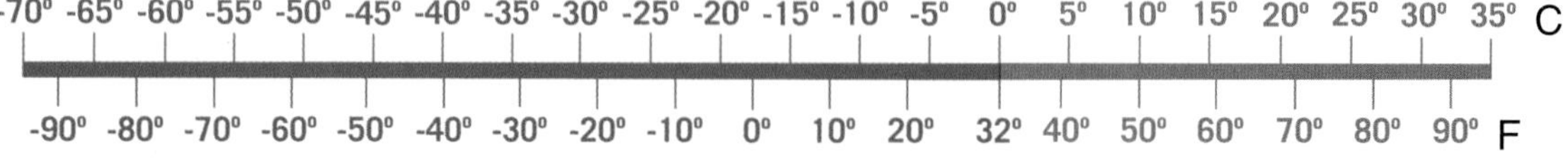

5.20 *(Continued)*

maps for the southern hemisphere in January and July. On these maps, the isotherms are nearly circular, decreasing to a center of low temperature on Antarctica near the south pole. The center is much colder in July, when most of the polar region is in perpetual night. We can also see this same general trend in the north polar maps, but the continents complicate the pattern. The general temperature gradient from the equator poleward is also evident on the Mercator maps.

2. **Large land masses located in the subarctic and arctic zones develop centers of extremely low temperatures in winter.** The two large land masses we have in mind are North America and Eurasia. The January north polar map shows these low-temperature centers very well. The cold center in Siberia, reaching –50°C (–58°F), is strong and well defined. The cold center over northern Canada is also quite cold (–35°C, –32°F) but is not as well defined. Both features are visible on the January Mercator map. Greenland shows a low-temperature center as well, but it has a high dome of glacial ice, as discussed below in point 6. An important factor in keeping winter temperatures low in these regions is the high albedo of snow cover, which reflects much of the winter insolation back to space.
3. **Temperatures in equatorial regions change little from January to July.** Note the broad space between 25°C (77°F) isotherms on both January and July Mercator maps. In this region, the temperature is greater than 25°C (77°F) but less than 30°C (86°F). Although the two isotherms move a bit from winter to summer, the equator always falls between them. This demonstrates the uniformity of equatorial temperatures. (An exception is the northern end of the Andes Mountains in South America, where high elevations, and thus cooler temperatures, exist at the equator.) Equatorial temperatures are uniform primarily because insolation at the equator does not change greatly with the seasons.
4. **Isotherms make a large north-south shift from January to July over continents in the midlatitude and subarctic zones.** Figure 5.21 demonstrates this principle. In the winter, isotherms dip equatorward, while in the summer, they arch poleward. This effect is shown in North America and Eurasia in the January and July Mercator maps. In January the isotherms drop down over these continents, and in June they curve upward. For example, the 15°C (59°F) isotherm lies over central Florida in January. But by July this same isotherm has moved far north, cutting the southern shore of Hudson Bay and then looping far up into northwestern Canada. In contrast are the isotherms over oceans, which shift much less. This striking difference is due to the contrast between oceanic and continental surface properties, which cause continents to heat and cool more rapidly than oceans.
5. **Highlands are always colder than surrounding lowlands.** You can see this by looking at the pattern of isotherms around the Rocky Mountain chain, in western North America, on the Mercator maps. In winter, the –5°C (23°F) and –10°C (14°F) isotherms dip down around the mountains, indicating that the center of the range is colder. In summer, the 20°C (68°F) and 25°C (77°F) isotherms also dip down, showing the same effect even though temperatures are much warmer. The Andes Mountains in South America show the effect even more strongly. The principle at work here is that temperatures decrease with an increase in elevation.
6. **Areas of perpetual ice and snow are always intensely cold.** Greenland and Antarctica contain our planet's two great ice sheets. Notice how they stand out on the polar maps as cold centers in both January and July. They are cold for two reasons. First, their surfaces are high in elevation, rising to over 3000 m (about 10,000 ft) in their centers. Second, the white snow surfaces reflect much of the insolation. Since little solar energy is absorbed, little is available to warm the snow surface and the air above it. The Arctic Ocean, bearing a cover of floating ice, also maintains its cold temperatures throughout the year. However, the cold is much less intense in January than on the Greenland Ice Sheet, since ocean water underneath the ice acts as a heat reservoir to keep the ice above from getting extremely cold.

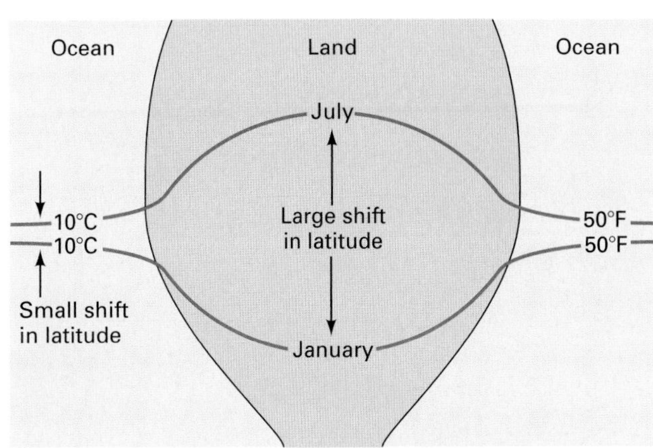

5.21 Seasonal migration of isotherms *Continental air temperature isotherms shift over a much wider latitude range from summer to winter than do oceanic air temperature isotherms. This difference occurs because oceans heat and cool much more slowly than continents.*

The Annual Range of Air Temperatures

Figure 5.22 is a world Mercator map showing the annual range of air temperatures. The lines, resembling isotherms, show the difference between the January and July monthly means. We can explain the features of this map by using the same effects of latitude, interior-maritime location, and elevation.

1. **The annual range increases with latitude, especially over northern hemisphere continents.** This trend is most clearly shown for North America and

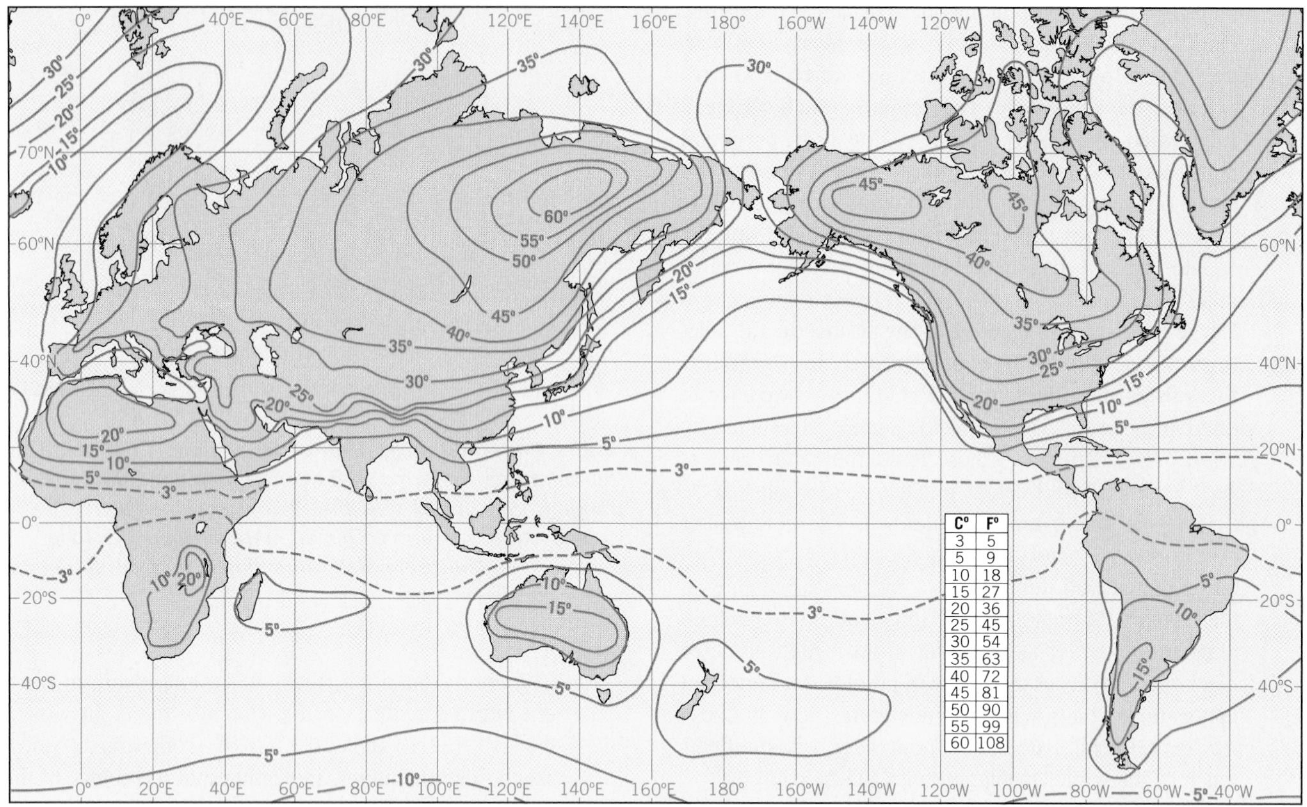

5.22 Annual range of air temperature in Celsius degrees *Data show differences between January and July means. The inset box shows the conversion of Celsius degrees to Fahrenheit for each isotherm value (Compiled by John E. Oliver.)*

Asia. This is due to the contrast between summer and winter insolation, which increases with latitude.

2. **The greatest ranges occur in the subarctic and arctic zones of Asia and North America.** The map shows two very strong centers of large annual range—one in northeast Siberia and the other in northwest Canada-eastern Alaska. In these regions, summer insolation is nearly the same as at the equator, while winter insolation is very low.
3. **The annual range is moderately large on land areas in the tropical zone, near the tropics of cancer and capricorn.** These are regions of large deserts—North Africa (Sahara), southern Africa (Kalahari), and Australia (interior desert) are examples. Dry air and the absence of clouds and moisture allow these continental locations to cool strongly in winter and warm strongly in summer, even though insolation contrasts with the season are not as great as at higher latitudes.
4. **The annual range over oceans is less than that over land at the same latitude.** This can be seen by following a parallel of latitude—40° N, for example. Starting from the right, we see that the range is between 5 and 10°C (9 and 18°F) over the Atlantic but increases to about 30°C (54°F) in the interior of North America. In the Pacific, the range falls to 5°C (9°F) just off the California coast and increases to 15°C (27°F) near Japan. In Central Asia, the range is near 35°C (63°F). Again, these major differences are due to the contrast between land and water surfaces. Since water heats and cools much more slowly than land, a narrower range of temperatures is experienced.
5. **The annual range is very small over oceans in the tropical zone.** As shown on the map, the range is less than 3°C (5°F) since insolation varies little with the seasons near the equator and water heats and cools slowly.

GLOBAL WARMING AND THE GREENHOUSE EFFECT

Tune in to the evening news, and one of the recurring scientific subjects you will hear about is global warming. Is our climate changing? Many scientists have concluded that the temperature of our planet is warming significantly because of human activities. Other scientists argue that the warming of the last few years is part of a natural global cycle.

You may also hear that carbon dioxide (CO_2), produced by human activities, is a major cause of concern in climate warming. This gas is released to the atmosphere in large quantities by fossil fuel burning, and as human energy consumption increases, so does atmospheric CO_2 concentration. The increase in CO_2 with time follows an *exponential growth* principle. This is the growth principle behind the compound interest that causes a savings account to increase in value. Many scientific phenomena show exponential growth. *Working It Out 5.3 • Exponential Growth* provides a closer look.

Working It Out 5.3

Exponential Growth

Exponential growth is a type of growth in which something grows by a constant percentage during each growth period. For example, the money in a bank account will grow according to the rate of interest paid, such as 5 percent per year. The important thing about this type of growth is that it is compounded. That is, a dollar will provide \$1.05 at the end of the first year, but the interest on the second year is paid on the \$1.05 accumulated from the first year, not on the original amount of \$1.

The figure on the right shows an example of exponential growth at two rates—3 percent and 6 percent per time period. The vertical axis shows the multiplier that is applied to the original amount. For example, after 20 time periods, the multiplier for the 6 percent growth rate is slightly more than 3, meaning the original quantity will have more than tripled in that time period at that growth rate. In contrast, the multiplier for the 3 percent growth rate is less than 2.

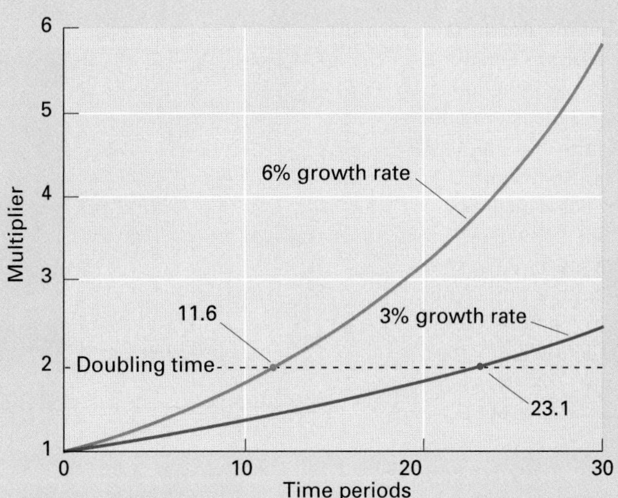

Exponential growth *The multiplier for a quantity growing at a constant percentage rate increases with time. The higher the growth rate, the more quickly the multiplier increases.*

Here is an approximate formula for exponential growth:

$$M = e^{(R \times T)}$$

where M is the multiplier; R is the rate, expressed as a decimal fraction of increase per time period (i.e., $R = 0.04$ for 4 percent per year); and T is the number of time periods (i.e., years) to elapse. The symbol e stands for the base of natural logarithms, which has a value of 2.718. To evaluate the expression, first find $(R \times T)$, then raise e to that power. Most scientific calculators will have a key labeled "exp" or "e^x" to evaluate this function.

Here's an example. Suppose the CO_2 concentration in the atmosphere is 360 ppm at present and is increasing at a rate of 4 percent per year. What concentration can we expect in 20 years if the present rate of growth continues?

$$M = e^{(R \times T)} = e^{(0.04 \times 20)} = 2.718^{0.80}$$
$$= 2.26$$

Thus, the concentration of 360 ppm will grow to about $360 \times 2.26 = 814$ ppm.

Sometimes it is convenient to think in terms of a doubling time—that is, the time it will take for a quantity to double given that it is growing exponentially at a fixed percentage rate. The doubling time is shown graphically for growth rates of 3 and 6 percent in the figure above. It will be the time associated with the multiplier 2. For 3 and 6 percent, doubling times are 11.6 and 23.1, respectively. There is a handy rule to figure the doubling time from the growth rate—divide the interest rate in percent into 70, and the result will be the doubling time. So, a 4 percent annual growth rate would lead to a doubling time of 70/4 = 17.5 years.

Why is the CO_2 buildup a problem? Recall from Chapter 4 that the greenhouse effect is caused by atmospheric absorption of outgoing longwave radiation, largely by carbon dioxide and water vapor, and that a portion of the absorbed energy is counterradiated back toward the Earth's surface. With more CO_2, more energy is absorbed and counterradiated, and the surface is warmed by an extra amount. For this reason, many scientists are concerned that the greenhouse effect is escalating and that global temperatures are rising—or will rise soon—in response. (See *Eye on Global Change 5.4 • Carbon Dioxide—On the Increase.*)

Also of concern are other gases that are normally present in much smaller concentrations—methane (CH_4), the chlorofluorocarbons, tropospheric ozone (O_3), and nitrous oxide (N_2O). (See also *Eye on Global Change 4.4 • The Ozone Layer—Shield to Life,* Chapter 4.) Taken together with CO_2, they are known as **greenhouse gases.** Though less abundant, these gases are better absorbers of longwave radiation than

Eye on Global Change | 5.4

Carbon Dioxide—On the Increase

In the centuries before global industrialization, carbon dioxide concentration in the atmosphere was at a level slightly less than 300 parts per million (ppm), or about $^3/_{100}$ths of a percent by volume. During the last hundred years or so, that amount has been substantially increased. Why? The answer is fossil fuel burning. When fuels like coal, oil, or natural gas are burned, they yield water vapor and carbon dioxide. The release of water vapor does not present a problem because a large amount of water vapor is normally present in the global atmosphere. But because the normal amount of CO_2 is so small, fossil fuel burning has raised the level to about 360 ppm. According to studies of bubbles of atmospheric gases trapped in glacial ice, this level is the highest attained in the last 420,000 years and nearly double the amount of CO_2 present during glaciations of the most recent Ice Age.

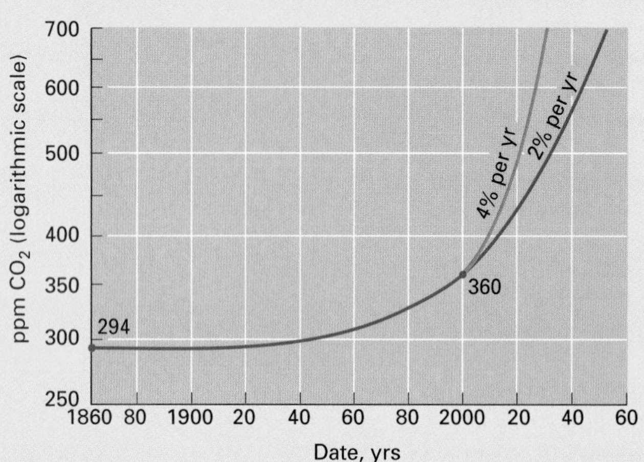

Increase in atmospheric carbon dioxide, observed to 2000 and predicted into the twenty-first century

The figure above shows how CO_2 has increased with time since 1860. Until 1940 or so, the level remained nearly stable. Even with future concerted global action to reduce CO_2 emissions, scientists estimate that levels will stabilize at a value not lower than about 550 ppm by the late twenty-first century. This will nearly double preindustrial levels.

Predicting the buildup of CO_2 is difficult because not all the carbon dioxide emitted into the air by fossil fuel burning remains there. Instead, a complex cycle moves

CO_2, and so are more effective, molecule per molecule, at enhancing the greenhouse effect.

Factors Influencing Climatic Warming and Cooling

Figure 5.23 shows how a number of important factors have influenced global warming since about 1850. The five bars on the left show the greenhouse gases, ranging from CO_2 to N_2O. Although carbon dioxide is the largest, note that the other four together contribute about the same amount of warming. Taken together, the total enhanced energy flow to the surface produced by greenhouse gases is about 3 W/m^2, which is about 1.25 percent of the solar energy absorbed by the Earth and atmosphere.

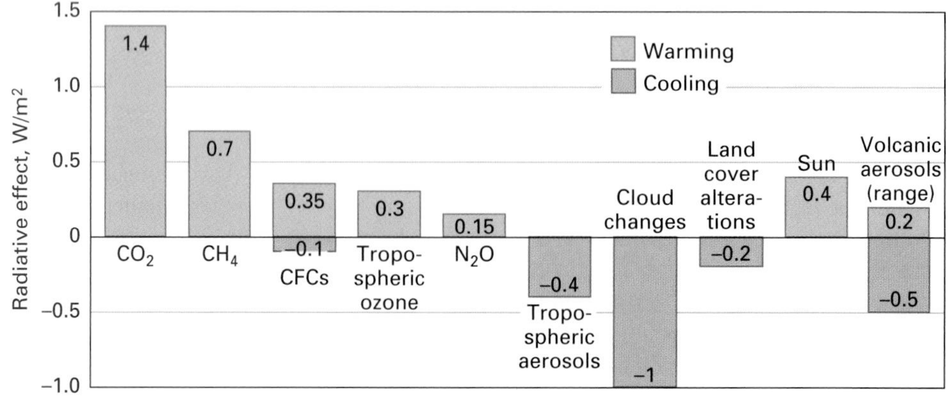

5.23 Factors affecting global warming and cooling *Greenhouse gases act primarily to enhance global warming, while aerosols, cloud changes, and land-cover alterations caused by human activity act to retard global warming. Natural factors may be either positive or negative. (After Hansen et al., 2000, Proc. Nat. Acad. Sci., used by permission.)*

CO_2 throughout the life layer. Photosynthesis is one part of this *carbon cycle.* Plants use CO_2 in photosynthesis, the process by which they use light energy to build their tissues. So, plants take in carbon dioxide, removing it from the atmosphere. When plants die, their remains are digested by decomposing organisms, which release CO_2. This process returns CO_2 to the atmosphere. Normally, these processes are in balance.

Human activity has tilted this balance by clearing land and burning the vegetation cover as new areas of forest are opened for development. This practice increases the amount of CO_2 in the air. When agricultural land is allowed to return to its natural forest state, CO_2 is removed from the air by growing trees. At present, scientists calculate that forests are growing more rapidly than they are being destroyed in the midlatitude regions of the northern hemisphere. This plant growth helps to counteract the buildup of CO_2 produced by fossil fuel burning. But it may be outweighed by the tropical deforestation that is taking place in South America, Africa, and Asia. (See *Eye on Global Change 24.1 • Exploitation of the Low-Latitude Rainforest Ecosystem.*)

Laboratory and field experiments have shown that plants exposed to increased concentrations of CO_2 will grow faster and better. The faster they grow, the more CO_2 they can take in, which helps to reduce the amount in the atmosphere. However, scientists are unsure whether increased CO_2 will stimulate plant growth under natural conditions. Early indications are that plants eventually adjust to the higher CO_2 levels and slow their growth rates.

Another part of the cycle involves the oceans. The ocean's surface layer contains microscopic plant life that takes in carbon dioxide. The CO_2 in the ocean water initially comes from the atmosphere and is mixed into the ocean by surface waves. When these microscopic floating plants die, their bodies sink to the ocean bottom. There they decompose and release CO_2, enriching waters near the ocean floor.

This CO_2 eventually returns to the surface through a system of global ocean current flows in which CO_2-poor waters sink in the northernmost Atlantic and CO_2-rich bottom waters rise to the surface in the northernmost Pacific. In fact, the ocean acts like a slow conveyor belt, moving CO_2 from the surface to ocean depths and releasing it again in a cycle lasting about 1500 years. (We return to this long-term ocean circulation pattern in Chapter 7.)

At present, scientists estimate that ocean surface waters absorb more CO_2 than they release, owing to increased atmospheric levels of CO_2. Therefore, carbon dioxide may be accumulating in ocean depths. However, current studies of global climate computer simulations indicate that the oceans may not be as effective in removing excess CO_2 as scientists previously thought.

Although there is a great deal of uncertainty about the movements and buildup rate of excess CO_2 released to the atmosphere by fossil fuel burning, one thing is certain. Without conversion to solar, nuclear, and hydroelectric power, fuel consumption will continue to release carbon dioxide, and its effect on climate will continue to increase.

Methane, CH_4, is naturally released by the decay of organic matter in wetlands. However, human activity generates about double that amount in rice cultivation, farm animal wastes, bacterial decay in sewage and landfills, fossil fuel extraction and transportation, and biomass burning. Chlorofluorocarbons (CFCs) are shown as having both a warming and cooling effect. The warming effect results because these compounds are very good absorbers of longwave energy. The cooling effect occurs because CFCs destroy ozone in the stratosphere, and ozone contributes to warming. (See *Eye on Global Change 4.4 • The Ozone Layer—Shield to Life.*) Ozone warming also occurs in the troposphere, where O_3 is created by air pollution. Nitrous oxide, N_2O, is released by bacteria acting on nitrogen fertilizers in soils and runoff water. Motor vehicles also emit significant amounts of N_2O.

The next three factors are primarily a result of human activity, and all tend to cool the Earth-atmosphere system. Tropospheric aerosols are produced largely by fossil fuel burning and are considered to be a potent form of air pollution. They include sulfate particles, fine soot, and organic compounds. Aerosols act to scatter solar radiation back to space, thus reducing the flow of solar energy available to warm the surface. In addition, they enhance the formation of low, bright clouds that also reflect solar radiation back to space. These, and other changes in cloud cover caused directly or indirectly by human activity, lead to a significant cooling effect. Land-cover alteration, which has also induced cooling, includes conversion of forested lands to cropland and pastures, which are brighter and reflect more solar energy.

The last two factors are natural in origin. Solar output has increased slightly, causing warming. Volcanic aerosols, however, have at times caused warming and at other times cooling.

Taken together, the warming effect of the greenhouse gases has outweighed the cooling effects of other factors, and the result is a net warming effect of about 1.6 W/m^2, which is about $^2/_3$ of 1 percent of the total solar energy flow absorbed by the Earth and atmosphere. Has this enhanced energy flow caused global temperatures to rise? If not, will it warm the Earth in the near future? Before addressing this important question, we need to review the record of the Earth's surface temperature over the last few centuries.

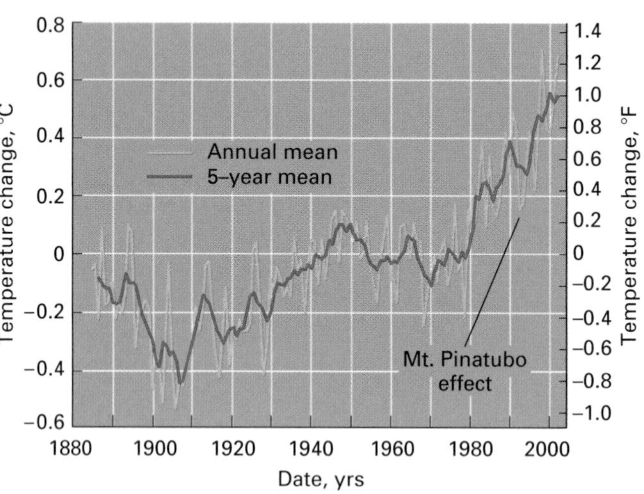

5.24 Mean annual surface temperature of the Earth, 1886–2002 *The vertical scale shows departures in degrees from a zero line of reference representing the average for the years 1951–1980. The yellow line shows the mean for each year. The red line shows a running five-year average. Note the effect of Mount Pinatubo in 1992–1993. (James Hansen/NASA Goddard Institute for Space Studies.)*

The Temperature Record

Figure 5.24 shows the Earth's mean annual surface temperature from 1866 to 2002 as obtained from surface air temperature measurements. The temperature is expressed as a difference from the average annual temperature for the period 1951–1980. Two curves are shown—yearly data and data smoothed over a five-year period. Although temperature has increased, and especially toward the end, there have been wide swings in the mean annual surface temperature.

Some of the variation is due to volcanic activity. We have already noted that volcanic activity can propel particles and gases—especially sulfur dioxide, SO_2—into the stratosphere, forming stratospheric aerosols. Strong winds spread the aerosols quickly throughout the entire layer. The aerosols have a cooling effect because they reflect incoming solar radiation. For example, the eruption of Mount Pinatubo in the Philippines in the spring of 1991 lofted 15 to 20 million tons of sulfuric acid aerosols into the stratosphere (Figure 5.25). The aerosol layer produced by the eruption reduced solar radiation reaching the Earth's surface between 2 and 3 percent for the year or so following the blast. In response, global temperatures fell about 0.3°C (0.5°F) in 1992 and 1993 (see Figure 5.24).

Although the period of direct air temperature measurement does not extend past the middle of the nineteenth century, the record can be extended further back by using tree-ring analysis. The principle is simple—each year, trees

5.25 Eruption of Mount Pinatubo, Philippine Islands, April 1991 *Volcanic eruptions like this can inject particles and gases into the stratosphere, influencing climate for several years afterwards.*

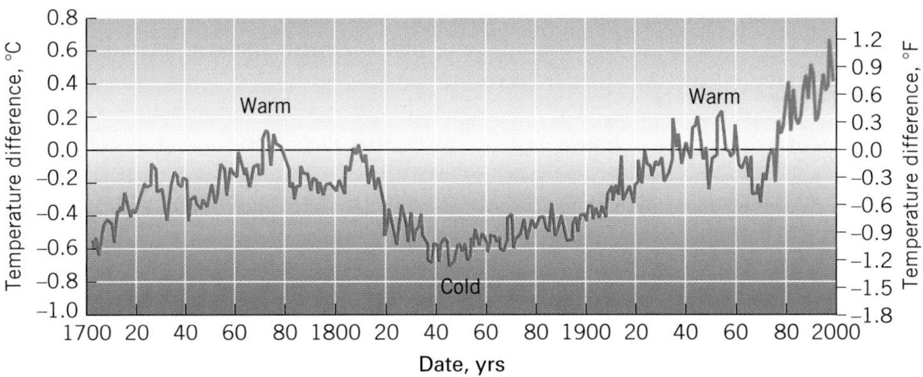

5.26 Three centuries of northern hemisphere temperatures *A reconstruction of the departures of northern hemisphere temperatures from the 1950–1965 mean, based on analyses of tree rings sampled along the northern tree limit of North America. Original data are from 1700 to 1975; for completeness, data from Figure 5.24 are shown from 1975 to the present in a different color. (Courtesy of Gordon C. Jacoby of the Tree-Ring Laboratory of the Lamont-Doherty Geological Observatory of Columbia University.)*

grow in diameter. In climates where the seasons are distinct, this growth can appear as an annual ring. If growing conditions are good, the annual ring is wide. If poor, the annual ring is narrow. For trees along the timberline in North America, the width is related to temperature—the trees grow better when temperatures are warmer. Since only one ring is formed each year, the date of each ring is easy to determine by counting backward from the present. Because these trees live a long time, the temperature record may be extended backward several centuries.

Figure 5.26 shows a reconstruction of northern hemisphere temperatures from 1700 to about 1975 using tree-ring analysis. Like Figure 5.24, it is expressed as a difference from a recent average. From 1880 to present, the temperatures reconstructed from tree-ring analysis seem to fit the observed temperatures (Figure 5.24) quite well. Tree-ring analysis of an earlier period shows us another cycle of temperature increases and decline. The low point, around 1840, marks a historic cold event during which European alpine glaciers became more active and advanced. Other evidence indicates that the two cycles in Figure 5.26 are part of a natural global cycle of temperature warming and cooling lasting about 150 to 200 years. These cycles have occurred regularly over the last thousand years.

What causes these cycles? Again, a number of theories have been offered, but no consensus has yet been reached as to the actual cause or causes. Still longer cycles of temperature change are evident in the fossil record, inducing the advance and retreat of continental and mountain glaciers during the Ice Age. These cycles are discussed in more detail in Chapter 20. Thus, the temperature record shows that the Earth's climate is naturally quite variable, responding to many different influences on many different time scales.

Future Scenarios

The year 1998 was the warmest year on record since the middle of the nineteenth century, with an average temperature of 14.7°C (58.5°F) compiled by NASA scientists at New York's Goddard Institute of Space Science. It was also the warmest year of the past thousand, according to reconstructions of past temperatures using tree rings and glacial ice cores by University of Massachusetts scientists. While a record year, 1998 was only very slightly warmer than 2002. The third-warmest year was 2001, followed by 1990 and 1995.

Is this recent warming an effect of human activity? In 1995 the Intergovernmental Panel on Climate Change, a United Nations-sponsored group of more than 2000 scientists, concluded for the first time that human activity has caused climatic warming. This judgment was based largely on computer simulations of global climate that account for the release of CO_2 and SO_2 from fossil fuel burning occurring since the turn of the century. The simulations agreed well with the patterns of warming observed over that period, leading to the conclusion that, in spite of natural variability, human influence has been felt in the climate record of the twentieth century. A new report, issued in 2001, cited increased confidence in this conclusion. *A Closer Look: Eye on Global Change 5.5 • The IPCC Report of 2001* provides a summary of observed climate change and predictions of climate change for the remainder of the twenty-first century.

Given that greenhouse gases are warming the planet, what will be the effect? Using various projections of the continuing release of greenhouse gases coupled with computer climate models, this same group of scientists projected that global temperatures will warm between 1.4°C (2.5°F) and 5.8°C (10.4°F) by the year 2100.

Why does this temperature rise cause concern? The problem is that many other changes may accompany a rise in temperature. One of these changes is a rise in sea level, as glaciers and sea ice melt in response to the warming. Current predictions call for a rise of 9 to 88 cm (3.5 to 34.6 in.) in sea level by the year 2100. This would place as many as 92 million people within the risk of annual flooding. Climate change could also promote the spread of insect-borne diseases such as malaria. Furthermore, climate boundaries may shift their positions, making some regions wetter while others become drier. Thus, agricultural patterns could shift, displacing large human populations as well as natural ecosystems.

A recently discovered effect of climatic warming is the enhancement of variability in climate. Events such as very high 24-hour precipitation—extreme snowstorms, rainstorms, sleet and ice storms, for example—appear to be occurring more frequently since 1980. More frequent and more intense spells of hot and cold weather may also be related to climatic warming.

The world has become widely aware of the problem of the buildup in CO_2 and other greenhouse gases. At the Rio de Janeiro Earth Summit in 1992, nearly 150 nations signed a treaty limiting emissions of greenhouse gases. Many details were ironed out at a subsequent meeting in Kyoto, Japan, in 1997. The Kyoto Protocol calls on a group of 38 industrial nations to reduce their emissions of CO_2 and other greenhouse gases to about 5 percent below 1990 levels. Nations of the European Union, the United States, and Japan are to reduce average emissions during 2008 and 2012 to levels that are 6, 7, and 8 percent lower, respectively, than 1990 levels. Achieving such reductions could have economic consequences ranging from minor to severe, depending on the analysis.

Meeting again in Buenos Aires in 1998, the participating nations set timetables for the implementation of plans for greenhouse gas reduction in 1999–2000. Important breakthroughs occurred when the conference adopted a plan to allow countries to trade emission rights, as well as a "clean development" mechanism that would permit industrial nations to invest in emissions-reducing enterprises in developing countries. For the first time, some developing countries pledged specific greenhouse gas reductions of their own.

The international effort to reduce releases of greenhouse gases suffered a setback in the spring of 2001, when President George W. Bush rejected American participation in the Kyoto Protocol. However, American resistance was not sufficient to keep 178 nations from pledging to join in the effort to reduce greenhouse gas emissions.

Although these efforts will slow the buildup of greenhouse gases, the ultimate solution will certainly involve greater reliance on solar, wind, and geothermal energy sources, which produce power without releasing CO_2. Energy conservation and development of new methods of utilizing fossil fuels that produce reduced CO_2 emissions will also be important.

A LOOK AHEAD In this chapter we have developed an understanding of both air temperatures and temperature cycles, along with the factors that influence them—insolation, latitude, surface type, continental-maritime location, and elevation. Air temperatures are a very important part of climate, and looking ahead, we will study the climates of the Earth in Chapters 9–11.

The other key ingredient of climate is precipitation—the subject covered in the next chapter. Here temperature plays a very important role, since moist air must cool before condensation forms water droplets and, eventually, precipitation.

Chapter Summary

- **Air temperature**—measured at 1.2 m (4 ft) above the surface—is influenced by insolation, latitude, surface type, location, and elevation. The energy balance of the ground surface is determined by **net radiation**, **conduction** to the soil, and **latent heat transfer** and **convection** to and from the atmosphere.
- Air temperature is measured using a *thermometer* or *thermistor*. Stations make daily minimum, maximum, and mean temperature measurements.
- The two major cycles of air temperature—daily and annual—are controlled by the cycles of insolation produced by the rotation and revolution of the Earth. These cycles induce cycles of net radiation at the surface. When net radiation in the daily cycle is positive, air temperatures increase; when negative, air temperatures decrease. This principle applies to both daily and annual temperature cycles.
- Temperatures of air and soil at or very close to the ground surface are more variable than air temperature as measured at standard height.
- Surface characteristics also affect temperatures. Rural surfaces are generally moist and slow to heat and cool, while urban surfaces are dry and absorb and give up heat readily. This difference creates an urban heat island effect.
- Air temperatures normally fall with altitude in the **troposphere**. The average value of decrease with altitude is the **environmental temperature lapse rate**, 6.4°C/1000 m (3.5°F/1000 ft). At the *tropopause*, this decrease stops. In the **stratosphere** above, temperatures increase slightly with altitude.
- Air temperatures observed at mountain locations are lower with higher elevation, and day-night temperature differences increase with elevation.
- When air temperature increases with altitude, a **temperature inversion** is present. This can develop on clear nights when the surface loses longwave radiation to space.
- Annual air temperature cycles are influenced by the annual pattern of net radiation, which depends largely on latitude.

- Maritime or continental location is another important factor. Ocean temperatures vary less than land temperatures because water heats more slowly, absorbs energy throughout a surface layer, and can mix and evaporate freely. Maritime locations that receive oceanic air therefore show smaller ranges of both daily and annual temperature.
- Global temperature patterns for January and July show the effects of latitude, maritime-continental location, and elevation. Equatorial temperatures vary little from season to season. Poleward, temperatures decrease with latitude, and continental surfaces at high latitudes can become very cold in winter. At higher elevations, temperatures are always colder.
- **Isotherms** over continents swing widely north and south with the seasons, while isotherms over oceans move through a much smaller range of latitude. The annual range in temperature increases with latitude and is greatest in northern hemisphere continental interiors.
- Our planet's global temperature changes from year to year. Within the last few decades, global temperatures have been increasing. CO_2 released by fossil fuel burning is important in causing warming, but so are the other **greenhouse gases**, CH_4, CFCs, O_3, and N_2O. Aerosols scatter sunlight back to space and induce more low clouds, so they tend to lower global temperatures. Solar output and volcanic activity also influence global temperatures.
- Most scientists agree that the human-induced buildup of greenhouse gases has begun to affect global climate. However, natural cycles, such as variations in the Sun's output, still provide strong influences. If we continue to release large quantities of greenhouse gases at increasing rates, we can expect a significant rise in global temperatures that will be accompanied by shifts of climate zones and increasing sea level.

Key Terms

air temperature
surface
net radiation
conduction
latent heat transfer
convection
transpiration
evapotranspiration
heat island
lapse rate
environmental temperature lapse rate
troposphere
stratosphere
temperature inversion
isotherm
greenhouse gases

Review Questions

1. Identify five important factors in determining air temperature and air temperature cycles.
2. What factors influence the temperature of a surface?
3. How are mean daily air temperature and mean monthly air temperature determined?
4. How does the daily temperature cycle measured within a few centimeters or inches of the surface differ from the cycle at normal air temperature measurement height?
5. Compare the characteristics of urban and rural surfaces and describe how the differences affect urban and rural air temperatures. Include a discussion of the urban heat island.
6. What are the two layers of the lower atmosphere? How are they distinguished? Name the two upper layers.
7. How and why are the temperature cycles of high mountain stations different from those of lower elevations?
8. Explain how latitude affects the annual cycle of air temperature through net radiation by comparing Manaus, Aswan, Hamburg, and Yakutsk.
9. Why do large water bodies heat and cool more slowly than land masses? What effect does this have on daily and annual temperature cycles for coastal and interior stations?
10. What three factors are most important in explaining the world pattern of isotherms? Explain how and why each factor is important, and what effect it has.
11. Turn to the January and July world temperature maps shown in Figure 5.20. Make six important observations about the patterns and explain why each occurs.
12. Turn to the world map of annual temperature range in Figure 5.22. What five important observations can you make about the annual temperature range patterns? Explain each.
13. Identify the important greenhouse gases and rank them in terms of their warming effect. What human-influenced factors act to cool global temperature? How?
14. Describe how global air temperatures have changed in the recent past. Identify some factors or processes that influence global air temperatures on this time scale.

Focus on Systems 5.1 • The Surface Energy Balance Equation

1. Consider the surface energy balance during the day. What happens if the surface falls under the shadow of a cloud? How will energy flows to and from the surface be affected? What would you expect to happen to the temperature of the surface?

2. Suppose that on a hot sunny day, a surface layer of soil dries out so that water is no longer available for evaporation. How will energy flows to and from the surface be affected? What will happen to the surface temperature?
3. Air temperature is measured at standard height (1.2 m, 4 ft) above the surface. Would you expect this air temperature to be warmer or colder than the surface during the day? Why? What about at night? Why?

Eye on Global Change 5.4 • Carbon Dioxide—On the Increase

1. Why has the atmospheric concentration of CO_2 increased in recent years?
2. How does plant life affect the level of atmospheric CO_2?
3. What is the role of the ocean in influencing atmospheric levels of CO_2?

A Closer Look:
Eye on Global Change 5.5 • The IPCC Report of 2001

1. What changes in global climate have been noted for the last half of the twentieth century in global temperature, snow and ice cover, precipitation, and greenhouse gas concentrations?
2. How is climate predicted to change by the end of the twenty-first century with respect to temperature, precipitation, snow and ice cover, and sea level?

Visualizing Exercises

1. Sketch graphs showing how insolation, net radiation, and temperature might vary from midnight to midnight during a 24-hour cycle at a midlatitude station such as Chicago.
2. Sketch a graph of air temperature with height showing a low-level temperature inversion. Where and when is such an inversion likely to occur?

Essay Questions

1. Portland, Oregon, on the north Pacific coast, and Minneapolis, Minnesota, in the interior of the North American continent, are at about the same latitude. Sketch the annual temperature cycle you would expect for each location. How do they differ and why? Select one season, summer or winter, and sketch a daily temperature cycle for each location. Again, describe how they differ and why.
2. Many scientists have concluded that human activities are acting to raise global temperatures. What human processes are involved? How do they relate to natural processes? Are global temperatures increasing now? What other effects could be influencing global temperatures? What are the consequences of global warming?

Problems

Working It Out 5.2 • Temperature Conversion

1. On a summer day, a Toronto radio station broadcasts a weather report forecasting a high of 38°C. What is this temperature in °F? On a winter day, a radio station in Buffalo broadcasts a weather report forecasting the same high (38°) but in degrees Fahrenheit. What is this temperature in °C?
2. A maximum-minimum thermometer records a high of 46°F and a low of 28°F. What is the temperature range in °C?
3. At what temperature will both Celsius and Fahrenheit thermometers display the same value? (*Hint:* Check your work using Figure 5.1.)

Working It Out 5.3 • Exponential Growth

1. If the present concentration of CO_2 is 360 ppm and it increases at a rate of 2 percent per year, what will the concentration be in 50 years? What will the concentration be if CO_2 is increasing at a rate of 3 percent?
2. The population of Singapore is about 2.8 million, and it is increasing at an annual rate of 1.3 percent. The Republic of Congo has a population of about 2.4 million, now increasing at an annual rate of 3.0 percent. What is the doubling time of each population? How large will the populations of these two countries be in 25 years if growth continues at the present rates?

Eye on the Landscape

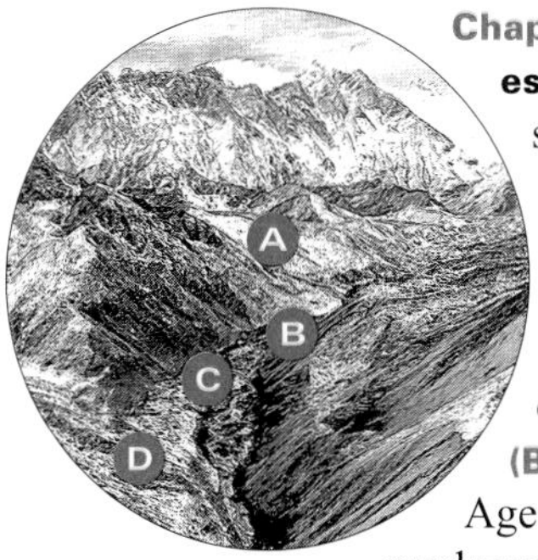

Chapter Opener Mount Everest This magnificent landscape was carved during the Ice Age by huge mountain glaciers, of which a remant can be seen at **(A)**. The Ice Age glaciers carved and deepened the valley at **(B)–(C)**, then retreated as the Ice Age ended. The melting glaciers produced vast volumes of water, laden with the rock debris that now fills the valley. In this process, fluvial action by meltwater created the terraces that now bear fallow fields **(C)**. This valley joins a larger one near **(D)**, where the glacier in the adjoining valley pushed debris into ridges known as lateral moraines. We describe the work of glaciers in shaping the landscape in Chapter 20, while the work of rivers is covered in Chapter 17.

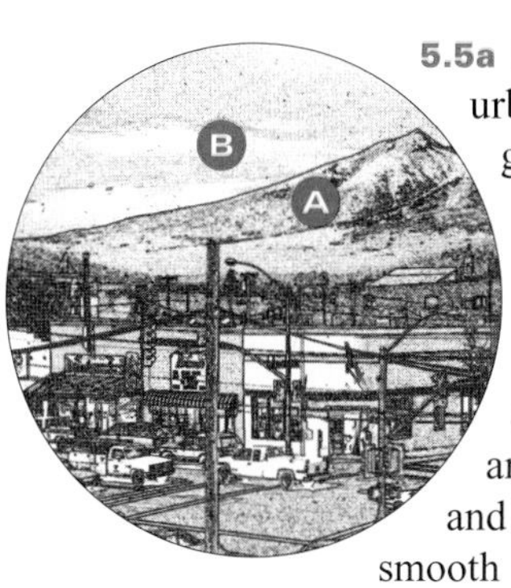

5.5a Urban surfaces Beyond the urban surfaces in the foreground is a distant view of a semiarid landscape in the western United States **(A)**. The sparse and low vegetation cover reflects the semiarid climate (see Chapters 9 and 11 for semiarid climate and vegetation). Note also the smooth profile of the land surface from the mountain peak to the base of the slope **(B)**. This shape indicates long-continued erosion by running water (fluvial erosion), which we describe in Chapter 17.

5.5b Rural surfaces The human influence on this agricultural landscape **(A)** is strong. A dam has created a small fire pond, providing water for both livestock and fire-fighting. The fields are mowed for hay, providing a groomed look by cutting all plants. Grazing by livestock would be more selective and give a more tufted appearance to the pasture. The young trees dotting the landscape are probably all planted and have been pruned to pleasing shapes. The split-rail fencing is attractively rustic but too expensive for most farm applications. In short, this is the spread of a gentleman farmer.

5.10 Clear air at high elevation—Mount Whitney The jagged topography of these high peaks **(A)** indicates glacial, rather than fluvial, erosion. Imagine a dome of glacial ice covering this entire scene, leaving only the tip of Mount Whitney uncovered. Ice flowing in and around the rock masses below the summit gouges the rock, tearing huge chunks loose along cracks and joints in the rock. Then the climate warms and the ice melts, revealing rugged slopes and sheer rock faces stripped by the ice action. We'll cover glacial landscapes in Chapter 20.

A CLOSER LOOK

Eye on Global Change 5.5 The IPCC Report of 2001

In 2001, the United Nations Intergovernmental Panel on Climate Change (IPCC), a body of over 600 scientists nominated by countries from all over the globe, issued three major reports on human-induced changes in global climate. The reports concerned the scientific basis for anticipated climate change, the projected impacts of the change, and strategies for mitigating the impacts.

To estimate the degree of climate change with time and over the globe, scientists of the IPCC used several complex global climate models—mathematical models that predict the state of the atmosphere and land and water surfaces at short time steps for long periods. The models were driven by predicted releases of greenhouse gases under different scenarios of global economic growth and social evolution through the end of the twenty-first century. The two graphs on page 133 show global temperature change and sea-level rise as modeled under these scenarios. Although there is variation between outcomes, it is clear that both global temperature and sea level will rise significantly by 2100 based on their analysis.

Here are some of the panel's more specific findings, taken from *Climate Change 2001: The Scientific Basis, A Report of the Working Group I of the Intergovernmental Panel on Climate Change,* Summary for Policymakers, IPCC, Geneva, Switzerland, 2001.

Recent Climate Change

- Global average surface temperature increased about 0.6°C (1.1°F) during the twentieth century. The 1990s was the warmest decade and 1998 the warmest year since 1861. Nighttime daily minimum air temperatures rose about twice as fast as daytime maximum temperatures. Since 1950, the frequency of extreme low temperatures has decreased. The frequency of extreme high temperatures has increased a smaller amount.
- The lowest 8 km (5 mi) of the atmosphere has warmed at a rate of about 0.05°C (0.09°F) per decade since at least 1979.
- Snow and ice cover has decreased by about 10 percent since the late 1960s. The duration of snow cover has been reduced by about two weeks in the mid- and high latitudes of the northern hemisphere. The extent of sea ice in spring and summer in the northern hemisphere has decreased by 10 to 15 percent since the 1950s. Summer and fall sea ice is thinner.
- Global average sea level rose between 10 and 20 cm (4 and 8 in.) during the twentieth century. Global ocean heat content has increased since at least the late 1950s.
- Precipitation increased by 0.5 to 1 percent per decade in the twentieth century over the mid- and high latitudes of the northern hemisphere. Rainfall decreased in subtropical regions. There was an increase of 2 to 4 percent in the frequency of heavy precipitation events in the latter half of the twentieth century. The frequency and intensity of droughts in some parts of Africa and Asia have also increased in recent decades.
- Cloud cover increased by about 2 percent in the mid- and high latitudes of the northern hemisphere during the twentieth century.
- Since 1970, El Niño episodes have been more frequent, intense, and persistent, compared to those in the previous 100 years.
- Some important aspects of climate have not changed. Some parts of the southern hemisphere oceans and parts of Antarctica have not warmed in recent decades. No significant changes have occurred in Antarctic sea ice coverage since 1978. Tropical and extratropical cyclone frequencies have not been shown to be changing.
- Concentrations of greenhouse gases in the atmosphere have increased as a result of human activities. CO_2 has increased by 31 percent since 1750. For the past two decades, CO_2 concentration has increased by about 1.5 parts per million (ppm) per year. Methane (CH_4) concentration has increased by 151 percent since 1750; nitrous oxide (N_2O) by about 17 percent; and ozone (O_3) by about 36 percent.
- There is newer and stronger evidence that human activities are responsible for most of the warming observed since 1950.

Climate Change in the Twenty-first Century

The IPCC reports also offer these predictions for the rest of the twenty-first century. Although they may not all come true, they are considered likely given the state of present knowledge and current levels of confidence in global climate model predictions, as well as reasonable scenarios for economic growth, development, and emission control.

- Atmospheric composition will continue to change. CO_2 from fossil fuel burning will continue to increase. As this occurs, land and ocean will take up a decreasing fraction of CO_2 released, accentuating the increase. By 2100, CO_2

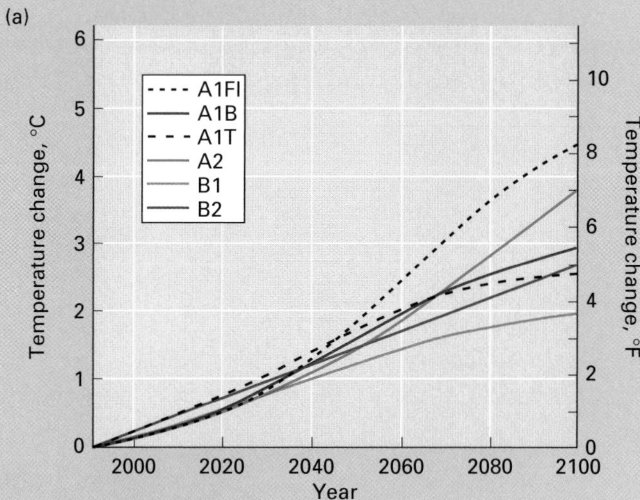

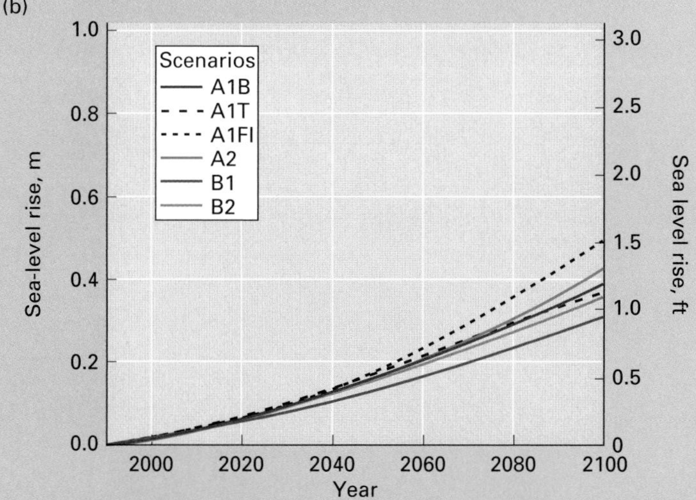

Increases in temperature and sea-level rise modeled by the IPCC *These graphs show how temperature (a) and sea level (b) are predicted to rise between now and 2100, based on global climate models. Curves show means for several models under six emissions scenarios. The A1 scenarios describe a world of rapid growth with economic and social convergence among regions leading to a more uniform world. Scenario A1F1 projects heavy reliance on fossil fuels; A1T on nonfossil energy sources; and A1B a balance across all sources. The A2 scenario is a more heterogeneous world with less convergence and greater regional isolation. The B scenarios are similar to A1 but move toward a service and information economy that emphasizes social and environmental sustainability. B1 assumes more global convergence, while B2 assumes more independence among regions. (From* IPCC, Climate Change Report 2001: Synthesis Report, *copyright IPCC 2001, used by permission of Cambridge University Press.*

concentration will have increased by 90 to 250 percent above the 1750 value, depending on the scenario of economic growth, development, and CO_2 release.

- Other greenhouse gas concentrations are likely to change, increasing under most scenarios. Aerosols could contribute significantly to warming if their release is unabated.
- Global average temperature will rise between 1.4 and 5.8°C (2.5 and 10.4°F) from 1990 to 2100. The projected rate of warming is much larger than that of the late twentieth century and is likely to be greater than any warming episode in the last 10,000 years.
- Land areas will warm more rapidly than the global average, especially northern latitudes in the winter.
- Precipitation will increase in northern mid- to high latitudes and in Antarctica in winter. Larger year-to-year variations in precipitation are likely. At low latitudes, there will be decreases in precipitation in some areas and increases in others.
- Over nearly all land areas, there will be higher maximum temperatures and more hot days; higher minimum temperatures, fewer cold days and frost days; reduced daily temperature range; more intense precipitation events; increased risk of summer drought; and an increase in peak wind and precipitation intensities of tropical cyclones.
- Snow and ice cover will decrease further, and ice caps and glaciers will continue their widespread retreat of the late twentieth century. The Antarctic Ice Sheet will gain mass from increased precipitation, while the Greenland Ice Sheet will thin.
- Sea level will rise from 9 to 88 cm (3.5 to 34.6 in.) depending on the scenario, due to thermal expansion and a gain in volume from melting ice.
- Human-generated climate change will persist for many centuries into the future. Even if concentrations of greenhouse gases are stabilized, oceans will continue to warm; sea level will continue to rise; and ice sheets and glaciers will continue to melt. Models suggest that a local warming of 3°C (5.4°F) over the Greenland ice cap, if sustained over 10 centuries, would completely melt the ice cap, raising sea level by about 7 m (9.8 ft).

6 ATMOSPHERIC MOISTURE AND PRECIPITATION

EYE ON THE LANDSCAPE

Storm over the Amazonian rainforest near Tefé, Amazonas, Brazil. In the equatorial rainforest, precipitation is largely in the form of convective showers. **What else would the geographer see?...Answers at the end of the chapter.**

Water plays a key role in the energy flows that shape our planet's climate and weather. In Chapter 4 we noted how ocean currents act to carry the annual surplus of solar energy from equatorial regions toward the poles. Another water-assisted part of this global energy flow is the flow of latent heat in the form of water vapor conducted poleward by atmospheric circulation. When water evaporates over warm tropical oceans, it takes up latent heat. When it condenses, often in distant colder regions, that latent heat is released in a different location. Precipitation, then, is an important part of this global energy flow process. Precipitation over land also provides water (and ice) that can move under the influence of gravity, thus carving the landforms and landscapes that provide the Earth's varied human habitats. We will return to this role of water in Chapters 16–18.

In this chapter, we focus on water in the air, both as vapor and as liquid and solid water. **Precipitation** is the fall of liquid or solid water from the atmosphere to reach the Earth's land or ocean surface. It is formed when moist air is cooled, causing water vapor to form liquid droplets or solid ice particles. If cooling is sufficiently long and intense, liquid and solid water particles will grow to a size too large to be carried in the air by air motion. They can then fall to Earth.

What causes the cooling of moist air? The answer is upward motion. Whenever air moves upward in the atmosphere, it is cooled. This is not simply because surrounding air temperatures tend to get lower as you go up, as we saw in Chapter 5. It is because of a simple physical principle stating that when air expands, it cools. And because atmospheric pressure decreases with altitude, air moving upward expands—and so at the same time it gets cooler.

What makes air move upward? There are three major causes. One occurs when winds move air up and over a mountain barrier. The second takes place when unequal heating at the ground surface creates a bubble of air that is warmer than the surrounding air. Since it is less dense, it is buoyed upward—like a cork under water bobbing to the surface. This chapter focuses on these two causes of uplift. The third occurs when a mass of cooler, denser air slides under a mass of warmer, lighter air, lifting the warmer air aloft. This type of upward movement occurs in weather systems, which we will cover in Chapter 8. But before we begin our study of atmospheric moisture and precipitation, we will briefly review the three states of water and the conversion of one state to another.

THREE STATES OF WATER

As shown in Figure 6.1, water can exist in three states—solid (ice), liquid (water), and gas (water vapor). A change of state from solid to liquid, liquid to gas, or solid to gas requires the input of heat energy. As we noted in Chapter 4, this energy is called *latent heat,* which is drawn in from the surroundings and stored within the water molecules. When the change goes the other way, from liquid to solid, gas to liquid, or gas to solid, this latent heat is released to the surroundings.

For each type of transition, there is a specific name, shown in the figure. Melting, freezing, evaporation, and condensation are all familiar terms. **Sublimation** is the direct transition from solid to vapor. Perhaps you have noticed that old ice cubes in your refrigerator's freezer seem to shrink away from the sides of the ice cube tray and get smaller with time. They shrink through sublimation, which is induced by the constant circulation of cold, dry air through the freezer. The ice cubes never melt, yet they lose bulk directly as vapor. **Deposition** is the reverse process, when water vapor crystallizes directly as ice.* Frost formed on a cold winter night is an example. The amounts of latent heat taken up and released with changes of state of water are presented in *Working It Out 6.1 • Energy and Latent Heat.*

6.1 Three states of water *Arrows show the ways that any one state of water can change into either of the other two states. Heat energy is absorbed or released, depending on the direction of change.*

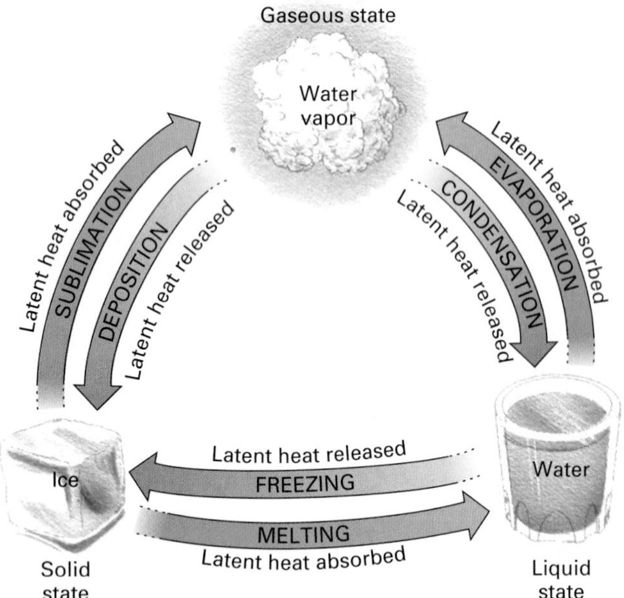

THE HYDROSPHERE AND THE HYDROLOGIC CYCLE

Let's now turn to the hydrosphere and the flows of water among ocean, land, and atmosphere. Recall from Chapter 2 that the hydrosphere includes water on the Earth in all its forms. About 97.2 percent of the hydrosphere consists of ocean saltwater, as shown in Figure 6.2. The remaining 2.8 percent is fresh water. The next largest reservoir is fresh water stored as ice in the world's ice sheets and mountain glaciers. This water accounts for 2.15 percent of total global water.

Fresh liquid water is found both on top of and beneath the Earth's land surfaces. Water occupying openings in soil and rock is called *subsurface water.* Most of it is held in deep storage as *ground water,* at a level where plant roots cannot access it. Ground water makes up 0.63 percent of the hydrosphere, leaving 0.02 percent of the water remaining.

The right-hand portion of Figure 6.2 shows how this small remaining proportion of the Earth's water is distributed. This proportion is important to us because it includes the water available for plants, animals, and human use. *Soil water,* which is held in the soil within reach of plant roots, comprises 0.005 percent of the global total. Water held in streams, lakes, marshes, and swamps is called *surface water.* Most of this surface water is about evenly divided between fresh water lakes and saline (salty) lakes. An extremely small proportion is held in the streams and rivers that flow toward the sea or inland lakes.

Note that the quantity of water held as vapor and cloud water droplets in the atmosphere is also very small—0.001 percent of the hydrosphere. Though small, this reservoir of water is of enormous importance. It provides the supply of precipitation that replenishes all freshwater stocks on land. And, as we will see in the next chapter, the flow of water vapor from warm tropical oceans to cooler regions provides a global flow of heat, in latent form, from low to high latitudes.

The movements of water among the great global reservoirs constitute the **hydrologic cycle.** In this cycle, water moves from land and ocean to the atmosphere as water vapor and returns as precipitation. Because precipitation over land exceeds evaporation, water also runs off the land to the oceans. We will return to the hydrologic cycle in more detail in Chapter 16.

---

* *Deposition* is a term used by chemists to describe the deposit of a solid directly from a vapor. Meteorologists generally use the term *sublimation* for both the change from vapor to solid and from solid to vapor, with the difference determined by the context. Since it is clearer to use a separate word for each process, we use the term *deposition* in this text.

Working It Out 6.1
Energy and Latent Heat

In the metric system, energy is measured in joules. The *joule* (J) is defined as one newton-meter, where a newton is the force produced by an acceleration of 1 m/s^2 applied to a 1 kg mass. The joule thus has units of force times distance, which is a measure of work or energy. The joule is related to the watt (W), since one watt is defined as the flow of one joule of energy per second. From this definition, the energy emitted by a 100-watt bulb in one second is 100 joules. In the English system, energy is measured by the British thermal unit (B.t.u.). However, this unit is little-used, and thus we do not indicate conversions of joules to B.t.u. in the text.

The amount of latent heat taken up or released by water when a change of state occurs depends on the change of state and the temperature of the water. For melting and freezing, the energy amounts to about 335 kilojoules per kilogram (kJ/kg) when water changes between a liquid at 0°C and a solid at the same temperature. For evaporation and condensation, the energy required or released is about 2260 kJ/kg at a temperature of 100°C. Note also that energy is required to heat the water from 0°C to 100°C. The amount required is 4.19 kJ/kg for each Celsius degree, or 419 kJ/kg for 100°. (The values given above depend partly on atmospheric pressure and are taken for sea-level conditions.)

To determine the amount of energy required for sublimation from solid to gas, we can add the amounts of heat required first for melting, then for warming to the boiling point, and finally, for evaporating. That is, $335 + 419 + 2260 = 3014$ kJ/kg. This will also be the amount of energy released when deposition occurs.

What quantity of energy is required to evaporate a kilogram of liquid water at 25°C? This quantity will be the sum of the energy required first to bring the water to the evaporation point and then the latent heat required to convert it to vapor. For the first quantity, each degree requires 4.19 kJ/kg, and there are $100 - 25 = 75$°C, so $4.19 \times 75 = 314$ kJ/kg are needed. Added to the 2260 kg necessary for the change of state, we have $314 + 2260 = 2574$ kJ/kg.

For now, we will summarize the main features of the hydrologic cycle in the global water balance, shown in Figure 6.3. The cycle begins with evaporation from water or land surfaces, in which water changes state from liquid to vapor and enters the atmosphere. In the case of land, this evaporation is enhanced by transpiration. Total evaporation is still about six times greater over oceans than land, however. This is because the oceans cover most of the planet and because land surfaces are not always wet enough to yield much evaporated water. Once in the atmosphere, water vapor can condense or deposit to form precipitation, which falls to Earth as rain, snow, or hail. Precipitation over the oceans is nearly four times greater than precipitation over land.

Upon reaching the land surface, precipitation has three fates. First, it can evaporate and return to the atmosphere as water vapor. Second, it can sink into the soil and then into the surface rock layers below. As we will see in later chapters, this subsurface water emerges from below to feed rivers, lakes, and even ocean margins. Third, precipitation can run off the land, concentrating in streams and rivers that eventually carry it to the ocean or to a lake in a closed inland basin. This flow of water is known as *runoff.*

Focus on Systems 6.2 • The Global Water Balance as a Matter Flow System provides a systems view of the hydrologic cycle. Evaporation, precipitation, and runoff are flow pathways for water in its various forms that are interconnected in this closed matter flow system.

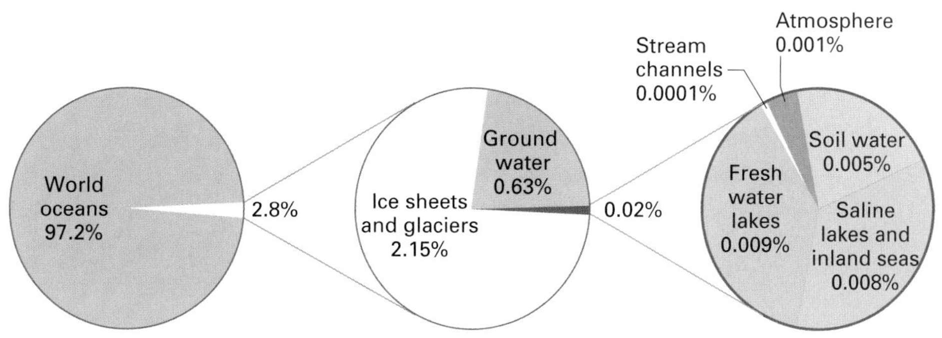

6.2 Volumes of global water in each reservoir of the hydrosphere *Nearly all the Earth's water is contained in the world ocean. Fresh surface and soil water make up only a small fraction of the total volume of global water.*

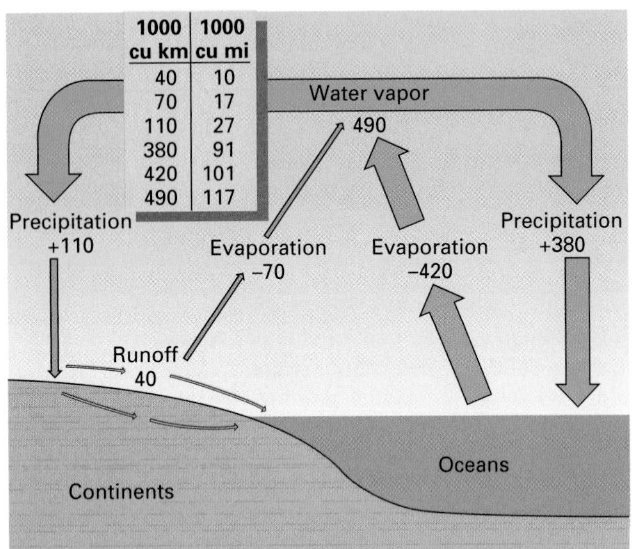

***6.3 The global water balance** Figures give average annual water flows in and out of world land areas and world oceans. Values are in thousands of cubic kilometers (cubic miles). Global precipitation equals global evaporation. (Based on data of John R. Mather.)*

For most of this chapter, we will be concerned with one aspect of the hydrologic cycle—the flow of water from the atmosphere to the surface in the form of precipitation. We will begin by examining how the water vapor content of air is measured and how clouds and precipitation form.

HUMIDITY

The amount of water vapor present in the air varies widely from place to place and time to time. It ranges from almost nothing in the cold, dry air of arctic regions in winter to as much as 4 or 5 percent of a given volume of air in the warm wet regions near the equator. The general term **humidity** refers to the amount of water vapor present in the air.

Understanding humidity and how the moisture content of air is measured involves an important principle—namely, that the maximum quantity of moisture that can be held at any time in the air is dependent on air temperature. Warm air can hold more water vapor than cold air—a lot more. Air at room temperature (20°C, 68°F) can hold about three times as much water vapor as air at freezing (0°C, 32°F).

Specific Humidity

The actual quantity of water vapor held by a parcel of air is its **specific humidity.** This measure is important because it describes how much water vapor is available for precipitation. Specific humidity is stated as the mass of water vapor contained in a given mass of air and is expressed as grams of water vapor per kilogram of air (g/kg). Figure 6.4 shows the relationship between air temperature and the maximum amount of water vapor that air can hold. We see, for example, that at 20°C (68°F), the maximum amount of water vapor that the air can hold—that is, the maximum specific humidity possible—is about 15 g/kg. At 30°C (86°F), it is nearly doubled—about 26 g/kg. For cold air, the values are quite small. At −10°C (14°F), the maximum is only about 2 g/kg.

Climatologists often use specific humidity to describe the moisture characteristics of a large mass of air. For example, extremely cold, dry air over arctic regions in winter may have a specific humidity as low as 0.2 g/kg. In comparison, the extremely warm, moist air of equatorial regions often holds as much as 18 g/kg. The total natural range on a worldwide basis is very wide. In fact, the largest values of specific humidity observed are from 100 to 200 times as great as the smallest.

Specific humidity is a geographer's yardstick for a basic natural resource—water. It is a measure of the quantity of water in the atmosphere that can be extracted as precipitation. Cold, moist air can supply only a small quantity of rain or snow, but warm, moist air is capable of supplying large amounts.

Figure 6.5 is a set of global profiles showing how specific humidity varies with latitude and how it relates to mean surface air temperature. Both humidity and temperature are measured at the same locations in standard thermometer shelters the world over, and also on ships at sea. Note that the horizontal axis of the graphs has an uneven scale of units—they get closer together toward both left and right edges, which represent the poles. In this case, the units are adjusted so that they reflect the amount of the Earth's surface present at that latitude. That is, since there is much more sur-

***6.4 Specific humidity and temperature** The maximum specific humidity of a mass of air increases sharply with rising temperature.*

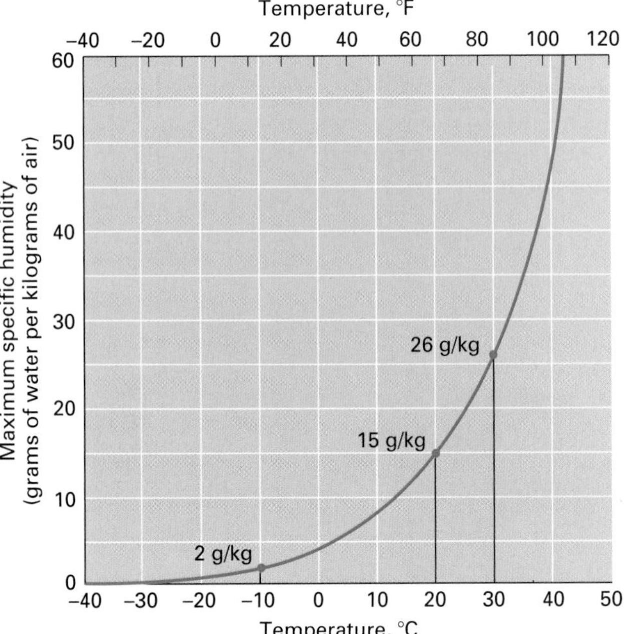

Focus on Systems 6.2

The Global Water Balance as a Matter Flow System

Since our planet contains only a fixed amount of water, a global balance must be maintained among flows of water to and from the lands, oceans, and atmosphere. Let's examine this idea in more detail. For our analysis, we will assume that the volume of ocean waters and the overall volume of fresh water in surface and subsurface water remains constant from year to year. This is probably quite reasonable, unless climate is changing rapidly.

The global water balance can be seen as a flow system diagrammed to the right, which is derived from Figure 6.3. In this flow system, the material that is flowing is water, primarily as liquid and solid water in precipitation and as water vapor in evaporation. For our analysis, we will consider the surface as consisting of two components—land and oceans. Water on the land portion of the Earth's surface is largely in the form of liquid water, but significant amounts are also present in solid form as snow and ice. The ocean-covered portion of the surface is largely liquid water, but the Arctic Ocean and some parts of the ocean near Antarctica have a cover of sea ice that is often topped with snow.

Flows of water link the atmosphere and the two surface components. Following Figure 6.3, these are marked with their values in units of thousands of cubic kilometers per year. For budgeting purposes, we adopt the convention that flows to the surface (land or oceans) are positive and flows leaving the surface are negative.

Let's first look at the flows into and out of the atmosphere. Since the cycle is in balance, they sum to zero. That is,

$$P_{LAND} + E_{LAND} + P_{OCEAN} + E_{OCEAN} = 0$$

$$(+110) + (-70) + (+380) + (-420) = 0$$

Here, P_{LAND} and P_{OCEAN} are precipitation flows from the atmosphere to the land and oceans, and E_{LAND} and E_{OCEAN} are evaporation flows from the land and oceans to the atmosphere. (Note that in the term *precipitation,* we include flows of water in both liquid (rain) and solid (snow) forms as well as direct deposition of water vapor as frost. Similarly, we take evaporation here to include sublimation of snow and ice to water vapor.)

In looking at these values, we also note that although the global balance for the atmosphere is zero, there is a positive balance for land (+40), indicating more precipitation than evaporation, and a negative balance for oceans (–40), indicating more evaporation than precipitation. Thus, each year land gains 40,000 km^3 of water while oceans lose 40,000 km^3 of water. Without a link between land and water, this would mean that the oceans would eventually empty and all the water would accumulate over land! Of course, this doesn't happen because a flow pathway connects land and oceans—runoff of water from rivers, glaciers, and ice sheets into the ocean. For the oceans to stay at the same level, and for the same amount of water to be stored on the land year after year, the flow in runoff must equal this difference—40 units, or 40,000 km^3.

Recall from earlier discussions of flow systems that all matter flow systems need a power source. What is the power source for this vast flow system of water in all its forms? The answer, of course, is the Sun. Solar energy evaporates water and moves it into the atmosphere, where it can later fall to Earth as precipitation. Gravity also plays a role in moving streams of water and ice off the lands and into the oceans as runoff.

The global water balance as a flow system

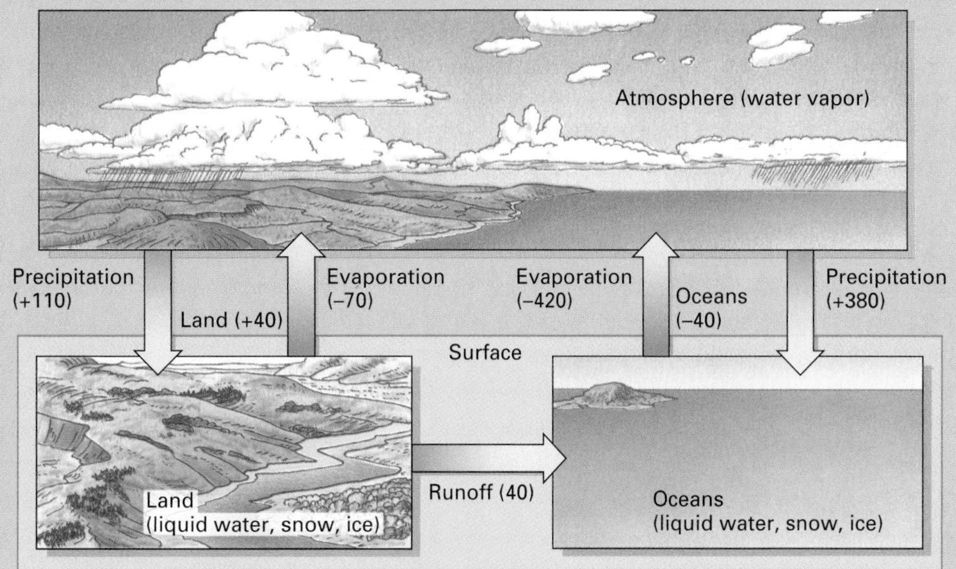

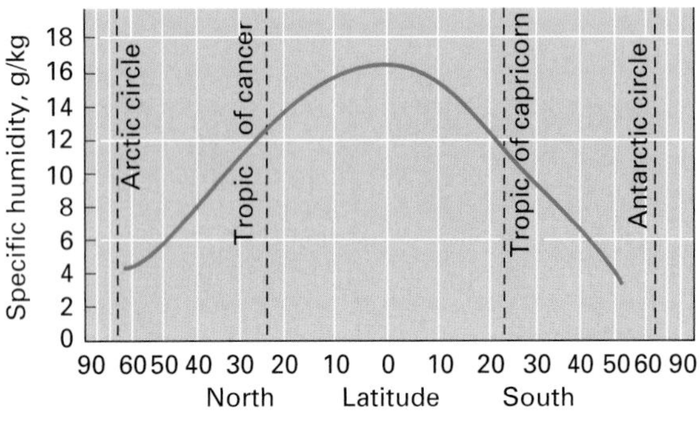

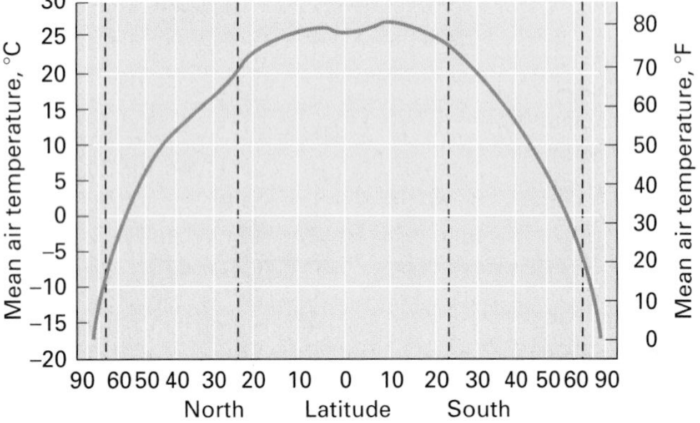

6.5 Global temperature and specific humidity *Pole-to-pole profiles of specific humidity (above) and temperature (below) show similar trends because the ability of air to hold water vapor (measured by specific humidity) is limited by temperature. (Data of J. von Hann, R. Süring, and J. Szava-Kovats as shown in Haurwitz and Austin, Climatology.)*

face area between 0° and 10° latitude than between 60° and 70°, the region between 0° and 10° is allocated more width on the graph.

Let's look first at specific humidity (upper graph). The curve clearly shows the largest values for the equatorial zones, with values falling off rapidly toward both poles. This curve follows the pattern of insolation quite nicely (see Figure 4.8). More insolation is available at lower latitudes, on average, to evaporate water in oceans or on moist land surfaces. Therefore, specific humidity values are higher at low latitudes than at high latitudes.

The global profile of mean (average) surface air temperature (lower graph) shows a similar shape to the specific humidity profile. We would expect this to be the case, since air temperature and maximum specific humidity vary together, as shown in Figure 6.4.

Another way of describing the water vapor content of air is by its **dew-point temperature.** If air is slowly chilled, it eventually will reach *saturation*. At this temperature, the air holds the maximum amount of water vapor possible. If further cooling continues, condensation will begin and dew will form. The temperature at which saturation occurs is therefore known as the dew-point temperature. Moist air will have a higher dew-point temperature than drier air.

Relative Humidity

The measure of humidity that we encounter every day is **relative humidity.** This measure compares the amount of water vapor present to the maximum amount that the air can hold at that temperature, expressed as a percentage. For example, if the air currently holds half the moisture possible at the present temperature, then the relative humidity is 50 percent. When the humidity is 100 percent, the air holds the maximum amount possible, is saturated, and will be at the dew-point temperature.

A change in relative humidity of the atmosphere can happen in one of two ways. The first is through direct gain or loss of water vapor. For example, if an exposed water surface or wet soil is present, additional water vapor can enter the air, raising the specific humidity. This process is slow because the water vapor molecules must diffuse upward from the surface into the air layer above.

The second way is through a change of temperature. Even though no water vapor is added, a lowering of temperature results in a rise of relative humidity. Recall that the capacity of air to hold water vapor is dependent on temperature. When the air is cooled, this capacity is reduced. The existing amount of water vapor then represents a higher percentage of the total capacity.

An example may help to illustrate these principles (Figure 6.6). Beginning in the early morning hours, at 4 A.M. the temperature is 5°C (41°F), and the relative humidity of the air is 100 percent. That is, the air is saturated and cannot hold any additional water vapor. By 10 A.M., the temperature has risen to 16°C (61°F). The relative humidity has dropped to 50 percent, even though the amount of water vapor in the air remains the same. By 3 P.M., the air has been warmed by the sun to 32°C (90°F). The relative humidity has dropped to 20 percent, which is very dry air. The same amount of water vapor is present in the air, but the capacity of the air to hold water vapor has greatly increased. As air temperature falls in the evening, relative humidity again rises, and the cycle repeats itself.

6.6 Relative humidity and air temperature *Relative humidity changes with temperature because the capacity of warm air to hold water vapor is greater than that of cold air. In this example, the amount of water vapor stays the same, and only the capacity to hold water changes.*

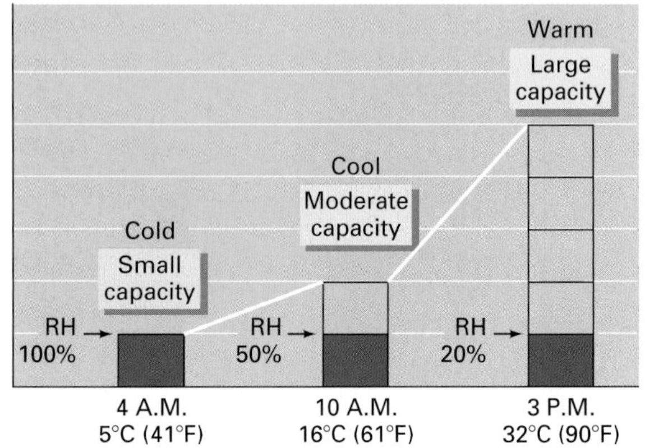

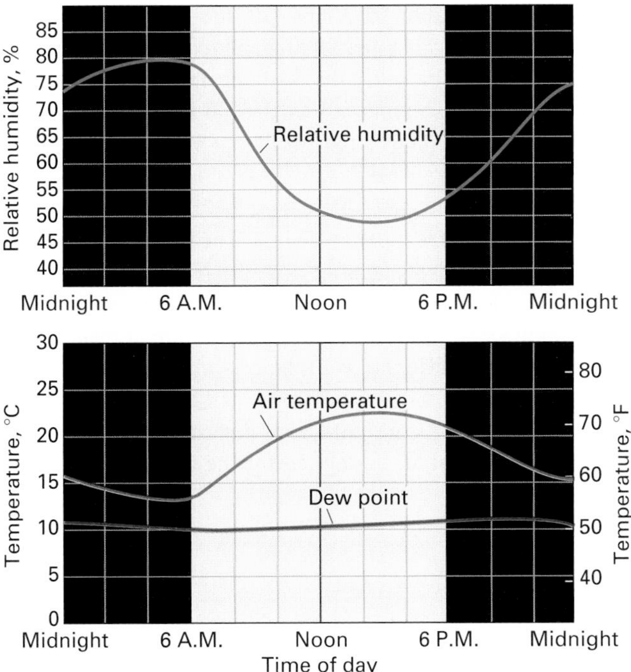

6.7 Daily cycle of relative humidity, air temperature, and dew point *Relative humidity (above) drops during the day as air temperature (below) rises in this example of hourly averages from Washington, D.C., for the month of May. Dew-point temperature remains nearly constant. (Data from National Weather Service.)*

Figure 6.7 illustrates the principle of relative humidity change caused by temperature change. Values shown are averages for May at Washington, D.C. As air temperature rises during the day, relative humidity falls, and vice versa. The dew-point temperature, which indicates the true moisture content, remains nearly constant.

How is humidity measured? A simple method uses two thermometers mounted together side by side in an instrument called a *sling psychrometer* (Figure 6.8). The wet-bulb thermometer bulb is covered with a sleeve of cotton fibers and is wetted. The dry-bulb thermometer remains dry. To operate the sling psychrometer, water is applied to the sleeve covering the wet bulb, and the two thermometers are then whirled in the open air. Evaporation cools the wet bulb. If the air is saturated, then evaporation cannot occur. In this case, no cooling results, and the temperature of the two thermometers is the same. If the air is not saturated, evaporation occurs and the wet-bulb thermometer is cooled. The drier the air, the greater is the cooling. To determine the relative humidity, a special sliding scale is used (Figure 6.8). Wet-bulb and dry-bulb temperatures are set on the scale, and relative humidity is read off directly. There are also instruments that read relative humidity directly. One such instrument uses a thin layer of a special material bonded to a metal film. The material absorbs water vapor in an amount depending on the relative humidity. The water vapor affects the ability of the metal film to hold an electric charge. This ability is sensed by an electronic circuit and converted to a direct reading of relative humidity.

THE ADIABATIC PROCESS

Given that ample water vapor is present in a mass of air, how is that related to precipitation? In other words, how is the water vapor turned into liquid or solid particles that can fall to Earth? The answer is by natural cooling of the air. When air is cooled to the dew point and below, some water vapor molecules must change state to form water droplets or ice crystals. Think about a moist sponge. To extract the water, you have to squeeze the sponge—that is, reduce its ability to hold water. Chilling the air is like squeezing the sponge. GEODISCOVERIES

How is air chilled sufficiently to produce precipitation? One mechanism for chilling air is nighttime cooling. As we saw in Chapter 5, the ground surface can become quite cold on a clear night through loss of longwave radiation. Thus, still air near the surface can be cooled below the condensation point, producing dew or frost. However, this mechanism is not sufficient to form precipitation. Precipitation is formed only when a substantial mass of air experiences a

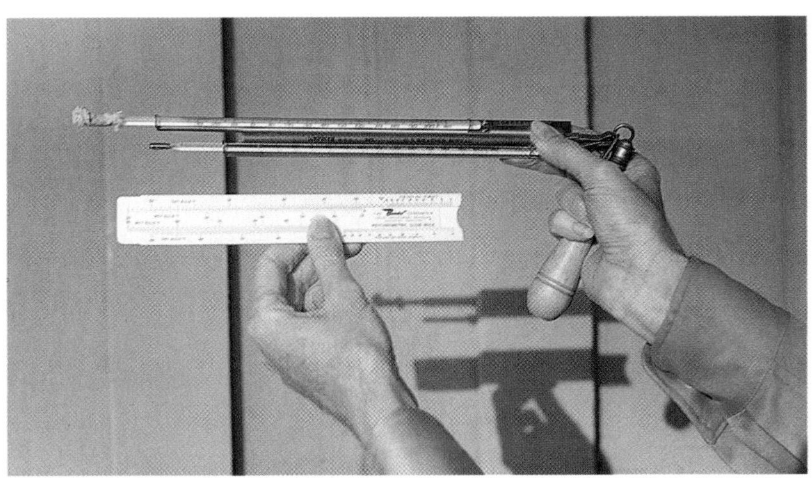

6.8 Sling psychrometer *This standard sling psychrometer uses paired thermometers. The cloth-covered wet-bulb thermometer projects beyond the dry-bulb thermometer. The handle is used to swing the thermometers in the free air. Evaporation lowers the wet-bulb temperature. Thus, the dryer the air, the greater the evaporation and the cooler the reading in relation to the dry-bulb thermometer. The sliding scale ruler, pictured below, enables rapid determination of relative humidity from wet- and dry-bulb readings.*

steady drop in temperature below the dew point. This happens when an air parcel is uplifted to a higher level in the atmosphere, as we will see in the following section.

Dry Adiabatic Rate

An important principle of physics is that when a gas is allowed to expand, its temperature drops. Conversely, when a gas is compressed, its temperature increases. If you have ever pumped up a bicycle tire using a hand pump, you have observed this latter effect yourself. As you pump vigorously, the metal pump gets hot. This occurs because air inside the pump is being compressed and therefore heated. In the same way, when a small jet of air escapes from a high-pressure hose, it feels cool. Perhaps you have flown on an airplane and noticed a small nozzle overhead that directs a stream of cool air toward you. When the air moves from higher pressure in the air hose that feeds the nozzle to lower pressure in the cabin, it expands and cools. Physicists use the term **adiabatic process** to refer to a heating or cooling process that occurs solely as a result of pressure change. That is, the change in temperature is not caused by heat flowing into or away from a volume of air, but only by a change in pressure on a volume of air.

Given that air cools or heats when the pressure on it changes, how does that relate to uplift and precipitation? The missing link is simply that atmospheric pressure decreases with an increase in altitude. So, if a parcel of air is uplifted, atmospheric pressure on the parcel will be lower, and it will expand and cool. Conversely, if a parcel of air descends, atmospheric pressure will be higher, and it will be compressed and warmed (Figure 6.9). The **dry adiabatic lapse rate** describes this behavior for a rising air parcel that has not reached saturation. This rate has a value of about 10°C per 1000 m (5.5°F per 1000 ft) of vertical rise. That is, if a parcel of air is raised 1 km, its temperature will drop by 10°C. Or, in English units, if raised 1000 ft, its temperature will drop by 5.5°F. This is the "dry" rate because condensation does not occur.

Note that in the previous chapter we encountered the environmental temperature lapse rate. The temperature lapse rate is simply an expression of how the temperature of still air varies with altitude. This rate will vary from time to time and from place to place, depending on the state of the atmosphere. It is quite different from the dry adiabatic lapse rate. The dry adiabatic rate applies to a mass of air in vertical motion. It is always constant and is determined by physical laws. No motion of air is implied for the temperature lapse rate.

Wet Adiabatic Rate

Let's continue examining the fate of a parcel of air that is moved upward in the atmosphere (Figure 6.10). We will assume that the parcel starts with a temperature of 20°C (68°F). As the parcel moves upward, its temperature drops at the dry adiabatic rate, 10°C/1000 m (5.5°F/1000 ft). At 500 m (1600 ft), the temperature has dropped by 5°C (9°F) to 15°C (59°F). At 1000 m (3300 ft), the temperature has fallen to 10°C (50°F). As the rising process continues, the air is cooled to its dew-point temperature, and condensation starts to occur. This is shown on the figure as the *lifting condensation level.*

Note, however, that the dew-point temperature changes slightly with elevation. Instead of remaining constant, it falls at the *dew-point lapse rate* of 1.8°C/1000 m (1.0°F/1000 ft). Suppose the dew-point temperature of the air mass is 11.8°C

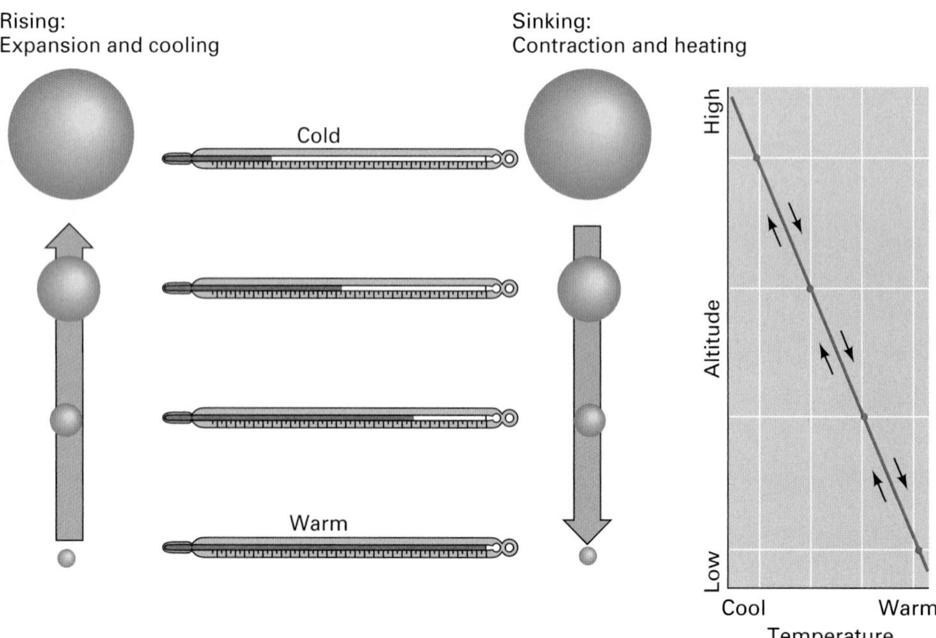

6.9 Adiabatic cooling and heating *A schematic diagram of adiabatic cooling and heating that accompanies the rising and sinking of a mass of air. When air is forced to rise, it expands and its temperature decreases. When air is forced to descend, its temperature increases. (A. N. Strahler.)*

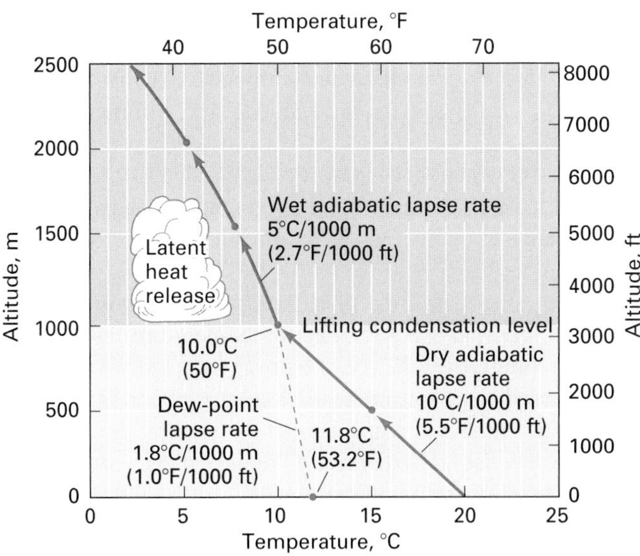

6.10 Adiabatic cooling *Adiabatic decrease of temperature in a rising parcel of air leads to condensation of water vapor into water droplets and the formation of a cloud. (Copyright © A. N. Strahler.)*

(53.2°F). Then at 1000 m (3300 ft), the dew-point temperature will be 10°C (50°F). That will also be the temperature of the parcel for the example above, so condensation will begin at that level. The lifting condensation level is thus determined by the initial temperature of the air and its dew point. *Working It Out 6.3 • The Lifting Condensation Level* shows how to calculate the lifting condensation level, given these two quantities.

If cooling continues, water droplets will form, producing a cloud. If the parcel of saturated air continues to rise, however, a new principle comes into effect—latent heat release. That is, when condensation occurs, latent heat is released and warms the uplifted air. In other words, two effects are occurring at once. First, the uplifted air is being cooled by the reduction in atmospheric pressure. Second, it is being warmed by the release of latent heat from condensation.

Which effect is stronger? As it turns out, the cooling effect is stronger, so the air will continue to cool as it is uplifted. But because of the release of latent heat, the cooling will occur at a lesser rate. This cooling rate for saturated air is called the **wet adiabatic lapse rate** and ranges between 4 and 9°C per 1000 m (2.2 and 4.9°F per 1000 ft).

Working It Out 6.3

The Lifting Condensation Level

When a parcel of air moves upward, it cools at the dry adiabatic rate of about 10°C/1000 m. This cooling occurs because the parcel is exposed to a lesser atmospheric pressure as it rises, and under the adiabatic principle, it expands and cools. The change in pressure also affects the dew-point temperature, which falls at a rate of about 1.8°C/1000 m. When the temperature of the cooling air parcel reaches the dew-point temperature, condensation will begin to occur. The elevation at which condensation begins to occur is referred to as the *lifting condensation level.*

Suppose an air parcel is at 20°C and is raised 500 m. What will its temperature be? Since the dry adiabatic rate is 10°C/1000 m, its temperature will drop by 500/1000 × 10 = 5°C. So, the temperature will be 20 − 5 = 5°C. Putting this in equation form,

$$T = T_0 - H \times R_{DRY}$$

where T is the temperature of the parcel, T_0 is the starting temperature at height 0 m, H is the height in meters, and R_{DRY} is the dry adiabatic rate, 10°C/1000 m.

Let's assume that the dew point of that same parcel of air is 11°C. What will the dew-point temperature be at 500 m? Since the dew-point lapse rate is 1.8°C/1000 m, the dew-point temperature will fall by 500/1000 × 1.8 = 0.9°C. So, the dew-point temperature will be 11 − 0.9 = 10.1°C. Similarly, we can write

$$T_D = T_{DEW} - H \times R_{DEW}$$

where T_D is the dew-point temperature of the parcel at height H, T_{DEW} is the starting dew-point temperature, and R_{DEW} is the dew-point lapse rate, 1.8°C/1000 m.

At what level will condensation occur? Condensation will occur when the parcel's temperature reaches the dew point—that is, at the height at which $T = T_D$. For this level, then, we can write

$$T_0 - H \times R_{DRY} = T_{DEW} - H \times R_{DEW} \quad (6.1)$$

After some algebraic rearrangement and substitution of the values for the dry adiabatic rate and the dew-point lapse rate, we have the formula

$$H = 1000 \times \frac{T_0 - T_{DEW}}{8.2} \quad (6.2)$$

which gives the lifting condensation level directly in meters.

What will be the lifting condensation level for the air parcel described above? Substituting,

$$H = 1000 \times \frac{20 - 11}{8.2} = 1000 \times \frac{9}{8.2} = 1098 \text{ m}$$

What will the temperature of the air parcel be at the lifting condensation level? By substituting in the first expression,

$$T = T_0 - H \times R_{DRY} = 20 - 198 \times \frac{10}{1000} = 9.0°\text{C}$$

Unlike the dry adiabatic lapse rate, which remains constant, the wet adiabatic lapse rate is variable because it depends on the temperature and pressure of the air and its moisture content. For most situations, however, we can use a value of 5°C/1000 m (2.7°F/1000 ft). The higher rates apply only to cold, relatively dry air that contains little moisture and therefore little latent heat. In Figure 6.10, the wet adiabatic rate is shown as a slightly curving line to indicate that its value increases with altitude.

Clouds

Clouds are frequent features of the atmosphere. Views of the Earth from space show that about half of the Earth is covered by clouds at any given time. In Chapter 2 we noted that low clouds reflect solar energy, thus cooling the Earth-atmosphere system, while high clouds absorb outgoing longwave radiation, thus warming the Earth-atmosphere system. In Chapter 8 we will return to this topic in more detail. In this chapter, however, we are concerned with clouds as the source of precipitation.

A **cloud** is made up of water droplets or ice particles suspended in air. These particles have a diameter in the range of 20 to 50 μm (0.0008 to 0.002 in.). Recall from Chapter 4 that μm denotes the *micrometer*, or one-millionth of a meter. Each cloud particle is formed on a tiny center of solid matter, called a *condensation nucleus*. This nucleus has a diameter in the range 0.1 to 1 μm (0.000004 to 0.00004 in.).

An important source of condensation nuclei is the surface of the sea. When winds create waves, droplets of spray from the crests of the waves are carried rapidly upward in turbulent air. Evaporation of sea water droplets leaves a tiny residue of crystalline salt suspended in the air. This aerosol strongly attracts water molecules. Another source of nuclei is the heavy load of dust carried by polluted air over cities, which substantially aids condensation and the formation of clouds and fog. But even very clean and clear air contains enough condensation nuclei for the formation of clouds.

In our everyday life at the Earth's surface, liquid water turns to ice when the surrounding temperature falls to the freezing point, 0°C (32°F), or below. However, when water is dispersed as tiny droplets in clouds, it can remain in the liquid state at temperatures far below freezing. Such water is described as *supercooled*. Clouds consist entirely of water droplets at temperatures down to about –12°C (10°F). As cloud temperatures grow colder, a mix of water droplets and ice crystals occurs. The coldest clouds, with temperatures below –40°C (–40°F), occur at altitudes of 6 to 12 km (20,000 to 40,000 ft) and are formed entirely of ice particles.

Cloud Forms

Clouds come in many shapes and sizes, from the small, white puffy clouds often seen in summer to the dark layers that produce a good, old-fashioned rainy day. Meteorologists classify clouds into four families, arranged by height: high, middle, and low clouds, and clouds with vertical development. These are shown in Figures 6.11 and 6.12. Some individual types with their names are also shown.

6.11 Cloud families and types *Clouds are grouped into families on the basis of height. Individual cloud types are named according to their form.*

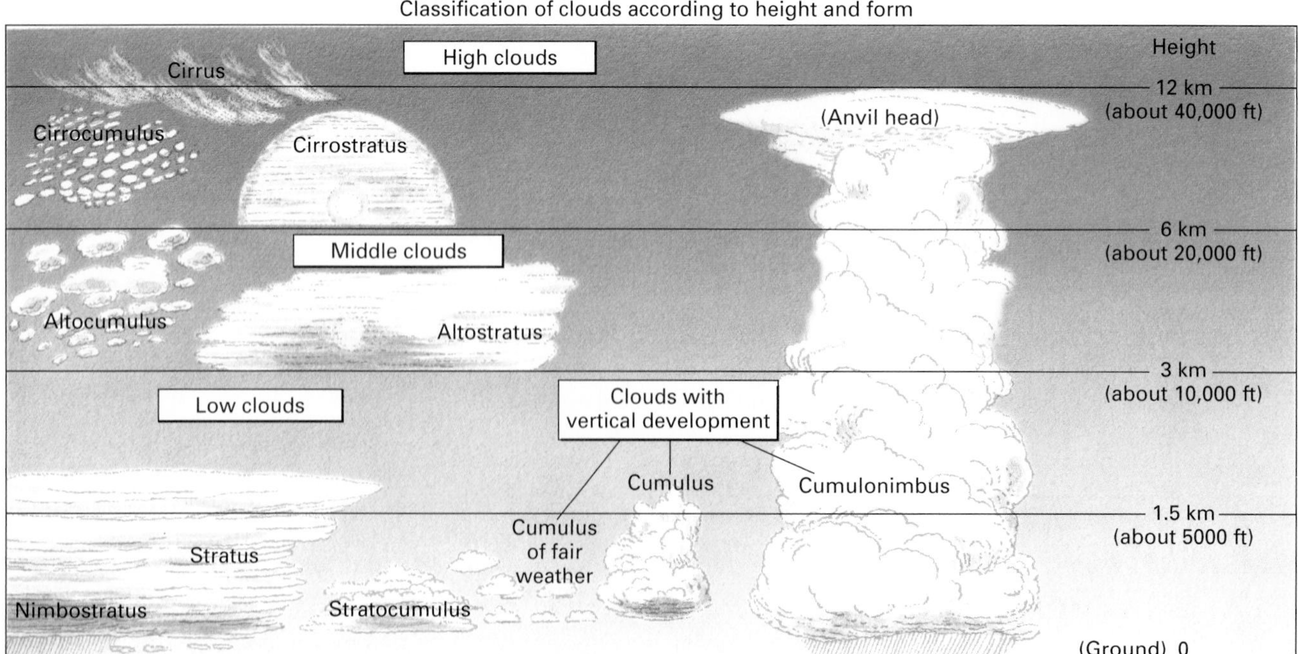

6.12 Cloud photos *(a) Rolls of altocumulus clouds, also known as a "mackerel sky." (b) High cirrus clouds lie above puffy cumulus clouds in this scene from Holland. (c) Fibrous cirrus clouds are drawn out in streaks by high-level winds to form "mare's tails." (d) Upper layer: lenticular clouds, a type of altostratus formed when moist air is uplifted as it crosses a range of hills. Lower layer: cumulus clouds, also triggered by the uplift.*

Clouds are grouped into two major classes on the basis of form—stratiform, or layered clouds, and cumuliform, or globular clouds. *Stratiform* clouds are blanket-like and cover large areas. A common type is *stratus,* which covers the entire sky. Stratus clouds are formed when air layers are forced to rise gradually over large regions. This can happen when one air layer overrides another. As the overriding layer rises and is cooled, condensation occurs over a large area, and a blanket-like cloud forms. If the overriding layer is quite moist and rising continues, dense, thick stratiform clouds result that can produce abundant rain or snow.

Cumuliform clouds are globular masses of cloud that are associated with small to large parcels of rising air. The air parcels rise because they are warmer than the surrounding air. Like bubbles in a fluid, they move upward. Of course, as they are buoyed upward, they are cooled by the adiabatic process. When condensation occurs, a cloud is formed. The most common cloud of this type is the *cumulus* cloud (Figure 6.12). Sometimes the upward movement yields dense, tall clouds that produce thunderstorms. This form of cloud is the *cumulonimbus.* ("Nimbus" is the Latin word for rain cloud, or storm.)

At the scale of an observer on Earth, cloud shapes and forms are visible clues to air motions that are otherwise invisible to the eye. At the continental or global scale, cloud patterns also serve to mark air motions and reveal major features of global air circulation. One of the most important tools for observing cloud patterns remotely is the GOES series of geostationary satellites. *Focus on Remote Sensing 6.4 • Observing Clouds from GOES* shows some examples of GOES images. GEODISCOVERIES

Fog

Fog is simply a cloud layer at or very close to the surface. In our industrialized world, fog is a major environmental hazard. Dense fog on high-speed highways can cause chain-reaction accidents, sometimes involving dozens of vehicles and causing injury or death to their occupants. At airports, landing delays and shutdowns because of fog bring economic losses to airlines and can inconvenience many thousands of travelers in a single day. For centuries, fog at sea has been a navigational hazard, increasing the danger of ship collisions and groundings. In addition, polluted fogs, like those of London in the early part of the twentieth century, can injure urban dwellers' lungs and take a heavy toll in lives.

One type of fog, known as *radiation fog,* is formed at night when the temperature of the air layer at the ground

Focus on Remote Sensing | 6.4

Observing Clouds from GOES

Some of the most familiar images of Earth acquired by satellite instruments are those of the Geostationary Operational Environmental Satellite (GOES) system. Images from the GOES series of satellites, and its pathfinding predecessors, Synchronous Meterological Satellite (SMS) –1 and –2, have been in constant use by meteorologists and weather forecasters since 1974. The primary mission of the GOES series is to view cloud patterns and track weather systems by providing frequent images of the Earth from a consistent viewpoint in space. This capability is of great economic value in forecasting storms and severe weather.

A key feature of the GOES series is its geostationary orbit. Recall from *Geographer's Tools 4.8 • Remote Sensing for Physical Geography* that a satellite in geostationary orbit rotates above the equator in the same direction as the Earth and at the same angular velocity—that is, it remains constantly above the same point on the equator. From this vantage point, an instrument can acquire a constant stream of images of the Earth beneath it. Thus, a geostationary imager is ideal for viewing clouds and tracking cloud patterns.

Altogether, 10 GOES satellites in three major generations have been placed in orbit since 1975. They have maintained a nearly continuous stream of images from two points on the equator, at latitudes 75° W and 135° W, that bracket North America. From 135° W (GOES-West), Pacific storm systems can be tracked as they approach the continent and move across the western states and provinces. From 75° W (GOES-East), weather systems are observed in the eastern part of the continent as they move in from the west. GOES-East also observes the tropical Atlantic, allowing identification of tropical storms and hurricanes as they form and move eastward toward the Caribbean Sea and southeastern United States.

The present generation of GOES platforms carries two primary instruments—the GOES Imager and the GOES Sounder. The imager acquires data in five spectral bands, ranging from the visible red to the thermal infrared. Images of clouds, water vapor, surface temperature, winds, albedo, and fires and smoke are routinely acquired. The sounder uses 19 spectral channels to sound the atmosphere and observe atmospheric profiles of temperature, moisture, ozone, and cloud height and amount.

The image at the right was acquired from the GOES-8 imager at 75° W lat. on September 2, 1994. In this near-noon image, the entire side of the globe nearest to the satellite is illuminated. The red and green primary colors in the image are derived from visible band data. Vegetated areas of the Earth's surface appear green, while semiarid and desert landscapes appear yellow-brown. The blue primary color is derived from a thermal infrared band but is scaled inversely so that cold areas are bright and warm areas are dark. Since clouds are bright white in visible wavelengths

level falls below the dew point. This kind of fog is associated with a low-level temperature inversion (Figure 5.12). Another fog type, *advection fog,* results when a warm, moist air layer moves over a cold surface. As the warm air layer loses heat to the surface, its temperature drops below the dew point, and condensation sets in. Advection fog commonly occurs over oceans where warm and cold currents occur side by side. When warm, moist air above the warm current moves over the cold current, condensation occurs. Fogs of the Grand Banks off Newfoundland are formed in this way because here the cold Labrador current comes in contact with the warmer waters of the Gulf Stream.

Sea fog is frequently found along the California coast. It forms within a cool marine air layer in direct contact with the colder water of the California current (Figure 6.13). Similar fogs are found on continental west coasts in the tropical latitude zones where cool, equatorward currents parallel the shoreline.

Precipitation

Clouds are the source of precipitation—the process that provides the fresh water essential for most forms of terrestrial life. Let's look in more detail at how precipitation is formed inside clouds. Precipitation can form in two ways. In warm clouds, fine water droplets condense, collide, and coalesce into larger and larger droplets that can fall as rain. In colder clouds, ice crystals form and grow in a cloud that contains a mixture of both ice crystals and water droplets.

The first process occurs when saturated air rises rapidly, and cooling forces additional condensation. Cloud droplets grow by added condensation and attain a diameter of 50 to 100 μm (0.002 to 0.004 in.). In collisions with one another, the droplets grow to about 500 μm in diameter (about 0.02 in.). This is the size of water droplets in drizzle. Further collisions and coalescence increase drop size and yield *rain.* Average raindrops have diameters of about 1000 to 2000 μm (about 0.04 to 0.1 in.), but they can reach a maximum diameter of about 7000 μm (about 0.25 in.). Above this value they and are colder than surface features, they appear white. The oceans appear blue because they are dark in the visible bands but are still somewhat warm.

For weather forecasting, an important tool of the latest generation of GOES imagers is the water vapor image. The image at the right shows a global water vapor image for August 11, 2000, constructed from GOES-East and -West images as well as those of two other geostationary satellites (MeteoSat and GMS). The brightest areas show regions of active precipitation—note the many spots of convective shower patches in the equatorial zone. Dark areas show low water vapor content—the desert band stretching across the Middle East to central Asia and the desert regions of southern Africa are examples. The dark lines moving across North America are frontal systems in which cool, dry air is pushing warmer, moister air southward and eastward. Note also dark tongues of dry air spiraling with light tongues of moist air in cyclonic systems in the southern hemisphere (Chapter 7). A bright patch in the Atlantic off the Carolina coast is Hurricane Alberto. The eye of the storm is faintly visible as a dark spot at the center.

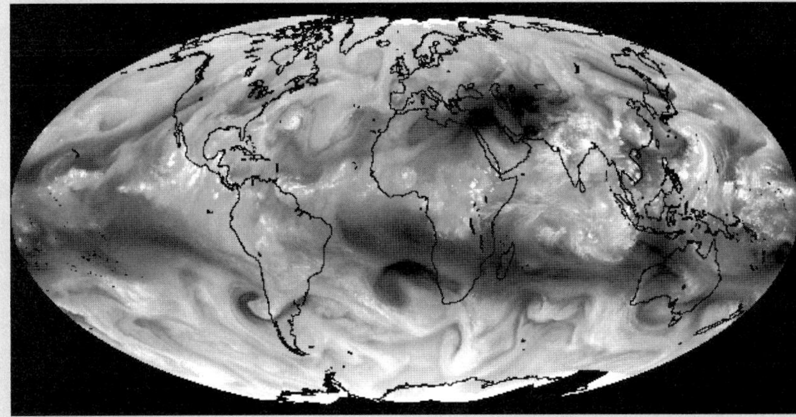

Water vapor composite image *Areas of high atmospheric water vapor content are bright in this image, prepared by merging data from four geostationary satellite instruments. (University of Wisconsin Space Science and Engineering Center.)*

Earth from GOES-8 *The GOES-8 geostationary satellite acquired this image on September 2, 1994. (Courtesy NASA. Image produced by M. Jentoft-Nilsen, F. Hasler, D. Chesters, and T. Nielsen.)*

6.13 **Eye on the Landscape**
Sea fog *A layer of sea fog along the Big Sur coast of California, south of San Francisco. A patch of fog has burned off in the center embayment, revealing the water surface and a pocket beach.* **What else would the geographer see? ... Answers at the end of the chapter.**

***6.14* Snowflakes** *These individual snow crystals, greatly magnified, were selected for their beauty. Most snowflakes are lumps of ice without such a delicate structure.*

become unstable and break into smaller drops while falling. This type of precipitation formation occurs in warm clouds typical of the equatorial and tropical zones.

Snow is produced by the second process in cool clouds that are a mixture of ice crystals and supercooled water droplets. As the ice crystals take up water vapor and grow by deposition, the supercooled water droplets lose water vapor by evaporation and shrink. When an ice crystal collides with a droplet of supercooled water, it induces freezing of the droplet. The ice crystals further coalesce to form snow particles, which can become heavy enough to fall from the cloud. Snowflakes formed entirely by deposition can have intricate crystal structures (Figure 6.14), but most particles of snow have endured collisions and coalescence and are simply tiny lumps of ice.

When the underlying air layer is below freezing, snow reaches the ground as a solid form of precipitation. Otherwise, it melts and arrives as rain. A reverse process, the fall of raindrops through a cold air layer, results in the freezing of rain and produces pellets or grains of ice. These are commonly referred to in North America as *sleet.* (Among the British, sleet refers to a mixture of snow and rain.)

Perhaps you have experienced an *ice storm.* This occurs when the ground is frozen and the lowest air layer is also below freezing. Rain falling through the layer is chilled and freezes onto ground surfaces as a clear glaze. Ice storms cause great damage, especially to telephone and power lines and to tree limbs pulled down by the weight of the ice. In addition, roads and sidewalks are made extremely hazardous by the slippery glaze. Actually, the ice storm is more accurately named an "icing" storm, since it is not ice that is falling but supercooled rain.

Hail, another form of precipitation, consists of lumps of ice ranging from pea- to grapefruit-sized—that is, with a diameter of 5 mm (0.2 in.) or larger. Most hail particles are roughly spherical in shape. The formation of hail will be explained in our discussion of the thunderstorm.

Precipitation is measured in units of depth of fall per unit of time—for example, centimeters or inches per hour or per day. A centimeter (inch) of rainfall would cover the ground to a depth of 1 cm (1 in.) if the water did not run off or sink into the soil. Rainfall is measured with a *rain gauge.* The simplest rain gauge is a straight-sided, flat-bottomed pan, which is set outside before a rainfall event. (An empty coffee can does nicely.) After the rain, the depth of water in the pan is measured.

A very small amount of rainfall, such as 2 mm (0.1 in.), makes too thin a layer to be accurately measured in a flat pan. To avoid this difficulty, a typical rain gauge is constructed from a narrow cylinder with a funnel at the top (Figure 6.15). The funnel gathers rain from a wider area than the

***6.15* Rain gauge** *This rain gauge consists of two clear plastic cylinders, here partly filled with red-tinted water. The inner cylinder receives rainwater from the funnel top. When it fills, the water overflows into the larger, outer cylinder.*

mouth of the cylinder, so the cylinder fills more quickly. The water level gives the amount of precipitation, which is read on a graduated scale.

Snowfall is measured by melting a sample column of snow and reducing it to an equivalent in rainfall. In this way, rainfall and snowfall records may be combined in a single record of precipitation. Ordinarily, a 10-cm (or 10-in.) layer of snow is assumed to be equivalent to 1 cm (or 1 in.) of rainfall, but this ratio may range from 30 to 1 in very loose snow to 2 to 1 in old, partly melted snow.

Precipitation Processes

So far, we have seen how air that is moving upward will be chilled by the adiabatic process to the saturation point and then to condensation, and how eventually precipitation will form. However, one key piece of the precipitation puzzle has been missing—what causes air to move upward? Air can move upward in three ways. First, it can be forced upward as a through-flowing wind. Consider the case of a mass of air moving up and over a mountain range, for example. If the mountain range is high enough and the air is moist enough, then precipitation will occur. This is called **orographic precipitation,** and we'll describe it more fully in the following section.

A second way for air to be forced upward is through *convection.* In this process, a parcel of air is heated, perhaps by a patch of warm ground, so that it is warmer and therefore less dense than the air around it. Like a bubble, it rises. If the air is quite moist and condensation sets in, then the release of latent heat of condensation will ensure that the parcel remains warmer than the surrounding air as it rises. Precipitation that results from this process is **convectional precipitation,** and it will be described after orographic precipitation.

A third way for air to be forced upward is through the movement of air masses. As we will see in Chapter 8, air masses are large bodies of air, each with a set of relatively uniform temperature and moisture properties. Air masses move normally from west to east in midlatitudes, and one can overtake another. When this happens, one of them will be forced aloft. Since this overtaking usually occurs in a common type of storm called a cyclone, this type of precipitation is known as *cyclonic precipitation.* We will treat the subject of cyclonic precipitation in Chapter 8.

Orographic Precipitation

In orographic precipitation, through-flowing winds move moist air up and over a mountain barrier. (The term *orographic* means "related to mountains.") Figure 6.16 shows

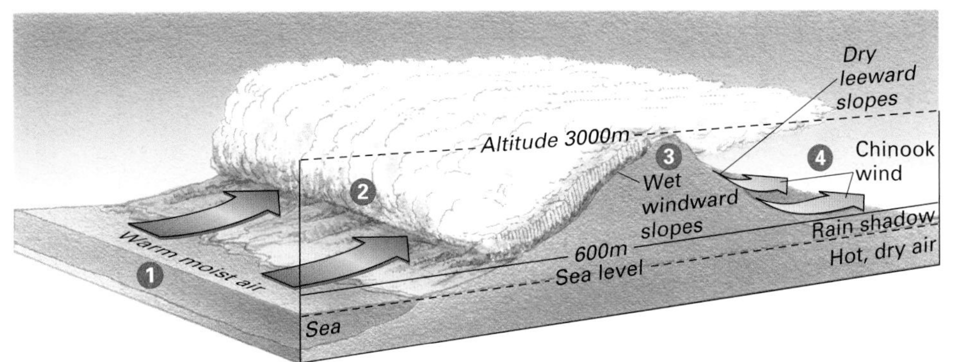

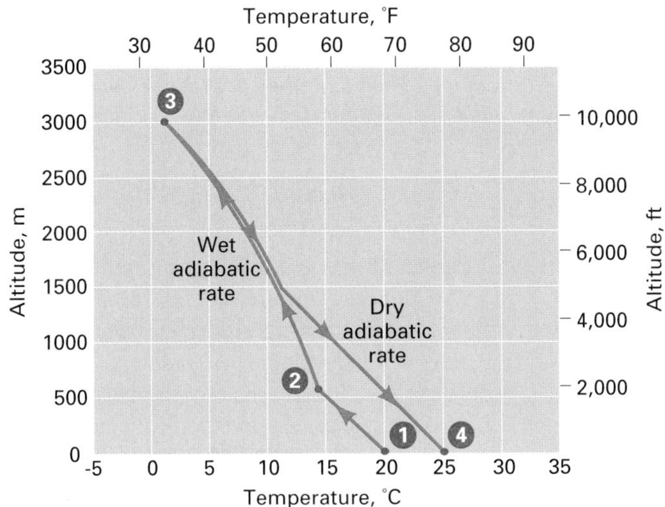

6.16 Orographic precipitation *The forced ascent of a warm, moist oceanic air mass over a mountain barrier produces precipitation and a rain shadow desert. As the air moves up the mountain barrier, it loses moisture through precipitation. As it descends the far slope, it is warmed.*

how this process works. Moist air arrives at the coast after passing over a large ocean surface (1). As the air rises on the windward side of the range, it is cooled at the dry adiabatic rate. When cooling is sufficient, the lifting condensation level is reached, condensation sets in, and clouds begin to form (2). Cooling now proceeds at the wet adiabatic rate. Eventually, precipitation begins. As the air continues up the slope, precipitation continues to fall. GEODISCOVERIES

After passing over the mountain summit, the air begins to descend down the leeward slopes of the range (3). However, as it descends, it is compressed. This means that it is warmed, according to the adiabatic principle. Since cooling is no longer occurring, cloud droplets and ice crystals evaporate or sublimate. Because evaporation and sublimation take up latent heat, at first the warming follows the wet adiabatic rate. Eventually the air clears, and as it continues to descend, it warms further at the dry adiabatic rate. At the base of the mountain on the far side (4), the air is now warmer—and drier, since much of its moisture has been removed in fallen precipitation. This effect creates a *rain shadow* on the far side of the mountain—a belt of dry climate that extends down the leeward slope and beyond. Several of the Earth's great deserts are of this type.

When strong and persistent, the descending flow of hot, dry air is termed a *chinook* wind. Channeled into valleys on the leeward side, it can raise local temperatures very rapidly. The dry air has great evaporating ability, and it can rapidly sublimate a snow cover in winter or dry a brush cover to tinder in summer. The chinook wind is one type of local wind, and we will discuss several other local wind types in Chapter 7.

California's rainfall patterns provide an excellent example of orographic precipitation and the rain shadow effect. A map of California (Figure 6.17) shows mean annual precipitation. It uses lines of equal precipitation, called *isohyets.* Focus now on central California and follow the arrow on the map. Prevailing westerly winds bring moist air in from the Pacific Ocean, first over the Coast Ranges of central and northern California, and then, after rain is deposited on the Coast Ranges, the air descends into the broad Central Valley. Here, precipitation is low, averaging less than 25 cm (10 in.) per year.

Next the air continues up and over the great Sierra Nevada, whose summits rise to 4200 m (about 14,000 ft) above sea level. Heavy precipitation, largely in the form of winter snow, falls on the western slopes of these ranges and nourishes rich forests. Passing down the steep eastern face of the Sierra Nevada, the air descends quickly into the Owens Valley, at about 1000 m (3300 ft) elevation. The adiabatic heating warms and dries the air, creating a rain shadow desert. In the Owens Valley, annual precipitation is less than 10 cm (4 in.). In this way, the orographic effect on air moving across the mountains of California produces a part of America's great interior desert zone, extending from eastern California and across Nevada.

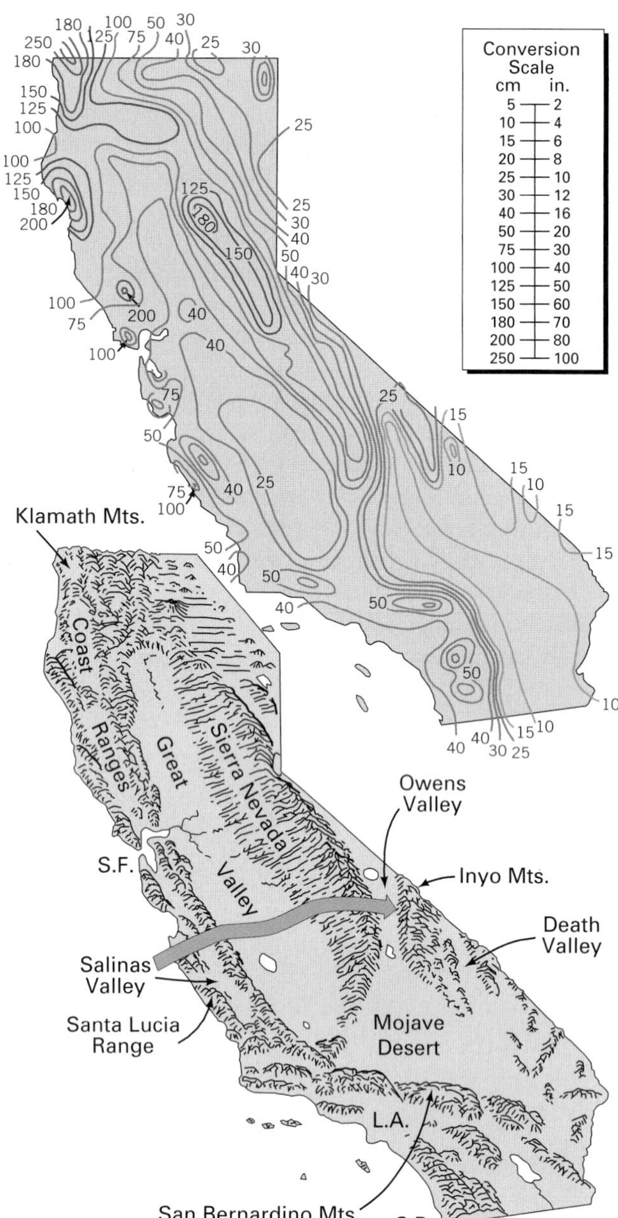

6.17 Orographic precipitation effects in California
The effect of mountain ranges on precipitation is strong in the state of California because of the prevailing flow of moist oceanic air from west to east. Centers of high precipitation coincide with the western slopes of mountain ranges, including the coast ranges and Sierra Nevada. To the east, in their rain shadows, lie desert regions.

Convectional Precipitation

The second process of inducing uplift in a parcel of air is *convection,* the upward motion of a parcel of heated air. In this process, strong updrafts occur within *convection cells*—vertical columns of rising air that are often found above warm land surfaces. Air rises in a convection cell because it is warmer, and therefore less dense, than the surrounding air. GEODISCOVERIES

The convection process starts when a surface is heated unequally. For example, consider an agricultural field surrounded by a forest. Since the field surface consists largely of bare soil and only a low layer of vegetation, it will be warmer under steady sunshine than the adjacent forest. This means that as the day progresses, the air above the field will grow warmer than the air above the forest.

Now, the density of air depends on its temperature—warm air is less dense than cooler air. The hot-air balloon operates on this principle. The balloon is open at the bottom, and in the basket below a large gas burner forces heated air into the balloon. Because the heated air is less dense than the surrounding air, the balloon rises. The same principle will cause a bubble of air to form over the field, rise, and break free from the surface. Figure 6.18 diagrams this process.

As the bubble of air rises, it is cooled adiabatically and its temperature will decrease as it rises. However, we know that the temperature of the surrounding air will normally decrease with altitude as well. Nonetheless, as long as the bubble is still warmer than the surrounding air, it will be less dense and will therefore continue to rise.

If the bubble remains warmer than the surrounding air and uplift continues, adiabatic cooling chills the bubble below the dew point. Condensation occurs, and the rising air column becomes a puffy cumulus cloud. The flat base of the cloud marks the lifting condensation level at which condensation begins. The bulging "cauliflower" top of the cloud is the top of the rising warm-air column, pushing into higher levels of the atmosphere. Normally, the small cumulus cloud will encounter winds aloft that mix it into the local air. After drifting some distance downwind, the cloud evaporates.

UNSTABLE AIR Sometimes, however, convection continues strongly, and the cloud develops into a dense cumulonimbus mass, or thunderstorm, from which heavy rain will fall. Two environmental conditions encourage the development of thunderstorms: (1) air that is very warm and moist, and (2) an environmental temperature lapse rate in which temperature decreases more rapidly with altitude than it does for either the dry or wet adiabatic lapse rates. Figure 6.19, which diagrams the convection process in unstable air, shows this type of lapse rate as having a flatter slope than the dry or wet adiabatic rates. Air with these characteristics is referred to as **unstable air.**

For the first condition, recall that there are two adiabatic rates—dry and wet (Figure 6.10). The wet rate is about half the dry rate. While condensation is occurring, the lesser wet rate applies. Thus, air in which condensation is occurring cools less rapidly with uplift. Furthermore, the wet rate is smallest for warm, moist air. This means that the temperature decrease experienced by warm, moist, rising, condensing air is quite small. And since the temperature decrease is small, the rising air is more likely to stay warmer than the surrounding air. Thus, uplift continues.

With regard to the second condition—an environmental temperature lapse rate value that exceeds the adiabatic rates—keep in mind that this rate describes the temperature of the surrounding, still air. Therefore, the temperature of the still air falls quite rapidly with altitude. This means that the condensing air in a rising parcel will stay warmer than the surrounding air. Again, uplift continues.

A simple example may make convection in unstable air clearer. Figure 6.19 shows how temperature changes with altitude for a parcel moving upward by convection in unstable air. The surrounding air is at a temperature of 26°C (78.8°F) at ground level and has a lapse rate of 12°C/1000 m (6.6°F/1000 ft). A parcel of air is heated by 1°C (1.8°F) to 27°C (80.6°F), and it begins to rise. At first, it cools at the dry adiabatic rate. At 500 m (1640 ft), the parcel is at 22°C (71.6°F), while the surrounding air is at 20°C (68°F). Since it is still warmer than the surrounding air, uplift continues. At 1000 m, the lifting condensation level is reached. The temperature of the parcel is 17°C (62.6°F).

As the parcel rises above the condensation level, the wet adiabatic rate applies, here using the value 5°C/1000 m (2.7°F/1000 ft). Now the parcel cools more slowly as it

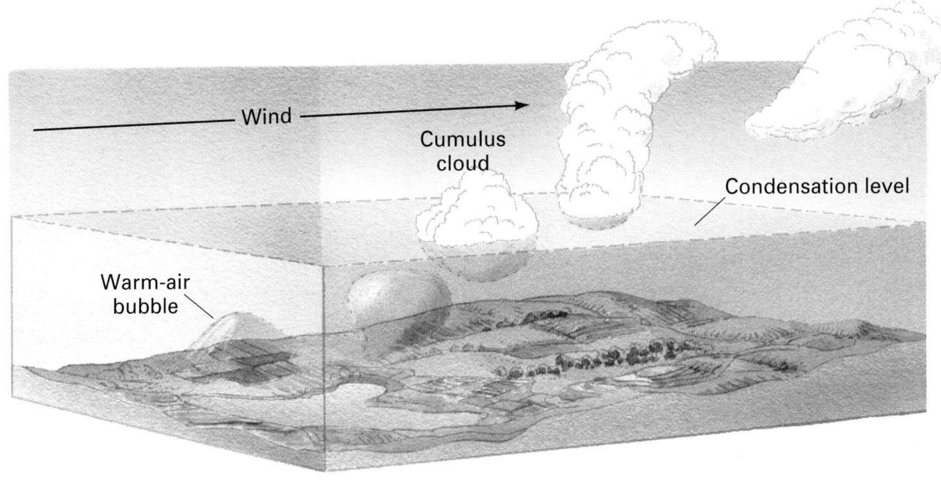

6.18 Formation of a cumulus cloud *A bubble of heated air rises above the lifting condensation level to form a cumulus cloud.*

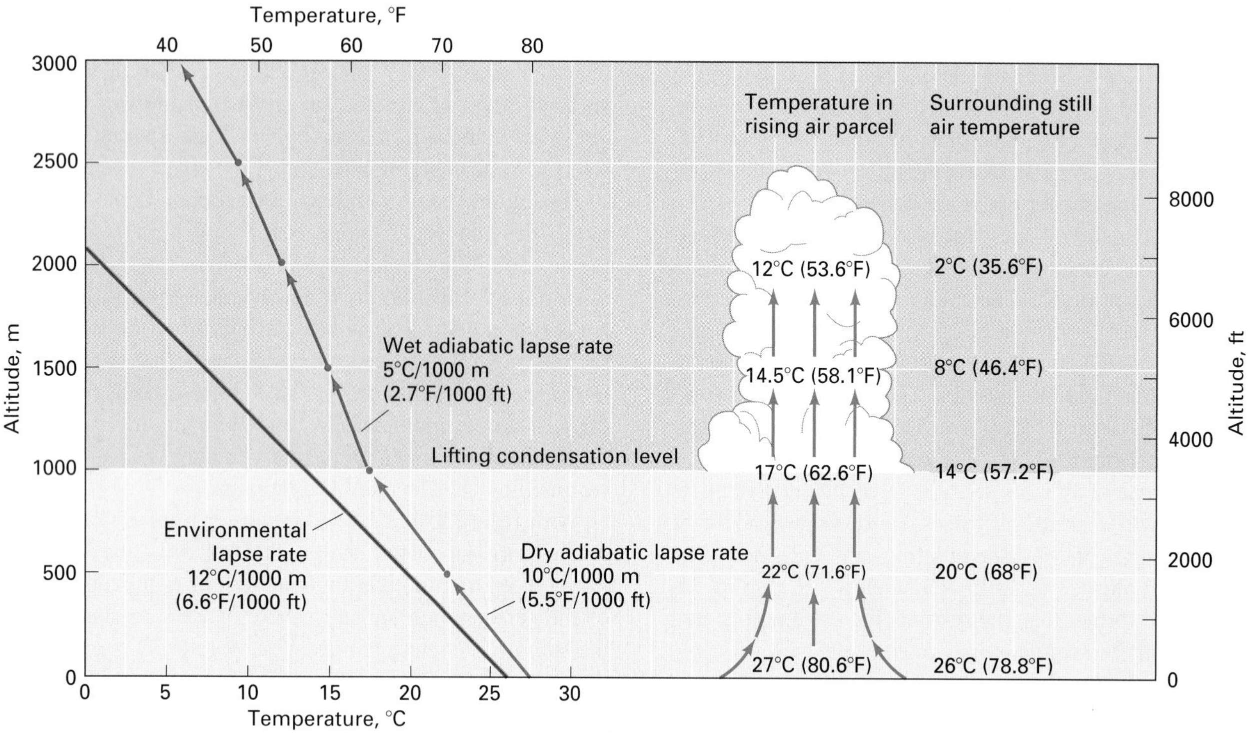

6.19 Convection in unstable air *When the air is unstable, a parcel of air that is heated sufficiently to rise will continue to rise to great heights.*

rises. At 1500 m (4920 ft), the parcel is 14.5°C (58.1°F), while the surrounding air is 8°C (46.4°F). Since the parcel is still warmer than the surrounding air, it continues to rise. Note that the difference in temperature between the rising parcel and the surrounding air now actually increases with altitude. This means that the parcel will be buoyed upward ever more strongly, forcing even more condensation and precipitation.

The key to the convectional precipitation process is latent heat. When water vapor condenses into cloud droplets or ice particles, it releases latent heat to the rising air parcel. By keeping the parcel warmer than the surrounding air, this latent heat fuels the convection process, driving the parcel ever higher. When the parcel reaches a high altitude, most of its water will have condensed. As adiabatic cooling continues, less latent heat will be released. As a result, the uplift will weaken. Eventually, uplift will stop, since the energy source, latent heat, is gone. The cell dies and dissipates into the surrounding air.

Unstable air is typical of summer air masses in the central and southeastern United States. As we will see later, summer weather patterns sweep warm, humid air from the Gulf of Mexico over the continent. Over a period of days, the intense summer insolation strongly heats the air layer near the ground, producing a steep environmental lapse rate. Thus, both of the conditions that create unstable air are present. As a result, thunderstorms are very common in these regions during the summer.

Unstable air is also likely to be found in the vast, warm and humid regions of the equatorial and tropical zones. In these regions, convective showers and thundershowers are common. At low latitudes, much of the orographic rainfall is actually in the form of heavy showers and thundershowers produced by convection. In this case, the forced ascent of unstable air up a mountain slope easily produces rapid condensation, which then triggers the convection process.

Thunderstorms

The thunderstorm is an awesome event (Figure 6.20). Intense rain and violent winds, coupled with lightning and thunder, renew our respect for nature's power. What is the anatomy of a thunderstorm?

A **thunderstorm** is an intense local storm associated with a tall, dense cumulonimbus cloud in which there are very strong updrafts of air. A single thunderstorm typically consists of several individual convection cells. A single convection cell is diagrammed in Figure 6.20. Air rises within the cell as a succession of bubble-like air parcels. Intense adiabatic cooling within the bubbles produces precipitation. It can be in the form of water at the lower levels, mixed water and snow at intermediate levels, and snow at high levels where cloud temperatures are coldest.

As the rising air parcels reach high levels, which may be 6 to 12 km (about 20,000 to 40,000 ft) or even higher, the rising rate slows. Strong winds typically present at such high altitudes drag the cloud top downwind, giving the thunderstorm cloud its distinctive shape—resembling an old-fashioned blacksmith's anvil.

6.20 Thunderstorms

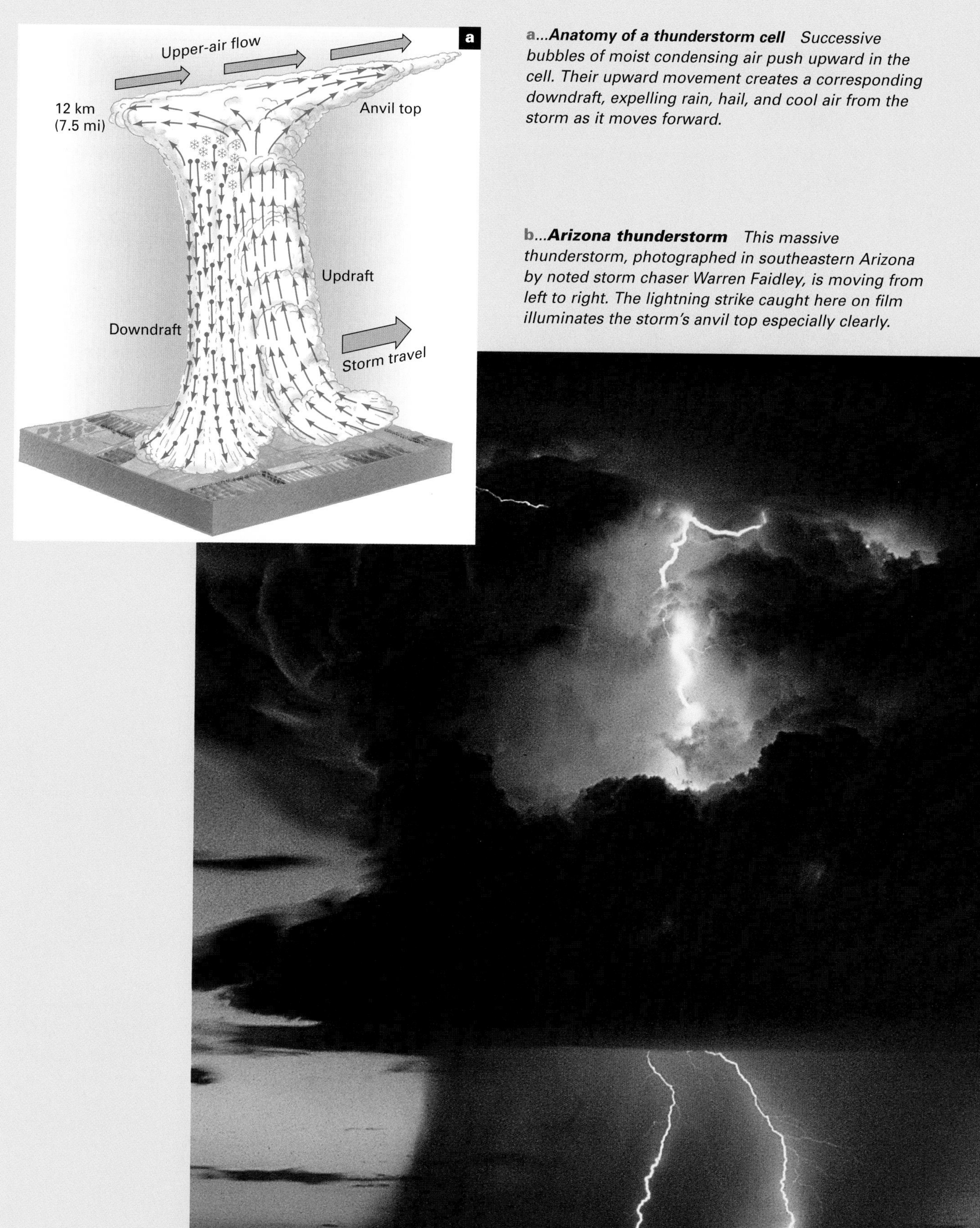

a...*Anatomy of a thunderstorm cell* *Successive bubbles of moist condensing air push upward in the cell. Their upward movement creates a corresponding downdraft, expelling rain, hail, and cool air from the storm as it moves forward.*

b...*Arizona thunderstorm* *This massive thunderstorm, photographed in southeastern Arizona by noted storm chaser Warren Faidley, is moving from left to right. The lightning strike caught here on film illuminates the storm's anvil top especially clearly.*

6.21 Hailstones *Fresh hailstones, up to about 1 cm (0.5 in.) in diameter.*

Ice particles falling from the cloud top act as nuclei for freezing and deposition at lower levels. Large ice crystals form and begin to sink rapidly. Melting, they coalesce into large, falling droplets. The rapid fall of raindrops adjacent to the rising air bubbles pulls the air downward and feeds a downdraft within the convection cell. Emerging from the cloud base, the downdraft of cool air, laden with precipitation, approaches the surface and spreads out in all directions. Part of the downdraft moves forward, causing more warm, moist surface air to rise and enter the updraft portion of the storm. This effect helps perpetuate the storm. The downdraft creates strong local winds. Wind gusts can sometimes be violent enough to topple trees and raise the roofs of weak buildings.

In addition to powerful wind gusts and heavy rains, thunderstorms can produce hail. Hailstones (Figure 6.21) are formed by the accumulation of ice layers on ice pellets that are suspended in the strong updrafts of the thunderstorm. In this process, they can reach diameters of 3 to 5 cm (1.2 to 2.0 in.). When they become too heavy for the updraft to support, they fall to Earth. Figure 6.22 shows a hailstorm in progress.

6.22 Eye on the Landscape Hailstorm in progress, Santa Cruz County, Arizona *The white streaks in this photo are marble-sized hailstones. Although the hail is falling from a cloud directly overhead, the low afternoon Sun still illuminates the foreground scene.* **What else would the geographer see?...Answers at the end of the chapter.**

A

B

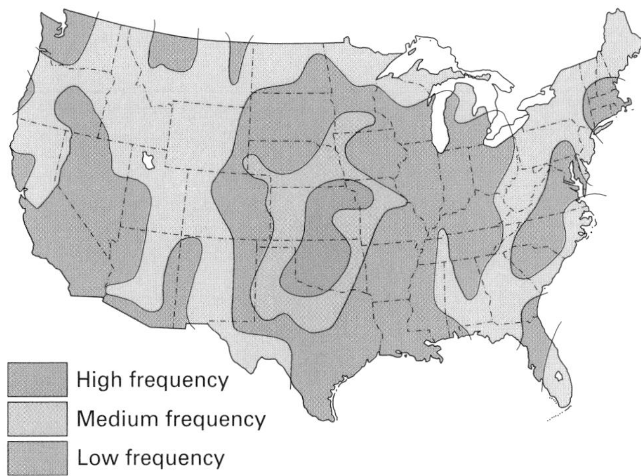

6.23 Frequency of severe hailstorms *As shown in this map of the 48 contiguous United States, severe hailstorms are most frequent in the midwestern plains states of Oklahoma and Kansas. A severe hailstorm is defined as a local convective storm producing hailstones equal to or greater than 1.9 cm (0.75 in.) in diameter. (From R. H. Skaggs, Proc. Assoc. American Geographers, vol. 6, Figure 2. Used by permission.)*

Annual losses from crop destruction caused by hailstorms amount to several hundred million dollars. Damage to wheat and corn crops is particularly severe in the Great Plains, running through Nebraska, Kansas, Missouri, Oklahoma, and northern Texas (Figure 6.23).

Another effect of convection cell activity is to generate lightning. This phenomenon occurs when updrafts and downdrafts cause positive and negative static charges to accumulate within different regions of the cloud. Lightning is a great electric arc—a series of gigantic sparks—passing between differently charged parts of the cloud mass or between cloud and ground (see Figure 6.20). During a lightning discharge, a current of as much as 60,000 to 100,000 amperes may develop. This current heats the air intensely, which makes it expand very rapidly—much like an explosion. This expansion sends out sound waves, which we recognize as a thunderclap. Most lightning discharges occur within the cloud, but a significant proportion strike land. In the United States, lightning causes a yearly average of about 150 human deaths and property damage of hundreds of millions of dollars, including loss by structural and forest fires set by lightning.

The thunderstorm is a good example of a flow system of matter and energy—one that is a bit more complicated than our previous examples. *Focus on Systems 6.5 • The Thunderstorm as a Flow System* views the thunderstorm from the systems perspective.

Microbursts

The downdraft that accompanies a thunderstorm can sometimes be very intense—so intense that it is capable of causing low-flying aircraft to crash. This type of intense downdraft is called a *microburst.* A diagram of a microburst is shown in Figure 6.24. The downward-moving air flows outward in all directions. It is often, but not always, accompanied by rain.

An aircraft flying through the microburst first encounters strong headwinds. They may cause a bumpy ride but do not interfere with the airplane's ability to fly. However, as it passes through the far side of the microburst, the aircraft encounters a strong tailwind. The lift of the airplane's wings depends on the speed of the air flowing across them, and the tailwind greatly reduces the air speed. This causes a loss of lift. If the tailwind is strong enough, the airplane cannot hold its altitude and may crash. Microbursts can be detected by special radar instruments that measure horizontal wind speeds. They are quite expensive, however, and are not installed at all airports.

Training procedures for pilots have reduced the incidence of aviation accidents in the United States attributed to microbursts and associated wind shear. However, in 1994 a jet aircraft attempting to land at Charlotte, North Carolina, crashed during a violent thunderstorm, killing 37 and injuring 20 others. A microburst was detected just as the airplane approached the airport, but the warning was not broadcast to the pilots on the radio channel they were following. Although they attempted to abort the landing, a pilot error occurred, and their aircraft was brought down by the severe tailwind of the microburst.

6.24 Anatomy of a microburst *(above) Schematic cross section showing a downdraft reaching the ground and producing a horizontal outflow. (below) Schematic profile of an aircraft taking off through a microburst, leading to a loss of lift and eventual crash. (Adapted from diagrams by Research Applications Program, National Center for Atmospheric Research, Boulder, Colorado.)*

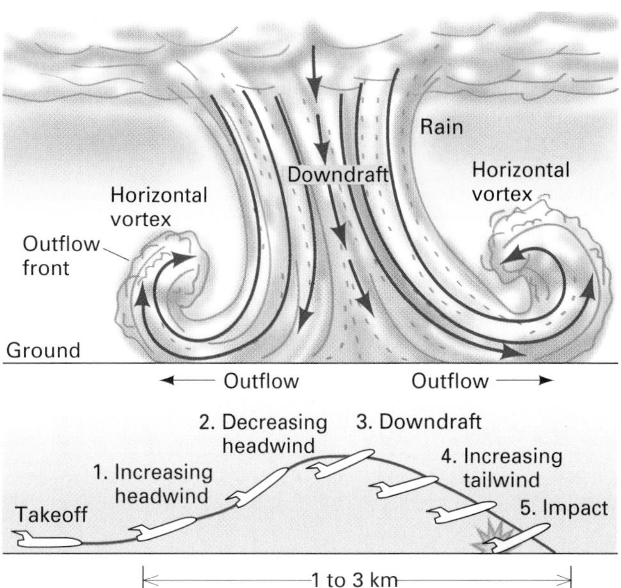

Geographers at Work | Bad Air Days and Other Mysteries

by Andrew C. Comrie ***University of Arizona***

While stratospheric ozone is a vital molecule that absorbs damaging solar ultraviolet radiation, it is a serious air pollutant when it occurs at low levels in the troposphere. Ozone is usually formed indirectly as a result of photochemical reactions in polluted air. It causes eyes to burn and tear, and is damaging to the lungs.

How and when does ozone form at a specific location? How is ozone formation related to local winds and turbulence? How is ozone, produced in cities and industrial areas, conveyed to rural areas? Which regions in a metropolitan area are likely to be threatened by high ozone levels and when?

These important questions are the research focus of Andrew Comrie at the University of Arizona. He has developed models to forecast daily pollution levels in urban environments; assessed the impact of ozone on the health of forests and modeled the sources and movement of polluted air into forested areas; and sampled turbulent air layers caused by sea breezes in Cape Town, South Africa, to assess the impact of a potential nuclear power plant leak.

Not content with such important achievements, Andrew has developed a climatology of the North American monsoon, which produces convectional rainfall in summer over the deserts of the American Southwest. He has also worked to develop the connection between outbreaks of valley fever, caused by a soil-dwelling fungus, and temperature, rainfall, and season.

Curious about how Andrew carries out his research and how he is helped by students? Read more about the work of this geographer-climatologist on our web site.

Andrew Comrie works to understand bad air days and other mysteries of the relationship between air quality and climate.

AIR QUALITY

One of the more important impacts of human activity on our planet is air pollution. Industrial processes and fossil fuel combustion release large quantities of unwanted and unhealthful substances into the air. Air quality is of great concern, not only for the health of ecosystems, but also for human health and comfort. Our interchapter feature, *A Closer Look: Eye on the Environment 6.6 • Air Pollution* provides an overview of air pollution, identifying the most important pollutants and describing some climatic effects of air pollution.

A LOOK AHEAD Precipitation has been the major focus of this chapter. Keep in mind that no matter what the cause of precipitation, latent heat is always released. Because a significant amount of oceanic evaporation results in precipitation over land, there is a very significant flow of latent heat from oceans to land. And, as we will see in the following chapter, global air circulation patterns move this latent heat poleward, helping to warm the continents and creating the distinctive pattern of climates that differentiates our planet.

Focus on Systems | 6.5

The Thunderstorm as a Flow System

A thunderstorm is a dramatic and spectacular example of a system in which matter, in the form of air and water, moves from one location to another under the influence of a coupled flow of energy. Let's look at the energy and matter flows in a thunderstorm in more detail. A diagram of these flows is below, with part *(a)* showing the structure of the storm, similar to Figure 6.20; part *(b)* showing how water flows and changes state within the storm; and part *(c)* diagramming the flow of energy within the storm.

Examining the flow of water in the thunderstorm system *(b)*, we find that cells of warm moist air rise in the forward part of the storm. They serve to bring in water vapor at low altitude, which condenses to liquid and solid forms as it moves upward. At the top of the cloud, some ice particles are carried away in throughflowing winds at high altitude. The remainder are joined by water droplets and descend on the rearward side of the storm, striking the ground as rain, hail, or even snow. Thus, the water flow forms a loop upward as vapor, then downward as precipitation.

Part *(c)*, as noted, shows the flow of energy within the storm. As the convection cells within the storm cloud move moist air aloft, condensation converts latent heat to sensible heat, enhancing the convective uplift. At the top of the storm, high-level winds carry a portion of the sensible heat forward and away from the storm. As liquid and solid water particles are carried *downward* on the trailing side of the storm, some evaporation and sublimation occur, shifting a portion of the sensible heat released to latent heat. At the surface, the net effect is that a larger quantity of latent heat is lost than gained, resulting in generally cooler air. Aloft, the air remains warmer than before the storm, owing to the gain in sensible heat.

The power source for the storm is the latent heat in the water vapor that moves aloft. By changing from the vapor to liquid or solid state, the vapor releases the energy that powers the storm. Some of that energy is recovered when solid or liquid water is converted to vapor in the downdraft cell, but a larger portion is dissipated at higher altitudes. Thus, the storm acts like a heat pump, moving heat from the surface to upper altitudes. Of course, the ultimate power source for the storm is really the Sun, which heated the surface and powered the evaporation that made the air moist to begin with.

The thunderstorm as a flow system of matter and energy

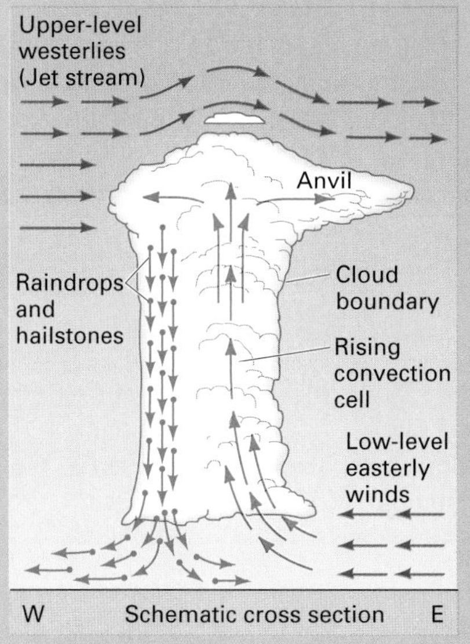

(a)

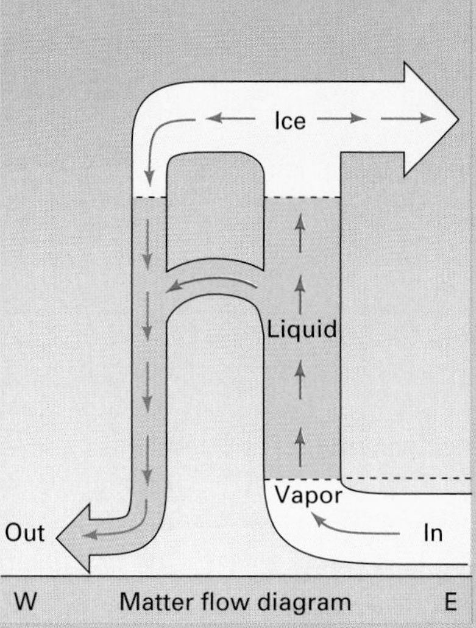

(b)

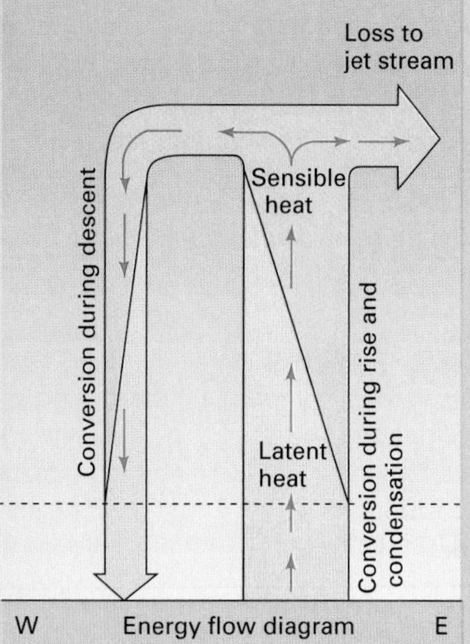

(c)

Chapter Summary

- **Precipitation** is the fall of liquid or solid water from the atmosphere to reach the Earth's land or ocean surface. Evaporating, condensing, melting, freezing, **sublimation**, and **deposition** describe the changes of state of water.
- Water moves freely between ocean, atmosphere, and land in the **hydrologic cycle**. The global water balance describes these flows. The fresh water in the atmosphere and on land in lakes, streams, rivers, and *ground water* is only a very small portion of the total water in the **hydrosphere**.
- **Humidity** describes the amount of water vapor present in air. The ability of air to hold water vapor depends on temperature. Warm air can hold much more water vapor than cold air.
- **Specific humidity** measures the mass of water vapor in a mass of air, in grams of water vapor per kilogram of air. **Relative humidity** measures water vapor in the air as the percentage of the maximum amount of water vapor that can be held at the given air temperature. The **dew-point temperature**, at which condensation occurs, also measures the moisture content of air.
- The *adiabatic principle* states that when a gas is compressed, it warms, and when a gas expands, it cools. When an air parcel moves upward in the atmosphere, it encounters a lower pressure and so expands and cools.
- The **dry adiabatic rate** describes the rate of cooling with altitude. If the air is cooled below the dew point, condensation or deposition occurs, and latent heat is released. This heat reduces the parcel's rate of cooling with altitude. When condensation or deposition is occurring, the cooling rate is described as the **wet adiabatic rate**.
- Clouds are composed of droplets of water or crystals of ice that form on *condensation nuclei*. Clouds typically occur in layers, as *stratiform* clouds, or in globular masses, as *cumuliform* clouds. *Fog* occurs when a cloud forms at ground level.
- Precipitation from clouds occurs as *rain, hail, snow,* and *sleet*. When *supercooled* rain falls on a surface below freezing, it produces an *ice storm*.
- There are three types of precipitation processes—**orographic, convectional,** and *cyclonic*. In orographic precipitation, air moves up and over a mountain barrier. As it moves up, it is cooled adiabatically and rain forms. As it descends the far side of the mountain, it is warmed, producing a *rain shadow* effect.
- In convectional precipitation, unequal heating of the surface causes an air parcel to become warmer and less dense than the surrounding air. Because it is less dense, it rises. As it moves upward, it cools, and condensation with precipitation may occur. Under conditions of **unstable air, thunderstorms** can form, yielding hail and lightning. (Cyclonic precipitation is described in Chapter 8.)

A Closer Look:
Eye on the Environment 6.6 • Air Pollution

- *Air pollution* is defined as unwanted gases, aerosols, and particulates injected into the air by human and natural activity. Polluting *gases, aerosols,* and *particulates* are generated largely by fuel combustion. **Smog,** a common form of air pollution, contains nitrogen oxides, volatile organic compounds, and ozone. *Inversions* can trap smog and other pollutant mixtures in a layer close to the ground, creating unhealthy air. *Acid deposition* of sulfate and nitrate particles can acidify soils and lakes, causing fish and tree mortality.

Key Terms

precipitation
sublimation
deposition
hydrologic cycle
humidity
specific humidity
dew-point temperature
relative humidity
adiabatic process
dry adiabatic lapse rate
wet adiabatic lapse rate
cloud
orographic precipitation
convectional precipitation
unstable air
thunderstorm
smog

Review Questions

1. Identify the three states of water and the six terms used to describe possible changes of state.
2. What is the hydrosphere? Where is water found on our planet? in what amounts? How does water move in the hydrologic cycle?
3. Define specific humidity. How is the moisture content of air influenced by air temperature?
4. Define relative humidity. How is relative humidity measured? Sketch a graph showing relative humidity and temperature through a 24-hour cycle.
5. Use the terms *saturation, dew point,* and *condensation* in describing what happens when an air parcel of moist air is chilled.
6. What is the adiabatic process? Why is it important?

7. Distinguish between dry and wet adiabatic lapse rates. In a parcel of air moving upward in the atmosphere, when do they apply? Why is the wet adiabatic lapse rate less than the dry adiabatic rate? Why is the wet adiabatic rate variable in amount?
8. How are clouds classified? Name four cloud families, two broad types of cloud forms, and three specific cloud types.
9. What is fog? Explain how radiation fog and advection fog form.
10. How is precipitation formed? Describe the process for warm and cool clouds.
11. Describe the orographic precipitation process. What is a rain shadow? Provide an example of the rain shadow effect.
12. What is unstable air? What are its characteristics?
13. Describe the convectional precipitation process. What is the energy source that powers this source of precipitation? Explain.

Focus on Systems 6.2 • The Global Water Balance as a Matter Flow System

1. Sketch a diagram showing five pathways of matter flow for water in the global water balance flow system. Indicate their magnitudes on the diagram.
2. Suppose that global climate warming increases evaporation over land and oceans. Provided that the cycle remains in balance (that is, $P_{LAND} + E_{LAND} + P_{OCEAN} + E_{OCEAN} = 0$), what will be the effect on precipitation?
3. At present, oceans cover 71 percent of the Earth's surface. Suppose this value were 50 percent. How do you think that the global water balance might be affected?

Focus on Systems 6.5 • The Thunderstorm as a Flow System

1. Compare and contrast the energy and matter flows in updraft cells and downdraft cells within a convective storm.
2. How would the convective precipitation process be affected by stable surrounding air?

A Closer Look:
Eye on the Environment 6.6 • Air Pollution

1. What are the most abundant air pollutants, and what are their sources?
2. What is smog? What important pollutant forms within smog, and how does this happen?
3. Distinguish between low-level and high-level inversions. How are they formed? What is their effect on air pollution? Give an example of a pollution situation of each type.
4. To what type of pollution does the term *acid deposition* refer? What are its causes? What are its effects?

Visualizing Exercises

1. Lay out a simple diagram showing the main features of the hydrologic cycle. Include flows of water connecting land, ocean, and atmosphere. Label the flow paths.
2. Graph the temperature of a parcel of air as it moves up and over a mountain barrier, producing precipitation.
3. Sketch the anatomy of a thunderstorm cell. Show rising bubbles of air, updraft, downdraft, precipitation, and other features.

Essay Questions

1. Water in the atmosphere is a very important topic for understanding weather and climate. Organize an essay or oral presentation on this topic, focusing on the following questions: What part of the global supply of water is atmospheric? Why is it important? What is its global role? How does the capacity of air to hold water vapor vary? How is the moisture content of air measured? Clouds and fog visibly demonstrate the presence of atmospheric water. What are they? How do they form?
2. Compare and contrast orographic and convectional precipitation. Begin with a discussion of the adiabatic process and the generation of precipitation within clouds. Then compare the two processes, paying special attention to the conditions that create uplift. Can convectional precipitation occur in an orographic situation? under what conditions?

Problems

Working It Out 6.1 • Energy and Latent Heat

1. Figure 6.3 shows that about 490 km^3 of water are evaporated from land and water surfaces each year. If all this evaporation occurs from a water surface at 15°C, how much energy is required? (1 m^3 of water has a mass of about 10^3 kg.)
2. U.S. energy consumption is about 8.5×10^{13} kJ/yr. How does this value compare with the annual energy of evaporation computed in problem 1?

Working It Out 6.3 • The Lifting Condensation Level

1. What is the lifting condensation level for a parcel of air at an initial temperature of 25°C and a dew-point temperature of 18°? What will be the temperature of the air parcel at that level?
2. Suppose that the same parcel of air is warmed to 30°C before it begins its ascent. What will be the lifting condensation level for the warmed parcel? What will be its temperature at that level?
3. Using algebra, show how expression (2) is derived from expression (1).

Eye on the Landscape

Chapter Opener Amazonian rainforest The structure of the rainforest canopy is readily visible in this striking photo. The diversity of this forest type is shown in the many different colors and forms of the tree crowns in view. Also interesting is the variety in height, with the crowns of some species clearly emerging from above the general level **(A)**, while the crowns of the others form a nearly closed canopy underneath. Holes in the canopy, where sunlight penetrates to the ground, are also visible. The equatorial rainforest is a subtype of the forest biome (Chapter 24).

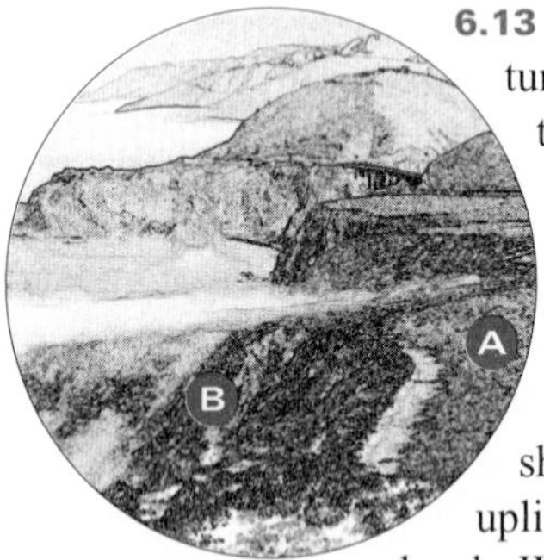

6.13 Sea fog An interesting feature of this dramatic seascape is the presence of gently sloping terrace surfaces **(A)**, used as hay fields, that lie above the marine cliffs. These terrace surfaces were formed by wave abrasion in the nearshore zone and have now been uplifted to well above present sea level. Wave abrasion is covered in Chapter 19; Figure 19.4 shows an abrasion platform that is currently being formed. The uplift is the result of tectonic activity just off the coast, where the Pacific crustal plate and the American plate are moving in different directions (see Chapter 13 for plate tectonics). Another interesting feature of this photo is the gullying and rapid erosion of the cliff **(B)**. The coast ranges of Big Sur receive abundant precipitation, as moist, eastward-moving air is raised over the mountain barrier. (See Orographic Precipitation, this chapter.) The soft sediments are easily eroded by heavy rains, keeping vegetation from stabilizing the slope. Loose sediment falls to the shoreline and is carried offshore by coastal currents. Erosion and gullying are subjects of Chapter 17; Figure 17.4 shows gullies developing on a flat landscape. Coastal erosion is covered in Chapter 19.

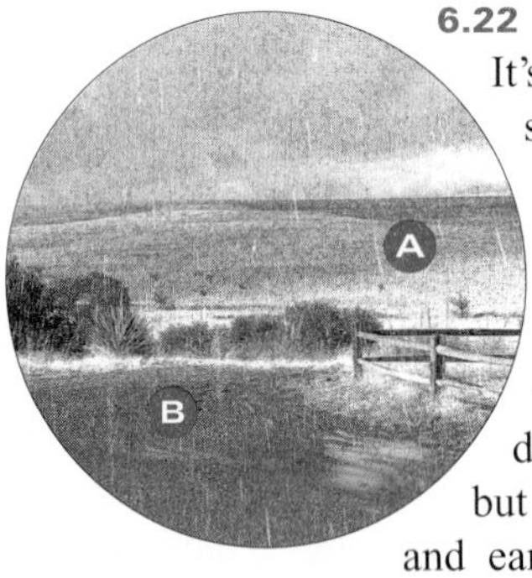

6.22 Hailstorm in progress It's easy to tell from this landscape that it's the dry season. The dry, brown grass has gone dormant **(A)**, leaving only green agaves and tough evergreen oaks to punctuate the sea of amber. The thundershower seems out of place, but after a nearly rainless spring and early summer, Arizona experiences a local monsoon in July and August as moister air brings convectional precipitation to the region. The soil in the road **(B)** is an aridisol—a desert soil of light reddish to gray color. Although lacking in organic matter, it can be quite productive if irrigated. Semiarid grasslands occur in the steppe climate, explained in Chapters 10 and 11. Soils are the subject of Chapter 21.

A CLOSER LOOK

Eye on the Environment 6.6 Air Pollution

Most people living in or near urban areas have experienced air pollution first hand. Perhaps you've felt your eyes sting or your throat tickle as you drive an urban freeway. Or you've noticed black dust on window sills or window screens and realized that you are breathing in that dust as well. Air pollution is an atmospheric phenomenon that is largely the result of human activity. In this *Eye on the Environment* feature, we provide a short introduction to air pollution.

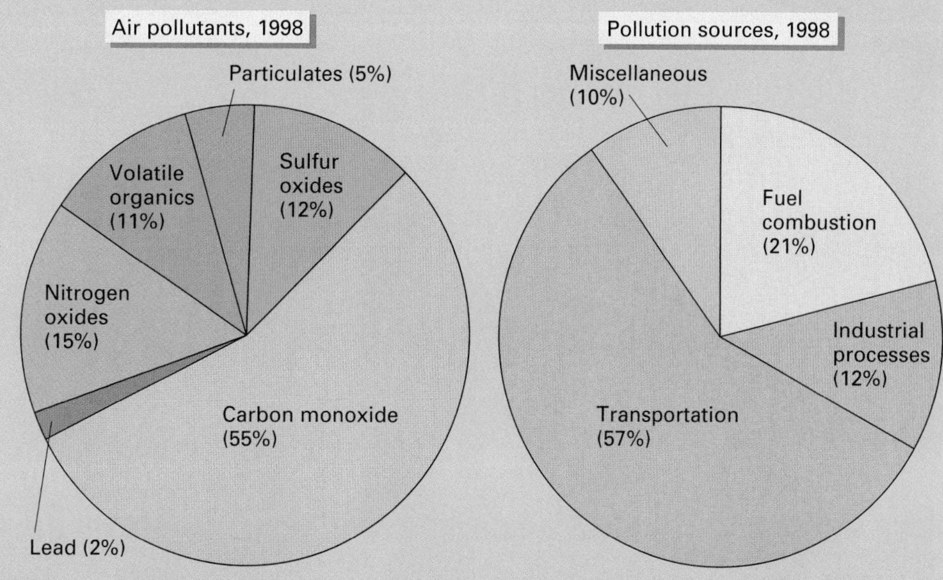

E6.1 Air pollutants and sources *These data are for the United States in 1998. (Data from EPA.)*

In this and preceding chapters, we have discussed two kinds of substances in air—aerosols and gases. Aerosols are small bits of matter in the air, so small that they float freely with normal air movements. Gases are molecular compounds that are mixed together to form the main body of air. To these two categories, we add a third category, *particulates*—larger, heavier particles that sooner or later fall back to Earth.

An *air pollutant* is an unwanted substance injected into the atmosphere from the Earth's surface by either natural or human activities. Air pollutants come as aerosols, gases, and particulates. Most pollutants generated by human activity arise in two ways. The first is through the day-to-day activities of large numbers of people, for example, in driving automobiles within urban centers. The second is through industrial activities, such as fossil fuel combustion or the smelting of mineral ores to produce metals.

Figure E6.1 shows the relative proportions of the most common air pollutants and their sources for the United States in 1998. The gases are carbon monoxide (CO), sulfur oxides, and nitrogen oxides. The nitrogen oxides can be NO, NO_2, or NO_3; this mixture is usually referred to as NO_X. Similarly, the sulfur oxides SO_2 and SO_3 are referred to as SO_X. Volatile organic compounds include evaporated gasoline, dry cleaning fluids, incompletely combusted fossil fuels, and similar pollutants. They occur as both gases and aerosols. The remaining pollutants shown in the chart are in the form of lead and particulates. Here, particulates include pollution particles that are fine enough to remain in the air as aerosols.

Combustion of fossil fuel, in transportation or as stationary fuel combustion, is the most important source of all these pollutants. As shown in the right half of the figure, it accounts for 78 percent of the emissions. Exhausts from gasoline and diesel engines contribute most of the carbon monoxide, half the volatile organic compounds, and about a third of the nitrogen oxides.

Stationary sources of fuel combustion include power plants that generate electricity and various industrial plants that also burn fossil fuels. They contribute most of the sulfur oxides because they burn coal or lower-grade fuel oils. These fuels are richer in sulfur than gasoline and diesel fuel. Stationary sources also supply most of the particulate matter. Some of it is fly ash—coarse soot particles emitted from smokestacks of power plants. These particles settle out quite quickly within close range of the source. Smaller and finer carbon particles, however, can remain suspended almost indefinitely and become aerosols. Industrial processes, such as the smelting of ore, are also important sources of air pollutants. They are the main source of lead (Figure E6.2).

Smog and Haze

When aerosols and gaseous pollutants are present in considerable density over an urban area, the resultant mixture is known as ***smog*** (Figure E6.3). This term was coined by combining the words "smoke" and "fog." Typically, smog allows hazy sunlight to reach the ground, but it may also be dense enough to hide aircraft

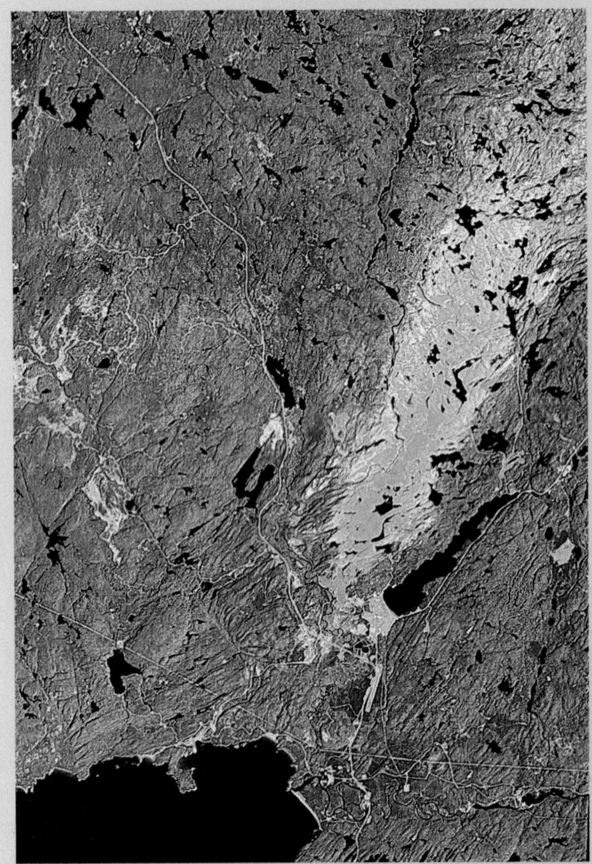

E6.2 Air pollution from smelters *This satellite image shows the devastating effects of pollutants emitted from smelting operations near Wawa, Ontario, Canada. Healthy vegetation is shown in red to pink tones. The blue streak on the right-hand side of the image is barren of vegetation. The smelting facilities are located at the narrowest point of the blue plume.*

E6.3 Smog in the city *Downtown Los Angeles projects skyward into a layer of smog.*

flying overhead from view. Smog irritates the eyes and throat, and it can corrode structures over long periods of time.

Modern urban smog has three main toxic ingredients: nitrogen oxides, volatile organic compounds, and ozone. Nitrogen oxides and volatile organic compounds, are largely automobile pollutants, although significant amounts can be released by industrial power plants and petroleum processing and storage facilities. Ozone is not normally produced directly by pollution sources in the city but forms through a photochemical reaction in the air. In this reaction, nitrogen oxides react with volatile organic compounds in the presence of sunlight to form ozone. Ozone in urban smog can harm plant tissues and eventually kill sensitive plants. It is also harmful to human lung tissue, and it aggravates bronchitis, emphysema, and asthma. This is the same compound that absorbs ultraviolet radiation in the ozone layer within the stratosphere. So, ozone in the stratosphere is desirable, but if close to Earth, it is undesirable indeed. Photochemical reactions can also produce other toxic compounds in smog.

Haze is a condition of the atmosphere in which aerosols. obscure distant objects. Haze builds up naturally in stationary air as a result of human and natural activity. When the air is humid and abundant water vapor is available, water films grow on suspended nuclei. This creates aerosol particles large enough to obscure and scatter light, reducing visibility. Natural haze aerosols include soil dust, salt crystals from the sea surface, hydrocarbon compounds from plants, plant pollen, and smoke from forest and grass fires. Some regions are noted for naturally occurring haze—the Great Smoky Mountains are an example.

Fallout and Washout

Pollutants generated by a combustion process are contained within hot exhaust air emerging from a factory smokestack or an auto tailpipe. Since hot air rises, the pollutants are at first carried aloft by convection. However, the larger particulates soon settle under gravity and return to the surface as *fallout.* Particles too small to settle out are later swept down to Earth by precipitation in a process called *washout.* Through a combination of fallout and washout, the atmosphere tends to be cleaned of pollutants. Although a balance between input and output of pollutants is achieved in the long run, the quantities stored in the air at a given time fluctuate widely.

Pollutants are also eliminated from the air over their source areas by wind. Strong, through-flowing winds will disperse pollutants into large volumes of cleaner air in the downwind direction. Strong winds can quickly sweep away most pollutants from an urban area, but during periods when winds are light or absent, the concentrations can rise to high values.

Inversion and Smog

The concentration of pollutants over a source area rises to its highest levels when vertical mixing (convection) of the air is inhibited. This happens in an inversion—a condition in which the temperature of the air increases with altitude.

Why does the inversion inhibit mixing? Recall that a heated air parcel, perhaps emerging from a smokestack or chimney, will rise as long as it is warmer than the surrounding air. Recall, too, that as it rises, it is cooled according to the adiabatic principle. In an inversion, however, the surrounding air gets warmer, not colder, with altitude. So, the parcel will cool as it rises, while the surrounding air becomes warmer. Thus, the parcel will quickly arrive at the temperature of the surrounding air, and uplift will stop. Under these conditions, heated air will move only a short distance upward. Pollutants in the air parcel will disperse at low levels, keeping concentrations high near the ground.

Two types of inversions are important in causing high air pollutant concentrations—low-level and high-level inversions. When a *low-level temperature inversion* develops over an urban area with many air pollution sources, pollutants are trapped under the "inversion lid." Heavy smog or highly toxic fog can develop. An example is the tragedy that occurred in Donora, Pennsylvania, in late October 1948, when a persistent low-level inversion developed. The city occupies a valley floor hemmed in by steeply rising valley walls. The walls prevented the free mixing of the lower air layer with that of the surrounding region (Figure E6.4) and helped keep the inversion intact. Industrial smoke and gases from factories poured into the inversion layer for five days, increasing the pollution level. A poisonous fog formed and began to take its toll. Twenty people died, and several thousands were stricken before a change in weather patterns dispersed the polluted layer.

Another type of inversion is responsible for the smog problem experienced in the Los Angeles Basin and other California coastal regions, ranging north to San Francisco and south to San Diego. Here special climatic conditions produce prolonged inversions and smog accumulations (Figure E6.5). Off the California coast is a persistent fair-weather system that is especially strong in the summer. This system produces a layer of warm, dry air at upper elevations. However, a cold current of upwelling ocean bottom water runs along the coast, just offshore. Moist ocean air moves across this cool current and is chilled, creating a cool, marine air layer.

Now, the Los Angeles Basin is a low, sloping plain lying between the Pacific Ocean and a massive mountain barrier on the north and east sides. Weak winds from the south and southwest move the cool, marine air inland over the basin. Further landward movement is blocked by the mountain barrier. Since there is a warm layer above this cool marine air layer, the result is an inversion. Pollutants accumulate in the cool air layer and produce smog. The upper limit of the smog stands out sharply in contrast to the clear air above it, filling the basin like a lake and extending into valleys in the bordering mountains

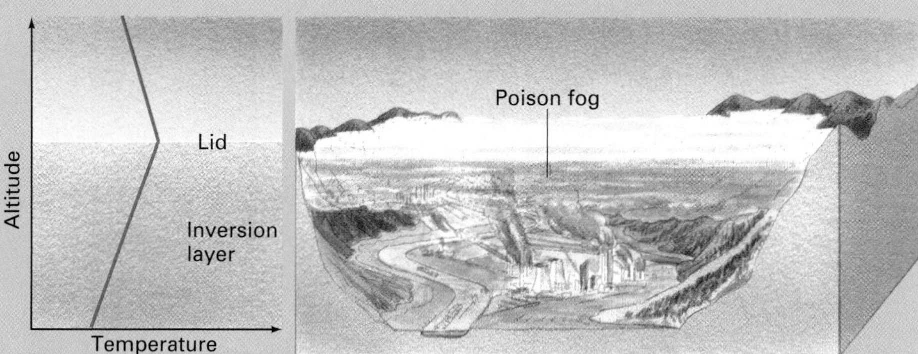

E6.4 Low-level temperature inversion *An inversion held air pollutants close to the ground, creating a poison fog accumulation at Donora, Pennsylvania, in October 1948.*

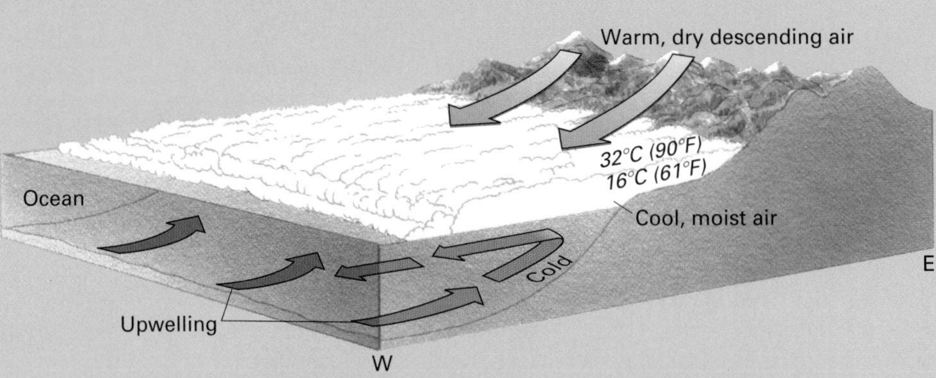

E6.5 High-level temperature inversion on the California coast *A layer of warm, dry, descending air from a persistent fair-weather system rides over a cool, moist marine air layer at the surface to create a persistent temperature inversion.*

(Figure E6.6). Since this type of inversion persists to a higher level in the atmosphere, we can describe it as a *high-level temperature inversion.*

An actual temperature inversion, in which temperature close to the surface increases with altitude, is not essential for building a high concentration of pollutants above a city. Light or calm winds and stable air are all that are required. *Stable air* has a temperature profile that decreases with altitude but at a slow rate. Some convectional mixing occurs in stable air, but the convectional precipitation process is inhibited. At certain times of the year, slow-moving masses of dry, stable air occupy the central and eastern portions of the North American continent. Under these conditions, a broad *pollution dome* can form over a city or region, and air quality will suffer (Figure E6.7). When there is a regional wind, the pollution from a large city will be carried downwind to form a *pollution plume.*

Climatic Effects of Urban Air Pollution

Urban air pollution reduces visibility and illumination. Specifically, a smog layer can cut illumination by 10 percent in summer and 20 percent in winter. Ultraviolet radiation is absorbed by ozone in smog. At

E6.6 Smog and haze at Los Angeles *A dense layer of smog and marine haze fills the Los Angeles Basin. The view is from a point over the San Gabriel Mountains, looking southwest.*

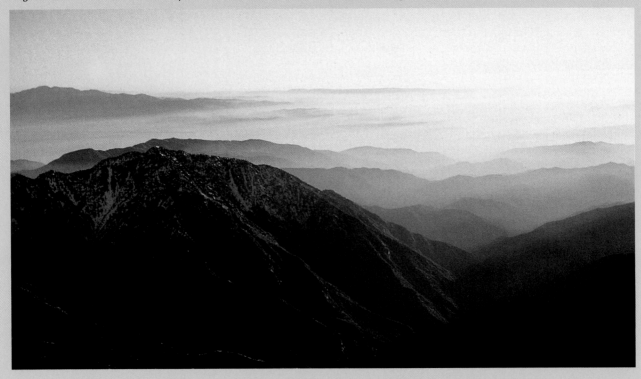

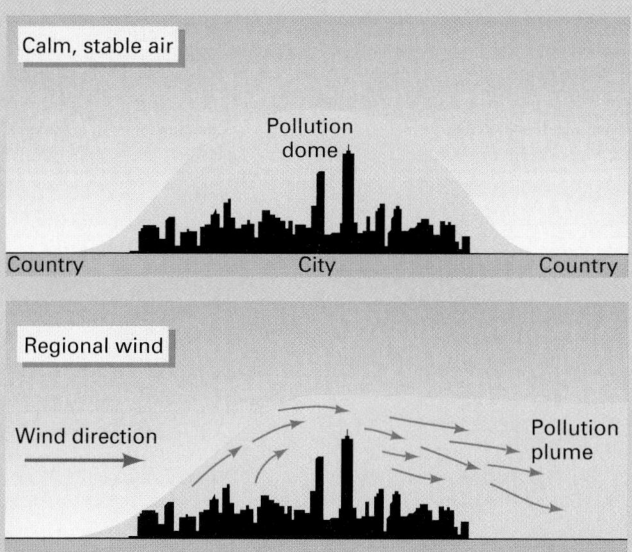

E6.7 Pollution dome and pollution plume *If calm, stable air overlies a major city, a pollution dome can form. When a wind is present, pollutants are carried away as a pollution plume.*

times, ultraviolet radiation is completely prevented from reaching the ground. Although this reduces the risk of human skin cancer, it may also permit increased viral and bacterial activity at ground level.

Winter fogs are much more frequent over cities than over the surrounding countryside. The fog is enhanced by the abundance of urban aerosols and particulates. Coastal airports, such as those of New York City, Newark, and Boston, suffer from an increased frequency of fog resulting from urban air pollution. Cities also show an increase in cloudiness and precipitation as compared to the surrounding countryside. This increase results from intensified convection generated when the lower air is heated by human activities.

Acid Deposition

Yet another air pollution problem is *acid deposition.* Perhaps you've heard about acid rain killing fish and poisoning trees. Acid rain is part of the phenomenon of acid deposition, which includes not only acid rain but also dry acidic dust particles. Acid rain simply consists of raindrops that have been acidified by air pollutants. Dry acidic particles are dust particles that are acidic in nature. They fall to Earth and coat the surface in a thin dust layer. When wetted by rain or fog, they acidify the water on leaves and soils. In winter, acid particles can mix with snow as it accumulates. In spring, when the snow melts, a surge of acid water is then released to soils and streams.

Acid deposition is produced by the release of sulfur dioxide (SO_2) and nitric oxide (NO_2) into the air. This occurs when fossil fuels are burned. Once in the air, SO_2 and NO_2 readily combine with oxygen and water in the presence of sunlight and dust particles to form sulfuric and nitric acid aerosols. These aerosols serve as condensation nuclei and acidify the tiny water droplets created. When the droplets coalesce in precipitation, acid raindrops or ice crystals result. Sulfuric and nitric acids can also be formed on dust particles, creating dry acid particles. These can be as damaging to plants, soils, and aquatic life as acid rain.

What are the effects of acid deposition? A primary effect is acidification of lakes and streams, injuring aquatic plants and animals and perhaps eventually sterilizing a lake, pond, or stream. Another problem is soil damage. Through a series of chemical effects, acid deposition can cause soils to lose nutrients, affecting the natural ecosystem within a region.

To measure acidity, scientists use the pH scale. Values on this scale range from 1 to 14, with high values indicating alkalinity and the low values acidity. Pure water has a pH of 7. Normally, rainwater is slightly acid, showing pH values of 5 to 6. In the 1960s European water chemists observed that the pH of rain in northwestern Europe had dropped to values as low as 3 in some samples, similar to household vinegar. Because pH numbers are on a logarithmic scale, these values mean that rain in these samples was 100 to 1000 times more acid than normal.

In North America, scientists studying the chemical quality of rainwater reported in the 1970s and 1980s that rainwater over a large area of the northeastern United States had an average pH of about 4. Values less than pH 4 occurred at times over many heavily industrialized U.S. cities, including Boston, New York, Philadelphia, Birmingham, Chicago, Los Angeles, and San Francisco.

Figure E6.8 is a map of the United States showing the acidity of rainwater averaged over 1999. There has been a significant improvement in the last decade, but values are still low in the Ohio Valley and in the northeast. Compare this with Figure E6.9, showing emissions of sulfur dioxide and nitrogen oxides by states. Although the emissions data are for an earlier year, a strong relationship, especially between SO_2 and the average acidity of rainwater, is clear.

An important factor in the level of impact of acid deposition on the

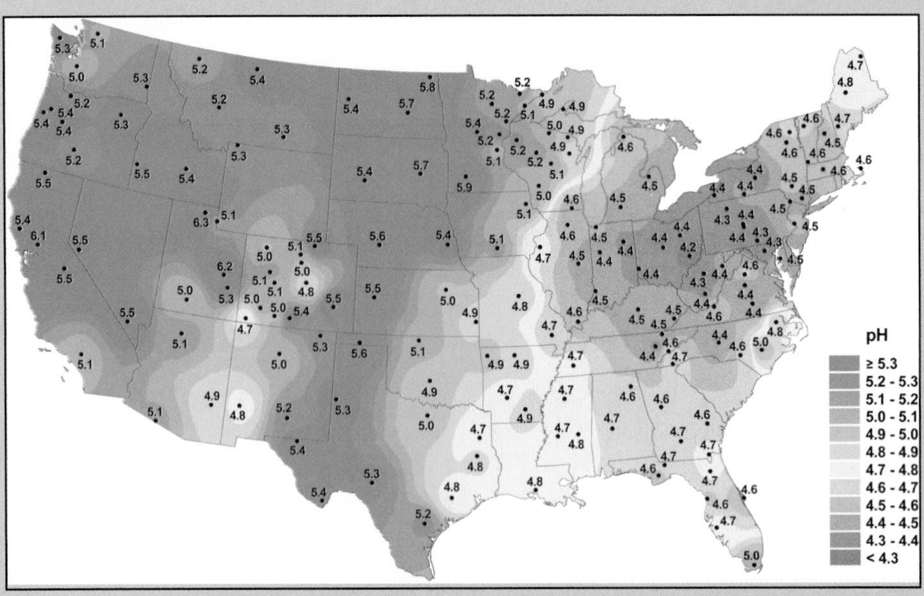

E6.8 Acidity of rainwater for the United States in 1999 *Values are pH units. The northeastern United States shows the lowest, most acid values. (National Atmospheric Deposition Program (NRSP-3)/National Trends Network, Illinois State Water Survey.)*

environment is the ability of the soil and surface water to absorb and neutralize acid. This factor ranges widely. In dry climates, acids can be readily neutralized because surface waters are normally somewhat alkaline. Areas where soil water is naturally somewhat acidic are most sensitive to acid deposition. Such areas are associated with moist climates generally and include the eastern United States and Canada, high-mountain regions of the western states and provinces, and the Pacific Northwest.

Acid deposition in Europe and North America has severely

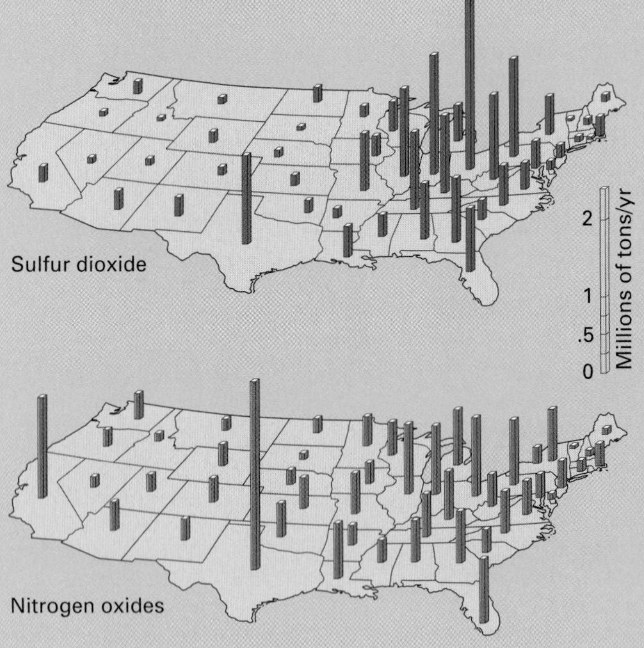

E6.9 Emissions of sulfur dioxide and nitrogen oxides by states for the year 1993 *Sulfur dioxide releases are especially heavy in the states that burn Appalachian coal. Releases of nitrogen oxides are high in Texas, due to oil processing, and in California, due to auto emissions. (Data from EPA, 1994.)*

E6.10 Acid deposition impact *A view from the Appalachian Trail, near Klingman's Dome, in Great Smoky Mountains National Park, Tennessee. Death of the many leafless firs in the scene has been attributed to stress caused by acid deposition.*

impacted some ecosystems. In Norway, acidification of stream water has virtually eliminated many salmon runs by inhibiting salmon egg development. Increased fish mortality, attributed to acidification, has been observed in many lakes in eastern Canada. In 1990, American scientists estimated that 14 percent of Adirondack lakes were heavily acidic, along with 12 to 14 percent of the streams in the Mid-Atlantic states.

Forests, too, have been damaged by acid deposition (Figure E6.10). In western Germany, the impact has been especially severe in the Harz Mountains and the Black Forest. In the 1990s, Europe continued to experience heavy losses in timber, resulting from fallout of sulfur dioxide in combination with nitrogen oxide emissions from vehicles, industry, and farm wastes. Fallout of sulfur oxides was most heavily concentrated in Germany, the Czech Republic, Slovakia, Poland, and Hungary.

During the 1990s, releases of sulfur oxides, nitrogen oxides, and volatile organic compounds were significantly reduced in the United States. These effects were due primarily to improved industrial emission controls. Although the situation is improving, acid deposition is still a very important problem in many parts of the world—especially Eastern Europe and the states of the former Soviet Union. There, air pollution controls have been virtually nonexistent for decades. Reducing pollution levels and cleaning up polluted areas will be a major task for these nations in the next decades.

Air Pollution Control

The United States and Canada have some of the strictest laws in the world limiting air pollution. As a result, many strategies have been developed to reduce emissions. Sometimes these strategies involve trapping and processing pollutants after they are generated. The "smog controls" on automobiles are an example. Sometimes the use of alternative, nonpolluting technology is in order—for example, substituting solar, wind, or geothermal power for coal burning to generate electricity.

In any event, both as individuals and as a society we can do much to preserve the quality of the air we breathe. Reducing fossil fuel consumption is an example. It will be a considerable challenge to our society and others to preserve and increase global air quality in the face of an expanding human population and increasing human resource consumption.

7 | WINDS AND THE GLOBAL CIRCULATION SYSTEM

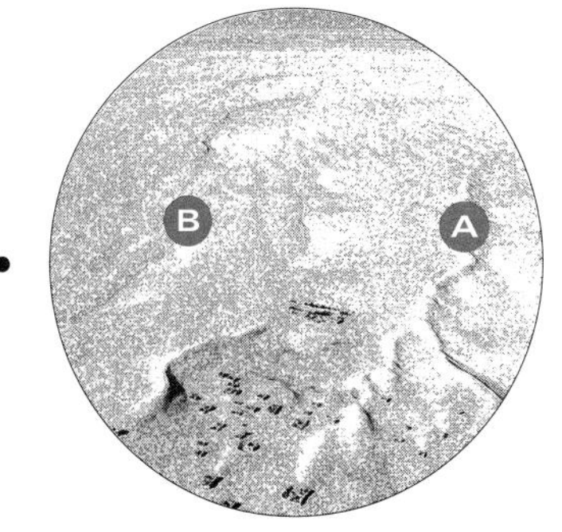

EYE ON THE LANDSCAPE
Village of Araouane, Mali. This ancient village in the southern Sahara, once a thriving stop on the great caravan route to Mauritania, is now slowly being buried by wind-driven sand. **What else would the geographer see?...Answers at the end of the chapter.**

The air around us is always in motion. From a gentle breeze on a summer's day to a cold winter wind, we experience moving air much of the time. Why does the air move? What are the forces that cause winds to blow? Why do winds blow more often in some directions than others? What is the pattern of wind flow in the upper atmosphere, and how does it affect our weather? How does the global wind pattern produce ocean currents? These are some of the questions we will answer in this chapter.

Winds are caused by unequal heating of the Earth's atmosphere. When the atmosphere in one location is heated to a temperature that is warmer than that in another location, a difference in pressure, or pressure gradient, results. As we will see later in more detail, this pressure gradient causes air at the surface to move toward the warmer location and air at upper levels to move away from the warmer location. This process directly explains many types of local winds, such as the sea breeze you might experience on an afternoon at the beach. It is also the cause of global wind motions, since the equatorial and tropical atmosphere is heated more intensely by the Sun than is the atmosphere at mid- and higher latitudes. In response to this difference in heating, global-scale pressure gradients move vast bodies of warm air poleward, and huge pools of cool air shift equatorward, strongly influencing our day-to-day weather. The direction of these global wind motions is also influenced by the Earth's rotation—a factor that we also discuss in this chapter.

Another feature of the global circulation system is ocean currents, which are largely driven by surface winds. Ocean currents, like global wind motions, act to move heat energy across parallels of latitude and are also affected by the Earth's rotation.

Before we turn to local and global wind patterns, however, we will need to examine atmospheric pressure and how wind develops in response to pressure differences. These are the subjects for the next few pages.

Atmospheric Pressure

The Earth's land surface is actually located at the bottom of a vast ocean of air—the Earth's atmosphere. Like the water in the ocean, the air in the atmosphere is constantly pressing on the solid or liquid surface beneath it. This pressure is exerted in all directions.

Atmospheric pressure exists because air molecules have mass and are constantly being pulled toward the Earth by gravity. As we all know and experience every day, the acceleration of mass by gravity creates a force that we refer to as weight. Pressure is a force per unit area. For the atmosphere, this force is thus produced by the weight of a column of air above a unit area of surface. At sea level, the force is the weight of about 1 kilogram of mass of air that lies above each square centimeter of surface (1 kg/cm^2) (Figure 7.1). In the English system, the force is about 15 pounds of weight for each square inch of surface (15 $lb/in.^2$).

***7.1 Atmospheric pressure** This figure depicts atmospheric pressure as the weight of a column of air. (a) Metric system. The weight of a column of air 1 cm on a side is balanced by the weight of a mass of about 1 kg. (b) English system. The weight of a column of air 1 in. on a side is balanced by a weight of about 15 pounds.*

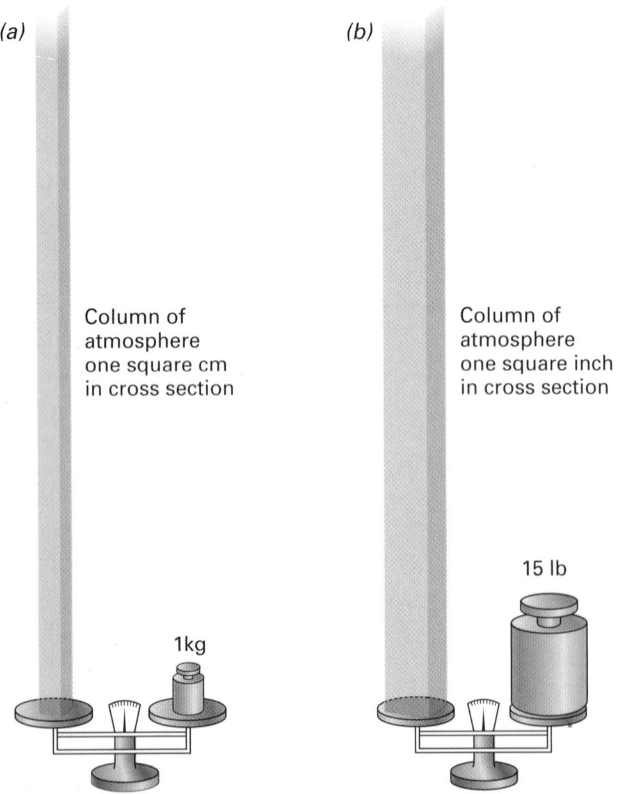

In the metric system, the unit of pressure is the *pascal* (Pa). At sea level, the average pressure of air is 101, 320 Pa. Although the pascal is the proper metric unit for pressure, many atmospheric pressure measurements are based on the *bar* and the **millibar** (mb) (1 bar = 1000 mb = 10,000 Pa), two former metric units that are still in wide use. In this text, we will use the millibar as the metric unit of atmospheric pressure, following this long tradition. Standard sea-level atmospheric pressure is 1013.2 mb.

Measuring Atmospheric Pressure

Perhaps you know that atmospheric pressure is measured by a **barometer.** But how does a barometer work? The principle is simple, and you have experienced it often. Visualize sipping a soda through a straw. By lowering your jaw and moving your tongue, you expand the volume of your mouth. This creates a partial vacuum, and air pressure forces the soda up through the straw and into your mouth.

A *mercury barometer* works on the same principle (Figure 7.2)—that atmospheric pressure will force a liquid up into a tube when a vacuum is present. To construct the barometer, a glass tube about 1 m (3 ft) long, sealed at one end, is completely filled with mercury. The open end is temporarily held closed. Then the tube is inverted, and the open end is immersed into a dish of mercury. When the opening of the tube is uncovered, the mercury in the tube falls some distance, but then it remains fixed at a level about 76 cm (30 in.) above the surface of the mercury in the dish.

What is happening here? At first, mercury flows out of the tube, under the force of gravity, and into the dish. Since there is no air in the top of the tube, a vacuum is formed.

***7.2 Mercury barometer** Atmospheric pressure pushes the mercury upward into the tube, balancing the pressure exerted by the weight of the mercury column. As atmospheric pressure changes, the level of mercury in the tube changes. At sea level, the average height of the mercury level is about 76 cm (30 in.).*

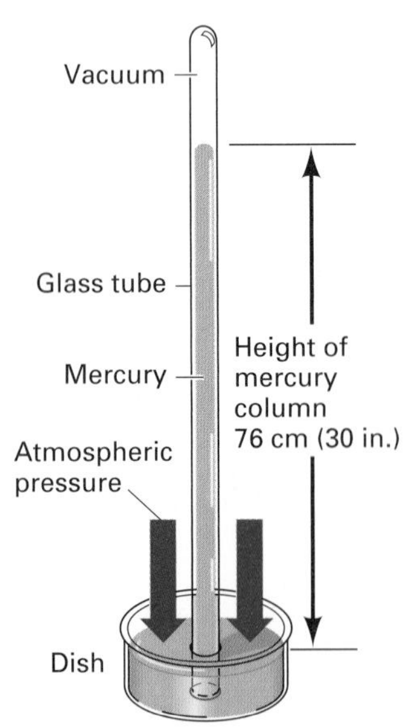

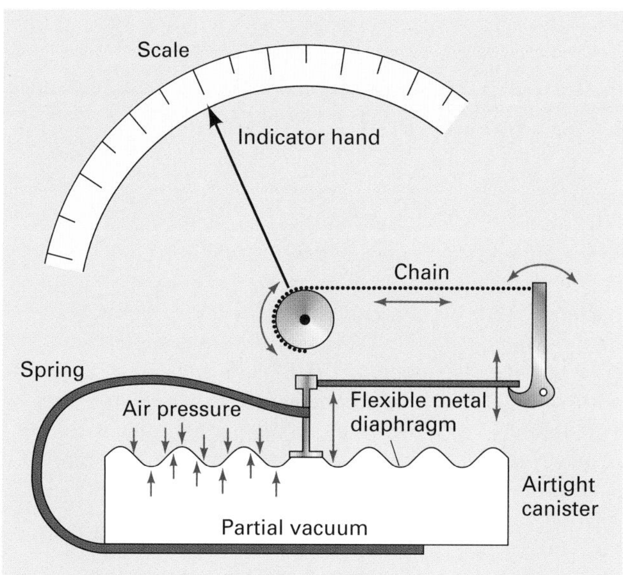

7.3 Aneroid barometer *As air pressure varies, the diaphragm moves up and down. This motion is transmitted mechanically to move an indicator hand along a scale.*

At the same time, atmospheric pressure pushes on the surface of the mercury in the dish, pressing the mercury upward into the tube. Since there is no air in the tube, there is no air pressure to push back against it. The pull of gravity on the mercury column then balances the pressure of the air, and the mercury level stays at a stable height. If air pressure changes, so will the height of the mercury column. So, this device will measure air pressure directly—but as a height in centimeters or inches rather than a pressure. The mercury barometer has become the standard instrument for weather measurements because of its high accuracy.

The widespread use of the mercury barometer has led to the common measurement of air pressure in heights of mercury, as centimeters or inches. On this scale, standard sea-level pressure is 76 cm Hg (29.92 in. Hg). The letters Hg are the chemical symbol for mercury. In this text, we will use in. Hg as the English unit for atmospheric pressure.

A more common type of barometer is the *aneroid barometer* (Figure 7.3). It uses a sealed canister of air, much like a small, flat can you might see in a supermarket. The canister has a slight partial vacuum, tensing the lid, or diaphragm, against a spring attached to the lid. The diaphragm flexes as air pressure changes, and through a mechanical linkage this fine movement drives a hand indicator that moves along a scale.

Atmospheric pressures at a location vary from day to day by only a small proportion. On a cold, clear winter day, the barometric pressure at sea level might be as high as 1030 mb (30.4 in. Hg), while in the center of a storm system, pressure might drop to 980 mb (28.9 in. Hg), a difference of about 5 percent. Pressure changes are associated with traveling weather systems, a topic that we will discuss in Chapter 8.

How Air Pressure Changes with Altitude

Figure 7.4 shows how atmospheric pressure decreases with altitude. Moving upward from the Earth's surface, the decrease of pressure with altitude is initially quite rapid. At higher altitudes, the decrease is much slower. Because atmospheric pressure decreases rapidly with altitude near the surface, a small change in elevation will often produce a significant change in air pressure. For example, you may have noticed that your ears sometimes "pop" from a pressure change during an elevator ride in a tall building. You may also have noticed your ears popping on an airplane flight. Aircraft cabins are pressurized to about 800 mb (24 in. Hg), which corresponds to an elevation of about 1800 m (5900 ft).

More serious effects on human physiology occur at greater elevations. Perhaps you've felt out of breath, or tired more easily, while walking along a high-mountain trail. With decreased air pressure, oxygen moves into lung tissues more slowly, producing shortness of breath and fatigue. These symptoms, along with headache, nosebleed, or nausea, are referred to as mountain sickness and are likely to occur at altitudes of 3000 m (about 10,000 ft) or higher. Most people

7.4 Atmospheric pressure and altitude *Atmospheric pressure decreases with increasing altitude above the Earth's surface.*

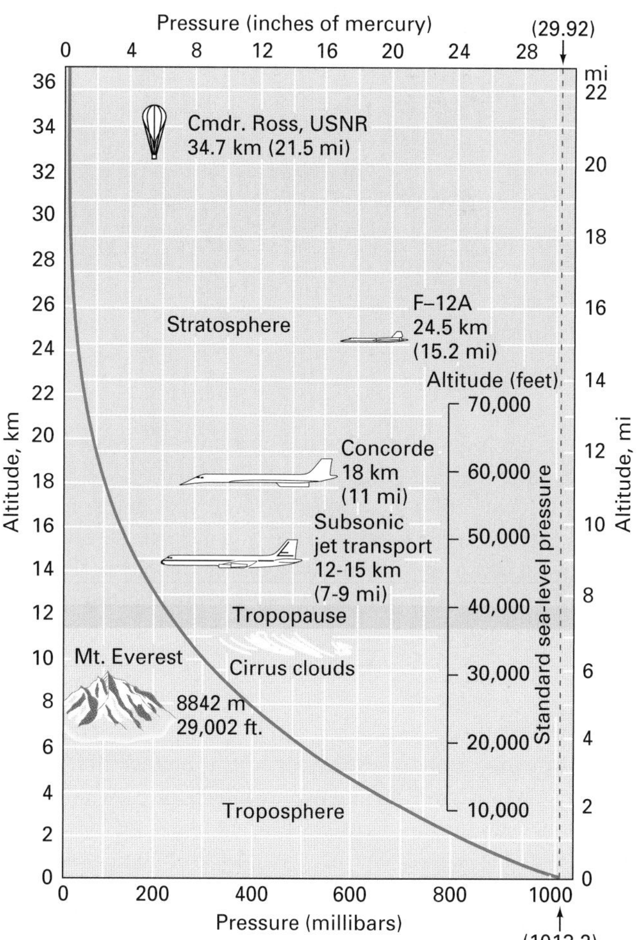

Working It Out 7.1
Pressure and Density in the Oceans and Atmosphere

Although it is sometimes useful to think of the Earth's surface as the bottom of a vast "ocean of air," the atmosphere is quite different from the oceans because the oceans are composed of a nearly incompressible liquid—water with dissolved compounds—and the atmosphere is composed of a mixture of readily compressible gases. Recall that both liquids and gases are classed as fluids.

The pressure at any level in a fluid is created by the weight of the mass of the fluid above that level. In the oceans, it's relatively simple to determine that weight. Let's consider a square meter (1 m × 1 m) of horizontal area at depth *D* within the ocean. Pressing down on that square meter is the weight of $D \times 1 \times 1$ m^3 of water. Each cubic meter of water weighs about 1000 kg or 1 metric ton (t). So as a quick rule of thumb, we can regard the pressure at any depth in the ocean as about $1 \times D = D$ t/m^2. That is,

$$P_{OCEAN} = D$$

where *P* is the pressure in t/m^2 and *D* is the depth in m. For example, the pressure at 500 m depth is about 500 t/m^2; at 5 km = 5000 m, it is about 5000 t/m^2.

These values are based on a constant density of 1 t/m^3. To be precise, there is a slight variation in the density of sea water with pressure, temperature, and salinity. That is, colder, saltier, and deeper water will be more dense. At the surface, the density of ocean water is 1.028 t/m^3, while at 5000 km depth, the density is 1.051 t/m^3, an increase of about 2.2 percent. The left figure on the facing page, which plots pressure with depth in the ocean, uses a constant density value of 1.035 t/m^3.

Note that at any depth in the ocean, there will be not only the weight of the water above, but also the weight of the atmosphere above the water. Sea-level pressure is about 1000 millibars, which converts to 1 t/m^2. Thus, the atmosphere only adds the pressure of about one more meter of depth.

Though small, the density differences in water with temperature and salinity can vary the pressure at a given depth enough to create pressure gradients and therefore induce movement of water from higher to lower pressure. An example is the convection loop circulation that occurs in a container of water when it is heated from below. This loop is described in more detail in *Focus on Systems 7.3 • The Convection Loop as an Energy Flow System.*

In contrast to the oceans, the atmosphere is readily compressible. What is the weight of a cubic meter of air? There is no rule-of-thumb answer for that question because density varies freely with pressure. When compressed, more molecules of the gases that compose air are present in a given volume, and so the volume will have a greater mass and weigh more. At the Earth's surface, the density of a cubic meter of air is about 1.225 kg/m^3, while at 11 km (about 36,000 ft), the height of a cruising jet aircraft, it is about 0.364 kg/m^3, or a little less than one-third of the surface value.

The difference is nearly all due to pressure, but temperature is also important. At a given pressure, warmer air will be less dense, while colder air will be more dense. In the chapter text, we have seen how warming a column of air creates a pressure gradient force that induces movements of air. In fact, the contrast between a warm atmosphere near the equator and a cold atmosphere near the poles sus-

adjust to the reduced air pressure of a high-mountain environment after several days.

The decrease of atmospheric pressure with height above the surface, as well as the increase of water pressure in the ocean with depth, are explained by the simple physical principles given in *Working It Out 7.1 • Pressure and Density in the Oceans and Atmosphere.*

WIND

From a gentle, cooling breeze on a warm summer afternoon to the howling gusts of a major winter storm, wind is a familiar phenomenon. **Wind** is defined as air motion with respect to the Earth's surface. Wind movement is dominantly horizontal. Air motions that are dominantly vertical are not called winds but are known by other terms, such as updrafts or downdrafts.

Measurement of Wind

The motion of air in wind, like all motion, is characterized by a direction and velocity (Figure 7.5). Wind direction can be determined by a *wind vane,* the most common of the weather instruments. The design of the wind vane keeps it facing into the wind. Wind direction is always given as the direction from which the wind is coming. So, a west wind is one that comes from the west and moves to the east (Figure 7.6).

Wind speed is measured by an *anemometer.* The most common type is the cup anemometer, which consists of three funnel-shaped cups mounted on the ends of spokes of a horizontal wheel. The cups rotate with a speed proportional to that of the wind. One version of the anemometer turns a small electric generator. The more rapidly the wheel rotates, the more current is generated. The current is measured by a meter, which is calibrated in units of wind speed. Units are either meters per second or miles per hour.

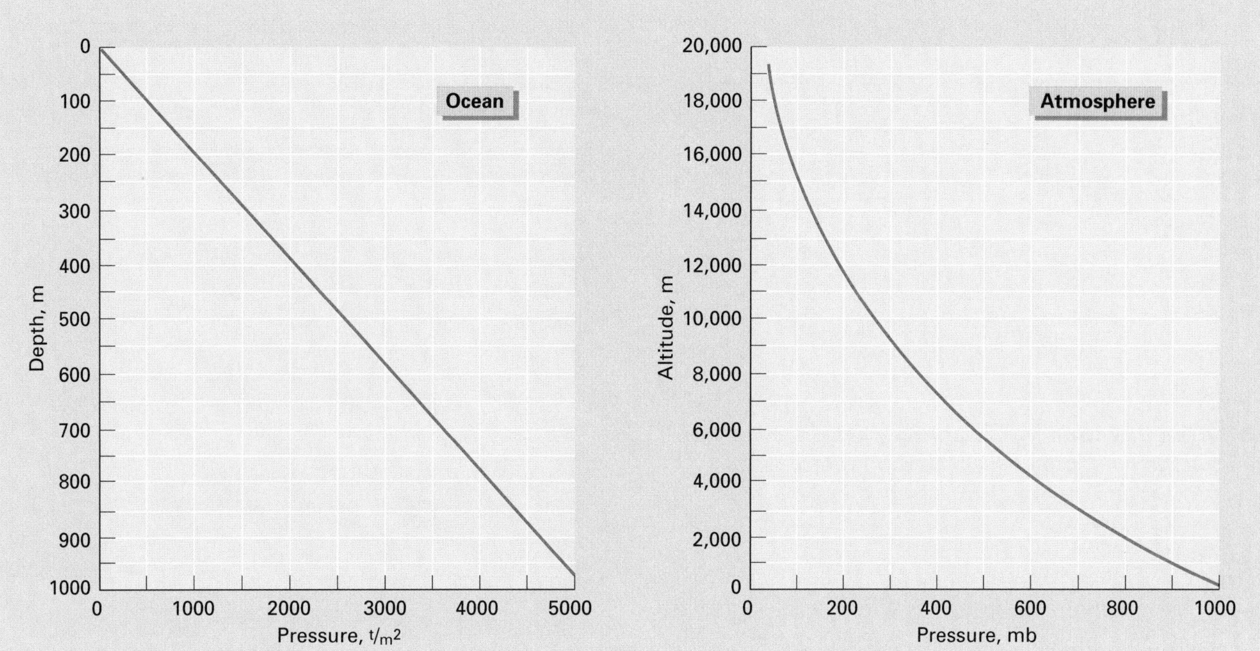

Pressure graphs *(left) Graph of pressure with depth in the oceans. (right) Graph of pressure with altitude in the atmosphere.*

tains the winds that constantly move and mix the troposphere.

Atmospheric pressure is plotted against altitude in the figure on the right. Note that the graph curves smoothly, whereas the graph of pressure against depth for the oceans is straight. The curve of the atmospheric pressure graph is a result of the fact that the density of air decreases with altitude.

Atmospheric pressure at any given altitude can be approximated by the following formula:

$$P_Z = 1014[1 - 0.0226Z]^{5.26}$$

where P_Z is the pressure in millibars at height Z in km. This formula takes into account the fact that both temperature and pressure decrease with altitude. This formula, which is derived for certain standard atmospheric conditions, is used in plotting the figure. Note that the value of 1014 is the surface atmospheric pressure in millibars for these standard conditions.

As an example, let's find the pressure at 12 km, which is approximately the altitude of a passenger jet on a transcontinental flight (about 39,000 ft). Substituting for Z, we have

$$P_Z = 1014 \times [1 - (0.0226 \times 12)]^{5.26}$$

$$= 1014 \times (1 - 0.271)^{5.26}$$

$$= 1014 \times (0.729)^{5.26}$$

$$= 1014 \times 0.189 = 192 \text{ mb}$$

Winds and Pressure Gradients

Wind is caused by differences in atmospheric pressure from place to place. Air tends to move from high to low pressure until the air pressures are equal. This follows the simple physical principle that any flowing fluid (such as air) subjected to gravity will move until the pressure at every level is uniform.

7.5 Combination cup anemometer and wind vane *The anemometer and wind vane observe wind speed and direction, which are displayed on the meter below. The wind vane and anemometer are mounted outside, with a cable from the instruments leading to the meter, which is located indoors. Also shown are a barometer and maximum-minimum thermometer. (Courtesy of Taylor Instrument Company and Ward's Natural Science Establishment, Rochester, N.Y.).*

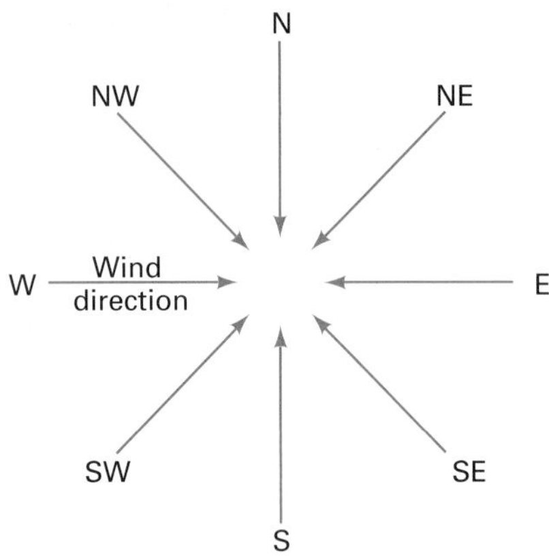

7.6 Wind direction *Winds are designated according to the compass point from which the wind comes. A west wind comes from the west, but the air is moving eastward.*

Figure 7.7 illustrates this principle. It is a simple map showing two locations, taken here as Wichita and Columbus. At Wichita, the pressure is high (H), and the barometer (adjusted to sea-level pressure) reads 1028 mb (30.4 in. Hg). At Columbus, the barometer is low (L) and reads 997 mb (29.4 in. Hg). Thus, there is a **pressure gradient** between Wichita and Columbus. The lines on the map around Wichita and Columbus connect locations of equal pressure and are called **isobars.** As you move between Wichita and Columbus, you cross these isobars on the map, encountering changing pressures as a result of the drop from 1028 to 997 mb (30.4 to 29.4 in. Hg).

Because atmospheric pressure is unequal at Wichita and Columbus, a *pressure gradient force* will push air from Wichita toward Columbus. This pressure gradient will produce a wind. The greater the pressure difference between the two locations, the greater this force will be and the stronger the wind. However, the wind will not necessarily move directly from Wichita to Columbus. It will be deflected by the Coriolis effect, as we will see later in this chapter.

A Simple Convective Wind System

Pressure gradients develop because of unequal heating of the atmosphere. To understand this process, we will look at a simple convective wind system, shown in Figure 7.8. Part

7.7 Isobars and a pressure gradient *High pressure is centered at Wichita, and low pressure is centered at Columbus.*

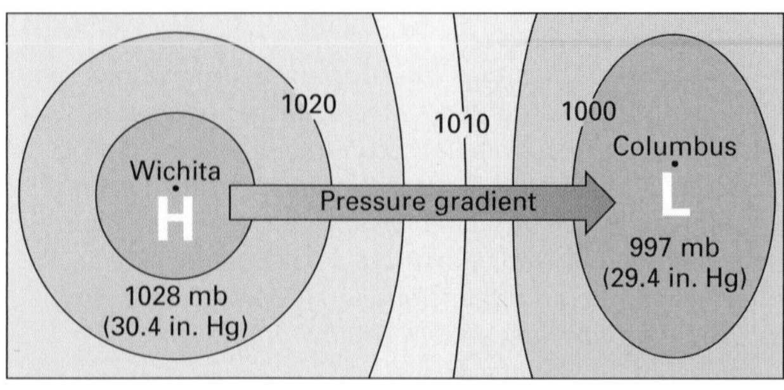

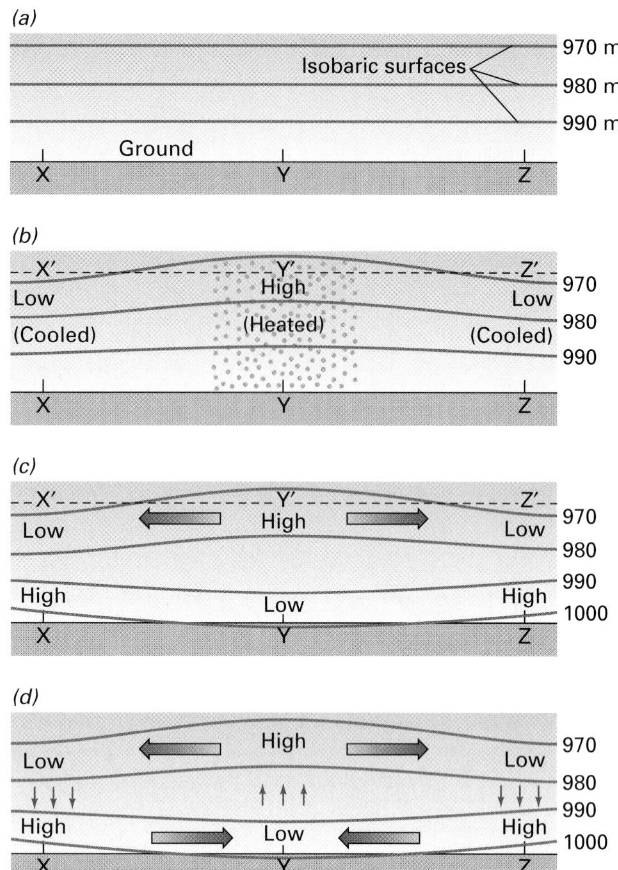

7.8 Formation of a simple convection loop *Heating of the atmosphere over point Y creates a pressure gradient that causes a convection loop.*

(a) depicts a uniform atmosphere above a ground surface. The lines mark isobaric surfaces, seen in cross section, and points *X, Y,* and *Z* mark the left, middle, and right ends of the cross section. Let's now imagine that the air is heated over *Y* (perhaps by sunlight) and cooled over *X* and *Z* (perhaps by radiation loss), as in part *(b).* As the air at *Y* expands in response to heating, it pushes the isobaric surfaces upward. At a constant height in the atmosphere, for example, along line *X′-Y′-Z′,* a pressure gradient is induced. We can see this by noting that at *Y′,* the pressure is between 980 and 970 mb, while at *X′* and *Z′,* it is less than 970 mb. Thus, there will be a pressure gradient from *Y′* to *X′* and *Y′* to *Z′.*

The pressure gradient next causes air to move away from *Y′,* as shown in part *(c).* However, as soon as this air moves, it changes the surface pressure at *X, Y,* and *Z.* Since there is less air above *Y,* pressure at the ground falls. Meanwhile, there is more air above points *X* and *Z,* so pressure there rises. This creates a new pressure gradient at the surface that moves air from *X* and from *Z* to *Y* as shown in part *(d).* The resulting circulation has produced two convection loops in which air at the surface converges at the location of heating and moves away, or diverges, at higher altitude. The principle that heating of the air produces a pressure gradient that causes air to move is an important one. We will return to it later in the chapter when we consider upper-level winds.

Sea and Land Breezes

Sea and *land breezes* are simple examples of how heating and cooling produce convection loops to create local winds. If you've ever vacationed at the beach in the summer, you may have noticed that in the afternoons a *sea breeze* often sets in. This wind brings cool air off the water, dropping temperatures and refreshing beachgoers and residents living close to the beach. Late at night, a *land breeze* may develop. This wind moves cooler air, which is chilled over land by nighttime radiant cooling, toward the water.

Figure 7.9 shows how these breezes work in more detail. Recall from Chapter 5 that land surfaces heat and cool more rapidly than water, so that the coastline is a location where temperature contrasts can easily develop. During the day, the air over the land is warmed by the land surface, and a convection loop forms. Air moves oceanward aloft, while surface winds bring cool marine air landward at the surface (part *b*). At night, radiation cooling over land creates a reversed convection loop, and an offshore breeze develops (part *c*).

The sea breeze–land breeze convection loop provides an example of surface thermal high- and low-pressure zones. That is, low surface pressure is associated with warm air, and high surface pressure with cool air. We will see this later in the chapter at a more global scale, when we note the tendency for deserts to have low atmospheric pressure and persistent regions of ice and snow to have high pressure.

Local Winds

Sea and *land* breezes are examples of *local wind* systems that are generated by local effects, such as a difference in heating of the atmosphere by different land surface types. Other types of local winds include mountain and valley winds, drainage winds, and Santa Ana winds.

Mountain winds and *valley winds* are local winds that alternate in direction in a manner similar to the land and sea breezes. During the day, mountain hill slopes are heated intensely by the Sun, causing a convection loop to form. As the surface part of the loop, an air current moves up valleys from the plains below—upward over rising mountain slopes, toward the summits. At night, the hill slopes are chilled by radiation, setting up a reversed convection loop. The cooler, denser hill slope air then moves valleyward, down the hill slopes, to the plain below.

Another group of local winds are known as *drainage winds,* in which cold, dense air flows under the influence of gravity from higher to lower regions. In a typical situation, cold, dense air accumulates in winter over a high plateau or high interior valley. Under favorable conditions, some of this cold air spills over low divides or through passes, flowing out on adjacent lowlands as a strong, cold wind.

Drainage winds occur in many mountainous regions of the world and go by various local names. The *mistral* of the Rhône Valley in southern France is a well-known example—it is a cold, dry local wind. On the ice sheets of Greenland and Antarctica, powerful drainage winds move down the gradient of the ice surface and are funneled through coastal valleys. Picking up loose snow, these winds produce blizzardlike conditions that last for days at a time.

Another type of local wind occurs when the outward flow of dry air from an anticyclone is combined with the local effects of mountainous terrain. An example is the *Santa Ana*—a hot, dry easterly wind that sometimes blows from the interior desert region of southern California across coastal mountain ranges to reach the Pacific coast. This wind is funneled through local mountain gaps or canyon floors where it gains great force. At times the Santa Ana wind carries large amounts of dust. Because this wind is dry, hot, and strong, it can easily fan wildfires in brush or forest out of control (Figure 7.10).

Still another type of local wind—the *chinook,* which we encountered in Chapter 6—results when strong regional winds pass over a mountain range and descend on the lee side. (See Figure 6.16.) As we know from studying orographic precipitation in Chapter 6, the descending air is warmed and dried. A chinook wind can sublimate snow or dry out soils very rapidly.

7.9 Sea and land breezes *The contrast between the land surface, which heats and cools rapidly, and the ocean surface, which has a more uniform temperature, induces pressure gradients to create sea and land breezes.*

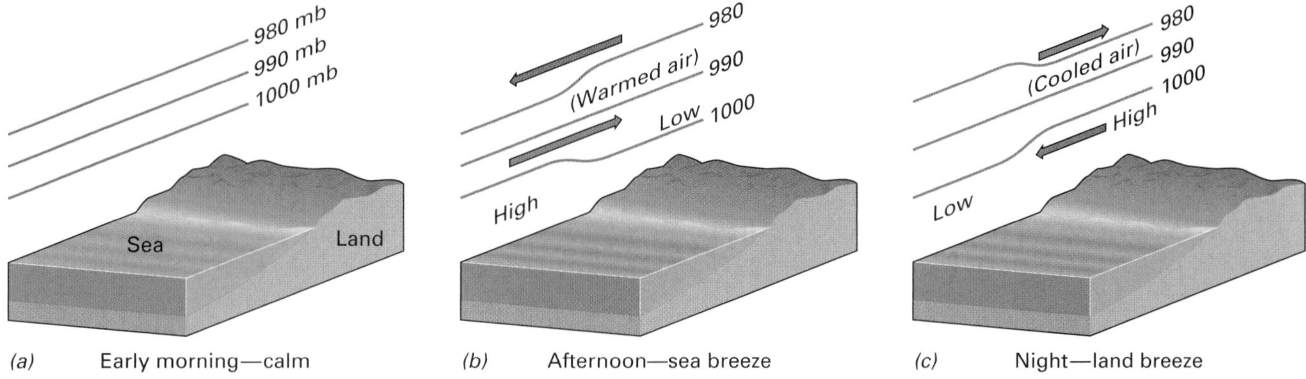

Geographers at Work | The Power of Wind

by Katherine Klink ***The University of Minnesota***

On a walk across a college campus you might feel quite a range of wind speeds and directions. Hurrying down a path between two buildings, a strong gust might push you from behind, while a few moments later, crossing a plaza, you might feel a more gentle wind coming from ahead. Why does the wind direction and speed vary so much from place to place? Clearly, buildings can channel wind, but what about trees? or topographic features such as valleys? Is it windier in the city? or in the country? Windier over forests? or croplands? What about wind direction? How is it that two America's Cup sailboats a few tens of meters away from each other can set their sails to different angles to sail the same course?

These are just the kind of questions that interest Katherine Klink, a geographer at the University of Minnesota who studies the climatology of wind. She is curious about the spatial and temporal variability of surface wind speed and direction. How is the wind at one location related to the wind at a nearby site? How is the wind field related to temperature? These are questions that Katherine seeks to answer by analyzing wind measurements acquired in many settings under many different situations.

Katherine Klink (shown with daughter Lindsey) studies the climatology of winds and the implications for generating power from wind.

Katherine is also interested in harvesting wind power as a renewable energy source. For this purpose, the variability of direction, speed, and duration of wind at a location is a key factor.

Another area of interest for Katherine is long-term change in wind. She has analyzed 30 years of wind speed data from about 175 locations in the continental United States to see if wind speeds are changing as human impact on our climate grows. Has she detected a significant change? Yes. . . .But you'll have to visit our book's web site to find out what she has concluded!

7.10 Eye on the Landscape ***Brushfire*** *Brushfires, such as this one in Oakland, California, are often fanned by strong downslope winds of hot, dry air (Santa Ana winds).* **What else would the geographer see? ... Answers at the end of the chapter.**

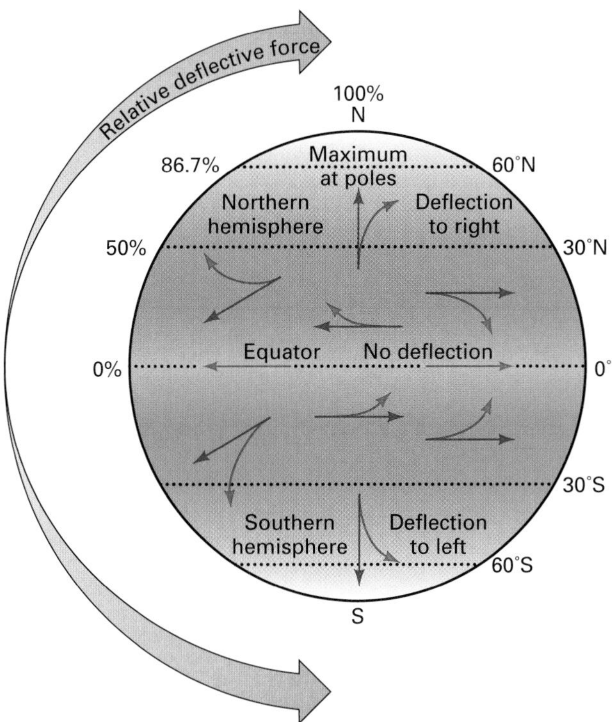

7.11 Coriolis effect direction and strength *The Coriolis effect acts to deflect the paths of winds or ocean currents to the right in the northern hemisphere and to the left in the southern hemisphere as viewed from the starting point. Blue arrows show the direction of initial motion, and red arrows show the direction of motion apparent to the Earth observer.*

Surface winds are generated when pressure gradient forces cause air to move. Energy of motion—kinetic energy—is stored in the air motion. The stored energy can be extracted by windmills or wind turbines to do other work, such as generate electric power. *Eye on the Environment 7.2 • Wind Power, Wave Power, and Current Power* describes this source of energy as well as those of waves and ocean currents.

The Coriolis Effect and Winds

The pressure gradient force tends to move air from high pressure to low pressure. For sea and land breezes, which are local in nature, this push produces a wind motion in about the same direction as the pressure gradient. For wind systems on a global scale, however, the direction of air motion will be somewhat different. The difference is due to the Earth's rotation, through the Coriolis effect. What is this effect, and how does it come about? GEODISCOVERIES

Through the **Coriolis effect,** an object in motion on the Earth's surface always appears to be deflected away from its course. The object moves as though a force were pulling it sideward. The apparent deflection is to the right in the northern hemisphere and to the left in the southern hemisphere. Deflection is strongest near the poles and decreases to zero at the equator. Figure 7.11 diagrams the effect. Note that the apparent force does not depend on direction of motion—the deflection occurs whether the object is moving toward the north, south, east, or west.

The Coriolis effect is a result of the Earth's rotation. Imagine that a rocket is launched from the north pole toward New York, following the 74° W longitude meridian

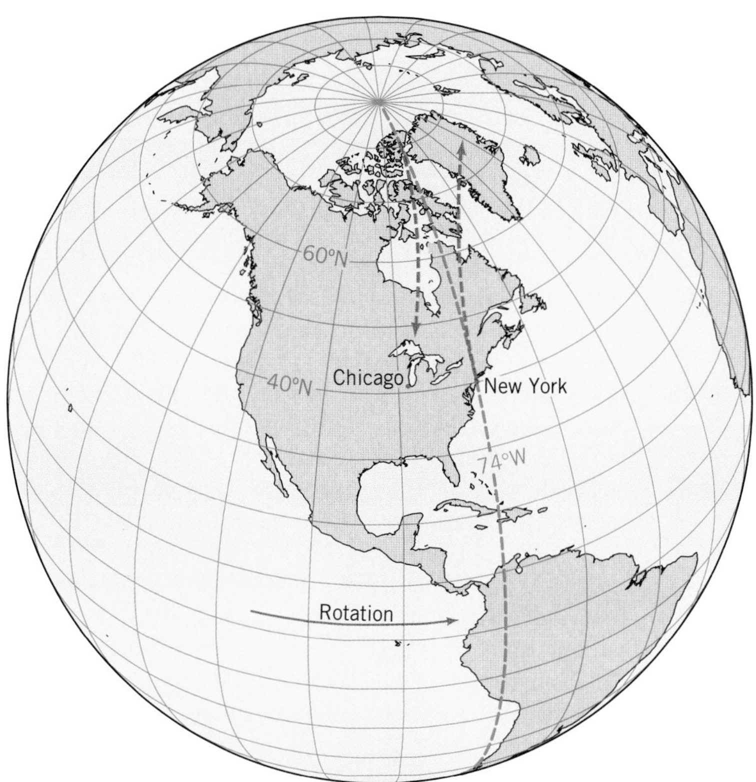

7.12 The Coriolis effect *The path of a rocket launched from the north pole toward New York will apparently be deflected to the right by the Earth's rotation and might land near Chicago. If launched from New York toward the north pole, its path will appear to be deflected to the right as well.*

Eye on the Environment 7.2

Wind Power, Wave Power, and Current Power

Wind power is an indirect form of solar energy that has been used for centuries. In the Low Countries of Europe, the windmill played a major role in pumping water from the polders as they were reclaimed from tidal land. In these low, flat areas, streams could not be adapted to waterpower, and the windmill was also used to grind grain.

The design of new forms of windmills has intrigued inventors for many decades. The total supply of wind energy is enormous. The World Meteorological Organization has estimated that the combined electrical-generating wind power of favorable sites throughout the world comes to about 20 million megawatts, a figure about 100 times greater than the total electrical generating capacity of the United States. Many problems must be solved, however, before wind power can become a major resource.

Small wind turbines are a promising source of supplementary electric power for individual farms, ranches, and homes. The Darrieus rotor, with circular blades turning on a vertical axis, is well adapted to small generators—those with less than 50 kilowatts output. Wind turbines presently in operation and being developed are adjusted in scale according to the purposes they serve. Turbines capable of producing power in the range of 50 to 200 kilowatts are already in operation to serve small communities.

Wind turbines with a generating capacity in the range of 50 to 100 kilowatts have been assembled in large numbers at favorable locations to form "wind

Modern windmills *A wind farm in San Gorgonio Pass, near Palm Springs, California. Wind speeds here average 7.5 m/s (17m/hr).*

(Figure 7.12). However, as it travels toward New York, the Earth rotates beneath its trajectory, moving from west to east. The rocket continues on its straight path, but to an observer at the launch point on the rotating Earth below, the rocket will appear to curve to the right, away from New York and toward Chicago. For the rocket to reach New York, its flight path would have to be adjusted to take the Earth's rotation into account.

What would happen if the rocket were launched from New York heading north along the 74° W meridian toward the pole? Because of the Earth's rotation, a point on the Earth's surface at the latitude of New York (about 40° N) moves eastward at about 1300 km/hr (about 800 mi/hr). Although the rocket is aimed properly along the meridian, at its launch its motion will have an eastward component of 1300 km/hr. However, as the rocket continues northward, it passes over land that is moving less swiftly eastward. At 60° N latitude, the eastward velocity of the Earth's surface is only about 800 km/hr (about 500 mi/hr). So, the rocket will appear to move more quickly eastward than the land underneath, and its path will appear deflected to the right.

From this analysis, we can see that the Coriolis effect is actually produced because observers on the ground and the geographic grid around them are not fixed but are in motion along with the Earth. Figure 7.13 illustrates this point. Imagine yourself fixed in space above the north pole *(a)*. As you watch, the geographic grid slowly turns underneath you. A straight motion, such as that of a rocket, will trace a curving path across the turning grid. Now imagine yourself fixed above New York, viewing the Earth from about the same point in space as in Figure 7.12. As you watch the Earth turn to the right, the intersections of meridians and parallels near New York will appear to rotate counterclockwise. Again, a straight path will appear to curve on this turning grid. Now imagine yourself at the equator *(c)*. Here the grid will move eastward but will not rotate. So a straight path does not curve at the equator. This is why the Coriolis effect is not felt at the equator.

farms." Arranged in rows along ridge crests, the turbines intercept local winds of exceptional frequency and strength. One such locality is the Tehachapi Pass in California, where winds moving southeastward in the San Joaquin Valley are funneled through a narrow mountain gap before entering the Mojave desert. The San Gorgonio Pass, near Palm Springs, presents a similar situation favorable to the development of wind farms. Another important wind farm locality lies about 80 km (50 mi) east of San Francisco in the Altamont Pass area of Alameda and Contra Costa counties. Here, daytime westerly winds of great persistence develop in response to the buildup of surface low pressure in the Great Valley to the east. A group of wind farms built here with a total of about 3000 turbines provides the Pacific Gas and Electric Company with a yield of over 500 million kilowatt-hours of electricity per year. Larger wind turbines, each capable of generating 2 to 4 megawatts, have been constructed at a number of sites.

Wave energy is another indirect form of solar energy. Nearly all ocean waves are produced by the stress of wind blowing over the sea surface. Water waves are characterized by motion of water particles in vertical orbits. Energy is extracted from wave motion by use of a floating object tethered to the seafloor. As the floating mass rises and falls, a mechanism is operated and drives a generator. Pneumatic systems use the principle of the bellows, operated by the changing pressure of the surrounding water as the water level rises and falls. A number of such devices are in the planning and testing stages.

Harnessing the vast power of a great current stream, such as the Gulf Stream or the Kuroshio, is a prospect that has not gone unnoticed. One rather grandiose plan proposes a large number of current-driven turbines to be tethered to the ocean floor so as to operate below the ocean surface. Each turbine would be 170 m (558 ft) in diameter and capable of generating 83 megawatts of power. An array of 242 such turbines could deliver 10,000 megawatts—a large part of Florida's power requirement and equivalent to the use of about 130 million barrels of crude oil per year.

Vertical wind turbines *These vertical wind turbines generate power without needing to rotate to face the wind. Altamont Pass, California.*

7.13 Rotation of the geographic grid *The Coriolis effect is produced by Earth's rotation, which turns the geographic grid. This makes straight motion appear curved. Apparent rotation is greatest at the poles and decreases to zero at the equator.*

(*a*) North Pole

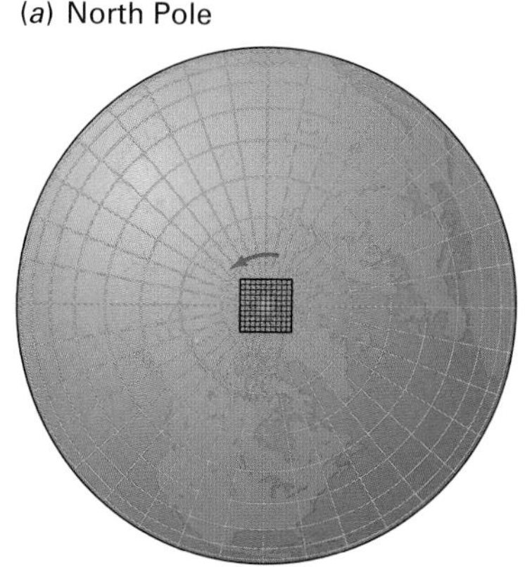

(*b*) Midlatitude location

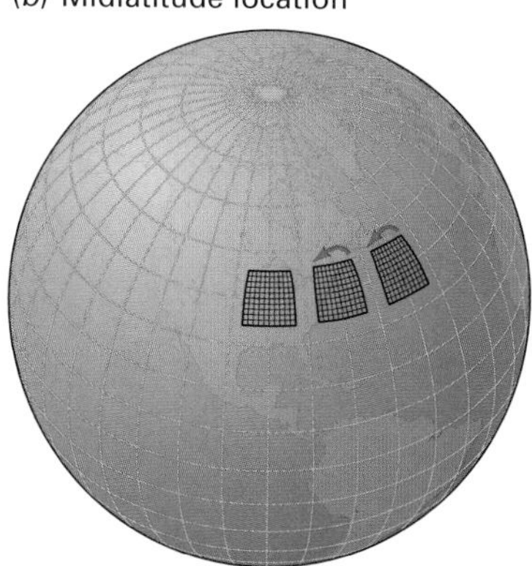

(*c*) Equator

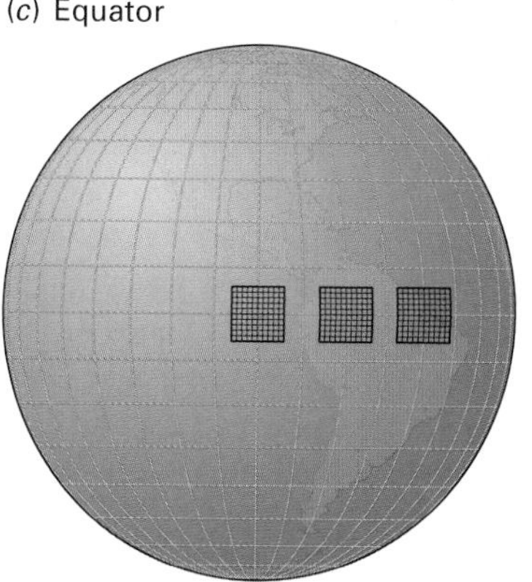

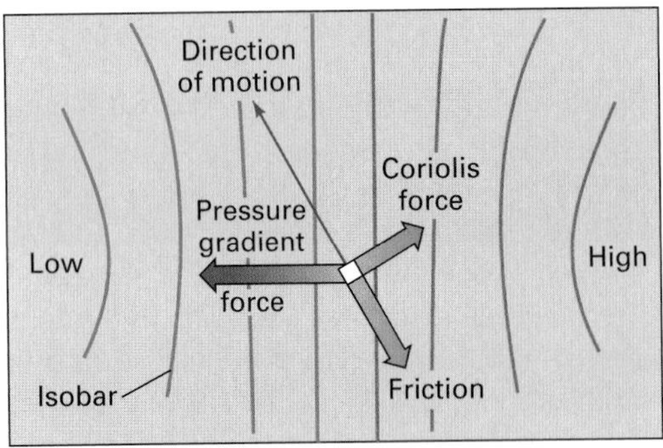

7.14 Balance of forces on a parcel of surface air
Although the pressure gradient will push a parcel of air toward low pressure, it will be deflected by the Coriolis force and slowed by friction with the surface. The direction of air motion will be the result of the three forces acting together.

In most real situations, we are concerned with analyzing the motions of ocean currents or air masses not from the viewpoint of space, but from the viewpoint of an Earth observer on the geographic grid. As a shortcut, we can treat the Coriolis effect as a sideward-turning force that always acts at right angles to the direction of motion. The strength of the Coriolis force increases with the speed of motion but decreases with latitude. If we treat the Coriolis effect this way, it will properly describe motion within the geographic grid.

Considering the Coriolis effect as a sideward-turning force acting on Earthbound motion, how does the Coriolis force effect wind patterns? A parcel of air in motion near the surface is subjected to three forces (Figure 7.14). First is the pressure gradient force that pushes the parcel toward low pressure. Second is the Coriolis force, which always acts at right angles to the direction of motion. Third is a frictional force exerted by the ground surface, which is proportional to the wind speed and always acts in the opposite direction to the direction of motion. The sum of these three forces produces a direction of motion that is toward low pressure but at an angle to the pressure gradient.

Cyclones and Anticyclones

Low- and high-pressure centers are common features of the daily weather map. Think of them as marking vast whirls of air in spiraling motion. In low-pressure centers, known as **cyclones,** air spirals inward and upward. This inward spiraling motion is *convergence.* In high-pressure centers, known as **anticyclones,** air spirals downward and outward. This outward spiraling motion is *divergence.*

How does the inward and outward motion of air in cyclones and anticyclones come about? Figure 7.15 shows surface wind patterns around cyclones and anticyclones in northern and southern hemispheres. When low pressure is at the center (cyclone), the pressure gradient is straight inward (left side of figure). When high pressure is at the center (anticyclone), the gradient is straight outward (right side). But because of the Coriolis force and friction with the surface, the surface air moves at an angle across the gradient. This creates the spiraling motion. In the northern hemisphere (upper portion), the cyclonic spiral will be counterclockwise because the Coriolis force acts to the right. In the southern hemisphere (lower portion), the cyclonic spiral will be clockwise because the Coriolis force acts to the left. For anticyclones, the situation is reversed.

We know from experience that low-pressure centers (cyclones) are often associated with cloudy or rainy weather and that high-pressure centers (anticyclones) are often associated with fair weather. Why? Recall from Chapter 6 that when air is forced aloft, it is cooled according to the adiabatic principle, and condensation and precipitation can occur. Thus, cloudy and rainy weather is often associated with the inward and upward air motion of cyclones. In the anticyclone, diverging air sinks and spirals outward. When air descends, we know from Chapter 6 that it is warmed according to the adiabatic principle, and thus condensation cannot occur. So anticyclones are often associated with fair weather.

Cyclones and anticyclones are typically large features of the lower atmosphere—perhaps a thousand kilometers (about 600 mi) across, or more. For example, a fair weather system, which is an anticyclone, may stretch from the Rockies to the Appalachians. Cyclones and anticyclones can remain more or less in one location, or they can move, sometimes rapidly, to create weather disturbances.

7.15 Air motion in cyclones and anticyclones
Surface winds spiral inward toward the center of a cyclone but outward and away from the center of an anticyclone. Because the Coriolis effect deflects moving air to the right in the northern hemisphere and to the left in the southern hemisphere, the direction of inspiraling and outspiraling reverses from one hemisphere to the other.

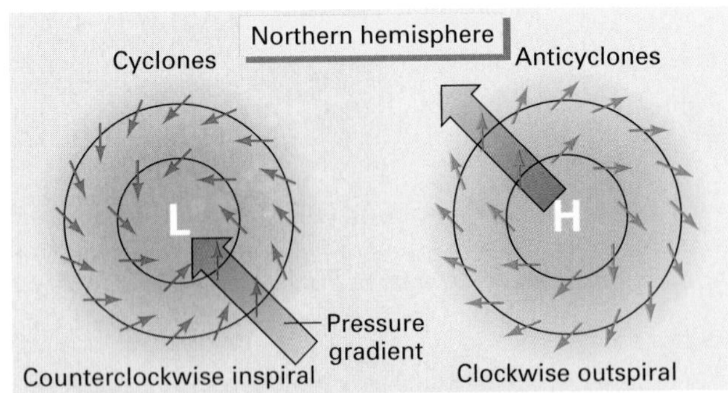

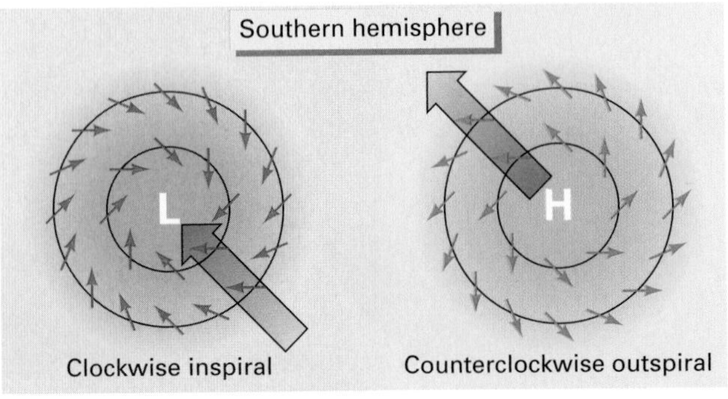

The vertical motions of air in adjacent cyclones and anticyclones are often connected in a convection loop. This loop can be viewed as an energy flow system and is described in more detail in *Focus on Systems 7.3 • The Convection Loop as an Energy Flow System.* Because a spiraling motion is involved, this loop is more complex than the simple ones we have discussed in this chapter and in Chapter 6.

Surface Winds on an Ideal Earth

Consider wind and pressure patterns on an ideal Earth—one without a complicated pattern of land and water and no seasonal changes. Pressure and winds for such an ideal Earth are shown in Figure 7.16. The surface wind patterns are shown on the globe's surface, and the wind patterns aloft are shown in the cross section on the right. GEODISCOVERIES

The most important features for understanding the wind pattern are Hadley cells, shown on the cross section. Recall that when air is heated, a convection loop can be formed. Since insolation is strongest when the Sun is directly overhead, the surface and atmosphere at the equator will be heated more strongly than other places on this featureless Earth. The result will be two convection loops, the **Hadley cells,** which form in the northern and southern hemispheres. In each Hadley cell, air rises over the equator and is drawn poleward by the pressure gradient. Of course, the air is turned by the Coriolis force and so heads westward as well as poleward. Eventually, it descends at about a 30° latitude, completing a convection loop. (The Hadley cell is named for George Hadley, who first proposed its existence in 1735.)

Since air rises at the equator, a zone of surface low pressure will result. This zone is known as the *equatorial trough.* As shown by the wind arrows, air in both hemispheres moves toward the equatorial trough, where it converges and moves aloft as part of the Hadley cell circulation. Convergence occurs in a narrow zone, named the **intertropical convergence zone (ITCZ).** Because the air is generally rising, winds in the ITCZ are light and variable. In earlier centuries mariners on sailing vessels referred to this region as the *doldrums,* where they were sometimes becalmed for days at a time.

On the poleward side of the Hadley cell circulation, air descends and surface pressures are high. This produces two **subtropical high-pressure belts,** each centered at about 30° latitude. Within the belts, two, three, or four very large and stable anticyclones are formed. At the centers of these anticyclones, air descends, and winds are weak. Calm prevails as much as one-quarter of the time. Because of the high frequency of calms, mariners named this belt the *horse latitudes.* The name is said to have originated in colonial times when New England traders carried cargoes of horses to the West Indies. When their ships were stranded for long

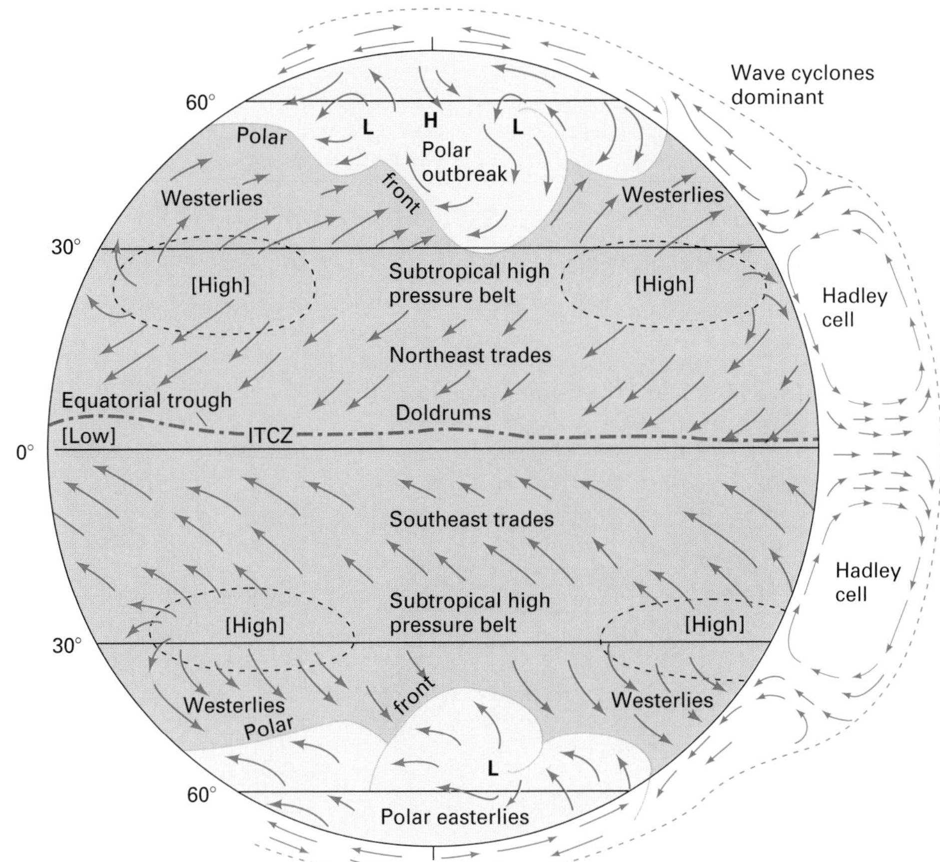

7.16 Global surface winds on an ideal Earth *This schematic diagram of global surface winds and pressures shows the features of an ideal Earth, without the disrupting effect of oceans and continents and the variation of the seasons. Surface winds are shown on the disk of the Earth, while the cross section at the right shows winds aloft.*

Focus on Systems | 7.3

The Convection Loop as an Energy Flow System

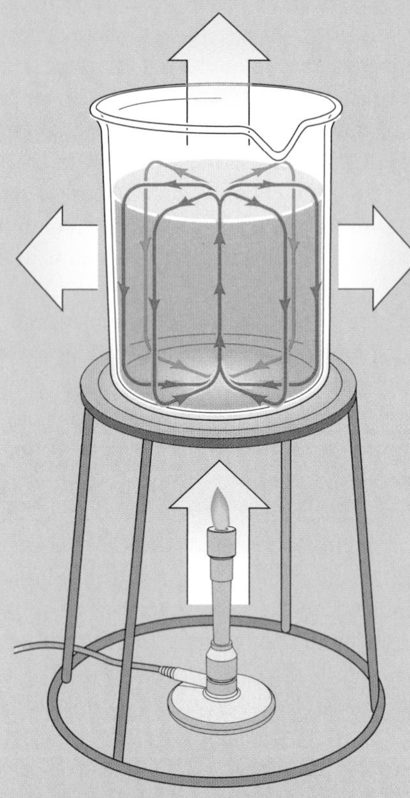

Convection in a heated beaker *The convectional circulation in a beaker of water heated by a low flame constitutes a flow system in which the fluid cycles in a three-dimensional loop and heat (broad arrows) is transported from the bottom center of the vessel to the water surface and sides of the container, where it warms the surrounding air.*

In Chapters 4 and 6, we established the principle that when an air column is heated, it creates a pressure gradient force that sets up a convection loop. The convection loop is actually a flow system in which a closed system of matter flow is linked with an open system of energy flow. To make this clearer, think of a laboratory beaker of water gently heated by a single, low flame, as sketched above. The flame heats the water directly above it, which expands and becomes less dense than the surrounding fluid. In the same way that this sets up a pressure gradient and convection loop in the atmosphere (Figure 7.8), a convection loop forms in the beaker. Warm water ascends in a cylinder at the center of the beaker, spreads out across the top of the liquid, descending along the sides of the beaker, and finally converges above the flame to be heated and rise again.

What happens to the sensible heat that was absorbed by the water? Some of the heat will be lost to the air above the beaker when the warm liquid spreads out over the top surface. Another portion will be lost from the sides of the beaker as the water descends and warms the glass and the air that surrounds the glass. Yet another portion will go to warm the liquid as a whole, since some mixing will occur even though the flow in the loop may be fairly smooth.

If we wait for a while, the system will come to a steady state in which the liquid constantly moves in a three-dimensional convection loop, absorbing heat at the bottom of the beaker and releasing heat at the top surface and sides. The net effect is then a continuous cycling of matter while energy flows in at one location and out at another.

When large convection loops are set up in the Earth's atmosphere by unequal surface heating, they are not so simple as the example of water heating in a beaker. The rotation of the Earth, through the Coriolis effect, induces a spiraling motion to the flows. The figure below shows an anticyclone and a cyclone linked together in a convection loop, as they might be in a sequence of eastward-moving weather systems. In the anticylone (high-pressure center), air converges aloft and descends in a spiraling motion. Near the surface, the flow diverges, fanning out and moving away from the center of the anticyclone. In the cyclone, air converges near the surface and spirals upward. Aloft, the air diverges and spirals outward.

The upward and downward flows are linked as a portion of the moving air flows from the anticyclone to the cyclone near the surface, and another portion flows from the cyclone to the anticyclone aloft. The loop is not perfect. Air escapes and enters from the sides of the flow, as shown by the arrows.

Dynamic convection loop *A cyclone and anticyclone linked together in a convection loop.*

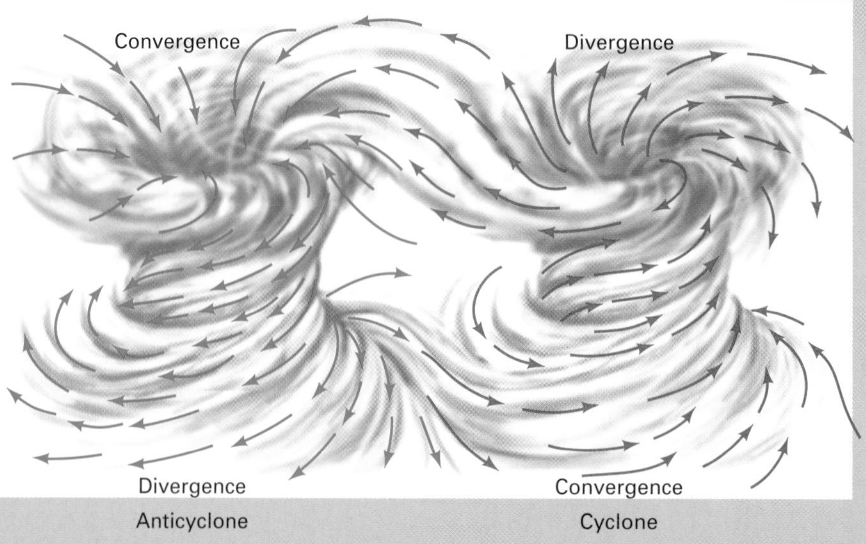

periods of time and the fresh water supplies ran low, the horses were thrown overboard.

Winds around the subtropical high-pressure centers are outspiraling and move toward equatorial as well as middle latitudes. The winds moving equatorward are the strong and dependable *trade winds*. North of the equator, these are from the northeast and are referred to as the *northeast trades*. To the south of the equator, they are from the southeast and are the *southeast trades*. Poleward of the subtropical highs, air spiraling outward produces southwesterly winds in the northern hemisphere and northwesterly winds in the southern hemisphere.

Between about 30° and 60° latitude, the pressure and wind pattern becomes more complex. This latitudinal belt is a zone of conflict between air bodies with different characteristics. Masses of cool, dry air move into the region, heading eastward and equatorward. These form *polar outbreaks*. The border of the polar outbreak is known as the *polar front*. (We'll have much more to say about air masses and the polar front in Chapter 8.) As a result of this activity, pressures and winds can be quite variable in the midlatitudes from day to day and week to week. On average, however, winds are more often from the west, so the region is said to have *prevailing westerlies*.

At the poles, the air is intensely cold. As a result, high pressure occurs. Outspiraling of winds around a polar anticyclone should create surface winds from a generally easterly direction, known specifically as *polar easterlies*. As we will see in discussing actual pressure and wind patterns, this situation exists only in the south polar region. In the north polar region, winds tend to have an eastward component, but there is too much variation in direction to consider polar easterlies as dominant winds there.

Global Wind and Pressure Patterns

So far, we've discussed the wind pattern for a seasonless, featureless Earth. Let's turn now to actual global surface wind and pressure patterns. Here, we will use the global maps of wind and pressure for January and July, shown in Figure 7.17. The pressures and winds shown are averages over many years for all daily observations in either January or July. They are corrected for the elevation of the recording station, so that pressures are shown for sea level. Average barometric pressure is 1013 mb (29.2 in. Hg). Values greater than this value are "high" (red lines), while values lower are "low" (green lines). The maps make use of both Mercator and polar projections. As we discuss the various features of global pressure and winds, be sure to examine both sets of maps. GEODISCOVERIES

Subtropical High-Pressure Belts

The largest and most prominent features of the maps are the subtropical high-pressure belts, created by the Hadley cell circulation. The high-pressure belt in the southern hemisphere conforms well to the pattern of Figure 7.16. It has three large high-pressure cells, each developed over oceans, that persist year round. A fourth, weaker high-pressure cell forms over Australia in July, as the continent cools during the southern hemisphere winter.

In the northern hemisphere, the situation is somewhat different. The subtropical high-pressure belt shows two large anticyclones centered over oceans—the Hawaiian High in the Pacific and the Azores High in the Atlantic. From January to July, these intensify and move northward. From their positions off the east and west coasts of the United States, they have a dominant influence on summer weather in North America.

Figure 7.18 is a schematic map of two large high-pressure cells bounded by continents, showing the features of circulation around the cells. As we noted earlier, the outspiraling circulation produces the easterly trade winds in the tropical and equatorial zones, and the westerlies in the subtropical zones and poleward. In these high-pressure cells, air on the east side subsides more intensely. This means that winds spiraling outward on the eastern side are drier. On the west side, subsidence is less strong. In addition, these winds travel long distances across warm, tropical ocean surfaces before reaching land, and so they pick up moisture and heat on their journey.

When the Hawaiian and Azores Highs intensify and move northward in the summer, our east and west coasts feel their effects. On the west coast, dry subsiding air from the Hawaiian High dominates, so fair weather and rainless conditions prevail. On the east coast, warm, moist air from the Azores High flows across the continent from the southeast, producing generally hot, humid weather for the central and eastern United States. In winter, these two anticyclones weaken and move to the south—leaving our weather to the mercy of colder winds and air masses from the north and west.

The ITCZ and the Monsoon Circulation

Recall from Chapter 4 that insolation is most intense when the Sun is directly overhead. Remember, too, that the latitude at which the Sun is directly overhead changes with the seasons, migrating between the tropics of cancer and capricorn. Since the Hadley cell circulation is driven by this heating, we can expect that the elements of the Hadley cell circulation—ITCZ and subtropical high-pressure belts—will also shift with the seasons. We have already noted this shift for the Hawaiian and Azores Highs, which intensify and move northward as July approaches.

In general, the ITCZ also shifts with the seasons, as we might expect. The shift is moderate in the western hemisphere. The ITCZ moves a few degrees north from January to July over the oceans. In South America, the ITCZ lies across the Amazon in January and swings northward by about 20° to the northern part of the continent. But what happens in Africa and Asia? By comparing the two Mercator maps of Figure 7.17, you can see that there is a huge shift indeed! In January, the ITCZ runs south across eastern Africa and crosses the Indian Ocean to northern Australia at

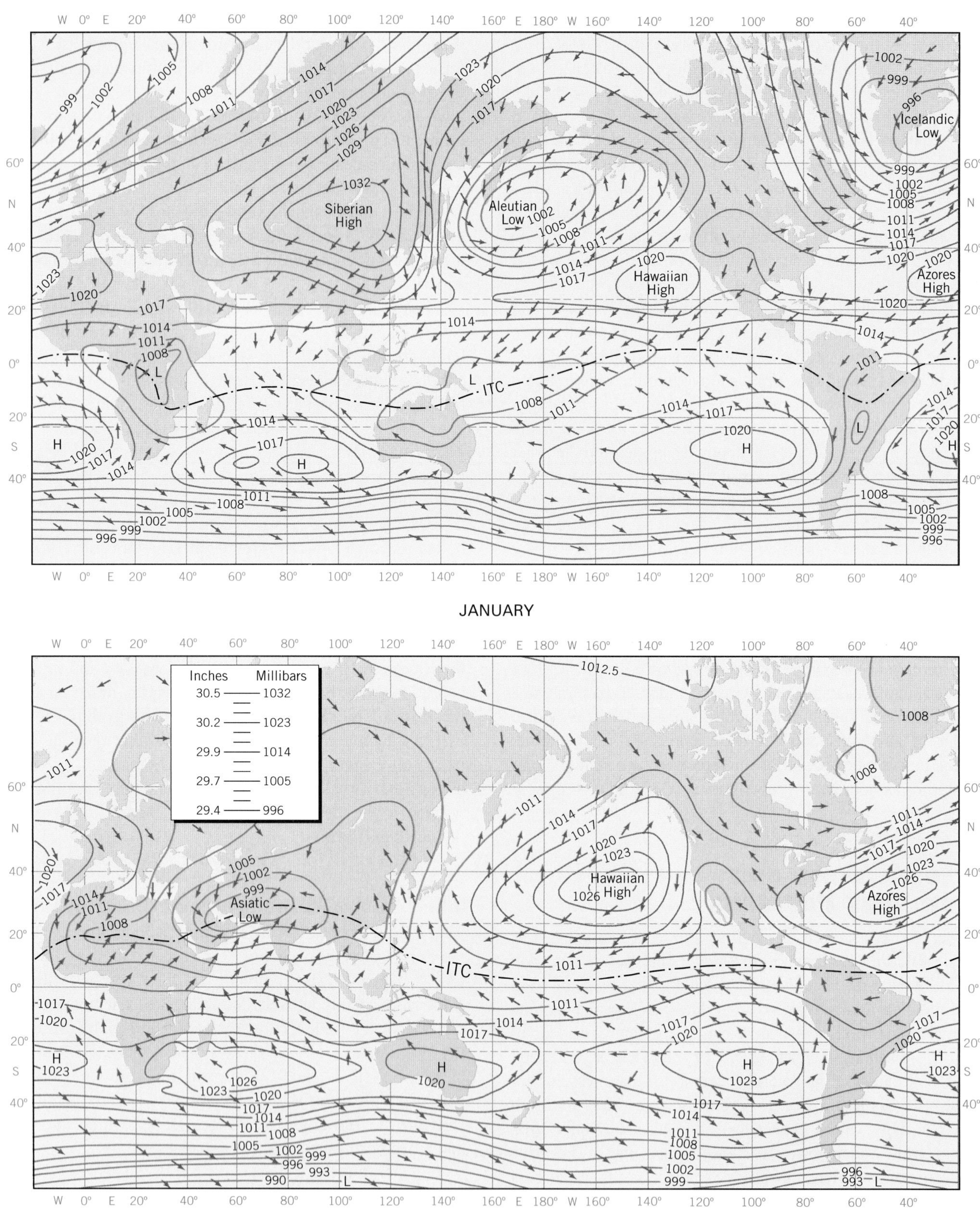

7.17 Atmospheric pressure maps *The maps show mean monthly atmospheric pressure and prevailing surface winds for January and July. Pressure units are millibars reduced to sea level. Many of the wind arrows are inferred from isobars. (Data compiled by John E. Oliver.) (Figure continues)*

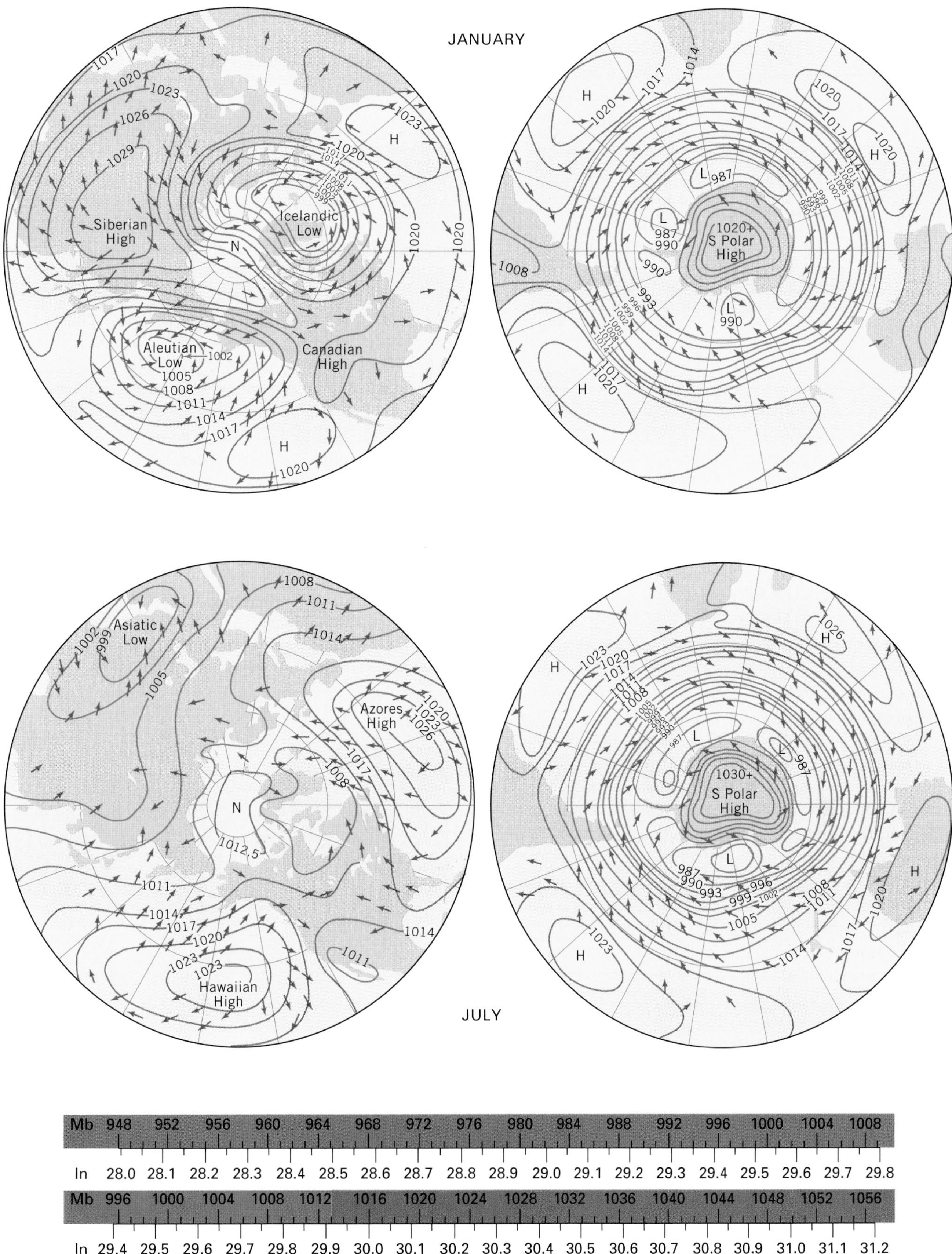
JANUARY
Siberian High
Icelandic Low
N
Aleutian Low
Canadian High
H
L
S Polar High
1020+
JULY
Asiatic Low
Azores High
Hawaiian High
1012.5
1030+
Mb 948 952 956 960 964 968 972 976 980 984 988 992 996 1000 1004 1008
In 28.0 28.1 28.2 28.3 28.4 28.5 28.6 28.7 28.8 28.9 29.0 29.1 29.2 29.3 29.4 29.5 29.6 29.7 29.8
Mb 996 1000 1004 1008 1012 1016 1020 1024 1028 1032 1036 1040 1044 1048 1052 1056
In 29.4 29.5 29.6 29.7 29.8 29.9 30.0 30.1 30.2 30.3 30.4 30.5 30.6 30.7 30.8 30.9 31.0 31.1 31.2

7.17 *(Continued)*

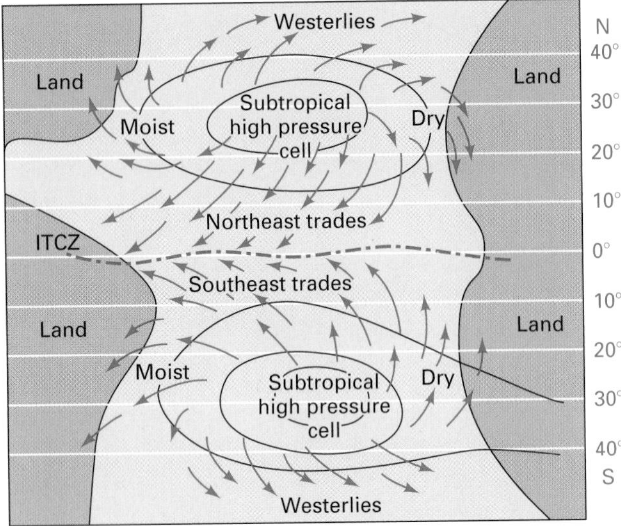

7.18 Subtropical high-pressure cells *Over the oceans, surface winds spiral outward from the subtropical high-pressure cells, feeding the trades and the westerlies. On the eastern side of the cells, air subsides more strongly, producing dry winds. On the western sides, subsidence is not as strong, and a long passage over oceans brings warm, moist air to the continents.*

a latitude of about 15° S. In July, it swings north across Africa along the southern side of the Sahara, and then it rises further north to lie along the south rim of the Himalayas, in India, at a latitude of about 25° N. This is a shift of about 40 degrees of latitude!

Why does such a large shift occur in Asia? To answer this question, look at the pressure and wind pattern over Asia. In January, an intense high-pressure system, the Siberian High, is found there. Recall from Chapter 5 that, in winter, temperatures in northern Asia are very cold, so this high pressure is to be expected. In July, this high-pressure center is absent, replaced instead by a low centered over the Middle Eastern desert region. The Asiatic low is produced by the intense summer heating of the landscape.

The movement of the ITCZ and the change in the pressure pattern with the seasons create a reversing wind pattern in Asia known as the **monsoon** (Figure 7.19). In the winter, there is a strong outflow of dry, continental air from the north across China, Southeast Asia, India, and the Middle East. During this *winter monsoon,* dry conditions prevail. In the summer, warm, humid air from the Indian Ocean and the southwestern Pacific moves northward and northwestward into Asia, passing over India, Indochina, and China. This airflow is known as the *summer monsoon* and is accompanied by heavy rainfall in southeastern Asia.

North America does not have the remarkable extremes of monsoon winds experienced in Asia. Even so, in summer there is a prevailing tendency for warm, moist air originating in the Gulf of Mexico to move northward across the central and eastern part of the United States. At times, moist air from the Gulf of California also invades the desert Southwest, causing widespread scattered thundershowers and creating the "Arizona monsoon." In winter, the airflow pattern across North America changes, and dry, continental air from Canada moves south and eastward.

Wind and Pressure Features of Higher Latitudes

The northern and southern hemispheres are quite different in their geography. As you can easily see from Figure 7.20, the northern hemisphere has two large continental masses, separated by oceans. An ocean is also at the pole. In the southern hemisphere, we find a large ocean with a cold, glacier-covered continent at the center. These differing land–water patterns strongly influence the development of high- and low-pressure centers with the seasons.

Let's examine the northern hemisphere first (Figure 7.17). Recall from Chapter 5 that continents will be cold in winter and warm in summer, as compared to oceans at the same latitude. But we know from our analysis earlier in the chapter that cold air will be associated with surface high pressure and warm air with surface low pressure. Thus, continents will show high pressure in winter and low pressure in summer.

7.19 Monsoon wind patterns *The Asiatic monsoon winds alternate in direction from January to July, responding to reversals of barometric pressure over the large continent. (Isobars labeled in millibars.)*

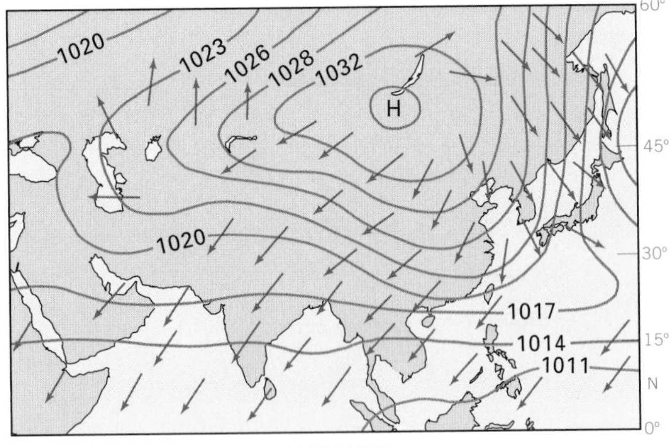

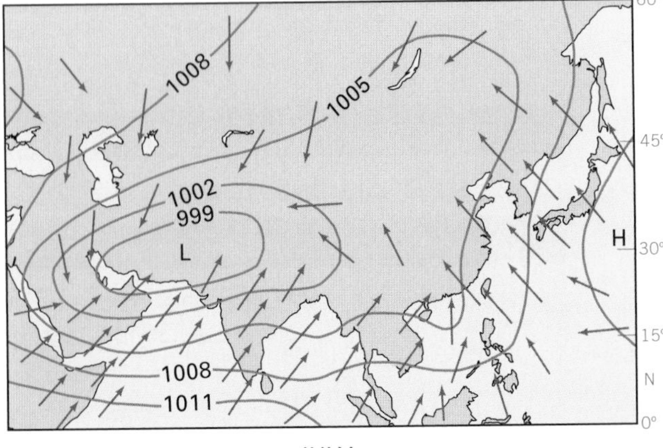

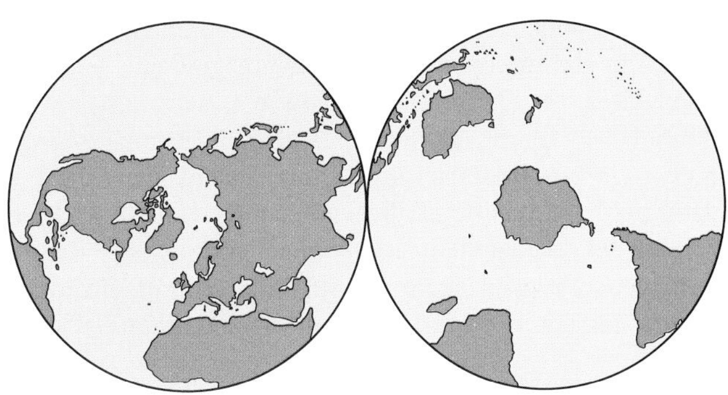

7.20 ***Land–ocean contrasts by hemisphere*** *The northern and southern hemispheres are quite different in their distributions of lands and oceans. One pole bears a deep ocean, and the other a great continental land mass.*

This pattern is very clear in the northern hemisphere polar maps of Figure 7.17. In winter, the strong Siberian High is found in Asia, and a weaker cousin, the Canadian High, is found in North America. From these high-pressure centers, air spirals outward, bringing cold air to the south. Over the oceans, two large centers of low pressure are striking—the Icelandic Low and the Aleutian Low. These two low-pressure centers are not actually large stable features that we would expect to find on every daily world weather map in January. Rather, they are regions of average low pressure where winter storm systems are spawned.

In summer, the pattern reverses. The continents show generally low surface pressure, while high pressure builds over the oceans. This pattern is easily seen on the northern hemisphere July polar map. The Asiatic Low is strong and intense. Inspiraling winds, forming part of the monsoon cycle, bring warm, moist Indian Ocean air over India and Southeast Asia. A lesser low forms over the deserts of the southwestern United States and northwestern Mexico. The two subtropical highs, the Hawaiian Highs and the Azores Highs, strengthen and dominate the Atlantic and Pacific Ocean regions. Outspiraling winds, following the pattern of Figure 7.18, keep the west coasts of North America and Europe warm and dry, and the east coasts of North America and Asia warm and moist (see Figure 7.17 also).

The higher latitudes of the southern hemisphere present a polar continent surrounded by a large ocean. Since Antarctica is covered by a glacial ice sheet and is cold at all times, a permanent anticyclone, the South Polar High, is centered there. Easterly winds spiral outward from the high-pressure center. Surrounding the high is a band of deep low pressure, with strong, inward-spiraling westerly winds. As early mariners sailed southward, they encountered this band, in which wind strength intensifies toward the pole. Because of the strong prevailing westerlies, they named these southern latitudes the "roaring forties," "flying fifties," and "screaming sixties." GEODISCOVERIES

WINDS ALOFT

The surface wind systems we have examined describe airflows at or near the surface. How does air move at the higher levels of the troposphere? Just as we saw for air near the surface, air aloft will move in response to pressure gradients and will be influenced by the Coriolis effect.

How do pressure gradients arise at upper levels? Before answering this question, we need to restate a simple principle that appeared in our discussion of the convection loop and is shown in Figure 7.8*b*. This principle is that pressure decreases less rapidly with height in warmer air than in colder air. The principle is derived from fundamental physical laws describing the temperature, density, and pressure relationships within a thick layer of gas. Also recall that the Earth's insolation is greatest near the equator and least near the poles. Thus, there will be a temperature gradient from the equator to the poles.

Figure 7.21 shows a diagram of this situation—a simple generalized cross section of the atmosphere from the pole on the left *(X)* to about 30° latitude on the right *(Y)*. The isobaric surfaces slope downward from the low latitudes to the pole. This creates a pressure gradient force. For example, at

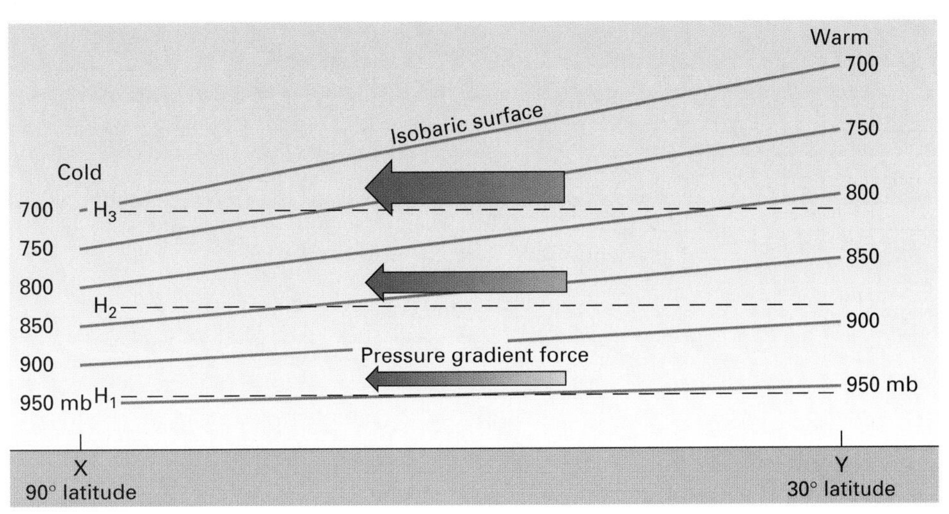

7.21 ***Upper-air pressure gradient*** *Because the atmosphere is warmer near the equator than at the poles, a pressure gradient force acts to push air poleward. The gradient force increases with altitude.*

height H_1, the pressure at X is about 940 mb, while at Y it is about 955 mb, creating a pressure gradient of 15 mb. But note how the pressure difference becomes greater with elevation. At H_2, the polar pressure is about 825, while the low-latitude pressure is about 890. Here the pressure gradient is 65 mb. At H_3, the gradient is about 100 mb. Of course, these values are just examples, but they show that the pressure gradient force increases with altitude, bringing strong winds at high altitudes.

The Geostrophic Wind

Given that a pressure gradient force exists pushing poleward, how will that gradient produce wind and what will be the wind direction? Recall that any wind motion will be subjected to the Coriolis force, which will turn it rightward in the northern hemisphere and leftward in the southern hemisphere. Thus, poleward air motion will be toward the east, creating west winds in both hemispheres.

Figure 7.22*a* shows how the Coriolis force acts on an upper air parcel. Compared to Figure 7.14, which diagrams the forces on a parcel moving close to the surface, we see that the upper air parcel moves without a friction force, since it is so far from the surface. Thus, there are only two forces on the parcel—the pressure gradient force and the Coriolis force. Imagine a still parcel of air that begins to move poleward in response to the pressure gradient force *(b)*. At first it travels in the direction of the pressure gradient, but as it accelerates, the Coriolis force pulls it increasingly toward the right. As its velocity increases, the parcel turns increasingly rightward until the Coriolis force just balances the gradient force. At that point, the sum of forces on the parcel is zero, so its speed and direction remain constant. We use the term *geostrophic wind* to identify this type of airflow. It occurs at upper levels in the atmosphere and is parallel to the isobars.

Figure 7.23 shows an upper-air map for North America on a late June day. The contours show the height of the 500-mb pressure level, and from our analysis of Figure 7.21 above, we can easily see that the 500-mb surface will dip down toward the ground where the upper-air pressure is lower and the air column is cooler. Similarly, the surface will rise up when the upper-air pressure is higher and the air column is warmer. So on this map, low height is low pressure, and high height is high pressure.

The map shows a large, upper-air low centered over the Great Lakes. Since low pressure aloft indicates cold air, this is a mass of cold, Canadian air that has moved southward to dominate the eastern part of the continent. A large, but weaker, upper-air high is centered over the southwestern desert. This will be a mass of warm air, heated by intense surface insolation on the desert below.

7.22 Geostrophic wind *(a) At upper levels in the atmosphere, a parcel of air is subjected to a pressure gradient force and a Coriolis force. (b) As a parcel of air moves in response to a pressure gradient, it is turned progressively sidewards until the gradient force and Coriolis force balance, producing the geostrophic wind.*

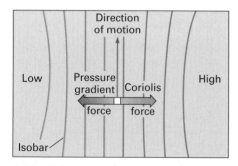

Direction of motion
Low
Pressure gradient force
Coriolis force
High
Isobar

(a)

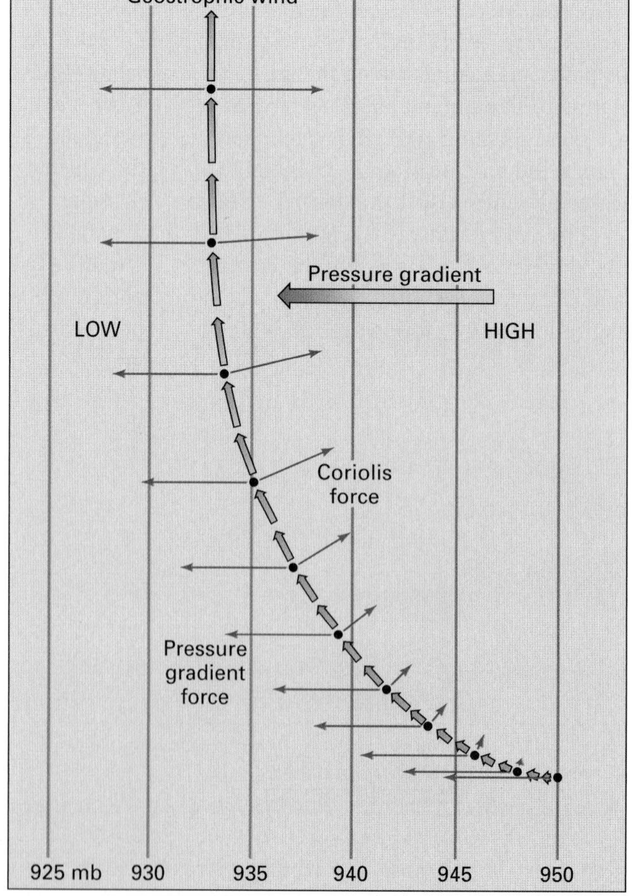

(b)

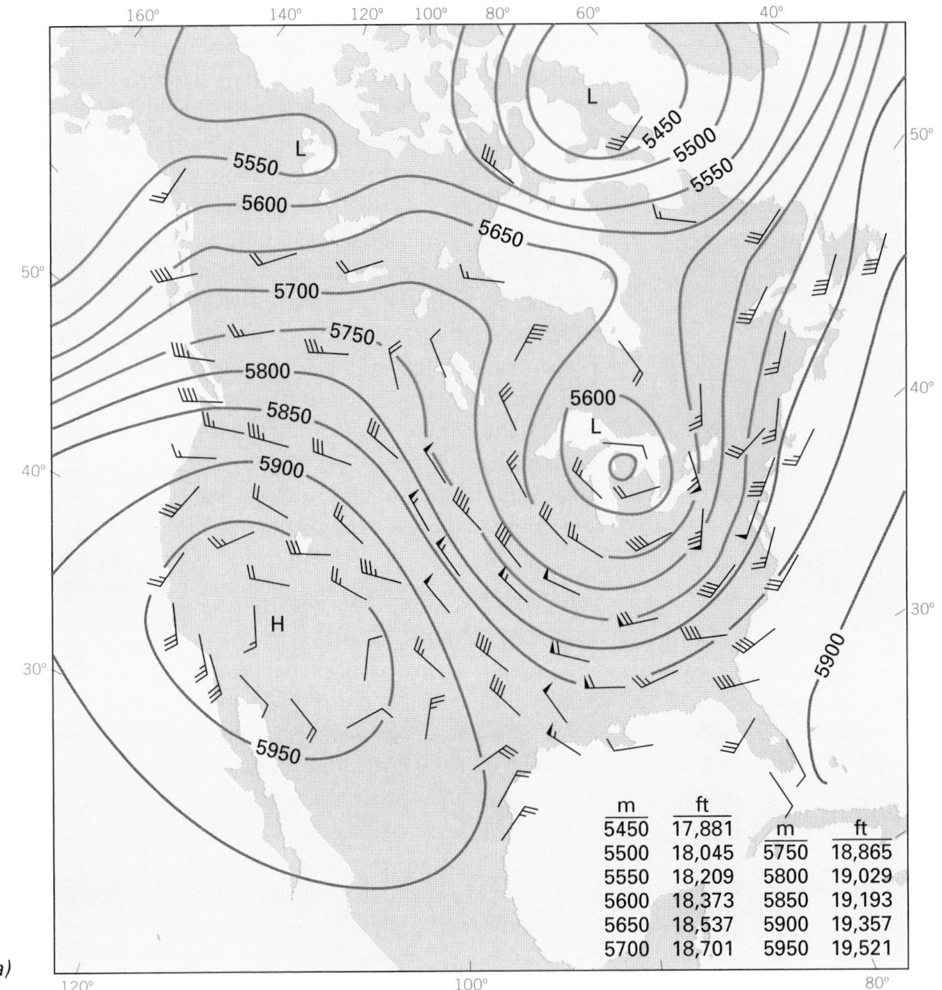

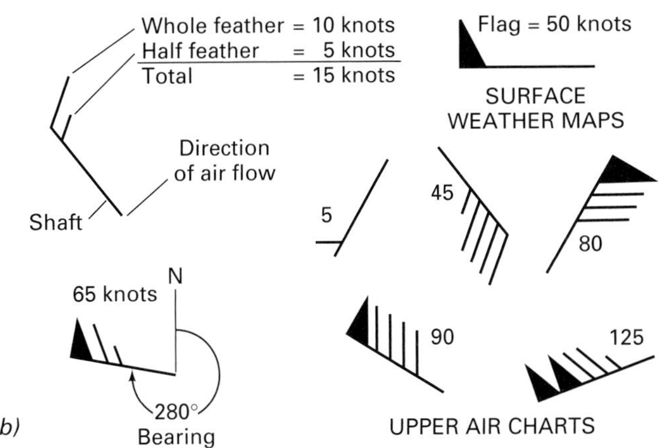

7.23 Upper-air wind map *(a) An upper-air map for a day in late June. Lines are height contours for the 500-mb surface. (b) Explanation of wind arrows. 1 knot (nautical mile per hour) = 0.514 meters per second.*

The figure also shows upper-air wind strengths and directions using the "arrow" symbol employed on scientific weather maps. Note that the winds generally follow the contours well, showing geostrophic flow. They are strongest where the contours are closest together because there the pressure gradient force will be strongest. Note also that the winds around the upper-air low tend to spiral inward and converge on the low. The converging winds descend to the surface, where surface high pressure is present, and then spiral outward. Similarly, the winds around the upper-air high spiral outward and diverge away from the center. At the surface, low pressure causes winds to spiral inward and upward.

Global Circulation at Upper Levels

Figure 7.24 sketches the general pattern of airflows at higher levels in the troposphere. The pattern has four major

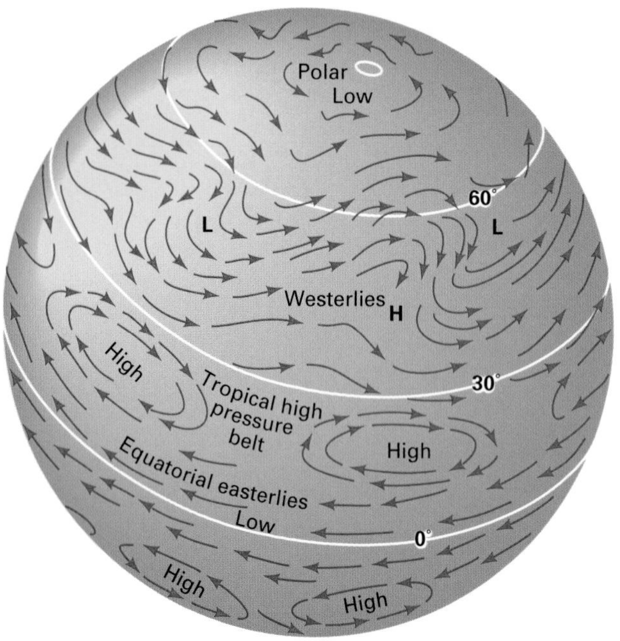

7.24 Global upper-level winds *In this generalized plan of global winds high in the troposphere, strong west winds dominate the mid- and high-latitude circulation. They often sweep to the north or south around centers of high and low pressure aloft. In the equatorial region, a weak easterly wind pattern prevails. (Copyright © A. N. Strahler.)*

features—weak equatorial easterlies, tropical high-pressure belts, upper-air westerlies, and a polar low.

We've noted above how the general temperature gradient from the tropics to the poles creates a pressure gradient force that generates westerly winds in the upper atmosphere. These *upper-air westerlies* blow in a complete circuit about the Earth, from about 25° lat. almost to the poles. They often exhibit large undulations from their westerly track, as we will describe shortly. At high latitudes the westerlies form a huge circumpolar spiral, circling a great polar low-pressure center. Toward lower latitudes, atmospheric pressure rises steadily, forming a *tropical high-pressure belt* at 15° to 20° N and S lat. This is the high-altitude part of the surface subtropical high-pressure belt, but it is shifted somewhat equatorward. Between the high-pressure ridges is a zone of lower pressure in which the winds are light but generally easterly. These winds are called the *equatorial easterlies.*

Thus, the overall picture of upper-air wind patterns is really quite simple—a band of weak easterly winds in the equatorial zone, belts of high pressure near the tropics of cancer and capricorn, and westerly winds, with some variation in direction, spiraling around polar lows.

Rossby Waves, Jet Streams, and the Polar Front

The smooth westward flow of the upper-air westerlies frequently forms undulations, called **Rossby waves.** Figure 7.25 shows how these waves develop and grow in the northern hemisphere. The waves arise in a zone of contact between cold polar air and warm tropical air, called the *polar front.* The number of Rossby waves ranges from three to seven.

For a period of several days or weeks, the flow may be fairly smooth. Then, an undulation develops, and warm air pushes poleward while a tongue of cold air is brought to the south. Eventually, the tongue is pinched off, leaving a pool of cold air at a latitude far south of its original location. This cold pool may persist for some days or weeks, slowly warming

7.25 Rossby waves *Rossby waves form in the westerlies of the northern hemisphere, marking the boundary between cold polar air and warm tropical air. This cycle shows the formation of large waves that are pinched off, leaving pools of polar air at lower latitudes. (Copyright © A. N. Strahler.)*

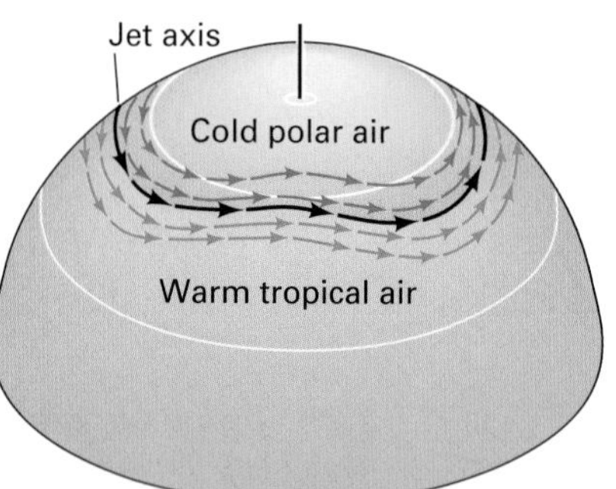

The jet stream begins to undulate.

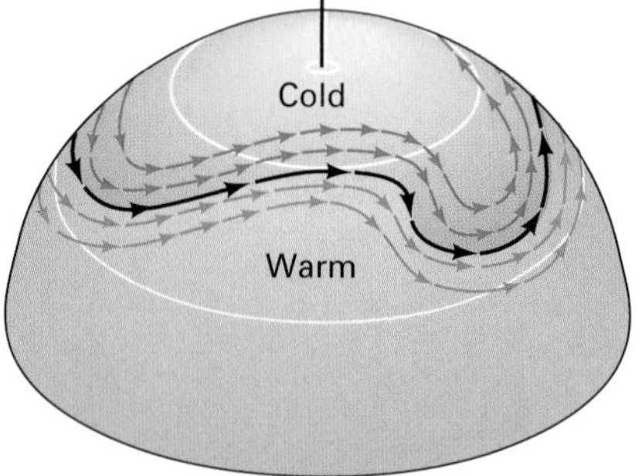

Rossby waves begin to form.

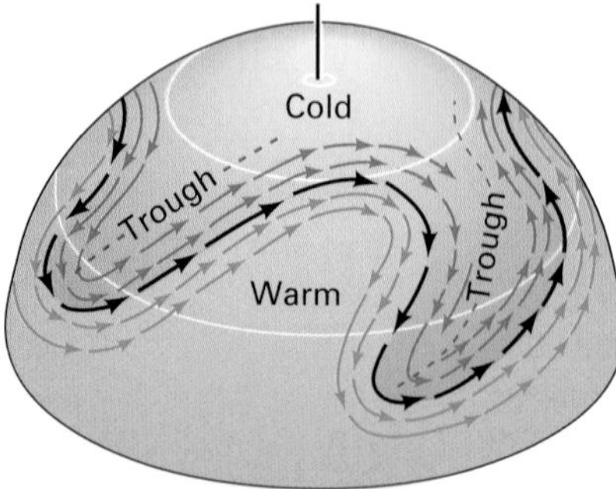

Waves are strongly developed. The cold air occupies troughs of low pressure.

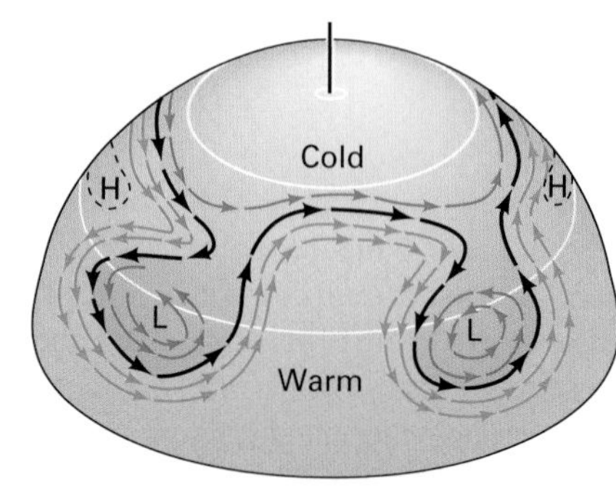

When the waves are pinched off, they form cyclones of cold air.

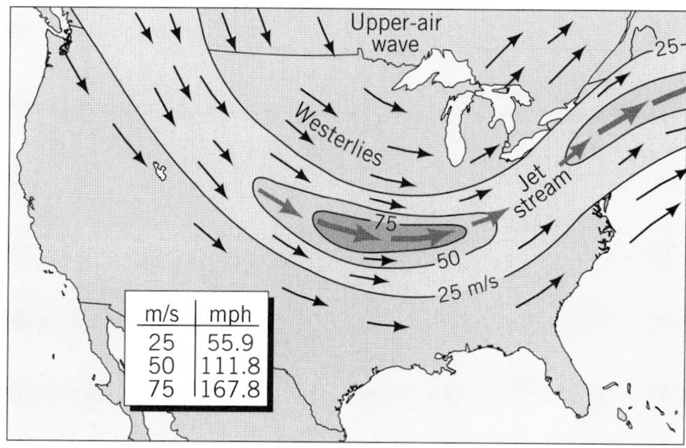

7.26 Polar jet stream *The polar jet stream is shown on this map by lines of equal wind speed. (National Weather Service.)*

with time. The low in Figure 7.23 is an example of such an air mass. Because of its cold center, it will show low pressure and convergence aloft. Air will descend in its core, and diverge at the surface, creating surface high pressure. Similarly, a warm air pool will contain rising air, with convergence and low pressure at the surface and divergence and high pressure aloft.

Because the Rossby wave circulation brings warm air poleward and cold air equatorward, it is a primary mechanism of poleward heat transport. It is also the reason weather in the midlatitudes is often so variable, as pools of warm, moist air and cold, dry air alternately invade midlatitude land masses.

Jet streams are important features of upper-air circulation. They are narrow zones at a high altitude in which wind streams reach great speeds. They occur where atmospheric pressure gradients are strong. Along a jet stream, pulse-like movements of air follow broadly curving tracks (Figure 7.26). The greatest wind speeds occur in the center of the jet stream, with velocities decreasing away from the center. There are three kinds of jet streams. Two are westerly wind streams, and the third is a weaker jet with easterly winds that develop in Asia as part of the summer monsoon circulation. These are shown in Figure 7.27.

The most poleward type of jet stream is located along the polar front (Figure 7.16). It is designated the *polar-front jet stream* (or simply, the "polar jet"). Located generally between 35° and 65° latitude, it is present in both hemispheres. The polar jet follows the edges of Rossby waves, since these mark the boundary between cold polar air and warm subtropical air. As we know from studying how atmospheric temperature affects pressure aloft, a strong temperature gradient will create a strong pressure gradient, which powers the jet stream.

The polar jet is typically found at altitudes of 10 to 12 km (about 33,000 to 40,000 ft), and wind speeds in the jet range from 75 to as much as 125 m/s (about 170 to 280 mi/hr). In the northern hemisphere, traveling aircraft often use the polar jet to increase ground speed when flying eastward. In the westward direction, flight paths are chosen to avoid the strong winds of the polar jet.

A second type of jet stream forms in the subtropical latitude zone—the *subtropical jet stream.* It occupies a position at the tropopause just above the subtropical high-pressure cells in northern and southern hemispheres (Figure 7.28). Here, westerly wind speeds reach maximum values of 100 to 110 m/s (about 215 to 240 mph). This jet is associated with the increase in velocity, induced by conservation of angular momentum, that occurs as an air parcel moves poleward from the equator. Rotating at the equator along with the Earth, an air parcel has a quantity of angular momentum based on the product of the turning rate of the Earth's rotation and the distance between the parcel and the axis of rotation. As it moves poleward, it gets closer to the Earth's axis of rotation, and by conservation of angular momentum, its turning rate must increase. That translates into an increase in the eastward velocity of the air parcel. This is the same principle that an ice skater utilizes when drawing his or her arms and legs inward to increase the speed of a spin. The result is

7.27 Upper-level circulation cross section *A schematic diagram of wind directions and jet streams along an average meridian from pole to pole. The four polar and subtropical jets are westerly in direction, in contrast to the single tropical easterly jet. (Copyright © A. N. Strahler.)*

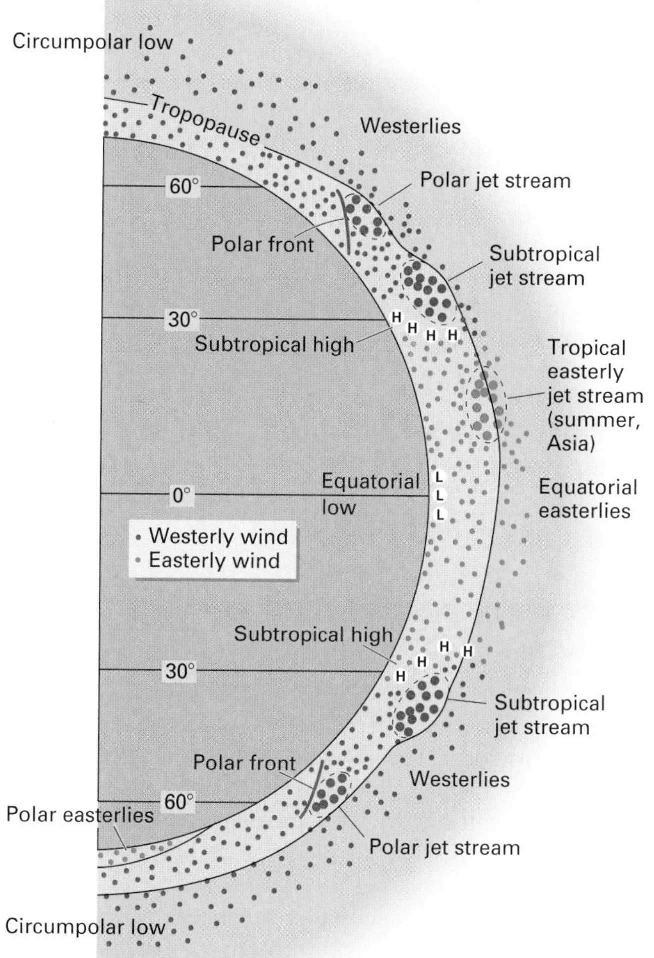

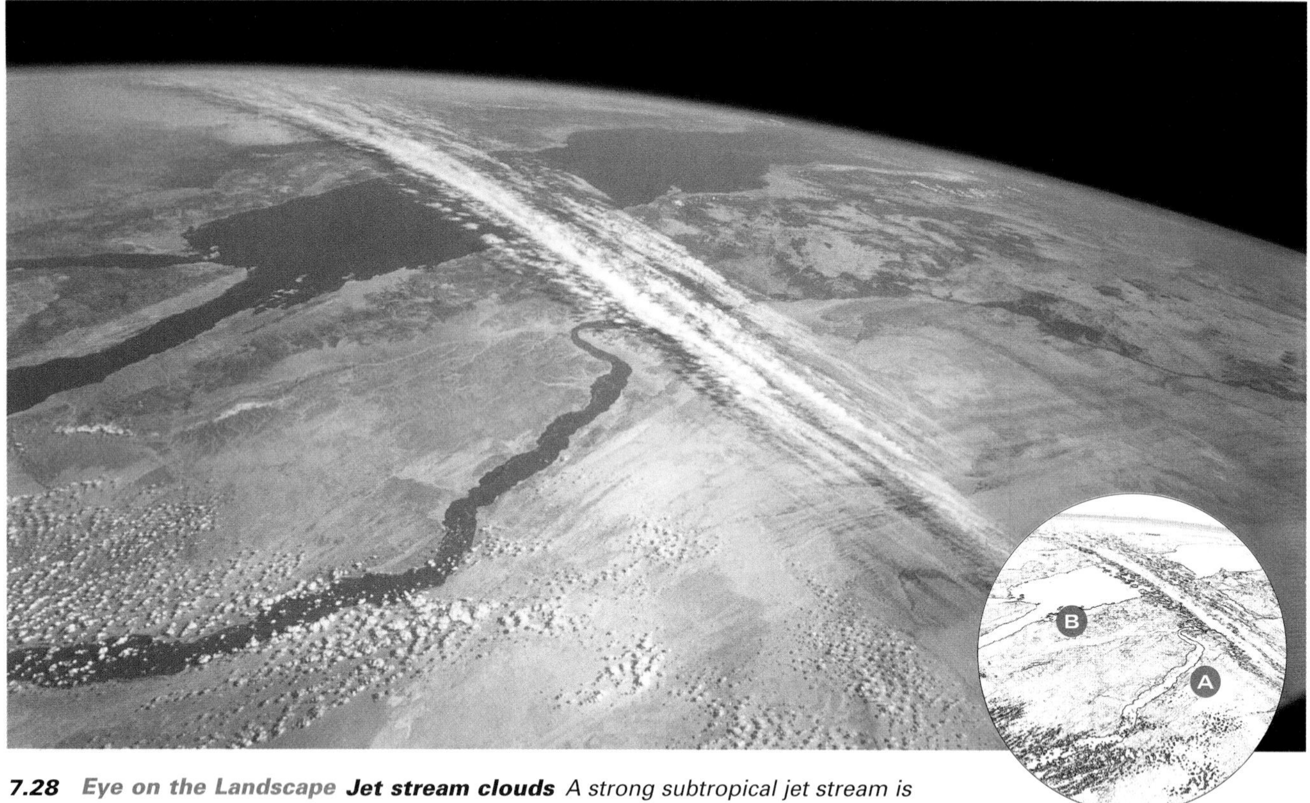

7.28 ***Eye on the Landscape*** ***Jet stream clouds*** *A strong subtropical jet stream is marked in this space photo by a narrow band of cirrus clouds that occurs on the equatorward side of the jet. The jet stream is moving from west to east at an altitude of about 12 km (40,000 ft). The cloud band lies at about 25° N lat. In this view, astronauts aimed their camera toward the southeast, taking in the Nile River Valley and the Red Sea. At the left is the tip of the Sinai peninsula.* **What else would the geographer see? ... Answers at the end of the chapter.**

a river of swiftly moving air that piles up as it descends over the subtropical highs. Cloud bands of the subtropical jet stream are pictured in Figure 7.28.

A third type of jet stream is found at even lower latitudes. Known as the *tropical easterly jet stream,* it runs from east to west—opposite in direction to that of the polar-front and subtropical jet streams. The tropical easterly jet occurs only in the summer season and is limited to a northern hemisphere location over Southeast Asia, India, and Africa.

Temperature Layers of the Ocean

As with the atmosphere, the ocean has a layered structure. Ocean layers are recognized in terms of temperature, with temperatures generally highest at the sea surface and decreasing with depth. This trend is not surprising, since the sources of heat that warm the ocean are solar insolation and heat supplied by the overlying atmosphere, both of which act to warm the water at or near the surface.

The ocean's layered temperature structure is shown in Figure 7.29. At low latitudes throughout the year and in middle latitudes in the summer, a warm surface layer develops. Here wave action mixes heated surface water with the water below it to form a warm layer that may be as thick as 500 m (about 1600 ft) with a temperature of 20° to 25°C (68° to 77°F) in oceans of the equatorial belt. Below the warm layer, temperatures drop rapidly in a zone known as the *thermocline.* Below the thermocline is a layer of very cold water extending to the deep ocean floor. Temperatures near the base of the deep layer range from 0°C to 5°C (32° to 41°F). In arctic and antarctic regions, the warm layer and thermocline are absent, as Figure 7.29 shows.

7.29 ***Ocean temperature structure*** *A schematic north-south cross section of the world ocean shows that the warm surface water layer disappears in arctic and antarctic latitudes, where very cold water lies at the surface. The thickness of the warm layer and the thermocline is greatly exaggerated.*

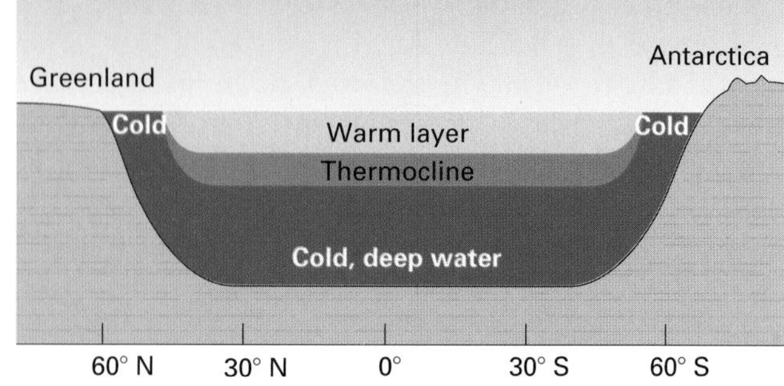

Ocean Currents

Just as there is a circulation pattern to the atmosphere, so there is a circulation pattern to the oceans. An *ocean current* is any persistent, dominantly horizontal flow of ocean water. Current systems act to exchange heat between low and high latitudes and are essential in sustaining the global energy balance. We can recognize both surface currents and deep currents. Surface currents are driven by prevailing winds. Deep currents are powered by changes in temperature and density occurring in surface waters that cause them to sink.

Surface Currents

The patterns of surface ocean currents are strongly related to prevailing surface winds. Energy is transferred from wind to water by the friction of the air blowing over the water surface. Because of the Coriolis effect, the actual direction of water drift is deflected about 45° from the direction of the driving wind.

Because ocean currents move warm waters poleward and cold waters toward the equator, they are important regulators of air temperatures. Warm surface currents keep winter temperatures in the British Isles from falling much below freezing in winter. Cold surface currents keep weather on the California coast cool, even in the height of summer.

Figure 7.30 shows the general features of the circulation of an idealized ocean between two continental masses. The circulation includes two large circular movements, called **gyres,** that are centered at latitudes of 20° to 30°. These gyres track the movements of air around the subtropical high-pressure cells. An *equatorial current* with westward flow marks the belt of the trade winds. Although the trades blow to the southwest and northwest, at an angle across the parallels of latitude, the water movement follows the parallels. The equatorial currents are separated by an equatorial countercurrent. A slow, eastward movement of water over the zone of the westerlies is named the *west-wind drift.* It covers a broad belt between lat. 35° and 45° in the northern hemisphere and between lat. 30° and 60° in the southern hemisphere.

Figure 7.31 presents a world map that shows surface ocean currents for the month of January. In the equatorial region, ocean currents flow westward, pushed by northeast and southeast trade winds. As these equatorial currents approach land, they are turned poleward along the west sides of the oceans, forming warm currents paralleling the coast. Examples are the Gulf Stream of eastern North America and the Kuroshio Current of Japan.

In the zone of westerly winds, a slow eastward motion of water is named the *west-wind drift.* As west-wind drift waters approach the western sides of the continents, they are deflected equatorward along the coast. The equatorward flows are cool currents. They are often accompanied by *upwelling* along continental margins. In this process, colder water from greater depths rises to the surface. Examples of cool currents with upwelling are the Humboldt (or Peru) Current, off the coast of Chile and Peru, the Benguela Current, off the coast of southern Africa, and the California Current.

West-wind drift water also moves poleward to join arctic and antarctic circulations. In the northeastern Atlantic Ocean, the west-wind drift forms a relatively warm current. This is the North Atlantic Drift, which spreads around the British Isles, into the North Sea, and along the Norwegian coast. The Russian port of Murmansk, on the arctic circle, remains ice-free year round because of this warm drift current.

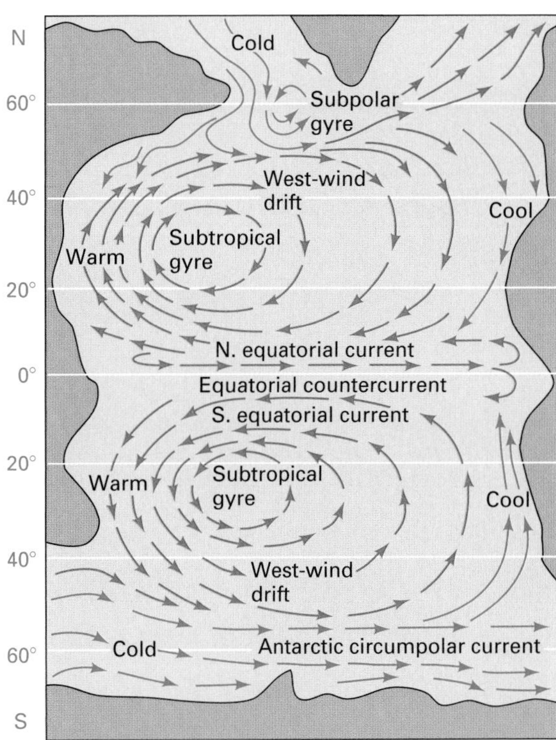

7.30 Ocean current gyres *Two great gyres, one in each hemisphere, dominate the circulation of shallow ocean waters.*

In the northern hemisphere, where the polar sea is largely landlocked, cold currents flow equatorward along the east sides of continents. Two examples are the Kamchatka Current, which flows southward along the Asian coast across from Alaska, and the Labrador Current, which flows between Labrador and Greenland to reach the coasts of Newfoundland, Nova Scotia, and New England.

The strong west winds around Antarctica produce an antarctic circumpolar current of cold water, shown in Figure 7.32. Some of this flow branches equatorward along the west coast of South America, adding to the Humboldt Current (Figure 7.31). Two small gyres lie close to the antarctic continent.

Figure 7.33 shows a satellite image of ocean temperature along the east coast of North America for a week in April. The Gulf Stream stands out as a tongue of red and yellow (warm) color, moving along the coast and heading off from North Carolina in a northeasterly direction. Cooler water from the Labrador Current, in green and blue tones, hugs the northern Atlantic coast. This current heads south and then turns to follow the Gulf Stream to the northeast. Instead of mixing, the two flows remain quite distinct. The boundary between them shows a wavelike flow, much like Rossby waves in the atmosphere. Warm and cold bodies of water are cut off to float freely, forming warm-core and cold-core rings. GEODISCOVERIES

Another phenomenon of ocean surface currents is *El Niño.* During an El Niño event, Pacific surface currents

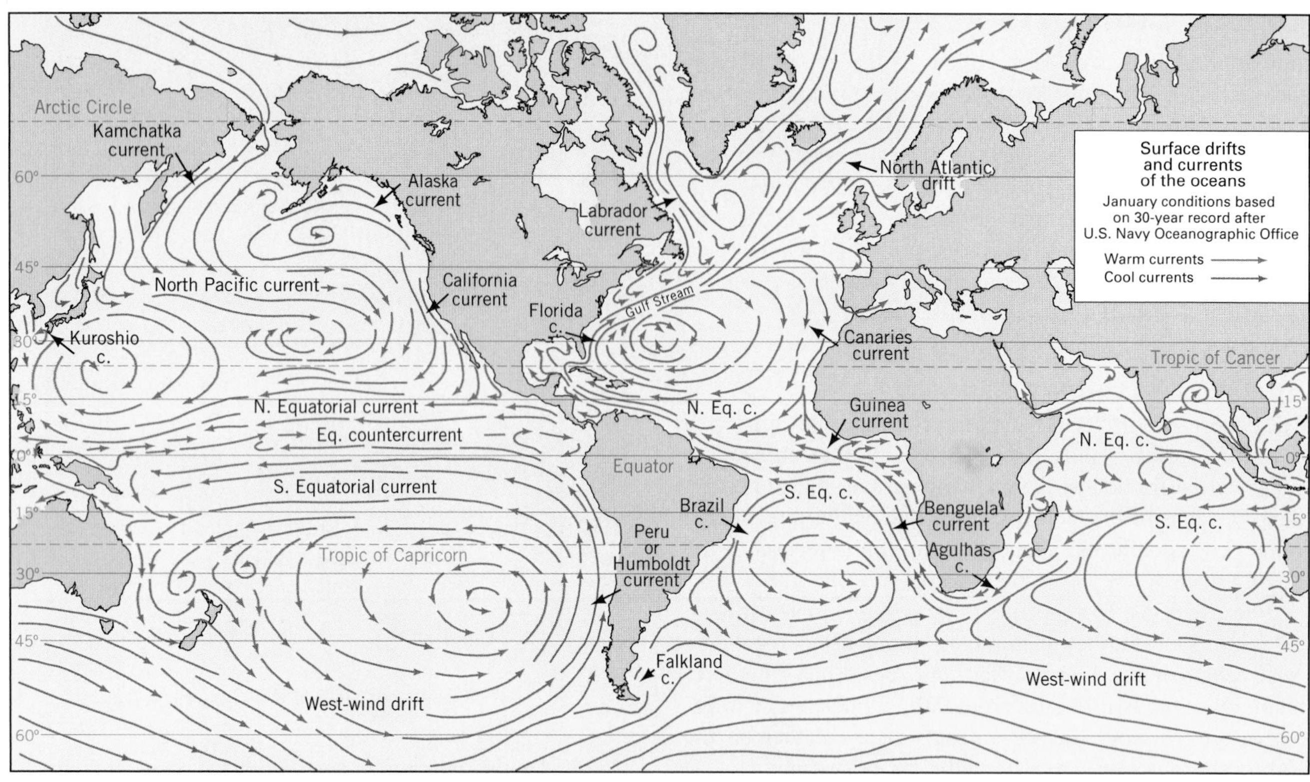

7.31 ***January ocean currents*** *Surface drifts and currents of the oceans in January. (Based on data from U.S. Navy Oceanographic Office. Redrawn and revised by A. N. Strahler.)*

shift into an unusual pattern. Pacific upwelling along the Peruvian coast ceases, trade winds weaken, and a weak equatorial eastward current develops. Global patterns of precipitation also change during El Niño events, bringing floods to some regions and droughts to others. In contrast to El Niño is *La Niña,* in which normal Peruvian coastal upwelling is enhanced, trade winds strengthen, and cool water is carried far westward in an equatorial plume.

Figure 7.34 shows two satellite images of sea-surface temperature observed during El Niño and La Niña years.

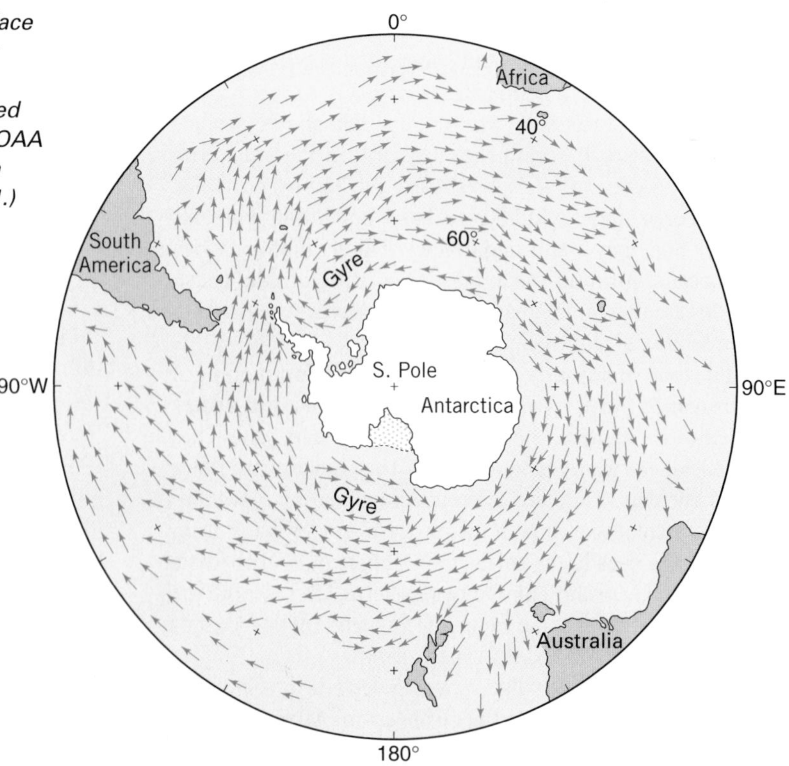

7.32 ***Currents of the Southern Ocean*** *Surface water of the Southern Ocean flows continuously from west to east around Antarctica as a broad circumpolar current. This map is greatly simplified from a computer-generated model. (Based on NOAA data as presented by D. Olbers and M. Wenzel in EOS, American Geophysical Union, vol. 71, no. 1.)*

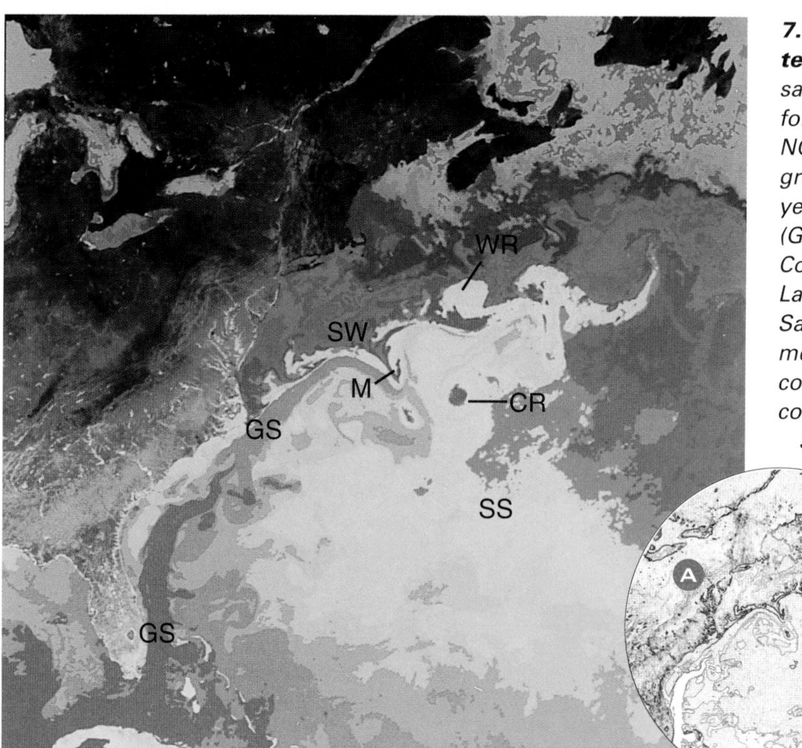

7.33** Eye on the Landscape* ***Sea-surface temperatures in the western Atlantic *A satellite image showing sea-surface temperature for a week in April from data acquired by the NOAA-7 orbiting satellite. Cold water appears in green and blue tones; warm water in red and yellow tones. The image shows the Gulf Stream (GS) and its interactions with cold water of the Continental Slope (SW), brought down by the Labrador Current, and the warm water of the Sargasso Sea (SS). Other features include a meander (M), a warm-core ring (WR), and a cold-core ring (CR). The image was made from data collected by the NOAA-7 orbiting satellite. Data of 35 satellite passes during a week in April were processed together to remove the effects of clouds on single images. (NASA. Courtesy of Otis B. Brown, Robert Evans, and M. Carle, University of Miami, Rosenstiel School of Marine and Atmospheric Science, Florida.)* **What else would the geographer see? ... Answers at the end of the chapter.**

7.34 La Niña and El Niño sea-surface temperatures *This striking image shows sea-surface temperatures in the eastern tropical Pacific during (a) La Niña and (b) El Niño years as measured by the NOAA-7 satellite. Green tones indicate cooler temperatures, while red tones indicate warmer temperatures. (Otis B. Brown/Univ. of Miami and Gene Carl Feldman, NASA/GSFC.)*

(a)

(b)

Eye on Global Change | 7.4

El Niño

At intervals of about three to eight years, a remarkable disturbance of ocean and atmosphere occurs. It begins in the eastern Pacific Ocean and spreads its effects widely over the globe. This disturbance lasts more than a year, bringing droughts, heavy rainfalls, severe spells of heat and cold, or a high incidence of cyclonic storms to various parts of the Pacific and its eastern coasts. This phenomenon is called *El Niño.* The expression comes from Peruvian fishermen, who refer to the *Corriente del Niño,* or the "Current of the Christ Child," in describing an invasion of warm surface water that occurs once every few years around Christmas time and greatly depletes their catch of fish. El Niño occurs at irregular intervals and with varying degrees of intensity. Notable El Niño events occurred in 1891, 1925, 1940–1941, 1965, 1972–1973, 1982–1983, 1989–1990, 1991–1992, 1994–1995, and 1997–1998 and 2002–2003. GEODISCOVERIES

Normally, the cool Humboldt (Peru) Current flows northward off the South American coast, and then at about the equator it turns westward across the Pacific as the south equatorial current (see Figure 7.31). The Humboldt Current is fed by upwelling of cold, deep water, bringing with it nutrients that serve as food for marine life. With the onset of El Niño, upwelling ceases, the cool water is replaced by warm, sterile water from the west, and the abundant marine life disappears.

In an El Niño year, a major change in barometric pressure occurs across the entire stretch of the equatorial zone as far west as southeastern Asia. Normally, low pressure prevails over northern Australia, the East Indies, and New Guinea, where the largest and warmest body of ocean water can be found (see figure below, left).

El Niño maps *Maps of pressures in the tropical Pacific and eastern Indian Ocean in November during normal and El Niño years. In a normal year (a), low pressure dominates in Malaysia and northern Australia. In an El Niño year (b), low pressure moves eastward to the central part of the western Pacific, and sea-surface temperatures become warmer in the eastern Central Pacific. (Copyright © A. N. Strahler.)*

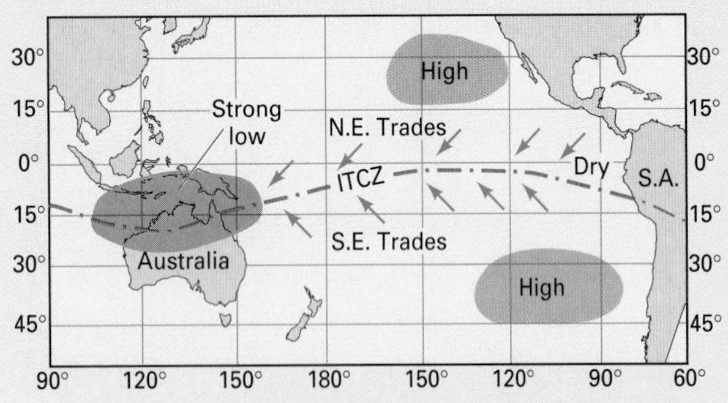

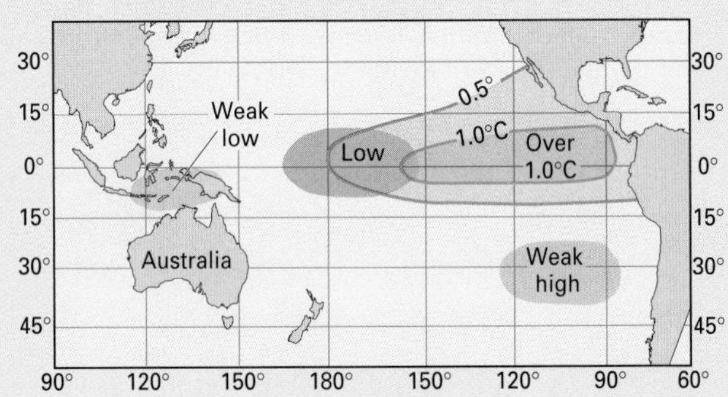

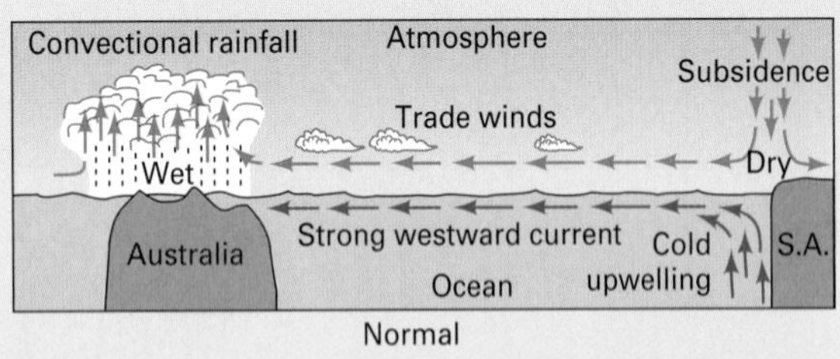

(*a*)

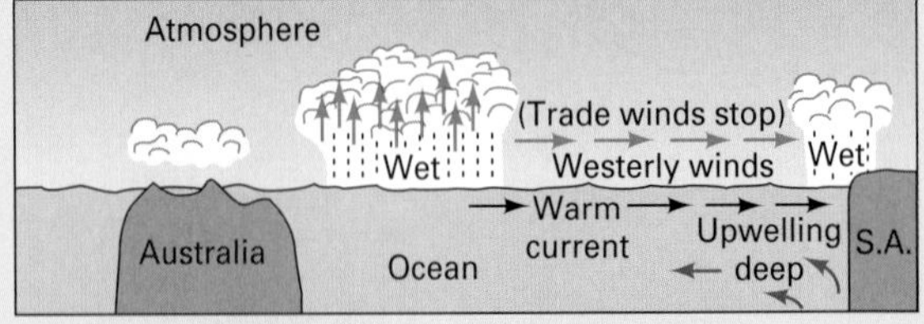

(*b*)

Abundant rainfall occurs in this area during December, which is the high-Sun period in the southern hemisphere.

During an El Niño event, the low-pressure system is replaced by weak high pressure and drought ensues. Pressures drop in the equatorial zone of the eastern Pacific, strengthening the equatorial trough. Rainfall is abundant in this new low-pressure region (right part of figure). The shift in barometric pressure patterns is known as the *Southern Oscillation.*

Surface winds and currents also change with this change in pressure. During normal conditions, the strong, prevailing trade winds blow westward, causing very warm ocean water to move to the western Pacific and to "pile up" near the western equatorial low. This westward motion causes the normal upwelling along the South American coast, as bottom water is carried up to replace the water dragged to the west. During an El Niño event, the easterly trade winds die with the change in atmospheric pressure. A weak westerly wind flow sometimes occurs, completely reversing the normal wind direction. Without the pressure of the trade winds to hold them back, warm waters surge eastward. Sea-surface temperatures and actual sea levels rise off the tropical western coasts of the Americas.

The major change in sea-surface temperatures that accompanies an El Niño can also shift weather patterns across large regions of the globe. The winter of 1997–1998 experienced one of the strongest El Niños of the century. Torrential rains drenched Peruvian and Ecuadorean coast ranges, producing mudflows, debris avalanches, and extensive river flooding. Large portions of Australia and the East Indies went rainless for months, and forest fires burned out of control in Sumatra, Borneo, and Malaysia. In east Africa, Kenya experienced rainfall 100 cm (80 in.) above normal. In North America, a series of powerful winter storms lashed the Pacific coast, doing extensive damage in California. Monstrous tornadoes ripped through Florida, killing over 40 people and destroying more than 800 homes. Meanwhile, mild winter conditions east of the Rockies saved vast amounts of fossil fuel while generating an ice storm that put 4 million people without power in Quebec and the northeastern United States. All told, the deranged weather of the El Niño of 1997–1998 did property damage estimated at $33 billion and killed an estimated 2100 people.

A somewhat rarer phenomenon, also capable of altering global weather patterns, is La Niña (the girl child), a condition roughly opposite to El Niño. During a La Niña period, sea-surface temperatures in the central and western Pacific Ocean fall to lower than average levels. This happens because the South Pacific subtropical high becomes very strongly developed during the high-Sun season. The result is abnormally strong southeast trade winds. The force of these winds drags a more-than-normal amount of warm surface water westward, which enhances upwelling along western continental coasts. Figure 7.34 provides a pair of satellite images of sea-surface temperature that contrast El Niño and La Niña conditions.

The El Niño of 1997–1998 was rapidly followed by the La Niña of 1998–1999. The result was heavier monsoon rains in India and more rain in Australia. In North America, winter conditions were colder than normal in the northwest and upper midwest. The eastern Atlantic region endured drought through spring and early summer. The hurricane season of 1998, spawning the monster storm Hurricane Mitch, was the deadliest in the past two centuries.

What causes the El Niño/Southern Oscillation phenomenon (also known as ENSO)? At first, meteorologists tried to explain the changes in winds and pressures by the changes in ocean surface currents, while oceanographers tried to explain the changes in surface currents by changes in winds and pressures. It wasn't until the 1950s that scientists realized that the two phenomena were linked. One view is that the cycle is simply a natural oscillation caused by the way in which the atmosphere and oceans are coupled by energy exchange. In any event, scientists now have good computer models that accept sea-surface temperature along with air temperature and pressure data and can predict El Niño events reasonably well some months before they occur.

El Niño and its alter ego La Niña show how dynamic our planet really is. As a grand-scale, global phenomenon, El Niño–La Niña shows how the circulation patterns of the ocean and atmosphere are linked and interact to provide teleconnections capable of producing extreme events affecting millions of people throughout the world.

During La Niña conditions (left), the cool water of the Humboldt Current, moving northward along the coast of Chile and Peru, is carried westward into the Pacific in a long plume. Much cold, upwelling water (dark green) is brought to the surface along the Peruvian coast. During an El Niño year (right), the eastward motion of warm water holds the Humboldt Current in check. Some slight upwelling occurs (yellow color), but the amount is greatly reduced compared to normal or La Niña years. *Eye on Global Change 7.4 • El Niño* provides a fuller description of the El Niño–La Niña phenomenon. GEODISCOVERIES

Deep Currents and Thermohaline Circulation

Deep currents move ocean waters in a slow circuit across the floors of the world's oceans. They are generated when surface waters become more dense and slowly sink downward. Coupled with these deep currents are very broad and slow surface currents on which the more rapid surface currents, described above, are superimposed. Figure 7.35 diagrams this slow flow pattern, which links all of the world's oceans. It is referred to as *thermohaline circulation,* since it depends on the temperature and salinity of North Atlantic Ocean waters. These factors act through changes to the density of sea water. Since warm water is slightly less dense than cold water, surface waters do not normally mix with cold bottom waters. Follow the figure carefully as we explain the process.

Beginning at *A* in the figure, warm Atlantic surface water slowly moves northward through the equatorial and tropical zones. As this surface layer is warmed, evaporation occurs and it becomes saltier. This increases the density of the layer slightly. As the water reaches the northern boundaries of the Atlantic, it loses heat to the atmosphere, and so it becomes colder and still more dense. Eventually, the surface layer becomes dense enough to sink *(B).* This sinking occurs all along the northern boundary of the North Atlantic.

7.35 Ocean circulation *Deep ocean currents, generated by the sinking of cold, salty water in the northern Atlantic, circulate sea water in slowly moving coupled loops involving the Atlantic, Pacific, Indian, and Southern oceans. Actual flows are much broader than the narrow flow paths shown here. (After A. J. Gordon, Nature 382:399–400, August, 1996. Used by permission.)*

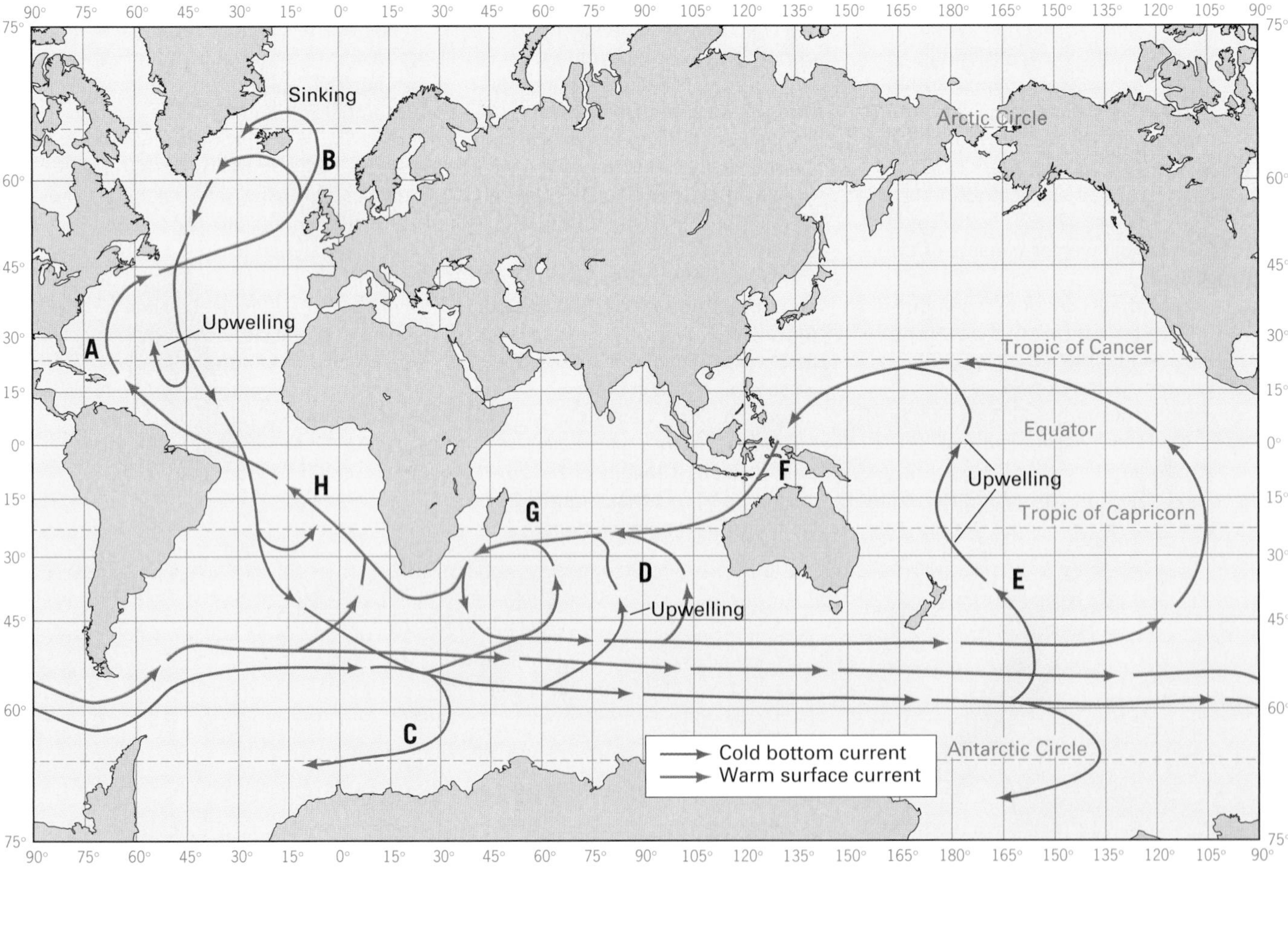

Carried along the bottom, the cold, dense water eventually reaches the Southern Ocean *(C)*. Here, upwelling and mixing occur, and the deep waters are carried into the surface waters of the Indian and southern Pacific oceans *(D, E)*. A coupled circulation loop moves surface water from the Pacific through the Indonesian seas *(F)* and into the Indian Ocean *(G)*. Most of this flow connects with a westward flow link along the Southern Ocean, poleward of the African and Australian continents. However, some of the flow escapes to the west around the southern tip of Africa to enter the South Atlantic *(H)*, completing the entire circuit.

Thermohaline circulation plays an important role in the carbon cycle by moving CO_2-rich surface waters into the ocean depths. As noted in Chapter 5 in *Eye on Global Change 5.4 • Carbon Dioxide—On the Increase,* deep ocean circulation provides a conveyor belt for storage and release of CO_2 in a cycle of about 1500 years' duration. This allows the ocean to moderate rapid changes in atmospheric CO_2 concentration, such as those produced by human activity through fossil fuel burning.

Some scientists have observed that thermohaline circulation could be slowed or stopped by inputs of fresh water into the North Atlantic. Such fresh water inputs could come from the sudden drainage of large lakes formed by melting ice at the close of the last Ice Age. The fresh water would decrease the density of the ocean water, keeping the water from becoming dense enough to sink. Without sinking, circulation would stop. In turn, this would interrupt a major flow pathway for the transfer of heat from equatorial regions to the northern midlatitudes. This mechanism could result in relatively rapid climatic change and is one explanation for periodic cycles of warm and cold temperatures experienced since the melting of continental ice sheets about 12,000 years ago.

A Look Ahead The global circulation of winds and currents paves the way for our next subject—weather systems. Recall from Chapter 6 that when warm, moist air rises, precipitation can occur. This happens in the centers of cyclones, where air converges. Although cyclones and anticyclones are generally large surface features, they move from day to day and are steered by the global pattern of winds. Thus, your knowledge of global winds and pressures will help you to understand how weather systems and storms develop and migrate. Your knowledge will also be very useful for the study of climate, which we take up in Chapter 9.

Chapter Summary

- The term *atmospheric pressure* describes the weight of air pressing on a unit of surface area. Atmospheric pressure is measured by a **barometer**. Atmospheric pressure decreases rapidly as altitude increases.
- Wind occurs when air moves with respect to the Earth's surface. Air motion is produced by **pressure gradients** that are formed when air in one location is heated to a temperature that is warmer than another. Heating creates high pressure aloft, which moves high-level air away from the area of heating. This motion induces low pressure at the surface, pulling surface air toward the area of heating, and a convection loop is formed.
- *Sea* and *land breezes* are examples of *convection loops* formed from unequal heating and cooling of the land surface as compared to a nearby water surface.
- *Local winds* are generated by local pressure gradients. The sea and land breezes, as well as *mountain* and *valley winds,* are examples caused by local surface heating. Other local winds include *drainage winds, Santa Ana,* and *chinook winds*.
- The Earth's rotation strongly influences atmospheric circulation through the **Coriolis effect**. The *Coriolis force* deflects wind motion, producing circular or spiraling flow paths around **cyclones** (centers of low pressure and *convergence*) and **anticyclones** (centers of high pressure and *divergence*).
- Because the equatorial and tropical regions are heated more intensely than the higher latitudes, two convection loops develop—the **Hadley cells**. These loops drive the *northeast* and *southeast trade winds*, the convergence and lifting of air at the **intertropical convergence zone (ITCZ)**, and the sinking and divergence of air in the **subtropical high-pressure belts**.
- The most persistent features of the global pattern of atmospheric pressure are the subtropical high-pressure belts, which are generated by the Hadley cell circulation. They intensify during their high-Sun season.
- The **monsoon** circulation of Asia responds to a reversal of atmospheric pressure over the continent with the seasons. A *winter monsoon* flow of cool, dry air from the northeast alternates with a *summer monsoon* flow of warm, moist air from the southwest.
- In the midlatitudes and poleward, westerly winds prevail. In winter, continents develop high pressure, and intense oceanic low-pressure centers are found off the Aleutian Islands and near Iceland in the northern hemisphere. In the summer, the continents develop low pressure as oceanic subtropical high-pressure cells intensify and move poleward.
- Winds aloft are dominated by a global pressure gradient force between the tropics and pole in each hemisphere that is generated by the hemispheric temperature gradient

from warm to cold. Coupled with the Coriolis force, the gradient generates strong westerly *geostrophic winds* in the upper air. In the equatorial region, weak easterlies dominate the upper-level wind pattern.

- **Rossby waves** develop in the *upper-air westerlies*, bringing cold, polar air equatorward and warmer air poleward. The *polar-front* and *subtropical* **jet streams** are concentrated westerly wind streams with high wind speeds. The *tropical easterly jet stream* is weaker and limited to Southeast Asia, India, and Africa.
- Oceans show a warm surface layer, a *thermocline*, and a deep cold layer. Near the poles, the warm layer and thermocline are absent.
- *Ocean* surface *currents* are dominated by huge **gyres** that are driven by the global surface wind pattern. *Equatorial currents* move warm water westward and then poleward along the east coasts of continents. Return flows bring cold water equatorward along the west coasts of continents.
- *El Niño* events occur when an unusual flow of warm water in the equatorial Pacific moves eastward to the coasts of Central and South America, suppressing the normal northward flow of the Humboldt Current. Upwelling along the Peruvian coast is greatly reduced. El Niño events normally occur on a three- to eight-year cycle and affect global patterns of precipitation in many regions. The causes of El Niño are not well understood.
- Slow, deep ocean currents are driven by the sinking of cold, salty water in the northern Atlantic. This *thermohaline circulation* pattern involves nearly all the Earth's ocean basins, and also acts to moderate the buildup of atmospheric CO_2 by moving CO_2-rich surface waters to ocean depths.

Key Terms

millibar
barometer
wind
pressure gradient
isobar
Coriolis effect
cyclone
anticyclone
Hadley cell
intertropical convergence zone (ITCZ)
subtropical high-pressure belts
monsoon
Rossby waves
jet stream
gyres

Review Questions

1. Explain atmospheric pressure. Why does it occur? How is atmospheric pressure measured and in what units? What is the normal value of atmospheric pressure at sea level? How does atmospheric pressure change with altitude?
2. Describe a simple convective wind system, explaining how air motion arises from a pressure gradient force induced by heating.
3. Describe land and sea breezes. How do they illustrate the concepts of pressure gradient and convection loop?
4. What is the Coriolis effect, and why is it important? What produces it? How does it influence the motion of wind and ocean currents in the northern hemisphere? in the southern hemisphere?
5. Define cyclone and anticyclone. How does air move within each? What is the direction of circulation of each in the northern and southern hemispheres? What type of weather is associated with each and why?
6. What is the Asian monsoon? Describe the features of this circulation in summer and winter. How is the ITCZ involved? How is the monsoon circulation related to the high- and low-pressure centers that develop seasonally in Asia?
7. Compare the winter and summer patterns of high and low pressure that develop in the northern hemisphere with those that develop in the southern hemisphere.
8. What are drainage winds? What local names are applied to them?
9. How does global scale heating of the atmosphere create a pressure gradient force that increases with altitude?
10. What is the geostrophic wind, and what is its direction with respect to the pressure gradient force?
11. Describe the basic pattern of global atmospheric circulation at upper levels.
12. What are Rossby waves? Why are they important?
13. Identify five jet streams. Where do they occur? In which direction do they flow?
14. What is the general pattern of ocean surface current circulation? How is it related to global wind patterns?
15. How does thermohaline circulation induce deep ocean currents?

Eye on the Environment 7.2 • Wind Power, Wave Power, and Current Power

1. Discuss wind power as a source of energy. Include indirect use in tapping waves and ocean currents for energy.

Focus on Systems 7.3 • The Convection Loop as an Energy Flow System

1. Diagram a simple convection loop and show how it uses a flow of energy to cycle matter. Show energy inputs and outputs.
2. Describe how the cyclone and anticylone are coupled in the atmospheric convection loop. In what respects does this loop differ from the simple convection loop described in 1. above?

Eye on Global Change 7.4 • El Niño

1. Compare the normal pattern of wind, pressure, and ocean currents in the equatorial Pacific with the pattern during an El Niño event.
2. What are some of the weather changes reported for El Niño events?
3. What is La Niña, and how does it compare with the normal pattern?

Visualizing Exercises

1. Sketch an ideal Earth (without seasons or ocean—continent features) and its global wind system. Label the following on your sketch: doldrums, equatorial trough, Hadley cell, ITCZ, northeast trades, polar easterlies, polar front, polar outbreak, southeast trades, subtropical high-pressure belts, and westerlies.
2. Draw four spiral patterns showing outward and inward flow in clockwise and counterclockwise directions. Label each as appropriate to cyclonic or anticyclonic circulation in the northern or southern hemisphere.

Essay Questions

1. An airline pilot is planning a nonstop flight from Los Angeles to Sydney, Australia. What general wind conditions can the pilot expect to find in the upper atmosphere as the airplane travels? What jet streams will be encountered? Will they slow or speed the aircraft on its way?
2. You are planning to take a round-the-world cruise, leaving New York in October. Your vessel's route will take you through the Mediterranean Sea to Cairo, Egypt, in early December. Then you will pass through the Suez Canal and Red Sea to the Indian Ocean, calling at Bombay, India, in January. From Bombay, you will sail to Djakarta, Indonesia, and then go directly to Perth, Australia, arriving in March. Rounding the southern coast of Australia, your next port of call is Auckland, New Zealand, which you will reach in April. From Auckland, you head directly to San Francisco, your final destination, arriving in June. Describe the general wind and weather conditions you will experience on each leg of your journey.

Problems

Working It Out 7.1 • Pressure and Density in the Oceans and Atmosphere

1. A diving pool is 5 m deep. What will be the pressure on a person swimming at the bottom of the pool, including atmospheric pressure? What fraction of the pressure is due to the water, and what fraction is due to the atmosphere? Suppose the swimmer is now a deep-sea diver at a depth of 100 m. What are the fractions for this case?
2. Mount Washington, in New Hampshire, has a summit elevation of 1917 m (6288 ft). Mount Whitney, in California, has a summit elevation of 4418 m (14,494 ft). Using either the pressure formula or reading directly from the figure, what barometric pressures would you expect to find at these two summits? What percentage of sea-level pressure is each value? (Use 1014 mb as surface atmospheric pressure.) (To obtain the fractional power 5.26, use the y^X button found on most calculators.)
3. In the course of the passage of a hurricane, a weather observer notes a change in barometric pressure from 1016 mb to 979 mb, a difference of 37 mb. Using the formula for atmospheric pressure with altitude, would a barometer reading 1014 mb at sea level change by at least that amount if taken up to an elevation of 350 m?

Eye on the Landscape

Chapter Opener Village of Araouane, Mali The Sahara has vast areas of wind-driven sand dunes and sheets. The ridges are longitudinal dunes **(A)**, which run parallel to the direction of the wind. However, these ridges seem to be influenced by winds coming from the upper left, crossing the ridges at an angle. The small dark dots in bands or lines **(B)** are individual plant crowns. In this part of the Sahara, ground water is close to the surface, and deep-rooted plants, called phreatophytes, are able to tap the water table. This underground water also supports the village through shallow wells. Chapter 19, The Work of Waves and Wind, provides more information on dunes and landforms made by wind. The adaptation of plants to differing water regimes is a subject of Chapter 23, Biogeographic Processes.

7.10 Brushfire The dense, shrubby vegetation on this slope is chaparral, a mix of tough woody shrubs with leathery leaves **(A)** that is often found on steep slopes in Mediterranean climates (covered in Chapter 10). To reduce moisture loss during the hot, dry summer, many plants of the chaparral have waxy, oily leaves that can prove very flammable under the right conditions. Chaparral is a vegetation type that depends on burning at regular intervals to maintain itself, so the fire in this picture is merely part of a natural process.

7.28 Jet stream clouds A desert landscape seen from space. Large expanses of the Middle East are essentially devoid of vegetation, and so a space photo such as this shows variation in soil and surface color. Many of the white areas **(A)** are wind-blown sand sheets (see Dunes, Chapter 19), while the medium tones are weathered rock materials that have not been carried as far by wind or water from their origin (see Chapters 17 and 19 for sediment transportation by water and wind). Ranges of hills and low mountains **(B)** appear darker due to a sparse vegetation cover. The Nile River Valley cuts a wide swath through the region, with its broad band of irrigated agriculture (see Chapter 16 on rivers). The overall blue tint is produced by atmospheric scattering (Chapter 4), which is more intense in the blue region of the spectrum.

7.33 Sea-surface temperatures This image shows a temperature pattern over the land **(A)** as well as the water. The general gradient from yellow and red tones (hot) to dark tones (cold) shows land surface temperatures decreasing strongly from Florida to the Maritime Provinces, as we might expect for a week in April. Industrial and residential regions, such as the Washington to New York corridor, Hudson River Valley, and Connecticut River Valley show warmer temperatures due to human activity (see Chapter 3 on urban-rural surface temperature contrasts). Note the dark color of the Appalachians, which are colder because of their elevation (see Chapter 5 for how surface temperature varies with elevation).

8 WEATHER SYSTEMS

Village on Guanaja, Islas de la Bahía, Honduras. The near-total destruction of this small village on an offshore island north of Honduras island was wrought by hurricane Mitch in October of 1998.

As we saw in Chapter 7, the Earth's atmosphere is in constant motion, driven by the planet's rotation and its uneven heating by the Sun. The horizontal motion of the wind moves air from one place to another, allowing air to acquire characteristics of temperature and humidity in one region and then carry those characteristics into another region. The vertical motion affects clouds and precipitation. When air is lifted, it is cooled, enabling clouds and precipitation to form. When air descends, it is warmed, retarding the formation of clouds and precipitation. In this way, the Earth's wind system influences the weather we experience from day to day—the temperature and humidity of the air, cloudiness, and the amount of precipitation.

Some patterns of wind circulation occur commonly and so present recurring patterns of weather. For example, traveling low-pressure centers (cyclones) of converging, inspiraling air often bring warm, moist air in contact with cooler, drier air, with clouds and precipitation as the result. We recognize these recurring circulation patterns and their associated weather as **weather systems.**

Weather systems range in size from a few kilometers, in the case of the tornado, to a thousand kilometers or more, in the case of a large traveling anticyclone. They may last for hours or weeks, depending on their size and strength. Some forms of weather systems—tornadoes and hurricanes, for example—involve high winds and heavy rainfall and can be very destructive to life and property.

AIR MASSES

Weather systems are often associated with the motion of air masses. An **air mass** is a large body of air with fairly uniform temperature and moisture characteristics. It can be several thousand kilometers or miles across, and it can extend upward to the top of the troposphere. A given air mass is characterized by a distinctive combination of surface temperature, environmental temperature lapse rate, and surface-specific humidity. Air masses range widely in temperature—from searing hot to icy cold—as well as in moisture content.

Air masses acquire their characteristics in *source regions.* In a source region, air moves slowly or stagnates, which allows the air to acquire temperature and moisture characteristics from the region's surface. For example, an air mass with warm temperatures and a high water vapor content develops over a warm equatorial ocean. Over a large tropical desert, slowly subsiding air forms a hot air mass with low humidity. A very cold air mass with a low water vapor content is generated over cold, snow-covered land surfaces in the arctic zone in winter.

Air masses move from one region to another under the influence of pressure gradients and upper-level wind patterns and are sometimes pushed or blocked by high-level jet stream winds. When an air mass moves to a new area, its properties will begin to change because it is influenced by the new surface environment. For example, the air mass may lose heat or take up water vapor.

Air masses are classified on the basis of the latitudinal position and the nature of the underlying surface of their source regions. Latitudinal position primarily determines surface temperature and the environmental temperature lapse rate of the air mass, while the nature of the underlying surface—continent or ocean—usually determines the moisture content. For latitudinal position, five types of air masses are distinguished, as shown in the following table.

Air Mass	Symbol	Source Region
Arctic	A	Arctic ocean and fringing lands
Antarctic	AA	Antarctica
Polar	P	Continents and oceans, lat. 50–60° N and S
Tropical	T	Continents and oceans, lat. 20–35° N and S
Equatorial	E	Oceans close to equator

For the type of underlying surface, two subdivisions are used:

Air Mass	Symbol	Source Region
Maritime	m	Oceans
Continental	c	Continents

Combining these two types of labels produces a list of six important types of air masses, shown in Table 8.1. The table also gives some typical values of surface temperature and specific humidity, although these can vary widely, depending on season. Air mass temperature can range from – 46°C (–51°F) for continental arctic (cA) air masses to 27°C (81°F) for maritime equatorial air masses (mE). Specific humidity shows a very high range—from 0.1 g/kg for the cA air mass to as much as 19 g/kg for the mE air mass. In other words, maritime equatorial air can hold about 200 times as much moisture as continental arctic air.

Figure 8.1 shows schematically the global distribution of source regions of these air masses. An idealized continent is shown in the center of the figure surrounded by ocean. Note from the figure that the polar air masses (mP, cP) originate in the subarctic latitude zone, not in the polar latitude zone. Recall that in Chapter 4 we defined "polar" as a region containing one of the poles. However, meteorologists use the word "polar" to describe air masses from the subarctic and subantarctic zones, and we will follow their usage when referring to air masses.

The maritime tropical air mass (mT) and maritime equatorial air mass (mE) originate over warm oceans in the tropical and equatorial zones. They are quite similar in temperature and water vapor content. With their high values of specific humidity, both are capable of very heavy yields of precipitation. The continental tropical air mass (cT) has its source region over subtropical deserts of the continents. Although this air mass may have a substantial water vapor content, it tends to be stable and has low relative humidity when highly heated during the daytime.

The maritime polar air mass (mP) originates over midlatitude oceans. Since the quantity of water vapor it holds is not as large as maritime tropical air masses, the mP air mass yields only moderate precipitation. Much of this precipita-

Table 8.1 | Properties of Typical Air Masses

Air Mass	Symbol	Source Region	Properties	Temperature °C	Temperature (°F)	Specific Humidity (gm/kg)
Maritime equatorial	mE	Warm oceans in the equatorial zone	Warm, very moist	27°	(81°)	19
Maritime tropical	mT	Warm oceans in the tropical zone	Warm, moist	24°	(75°)	17
Continental tropical	cT	Subtropical deserts	Warm, dry	24°	(75°)	17
Maritime polar	mP	Midlatitude oceans	Cool, moist(winter)	4°	(39°)	4.4
Continental polar	cP	Northern continental interiors	Cold, dry (winter)	–11°	(12°)	1.4
Continental arctic (and continental antarctic)	cA (cAA)	Regions near north and south poles	Very cold, very dry (winter)	–46°	(–51°)	0.1

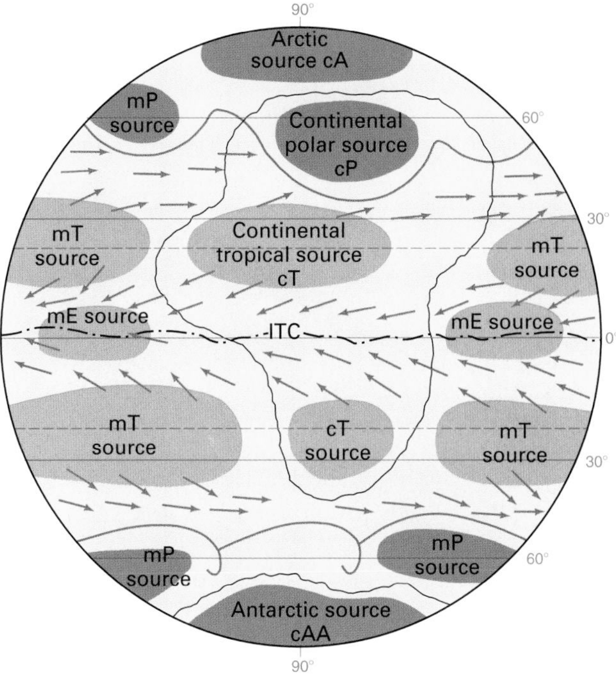

8.1 Global air masses and source regions *An idealized continent, producing continental (c) air masses, is shown at the center. It is surrounded by oceans, producing maritime air masses (m). Tropical (T) and equatorial (E) source regions provide warm or hot air masses, while polar (P), arctic (A), and antarctic (AA) source regions provide colder air masses of low specific humidity.*

tion is orographic and occurs over mountain ranges on the western coasts of continents. The continental polar air mass (cP) originates over North America and Eurasia in the subarctic zone. It has low specific humidity and is very cold in winter. Last is the continental arctic (and continental antarctic) air mass type (cA, cAA), which is extremely cold and holds almost no water vapor.

North American Air Masses

The air masses that form in and near North America have a strong influence on North American weather. The source regions of these air masses are shown in Figure 8.2. The continental polar (cP) air mass of North America originates over north-central Canada. This air mass forms tongues of cold, dry air that periodically extend south and east from the source region to produce anticyclones accompanied by cool or cold temperatures and clear skies. The arctic air mass (cA) that develops over the Arctic Ocean and its bordering lands of the arctic zone in winter is extremely cold and stable. When this air mass moves southward, it produces a severe cold wave.

The maritime polar air mass (mP) originates over the North Pacific and Bering Strait, in the region of the persistent Aleutian low-pressure center. This air mass is characteristically cool and moist, with a tendency in winter to become unstable, giving heavy precipitation over coastal ranges. Another maritime polar air mass of the North American region originates over the North Atlantic Ocean. It, too, is cool and moist, and is felt especially in Canada's maritime provinces in winter.

Of the tropical air masses, the most common visitor to the central and eastern United States is the maritime tropical air mass (mT) from the Gulf of Mexico. It moves northward, bringing warm, moist, unstable air over the eastern part of the country. In the summer, particularly, this air mass brings hot, sultry weather to the central and eastern United States. It also produces many thunderstorms. Closely related is a maritime tropical air mass from the Atlantic Ocean east of Florida, over the Bahamas.

Over the Pacific Ocean, a source region of another maritime tropical air mass (mT) lies in the cell of high pressure located to the southwest of lower California. Occasionally, in summer, this moist unstable air mass penetrates the southwestern desert region, bringing severe thunderstorms to southern California and southern Arizona. In winter, a tongue of mT air frequently reaches the California coast, bringing heavy rainfall that is intensified when forced to rise over coastal mountain ranges.

A hot, dry continental tropical air mass (cT) originates over northern Mexico, western Texas, New Mexico, and Arizona during the summer. This air mass does not travel widely but governs weather conditions over the source region.

8.2 North American air mass source regions and trajectories *Air masses acquire temperature and moisture characteristics in their source regions, then move across the continent. (Data from U.S. of Dept. Commerce.)*

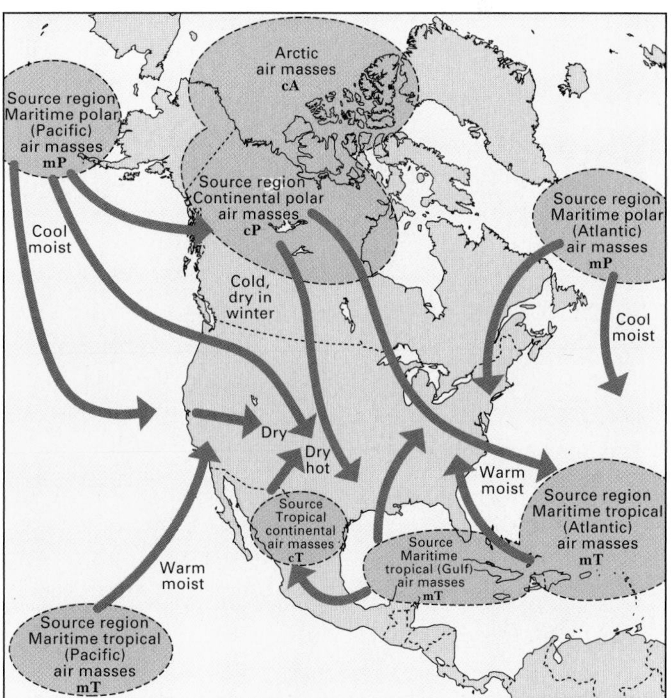

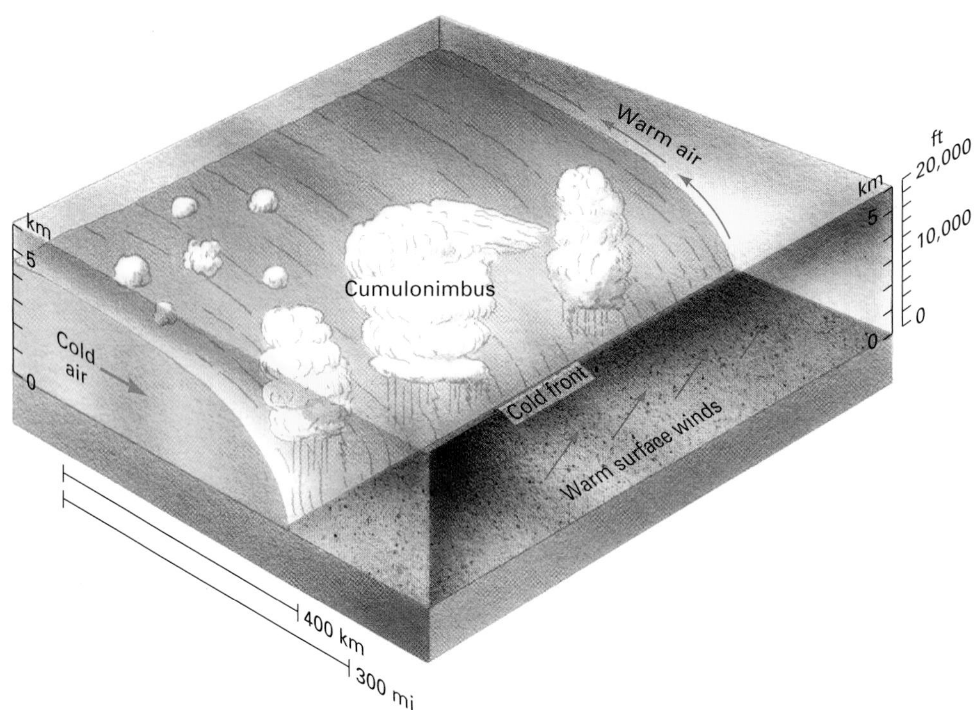

8.3 Cold front *At a cold front, a cold air mass lifts a warm air mass aloft. The upward motion sets off a line of thunderstorms. The frontal boundary is actually much less steep than is shown in this schematic drawing. (Drawn by A. N. Strahler.)*

Cold, Warm, and Occluded Fronts

A given air mass usually has a sharply defined boundary between itself and a neighboring air mass. This boundary is termed a **front.** We saw an example of a front in the contact between polar and tropical air masses, shown in Figures 7.16, 7.25, and 7.27. This feature is the *polar front,* and it is located below the axis of the jet stream in the upper-air waves. GEODISCOVERIES

Figure 8.3 shows the structure of a front along which a cold air mass invades a zone occupied by a warm air mass. A front of this type is called a **cold front.** Because the colder air mass is denser, it remains in contact with the ground. As it moves forward, it forces the warmer air mass to rise above it. If the warm air is unstable, severe thunderstorms may develop. Thunderstorms near a cold front often form a long line of massive clouds stretching for tens of kilometers (Figure 8.4).

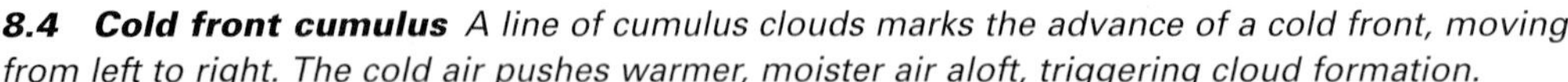

8.4 Cold front cumulus *A line of cumulus clouds marks the advance of a cold front, moving from left to right. The cold air pushes warmer, moister air aloft, triggering cloud formation.*

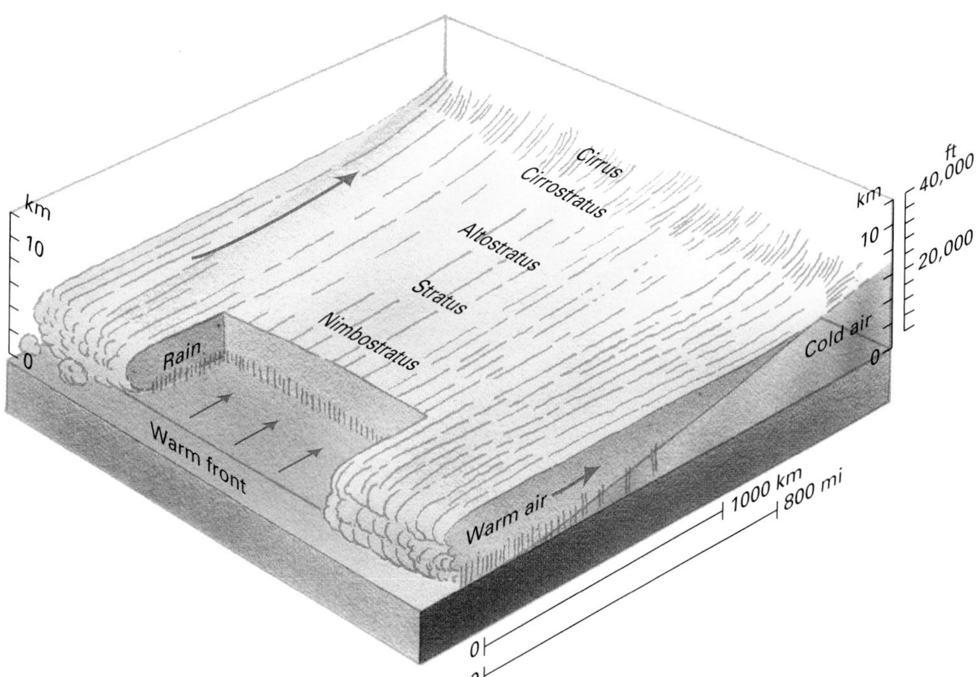

8.5 Warm front *In a warm front, warm air advances toward cold air and rides up and over the cold air. A notch of cloud is cut away to show rain falling from the dense stratus cloud layer. (Drawn by A. N. Strahler.)*

Figure 8.5 diagrams a **warm front** in which warm air moves into a region of colder air. Here, again, the cold air mass remains in contact with the ground because it is denser. The warm air mass is forced to rise on a long ramp over the cold air below. The rising motion causes stratus clouds to form, and precipitation often follows. If the warm air is stable, the precipitation will be steady (Figure 8.5). If the warm air is unstable, convection cells can develop, producing cumulonimbus clouds with heavy showers or thunderstorms (not shown in the figure).

Cold fronts normally move along the ground at a faster rate than warm fronts. Thus, when both types are in the same neighborhood, a cold front can overtake a warm front. The result is an **occluded front,** diagrammed in Figure 8.6. ("Occluded" means closed or shut off.) The colder air of the fast-moving cold front remains next to the ground, forcing both the warm air and the less cold air ahead to rise over it. The warm air mass is lifted completely free of the ground.

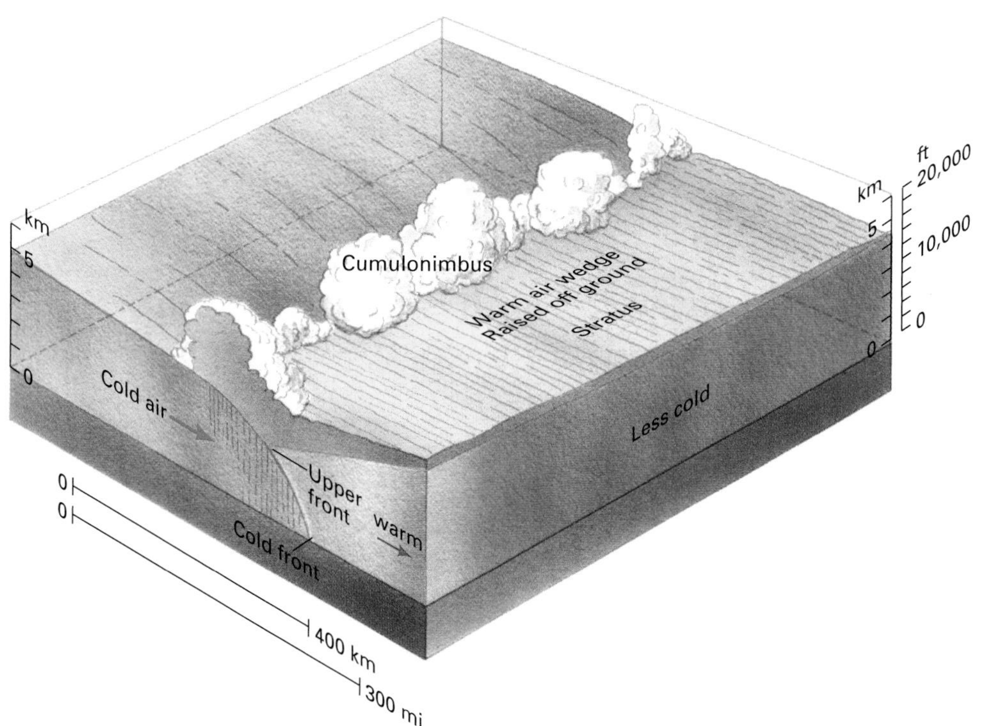

8.6 Occluded front *In an occluded front, a warm front is overtaken by a cold front. The warm air is pushed aloft, and it no longer contacts the ground. Abrupt lifting by the denser cold air produces precipitation. (Drawn by A. N. Strahler.)*

TRAVELING CYCLONES AND ANTICYCLONES

Air masses are set in motion by wind systems—typically, cyclones and anticyclones that involve masses of air moving in a spiraling motion. As we saw in Chapter 7, air spirals inward and converges in a *cyclone,* while air spirals outward and diverges in an *anticyclone.* Most types of cyclones and anticyclones are large features that move slowly across the Earth's surface, bringing changes in the weather as they move. These are referred to as *traveling cyclones* and *traveling anticyclones.*

In a cyclone, convergence and upward motion cause air to rise and be cooled adiabatically. If the air is moist, condensation or deposition can occur. This is **cyclonic precipitation.** Many cyclones are weak and pass overhead with little more than a period of cloud cover and light precipitation. However, when pressure gradients are steep and the inspiraling motion is strong, intense winds and heavy rain or snow can accompany the cyclone. In this case, the disturbance is called a **cyclonic storm.**

Traveling cyclones fall into three types. First is the wave cyclone of midlatitude, arctic, and antarctic zones. This type of cyclone ranges in intensity from a weak disturbance to a powerful storm. Second is the tropical cyclone of tropical and subtropical zones. This type of cyclone ranges in intensity from a mild disturbance to the highly destructive hurricane or typhoon. A third type is the tornado, a small, intense cyclone of enormously powerful winds. The tornado is much, much smaller in size than other cyclones, and it is related to strong convectional activity.

In an anticyclone, divergence and downward motion cause air to descend and be warmed adiabatically. Thus, condensation does not occur. Skies are fair, except for occasional puffy cumulus clouds that sometimes develop in a moist surface air layer. Because of these characteristics, anticyclones are often termed *fair-weather systems.* Toward the center of an anticyclone, the pressure gradient is weak, and winds are light and variable. Traveling anticyclones are found in the midlatitudes. They are typically associated with ridges or domes of clear, dry air that move eastward and equatorward. Figure 8.7 is a geostationary satellite image of eastern North America. A large anticyclone is centered over the area, bringing fair weather and cloudless skies.

Wave Cyclones

In middle and high latitudes, the dominant form of weather system is the **wave cyclone,** a large inspiral of air that repeatedly forms, intensifies, and dissolves along the polar front. Figure 8.8 shows a situation favorable to the formation of a wave cyclone. Two large anticyclones are in contact on the polar front. One contains a cold, dry polar air mass, and the other a warm, moist maritime air mass. Air-

8.7* Eye on the Landscape *Picture of an anticyclone *This geostationary satellite image of eastern North America shows a large anticyclone centered over the area, bringing fair weather and cloudless skies. The boundary between clear sky and clouds running across the Gulf of Mexico and the Florida peninsula delineates the leading edge of the cool, dry air mass. The cloud edge at the top of the photo marks the cold fronts of two air masses advancing eastward and southward.* **What else would the geographer see? ... Answers at the end of the chapter.**

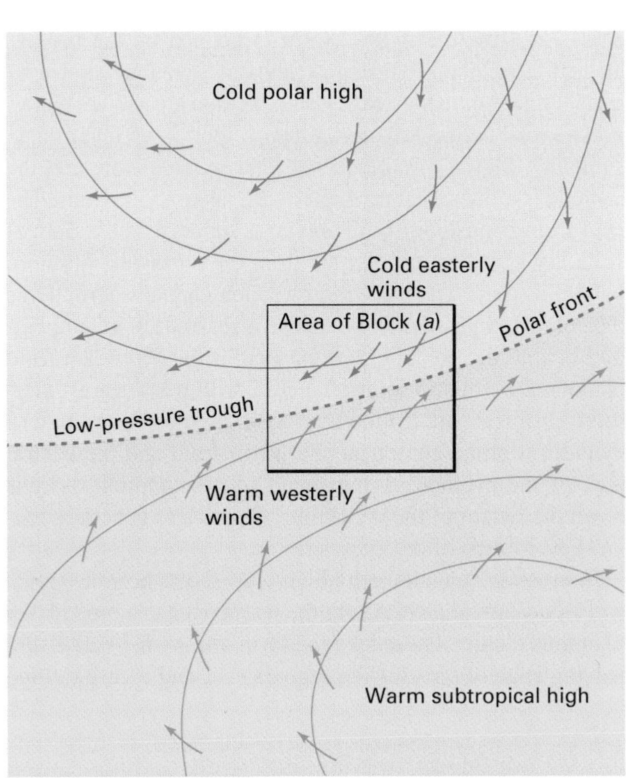

8.8 Conditions for formation of a wave cyclone *Two anticyclones, one with warm subtropical air and the other with cold polar air, are in contact on the polar front. The shaded area is shown in block (a) of the following figure as the early stage of development of a wave cyclone.*

flow converges from opposite directions on the two sides of the front, setting up an unstable situation. The wave cyclone will begin to form between the two high-pressure cells in a zone of lower pressure referred to as a *low-pressure trough.*

How does a wave cyclone form, grow, and eventually dissolve? Figure 8.9 shows the life history of a wave cyclone. In block *(a)* (early stage), the polar-front region shows a wave beginning to form. Cold air is turned in a southerly direction and warm air in a northerly direction, so that each advances on the other. As these frontal motions develop, precipitation will form.

In block *(b)* (open stage), the wave disturbance along the polar front has deepened and intensified. Cold air actively pushes southward along a cold front, and warm air actively moves northeastward along a warm front. The zones of precipitation along the two fronts are now strongly developed but are wider along the warm front than along the cold front.

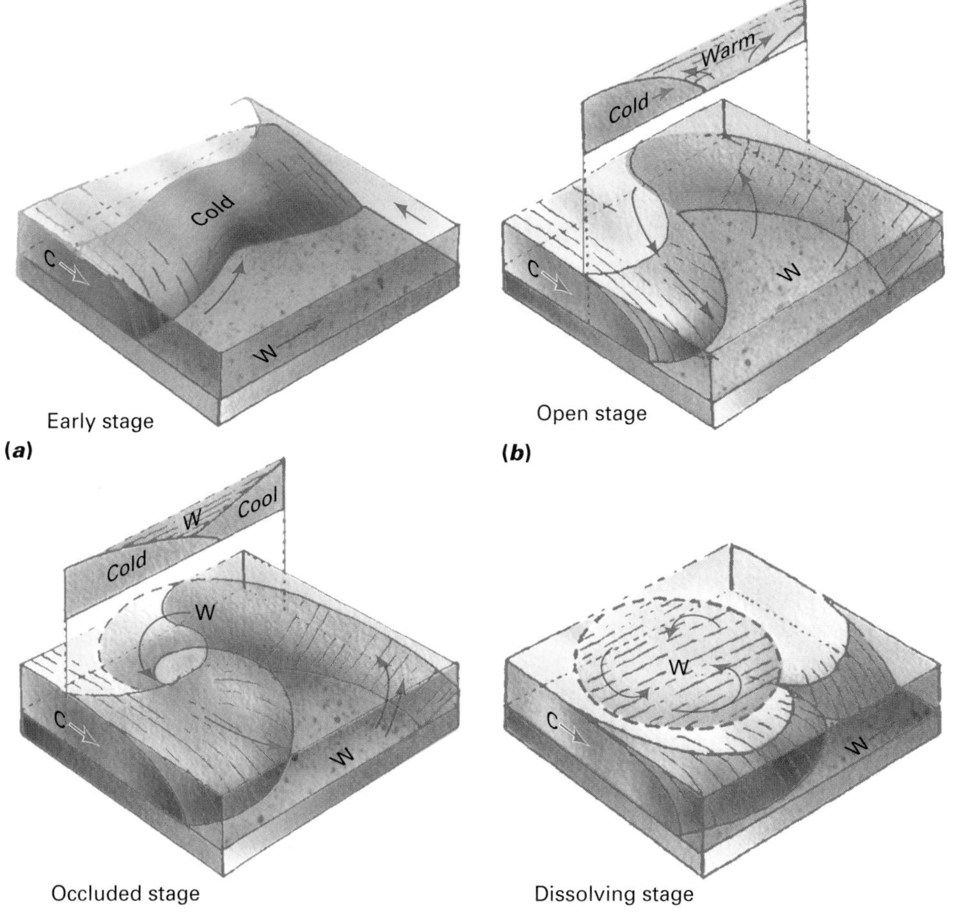

8.9 Development of a wave cyclone *In (a), a wave motion begins at a point along the polar front. The wave along the cold and warm fronts deepens and intensifies in (b). In (c), the cold front overtakes the warm front, producing an occluded front in the center of the cyclone. Later, the polar front is reestablished with a mass of warm air isolated aloft (d).*

In block *(c)* (occluded stage), the cold front has overtaken the warm front, producing an occluded front. The warm air mass at the center of the inspiral is forced off the ground, intensifying precipitation. Eventually, the polar front is reestablished (block *(d)*, dissolving stage), but a pool of warm, moist air remains aloft. As the moisture content of the pool is reduced, precipitation dies out, and the clouds gradually dissolve.

Keep in mind that a wave cyclone is quite a large feature—1000 km (about 600 mi) or more across. Also, the cyclone normally moves eastward as it develops, propelled by prevailing westerlies aloft. Therefore, blocks *(a)–(d)* are like three-dimensional snapshots taken at intervals along an eastbound track.

Weather Changes Within a Wave Cyclone

How does weather change as a wave cyclone passes through a region? Figure 8.10 shows two simplified weather maps of the eastern United States depicting conditions on successive days. The structure of the storm is defined by the isobars, labeled in millibars. The three kinds of fronts are shown by special line symbols. Areas of precipitation are shown in gray.

The map on the left shows the cyclone in an open stage, similar to Figure 8.9, block *(b)*. The isobars show that the cyclone is a low-pressure center with inspiraling winds. The cold front is pushing south and east, supported by a flow of cold, dry continental polar air from the northwest filling in behind it. Note that the wind direction changes abruptly as the cold front passes. There is also a sharp drop in temperature behind the cold front as cP air fills in. The warm front is moving north and somewhat east, with warm, moist maritime tropical air following. The precipitation pattern includes a broad zone near the warm front and the central area of the cyclone. A thin band of precipitation extends down the length of the cold front. Cloudiness generally prevails over much of the cyclone.

A cross section along the line A–A′ shows how the fronts and clouds are related. Along the warm front is a broad layer of stratus clouds. These take the form of a wedge with a thin leading edge of cirrus. (See Figures 6.11 and 6.12 for cloud

8.10 Simplified surface weather maps and cross sections through a wave cyclone *In the open stage (left), cold and warm fronts pivot around the center of the cyclone. In the occluded stage (right), the cold front has overtaken the warm front, and a large pool of warm, moist air has been forced aloft.*

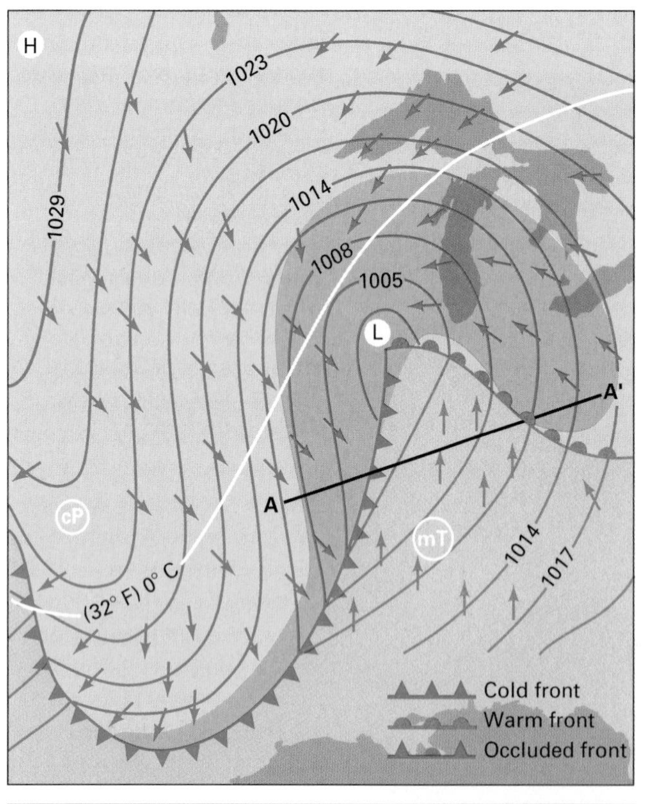

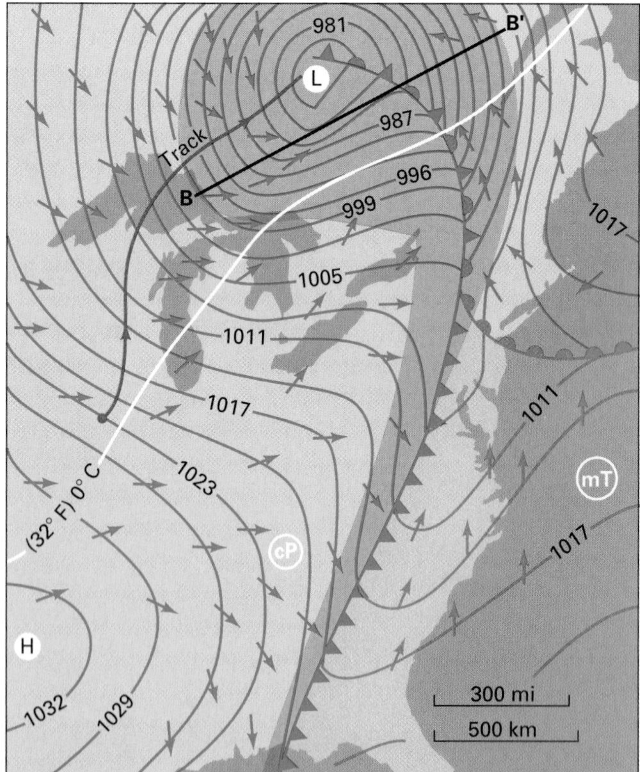

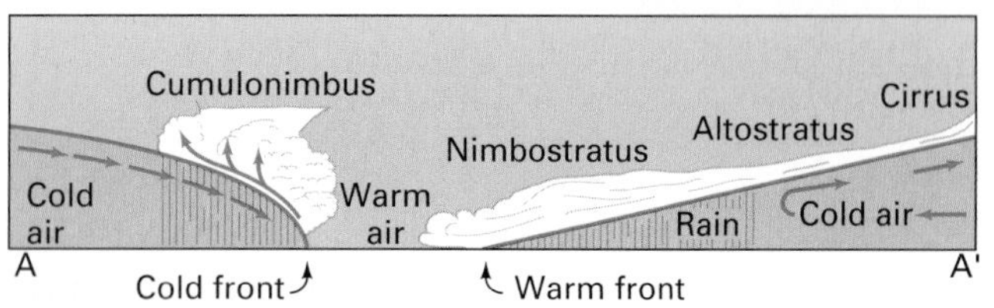

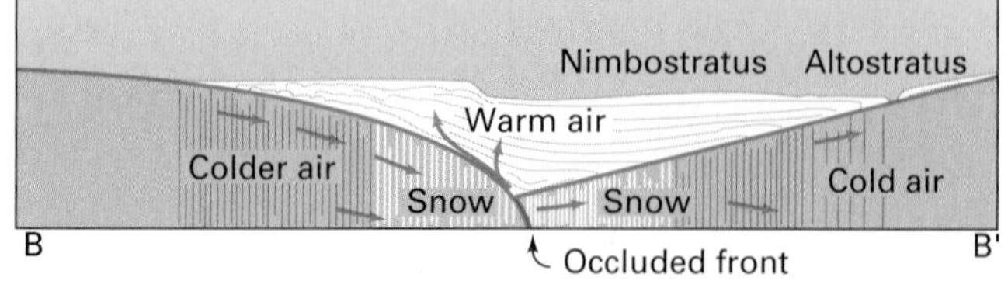

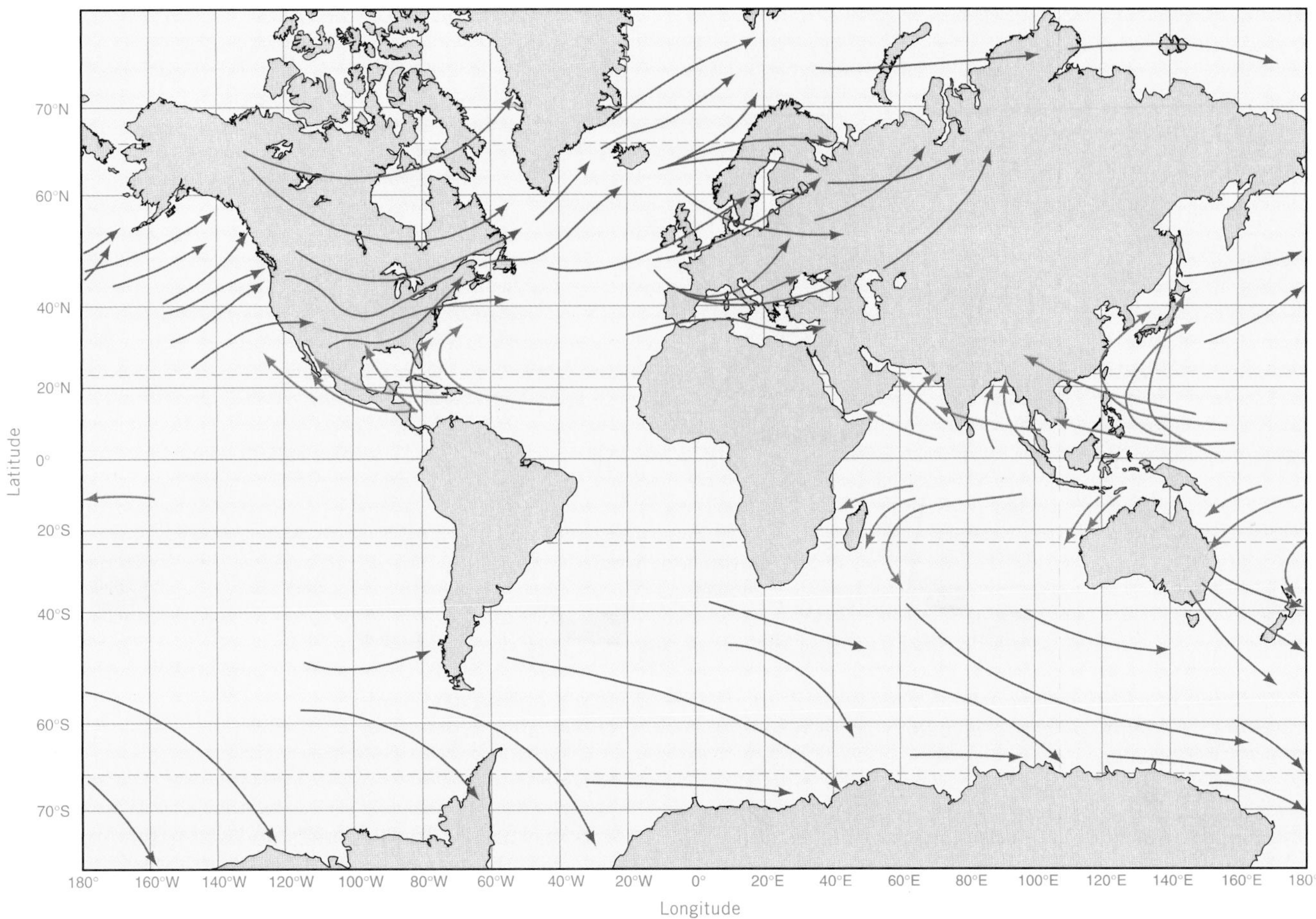

8.11 Paths of tropical cyclones and wave cyclones *This world map shows typical paths of tropical cyclones (red) and midlatitude wave cyclones (blue). (Based on data of S. Pettersen, B. Haurwitz and N. M. Austin, J. Namias, M. J. Rubin, and J-H. Chang.)*

types.) Westward, this wedge thickens to altostratus, then to stratus, and finally to nimbostratus with steady rain. Within the sector of warm air, the sky may partially clear with scattered cumulus. Along the cold front are cumulonimbus clouds associated with thunderstorms. These yield heavy rains but only along a narrow belt.

The weather map on the right shows conditions 24 hours later. The cyclone has moved rapidly northeastward, its track shown by the red line. The center has moved about 1600 km (1000 mi) in 24 hours—a speed of just over 65 km (40 mi) per hour. The cold front has overtaken the warm front, forming an occluded front in the central part of the disturbance. A high-pressure area, or tongue of cold polar air, has moved in to the area west and south of the cyclone, and the cold front has pushed far south and east. Within the cold air tongue, the skies are clear. A cross section below the map shows conditions along the line B–B′, cutting through the occluded part of the storm. Notice that the warm air mass is lifted well off the ground and yields heavy precipitation.

Cyclone Tracks and Cyclone Families

Wave cyclones tend to form in certain areas and travel common paths until they dissolve. Figure 8.11 is a world map showing common paths of wave cyclones and tropical cyclones. (We will discuss tropical cyclones in detail later in this chapter.) The western coast of North America commonly receives wave cyclones arising in the North Pacific Ocean. Wave cyclones also originate over land, shown by the tracks starting in Alaska, the Pacific Northwest, the south-central United States, and along the Gulf coast. Most of these tracks converge toward the northeast and pass into the North Atlantic, where they tend to concentrate in the region of the Icelandic Low.

In the northern hemisphere, wave cyclones are heavily concentrated in the neighborhood of the Aleutian and Icelandic Lows. These cyclones commonly form in a succession, traveling as a chain across the North Atlantic and North Pacific oceans. Figure 8.12, a world weather map, shows several such cyclone families. Each wave cyclone moves northeastward, deepening in low pressure and occluding to form an upper-air low. For this reason, intense cyclones arriving at the western coasts of North America and Europe are usually occluded.

In the southern hemisphere, storm tracks are more nearly along a single lane, following the parallels of latitude. Three such cyclones are shown in Figure 8.12. This track is more uniform because of the uniform pattern of ocean surface cir-

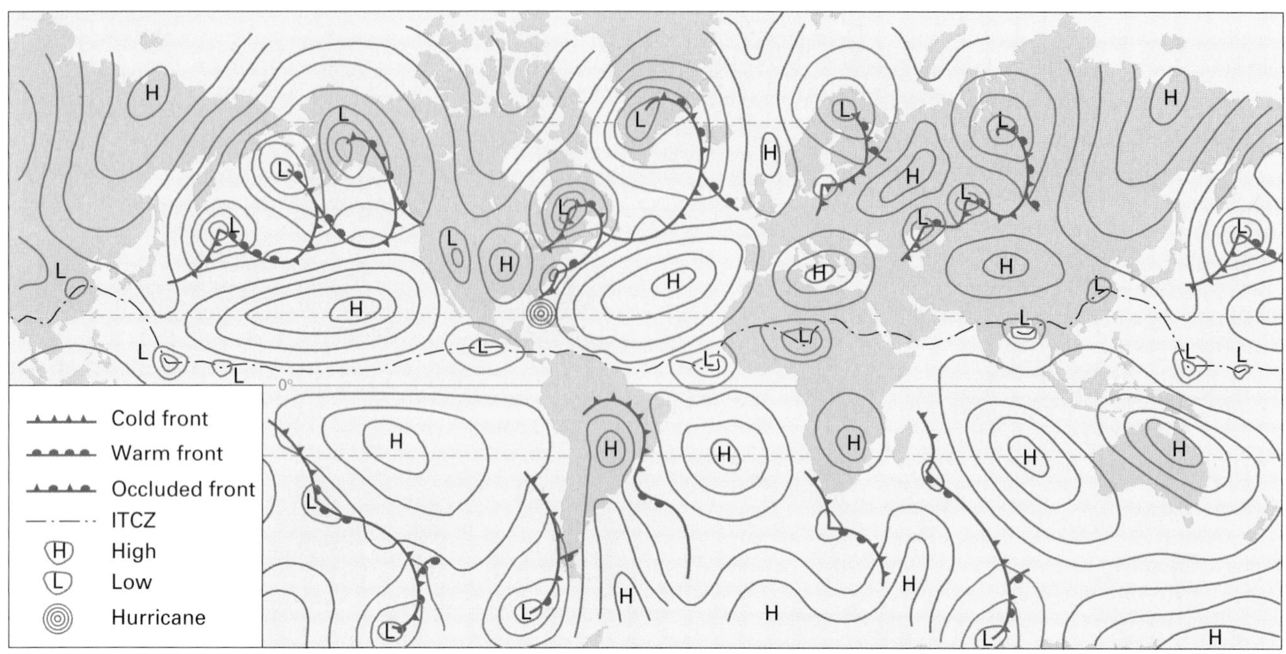

8.12 Daily world weather map *A daily weather map of the world for a given day during July or August might look like this map, which is a composite of typical weather conditions. (After M. A. Garbell.)*

cling the globe at these latitudes. Only the southern tip of South America projects southward to break the monotonous expanse of ocean.

The Tornado

A **tornado** is a small but intense cyclonic vortex in which air spirals at tremendous speed. It is associated with thunderstorms spawned by fronts in midlatitudes of North America. Tornadoes can also occur inside tropical cyclones (hurricanes).

The tornado appears as a dark funnel cloud hanging from the base of a dense cumulonimbus cloud (Figure 8.13). At its lower end, the funnel may be 100 to 450 m (about 300 to 1500 ft) in diameter. The base of the funnel appears dark because of the density of condensing moisture, dust, and debris swept up

8.13 Tornado *This tornado touched down near Clearwater, Kansas, on May 16, 1991. Surrounding the funnel is a cloud of dust and debris carried into the air by the violent winds.*

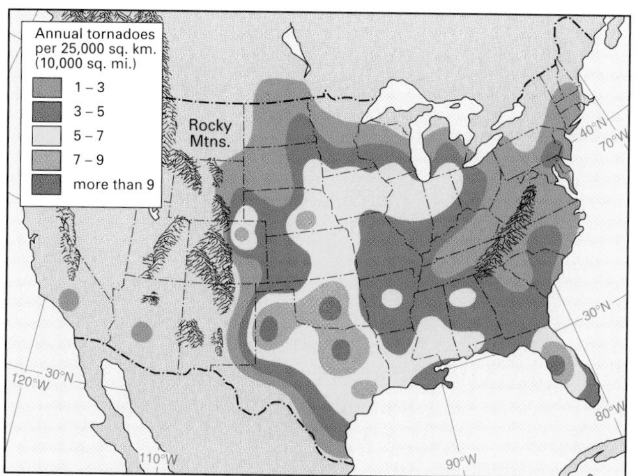

8.14 Frequency of occurrence of observed tornadoes in the conterminous United States and southern Canada *The data shown in this map span a 30-year record, 1960–1989. (Courtesy of Edward W. Ferguson, National Severe Storms Forecast Center, National Weather Service.)*

by the wind. Wind speeds in a tornado exceed speeds known in any other storm. Estimates of wind speed run as high as 100 m/s (about 225 mi/hr). As the tornado moves across the countryside, the funnel writhes and twists. Where it touches the ground, it can cause the complete destruction of almost anything in its path. GEODISCOVERIES

Tornadoes occur as parts of cumulonimbus clouds traveling in advance of a cold front. They seem to originate where turbulence is greatest. They are most common in the spring and summer but can occur in any month. Where a cold front of maritime polar air lifts warm, moist maritime tropical air, conditions are most favorable for tornadoes. As shown in Figure 8.14, they occur in greatest numbers in the central and southeastern states and are rare over mountainous and forested regions. They are almost unknown west of the Rocky Mountains and are relatively less frequent on the eastern seaboard. Tornadoes are a typically American phenomenon, being most frequent and violent in the United States. They also occur in Australia in substantial numbers and are occasionally reported from other midlatitude locations.

Devastation from a tornado is often complete within the narrow limits of its path (Figure 8.15). Only the strongest buildings constructed of concrete and steel can withstand the extremely violent winds. The National Weather Service maintains a tornado forecasting and warning system. Whenever weather conditions favor tornado development, the danger area is alerted, and systems for observing and reporting a tornado are set in readiness. GEODISCOVERIES

Tropical and Equatorial Weather Systems

So far, the weather systems we have discussed are those of the midlatitudes and poleward. Weather systems of the tropical and equatorial zones show some basic differences from those of the midlatitudes. Upper-air winds are often weak, so air mass movement is slow and gradual. Air masses are warm and moist, and tend to have similar characteristics. Thus, clearly defined fronts and large, intense wave cyclones are missing. On the other hand, intense convectional activity occurs because of the high moisture content of low-latitude maritime air masses. In these very moist air masses, even slight convergence and uplifting can be sufficient to trigger precipitation.

Clouds and precipitation in tropical and equatorial regions have recently become the object of a remote sensing mission. Of special interest is the intensity and duration of rainfall over oceans, which is not well monitored. *Focus on Remote Sensing 8.1 • TRMM—The Tropical Rainfall Monitoring Mission* provides more information.

8.15 Tornado destruction *A very powerful tornado swept this broad path through Midwest City, Oklahoma, on May 3, 1999.*

Focus on Remote Sensing | 8.1

TRMM—The Tropical Rainfall Monitoring Mission

As we saw when studying the Earth's global radiation balance in Chapter 4, the latitude belt between the Earth's two tropics receives the major portion of the Sun's energy. It is the region of the Hadley cell circulation, covered in Chapter 7, in which warm, moist air converges near the equator (ITCZ) and rises in a giant convection loop that moves heat from the equatorial region to the tropics. As the air rises, convectional precipitation occurs. In the condensation process, latent heat is released to the surrounding atmosphere. The effect is to gather up solar energy by evaporation of water from large areas of ocean or moist land surface and then release that energy to the atmosphere over a much smaller region as precipitation forms.

Mathematical models of atmospheric circulation have long identified this heating of middle levels of the atmosphere in the intertropical convergence zone (ITCZ) as an important process in atmospheric dynamics. However, scientists have lacked specific knowledge about where and when intertropical convectional precipitation occurs and how rainfall formation, and therefore energy release, occurs within rain clouds. Although weather radars have allowed the study of convectional precipitation over land, there has been no such source of data over oceans.

This lack of data, and its importance for weather and climate modeling, led NASA and the National Space Development Agency of Japan (NASDA) to devise and fly the Tropical Rainfall Measuring Mission (TRMM), which monitors clouds and precipitation in the intertropical zone. This satellite platform, launched from the Tanegashima, Japan, spaceport in November 1997, has a unique orbit. Instead of the polar Sun-synchronous orbit of a typical Earth imager (see *Geographer's Tools 4.8 • Remote Sensing for Physical Geography*), TRMM's orbit is strongly inclined so that the platform revolves from tropic to tropic, covering only the region from 35° N to 35° S latitudes. With a low altitude and a short orbital period, the orbit allows the platform's instruments to observe precipitation over land and ocean surfaces in this region at all times of day and night. In this way, scientists can accumulate accurate statistics about rainfall frequency, intensity, duration, and energy released for this portion of the globe. This, in turn, will allow mathematical models of atmospheric circulation to predict global energy transport, winds, and precipitation with much better accuracy.

The TRMM satellite platform carries five instruments, of which three are used for observations

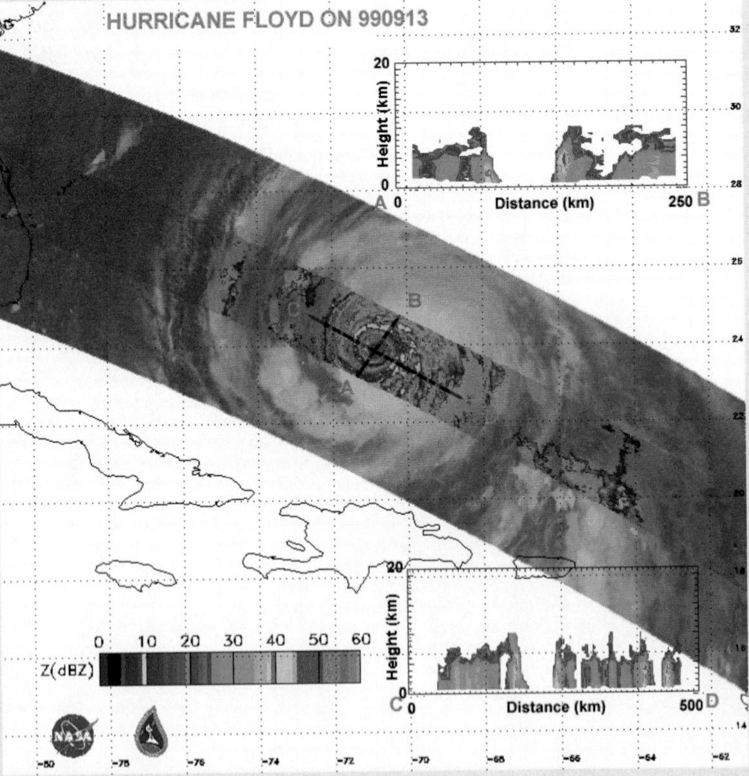

TRMM observes Hurricane Floyd *This data swath from TRMM instruments shows Hurricane Floyd as it passed north and east of Cuba and the Greater Antilles in September 1999. Cross sections show the concentration of raindrops and ice particles inside the clouds. (NASA)*

of clouds and precipitation: the precipitation radar, passive microwave imager, and visible and infrared scanner. The precipitation radar provides three-dimensional maps of storm structure, including the intensity and distribution of rain, precipitation type, height of storm clouds, and height at which snow melts into rain. It works by sending a pulse of radio waves toward the Earth and then listening for echoes that are scattered back by water droplets and ice particles. The timing and intensity of the echoes are used to reconstruct a profile of precipitation from the surface to the top of the storm cloud.

The passive microwave imager uses a different principle to monitor water vapor, cloud water, and rainfall intensity. The principle is that water droplets emit more energy in the microwave portion of the spectrum than does a water surface. This allows the measurement of precipitation over oceans. The emitted radiation is very small, however, and the measuring instrument must therefore be quite accurate. There are also methods for estimating precipitation over land using microwave emissions, but they are less effective.

The third imager is a visible and infrared scanner. It acquires images in five spectral regions, ranging from the visible to the thermal infrared. Its visible band tracks clouds by imaging them as bright objects. Thermal bands measure cloud-top temperature, which indicates cloud height. This imager is also quite similar to a number of instruments on other satellites, so it provides a way of linking the unique observations of the precipitation radar and microwave imager to data collected by other platforms at other times. The TRMM platform also carries a CERES instrument (see *Focus on Remote Sensing 4.5 • CERES—Clouds and the Earth's Radiant Energy System*) and a lightning imaging sensor.

The image on the facing page shows TRMM data obtained while passing over Hurricane Floyd on September 13, 1999. The inward-spiraling cloud pattern of this tropical cyclone is clearly shown by the wide track of the visible and infrared scanner. The narrow track shows precipitation as imaged by the other instruments. The two cross sections, A–B and C–D, show the structure of the storm. Precipitation is most intense in the yellow and red areas in the eye-wall of the storm, but some of the outer precipitation bands visible in cross section C–D also show very intense precipitation.

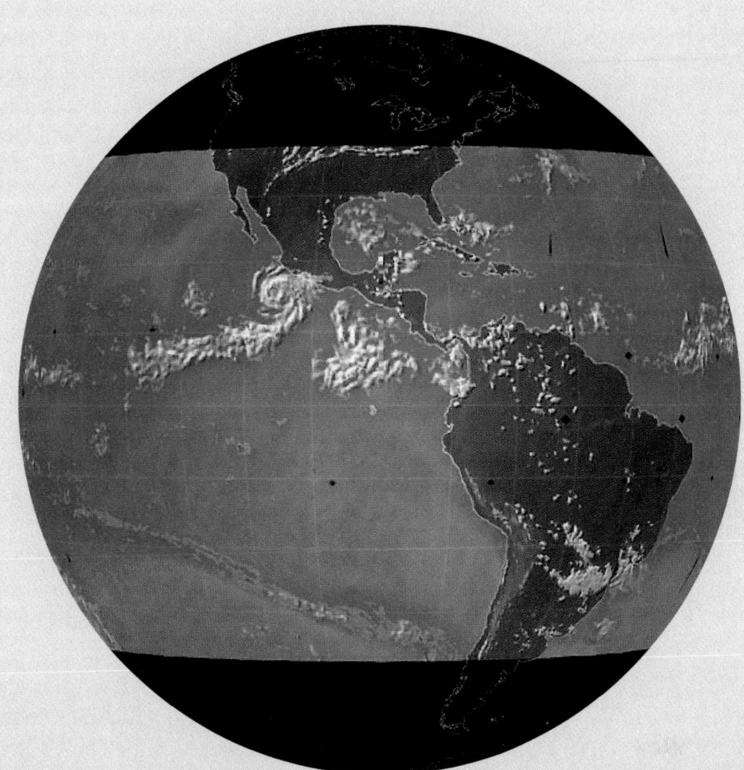

Global rainfall *This image shows precipitation as observed over a two-day period by TRMM instruments. (NASA)*

The image above is a composite of precipitation measured over a two-day period. Again, yellow and red tones indicate intense rainfall. Note the tropical cyclone present off the west coast of Mexico. Also visible are lines of frontal precipitation across the United States at the top of the image and across the southern Pacific from the equator southeast to southern Chile.

Like many other new orbiting platforms of instruments observing the Earth from space, TRMM is rapidly improving our scientific knowledge of the Earth as a system as well as our ability to model that system and to predict how human activity will alter it.

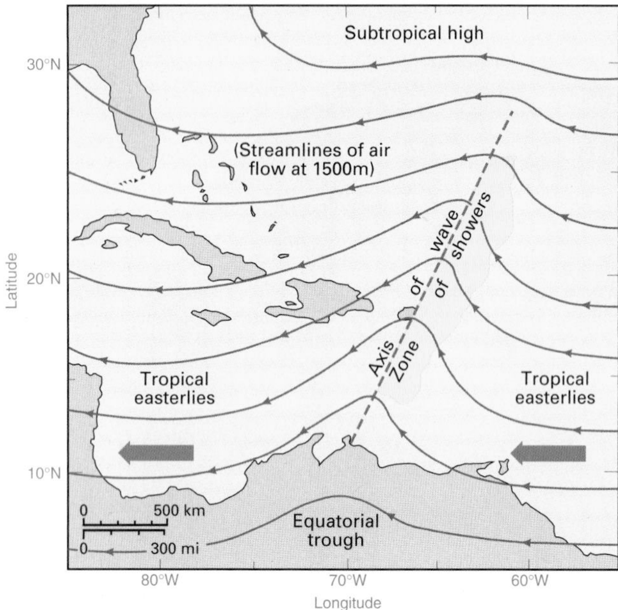

***8.16 Easterly wave** An easterly wave passing over the West Indies. (Data from H. Riehl, Tropical Meteorology, New York: McGraw-Hill.)*

Easterly Waves and Weak Equatorial Lows

One of the simplest forms of tropical weather systems is an *easterly wave,* a slowly moving trough of low pressure within the belt of tropical easterlies (trades). These waves occur in latitudes 5° to 30° N and S over oceans, but not over the equator itself. Figure 8.16 is a simplified upper-air map of an easterly wave showing wind patterns, the axis of the wave, isobars, and the zone of showers. At the surface, a zone of weak low pressure underlies the axis of the wave. The wave travels westward at a rate of 300 to 500 km (about 200 to 300 mi) per day. Surface airflow converges on the eastern, or rear, side of the wave axis. This convergence causes the moist air to be lifted, producing scattered showers and thunderstorms. The rainy period may last a day or two as the wave passes.

Another related weather system is the *weak equatorial low,* a disturbance that forms near the center of the equatorial trough. Moist equatorial air masses converge on the center of the low, causing rainfall from many individual convectional storms. Several such weak lows are shown on the world weather map (Figure 8.12), lying along the ITCZ. Because the map is for a day in July or August, the ITCZ is shifted well north of the equator. During this season, the rainy monsoon is in progress in Southeast Asia. It is marked by a weak equatorial low in northern India.

Polar Outbreaks

Another distinctive feature of low-latitude weather is the occasional penetration of powerful tongues of cold polar air from the midlatitudes into very low latitudes. These tongues are known as *polar outbreaks.* The leading edge of a polar outbreak is a cold front with squalls, which is followed by unusually cool, clear weather with strong, steady winds. The polar outbreak is best developed in the Americas. Outbreaks that move southward from the United States into the Caribbean Sea and Central America are called "northers" or "nortes," while those that move north from Patagonia into tropical South American are called "pamperos." One such outbreak is shown over South America on the world weather map (Figure 8.12). A severe polar outbreak may bring subfreezing temperatures to the highlands of South America and severely damage such essential crops as coffee.

Tropical Cyclones

The most powerful and destructive type of cyclonic storm is the **tropical cyclone,** which is known as the *hurricane* in the western hemisphere, the *typhoon* in the western Pacific off the coast of Asia, and the *cyclone* in the Indian Ocean. This type of storm develops over oceans in 8° to 15° N and S latitudes but not closer to the equator. The exact mechanism of formation is not known, but typically the tropical cyclone originates as an easterly wave or weak low, which then intensifies and grows into a deep, circular low. High sea-surface temperatures, over 27°C (81°F), are required for tropical cyclones to form. Once formed, the storm moves westward through the trade-wind belt, often intensifying as it travels. It can then curve northwest, north, and northeast, steered by westerly winds aloft. Tropical cyclones can penetrate well into the midlatitudes, as many residents of the southern and eastern coasts of the United States can attest. GEODISCOVERIES

An intense tropical cyclone is an almost circular storm center of extremely low pressure. Because of the very strong pressure gradient, winds spiral inward at high speed. Convergence and uplift are intense, producing very heavy rainfall (Figure 8.17). The storm gains its energy

***8.17 Hurricane weather map** A simplified weather map of a hurricane passing over the western tip of Cuba. Daily locations, beginning on September 3, are shown as circled numerals. The recurving path will take the storm over Florida. Shaded areas show dense rain clouds as seen in a satellite image.*

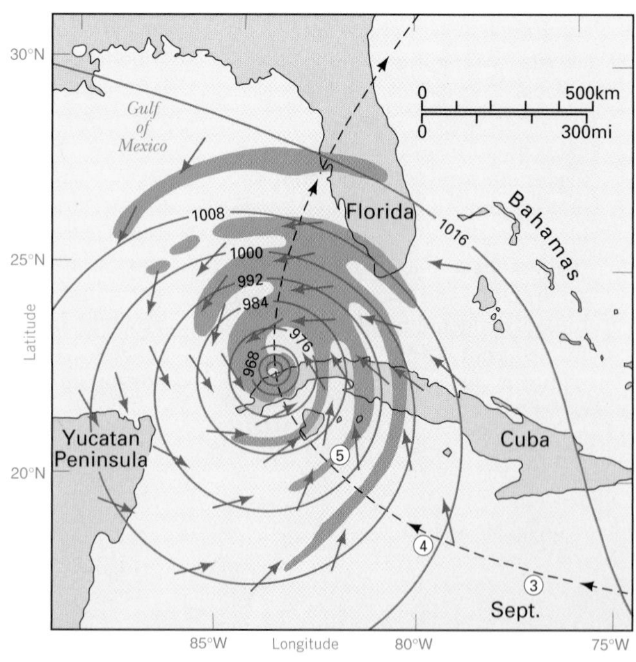

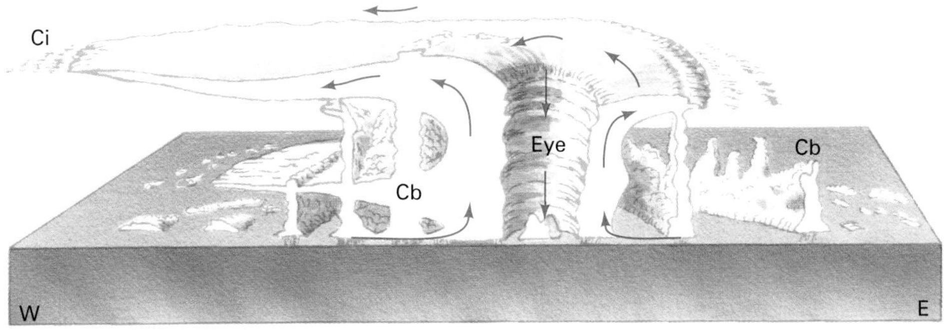

8.18 Anatomy of a hurricane *In this schematic diagram, cumulonimbus (Cb) clouds in concentric rings rise through dense stratiform clouds. Cirrus clouds (Ci) fringe out ahead of the storm. Width of diagram represents about 1000 km (about 600 mi). (Redrawn from NOAA, National Weather Service.)*

through the release of latent heat as the intense precipitation forms. The storm's diameter may be 150 to 500 km (about 100 to 300 mi). Wind speeds can range from 30 to 50 m/s (67 to 112 mi/hr) and sometimes much higher. Barometric pressure in the storm center commonly falls to 950 mb (28.1 in. Hg) or lower.

A characteristic feature of the tropical cyclone is its central eye, in which clear skies and calm winds prevail (Figure 8.18). The eye is a cloud-free vortex produced by the intense spiraling of the storm. In the eye, air descends from high altitudes and is adiabatically warmed. As the eye passes over a site, calm prevails, and the sky clears. Passage of the eye may take about half an hour, after which the storm strikes with renewed ferocity, but with winds in the opposite direction. Wind speeds are highest along the cloud wall of the eye.

The intensity of tropical cyclones is rated on the Simpson–Saffir scale, shown in Table 8.2. This scale ranks storms based on the central pressure of the storm, mean wind speed, and height of accompanying storm surge. Category 1 storms are weak, while category 5 storms are devastating.

Refer to Figure 8.11, presented earlier, which shows typical paths of both wave cyclones and tropical cyclones. As you can see, tropical cyclones always form over oceans. In the western hemisphere, hurricanes originate in the Atlantic off the west coast of Africa, in the Caribbean Sea, or off the west coast of Mexico. Curiously, tropical cyclones do not form in the South Atlantic or southeast Pacific regions. As a result, South America is never threatened by these severe storms. In the Indian Ocean, cyclones originate both north and south of the equator, moving north and east to strike India, Pakistan, and Bangladesh, as well as south and west to strike the eastern coasts of Africa and Madagascar. Typhoons of the western Pacific also form both north and south of the equator, moving into northern Australia, Southeast Asia, China, and Japan.

Tracks of tropical cyclones of the North Atlantic are shown in detail in Figure 8.19. Most of the storms originate at 10° to 20° N latitude, travel westward and northwestward through the trades, and then turn northeast at about 30° to 35° N latitude into the zone of the westerlies. Here their intensity lessens, especially if they move over land. In the trade-wind belt, the cyclones travel 10 to 20 km (6 to 12 mi) per hour. In the zone of the westerlies, their speed is more variable.

Tropical cyclones are now tracked using satellite images. They are often easy to identify from their distinctive pattern of inspiraling bands of clouds and a clear central eye. Figure 8.20 provides a gallery of satellite images of tropical cyclones.

Table 8.2 Simpson–Saffir Scale of Tropical Cyclone Intensity

Category	Central Pressure mb (in. Hg)	Storm Surge m (ft)	Mean Wind m/s (mph)
1 Weak	>980 (>29.0)	1.2–1.7 (4–5)	33–42 (74–95)
2 Moderate	965–979 (28.5–29.0)	1.8–2.6 (6–8)	43–49 (96–110)
3 Strong	945–964 (27.9–28.5)	2.7–3.8 (9–12)	50–58 (111–130)
4 Very Strong	920–944 (27.2–27.9)	3.9–5.6 (13–18)	59–69 (131–155)
5 Devastating	<920 (<27.2)	>5.6 (>18)	>69 (>155)

8.19 Tracks of typical hurricanes occurring during August *The storms arise in warm tropical waters and move northwest. On entering the region of prevailing westerlies, the storms change direction and move toward the northeast.*

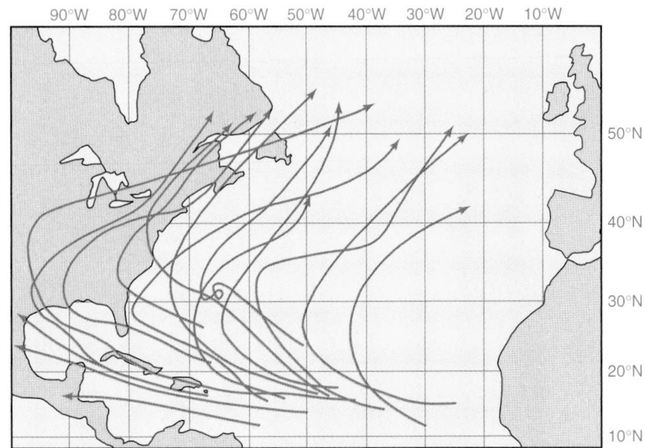

8.20 Tropical Cyclone Gallery

a...***Hurricane Andrew*** *approaching the Louisiana coast on August 25, 1992, as depicted in an enhanced image from NOAA's Advanced Very High Resolution Imaging Radiometer (AVHRR).*

b...***Hurricane Linda*** *approaching Baja California on 1997 September 12. Data are from the NOAA GOES-9 satellite. (NASA. Images and rendering by Marit Jentoft-Nilsen.)*

c...*Hurricane Mitch* *on October 26, 1998, off the coast of Honduras, as it approached the Yucatan Peninsula. Data from NOAA's AVHRR instrument. (NOAA.)*

d...*Hurricane Isabel* *This striking MODIS satellite image from September 18, 2003, shows Hurricane Isabel at about the time its eye crossed the coast of North Carolina. This major storm was responsible for at least 13 direct deaths and 17 others indirectly related to the storm. Early damage estimates centered on about a billion dollars.*

Eye on the Environment | 8.2

Hurricane Andrew—Killer Cyclone

Tropical cyclones, or hurricanes, are among the most destructive of natural phenomena. With sustained high winds and rain, they can wreak a pattern of destruction across vast areas of land. Coastal areas are particularly vulnerable. Since hurricanes develop and reach their maximum intensity over ocean waters, coastal regions experience the full brunt of the storm. Storm surges caused by strong onshore winds can raise sea levels and bring storm-ravaged surf inland to batter coastal structures well back from the beach. Low coastal islands can be completely submerged, drowning their inhabitants.

Hurricane Andrew, striking the south Florida Peninsula on August 23–24, 1992, and moving on to ravage the Louisiana coast on August 26, was the most damaging natural disaster ever to hit the United States. In the path of Andrew in south Dade Country, Florida, was a population of 355,000 people. By the time the storm was over, 80,000 dwellings were demolished or destroyed. Another 55,000 dwellings were seriously damaged. Estimates of property loss ranged between $20 and $30 billion. The death toll was small, however—the storm claimed only 43 victims. Evacuations and emergency warnings reduced the toll significantly. In Louisiana, Andrew struck a rural area of agricultural fields, marshlands, and small towns, so damage was less intense. Another 15 people died, and additional damage was estimated at about $2 billion.

Hurricane Andrew, like many other tropical cyclones, began as a tropical easterly wave of showers off the African coast near the Cape Verde Islands on August 14. Moving westward, it became a tropical depression with winds of 18 m/s (40 mi/hr) on August 17 and was named Andrew—the first tropical cyclone of the 1992 season. By August 20, the storm had moved westward but gained little strength. Opposing the westward surface motion of the storm was an upper air flow from west to east, creating a high-level wind shear that kept the storm from developing a strong vertical structure. However, by August 21, an unusual easterly upper air flow took hold, eliminating the wind shear. This allowed Andrew to gain strength. By 11 P.M. on the 21st, Andrew's winds had increased to 27 m/s (60mi/hr).

Andrew continued to intensify. On Saturday morning, August 22, the storm became a category 1 hurricane on the Simpson–Saffir scale, and later in the day, it attained category 2 status. The storm continued its rapid development, becoming a category 3, then a category 4 storm on Sunday, August 23. Bearing winds of 72 m/s (160 mi/hr) and a central low pressure of 924 mb (27.33 in. Hg), the hurricane swept through the southern Bahamas that evening, killing four and leaving 1700 homeless. Between 5:00 and 6:00 A.M. Monday morning, Andrew's eye crossed the coast of Florida just south of Miami. Sustained winds were in the range of 65 m/s (145 mi/hr). A few miles to the north of the eye, the anemometer atop the National Hurricane Center measured a peak of 73 m/s (164 mi/hr) before breaking at the height of the storm. Other anemometers

Andrew's wrath *Hurricane Andrew blew the roofs from these homes in the suburbs of Miami, Florida, exposing their contents to wind and water damage.*

Tropical cyclones occur only during certain seasons. For hurricanes of the North Atlantic, the season runs from May through November, with maximum frequency in late summer or early autumn. In the southern hemisphere, the season is roughly the opposite. These periods follow the annual migrations of the ITCZ to the north and south with the seasons, and correspond to periods when ocean temperatures are warmest.

For convenience, tropical cyclones are given names as they are tracked by weather forecasters. Male and female names are alternated in an alphabetical sequence renewed each season. Two sets of names are used—one for hurricanes of the Atlantic and one for typhoons of the Pacific. Names are reused, but the names of storms that cause significant damage or destruction are retired from further use. GEODISCOVERIES measured gusts of about 78 m/s (175 mi/hr). The storm surge brought sea level up to 5.2 m (16.9 ft) above normal.

Crossing Florida at a speed of 7–8 m/s (16–18 mi/hr) and entering the Gulf of Mexico, Andrew headed west and slightly north (see Figure 18.20). Intensifying over the warm Gulf water, the hurricane reattained category 4 status and struck the Louisiana coast on Wednesday, August 26. Low barrier islands were overtopped, eroded, and cut with new channels. Marshes and shrimping grounds were filled with sand and debris. Sugar cane crops were flattened. Many dwellings were severely damaged. Moving inland and weakening, Andrew headed northeast, finally disintegrating into a region of showers in the Ohio River Valley on August 28.

Although Andrew's winds in south Florida were measured in the 60–74 m/s (135–165 mi/hr) range, an analysis of the damage patterns afterward indicated that winds were still higher in some locations. These winds may have resulted from a particularly intense convectional cell that developed on the north side of the storm as it crossed the coastline.

Another possibility, according to Professor Ted Fujita of the University of Chicago, is a "suction vortex"—a swirl of faster winds that occurs within the storm's eye-wall. Although the vortex normally achieves wind speeds of only 9 to 13 m/s (20 to 30 mi/hr), the vortex can be sucked upward by a strong local updraft, causing it to narrow and spin more intensely. In fact, the vortex can attain wind speeds of up to 35 m/s (80 mi/hr). Where the swirling motion reinforces the overall storm wind of perhaps 55 m/s (120 mi/hr), total wind speed can approach 90 m/s (200 mi/hr).

Andrew's harvest *The entrance to Pinelands, in the Everglades National Park, is blocked by fallen trees brought down by the high winds of Hurricane Andrew.*

No normal structure can survive such winds for more than a minute or two. Fortunately, these winds are restricted to a narrow, tornado-like swath of up to 100 m (about 300 ft) in width.

The damage was not restricted to human settlements. Hurricane Andrew delivered a solid blow to the Everglades as well. Most "hammocks"—tree-covered islands in the watery grassland—were stripped bare of leaves, their trees broken and splintered. Mangrove swamps on the western edge of the Everglades were also defoliated.

Although the Everglades ecosystem has endured the periodic assaults of hurricanes for centuries and is well-adjusted to surviving them, ecologists are concerned about the regrowth process. A number of imported species threaten to take over, crowding out the native ones. It is not certain how the presence of these exotic plants will influence the recovery of the hammocks and other areas opened up by Andrew's high winds and waters.

Andrew is an example of the brutal power of the tropical cyclone to change the face of the Earth. Only a massive earthquake in a major urban area is capable of doing more damage. Although little can be done to stop such disasters from occurring, much can be done to reduce the damage they cause. Warnings and evacuations have reduced death rates to very low numbers. What is needed is better attention to building structures that are more wind-resistant and have such simple protections as storm shutters. Preserving shallow bays and mangrove swamps in natural states can also help by brunting the force of storm surges. Keeping beach-front development in check can reduce the number and size of structures exposed to the full force of hurricane winds and waves. Although Andrew was the first major hurricane to hit south Florida in nearly 30 years, the next hurricane will probably hit much sooner. Let's hope that this region will fare better next time, thanks to the lessons learned from Andrew.

Impacts of Tropical Cyclones

Tropical cyclones can be tremendously destructive storms. Islands and coasts feel the full force of the high winds and flooding as tropical cyclones move onshore. For example, Hurricane Andrew, striking the Florida coast near Miami, was the most devastating storm ever to occur in the United States, claiming as much as $25 billion in property damage and 43 lives. *Eye on the Environment 8.2 • Hurricane Andrew—Killer Cyclone* provides more information about this costly natural disaster.

The most serious effect of tropical cyclones is usually coastal destruction by storm waves and very high tides. Since the atmospheric pressure at the center of the cyclone is so low, sea level rises toward the center of the storm. High winds create a damaging surf and push water toward the coast, raising sea level even higher. Waves attack the shore at points far inland of the normal tidal range. Low pressure, winds, and the underwater shape of a bay floor can combine to produce a sudden rise of water level, known as a **storm surge.** During surges, ships are lifted by the high waters and can be stranded far inland.

If high tide accompanies the storm, the limits reached by inundation are even higher. This flooding can create enormous death tolls. At Galveston, Texas, in 1900, a sudden storm surge generated by a severe hurricane flooded the low coastal city and drowned about 6000 persons. Low-lying coral atolls of the western Pacific may be entirely swept over by wind-driven sea water, washing away palm trees and houses and drowning the inhabitants.

Also important is the large amount of rainfall produced by tropical cyclones. For some coastal regions, these storms provide much of the summer rainfall. Although this rainfall is a valuable water resource, it can also produce freshwater flooding, raising rivers and streams out of their banks. On steep slopes, soil saturation and high winds can topple trees and produce disastrous mudslides and landslides.

Tropical cyclone activity varies from year to year and decade to decade. In the Atlantic Basin, 13 strong hurricanes of category 3 or higher struck the eastern United States or Florida Peninsula from 1947 to 1969, while only one hurricane struck the same region in the period 1970–1987. Although a number of strong storms, including category 4 or 5 hurricanes Gilbert (1988), Hugo (1989), and Andrew (1992) occurred in 1988–1992, Atlantic hurricane activity remained depressed until 1994. However, the 1995 season provided 19 named tropical storms—two fewer than the record of 1933—and 11 hurricanes—only one short of the record set in 1969. During August 27 and 28, 1995, a record of five tropical cyclones were active simultaneously in the Atlantic Basin. The 1996 season brought a respite with only two moderate hurricanes—Bertha and Fran—striking nearly the same region of the North Carolina coast. The following year also showed reduced activity, with only weak Hurricane Danny making landfall on the Louisiana–Alabama coast.

But hurricanes returned with a vengeance in 1998, with two killer storms—Georges and Mitch. Hurricane Georges, a category 4 storm in the central Atlantic, weakened somewhat before making landfalls in the Lesser Antilles islands of Antigua, St. Kitts, and Nevis, and then proceeded along the Greater Antilles chain from Puerto Rico to Hispaniola and Cuba. After grazing Key West, Florida, Georges finally made continental landfall near Biloxi, Mississippi, on September 28 with winds of about 50 m/s (112 mi/hr).

About a month later, Hurricane Mitch, one of the deadliest Atlantic tropical cyclones in history, struck the heart of Central America. Claiming a toll of 9086, Mitch killed most of its victims with its torrential rains and their aftermath. The monster category 5 storm attained a central low pressure of 905 mb with winds of 77.2 m/s (172 mi/hr) in the western Caribbean on October 26, 1998. This central pressure is the fourth lowest ever measured in an Atlantic hurricane and ties Hurricane Mitch with Hurricane Camille of 1969. Moving at an average speed of only 2.2 m/s (5 mi/hr), Mitch hovered over the seven-nation Central American isthmus for nearly a week, dumping about 60 cm (24 in.) of rain at most locations. As much as 91 cm (36 in.) was reported at one mountain weather station, and it is likely that higher rainfalls went unrecorded. Countless mudflows and debris avalanches coursed down the rugged slopes of the western ranges of Honduras and Nicaragua, burying entire villages and villagers alike. Major rivers quickly flooded, sweeping floodplain settlements away on roiling waters. The damage was truly devastating. Honduras and Nicaragua suffered losses of about half their annual gross national products. Guatemala and El Salvador were also hard hit.

The most recent forecasts suggest that we may be entering an active period for Atlantic hurricanes, similar to that of 1947–1969. If so, we can expect the cost to human society to increase substantially, given the buildup in coastal development that has occurred in hurricane-prone regions.

POLEWARD TRANSPORT OF HEAT AND MOISTURE

Recall from Chapter 4 that vast global heat flows extend poleward from the equatorial and tropical regions, where the Sun's energy is concentrated and net radiation is positive, to the middle and higher latitudes, where net radiation is negative. These heat flows are referred to as *poleward heat transport.* Weather systems such as cyclones are byproducts of this heat transport, which sends warm, moist air poleward to mix with cooler, drier air, thus creating cyclonic precipitation. With our knowledge of the global circulation patterns of the atmosphere and oceans, and our knowledge of precipitation processes, we can now examine poleward heat transport in more detail.

Atmospheric Heat and Moisture Transport

Figure 8.21 is a schematic diagram of global heat and moisture flow within the atmosphere. An important feature is the Hadley cell circulation—a global convection loop in which moist air converges and rises in the intertropical convergence zone (ITCZ) while subsiding and diverging in the subtropical high-pressure belts. *Focus on Systems 8.3 • Hadley Cell Circulation as a Convection Loop Flow System* shows the anatomy of a Hadley cell in more detail.

The Hadley cell convection loop acts to pump heat from warm equatorial oceans poleward to the subtropical zone. Near the surface, air flowing toward the ITCZ picks up water vapor evaporated by sunlight from warm ocean surfaces, increasing the moisture and latent heat content of the air. With convergence and uplift of the air at the ITCZ, the

Hadley Cell Circulation as a Convection Loop Flow System

The atmosphere is a largely clear, fluid layer that transmits most of the Sun's energy to the surface of the Earth. Because solar energy does not warm the Earth's surface uniformly, the Earth's atmosphere and oceans exhibit a complex circulation pattern of air and water flows that acts to redistribute absorbed solar radiation more evenly across the globe. In this circulation, convection loops play an important role.

The Hadley cell is a good example of a global-scale convectional circulation pattern. The figure below shows the Hadley cell circulation on a day in December. The diagram uses lines of equal wind velocity, called isotachs. Two kinds of isotachs are shown. The upper diagram *(a)* shows wind speed in a horizontal (north-south) direction, while the lower diagram *(b)* shows wind speed in a vertical direction.

Looking first at the horizontal component of the wind *(a)*, a nest of centered ovals at the bottom of the figure shows a center of strong horizontal motion near the surface. This flow is southward, as shown by the direction of the small arrow. At the top of the figure is another nest of solid isotachs, indicating a strong wind flow in the northern direction. These two flows are the connecting legs of horizontal flow in the convection loop. The flows are strongest near 10° N latitude. The vertical component of the wind, shown in *(b)*, also shows two centers of wind strength. Near the equator, motion is upward. At about 15° N latitude, the motion is downward.

Taken together, the four flows shown by the isotachs provide a single convection loop, diagrammed by the broad arrows. In this loop, sensible heat is released from latent heat by condensation in rising air near the equator and is carried by the air in motion to the upper part of the troposphere. Here, a portion of the heat is lost by direct radiation to space. As the air descends, a portion leaks out of the cycle and heads poleward. This air delivers a flow of heat to higher latitudes. In its place, cooler air enters the loop near the surface.

Although the Hadley cell circulation may fit a simple convection loop fairly well overall, the motions of the atmosphere, induced by solar heating of the Earth's surface and influenced by the Earth's rotation, are generally quite complex. These motions also power the surface currents of the ocean and, ultimately, the deep currents as well. We can view all the motions of the Earth's oceans and atmosphere as a continuous cycle of interlinked fluid flows—a cycle powered by an energy transport system that moves heat from low to high latitudes.

A cross section of a Hadley cell *In this diagram, taken for a day in December, contour lines are isotachs, or lines of equal velocity. (a) North-south component of velocity, ranging up to about 4 m/s. (b) Upward-downward component, ranging up to about 1 cm/s. Note that the height of the cell is greatly exaggerated. It is really a very flat feature when its height is compared to its latitudinal width. (Copyright © A. N. Strahler.)*

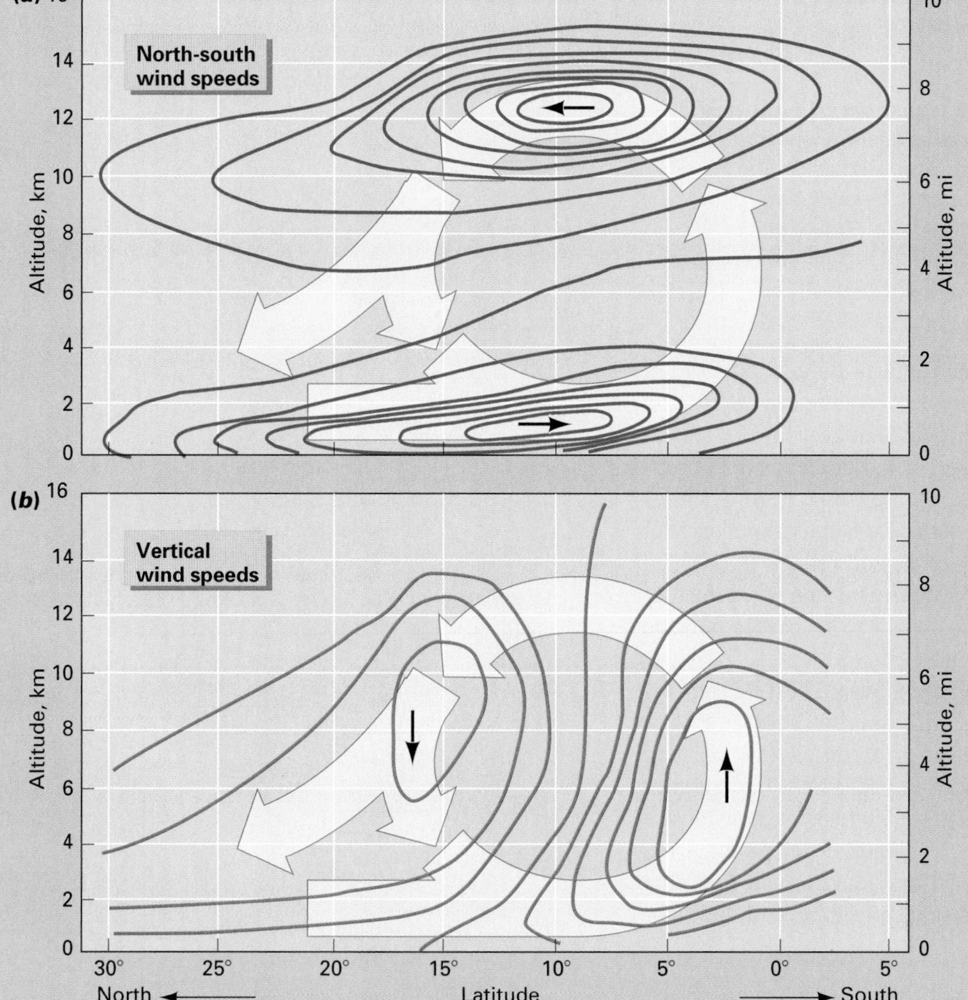

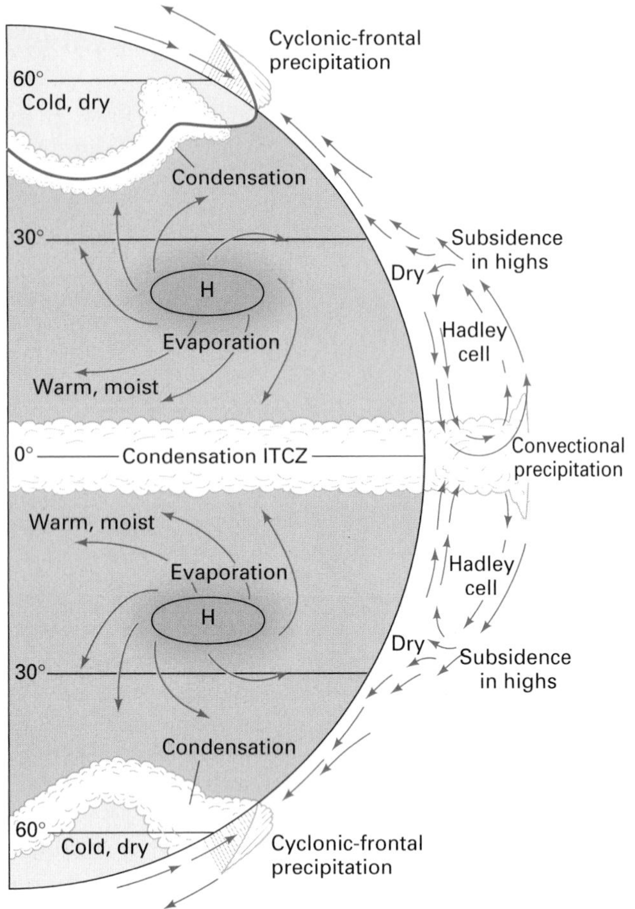

8.21 Global atmospheric transport of heat and moisture *Major mechanisms of heat transport in the atmosphere include intertropical convergence, the Hadley cell circulation, and cyclonic-frontal precipitation in the midlatitudes.*

latent heat is released as sensible heat in the condensation that occurs. Air traveling poleward in the return circulation aloft retains much of this heat, although some is lost to space by radiant cooling. When the air descends in the subtropical high-pressure belts, the sensible heat becomes available at the surface. The net effect is thus to gather heat from tropical and equatorial zones and release the heat in the subtropical zone, where it can be conveyed further poleward by the motion of mT and cT air masses into the midlatitudes.

In the mid- and high latitudes, poleward heat transport is also produced by the Rossby wave mechanism (see Figure 7.25). Lobes of cold, dry polar or arctic air (cP, cA, and cAA air masses) plunge toward the equator, while tongues of warmer, moister air (mT and mP air masses) flow toward the poles. At their margins, cyclones develop, releasing latent heat in precipitation and providing a heat flow that warms the mid- and higher latitudes well beyond the capabilities of the Sun.

Oceanic Heat Transport

Just as atmospheric circulation plays a role in moving heat from one region of the globe to another, so does oceanic circulation. Chapter 7 documented the thermohaline circulation of the ocean, in which the sinking of cold, salty water in the North Atlantic is linked to a slow-moving system of deep ocean currents, upwelling, and surface currents (Figure 7.35). Figure 8.22 diagrams the key features of this flow. At the surface, it involves slow drifting of huge volumes of Atlantic surface water from the South Atlantic to the North Atlantic. Exposed to intense insolation in the equatorial and tropical zones, the water is warmed. Evaporation also concentrates salt in the surface layer, making the water more dense. As it travels northward, the water cools, losing heat and warming the atmosphere above it. Finally, the water is chilled almost to its freezing point. Because of its cold temperature and saltiness, the North Atlantic water is dense enough to sink to the bottom of the Atlantic Basin. This cold, Atlantic bottom water then heads slowly southward.

Note that this circulation pattern is also a type of sinking convection loop powered by the loss of heat and increase in the

8.22 Thermohaline convection loop *Warm surface waters flow into the North Atlantic, cool, and sink to the deep Atlantic Basin, creating a circulation that warms westerly winds moving onto the European continent.*

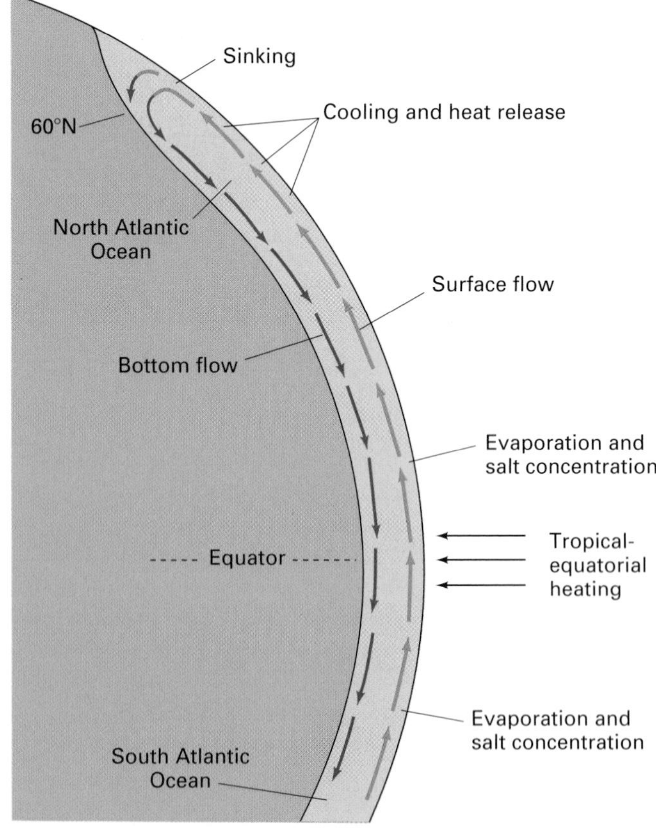

density of the surface water as it moves poleward. In effect, this loop acts like a heat pump in which sensible heat is acquired in tropical and equatorial regions and is moved northward into the North Atlantic, where it is transferred to the air. Since wind patterns move air eastward at higher latitudes, this heat ultimately warms Europe. The amount of heat released is quite large. In fact, a recent calculation shows that it accounts for about 35 percent of the total solar energy received by the Atlantic Ocean north of 40° latitude! This type of circulation does not occur in the Pacific or Indian oceans.

Cloud Cover, Precipitation, and Global Warming

Thus far in this text, we have examined climatic change from several different perspectives. Let's now focus on how global climate might be influenced by an increase in clouds and precipitation. Recall that global temperatures have been rising over the last 20 years. As part of this global warming, satellite data have detected a rise in temperature of the global ocean surface of about 1°C (1.8°F) over the past decade. Any rise in sea-surface temperature increases the rate of evaporation. And the increase in evaporation will raise the average atmospheric content of water vapor. What effect will this have on climate?

Water has several roles in global climate. First, in its vapor state it is one of the greenhouse gases. That is, it absorbs and emits longwave radiation, thus enhancing the warming effect of the atmosphere above the Earth's surface. In fact, water is more important than CO_2 in creating the greenhouse effect. Thus, one result of an increase in global water vapor in the atmosphere should enhance warming.

Second, water vapor can condense or deposit, forming clouds. Will more clouds increase or decrease global temperatures? That question is still being debated by the scientists who study and model the atmosphere. Clouds can have two different effects on the surface radiation balance. As large, white bodies, they can reflect a large proportion of incoming shortwave radiation back to space, thus acting to cool global temperatures. But cloud droplets and ice particles also absorb longwave radiation from the ground and return that emission as counterradiation. This absorption is an important part of the greenhouse effect, and it is much stronger for water as cloud droplets or ice particles than as water vapor. Thus, clouds also act to warm global temperatures by enhancing longwave reradiation from the atmosphere to the surface.

Which effect, longwave warming or shortwave cooling, will dominate? The best information at present, obtained from satellite measurements, is that the average flow of shortwave energy reflected by clouds back to space is about 50 W/m^2, while the greenhouse warming effect of clouds amounts to about 30 W/m^2. Thus, the net effect of clouds is a cooling of the planet by an energy flow of about 20 W/m^2.

What will happen when surface temperature increases, water vapor in the atmosphere increases, and more clouds form? Computer models of global climate generally agree that longwave warming will be enhanced more than shortwave cooling. This means that the net effect of clouds will be a positive feedback that enhances the greenhouse effect and so accentuates the surface warming. After a small increase, the effect of clouds will still be to cool the planet but not at so great a rate. Shortwave cooling will still prevail but to a lesser degree. Another way to put it is to say that with global warming, the cooling effects of clouds on climate will be reduced somewhat, leading to temperatures that are even warmer. *Focus on Systems 8.4 • Feedback Loops in the Global Climate System* provides a more detailed look at the feedback loops between surface warming and clouds.

What about precipitation? This is the third role of water in global climate. With more water vapor and more clouds in the air, more precipitation should result. Think, however, about what might happen if precipitation increases in arctic and subarctic zones. In this case, more of the Earth's surface could be covered by snow and ice. Since snow is a good reflector of solar energy, this would increase the Earth's albedo, thus tending to reduce global temperatures. Another effect might be to increase the depth of snow, thus tying up more water in snow packs and reducing runoff to the oceans. Reducing runoff would reduce the rate at which sea level has been rising, presumably as a result of climate warming (see Chapter 19).

At this time, scientists are unsure how the global climate system will respond to global warming induced by the CO_2 increases predicted for the twenty-first century. Perhaps increased cloud cover and enhanced precipitation will slow the warming trend—or perhaps they will enhance it. As time goes by, however, our understanding of global climate and our ability to predict its changes are certain to increase.

A Look Ahead With a knowledge of the global circulation patterns of the atmosphere and oceans, as well as an understanding of weather systems and how they produce precipitation, the stage is set for global climate—which is the topic of Chapters 9–11. As we will see, the annual cycles of temperature and precipitation that most regions experience are quite predictable, given the changes in wind patterns, air mass flows, and weather systems that occur with the seasons. Our description of the world's climates grows easily and naturally from the principles you have mastered thus far in your study of physical geography.

Focus on Systems | 8.4

Feedback Loops in the Global Climate System

An important feature of many flow systems is the feedback loop, which we discussed in Chapter 2. As we saw there, a feedback loop is a connection that either reduces or amplifies a change that occurs within a system. In the case of negative feedback, the change is reduced, while in the case of positive feedback, the change is amplified.

As an example of feedback, consider the linkage between global warming and atmospheric water in the form of water vapor, clouds, and precipitation, which is explained in the last section of this chapter. In this linkage, increased levels of atmospheric CO_2 and other greenhouse gases increase surface temperatures by enhancing the greenhouse effect. Increased surface temperatures in turn enhance evaporation, generating more water vapor in the atmosphere, more clouds, and greater precipitation.

The linkages discussed in the text are diagrammed in the figure below. Increased concentrations of CO_2 and other greenhouse gases boost the greenhouse effect, increasing planetary temperature and raising atmospheric specific humidity. Since water vapor is itself a greenhouse gas, the increase in specific humidity forms a positive feedback. Increased cloud cover also provides positive feedback, since liquid and solid water particles are more effective at greenhouse warming than water vapor alone. But increased clouds provide negative feedback as well when they reflect solar radiation back to space. Increased precipitation could provide a negative feedback if it resulted in greater snow and ice cover, which would reflect more solar radiation back to space. Note that these are just a few of the feedback loops within the global climate system. Climatologists and global climate modelers have proposed and documented many others.

As you can see from this example, feedback loops can be quite important in understanding how systems behave. Systems with negative feedback loops will tend to be more stable, so that when changes occur, the system moderates them. Systems with positive feedback loops will tend to be more dynamic, since changes are amplified by the structure of the system. It's hard to tell at this time whether negative or positive feedback loops dominate the global climate system. However, considering the cost of changing climate patterns to human society, let's hope that future changes are moderated by negative feedback mechanisms rather than accentuated by positive ones.

Feedback loops *Diagram of positive and negative feedback loops in a coupled system of surface temperature and cloud cover.*

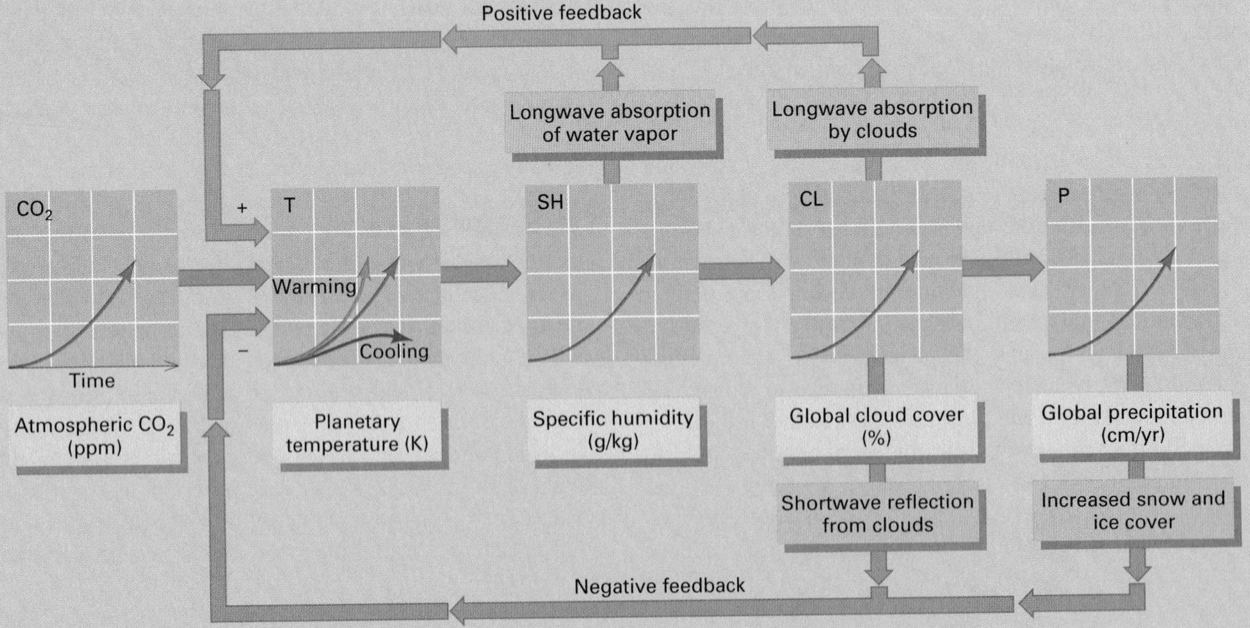

Chapter Summary

- A **weather system** is a recurring atmospheric circulation pattern with its associated weather. Weather systems include wave cyclones, traveling anticyclones, tornadoes, easterly waves, weak equatorial lows, and tropical cyclones.
- **Air masses** are distinguished by the latitudinal location and type of surface of their *source regions*. Air masses influencing North America include those of continental and maritime source regions, and of arctic, polar, and tropical latitudes.
- The boundaries between air masses are termed **fronts**. These include **cold** and **warm fronts**, where cold or warm air masses are advancing. In the **occluded front**, a cold front overtakes a warm front, pushing a pool of warm, moist air mass above the surface.
- *Traveling cyclones* include wave cyclones, tropical cyclones, and the tornado. The *traveling anticyclone* is typically a fair-weather system.
- **Wave cyclones** form in the midlatitudes at the boundary between cool, dry air masses and warm, moist air masses. In the wave cyclone, a vast inspiraling motion produces cold and warm fronts, and eventually an occluded front. Precipitation normally occurs with each type of front.
- **Tornadoes** are very small, intense cyclones that occur as a part of thunderstorm activity. Their high winds can be very destructive.
- Tropical weather systems include *easterly waves* and *weak equatorial lows*. Easterly waves occur when a weak low-pressure trough develops in the easterly wind circulation of the tropical zones, producing convergence, uplift, and shower activity. Weak equatorial lows occur near the intertropical convergence zone. In these areas of low pressure, convergence triggers abundant convectional precipitation.
- **Tropical cyclones** can be the most powerful of all storms. They develop over very warm tropical oceans and can intensify to become vast inspiraling systems of very high winds with very low central pressures. As they move onto land, they bring heavy surf and storm surges of very high waters. Tropical cyclones have caused great death and destruction in coastal regions.
- Global air and ocean circulation provides the mechanism for *poleward heat transport* by which excess heat moves from the equatorial and tropical regions toward the poles. In the atmosphere, the heat is carried primarily in the movement of warm, moist air poleward, which releases its latent heat when precipitation occurs. In the oceans, a global circulation moves warm surface water northward through the Atlantic Ocean. Heated in the equatorial and tropical regions, the surface water loses its heat to the air in the North Atlantic and sinks to the bottom. These heat flows help make northern and southern climates warmer than we might expect based on solar heating alone.
- Because global warming, produced by increasing CO_2 levels in the atmosphere, will increase the evaporation of surface water, atmospheric moisture levels will increase. This will tend to enhance the greenhouse effect. But more clouds are likely to form, and this should cool the planet. Increased moisture could also reduce temperatures by increasing the amount and duration of snow cover. Further research on the effect of global warming on climate is needed.

Key Terms

weather system	cold front	cyclonic precipitation	tornado
air mass	warm front	cyclonic storm	tropical cyclone
front	occluded front	wave cyclone	storm surge

Review Questions

1. Define air mass. What two features are used to classify air masses?
2. Compare the characteristics and source regions for mP and cT air mass types.
3. Describe a tornado. Where and under what conditions do tornadoes typically occur?
4. Identify three weather systems that bring rain in equatorial and tropical regions. Describe each system briefly.
5. Describe the structure of a tropical cyclone. What conditions are necessary for the development of a tropical cyclone? Give a typical path for the movement of a tropical cyclone in the northern hemisphere.
6. Why are tropical cyclones so dangerous?
7. How does the global circulation of the atmosphere and oceans provide poleward heat transport?
8. How does water, as vapor, clouds, and precipitation, influence global climate? How might water in these forms act to enhance or retard climatic warming?

Eye on Environment 8.2 • Hurricane Andrew—Killer Cyclone

1. Where did Hurricane Andrew first form, and how did it develop and move?
2. What damage was sustained by South Florida and coastal Louisiana from Hurricane Andrew?
3. What were the maximum wind speeds of Andrew? How were they achieved?
4. How did Hurricane Andrew affect the Everglades?

Focus on Systems 8.3 • Hadley Cell Circulation as a Convection Loop Flow System

1. Sketch a cross section of Hadley cell circulation, using arrows to indicate air flow direction. At what latitudes are northerly, southerly, upward, and downward flows strongest?

Focus on Systems 8.4 • Feedback Loops in the Global Climate System

1. What is a feedback loop? Contrast the effects of negative and positive feedback loops on a system.
2. Provide an example of a system with a feedback loop and explain how the loop affects the system.

Visualizing Exercises

1. Identify three types of fronts and draw a cross section through each. Show the air masses involved, the contacts between them, and the direction of air mass motion.
2. Sketch two weather maps, showing a wave cyclone in open and occluded stages. Include isobars on your sketch. Identify the center of the cyclone as a low. Lightly shade areas where precipitation is likely to occur.

Essay Questions

1. Compare and contrast midlatitude and tropical weather systems. Be sure to include the following terms or concepts in your discussion: air mass, convectional precipitation, cyclonic precipitation, easterly wave, polar front, stable air, traveling anticyclone, tropical cyclone, unstable air, wave cyclone, and weak equatorial low.
2. Prepare a description of the annual weather patterns that are experienced through the year at your location. Refer to the general temperature and precipitation pattern as well as the types of weather systems that occur in each season.

Eye on the Landscape

8.7 **Picture of an anticyclone** On this clear satellite image, it is easy to see some of the major features of the eastern North American landscape. At (A), note the difference in soil color along the Mississippi River. These are alluvial soils, deposited over millennia by the Mississippi during the Ice Ages. At (B), note the curving arc that marks the boundary between soft coastal plain sediments to the south and the weathered rocks and soils of the piedmont, to the north. (Figure 17.8 diagrams this location.) Actually, this boundary can be traced all the way to New York City. Further inland lie the Appalachians (C), showing puffy orographic cumulus clouds. (We covered orographic precipitation in Chapter 4.)

9 THE GLOBAL SCOPE OF CLIMATE

EYE ON THE LANDSCAPE
Drying dates in a palm grove south of Cairo, Egypt. Date palms are unique to the hot, dry climates of the world's deserts. The dates are red or yellow when freshly picked, and dry to a range of brown colors. **What else would the geographer see?...Answers at the end of the chapter.**

What do we mean by climate? In its most general sense, **climate** is the average weather of a region. To describe the average weather of a region, we could use many of the measures describing the state of the atmosphere that we have already encountered. These might include daily net radiation, barometric pressure, wind speed and direction, cloud cover and type, presence of fog, precipitation type and intensity, incidence of cyclones and anticyclones, frequency of frontal passages, and other such items of weather information. However, observations as detailed as these are not made regularly at most weather stations around the world. To study climate on a worldwide basis, we must turn to the two simple measurements that are made daily at every weather station—temperature and precipitation. We will be concerned with both their average values for each month of the year and their variation across the months of the year.

Note that temperature and precipitation strongly influence the natural vegetation of a region—for example, forests occur generally in moist regions, and grasslands in dry regions. The natural vegetation cover is often a distinctive feature of a climatic region and typically influences the human use of the area. Temperature and precipitation are also important factors in the cultivation of crop plants—a necessary process for human survival. And the development of soils, as well as the types of processes that shape landforms, are partly dependent on temperature and precipitation. For these reasons, we will find that climates defined on the basis of temperature and precipitation also help set apart many features of the environment, not just climate alone. This is why the study of global climates is such an important part of physical geography.

KEYS TO CLIMATE

A few simple principles discussed in earlier chapters are very helpful in understanding the global scope of climate. First, recall from Chapter 5 that two major factors influence the annual cycle of air temperature experienced at a station: latitude and coastal versus continental location.

- *Latitude.* The annual cycle of temperature at a station depends on its latitude. Near the equator, temperatures are warmer and the annual range is low. Toward the poles, temperatures are colder and the annual range is greater. These effects are produced by the annual cycle of insolation, which varies with latitude.
- *Coastal-continental location.* Coastal stations show a smaller annual variation in temperature, while the variation is larger for stations in continental interiors. This effect occurs because ocean surface temperatures vary less with the seasons than do temperatures of land surfaces.

Air temperature also has an important effect on precipitation. Recall this principle from Chapter 6:

- *Warm air can hold more moisture than cold air.* This means that colder regions generally have lower precipitation than warmer regions. Also, precipitation will tend to be greater during the warmer months of the temperature cycle.

Another key idea associated with climate is the *time cycle,* which we discussed in Chapter 2. The primary driving force for weather, as we have seen, is the flow of solar energy received by the Earth and atmosphere. Since that energy flow varies on daily cycles with the planet's rotation and on annual cycles with its revolution in orbit, it imposes these cycles on temperature and precipitation. Other time cycles appear in climate as well. *Focus on Systems 9.1 • Time Cycles of Climate* provides a closer look at time cycles in climate.

Keeping these key ideas in mind as you read this chapter and the two that follow will help make the climates easy to understand and explain.

Temperature Regimes

Let's look in more detail at the influence of latitude and location on the annual temperature cycle of a station. Figure 9.1 shows some typical patterns of mean monthly temperatures observed at stations around the globe. We can refer to these patterns as *temperature regimes*—distinctive types of annual temperature cycles related to latitude and location. In the figure, each regime has been labeled according to its latitude zone: equatorial, tropical, midlatitude, and subarctic. Some labels also describe the location of the station in terms of its position on a landmass—"continental" for a continental interior location, and "west coast" or "marine" for a location close to the ocean.

The equatorial regime (Douala, Cameroon, 4° N) is uniformly very warm. Temperatures are close to 27°C (81°F) year-round. There are no temperature seasons because insolation is nearly uniform through the year. In contrast, the tropical continental regime (In Salah, Algeria, 27° N) shows a very strong temperature cycle. Temperatures change from very hot when the Sun is high, near one solstice, to mild at the opposite solstice. However, the situation is quite different at Walvis Bay, Southwest Africa (23° S), which is at nearly the same latitude and so has about the same insolation cycle as In Salah. At Walvis Bay, we find the tropical west-coast regime, which has only a weak annual cycle and no extreme heat. The difference, of course, is due to the moderating effects of the maritime location of Walvis Bay. This moderating effect persists poleward, as shown in the two temperature graphs for the midlatitude west-coast regime—Monterey, California (36° N) and Sitka, Alaska (57° N).

In continental interiors, however, the annual temperature cycle remains strong. The midlatitude continental regime of Omaha, Nebraska (41° N), and the subarctic continental regime of Fort Vermilion, Alberta (58° N), show annual variations in mean monthly temperature of about 30°C (54°F) and 40°C (72°F), respectively. The ice sheet regime of Greenland (Eismitte, 71° N) is in a class by itself, with severe cold all year.

Other regimes can be identified, and, because they grade into one another, the list could be expanded indefinitely. However, what is important is that (1) annual variation in insolation, which is determined by latitude, provides the basic control on temperature patterns, and (2) the effect of location—maritime or continental—moderates that variation.

The temperature regimes in Figure 9.1 show the variation in mean monthly temperature through the year. Monthly temperature is simply the average of daily temperatures during the month, and daily temperature is the average of the daily maximum and minimum. Mean monthly temperature is the monthly temperature for a particular month of the year averaged over a long period of record, usually several decades. *Working It Out 9.2 • Averaging in Time Cycles* demonstrates the power of averaging to reveal distinctive time cycles in data, such as the temperature regimes presented above.

Global Precipitation

Global precipitation patterns are largely determined by air masses and their movements, which in turn are produced by global air circulation patterns. Before taking a detailed look at global precipitation patterns, let's examine the general patterns expected for a hypothetical supercontinent that has most of the features of the Earth's continents but is simplified (Figure 9.2). The map recognizes and defines five classes of annual precipitation: wet, humid, subhumid, semi-arid, and arid.

Beginning with the equatorial zone, the figure shows a wet band stretching across the continent. This band is produced by convectional precipitation in weak equatorial lows near the intertropical convergence zone, as described in

Focus on Systems 9.1

Time Cycles of Climate

Chapter 2 presented the concept of a time cycle—a rhythmic change that affects natural flow systems. One major class of time cycles evident in climate consists of the astronomical cycles. The two most important ones are described in Chapter 3—the daily cycle of the Earth's rotation on its axis, and the annual cycle of the Earth's revolution around the Sun. The daily cycle of rotation results in a daily temperature cycle, shown in *(a)*, which is produced by hourly changes in insolation and surface energy balance described in Chapter 5. The annual cycle of revolution produces an annual cycle of daylight duration, shown in *(b)*. Daylight duration is determined by latitude and solar declination, and is strongly related to the total daily energy flow received at the surface. In both curves, we've sketched the situation for a midlatitude continental location.

The standard climograph consists of annual cycles of temperature and precipitation averaged over several or many consecutive years (see Figure 9.8). Each of the red dots showing the mean air temperature of a given month is itself the average value of all of the monthly averages in the period of record. Suppose we were to plot the climographs of single years in sequence, as in *(c)*, which shows a sketch for temperature. Each year's curve is different from those of the other years. The same would hold true for the precipitation bars on the climographs. Nevertheless, there is an annual rhythm in the climate data, even though the "beats" in this rhythm are far from uniform.

Let's turn now to long strings of annual data of air temperature, covering two to three centuries. A good example is Figure 5.26, showing the record of annual tree-ring growth. Here, the annual temperature value is reduced to a single number, represented by a single dot. The dots are then connected by short, straight lines, giving the saw-tooth graph. You can see here a major time cycle with a length of about 140 years. Superimposed is a smaller, shorter cycle that is on the order of perhaps one decade. Now examine Figure 5.24, a similar graph, in which annual values are shown by the yellow line. An added feature here is a smooth red curve—a five-year running average that helps to reveal significant cycles of more moderate magnitudes. The general, overall rise in values from left to right may be a much larger cycle of which we are seeing only one part. The interpretation of this large cycle is, of course, the subject of the great ongoing debate about global warming. We will return to time cycles again in Chapter 20 in the context of the Ice Age.

To sum it up, climate is always changing through time, showing rhythms of rise and fall of temperature and precipitation. The cycles we observe in nature are nested in a hierarchy of sizes—smaller ones within larger ones. Scientists can plot the cycles of the past with some assurance, but prediction of trends to come is filled with uncertainty.

Time cycles in climate

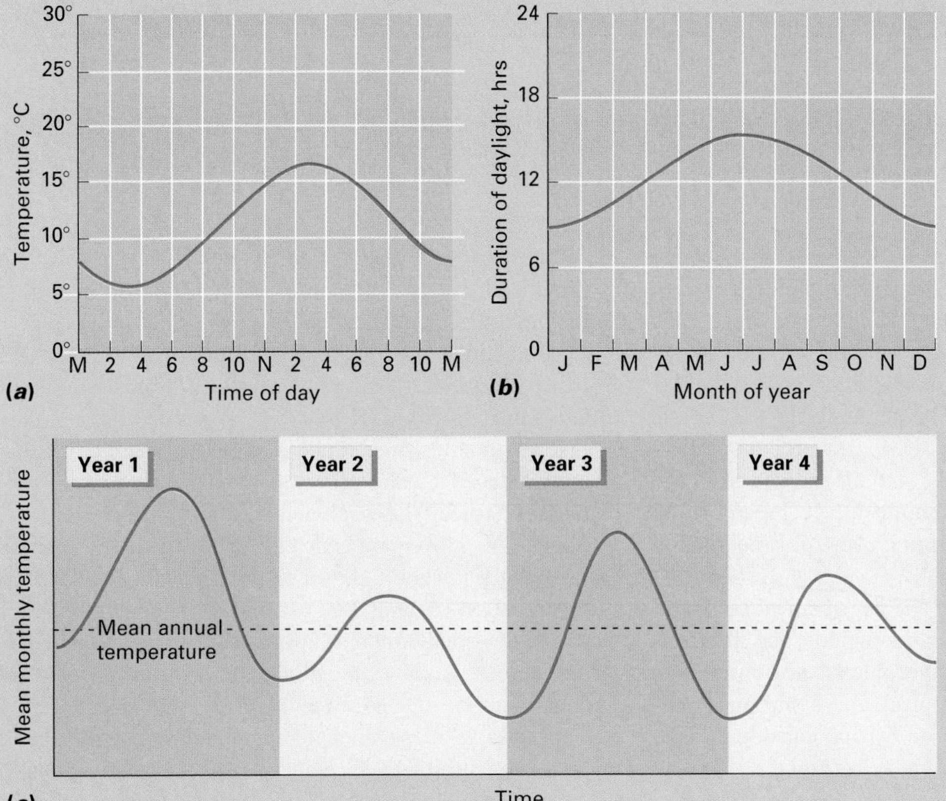

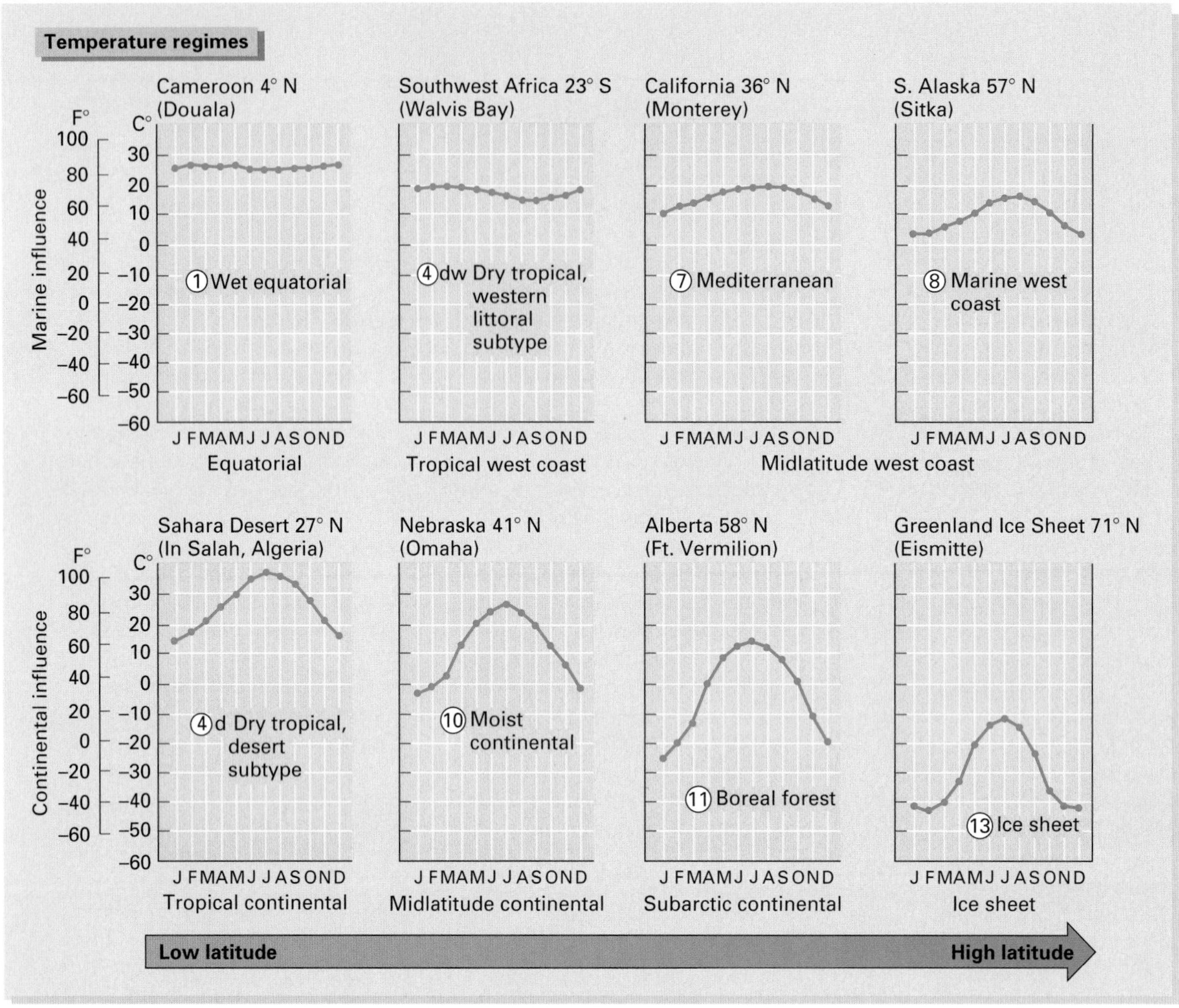

9.1 Temperature regimes *(above and right) Some important temperature regimes, represented by annual cycles of air temperature. (Based on the Goode Base Map.)*

Chapter 8. Note that the wet band widens and is extended poleward into the tropical zone along the continent's eastern coasts. This region is kept moist by the influence of the trade winds, which move warm, moist mT air masses and tropical cyclones westward onto the continental coast. Farther poleward, humid conditions continue along the east coasts into the midlatitude zones. In these regions, subtropical high-pressure cells tend to move mT air masses from the east onto the continent in the summer (see Figure 7.17), while, in winter, wave cyclones bring cyclonic precipitation from the west (see Figure 8.10).

Another important feature of the hypothetical continent is the pattern of arid and semiarid regions that stretches from tropical west coasts to subtropical and midlatitude continental interiors. In the tropical and subtropical latitudes, the arid pattern is produced by dry, subsiding air in persistent subtropical high-pressure cells (Figure 7.17). The aridity continues eastward and poleward into semiarid continental interiors (for example, see Figure 9.3). These regions remain relatively dry because they are far from source regions for moist air masses. Rain shadow effects provided by coastal mountain barriers are also important in maintaining inland aridity.

Yet another obvious feature of the supercontinent is the pair of wet bands along the west coasts of the midlatitude and subarctic zones. Figure 9.4 provides an example of the lush green landscapes found in these regions. The wet bands are produced by the eastward movement of moist mP air masses, typically as occluded wave cyclones, onto the continent. This movement is driven by the prevailing westerlies.

In the arctic zone, shown on the continent as arctic desert, precipitation remains low because air temperatures are low and only a small amount of moisture can be held in cold air.

These features of the supercontinent are echoed in the actual pattern of global precipitation, shown in Figure 9.5. This map of mean annual precipitation shows **isohyets**—lines drawn through all points having the same annual precipitation. Using the same logic that we used to explain the precipitation patterns of the hypothetical continent, we can recognize seven global precipitation regions as follows.

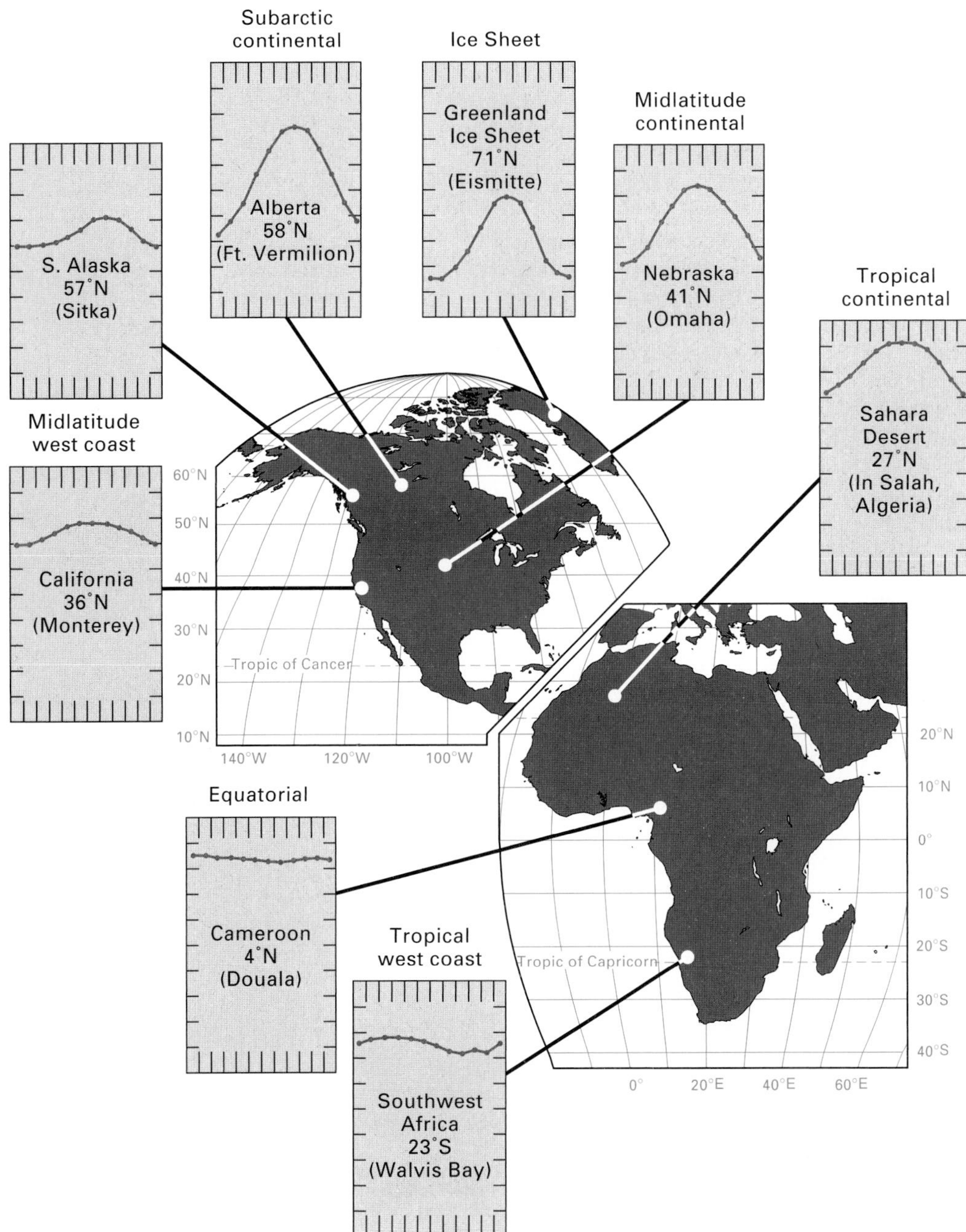

(Note that for regions where all or most of the precipitation is rain, we use the word "rainfall." For regions where snow is a significant part of the annual total, we use the word "precipitation.") Table 9.1 summarizes the observations below.

1. **Wet equatorial belt.** This zone of heavy rainfall, over 200 cm (80 in.) annually, straddles the equator and includes the Amazon River Basin in South America, the Congo River Basin of equatorial Africa, much of the African coast from Nigeria west to Guinea, and the East Indies. Here the prevailing warm temperatures and high-moisture content of the mE air masses favor abundant convectional rainfall. Thunderstorms are frequent year-round.
2. **Trade-wind coasts.** Narrow coastal belts of high rainfall, 150 to 200 cm (about 60 to 80 in.), and locally even more, extend from near the equator to latitudes of about 25° to 30° N and S on the eastern sides of every continent or large island. Examples include the eastern coast of Brazil, Central America, Madagascar, and northeastern Australia. The rainfall of these coasts is supplied by moist mT air masses from warm oceans, brought over the land by the trade winds. As they encounter coastal hills and mountains, these air masses produce heavy orographic rainfall.
3. **Tropical deserts.** In striking contrast to the wet equatorial belt astride the equator are the two zones of vast

Working It Out | 9.2

Averaging in Time Cycles

The weather at most stations can vary widely from day to day, or even from hour to hour. For the study of climate, however, we are concerned with the weather in a statistical, or average, sense. To discover that average, it is often necessary to observe such variables as daily temperature and precipitation over periods of decades.

Part *(a)* of the figure below shows monthly precipitation for a period of 20 years, from 1937 to 1994, as observed at St. Louis, Missouri. A first inspection of the graph shows that there is a rhythmic pattern to the precipitation, with sequences of wet months interspersed with sequences of drier months. This pattern matches that of the moist continental climate ⑩, with wet summers and dry winters. But beyond that general observation, it's hard to determine what the shape of the typical annual cycle of precipitation looks like for this station. In fact, it's difficult even to make a guess about which month is, on the average, the wettest or the driest.

To uncover the typical monthly cycle of precipitation, finding an average is necessary. The *average,* or *mean,* of a number of observations is simply the sum of the observations divided by the number of such observations. In commonly used algebraic notation, we can write

$$\bar{P} = \frac{1}{n} \sum_{i=1}^{n} P_i$$

where $\bar{P}$ (pronounced *P*-bar) is the mean precipitation for all observations in a particular month; P_i is the precipitation value associated with that month in the *i*th year; *n* is the number of years for which we have

St. Louis precipitation *Precipitation records and statistics for St. Louis, Missouri. (Data of NOAA/NCDC.)*

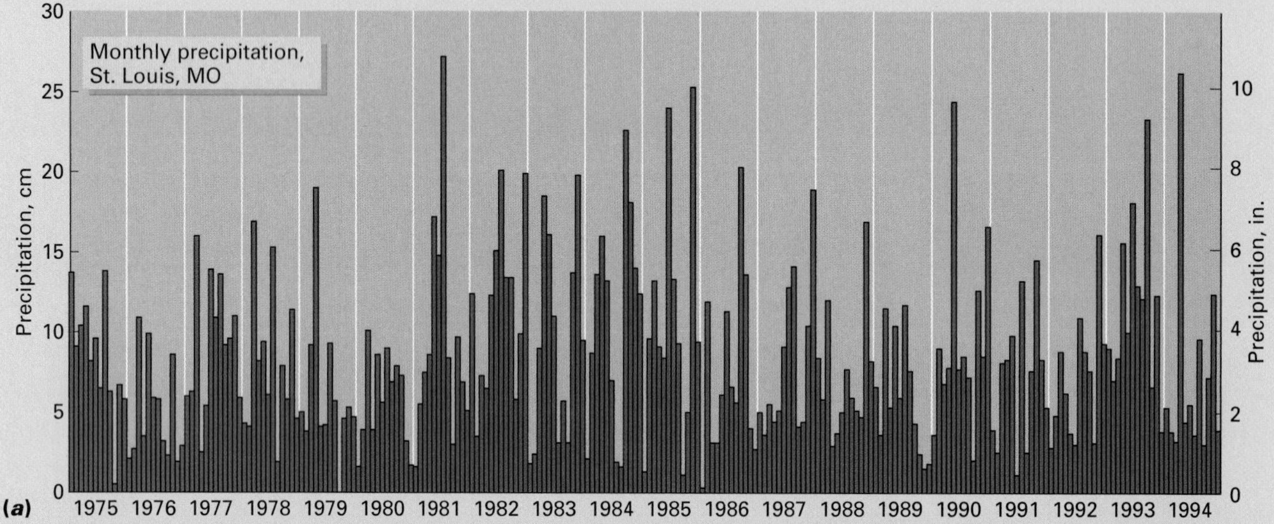

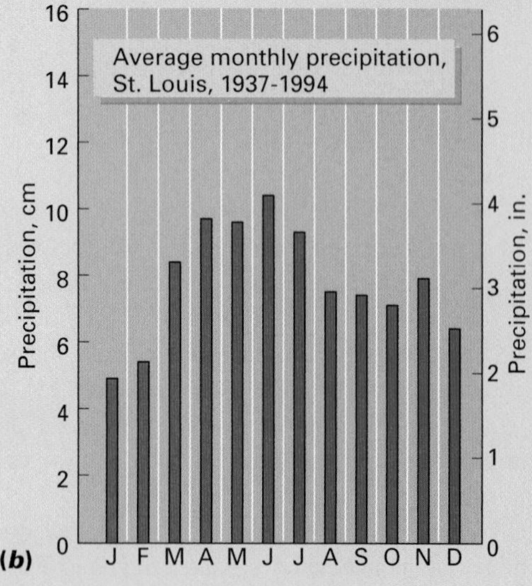

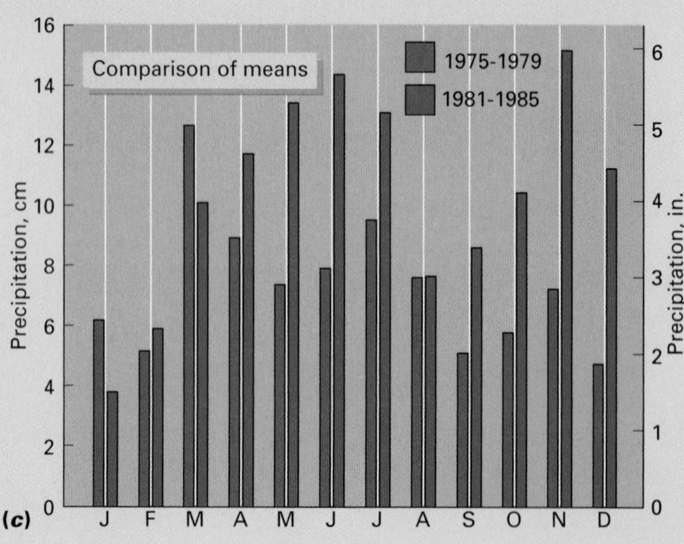

data for that particular month; and the subscript *i* denotes the number of the yearly observation, which ranges from *i*=1 to *n*. The Greek letter (capital) sigma, Σ, is used to denote the sum, with the lower and upper notations showing that the sum is to be formed for each observation in the sequence from *i*=1 to *n*. This is simply a rather formal way of saying, "Find the average by adding up all the values and then dividing their sum by the number of values."

The monthly averages for the 58 years of record (1937–1994) at this station are shown in graph *(b)* of the figure. Note that the monthly cycle has a fairly smooth shape. Maximum precipitation occurs in the spring months of April, May, and June, decreasing to a low point in January. By taking the average, the year-to-year variability is eliminated and the precipitation cycle appears.

Consider what the time cycle might look like with only a few years of record—say five years. By looking at graph *(a)*, you can see that if we chose the five years from 1975 to 1979, we would get a different average than for the years 1981–1985. Part *(c)* of the figure compares the means of these two five-year sequences. Clearly, 1975–1979 is a sequence of drier years, while the 1981–1985 period is wetter. This is true for all months except April, which has more rainfall during the years 1975–1979 than 1981–1985.

In general, the longer the period of averaging, the closer the average will come to the true long-term average, and the smoother the cycle will be. Thus, averaging is a very useful tool for revealing patterns and time cycles in data sequences.

tropical deserts lying approximately on the tropics of cancer and capricorn. These are hot, barren deserts, with less than 25 cm (10 in.) of rainfall annually and in many places with less than 5 cm (2 in.). They are located under and are caused by the large, stationary subtropical cells of high pressure, in which the subsiding cT air mass is adiabatically warmed and dried. These deserts extend off the west coasts of the lands and out over the oceans. Rain here is largely convectional and extremely unreliable.

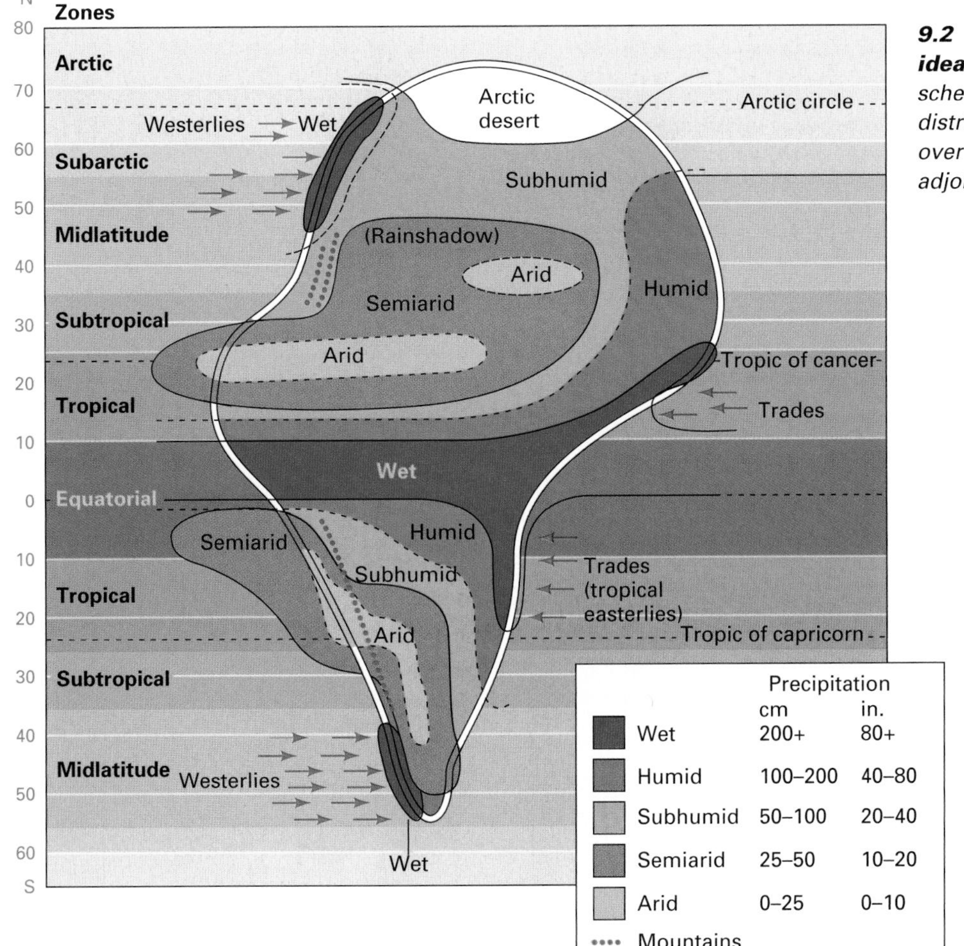

9.2 Precipitation on an idealized supercontinent *A schematic diagram of the distribution of annual precipitation over an idealized continent and adjoining oceans.*

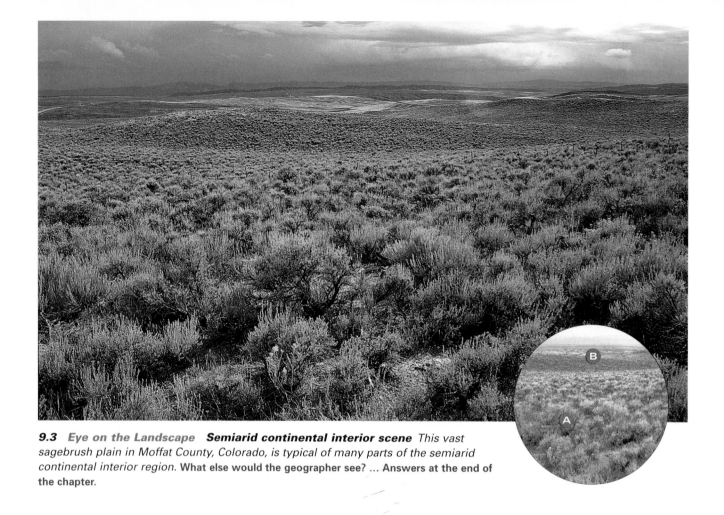

9.3 **Eye on the Landscape** ***Semiarid continental interior scene*** *This vast sagebrush plain in Moffat County, Colorado, is typical of many parts of the semiarid continental interior region.* **What else would the geographer see? ... Answers at the end of the chapter.**

9.4 **Eye on the Landscape** ***Midlatitude west-coast scene*** *West-coast midlatitude climates are moist and cool most of the year. The Irish coast, shown in this photo from the Dingle Peninsula, is an example.* **What else would the geographer see? ... Answers at the end of the chapter.**

Table 9.1 | **World Precipitation Regions**

Name	Latitude Range	Continental Location	Prevailing Air Mass	Annual Precipitation	
				Centimeters	Inches
1. Wet equatorial belt	10° N to 10° S	Interiors, coasts	mE	Over 200	Over 80
2. Trade-wind coasts (windward tropical coasts)	5–30° N and S	Narrow coastal zones	mT	Over 150	Over 60
3. Tropical deserts	10–35° N and S	Interiors, west coasts	cT	Under 25	Under 10
4. Midlatitude deserts and steppes	30–50° N and S	Interiors	cT, cP	10–50	4–20
5. Moist subtropical regions	25–45° N and S	Interiors, coasts	mT (summer)	100–150	40–60
6. Midlatitude west coasts	35–65° N and S	West coasts	mP	Over 100	Over 40
7. Arctic and polar deserts	60–90° N and S	Interiors, coasts	cP, cA	Under 30	Under 12

4. **Midlatitude deserts and steppes.** Farther northward, in the interiors of Asia and North America between lat. 30° and lat. 50°, are great deserts, as well as vast expanses of semiarid grasslands known as **steppes.** Annual precipitation ranges from less than 10 cm (4 in.) in the driest areas to 50 cm (20 in.) in the moister steppes. Dryness here results from remoteness from ocean sources of moisture.

   Located in regions of prevailing westerly winds, these arid lands typically lie in the rain shadows of coastal mountains and highlands. For example, the Cordilleran Ranges of Oregon, Washington, British Columbia, and Alaska shield the interior of North America from moist mP air masses originating in the Pacific. Upon descending into the intermountain basins and interior plains, the mP air masses are warmed and dried. Similarly, mountains of Europe and the Scandinavian Peninsula obstruct the flow of moist mP air masses from the North Atlantic into western Asia. The great southern Asiatic ranges also prevent the entry of moist mT and mE air masses from the Indian Ocean.

   The southern hemisphere has too little land in the midlatitudes to produce a true continental desert, but the dry steppes of Patagonia, lying on the lee side of the Andean chain, are roughly the counterpart of the North American deserts and steppes of Oregon and northern Nevada.
5. **Moist subtropical regions.** On the southeastern sides of the continents of North America and Asia, in lat. 25° to 45° N, are the moist subtropical regions, with 100 to 150 cm (about 40 to 60 in.) of rainfall annually. Smaller areas of the same kind are found in the southern hemisphere in Uruguay, Argentina, and southeastern Australia. These regions are positioned on the moist western sides of the oceanic subtropical high-pressure centers. As a result, the lands receive moist mT air masses from the tropical ocean that are carried poleward over the adjoining land. Commonly, too, these areas receive heavy rains from tropical cyclones.
6. **Midlatitude west coasts.** Another distinctive wet location is on midlatitude west coasts of all continents and large islands lying between about 35° and 65° in the region of prevailing westerly winds. In these zones, abundant orographic precipitation occurs as a result of forced uplift of mP air masses. Where the coasts are mountainous, as in Alaska and British Columbia, southern Chile, Scotland, Norway, and South Island of New Zealand, the annual precipitation is over 200 cm (79 in.). During the Ice Age, this precipitation fed alpine glaciers that descended to the coast, carving the picturesque deep bays (fiords) that are so typically a part of the scenery there.
7. **Arctic and polar deserts.** A seventh precipitation region is formed by the arctic and polar deserts. Northward of the 60th parallel, annual precipitation is largely under 30 cm (12 in.), except for the west-coast belts. Cold cP and cA air masses cannot hold much moisture, and consequently, they do not yield large amounts of precipitation. At the same time, however, the relative humidity is high and evaporation rates are low.

Seasonality of Precipitation

Total annual precipitation is a useful quantity in establishing the character of a climate type, but it does not account for the seasonality of precipitation. The variation in monthly precipitation through the annual cycle is a very important factor in climate description. If there is a pattern of alternating dry and wet seasons instead of a uniform distribution of precipitation throughout the year, we can expect that the natural vegetation, soils, crops, and human use of the land will all be different. It also makes a great deal of difference whether the wet season coincides with a season of higher temperatures or with a season of lower temperatures. If the warm season is also wet, growth of both native plants and crops will be enhanced. If the warm season is dry, the stress on growing plants will be great and irrigation will be required for crops.

Monthly precipitation patterns can be described largely by three types: (1) uniformly distributed precipitation; (2) a

9.5 World precipitation *Mean annual precipitation of the world. Isohyets are labeled in cm (in.)*

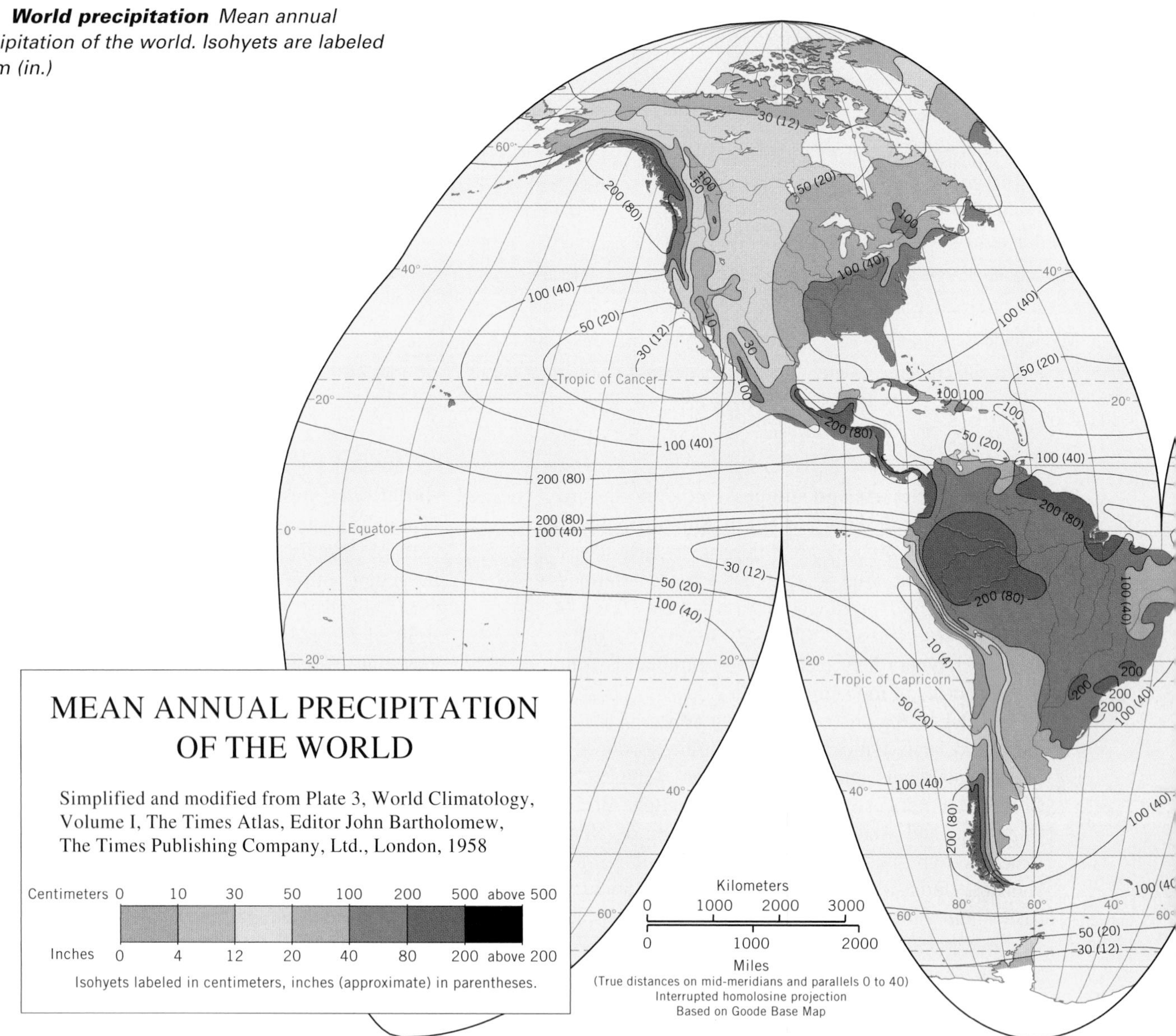

precipitation maximum during the summer (or season of high Sun), in which insolation is at its peak; and (3) a precipitation maximum during the winter or cooler season (season of low Sun), when insolation is least. Note that the uniform type of pattern can include a wide range of possibilities from little or no precipitation in any month to abundant precipitation in all months.

Figure 9.6 shows a set of monthly precipitation diagrams selected to illustrate the major types that occur over the globe. Two stations show the uniformly distributed pattern described above—Singapore, a wet equatorial station near the equator ($1^1/_2$° N), and Tamanrasset, Algeria, a tropical desert station very near the tropic of cancer at 23° N. At Singapore, rainfall is abundant in all months, but some months have somewhat more than others. Tamanrasset has so little rain in any month that it scarcely shows on the graph.

Chittagong, Bangaladesh ($22^1/_2$° N), and Kaduna, Nigeria ($10^1/_2$° N), both show patterns of the second type—that is, a wet season at the time of high Sun (summer solstice) and a dry season at the time of low Sun (winter solstice). Chittagong is an Asian monsoon station, with a very large amount of precipitation falling during the high Sun season. Kaduna, an African station with about half the total annual precipitation, shows a similar pattern and is also of the wet-dry tropical type. Both of these stations experience their wet season when the intertropical convergence zone (ITCZ) is nearby and their dry season when the ITCZ has retreated to the other hemisphere.

The summer precipitation maximum also occurs at higher latitudes on the eastern sides of continents. Shanghai, China (31° N), shows this pattern nicely in the subtropical zone. The same summer maximum persists into the midlatitudes. For example, Harbin, in eastern China (46° N), has a long, dry winter with a marked period of summer rain.

In contrast to these patterns are cycles with a winter precipitation maximum. Palermo, Sicily (38° N), is an example of the Mediterranean type, named for its prevalence in the lands surrounding the Mediterranean Sea. This type experi-

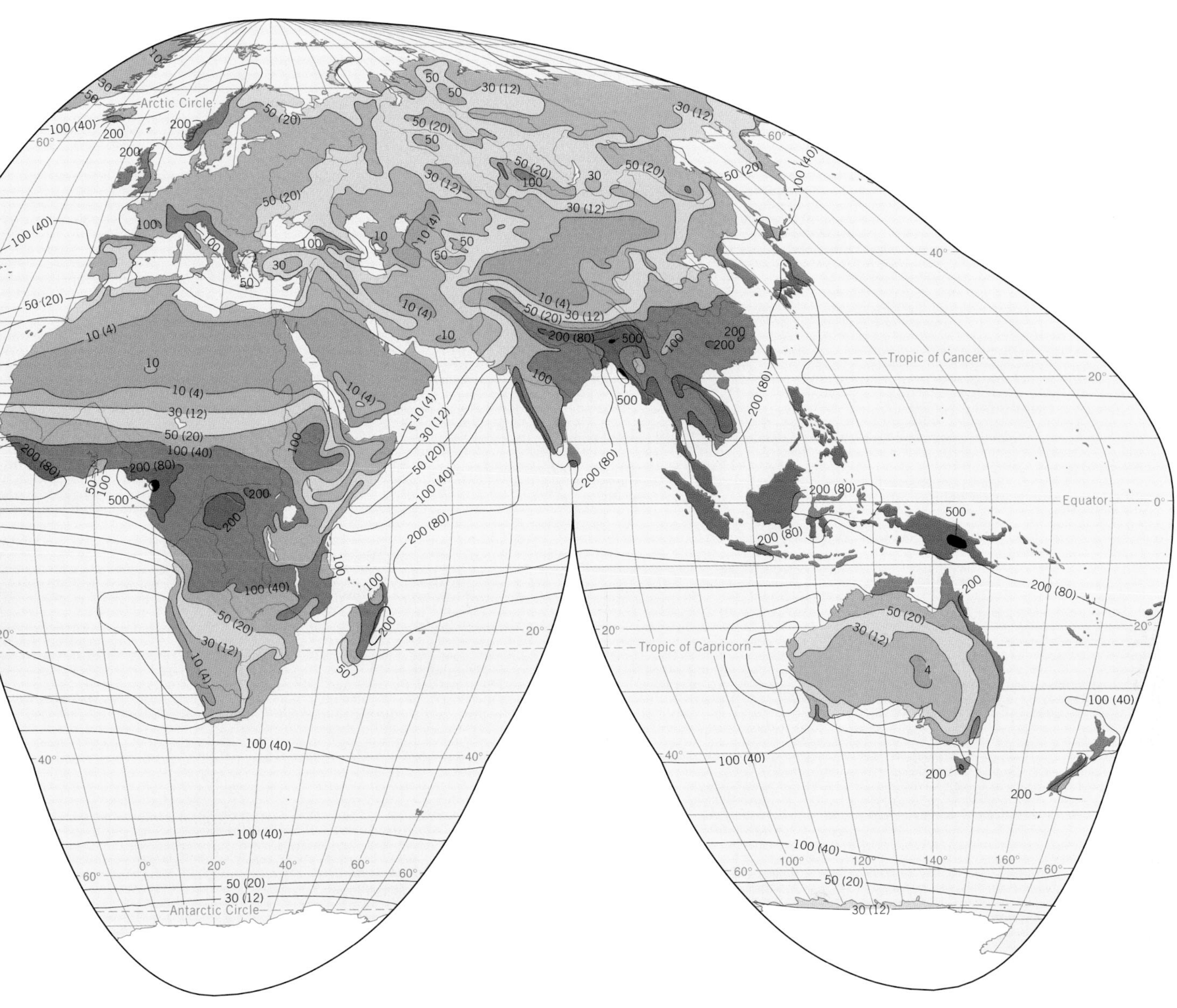

ences a very dry summer but has a moist winter. Southern and central California are also regions of this climate type. In Mediterranean climates, summer drought is produced by subtropical high-pressure cells, which intensify and move poleward during the high-Sun season. They extend into the regions of Mediterranean climate, providing the hot dry weather of their associated cT air masses and blocking the passage of other, moister, air masses. In the low-Sun season, the subtropical high-pressure cells move equatorward and weaken, allowing frontal and cyclonic precipitation to penetrate Mediterranean climate regions.

The dry-summer, moist-winter cycle is carried into higher midlatitudes along narrow strips of west coasts. Shannon Airport, Ireland (53° N), shows this marine west-coast type, although the difference between summer and winter rainfall is not as marked. Summers have less rainfall here for two reasons. First, the blocking effects of subtropical high-pressure cells tend to extend poleward into these regions, keeping moister air masses and cyclonic storms away. Second, the cyclonic storms that produce much of the winter precipitation are reduced in intensity during the high-Sun season. This occurs because the temperature and moisture contrasts between polar and arctic air masses and tropical air masses are weaker in summer, owing to increased high-latitude insolation.

Climate Classification

Mean monthly values of air temperature and precipitation can describe the climate of a weather station and its nearby region quite accurately. To study climates from a global viewpoint, climatologists classify these values into distinctive climate types. This requires developing a set of rules to use in examining monthly temperature and precipitation values. By applying the rules, the climatologist can use each station's data to determine the climate to which it belongs. This textbook recognizes 13 distinctive

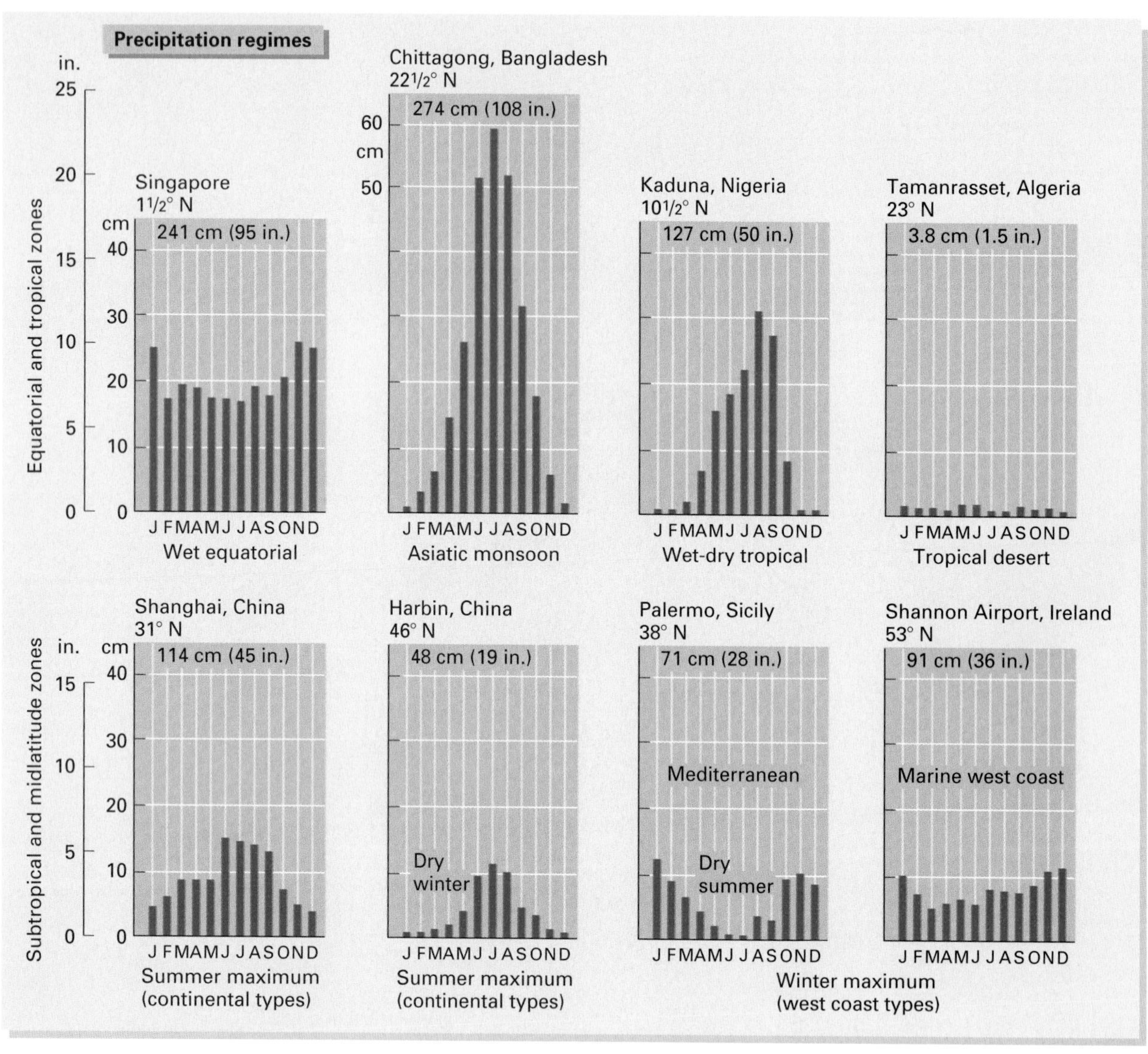

***9.6 Seasonal precipitation patterns** (above and right) Eight precipitation types selected to show various seasonal patterns. (Based on the Goode Base Map.)*

climate types that are designed to be understood and explained by air mass movements and frontal zones. The rules that actually define these types, shown in Appendix 2, are based on an analysis of how the amount of moisture held in the soil varies throughout the year as determined by air temperature and rainfall. This is an important topic that we will take up in Chapters 16 and 21. The specific climate rules are not important to our discussion here, which is to show that the classification follows quite naturally from the understanding of global temperature and precipitation processes that you have acquired in prior chapters. GEODISCOVERIES

The system of climates presented in this book and defined in Appendix 2 is designed to flow quite naturally from the principles governing temperature and precipitation that we have discussed above. An alternative classification is that devised by the Austrian climatologist Vladimir Köppen in 1918 and modified by Geiger and Pohl in 1953. It uses a system of letters to label climates. It is presented in a special supplement located as an interchapter feature following the end of this chapter.

Recall from Chapter 8 that air masses are classified according to the general latitude of their source regions and their surface type—land or ocean—within that region (see Figure 8.1). The latitude determines the temperature of the air mass, which can also depend on the season. The kind of surface, land or ocean, controls the moisture content. Since the air mass characteristics control the two most important climate variables—temperature and precipitation—we can explain climates using air masses as a guide.

We also know from Chapter 8 that frontal zones are regions in which air masses are in contact. And, when unlike air masses are in contact, cyclonic precipitation is likely to develop. However, the position of frontal zones changes with the seasons. For example, the polar-front zone lies generally across the middle latitudes of the

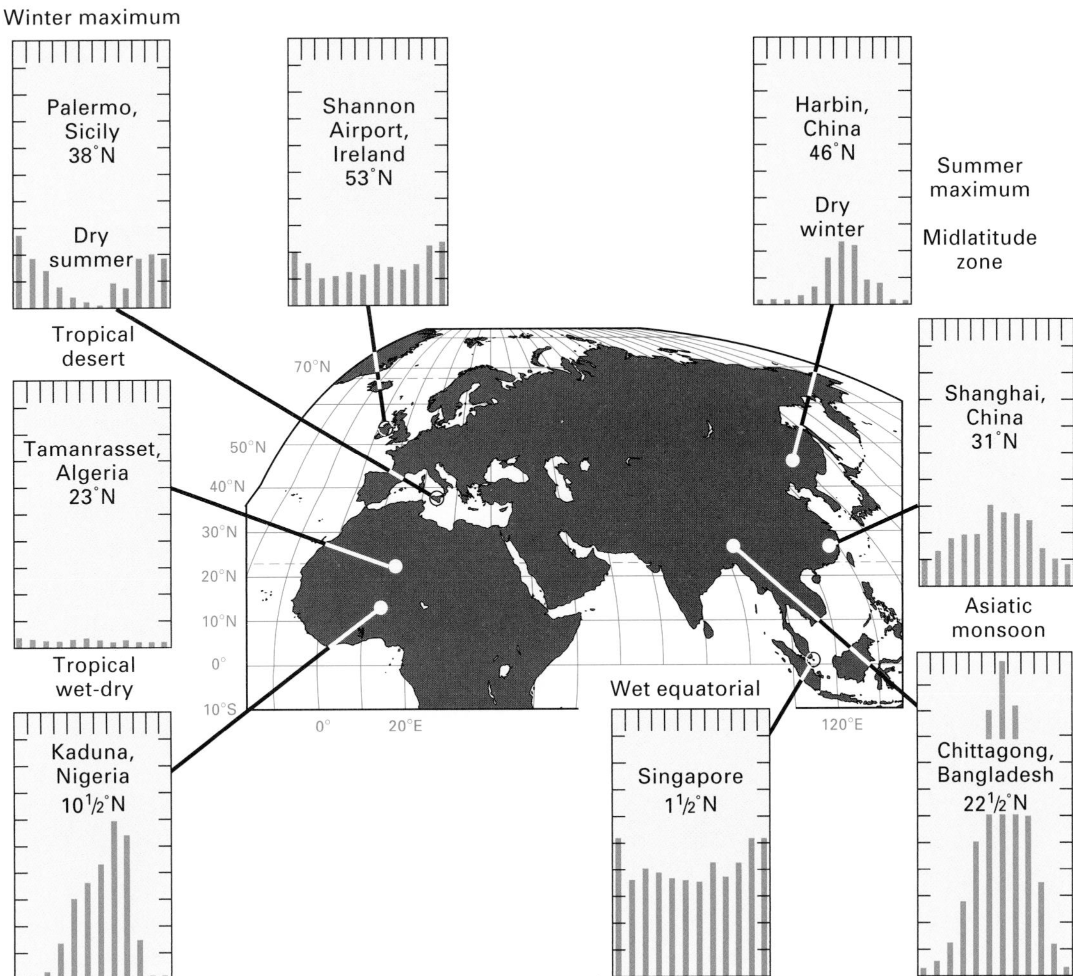

United States in winter, but it retreats northward to Canada during the summer. The seasonal movements of frontal zones therefore influence annual cycles of temperature and precipitation.

Figure 9.7 shows our schematic diagram of air mass source regions. We have subdivided this diagram into global bands that contain three broad groups of climates: low-latitude (Group I), midlatitude (Group II), and high-latitude (Group III). They are described briefly as follows.

- *Group I: Low-Latitude Climates.* The region of low-latitude climates (Group I) is dominated by the source regions of continental tropical (cT), maritime tropical (mT), and maritime equatorial (mE) air masses. These source regions are related to the three most obvious atmospheric features that occur within their latitude band—the two subtropical high-pressure belts and the equatorial trough at the intertropical convergence zone (ITCZ). Air of polar origin occasionally invades regions of low-latitude climates. Easterly waves and tropical cyclones are important weather systems in this climate group.
- *Group II: Midlatitude Climates.* The region of midlatitude climates (Group II) lies in the polar-front zone—a zone of intense interaction between unlike air masses. Here tropical air masses moving poleward and polar air masses moving equatorward are in conflict. Wave cyclones are normal features of the polar front, and this zone may contain as many as a dozen wave cyclones around the globe.
- *Group III: High-Latitude Climates.* The region of high-latitude climates (Group III) is dominated by polar and arctic (including antarctic) air masses. In the arctic belt of the 60th to 70th parallels, continental polar air masses meet arctic air masses along an *arctic-front zone,* creating a series of eastward-moving wave cyclones. In the southern hemisphere, there are no source regions in the subantarctic belt for continental polar air—just a great single oceanic source region for maritime polar (mP) air masses. The pole-centered continent of Antarctica provides a single great source of the extremely cold, dry antarctic air mass (cAA). These two air masses interact along the *antarctic-front zone.*

Within each of these three climate groups are a number of climate types (or simply, climates)—four low-latitude

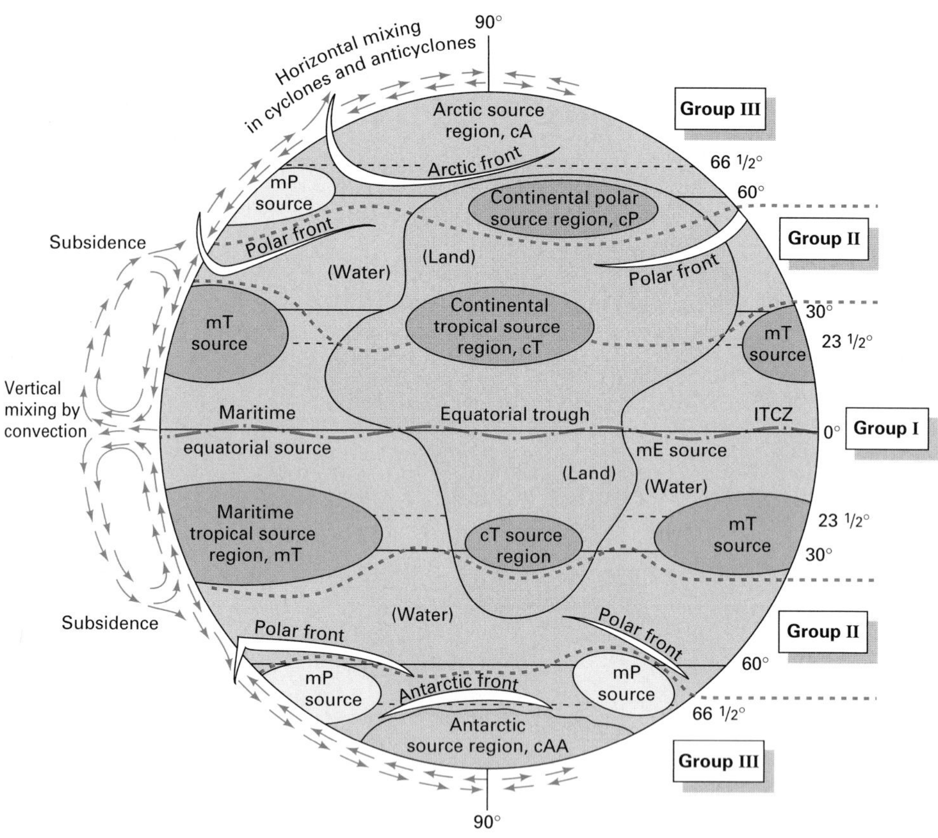

9.7 Climate groups and air mass source regions *Using the map of air mass source regions, we can identify five global bands associated with three major climate groups. Within each group is a set of distinctive climates with unique characteristics that are explained by the movements of air masses and frontal zones.*

climates (Group I), six midlatitude climates (Group II), and three high-latitude climates (Group III)—for a total of 13 climate types. In this textbook, the climates are numbered for ease of identification on maps and diagrams. We will refer to each climate by name because the names describe the general nature of the climate and also suggest its global location. For convenience, we include the climate number next to its name in the text.

In presenting the climates, we will make use of a pictorial device called a **climograph.** It shows the annual cycles of monthly mean air temperature and monthly mean precipitation for a location, along with some other useful information. Figure 9.8 is an example of a climograph. The data plotted are for Kayes, a station in the African country of Mali, at 14° N. At the top of the climograph, the mean monthly temperature is plotted as a line graph. At the bottom, the mean monthly precipitation is shown as a bar graph. The annual range in temperature and the total annual precipitation are stated on every climograph as well. Most climographs also display dominant weather features, which are shown using picture symbols. For Kayes, the two dominant features are the subtropical high and equatorial trough (ITCZ). Many of our climographs also include a small graph of the Sun's declination in order to help show when solstices and equinoxes occur.

The world map of climates, Figure 9.9, shows the actual distribution of climate types on the continents. This map, based on data collected at a large number of observing stations, is simplified because the climate boundaries are uncertain in many areas where observing stations are thinly distributed.

Overview of the Climates

Although Chapters 10 and 11 will provide a thorough examination of the 13 climates in our system, here is a quick and informal description of each.

LOW-LATITUDE CLIMATES (GROUP I)

- *Wet equatorial* ①. Warm to hot with abundant rainfall, this is the steamy climate of the Amazon and Congo basins.
- *Monsoon and trade-wind coastal* ②. This warm to hot climate has a very wet rainy season. It occurs in coastal regions that are influenced by trade winds or a monsoon circulation. The climates of Vietnam and Bangladesh are good examples.
- *Wet-dry tropical* ③. A warm to hot climate with very distinct wet and dry seasons. The monsoon region of India falls into this type, as does much of the Sahel region of Africa.
- *Dry tropical* ④. The climate of the world's hottest deserts—extremely hot in the high-Sun season, a little cooler in the low-Sun season, with little or no rainfall. The Sahara desert, Saudi Arabia, and the central Australian desert are of this climate.

MIDLATITUDE CLIMATES (GROUP II)

- *Dry subtropical* ⑤. Another desert climate, but not quite as hot as the dry tropical climate ④, since it is found farther poleward. This type includes the hottest part of the American southwest desert.

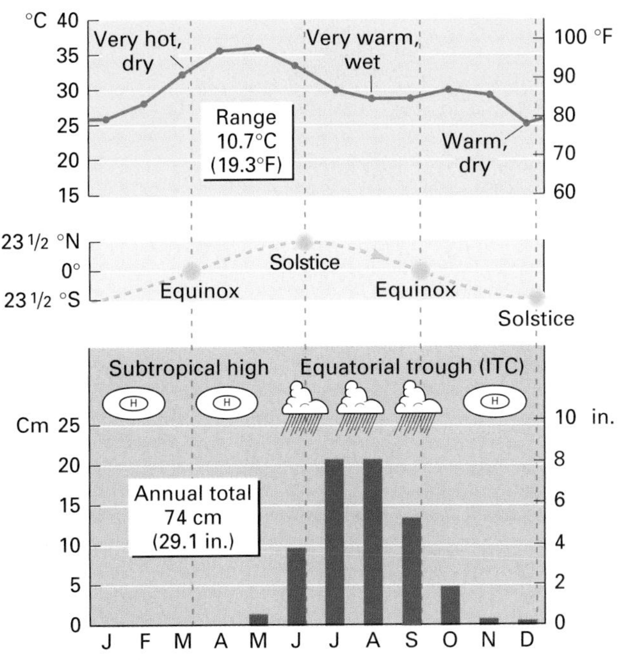

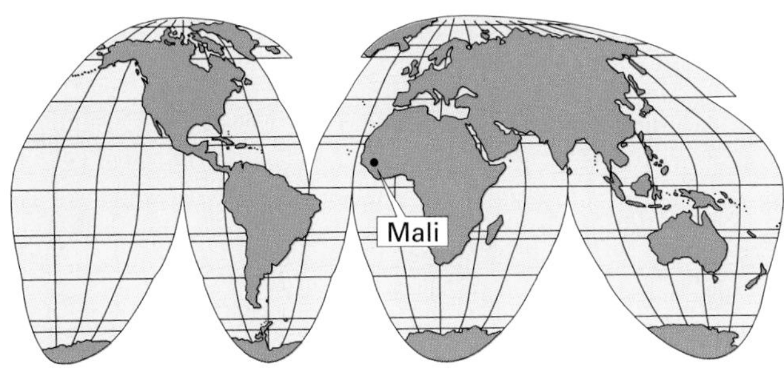

9.8 Typical climograph *Kayes, Mali, lat. 14° N, is located in western Africa, just south of the Sahara desert. Here, the climate is nearly rainless for seven months of the year. During these months, dry subsiding air from subtropical high-pressure cells dominates. Rainfall occurs when the intertropical convergence zone moves northward and reaches the vicinity of Kayes in June or July. Temperatures are hottest in April and May, before the rainy season begins.*

- *Moist subtropical* ⑥. The climate of the southeastern regions of the United States and China—hot and humid summers, with mild winters and ample rainfall year-round.
- *Mediterranean* ⑦. Hot, dry summers and rainy winters mark this climate. Southern and central California, as well as the lands of the Mediterranean region—Spain, southern Italy, Greece, and the coastal regions of Lebanon and Israel—are prime examples.
- *Marine west-coast* ⑧. The climate of the Pacific Northwest—coastal Oregon, Washington, and British Columbia. Warm summers and cool winters, with more rainfall in winter, are characteristics of this climate.
- *Dry midlatitude* ⑨. This dry climate is found in midlatitude continental interiors. The steppes of central Asia and the Great Plains of North America are familiar locales with this climate—warm to hot in summer, cold in winter, and with low annual precipitation.
- *Moist continental* ⑩. This is the climate of the eastern United States and lower Canada—cold in winter, warm in summer, with ample precipitation through the year.

HIGH-LATITUDE CLIMATES (GROUP III)

- *Boreal forest* ⑪. Short, cool summers and long, bitterly cold winters characterize this snowy climate. Northern Canada, Siberia, and central Alaska are regions of boreal forest climate.
- *Tundra* ⑫. Although this climate has a long, severe winter, temperatures on the tundra are somewhat moderated by proximity to the Arctic Ocean. This is the climate of the coastal arctic regions of Canada, Alaska, Siberia, and Scandinavia.
- *Ice sheet* ⑬. The bitterly cold temperatures of this climate, restricted to Greenland and Antarctica, can drop below –50°C (–58°F) during the sunless winter months. Even during the 24-hour days of summer, temperatures remain well below freezing.

Dry and Moist Climates

All but two of the 13 climate types introduced above are classified as either dry climates or moist climates. *Dry climates* are those in which total annual evaporation of moisture from the soil and from plant foliage exceeds the annual precipitation by a wide margin. Generally speaking, the dry climates do not support permanently flowing streams. The soil is dry much of the year, and the land surface is clothed with sparse plant cover—scattered grasses or shrubs—or simply lacks a plant cover. *Moist climates* are those with sufficient rainfall to maintain the soil in a moist condition through much of the year and to sustain the year-round flow of the larger streams. Moist climates support forests or prairies of dense tall grasses.

Within the dry climates there is a wide range of degree of aridity, ranging from very dry deserts nearly devoid of plant life to moister regions that support a partial cover of grasses or shrubs. To recognize this diversity, we will refer to two dry climate subtypes: (1) *semiarid* (or steppe) and (2) *arid*. The semiarid (steppe) subtype, designated by the letter *s,* has enough precipitation to support sparse grasses and shrubs. This subtype is found adjacent to moist climates. The arid subtype, indicated by the letter *a,* ranges from extremely dry to transitional with semiarid.

Two of our 13 climates cannot be accurately described as either dry or moist climates. These are the wet-dry tropical ③ and Mediterranean ⑦ climate types. Instead, they show a seasonal alteration between a very wet season and a very dry season. This striking contrast in seasons gives a special character to the two climates, and thus, we have singled them out for special recognition as *wet-dry climates*. In Figure 9.6, they are designated as having the wet-dry tropical and the Mediterranean precipitation patterns. Table 9.2 summarizes moist, dry, and wet-dry climates as they occur within the three main climate groups.

9.9 Climates of the world
Compiled from station data by A. N. Strahler.

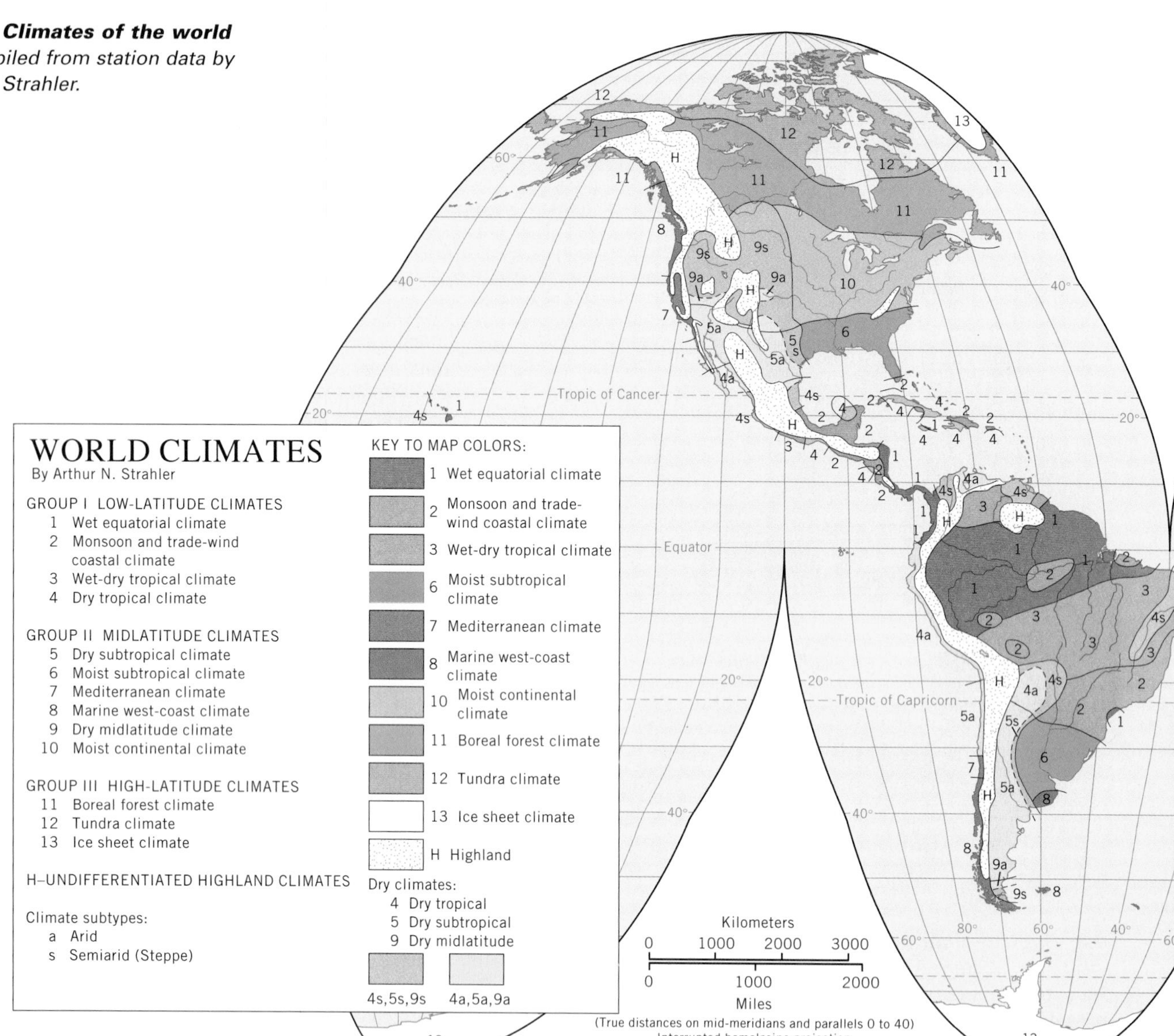

A LOOK AHEAD This introduction to global climate has stressed the relationship between climate and the factors that influence annual cycles of temperature and precipitation. Temperature cycles may be uniform, seasonal (continental), or moderated by oceanic influences (marine). Precipitation may be uniform (ranging from scarce in all months to abundant in all months), may have a maximum at the time of high Sun, or may have a maximum at the time of low Sun. These temperature and precipitation cycles in turn are produced by the annual cycle of insolation as it varies with latitude and by the global patterns of atmospheric circulation and air mass movements. The next two chapters examine the climates in more detail, including descriptions of key environmental characteristics that make each climate unique.

Table 9.2 | **Moist, Dry, and Wet-Dry Climate Types**

	Climate Type		
Climate Group	**Moist**	**Dry**	**Wet-Dry**
I: Low-latitude Climates	Wet equatorial ① Monsoon and trade-wind coastal ②	Dry tropical ④ (*s*, steppe; *a*, arid)	Wet-dry tropical ③
II: Midlatitude Climates	Moist subtropical ⑥ Marine west-coast ⑧ Moist continental ⑩	Dry subtropical ⑤ (*s*, *a*) Dry midlatitude ⑨ (*s*, *a*)	Mediterranean ⑦
III: High-latitude Climates	Boreal forest ⑪ Tundra ⑫	Ice sheet ⑬	

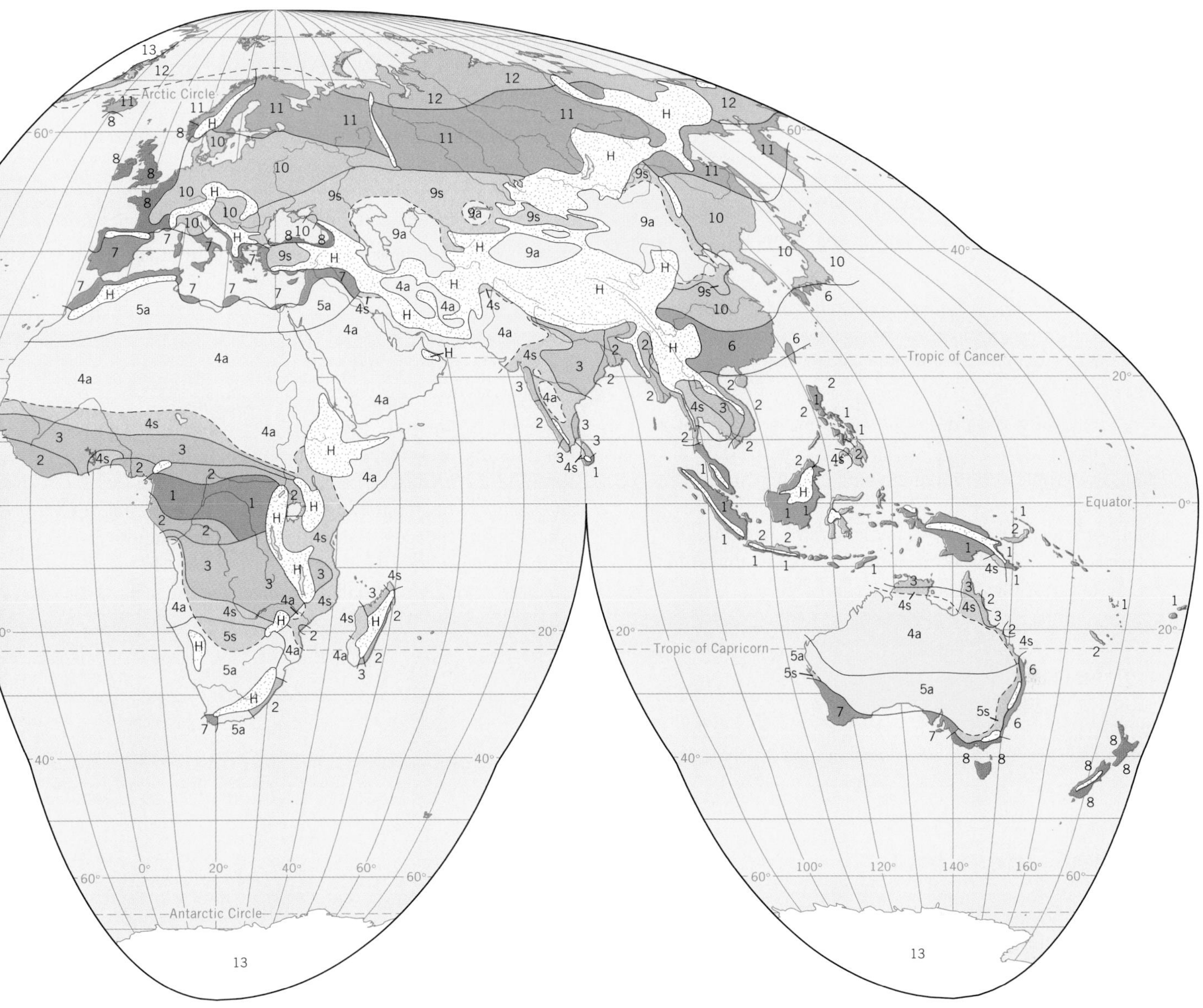

Chapter Summary

- **Climate** is the average weather of a region. Because temperature and precipitation are measured at many stations worldwide, we can use the combined annual patterns of monthly averages of temperature and precipitation to assign climate types.
- *Temperature regimes* are typical patterns of annual variation in temperature. They depend on latitude, which determines the annual pattern of insolation, and on location—continental or maritime—which enhances or moderates the annual insolation cycle.
- Global precipitation patterns are largely determined by air masses and their movements, which in turn are produced by global air circulation patterns. The main features of the global pattern of rainfall are:
  1. a wet equatorial belt produced by convectional precipitation around the ITCZ;
  2. trade-wind coasts that receive moist flows of mT air from trade winds as well as tropical cyclones;
  3. tropical deserts located under subtropical high-pressure cells;
  4. midlatitude deserts and **steppes**, which are dry because they are far from maritime moisture sources;
  5. moist subtropical regions that receive westward flows of moist mT air in the summer and eastward-moving wave cyclones in winter;
  6. midlatitude west coasts, which are subjected to eastward flows of mP air and occluded wave cyclones by prevailing westerly winds; and
  7. polar and arctic deserts, where little precipitation falls because the air is too cold to hold much moisture.
- Annual patterns of precipitation fall into three patterns: uniform (ranging from abundant to scarce); high-Sun (summer) maximum; and low-Sun (winter) maximum.
- There are three groups of climate types, arranged by latitude. *Low-latitude climates (Group I)* are dominated by mE, mT, and cT air masses, and are largely related to the

global circulation patterns that produce the ITCZ, trade winds, and subtropical high-pressure cells. They include:

Wet equatorial ① (warm to hot with abundant rainfall).
Monsoon and trade-wind coastal ② (warm to hot with a very wet rainy season at high Sun).
Wet-dry tropical ③ (warm to hot with very distinct wet and dry seasons).
Dry tropical ④ (extremely hot in the high-Sun season, a little cooler in the low-Sun season, with little or no rainfall).

- *Midlatitude climates (Group II)* lie in the polar-front zone and are strongly influenced by eastward-moving wave cyclones in which mT, mP, and cP air masses interact. They include:

Dry subtropical ⑤ (a desert climate, not quite as hot as dry tropical ④.
Moist subtropical ⑥ (hot and humid summers, mild winters, ample rainfall).
Mediterranean ⑦ (hot, dry summers, rainy winters).
Marine west-coast ⑧ (warm summers, cool winters, more rainfall in winter).
Dry midlatitude ⑨ (warm in summer, cold in winter, low annual precipitation).
Moist continental ⑩ (warm in summer, cold in winter, ample precipitation through the year).

- *High-latitude climates (Group III)* are dominated by polar and arctic (antarctic) air masses. Wave cyclones mixing mP and cP air masses along the *arctic-front zone* provide precipitation in this region. Climates of high latitudes include:

Boreal forest ⑪ (short, cool summers, with bitterly cold, snowy winters).
Tundra ⑫ (very short or nonexistent summers with cold winters).
Ice sheet ⑬ (bitterly cold, even in summer).

- *Dry climates* are those in which precipitation is largely evaporated from soil surfaces and transpired by vegetation, so that permanent streams cannot be supported. Within dry climates, there are two subtypes: *arid* (driest) and *semiarid* or steppe (a little wetter). In *moist climates*, precipitation exceeds evaporation and transpiration, providing for sustained year-round stream flow. In *wet-dry climates*, strong wet and dry seasons alternate.

Key Terms

climate
isohyet
steppe
climograph

Review Questions

1. Discuss the use of monthly records of average temperature and precipitation to characterize the climate of a region. Why are these measures useful?
2. Why are latitude and location (maritime or continental) important factors in determining the annual temperature cycle of a station?
3. How does air temperature, as a climatic variable, influence precipitation?
4. Describe three temperature regimes and explain how they are related to latitude and location.
5. Identify seven important features of the global map of precipitation and describe the factors that produce each.
6. The seasonality of precipitation at a station can generally be described as following one of three patterns. Identify them, explaining how each pattern can arise and providing an example.
7. What are the important global circulation patterns and air masses that influence low-latitude (Group I) climates? What are their effects?
8. What air masses and circulation patterns influence the midlatitude (Group II) climates and how?
9. Identify the air masses and frontal zones that are important in determining high-latitude climates (Group III) and explain their effects.
10. Name the climates by number (1–13) and group (I–III).

Focus on Systems 9.1 • Time Cycles of Climate

1. What are the two primary time cycles evident in climate data? How do they influence weather at a location?
2. Provide an example of another time cycle evident in temperature-records of the past few centuries.

Visualizing Exercises

1. Given that Santa Barbara, California, enjoys a midlatitude west-coast location, sketch a possible climograph showing monthly temperature and precipitation for Santa Barbara using the temperature and precipitation graphs in Figures 9.1 and 9.6.
2. Do the same for St. Louis, Missouri, noting that it has a midlatitude continental location.

Essay Question

1. Sketch a hypothetical supercontinent with a shape and features of your own choosing. It should stretch from about 70° N to 40° S latitude. Add some north-south mountain ranges, positioned where you like. Then select four locations on the supercontinent, describing and explaining the annual cycles of temperature and precipitation you would expect at each.

Problems

Working It Out 9.2 • Averaging in Time Cycles

1. The table below shows monthly precipitation in cm at St. Louis for 1990–1994. Find the average for each month for this five-year period.

Yr	J	F	M	A	M	J	J	A	S	O	N	D
1990	3.6	9.0	6.8	7.8	24.4	7.7	8.5	7.2	2.0	12.6	8.5	16.6
1991	3.9	2.5	8.1	8.3	9.8	1.1	13.2	2.5	7.6	14.5	8.3	5.3
1992	2.8	4.8	8.8	6.2	3.7	3.0	10.9	8.8	7.6	3.1	16.1	9.3
1993	9.0	7.0	8.4	15.6	10.0	18.1	12.9	12.1	23.3	6.6	12.3	3.8
1994	5.3	3.8	3.2	26.2	4.4	5.5	3.6	9.6	3.0	7.2	12.4	3.9

2. Using graph paper, plot the monthly averages for 1990–1994 in a similar fashion to graph *b* of the figure. Then plot the 58-year monthly means, using the monthly values shown on graph (*b*). How do the two graphs compare?

Eye on the Landscape

Chapter Opener Drying dates south of Cairo Date palms (A) thrive in hot, dry climates where ample water is available. They are a staple of Saharan oases—areas where fresh ground water is close to the surface and can be reached easily by plant roots. Date palms can also be grown in other desert regions if irrigated. We touch on oases in Chapter 10, and Figure 10.19 shows an oasis in the dry western desert of China. Water limits to plants are a subject of Chapter 23, Biogeographic Processes.

9.3 Semiarid continental interior scene The tough and wiry but fragrant sagebrush (A) dominates the vegetation in this photo and is found widely throughout the semiarid southwest. Where grazing is more intensive, the unpalatable sagebrush gains ground on grasses and forbs, as it has in this scene. Rolling hills (B) are typical of vast areas in the intermountain region. These smooth erosion features are carved and maintained largely by infrequent but intense rainfall events that move bare soil downslope by overland flow. Disturbance is covered in Chapter 22, while we treat overland flow in Chapter 16.

9.4 Midlatitude west coast scene These marine cliffs (A) indicate a landscape that has recently been drowned by ocean waters. During the Ice Age, sea level was as much as 100 m (330 ft) lower than it is today. As the continental ice sheets melted, sea level rose and waves began to attack the shoreline carving the cliffs. Another interesting feature of this photo is the lack of natural vegetation (B). Before human settlement, this area would have been forested with a mix of deciduous and evergreen trees, but no trace of that former landscape remains. Many European landscapes, though pleasing to the eye, are entirely dependent on human activities. Coastlines of submergence are a topic of Chapter 19; human influence on natural landscapes is a subject of Chapter 23.

A CLOSER LOOK

Special Supplement 9.3 The Köppen Climate System

Air temperature and precipitation data have formed the basis for several climate classifications. One of the most important of these is the Köppen climate system, devised in 1918 by Dr. Vladimir Köppen of the University of Graz in Austria. For several decades, this system, with various later revisions, was the most widely used climate classification among geographers. Köppen was both a climatologist and plant geographer, so that his main interest lay in finding climate boundaries that coincided approximately with boundaries between major vegetation types.

Under the Köppen system, each climate is defined according to assigned values of temperature and precipitation, computed in terms of annual or monthly values. Any given station can be assigned to its particular climate group and subgroup solely on the basis of the records of temperature and precipitation at that place.

Note that mean annual temperature refers to the average of 12 monthly temperatures for the year. Mean annual precipitation refers to the average of the entire year's precipitation as observed over many years.

The Köppen system features a shorthand code of letters designating major climate groups, subgroups within the major groups, and further subdivisions to distinguish particular seasonal characteristics of temperature and precipitation. Five major climate groups are designated by capital letters as follows:

A ***Tropical rainy climates***
Average temperature of every month is above 18°C (64.4°F). These climates have no winter season. Annual rainfall is large and exceeds annual evaporation.

B ***Dry climates***
Evaporation exceeds precipitation on the average throughout the year. There is no water surplus; hence, no permanent streams originate in B climate zones.

C ***Mild, humid (mesothermal) climates***
The coldest month has an average temperature of under 18°C (64.4°F) but above –3°C (26.6°F); at least one month has an average temperature above 10°C (50°F). The C climates thus have both a summer and a winter.

D ***Snowy-forest (microthermal) climates***
The coldest month has an average temperature of under –3°C (26.6°F). The average temperature of the warmest month is above 10°C (50°F). (Forest is not generally found where the warmest month is colder than 10°C (50°F).)

E ***Polar climates***
The average temperature of the warmest month is below 10°C (50°F). These climates have no true summer.

Note that four of these five groups (*A, C, D,* and *E*) are defined by temperature averages, whereas one (*B*) is defined by precipitation-to-evaporation ratios. Groups *A, C,* and *D* have sufficient heat and precipitation for the growth of forest and woodland vegetation. Figure S9.1 shows the boundaries of the five major climate groups, and Figure S9.2 is a world map of Köppen climates.

Subgroups within the five major groups are designated by a second letter according to the following code.

S ***Semiarid (steppe)***
W ***Arid (desert)***

(The capital letters S and W are applied only to the dry B climates.)

- **f** Moist, adequate precipitation in all months, no dry season. This modifier is applied to A, C, and D groups.
- **w** Dry season in the winter of the respective hemisphere (low-Sun season).
- **s** Dry season in the summer of the respective hemisphere (high-Sun season).
- **m** Rainforest climate, despite short, dry season in monsoon type of precipitation cycle. Applies only to A climates.

From combinations of the two letter groups, 12 distinct climates emerge:

A ***Tropical rainforest climate***
The rainfall of the driest month is 6 cm (2.4 in.) or more.

Am ***Monsoon variety of Af***
The rainfall of the driest month is less than 6 cm (2.4 in.). The dry season is strongly developed.

Aw ***Tropical savanna climate***
At least one month has rainfall less than 6 cm (2.4 in.). The dry season is strongly developed.

Figure S9.3 shows the boundaries between Af, Am, and Aw climates as determined by both annual rainfall and rainfall of the driest month.

BS ***Steppe climate***
A semiarid climate characterized by grasslands, it occupies an intermediate position between the

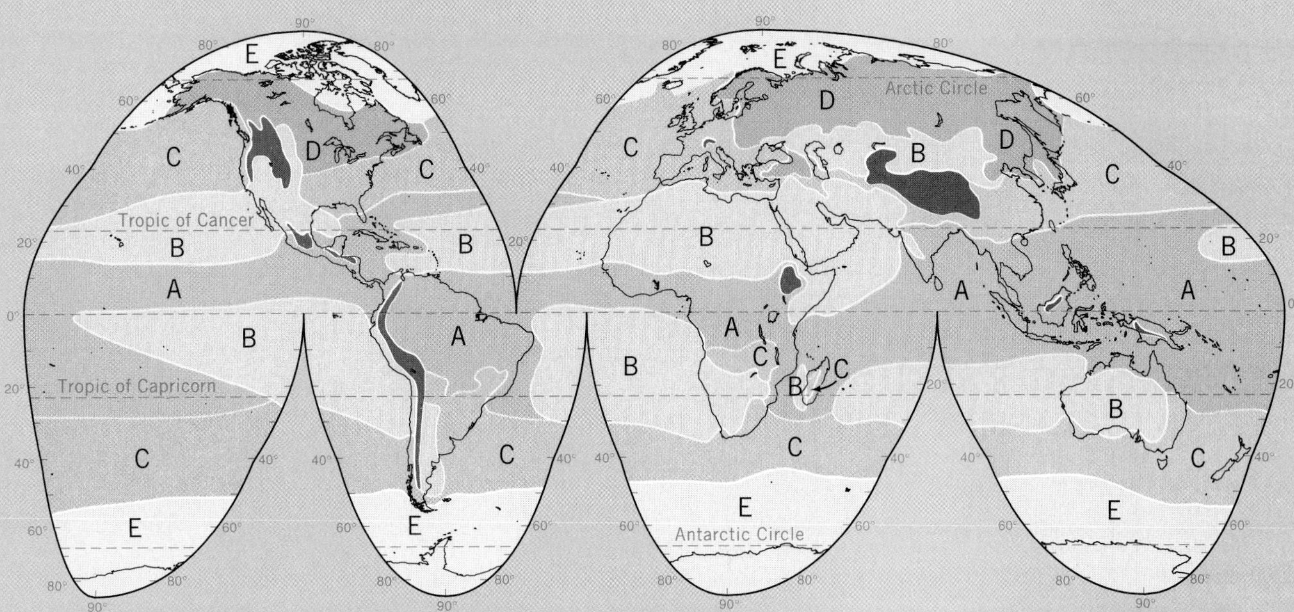

S9.1 Generalized Köppen climate map *Highly generalized world map of major climate regions according to the Köppen classification. Highland areas are in black. (Based on Goode Base Map.)*

desert climate (BW) and the more humid climates of the A, C, and D groups. Boundaries are determined by formulas given in Figure S9.4.

BW Desert climate
Desert has an arid climate with annual precipitation of usually less than 40 cm (15.7 in.). The boundary with the adjacent steppe climate (BS) is determined by formulas given in Figure S9.4.

C *Mild humid climate with no dry season*
Precipitation of the driest month averages more than 3 cm (1.2 in.).

Cw *Mild humid climate with a dry winter*
The wettest month of summer has at least 10 times the precipitation of the driest month of winter. (Alternative definition: 70 percent or more of the mean annual precipitation falls in the warmer six months.)

Cs *Mild humid climate with a dry summer*
Precipitation of the driest month of summer is less than 3 cm (1.2 in.). Precipitation is at least three times as much as the driest month of summer. (Alternative definition: 70 percent or more of the mean annual precipitation falls in the six months of winter.)

Df *Snowy-forest climate with a moist winter*
No dry season.

Dw *Snowy-forest climate with a dry winter*

ET *Tundra climate*
The mean temperature of the warmest month is above 0°C (32°F) but below 10°C (50°F).

EF *Perpetual frost climate*
In this ice sheet climate, the mean monthly temperatures of all months are below 0°C (32°F).

To denote further variations in climate, Köppen added a third letter to the code group. The meanings are as follows:

- ***a*** With hot summer; warmest month is over 22°C (71.6°F); C and D climates.
- ***b*** With warm summer; warmest month is below 22°C (71.6°F); C and D climates.
- ***c*** With cool, short summer; less than four months are over 10°C (50°F); C and D climates.
- ***d*** With very cold winter; coldest month is below –38°C (–36.4°F); D climates only.
- ***h*** Dry-hot; mean annual temperature is over 18°C (64.4°F); B climates only.
- ***k*** Dry-cold; mean annual temperature is under 18°C (64.4°F); B climates only.

As an example of a complete Köppen climate code, BWk refers to a cool desert climate, and Dfc refers to a cold, snowy-forest climate with cool, short summer.

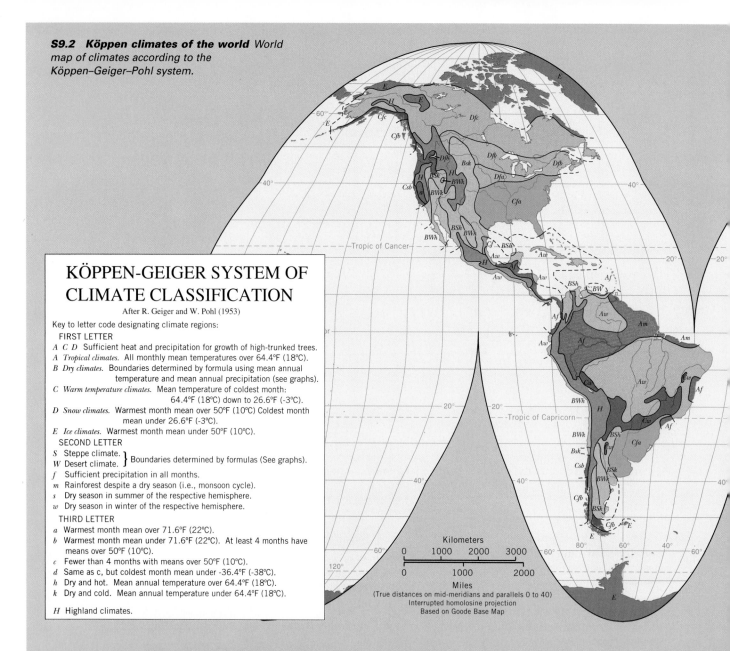

S9.2 Köppen climates of the world *World map of climates according to the Köppen–Geiger–Pohl system.*

KÖPPEN-GEIGER SYSTEM OF CLIMATE CLASSIFICATION

After R. Geiger and W. Pohl (1953)

Key to letter code designating climate regions:

FIRST LETTER

A C D Sufficient heat and precipitation for growth of high-trunked trees.
A Tropical climates. All monthly mean temperatures over 64.4°F (18°C).
B Dry climates. Boundaries determined by formula using mean annual temperature and mean annual precipitation (see graphs).
C Warm temperature climates. Mean temperature of coldest month: 64.4°F (18°C) down to 26.6°F (-3°C).
D Snow climates. Warmest month mean over 50°F (10°C) Coldest month mean under 26.6°F (-3°C).
E Ice climates. Warmest month mean under 50°F (10°C).

SECOND LETTER

S Steppe climate.
W Desert climate. } Boundaries determined by formulas (See graphs).
f Sufficient precipitation in all months.
m Rainforest despite a dry season (i.e., monsoon cycle).
s Dry season in summer of the respective hemisphere.
w Dry season in winter of the respective hemisphere.

THIRD LETTER

a Warmest month mean over 71.6°F (22°C).
b Warmest month mean under 71.6°F (22°C). At least 4 months have means over 50°F (10°C).
c Fewer than 4 months with means over 50°F (10°C).
d Same as c, but coldest month mean under -36.4°F (-38°C).
h Dry and hot. Mean annual temperature over 64.4°F (18°C).
k Dry and cold. Mean annual temperature under 64.4°F (18°C).

H Highland climates.

Kilometers 0 1000 2000 3000
0 1000 2000 Miles
(True distances on mid-meridians and parallels 0 to 40)
Interrupted homolosine projection
Based on Goode Base Map

S9.3 Boundaries of the A climates

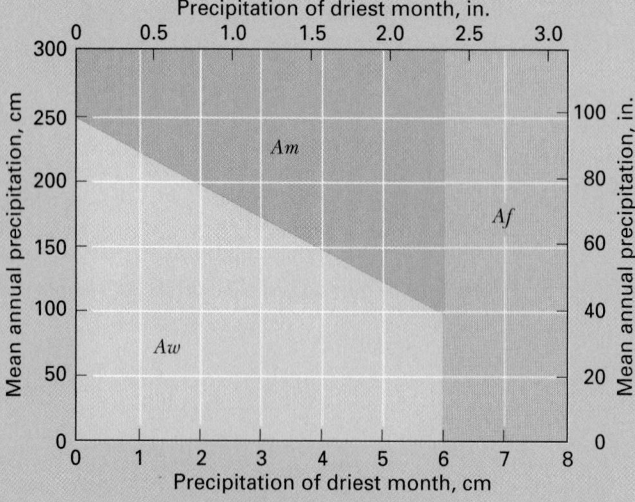

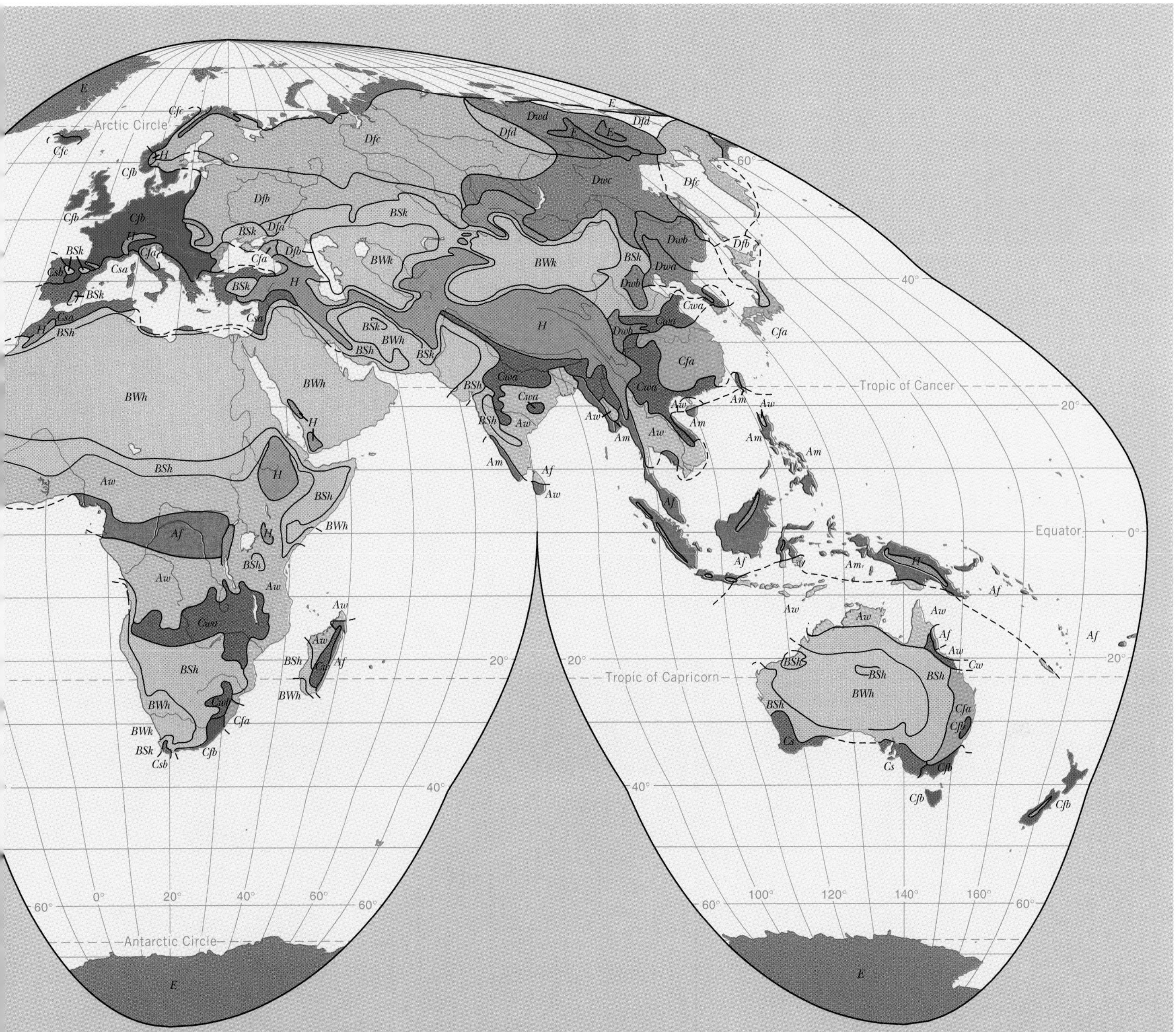

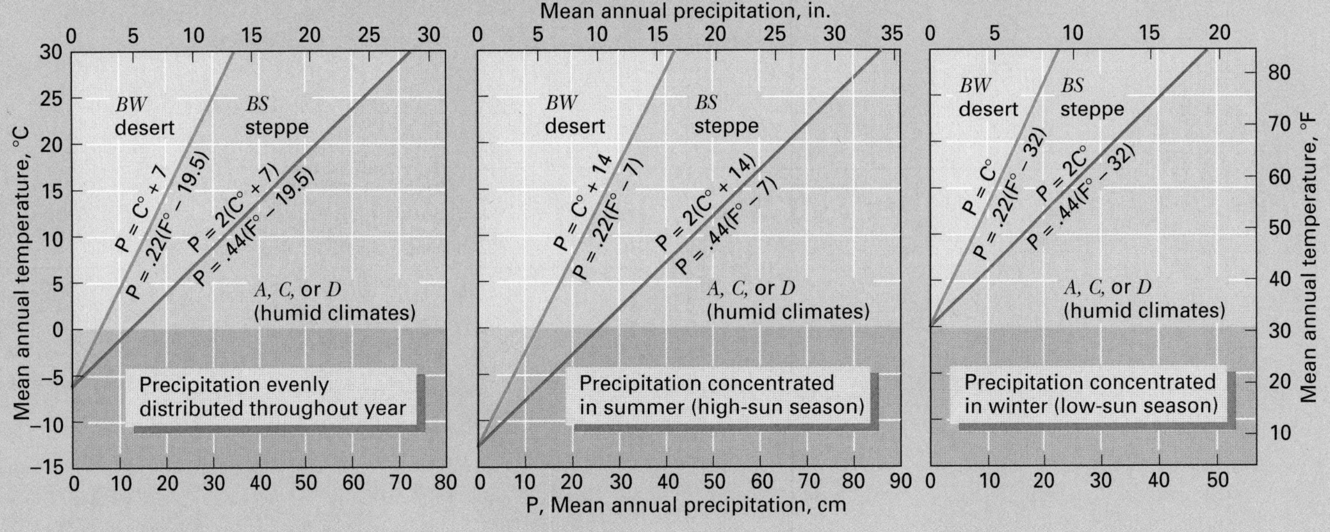

S9.4 Boundaries of the B climates *Upper figures: metric system. Lower figures: English system*

10 LOW-LATITUDE CLIMATES

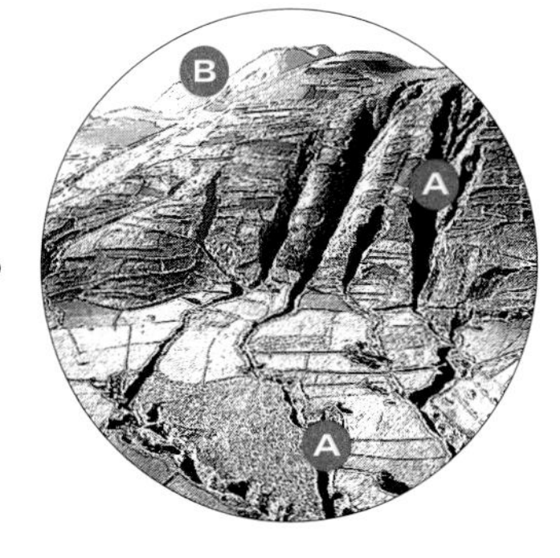

EYE ON THE LANDSCAPE
Tropical agriculture on volcanic mountain slopes near Ankisabe, Madagascar. The monsoon and trade-wind coastal climate of the west coast of Madagascar provides abundant rainfall for multiple crops. **What else would the geographer see?...Answers at the end of the chapter.**

With our introduction to global climates now complete in Chapter 9, we can describe each of the climates and their associated environments in turn, beginning with the low-latitude climates of this chapter. Keep in mind that the annual cycles of temperature and precipitation for each climate region can be easily related to the latitude and location (coastal or interior) of the region, as well as to the characteristic movements of air masses and fronts through the region as they are controlled by global-scale patterns of atmospheric circulation.

Our description of the environments associated with the climate types will stress the natural vegetation covers, agricultural practices, and other climate-related human activities that occur in the regions of each climate type. We will also touch on the nature of soils in some climate regions. Both soils and vegetation are covered more thoroughly in Part 5.

OVERVIEW OF THE LOW-LATITUDE CLIMATES

The *low-latitude climates* lie for the most part between the tropics of cancer and capricorn. In terms of world latitude zones, the low-latitude climates occupy all of the equatorial zone (10° N to 10° S), most of the tropical zone (10–15° N and S), and part of the subtropical zone. In terms of prevailing pressure and wind systems, the region of low-latitude climates includes the equatorial trough of the intertropical convergence zone (ITCZ), the belt of tropical easterlies (northeast and southeast trades), and large portions of the oceanic subtropical high-pressure belt (Figure 9.7).

Figure 10.1 shows climographs for the four low-latitude climates. These climates range from extremely moist—the

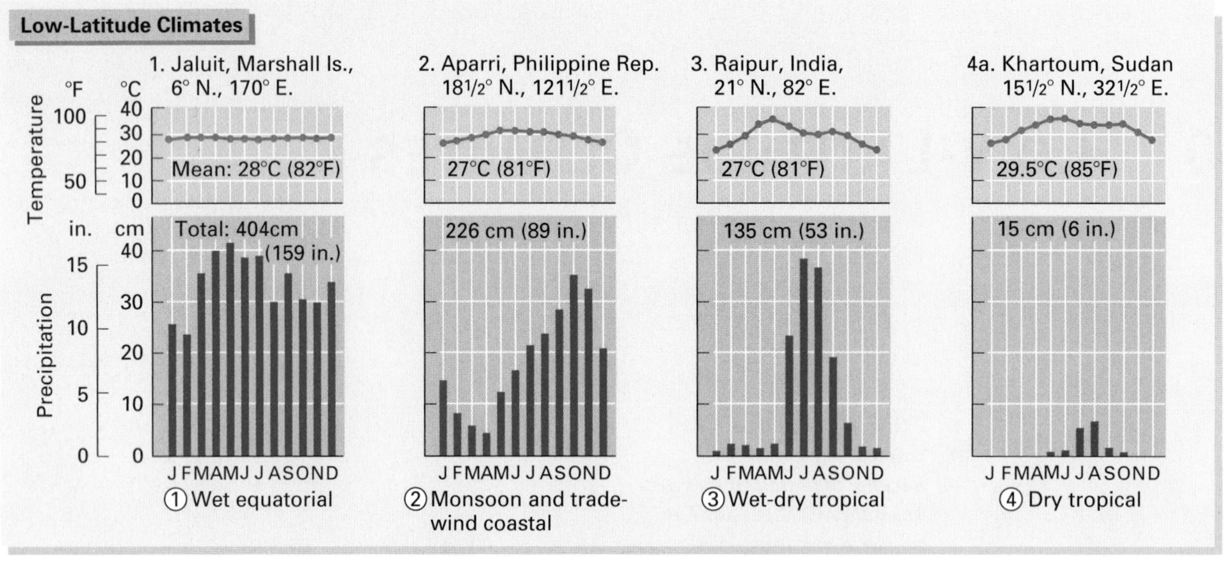

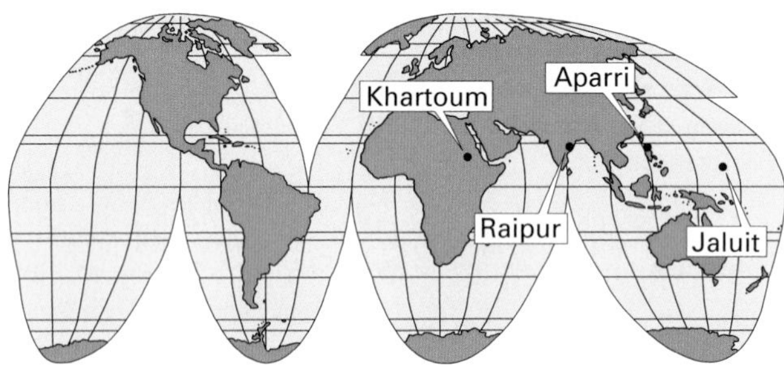

10.1 Low-latitude climographs *These four climographs show the key features of the four low-latitude climates.*

wet equatorial climate ①—to extremely dry—the dry tropical climate ④. They also vary strongly in the seasonality of their rainfall. In the wet equatorial climate ①, rainfall is abundant year-round. But in the wet-dry tropical climate ③, rainfall is abundant for only part of the year. During the remainder of the year, little or no rain falls. The seasonal temperature cycle also varies among these climates. In the wet equatorial climate ①, temperatures are nearly uniform throughout the year. In the dry tropical climate ④, there is a strong annual temperature cycle. Table 10.1 summarizes some of the characteristics of these climates.

The Wet Equatorial Climate ① (Köppen: *Af*)

The **wet equatorial climate** ① is a climate of the intertropical convergence zone (ITCZ), which is nearby for most of the year. The climate is dominated by warm, moist maritime equatorial (mE) and maritime tropical (mT) air masses that yield heavy convectional rainfall. Precipitation is plentiful in all months, and the annual total often exceeds 250 cm (about 100 in.). However, there is usually a seasonal pattern to the rainfall, so that rainfall is greater during some part of the year. This period of heavier rainfall occurs when the ITCZ migrates into the region. Remarkably uniform temperatures prevail throughout the year. Both mean monthly and mean annual temperatures are typically close to 27°C (81°F).

Figure 10.2 shows the world distribution of the wet equatorial climate ①. This climate is found in the latitude range 10° N to 10° S. Its major regions of occurrence include the Amazon lowland of South America, the Congo Basin of equatorial Africa, and the East Indies, from Sumatra to New Guinea.

Figure 10.3 is a climograph for Iquitos, Peru (located in Figure 10.2), a typical wet equatorial station situated close to the equator in the broad, low basin of the upper Amazon River. Notice the very small annual range in temperature and the very large annual rainfall total. Monthly air temperatures are extremely uniform in the wet equatorial climate ①. Typically, mean monthly air temperature will range between 26° and 29°C (79° and 84°F) for stations at low elevation in the equatorial zone.

The Monsoon and Trade-Wind Coastal Climate ② (Köppen: *Af, Am*)

Like the wet equatorial climate ①, the **monsoon and trade-wind coastal climate** ② has abundant rainfall. But unlike

Table 10.1 | Low-Latitude Climates

Climate	Temperature	Precipitation	Explanation
Wet equatorial ①	Uniform temperatures, mean near 27 °C (81°F).	Abundant rainfall, all months, from mT and mE air masses. Annual total may exceed 250 cm (100 in.).	The ITCZ dominates this climate, with abundant convectional precipitation generated by convergence in weak equatorial lows. Rainfall is heaviest when the ITCZ is nearby.
Monsoon and trade-wind coastal ②	Temperatures show an annual cycle, with warmest temperatures in the high-Sun season.	Abundant rainfall but with a strong seasonal pattern.	Trade-wind coastal: Rainfall from mE and mT air masses is heavy when the ITCZ is nearby, lighter when the ITCZ moves to the opposite hemisphere. Asian monsoon coasts: dry air flowing southwest in low-Sun season alternates with moist oceanic air flowing northeast, producing a seasonal rainfall pattern on west coasts.
Wet-dry tropical ③	Marked temperature cycle, with hottest temperatures before the rainy season.	Wet high-Sun season alternates with dry low-Sun season.	Subtropical high pressure moves into this climate in the low Sun-season, bringing very dry conditions. In the high-Sun season, the ITCZ approaches and rainfall occurs. Asian monsoon climate: alternation of dry continental air in low-Sun season, with moist oceanic air in low-Sun season with moist oceanic air in high-Sun season brings a strong pattern of dry and wet seasons.
Dry tropical ④	Strong temperature cycle, with intense hot temperatures during high-Sun season.	Low precipitation. Sometimes rainfall occurs when the ITCZ is near.	This climate is dominated by subtropical high pressure, which provides clear, stable air for much or all of year. Insolation is intense during high-Sun period.

the wet equatorial climate ①, the rainfall of the monsoon and trade-wind coastal climate ② always shows a strong seasonal pattern. This seasonal pattern is due to the migration of the intertropical convergence zone (ITCZ). In the high-Sun season ("summer," depending on the hemisphere), the ITCZ is nearby and monthly rainfall is greater. In the low-Sun season, when the ITCZ has migrated to the other hemisphere, subtropical high pressure dominates and monthly rainfall is less. Figure 10.2 shows the global distribution of the monsoon and trade-wind coastal climate ②. The climate occurs over latitudes from 5° to 25° N and S.

As the name of the monsoon and trade-wind coastal climate ② suggests, this climate type is produced by two somewhat different situations. On trade-wind coasts, rainfall is produced by moisture-laden maritime tropical (mT) and maritime equatorial (mE) air masses. These are moved onshore onto narrow coastal zones by trade winds or by monsoon circulation patterns. As the warm, moist air passes over coastal hills and mountains, the orographic effect touches off convectional shower activity. Shower activity is also intensified by easterly waves, which are more frequent when the ITCZ is nearby. The east coasts of land masses experience this trade-wind effect because the trade winds blow from east to west. Trade-wind coasts are found along the east sides of Central and South America, the Caribbean Islands, Madagascar (Malagasy), Southeast Asia, the Philippines, and northeast Australia (see Figure 10.2).

The coastal precipitation effect also applies to the summer monsoon of Asia, when the monsoon circulation brings mT air onshore. However, the onshore monsoon winds blow from southwest to northeast, so it is the western coasts of land masses that are exposed to this moist airflow. Western India and Myanmar (formerly Burma) are examples. Moist air also penetrates well inland in Bangladesh, providing the very heavy monsoon rains for which the region is well known.

10.2 World map of wet equatorial ① and monsoon and trade-wind coastal climates ② *(Based on Goode Base Map.)*

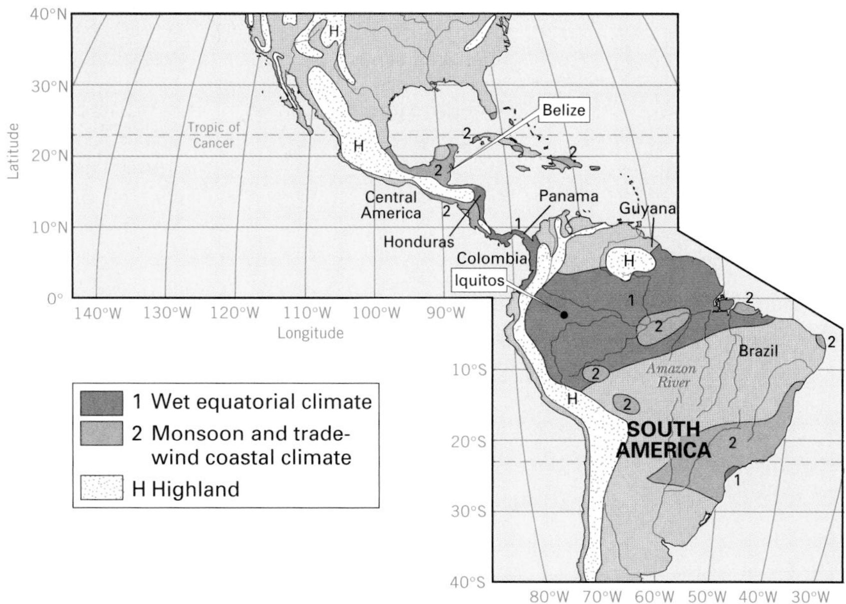

In central and western Africa, and southern Brazil, the monsoon pattern shifts the intertropical convergence zone over 20° of latitude, or more (see Figure 7.18). Here, heavy rainfall occurs in the high-Sun season, when the ITCZ is nearby. Drier conditions prevail in the low-Sun season, when the ITCZ is far away.

Temperatures in the monsoon and trade-wind coastal climate ②, though warm throughout the year, also show an annual cycle. Warmest temperatures occur in the high-Sun season, just before arrival of the ITCZ brings clouds and rain. Minimum temperatures occur at the time of low Sun.

Figure 10.4 is a climograph for the city of Belize, in the Central American country of Belize (Figure 10.2). This east coast city, located at lat. 17° N, is exposed to the tropical

10.3 Wet equatorial climate ① *Iquitos, Peru, lat. 3° S, is located in the upper Amazon lowland, close to the equator. Temperatures differ very little from month to month, and there is abundant rainfall throughout the year.*

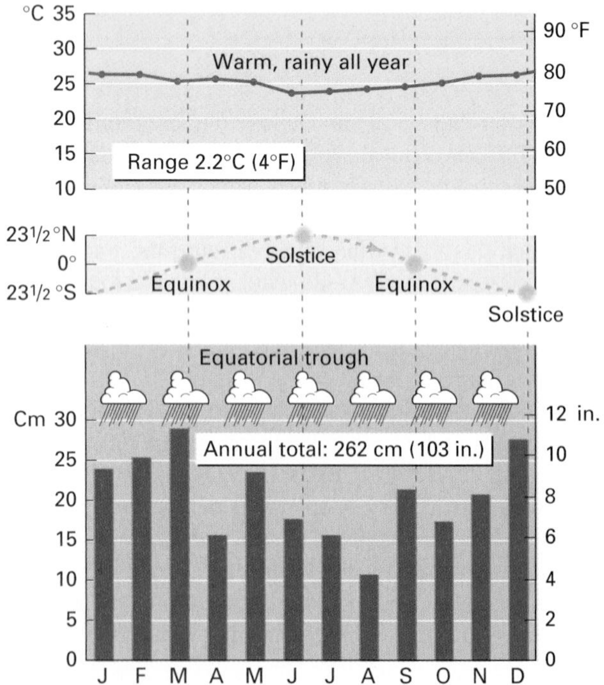

10.4 Trade-wind coastal climate ② *This climograph for Belize, a Central American east-coast city at lat. 17° N, shows a marked season of low rainfall following the period of low Sun. For the remainder of the year, precipitation is high, produced by warm, moist northeast trade winds.*

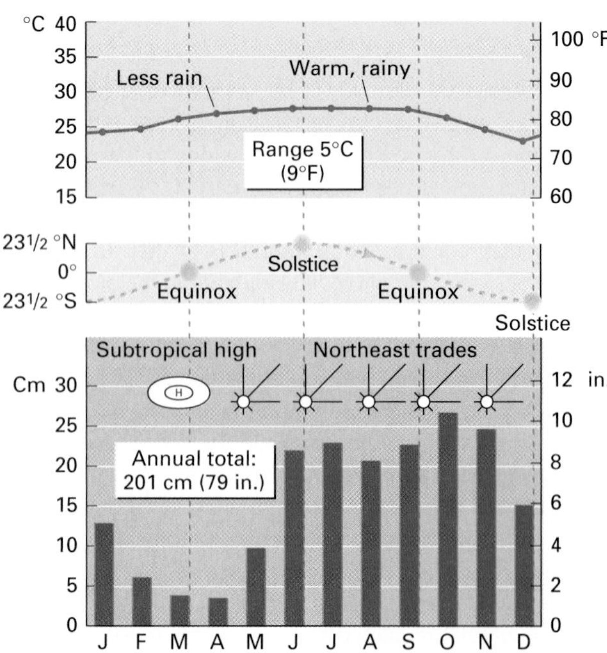

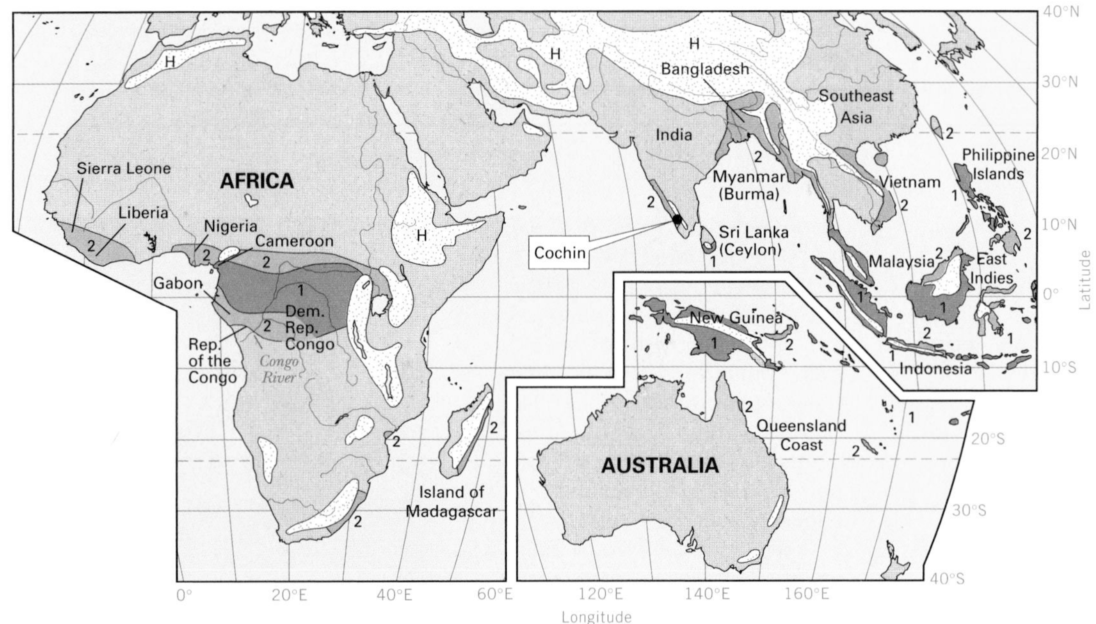

easterly trade winds. Rainfall is abundant from June through November, when the ITCZ is nearby. Easterly waves are common in this season, and on occasion a tropical cyclone will bring torrential rainfall. Following the December solstice, rainfall is greatly reduced, with minimum values in March and April. At this time, the ITCZ lies farthest away, and the climate is dominated by subtropical high pressure. Air temperatures show an annual range of 5° C (9° F) with maximum in the high-Sun months.

The Asiatic monsoon shows a similar pattern, but there is an extreme peak of rainfall during the high-Sun period and a well-developed dry season with two or three months of only small rainfall amounts. The climograph for Cochin, India, provides an example (Figure 10.5). Located at lat. 10° N on the west coast of lower peninsular India (see Figure 10.2), Cochin receives the warm, moist southwest winds of the summer monsoon. In this season, monthly rainfall is extreme in both June and July. A strongly pronounced season of low rainfall occurs at time of low Sun—December through March. Air temperatures show only a very weak annual cycle, cooling a bit during the rains, so the annual range is small at this low latitude.

The Low-Latitude Rainforest Environment

Our first two climates—wet equatorial ① and monsoon and trade-wind coastal ②—are quite uniform in temperature and have a high annual rainfall. These factors create a

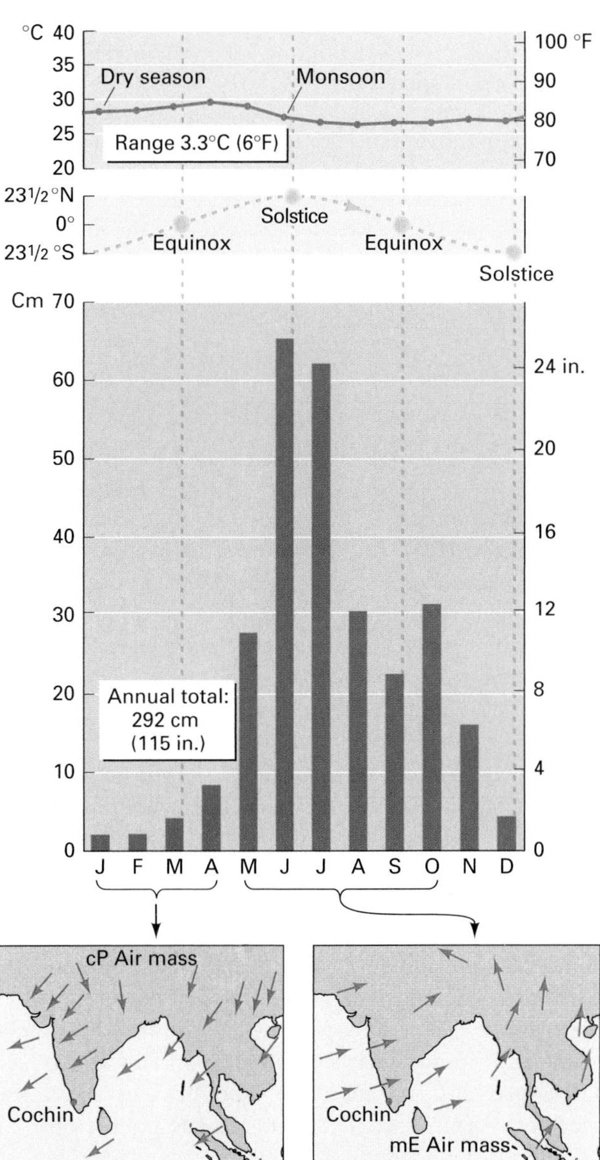

10.5 Monsoon coastal climate ② *Cochin, India, on a windward coast at lat. 10° N, shows an extreme peak of rainfall during the rainy monsoon, contrasting with a short dry season at time of low Sun.*

Geographers at Work

Conserving Tropical Biodiversity Outside of National Parks

by Lisa Naughton ***University of Wisconsin, Madison***

As the vast rainforests of the equatorial and tropical zones shrink under the pressure of commercial exploitation for timber and cash-crop agriculture, many countries have created national parks, game parks, and conservation areas to protect the natural environment and its biodiversity. In these reserves, human presence and activity is strictly controlled. Yet the natural environment also includes the local peoples who reside in or near the parks.

At the borders of the reserve, the economic needs of local people often clash with wildlife survival. When an elephant enters a farmer's field to feast on corn, its behavior is viewed as unnatural to the ecologist and highly undesirable by the farmer. However, the wildlife and farmer have coexisted for millenia in an uneasy relationship in which the wildlife exploits the hard work of the farmer but is often hunted in return. To avoid the loss of endangered species as well as the privations of their destructive interactions with humans, it is essential to study the behavior of both humans and animals at the fringe of the reserve.

That's where Lisa Naughton comes in. She is fascinated by human–wildlife interactions at park edges, and strives to understand the feeding strategies of wildlife and their survival in agroforestry systems in order to balance farming aims with conservation goals. Lisa has worked in forests of both Latin America and Africa, and has had thrilling encounters with wildlife ranging from elephants and chimpanzees in Uganda to fresh water dolphins in the Equadorian Amazon.

Lisa's present work takes her to the southeastern corner of Peru, near the headwaters of the Amazon system. Here, a transoceanic road linking Brazil with Peru has opened up to farms, gold miners, and loggers who are settling near the edge of a national park. Read more about how she and her team of Peruvian biologists and geographers systematically sample farms and interview farmers to monitor wildlife and human behavior in this critical zone of interaction by visiting our web site.

Lisa Naughton with Ecuadorian colleagues studying the impacts of gold mining in a cloud forest park.

special environment—the *low-latitude rainforest environment* (Figure 10.6). In the rainforest, streams flow abundantly throughout most of the year, and river channels are lined along the banks with dense forest vegetation. Indigenous peoples of the rainforest once traveled the rivers in dugout canoes, although today most enjoy easier travel powered by outboard motor. Over a century ago, larger shallow-draft river craft turned the major waterways into the main arteries of trade, connecting towns and cities on the river banks. Aircraft have added a new dimension in mobility, crisscrossing the almost trackless green sea of forest to find landings in clearings or on broad reaches of the larger rivers.

In the low-latitude rainforest environment, the abundant rainfall and prevailing warm soil temperatures promote the decay and decomposition of rock to great depths, so that a thick soil layer is usually present. The soil is typically rich in oxides of iron and has a deep red color. This kind of soil has largely lost its ability to hold nutrient substances needed by plants such as the grasses and grain crops that are important in modern agriculture practiced in higher latitudes. There are, however, many kinds of native plants adapted to these soils. The most conspicuous of these plants are the great forest trees with broad leaves that comprise the majority of tree species within the evergreen rainforest. The key to the success of this kind of forest lies in the ability of the trees to quickly reuse (recycle) the essential plant nutrients that are released by the decay of fallen leaves and branches.

Low-latitude rainforests, unlike higher latitude forests, possess a great diversity of plant and animal species. The low-latitude rainforest can contain as many as 3000 different tree species in an area of only a few square kilometers, whereas midlatitude forests possess fewer than one-tenth that number. The number of types of animals found in the rainforest is also very large. A 16 km^2 (6 mi^2) area in Panama near the Canal, for example, contains about 20,000 species of insects, while all of France has only a few hundred.

Animal life of the rainforest is most abundant in the upper layers of the rainforest. Above the canopy, birds and bats are important predators and feed on insects above and

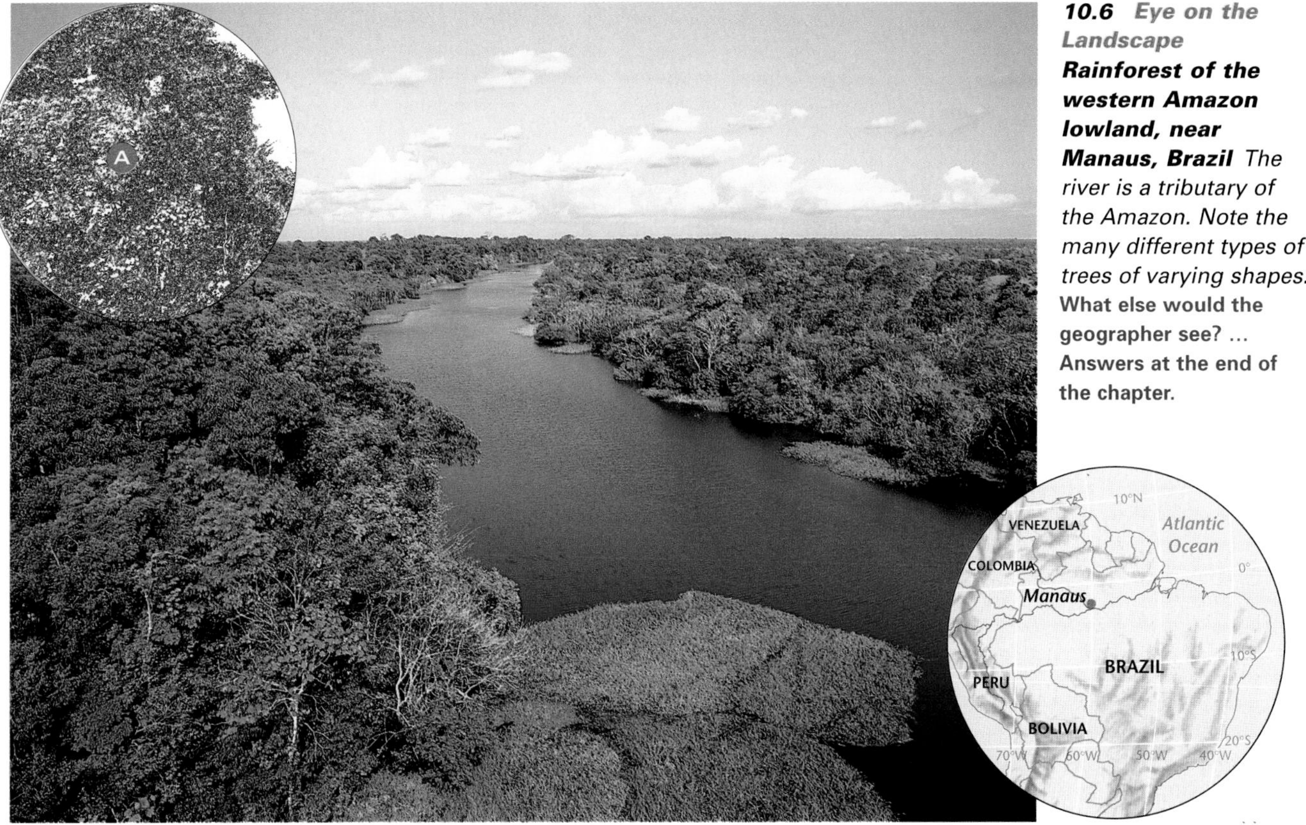

***10.6 Eye on the Landscape* Rainforest of the western Amazon lowland, near Manaus, Brazil** *The river is a tributary of the Amazon. Note the many different types of trees of varying shapes.* **What else would the geographer see? ... Answers at the end of the chapter.**

within the topmost leaf canopy. Below this level live a wide variety of birds, mammals, reptiles, and invertebrates. These animals feed on the leaves, fruit, and nectar that are abundantly available in the main part of the canopy. Ranging between the canopy and ground are small climbing animals—monkeys, for example—that forage in both layers. At the surface are the large ground animals, including herbivores—plant-eating animals—that graze the low leaves and fallen fruits, and carnivores—meat eaters—that prey on the abundant animals.

PLANT PRODUCTS AND FOOD RESOURCES OF THE RAINFOREST

Many products of the rainforest have economic value. Rainforest lumber, such as mahogany, ebony, or balsawood, is an important export. Quinine, cocaine, and other drugs come from the bark and leaves of tropical plants. Cocoa is derived from the seed kernel of the cacao plant. Natural rubber is made from the sap of the rubber tree. The tree comes from South America, where it was first exploited. Rubber trees also are widely distributed through the rainforest of Africa. Today, the principal production is from plantations in Indonesia, Malaysia, Thailand, Vietnam, and Sri Lanka (Ceylon).

In many areas of the rainforest and its fringes, agriculture is practiced by the slash-and-burn method, in which small patches of forest are cut down and then burned on the site. The wood ashes contain nutrients, and when spread over the soil, they serve as a simple fertilizer. Crops are then grown in the soil for a few years, until the nutrients are gone. The now unproductive plot is abandoned and eventually returns to rainforest. *Eye on Global Change 24.1 • Exploitation of the Low-Latitude Rainforest Environment* provides more detail on this process and discusses the effects of large-area clearing of the rainforest for mechanized agriculture and grazing.

An important class of food plants native to the wet low-latitude environment are starchy staples. Some of these are root crops, while others are fruits. Manioc, also known as cassava, has a tuberous root—something like a sweet potato—that reaches a length over 0.3 m (1 ft) and may weigh several kilograms. Yams, like maniocs, are large underground tubers. They are a major source of food in West Africa and the Caribbean. The taro plant possesses a starchy enlarged underground stem that has considerable food value. Visitors to Hawaii know taro through its transformation into poi, a fermented paste. Taro was imported into Africa from Southeast Asia and eventually reached the Caribbean region. Among the starchy fruits of the wet low-latitude environment are the banana and plantain. The banana, first cultivated for food in Southeast Asia, then spread to Africa and the Americas. The plantain is a coarse variety of the banana that is starchy, has little sugar, and requires cooking.

The plant that has perhaps been most important to humans in the low latitudes is the coconut palm. Besides being a staple food, it provides a multitude of useful products in the form of fiber and structural materials. The coconut palm flourishes on islands and coastal fringes of the wet equatorial and tropical climates (Figure 10.7). Copra, the dried meat of the coconut, is a valuable source of vegetable oil. Copra and

10.7 Coconut palms *This coastal grove of coconut palms in City of Refuge, Kona Coast, Island of Hawaii, illustrates the distinctive habit of this useful plant. Coconut palms provide many useful products, from food to fiber. (Arthur N. Strahler.)*

coconut oil are major products of Indonesia, the Philippines, and New Guinea. Palm oil and palm kernels of other palm species are important products of the equatorial zone of West Africa and the Congo River Basin.

THE WET-DRY TROPICAL CLIMATE ③ (KÖPPEN: *AW, CWA*)

In the monsoon and trade-wind coastal climate ②, we noted that the movements of the intertropical convergence zone into and away from the climate region produce a seasonal cycle of rainfall and temperature. As we move farther poleward, this cycle becomes stronger, and the monsoon and trade-wind coastal climate ② grades into the **wet-dry tropical climate** ③.

The wet-dry tropical climate ③ is distinguished by a very dry season at low Sun that alternates with a very wet season at high Sun. During the low-Sun season, when the equatorial trough is far away, dry continental tropical (cT) air masses prevail. In the high-Sun season, when the ITCZ is nearby, moist maritime tropical (mT) and maritime equatorial (mE) air masses dominate. Cooler temperatures accompany the dry season but give way to a very hot period before the rains begin.

Figure 10.8 shows the global distribution of the wet-dry tropical climate ③. It is found at latitudes of 5° to 20° N and S in Africa and the Americas, and at 10° to 30° N in Asia. In Africa and South America, the climate occupies broad bands poleward of the wet equatorial and monsoon and trade-wind coastal climates. Because these regions are farther away from the ITCZ, less rainfall is triggered by the ITCZ during

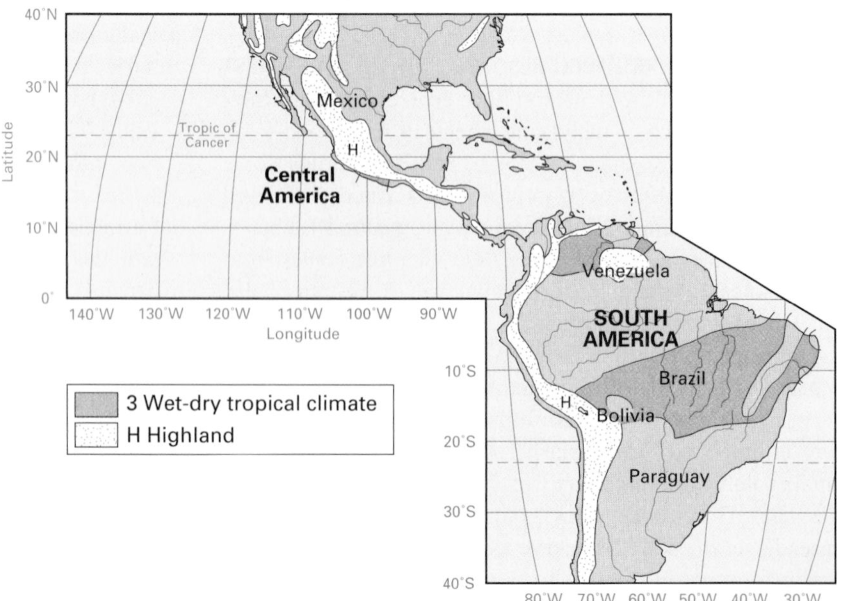

10.8 World map of the wet-dry tropical climate ③ *(Based on Goode Base Map.)*

the rainy season, and subtropical high pressure can dominate more strongly during the low-Sun season. In central India and Indochina, the regions of wet-dry tropical climate ③ are somewhat protected by mountain barriers from the warm, moist mE and mT air flows provided by trade and monsoon winds. These barriers create a rain shadow effect, so that even less rainfall occurs during the rainy season and the dry season is drier still.

Figure 10.9 is a climograph for Timbo, Guinea, at lat. 10° N in West Africa (Figure 10.8). Here the rainy season begins just after the March equinox and reaches a peak in August, about two months following the June solstice. At this time, the ITCZ has migrated to its most northerly position, and moist mE air masses flow into the region from the ocean lying to the south. Monthly rainfall then decreases as the low-Sun season arrives and the ITCZ moves to the south. Three months—December through February—are practically rainless. During this season, subtropical high pressure dominates the climate, and stable, subsiding continental tropical (cT) air pervades the region. The temperature cycle is closely linked to both the solar cycle and the precipitation pattern. In February and March, insolation increases, and air temperature rises sharply. A brief hot season occurs. As soon as the rains set in, the effect of cloud cover and evaporation of rain causes the temperature to decline. By July, temperatures have resumed an even level.

A characteristic of the tropical wet-dry climate ③ is its large year-to-year variability in precipitation. In this climate, rainfall occurs when the ITCZ migrates into the region. In some years, this migration fails and rainfall is greatly reduced. *Working It Out 10.1 • Cycles of Rainfall in the Low Latitudes* shows how interannual variation can be measured, and looks at multiyear cycles in some tropical wet-dry climate ③ stations.

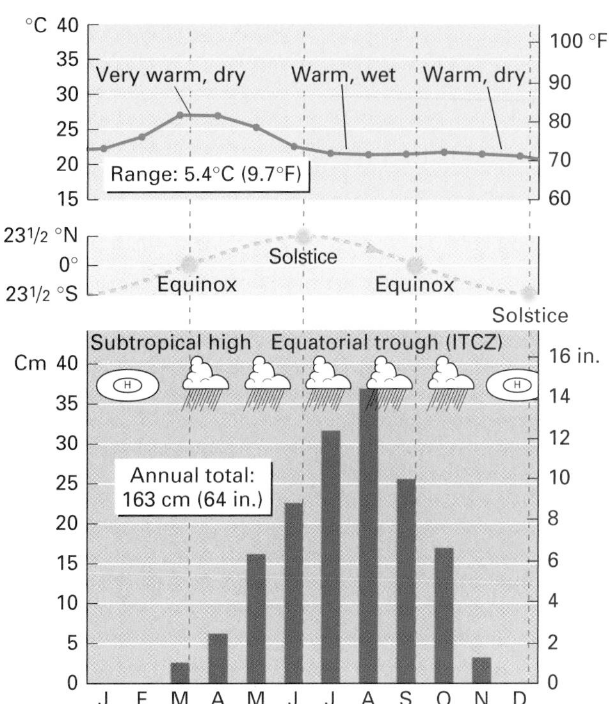

10.9 Wet-dry tropical climate ③ *Timbo, Guinea, at lat. 10° N, is located in West Africa. A long wet season at time of high Sun alternates with an almost rainless dry season at time of low Sun.*

The Savanna Environment

The wet-dry tropical climate ③ is the home of the savanna environment. The native vegetation of this climate must survive alternating seasons of very dry and very wet weather. Most plants enter a dormant phase during the dry period, then burst forth into leaf and bloom with the coming of the

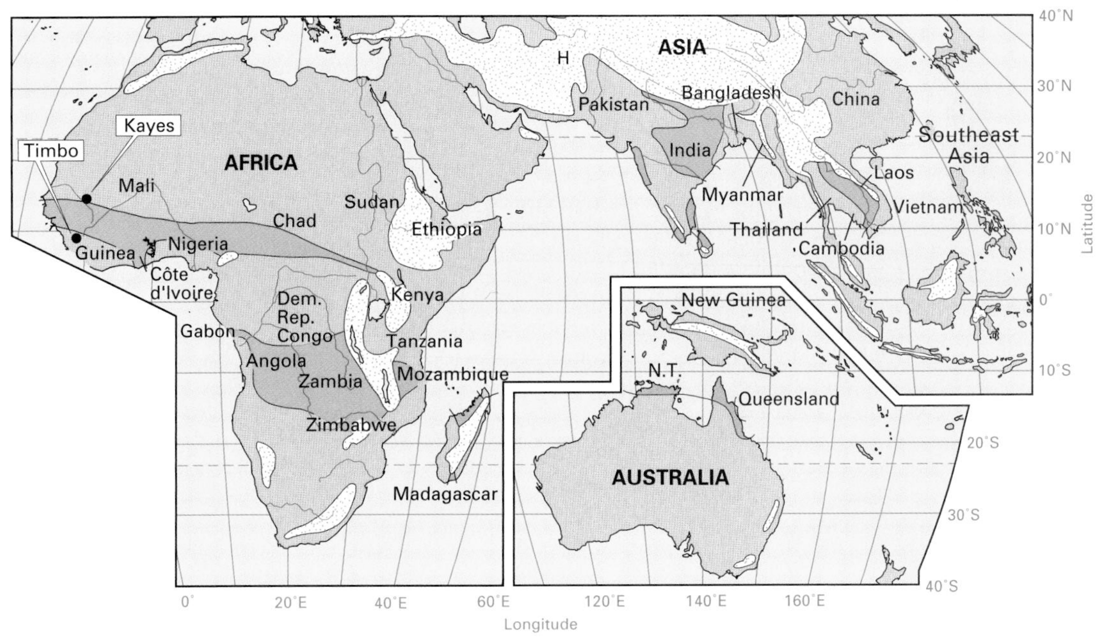

Working It Out 10.1

Cycles of Rainfall in the Low Latitudes

The cycle of monthly precipitation shown on the climographs presented in Chapters 9–11 is basically controlled by the Sun's declination cycle, which rhythmically varies the insolation received at a location according to its latitude and the time of the year. Through the mechanisms described in Chapters 6–9, this cycle produces an annual cycle of precipitation that is revealed by taking monthly averages over a long period of record. (See *Working It Out 9.2 • Averaging in Time Cycles,* Chapter 9.)

Although this cycle is a strong one at most locations, there are also other cycles that influence rainfall over a sequence of years rather than from month to month. However, they are not as strong or predictable as the annual cycle. But when we have several "wet" years followed by several "dry" years in a repeating sequence, another cycle may be at work.

Let's examine rainfall cycles that span several decades. These are shown in the three graphs to the right. Each graph presents annual rainfall totals for a sequence of 46 years at three low-latitude stations: (*a*) Padang, Sumatra, in the wet equatorial climate ①; (*b*) Bombay, India, in the wet-dry tropical climate ③; and (*c*) Abbassia, near Cairo, Egypt, which is in the dry tropical climate ④. (Notice the difference in vertical scale of the Abbassia record.) These data are selected for that month which, on the average, is the rainiest month at the recording station.

The data are presented in two ways: annual bars and a superimposed continuous curved line that smooths those annual values. From a close look at the graph, it seems that many of the peak bars are followed by either two or three lower bars before reversal to a higher bar. This suggests that there is a rainfall cycle with a length, or period, of between two and three years. The smoothed curve shows this cycle rather well for Padang and Bombay, especially in the early part of the sequence.

Another type of information about a cycle is its *amplitude.* For a smooth wavelike curve, amplitude is the difference in height between a crest and the adjacent trough. That difference is expressed here in centimeters of rainfall. Using a different vocabulary that better suits the description of climate, we find that the varying amplitude of the curve is a measure of the *variability* of the precipitation. From an inspection of the amplitude of the cycles at Padang and Bombay, it appears that rainfall at Bombay has somewhat more variability.

One way to measure the variability is to examine the *average deviation* of each yearly value from the mean value. The deviation is simply the difference between a single value and the mean of all the values, taken without respect to sign. Thus, we can define the deviation as

$$D_i = | P_i - \overline{P} |$$

where D_i is the deviation of the *i*th observation, P_i; $\overline{P}$ is the mean (see *Working It Out 9.2 • Averaging in Time Cycles*); and the two vertical bars $||$ indicate absolute value. (The absolute value is simply the value taken without respect to sign. It is always positive.) The average deviation $\overline{D}$ is then

$$\overline{D} = \frac{1}{n}\sum_{i=1}^{n} D_i = \frac{1}{n}\sum_{i=1}^{n} | P_i - \overline{P} |$$

where there are *n* observations.

Calculation of the mean and mean deviation for the three sets of observations yields the following results:

Station	Mean, cm	Mean deviation, cm
Padang	51.4	12.9
Bombay	61.0	20.7
Abbassia	0.67	0.72

Looking first at the means, we find that Bombay has the largest value, with 61 cm (24 in.) of July rainfall. Padang is next, with an average of 51 cm (20 in.) of rainfall during November. Compared to these two stations, Abbassia is very dry, with an average of less than 1 centimeter of rainfall in its rainiest month, January. Considering the mean deviations of these stations, we see that Bombay has the largest year-to-year fluctuation. Padang is next, followed by Abbassia with a very small value.

Note that if we look only at the mean deviation, we would conclude that January rainfall is very constant at Abbassia. However, we can see from the graph that in many years no rainfall at all occurs at Abbassia, while in others more than 3 cm falls. For the residents of this location, the variability is great, indeed! This suggests that we need to compare the mean deviation with the mean as a measure of *relative variability.* Taking the ratio of the mean deviation to the mean, or $\overline{D} \div \overline{P}$, we have

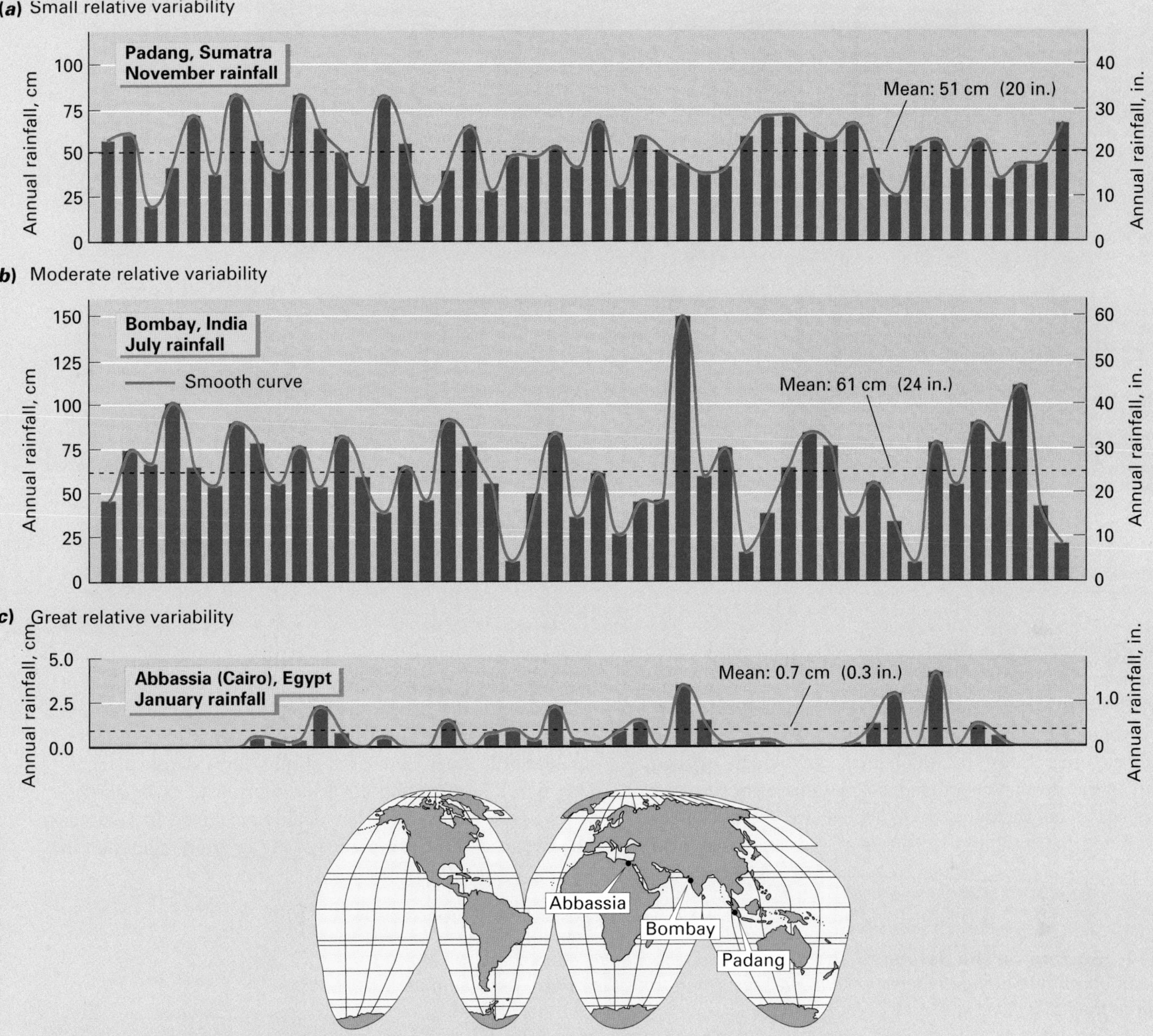

Wettest-month rainfall at three tropical wet-dry climate ③ stations *These data for Padang, Sumatra, Bombay, India, and Abbassia, Egypt, show a tendency for wettest-month rainfall to rise and fall in a two- to three-year cycle.*

Station	Relative Variability
Padang	12.9 ÷ 51.4 = 0.25
Bombay	20.7 ÷ 61.0 = 0.34
Abbassia	0.72 ÷ 0.67 = 1.07

Another way to put this is to observe that the average deviation is 25 percent of the mean for Padang, 34 percent for Bombay, and 107 percent for Abbassia. By this measure, Abbassia's January rainfall is the most variable, while Padang's is least variable.

This exercise not only presents a way to measure the variability of precipitation, but it also demonstrates the large variability in rainfall that is characteristic of monthly rainfall in many of the low-latitude climates.

10.10 Savanna woodland of the Serengeti Plains, Tanzania, East Africa *Acacia trees with flattened crowns are scattered on this grassy landscape.*

rains. For this reason, the native plant cover can be described as *rain-green vegetation.*

Rain-green vegetation consists of two basic types. First is *savanna woodland* (Figure 10.10). By *woodland* we mean an open forest in which trees are widely spaced apart. In the savanna woodland, coarse grasses occupy the open space between the trees, which are often coarse-barked and thorny. Large expanses of grassland may also be present (Figure 10.11). In the dry season, the grasses turn to straw, and many of the tree species shed their leaves to cope with the drought. The second vegetation type is found in the more arid parts of the climate region. Here, small thorny trees and large shrubs form dense patches. This type of vegetation cover is referred to as *thorntree-tall-grass savanna.* (Further details are given in Chapter 24.) Because of the prevalence of the savanna vegetation in

10.11 Grazing on the Serengeti *Vast herds of wildebeest and zebra graze the lush vegetation of the Serengeti Plains in Tanzania, still green from the rains. These animals migrate long distances in search of food and water.*

the wet-dry tropical climate ③, it can be identified as the savanna environment.

In the savanna environment, river channels that are not fed by nearby moist mountain regions are nearly or completely dry in the low-Sun dry season. In the rainy season, these river channels become filled to their banks with swiftly flowing, turbid water. The rains are not reliable, and agriculture without irrigation is hazardous at best. When the rains fail, a devastating famine can ensue. The Sahel region of Africa, discussed further in *Eye on Global Change 10.2 • Drought and Land Degradation in the African Sahel,* is a region well-known for such droughts and famines.

Soils of the savanna environment are similar in their physical characteristics and fertility to those of the rainforest environment—that is, they are largely soils of low fertility and red color. However, substantial areas of the savanna environment have fertile soils developed and sustained by the slow accumulation of windblown dust from adjacent deserts. Equally important are highly fertile soils that occur along major through-flowing rivers. Annual flooding of these rivers leaves lowland deposits of fertile silt carried down from distant mountain ranges.

ANIMAL LIFE OF THE AFRICAN SAVANNA The natural animal life of the savanna grasslands and woodlands is closely adapted to the vegetation and climate. These are the regions of the carnivorous game animals and the vast multitudes of grazing animals on which they prey (Figure 10.11). The savanna of Africa is the natural home of large herbivores, such as wildebeest, gazelle, deer, antelope, buffalo, rhinoceros, zebra, giraffe, and elephant. Their predators are the lion, leopard, hyena, wild dog, and jackal. Some of the herbivores depend on fleetness of foot to escape the predators. Others, such as the rhinoceros, buffalo, and elephant, defend themselves by their size, strength, or armor-thick hide. The giraffe is peculiarly well-adapted to savanna woodlands. Its long neck permits browsing on the higher foliage of the scattered trees.

The dry season brings a severe struggle for existence to animals of the African savanna. As streams and hollows dry up, the few muddy water holes must supply all drinking water. Danger of attack by carnivores is greatly increased.

The savanna ecosystem in Africa currently faces the prospect of widespread destruction. The primary threat is the loss of habitat that occurs as human activities claim ever-larger portions of the native savanna. Parks set aside to preserve the savanna ecosystem confine the grazing animals to a narrow range and prevent their seasonal migrations in search of food. In some instances, the growth in animal populations has been phenomenal because they have been protected from hunting. In confined preserves they rapidly consume all available vegetation in futile attempts to survive. Rapidly growing human populations are bringing increasing pressure on management agencies to allow encroachment on game preserves in order to expand cattle grazing and agriculture. Poaching on a large scale is now threatening the extinction of the elephant and rhinoceros in many areas.

AGRICULTURAL RESOURCES OF THE SAVANNA ENVIRONMENT The agricultural resources and practices of the savanna environment differ significantly between Asia and Africa. In Southeast Asia, the stronger Asiatic monsoon system is a key factor. There are also vastly larger human populations in the Asiatic monsoon nations than elsewhere in low latitudes, and these peoples have evolved their own unique culture patterns.

Of the staple foods grown widely in Southeast Asia, rice is dominant. About one-third of the human race subsists on rice, and most of this vast population is crowded into the arable lands of Southeast Asia. Intensive rice cultivation in this part of the world actually spans three climate types: monsoon and trade-wind coastal climate ②, wet-dry tropical climate ③, and moist subtropical climate ⑥. Rice requires flooding of the ground at the time the seedling plants are cultivated, and this activity has been traditionally timed to coincide with the peak of the rainy monsoon season (Figure 10.12). The crop matures and is harvested in the dry season. Sugar cane is another important crop that grows rapidly during the rainy season and is harvested in the dry season.

10.12 Rice cultivation *Much rice cultivation in the monsoon belt of Asia occurs in terraced fields such as these on the island of Bali, Indonesia. Terraces are needed because young rice plants require flooded paddies. Terraces also act to reduce soil erosion when steep slopes are brought into agriculture.*

Eye on Global Change | 10.2

Drought and Land Degradation in the African Sahel

The wet-dry tropical climate ③ is subject to years of devastating drought as well as to years of abnormally high rainfall that can result in severe floods. Climate records show that two or three successive years of abnormally low rainfall (a drought) typically alternate with several successive years of average or higher than average rainfall. This variability is a permanent feature of the wet-dry tropical climate ③, and the plants and animals inhabiting this region have adjusted to the natural variability in rainfall, with one exception: the human species.

The wet-dry tropical climate ③ of West Africa, including the adjacent semiarid southern belt of the dry tropical climate ④ to the north, provides a lesson on the human impact on a delicate ecological system. Countries of this perilous belt, called the *Sahel,* or *Sahelian zone,* are shown in the figure at the right. From 1968 through 1974, all these countries were struck by a severe drought. Both nomadic cattle herders and grain farmers

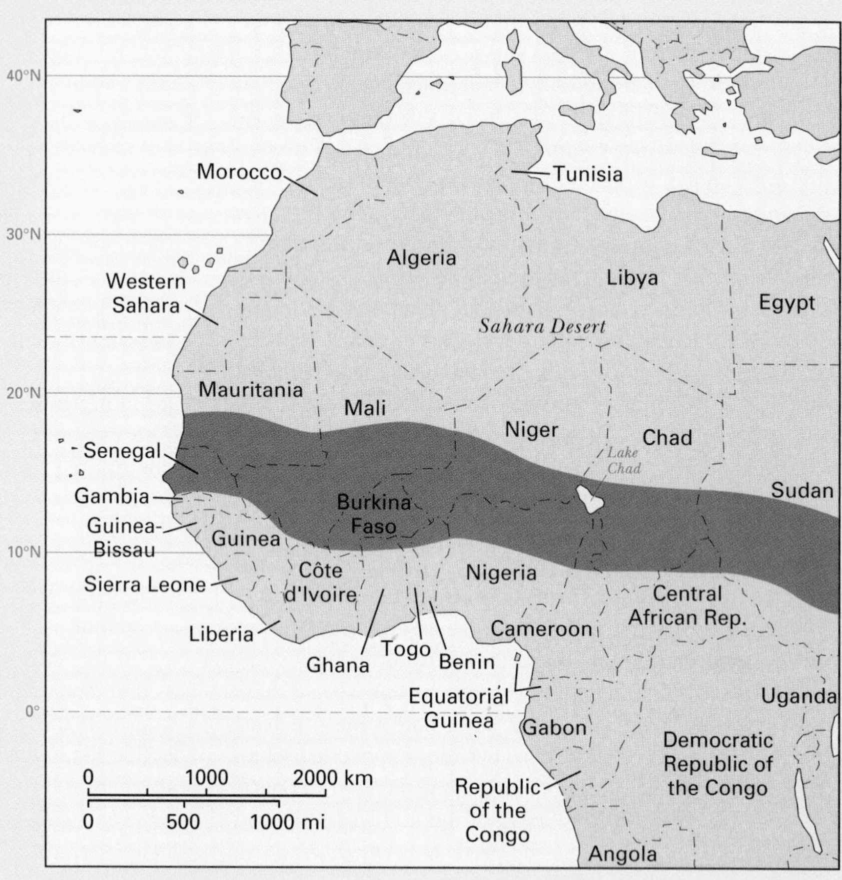

The African Sahel *The Sahel, or Sahelian zone, shown in color, lies south of the great Sahara desert of North Africa.*

Sahelian drought *At the height of the Sahelian drought, vast numbers of cattle had perished and even the goats were hard pressed to survive. Trampling of the dry ground prepared the region for devastating soil erosion in the rains that eventually ended the drought.*

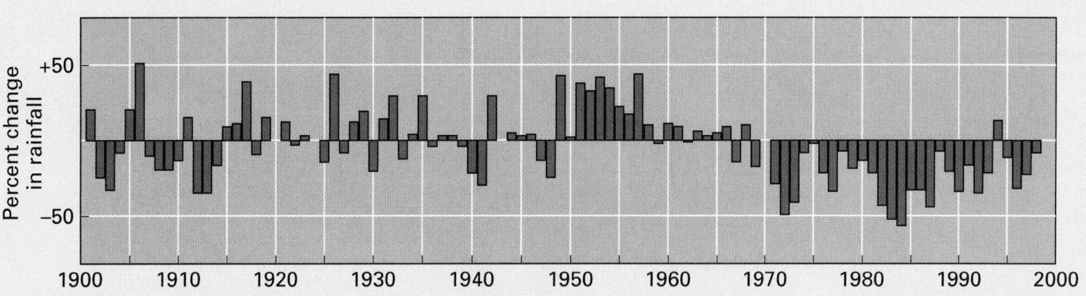

Rainfall in the Sahel *Rainfall fluctuations for stations in the western Sahel, 1901–1998, expressed as a percent departure from the long-term mean. (Courtesy of Sharon E. Nicholson, Department of Meteorology, Florida State University, Tallahassee.)*

share this zone. During the drought, grain crops failed and foraging cattle could find no food to eat. In the worst stages of the Sahel drought, nomads were forced to sell their remaining cattle. Because the cattle were their sole means of subsistence, the nomads soon starved. Some 5 million cattle perished, and it has been estimated that 100,000 people died of starvation and disease in 1973 alone.

The Sahelian drought of 1968–1974 was associated with a special phenomenon, which at that time was called *desertification*—the permanent transformation of the land surface by human activities to resemble a desert, largely through the destruction of grasses, shrubs, and trees by grazing animals and fuel wood harvesting. That term has now been abandoned in favor of *land degradation.* This degradation accelerates the effects of soil erosion, such as gullying of slopes and accumulations of sediment in stream channels. The removal of soil by wind is also intensified.

Periodic droughts throughout past decades are well documented in the Sahel, as they are in other world regions of wet-dry tropical climate ③. Shown above is a graph showing the percentage of departures from the long-term mean of each year's rainfall in the western Sahel from 1901 through 1995. Note the wide year-to-year variation. Since about 1950, the durations of periods of continuous departures both above and below the mean seem to have increased substantially. The period of sustained high-rainfall years in the 1950s contrasts sharply with a series of severe drought episodes starting in 1971. To obtain an earlier record, scientists have examined fluctuations in the level of Lake Chad. In times of drought, the lake's shoreline retreats, while in times of abundant rainfall, it expands. The changes document periods of rainfall deficiency and excess—1820–1840, below normal; 1870–1895, above normal; 1895–1920, below normal.

The sustained drought periods of the past show that droughts and wet periods are a normal phenomenon in the Sahel. In prior eras, the landscape was able to recover from the droughts during periods of abundant rainfall. Today, the pressure of increased populations of humans and cattle keeps the land degraded. As long as these populations remain high, the land degradation will be permanent.

The effect of an accelerated global warming in the coming decades on regions of wet-dry tropical climate ③ is difficult to predict. Some regions may experience greater swings between drought and surplus precipitation. At the same time, these regions may migrate poleward as a result of intensification of the Hadley cell circulation. Thus, the Sahel region may move northward into the desert zone of the Sahara. However, such changes are highly speculative. At this time, there is no scientific consensus on how climatic change will affect the Sahel or other regions of wet-dry tropical climate ③.

FOOD AND WOODLAND RESOURCES OF THE AFRICAN SAVANNA

Because the savanna environment in Africa is a transition zone between rainforest and desert environments, there is a corresponding gradation of plant resources and the ways in which they are used to support human life. In the moister zones, those of the savanna woodland with a long wet season, agriculture follows a pattern known as *bush-fallow farming.* This practice is related in some ways to the slash-and-burn agriculture of the rainforest. Trees are cut from a small area, piled up, and burned. The ash provides fertilizer for cultivated plots. After a few years, the land is allowed to revert to shrub and tree growth. Some of the same crops that are grown in rainforest agriculture are grown in wetter areas of the bordering savanna. Crops include grains, yams, soybeans, and sugar cane. Tobacco and cotton are also grown.

In the drier savanna grassland and thorntree-savanna, where the wet season is short, the main subsistence crops of

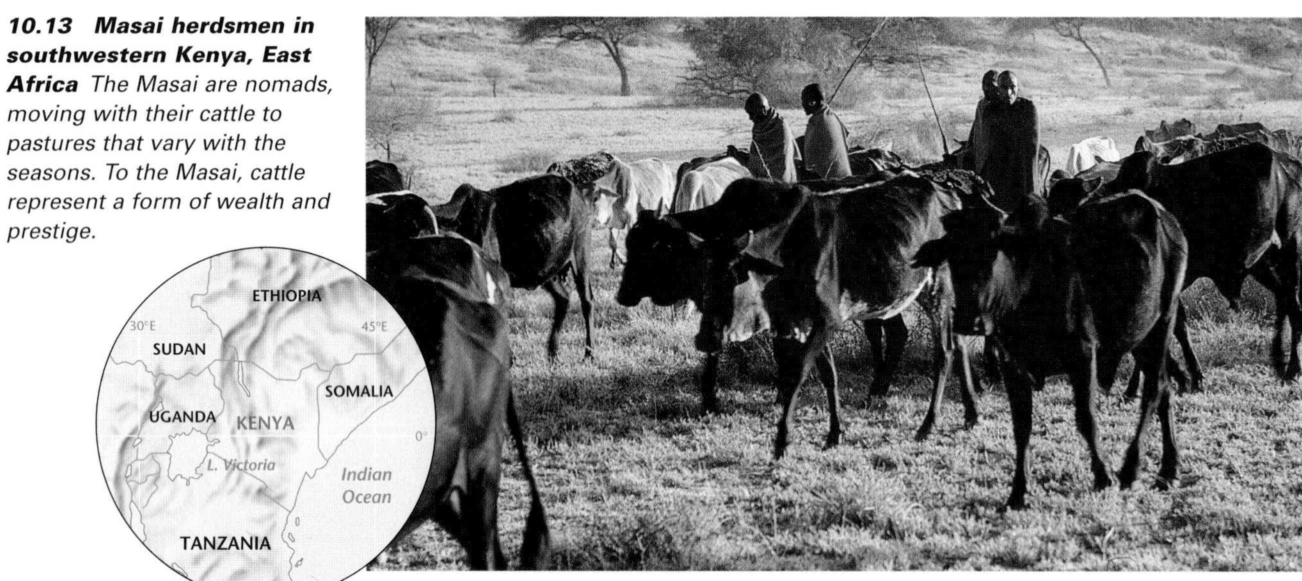

10.13 Masai herdsmen in southwestern Kenya, East Africa *The Masai are nomads, moving with their cattle to pastures that vary with the seasons. To the Masai, cattle represent a form of wealth and prestige.*

uplands are corn, millet (a kind of grain), sorghum, and peanuts. Sorghum—a tall, leafy plant somewhat resembling corn—is also an important food crop of the savanna environment. It is well-adapted to conditions of a short wet season and a long, hot dry season. The peanut is another major food crop of the savanna environment. Introduced from India, it is now grown in West Africa. Cotton, along with peanuts, is an important cash crop for export.

Besides the agricultural system of the permanent farmers, there exists a nomadic cattle culture, in which large numbers of cattle are maintained as a display of wealth (Figure 10.13). The cattle also provide food in the form of milk, butter, and blood. Following a seasonal pattern, the cattle are moved into the semidesert zone in the rainy season to graze on grasses. They are then returned to the savanna grassland zone in the dry season, where they graze on fallow croplands and rely on water holes for survival.

The savanna woodland belt of Africa provides a number of other plant resources besides cultivated food crops. Trees are cut for firewood, which is the only fuel for cooking available to most inhabitants. Trees also provide construction poles for dwellings. Among the savanna woodland trees that furnish important export products are the cashew-nut tree and the kapok tree. Gum arabic is taken from acacia trees. Most native trees of the savanna woodland have little commercial value as export lumber (an exception is black teak), but plan-

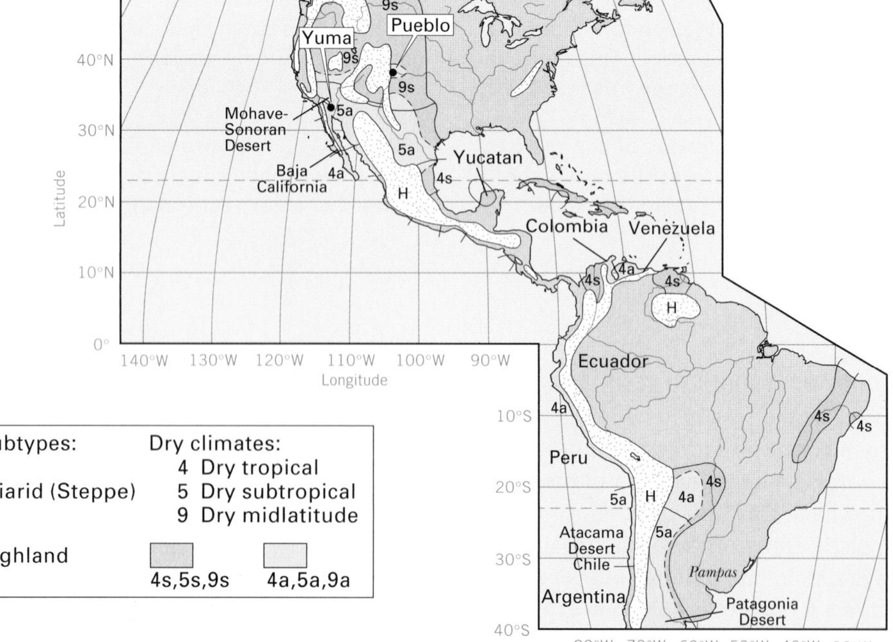

10.14 World map of the dry tropical ④, dry subtropical ⑤, and dry midlatitude ⑨ climates *The latter two climates are poleward and eastward extensions of the dry tropical climate ④ with cooler temperatures. (Based on Goode Base Map.)*

tations of exotic tree species have been introduced in some areas. Both pine and eucalyptus have been successfully introduced as sources of pulpwood. These trees grow very rapidly, as compared with pulpwood trees in high latitudes.

The Dry Tropical Climate ④ (Köppen: *BWh, BSh*)

The **dry tropical climate** ④ is found in the center and east sides of subtropical high-pressure cells. Here, air strongly subsides, warming adiabatically and inhibiting condensation. Rainfall is very rare and occurs only when unusual weather conditions move moist air into the region. Since skies are clear most of the time, the Sun heats the surface intensely, keeping air temperatures high. During the high-Sun period, heat is extreme. During the low-Sun period, temperatures are cooler. Given the dry air and lack of cloud cover, the daily temperature range is very large.

The driest areas of the dry tropical climate ④ are near the tropics of cancer and capricorn. As we travel from the tropics toward the equator, we find that somewhat more rain falls. Continuing in this direction, we encounter a short rainy season at the time of year when the ITCZ is near, and the climate grades into the wet-dry tropical climate ③.

Figure 10.14 shows the global distribution of the dry tropical climate ④. Nearly all of the dry tropical climate ④ areas lie in the latitude range 15° to 25° N and S. The largest region is the Sahara–Saudi Arabia–Iran–Thar desert belt of North Africa and southern Asia. This vast desert expanse includes some of the driest regions on Earth. Another large region of dry tropical climate ④ is the desert of central Australia. The west coast of South America, including portions of Ecuador, Peru, and Chile, also exhibits the dry tropical climate ④. However, temperatures there are moderated by a cool marine air layer that blankets the coast.

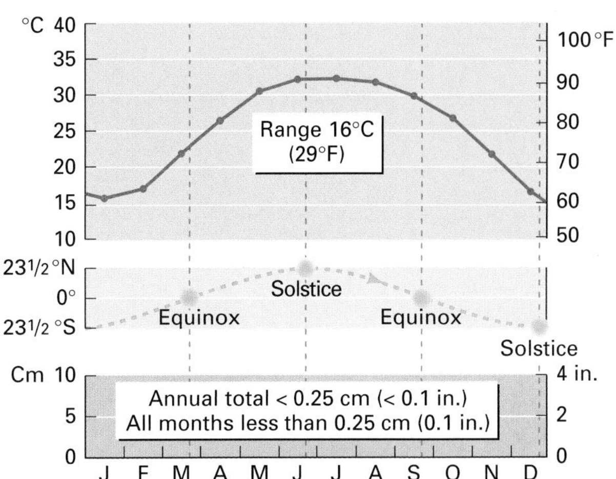

10.15 Dry tropical climate ④, dry desert *Wadi Halfa is a city on the Nile River in Sudan at lat. 22° N, close to the Egyptian border. Too little rain falls to be shown on the graph. Air temperatures are very high during the high-Sun months.*

Figure 10.15 is a climograph for a dry tropical station in the heart of the North African desert. Wadi Halfa, Sudan (Figure 10.14), lies at lat. 22° N, almost on the tropic of cancer. The temperature record shows a strong annual cycle with a very hot period at the time of high Sun, when three consecutive months average 32°C (90°F). Daytime maximum air temperatures are frequently between 43° and 48°C

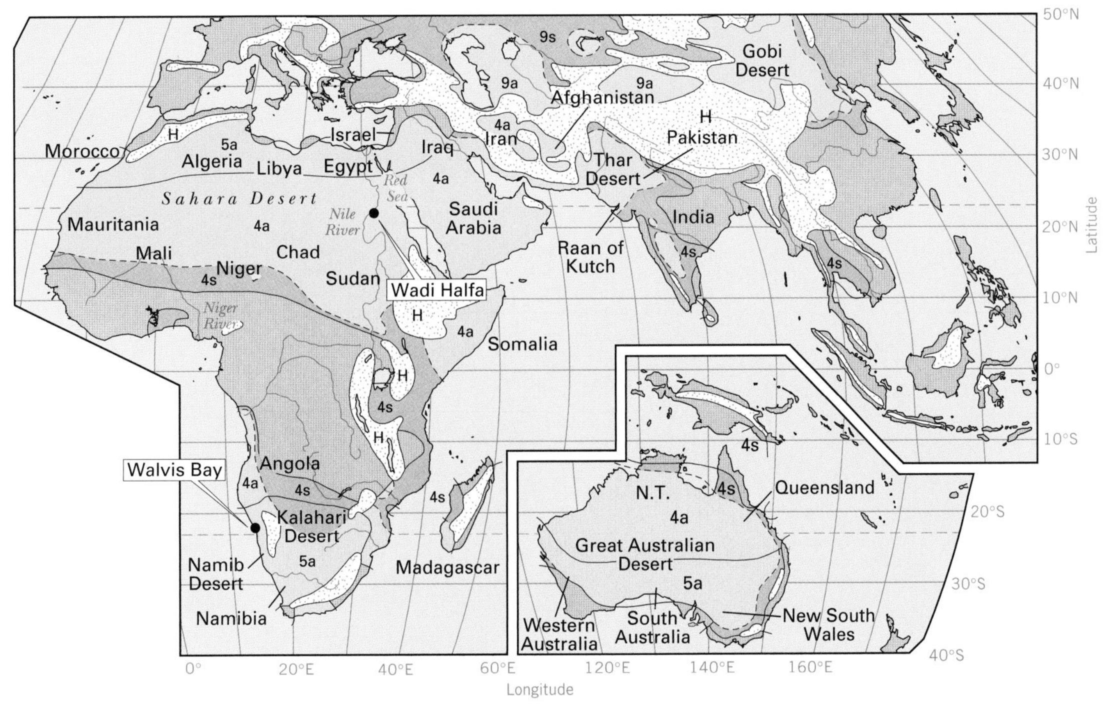

10.16 Desert Landscapes

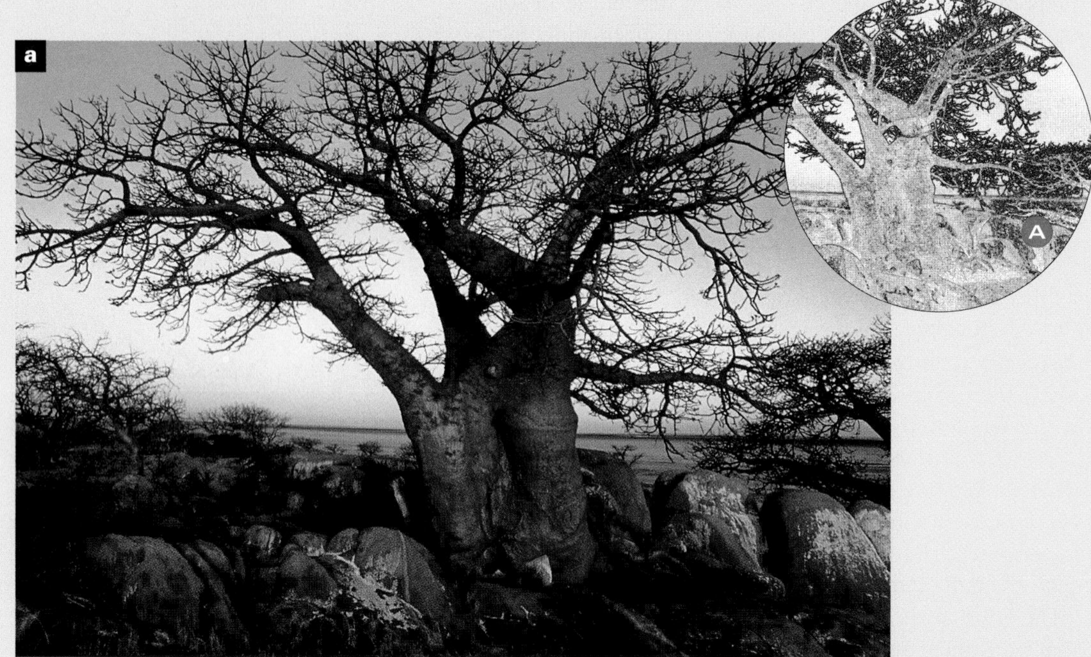

a...*Eye on the Landscape* ***The baobab*** *This strange-looking baobab tree is a common inhabitant of the thorntree semidesert of Botswana, in the Kalahari region of southern Africa.* **What else would the geographer see?...Answers at the end of the chapter.**

b...***Great Australian desert*** *Red colors dominate this desert scene from the Rainbow Valley, south of Alice Springs.*

c...***Sahara Desert*** *This sandy plain in Algeria is dotted with date palms that tap ground water near the surface.*

d...*Eye on the Landscape*
Gobi desert *Dry, wind-swept basins alternate with rock outcrops and ranges in the high Gobi desert of Mongolia.* **What else would the geographer see?...Answers at the end of the chapter.**

Sahara
Gobi
Baja
Kalahari
Australian desert

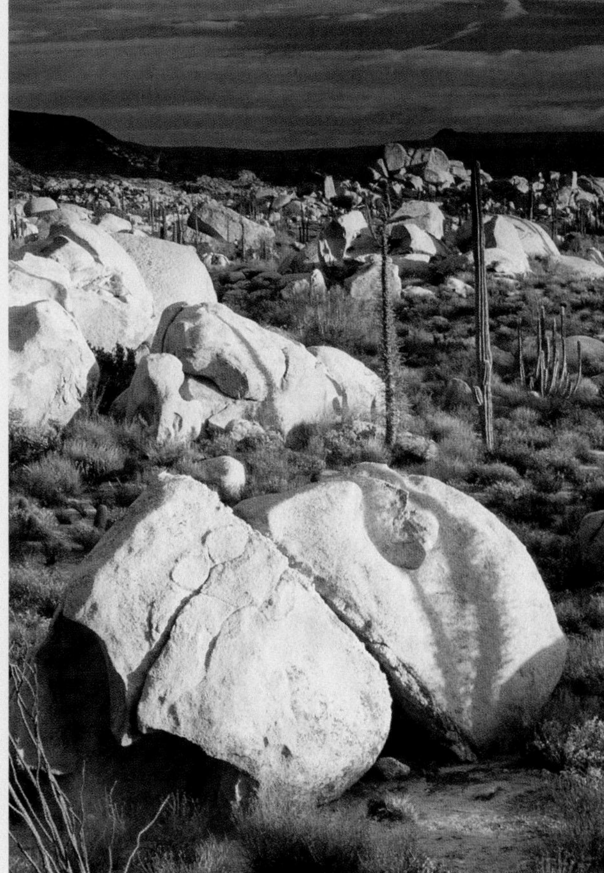

e...***Baja California desert***
Rounded granite boulders and many plants adapted to dry desert conditions are visible in this scene from the Catavina desert of the northern Baja California Pennisula, Mexico.

(about 110° to 120°F) in the warmer months. There is a comparatively cool season at the time of low Sun, but the coolest month averages a mild 16°C (61°F), and freezing temperatures are rarely recorded. No rainfall bars are shown on the climograph because precipitation averages less than 0.25 cm (0.1 in.) in all months. Over a 39-year period, the maximum rainfall recorded in a 24-hour period at Wadi Halfa was only 0.75 cm (0.3 in.).

Although the world's dry climates consist largely of extremely *arid* deserts, there are in addition broad zones at the margins of the desert that are best described as *semiarid*. These zones have a short wet season that supports the growth of grasses on which animals (both wild and domestic) graze. Geographers also call these semiarid regions **steppes.** Nomadic tribes and their herds of animals visit these areas during and after the brief moist period. In Figures 9.9 (world climate map) and 10.14 (world map of dry climates), the two subdivisions of dry climates are distinguished with the letters *a* (arid) and *s* (semiarid).

An example of the semiarid dry tropical climate ④ is that of Kayes, Mali, which we presented in Chapter 9 as a sample climograph (Figure 9.8). Located in the Sahel region of Africa, this station has a distinct rainy season that occurs when the intertropical convergence zone moves north in the high-Sun season. This precipitation pattern shows the semiarid subtype as a transition between the arid dry tropical climate ④ and the wet-dry tropical climate ③.

The Earth's desert landscapes are actually quite varied (Figure 10.16). Much of the arid desert consists of barren areas of drifting sand or sterile salt flats. However, in semiarid regions, thorny trees and shrubs are often abundant, since the climate includes a small amount of regular rainfall. Figure 10.16a shows a scene from the semiarid part of the Kalahari desert in southwest Africa (Figure 10.14), featuring the strange-looking baobab tree. Baja California also supports a sparse collection of plants especially adapted to semiarid desert conditions.

An important variation of the dry tropical climate ④ occurs in narrow coastal zones along the western edge of continents. These regions are strongly influenced by cold ocean currents and the upwelling of deep, cold water, which occurs just offshore. The cool water moderates coastal zone temperatures and reduces the seasonality of the temperature cycle. Figure 10.17 shows a climograph for Walvis Bay, a port city on the west coast of Namibia (South-West Africa), at lat. 23° S (Figure 10.14). (In the figure, the yearly cycle begins with July because this is a southern hemisphere station.) For a location nearly on the tropic of capricorn ($23^1/_2$° S), the monthly temperatures are remarkably cool, with the warmest monthly mean only 19°C (66°F) and the coolest monthly mean 14°C (57°F). This provides an annual range of only 5°C (9°F). Coastal fog is a persistent feature of this climate. Another important occurrence of this western coastal desert subtype is along the western coast of South America in Peru and Chile. Figure 10.18 shows a stretch of this coastline in northern Chile, where it is known as the Atacama desert.

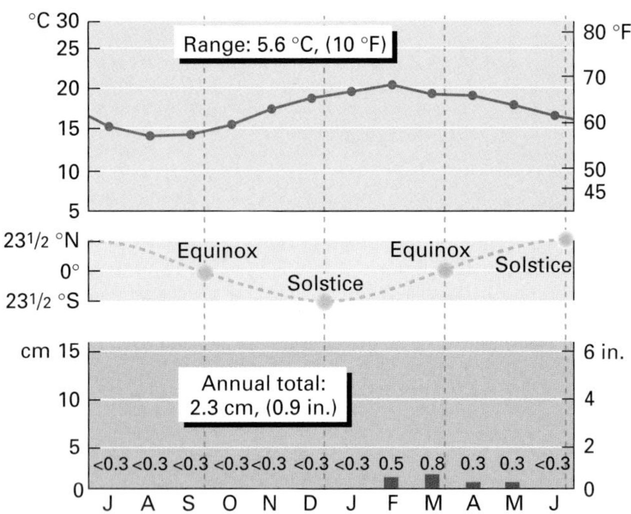

10.17 Dry tropical climate ④, western coastal desert subtype *Walvis Bay, Namibia (South-West Africa), is a desert station on the west coast of Africa at lat. 23° S. Air temperatures are cool and remarkably uniform throughout the year.*

The Tropical Desert Environment

The tropical deserts and their bordering semiarid zones comprise a global environmental region sustained by subsiding air masses of the continental high-pressure cells. Because desert rainfall is so infrequent, river channels and the beds of smaller streams are dry most of the time. However, a sudden and intense desert downpour can cause local flooding of brief duration that transports large amounts of silt, sand, gravel, and boulders. These events are termed *flash floods* (see Chapter 17). Major river channels often end in flat-floored basins having no outlet. Here clay and silt are deposited and accumulate, along with layers of soluble salts. Shallow salt lakes occupy some of these basins. The action of desert streams and the deposition of their salts by evaporation are covered in some detail in Chapters 15 and 17.

Large expanses of the very dry low-latitude desert appear to be entirely devoid of plant life except, perhaps, along the banks of some of the larger dry stream channels. Large barren areas may be covered with a layer of loosely fitted rock particles (sometimes called "desert pavement"), while loose drifting sand completely mantles other large areas. These kinds of surfaces are described in Chapter 19.

Many desert plants survive the harsh environment because they can quickly take advantage of a rare rainfall that may arrive only once in several years. These include fast-growing annuals, which survive the dry periods as dormant seeds, then sprout, grow, flower, and set seed before soil moisture is exhausted. Hard-leafed or spiny shrubs, which resist water loss by transpiration, are also common. Some desert-dwellers are succulent plants and store water in their spongy tissues—cacti are an example. Still others, like the tamarisk (or salt cedar), grow along the dry beds of major watercourses and

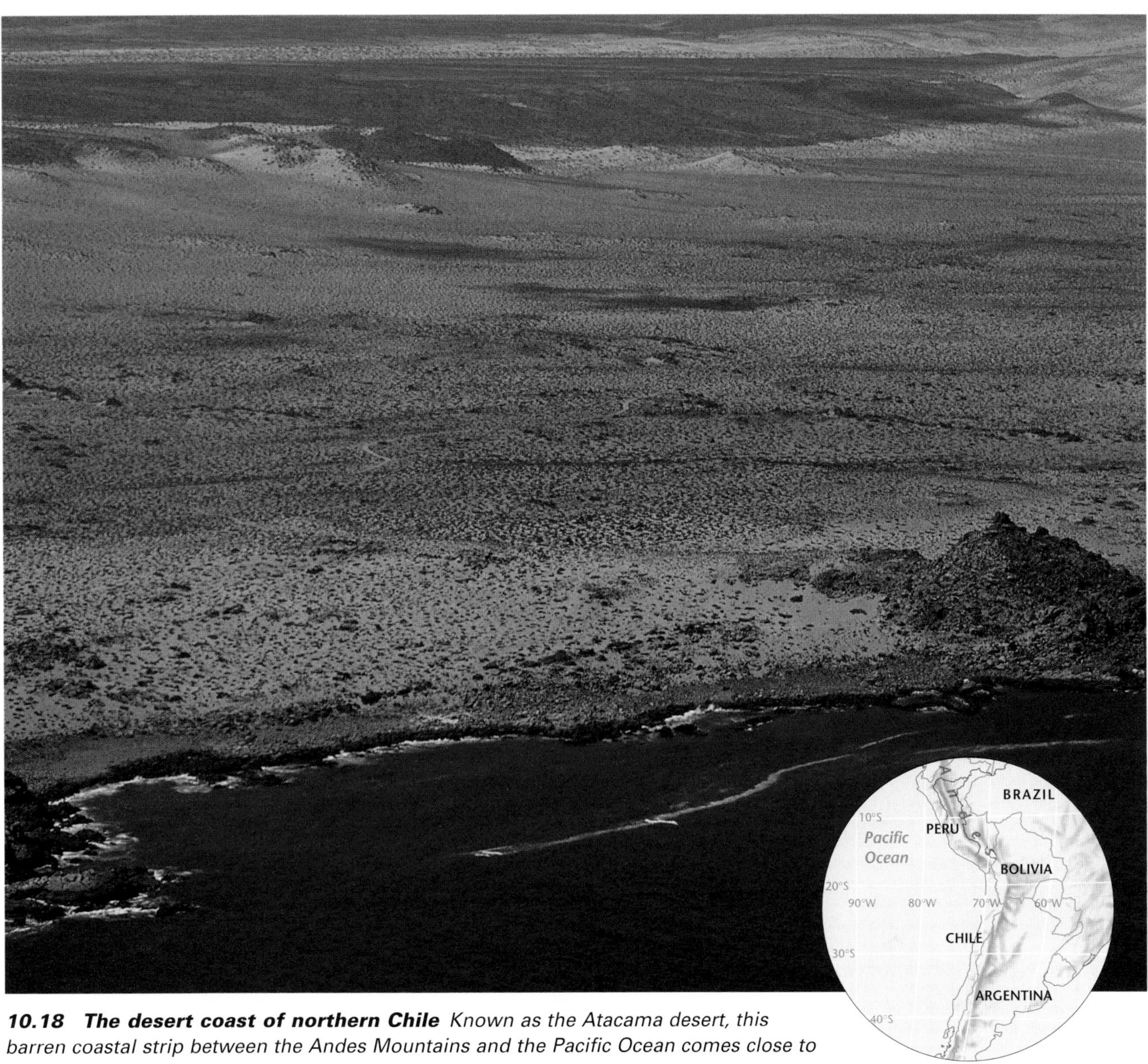

10.18 The desert coast of northern Chile *Known as the Atacama desert, this barren coastal strip between the Andes Mountains and the Pacific Ocean comes close to being a truly rainless zone.*

send their roots deep below the riverbed to reach stores of water that have infiltrated during river floods.

In various low places in the dry desert, water can be reached by digging or drilling wells that tap the ground water zone, where porous rock material is saturated with fresh water. Where such water supplies are available, they can be used to irrigate agricultural plots, thus creating an *oasis* (Figure 10.19).

Highland Climates of Low Latitudes

Highland climates, shown in white on the world climate map, are cool to cold, usually moist, climates that occupy mountains and high plateaus. Generally, the higher the location, the colder and wetter is its climate. Temperatures are lower since air temperatures in the atmosphere normally decrease with altitude (see Chapter 5). Rainfall increases because orographic precipitation tends to be induced when air masses ascend to higher elevations (Chapter 6). Highland climates are not usually included in the broad schemes of climate classification. Many small highland areas are simply not shown on a world map.

The character of the climate of a given highland area is usually closely related to that of the climate of the surrounding lowland, particularly the form of the annual temperature cycle and the times of occurrence of wet and dry seasons. An example of this effect in the tropical zone is shown by climographs for two Indian stations in close geographical proximity (Figure 10.20). New Delhi, the capital city, lies in the Ganges lowland. Simla, a mountain refuge from the hot weather, is located nearby at about 2000 m (about 6500 ft) in altitude in the foothills of the Himalayas. When the hot-season temperature averages over 32°C (90°F) in New Delhi, Simla is enjoying a pleasant 18°C (64°F), which is a full 14°C

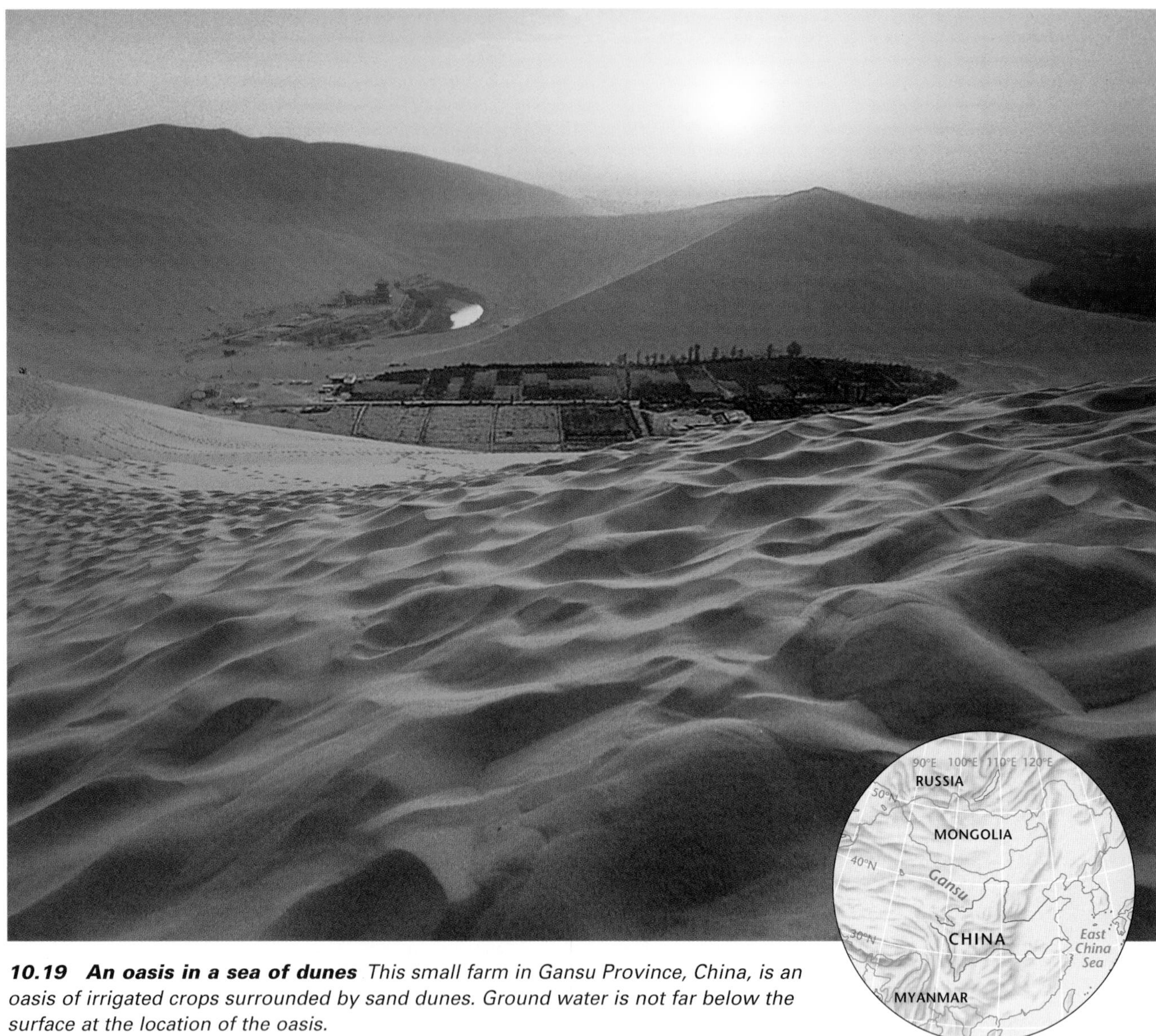

10.19 An oasis in a sea of dunes *This small farm in Gansu Province, China, is an oasis of irrigated crops surrounded by sand dunes. Ground water is not far below the surface at the location of the oasis.*

(25°F) cooler. Notice, however, that the two temperature cycles are quite similar in phase, with the minimum month being January for both. The annual rainfall cycles are also similar. New Delhi shows the typical rainfall pattern of the wet-dry tropical climate of Southeast Asia, with monsoon rains peaking in July and August. Simla has the same pattern, but the amounts are larger in every month, and the monsoon peak is very strong. Simla's annual total rainfall is well over twice that of New Delhi.

At high altitudes in tropical regions, native peoples have practiced agriculture in spite of a hostile environment that includes cold temperatures and intense solar radiation. The equatorial Andes provides an example. The high intermountain basins of the Bolivian Plateau, or Altiplano, are in the altitude range 3200 to 4300 m (about 10,500 to 14,000 ft) and are above the upper limits of the rainforest. Here, in small plots, corn, wheat, barley, and potatoes can be cultivated on a limited scale, and nearby mountain pastures provide grazing for domesticated animals (Figure 10.21).

A LOOK AHEAD Low-latitude climates are a study in contrasts, from the wettest of climates to the driest and from the hottest to the most uniform in temperature. They are dominated by two important features of atmospheric circulation—the intertropical convergence zone and the subtropical high-pressure cells. In our next chapter, we turn to the climates of the midlatitudes and high latitudes. Although this group cannot claim the wettest or hottest of climates, it includes climates that are still very wet, very dry, and very hot, at least at certain times of the year. As we will see in the next chapter, it is the polar-front boundary and the contrast of warm, moist air masses of the subtropics with colder, drier air masses of the polar regions that is the most important factor in influencing the temperature and precipitation cycles of midlatitude and high-latitude climates.

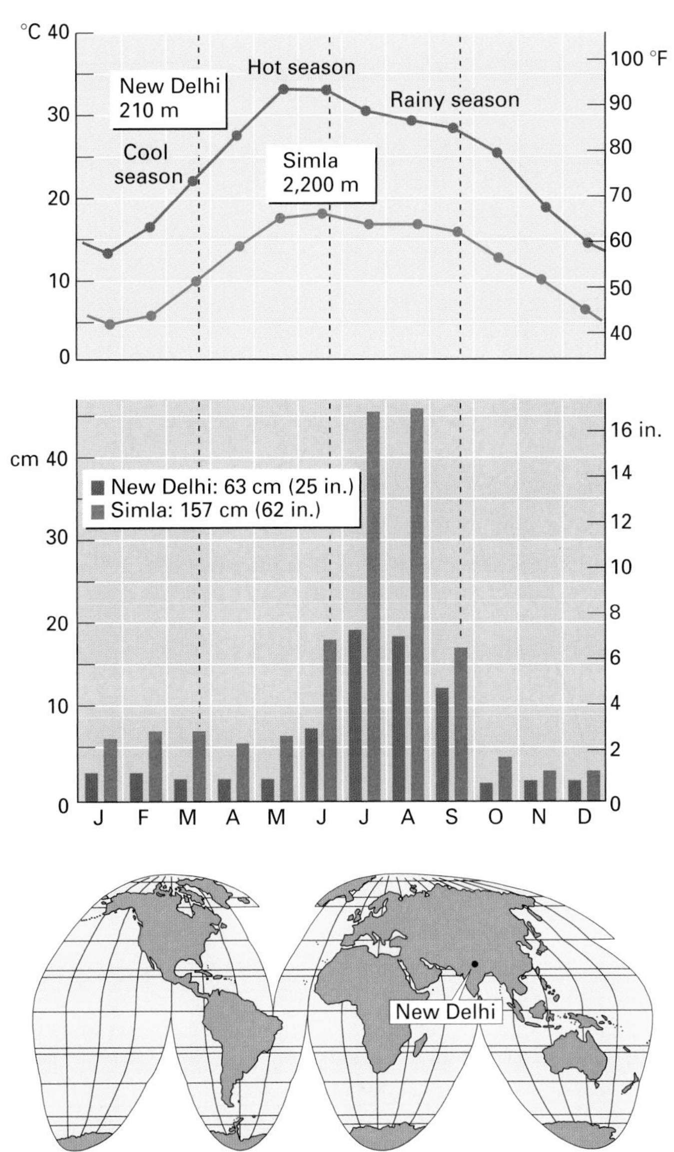

10.20 Climographs for New Delhi and Simla *Both of these stations are in northern India but at different elevations. Simla is a welcome refuge from the intense heat of the Gangetic Plain, where New Delhi is located, in May and June. But in July and August, Simla is much wetter.*

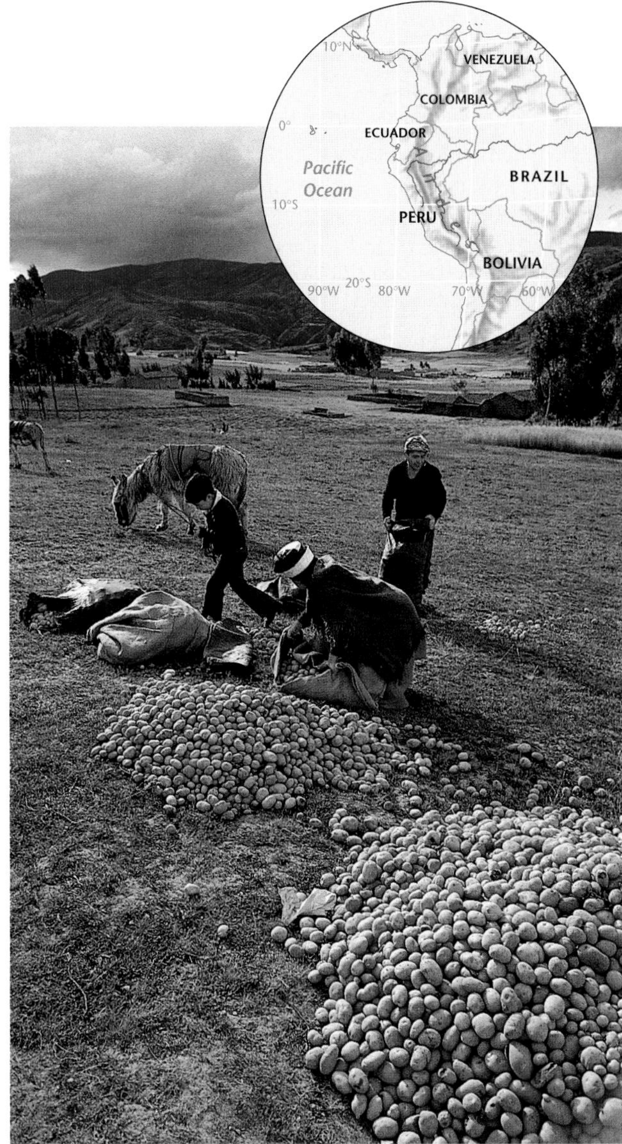

10.21 High-altitude agriculture *Potatoes are a mainstay of Peruvian agriculture in the high Andes. Here, a potato crop is being sorted.*

Chapter Summary

- *Low-latitude climates* are located largely between 30° N and S latitudes, and are controlled by the characteristics and annual movements of the ITCZ and the subtropical highs. These climates range from very moist to very dry and from quite uniform to very seasonal.
- In the **wet equatorial climate** ①, the ITCZ is always nearby, and so rainfall is abundant throughout the year. The annual temperature cycle is very weak, so that the daily range greatly exceeds the annual range.
- The **monsoon and trade-wind coastal climate** ② shows a dry period when the ITCZ has migrated toward the opposite tropic. With the return of the ITCZ, monsoon circulation and enhanced easterly waves provide increased rainfall in this climate type. Temperatures are highest in the dry weather before the onset of the wet season.
- The wet equatorial climate ① and monsoon and trade-wind coastal climate ② provide the *low-latitude rainforest environment*, which is one of tall, broadleaved evergreen trees and great species diversity. The abundant rainfall yields freely flowing streams and rivers, and also acts to wash nutrients from the soil. The rainforest provides many products of economic value, including hardwoods, pharmaceuticals, and rubber. Manioc, yams, taro, banana, and coconut are native foodstuffs.

- The **wet-dry tropical climate** ③ has a very dry period at the time of low Sun and a wet season at the time of high Sun, when the ITCZ is near. Temperatures peak strongly just before the onset of the wet season.
- Associated with this climate type is the *savanna environment*, including *savanna woodland* and *thorntree-tall-grass savanna*. The savanna is the home of the large grazing herbivores and carnivores that prey on them, but human pressure on the savanna environment is diminishing their habitat. In Asia, where the monsoon is strong, much of this climate zone is in rice cultivation. In Africa, the wetter savanna regions are in *bush-fallow farming*, yielding subsistence food crops and some exports. Cattle, raised for food, graze savanna grasses and fallow cropland.
- Extreme year-to-year variability in rainfall in the tropical wet-dry climate ③ provides a constant threat of extreme drought, especially in the Sahel region of western and north-central Africa. The drought not only yields famine and disease, but also produces land degradation brought about by overgrazing and fuel wood harvesting.
- The **dry tropical climate** ④ is dominated by subtropical high pressure and is often nearly rainless. Temperatures are very high during the high-Sun season. While the *semiarid* subtype of this climate has a short wet season and supports **steppes** of sparse grasses, the *arid* subtype provides the tropical desert environment. In the desert, streams flow only after rare heavy rain showers, producing *flash floods*.
- Plants have adapted many strategies to survive the harsh environment of the dry tropical climate ④, ranging from an annual habit to deep roots that tap groundwater. Human habitation is only possible where water is available.
- **Highland climates** of low latitudes normally show a similar seasonality to lowland climates of nearby regions, but are cooler and wetter. Even on high plateaus well above the range of rainforest, agriculture is practiced.

Key Terms

wet equatorial climate ①
monsoon and trade-wind coastal climate ②
wet-dry tropical climate ③
dry tropical climate ④
steppes
highland climates

Review Questions

1. Why is the annual temperature cycle of the wet equatorial climate ① so uniform?
2. What are the main features of the low-latitude rainforest environment? Name some typical plant products of the rainforest. Which plants provide staple foods for inhabitants of the moist low-latitude environments?
3. The wet-dry tropical climate ③ has two distinct seasons. What factors produce the dry season? the wet season?
4. Describe the two main types of natural vegetation that occur in the savanna environment. Identify some typical animals of the African savanna.
5. How do native peoples of the African savanna utilize their environment?
6. Why are the dry tropical ④, dry subtropical ⑤, and dry midlatitude ⑨ climates dry? How do they differ in temperature and precipitation cycles?
7. How do the arid (a) and semiarid (s) subtypes of the dry climates differ?
8. Identify the key features of the tropical desert environment.
9. How do low-latitude highland climates differ from their counterparts at low elevation?

Eye on Global Change 10.2 • Drought and Land Degradation in the African Sahel

1. What is meant by the term *land degradation?* Provide an example of a region in which land degradation has occurred.
2. Examine the Sahel rainfall graph carefully. Compare the pattern of rainfall fluctuations for the periods 1901–1950 and 1951–present. How do they differ?

Visualizing Exercise

1. Sketch the temperature and rainfall cycles for a typical station in the monsoon and trade-wind coastal climate ②. What factors contribute to the seasonality of the two cycles?

Essay Questions

1. The intertropical convergence zone (ITCZ) moves north and south with the seasons. Describe how this movement affects the four low-latitude climates.
2. Compare and contrast the low-latitude environments of Africa.

Problems

Working It Out 10.1 • Cycles of Rainfall in the Low Latitudes

1. The data below are measurements of total precipitation for November for the years 1985–1994 at San Juan, Puerto Rico. (November is the wettest month during this 10-year period.) Find the mean deviation of these data.

Year	Precipitation, cm
1985	11.5
1986	14.9
1987	19.0
1988	14.4
1989	12.6
1990	13.5
1991	15.6
1992	30.4
1993	11.0
1994	21.1

2. Calculate the relative variability by taking the ratio of the mean deviation to the mean. How does the result compare to those of the wettest months of Padang, Bombay, and Abbassia?

Eye on the Landscape

Chapter Opener Tropical agriculture, Madagascar The steep slopes of this volcanic terrain are being rapidly eroded by huge gullies **(A)**. Slash-and-burn agriculture in preexisting forests, followed by wall-to-wall logging, removed the vegetation cover that was essential to stabilize the soils. Now partitioned into small fields, the volcanic slopes are still wasting away rapidly, shown by the sharp edges of the gully to the right of the upper **(A)** label. Adjacent volcanic domes **(B)** have been stripped of vegetation cover, but are not yet laid out for agriculture. Gullying is a type of erosion we describe in Chapter 17, while volcanic domes are covered in Chapters 14 and 18.

10.6 Amazon rainforest The rainforest is a very diverse and complex environment **(A)**. In this aerial view, you can distinguish many different types of trees based on leaf color and texture, height, and branching pattern. Note that some very large tree crowns project above the others—these are *emergents,* which are characteristic of mature rainforest and play a special role in the ecology of the rainforest. See Chapter 24, Global Econsystems for more information.

10.16a The baobab Note the rounded rocks around the base of the baobab **(A)**. In desert climates, chemical weathering breaks apart some types of crystalline rocks grain by grain, producing the rounded forms such as these. See Chapter 15, granular disintegration and chemical weathering.

10.16d Gobi desert This scene gives the impression of ancient weathered rocks **(A)** emerging from a sea of rock fragments **(B)**. The gently sloping surfaces are fans built of weathered rock material that has been carried away from the mountains in runoff from infrequent, but severe, rainstorms (see Chapter 17, Alluvial Fans). In some cases, the sloping surface may be underlain by eroded rock with a thin veneer of gravel (Chapter 17, pediments). The surfaces are smoothed by wind action, which removes fine material and leaves coarse sand and gravel mantling the surface (see Chapter 19, desert pavement).

11 Midlatitude and High Latitude Climates

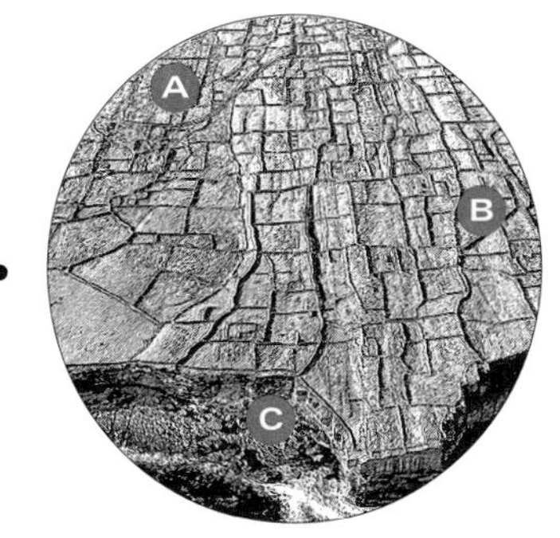

Eye on the Landscape

Cliffs of Inishmore, Aran Islands, Ireland. The abundant precipitation of the marine west-coast ⑧ has nourished agriculture here for centuries. The fields are surrounded by stone walls that protect the plots from wind erosion.

What else would the geographer see?...Answers at the end of the chapter.

This chapter continues our examination of the global climates, focusing on midlatitude and high-latitude climates. As in Chapter 10, we will find that the temperature and precipitation cycles of the individual climates can be easily explained by the effects of latitude, position on the land mass, and movements of air masses and fronts.

Overview of the Midlatitude Climates

The *midlatitude climates* almost fully occupy the land areas of the midlatitude zone and a large proportion of the subtropical latitude zone. Along the western fringe of Europe, they extend into the subarctic latitude zone as well, reaching to the 60th parallel. Unlike the low-latitude climates, which are about equally distributed between northern and southern hemispheres, nearly all of the midlatitude climate area is in the northern hemisphere. In the southern hemisphere, the land area poleward of the 40th parallel is so small that the climates are dominated by a great southern ocean and do not develop the continental characteristics of their counterparts in the northern hemisphere.

The midlatitude climates of the northern hemisphere lie in a broad zone of intense interaction between two groups of very unlike air masses (Figure 9.7). From the subtropical zone, tongues of maritime tropical (mT) air masses enter the midlatitude zone. There, they meet and conflict with tongues of maritime polar (mP) and continental polar (cP) air masses along the discontinuous and shifting polar-front zone.

In terms of prevailing pressure and wind systems, the midlatitude climates include the poleward halves of the

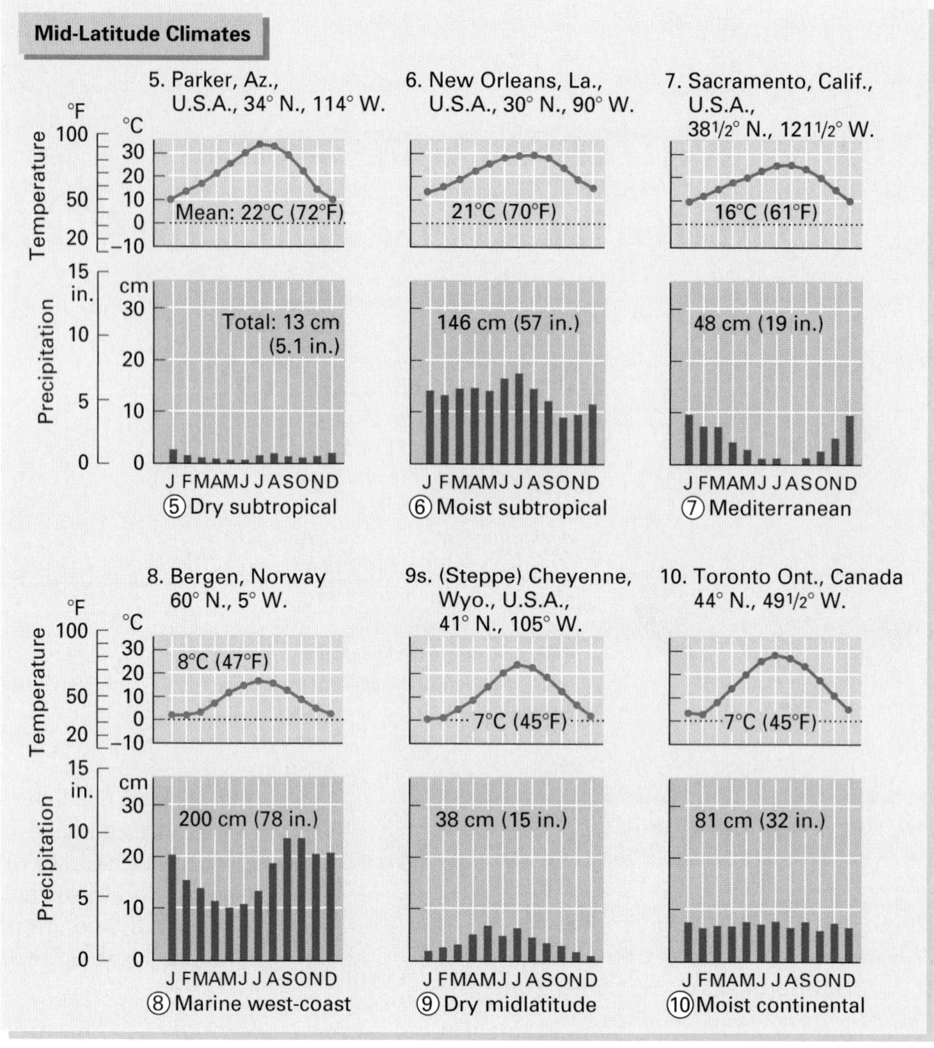

11.1 Climographs for the six midlatitude climates

great subtropical high-pressure systems and much of the belt of prevailing westerly winds (see Figure 7.15). As a result, weather systems, such as traveling cyclones and their fronts, characteristically move from west to east. This dominant global eastward airflow influences the distribution of climates from west to east across the North American and Eurasian continents.

The influence of traveling cyclones and the interaction between warm, moist air masses and cooler, drier air masses in the midlatitude climate zone makes for climates that are quite variable from day to day and month to month. *Working It Out 11.1 • Standard Deviation and Coefficient of Variation* uses monthly precipitation from stations in two different midlatitude climates to show how variability is measured quantitatively.

The six midlatitude climate types range from two that are very dry to three that are extremely moist (Figure 11.1). The midlatitude climates span the range from those with strong wet and dry seasons—the Mediterranean climate ⑦—to those with precipitation that is more or less uniformly distributed through the year. Temperature cycles are also quite varied. Low annual ranges are seen along the windward west coasts. In contrast, annual ranges are large in the continental interiors. Table 11.1 summarizes the important features of these climates.

The Dry Subtropical Climate ⑤ (Köppen: *BWh, BWk, BSh, BSk*)

The **dry subtropical climate** ⑤ is simply a poleward extension of the dry tropical climate ④, caused by somewhat similar air mass patterns. A point of difference is in the annual temperature range, which is greater for the dry subtropical climate ⑤. The lower latitude portions have a distinct cool season, and the higher latitude portions, a cold season. The cold season, which occurs at a time of low Sun, is due in part to invasions of cold continental polar (cP) air masses from higher latitudes. Precipitation occurring in the low-Sun season is produced by midlatitude cyclones that occasionally move into the subtropical zone. As in the dry tropical climate ④, both arid and semiarid subtypes are recognized.

Table 11.1 | **Midlatitude Climates**

Climate	Temperature	Precipitation	Explanation
Dry subtropical ⑤	Distinct cool or cold season at low-Sun period.	Precipitation is low in nearly all months.	This climate lies poleward of the subtropical high-pressure cells and is dominated by dry cT air most of the year. Rainfall occurs when moist mT air reaches the region, either in summer monsoon flows or in winter frontal movements.
Moist subtropical ⑥	Temperatures show strong annual cycle, but with no winter month below freezing.	Abundant rainfall, cyclonic in winter and convectional in summer. Humidity generally high.	The flow of mT air from the west sides of subtropical high-pressure cells provides moist air most of the year. cP air may reach this region during the winter.
Mediterranean ⑦	Temperature range is moderate, with warm to hot summers and mild winters.	Unusual pattern of wet winter and dry summer. Overall, drier when nearer to subtropical high pressure.	The poleward migration of subtropical high-pressure cells moves clear, stable cT air into this region in the summer. In winter, cyclonic storms and polar frontal precipitation reach the area.
Marine west-coast ⑧	Temperature cycle is moderated by marine influence.	Abundant precipitation but with a winter maximum.	Moist mP air, moving inland from the ocean to the west, dominates this climate most of the year. In the summer, subtropical high pressure reaches these regions, reducing precipitation.
Dry midlatitude ⑨	Strong temperature cycle with large annual range. Summers warm to hot, winters cold to very cold.	Precipitation is low in all months but usually shows a summer maximum.	This climate is dry because of its interior location, far from mP source regions. In winter, cP dominates. In summer, a local dry continental air mass develops.
Moist continental ⑩	Summers warm, winters cold with three months below freezing. Very large annual temperature range.	Ample precipitation, with a summer maximum.	This climate lies in the polar-front zone. In winter, cP air dominates, while mT invades frequently in summer. Precipitation is abundant, cyclonic in winter and convectional in summer.

Figure 10.14 shows the global distribution of the dry subtropical climate ⑤. A broad band of this climate type is found in North Africa, connecting with the Near East. Southern Africa and southern Australia also contain regions of dry subtropical climate ⑤ poleward of the dry tropical climate ④. In South America, a band of dry subtropical climate ⑤ occupies Patagonia, a region east of the Andes in Argentina. In North America, the Mojave and Sonoran deserts of the American Southwest and northwest Mexico are of the dry subtropical type.

Figure 11.2 is a climograph for Yuma, Arizona, a city within the arid subtype of the dry subtropical climate ⑤, close to the Mexican border at lat. 33° N. The pattern of monthly temperatures shows a strong seasonal cycle with a dry, hot summer. A cold season brings monthly means as low as 13°C (55°F). Freezing temperatures—0°C (32°F) and below—can be expected at night in December and January. The annual range is 20°C (36°F). Precipitation, which totals about 8 cm (3 in.), is small in all months but has peaks in late winter and late summer. The August maximum is caused by the invasion of maritime tropical (mT) air masses, which bring thunderstorms to the region. Higher rainfalls from December through March are produced by midlatitude wave cyclones following a southerly path. Two months, May and June, are nearly rainless.

Working It Out 11.1

Standard Deviation and Coefficient of Variation

Although the mean, or average, of such variables as monthly precipitation or temperature is an important determiner of climate, the variability from one year to the next is also quite important. In *Working It Out 10.1 • Cycles of Rainfall in the Low Latitudes* (Chapter 10), we introduced the mean deviation as a useful measure of variation around a mean. We also provided the ratio of the mean deviation to the mean as a measure of relative variability.

A more commonly used measure of variation is the *sample standard deviation,* which is defined for a sample of *n* precipitation values P_i, i = 1, ..., *n,* as

$$s_P = \sqrt{\frac{1}{n-1}\sum_{i=1}^{n} D_i^2}$$

where s_P is the sample standard deviation of *P* and D_i is the deviation of the *i*th observation from the sample mean, which was defined in Chapter 9 as $D_i = P_i - \overline{P}$. ($\overline{P}$ is the mean of the sample.) To find the sample standard deviation, simply take each deviation and square it. Note that any deviations that are negative will become positive after squaring. Thus, there is no need to take the absolute value of the deviation, as we did in *Working It Out 9.2 • Averaging in Time Cycles.* Next, form the sum of the squared deviations, and divide that sum by $n - 1$—that is, one less than the number of samples. This is a bit like taking the average of the squared values but dividing by $n - 1$ instead of *n* for reasons of statistical theory. Last, take the square root of the result.

As an example of the sample standard deviation, examine the set of graphs shown to the right. Graphs (*a*) and (*b*) present monthly precipitation for two stations for a 10-year period from 1985 to 1994. San Jose, California, in the Mediterranean climate ⑦, has dry summers and wet winters and a low overall annual average precipitation of 36 cm (14 in.). Many months in the 10-year record are completely dry. Baton Rouge, Louisiana, in the moist subtropical climate ⑥, is a much wetter location, with an average annual precipitation of 173 cm (68 in.).

What about the variation in precipitation experienced at these stations? When comparing the two graphs visually, the Baton Rouge graph gives the impression of greater variation, largely because many months are much wetter than those of San Jose. Graph (*c*) compares the monthly standard deviations for each sample. In all months, the value is larger for Baton Rouge than for San Jose, in keeping with the impression gained from comparing graphs (*a*) and (*b*). The standard deviation for the sample of annual totals, shown as the last pair of bars, is larger than the monthly values for both locations since the annual total itself is always larger than those of individual months.

However, the standard deviation does not measure the relative variation—that is, the variation with respect to the mean. Graph (*d*) shows the *coefficient of variation* for each location by month. The coefficient of variation for a sample is defined as the ratio of the standard deviation to the mean. That is,

$$CV_P = \frac{s_P}{\overline{P}}$$

where CV_P is the coefficient of variation for precipitation. The graph shows clearly that the relative variation of monthly rainfall at San Jose is larger than that of Baton Rouge in all months. Note that the coefficient of variation is highest in San Jose in the months of September and May, which are just before and just after the wet season. These are normally dry months, but in some years substantial rainfall occurs. Note also that the coefficients of variation for the annual totals (rightmost bars in graph *d*) are smaller than those of the monthly averages, showing that the relative variation of the total is less. This stands to reason because a deficiency in one month can be made up in the next.

The mean, standard deviation, and coefficient of variation of a sample are values that give basic information about the sample and its variability. They are referred to as *sample statistics* and are widely used in many branches of science.

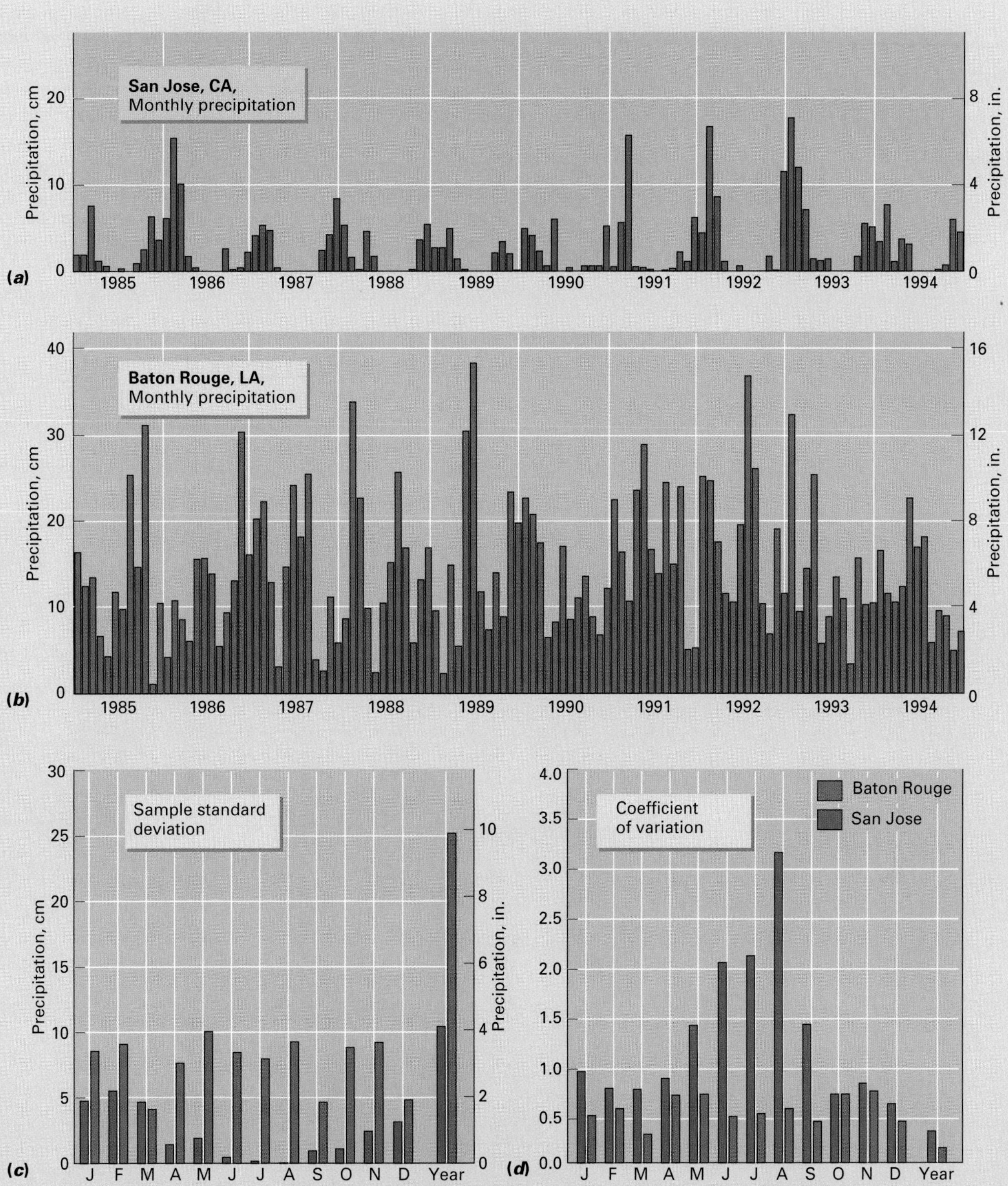

Precipitation data and statistics for San Jose, California, and Baton Rouge, Louisiana.
(Data of NOAA/NCDC.)

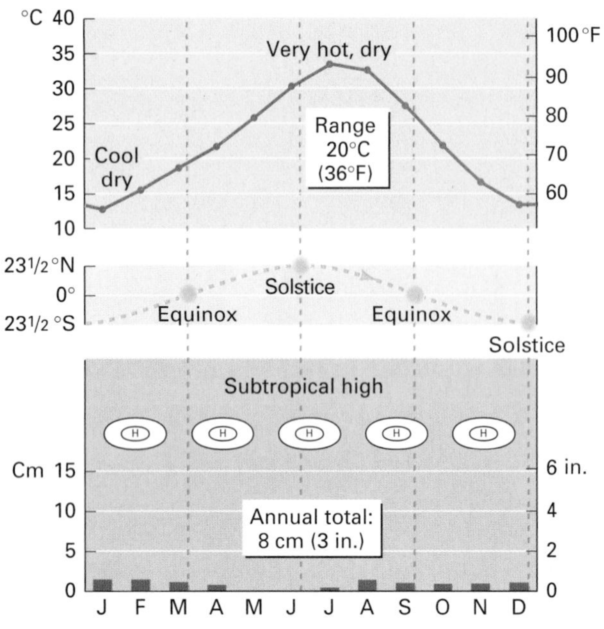

11.2 Dry subtropical climate ⑤ *Yuma, Arizona, lat. 33° N, has a strong seasonal temperature cycle. Compare with Wadi Halfa (Figure 10.15).*

The Subtropical Desert Environment

The environment of the dry subtropical climate ⑤ is similar to that of the dry tropical climate ④ in that both are very dry. The boundary between these two climate types is gradational. But if we were to travel northward in the subtropical climate zone of North America, arriving at about 34° N in the interior Mojave desert of southeastern California, we would encounter environmental features significantly different from those of the low-latitude deserts of tropical Africa, Arabia, and northern Australia. Although the great summer heat of the low-elevation regions of the Mojave desert is comparable to that experienced in the Sahara desert, the low Sun brings a winter season that is not found in the tropical deserts. Here, cyclonic precipitation can occur in most months, including the cool low-Sun months.

In the Mojave desert and adjacent Sonoran desert, plants are often large and numerous, in some places giving the appearance of an open woodland. One example is the occurrence of forest-like stands of the tall, cylindrical saguaro cactus. Another is the woodland of Joshua trees found in higher parts of the Mojave desert (Figure 11.3). Other distinctive plants include the prickly pear cactus, the ocotillo plant, and the creosote bush.

11.3 A Mojave desert landscape *The strange-looking Joshua tree, shown in the foreground, is abundant at higher elevations in the Mojave desert.*

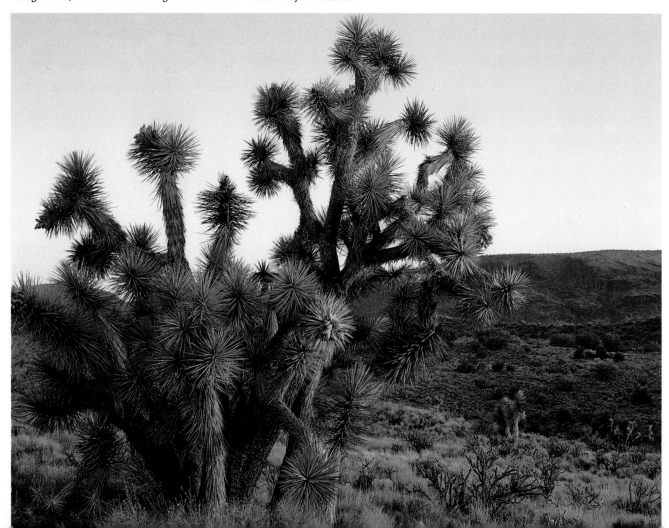

Both plants and animals of deserts are adapted to the dry environment. As in the dry tropical environment, many annual plants remain dormant as seeds during long dry periods, then spring to life, flower, and bloom very quickly when rain falls. Certain invertebrate animals adopt the same life pattern. For example, the tiny brine shrimp of the North American desert may wait many years in dormancy until normally dry lakebeds fill with water, an event that occurs perhaps three or four times per century. The shrimp then emerge and complete their life cycles before the lake evaporates.

The human species adapts to the American subtropical desert environment by the application of technology. Irrigation projects import water in abundance, and growers apply it to croplands, where much of it is lost by intense evapotranspiration. Pipelines and highways provide fuels and building materials, machinery, and home appliances (the air conditioner). Transmission lines bring electricity from distant dams and coal-fired power plants. These human technological adaptations have run into serious environmental difficulties, one of which we will present in depth in Chapter 21. It is the deterioration of croplands by deposition of salts left by evaporation of irrigation waters and the accompanying waterlogging of the soil.

The Moist Subtropical Climate ⑥ (Köppen: *Cfa*)

Recall that circulation around the subtropical high-pressure cells provides a flow of warm, moist air onto the eastern side of continents (see Figure 7.18). This flow of maritime tropical (mT) air dominates the **moist subtropical climate** ⑥. Summer in this climate sees abundant rainfall, much of it convectional. Occasional tropical cyclones further enhance summer precipitation. In Southeast Asia, this climate is characterized by a strong monsoon effect, with summer rainfall much increased above winter rainfall. Summer temperatures are warm, with persistent high humidity.

Winter precipitation in the moist subtropical climate ⑥ is also plentiful, produced in midlatitude cyclones. Invasions of continental polar (cP) air masses are frequent in winter, bringing spells of subfreezing weather. No winter month has a mean temperature below 0°C (32°F).

Figure 11.4 presents a global map of the moist subtropical climate ⑥. It is found on the eastern sides of continents in the latitude range 20° to 35° N and S. In South America, it includes parts of Uruguay, Brazil, and Argentina. In Australia, it consists of a narrow band between the eastern coastline and the eastern interior ranges. Southern China, Taiwan, and southernmost Japan are regions of the moist subtropical climate ⑥ in Asia. In the United States, the moist subtropical climate ⑥ covers most of the Southeast, from the Carolinas to east Texas.

Figure 11.5 is a climograph for Charleston, South Carolina, located on the eastern seaboard at lat. 33° N. In this region, a marked summer maximum of precipitation is typical. Total annual rainfall is abundant—120 cm (47 in.)—and ample precipitation falls in every month. The annual temperature cycle is strongly developed, with a large annual range of 17°C (31°F). Winters are mild, with the January mean temperature well above the freezing mark.

The Moist Subtropical Forest Environment

The abundant rainfall of the moist subtropical climate ⑥ provides sufficient soil water for crops without irrigation in most years. Rivers and streams flow copiously through much of the year. Flooding can be severe at times from tropical cyclones that come inland and produce torrential rains in the high-Sun months. The large water surplus of the moist subtropical climate has important implications in terms of economic development. The large flows of rivers can furnish abundant fresh water resources for urbanization and industry without competition from irrigation demands. Evaporative losses from reservoirs are much less important than those in arid lands. The maintenance of ample stream flows tends to reduce the dangers of severe water pollution and its adverse effects on ecosystems of streams and estuaries.

Much of the natural vegetation in the North American area of this moist climate type consists of *broadleaf deciduous forest.* (In a deciduous forest, leaf shedding occurs annually, and the trees are bare throughout the cold or dry season.) The oak tree is a typical representative of this forest. Broadleaf deciduous forest can be found on the interior uplands of the Carolinas and across Tennessee. Today, of course, large parts of this formerly forested zone have been replaced by agricultural croplands. An entirely different forest type occurs farther south and east on the low, sandy coastal plain of the Gulf states, on the Florida Peninsula, and on the Atlantic coastal plain of Georgia and the Carolinas. It is the *southern pine forest,* which is well adapted to sandy soils. (See world vegetation map, Figure 24.3.) Pines belong to the class of needleleaf trees, and they retain green foliage throughout the year.

Over a large part of southern China and the south island of Japan, the native vegetation was formerly a broadleaf forest of the evergreen type, in which the leaf canopy remains green throughout the year. It is called the *broadleaf evergreen forest* (Chapter 24). Today this forest is gone from China, but some small areas in Japan remain, allowing us to reconstruct its composition. It contained a large number of species of evergreen oaks, as well as species of the magnolia tree and the camphor tree, all peculiar to that part of the world. On the North American continent, in Louisiana and farther east on the Gulf coast, the evergreen broadleaf forest is represented by a few small areas of Evangeline oak and magnolia (Figure 11.6).

The warm summer climate and high rainfall of the moist subtropical environment tend to wash nutrients from the soil layer. As in the soils of the moist low-latitude climates, iron oxides accumulate in the soil, giving it colors ranging from yellow to red. For agricultural crops, especially grains or cereals, the soils of the moist subtropical

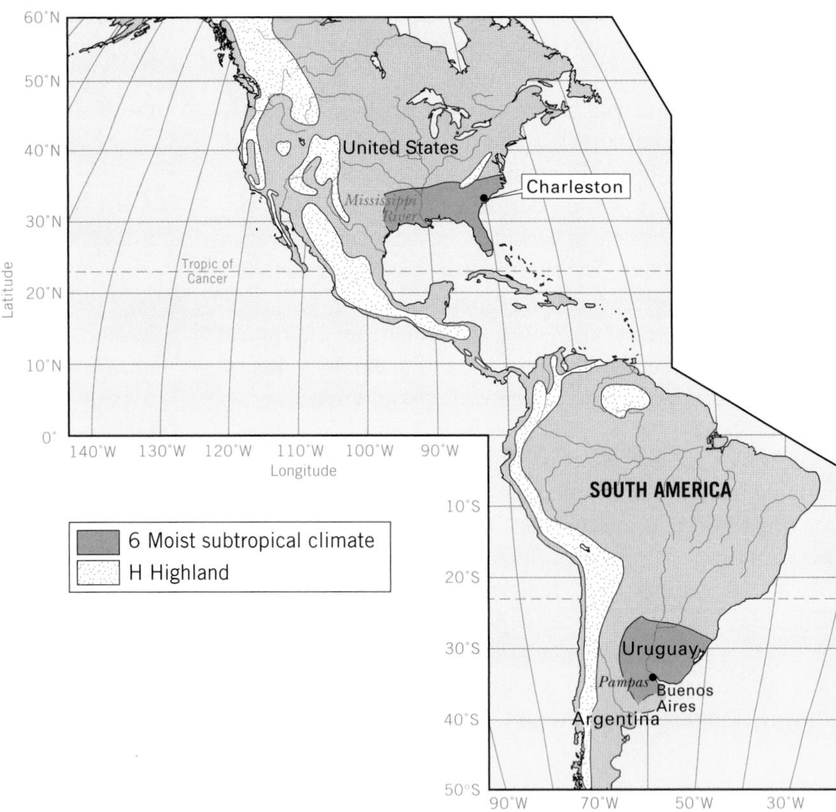

11.4 World map of the moist subtropical climate *⑥ (Based on Goode Base Map.)*

environment rate as low in fertility and require large applications of fertilizers. Another unfavorable factor is the susceptibility of these soils to severe erosion and gullying when exposed by forest removal and intensive cultivation.

11.5 Moist subtropical climate *⑥ Charleston, South Carolina, lat. 33° N, has a mild winter and a warm summer. There is ample precipitation in all months but a definite summer maximum.*

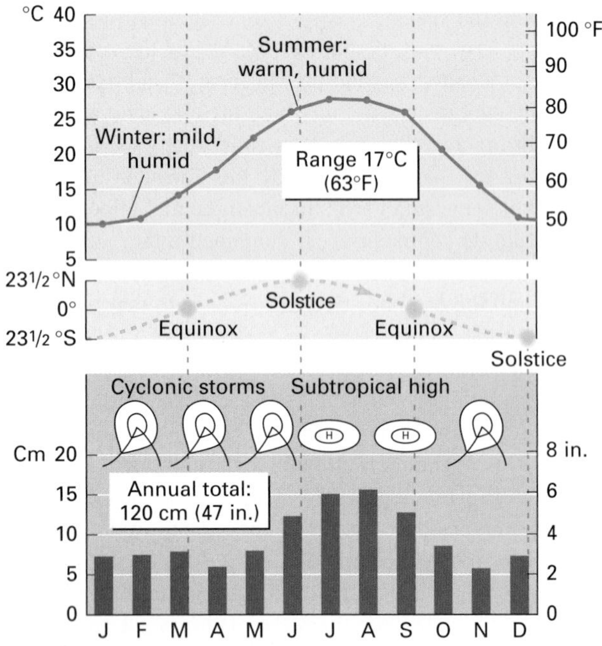

We will discuss these problems of soil fertility and erodibility in Chapter 21.

AGRICULTURAL RESOURCES OF THE MOIST SUBTROPICAL FOREST ENVIRONMENT In the two major regions of moist subtropical climate ⑥—North America and Southeast Asia—agricultural use of the land follows quite different patterns. The differences are partly historical and cultural, but also reflect the stronger monsoon effect in Asia, which causes a stronger concentration of precipitation in the summer. The human population is vastly denser in Southeast Asia than in the New World and subsists largely on rice, the dominant staple food crop. Two and even three rice crops are harvested annually in southern China. The rice crop is often followed by planting of wheat, winter legumes, peas, or green fertilizer crops. Except in the Mississippi delta region, little rice is produced in the southern United States. Both American and Asiatic regions produce sugar cane, peanuts, tobacco, and cotton, although not on an equally intense basis in both regions. One striking difference is that tea is widely cultivated in Southeast Asia but not at all in the southern United States. Corn is a major crop in the southern United States that is becoming more important in southern China.

The potential of the moist subtropical climate to produce more food rests in more intensive land use instead of expansion of the area now under cultivation. The best land is already in use, and, in Southeast Asia, elaborate terrace systems have been used for centuries to allow the farming of steep hill slopes. New genetic strains of rice and corn offer promise of increased yields when the necessary

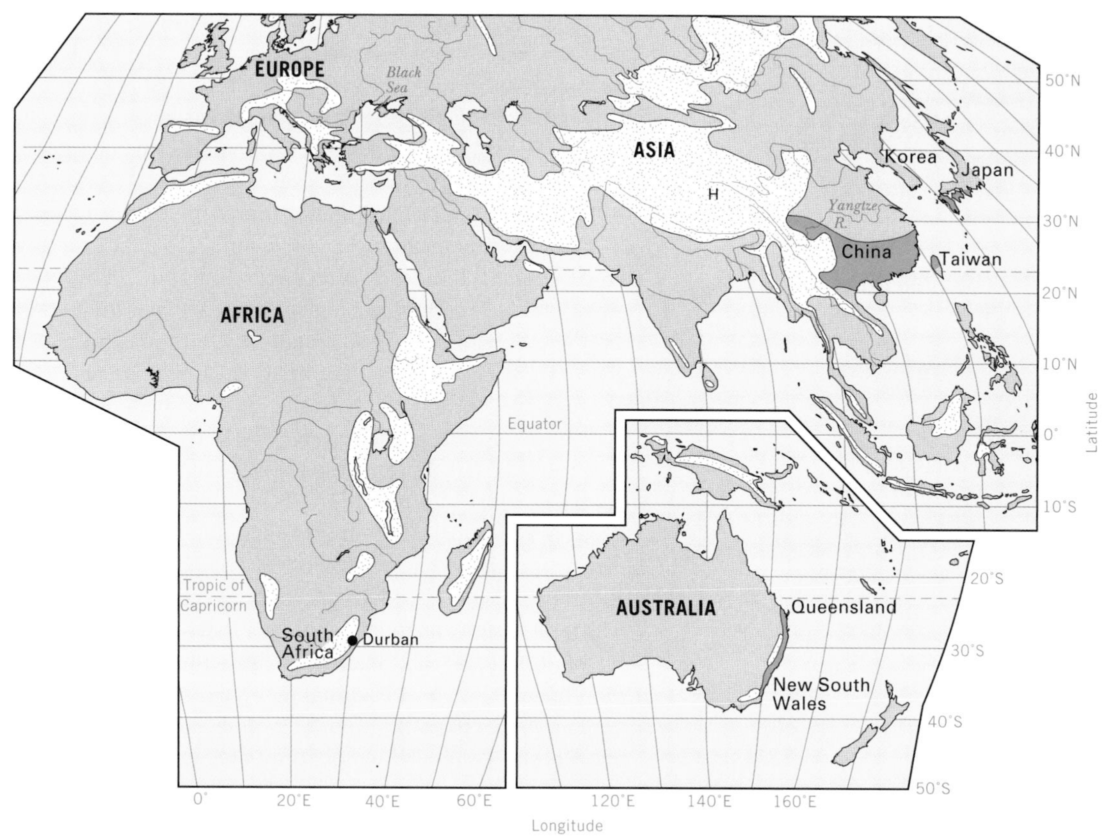

fertilizers are used. Japan and Taiwan apply very high levels of fertilizers and achieve high rice yields. The People's Republic of China is only now in the process of sharply increasing its production and use of fertilizers from comparatively low levels of the recent past. China is also developing independently some new high-yielding strains of rice and wheat.

In the southern United States, cattle production is another source of increased food production and makes use of soils too sandy for field crops. With soil water frequently replenished through the long, warm summer, pasture and range land can be continuously productive. Tree farming is also an important use of sandy soils. Pines are well adapted to rapid growth on sandy soils and thrive where nutrient bases are in short supply.

THE MEDITERRANEAN CLIMATE ⑦ (KÖPPEN: *CSA, CSB*)

The **Mediterranean climate** ⑦ is unique among the climate types because its annual precipitation cycle has a wet winter and a very dry summer. The reason for this precipitation

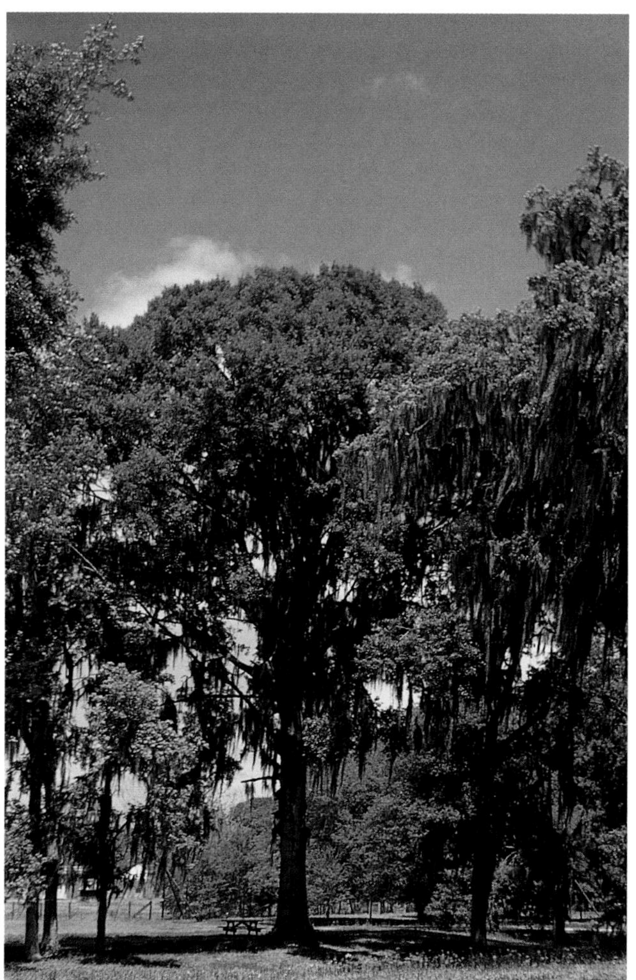

11.6 Broadleaf evergreen forest of the Gulf coast
A typical species of the broadleaf evergreen forest is the Evangeline oak, here bearing Spanish "moss"—an epiphyte that forms long beard-like streamers. The ground beneath is maintained as a lawn. Evangeline State Park, Bayou Teche, Louisiana. (A. N. Strahler.)

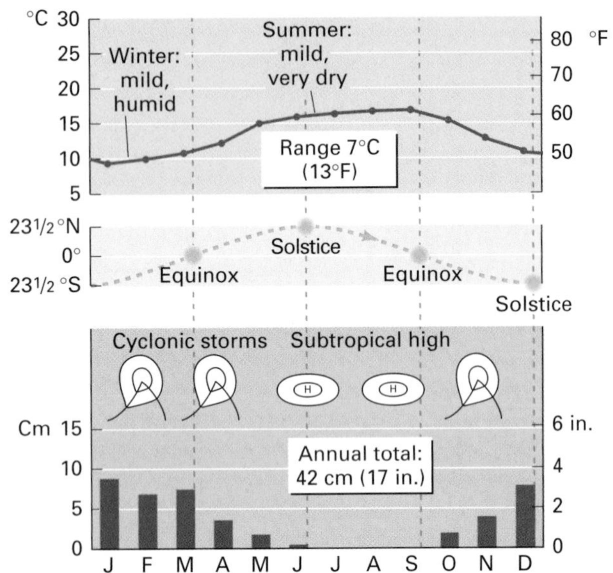

11.7 Mediterranean climate ⑦ *Monterey, California, lat. 36° N, has a very weak annual temperature cycle because of its closeness to the Pacific Ocean. The summer is very dry.*

cycle lies in the poleward movement of the subtropical high-pressure cells during the summer season. The Mediterranean climate ⑦ is located along the west coasts of continents, just poleward of the dry, eastern side of the subtropical high-pressure cells (see Figure 7.18). When the subtropical high-pressure cells move poleward in summer, they enter the region of this climate. Dry continental tropical (cT) air then dominates, producing the dry summer season. In winter, the moist mP air mass invades with cyclonic storms and generates ample rainfall.

In terms of total annual rainfall, the Mediterranean climate ⑦ spans a wide range from arid to humid, depending on location. Generally, the closer an area is to the tropics, the stronger the influence of subtropical high pressure will be, and thus the drier the climate. The temperature range is moderate, with warm to hot summers and mild winters. Coastal zones between lat. 30° and 35° N and S, such as southern California, show a smaller annual range, with very mild winters.

Figure 11.7 is a climograph for Monterey, California, a Pacific coastal city at lat. 36° N. The annual temperature cycle is very weak. The small annual range and cool summer reflect the strong control of the cold California current and its cool, marine air layer. Fogs are frequent. This temperature regime is found only in a narrow coastal zone. Rainfall drops to nearly zero for four consecutive summer months but rises to substantial amounts in the rainy winter season. In California's Central Valley, the dry summers persist, but the annual temperature range is greater, so that daily high temperatures match those of the adjacent dry climates.

A global map of the Mediterranean climate ⑦ is shown in Figure 11.8. It is found in the latitude range 30° to 45° N

11.8 World map of the Mediterranean ⑦ and marine west-coast ⑧ climates *(Based on Goode Base Map.)*

and S. In the southern hemisphere, it occurs along the coast of Chile, in the Cape Town region of South Africa, and along the southern and western coasts of Australia. In North America, it is found in central and southern California. In Europe, this climate type surrounds the Mediterranean Sea, which gives the climate its distinctive name.

The Mediterranean Climate Environment

The Mediterranean climate ⑦ is a very attractive one for human habitation. The attraction lies in the thermal cycle of year-round pleasant temperatures, especially where moderated by coastal influences. The mild winters with abundant sunshine (despite periods of substantial rainfall) are a welcome refuge from the severe winters of the midlatitude continental interiors of Eurasia and North America. However, the low annual precipitation with dry summers makes fresh water scarce, requiring a large investment in technology to deliver enough water to support the heavy load of humanity. Figure 11.9 shows some typical landscapes and features of the Mediterranean climate ⑦.

Soil fertility in valley and lowland areas of the Mediterranean climate is naturally high, as it usually is in semiarid climates of the midlatitudes. Because overall rainfall is low, the nutrients that are essential to forage grasses and grains, fruit trees, vegetables, and many other varieties of plants are not washed out of the soil. The soils native to the Mediterranean environment belong to a special category of subclasses, explained in Chapter 21, and are not easily described in a few words.

The native vegetation of the Mediterranean climate environment is adapted to survival through the long summer drought. Shrubs and trees are typically equipped with small, hard or thick leaves that resist water loss through transpiration. These plants are called *sclerophylls;* the prefix scler, from the Greek for "hard," is combined with phyllo, which is Greek for "leaf." Sclerophylls of the Mediterranean environment are typically evergreen, retaining their leaves through the entire yearly cycle. Examples are the evergreen oaks, of which there are several common species in California, and the cork oak of the Mediterranean lands (Figure 11.9*a*). Another example is the olive tree, native to the Mediterranean lands. In Australia, the thick-leafed eucalyptus tree is the dominant sclerophyll. Oak woodland of California bears a ground cover of grasses that turn to straw in the summer (Figure 11.9*b*). At higher elevations, drought-resistant conifers often occur.

Another form of native vegetation in this environment is a cover of drought-resistant shrubs, including sclerophylls and spiny-leafed species. In the Mediterranean lands, this scrub vegetation goes under the name of maquis or garrigue. In California, where it is called *chaparral,* it clothes steep hill and mountain slopes too dry to support oak woodland or oak forest (Figure 11.9*c*).

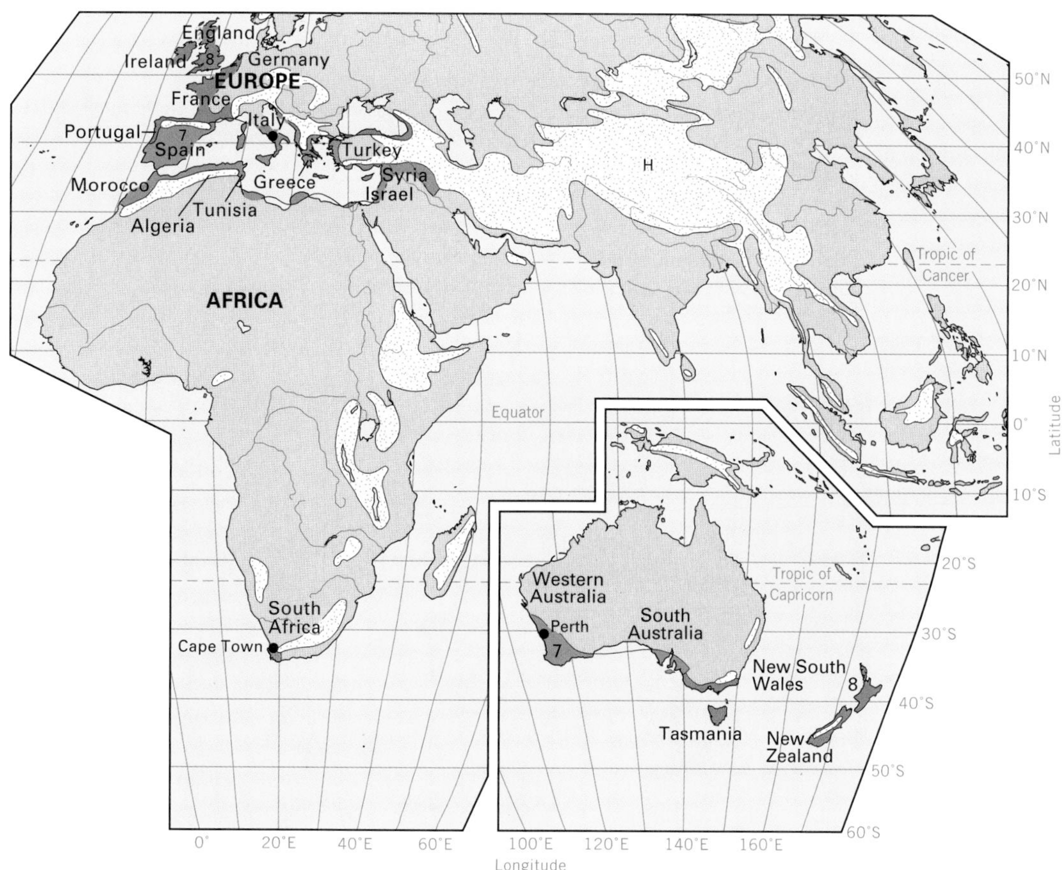

11.9 Mediterranean Landscapes

a...***Cork oak*** *The cork oak, Quercus suber, has tough, thick outer bark that is harvested for wine corks. The underlying live bark is reddish in color. Algarve Province, Portugal.*

b...***California oak woodland in summer*** *Blue oaks (Quercus douglasii) are scattered over a cover of dry grasses in the inner Coast Ranges, near Williams, California.*

c...*Eye on the Landscape* ***California chaparral*** *Wiry shrubs cloak the steep slopes of a coast range near Acton, California.* **What else would the geographer see?... Answers at the end of the chapter.**

d...*Orchard agriculture* *Groves of lemon, orange, and avocado trees surround these homes and stables in Montecito, California. Chaparral covers the steep slopes of the Santa Ynez Mountains in the background. (A. N. Strahler.)*

Wildfire is an integral part of the Mediterranean environment of California. Chaparral is extremely flammable during the long summer fire season (Figure 7.20). Brushfires rage through chaparral and oak forests and leave the soil surface bare and unprotected. When torrential rains occur in winter, large quantities of coarse mineral debris are swept downslope and carried long distances by streams in flood. Mudflows and debris floods (usually called "mudslides" in the news media) are particularly destructive to communities on canyon floors (see Chapter 15).

Throughout the Mediterranean lands of Europe, North Africa, and the Near East, devastating soil erosion, induced by human activity over the past 2000 years or longer, has left its scars on the landscape. Many hillsides have been denuded of their soils and are now rocky and barren. Sediment, representing the displaced soil, has formed thick layers of sand and silt in adjacent valley floors.

AGRICULTURE IN THE MEDITERRANEAN ENVIRONMENT

Lands bordering the Mediterranean Sea produce cereals—wheat, oats, and barley—where arable soils are extensive enough to be cultivated. However, we usually think of that region as an important source of citrus fruits, grapes, and olives for European markets. Cork from the bark of the cork oak is also a product of economic value (Figure 11.9*a*). In central and southern California, citrus, grapes, avocados, nuts (almond, walnut), and other fruits are grown extensively (Figure 11.9*d*). Irrigated lowland soils are also highly productive of vegetable crops, such as carrots, lettuce, cauliflower, broccoli, artichokes, and strawberries, as well as sugar beets and forage crops (alfalfa). Cattle ranching and sheep grazing are of major importance on grassy hill slopes unsuited to field crops and orchards and on irrigated lowland pastures.

Because the Mediterranean environment is limited in global extent to comparatively small land areas, it offers little prospect for the expansion of croplands to provide major additions to the world's food supply. Irrigation is essential for high productivity, but heavy irrigation of lowland soils poses some hazards: salt accumulation and waterlogging (see Chapter 21). Urbanization and industrial development also face major problems of obtaining additional water supplies through importation over long distances by aqueduct. Nevertheless, the mild, sunny climate of southern California has proved a powerful population magnet, and water importation has already been developed on a mammoth scale. Expansion of suburban housing has, however, begun to take over rich, flat croplands, reducing the agricultural potential in a number of areas.

Rainfall in the Mediterranean climate ⑦ can be quite variable from year to year. Sometimes the weather patterns that provide winter precipitation fail to appear, or appear infrequently, leading to drought. In other years, precipitation may be much greater than normal. *Focus on Systems 11.2 • California Rainfall Cycles and El Niño* provides an example—rainfall in Santa Barbara over a 22-year period—and its possible relation to such global climate events as El Niño cycles and volcanic eruptions.

THE MARINE WEST-COAST CLIMATE ⑧ (KÖPPEN: *CFB, CFC*)

The **marine west-coast climate** ⑧ occupies midlatitude west coasts. These locations receive the prevailing westerlies from over a large ocean and experience frequent cyclonic storms involving cool, moist mP air masses. Where the coast is mountainous, the orographic effect causes a very large annual precipitation. In this moist climate, precipitation is plentiful in all months, but there is often a distinct winter maximum. In summer, subtropical high pressure extends poleward into the region, reducing rainfall. The annual temperature range is comparatively small for midlatitudes. The marine influence keeps winter temperatures mild, as compared with inland locations at equivalent latitudes.

The global map of the marine west-coast climate ⑧ (Figure 11.8) shows the areas in which this climate occurs. In North America, the climate occupies the western coast from Oregon to northern British Columbia. In Western Europe, the British Isles, Portugal, and much of France fall into the marine west-coast climate. New Zealand and the southern tip of Australia, as well as the island of Tasmania, are marine west-coast climate regions found in the southern hemisphere, as is the Chilean coast south of 35° S. The general latitude range of this climate is 35° to 60° N and S.

Figure 11.10 is a climograph for Vancouver, British Columbia, just north of the U.S.-Canadian border. The annual precipitation is very great, and most of it falls during the winter months. Notice the greatly reduced rainfall in the summer months. The temperature cycle shows a remarkably

11.10 Marine west-coast climate ⑧ *Vancouver, British Columbia, lat. 49° N, has a large annual total precipitation but with greatly reduced amounts in the summer. The annual temperature range is small, and winters are very mild for this latitude.*

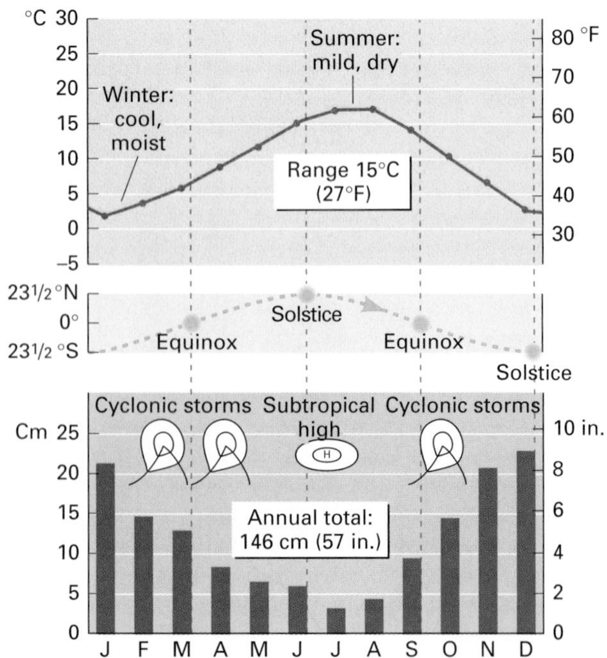

small range for this latitude. Even the winter months have averages above the freezing mark.

The Marine West-Coast Environment

Because of the abundant precipitation, lowland soils of marine west-coast climate ⑧ regions show a loss of nutrients but in Europe have still retained moderate fertility. Applications of fertilizers and lime are needed for bountiful crop production, and in Europe these soils have been successfully cultivated for centuries. Much of the land surface within the marine west-coast environment in northern Europe, British Columbia, southern Chile, and the South Island of New Zealand is on mountainous slopes that have been heavily scoured by the ice sheets and mountain glaciers of the recent Ice Age. Soils of these glaciated areas are extremely young and are poorly developed.

Forest is the native vegetation of this environment. Dense needleleaf forests of fir, cedar, hemlock, and spruce (Figure 11.11) flourish in the wet mountainous areas of the northern Pacific coast. Under the lower precipitation regime of Ireland, southern England, France, and the Low Countries, a broadleaf deciduous forest was the native vegetation, but much of it disappeared many centuries ago under cultivation. Only scattered forest plots or groves remain (see Figure 11.12). Sometimes called "summergreen" deciduous forest, it is dominated by tall broadleaf trees that provide a continuous and dense canopy in summer but shed their leaves completely in winter. Dominant tree species of this forest type in western Europe are oak and ash, with beech in the cooler and moister areas. Figure 24.11 is a map showing the distribution of midlatitude deciduous forests.

AGRICULTURE AND WATER RESOURCES The marine west-coast environment of western Europe and the British Isles has been intensively developed for centuries for such diverse uses as crop farming, dairying, orchards, and forest (Figure 11.12). It is an environment in most respects similar in agricultural character and productivity to the moist continental climate environment with which it merges on the east.

In North America, the mountainous terrain of the coastal belt offers only limited valley floors for agriculture, which is generally diversified farming. Forests are the primary plant resource, and here they constitute perhaps the greatest structural and pulpwood timber resource on Earth. Douglas fir, western cedar, and western hemlock are the principal lumber trees of the Pacific Northwest (Figure 11.13). The same mountainous terrain that limits agriculture in the Pacific Northwest is a producer of enormous water surpluses that run to the sea in rivers. In many cases, these flows have been dammed to provide hydroelectric power, as for example in the Columbia River basin. However, the dams have not been without environmental consequences, such as limiting salmon runs and changing coastal estuaries.

11.11 ***Needleleaf forest*** *Sitka spruce and hemlock populate the Hoh rainforest, Olympic National Park, Washington. Ferns and mosses provide a lush ground cover in this very wet environment.*

Focus on Systems | 11.2

California Rainfall Cycles and El Niño

California's coastal zone from the San Francisco Bay area as far south as San Diego lies in a coastal subtype of the Mediterranean climate ⑦. It has a long, nearly rainless summer season, often cursed by disastrous wildfires. Relief from the summer drought comes in a few winter months that can bring heavy rainfalls, and with them disastrous stream floods and mudflows. Substantial amounts of rain can fall in November, or even earlier, but the months of heavy rainfall totals are December through March. In some years, heavy rainfall may continue well into April.

Our upper graph *(a)* shows the year's total rainfall for Santa Barbara, a coastal city lying about midway between San Francisco and San Diego, for the period 1963–1994. The rain gauge is positioned at low elevation not far from the coast. Values are shown for the water year. This period begins on October 1 and ends on September 30 in order to place all winter rainfall during the same recording period. The year-to-

Rainfall in Santa Barbara compared with the Southern Oscillation Index *Note that the scale of the index is reversed for easier comparison with the rainfall data. (Data of NOAA/NCDC.)*

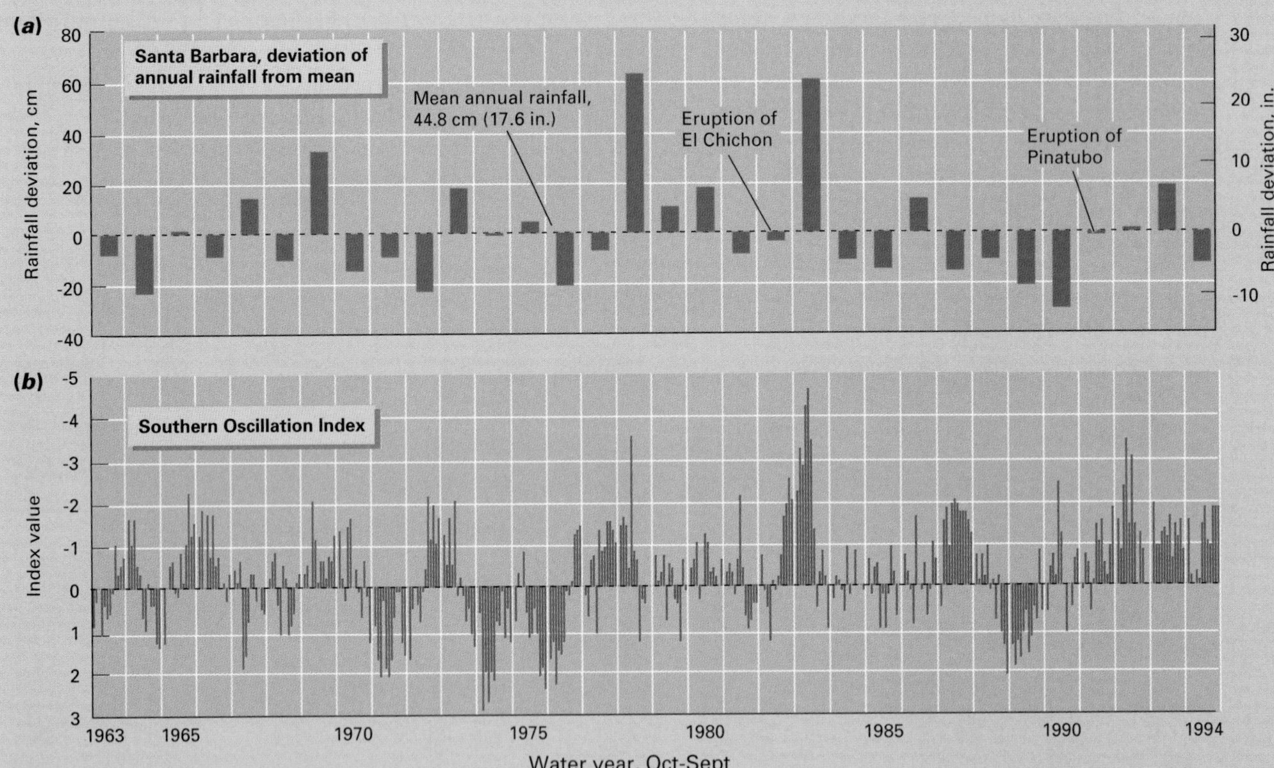

year relative variability is obviously great, with high peaks rising far above the mean value of 44.8 cm (17.6 in.). The time cycle of these annual accumulations seems to be irregular in both its period and its amplitude. (See *Working It Out 10.1 • Cycles of Rainfall in the Low Latitudes,* Chapter 10.) Forecasting ahead into the winter rain season from one year to the next is a most unreliable undertaking with no obvious repeating period to rely on.

On the other hand, rainfall forecasting for Santa Barbara could stand a good chance of being reasonably reliable if climatologists should discover a phenomenon with a similar pattern that is displayed several months in advance of the rain season. We would call it a "precursory event"—one that precedes and predicts an event to come. What precursory weather phenomenon might this be? Surely you can guess the messenger, for it is described in an earlier chapter. El Niño, in combination with La Niña, might serve as the messenger. We can read these warning signals from a graph of the Southern Oscillation, which is under continuous observation. For the Southern Oscillation signal to be effective in forecasting, it must occur at least a few months in advance of the start of the rain season.

Our second graph *(b)* shows the recorded monthly values of the Southern Oscillation Index for the same period of years as the Santa Barbara rainfall graph. The index is the difference between the monthly barometric pressure at Darwin, Australia, and at the Island of Tahiti. Note that the *y*-axis is reversed, so that negative values of the index (i.e., El Niños) can be correlated more easily with rainfall peaks. A persistent change in the upward (negative) direction signals El Niño, while such a change in the downward (positive) direction signals La Niña.

We have aligned the two graphs so that you can see if there is any consistent association of the El Niño/La Niña pressure signals with strong highs and lows in the rainfall graph. The two wettest years—1978 and 1983—seem nicely related to El Niños. The Southern Oscillation Index begins dropping some months before the onset of the wet year, thus signaling the excess rainfall. However, the third wettest year, 1969, is not well signaled and is only associated with weak El Niño conditions. Further, strong El Niños in 1965–1966 and 1987 are not associated with abnormally high rainfalls.

What about the relation between dry years and La Niña? Obvious La Niña events occur in 1971, 1974, 1975–1976, and 1989. Although these do not seem strongly related to individual low-rainfall years, at least we can say the annual rainfall during La Niña periods is always below the mean to some degree.

From these comparisons, we can see that while there is some relationship between Santa Barbara rainfall and the Southern Oscillation Index, it is not all that reliable. What other signals might predict a wet year for Santa Barbara? One possibility would be a rise in sea-surface temperature levels in the northern hemisphere of the eastern Pacific. Warmer sea-surface temperatures might act to add more moisture to maritime air masses that are drawn into occluded cyclones, which in turn provide much of the heaviest coastal rainfall to southern California. Although there is no sea-surface temperature index for the northeastern Pacific, sea-surface temperature is now continuously monitored by satellite instruments and such an index might be developed.

Another possible controlling factor is volcanic eruptions. Marked on the upper graph are the eruptions of El Chichon (1982) and Pinatubo (1991). Both of these events propelled significant volumes of aerosols into the stratosphere, although the volume of Pinatubo aerosols was three times greater. (Recall from Chapter 5 that volcanic aerosols reflect more sunlight back to space, thus cooling the Earth's surface and lower atmosphere, and so impacting global climate.) Whereas El Chichon has a perfect correspondence with a high rainfall at Santa Barbara in the following winter, Pinatubo shows no rainfall response in the first following year. Again we are frustrated!

Whatever the precursor event may prove to be, uncovering such teleconnections between weather and climate phenomena in different parts of the world is a challenging and fascinating pastime that can be of real value to human populations at the mercy of the vagaries of climate.

11.12 Marine west-coast landscape *This lush landscape in Devon, England, is located in the domain of the summergreen deciduous forest. After many centuries of human occupation and cultivation, the forest is now reduced to small patches and scattered individual trees in hollows and long fence rows.*

11.13 Needleleaf forest *The marine west-coast climate ⑧ of North America is the natural home of vast forests of evergreen needleleaf conifers. This photo shows clear-cut patches of forest in British Columbia, on Vancouver Island.*

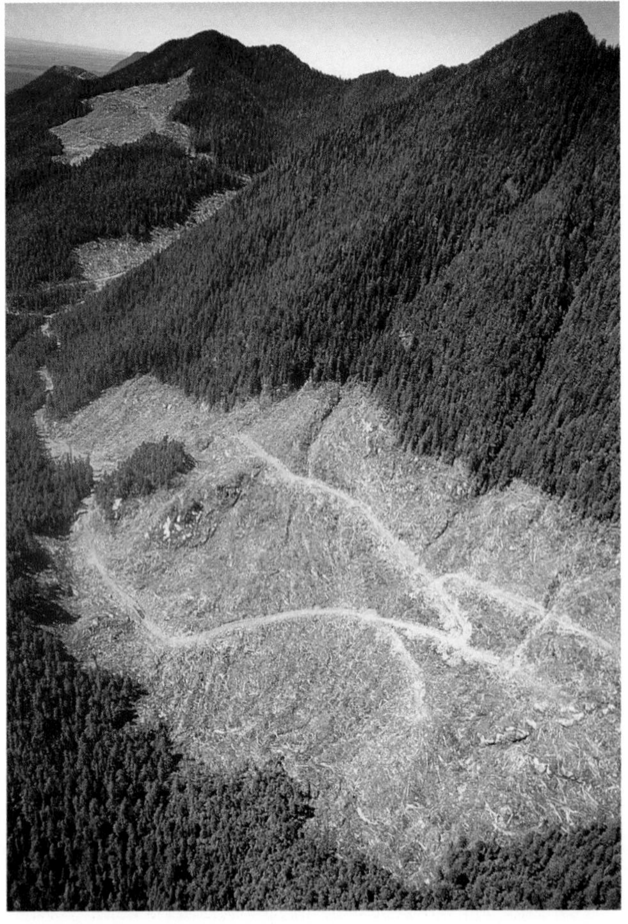

THE DRY MIDLATITUDE CLIMATE ⑨ (KÖPPEN: *BWk, BSk*)

The **dry midlatitude climate** ⑨ is limited almost exclusively to interior regions of North America and Eurasia, where it lies within the rain shadow of mountain ranges on the west or south. Maritime air masses are effectively blocked out much of the time, so that the continental polar (cP) air mass dominates the climate in winter. In summer, a dry continental air mass of local origin is dominant. Summer rainfall is mostly convectional and is caused by occasional invasions of maritime air masses. The annual temperature cycle is strongly developed, with a large annual range. Summers are warm to hot, but winters are cold to very cold.

The largest expanse of the dry midlatitude climate ⑨ is in Eurasia, stretching from the southern republics of the former Soviet Union to the Gobi desert and northern China (Figure 10.14). In the central portions of this region lie true deserts of the arid climate subtype, with very low precipitation. Extensive areas of highlands occur here as well. In North America, the dry western interior regions, including the Great Basin, Columbia Plateau, and the Great Plains, are of the semiarid subtype. A small area of dry midlatitude climate ⑨ is found in southern Patagonia, near the tip of South America. The latitude range of this climate is 35° to 55° N.

Figure 11.14 is a climograph for Pueblo, Colorado, a semiarid station located at lat. 38° N, just east of the Rocky Mountains. Total annual precipitation is 31 cm (12 in.). Most of this precipitation is in the form of convectional summer rainfall, which occurs when moist maritime tropical (mT) air masses invade from the south and produce thun-

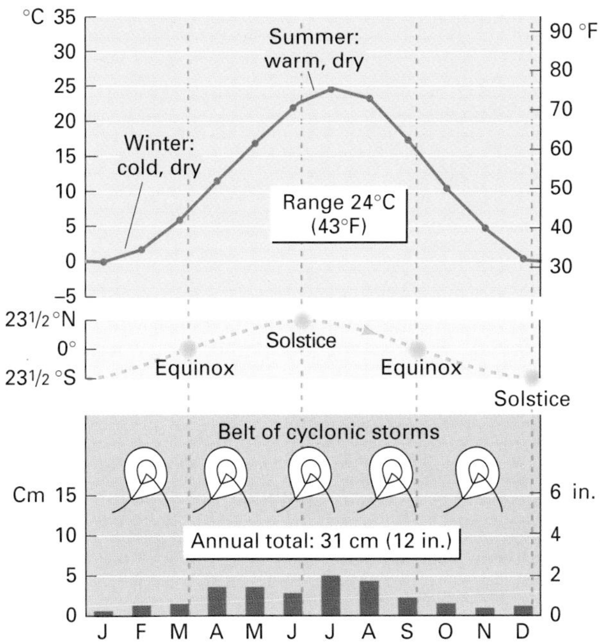

11.14 Dry midlatitude climate ⑨ *Pueblo, Colorado, lat. 38° N, shows a marked summer maximum of rainfall in the summer months. Note also the cold, dry winter season. Figure 10.14 shows the location of this station.*

derstorms. In winter, snowfall is light and yields only small monthly precipitation averages. The temperature cycle has a large annual range, with warm summers and cold winters. January, the coldest winter month, has a mean temperature just below freezing.

Annual precipitation in the dry midlatitude climate ⑨ is quite variable and there may be large differences in precipitation from year to year. Sequences of overly dry and overly wet years are often the rule. When dry conditions prevail for several years, the drought may be intense. *Eye on the Environment 11.3 • Drought and the Dust Bowl* documents the famous drought of the North American west in the 1930s.

The Dry Midlatitude Environment

The arid subtype of the dry midlatitude climate ⑨a is restricted to the driest of midlatitude continental interiors. As an extension of the arid subtype of the dry subtropical climate ⑤a, this cold desert environment supports only desert vegetation, or in North America, a cover of sagebrush and associated low woody shrubs.

Much more extensive is the steppe subtype (⑨s), which is found in large regions of interior North America and central Asia. In this subtype, low annual precipitation combined with a strongly continental thermal regime has produced soils of high natural fertility that retain large supplies of the nutrient elements, including calcium, magnesium, potassium, and sodium. These soils are moderately to strongly alkaline, in contrast to soils of the moister midlatitude climates that are acid in chemical balance. (See Chapter 21 for details.)

Grasses thrive on such nutrient-rich soils in climates of low precipitation, and thus the native vegetation of the dry midlatitude steppe consists principally of hardy perennial short grasses capable of enduring severe summer drought. In North America, this cover is termed *short-grass prairie* rather than steppe. Figure 24.19 maps the areas of short-grass prairie.

AGRICULTURAL RESOURCES OF THE SHORT-GRASS PRAIRIE

Wheat is perhaps the most important single crop produced in unirrigated areas of prairie bordering on moister climate zones. One such important wheat-producing region lies in southern Alberta and Saskatchewan and in the northern border region of Montana. Another is the Palouse Hills region of southeastern Washington state and western Idaho (Figure 11.15). Here the crop is spring wheat, which is planted in the spring of the year. Using the water remaining in the soil from winter rains and snows, and the precipitation that falls in late spring and early summer, the crop is able to reach maturity for midsummer harvesting. In Russia, the rich wheat region of the Ukraine continues in a narrow zone far eastward across the steppes of Kaza-

11.15 Dry midlatitude agriculture *Wheat farming in the rolling Palouse Hills region of eastern Washington.*

Eye on the Environment 11.3

Drought and the Dust Bowl

Steppe grasses do not form a complete sod cover—loose, bare soil is exposed between the grass clumps. For this reason, overgrazing during a series of dry years can easily reduce the hold of grasses enough to permit destructive deflation (wind erosion), followed by water erosion and gullying.

On the Great Plains of the United States and Canada, deflation and soil drifting reached disastrous proportions during a series of drought years in the middle 1930s, following a great expansion of wheat cultivation. During the drought a sequence of exceptionally intense dust storms occurred (see photo below). Within their formidable black clouds, visibility declined to nighttime darkness, even at noonday. The affected area, which also included part of the adjacent moist continental climate, became known as the *Dust Bowl.* Many centimeters of soil were removed from fields and transported out of the region as suspended dust, while the coarser silt and sand particles accumulated in drifts along fence lines and around buildings. The combination of environmental degradation and repeated crop failures caused widespread abandonment of farms and a general exodus of farm families.

Geographers who have studied the Dust Bowl phenomenon do not agree on how great a role soil cultivation and livestock grazing played in inducing deflation. The drought was a natural event over which humans had no control, but it seems reasonable that the natural grassland would have sustained far less soil loss and drifting if it had not been destroyed by the plow.

Although we cannot prevent cyclic occurrences of drought over the Great Plains, measures can be taken to minimize the deflation and soil drifting occurring in periods of dry soil conditions. Improved farming practices include use of deeply carved furrows that act as traps to soil movement. Leaving the stubble from harvested grain crops reduces deflation when land is lying fallow, and tree belts may have significant effect in reducing the intensity of wind at ground level.

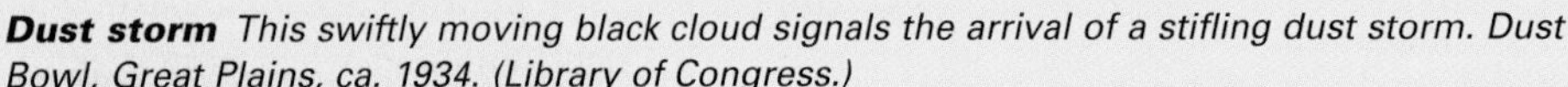

Dust storm *This swiftly moving black cloud signals the arrival of a stifling dust storm. Dust Bowl, Great Plains, ca. 1934. (Library of Congress.)*

khstan. In northern China, wheat is grown within a steppe region bordering the moist continental climate.

Wheat production of the midlatitude steppes is very much at the mercy of variations in seasonal rainfall. Good years and poor years follow cyclic variations. Soil water, not soil fertility, is the key to wheat production over these vast steppe lands. In the western Great Plains, there has been a great increase in recent decades in the use of ground water pumped to the surface and distributed by irrigation systems. This ground water source is rapidly being depleted and will ultimately fail.

Semiarid steppes form the great sheep and cattle ranges of the world. The steppes of central Asia have for centuries supported a nomadic population whose sheep and goats find subsistence on the sparse grassland. On the vast expanses of the western Great Plains, the American bison lived in great numbers until being almost exterminated by hunters. The short-grass veldt of South Africa also supported much game at one time.

The Moist Continental Climate ⑩ (Köppen: *Dfa, Dfb, Dwa, Dwb*)

The **moist continental climate** ⑩ is located in central and eastern parts of North America and Eurasia in the midlatitudes. This climate lies in the polar-front zone—the battleground of polar and tropical air masses. Seasonal temperature contrasts are strong, and day-to-day weather is highly variable. Ample precipitation throughout the year is increased in summer by invading maritime tropical (mT) air masses. Cold winters are dominated by continental polar (cP) and continental arctic (cA) air masses from subarctic source regions.

In eastern Asia—China, Korea, and Japan—the seasonal precipitation pattern shows more summer rainfall and a drier winter than in North America. This is an effect of the monsoon circulation, which moves moist maritime tropical (mT) air across the eastern side of the continent in summer and dry continental polar southward through the region in winter. In Europe, the moist continental climate ⑩ lies in a higher latitude belt (45° to 60° N) and receives precipitation from mP air masses coming from the North Atlantic.

Madison, Wisconsin, lat. 43° N, in the American Midwest (Figure 11.16), provides an example of the moist continental climate ⑩. The annual temperature range is very large. Summers are warm, but winters are cold, with three consecutive monthly means well below freezing. Precipitation is ample in all months, and the annual total is large. There is a summer maximum of precipitation when the maritime tropical (mT) air mass invades, and thunderstorms are formed along moving cold fronts and squall lines. Much of the winter precipitation is in the form of snow, which remains on the ground for long periods.

Figure 11.17 shows the global locations of the moist continental climate ⑩. It is restricted to the northern hemisphere, occurring in latitudes 30° to 55° N in North America and Asia, and in latitudes 45° to 60° N in Europe. In Asia, it is found in northern China, Korea, and Japan. Most of central and eastern Europe has a moist continental climate, as does most of the eastern half of the United States from Tennessee to the north, as well as the southernmost strip of eastern Canada.

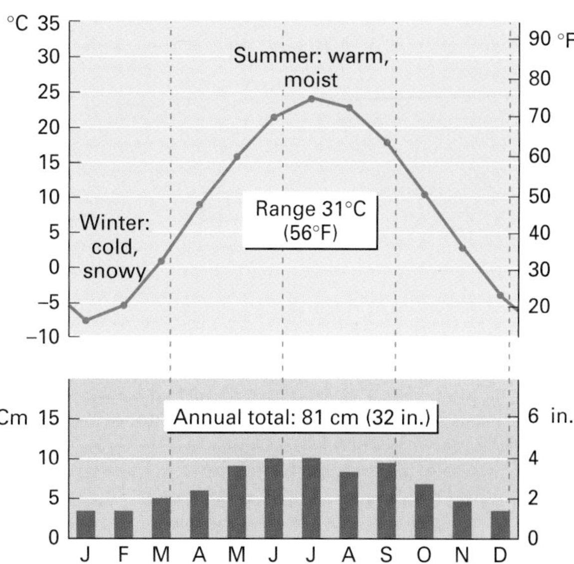

11.16 Moist continental climate ⑩ *Madison, Wisconsin, lat. 43° N, has cold winters and warm summers, making the annual temperature range very large.*

The Moist Continental Forest and Prairie Environment

With ample precipitation throughout the year and only a short period of summer dryness, the moist continental climate supports forests as the native vegetation. Soils beneath these forests show the effects of the moist environment through the washing out of soil nutrients and other soil components and a strong tendency to soil acidity. These effects are most severe in the colder, more northerly parts of the climate zone, where strongly acidic soils are found on sandy surface layers. Here the evergreen needleleaf forest dominates—the pine forest of the Great Lakes region is an example. Throughout much of the northeastern United States and southeastern Canada, mixed coniferous and deciduous forest is the native type. It grades southward into broadleaf deciduous forest, which is found in a large area of the eastern United States. Here, soils are more fertile and well suited to crop cultivation. Deciduous forest remnants are also found in this climate in central and eastern Europe, and in a narrow belt penetrating far eastward into Siberia. Deciduous forests are also found in north-central China and in Korea.

An important environmental factor has been the activity of the great ice sheets that recently covered the northern parts of this climate region. The ice not only carved many landforms, but also left behind deposits of fresh soil materials in many places. These effects are explained in Chapter 20.

When traced westward into the continental interior of North America, the moist continental climate becomes progressively drier. This gradient influences both soils and vegetation. There is a large region of the Middle West, starting in Illinois and continuing west through Iowa and into Nebraska, that formerly supported a natural cover of tall, dense grasses—*tall-grass prairie.* This kind of prairie once extended from about the United States–Canada border southward to the Gulf coast, but today only a few small remnants can be found (see Figure 24.19).

AGRICULTURAL RESOURCES OF THE MOIST CONTINENTAL CLIMATE Because of the availability of soil water through a

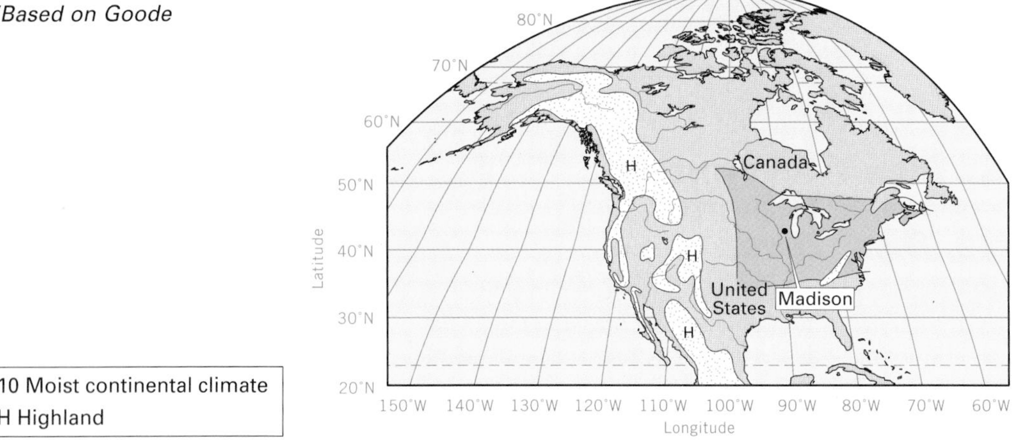

***11.17* World map of the moist continental climate** *⑩ (Based on Goode Base Map.)*

warm summer growing season, the moist continental environment has an enormous potential for food production. Throughout Europe, large areas of the moist continental forest environment have been under field crops, pastures, and vineyards for centuries (Figure 11.18). The cooler, more northerly sections in North America and Europe support dairy farming on a large scale. A combination of acid soils and unfavorable glacial terrain in the form of bogs and lakes, rocky hills, and stony soils has deterred crop farming in many parts of these northern regions.

***11.18* Eye on the Landscape Moist continental landscape** *The Mosel River in its winding, entrenched meandering gorge through the Rhineland-Pfalz province of western Germany. On the steep, undercut valley wall at the right are vineyards and forested land. Abundant rainfall, fertile soils, and mild summer temperatures provide an environment well suited to agriculture.* **What else would the geographer see?...Answers at the end of the chapter.**

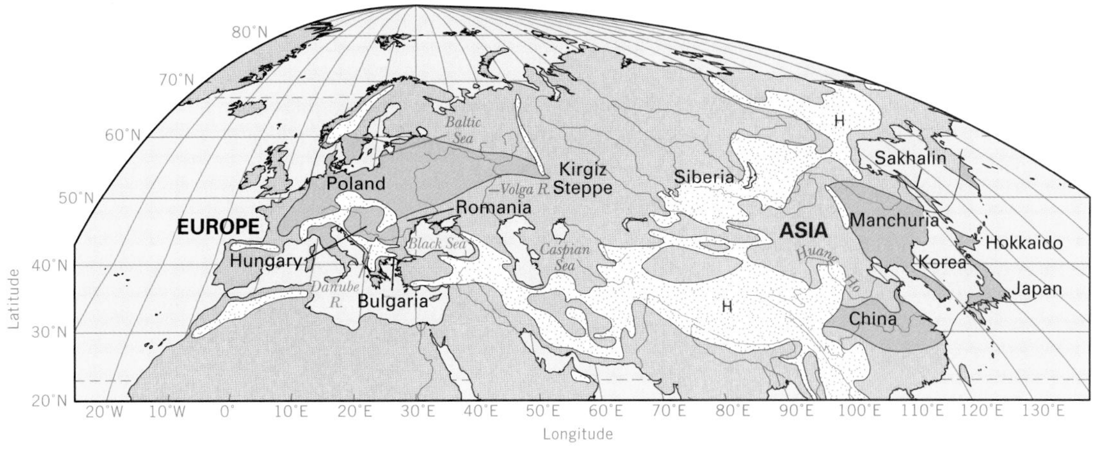

Farther south, plains formed on former lake floors and on undulating uplands are ideally suited to crop farming. Here, soils are of high fertility. Cereals grown extensively in North America and Europe include corn, wheat, rye (especially in Europe), oats, and barley. Corn is also an important crop in Hungary and Romania. Beet sugar is an important product of this environmental region in Europe but not in North America. On the other hand, soybeans are intensively cultivated in the midwestern United States and in northern China and Manchuria, but very little in Europe. Rice is a dominant crop in both South Korea and Japan, much farther poleward than elsewhere in Asia. The rice seedlings can be planted in paddies flooded during the brief but copious rains of midsummer, then harvested in the dry autumn. (Among geographers, this northern rice area is often included in the region called Monsoon Asia.) Agricultural productivity of the tall-grass prairie lands in the United States is now legendary under the name of the "corn belt." Corn production is concentrated most heavily in the prairie plains of Illinois, Iowa, and eastern Nebraska. Wheat is also a major crop near the western limits of the tall-grass prairie in Kansas and Oklahoma.

Overview of the High-Latitude Climates

By and large, the *high-latitude climates* are climates of the northern hemisphere (Figure 11.19). They occupy the northern subarctic and arctic latitude zones, but they also extend southward into the midlatitude zone as far south as about the 47th parallel in eastern North America and eastern Asia. One of these, the ice sheet climate ⑬, is present in both hemispheres in the polar zones. Table 11.2 provides an overview of the three high-latitude climates.

The high-latitude climates coincide closely with the belt of prevailing westerly winds that circles each pole (see Figure 7.15). In the northern hemisphere, this circulation

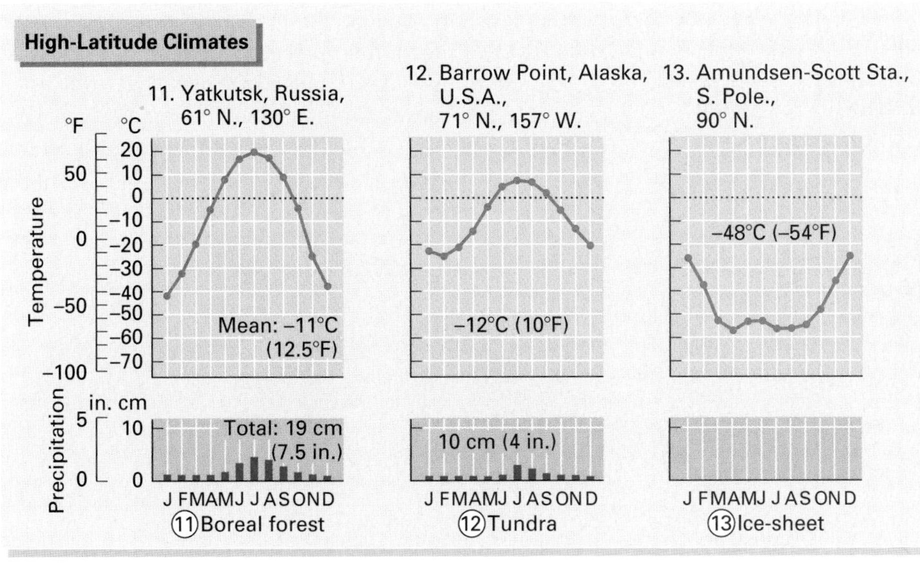

11.19 Climographs of the three high-latitude climates

Table 11.2 | High-Latitude Climates

Climate	Temperature	Precipitation	Explanation
Boreal forest ⑪	Short, cool summers and long, bitterly cold winters. Greatest annual range of all climates.	Annual precipitation small, falling mostly in summer months.	This climate falls in the source region for cold, dry stable cP air masses. Traveling cyclones, more frequent in summer, bring precipitation from mP air.
Tundra ⑫	No true summer but a short mild season. Otherwise, cold temperatures.	Annual precipitation small, falling mostly during mild season.	The coastal arctic fringes occupied by this climate are dominated by cP, mP, and cA air masses. The maritime influence keeps winter temperatures from falling to the extreme lows of interiors.
Ice sheet ⑬	All months below freezing, with lowest global temperatures on Earth experienced during Antarctic winter.	Very low precipitation, but snow accumulates since temperatures are always below freezing.	Ice sheets are the source regions of cA and cAA air masses. Temperatures are intensely cold.

sweeps maritime polar (mP) air masses, formed over the northern oceans, into conflict with continental polar (cP) and continental arctic (cA) air masses on the continents. Rossby waves form in the westerly flow, bringing lobes of warmer, moister air poleward into the region in exchange for colder, drier air moved equatorward. The result of these processes is frequent wave cyclones, produced along a discontinuous and constantly fluctuating arctic-front zone. In summer, tongues of maritime tropical air masses (mT) reach the subarctic latitudes to interact with polar air masses and yield important amounts of precipitation.

THE BOREAL FOREST CLIMATE ⑪ (KÖPPEN: *DFC, DFD, DWC, DWD*)

The **boreal forest climate** ⑪ is a continental climate with long, bitterly cold winters and short, cool summers. It occupies the source region for cP air masses, which are cold, dry, and stable in the winter. Invasions of the very cold cA air mass are common. The annual range of temperature is greater than that of any other climate and is greatest in Siberia. Precipitation increases substantially in summer, when maritime air masses penetrate the continent with traveling cyclones, but the total annual precipitation is small. Although much of the boreal forest climate is moist, large areas in western Canada and Siberia have low annual precipitation and are therefore cold and dry.

The global extent of the boreal forest climate ⑪ is presented in Figure 11.20. In North America, it stretches from central and western Alaska, across the Yukon and Northwest Territories to Labrador on the Atlantic coast. In Europe and Asia, it reaches from the Scandinavian Peninsula eastward across all of Siberia to the Pacific. In latitude, this climate type ranges from 50° to 70° N.

Figure 11.21 is a climograph for Fort Vermilion, Alberta, at lat. 58° N. The very great annual temperature range shown here is typical for North America. Monthly mean air temperatures are below freezing for seven consecutive months. The summers are short and cool. Precipitation shows a marked annual cycle with a summer maximum, but the total annual precipitation is small. Although precipitation in winter is small, a snow cover remains over solidly frozen ground through the entire winter. On the same climograph, temperature data are shown for Yakutsk, a Siberian city at lat. 62° N. The enormous annual range is evident, as well as the extremely low means in winter months. January reaches a mean of about –42°C (–44°F). Except for the ice sheet inte-

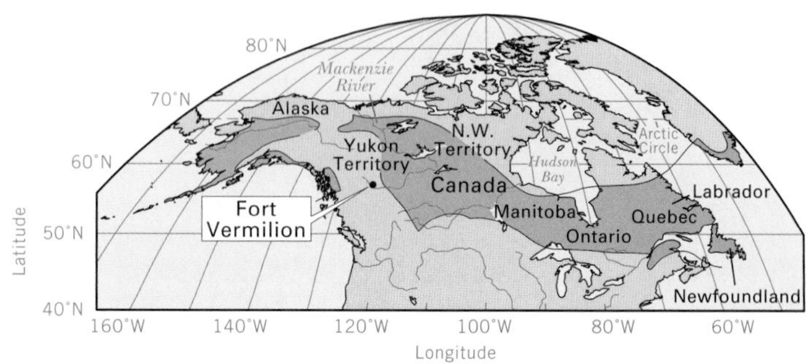

11.20 World map of the boreal forest climate ⑪ *(Based on Goode Base Map.)*

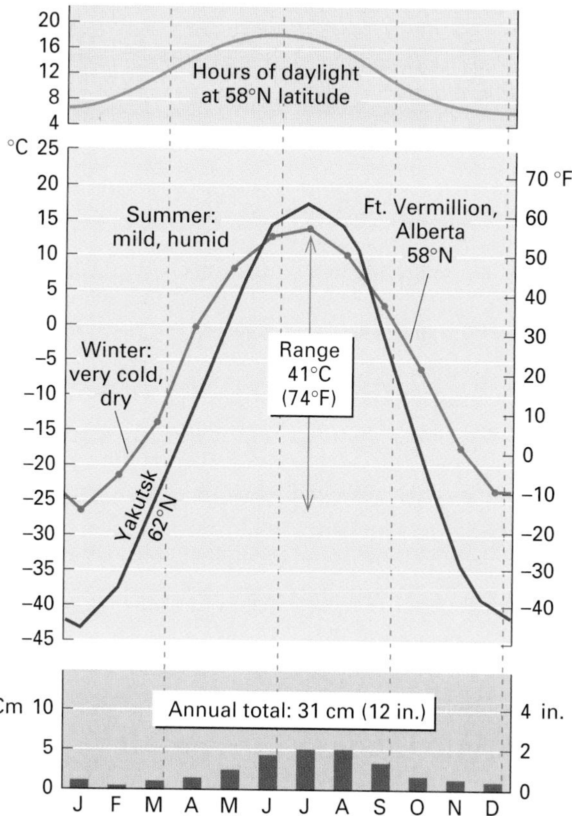

11.21 Boreal forest climate *⑪ Extreme winter cold and a very great annual range in temperature characterize the climate of Fort Vermilion, Alberta. The temperature range of Yakutsk, Siberia, is even greater.*

riors of Antarctica and Greenland, this region is the coldest on Earth. Precipitation is not shown for Yakutsk, but the annual total is very small.

The Boreal Forest Environment

The land surface features of much of the region of boreal forest climate were shaped beneath the great ice sheets of the last ice age. Severe erosion by the moving ice exposed hard bedrock over vast areas and created numerous shallow rock basins. Bouldery rock debris mantles the rock surface in many places. Many of the shallower rock basins have been filled by organic bog materials. GEODISCOVERIES

The dominant upland vegetation of the boreal forest climate ⑪ region is boreal forest, consisting of *needleleaf trees.* In North America and Europe, these are evergreen needleleaf trees, mostly pine, spruce, and fir. Figure 24.15 is a composite map showing the extent of the boreal forest. In central and eastern Siberia, the boreal forest is dominated by the larch, which sheds its needles in winter and is thus a deciduous tree. Associated with the needleleaf trees are stands of aspen, balsam poplar, willow, and birch. Along the northern fringe of boreal forest lies a zone of woodland in which low trees, such as black spruce, are spaced widely apart. The open areas are covered by a surface layer of lichens and mosses (Figure 11.22). This cold woodland is referred to as *taiga.*

Although the growing season in the boreal forest climate ⑪ is short, crop farming is still possible. It is largely limited to lands surrounding the Baltic Sea, bordering Finland and Sweden. Crops grown in this area include barley, oats, rye, and wheat. Along with dairying, these crops primarily supply food for subsistence. The principal nonmineral economic products throughout the subarctic lands of eastern Canada are pulpwood and lumber from the needleleaf forests. Logs are carried down the principal rivers to pulp mills and lumber mills. Forests of pine and fir in Sweden, Finland, and European Russia are the primary plant resource. Their wood products are exported in the form of paper, pulp, cellulose, and construction lumber.

THE TUNDRA CLIMATE ⑫ (KÖPPEN: *ET*)

The **tundra climate** ⑫ occupies arctic coastal fringes and is dominated by polar (cP, mP) and arctic (cA) air masses. Winters are long and severe. A moderating influence of the nearby ocean water prevents winter temperatures from falling to the extreme lows found in the continental interior. There is a very short mild season, which many climatologists do not recognize as a true summer.

The world map of the tundra climate ⑫ (Figure 11.23) shows the tundra ringing the Arctic Ocean and extending across the island region of northern Canada. It includes the Alaskan north slope, the Hudson Bay region, and the Greenland coast in North America. In Eurasia, this climate type occupies the northernmost fringe of the Scandinavian

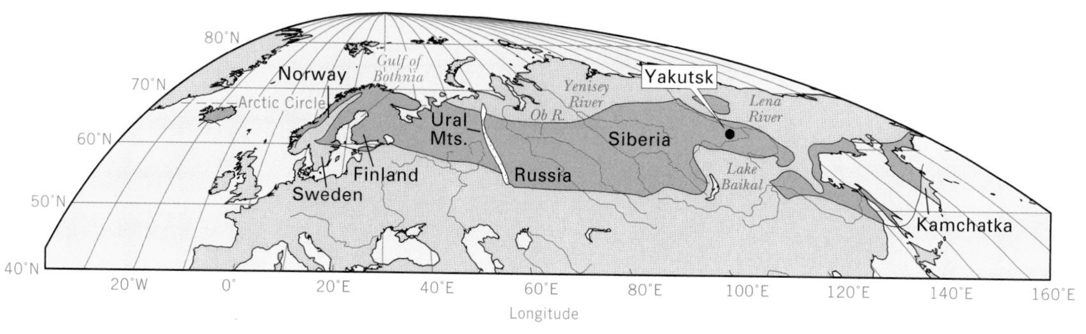

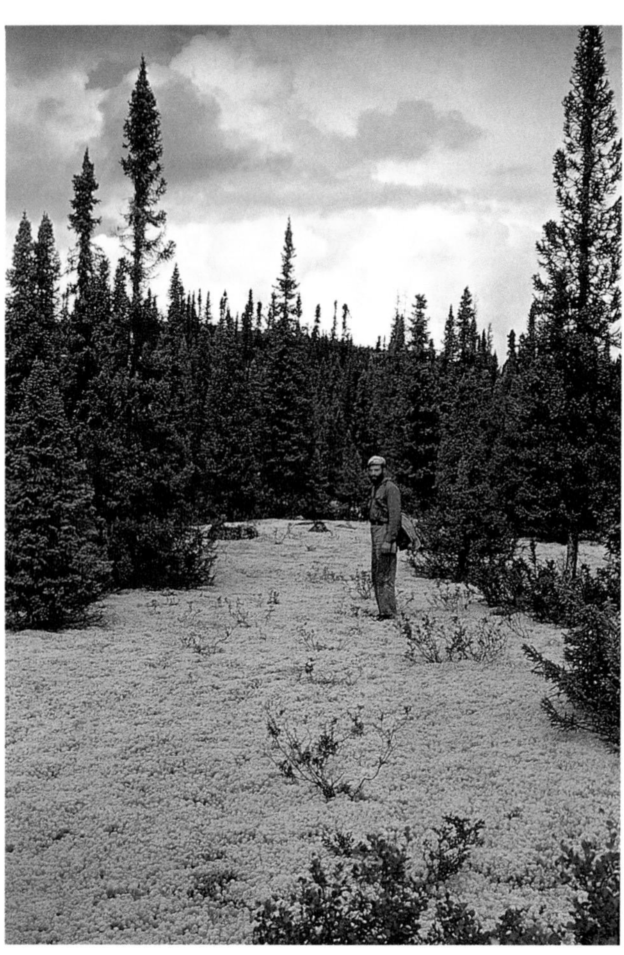

11.22 Boreal forest *Lichen woodland near Fort McKenzie, lat. 57° N, in northern Quebec. The trees are black spruce. Between the trees is a carpet of lichen. (R. N. Drummond.)*

Peninsula and Siberian coast. The Antarctic Peninsula (not shown in Figure 11.23) belongs to the tundra climate ⑫. The latitude range for this climate is 60° to 75° N and S, except for the northern coast of Greenland, where tundra occurs at latitudes greater than 80° N.

Figure 11.24 is a climograph for Upernavik, located on the west coast of Greenland at lat. 73° N. A short mild period, with above-freezing temperatures, is equivalent to a summer season in lower latitudes. The long winter is very cold, but the annual temperature range is not as large as that for the boreal forest climate to the south. Total annual precipitation is small. Increased precipitation beginning in July is explained by the melting of the sea-ice cover and a warming of ocean water temperatures. This increases the moisture content of the local air mass, allowing more precipitation.

The Arctic Tundra Environment

The term *tundra* describes both an environmental region and a major class of vegetation (Figure 11.25). (An equivalent climatic environment—called *alpine tundra*—prevails in many global locations in high mountains above the timberline.) Soils of the arctic tundra are poorly developed and

11.23 World map of the tundra climate ⑫

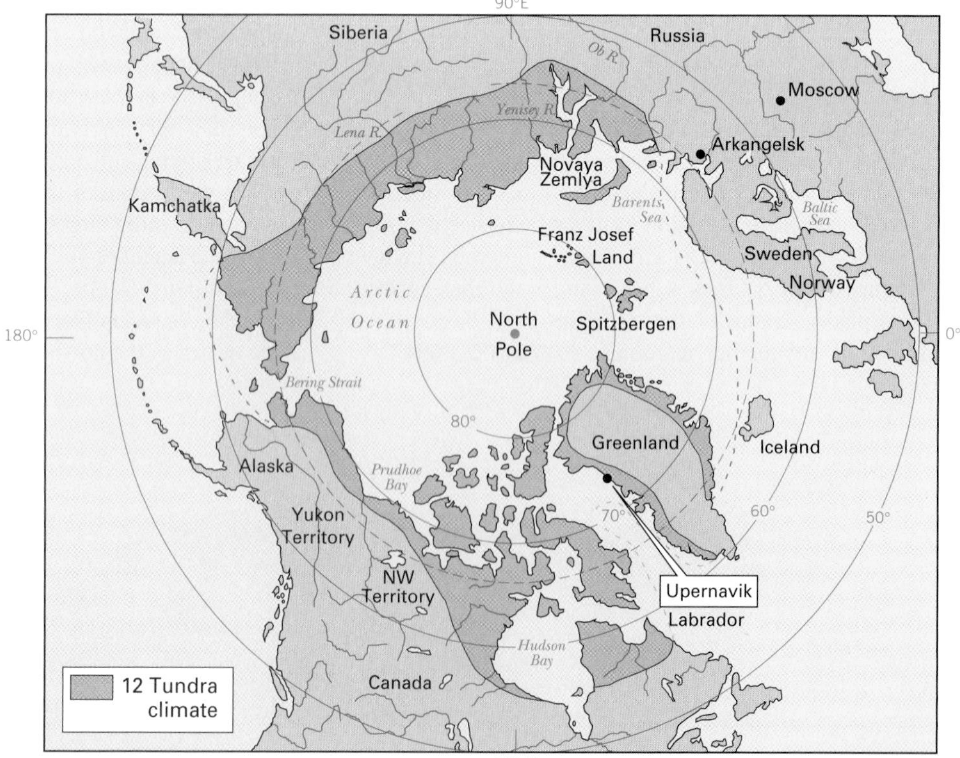

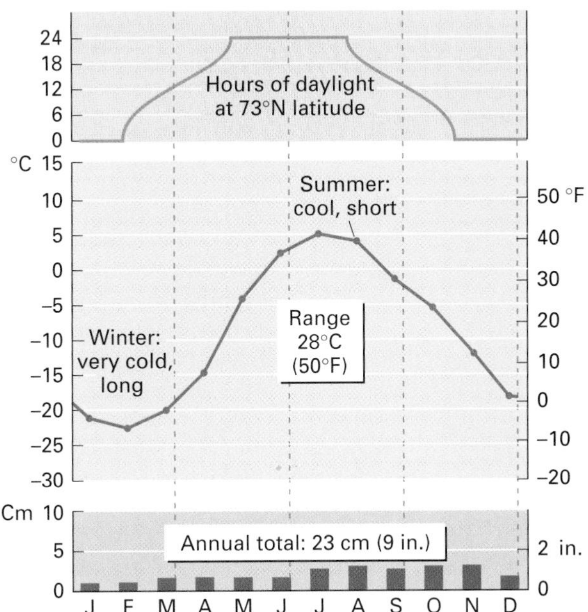

11.24 Tundra climate *⑫ Upernavik, Greenland, lat. 73° N, shows a smaller annual range than Fort Vermilion (Figure 11.21).*

consist of freshly broken mineral particles and varying amounts of humus (finely divided, partially decomposed plant matter). Peat bogs are numerous. Because soil water is solidly and permanently frozen not far below the surface, the summer thaw brings a condition of water saturation to the soil.

Vegetation of the tundra consists of a cover of scattered grasses, sedges, and lichens, along with shrubs of willow. Vegetation is scarce on dry, exposed slopes and summits. Here the surface cover is often a rocky pavement of angular rock fragments. Trees exist in the tundra only as small, shrub-like plants. They are stunted because of the seasonal damage to roots by freeze and thaw of the soil layer and to branches exposed to the abrading action of wind-driven snow. In some places, a distinct tree line separates the forest and tundra. It coincides approximately with the 10°C (50°F) isotherm of the warmest month and has been used by geographers as a boundary between boreal forest and tundra.

The number of species in the tundra environment is small, but the abundance of individuals is high. Among the animals, vast herds of caribou in North America or reindeer (their Eurasian relatives) roam the tundra, lightly grazing the lichens and plants and moving constantly (Figure 11.25). A smaller number of musk-oxen are also consumers of the

11.25 Tundra landscape *Caribou migrating across the arctic tundra of northern Alaska.*

Geographers at Work | Unearthing the Secrets of the Spanish Flu

by Kirsty Duncan ***The University of Toronto, Ontario***

What is the deadliest infectious disease in recorded history? Although the bubonic plagues of the Middle Ages and AIDS, of our own generation, come to mind, the influenza called the Spanish flu killed 20 to 40 million people worldwide—more than all the casualties in the First World War. And by 1919 it had all but disappeared.

Why was this disease so deadly? Where did it come from and why did it disappear so fast? These fascinating questions are within the province of medical geography, a new and emerging field of geography that deals with geographical epidemiology, ethnomedicine, and the spatial aspects of health care delivery, policy, and politics.

The Spanish flu of 1918 is particularly fascinating to Kirsty Duncan, a young medical geographer at the University of Toronto. She reasoned that perhaps the flu had ranged so far into the arctic zone that the bodies of some of its victims might have been preserved in graves sunk into permanently frozen ground. And, by sampling the tissues of the frozen victims, perhaps the virus could be unearthed for further study.

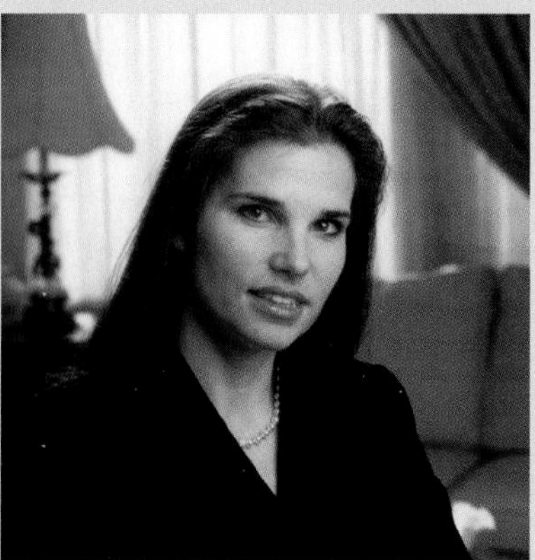

Kirsty Duncan specializes in medical geography. She hopes to connect the Spanish flu of 1918 to encephalitis lethargica, which claimed 5 million victims between 1915 and 1930.

After three years of investigation, she discovered the graves of six miners on the Norwegian islands of Svalbard, nearly 100 km (62 mi) north of the mainland of Norway, who died of the Spanish flu. Putting together a team of experts in geography, geology, geophysics, medical archeology, medicine, microbiology, pathology and virology from Canada, Norway, the United States, and England, she embarked on a project to disinter the miners and sample their remains for the influenza virus. To read her story and find out what she and her team discovered, visit our web site.

tundra vegetation. Wolves and wolverines, arctic foxes, and polar bears are predators. Among the smaller mammals, snowshoe rabbits and lemmings are important herbivores. Invertebrates are scarce in the tundra, except for a small number of insect species. Black flies, deerflies, mosquitoes, and "no-see-ums" (tiny biting midges) are all abundant and can make July on the tundra most uncomfortable for humans and animals. Reptiles and amphibians are also rare. The boggy tundra, however, offers an ideal summer environment for many migratory birds such as waterfowl, sandpipers, and plovers.

Arctic Permafrost

Because of the cold temperatures experienced in the tundra and northern boreal forest climate zones, the ground is typically frozen to great depth. This perennially frozen ground, or **permafrost,** prevails over the tundra region and a wide bordering area of boreal forest climate. Normally, a top layer of the ground will thaw each year during the mild season. This active layer of seasonal thaw is from 0.6 to 4 m (2 to 13 ft) thick, depending on latitude and the nature of the ground.

Continuous permafrost, which extends without gaps or interruptions under all surface features, coincides largely with the tundra climate, but also includes a large part of the boreal forest climate in Siberia. *Discontinuous permafrost,* which occurs in patches separated by frost-free zones under lakes and rivers, occupies much of the boreal forest climate zone of North America and Eurasia. Sporadic occurrence of permafrost in small patches extends into the southern limits of the boreal forest climate. Chapter 15 provides more detail on permafrost in arctic and boreal regions.

THE ICE SHEET CLIMATE ⑬ (KÖPPEN: *EF*)

The **ice sheet climate** ⑬ coincides with the source regions of arctic (A) and antarctic (AA) air masses, situated on the vast, high ice sheets of Greenland and Antarctica and over polar sea ice of the Arctic Ocean. Mean annual temperature is much lower than that of any other climate, with no monthly mean above freezing. Strong temperature inversions, caused by radiation loss from the surface, develop over the ice sheets. In Antarctica and Greenland, the high surface altitude of the ice sheets intensifies the cold. Strong cyclones with blizzard winds are frequent. Precipitation, almost all occurring as snow, is very low but accumulates because of the continuous cold. The latitude range for this climate is 65° to 90° N and S.

Figure 11.26 shows temperature graphs for several representative ice sheet stations. The graph for Eismitte, a research station on the Greenland ice cap, shows the northern hemisphere temperature cycle, whereas the other four examples are all from Antarctica. Temperatures in the interior of Antarctica have proved to be far lower than those at any other place on Earth. A Russian meteorological station at Vostok, located about 1300 km (about 800 mi) from the south pole at an altitude of about 3500 m (11,500 ft), may be the world's coldest spot. Here a low of −88.3°C (−127°F) was observed in 1958. At the pole itself (Amundsen-Scott Station), July, August, and September of 1957 had averages of about −60°C (−76°F). Temperatures are considerably higher, month for month, at Little America in Antarctica because it is located close to the Ross Sea and is at a low altitude.

11.26 Ice sheet climate ⑬ *Temperature graphs for five ice sheet stations.*

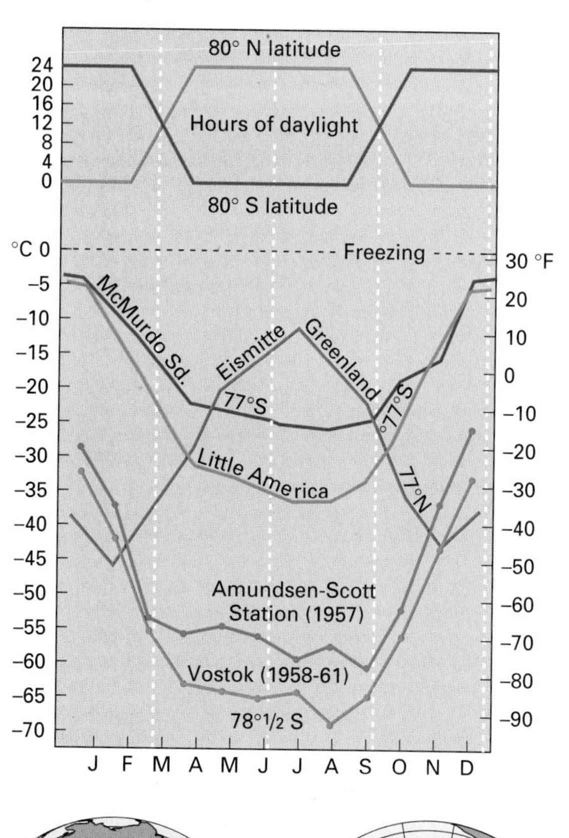

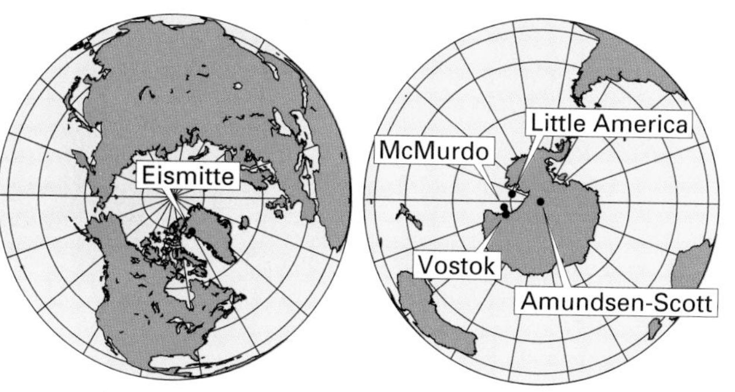

The Ice Sheet Environment

Because of low monthly mean temperatures throughout the year over the ice sheets, this environment is devoid of vegetation and soils. The few species of animals found on the ice margins are associated with a marine habitat. In terms of habitation by humans, the ice sheet environment is extremely hostile because of extreme cold, high winds, and a total lack of food and fuel resources. Enormous expenditures of energy are required to import these necessities of life and to provide shelter. These efforts are justified for scientific research, but in the foreseeable future there is little prospect that this icy environment will provide useful supplies of energy or minerals.

Our Changing Climate

In Chapter 9, we defined climate as the average weather of a region, described by the average values of monthly temperature and precipitation observed at weather stations. These average values are obtained over significant periods of time—typically decades. However, it is important to realize that human activities have now modified the weather we experience. Moreover, that modification will continue into the future, with increasingly stronger effects. In this way, the climate of every location is changing, and changing because of human modification of the atmosphere, land, and oceans.

As we noted in our interchapter feature, *A Closer Look: Eye on Global Change 5.5 • The IPCC Report of 2001,* recent human activity has raised global temperatures, which in turn have reduced global snow and ice cover and raised sea levels. Precipitation, enhanced by greater evaporation, has increased in mid- and high-latitude regions, but decreased in subtropical regions. The variability of weather has also increased, with extreme precipitation events more frequent. These trends will continue as we continue to release greenhouse gases into the atmosphere, clear forests for timber and agriculture, build dams and channel rivers, and pollute the air with soot and aerosols.

How will climate change affect you and your life? That depends on where you live. For most of North America, temperatures of the twenty-first century will rise significantly, bringing warmer winters and hotter summers. Although precipitation will increase in many regions, warmer temperatures will cause more evapotranspiration, and the result will be more summer drought and higher heat indexes. Meanwhile, more frequent extremes of precipitation will enhance flooding and storm damage. If you live on the coast, you may see major changes in wetlands and estuaries from a rise in sea level, and you may also experience economic losses in greater and more frequent storm surges of hurricanes and coastal storms. Figure 11.27 illustrates some of these effects. Our interchapter feature, *A Closer Look: Eye on Global Change 11.4 • Regional Impacts of Climate Change on North America*, provides summaries of climate change predictions and impacts for nine major North American regions. Take the time to review them carefully to see what human modification of nature holds in store for you.

11.27 Effects of Global Climate Change

a...**Dead and dying trees** *Under insect attack, this stand of white firs in Kings Canyon National Park has been decimated by tussock moth caterpillars. Insect attacks like this are often brought on by drought stress, which will be enhanced by global warming in many midlatitude regions.*

b...**Flash flooding** *Heavy rainfalls in August 2002 quickly caused the Rodl river in Rottenegg, Austria, to overrun its banks. Frequent and more intense summer rainfall in midlatitudes is predicted by global climate change models.*

c...**Severe storms** *Huge waves generated by a winter storm in January, 2003, crash into a sea wall at Winthrop, Massachusetts. Global climate change models predict that strong storms and waves will be more frequent in the midlatitudes in the future.*

A LOOK AHEAD The climates of the Earth are remarkably diverse, ranging from the hot, humid wet equatorial climate ① at the equator to the bitterly cold and dry ice sheet climate ⑬ at the poles. Between these extremes are the other climates, each with distinctive features—such as dry summers or dry winters, or uniform or widely varying temperatures. The global environments associated with each climate type are also highly varied, from the lush, equatorial rainforest to the stunted willows of the tundra. Some of these environmental regions are more hospitable to human habitation and use than are others. As we have shown in this chapter and its predecessor, climate exerts strong controls on vegetation and soils, especially at the global level. We will return to a more detailed look at vegetation and soils in Part 5.

The next set of topics in our study in physical geography concerns the lithosphere—the realm of the solid Earth, as we described in Chapter 1. Our survey of the lithosphere, in Part 3, *Systems and Cycles of the Solid Earth,* will begin with the nature of Earth materials and then move to a description of how continents and ocean basins are formed and how they are continually changing, even today. Last, we turn to landforms occurring within continents that are produced by such lithospheric processes as volcanic activity and earthquake faulting.

Chapter Summary

- *Midlatitude climates* are quite varied, since they lie in a broad zone of intense interaction between tropical and polar air masses.
- The **dry subtropical climate** ⑤ is dominated by subtropical high pressure and resembles the dry tropical climate ④, but it has a larger annual temperature range and a distinct cool season. The Mojave and Sonoran deserts, which occur in this climate type, are distinguished by a cover of specialized plants, such as the Joshua tree and saguaro cactus.
- Abundant rainfall with a summer maximum is a characteristic of the **moist subtropical climate** ⑥. The temperature cycle of this type includes cool winters with spells of subfreezing weather and warm, humid summers. The natural vegetation cover is forest—*broadleaf deciduous* in most regions, with *southern pine* forest in regions of sandy soils. However, agriculture has largely replaced natural forests throughout this climate type. In Asia, rice culture is dominant, while in North America, sugar cane, peanuts, tobacco, and cotton are more typical.
- The **Mediterranean climate** ⑦ is unique because its annual precipitation cycle has a wet winter and a dry summer. The temperature range is moderate, with warm to hot summers and mild winters. The vegetation of the Mediterranean climate environment includes many evergreen *sclerophylls*, such as the cork oak, olive tree, and eucalyptus. Oak woodlands are also typical. Steep slopes are clothed with the drought-resistant shrubs of the *chaparral*. Due to a long history of human habitation, little natural vegetation remains in the Mediterranean climate environment of Europe.
- The **marine west-coast climate** ⑧, like the Mediterranean climate ⑦, has a winter precipitation maximum. The marine influence keeps temperature mild with a narrowed annual range. Forest is the native vegetation of the marine west-coast environment—dense needleleaf forest on the northern Pacific coast and broadleaf deciduous forest on the coast of western Europe. In Europe, lands of this climate have been under intensive cultivation for many centuries, and little forest remains. In North America, this climate zone is a source of many diverse forest products.
- The **dry midlatitude climate** ⑨ has both *arid* (⑨a) and *semiarid* (⑨s) subtypes. Both types have warm to hot summers and cold to very cold winters. The semiarid subtype typically has fertile soils with a sparse cover of grasses, which is termed *short-grass prairie* in North America and steppe in Eurasia. Wheat is the dominant crop, with cattle grazing an important commercial activity. Because rainfall is highly variable, drought is a recurring event. When combined with overgrazing, drought can create dust bowl conditions.
- The **moist continental climate** ⑩ has ample precipitation with a summer maximum. Summers are warm and winters cold. In its northern regions, the moist continental environment sustains a cover of evergreen needleleaf forest, while farther south, the forest is broadleaf deciduous. In North America, a region of this climate in the Middle West once supported *tall-grass prairie*. This region is now the corn belt, an area of legendary crop productivity.
- *High-latitude climates* have low precipitation since air temperatures are low. The precipitation generally occurs during the short warm period.
- The **boreal forest climate** ⑪ has long, bitterly cold winters. The boreal forest consists of needleleaf trees of pine, spruce, fir, and larch. Patches of deciduous aspen, poplar, willow, and birch also occur. Dairying, limited crop farming, and timber harvesting for pulp and lumber are economic activities of the boreal forest environment.
- The **tundra climate** ⑫ occupies arctic coastal fringes. Because of the marine influence, winter temperatures are not as bitterly cold as those of the boreal forest climate ⑪. The tundra is a vegetation cover of scattered grasses, sedges, lichens, and dwarf shrubs, with patches of rock fragments devoid of plants. Although the environment is

harsh, wildlife is abundant, especially in the warm months. **Permafrost** underlies much of the tundra, which presents special problems for construction of buildings and roads.

- The **ice sheet climate** ⑬ is the coldest of all climates, with no monthly mean temperature above freezing. The severity of the climate prohibits nearly all human habitation.

- Global climate is changing, largely because of human activities. Temperatures are rising, and precipitation is increasing in mid- and northern latitudes while decreasing in the subtropics. Extreme weather events are more frequent and more severe. Sea level is rising. The impacts of these changes include drought, heat waves, greater flood and storm damage, and changes in coasts and wetlands.

Key Terms

dry subtropical climate ⑤
moist subtropical climate ⑥
Mediterranean climate ⑦
marine west-coast climate ⑧
dry midlatitude climate ⑨
moist continental climate ⑩
boreal forest climate ⑪
tundra climate ⑫
permafrost
ice sheet climate ⑬

Review Questions

1. What climate type is associated with the Mojave desert? Describe some of the features of the Mojave desert environment.
2. Both the moist subtropical ⑥ and moist continental ⑩ climates are found on eastern sides of continents in the midlatitudes. What are the major factors that determine their temperature and precipitation cycles? How do these two climates differ?
3. What natural vegetation types are associated with the moist subtropical forest environment? How does human use of this environment differ between eastern North America and eastern Asia?
4. Both the Mediterranean ⑦ and marine west-coast ⑧ climates are found on the west coasts of continents. Why do they experience more precipitation in winter than in summer? How do the two climates differ?
5. Identify the characteristic features of the natural vegetation of the Mediterranean climate environment. Describe human use of this environment for agriculture and the limitations the climate places on these uses.
6. Contrast the natural vegetation types and human use of the land in the marine west-coast environments of North America and Europe.
7. The dry midlatitude environment is one of great agricultural importance. Why?
8. The moist continental climate ⑩ region supports at least two important types of forest. What are they?
9. Identify the key agricultural products of the moist continental climate region. What is the role of tall-grass prairie lands?
10. The subtropical high-pressure cells influence several climate types in the low latitudes and midlatitudes. Identify the climates and describe the effects of the subtropical high-pressure cells on them.
11. Both the boreal forest ⑪ and tundra ⑫ climate are climates of the northern regions, but the tundra is found fringing the Arctic Ocean and the boreal forest is located further inland. Compare these two climates from the viewpoint of coastal-continental effects.
12. The boreal forest environment includes vast areas of land that are little influenced by human activity. Describe the natural vegetation of the boreal forest environment.
13. What are the common plants of the tundra environment? What animals inhabit this climate zone?
14. Distinguish between two types of permafrost. What challenges does the permafrost environment provide for human habitation?
15. What is the coldest climate on Earth? How is the annual temperature cycle of this climate related to the cycle of insolation?

Focus on Systems 11.2 • California Rainfall Cycles and El Niño

1. What is the relationship between annual rainfall at Santa Barbara and the Southern Oscillation Index?
2. Is the relation between annual rainfall and the Southern Oscillation Index reliable enough to predict high-rainfall years? Defend your answer by discussing the two graphs in detail.

Eye on the Environment 11.3 • Drought and the Dust Bowl

1. What environmental disaster occurred on the Great Plains of North America in the 1930s? Explain.

A Closer Look:
Eye on Global Change 11.4 • Regional Impacts of Climate Change on North America

1. Identify six major changes in North American climate that are anticipated for the remainder of the twenty-first century and describe their general impact.
2. Summarize climate changes and effects on North America for the climate pairs Mediterranean ⑦ and marine

west-coast ⑧; boreal forest ⑪ and tundra ⑬; dry subtropical ⑤ and dry midlatitude ⑨; moist subtropical ⑥ and moist continental ⑩.

3. Find your own region in Table E11.1 and compare it to another region that you have visited or are familiar with in some way. How do predicted climate changes and impacts differ? How would a farmer or rancher fare in the two regions? an urban dweller?
4. How do climate change impacts on the water cycle vary from region to region in North America? Which region will be influenced most, in your opinion? least?

Visualizing Exercises

1. Sketch climographs for the Mediterranean climate ⑦ and the dry midlatitude climate ⑨. What are the essential differences between them? Explain why they occur.
2. Suppose South America were turned over. That is, imagine that the continent was cut out and flipped over end-for-end so that the southern tip was at about 10° N latitude and the northern end (Venezuela) was positioned at about 55° S. The Andean chain would still be on the west side, but the shape of the land mass would now be quite different. Sketch this continent and draw possible climate boundaries, using your knowledge of global air circulation patterns, frontal zones, and air mass movements.

Essay Questions

1. Discuss the role of the polar front and the air masses that come in conflict in the polar-front zone in the temperature and precipitation cycles of the midlatitude and high-latitude climates.
2. Compare and contrast the differing environments of the United States and Canada by constructing a transect linking five major cities in these countries and then describing the typical natural vegetation and human use of the land you would encounter in traveling along the transect.

Problems

Working It Out 11.1 • Standard Deviation and Coefficient of Variation

1. The following data are monthly rainfall values for Baton Rouge's wettest month (June) and driest month (September). For each month, find the sample standard deviation and the coefficient of variation. Which month has the largest standard deviation? Which has the highest relative variation?

Year	June	September
1985	11.8	14.7
1986	15.7	9.4
1987	24.2	3.9
1988	10.5	16.9
1989	38.4	14.0
1990	17.1	13.6
1991	16.7	15.0
1992	19.6	10.4
1993	8.9	3.4
1994	17.0	9.6

Eye on the Landscape

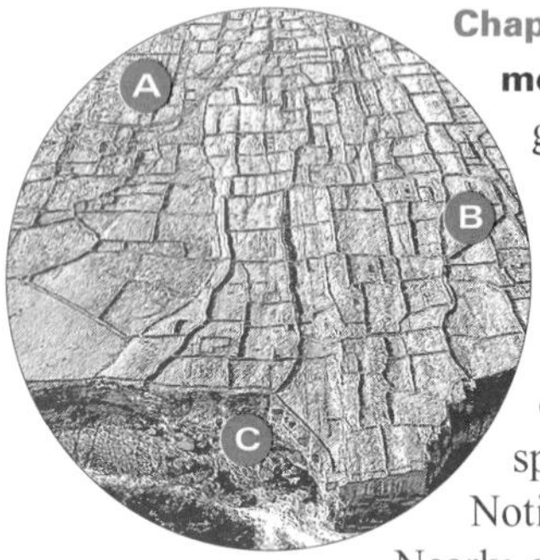

Chapter Opener Cliffs of Inishmore, Ireland The distinctive grain of this landscape comes from its limestone bedrock **(A)**, which dips shallowly to the left. The outcropping of individual limestone beds creates the linear pattern that spans the width of the island. Notice how thin the soils are. Nearly every field has rock outcrops within it **(B)**. The thin soils are a result of the dissolution of the pure limestone by abundant rainwater, leaving little residual material behind to form soil. Notice also the huge, rectangular blocks of limestone **(C)**. Limestone is a sedimentary rock that is described in more detail in Chapter 12. Its dissolution is a form of chemical weathering, discussed in Chapter 15.

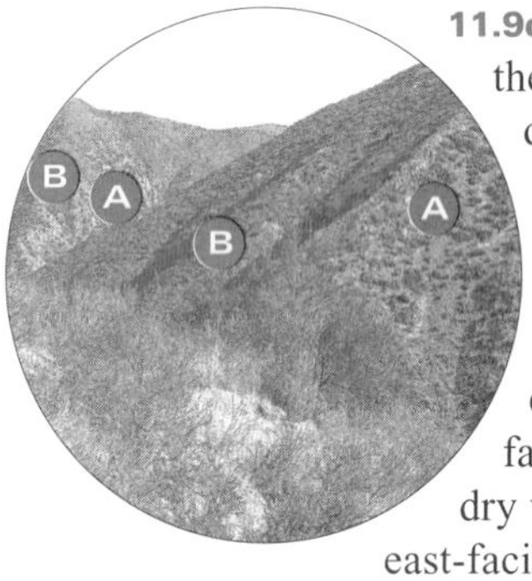

11.9c California chaparral Note the differences in vegetation density present on these chaparral slopes. In Mediterranean climates ⑦, slopes often show marked asymmetry of their vegetation cover, with south- and west-facing slopes showing sparse, dry vegetation **(A)** and north- and east-facing slopes showing lusher, denser plant covers **(B)**. The effect is usually due to differing insolation, which is stronger on south- and west-facing slopes. This creates more evaporation from the soil, which tends to limit plant growth. The effect can also be seen in Figure 11.9*b* where blue oaks are more frequent on the slopes facing to the left. Water limitations to plants are discussed in Chapter 23, Biogeographic Processes.

11.18 Moist continental landscape The Mosel, like many larger rivers, carves large loops in its path that are termed *meanders* **(A)**. As the loop expands outward, it cuts strongly into the adjacent terrain, creating a steep slope **(B)**. As it migrates outward, the loop leaves gentle slopes mantled by river sediments **(C)**. River meanders are discussed in Chapter 17.

A CLOSER LOOK

Eye on Global Change 11.4 Regional Impacts of Climate Change on North America

By nearly all accounts, the Earth is warming under the influence of human-induced releases of carbon dioxide into the atmosphere. Will the warming continue? How will the effects be felt in different regions? To answer these questions, we must rely on global climate models—mathematical models that run on supercomputers and model daily weather for long periods under assumptions of an increasing greenhouse effect. Though imperfect, the models point the way to possible effects that have great implications for human society as well as natural ecosystems.

Eye on Global Change 5.5 • The IPCC Report of 2001 has described the changes in global climate we can expect for the balance of the twenty-first century, based on runs of several global climate models under a wide range of scenarios of forcing variables. Recall that these predictions included:

- Global average temperatures rising 1.4–5.8°C (2.5–10.4°F), with temperatures of land areas, and especially northern-latitude and winter temperatures, rising even more.
- Increased precipitation in northern mid- and high latitudes.
- Over land, more hot days and increased summer drought, coupled with increased precipitation in extreme events.
- Continued melting and retreat of ice caps and glaciers.
- Continued rise in sea level by 9–88 cm (3.5–35.6 in.).

Because the implications of these predictions are so far-reaching and have so many important consequences, the United States and Canada recently completed large studies predicting climate change and analyzing the effects by geographic region.* Table E11.1 reports these regional changes and their impacts under headings of agriculture, forests and ecosystems, the water cycle, urban and human impact, transportation and infrastructure, coastal and marine environments, and natural hazards. From the table it is clear that every region will suffer change and disruption, sometimes with very far-reaching consequences. The accompanying diagram and photo display illustrate some of the impacts and stresses that will occur.

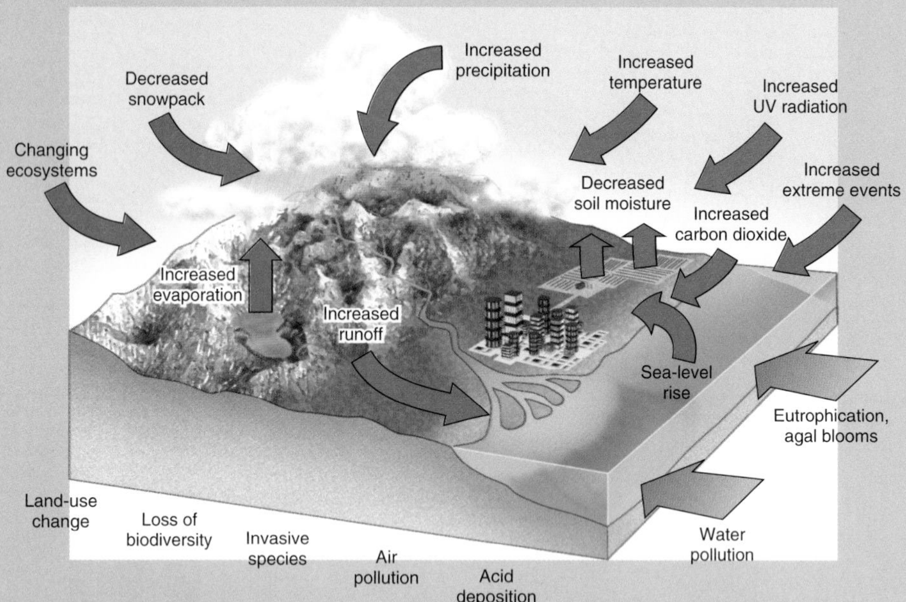

E11.4.1 Multiple stresses of a changing climate. *Natural ecosystems and human activities will be stressed in many ways, as diagrammed here, by climate change. (After National Assessment Synthesis Team, U.S. Global Change Research Program.)*

The principal changes in climate for North America forecast by the global climate models in the American and Canadian reports include:

- *General air temperature increases, with more northerly locations showing greater increases than southerly locations.* These will lead to major shifts in ecosystems, as forests migrate farther north, grasslands and savannas expand, and tundra shrinks.

*For the United States, see National Assessment Synthesis Team, *Climate Change Impacts on the United States: The Potential Consequences of Climate Variability and Change,* U.S. Global Change Research Program, Washington, DC, 2000. Canada has published a series of eight detailed reports on climate change within its regions as part of the Canada Country Study. They are summarized in two documents, *The National Summary for Policy Makers* and *Highlights for Canadians,* both issued in 1997 and available at http://www.ec.gc.ca/climate/ccs.

- *Warmer nighttime temperatures and warmer winter low temperatures.* These will increase growing seasons in many areas as well as moderate extreme cold weather events. Frozen rivers, lakes, and sea ice in Canada will thaw sooner and freeze later, facilitating water-based transportation.
- *Higher summer temperatures and heat indexes in most regions.* Extreme heat events will become more frequent, severely stressing crops, livestock, humans, and water supplies. Air quality in cities will be lowered by heat and smog.
- *More precipitation in most regions.* Although precipitation will increase, rising temperatures will stimulate more evaporation, causing summer drought and water shortages in many regions.
- *More frequent extreme precipitation events.* These will lead to more flooding, both locally in thundershowers and squall lines and regionally in hurricanes and extratropical cyclones.
- *Enhanced effects of El Niño and La Niña.* These cycles influence the frequency and magnitude of Pacific storms and Atlantic hurricanes, as well as wet and dry years in the southeast.

Considering specific climates, the changes and impacts can be highlighted as follows:

- *Mediterranean ⑦ and Marine West-Coast Climates ⑧:* Pacific subtropical high pressure will move northward and intensify during the summer, increasing air temperatures and reducing precipitation in these climate types. In the winter, the pressure gradient between Aleutian low pressure and Pacific subtropical high pressure will intensify, bringing more frequent and more severe storms to the coast and inland regions. These effects will accentuate the wet-winter–dry-summer contrast in west-coast climates. Contention over water resources between urban and agricultural uses, as well as between political divisions, will be enhanced.
- *Boreal Forest ⑪ and Tundra Climates ⑬:* Temperatures will moderate, particularly in winter, and the growing season will increase in length. Boreal forests will move northward, displacing tundra, and will in turn be replaced by mixed forests on their southern boundary. Fires, disease, and insect damage during prolonged summer water stress will reduce forest productivity. Extensive thawing of discontinuous permafrost will cause erosion, landslides and mudflows, sinking of the ground surface, and damage to forests and structures. Oxidation of organic matter in soils will release CO_2, enhancing the greenhouse effect and exacerbating global warming
- *Dry Subtropical ⑤ and Dry Midlatitude Climates ⑨:* Winter precipitation in the arid southwest will increase, benefiting urban areas and agriculture. Deserts will retreat, and some will be replaced by grasslands and savannas. Rain shadow effects in the intermountain region and east of the rockies will be enhanced, reducing soil moisture and stressing irrigation systems. While precipitation will increase in more northern regions, it will not keep pace with enhanced evaporation from rising temperatures, and summer soil moisture will decrease, impacting dry-land farming.
- *Moist Subtropical ⑥ and Moist Continental Climates ⑩:* Although precipitation will increase in the midwest, higher temperatures will reduce soil moisture, lowering water levels in lakes and rivers in summer. This will degrade water quality and affect river and lake transportation in the midwest. In the southeast, warming is predicted, but models disagree on precipitation change. More frequent extreme precipitation events here will cause more frequent flooding. Sea-level rise will lead to increased damage to barrier beaches and structures from storm surges, as well as to salt-water encroachment on coastal ground water tables. In the northeast, temperature increases will be more moderate, but still cause migration of forest species northward. The maritime provinces will warm only slightly, if at all, but will be affected by sea-level rise. In northern Ontario and Quebec, warming will be more intense, leading to species migration and ecosystem disruption.

From these observations it should be obvious that global warming and climate change will have major impacts on North Americans. To reduce the costs to society and the environment, we need to follow a two-pronged strategy. First, abatement is needed. We need to reduce emissions of CO_2 and other greenhouse gases by reducing our dependence on fossil fuels. This will require both active energy conservation and shifting to alternative energy sources such as solar, wind, and wave power. Water conservation will also be important in many regions, as water supplies are reduced and demand increases.

Second, mitigation will be required. Substantial investments in infrastructure will be needed, for example, in construction of new coastal and maritime structures where sea levels are rising or water levels are falling. In agriculture, new crop varieties and cropping practices will have to be developed. In nearly all economic endeavors, we will have to factor in the costs of dealing with a greater risk of extreme events and their impact.

Whatever the course of global warming, human society will be best served anticipating and planning for a changing future, rather than by reacting to it as it occurs.

Table E11.1 | Impact of Global Climate Change on Regions of the United States and Canada

Region	Observed Climate Trends, Past Century	Future Climate Trends, Next Century	Agriculture	Forests and Ecosystems
Alaska, British Columbia, and Yukon	• Substantial warming—2°C (4°F) since 1950s; especially interior in winter, 4°C (7°F). • Growing season lengthened 14 days since 1950s • Precipitation increased 30% since 1968	• Warming to continue, 3–10°C (7–18°F) • Precipitation to increase 20–25% with 10% decrease along south coast	• Lengthened growing season will enhance productivity • Summer drought on south coast may reduce productivity	• Warming will increase forest productivity on moist southern coast but reduce productivity in dry interior • Increased disturbance, including insects, blow-downs, and fire will reduce forest productivity • Warming will enhance decay of soil organic matter, adding CO_2 to carbon cycle • Potential shift of boreal forest northward into tundra zone
Canadian Arctic	• Mackenzie River district has warmed by 1.5°C (3°F); arctic tundra by 0.5°C (1°F) over last 100 years • Arctic mountains and fiords of eastern Arctic have cooled slightly	• Future winter temperature increases of 5–7°C (9–13°F) predicted over mainland and arctic islands • Modest cooling in extreme eastern arctic • Summer temperatures to increase up to 5°C (9°F) on mainland, 1–2°C (2–4°F) over marine areas • Annual precipitation to increase up to 25%	• Agricultural opportunities (e.g., irrigated wheat) will arise in central and upper Mackenzie River basin, but will be restricted by availability of suitable soils	• Tundra and taiga/tundra ecosystems reduced by as much as 2/3 in size • Freshwater species to migrate northward about 150 km (50 mi) per °C (°F) increase in temperature • Seal, sea lion, and walrus populations will decline through pack ice recession • Muskoxen and high-arctic Peary caribou may become extinct
Pacific Northwest	• Average annual temperature has increased 0.5–1.5°C (1–3°F) over most of the region • Annual precipitation has increased by 10%; to 30–40% east of the Cascades • Warm-dry and cool-wet years correlate with ENSO and Pacific Decadal oscillation cycles	• Annual temperatures to increase steadily, reaching +4–4.5°C (+7–8°F) by 2090 • Winter temperatures to rise by 4.5–6°C (8–11°F) • Precipitation to rise, possibly as much as 50% by 2090, with summer precipitation unchanged or slightly decreasing • Extreme precipitation events increase substantially	• Dry-land farming cycle shifted by wetter and warmer winters • Summer droughts will impact irrigated agriculture in conflict with urban usage	• Salmon populations reduced by increased winter flooding, reduced summer and fall flows, rising stream and estuary temperatures • Conifer forest will be stressed by warm, dry summers leading to pest infestations and fires • Conifer recruitment reduced by summer stress on seedlings

Water Cycle	Urban/Human Impact	Transportation and Infrastructure	Coastal/Marine Environments	Natural Hazards
• Spring flooding enhanced by warm temperatures and rainfall • Flood protection works on south coast rivers and streams may not be adequate	• Summer drought along south coast and southern interior will place water supplies for urban areas in contention with agriculture • Possible urban air quality impacts of summer drought	• Melting permafrost will damage roads, pipelines, and structures • Sea-level rise will increase coastal flooding, placing docks and port facilities at risk	• Sea-level rise could be as much as 30–50 cm (12–20 in.) • Marine ecosystems of Gulf of Alaska and Bering Sea already show large fluctuations with causes unknown; climate change effect likely to be large and unpredictable	• Retreat of glaciers will cause local flooding and enable landslides and mudflows, causing loss of life and infrastructure • Permafrost will thaw in discontinuous permafrost regions, causing erosion, landslides, sinking of ground surface, damage to forests, buildings, infrastructure • Sea ice will retreat, allowing coastal erosion and storm surges
• Increased evaporation in arctic regions from warmer atmosphere and longer thaw period will decrease flows and levels of northward-flowing rivers • River ice season reduced by 1 month by 2050; ice season for large lakes by 2 weeks	• Subsistence of native peoples likely to be made more difficult as populations of mammals, fish, and sea birds fluctuate, and lack of snow and ice makes hunting more difficult	• Reduced sea ice will benefit offshore oil and gas operations • Melting permafrost will negatively affect pipelines • Shipping season extended, with easy transit through northwest passage at times	• Flooding of coastal ecosystems, e.g. Mackenzie Delta, from sea-level rise • Reduced sea ice and increased water temperatures with unknown effects on marine biota	• Melting of permafrost will cause land subsidence, forest loss, damage to structures • Retreat of eastern mountain glaciers will produce soil instability with landslides and mudflows
• Warmer, wetter winters will increase flooding in rainfed rivers and trade snow pack for runoff; snow pack reduced by 75–125 cm (30–50 in.) • Summer shortages will be more severe because of reduced snow pack and earlier melting; summer soil moisture reduced 10–25% • Allocation conflicts and conflicting authorities may enhance vulnerability to drought	• Sea-level rise to require substantial investments to control coastal flooding, especially in southern Puget Sound where subsidence is occurring • Summer drought to impact urban areas through water supply, air quality	• New investments needed for water resource management for winter floods and summer droughts	• Increased frequency of severe storms will enhance storm surge flooding and coastal erosion, especially during El Niño events	• Higher precipitation and more extreme events will increase soil saturation, mudflows and mass movements • Winter flooding enhanced, with large populations at risk

(table continues)

Table E11.1 | (continued)

Region	Observed Climate Trends, Past Century	Future Climate Trends, Next Century	Agriculture	Forests and Ecosystems
American West	• Temperatures have risen 1–3°C (2–5°F) • Precipitation has generally increased; some areas by greater than 50%, but Arizona drier • Extreme precipitation events more frequent • Snow season has decreased by 16 days since 1951 in California and Nevada • Exceptionally wet and dry periods experienced	• Temperature to increase 1.5–2°C (3–4°F) by 2030s; 4.5–6°C (8–11°F) by 2090s • Precipitation to increase, especially winter precipitation over California • Rockies to get drier • More extreme wet and dry years	• Increased precipitation will increase crop yields and decrease water demands • Milder winter temperatures will lengthen the growing season and shift cropping areas northward • Higher temperatures will increase heat stress, weeds, pathogens, etc. • Forage production increased, growing and grazing season lengthened, with more precipitation and warmer temperatures	• Grassland, woodlands, and forests to increase, with a loss of desert vegetation • In fragmented western landscape, species can't migrate easily in response to warming; alpine ecosystems could become extinct • Increases in fires during dry cycles and air pollution damage could reduce productivity
Great Plains and Canadian Prairies	• Temperatures have risen more than 1°C (2°F) in past century; increases up to 3°C (5.5°F) in northern Great Plains and Canadian prairies • Annual precipitation decreased 10% in northern section, increased 10% in eastern section • Texas has had more high-intensity rainfall events • Snow season ends earlier in spring	• Temperatures increase and soil moisture decreases, even with increased rainfall • More frequent heat events, with impact on people and livestock • Precipitation increases 10–20 percent in northern area and decreases 10–20% in southern area and in Rocky Mountain rain shadow	• Higher temperatures and lower soil moisture will decrease crop yields in Canadian Prairies • Longer growing season may increase crop production at northern limits • Southern area and Rocky Mountain rain shadow to suffer reduced soil moisture and increased crop stress	• Decreased severity of winters and changed habitats will alter distributions of big game, waterfowl, and game birds • Some freshwater fish species may extend ranges north • Some seasonal and semipermanent wetlands will dry up • Some native species to be displaced by undesirable invasive species as climate changes
Ontario and Midwest	• Northern portion has warmed by 2°C (4°F), southern portion (Ohio River Valley) has cooled by 0.5°C (1°F) • Precipitation has increased 10–20% in 20th century • Substantial rise in number of heavy and very heavy precipitation events	• Temperature to rise 3–6°C (5–10°F) • Average minimum temperature to rise as much as 0.5–1°C (2–2°F) more than maximum temperature • Precipitation to increase 10-30% • Precipitation increase will not keep pace with temperature increase, creating soil moisture deficits, reduced lake and river levels, more frequent drought	• With longer growing season, double cropping will become more prevalent • CO_2 fertilization will enhance crop yields • Reduced soil moisture will reduce yields in some areas of the corn belt	• Boreal forest to shift north and suffer more fire, insects, disease; transitional forests to shift north as well • Changes in water temperature will affect freshwater ecosystems, changing species compositions • Runoff of excess nutrients and wastes into lakes and rivers will be enhanced by heavy precipitation events • Decreasing lake levels will impact current wetlands • Birds and wildlife populations will be impacted by changing habitats and food sources

Water Cycle	Urban/Human Impact	Transportation and Infrastructure	Coastal/Marine Environments	Natural Hazards
• Increased precipitation will increase water supplies, reduce demand, and ease competition among competing uses • Greater runoff will increase hydropower production and ease some water quality problems • But in drier regions, expect reduced supply, power production, and more contention among users	• Sea-level rise will impact coastal communities with increased lowland flooding and cliff erosion • Warmer temperatures will increase heat stress on urban populations during dry summers and lower air quality in dry summers • For tourism, more backpacking and less skiing • Less water available for recreation during dry years	• Flood control required for enhanced winter rains • New waterworks for supply during dry years	• Rising sea levels will threaten coastal wetlands, e.g., San Francisco Bay, and reduce diversity	• More extreme rainfall events will cause more mass movements of soil • In wet years, floods will be more damaging
• More competition among agriculture, ecosystem, urban, industrial, and recreational users for water in American Great Plains • Canadian hydroelectric power will compete with agriculture and urban uses of water • Little hope for recharge of declining ground water reserves in southern Great Plains	• Heat index to increase by as much as 25% in northern Texas, Oklahoma, Kansas region, along with more frequent heat events • Changing climate will impact small family farmers and ranchers harder than commercial agriculture	• River transportation affected by reduced water levels during summer droughts		• River flooding from extreme events will be enhanced • Sand Hills (dunes) may become mobile again if vegetation cover declines
• Great Lakes levels to fall 0.3–1.5 m (1–5 ft); outflow to St. Lawrence Seaway reduced 20-40% • Increased national and international tension over lake and river waters • Reduced river levels will reduce water quality	• Reduced risk of life-threatening cold in winter but increased risk of life-threatening heat in summer • Increased frequency of urban air pollution episodes • River flooding more frequent	• Reduced water levels on rivers and lakes will make water-based transportation more difficult and expensive; will require new docks and locks • Reduced ice cover lengthens lake/river shipping seasons		• Flash flooding from more frequent heavy rains will wash sediment, agricultural pollutants into rivers and lakes • Sustained heavy precipitation events may cause more catastrophic river flooding

(table continues)

Table E11.1 (continued)

Region	Observed Climate Trends, Past Century	Future Climate Trends, Next Century	Agriculture	Forests and Ecosystems
Southeast	• Temperatures were warm during the 1920s-1940s; cooled through the 1960s; warmed again starting in 1970s to levels of the 1920s and 1930s • Coastal regions warmed by 2°C (4°F) • Rainfall increased 20–30% or more over much of the region • Strong El Niño and La Niña effects provide large seasonal and interannual variation in temperature and precipitation, with hurricanes more frequent in La Niñas	• Models predict warming from about 3–6°C (5–10°F), depending on model and region • Models disagree on precipitation predictions, ranging from neutral or slight increase to 25% increase • El Niño rainfall and La Niña droughts may intensify as atmospheric CO_2 increases • Heat indexes rise dramatically	• Models disagree on soil moisture trends; could increase or decrease depending on amount of warming and change in precipitation • Models predict different crop yield scenarios ranging from increases in interior regions to decreases for dry-land crops • Significant decreases in yields of corn, peanuts, sorghum along Gulf Coast	• Models disagree on forest productivity; generally it increases, but decreases in some places for some ecosystems in some scenarios • If soil moisture drops, pine forests are replaced by savannas and grasslands
Northeast and Southern Quebec	• Temperatures have increased as much as 2°C (4°F) in last 100 years along coastal margins • Precipitation has increased greater than 20% over much of the region • Precipitation extremes are increasing, drought decreasing • Period between first and last dates of snow on the ground has decreased by 7 days in the last 50 years	• Temperatures will increase 3–5°C (5–9°F) depending on area and model • Winter minimum temperatures show greatest increases • Precipitation projections range from neutral or modest increases to nearly 25% • Winter snowfalls and periods of extreme cold will decrease • Ice storms, rains over frozen ground, and rapid snow melting events will increase • Heavy precipitation events will increase; hurricane frequency could increase	• Temperature and rainfall increases may enhance crop productivity • Earlier greenup will extend growing season • More frequent extreme rainfalls and storms will affect local agriculture	• Forest composition will change; northern hardwoods move north as oak-hickory forests replace them; sugar maple to migrate out of New England • Lobster populations will move northward • Migratory bird habitat will be reduced by sea-level rise • Trout populations reduced in Pennsylvania • Muting of fall foliage as species composition changes with added warmth • Milder winters will favor insect vectors for human and animal disease
Canadian Atlantic Provinces	• Slight cooling experienced in past 50 years • Higher frequency of extreme events observed	• No strong evidence for significant warming • Increased intensity and/or frequency of storms anticipated • More temperature extremes, leading to milder winters, early extended thaws, late springs, early frosts	• Increased frequency of storms could impact crop productivity	• Fish habitat could be lost; distribution of fish species and migration patterns altered, with life cycles interrupted • Winter ranges of terrestrial birds could shift; seabirds could change in range, distribution, and breeding success • Reduced snow cover could increase deer populations but reduce forest regeneration and species diversity • Increased forest blow-downs and mortality with increased storms and insect outbreaks

Water Cycle	Urban/Human Impact	Transportation and Infrastructure	Coastal/Marine Environments	Natural Hazards
• Stresses on water quality are associated with intensive agricultural practices, urban development, coastal processes, and mining activities • Water quality is impacted by higher temperatures that reduce dissolved oxygen and by contamination from agricultural runoff, untreated sewage, and chemical releases in catastrophic rainfall events	• Hurricanes, floods, and heat waves have major impacts on human population; climate change could enhance their frequency and severity • Low air quality episodes from increased temperatures and pollution releases accompany heat waves	• Flooding from extreme events may create transportation difficulties on roads and rivers	• Rising sea level, coupled with subsidence due to ground water withdrawal, sediment compaction, wetland drainage, and levee construction, will cause continued loss of coastal wetlands • Salt water intrusion of ground water from rising sea level will continue to kill coastal forests • Rising sea level increases frequency of storm surge and flooding of low-lying areas by storms and causes retreat of barrier beaches	• Coastal flooding in hurricanes can produce catastrophic loss of life
• Variations in water supply in Quebec may adversely affect power generation	• Climate change will add significantly to stresses on major urban areas, such as electric power consumption, summer air quality, heat waves, etc. • Summer recreation season to increase, but winter ski season to be shorter • Loss of beachfront property and recreation areas from sea level rise	• Low water levels in the St. Lawrence River will impact shipping and the marine environment • Sea-level rise may require engineering structures for coastal protection	• Estuarine water quality will fall as increased temperatures reduce dissolved oxygen and extreme rainfall events provide more polluted runoff and reduce salinity • Sea-level rise will substantially increase loss of wetlands and marshes	• Enhanced coastal flooding in hurricanes and northeasters from sea-level rise
• Less winter snow cover is predicted • More extremes of excessive moisture and drought • Declining runoff would reduce hydroelectric power generation	• Sea-level rise will increase danger of flooding of urban areas	• Changes in ice-free days would affect marine transportation and offshore oil and gas industry.	• Sea-level rise will affect coastlines from Bay of Fundy to Newfoundland; threaten Acadian dykeland agriculture in Nova Scotia	• Sea-level rise will enhance coastal flooding during severe storms

Chapters in Part 3

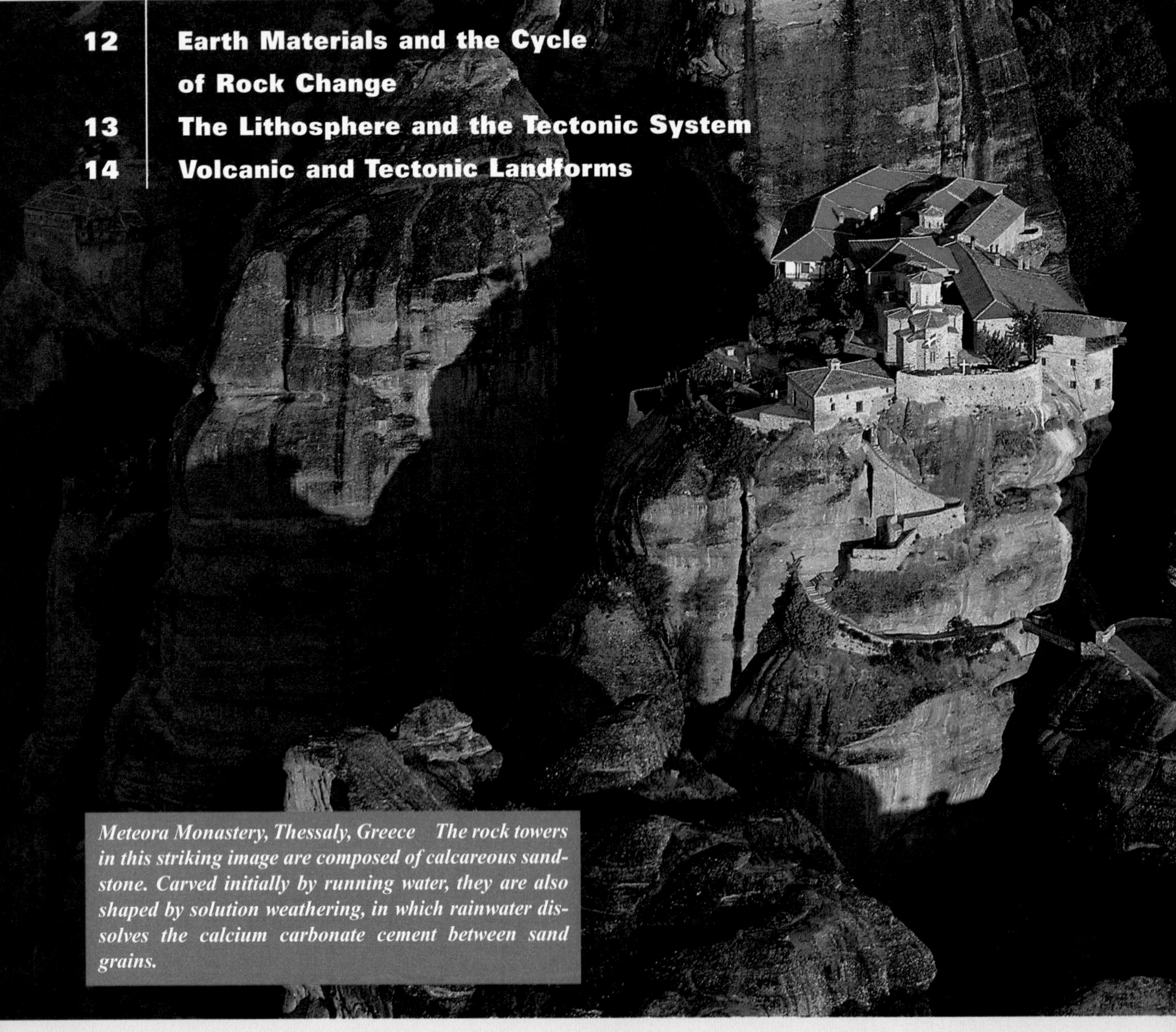

Meteora Monastery, Thessaly, Greece *The rock towers in this striking image are composed of calcareous sandstone. Carved initially by running water, they are also shaped by solution weathering, in which rainwater dissolves the calcium carbonate cement between sand grains.*

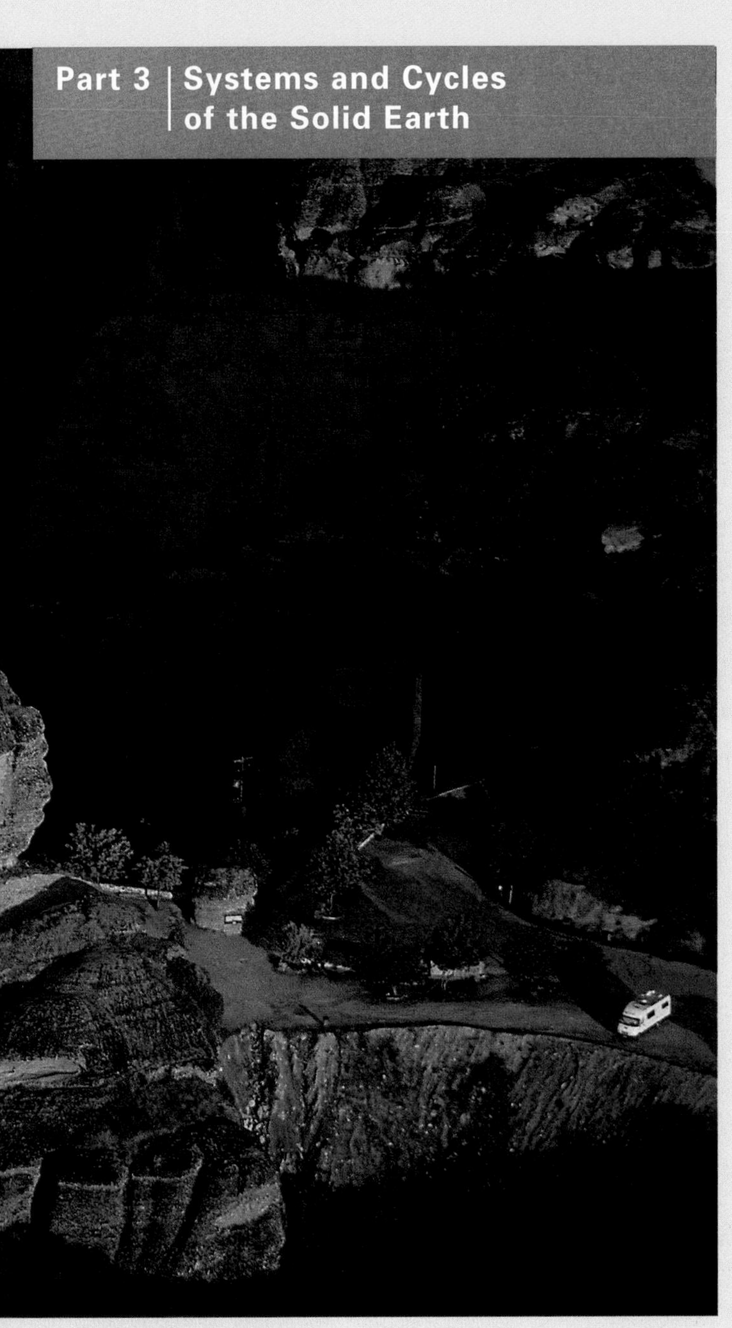

Part 3 | Systems and Cycles of the Solid Earth

In Part 3, we move from examining the flows of energy and matter at the Earth's surface to the systems and cycles of the solid Earth. These systems differ in two very important respects. One is their power sources—surface energy and matter flows are driven by solar radiation, while solid Earth systems are driven by the Earth's internal heat. These two great systems of energy and matter flow also operate at very different time scales. For surface energy flow systems, the most important rhythms are daily and annual, responding to the rotation of the Earth on its axis and the revolution of the Earth around the Sun. For Earth systems, on the other hand, cycles are typically measured in thousands to millions of years. ■ Chapter 12 begins Part 3 by presenting the basic materials of the solid Earth—rocks and minerals—and some principles of their formation. These are linked in the cycle of rock change, which describes how Earth materials are constantly being cycled and recycled over geologic time. In Chapter 13, we introduce the grandest cycle of them all—that of plate tectonics, in which the Earth's surface configuration of continents and ocean basins slowly changes as forces deep within the Earth cause vast tectonic plates to converge, collide, split, and separate. The present pattern of plates and their motions explains the location of many geologic surface phenomena, such as earthquakes and volcanoes. These and other manifestations of geologic activity, such as folds and faults, are the subject of Chapter 14. ■ Physical geographers are concerned with the solid Earth because the Earth's outer layer serves as a platform, or base, for life on the lands. This platform provides the continental surfaces that are carved into landforms by moving water, wind, and glacial ice. Landforms, in turn, influence the distribution of ecosystems and exert strong controls over human occupation of the lands. Landforms made by water, wind, gravity, and ice are the subject of the chapters in Part 4 of this book.

12 EARTH MATERIALS AND THE CYCLE OF ROCK CHANGE

EYE ON THE LANDSCAPE

Village of Bacolor, Luzon, Philippines. The nearby eruption of Mount Pinatubo in 1991 yielded a huge volume of volcanic ash that was subsequently carried downslope in hurricane rains, burying this village in many meters of sediment. **What else would the geographer see?... Answers at the end of the chapter.**

The solid Earth is a platform for both life and landform-making processes. We begin Part 3, our study of solid Earth, by examining the minerals and rocks that are found at or near the Earth's surface. There are a great number of minerals and rocks, but we will focus only on the few that are most important for a broad understanding of the continents, ocean basins, and their varied features. We will also examine the rock cycle, in which Earth materials are constantly formed and reformed over geologic time in a cycle powered largely by heat from the Earth's interior.

THE CRUST AND ITS COMPOSITION

The thin, outermost layer of our planet is the *Earth's crust.* This skin of varied rocks and minerals ranges from about 8 to 40 km (about 5 to 25 mi) thick and contains the continents and ocean basins. It is the source of soil on the lands, of salts of the sea, of gases of the atmosphere, and of all the water of the oceans, atmosphere, and lands.

Figure 12.1 displays the eight most abundant elements of the Earth's crust in terms of percentage by weight. Oxygen, the predominant element, accounts for a little less than half the total weight. Second is silicon, which accounts for a little more than a quarter. Together they account for 75 percent of the crust, by weight.

Aluminum accounts for approximately 8 percent and iron for about 5 percent of the Earth's crust. These metals are very important to our industrial civilization, and, fortunately, they are relatively abundant. Four other metallic elements—calcium, sodium, potassium, and magnesium—make up the remaining 12 percent. All four occur at about the same order of abundance (2 to 4 percent). These elements are essential

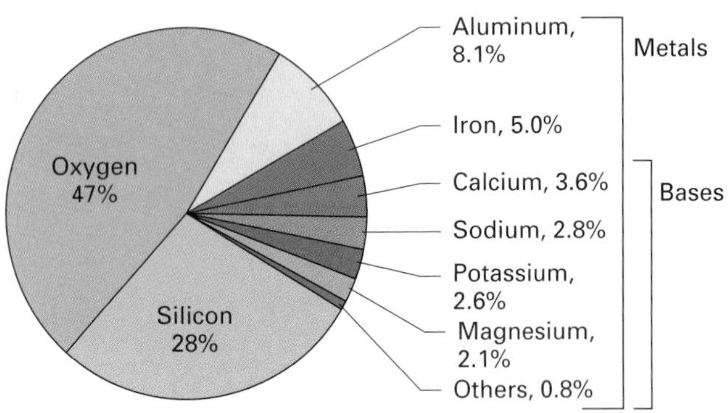

12.1 Crustal composition *The eight most abundant elements in the Earth's crust, measured by percentage of weight. Oxygen and silicon dominate, with aluminum and iron following.*

nutrients for plant and animal life and are described in Chapter 22 as plant nutrients. Because of this role, they are also important determiners of soil fertility (Chapter 21).

Of the remaining chemical elements composing the Earth's crust, a few are very important from the viewpoint of powering the cycles of the solid Earth. These are the radioactive forms of elements that, through slow radioactive decay, provide a nearly infinite source of heat that seeps slowly outward from the Earth's interior. Some of these are introduced in *Working It Out 12.1 • Radioactive Decay.* We will return to the topic of radioactive decay and radiogenic heating in Chapter 13.

Rocks and Minerals

The elements of the Earth's crust are usually bonded with other elements to form chemical compounds. We recognize these compounds as minerals. A **mineral** is a naturally occurring, inorganic substance that usually possesses a definite chemical composition and characteristic atomic structure. Most minerals have a crystalline structure—gemstones such as diamonds or rubies are examples, although their crystalline shape is enhanced by the stonecutter. Quartz is an example of a very common crystalline mineral. It usually occurs as a clear, six-sided prism (Figure 12.2). GEODISCOVERIES

Minerals are combined into **rock,** which we can broadly define as an assemblage of minerals in the solid state. Rock comes in a very wide range of compositions, physical characteristics, and ages. A given variety of rock is usually composed of two or more minerals, and often many different minerals are present. However, a few rock varieties consist almost entirely of one mineral. Most rock of the Earth's crust is extremely old by human standards, with the age of formation often ranging back many millions of years. But rock is also being formed at this very hour as active volcanoes emit lava that solidifies on contact with the atmosphere or ocean.

Rocks of the Earth's crust fall into three major classes. (1) **Igneous rocks** are solidified from mineral matter in a high-temperature molten state. (2) **Sedimentary rocks** are formed from layered accumulations of mineral particles derived mostly by weathering and erosion of preexisting

12.2 Quartz crystals *These large crystals of quartz form six-sided, translucent columns.*

rocks. (3) **Metamorphic rocks** are formed from igneous or sedimentary rocks that have been physically or chemically changed, usually by application of heat and pressure during crustal mountain-making.

The three classes of rocks are constantly being transformed from one to another in a continuous process through which the crustal minerals have been recycled during many millions of years of geologic time. Figure 12.3 diagrams these transformations. In the process of melting, preexisting

12.3 Rock transformation *The three classes of rock are transformed into one another by weathering and erosion, melting, and exposure to heat and pressure.*

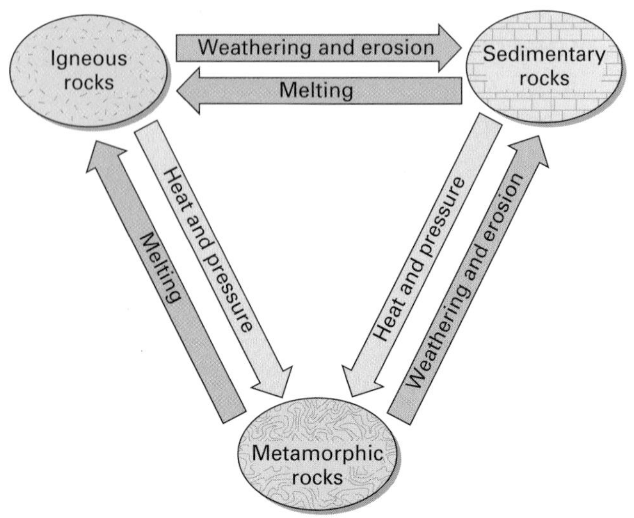

Radioactive Decay

The Earth's interior is largely heated by the spontaneous decay of naturally occurring radioactive isotopes of certain elements. You may recall from basic chemistry that the properties of an element are determined by the number of positively charged particles, or *protons,* contained in its nucleus. The number of protons, called the *atomic number,* identifies the chemical element to which the nucleus belongs. For example, the element uranium has 92 protons. Nuclei of atoms also contain neutrons, which are rather similar to protons but have no charge. The total of protons and neutrons within the nucleus of an atom is known as the *atomic mass number.*

Some elements are found in forms with different mass numbers. These forms are known as *isotopes.* For example, the most common isotope of uranium (U) is uranium-238, a form with an atomic mass of 238. Another form is uranium-235, which has an atomic mass number of 235. Chemists distinguish these two forms by writing them as ^{238}U and ^{235}U.

A key to understanding radioactivity is that certain isotopes are *unstable,* meaning that the composition of the nucleus of the isotope can experience an irreversible change. This change process is known as *radioactive decay.* When a nucleus decays, it emits matter and energy. The energy is absorbed by the surrounding matter and thus ultimately takes the form of sensible heat. In some types of decay protons may be lost or gained. This means that an atom of one element may be transformed into an atom of another. For example, the uranium isotope ^{238}U decays to form ^{234}Th, an isotope of the element thorium. This new isotope is known as a *daughter product.* Often the new isotope created will be unstable as well and will decay into yet another isotope of a different element. As this process continues, the result is a decay chain of daughter products that eventually ends in the formation of a stable isotope. For example, ^{238}U ultimately forms the stable isotope lead-206, ^{206}Pb.

The significance of radioactive decay is that it provides an internal source of heat for the Earth—a source that accounts for the melting of solid rock to form magma, and thus creates igneous rocks. This heating effect is described more fully in *Focus on Systems 12.4 • Powering the Cycle of Rock Change.* The decay of ^{238}U is only one of several radioactive decay chains that are important in heating the Earth from within. Other chains begin with ^{235}U, thorium-232 (^{232}Th), and potassium-40 (^{40}K).

The time rate at which unstable isotopes decay spans a very large magnitude. Some isotopes decay in a matter of milliseconds, while others decay over a period of billions of years. The rate of decay of an unstable isotope is measured by its *half-life*—the period of time in which a number of atoms of the isotope will be reduced by half. For example, the half-life of ^{238}U is 4.47 billion years, meaning that one gram of ^{238}U will be reduced to one-half of a gram after that length of time. In another 4.47 billion years, this half-gram will be reduced to a quarter-gram, and so forth.

The graph below shows an example of the decay of potassium-40 (^{40}K). This unstable isotope has a half-life of 1.28 billion years. The *y*-axis shows the proportion of the original quantity remaining after the elapsed time shown on the *x*-axis. This curve has a *negative exponential* form—that is, a curve that decreases as a negative exponential function of e, the base of natural logarithms. For this type of function, we can write the proportion P remaining at time t as

$$P(t) = e^{-kt},$$

where k is a constant related to the half-life by the expression

$$k = \frac{0.693}{H}$$

Here, H is the half-life and 0.693 is the natural logarithm of 2. For the case of ^{40}K, $k = 0.693/1.28 = 0.542$ and therefore

$$P(t) = e^{-0.542t}$$

Since the half-life for ^{40}K is measured in billions of years, t in this expression will also be in billions of years. For example, the proportion of ^{40}K atoms remaining after 500 million years (=0.5 billion years) will be

$$P(0.5) = e^{-0.542 \times 0.5} = e^{-0.271} = 0.763$$

When ^{40}K decays, one of two products is produced. In the first type of decay, calcium-40 (^{40}Ca) is created. In the second type, which occurs less frequently, argon-40 (^{40}Ar) is created. Both ^{40}Ca and ^{40}Ar are stable isotopes and undergo no further decay. The graph shows how these products will slowly build as they are formed by the decay of ^{40}K.

The phenomenon of radioactive decay is exploited in the application of one of the most important tools available to geologists—radiometric dating, which we explore further in *Working It Out 13.1 • Radiometric Dating.*

Exponential decay and growth curves for ^{40}K, ^{40}Ca, and ^{40}Ar *(Copyright © A. N. Strahler.)*

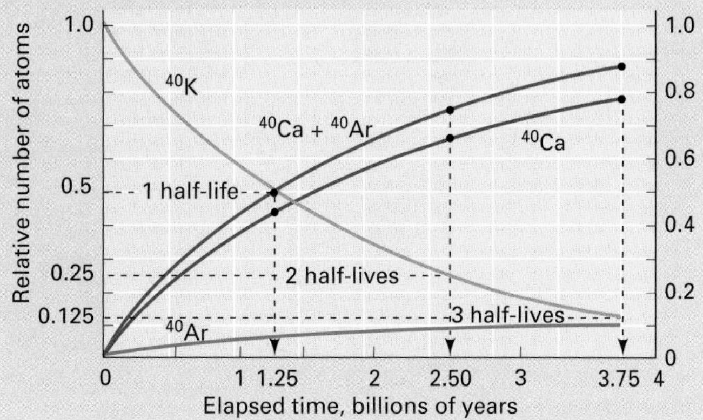

Table 12.1 | **Some Common Igneous Rock Types**

Subclass	Rock Type	Composition
Intrusive (Cooling at depth, producing coarse crystal texture)	Granite	Felsic minerals, typically quartz, feldspars, and mica
	Diorite	Felsic minerals without quartz, usually including plagioclase feldspar and amphibole
	Gabbro	Mafic minerals, typically plagioclase feldspar, pyroxene, and olivine
	Peridotite	An ultramafic rock of pyroxene and olivine
Extrusive (Cooling at the surface, producing fine crystal texture)	Rhyolite	Same as granite
	Andesite	Same as diorite
	Basalt	Same as gabbro

12.4 Silicate minerals and igneous rocks *Only the most important silicate mineral groups are listed, along with four common igneous rock types. The patterns shown for mineral grains indicate their general appearance through a microscope, and do not necessarily indicate the relative amounts of the minerals within the rocks.*

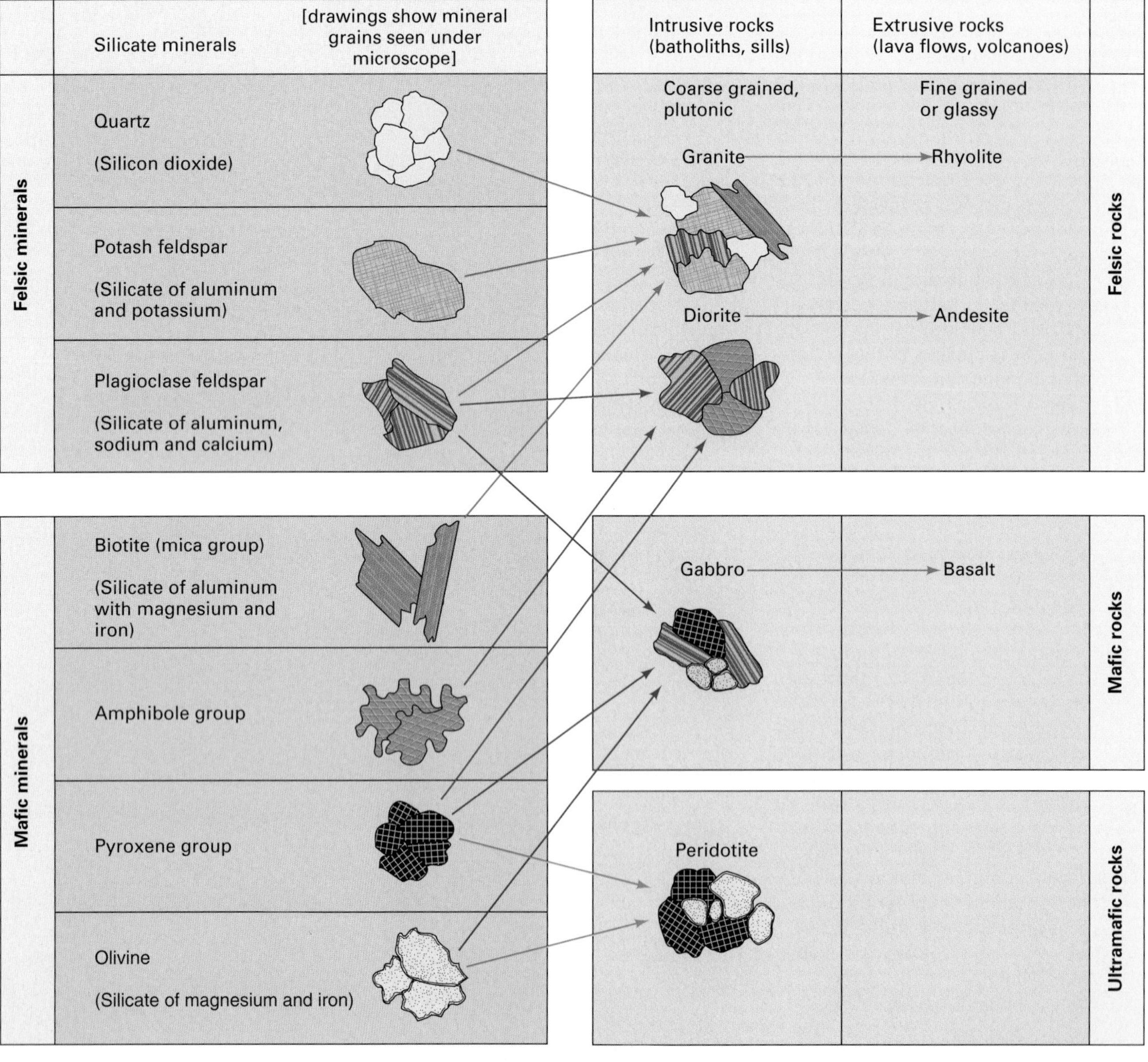

rock of any class is melted and then later cools to form igneous rock. In weathering and erosion, preexisting rock is broken down and accumulated in layers that become sedimentary rock. Heat and pressure convert igneous and sedimentary rocks to metamorphic rock. We will return to this cycle at the end of the chapter, after we have taken a more detailed look at rocks, minerals, and their formation processes.

Igneous Rocks

Igneous rocks are formed when molten material moves from deep within the Earth to a position within or atop the crust. There the molten material cools, forming rocks composed of mineral crystals. Table 12.1 presents the more important igneous rock types that we will refer to in this chapter.

Most igneous rock consists of *silicate minerals,* chemical compounds that contain silicon and oxygen atoms. Most of the silicate minerals also have one, two, or more of the metallic elements listed in Figure 12.1—that is, aluminum, iron, calcium, sodium, potassium, and magnesium. Although there are many silicate minerals, we will only be concerned with seven of them, which are shown in Figure 12.4.

Among the most common minerals of all rock classes is **quartz,** mentioned earlier, which is silicon dioxide (SiO_2) (Figure 12.2). It is quite hard and resists chemical breakdown. Two silicate-aluminum minerals called *feldspars* follow. One type, *potash feldspar,* contains potassium as the dominant metal besides aluminum. A second type, *plagioclase feldspar,* is rich in sodium, calcium, or both. Quartz and feldspar form a silicate mineral group described as **felsic** ("fel" for feldspar; "si" for silicate).

Quartz and the feldspars are light in color (white, pink, or grayish) and lower in density than the other silicate minerals. By *density,* we mean the mass of matter contained in a unit volume. Figure 12.5 shows the density of four Earth materials: water, quartz, the mineral olivine (discussed below), and pure iron.

The next three silicate minerals are actually groups of minerals, with a number of mineral varieties in each group. They are the *mica, amphibole,* and *pyroxene* groups. All three are silicates containing aluminum, magnesium, iron, and potassium or calcium. The seventh mineral, *olivine,* is a silicate of only magnesium and iron that lacks aluminum. Altogether, these minerals are described as **mafic** ("ma" for magnesium; "f" from the chemical symbol for iron, Fe). The mafic minerals are dark in color (usually black) and are denser than the felsic minerals. GEODISCOVERIES

Common Igneous Rocks

Igneous rocks solidify from rock in a hot, molten state, known as **magma.** From pockets a few kilometers below the Earth's surface, magma makes its way upward through fractures in older solid rock and eventually solidifies as igneous rock. No single igneous rock is made up of all seven silicate minerals listed in Figure 12.4. Instead, a given rock variety contains three or four of those minerals as the major ingredients. The mineral grains in igneous rocks are very tightly interlocked, and the rock is normally very strong.

The column in the center of Figure 12.4 shows four common igneous rocks. Each rock is connected by arrows to the principal minerals it contains. These four rocks are carefully selected to be used in our later explanation of features of the Earth's crust.

The first igneous rock is **granite.** The bulk of granite consists of quartz (27 percent), potash feldspar (40 percent), and plagioclase feldspar (15 percent). The remainder is mostly biotite and amphibole. Because most of the volume of granite is of felsic minerals, we classify granite as a *felsic igneous rock.* Granite is a mixture of white, grayish or pinkish, and black grains, but the overall appearance is a light gray or pink color (Figure 12.6*a*).

Diorite, the second igneous rock on the list, lacks quartz. It consists largely of plagioclase feldspar (60 percent) and secondary amounts of amphibole and pyroxene. Diorite is a light-colored felsic rock that is only slightly denser than granite. When its grains are coarse, it may have a speckled appearance (Figure 12.6*b*).

The third igneous rock is *gabbro,* in which the major mineral is pyroxene (60 percent). A substantial amount of plagioclase feldspar (20 to 40 percent) is present, and, in addition, there may be some olivine (0 to 20 percent). Since the mafic minerals pyroxene and olivine are dominant, gabbro is classed as a *mafic igneous rock.* It is dark in color and denser than the felsic rocks.

12.5 Density *The concept of density is illustrated by several cubes of the same size, but of different materials, hung from coil springs under the influence of gravity. The stretching of the coil spring is proportional to the density of each material, shown in thousands of kilograms per cubic meter. Quartz and olivine are common minerals. (Copyright © A. N. Strahler.)*

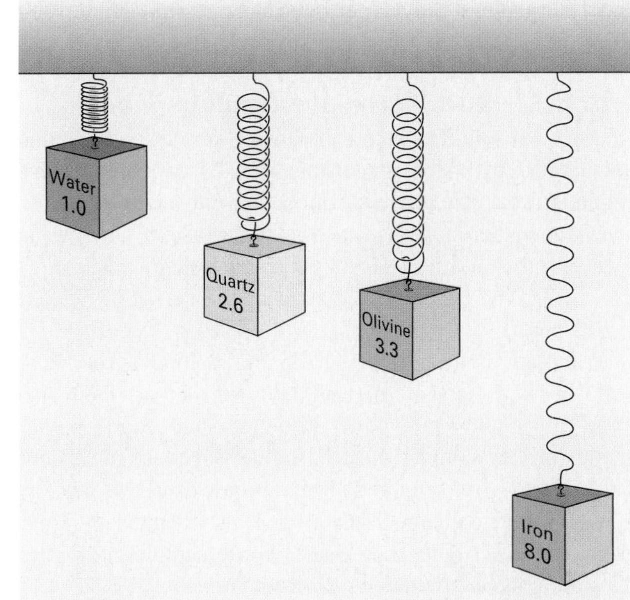

12.6 Intrusive igneous rock samples *(a) A coarse-grained granite. Dark grains are amphibole and biotite; light grains are feldspars; clear grains are quartz. (b) A coarse-grained diorite. Feldspar grains are light; amphibole and pyroxene grains are dark. Compare with granite in (a)—clear quartz grains are lacking.*

The fourth igneous rock, *peridotite,* is dominated by olivine (60 percent). The rest is mostly pyroxene (40 percent). Peridotite is classed as an *ultramafic igneous rock,* denser even than the mafic types.

These four common varieties of igneous rock show an increasing range of density, from felsic, through mafic, to ultramafic types. This arrangement according to density is duplicated on a grand scale in the principal rock layers that comprise the solid Earth, with the least dense layer (mostly felsic rocks) near the surface and the densest layer (ultramafic rocks) deep in the Earth's interior. We will stress this layered arrangement again in describing the Earth's crust and the deeper interior zones in Chapter 13.

Intrusive and Extrusive Igneous Rocks

Magma that solidifies below the Earth's surface and remains surrounded by older, preexisting rock is called **intrusive igneous rock.** The process of injection into existing rock is *intrusion.* Where magma reaches the surface, it emerges as **lava,** which solidifies to form **extrusive igneous rock** (Figure 12.7). The process of release at the surface is called *extrusion.*

Although both intrusive rock and extrusive rock can solidify from the same original body of magma, their outward appearances are quite different when you compare freshly broken samples of each. Intrusive igneous rocks cool very slowly—over hundreds or thousands of years—and, as a result, develop large mineral crystals—that is, they are *coarse-textured.* The granite and diorite samples in Figure 12.6 are good examples of coarse texture.

12.7 Fresh lava *This fresh black lava shows a glassy surface and flow structures. Chain of Craters Road, Hawaii Volcanoes National Park, Hawaii.*

12.8 Extrusive igneous rock samples *(a) Obsidian, or volcanic glass. The smooth, glassy appearance is acquired when a gas-free lava cools very rapidly. This sample shows red and black streaks caused by minor variations in composition. (b) A specimen of scoria, a form of lava containing many small holes and cavities produced by gas bubbles.*

In an intrusive igneous rock, individual mineral crystals are visible with the unaided eye or with the help of a simple magnifying lens. In an extrusive rock, which cools very rapidly, the individual crystals are very small. They can only be seen through a microscope. Rocks with such small crystals are termed *fine-textured.* If the lava contains dissolved gases, the gases expand as the lava cools, forming a rock with a frothy, bubble-filled texture, called *scoria* (Figure 12.8*a*). Sometimes a lava cools to form a shiny natural volcanic glass (Figure 12.8*b*). Most lava solidifies simply as a dense, uniform rock with a dark, dull surface.

Since the outward appearance of intrusive and extrusive rocks formed from the same magma are so different, they are named differently. The igneous rock types we have discussed so far—granite, diorite, gabbro, and peridotite—are intrusive rocks. Except for peridotite, all have counterparts as extrusive rocks. They are named in the right column of Figure 12.4. Each one has the same mineral composition as the intrusive rock named at the left. *Rhyolite* is the name for lava of the same composition as granite; **andesite** is lava with the mineral composition of diorite; and **basalt** is lava of the composition of gabbro. Andesite and basalt are the two most common types of lavas. Rhyolite and andesite are pale grayish or pink in color, whereas basalt is black. Lava flows, along with particles of solidified lava blown explosively from narrow vents, often accumulate as isolated hills or mountains that we recognize as volcanoes. They are described in Chapter 14.

A body of intrusive igneous rock is called a **pluton.** Granite typically accumulates in enormous plutons, called *batholiths.* Figure 12.9 shows the relationship of a batholith to the overlying rock. As the hot fluid magma rises, it melts and incorporates the older rock lying above it. A single batholith extends down several kilometers and may occupy an area of several thousand square kilometers.

Figure 12.9 shows other common forms of plutons. One is a *sill,* a plate-like layer formed when magma forces its way between two preexisting rock layers. In the example shown, the sill has lifted the overlying rock layers to make room. A second kind of pluton is the *dike,* a wall-like body formed when a vertical rock fracture is forced open by magma. Commonly, these vertical fractures conduct magma to the land surface in the process of extrusion. Figure 12.10 shows a dike of mafic rock cutting across layers of older rock. The dike rock is fine-textured because of its rapid cooling. Magma entering small, irregular, branching fractures in the surrounding rock solidifies in a branching network of thin *veins.*

The formation of extrusive igneous rock can be witnessed today where volcanic processes are active. *Eye on the Environment 12.2 • Battling Iceland's Heimaey Volcano* describes the struggle of courageous Icelanders to save their volcanic island village and harbor from the onslaught of a volcanic eruption.

12.9 Volcanic rock formations *This block diagram illustrates various forms of intrusive igneous rock plutons as well as an extrusive lava flow.*

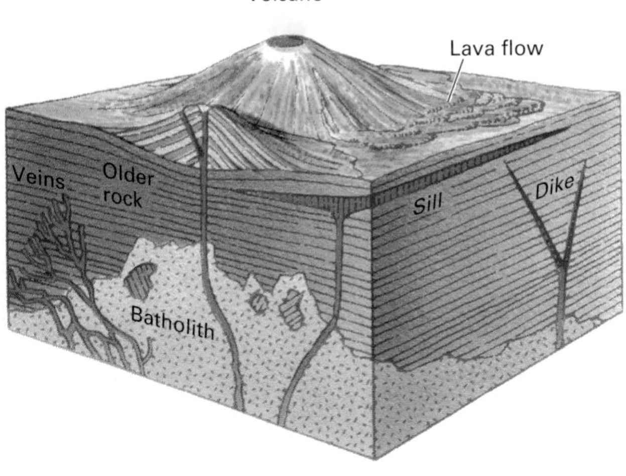

Eye on the Environment | 12.2

Battling Iceland's Heimaey Volcano

Normally the processes of rock formation are very slow, but when a volcanic eruption occurs, large volumes of fresh rock can be created very quickly. In 1973, inhabitants of Heimaey, a small island very near the southern coast of Iceland, found themselves witnessing the birth of new rock from a vantage point that was a little too close.

It was around 2:00 A.M., early on a January morning, that the eruption began. First, a fissure split the eastern side of the island from one coast to the other, sending up a curtain of fire in a pyrotechnical display nearly 2 km (1.2 mi) long. Soon, however, the spraying fountains of volcanic debris became restricted to a small area not far from Helgafell, an older volcanic cone. The lava and lava fragments of ash and cinders, called *tephra,* poured out at a rate of 100 m^3/sec (about 3500 ft^3/sec), building a cone that soon reached a height of about 100 m (about 300 ft) above sea level. It was dubbed Kirkjufell after a farmstead, Kirkjubaer, which lay beneath the debris.

It wasn't long before strong easterly winds set in, and Iceland's

Vestmannaeyjar after the eruption *Lava (right) invaded the village of Vestmannaeyjar, while fine black cinders (now swept from roofs and roads) covered the town.*

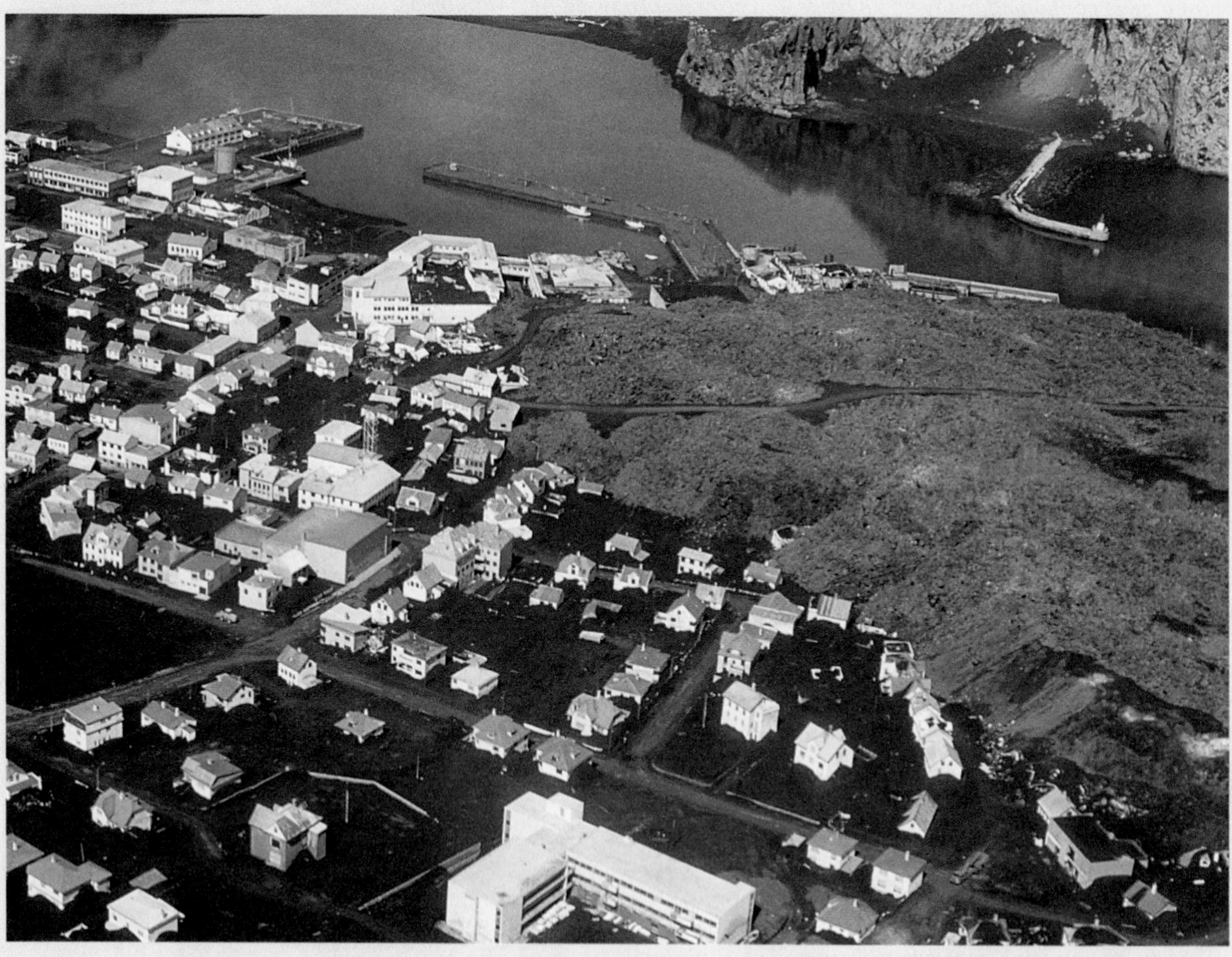

main fishing port, Vestmannaeyjar, located within a kilometer (about a half-mile) from the eruption, began to receive a "snowfall" of fine, black cinders. Houses on the east side of the town were buried under the tephra. Many of the houses collapsed under the weight of the hot ash, while others were set afire. Lava flows soon reached the village, burying and burning still other buildings (see photo).

Over the passing weeks, the emissions of tephra ceased, but lava continued to flow from the cone, which soon reached 200 m (about 600 ft) in height. Unfortunately, the lava began to flow northward, narrowing the harbor and threatening the future livelihood of the town and its evacuated inhabitants. Fishing is a major industry in Iceland, accounting for nearly 80 percent of Iceland's foreign exchange. Vestmannaeyjar normally lands and processes 20 percent of the nation's fish catch.

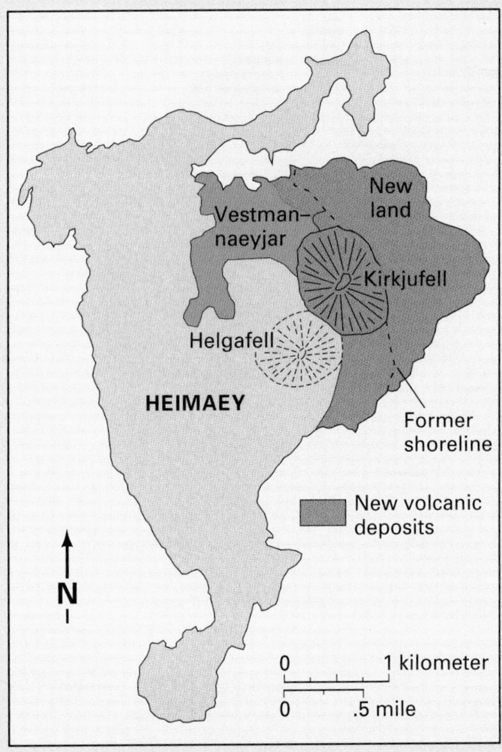

Map of Heimaey Island, Iceland *The map shows the extent of lava flows from the eruptions of the Kirkjufell crater. (Modified from U.S. Geological Survey.)*

Because the harbor and fishing port are important to the nation's economy, Icelanders embarked on a bold plan to save the harbor by altering the course of the lava, diverting the flow eastward. This was to be accomplished by cooling the flow on its north edge with water streams, creating a natural wall to channel the flow alongside the harbor instead of across it. Within a few weeks, the first pumping of water onto the flowing lava began. By early March, a pump ship came into operation in the harbor, providing a steady flow of sea water to cool the slowly moving flows. It was joined in April by as many as 47 high-capacity pumps floating on barges, delivering in total as much as 1 cu m of sea water per second (35 ft^3/sec).

The most effective technique for slowing the lava at a particular location began with cooling the edge and nearby surface of the flow with water from hoses. This allowed bulldozers to build a crude road up and over the slowly moving flow, using nearby tephra as the road material. Large plastic pipes were then laid along the road and across the flow, with small holes spraying water on hot spots. As long as sea water was flowing in the plastic pipes, they remained cool enough not to melt. After a day or so, the flow typically began to slow. Pumping usually continued for about two weeks, until the lava stopped steaming at spray points. The result of the water applications was to build a broad wall of cool, lava rubble with thickening lava behind it.

The huge undertaking successfully stabilized the northern front of the lava flow, keeping the harbor from becoming closed (see map). Although the lava had indeed reached the harbor, it had merely narrowed its entrance. In fact, the new flow improved the harbor's ability to shelter the boats within it (see map). Within about five months, the eruption was over, and the digging out began in earnest. Within a year, life was back to normal on Heimaey.

The diversion of the lava and reconstruction of the village of Vestmannaeyjar were extremely costly. To pay for the cost, Iceland passed a special tax increase, requiring the average Icelandic family to pay about 10 percent of its annual income for one year. Generous foreign aid also helped cover the enormous expenses borne by the tiny nation.

12.10 Exposure of a dike *A dike of mafic igneous rock with nearly vertical parallel sides, cutting across flat-lying sedimentary rock layers. Arrows mark the contact between igneous rock and sedimentary rock. Spanish Peaks region, Colorado.*

Chemical Alteration of Igneous Rocks

The minerals in igneous rocks are formed at high temperatures, and often at high pressures, as magma cools. When igneous rocks are exposed at or near the Earth's surface, the conditions are quite different. Temperatures and pressures are low. Also, the rocks are exposed to soil water and ground water solutions that contain dissolved oxygen and carbon dioxide. In this new environment, the minerals within an igneous rock may no longer be stable. Instead, most of these minerals undergo a slow chemical change that weakens their structure. Chemical change in response to this alien environment is called **mineral alteration.** It is a process of *weathering,* which we will discuss further in Chapter 15.

Weathering also includes the physical forces of disintegration that break up igneous rock into small fragments and separate the component minerals, grain by grain. This breakup, or fragmentation, is essential for the chemical reactions of mineral alteration. The reason is that fragmentation results in a great increase in the mineral surface area that is exposed to chemically active solutions in soils. (We will take up the processes of physical disintegration of rocks in Chapter 15.)

Another chemical alteration process is *oxidation.* This process occurs when oxygen dissolved in soil or ground water reacts with minerals. Oxidation is the normal fate of most silicate minerals exposed at the surface. With oxidation, the silicate minerals are converted to *oxides,* in which silicon and the metallic elements—such as calcium, magnesium, and iron—each bond completely with oxygen. Oxides are very stable. Quartz, with the composition silicon dioxide (SiO_2), is a common mineral oxide. It is very long-lasting and is found abundantly in many types of rocks and sediments. As we will see shortly, quartz is a major constituent of sedimentary rocks.

Silicate minerals do not generally dissolve in water, but some react chemically with water in a chemical alteration process called *hydrolysis.* This process is not merely a soaking or wetting of the mineral, but a true chemical change that produces a different mineral compound. The products of hydrolysis are stable and long-lasting, as are the products of oxidation.

Some of the alteration products of silicate minerals are clay minerals, described in Chapter 21. These minerals are produced by a combination of oxidation and hydrolysis acting on silicates. A **clay mineral** is one that has plastic properties when moist, because it consists of very small, thin flakes that become lubricated by layers of water molecules. Clay minerals formed by mineral alteration are abundant in common types of sedimentary rocks.

Solution is another mechanism by which minerals are altered. Most minerals do not dissolve directly in water. One important exception is rock salt (sodium chloride), which is a constituent of sea water. When carbon dioxide dissolves in water, a weak acid—*carbonic acid*—is formed. Carbonic acid can dissolve certain minerals, especially calcium carbonate. In addition, where decaying vegetation is present, soil water contains many complex organic acids that are capable of reacting with minerals.

SEDIMENTS AND SEDIMENTARY ROCKS

We can now turn to the second great rock class, the *sedimentary rocks.* The mineral particles in sedimentary rocks can be derived from preexisting rock of any of the three rock classes as well as from newly formed organic matter. However, igneous rock is the most important original source of the inorganic mineral matter that makes up sedimentary rock. For example, a granite can weather to yield grains of quartz and particles of clay minerals derived from feldspars, thus contributing sand and clay to a sedimentary rock. Sedimentary rocks include rock types with a wide range of physical and chemical properties. We will only touch on a few of the most important kinds of sedimentary rocks, which are shown in Table 12.2.

In the process of mineral alteration, solid rock is weakened, softened, and fragmented, yielding particles of many sized and mineral compositions. When transported by a fluid medium—air, water, or glacial ice—these particles are known collectively as **sediment.** Used in its broadest sense, sediment includes both inorganic and organic matter. Dissolved mineral matter in solution must also be included.

Streams and rivers carry sediment to lower land levels, where sediment can accumulate. The most favorable sites of sediment accumulation are shallow sea floors bordering continents. But sediments also accumulate in inland valleys, lakes, and marshes. Thick accumulations of sediment may become deeply buried under newer (younger) sediments. Wind and glacial ice also transport sediment but not necessarily to lower elevations or to places suitable for accumulation. Over long spans of time, the sediments can undergo physical or chemical changes, becoming compacted and hardened to form sedimentary rock.

Table 12.2 | **Some Common Sedimentary Rock Types**

Subclass	Rock Type	Composition
Clastic (Composed of rock and/or mineral fragments)	Sandstone	Cemented sand grains
	Siltstone	Cemented silt particles
	Conglomerate	Sandstone containing pebbles of hard rock
	Mudstone	Silt and clay, with some sand
	Claystone	Clay
	Shale	Clay, breaking easily into flat flakes and plates
Chemically precipitated (Formed by chemical precipitation from sea water or salty inland lakes)	Limestone	Calcium carbonate, formed by precipitation on sea or lake floors
	Dolomite	Magnesium and calcium carbonates, similar to limestone
	Chert	Silica, a microcrystalline form of quartz
	Evaporites	Minerals formed by evaporation of salty solutions in shallow inland lakes or coastal lagoons
Organic (Formed from organic material)	Coal	Rock formed from peat or other organic deposits; may be burned as a mineral fuel
	Petroleum (mineral fuel)	Liquid hydrocarbon found in sedimentary deposits; not a true rock but a mineral fuel
	Natural gas (mineral fuel)	Gaseous hydrocarbon found in sedimentary deposits; not a true rock but a mineral fuel

There are three major classes of sediment. First is **clastic sediment,** which consists of inorganic rock and mineral fragments, called *clasts.* Examples are the materials in a sand bar of a river bed or on a ocean beach. Second is **chemically precipitated sediment,** which consists of inorganic mineral compounds precipitated from a saltwater solution or as hard parts of organisms. In the process of chemical precipitation, ions in solution combine to form solid mineral matter separate from the solution. A layer of rock salt, such as that found in dry lake beds in arid regions, is an example. A third class is **organic sediment** which consists of the tissues of plants and animals, accumulated and preserved after the death of the organism. An example is a layer of peat in a bog or marsh.

Sediment accumulates in more-or-less horizontal layers, called **strata,** or simply "beds" (Figure 12.11). Individual strata are separated from those below and above by surfaces called stratification planes or bedding planes. These separation surfaces allow one layer to be easily removed from the next. Strata of widely different compositions can occur alternately, one above the next. GEODISCOVERIES

12.11 *Eye on the Landscape* ***Sandstone strata*** *This photo of an eroded sandstone formation on the Colorado Plateau shows individual layers, or strata, that make up the rock. The layers were originally deposited within a field of moving sand dunes.* **What else would the geographer see? ... Answers at the end of the chapter.**

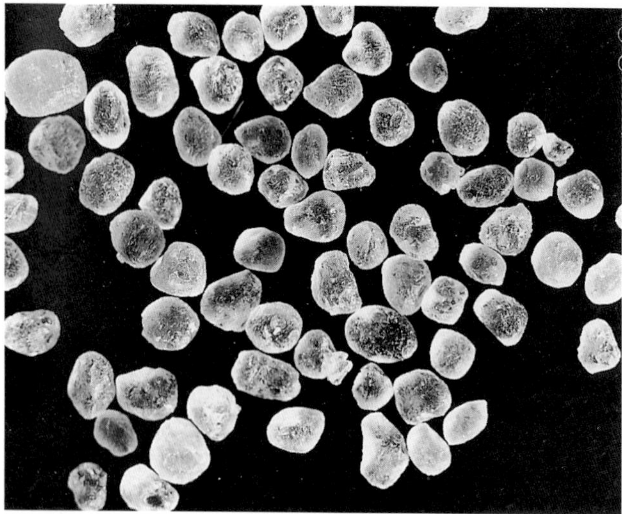

12.12 Rounded quartz grains from an ancient sandstone *The grains average about 1 mm (0.039 in.) in diameter.*

Clastic Sedimentary Rocks

Clastic sediments are derived from clasts of any and all of the rock groups—igneous, sedimentary, metamorphic—and thus may include a very wide range of minerals for sedimentary rock formation. Silicate minerals are the most important, both in original form and as altered by oxidation and hydrolysis. Quartz and feldspar usually dominate. Because quartz is hard and is immune to alteration, it is the most important single component of the clastic sediments (Figure 12.12). Second in abundance are fragments of unaltered fine-grained parent rocks, such as tiny pieces of lava rock. Feldspar and mica are also commonly present. Clay minerals are major constituents of the finest clastic sediments.

The range of particle sizes in a clastic sediment determines how easily and how far the particles are transported by water currents. The finer the particles, the more easily they are held suspended in the fluid. On the other hand, the coarser particles tend to settle to the bottom of the fluid layer. In this way, a separation of size grades, called *sorting,* occurs. Sorting determines the texture of the sediment deposit and of the sedimentary rock derived from that sediment. The finest clay particles do not settle out unless they are made to clot together into larger clumps. This clotting process, called flocculation, normally occurs when river water carrying clay mixes with the saltwater of the ocean.

When sediments accumulate in thick sequences, the lower strata are exposed to the pressure produced by the weight of the sediments above them. This pressure compacts the sediments, squeezing out excess water. Cementation occurs as dissolved minerals recrystallize where grains touch and in the spaces between mineral particles. Silicon dioxide (quartz, SiO_2) is very slightly soluble in water, and so the cement is often a form of quartz, called *silica,* which lacks a true crystalline form. Calcium carbonate ($CaCO_3$) is another common material that cements clastic sedimentary rocks. Compaction and cementation produce sedimentary rock.

The important varieties of clastic sedimentary rock are distinguished by the size of their particles. They include sandstone, conglomerate, mudstone, claystone, and shale. **Sandstone** is formed from fine to coarse sand (Figure 12.13). The cement may be silica or calcium carbonate. The sand grains are commonly of quartz, such as those shown in Figure 12.12. (Refer to Figure 21.2 for the names and diameters of the various grades of sediment particles.) Sandstone containing numerous rounded pebbles of hard rock is called *conglomerate* (Figure 12.14).

A mixture of water with particles of silt and clay, along with some sand grains, is called *mud.* The sedimentary rock hardened from such a mixture is called *mudstone.* Compacted and hardened clay layers become *claystone.* Sedimentary rocks of mud composition are commonly

12.13 **Eye on the Landscape** ***Massive sandstone*** *A thick formation of massive sandstone forms the sheer cliff in this view of rock formations on the North Rim of the Grand Canyon, Arizona. Termed the Coconino Formation, the sandstone is an ancient dune deposit.* **What else would the geographer see? ... Answers at the end of the chapter.**

12.14 Conglomerate section *A piece of quartzitic conglomerate, cut through and polished, reveals rounded pebbles of quartz (clear and milky colors) and chert (grayish). It is about 12 cm (4.7 in.) in diameter.*

layered in such a way that they easily break apart into small flakes and plates. The rock is then described as being *fissile* and is given the name **shale.** Shale, the most abundant of all sedimentary rocks, is formed largely of clay minerals. The compaction of the mud to form mudstone and shale involves a considerable loss of volume as water is driven out of the clay. Figure 12.15 shows an eroded shale formation in southern Utah. Here some of the shale beds have distinctive colors, so the bedding is especially prominent.

Chemically Precipitated Sedimentary Rocks

Under favorable conditions, mineral compounds are deposited from the salt solutions of seawater and of salty inland lakes in desert climates. One of the most common sedimentary rocks formed by chemical precipitation is **limestone,** composed largely of the mineral calcite. *Calcite* is calcium carbonate ($CaCO_3$). Marine limestones—limestone strata formed on the sea floor—accumulated in thick layers in many ancient seaways in past geologic eras (Figure 12.16). A closely related rock, also formed by chemical precipitation, is *dolomite,* composed of calcium-magnesium carbonate. Limestone and dolomite are grouped together as the *carbonate rocks.* They are dense rocks with white, pale gray, or even black color.

Sea water also yields sedimentary layers of silica in a hard, noncrystalline form called *chert.* Chert is a variety of sedimentary rock, but it also commonly occurs combined with limestone (cherty limestone).

Shallow water bodies acquire a very high level of salinity where evaporation is sustained and intense. One type of shallow water body is a bay or estuary in a coastal desert region. Another type is the salty lake of inland desert basins (see Figure 17.30). Sedimentary minerals and rocks deposited from such concentrated solutions are called **evaporites.** Ordinary rock salt, the mineral halite (sodium chloride), has accumulated in this way in thick sedimentary rock layers. These layers are mined as major commercial sources of halite.

Hydrocarbon Compounds in Sedimentary Rocks

Hydrocarbon compounds (compounds of carbon, hydrogen, and oxygen) form a most important type of organic sediment—one on which human society increasingly depends for fuel. These substances occur both as solids (peat and coal) and as liquids and gases (petroleum and natural gas). Only coal qualifies physically as a rock. *Peat* is a soft, fibrous substance of brown to black color. (See Figure 21.20.) It accumulates in a bog environment where the continual presence of water inhibits the decay of plant remains.

At various times and places in the geologic past, plant remains accumulated on a large scale, accompanied by sinking of the area and burial of the compacted organic matter under thick layers of inorganic clastic sediments. *Coal* is the

12.15 Shale butte *These flat-lying strata consist of layers of different colored shales that are easily eroded. Near Glen Canyon City in southern Utah. (A. N. Strahler.)*

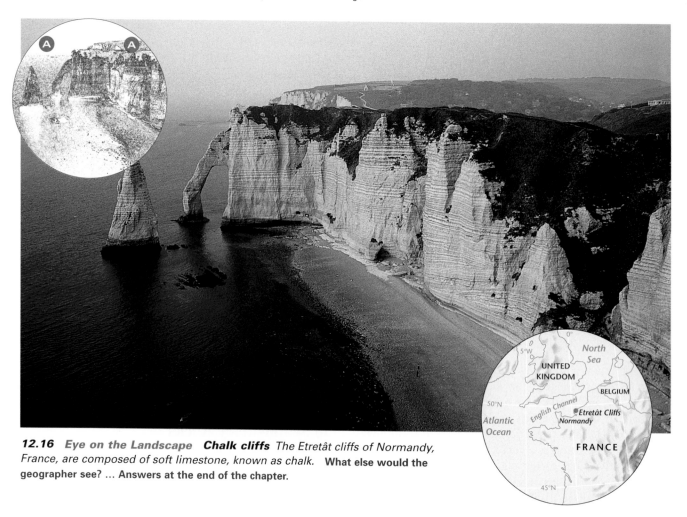

12.16 **Eye on the Landscape** **Chalk cliffs** *The Etretât cliffs of Normandy, France, are composed of soft limestone, known as chalk.* **What else would the geographer see? ... Answers at the end of the chapter.**

end result of this process (Figure 12.17). Individual coal seams are interbedded with shale, sandstone, and limestone strata.

Petroleum (or *crude oil,* as the liquid form is often called) includes many hydrocarbon compounds. *Natural gas,* which is found in close association with accumulations of liquid petroleum, is a mixture of gases. The principal gas is methane (marsh gas, CH_4). Geologists generally agree that petroleum and natural gas are of organic origin, but the nature of their formation is not fully understood. They are classed not as minerals but rather separately as mineral fuels.

Natural gas and petroleum commonly occupy open interconnected pores in a thick sedimentary rock layer—a porous

12.17 **Strip mine** *A coal seam near Sheridan, Wyoming, being strip mined by heavy equipment.*

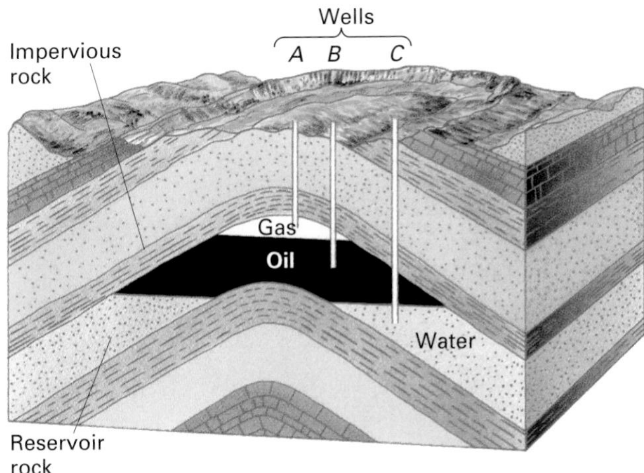

12.18 Trapping of oil and gas *Idealized cross section of an oil pool on a dome structure in sedimentary strata. Well A will draw gas; well B will draw oil; and well C will draw water. The cap rock is shale; the reservoir rock is sandstone. (Copyright © A. N. Strahler.)*

Table 12.3 Some Common Metamorphic Rock Types

Rock Type	Description
Slate	Shale exposed to heat and pressure that splits into hard flat plates
Schist	Shale exposed to intense heat and pressure that shows evidence of shearing
Quartzite	Sandstone that is "welded" by a silica cement into a very hard rock of solid quartz
Marble	Limestone exposed to heat and pressure, resulting in larger, more uniform crystals
Gneiss	Rock resulting from the exposure of clastic sedimentary or intrusive igneous rocks to heat and pressure

sandstone, for example. The simplest arrangement of strata favorable to trapping petroleum and natural gas is an uparching of the type shown in Figure 12.18. Shale forms an impervious cap rock. A porous sandstone beneath the cap rock serves as a reservoir. Natural gas occupies the highest position, with the oil below it, and then water, in increasing order of density.

Yet another form of occurrence of hydrocarbon fuels is *bitumen,* a variety of petroleum that behaves much as a solid, although it is actually a highly viscous liquid. Bitumen goes by other common names, such as tar, asphalt, or pitch. In some localities, bitumen occupies pore spaces in layers of sand or porous sandstone. It remains immobile in the enclosing sand and will flow only when heated. Outcrops of *bituminous sand (oil sand)* exposed to the Sun will show bleeding of the bitumen. Perhaps the best known of the great bituminous sand deposits are those occurring in Alberta, Canada. Where exposed along the banks of the Athabasca River, the oil sand is extracted from surface mines.

Hydrocarbon compounds in sedimentary rocks are important because they provide an energy resource on which modern human civilization depends. These **fossil fuels,** as they are called collectively, have required millions of years to accumulate. However, they are being consumed at a very rapid rate by our industrial society. These fuels are nonrenewable resources. Once they are gone, there will be no more because the quantity produced by geologic processes even in a thousand years is scarcely measurable in comparison to the quantity stored through geologic time.

GEODISCOVERIES

Metamorphic Rocks

Any type of igneous or sedimentary rock may be altered by the tremendous pressures and high temperatures that accompany the mountain-building processes of the Earth's crust. The result is a rock so changed in texture and structure as to be reclassified as **metamorphic rock.** Mineral components of the parent rock are, in many cases, reconstituted into different mineral varieties. Recrystallization of the original minerals can also occur. Our discussion of metamorphic rocks will mention only five common types—slate, schist, quartzite, marble, and gneiss (Table 12.3).

Slate is formed from shale that is heated and compressed by mountain-making forces. This fine-textured rock splits neatly into thin plates, which are familiar as roofing shingles and as patio flagstones. With application of increased heat and pressure, slate changes into **schist,** representing the most advanced stage of metamorphism. Schist has a structure called foliation, consisting of thin but rough, irregularly curved planes of parting in the rock (Figure 12.19). These are evidence of *shearing*—a stress that pushes the layers sideways, like a deck of cards pushed into a fan with the sweep of a palm. Schist is set apart from slate by the coarse texture of the mineral grains, the abundance of mica, and occasionally the presence of scattered large crystals of newly formed minerals, such as garnet.

12.19 Schist sample *A specimen of foliated schist.*

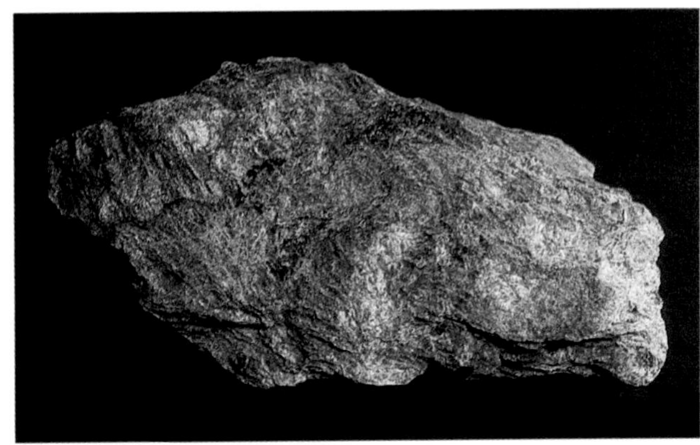

12.20 Banded gneiss *This surf-washed rock surface exposes banded gneiss, Pemaquid Point, Maine.*

The metamorphic equivalent of conglomerate, sandstone, and siltstone is **quartzite,** which is formed by the addition of silica to fill completely the open spaces between grains. This process is carried out by the slow movement of underground waters carrying silicate into the rock, where it is deposited.

Limestone, after undergoing metamorphism, becomes *marble,* a rock of sugary texture when freshly broken. During the process of internal shearing, calcite in the limestone reforms into larger, more uniform crystals than before. Bedding planes are obscured, and masses of mineral impurities are drawn out into swirling bands.

Finally, the important metamorphic rock **gneiss** may be formed either from intrusive igneous rocks or from clastic sedimentary rocks that have been in close contact with intrusive magmas. A single description will not fit all gneisses because they vary considerably in appearance, mineral composition, and structure. One variety of gneiss is strongly banded into light and dark layers or lenses (Figure 12.20), which may be bent into wavy folds. These bands have differing mineral compositions. They are thought to be relics of sedimentary strata, such as shale and sandstone, to which new mineral matter has been added from nearby intrusive magmas.

One of the more exciting developments in remote sensing in the past few years has been the launch of the ASTER instrument (Advanced Spaceborne Thermal Emission and Reflection Spectroradiometer) on NASA's Terra satellite platform. It has a unique ability to aid in geologic mapping and in identification of areas with potentially valuable ore deposits. *Focus on Remote Sensing 12.3 • Geologic Mapping with ASTER* tells this interesting story. GEODISCOVERIES

The Cycle of Rock Change

The processes that form rocks, when taken together, constitute a single system that cycles and recycles Earth materials over geologic time from one form to another. The **cycle of rock change,** shown in Figure 12.21, describes this system. There are two environments—a surface environment of low pressures and temperatures and a deep environment of high pressures and temperatures. The surface environment is the site of rock alteration and sediment deposition. In this environment, igneous, sedimentary, and metamorphic rocks are uplifted and exposed to air and water. Their minerals are altered chemically and broken free from the parent rock, yielding sediment. The sediment accumulates in basins, where deeply buried sediment layers are compressed and cemented into sedimentary rock.

Sedimentary rock, entering the deep environment, is heated by the slow radioactive decay of elements and comes into a zone of high confining pressure. Here, it is transformed into metamorphic rock. Pockets of magma are formed in the deep environment and move upward, melting and incorporating surrounding rock as they rise. Upon reaching a higher level, magma cools and solidifies, becoming intrusive igneous rock. Or it may emerge at the surface to form extrusive igneous rock. Either way, the cycle is completed.

The cycle of rock change has been active since our planet became solid and internally stable, continuously forming and reforming rocks of all three major classes. Not even the oldest igneous and metamorphic rocks found thus far are the "original" rocks of the Earth's crust, for they were recycled eons ago.

The loops in the cycle of rock change are powered by a number of sources, ranging from solar energy to radiogenic

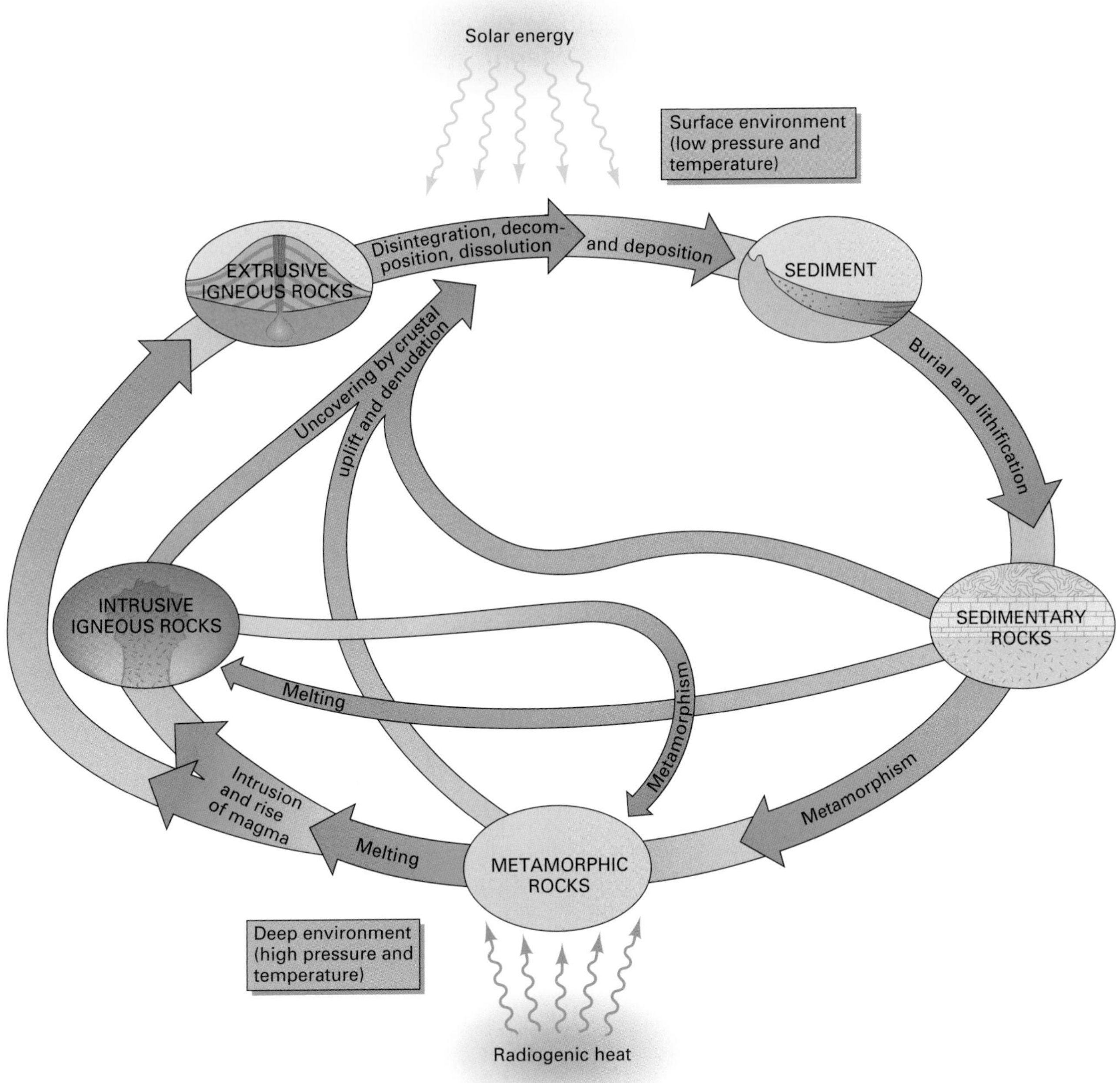

12.21 The cycle of rock change *This cycle links sediment with sedimentary, metamorphic, intrusive igneous, and extrusive igneous rocks in processes of rock formation and destruction.*

heat. *Focus on Systems 12.4 • Powering the Cycle of Rock Change* describes these sources in more detail.

A LOOK AHEAD This chapter has described the minerals and rocks of the Earth's surface and the processes of their formation. As we look further into the Earth's outermost layers in the next chapter, we will see that these processes do not occur everywhere. Instead, there is a grand plan that organizes the formation and destruction of rocks and distributes the processes of the cycle of rock change in a geographic pattern. This pattern is controlled by the pattern in which the solid Earth's brittle outer layer is fractured into great plates that split and separate and also converge and collide. The grand plan is plate tectonics, the scheme for understanding the dynamics of the Earth's crust over millions of years of geologic time.

Focus on Remote Sensing | 12.3

Geologic Mapping with ASTER

A unique imaging instrument built by Japan that is of special interest to geologists is aboard NASA's Terra satellite platform, launched in December 1999. ASTER, the Advanced Spaceborne Thermal Emission and Reflection Spectroradiometer, acquires multispectral images in three different waveband regions—visible and near-infrared, shortwave infrared, and thermal infrared. Recall from *Geographer's Tools 4.8 • Remote Sensing for Physical Geography* that human eyes are sensitive only to the visible light portion of the spectrum, from about 0.4 to 0.65 μm, while the Sun provides enough energy for good imaging from about 0.3 to 3.0 μm. This means that surface substances can be imaged in these wavelengths and have "colors" that are unknown in human experience. We refer to these colors as spectral signatures. As it happens, minerals and rocks often have distinctive spectral signatures in near- and shortwave infrared domains, and so near- and shortwave infrared images are particularly useful for geologic mapping.

Another spectral region of particular interest is the thermal infrared, especially the wavelengths from 8.0 to 12.0 μm. Here the energy for sensing is the radiant energy emitted by substances as a function of their temperature. However, some type of rocks and minerals emit more energy of a given wavelength at a given temperature than others. In this way, we can think of these substances as having spectral signatures in the thermal infrared, too.

The two graphs below illustrate this principle. They show reflectance plotted against wavelength for two common minerals—quartz and calcite. Also shown are the spectral locations of the ASTER wavebands. In the left graph, we can see that both quartz and calcite are quite bright in ASTER's two visible bands (bands 1 and 2, green and red). They appear white to the eye, with calcite, which is slightly brighter in the red band, appearing to be a "warmer" white. In the shortwave infrared, however, calcite is significantly darker in ASTER band 8 than in surrounding bands 7 and 9. The right graph shows spectral reflectance in the thermal infrared spectral region. Calcite is brightest in ASTER band 13 while being darkest in the adjacent band 14. Quartz has a distinctive signature as well, being about 40 percent brighter in bands 10–12 than in 13–14. Thus, both minerals have unique spectral signatures in this spectral region, too.

Color images in the three waveband regions, shown at right, illustrate how ASTER can help identify different types of rocks and minerals by their spectral signatures. The images are from the Saline Valley, a dry semidesert region in southeastern California near Death Valley, and were acquired on March 30, 2000. Trending diagonally across the center of the image is the valley itself, which is the now-nearly-dry bed of a lake that was filled with water during the Ice Age. Lying to the southwest is the Inyo mountain chain, with a number of steep valleys cutting into its flanks. In the lower left corner, we see a small part of the Owens Valley. To the northeast of Saline Valley lies a complex geologic region at the northern end of the Panamint Range.

Image *(a)* is a color infrared display in which true colors of green, red, and near-infrared are shown

Reflectance spectra for quartz and calcite *Wavebands in which ASTER acquires images are identified as numbered bands. Left, the visible and near-infrared spectral region; right, the thermal infrared spectral region.*

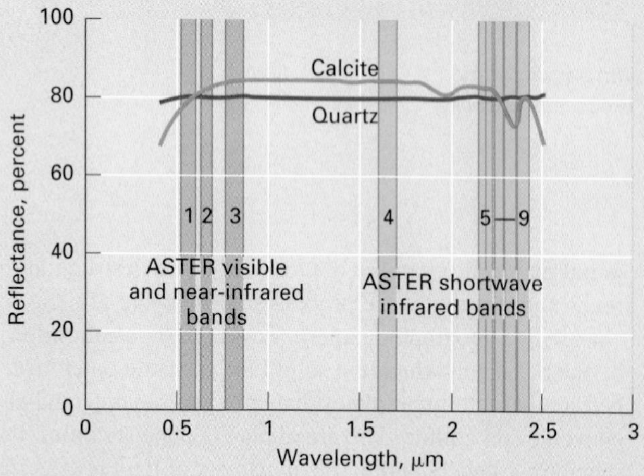

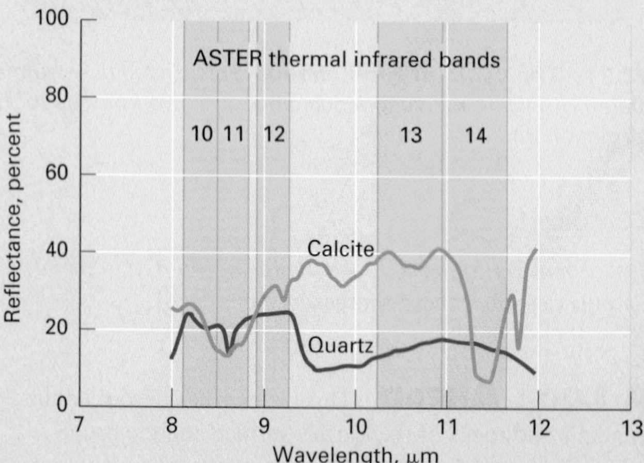

Multispectral ASTER images of Saline Valley, California *(a) visible and near-infrared image; (b) shortwave infrared image; (c) thermal infrared image. (NASA/GSFC/MITI/ERSDAC/JAROS and U.S./Japan ASTER Science Team.)*

as blue, green, and red in the image. The forests of the Inyos appear in a distinctive red color because vegetation is very bright in the near-infrared region. Snow along the summit of the Inyos is white, as are the salty dry basins of the valleys. Many different rock types appear, in tones ranging from brown, gray, and yellow to blue. Image *(b)* displays ASTER shortwave infrared bands 8, 6, and 4 as blue, green and red. These bands distinguish carbonate, sulfate, and clay minerals. Limestones are yellow-green, while the clay mineral kaolinite appears in purplish tones. In *(c)*, ASTER bands 10, 12, and 13 are shown in blue, green, and red, respectively. Here, rocks rich in quartz appear in red tones, while carbonate rocks are green. Mafic volcanic rocks are visible in purple tones.

By comparing the three images, it is easy to see how the spectral information acquired by ASTER has the ability to make geologic mapping faster and easier. By using ASTER data, geologists can not only come to understand better the geologic history of a region, but also identify the geologic settings in which valuable mineral ores may be found.

Focus on Systems | 12.4

Powering the Cycle of Rock Change

In the cycle of rock change, the materials of the lithosphere are constantly being formed and transformed in both their physical and mineral composition. Deep within the crust, rocks and sediments are compressed, consolidated, baked, sheared, and sometimes melted. They eventually emerge at the surface either by extrusion, in the case of lava, or by the stripping off of the overlying rock layers by erosion. Once at the surface, the rocks are physically and chemically weathered into sediment. The sediment then moves to low places where it accumulates and can be buried, thus completing the cycle.

What energy sources power the rock cycle? Let's examine the underground part of the cycle first. For this part of the cycle, the main power source is *radiogenic heat.* This term refers to heat that is slowly released by the radioactive decay of unstable isotopes that were originally formed early in the history of the solar system. (See *Working It Out 12.1 • Radioactive Decay,* for more details.) Isotopes of uranium (^{238}U, ^{235}U), thorium (^{232}Th), and potassium (^{40}K), along with the daughter products generated by their decay, are the source of nearly all of this heating.

The upper figure below plots temperature with depth below the Earth's surface. Two curves are shown—one for temperatures beneath continents and one for temperatures beneath ocean basins. Although the curves are

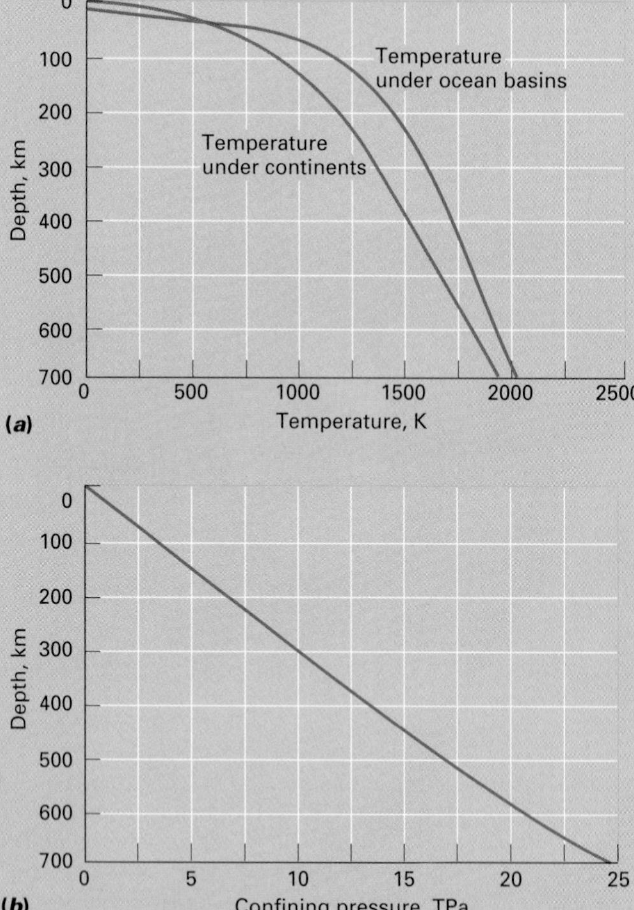

Change in temperature (a) and pressure (b) with depth from the Earth's surface *(Copyright © A. N. Strahler)*

slightly different, they show that the temperature increases rapidly at first and then increases only slowly with depth. Most of the radiogenic heat is liberated in the rock beneath the continents, within the uppermost 100 km (about 60 mi) or so. This observation fits with chemical analyses of continental rocks, which show that isotopes of uranium, thorium, and potassium are most abundant in upper layers of continents. Radiogenic heat is sufficient to keep Earth layers below the crust close to the melting point and thus provides the power source for the heating by which metamorphic and igneous rocks are formed from preexisting rocks.

A secondary power source is gravity. As sedimentary layers are buried more and more deeply, the pressure on the layers increases. The lower graph on the left plots pressure with depth below the Earth's surface. This pressure acts to bring mineral grains or plates into very close contact, where they can bond to one another. Water, often containing dissolved silica or calcium carbonate, is forced out of the sediments while leaving deposits of these minerals to cement the grains in place. Thus, the rock density increases with time.

Gravity and radiogenic heat are power sources that alter rock composition and structure. But what causes the motions of rocks, in which sediments and preexisting rocks are buried deep with the Earth and then later brought to the surface? These motions are part of the plate tectonic cycle, which we will examine in the following chapter.

The above-ground portion of the cycle of rock change is driven by several power sources, of which the Sun and gravity are the most important. Solar energy, and the unequal heating of the Earth's surface that it provides, is the primary energy source for the motions of the atmosphere and oceans, as we have seen in prior chapters. A lesser, but still important, power source for atmospheric and oceanic movements is the Earth's rotation, which transfers momentum to the fluid atmosphere and ocean. The tidal forces of the Sun and Moon also act to keep the ocean and atmosphere in motion.

The circulation of the atmosphere and oceans produces moving fluids—air, water, and glacial ice—that transport weathered rock particles while reducing them to ever-finer sizes. In this way, the particles are also directly exposed to oxygen and water, allowing alteration of their chemical composition. The action of gravity on the particles eventually brings them to sedimentary basins, where they accumulate and the underground portion of the cycle of rock change begins. The figure above shows a sketch of the flows of energy that power the cycle of rock change.

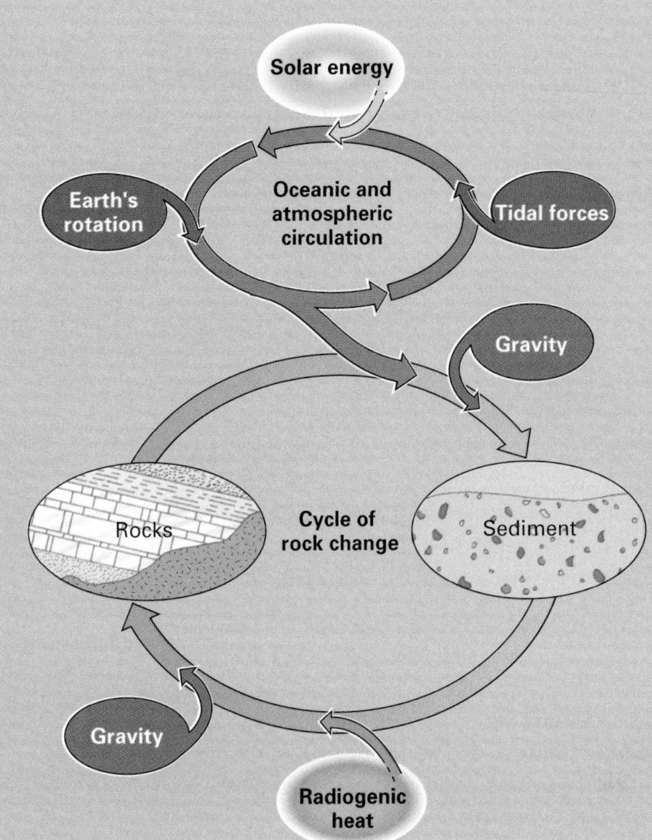

Flows of energy that drive the cycle of rock change

Chapter Summary

- The elements oxygen and silicon dominate the *Earth's crust.* Metallic elements, which include aluminum, iron, and the base elements, account for nearly all the remainder.
- **Minerals** are naturally occurring, inorganic substances. Each has an individual chemical composition and atomic structure. Minerals are combined into **rock**.
- *Silicate minerals* make up the bulk of **igneous rocks**. They contain silicon and oxygen together with some of the metallic elements.
- There are three broad classes of igneous rocks, depending on their mineral content. **Felsic** rocks contain mostly felsic minerals and are least dense; **mafic** rocks, containing mostly mafic minerals, are denser; and *ultramafic* rocks are most dense. Because felsic rocks are least dense, they are generally found in the upper layers of the Earth's crust. Mafic and ultramafic rocks are more abundant in the layers below.
- If **magma** erupts on the surface to cool rapidly as lava, the rocks formed are **extrusive** and have a fine crystal texture. If the magma cools slowly below the surface as a **pluton**, the rocks are **intrusive** and the crystals are larger. **Granite** (felsic, intrusive), **andesite** (felsic, extrusive), and **basalt** (mafic, extrusive), are three very common igneous rock types.
- Most silicate minerals found in igneous rocks undergo **mineral alteration** when exposed to air and moisture at the Earth's surface. Mineral alteration occurs through *oxidation, hydrolysis*, or *solution*. **Clay minerals** are commonly produced by mineral alteration.
- **Sedimentary rocks** are formed in layers, or **strata**. **Clastic sedimentary** rocks are composed of fragments of rocks and minerals that usually accumulate on ocean floors. As the layers are buried more and more deeply, water is pressed out and particles are cemented together. **Sandstone** and **shale** are common examples.
- **Chemical precipitation** also produces sedimentary rocks, such as **limestone**. *Coal, petroleum,* and *natural gas* are hydrocarbon compounds occurring in sedimentary rocks that are used as mineral fuels.
- **Metamorphic rocks** are formed when igneous or sedimentary rocks are exposed to heat and pressure. Shale is altered to *slate* or **schist**, sandstones become **quartzite**, and intrusive igneous rocks or clastic sediments are metamorphosed into **gneiss**.
- In the **cycle of rock change**, rocks are exposed at the Earth's surface, and their minerals are broken free and altered to form sediment. The sediment accumulates in basins, where the layers are compressed and cemented into sedimentary rock. Deep within the Earth, the heat of radioactive decay melts preexisting rock into magma, which can move upward into the crust to form igneous rocks that cool at or below the surface. Rocks deep in the crust are exposed to the heat and pressure, forming metamorphic rock. Mountain-building forces move deep igneous, sedimentary, and metamorphic rocks upward to the surface, providing new material for surface alteration and breakup and completing the cycle.

Key Terms

mineral
rock
igneous rocks
sedimentary rocks
metamorphic rocks
quartz
felsic
mafic
magma
granite
intrusive igneous rock
lava
extrusive igneous rock
andesite
basalt
pluton
mineral alteration
clay mineral
sediment
clastic sediment
chemically precipitated sediment
organic sediment
strata
sandstone
shale
limestone
evaporites
fossil fuels
metamorphic rock
schist
quartzite
gneiss
cycle of rock change

Review Questions

1. What is the Earth's crust? What elements are most abundant in the crust?
2. Define the terms *mineral* and *rock*. Name the three major classes of rocks.
3. What are silicate minerals? Describe two classes of silicate minerals.
4. Name four types of igneous rocks and arrange them in order of density.
5. How do igneous rocks differ when magma cools (a) at depth and (b) at the surface?
6. How are igneous rocks chemically altered? Identify and describe three processes of chemical alteration.
7. What is sediment? Define and describe three types of sediments.
8. Describe two processes that produce sedimentary rocks, and identify at least three important varieties of clastic sedimentary rocks.
9. How are sedimentary rocks formed by chemical precipitation?
10. What types of sedimentary deposits consist of hydrocarbon compounds? How are they formed?
11. What are metamorphic rocks? Describe at least three types of metamorphic rocks and how they are formed.

Eye on the Environment 12.2 • Battling Iceland's Heimaey Volcano

1. What were the initial effects of the eruption of Heimaey's volcano?
2. What important industry was threatened by the volcanic activity and why?
3. How was sea water used to keep lava from blocking the harbor?

Focus on Systems 12.4 • Powering the Cycle of Rock Change

1. How do pressure and temperature change with depth below the Earth's surface? Sketch a graph of temperature change with depth and explain its shape.
2. Identify and describe the sources of energy that power the cycle of rock change.

Visualizing Exercises

1. Sketch a cross section of the Earth showing the following features: batholith, sill, dike, veins, lava, and volcano.
2. Sketch the cycle of rock change and describe the processes that act within it to form igneous, sedimentary, and metamorphic rocks.

Essay Question

1. A granite is exposed at the Earth's surface, high in the Sierra Nevada mountain range. Describe how mineral grains from this granite might be released, altered, and eventually become incorporated in a sedimentary rock. Trace the route and processes that would incorporate the same grains in a metamorphic rock.

Problems

Working It Out 12.1 • Radioactive Decay

1. The half-life of potassium-40 (^{40}K) is 1.28 billion years. What percent of a quantity of ^{40}K will remain after 1 billion years? 3 billion years? (*Hint:* check your work by referring to the figure in the box.)
2. The half-life of thorium-232 (^{232}Th) is 14.1 billion years. What percent of a quantity of pure ^{232}Th will remain after 5 billion years? 10 billion years? 15 billion years?
3. The unstable isotope carbon-14 (^{14}C) has a half-life of 5730 years. How many years will be required for the proportion of a pure sample of ^{14}C to be reduced to 0.1 (10 percent)?

Eye on the Landscape

Chapter Opener Village of Bacolor, Luzon, Philippines The work of flowing water in carving the landscape is evident in the large expanses of fresh sediment covering this landscape. At **(A)**, gullies are forming, carving small valleys and sending sediment downstream. At **(B)**, the stream shows a braided pattern of channels that merge and separate, weaving their way downslope. This braided pattern is characteristic of streams carrying large volumes of sediment. Gullies and braided streams are discussed in Chapter 17.

12.11 Sandstone strata Note how these fine sedimentary beds **(A)** are crossed at angles by other beds. This cross-bedding is characteristic of dune sands. As the sand dune moves forward, sand layers accumulate on its sloping forward surface. Later, the wind erodes the dune, forming a new surface across the beds that cuts them at an angle. That surface is then covered by more sand layers. The result is the distinctive pattern of sloping sand beds cut by other beds at low angles that we see in this rock formation. Sand dune formation and movement is covered in Chapter 19.

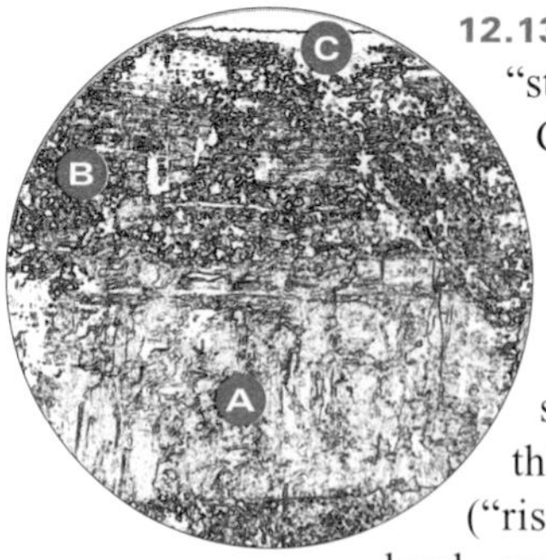

12.13 Massive sandstone The "stairstep" topography of the Grand Canyon is a unique product of uplift of the Colorado Plateau and deep downcutting by the through-flowing Colorado River. The steep canyon walls carved by this process form vertical faces ("risers") **(A)** where massive, hard, rock formations, such as the Coconino Formation, are exposed. Weaker beds of sandstone and shale form sloping surfaces ("steps") **(B)** with a cover of rock fragments and sediment. Although the region is very dry, there is enough rainfall at higher elevations to support an open stand of pinyon pine and juniper forest atop the plateau **(C)** as well as a sparse cover on the slopes of the canyon walls. You can read more about landforms of horizontal rock strata in Chapter 18. Altitudinal zonation of vegetation in the Grand Canyon region is discussed in Chapter 23.

12.16 Chalk cliffs Note the hazy air **(A)** that is apparent in this photo. Marine aerosols are particularly rich in tiny particles of sea salt. Like ordinary table salt during humid weather, these particles attract water vapor. The vapor accumulates around the salt particles until tiny droplets are formed. These droplets absorb and scatter sunlight, creating the marine haze that is so evident in this photo. Recall that haze was covered in Chapter 6 in our interchaper feature, *A Closer Look: Eye on the Environment 6.6 • Air Pollution.*

13 The Lithosphere and the Tectonic System

Þingvellir Fault east of Reykjavik, Iceland. Iceland lies on the boundary of the Eurasian and American plates, which are spreading apart at the rate of 2 cm (1 in.) per year.

On the globes and maps we've seen since early childhood, the outline of each continent is so unique that we would never mistake one continent for another. But why are no two continents even closely alike? As we will see in this chapter, the continents have had a long history. In fact, in each of the continents some regions of metamorphic rocks date back more than 2 billion years. As part of that history, the continents have been fractured and split apart, as well as pushed together and joined.

Perhaps you've visited an old New England farmhouse that was constructed over hundreds of years—first the small two-room house, to which the kitchen shed was added at the back, and then the parlor wing at the side. Later the roof was raised for a second story, the tool house and barn were joined to the main house, the carriage house was built, and so it went. The Earth's continents have that kind of history. They are composed of huge masses of continental crust that have been assembled at different times in each continent's history. The theory describing the motions and changes through time of the continents and ocean basins, and the processes that fracture and fuse them, is called *plate tectonics.*

Briefly stated, plate tectonic theory maintains that the Earth's outermost solid layer consists of huge plates that are in constant but slow motion, powered by energy sources deep within the Earth. Floating on a layer of plastic rock below, these rigid plates can both collide with other plates and split apart to form new plates. When crustal plates collide, one plate may be forced under the other in a process called *subduction.* In this case, part of the subducted plate can melt, creating pockets of magma that move upward to the surface and create volcanic mountain chains and island arcs. Plates can also collide without subduction, forming mountain chains of rocks that are fractured, folded, and compressed into complex mountain structures. Where crustal plates split apart, creating rift valleys and ultimately ocean basins, lava moves upward to fill the gap. Thus, the Earth's surface arrangement of continents and oceans is constantly but slowly changing. Ocean basins expand with rifting and contract with subduction, while continents grow by collision and lose parts by splitting.

Because plate motions are powered by internal Earth forces, they are independent of surface conditions. So we find volcanoes erupting in the cold desert of Antarctica as well as near the equator in African savannas. An alpine mountain range has been pushed up in the cold subarctic zone of Alaska, where it runs east-west. Yet another range lies astride the equator in South America and runs north-south. Both ranges lie in belts of crustal collision, where many strong earthquakes occur.

Although internal tectonic processes operate independently, the processes that determine climate, vegetation, and soils are dependent on the major relief features and Earth materials provided by the tectonic setting. Thus, an understanding of plate tectonics is important to our understanding of the global patterns of the Earth's landscapes—including its climate, soils, vegetation, and, ultimately, human activity.

In this chapter we will survey the major geologic features of our planet, starting with the layered structure of its deep interior. We will then examine the crust and compare the crust of the continents with the crust of the ocean basins. Lastly, we will turn to plate tectonics and describe how plate movements have created broad regions of igneous, sedimentary, and metamorphic rocks. The motions and histories of these vast rock plates, splitting and separating to form ocean basins, closing and colliding to form mountain ranges, can be taken as an overarching framework to organize a great body of geologic and geographic knowledge. That is why this revolutionary framework—plate tectonic theory—ranks with the theory of evolution as one of the great milestones in scientific study of the Earth.

THE STRUCTURE OF THE EARTH

What lies deep within the Earth? From studies of earthquake waves, reflected from deep Earth layers, scientists have discovered that our Earth is far from uniform from its outer crust to its center. Instead, it consists of a central core with several layers, or shells, surrounding it. The densest matter is at the center, and each layer above it is increasingly less dense. This structure is inherited from the Earth's earliest history as it was formed by accretion from a mass of gas and dust orbiting the Sun. We will begin our examination of the Earth's inner structure at the center and then work outward.

The Earth's Interior

Figure 13.1 is a cutaway diagram of the Earth showing its interior. The Earth as a whole is an almost spherical body approximately 6400 km (about 4000 mi) in radius. The center is occupied by the **core,** which is about 3500 km (about 2200 mi) in radius. Because earthquake waves suddenly change behavior upon reaching the core, scientists have concluded that the outer core has the properties of a liquid. However, the innermost part of the core is in the solid state. Based on earthquake waves (and other kinds of data), it has

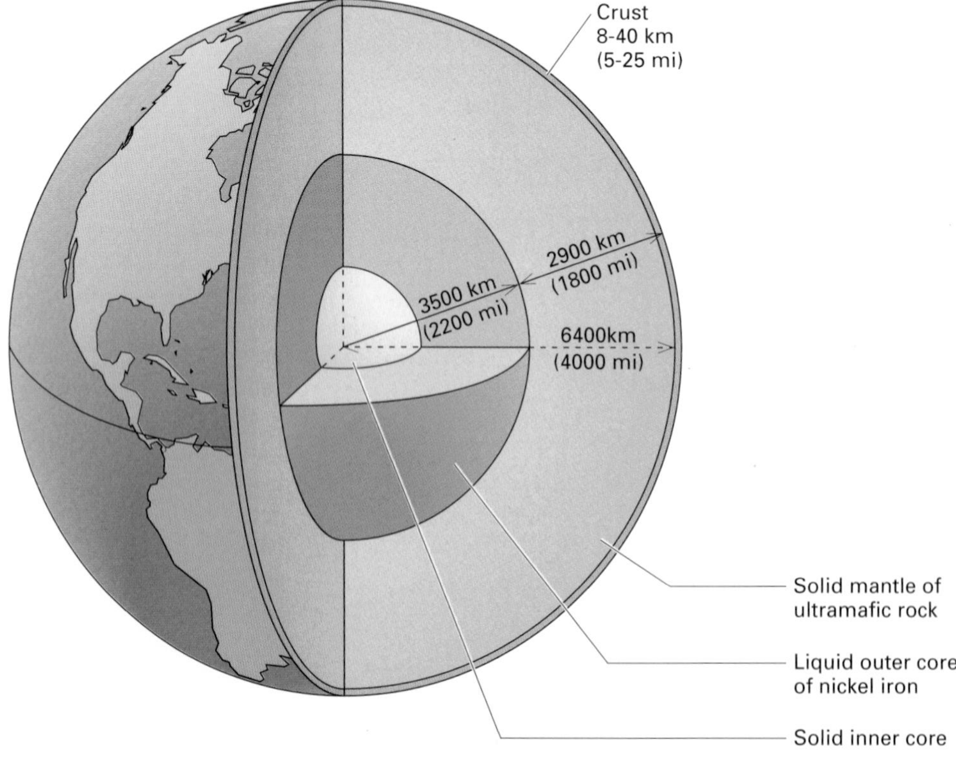

13.1 The Earth's interior *This cutaway diagram of the Earth shows the inner core of iron, which is solid, surrounded by a liquid outer core. The mantle, which surrounds the core, is a thick inner layer of ultramafic rock. The crust is too thin to show to correct scale. Thicknesses of layers are rounded to the nearest 100 units.*

long been inferred that the core consists mostly of iron, with some nickel. The core is very hot, but its temperature is not known very well. It is estimated to lie somewhere between 3000°C and 5000°C (about 5400°F to 9000°F).

Enclosing the metallic core is the **mantle,** a rock shell about 2900 km (about 1800 mi) thick. Earthquake waves indicate that mantle rock is composed of mafic minerals similar to olivine (a silicate of magnesium and iron) and may resemble the ultramafic igneous rock peridotite, which is found exposed here and there on the continental surface (see Figure 12.4 and Table 12.1). Temperatures in the mantle range from about 2800°C (about 5100°F) near the core to about 1800°C (about 3300°F) near the crust.

The outermost and thinnest of the Earth shells is the **crust,** a layer normally about 8 to 40 km (about 5 to 25 mi) thick (Figure 13.2). It is formed largely of igneous rock, but it also contains substantial proportions of metamorphic rock and a comparatively thin upper layer of sedimentary rock. The base of the crust is sharply defined where it contacts the mantle. This contact is detected by the way in which earthquake waves abruptly change velocity at that level. The boundary surface between crust and mantle is called the *Moho,* a simplification of the name of the scientist, Andrija Mohorovicic, who discovered it in 1909.

The continental crust is quite different from the crust beneath the oceans. From studies of earthquake waves, geologists have concluded that the **continental crust** consists of two continuous zones—a lower, continuous rock zone of mafic composition, which is more dense, and an upper, continuous zone of felsic rock, which is less dense (Figure 13.2). Because the felsic portion has a chemical composition similar to that of granite, it is commonly described as being *granitic rock.* Much of the granitic rock is metamorphic rock. There is no sharply defined separation between the felsic and mafic zones.

While continental crust typically has two layers of different composition, **oceanic crust** has only one. It consists almost entirely of mafic rocks with the composition of basalt and gabbro. Basalt, as lava, forms an upper zone, whereas gabbro, an intrusive rock of the same composition, lies beneath the basalt.

Another key distinction between continental and oceanic crust is that the crust is much thicker beneath the continents than beneath the ocean floors (Figure 13.2). While 35 km (22 mi) is a good average figure for crustal thickness beneath the continents, 7 km (4 mi) is a typical figure for thickness of the basalt and gabbro crust beneath the deep ocean floors. Later in this chapter we will relate this difference in thickness to the processes that form the two types of crust.

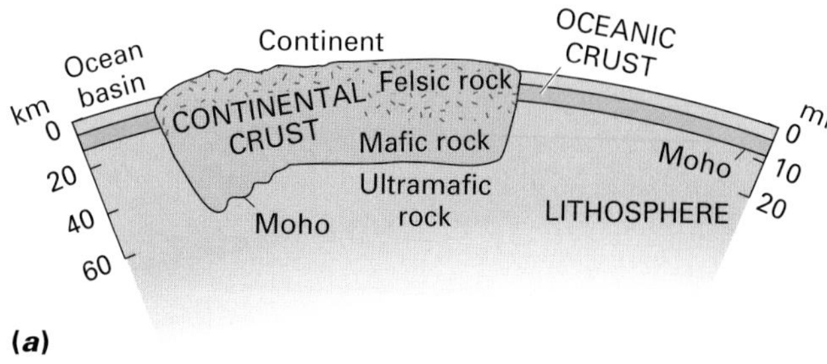

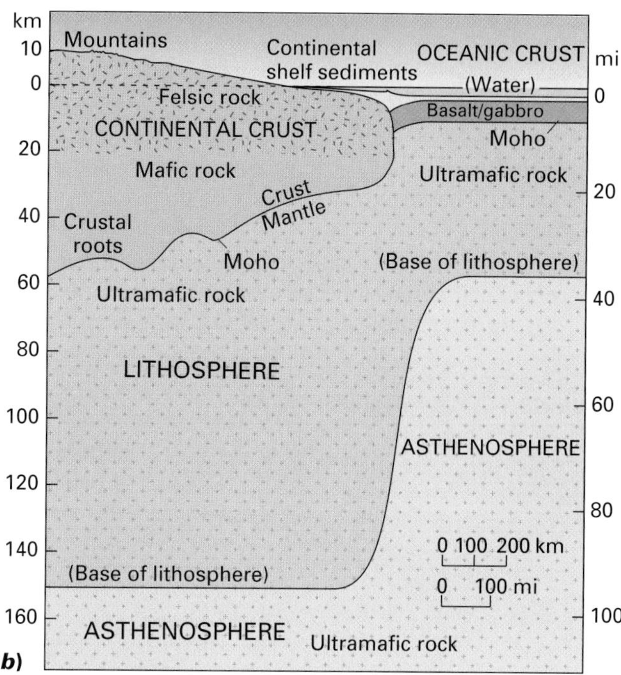

13.2 Outer layers of the Earth *(a) Idealized cross section of the Earth's crust and upper mantle. (b) Details of the crust and mantle at the edge of a continent, including the types of rocks found there. Also shown are the lithosphere and asthenosphere. (Copyright © A. N. Strahler.)*

The Lithosphere and Asthenosphere

Geologists use the term **lithosphere** to mean an outer Earth zone, or shell, of rigid, brittle rock. The lithosphere includes not only the crust, but also the cooler, upper part of the mantle that is composed of brittle rock. The lithosphere ranges in thickness from 60 to 150 km (40 to 95 mi). It is thickest under the continents and thinnest under the ocean basins.

Some tens of kilometers deep in the Earth, the brittle condition of the lithospheric rock gives way gradually to a plastic, or "soft," layer named the **asthenosphere** (Figure 13.2). (This word is derived from the Greek root *asthenes,* meaning "weak.") However, at still greater depth in the mantle, the strength of the rock material again increases. Thus, the asthenosphere is a soft layer sandwiched between the "hard" lithosphere above and a "strong" mantle rock layer below. In terms of states of matter, the asthenosphere is not a liquid, even though its temperature reaches 1400°C (about 2600°F).

The lithosphere resting atop the asthenosphere can be thought of as a hard, brittle shell resting on a soft, plastic underlayer. Because the asthenosphere is soft and plastic, the rigid lithosphere can easily move over it. The lithospheric shell consists of large pieces called **lithospheric plates.** A single plate can be as large as a continent and can move independently of the plates that surround it. Like great slabs of floating ice on the polar sea, lithospheric plates can separate from one another at one location, while elsewhere they may collide in crushing impacts that raise great ridges. Along these collision ridges, one plate can be found diving down beneath the edge of its neighbor. These varied sorts of plate movements will be described shortly.

THE GEOLOGIC TIME SCALE

To place the great movements of lithospheric plates in their correct positions in historical sequence, we will need to refer to some major units in the scale of geologic time. Table 13.1 lists the major geologic time divisions. All time older than 570 million years (m.y.) before the present is *Precambrian time.* Three *eras* of time follow: *Paleozoic, Mesozoic,* and *Cenozoic.* These eras saw the evolution of life-forms in the oceans and on the lands. The geologic eras are subdivided into *periods.* Their names, ages, and durations are also given in Table 13.1.

The Cenozoic Era is particularly important in terms of the continental surfaces because nearly all landscape features seen today have been produced in the 65 million years since that era began. The Cenozoic Era is comparatively short in duration, scarcely more than the average duration of a single period in older eras. It is subdivided directly into seven lesser time units called *epochs.* (Note that the terms *Tertiary Period* and *Quaternary Period* are sometimes applied to the Cenozoic Epochs Paleocene through Pliocene, and Pleistocene, respectively.) Details of the Pleistocene and Holocene epochs are given in Chapter 20.

The human genus *Homo* evolved during the late Pliocene Epoch and throughout the Pleistocene Epoch. As you can see, the period of human occupation of the Earth's surface—a few million years at best—is but a fleeting moment in the vast duration of our planet's history.

Geologists use a variety of techniques to establish the ages of rocks within the geologic time scale. One of the most important tools is *radiometric dating,* which establishes the age of certain minerals within rocks using principles of radioactive decay. This process is described in more detail in *Working It Out 13.1 • Radiometric Dating.*

Table 13.1 | Geologic Time Scale

Era	Period	Epoch	Duration (millions of years)	Age (millions of years)	Orogenies	Evolution of Life-Forms
Cenozoic		Holocene	(10,000 yr)			
		Pleistocene	2	2		Human genus
		Pliocene	3	5		
		Miocene	19	24		Hominoids
		Oligocene	13	37		Whales
		Eocene	21	58		Bats
		Paleocene	8	66	Cordilleran	Mammals
Mesozoic	Cretaceous		78	144		Flowering plants
	Jurassic		64	208		Dinosaurs (extinct)
	Triassic		37	245	Allegheny, or Hercynian	Turtles
Paleozoic	Permian		41	286		Frogs
	Carboniferous		74	360		Conifers; higher fishes
	Devonian		48	408	Caledonian	Vascular plants; primitive fishes
	Silurian		30	438		
	Ordovician		67	505		
	Cambrian		65	570		Invertebrates

Precambrian Time (Extends to oldest known rocks, about 4 billion years)
Age of Earth as a planet: 4.6 to 4.7 billion years.
Age of universe: 17 to 18 billion years.

Working It Out | 13.1
Radiometric Dating

How old is the Earth? How old are the oldest rocks? These questions have interested geologists ever since systematic study of the Earth began. However, it was not until the early part of the twentieth century, with the discovery of radioactive decay, that scientists uncovered a principle that would provide accurate answers to these questions. And it was not until the middle of the century, when technology provided methods to measure minute amounts of trace elements in rocks, that it became possible to apply this principle to the practical problem of determining the ages of rocks.

As noted in *Working It Out 12.1 • Radioactive Decay,* certain isotopes of elements undergo a spontaneous decay process in which their atomic composition changes. Recall from that feature that ^{238}U decays to form the daughter product ^{234}Th. The half-life for this decay is 4.47×10^9 yr. However, ^{234}Th is itself unstable, and decays to form another isotope, which is also unstable, and decays to form yet another isotope, and so on, until the stable isotope lead-206 (^{206}Pb) is ultimately formed. Recall also that another naturally occurring isotope of uranium, ^{235}U, has a similar decay sequence and yields the stable isotope lead-207 (^{207}Pb) with a half-life of 7.04×10^8 years.

Suppose that an initial quantity of ^{238}U is included in the formation of an igneous rock early in the Earth's history. Locked in place within the crystalline structure of minerals within the rock, the atoms of ^{238}U will slowly transform themselves to ^{206}Pb at a rate that depends only on the half-life of ^{238}U. This is the principle behind *radiometric dating*—using the ratios of particular elemental isotopes to determine the time of formation or metamorphism of a rock.

How exactly does radiometric dating work? Recall from *Working It Out 12.1* that

$$P(t) = e^{-kt}$$

where $P(t)$ is the proportion of an unstable isotope remaining after time t, and $k = 0.693/H$, with H the half-life of the isotope. From this relation, we can derive the formula

$$t = \frac{1}{k} \ln \left[\frac{D}{M} + 1 \right]$$

which gives the time of decay, t, as a function of k and D/M, the ratio of daughter to mother atoms in the sample. Suppose, for example, that a chemical analysis of a mineral shows that for every two atoms of ^{238}U, one atom of ^{206}Pb is present. Then $D/M = 1/2 = 0.5$, and since $k = 0.693/4.47 = 0.155$ using units of billions of years (1 b.y. = 10^9 yr), we have

$$t = \frac{1}{0.155} \ln[0.5 + 1] = \frac{0.405}{0.155} = 2.61 \text{ b.y.}$$

The practice of radioactive dating goes well beyond this simple example, but the principle is basically the same—that by analyzing the concentration of certain radioactive isotopes in certain minerals carefully and exactly, it is possible to date them with certainty.

How old are the oldest rocks? Among the oldest rocks known are ancient shield rocks from Greenland, which have been dated at 3.80 b.y. Even older are some shield rocks from Antarctica, which date to 3.93 b.y. But the oldest rocks so far discovered are ancient gneisses in northwestern Canada. They contain zircon crystals with ages of 3.96 b.y.

How old is the Earth? During the earliest eras of the Earth's history, continental crust was being vigorously formed and reformed, so rocks and minerals from these early times are long gone. However, some types of meteorites have radiometric ages of about 4.6 b.y., and these also have the same mixture of naturally occurring lead isotopes as the Earth's rocks. This latter fact leads geochemists to suspect strongly that these meteorites were formed from the same primordial matter as the Earth itself, thus dating the formation of the Earth at about 4.6 billion years ago.

MAJOR RELIEF FEATURES OF THE EARTH'S SURFACE

Before we begin our discussion of plate tectonics, it is important to examine the major relief features of the Earth—its continents and ocean basins—since they are produced by plate tectonic processes. Detailed global maps show that about 29 percent of the Earth's surface is land and 71 percent oceans (Figure 13.3). If the seas were to drain away, however, we would see that broad sloping areas lie close to the continental shores. These continental shelves are covered by shallow water, less than 150 m (500 ft) deep. From these relatively shallow continental shelves, the ocean floor drops rapidly to depths of thousands of meters. In other words, the ocean basins seem to be more than "brim full" of water—the oceans have spread over the margins of ground that would otherwise be assigned to the continents. If the ocean level were to drop by 150 m (about 500 ft), the shelves would be exposed, adding about 6 percent to the area of the continents. The surface area of continents would then be increased to 35 percent, and the ocean basin area decreased to 65 percent. These revised figures represent the true relative proportions of continents and oceans.

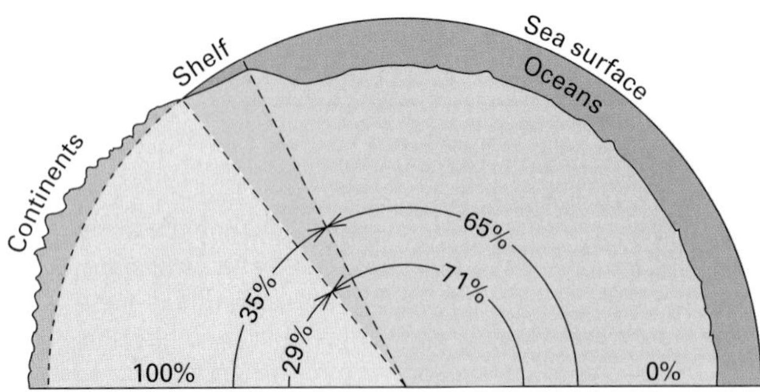

13.3 Continent and ocean areas compared *Actual global percentages of land and ocean areas compared with percentages if sea level were to drop 180 m (about 600 ft), exposing the continental shelves.*

Figure 13.4 is a map showing some of the major relief features of continents and oceans, including mountain arcs, island arcs, ocean trenches, and the midoceanic ridge. We will refer to this map as we discuss the relief features of continents and oceans in turn.

Relief Features of the Continents

Broadly viewed, the continents consist of two basic subdivisions: active belts of mountain-making and inactive regions of old, stable rock. The mountain ranges in the active belts grow through one of two very different geologic processes. First is *volcanism,* the formation of massive accumulations of volcanic rock by extrusion of magma. Many lofty mountain ranges consist of chains of volcanoes built of extrusive igneous rocks.

The second mountain-building process is *tectonic activity,* the breaking and bending of the Earth's crust under internal Earth forces. This tectonic activity usually occurs when great lithospheric plates come together in titanic collisions. (We will return to this topic later in this chapter.) Crustal masses that are raised by tectonic activity form mountains and plateaus. In some instances, both volcanism and tectonic activity have combined to produce a mountain range. Tectonic activity can also lower crustal masses to form depressions that may be occupied by ocean embayments or inland seas. Landforms produced by volcanic and tectonic activity are the subject of Chapter 14.

ALPINE CHAINS Active mountain-making belts are narrow zones that are usually found along the margins of lithospheric plates. These belts are sometimes referred to as *alpine chains* because they are characterized by high, rugged mountains, such as the Alps of Central Europe. These mountain belts were formed in the Cenozoic Era by volcanism or tectonic activity or a combination of both. Alpine mountain-building continues even today in many places.

13.4 Tectonic features of the world *Principal mountain arcs, island arcs, and trenches of the world and the midoceanic ridge. (Midoceanic ridge map copyright © A. N. Strahler.)*

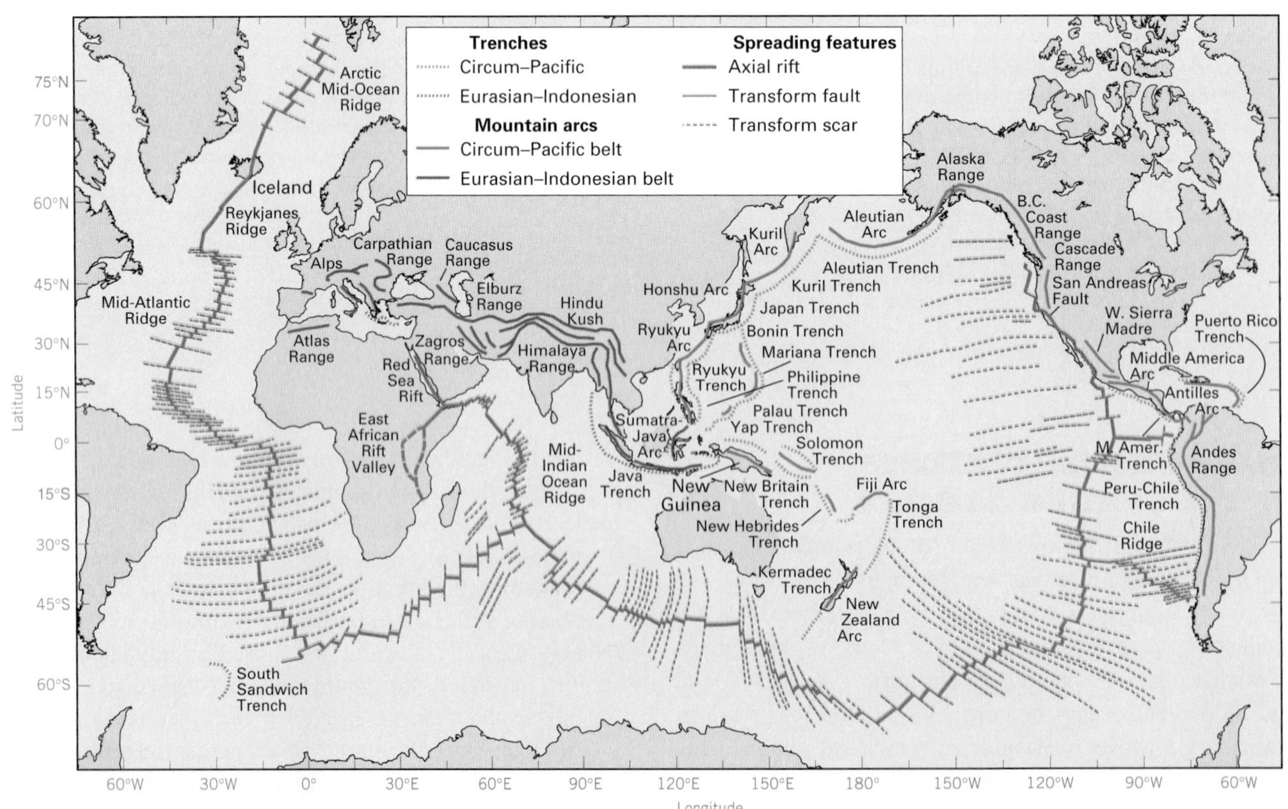

The alpine chains are characterized by broadly curved lines on the world map (Figure 13.4). Each curved section of an alpine chain is referred to as a **mountain arc.** The arcs are linked in sequence to form two principal mountain belts. One is the *circum-Pacific belt* (shown in green), which rings the Pacific Ocean Basin. In North and South America, this belt is largely on the continents and includes the Andes and Cordilleran ranges. In the western part of the Pacific Basin, the mountain arcs lie well offshore from the continents and take the form of **island arcs.** Partly submerged, they join the Aleutians, Kurils, Japan, the Philippines, and other smaller islands. These island arcs are the result of volcanic activity. Between the larger islands, the arcs are represented by volcanoes rising above the sea as small, isolated islands.

The second chain of major mountain arcs forms the *Eurasian-Indonesian belt,* shown in blue in Figure 13.4. It starts in the west at the Atlas Mountains of North Africa and continues through the European Alps and the ranges of the Near East and Iran to join the Himalayas. The belt then continues through Southeast Asia into Indonesia, where it abruptly meets the circum-Pacific belt. Later we will return to these active belts of mountain-making and explain them in terms of subduction of lithospheric plates.

CONTINENTAL SHIELDS Belts of recent and active mountain-making account for only a small portion of the continental crust. The remainder consists of comparatively inactive regions of much older rock. Within these stable regions we can recognize two types of crustal structure—continental shields and mountain roots. **Continental shields** are low-lying continental surfaces beneath which lie igneous and metamorphic rocks in a complex arrangement. Figure 13.5 is a generalized map showing the shield areas of the continents. Two classes of shields are shown: exposed shields and covered shields.

Exposed shields include very old rocks, mostly from Precambrian time, and have had a very complex geologic history. An example of an exposed shield is the Canadian Shield of North America. Exposed shields are also extensive in Scandinavia, South America, Africa, Asia, peninsular India, and Australia. The exposed shields are largely regions of low hills and low plateaus, although there are some exceptions where large crustal blocks have been recently uplifted. Many thousands of meters of rock have been eroded from these shields throughout the past half-billion or more years.

Covered shields are areas of the continental shields that are covered by younger sedimentary layers, ranging in age from Paleozoic through Cenozoic eras. These strata accumulated at times when the shields were inundated by shallow seas. Marine sediments were laid down on the ancient shield rocks in thicknesses ranging from hundreds to thousands of meters. These shield areas were then broadly arched upward to become land surfaces again. Erosion has since

13.5 Continental shields *Generalized world map of continental shields, exposed and covered. Continental centers of early Precambrian age (Archean) lie with the areas encircled by a broken red line. Heavy brown lines show mountain roots of Caledonian and Hercynian (Appalachian) orogenies. (Based in part on data of R. E. Murphy, P. M. Hurley, and others. Copyright © A. N. Strahler.)*

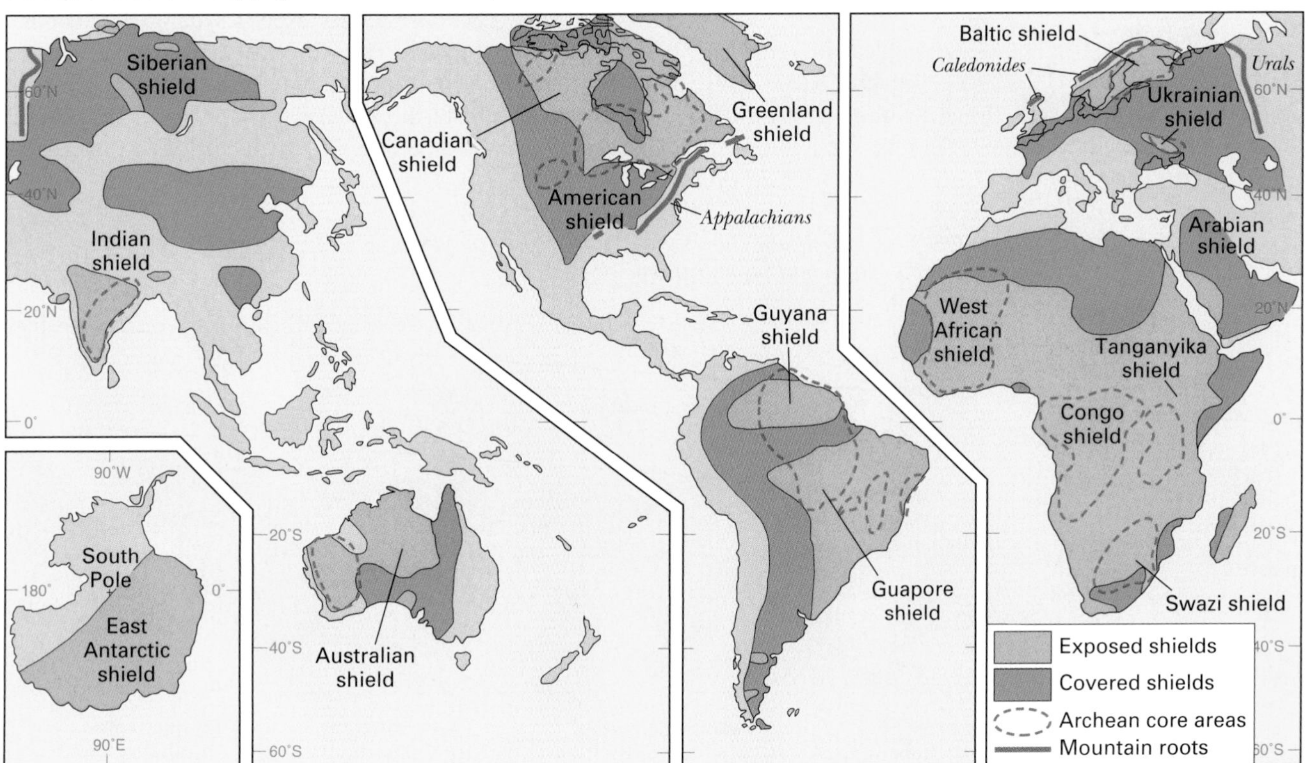

removed large sections of their sedimentary cover, but it still remains intact over vast areas. The covered shields are shown in Figure 13.5.

Some core areas of the shields are composed of rock as old as early Precambrian time, dating back to a time period called the Archean Eon, 2.5 to 3.5 billion years ago. On our map, these ancient areas are shown encircled by bold lines. The ancient cores are exposed in some areas but covered in others.

ANCIENT MOUNTAIN ROOTS Remains of older mountain belts lie within the shields in many places. These *mountain roots* are mostly formed of Paleozoic and early Mesozoic sedimentary rocks that have been intensely bent and folded, and in some locations changed into metamorphic rocks—slate, schist, and quartzite, for example. Thousands of meters of overlying rocks have been removed from these old tectonic belts, so that only the lowermost structures remain. Roots appear as chains of long, narrow ridges, rarely rising over a thousand meters above sea level.

One important system of mountain roots was formed in the Paleozoic Era, during a great collision between two enormous lithospheric plates that took place about 400 million years ago. This collision created high alpine mountain chains that have since been worn down to belts of subdued mountains and hills. Today these roots, called Caledonides, form a highland belt across the northern British Isles and Scandinavia. They are also present in the Maritime Provinces of eastern Canada and in New England. A second, but younger, root system was formed during another great collision of plates near the close of the Paleozoic Era, about 250 million years ago. In North America, this highland system is represented by the Appalachian Mountains. The Caledonides and Appalachians are shown as mountain roots in Figure 13.5.

Relief Features of the Ocean Basins

The relief features of ocean basins are quite different from those of the continents. Crustal rock of the ocean floors consists almost entirely of basalt, which is covered over large areas by a comparatively thin accumulation of sediments. Age determinations of the basalt and its sediment cover show that the oceanic crust is quite young, geologically speaking. Much of the oceanic crust is less than 60 million years old, although some large areas have ages of about 65 to 135 million years. When we consider that the great bulk of the continental crust is of Precambrian age—mostly over 1 billion years old—the young age of the oceanic crust is quite remarkable. However, we will soon see how plate tectonic theory explains this young age.

THE MIDOCEANIC RIDGE AND OCEAN BASIN FLOOR Figure 13.6 shows the important relief features of ocean basins. The ocean basins are characterized by a central ridge structure that divides the basin about in half. This *midoceanic ridge* consists of submarine hills that rise gradually to a rugged central zone. Precisely in the center of the ridge, at its highest point, is the *axial rift,* which is a narrow, trench-like feature. The location and form of this rift suggest that the crust is being pulled apart along the line of the rift.

The midoceanic ridge and its principal branches can be traced through the ocean basins for a total distance of about 60,000 km (about 37,000 mi). Figure 13.4 shows the extent of the ridge. Beginning in the Arctic Ocean, it then divides the Atlantic Ocean Basin from Iceland to the South Atlantic. Turning east, it enters the Indian Ocean, where one branch penetrates Africa. The other branch continues east between Australia and Antarctica, and then swings across the South Pacific. Nearing South America, it turns north and reaches North America at the head of the Gulf of California.

On either side of the midoceanic ridge are broad, deep plains that belong to the *ocean basin floor* (Figure 13.6). Their average depth below sea level is about 5000 m (about 16,000 ft). These large, flat expanses of ocean floor are called *abyssal plains.* They are extremely smooth because they have been built up of fine sediment that has settled slowly and evenly from ocean water above. Some abyssal regions include *abyssal hills,* small hills rising to heights

13.6 Ocean basins *This schematic block diagram shows the main features of ocean basins. It applies particularly well to the North and South Atlantic oceans.*

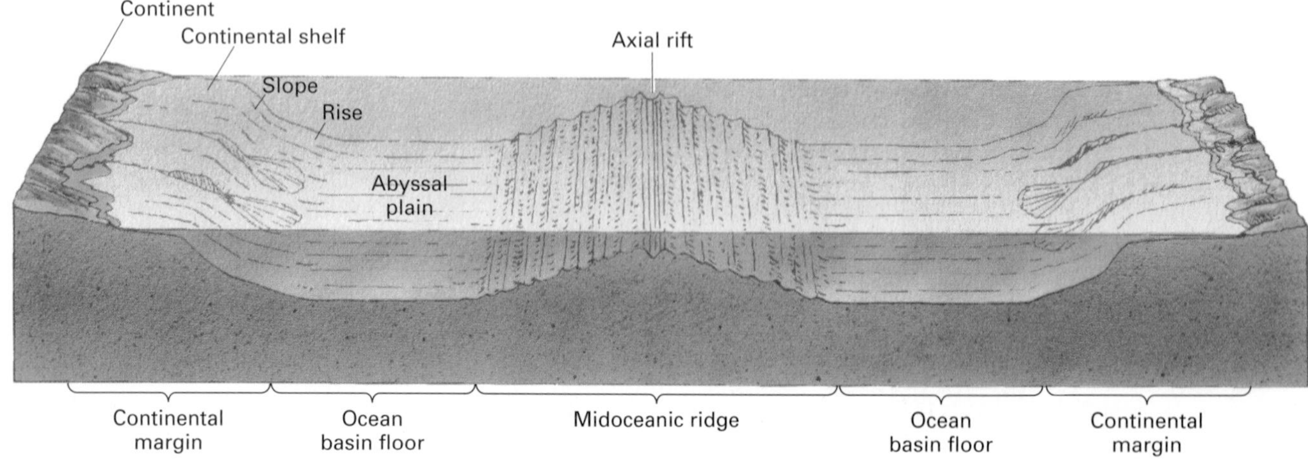

from a few tens of meters to a few hundred meters above the ocean floor.

Many details of the ocean basins and their submarine landforms are shown in Figure 13.7. This image of the ocean floor was constructed from data of a U.S. Navy satellite that measured the surface height of the ocean very precisely using radar. Because small variations in gravity caused by undersea ridges and valleys produce corresponding small variations in the ocean height, it is possible to infer the undersea topography from ocean height measurement. For example, the mass of a submerged seamount may attract enough sea water to raise sea levels above it by 1 to 2 m (3 to 6 ft), depending on its bulk. Similarly, the ocean surface level drops over a deep ocean trench, where there is less mass to attract sea water. The right page of the figure shows the North Atlantic Basin. The prominent central feature is the Mid-Atlantic Ridge. The left page shows the deep trenches of the western Pacific that mark the positions of subduction arcs—features of plate motion that we will discuss shortly.

CONTINENTAL MARGINS The *continental margin,* shown on the left and right sides of Figure 13.6, is the narrow zone in which oceanic lithosphere is in contact with continental lithosphere (see Figure 13.2*b*). As the continental margin is approached from deep ocean, the ocean floor begins to slope gradually upward, forming the *continental rise* (Figure 13.6). The floor then steepens greatly on the *continental slope.* At the top of this slope we arrive at the edge of the *continental shelf,* a gently sloping platform that is about 120 to 160 km (about 75 to 100 mi) wide along the eastern margin of North America. Water depth is about 150 m (about 500 ft) at the outer edge of the shelf. These three features—rise, slope, and shelf—form the continental margin.

Figure 13.6 illustrates a symmetrical ocean floor model—a midoceanic ridge with ocean basin floors on either side. This model nicely fits the North Atlantic and South Atlantic Ocean basins. It also applies rather well to the Indian Ocean and Arctic Ocean basins. The margins of these symmetrical basins are described as *passive continental margins* because they have not been subjected to strong tectonic and volcanic activity during the last 50 million years. This is because the continental and oceanic lithosphere that join at a passive continental margin are part of the same lithospheric plate and move together, away from the axial rift.

Many passive continental margins have accumulated great thicknesses of continental sediments. Near the continent, the strata of the shelf form a wedge-shaped deposit, thickening oceanward. A block diagram, Figure 13.8, shows details of this deposit. The river-borne sediments brought from the land are spread over the shallow sea floor by currents. Beneath the continental rise and its adjacent abyssal plain is another thick sediment deposit. It is formed from deep-sea sediments carried down the continental slope by swift muddy currents. The weight of these accumulated sediments causes them to sink, depressing the oceanic crust at its margin with the continental crust.

Deltas built by rivers contribute a great deal of the shelf sediment. Tongue-like turbid currents thickened with fine silt and clay from these deltas carve submarine canyons into the continental shelf. Where these currents emerge from the canyons, they deposit their sediment and build *deep-sea cones* (Figure 13.9).

The margins of the Pacific Ocean Basin are very different from those of the Atlantic. Although the Pacific has a midoceanic ridge with ocean basin floors on either side, its margins are characterized by mountain arcs or island arcs with deep offshore *oceanic trenches* (see Figure 13.7). Geologists refer to these trenched ocean-basin edges as *active continental margins.* As we will describe in more detail in the next section, oceanic crust is here being bent downward and forced under continental crust, creating trenches and inducing volcanic activity. The locations of the major trenches are shown in Figure 13.4. Trench floors can reach depths of nearly 11,000 m (about 36,000 ft), although most range from 7000 to 10,000 m (about 23,000 to 33,000 ft).

PLATE TECTONICS

With the structures and features of continents and ocean basins as a background, we can now turn to the motion of lithospheric plates and their interactions at boundaries—that is, to **plate tectonics. Tectonics** is a noun meaning "the study of tectonic activity." It refers here to all forms of breaking and bending of the entire lithosphere, including the crust. Breaking and bending occur on a very large scale when lithospheric plates collide. Sedimentary strata that lay flat and undisturbed for tens of millions of years on ocean floors and continental margins are crumpled and sheared into fragments by such collisions. The collision process itself may take millions of years to complete, but in the perspective of geologic time, it's only a brief episode.

Tectonic Processes

Generally speaking, prominent mountain masses and mountain chains (other than volcanic mountains) are elevated by one of two basic tectonic processes: *compression* and *extension* (Figure 13.10). Extensional tectonic activity—"pulling apart"—occurs where oceanic plates are separating or where a continental plate is undergoing breakup into fragments. As the crust thins, it is fractured and pushed upward, producing block mountains. We will return to a more detailed description of this process in a later section on continental rupture.

Compressional tectonic activity—"squeezing together" or "crushing"—acts at converging plate boundaries. The result is often an alpine mountain chain consisting of intensely deformed strata of marine origin. The strata are tightly compressed into wavelike structures called *folds.*

A simple experiment you can do will demonstrate how folding occurs in mountain-building. Take an ordinary towel of thick, limp cloth and lay it out on a smooth table top. Using both hands, palms down, bring the ends of the cloth slowly together. First, a simple up-fold will develop and

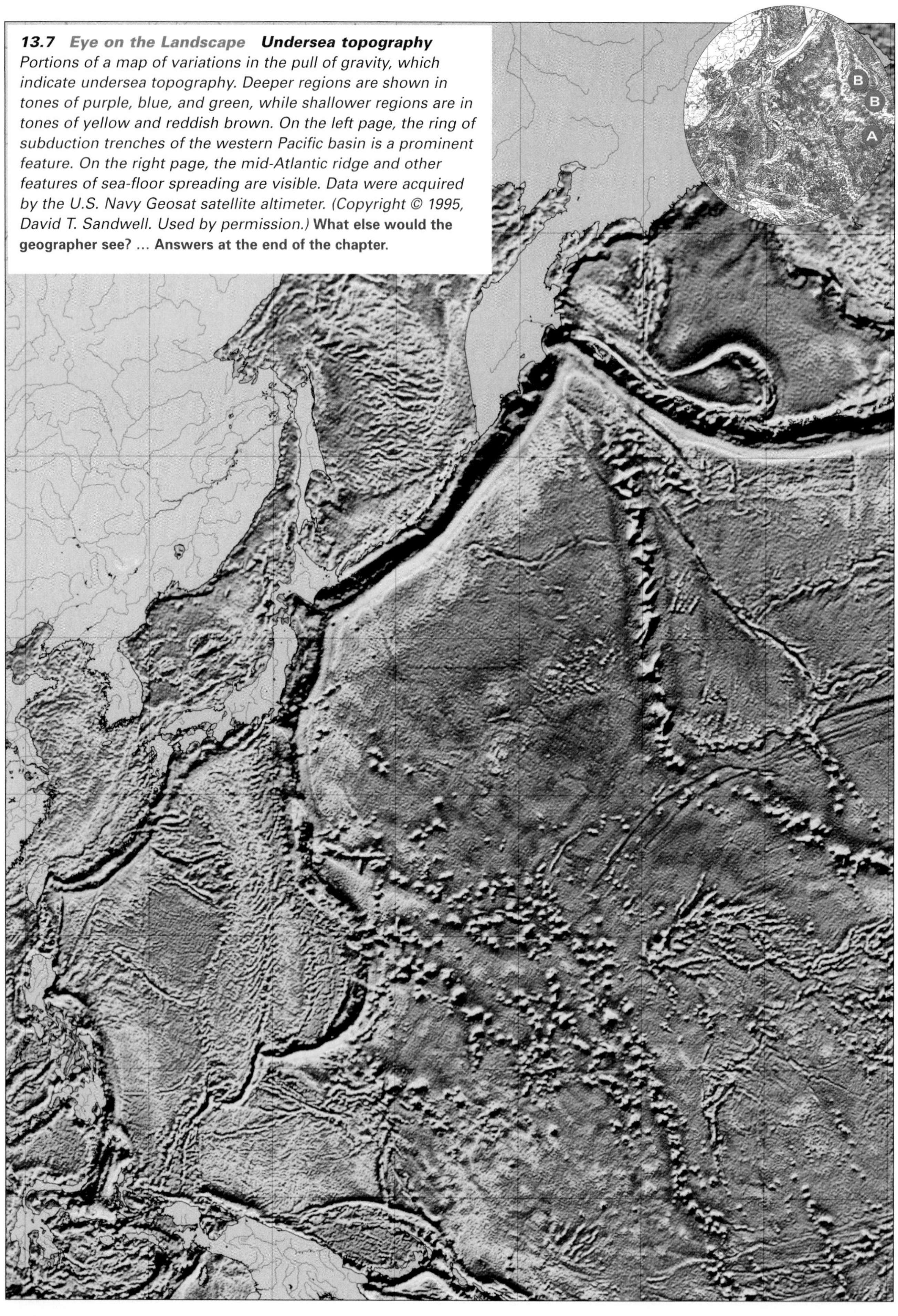

13.7 Eye on the Landscape **Undersea topography**
Portions of a map of variations in the pull of gravity, which indicate undersea topography. Deeper regions are shown in tones of purple, blue, and green, while shallower regions are in tones of yellow and reddish brown. On the left page, the ring of subduction trenches of the western Pacific basin is a prominent feature. On the right page, the mid-Atlantic ridge and other features of sea-floor spreading are visible. Data were acquired by the U.S. Navy Geosat satellite altimeter. (Copyright © 1995, David T. Sandwell. Used by permission.) **What else would the geographer see? ... Answers at the end of the chapter.**

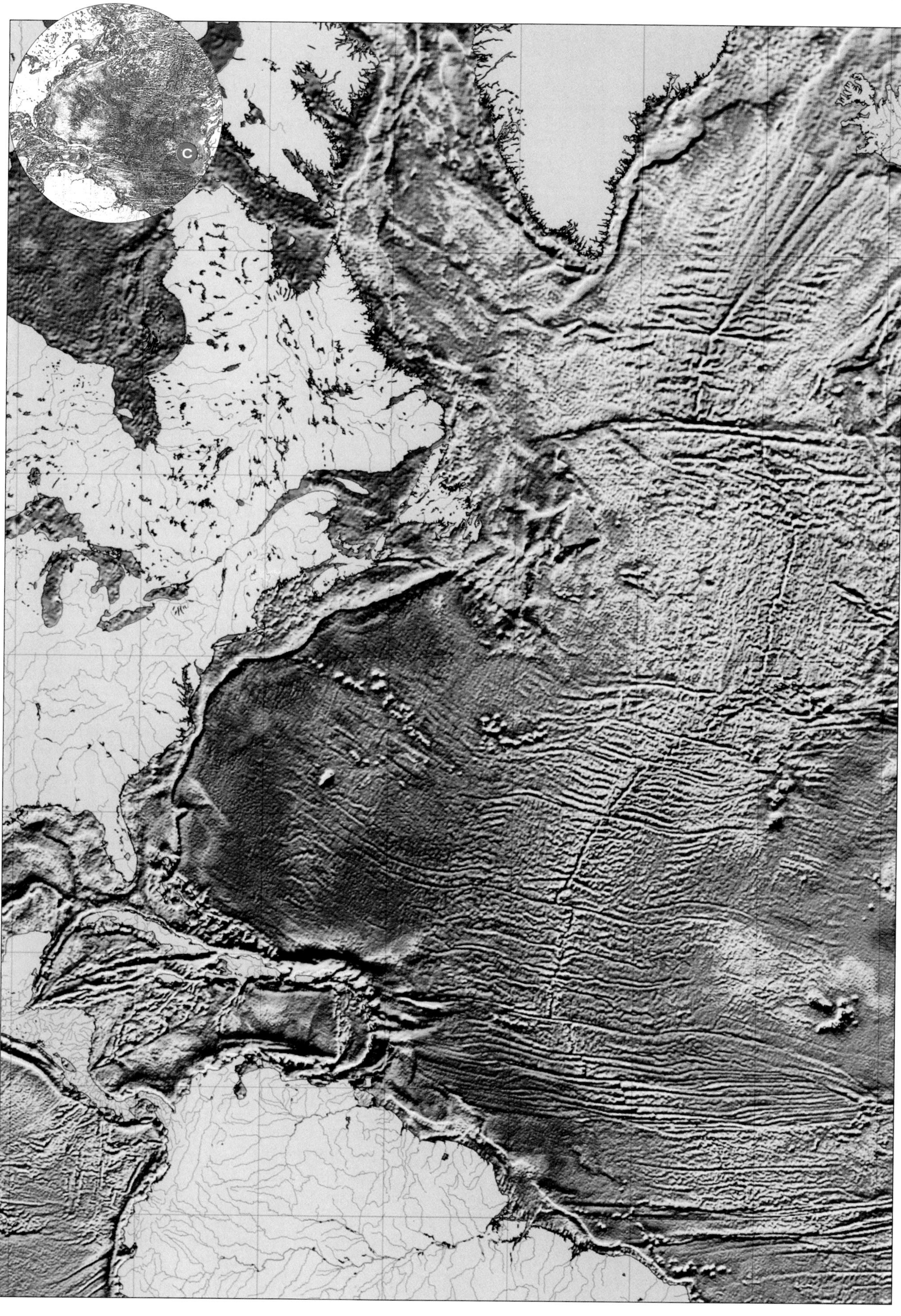
C

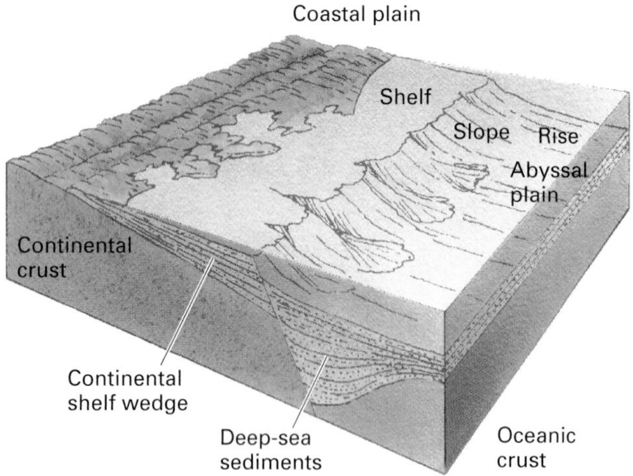

13.8 Structure of the continental shelf *This block diagram shows an inner wedge of sediments beneath the continental shelf and an outer wedge of deep-sea sediments beneath the continental rise and abyssal plain.*

grow, until it overturns to one side or the other. Then more folds will form, grow, and overturn, giving you a simple model of a new mountain range.

Typically, alpine folds can be traced through a history. First, they are overturned, and then they become *recumbent* as they are further overturned upon themselves (Figure 13.11). Accompanying the folding is a form of faulting in which slices of rock move over the underlying rock on fault surfaces with gentle inclination angles. These are called *overthrust faults.* The individual rock slices, called *thrust sheets,* are carried many tens of kilometers over the underlying rock. In the European Alps, thrust sheets of this kind were named *nappes* (from the French word meaning "cover sheet" or "tablecloth"). Nappes may be thrust one over the other to form a great pile. The entire deformed rock mass produced by such compressional mountain-making is called an *orogen,* and the event that produced it is an *orogeny.*

13.9 Deep-sea cone *A deep-sea cone of sediment forms at the base of a submarine canyon cut into the continental shelf. (Copyright © A. N. Strahler.)*

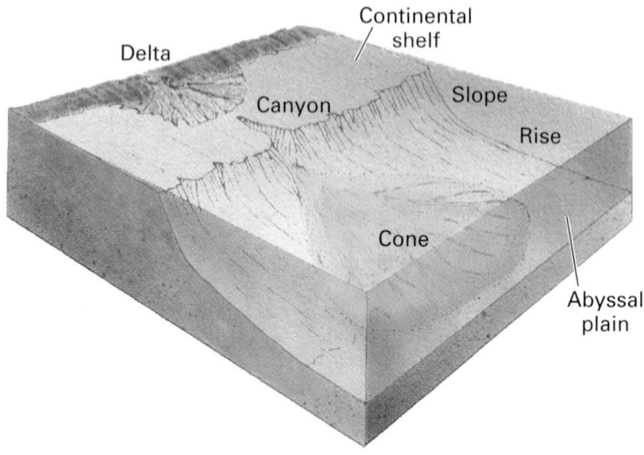

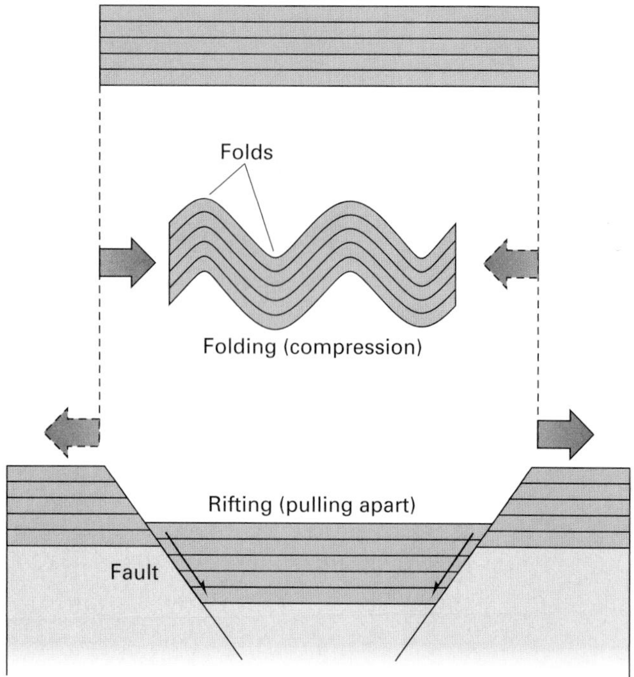

13.10 Two basic forms of tectonic activity *Flat-lying rock layers may be compressed to form folds or pulled apart to form faults by rifting.*

Plate Motions and Interactions

Figure 13.12 shows the major features of plate interactions. The vertical dimension of the block diagram *(a)* is greatly exaggerated, as are the landforms. A true-scale cross section *(b)* shows the correct relationships between crust and lithosphere, but surface relief features are too small to be shown. Diagram *(c)* is a sketch of plate motions on a spherical Earth.

Note that the figure shows two different kinds of lithospheric plates (see Figure 13.12*b*). Plates that lie beneath the ocean basins consist of *oceanic lithosphere.* This form of lithosphere is comparatively thin (about 50 km or 30 mi thick). The diagrams show two plates, X and Y, both made up of oceanic lithosphere. Plate Z bears continental crust and is made up of *continental lithosphere,* which is much thicker (about 150 km or 95 mi).

The lithosphere can be thought of as "floating" on the soft asthenosphere, but there is a difference in the relative surface heights of the two kinds of floating lithosphere. Consider two blocks of wood, one thicker than the other, floating in a pan of water. The surface of the thick block will ride higher above the water surface than that of the thin block. This principle explains why the continental surfaces rise high above the ocean floors.

As shown in Figure 13.12, plates X and Y are pulling apart along their common boundary, which lies along the axis of a midoceanic ridge. This pulling apart tends to create a gaping crack in the crust, but magma continually rises from the mantle beneath to fill it. The magma appears as basaltic lava in the floor of the rift and quickly congeals. At

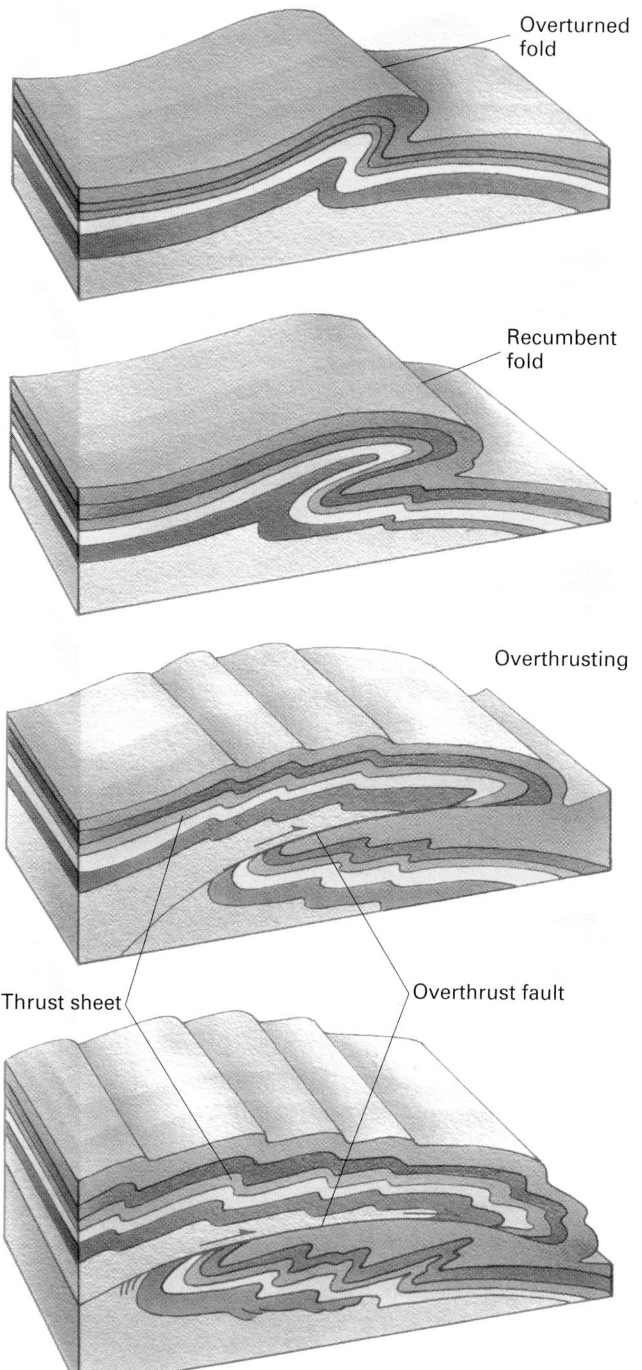

13.11 Folding in compressional tectonic activity *These schematic diagrams show the development of a recumbent fold, broken by a low-angle overthrust fault to produce a thrust sheet, or nappe, in alpine structure. (Based on diagrams by A. Heim, 1922, Geologie der Schweiz, vol. II-1, Tauschnitz, Leipzig.)*

greater depth under the rift, magma solidifies into *gabbro,* an intrusive rock of the same composition as basalt. Together, the basalt and gabbro continually form new oceanic crust. This type of boundary between plates is termed a *spreading boundary.*

At the right, the oceanic lithosphere of plate Y is moving toward the thick mass of continental lithosphere that comprises plate Z. Where these two plates collide, they form a *converging boundary.* Because the oceanic plate is comparatively thin and dense, in contrast to the thick, buoyant continental plate, the oceanic lithosphere bends down and plunges into the soft layer, or asthenosphere. The process in which one plate is carried beneath another is called **subduction.**

The leading edge of the descending plate is cooler and therefore denser than the surrounding hot, soft asthenosphere. As a result, the slab "sinks under its own weight," once subduction has begun. However, the slab is gradually heated by the surrounding hot rock and thus eventually softens. The underportion, which is mantle rock in composition, simply reverts to mantle rock as it softens.

The descending plate is covered by a thin upper layer of less dense mineral matter derived from oceanic and continental sediments. This material can melt and become magma. The magma tends to rise because it is less dense than the surrounding material. Figure 13.12*b* shows some magma pockets formed from the upper edge of the slab. They are pictured as rising like hot-air balloons through the overlying continental lithosphere. When they reach the Earth's surface, they form a chain of volcanoes lying about parallel with the deep oceanic trench that marks the line of descent of the oceanic plate.

Viewing plate Y as a unit in Figure 13.12, we see that this single lithospheric plate is simultaneously undergoing both *accretion* (growth by addition) and *consumption* (loss by softening and melting in subduction zones). If rates of accretion and consumption are equal, the plate will maintain its overall size. If consumption is greater, the area of the plate will decrease. If accretion is greater, the area of the plate will expand.

We have yet to consider a third type of lithospheric plate boundary. Two lithospheric plates may be in contact along a common boundary on which one plate merely slides past the other with no motion that would cause the plates either to separate or to converge (Figure 13.13). This is a *transform boundary.* The plane along which motion occurs is a nearly vertical fracture extending down through the entire lithosphere, and it is called a *transform fault.* A *fault* is a rock plane along which there is motion of the rock mass on one side with respect to that on the other. (More about faults will appear in Chapter 14.) Transform boundaries are often associated with midoceanic ridges and are shown in Figures 13.7 and 13.12. GEODISCOVERIES

PLATE BOUNDARIES SUMMARIZED In summary, there are three major kinds of active plate boundaries:

- *Spreading boundaries.* New lithosphere is being formed by accretion. Example: Sea-floor spreading along the axial rift.

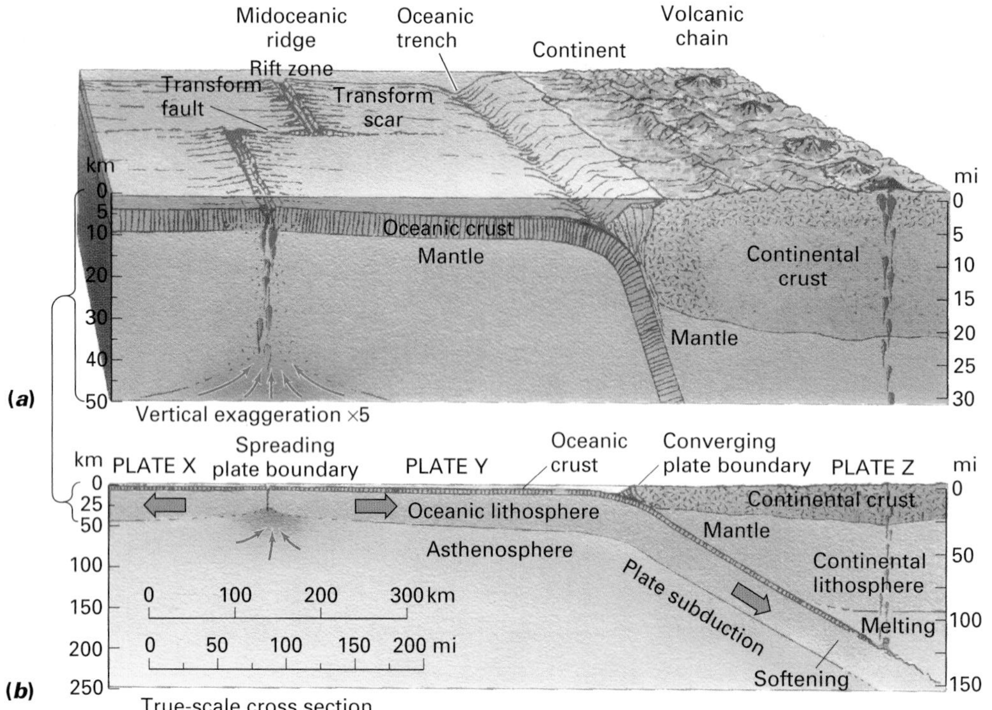

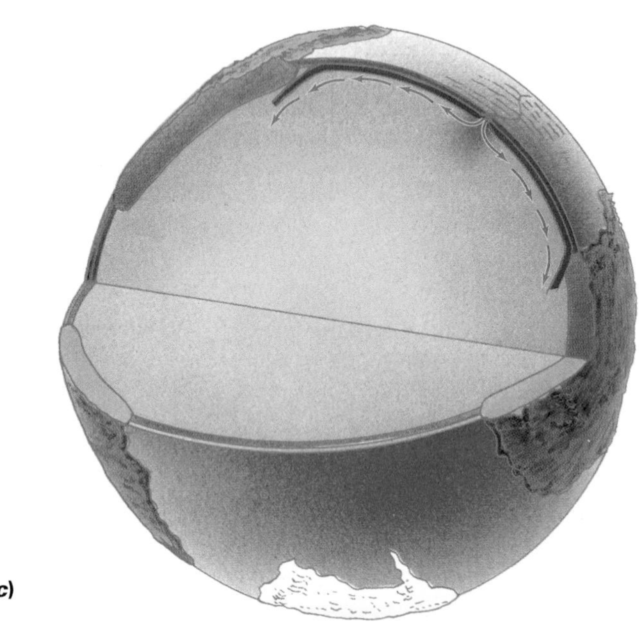

13.12 *Schematic cross sections showing some of the important elements of plate tectonics* *Diagram (a) is greatly exaggerated in vertical scale, and emphasizes surface and crustal features. Only the uppermost 30 km (20 mi) is shown. Diagram (b) is drawn to true scale and shows conditions to a depth of 250 km (155 mi). Here the actual relationships between lithospheric plates can be examined, but surface features are too small to be shown. Diagram (c) is a pictorial rendition of plates on a spherical Earth and is not to scale. (Copyright © A. N. Strahler.)*

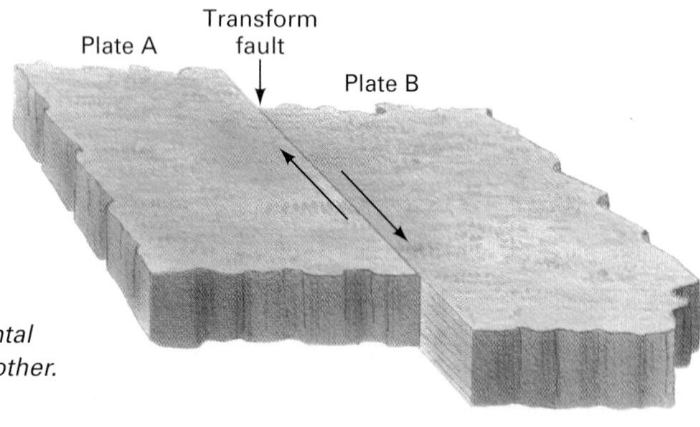

13.13 *Transform fault* *A transform fault involves the horizontal motion of two adjacent lithospheric plates, one sliding past the other. (Copyright © A. N. Strahler.)*

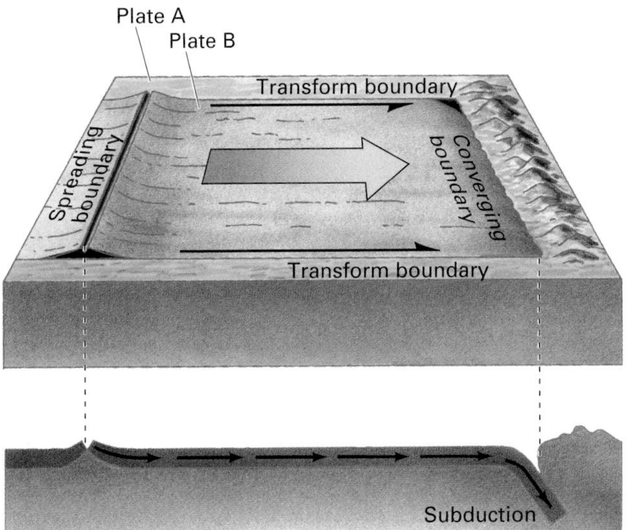

13.14 A simple lithospheric plate *A schematic diagram of a single rectangular lithospheric plate with two transform boundaries. (Copyright © A. N. Strahler.)*

- *Converging boundaries.* Subduction is in progress, and lithosphere is being consumed. Example: Active continental margin.
- *Transform boundaries.* Plates are gliding past one another on a transform fault. Example: Transform boundary associated with midoceanic ridge.

Let us put these three boundaries into a pattern to include an entire lithospheric plate (Figure 13.14). Visualize the sunroof of an automobile, in which a portion of the roof slides to open. Just as the sunroof can move by sliding past the fixed portion of the roof at its sides, so a lithospheric plate can move by sliding past other plates on transform faults. Where the sunroof opens, the situation is similar to a spreading boundary. Where the sunroof slides under the rear part of the roof, the situation is similar to a converging boundary. Boundaries of lithospheric plates can be curved as well as straight, and individual plates can pivot as they move. There are many geometric variations in the shapes and motions of individual plates.

The motions of lithospheric plates follow a time cycle in which a single supercontinent is split apart, then joined back together, and then split apart again. Named the Wilson Cycle, it is described in more detail in *Focus on Systems 13.2 • The Wilson Cycle and Supercontinents.*

The Global System of Lithospheric Plates

The global system of lithospheric plates consists of six great plates. These are listed in Table 13.2 and are shown on a world map in Figure 13.15. Several lesser plates and subplates are also recognized. They range in size from intermediate to comparatively small. Plate boundaries are shown by symbols, explained in the key accompanying the map.

The great Pacific plate (Figure 13.15*a*) occupies much of the Pacific Ocean Basin and consists almost entirely of oceanic lithosphere. Its relative motion is northwesterly, so that it has a subduction boundary along most of the western and northern edge. The eastern and southern edge is mostly a spreading boundary. A sliver of continental lithosphere is included and makes up the coastal portion of California and all of Baja California. The California portion is bounded by an active transform fault (the San Andreas Fault).

The American plate (Figure 13.15*a*) includes most of the continental lithosphere of North and South America as well as the entire oceanic lithosphere lying west of the midoceanic ridge that divides the Atlantic Ocean Basin down the middle. For the most part, the western edge of the American plate is a subduction boundary, with oceanic lithosphere diving beneath the continental lithosphere. The eastern edge is a spreading boundary. (Some classifications recognize separate North American and South American plates.)

The Eurasian plate (Figure 13.15*b*) is mostly continental lithosphere, but it is fringed on the west and north by a belt of oceanic lithosphere. The African plate has a central core of continental lithosphere nearly surrounded by oceanic lithosphere.

The Austral-Indian plate takes the form of a long rectangle. It is mostly oceanic lithosphere but contains two cores of continental lithosphere—Australia and peninsular India. Recent evidence shows that these two continental masses are moving independently and may actually be considered to be parts of separate plates. The Antarctic plate (Figure 13.15*c*) has an elliptical shape and is almost completely enclosed by a spreading plate boundary. This means that the other plates are moving away from the pole. The continent of Antarctica forms a central core of continental lithosphere completely surrounded by oceanic lithosphere.

Of the nine lesser plates, the Nazca and Cocos plates of the eastern Pacific are rather simple fragments of oceanic lithosphere bounded by the Pacific midoceanic spreading boundary on the west and by a subduction boundary on the east. The Philippine plate is noteworthy as having subduction boundaries on both east and west edges. Two small but distinct lesser plates—Caroline and Bismarck—lie to the southeast of the Philippine plate. The Arabian plate has two transform fault boundaries, and its relative motion is northeasterly. The Caribbean plate also has important transform fault boundaries. The tiny Juan de Fuca plate is steadily diminishing in size and will eventually disappear by subduction beneath the American plate. Similarly, the Scotia

Table 13.2 | **The Lithospheric Plates**

Great plates	Lesser plates
Pacific	Nazca
American (North, South)	Cocos
Eurasian	Philippine
Persian subplate	Caribbean
African	Arabian
Somalian subplate	Juan de Fuca
Austral-Indian	Caroline
Antarctic	Bismark
	Scotia

Focus on Systems | 13.2

The Wilson Cycle and Supercontinents

When geoscientists think about time cycles, they have in mind spans of time measuring hundreds of millions of years from beginning to end of just a single cycle. The grand cycle of tectonic events bears the name of a distinguished Canadian geophysicist—Professor J. Tuzo Wilson of the University of Toronto. He made important new discoveries and interpretations that helped lead to a full-blown theory of plate tectonics. By the time he formulated his theory, all of the parts were known—continental rupture to start the opening of a new ocean basin, plate subduction to generate island arcs, and collisions to weld continents together. Using all of these separate lithospheric activities, Wilson forged a logical time chain of plate tectonic events, and it quickly gained wide acceptance as the *Wilson cycle.*

The Wilson cycle *Schematic diagram depicting the six stages of the Wilson cycle. The diagrams are not to true scale. (From Plate Tectonics, copyright © 1998 by Arthur N. Strahler. Used by permission.)*

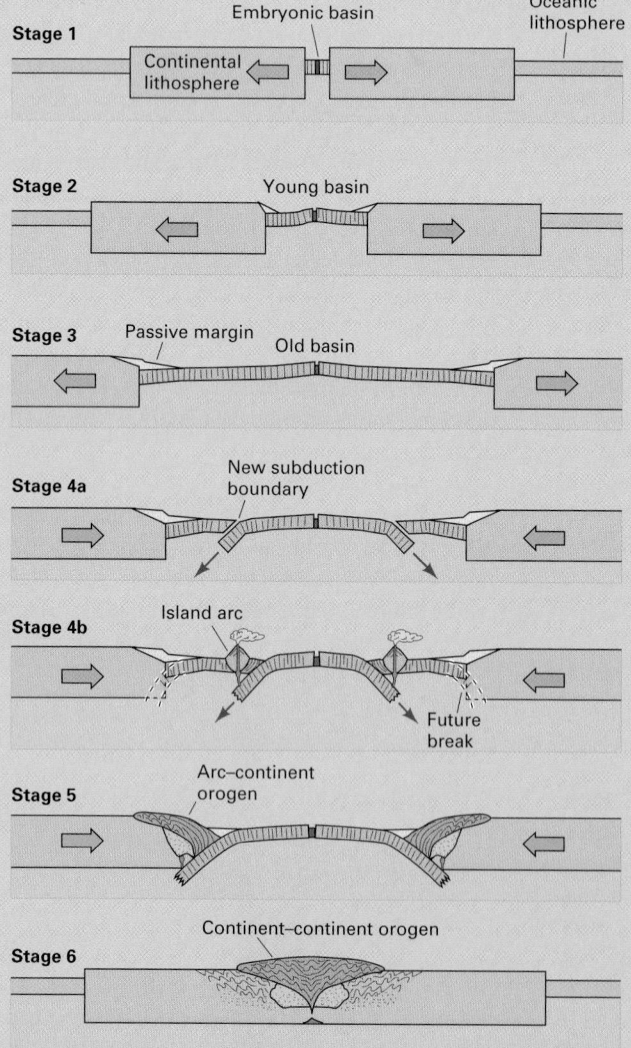

The Wilson cycle can best be appreciated by use of a cartoon-like presentation of the essential elements of plate tectonics. In our illustration below, these elements are shown arranged throughout six stages and are listed at left with references to our text diagrams and examples found on the planet today. This ideal cycle requires some 300 to 500 million years to complete.

- *Stage 1.* Embryonic ocean basin. (See also Figure 13.19*b.*) The Red Sea separating the Arabian Peninsula from Africa is an active example.
- *Stage 2.* Young ocean basin. The Labrador Basin, a branch of the North Atlantic lying between Labrador and Greenland, is a fair example of this stage.
- *Stage 3.* Old ocean basin. (Figure 13.19*c.*) Includes all of the vast expanse of the North and South Atlantic oceans and the Antarctic Ocean. Passive margin sedimentary wedges have become wide and thick.
- *Stage 4a.* The ocean basin begins to close as continental plates collide with it. New subduction boundaries begin to form.
- *Stage 4b.* Island arcs have risen and grown into great volcanic island chains. (Figure 13.18*a.*) These are found surrounding the Pacific plate, with the Aleutian arc as a fine example.
- *Stage 5.* Closing continues. Formation of new subduction margins close to the continents is followed by arc–continent collisions. The Japanese Islands represent this stage.
- *Stage 6.* The ocean basin has finally closed with a collision orogen, forming a continental suture. (Figure 13.18*c.*) The Himalayan orogen is a fine recent example, with activity continuing today.

Note that the continent of the final stage is wider than the conti-

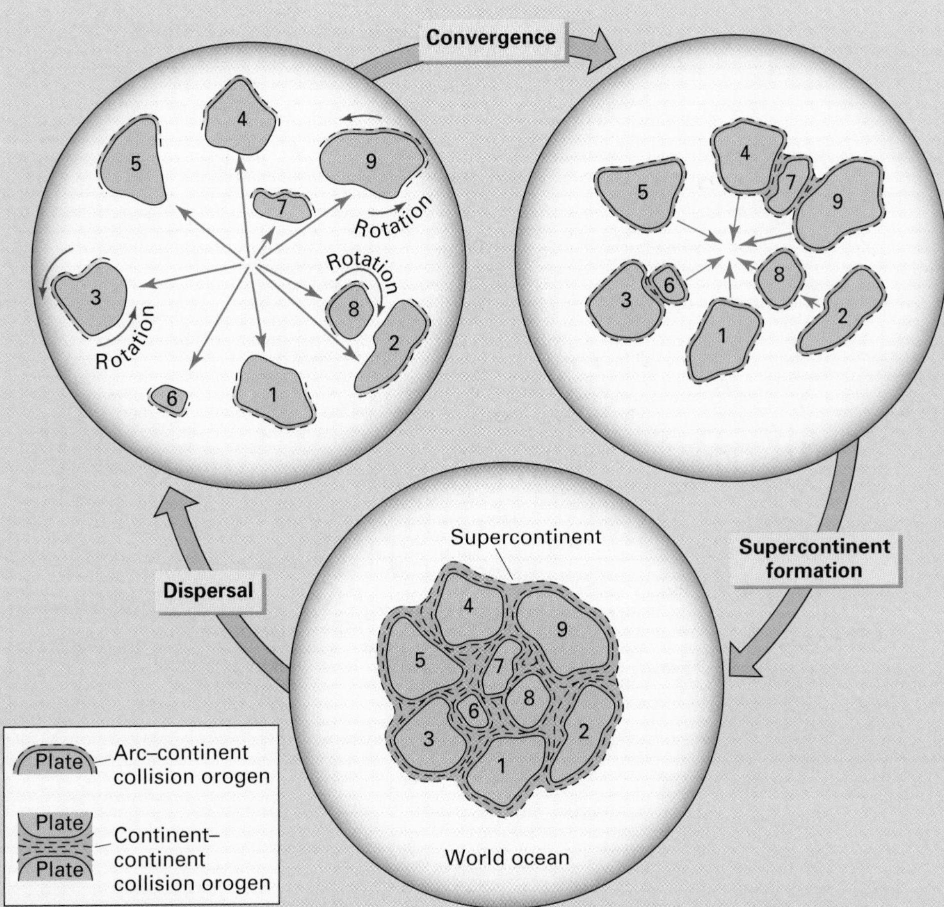

Supercontinent cycle *Schematic diagram of the supercontinent cycle in three phases. A pattern of wiggly dashes shows collision orogens that accumulate around and between the original continents. (From Plate Tectonics, copyright © 1998 by Arthur N. Strahler. Used by permission.)*

nent of the first stage because of the additional continental lithosphere formed during the intervening collisions. This is how continents can grow over geologic time.

Our stage diagrams show a single continent splitting into two continents, then rejoining to form one. Suppose that over the entire globe several continents are splitting apart and reclosing at about the same time. Expanding on this idea, imagine that these fragments become detached from a single world continent that can be called a *supercontinent.*

Several powerful lines of evidence show that a supercontinent actually came into existence, starting about 200 million years ago. Called *Pangea,* it is described in our chapter text. Good evidence has now been found that an earlier supercontinent, dubbed *Rodinia,* was fully formed about 700 million years ago. It consisted of early representatives of the same continents that later made up Pangea. Rodinia broke apart and its fragments were carried away in different directions. Then they reversed their motions and headed back toward a common center, where many continent–continent collisions bonded them together by sutures to comprise Pangea. Some interesting evidence has also pointed to the former existence of a supercontinent even older than Rodinia.

The illustration above presents the supercontinent cycle as a loop with three stages. Analogous to Stages 1–3 of the Wilson cycle is the dispersal phase, followed by a convergence phase that corresponds to Wilson stages 4 through 6. The cycle ends in a complete new supercontinent. Many new collision orogens are formed between the original continental fragments. Given, say, 3000 million years ago (Middle Archean time) as the first occurrence of a full-blown supercontinent cycle, there could have been some 6 to 10 such cycles. The hypothesis of a time cycle of supercontinents, repeating the Wilson cycle over and over again, now holds its place as the basic theme of the geologic evolution of our planet.

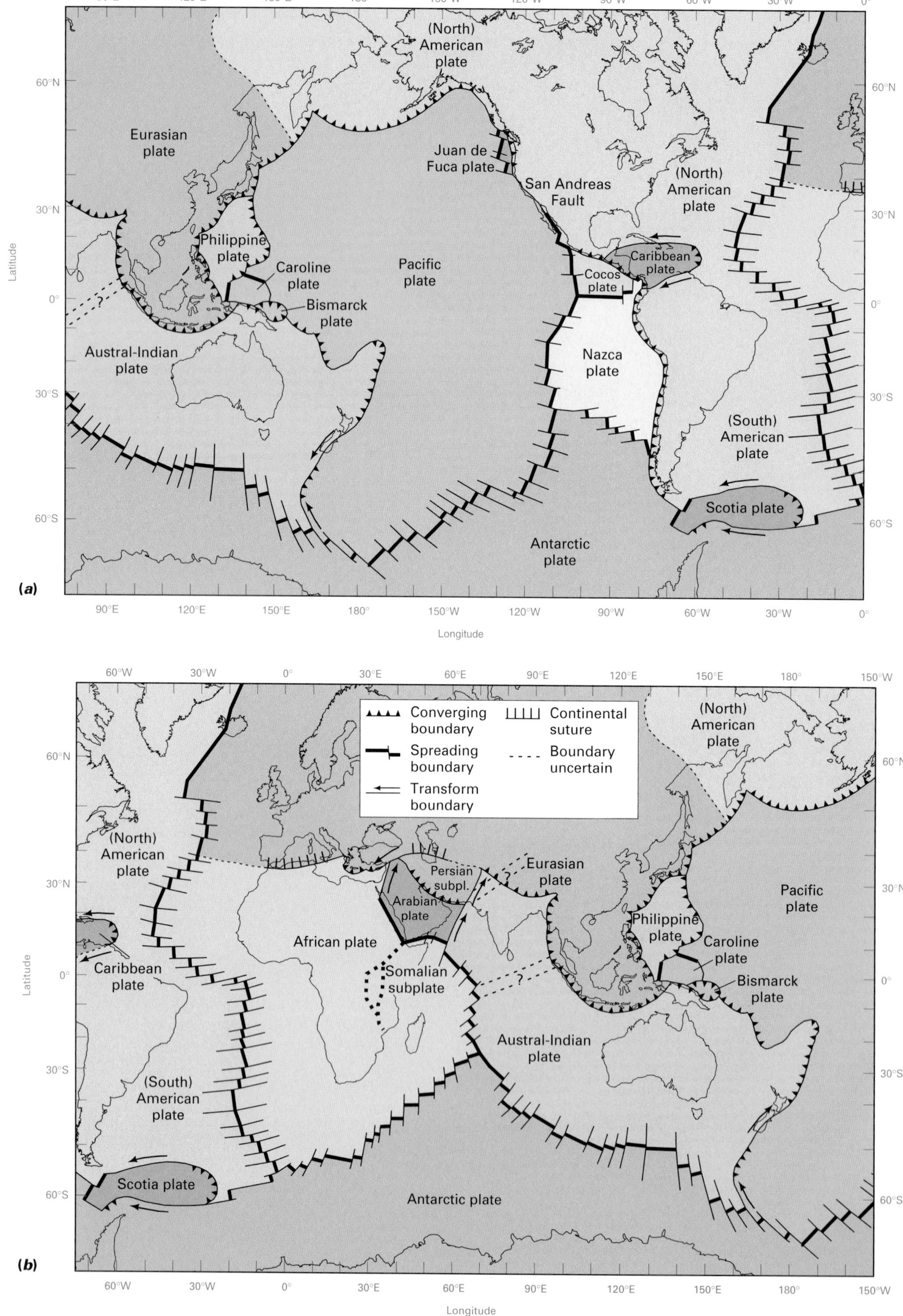
(North) American plate
Eurasian plate
Juan de Fuca plate
San Andreas Fault
(North) American plate
Philippine plate
Caroline plate
Pacific plate
Caribbean plate
Cocos plate
Bismarck plate
Austral-Indian plate
Nazca plate
(South) American plate
Scotia plate
Antarctic plate
(a)
Latitude
Longitude
Converging boundary
Spreading boundary
Transform boundary
Continental suture
Boundary uncertain
(North) American plate
Eurasian plate
Persian subpl.
Arabian plate
African plate
Caribbean plate
Somalian subplate
Philippine plate
Caroline plate
Bismarck plate
Pacific plate
Austral-Indian plate
(South) American plate
Scotia plate
Antarctic plate
(b)

13.15 World maps of lithospheric plates

(South) American plate
African plate
MADAGASCAR
Scotia Plate
Antarctic plate
Nazca plate
South Pole
Austral–Indian plate
Pacific plate
AUSTRALIA
Polar stereographic projection
Longitude

(c)

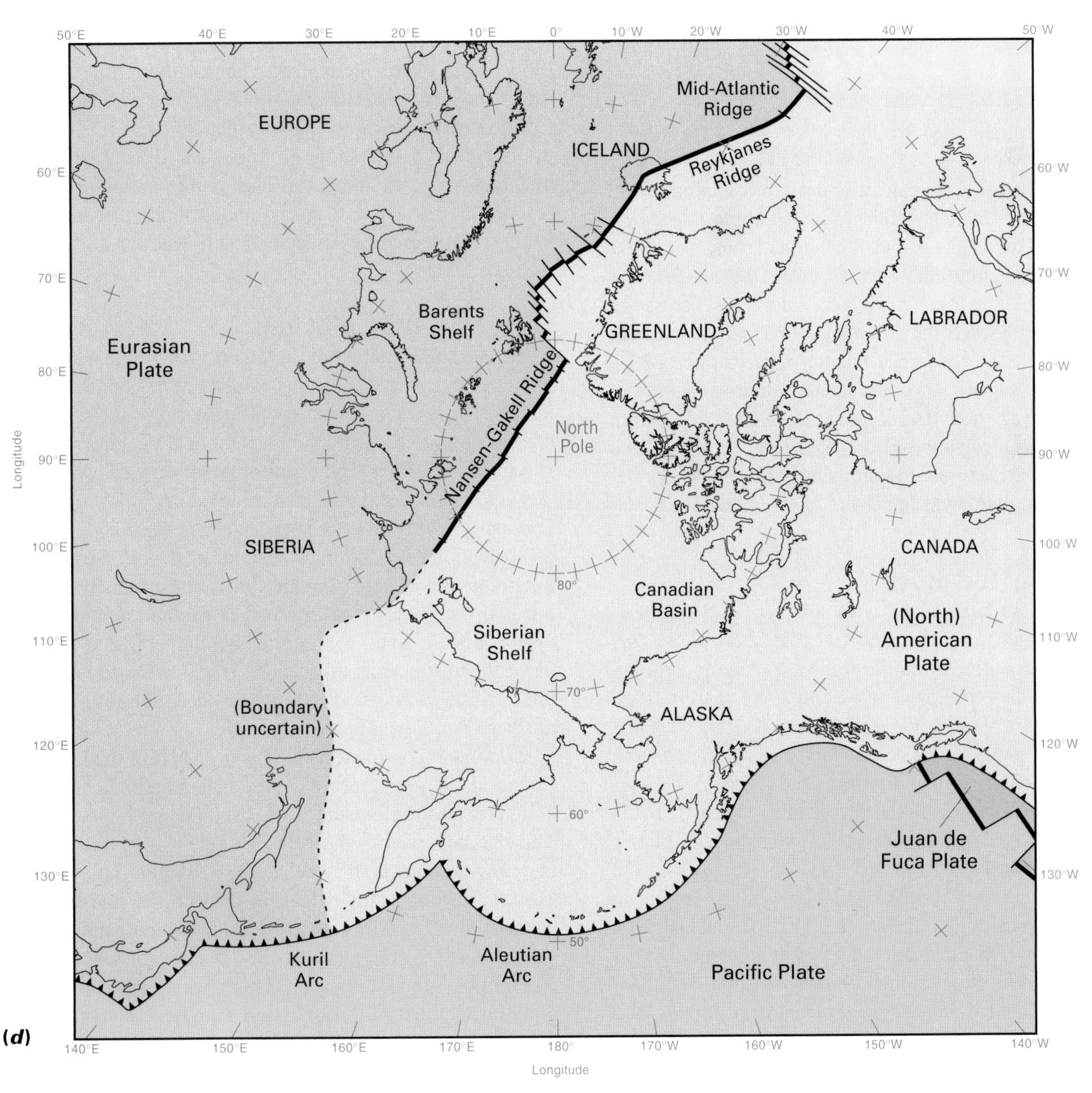

(d)

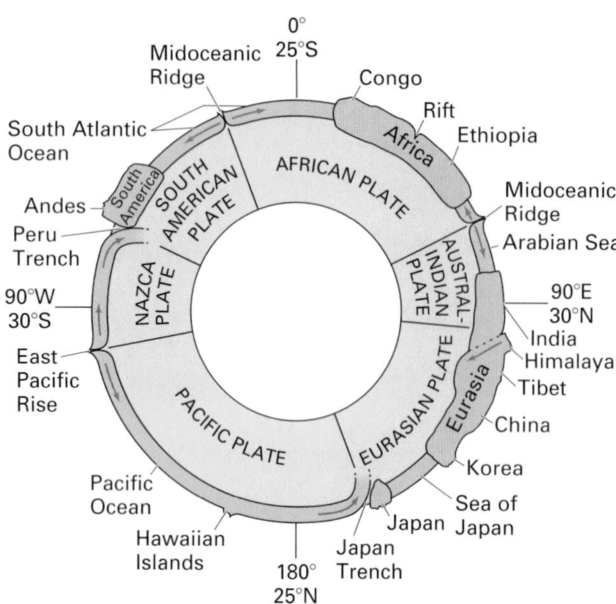

13.16 Arrangement of lithospheric plates *Schematic circular cross section of the major plates on a great circle tilted about 30 degrees with respect to the equator. (A. N. Strahler.)*

plate is being consumed by the American and Antarctic plates.

Figure 13.16 is a schematic circular cross section of the lithosphere along a great circle in low latitudes. It shows several of the great plates and their boundaries. The great circle is tilted by about 30° with respect to the equator. The section crosses the African plate, heads northeast across the Eurasian plate through the Himalayas to Japan and Korea, then dips southeast across the Pacific plate, cutting across the South American plate, and finally returns to Africa. Three spreading boundaries at midoceanic ridges are encountered, with two subduction zones (the Japan and Peru trenches) and a continent-to-continent collision, where the Austral-Indian plate dives under the Eurasian plate.

Subduction Tectonics

Converging plate boundaries, with subduction in progress, are zones of intense tectonic and volcanic activity. The narrow zone of a continent that lies above a plate undergoing subduction is therefore an active continental margin. (See plates Y and Z in Figure 13.12.) Figure 13.17 shows some details of the geologic processes that are associated with plate subduction. Two diagrams are used. Part *(a)* is exaggerated to show crustal and surface details, and part *(b)* is drawn to true scale to show the lithospheric plates. GEODISCOVERIES

The oceanic trench receives sediment coming from two sources. Carried along on the moving oceanic plate is deep-ocean sediment—fine clay and ooze—that has settled to the ocean floor. From the continent comes terrestrial sediment in the form of sand and mud brought by streams to the shore and then swept into deep water by currents. In the bottom of the trench, both types of sediment are intensely deformed and are dragged down with the moving plate. The deformed sediment is then scraped off the plate and shaped into wedges that ride up, one over the other, on steep fault planes. The wedges accumulate at the plate boundary, forming an *accretionary prism* in which metamorphism takes place. In this way, new continental crust of metamorphic rock is formed, and the continental plate is built outward.

The accretionary prism is of relatively low density and tends to rise, forming a *tectonic crest.* The tectonic crest is shown to be submerged in the figure, but in some cases it forms an island chain paralleling the coast—that is, a *tectonic arc.* Between the tectonic crest and the mainland is a shallow trough, the *forearc trough* (Figure 13.17*b*). This trough traps a great deal of terrestrial sediment, which accumulates in a basin-like structure. The bottom of the forearc trough continually subsides under the load of the added sediment. In some cases, the seafloor of the trough is flat and shallow, forming a type of continental shelf. Sediment carried across the shelf moves down the steep outer slope of the accretionary prism in tongue-like flows of turbidity currents.

The lower diagram of Figure 13.17 shows the descending lithospheric plate entering the asthenosphere. Intense heating of the upper surface of the plate melts the oceanic crust, forming basaltic magma. As this magma rises, it is changed in chemical composition at the base of the crust and becomes andesite magma. The rising andesite magma reaches the surface to form volcanoes of andesite lava, such as those we see in the Andes of South America.

Orogens and Collisions

Visualize, now, a situation in which two continental lithospheric plates converge along a subduction boundary. Ultimately, the two masses must collide because the impacting masses are too thick and too buoyant to allow either plate to slip under the other. The result is an orogeny in which various kinds of crustal rocks are crumpled into folds and sliced into nappes. This process has been called "telescoping" after the behavior of a folding telescope that collapses from a long tube into a short cylinder. Collision permanently unites the two plates, terminating further tectonic activity along that collision zone. Appropriately, the collision zone is named a **continental suture.** Our world map of plates, Figure 13.15, uses a special symbol for these sutures, shown in northern Africa and near the Aral Sea.

Continent–continent collisions occurred in the Cenozoic Era along a great tectonic line that marks the southern boundary of the Eurasian plate. (See Figure 13.15, map *b.*) The line begins with the Atlas Mountains of North Africa, and it runs across the Aegean Sea region into western Turkey. Beyond a major gap in Turkey, the line takes up again in the Zagros Mountains of Iran. Jumping another gap in southeastern Iran and Pakistan, the collision line sets in again in the great Himalayan Range, where it is still active.

Each segment of this collision zone represents the collision of a different north-moving plate against the single and relatively immobile Eurasian plate. A European seg-

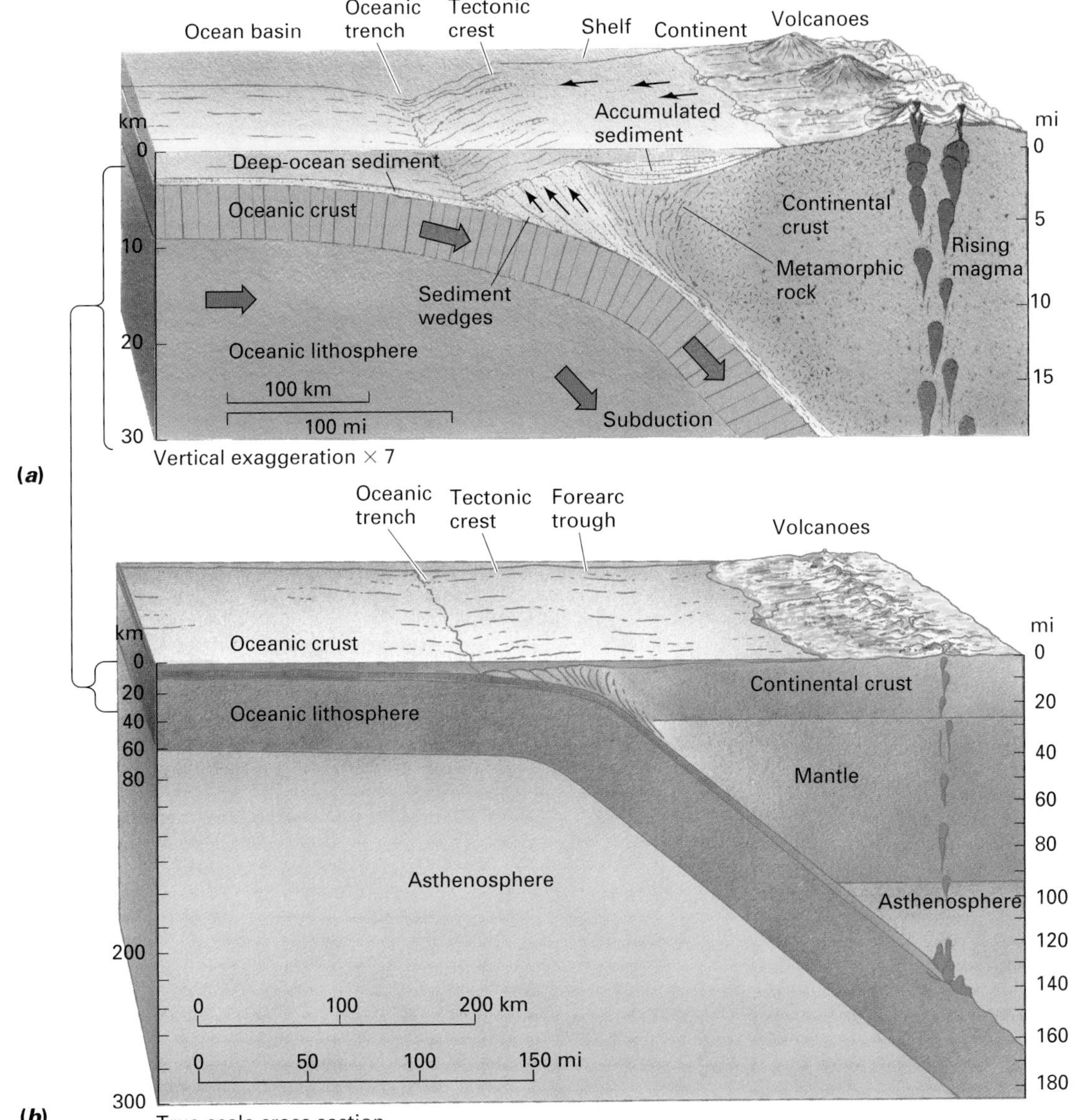

13.17 Typical features of an active subduction zone *The upper diagram (a) uses a great vertical exaggeration to show surface and crustal details. Sediments scraped off the moving plate form tilted wedges that accumulate in a rising tectonic mass. Near the mainland is a shallow trough in which sediment brought from the land accumulates. Metamorphic rock forms above the descending plate. Magma rising from the top of the descending plate reaches the surface to build a chain of volcanoes. Diagram (b) is a true-scale cross section showing the entire thickness of the lithospheric plates. (Copyright © A. N. Strahler.)*

ment containing the Alps was formed when the African plate collided with the Eurasian plate in the Mediterranean region. A Persian segment resulted from the collision of the Arabian plate with the Eurasian plate. A Himalayan segment represents the collision of the Indian continental portion of the Austral-Indian plate with the Eurasian plate.

Figure 13.18 is a series of cross sections in which the tectonic events of a typical continent–continent collision are reconstructed. Diagram *a* shows a passive margin at the left and an active subduction margin at the right. As the ocean between the converging continents is eliminated, a succession of overlapping thrust faults cuts through the oceanic crust in diagram *b*. The thrust slices ride up, one over the other, telescoping the oceanic crust and the sediments above it. As the slices become more and more tightly squeezed, they are forced upward. The upper part of each thrust sheet assumes a horizontal attitude to form a nappe in diagram *c*, which then glides forward under gravity on a low downgrade. A mass of metamorphic rock is formed between the joined continental plates, welding them together. This new rock mass is the continental suture. It is a distinctive type of orogen.

Continent–continent collisions have occurred many times since the late Precambrian time. Several ancient sutures have been identified in the continental shields. The Ural Mountains, which divide Europe from Asia, are one such suture, formed near the end of the Paleozoic Era.

Another type of collision is the arc–continent collision. Recall that island arcs often form in subduction zones as sediments of lesser density are carried into the mantle, then melt and rise as magma. In some situations, an ancient island arc can be carried toward a continent and collide with a passive continental margin. This creates a complex orogen that includes folded, faulted, and metamorphosed rocks derived from both the island arc and the sediments of the forearc trough (Figure 13.17*b*). Examples of orogens resulting from arc-continent collisions can be found on both eastern and western sides of North America. They include the ancient Appalachian and Ouachita mountain belts as well as the younger Cordilleran Ranges of western North America.

Continental Rupture and New Ocean Basins

We have already noted that the continental margins bordering the Atlantic Ocean Basin on both its eastern and western sides are very different from the active margin of a subduction zone. At present, the Atlantic margins have no important tectonic activity and are passive continental margins. Even so, they represent the contact between continental lithosphere and oceanic lithosphere, with continental crust meeting oceanic crust, as shown in Figure 13.8. The formation of passive margins begins when a single plate of continental lithosphere is rifted apart. This process is called *continental rupture.*

Figure 13.19 uses three schematic block diagrams to show how continental rupture takes place and leads to the development of passive continental margins. At first, the crust is both lifted and stretched apart as the lithospheric plate is arched upward. The crust fractures and moves along faults—upthrown blocks form mountains, while down-dropped blocks form basins.

Eventually a long narrow valley, called a *rift valley,* appears (block *a*). The widening crack in its center is continually filled in with magma rising from the mantle below. The magma solidifies to form new crust in the floor of the rift valley. Crustal blocks on either side slip down along a succession of steep faults, creating a mountainous landscape. The Rift Valley system of East Africa—described in more detail in Chapter 14—is a notable example of this stage of continental rupture. As separation continues, the rift valley widens and opens to the ocean. The ocean enters the narrow valley with a spreading plate boundary running down its center (block *b*). Rising magma in the central rift produces new oceanic crust and lithosphere.

The Red Sea is a narrow ocean formed by a continental rupture. Its straight coasts mark the edges of the rupture. The widening of such an ocean basin can continue until a large ocean has formed and the continents are widely separated (Figure 13.19, block *c*).

Figure 13.20 is an astronaut photo of the Red Sea where it joins the Gulf of Aden. As shown in the inset map, this is

13.18 Continent–continent collision *Schematic cross sections showing continent–continent collision and the formation of a suture zone with nappes. (Copyright © A. N. Strahler.)*

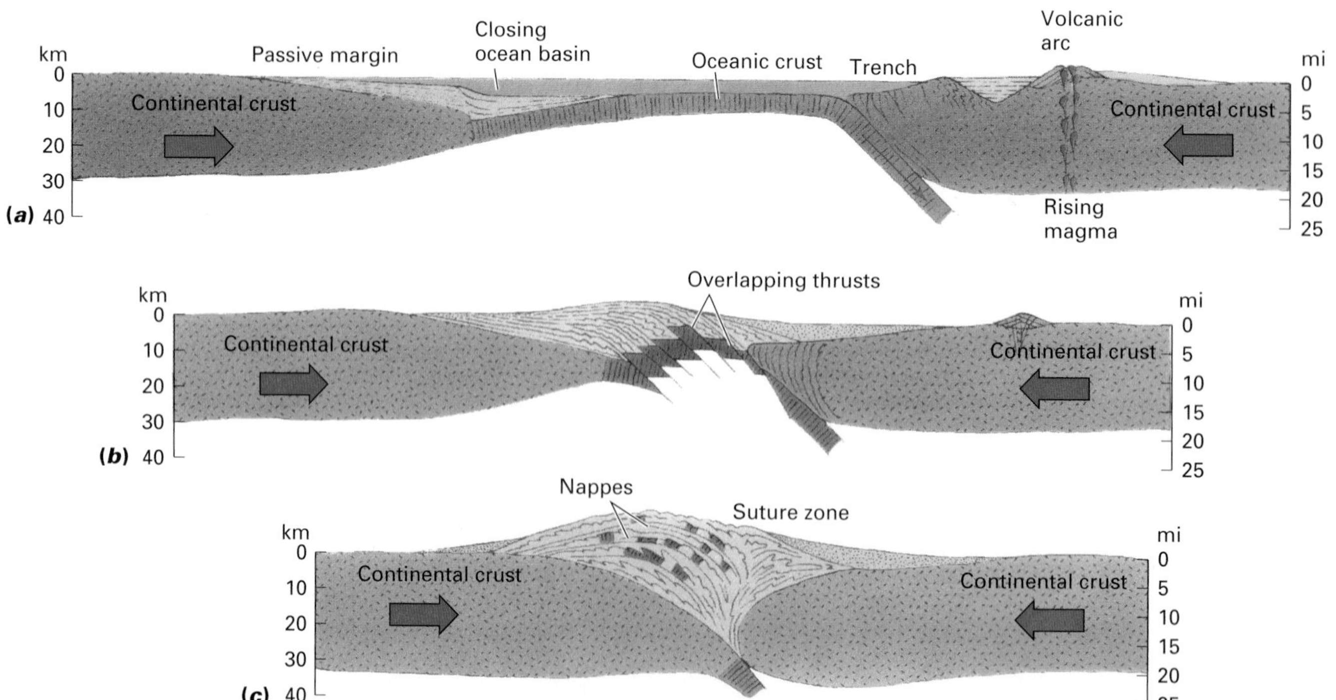

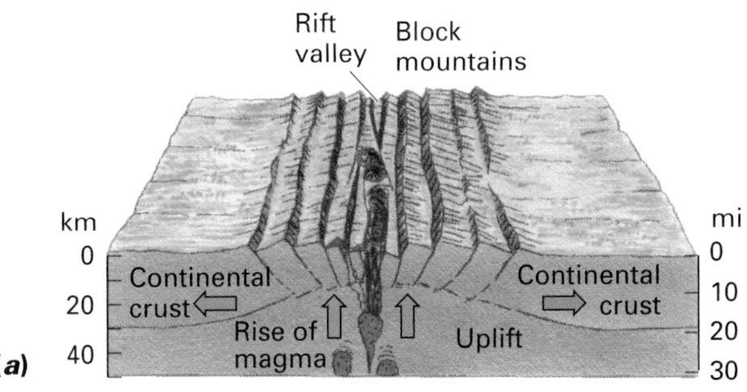

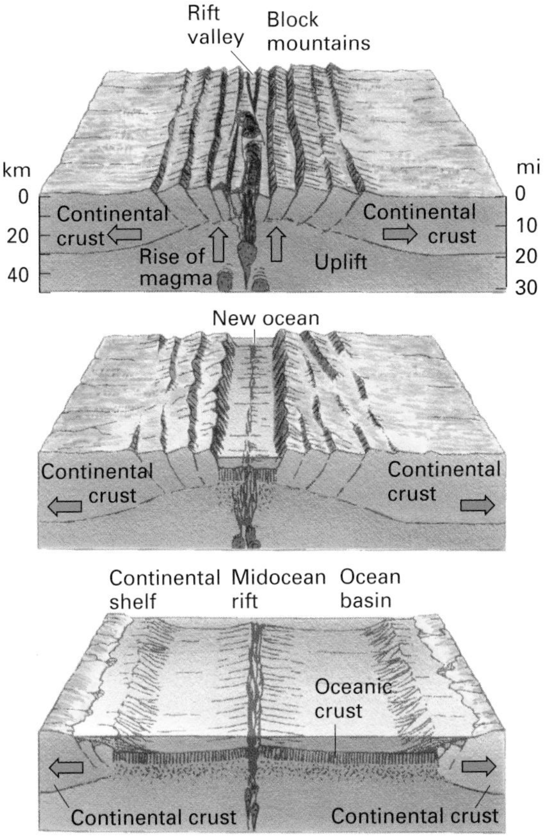

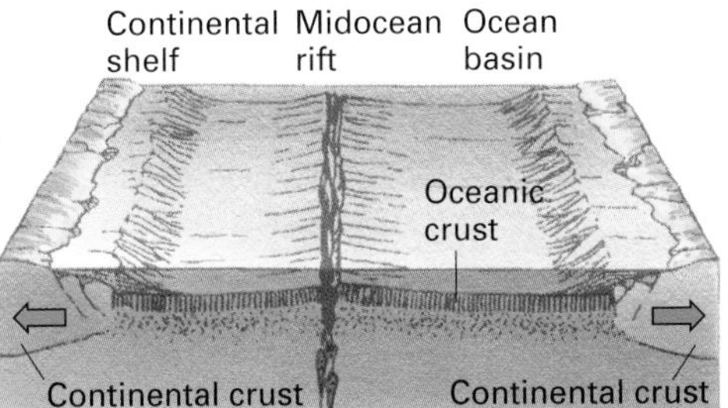

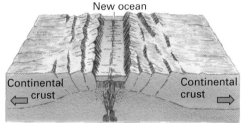

13.19 Continental rupture and spreading *Schematic block diagrams showing stages in continental rupture and the opening up of a new ocean basin. The vertical scale is greatly exaggerated to emphasize surface features. (a) The crust is uplifted and stretched apart, causing it to break into blocks that become tilted on faults. (b) A narrow ocean is formed, floored by new oceanic crust. (c) The ocean basin widens, while the passive continental margins subside and receive sediments from the continents. (Copyright © A. N. Strahler.)*

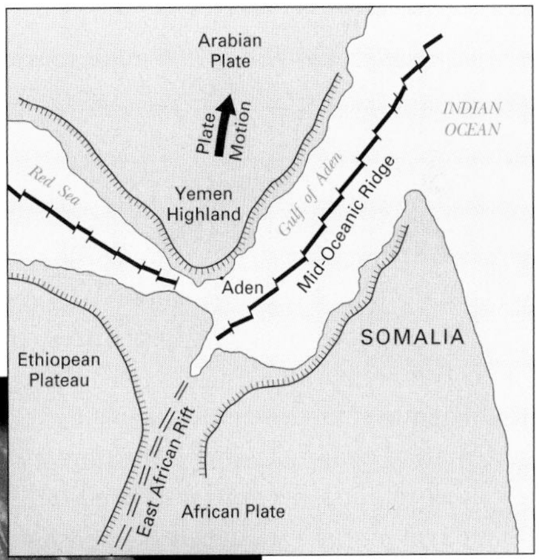

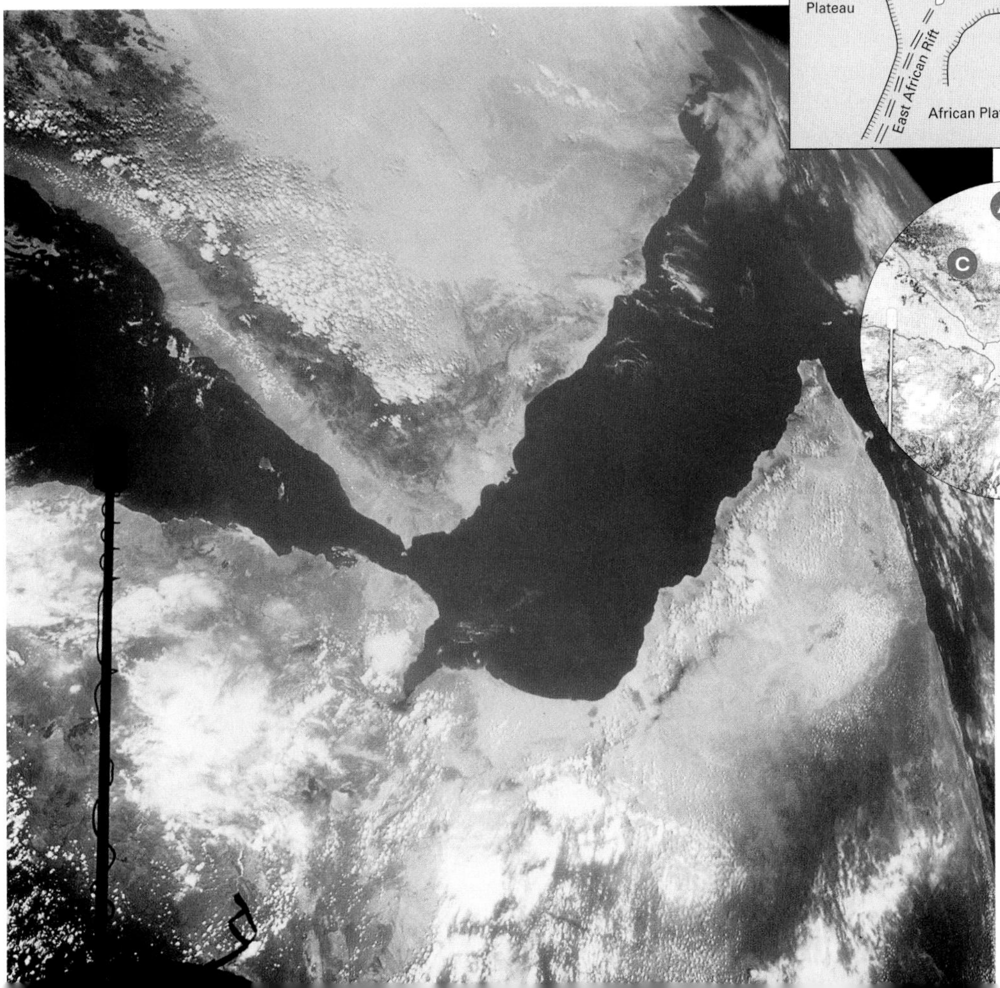

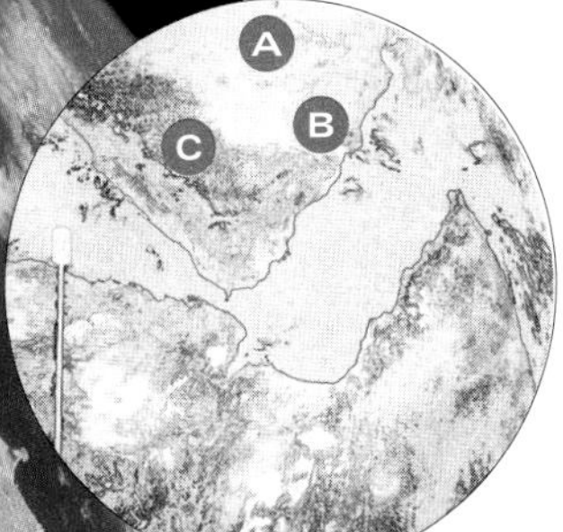

13.20 Eye on the Landscape The Red Sea and Gulf of Aden from orbit *This spectacular photo, taken by astronauts on the Gemini XI mission, shows the southern end of the Red Sea and the southern tip of the Arabian Peninsula.* **What else would the geographer see?...Answers at the end of the chapter.**

a triple junction of three spreading boundaries created by the motion of the Arabian plate pulling away from the African plate. It is easy to visualize how the two plates have split apart, allowing the ocean to enter.

During the process of opening of an ocean basin, the spreading boundary develops a series of offset breaks, one of which is shown in the upper left-hand part of diagram *a* of Figure 13.12. The offset ends of the axial rift are connected by an active transform fault. As spreading continues, the offsets slide past each other, and a scar-like feature is formed on the ocean floor as an extension of the transform fault. These *transform scars* take the form of narrow ridges or cliff-like features and may extend for hundreds of kilometers across the ocean floors. Before the true nature of these scars was understood, they were called "fracture zones." That name still persists and can be seen on maps of the ocean floor. The scars are not, for the most part, associated with active faults.

What is the power source for these slow motions of vast lithospheric plates? Geophysicists generally agree that the heat produced by radioactive decay of unstable isotopes in the crust and mantle powers plate motions. *Focus on Systems 13.3 • The Power Source for Plate Movements* provides more information.

Geologists could show by use of established principles of physics that this proposed mechanism was impossible. In 1930, Wegener perished of cold and exhaustion while on an expedition to the Greenland ice sheet. There followed three decades in which only a small minority of scientists promoted his hypothesis.

In the 1960s, however, seismologists showed beyond doubt that thick lithospheric plates are in motion, both along the midoceanic ridge and beneath the deep offshore trenches of the continents. Other geophysicists used magnetic data "frozen" into crustal rock to conclude that the continents had moved great distances apart. Within only a few years, Wegener's scenario was validated, but only by applying a mechanism for the process that was never dreamed of in his time.

The continents are moving today. Data from orbiting satellites have shown that rates of separation, or of convergence, between two plates are on the order of 5 to 10 cm (about 2 to 4 in.) per year, or 50 to 100 km (about 30 to 60 mi) per million years. At that rate, global geography must have been very different in past geologic eras than it is today. Many continental riftings and many plate collisions have taken place over the past two billion years. Single

Continents of the Past

Although modern plate tectonic theory is only a few decades old, the concept of a breakup of an early supercontinent into fragments that drifted apart dates back into the nineteenth century and beyond. Almost as soon as good navigational charts became available to show the continental outlines, geographers became intrigued with the close correspondence in outline between the eastern coast of South America and the western coastline of Africa. Credit for the first full-scale scientific hypothesis of the breakup of a single large continent belongs to Alfred Wegener, a German meteorologist and geophysicist, who offered geologic evidence as early as 1915 that the continents had once been united and had drifted apart (Figure 13.21). He reconstructed a supercontinent named *Pangea,* which existed intact as early as about 300 million years ago, in the Carboniferous Period.

A storm of controversy followed, and many American geologists denounced Wegener's "continental drift" hypothesis. However, he had some loyal supporters in Europe, South Africa, and Australia. Several lines of hard scientific evidence presented by Wegener strongly favored the former existence of Pangea. The prominent geographer and climatologist, Vladimir Köppen collaborated with Wegener in introducing favorable evidence from distribution patterns of fossil and present-day plant species. But Wegener's explanation of the physical process that separated the continents was weak and was strongly criticized on valid physical grounds. He had proposed that a continental layer of less dense rock had moved like a great floating "raft" through a "sea" of denser oceanic crustal rock.

13.21 Wegener's Pangea *Alfred Wegener's 1915 map fits together the continents that today border the Atlantic Ocean basin. The sets of dashed lines show the fit of Paleozoic tectonic structures between Europe and North America and between southernmost Africa and South America. (From A. Wegener, 1915, Die Entstehung der Kontinente und Ozeane, F. Vieweg, Braunschweig.)*

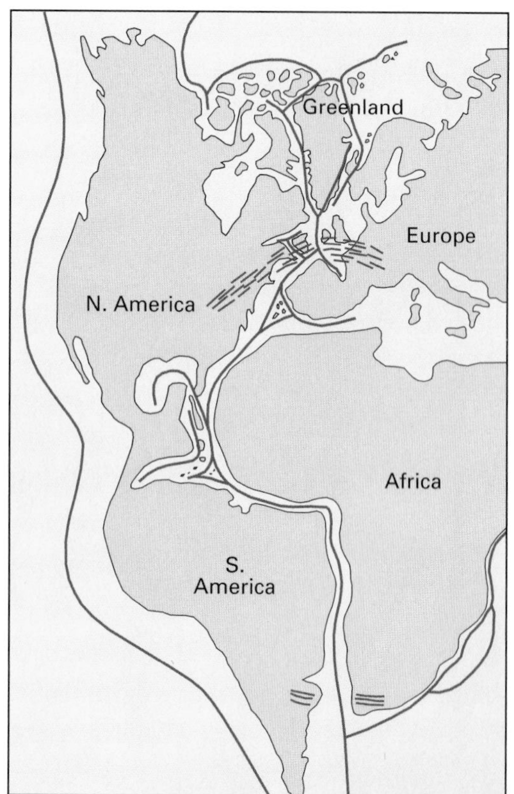

Focus on Systems | 13.3

The Power Source for Plate Movements

The system of lithospheric plates in motion represents a huge flow system of dense mineral matter. Its operation requires enormous power within an internal energy system. It is generally agreed that the source of this energy system lies in the phenomenon of radioactivity. (See *Working It Out 12.1 • Radioactive Decay,* and *Focus on Systems 12.4 • Powering the Cycle of Rock Change.*)

Radioactive elements in the crust and upper mantle constantly give off heat in the spontaneous process of radioactive decay. As the temperature of mantle rock rises, the rock expands. Recall that in the case of the atmosphere, upward motion of warmer, less dense air takes place by convection. It is thought that, in a somewhat similar type of convection, mantle material rises steadily beneath spreading plate boundaries. Geologists do not yet have a full understanding of how this rise of heated rock causes plates to move. The favored hypothesis states that as the rising mantle lifts the oceanic lithospheric plate to a higher elevation, the plate tends to move horizontally away from the spreading axis under the influence of gravity (see broad arrows on accompanying figure below). At the same time, of course, the plate undergoes cooling, becomes denser, and sinks steadily lower.

At the far edge of the plate, subduction occurs because the oceanic plate is colder and denser than the asthenosphere through which it sinks. The motion of the downgoing oceanic plate exerts a drag on the surrounding asthenosphere. This drag on the overlying asthenosphere sets in motion a horizontal flow current moving toward the sinking slab. The current then turns abruptly down to drag against the upper surface of the slab. At this downturn, the mantle beneath the continental lithosphere becomes highly heated and causes the upper surface of the slab to melt. Here, magma is released and rises in bubble-like bodies to penetrate the thick continental lithosphere. Upon cooling, this magma forms plutons. Magma that reaches the surface builds volcanoes. The lower end of the descending slab is also heated and softened, and is carried slowly away in the mantle at great depth, moving back toward the spreading boundary, where some of it can perhaps be recycled.

So we have here a power system that generates a vertical loop of material flow bringing hot asthenosphere near the Earth's surface, where heat is passed up into the oceanic and atmospheric layers. It is a one-way power system that depends on a radiogenic heat source that is being slowly depleted. As time passes, the plate tectonic system will lose power and steadily slow its rates of plate motion. Looming in the far, far distant future is a "heat death" of the Earth's tectonic system, when there will be no more great earthquakes and no more violent volcanic eruptions.

Mantle flow system and heat exchange *A schematic diagram showing a system of mantle flow (thin arrows) that transports heat to the Earth's surface where it is lost through volcanic action. (Copyright © A. N. Strahler.)*

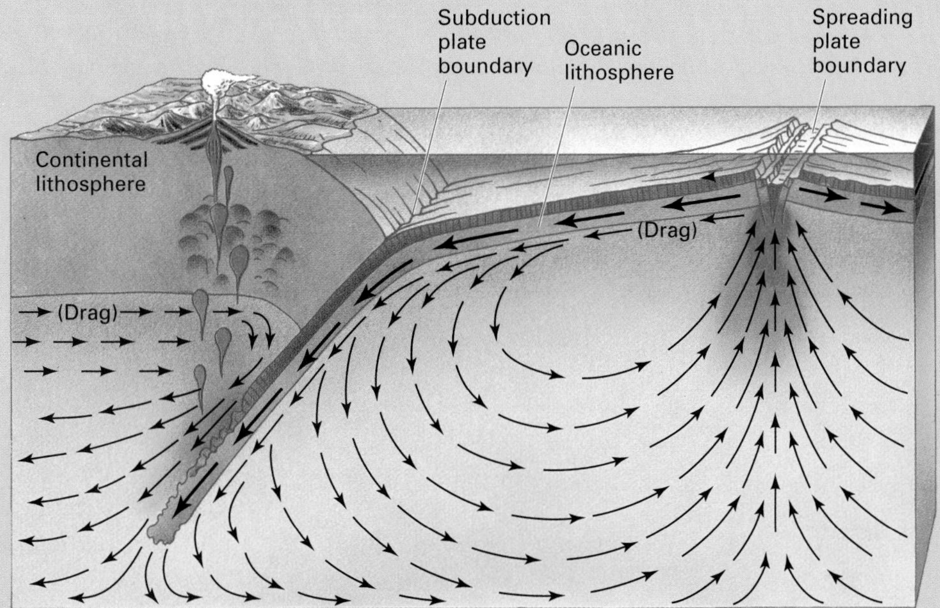

Geographers at Work | Paleogeography: Clues to the Causes of "Snowball Earth"

by Joe Meert ***University of Florida***

As we enter an era of global warming, it is hard to imagine an Earth entirely covered with ice—a "Snowball Earth" experiencing widespread glaciation at all latitudes. However, there is strong evidence that a Snowball Earth existed long ago in the Earth's history. The evidence lies in ancient rocks, formed from glacial deposits, that have been the object of geological interest for nearly half a century. These rocks—called tillites—were formed of a distinctive mixture of coarse and broken rock particles and chunks plastered together with silt and clay underneath moving ice. They are found today in very old rocks of the continental shields, far from the locations in which they were deposited.

Imagine a time when the Earth's geography was very different. Around 800 million years ago, lithospheric plates were clustered into a supercontinent, Rodinia, located largely near the equator. Rodinia then began to break apart.

Joe Meert in front of a sequence of Precambrian rocks in southern Norway that include glacial deposits from the Snowball Earth.

Over the next 200 million years, as its fragments migrated away and assumed their own paths, climate cooled and glaciation became widespread, even at latitudes near the equator. By 600 million years ago, the glacial epochs were over.

What was the cause of this event? While we know that the Earth's temperature is related to such factors as solar radiation and greenhouse gases, the physical landscape of the continents on the planet can also influence global temperature by altering global circulation patterns. Perhaps the breakup of Rodinia led to an overall cooling trend that, in turn, led to an icehouse planet.

The puzzle of Snowball Earth and the paleogeography of lithospheric plates has been a passion of Joe Meert for the last decade. His research specialty, paleomagnetism, provides the tools to determine the location of ancient rocks with respect to the magnetic poles, as well as to establish the wanderings of the poles themselves over geologic time. To find out how Joe uses these tools and what he has learned about the Snowball Earth, visit our web site at http://www.wiley.com/college/Strahler/stuff/morestuff.

continents have fragmented into smaller ones, while at other times, small continents have merged to form large ones. (See *Focus on Systems 13.2 • The Wilson Cycle and Supercontinents.*)

A brief look at the changing continental arrangements and locations over the past 250 m.y. is worthwhile in our study of physical geography because those changes brought with them changes in climate, soils, and vegetation on each continent as the continents moved across the parallels of latitude. Geography of the geologic past bears the name "paleogeography." If you decide to become a paleogeographer, you will need to carry with you into the past all you have learned about the physical geography of the present.

Wegener had made a crude map showing stages in the breakup of Pangea, but his timetable had been in part incorrect. Modern reconstructions of the global arrangements of past continents have been available since the mid-1960s. Global maps drawn for each geologic period have been repeatedly revised in the light of new evidence, but differences of interpretation affect only the minor details.

Figure 13.22 shows stages in the separations and travels of our continents, starting in the Permian Period of the Mesozoic Era, about 250 m.y. ago. In the first map, Pangea lies astride the equator and extends nearly to the two poles. Regions that are now North America and western Eurasia lie in the northern hemisphere. Jointly they are called *Laurasia.* Regions that are now South America, Africa, Antarctica, Australia, New Zealand, Madagascar, and peninsular India lie south of the equator. Jointly, they go by the name *Gondwana.* Subsequent maps (*b* through *d*) show the breaking apart and dispersal of the Laurasia and Gondwana plates to yield their modern components and locations, as seen in map *(e).*

Note in particular that North America traveled from a low-latitude location into high latitudes, finally closing off

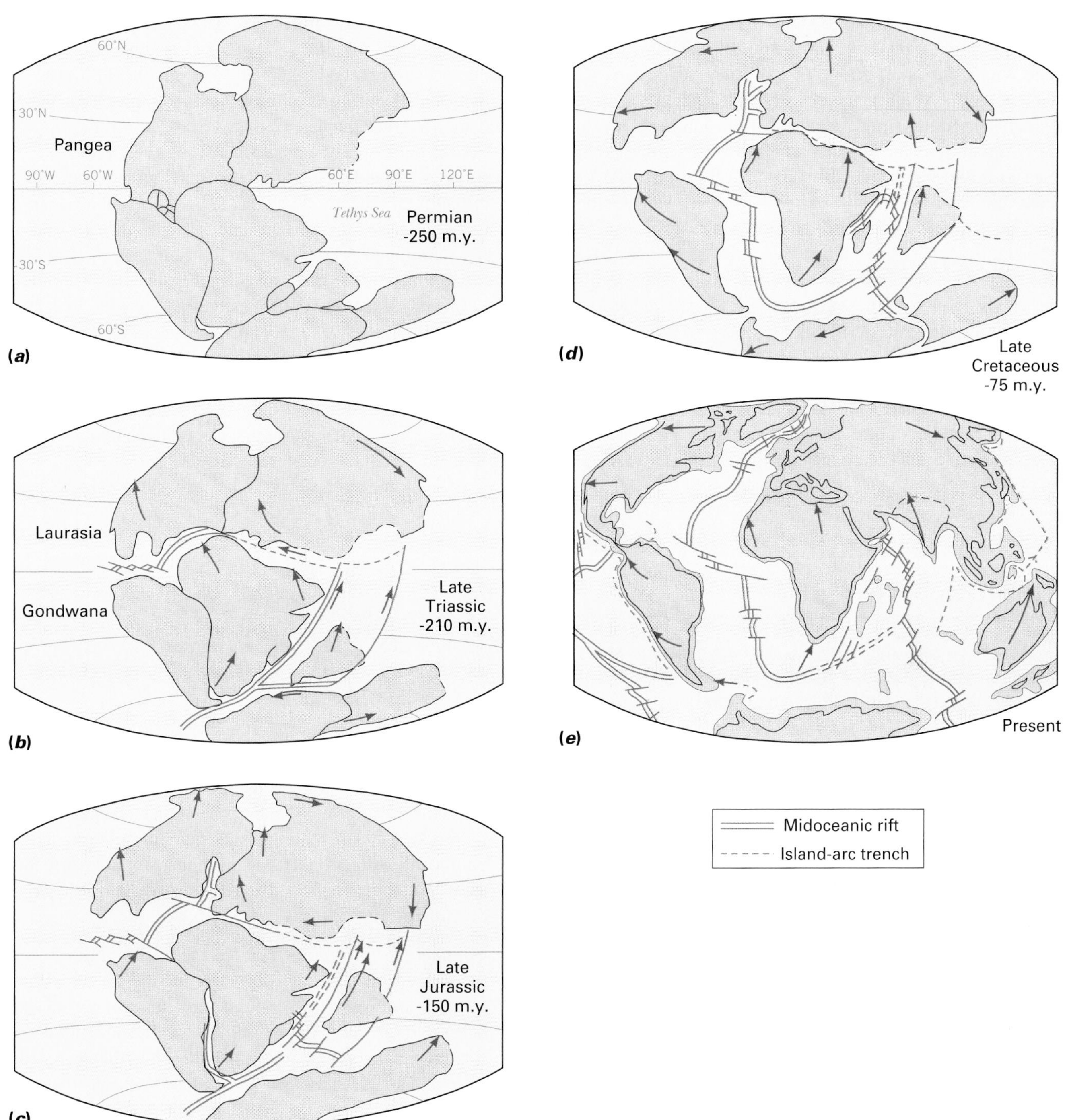

13.22* *Pangea's breakup *The breakup of Pangea is shown in five stages. Inferred motion of lithospheric plates is indicated by arrows. (Redrawn and simplified from maps by R. S. Dietz and J. C. Holden, Jour. Geophysical Research, vol. 75, pp. 4943–4951, Figures 2 to 6. Copyrighted by the American Geophysical Union. Used by permission.)*

the Arctic Ocean as a largely landlocked sea. This change may have been a major factor in bringing on the Ice Age in late Cenozoic time, about two million years ago. (See Chapter 20.) Another traveler was the Indian Peninsula, which started out from a near-subarctic southerly location in Permian time and streaked northeast across the Thethys Sea to collide with Asia in the northern tropical savanna zone. India's bedrock still bears the grooves and scratches of a great glaciation in Permian time, when it was nearer to the south pole. For the future, we may safely predict that the Atlantic Ocean basins will become much wider, and the westward motion of the Americas will cause a reduction in the width of the Pacific Basin.

A Look Ahead Over the long spans of geologic time, our planet's surface is shaped and reshaped endlessly. Oceans open and close, mountains rise and fall. Rocks are folded, faulted, and fractured. Arcs of volcanoes spew lava to build chains of lofty peaks. Deep trenches consume sediments that are carried beneath adjacent continents. These activities are byproducts of the motions of lithospheric plates and are described by plate tectonics.

For geographers, the great value of plate tectonics is that it provides a grand scheme for understanding the nature and distribution of the largest and most obvious features of our planet's surface—its continents and ocean basins and their major relief features. However, large features, such as mountain ranges, are made up of many smaller features—individual peaks, for example. In the next chapter, we will look at the continental surface in more detail, examining the volcanic and tectonic landforms that result when volcanoes erupt and rock layers are folded and faulted. As we will see, the formation of these landforms is often marked by powerful earthquakes—phenomena that can produce major impacts, indeed, on modern technological society.

Chapter Summary

- At the center of the Earth lies the **core**—a dense mass of liquid iron and nickel that is solid at the very center. Enclosing the metallic core is the **mantle**, composed of ultramafic rock. The outermost layer is the **crust**. **Continental crust** consists of two zones—a lighter zone of felsic rocks atop a denser zone of mafic rocks. **Oceanic crust** consists only of denser, mafic rocks.
- The **lithosphere**, the outermost shell of rigid, brittle rock, includes the crust and an upper layer of the mantle. Below the lithosphere is the **asthenosphere**, a region of the mantle in which mantle rock is soft or plastic.
- Geologists trace the history of the Earth through the geologic time scale. *Precambrian time* includes the Earth's earliest history. It is followed by three major divisions—the *Paleozoic, Mesozoic*, and *Cenozoic* eras.
- Continental masses consist of active belts of mountain-making and inactive regions of old, stable rock. Mountain-building occurs by *volcanism* and *tectonic activity*. *Alpine chains* include **mountain arcs** and **island arcs**. They occur in two principal mountain belts—the circum-Pacific and Eurasian-Indonesian belts.
- **Continental shields** are regions of low-lying igneous and metamorphic rocks. They may be *exposed* or *covered* by layers of sedimentary rocks. Ancient *mountain roots* lie within some shield regions.
- The ocean basins are marked by a *midoceanic ridge* with its central *axial rift*. This ridge occurs at the site of crustal spreading. Most of the *ocean basin floor* is *abyssal plain*, covered by fine sediment. As *passive continental margins* are approached, the *continental rise*, *slope*, and *shelf* are encountered. At *active continental margins*, deep *oceanic trenches* lie offshore.
- The two basic tectonic processes are extension and compression. Both processes can lead to the formation of mountains. *Extension* occurs in the splitting of plates, when the crust thins, is fractured, and then pushed upward to produce block mountains. When lithospheric plates collide, *compression* occurs, shaping rock layers into *folds* that then break and move atop one another along *overthrust* faults.
- *Continental lithosphere* includes the thicker, lighter continental crust and a rigid layer of mantle rock beneath. *Oceanic lithosphere* is comprised of the thinner, denser oceanic crust and rigid mantle below. The lithosphere is fractured and broken into a set of **lithospheric plates**, large and small, that move with respect to each other.
- Where plates move apart, a *spreading boundary* occurs. At *converging boundaries*, plates collide. At *transform boundaries*, plates move past one another on a *transform fault*. There are six major lithospheric plates.
- When oceanic lithosphere and continental lithosphere collide, the denser oceanic lithosphere plunges beneath the continental lithospheric plate, a process called **subduction**. A trench marks the site of downplunging. Some subducted oceanic crust melts and rises to the surface, producing volcanoes. Under the severe compression that occurs with continent–continent collision, the two continental plates are welded together in a zone of metamorphic rock named a **continental suture**.

- In *continental rupture*, extensional tectonic forces move a continental plate in opposite directions, creating a *rift valley*. Eventually, the rift valley widens and opens to the ocean, and new oceanic crust forms as spreading continues.
- Plate movements are thought to be powered by *convection currents* in the plastic mantle rock of the asthenosphere.
- During the Permian Period, the continents were joined in a single, large supercontinent—Pangea—that broke apart, leading eventually to the present arrangement of continents and ocean basins.

Key Terms

core
mantle
crust
continental crust
oceanic crust
lithosphere
asthenosphere
lithospheric plates
mountain arc
island arcs
continental shields
plate tectonics
tectonics
subduction
continental suture

Review Questions

1. Describe the Earth's inner structure, from the center outward. What types of crust are present? How are they different?
2. How do geologists use the term *lithosphere?* What layer underlies the lithosphere, and what are its properties? Define the term *lithospheric plate.*
3. More recent geologic time is divided into three eras. Name them in order from oldest to youngest. How do geologists use the terms *period* and *epoch?* What age is applied to time before the earliest era?
4. What proportion of the Earth's surface is in ocean? in land? Do these proportions reflect the true proportions of continents and oceans? If not, why not?
5. What are the two basic subdivisions of continental masses?
6. What term is attached to belts of active mountain-making? What are the two basic processes by which mountain belts are constructed? Provide examples of mountain arcs and island arcs.
7. What is a continental shield? How old are continental shields? What two types of shields are recognized?
8. Describe how compressional mountain-building produces folds, faults, overthrust faults, and thrust sheets (nappes).
9. Name the six great lithospheric plates. Identify an example of a spreading boundary by general geographic location and the plates involved. Do the same for a converging boundary.
10. Describe the process of subduction as it occurs at a converging boundary of continental and oceanic lithospheric plates. How is the continental margin extended? How is subduction related to volcanic activity?
11. How does continental rupture produce passive continental margins? Describe the process of rupturing and its various stages.
12. What are transform faults? Where do they occur?
13. Provide a brief history of the idea of "drifting continents."
14. What was Wegener's theory about "continental drift?" Why was it opposed at the time?
15. Briefly summarize the history of our continents that geologists have reconstructed beginning with the Permian period.

Focus on Systems 13.2 • The Wilson Cycle and Supercontinents

1. What is the Wilson cycle of plate tectonics? What is the net effect of a Wilson cycle on the continental mass?
2. Using a sketch, describe the cycle by which supercontinents are formed and reformed.

Focus on Systems 13.3 • The Power Source for Plate Movements

1. How is the principle of convection thought to be related to the power source for plate tectonic motions?
2. What role does gravity play in the motion of lithospheric plates?

Visualizing Exercises

1. Sketch a cross section of an ocean basin with passive continental margins. Label the following features: midoceanic ridge, axial rift, abyssal plain, continental rise, continental slope, and continental shelf.
2. Identify and describe two types of lithospheric plates. Sketch a cross section showing a collision between the two types. Label the following features: oceanic crust, continental crust, mantle, oceanic trench, and rising magma. Indicate where subduction is occurring.
3. Sketch a continent–continent collision and describe the formation process of a continental suture. Provide a present-day example where a continental suture is being formed, and give an example of an ancient continental suture.
4. Figure 13.15 is a Mercator map of lithospheric plates, whereas Figure 13.16 presents a cross section of plates on a great circle. Construct a similar cross section of plates on the 30° S parallel of latitude. As in Figure 13.16, label the plates and major geographic features. Refer to Figure 13.4 for arcs, trenches, and midocean ridges.

Essay Question

1. Suppose astronomers discover a new planet that, like Earth, has continents and oceans. They dispatch a reconnaissance satellite to photograph the new planet. What features would you look for, and why, to detect past and present plate tectonic activity on the new planet?

Problems

Working It Out 13.1 • Radiometric Dating

1. A geochemist analyzes a sample of zircon obtained from an igneous rock and observes that the ratio of ^{206}Pb to ^{238}U atoms is 0.448. Using the formula, what is the age of the sample?
2. The geochemist also determines the atomic ratio of ^{207}Pb to ^{235}U as 9.56 for the same sample. For this decay process, the half-life is $H = 7.04 \times 10^8$ yr. Calculate k and then determine the age from the D/M ratio of 9.56. Is it consistent with the age obtained from the $^{206}Pb/^{238}U$ ratio?

Eye on the Landscape

13.7 Undersea topography Notice the Hawaiian Islands **(A)** and the trail of seamounts **(B)** leading up to them. The islands are outpourings of lava from a "hotspot" in the mantle. The hotspot has remained fixed, however, as the Pacific lithospheric plate has moved across it. The seamounts thus mark the path of motion of the Pacific plate. Note that the direction of motion changed abruptly at one point in time. Some hotspots are relatively recent and don't show a trail of ancient eruptions. The Cape Verde Islands **(C)** are an example. Hotspot volcanoes are described in Chapter 14.

13.20 Red Sea and Gulf of Aden from orbit The pinkish tones of the interior Arabian desert **(A)** are caused by iron oxides in the desert rocks and soils; note the absence of any dark vegetation (see Chapter 15 for weathering and oxide formation). At **(B)**, a dark plateau has been dissected by a stream system leading down to the low, central desert; fluvial dissection is covered in Chapter 17. The cumulus clouds at **(C)** are the result of the lifting of moist air over the coastal mountain ranges. Recall that we covered orographic precipitation in Chapter 6.

14 VOLCANIC AND TECTONIC LANDFORMS

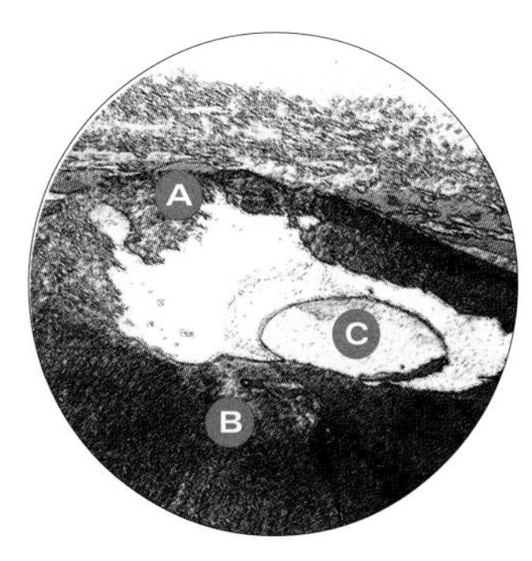

EYE ON THE LANDSCAPE

Lake in a crater of Mutnovsky Volcano, Kamchatka Peninsula, Russia. Located above the subduction zone where the Pacific plate plunges beneath the North American plate, the Kamchatka Peninsula is the site of recent volcanic activity. The last eruption here occurred in 1961. **What else would the geographer see?...Answers at the end of the chapter.**

The two previous chapters in Part 3—on Earth materials and plate tectonics—have now set the stage for the study of volcanic and tectonic landforms, which is the subject of this chapter. These features of the landscape are created by internal Earth forces. Upwelling magma spews forth explosively from vents and fissures, creating steep-sided volcanoes or flooding vast areas with molten rock. Compressional stresses generated by plate motions fold flat-lying rocks into wavelike forms, producing long ridges from upfolds and long valleys from downfolds. Earthquake faults mark the lines on which rock layers break and move past one other, producing huge uplifted mountain blocks next to deep, downdropped valleys.

By looking in detail at landforms created by volcanic and tectonic activity, we are moving down from the global perspective of enormous lithospheric plates that collide to form vast mountain arcs or separate to form yawning ocean basins. We are now "zooming in" to examine landscape features that are actually small enough to view from a single vantage point. As we will see, the global perspective of plate tectonics helps explain how these landscape features are created and where they occur.

LANDFORMS

Landforms are the surface features of the land—for example, mountain peaks, cliffs, canyons, plains, beaches, and sand dunes. Landforms are created by many processes, and much of the remainder of this book describes landforms and how they are produced. **Geomorphology** is the scientific study of the processes that shape landforms. In this chapter, we will examine landforms produced directly by volcanic and tectonic processes.

The shapes of continental surfaces reflect the "balance of power," so to speak, between two sets of forces. Internal Earth forces act to move crustal materials upward through volcanic and tectonic processes, thus bringing fresh rock to the Earth's surface. In opposition are processes and forces that act to lower continental surfaces by removing and transporting mineral matter through the action of running water, waves and currents, glacial ice, and wind. We can refer to these processes of land sculpture collectively as **denudation.**

Seen in this perspective, landforms in general fall into two basic groups—initial landforms and sequential landforms. *Initial landforms* are produced directly by volcanic and tectonic activity. They include volcanoes and lava flows, as well as downdropped rift valleys and elevated mountain blocks in zones of recent crustal deformation. Figure 14.1*a* shows a mountain block uplifted by crustal activity. The energy for lifting molten rock and rigid crustal masses to produce the initial landforms comes from an internal heat source. As we explained in Chapter 13, this heat energy is generated largely by natural radioactivity in rock of the Earth's crust and mantle. It is the fundamental energy source for the motions of lithospheric plates.

Landforms shaped by processes and agents of denudation belong to the group of *sequential landforms.* The word "sequential" means that they follow in sequence after the initial landforms have been created, and a crustal mass has been raised to an elevated position. Figure 14.1*b* shows an uplifted crustal block (an initial landform) that has been attacked by agents of denudation and carved up into a large number of sequential landforms.

14.1 Initial and sequential landforms *An initial landform is created, here by tectonic activity, then carved into sequential landforms. (Drawn by A. N. Strahler.)*

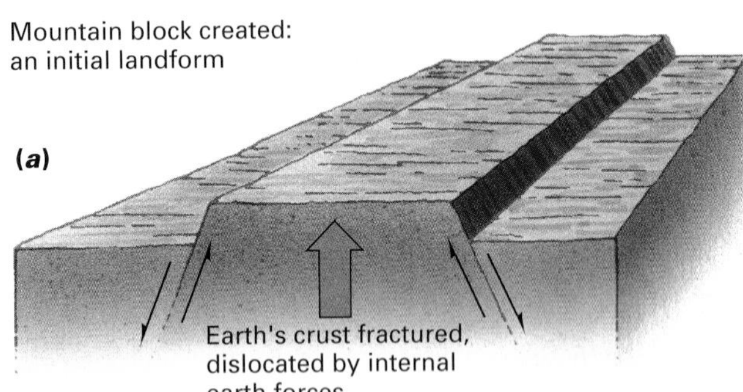

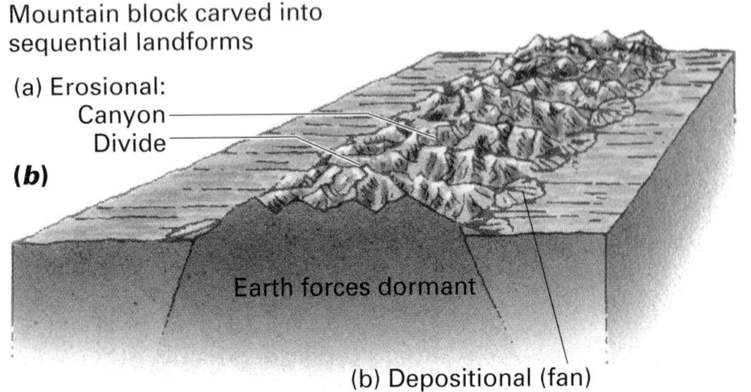

You can think of any landscape as representing the existing stage in a great contest of opposing forces. As lithospheric plates collide or pull apart, internal Earth forces periodically elevate parts of the crust to create initial landforms. The external agents of denudation—running water, waves, wind, and glacial ice—persistently wear these masses down and carve them into vast numbers of smaller sequential landforms.

Today we can observe the many stages of this endless struggle between internal and external forces by traveling to different parts of the globe. Where we find high alpine mountains and volcanic chains, internal Earth forces have recently dominated the contest. In the rolling low plains of the continental interiors, the agents of denudation have won a temporary victory. At other locations, we can find many intermediate stages. Because the internal Earth forces act repeatedly and violently, new initial landforms keep coming into existence as old ones are subdued.

This chapter focuses on initial landforms that are created directly by volcanic activity and tectonic activity. After we have described the most important agents of denudation in Chapters 15 and 17, we will return in Chapter 18 to the sequential landforms that develop on volcanic and tectonic landforms.

Volcanic Activity

We have already identified *volcanism,* or volcanic activity, as one of the forms of mountain-building. The extrusion of magma builds landforms, and these landforms can accumulate in a single area both as volcanoes and as thick lava flows. Through these volcanic activities, imposing mountain ranges are constructed.

A **volcano** is a conical or dome-shaped initial landform built by the emission of lava and its contained gases from a constricted vent in the Earth's surface (Figure 14.2). The magma rises in a narrow, pipe-like conduit from a magma reservoir lying beneath. Upon reaching the surface, igneous material may pour out in sluggish, tongue-like lava flows. Magma may also be violently ejected in the form of solid fragments driven skyward under pressure of confined gases. Ejected fragments, ranging in size from boulders to fine dust, are collectively called *tephra.* Forms and dimensions of a volcano are quite varied, depending on the type of lava and the presence or absence of tephra.

Stratovolcanoes

The nature of volcanic eruption, whether explosive or quiet, depends on the type of magma. Recall from Chapter 12 that there are two main types of igneous rocks: felsic and mafic. The felsic lavas (rhyolite and andesite) have a high degree of viscosity—that is, they are thick and gummy, and resist flow. So, volcanoes of felsic composition typically have steep slopes, and lava usually does not flow long distances from the volcano's vent. When the volcano erupts, tephra falls on the area surrounding the crater and contributes to the structure of the cone. Volcanic

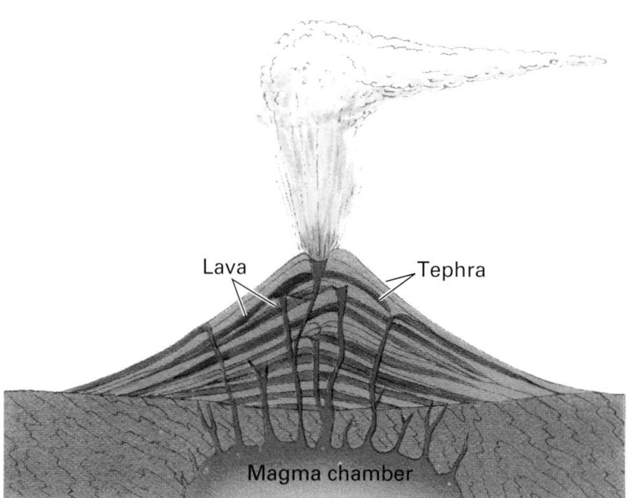

14.2 Anatomy of a stratovolcano *Idealized cross section of a stratovolcano with feeders from magma chamber beneath. The steep-sided cone is built up from layers of lava and tephra. (Copyright © A. N. Strahler.)*

bombs are also included in the tephra. These solidified masses of lava range up to the size of large boulders and fall close to the crater.  GEODISCOVERIES

The interlayering of sluggish streams of felsic lava and falls of tephra produces a **stratovolcano.** This tall, steep-sided cone usually steepens toward the summit, where a bowl-shaped depression—the *crater*—is located. The crater is the principal vent of the volcano. Felsic lavas usually hold large amounts of gas under high pressure. As a result, these lavas can produce explosive eruptions. The eruption of Mount St. Helens in 1980 is an example of an explosive eruption of felsic lavas (Figure 14.3). Very fine volcanic dust from such eruptions can rise high into the troposphere and stratosphere, traveling hundreds or thousands of kilometers before settling to the Earth's surface.

Lofty, conical stratovolcanoes are well known for their scenic beauty. Fine examples are Mount Hood and Mount St. Helens in the Cascade Range, Mount Fuji in Japan, Mount Mayon in the Philippines (Figure 14.3), and Mount Shishaldin in the Aleutian Islands. Stratovolcanoes are sometimes referred to as *composite volcanoes,* or composite cones, since they are composed of composites of ash and lava.

Another important form of emission from explosive stratovolcanoes is a cloud of white-hot gases and fine ash. This intensely hot cloud, or "glowing avalanche," travels rapidly down the flank of the volcanic cone, searing everything in its path. On the Caribbean island of Martinique, in 1902, a glowing cloud issued without warning from Mount Pelée. It swept down on the city of St. Pierre, destroying the city and killing all but two of its 30,000 inhabitants. More recently, glowing avalanches spawned by a 1997 eruption of the Soufrière Hills volcano on Montserrat, another island of the Lesser Antilles, killed about 20 people in small villages. Plymouth, the island's capital, was flooded with hot ash and debris, causing extensive fires that devastated the evacuated city (Figure 14.4). The southern two-thirds of the small island was left uninhabitable.

CALDERAS One of the most catastrophic of natural phenomena is a volcanic explosion so violent that it destroys the entire central portion of the volcano. Vast quantities of ash and dust are emitted and fill the atmosphere for many hundreds of square kilometers around the volcano. Only a great central depression, named a *caldera,* remains after the explosion. Although some of the upper part of the volcano is blown outward in fragments, most of it settles back into the cavity formed beneath the former volcano by the explosion.

Krakatoa, a volcanic island in Indonesia, exploded in 1883, leaving a huge caldera. Great seismic sea waves generated by the explosion killed many thousands of persons living in low coastal areas of Sumatra and Java. About 75 km^3 (18 mi^3) of rock was blown out of the crater during the explosion. Vast quantities of gas and fine particles of dust were carried into the stratosphere, where they contributed to a rosy glow of sunrises and sunsets that was seen around the world for several years afterward.

A classic example of a caldera produced in prehistoric times is Crater Lake, Oregon (Figure 14.3). The former volcano, named Mount Mazama, is estimated to have risen 1200 m (about 4000 ft) higher than the present caldera rim. The great explosion and collapse occurred about 6600 years ago.

STRATOVOLCANOES AND SUBDUCTION ARCS Most of the world's active stratovolcanoes lie within the circum-Pacific mountain belt. Here, subduction of the Pacific, Nazca, Cocos, and Juan de Fuca plates is active. In Chapter 13, we explained how andesitic magmas rise beneath volcanic arcs of active continental margins and island arcs (see Figure 13.12). One good example is the volcanic arc of Sumatra and Java, lying over the subduction zone between the Australian plate and the Eurasian plate. Another is the Aleutian volcanic arc, located where the Pacific plate dives beneath the North American plate. The Cascade Mountains of northern California, Oregon, and Washington form a similar chain that continues into the Garibaldi volcanic belt of southern British Columbia. Important segments of the Andes Mountains in South America consist of stratovolcanoes.

Shield Volcanoes

In contrast to thick, gassy felsic lava, mafic lava (basalt) is often highly fluid. It typically has a low viscosity and holds little gas. As a result, eruptions of basaltic lava are usually quiet, and the lava can travel long distances to spread out in thin layers. Typically, then, large basaltic volcanoes are broadly rounded domes with gentle slopes. They are referred to as **shield volcanoes.** Hawaiian volcanoes are of this type.

The shield volcanoes of the Hawaiian Islands are characterized by gently rising, smooth slopes that flatten near the top, producing a broad-topped volcano (Figure 14.5). Domes on the island of Hawaii rise to summit elevations of about 4000 m (about 13,000 ft) above sea level. Including the basal portion lying below sea level, they are more than twice that high. In width they range from 16 to 80 km (10 to 50 mi) at sea level and up to 160 km (about 100 mi) at the submerged base. The basalt lava of the Hawaiian volcanoes

14.3 Volcano Gallery

a

a...***Mount St. Helens*** *This stratovolcano of the Cascade Range in southwestern Washington violently erupted on the morning of May 18, 1980, emitting a great cloud of condensed steam, heated gases, and ash from the summit crater. Within a few minutes, the plume had risen to a height of 20 km (12 mi), and its contents were being carried eastward by stratospheric winds. The eruption was initiated when an explosion demolished the northern portion of the cone, which is concealed from this viewpoint.*

20°N
Philippine Sea
Luzon
South China Sea
Mount Mayon
PHILIPPINES
10°N
120°E
130°E

b

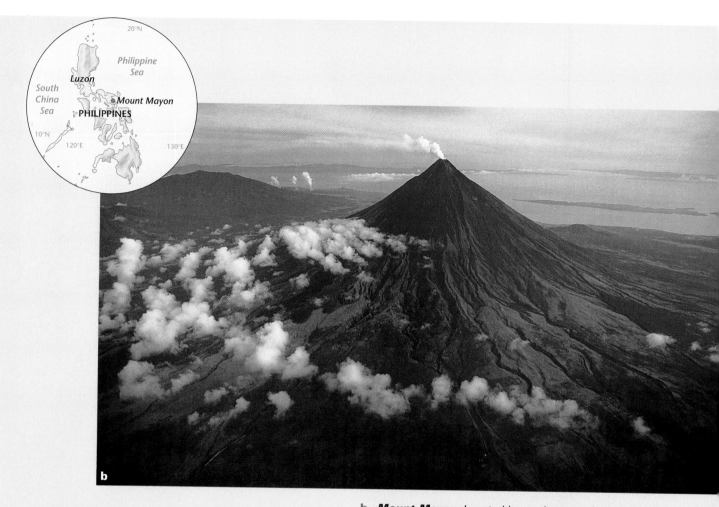

b...*Mount Mayon* *Located in southeastern Luzon, the Philippines, Mount Mayon is often considered the world's most nearly perfect stratovolcanic cone. Its summit rises to an altitude of nearly 2400 m (about 8000 ft). Mount Mayon has erupted at least 40 times since 1616. Its most recent eruption was in March, 2000, causing the evacuation of a local population of about 55,000 people.*

c...*Crater Lake* *Crater Lake, Oregon, is a water-filled caldera marking the remains of the summit of Mount Mazama, which exploded about 6600 years ago. Wizard Island (center foreground) was built on the floor of the caldera after the major explosive activity had ceased. It is an almost perfectly shaped cone of cinders capping small lava flows.*

14.4 Ash cloud *A cloud of hot, dense volcanic ash, emitted by the Soufrière Hills volcano, courses down this narrow valley on the island of Monserrat in the Lesser Antilles.*

14.5 Basaltic shield volcanoes of Hawaii *At lower left is the now-cold Halemaumau pit crater, formed in the floor of the central depression of Kilauea volcano. On the distant skyline is the snow-capped summit of Mauna Kea volcano, its elevation over 4000 m (about 13,000 ft).*

is highly fluid and travels far down the gentle slopes. Most of the lava flows issue from fissures (long, gaping cracks) on the flanks of the volcano.

Hawaiian lava domes have a wide, steep-sided central depression that may be 3 km (2 mi) or more wide and several hundred meters deep. These large depressions are a type of collapsed caldera. Molten basalt is sometimes seen in the floors of deep pit craters that occur on the floor of the central depression or elsewhere over the surface of the lava dome.

HOTSPOTS, SEA-FLOOR SPREADING, AND SHIELD VOLCANOES

The chain of Hawaiian volcanoes were created by the motion of the Pacific plate over a *hotspot*—a plume of upwelling basaltic magma deep within the mantle, arising far down in the asthenosphere. As the hot mantle rock rises, magma forms in bodies that melt their way through the lithosphere and reach the sea floor. Each major pulse of the plume sets off a cycle of volcano formation. However, the motion of the oceanic lithosphere eventually carries the volcano away from the location of the deep plume, and so it becomes extinct. Erosion processes wear the volcano away, and ultimately it becomes a low island. Continued attack by waves and slow settling of the island reduces it to a coral-covered platform. Eventually only a sunken island, or *guyot,* exists. *Focus on System 14.1 • The Life Cycle of a Hotspot Volcano* describes this process in more detail.

A few basaltic volcanoes also occur along the midoceanic ridge, where sea floor spreading is in progress. Perhaps the outstanding example is Iceland, in the North Atlantic Ocean. Iceland is constructed entirely of basalt. Basaltic flows are superimposed on older basaltic rocks as dikes and sills formed by magma emerging from deep within the spreading rift. Mount Hekla, an active volcano on Iceland, is a shield volcano somewhat similar to those of Hawaii. In Chapter 12, *Eye on the Environment 12.2 • Battling Iceland's Heimaey Volcano* documents how valiant Icelanders coped with a recent eruption. Other islands consisting of basaltic volcanoes located along or close to the axis of the Mid-Atlantic Ridge are the Azores, Ascension, and Tristan da Cunha.

Where a mantle plume lies beneath a continental lithospheric plate, the hotspot may generate enormous volumes of basaltic lava that emerge from numerous vents and fissures and accumulate layer upon layer. The basalt may ultimately attain a thickness of thousands of meters and cover thousands of square kilometers. These accumulations are called *flood basalts.*

An important American example is found in the Columbia Plateau region of southeastern Washington, northeastern Oregon, and westernmost Idaho. Here, basalts of Cenozoic age cover an area of about 130,000 km^2 (about 50,000 mi^2)—nearly the same area as the state of New York. Individual basalt flows are exposed along the walls of river gorges as cliffs in which vertical joint columns are conspicuous (Figure 14.6).

Associated with flood basalts, shield volcanoes, and scattered occurrences of basaltic lava flows is a small vol-

14.6 *Eye on the Landscape* ***Flood basalts*** *Basalt lava flows exposed in cliffs bordering the Columbia River in Washington. Each set of cliffs is a major lava flow. In cooling, vertical cracks form in the lava, creating tall columns.* **What else would the geographer see? ... Answers at the end of the chapter.**

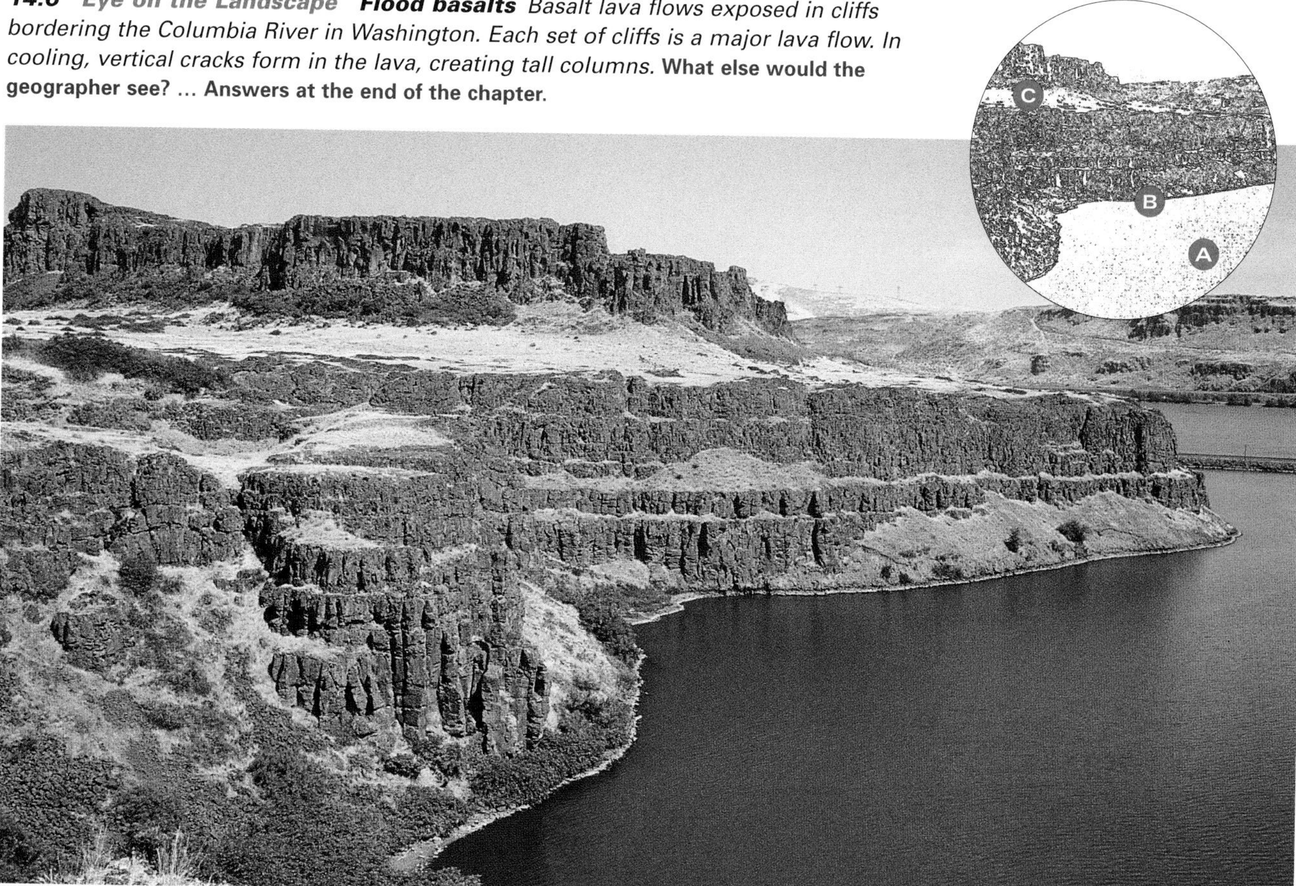

Focus on Systems 14.1

The Life Cycle of a Hotspot Volcano

The time cycles we have examined so far in earlier chapters are of the repeating type. That is, each individual cycle is followed by the next, in an endless procession. The astronomical cycles of Earth–Sun relationships are examples, generating repeating annual cycles of such variable quantities as daily air temperature or monthly rainfall amounts.

In this Focus on Systems feature, we turn to what can be called a *life cycle.* This type of cycle does not repeat, but rather it consists of a continuous progression that can be seen as having a number of stages from a beginning to an end. The higher organisms have life cycles, expressed in the various stages of growth, development, and aging from conception to death.

Nonorganic life cycles can also be identified in nature. For example, the life of a raindrop starts with a condensation of free gaseous water molecules, forming a tiny cloud droplet that in turn coalesces by collisions with other droplets to become a large drop and ends its cycle by collision with the surface of the Earth. Volcanoes fall into this kind of life cycle. A volcano can appear without warning at a location where no other volcanic activity has been seen. The volcanic cone grows to a great mountain, then seemingly becomes dormant, and may end its life in a gigantic explosion that leaves a deep caldera.

The life cycle of a basaltic shield volcano that grows upward from the abyssal ocean floor is particularly interesting. This growth may take it from a depth greater than 5000 m (about 16,000 ft) below the sea surface to a height greater than 4000 m (about 13,000 ft) above the surface, thus building a huge mass of igneous rock atop the sea floor. A unique feature of this type of volcano is that it rests on an oceanic lithosphere that is in steady horizontal motion over a soft asthenosphere. You might say that the volcano rides on a conveyor belt.

The chain of Hawaiian volcanic domes is the best example of this type of volcano. However, this chain lies far away from the spreading boundary of the Pacific plate, which would be a natural source of magma. What, then, is the magma source for these islands? Based on geophysical evidence, geologists have concluded that there is a *hotspot* beneath the islands—a persistent source of heat providing pulses or bubbles of rising magma that built the Hawaiian chain, island by island, as the Pacific plate moved over the hotspot. The upper figure to the right *(a)* diagrams this process.

The life cycle of these basaltic shield volcanoes was worked out by two geologists who became leading authorities on the Hawaiian volcanoes, Gordon Macdonald and Agatin Abbot. The typical stages of the life cycle are shown in the figure to the right. In the first stage (1), magma rising from the hotspot on the ocean floor forms a low basaltic lava dome. After the dome is constructed to full height (2), a central caldera is formed (3). A post-caldera stage (4) then follows in which large cinder cones fill the caldera and renew some of the original mass. Eventually, dormancy sets in and erosional processes lower the mountain while waves cut back the

(a)

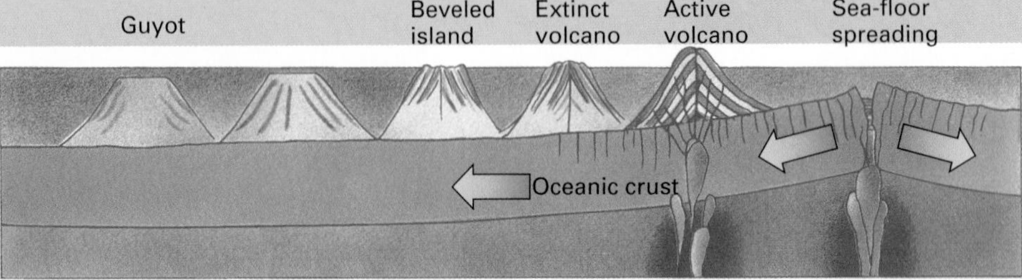

(b)

Volcanic chain *A chain of volcanoes is formed by an oceanic plate moving over a hotspot. (Copyright © by A. N. Strahler.)*

Life cycle of a typical Hawaiian volcano *(Adapted by permission from Macdonald and Abbott, 1970, Volcanoes in the Sea: The Geology of Hawaii, Honolulu, University of Hawaii Press, p. 138, Figure 92.)*

1. Deep marine stage

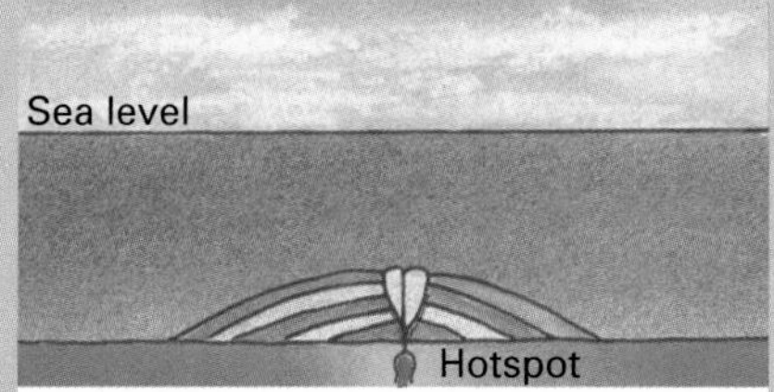

2. Shield-building stage

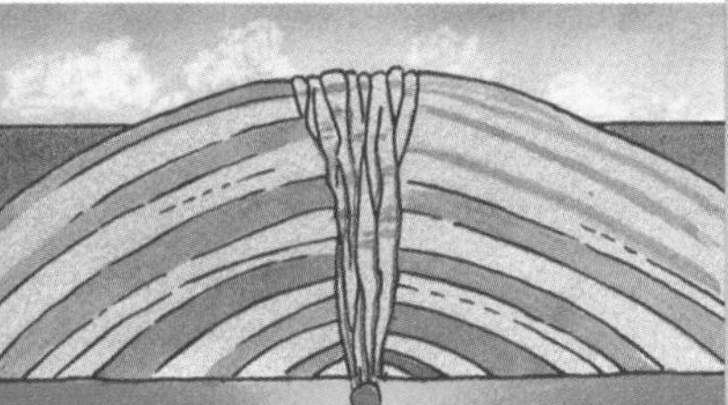

3. Caldera stage

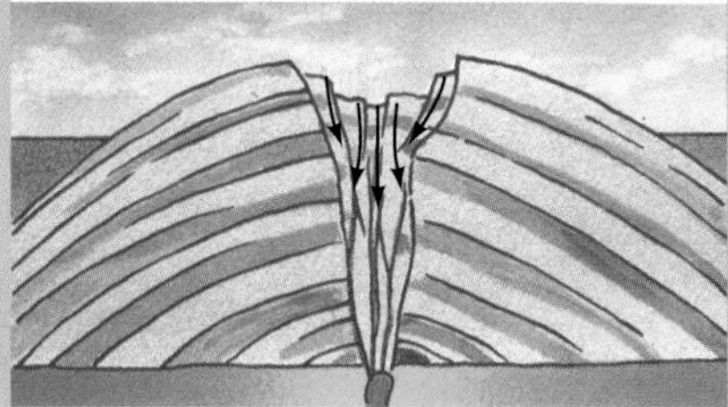

4. Cinder cone stage

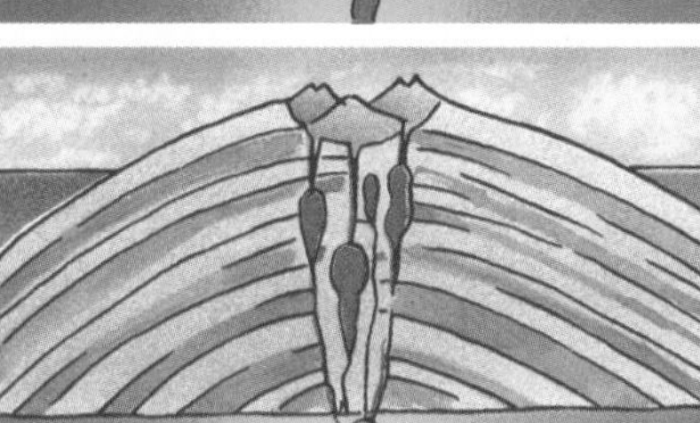

5. Erosional stage

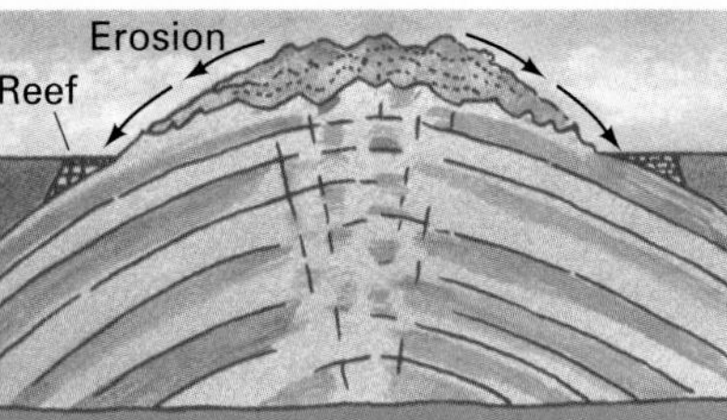

6. Atoll stage

7. Guyot stage

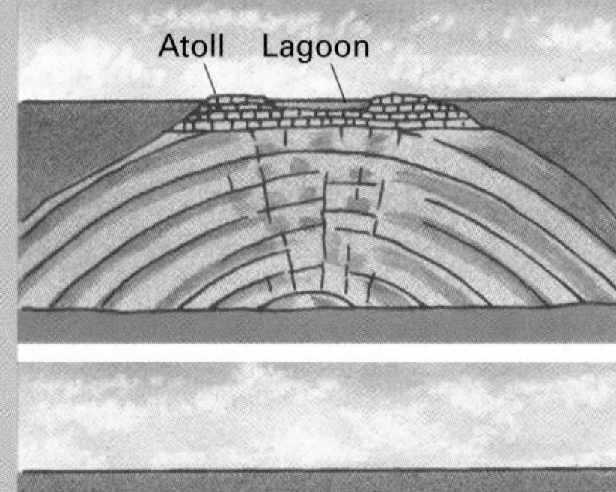

coast (5). Fringing coral reefs that form on the fringe of the dome become broader. The volcano is now fully extinct. No longer pushed upward by upwelling magma, the oceanic crust holding the volcano steadily subsides, lowering the volcano. Erosional beveling is completed at the atoll stage (6), when a thick layer of reef corals and related lagoon sediments has formed. In a final stage (7), continued crustal subsidence drowns the reef corals and only a sunken island, called a *guyot,* remains. That name honors a Swiss scholar, Arnold Henry Guyot, who in 1854 was appointed to Princeton University as a professor of physical geography.

It is now agreed that the hotspot that generated the entire chain of islands is produced by a *mantle plume,* arising far down in the asthenosphere. As the hot mantle rock rises, magma forms in bodies that melt their way through the lithosphere and reach the sea floor. Each major pulse of the plume sets off a volcanic cycle. As the volcano moves away from the location of the deep plume, it undergoes the middle and late stages (5–7) of the cycle. This lithospheric motion has produced a long trail of sunken islands and guyots, shown on the map below. (See Figure 13.7 as well.) The Hawaiian trail, trending northwestward, is 2400 km (about 1500 mi) long and includes a sharp bend to the north caused by a sudden change of direction of the Pacific plate. This distant leg consists of the Emperor Seamounts. Several other long trails of volcanic seamounts cross the Pacific Ocean Basin. They, too, follow parallel paths that reveal the plate motion.

Hawaiian seamount chain in the northwest Pacific Ocean Basin *Dots are summits; the enclosing colored area marks the base of the volcano at the ocean floor. (Copyright © by A. N. Strahler.)*

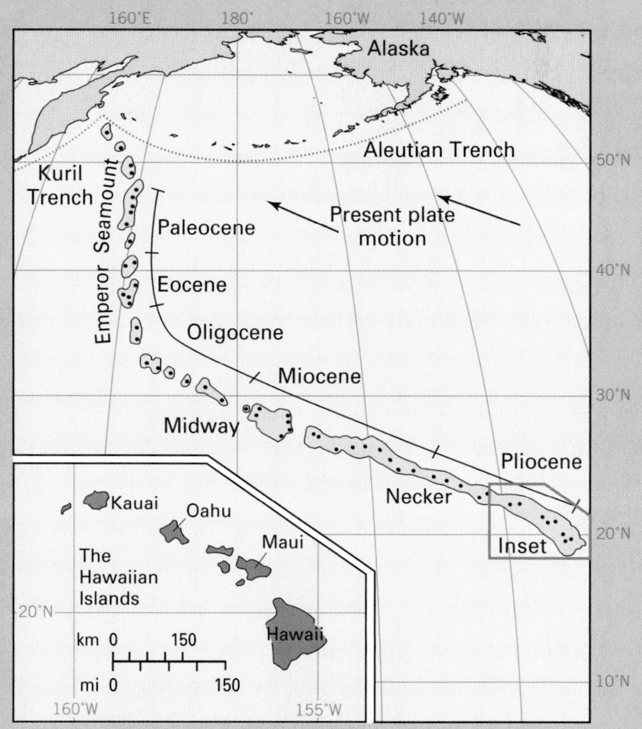

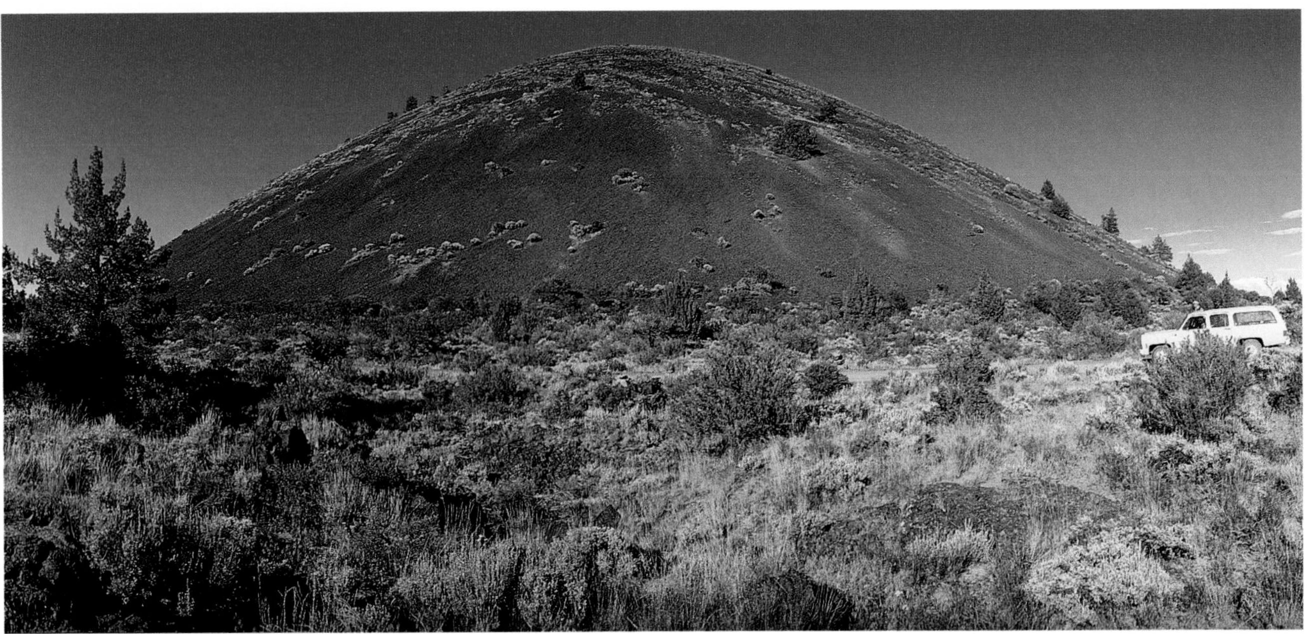

14.7 Cinder cone *This young cinder cone is built atop rough-surfaced basalt lava flows. Lava Beds National Monument, northern California.*

cano known as a *cinder cone* (Figure 14.7). Cinder cones form when frothy basalt magma is ejected under high pressure from a narrow vent, producing tephra. The rain of tephra accumulates around the vent to form a roughly circular hill with a central crater. Cinder cones rarely grow to heights of more than a few hundred meters. An exceptionally fine example of a cinder cone is Wizard Island, built on the floor of Crater Lake long after the caldera was formed (Figure 14.3).

HOT SPRINGS AND GEYSERS Where hot rock material is near the Earth's surface, it can heat nearby groundwater to high temperatures. When the groundwater reaches the surface, it provides *hot springs* at temperatures not far below the boiling point of water (Figure 14.8). At some places, jetlike emissions of steam and hot water occur at intervals from small vents—producing *geysers* (Figure 14.9). Since the water that emerges from hot springs and geysers is largely groundwater that has been heated in contact with hot rock, this water is recycled surface water. Little, if any, is water that was originally held in rising bodies of magma.

The heat from masses of lava close to the surface in areas of hot springs and geysers provides a source of energy for electric power generation. *Eye on the Environment 14.2 • Geothermal Energy Sources* provides more information about this application.

14.8 Mammoth Hot Springs *Small terraces ringed by mineral deposits hold steaming pools of hot water as the spring cascades down the slope. This example of geothermal activity is from Yellowstone National Park, Wyoming.*

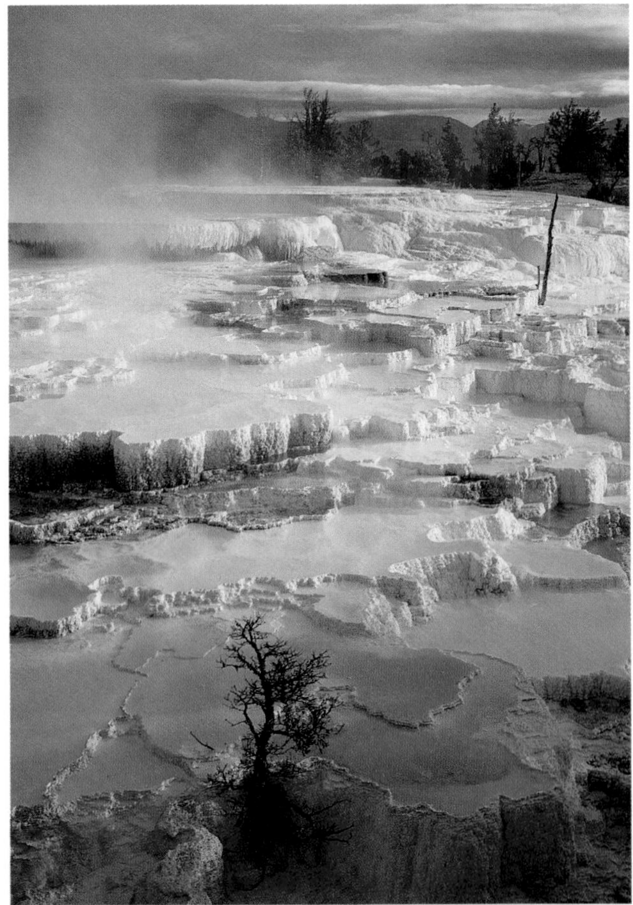

Volcanic Activity over the Globe

Figure 14.10 shows the locations of volcanoes that have been active within the last 12,000 years. By comparing this map with that of Figure 13.4, it is easy to see that many volcanoes are located along subduction boundaries. In fact,

Eye on the Environment | 14.2

Geothermal Energy Sources

Geothermal energy is energy in the form of sensible heat that originates within the Earth's crust and makes its way to the surface by conduction. Heat may be conducted upward through solid rock or carried up by circulating ground water that is heated at depth and returns to the surface. Concentrated geothermal heat sources are usually associated with igneous activity, but there also exist deep zones of heated rock and ground water that are not directly related to igneous activity.

Observations made in deep mines and bore holes show that the temperature of rock increases steadily with depth. Although the rate of increase falls off quite rapidly with increasing depth, temperatures attain very high values in the upper mantle, where rock is close to its melting point. Heat within the Earth's crust and mantle is produced largely by radioactive decay. (See *Focus on Systems 12.4 • Powering the Cycle of Rock Change.*) Slow as this internal heat production is, it scarcely diminishes with time, and the basic energy resource it provides can be regarded as limitless on a human scale.

It might seem simple enough to obtain all our energy needs by drilling deep holes at any desired location into the crust and letting the hot rock turn injected fresh water into steam, which we could use to generate electricity as our primary energy resource. Unfortunately, at the depths usually required to furnish the needed heat intensity, crustal rock tends to close any cavity or opening by rupture and slow flowage. This phenomenon would either prevent the holes from being drilled or would close them in short order. Generally, then, we must look for geothermal localities, where special conditions have caused hot rock and hot ground water to lie within striking distance of conventional drilling methods. Areas of hot springs and geysers are primary candidates.

Natural hot-water and steam localities were the first type of geothermal energy source to be developed and at present account for nearly all production of geothermal electrical power. Wells are drilled to tap the hot water. When it reaches the surface, the water flashes into steam under the reduced pressure of the atmosphere. The steam is fed into generating turbines to produce electricity, then condensed in large cooling towers (see photo below). The resulting hot water is usually released into surface stream flow, where it may create a thermal pollution problem. The larger steam fields have sufficient energy to generate at least 15 megawatts of electric power, and a few can generate 200 megawatts or more.

In certain areas, the intrusion of magma has been sufficiently recent that solid igneous rock of a batholith is still very hot in a depth range of perhaps 2 to 5 km (about 1 to 3 mi). At this depth, the rock is strongly compressed and contains little, if any, ground water. Rock in this zone may be as hot as 300°C (about 575°F) and could supply an enormous quantity of heat energy. The planned development of this resource includes drilling into the hot zone and then shattering the surrounding rock by hydrofracture—a method using water under pressure that is widely used in petroleum development. Surface water would then be pumped down one well into the fracture zone and heated water pumped up another well. Although some experiments have been conducted, this heat source has not yet been exploited in any practical way.

Geothermal power plant *This electricity-generating power plant at the Geysers, California, runs on steam produced by superheated ground water. Steam pipes in the foreground lead to the plant. After use in generating turbines, the steam is condensed in large cylindrical towers.*

14.9 Old Faithful Geyser *An eruption of Old Faithful Geyser in Yellowstone National Park, Wyoming.*

the "ring of fire" around the Pacific Rim is the most obvious feature on the map. Other volcanoes are located on or near oceanic spreading centers. Iceland is an example. In continental regions, spreading in East Africa has also produced volcanoes. (We'll return to the East African rift later in this chapter.) Hotspot activity, as in the location of the Hawaiian Islands, is also present.

Volcanic Eruptions as Environmental Hazards

The eruptions of volcanoes and lava flows are environmental hazards of the severest sort, often taking a heavy toll of plant and animal life and devastating human habitations. What natural phenomenon can compare with the Mount Pelée disaster in which thousands of lives were snuffed out in seconds? Perhaps only an earthquake or storm surge of a tropical cyclone is equally disastrous.

Wholesale loss of life and destruction of towns and cities are frequent in the history of peoples who live near active volcanoes. Loss occurs principally from sweeping clouds of incandescent gases that descend the volcano slopes like great avalanches, from lava flows whose relentless advance engulfs whole cities, from showers of ash, cinders, and bombs, and from violent earthquakes associated with volcanic activity. For habitations along low-lying coasts there is the additional peril of great seismic sea waves, generated

14.10 Volcanic activity of the Earth *Dots show the locations of volcanoes known or believed to have erupted within the past 12,000 years. Each dot represents a single volcano or cluster of volcanoes. (After data of NOAA. Copyright © A. N. Strahler.)*

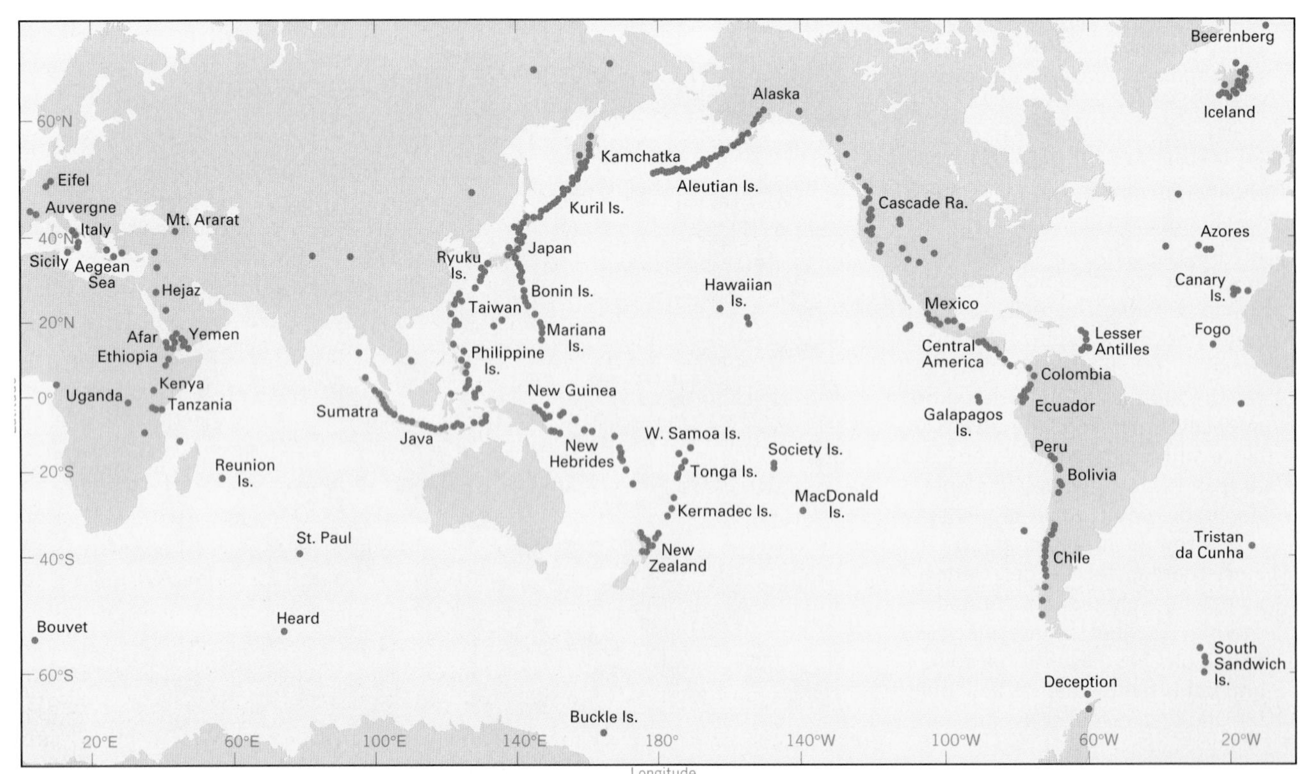

elsewhere by explosive destruction of undersea or island volcanoes.

In 1985, an explosive eruption of Ruiz Volcano in the Colombian Andes caused the rapid melting of ice and snow in the summit area. Mixing with volcanic ash, the water formed a variety of mudflow known as a *lahar.* Rushing downslope at speeds up to 145 km (90 mi) per hour, the lahar became channeled into a valley on the lower slopes, where it engulfed a town and killed more than 20,000 persons.

Scientific monitoring techniques are reducing the toll of death and destruction from volcanoes. By analyzing the gases emitted from the vent of an active volcano, as well as the minor earthquakes and local land tilting that precede a major quake, scientists have successfully predicted periods of volcanic activity. Extensive monitoring of Mount Mayon and the Mexican volcano Popocatepetl to predict recent eruptions has led to evacuations that saved hundreds or thousands of lives. However, not every volcano is well monitored or predictable.

Despite their potential for destructive activity, volcanoes are a valuable natural resource in terms of recreation and tourism. Few landscapes can rival in beauty the mountainous landscapes of volcanic origin. National parks have been made of Mount Rainier, Mount Lassen, and Crater Lake in the Cascade Range, a mountain mass largely of volcanic construction. British Columbia's Garibaldi Provincial Park preserves volcanic mountain vistas in a vast wilderness area of snow-covered peaks and swift rivers. Hawaii Volcanoes National Park recognizes the natural beauty of Mauna Loa and Kilauea—their breathtaking displays of molten lava are a living textbook of igneous processes. *Focus on Remote Sensing 14.3 • Remote Sensing of Volcanoes* shows Mount Vesuvius, Mount Fuji, and Popocatepetl, imaged by several different remote sensing systems.

Landforms of Tectonic Activity

Recall from Chapter 13, our introduction to global plate tectonics, that there are two basic forms of tectonic activity: compression and extension (see Figure 13.10). Along converging lithospheric plate boundaries, tectonic activity is primarily compression. In subduction zones, sedimentary layers of the ocean floor are compressed within a trench as the descending plate forces them against the overlying plate. In continental collision, compression is of the severest kind. In zones of rifting of continental plates, however, the brittle continental crust is pulled apart and yields by faulting. This motion is extensional and produces faults. We begin our examination of landforms of tectonic activity with folding produced by compression.

Fold Belts

In Chapter 13, we saw that two plates of continental lithosphere can collide, causing severe compressional stress. When flat-lying strata from a continental shelf or margin are caught in the collision, they can experience **folding** (see Figure 13.10). The wavelike shapes imposed on the strata consist of alternating arch-like upfolds, called *anticlines,* and trough-like downfolds, called *synclines.* Thus, the initial landform associated with an anticline is a broadly rounded mountain ridge, and the landform corresponding to a syncline is an elongate, open valley.

An example of open folds of comparatively young geologic age that has long attracted the interest of geographers is the Jura Mountains of France and Switzerland. Figure 14.11 is a block diagram of a small portion of that fold belt. The rock strata are mostly limestone layers and were capable of being deformed by bending with little brittle fracturing. Folding occurred in late Cenozoic (Miocene) time. Notice that each mountain crest is associated with the axis of an anticline, while each valley lies over the axis of a syncline. Some of the anticlinal arches have been partially removed by erosion processes. The rock structure can be seen clearly in the walls of the winding gorge of a major river that crosses the area. The Jura folds lie just to the north of the main collision orogen of the Alps. Because of their location near a mountain mass, they are called *foreland folds.* The ridge and valley region of Pennsylvania, Maryland, and Virginia is another example but of much greater age.

Faults and Fault Landforms

A **fault** in the brittle rocks of the Earth's crust occurs when rocks suddenly yield to unequal stresses by fracturing. Faulting is accompanied by a displacement—a slipping motion—along the plane of breakage, or *fault plane.* Faults are often of great horizontal extent, so that the surface trace, or fault line, can sometimes be followed along the ground for many kilometers. Most major faults extend down into the crust for at least several kilometers.

Faulting occurs in sudden slippage movements that generate earthquakes. A single fault movement may result in slippage of as little as a centimeter or as much as 15 m (about 50 ft). Successive movements may occur many years or decades apart, even several centuries apart. Over long time spans, the accumulated displacements can amount to tens or hundreds of kilometers. In some places, clearly recognizable sedimentary rock layers are offset on opposite sides of a fault, allowing the total amount of displacement to be measured accurately.

NORMAL FAULTS One common type of fault associated with crustal rifting is the **normal fault** (Figure 14.12*a*). The plane of slippage, or fault plane, is steeply inclined. The crust on one side is raised, or upthrown, relative to the other, which is downthrown. A normal fault results in a steep, straight, cliff-like feature called a *fault scarp* (Figure 14.12). Fault scarps range in height from a few meters to a few hundred meters (Figure 14.13). Their length is usually measurable in kilometers. In some cases they attain lengths as great as 300 km (about 200 mi).

Focus on Remote Sensing | 14.3

Remote Sensing of Volcanoes

Volcanoes are always attractive subjects for remote sensing. As monumental landforms, they have distinctive shapes and appearances that are easy to recognize in satellite imagery. They are also very dynamic subjects with the capacity to erupt in spectacular fashion, providing smoke plumes, ash falls, and lava flows. Here's a gallery of images of volcanoes from a suite of remote imagers with different characteristics.

Mount Vesuvius. This famous volcano erupted in A.D. 79, ejecting a huge cauliflower-shaped cloud of ash and debris that rapidly settled to the Earth. The nearby Roman city of Pompeii was buried in as much as 30 m (100 ft) of ash and dust, leaving an archeological treasure that was not discovered until 1748. In recent history, major eruptions were recorded in 1631, 1794, 1872, 1906, and 1944.

The image of Mount Vesuvius was acquired by the Advanced Spaceborne Thermal Emission and Reflection Radiometer (ASTER) (see *Focus on Remote Sensing 12.3 • Geologic Mapping with ASTER*) on September 26, 2000. Spatial resolution is 15 by 15 m (49 by 49 ft). Red, green, and blue colors in the image have been assigned to near-infrared, red, and green spectral bands, respectively. Vegetation appears bright red, with urban areas in blue and green tones. The magnitude of development around the volcano shows that the impact of a major eruption would be particularly catastrophic. If sudden, thousands of deaths would be expected.

Mount Vesuvius imaged by ASTER *(Image courtesy NASA/GSFC/MITI/ERSDAC/JAROS and U. S./Japan ASTER Science Team.)*

Mount Fuji. Like Mount Vesuvius, Mount Fuji lies within striking distance of a large population center—Tokyo, Japan, located about 100 km (about 60 mi) to the northeast. Although it has the symmetry of a simple cone, it is actually a complex structure with two former volcanic cones buried within its outer form. Mount Fuji is considered an active volcano, with 16 eruptions since A.D. 781. Its last eruption, in 1707, darkened the noontime sky and shed dust on the present-day Tokyo region.

This striking image of Mount Fuji, showing the

Mount Fuji imaged by the Shuttle Radar Topography Mission *(Image courtesy NASA/JPL/NIMA.)*

Tokyo region in the foreground, was acquired by the Shuttle Radar Topography Mission interferometric synthetic aperture radar on February 21, 2000. This type of radar sends simultaneous pulses of radio waves toward the ground from two antennas spaced 60 m (about 200 ft) apart. Very slight differences in the return signals can be related to the ground height. In this way, the radar can map elevations very precisely. The image is constructed by computer, simulating a viewpoint above and to the east of Tokyo. A color scale is assigned to elevation, ranging from white to green to brown. Vertical scale is doubled for visualization, so Mount Fuji and surrounding peaks appear twice as steep as they actually are.

Popocatepetl. On December 18, 2000, Popocatepetl came to life, spewing molten lava in plumes and rivers from its summit of 5470 m (17,946 ft). Located only 65 km (40 mi) from Mexico City, the "smoking mountain" of the Nahuatl language is located within view of nearly 30 million people. Following its last eruption in 1994, the volcano was under constant monitoring, and as a result, scientists were able to anticipate the eruption. About half the population in the valleys directly below the volcano were evacuated, and no injuries were reported.

The close-up image of Popocatepetl was acquired by Landsat-7 on January 4, 1999, at a spatial resolution of 30 by 30 m (98 by 98 ft). Snow and ice flank the summit crater. Canyons carved into the volcano lead away from the summit. The lower slopes are thickly covered with vegetation, which appears green in this image. The surrounding plain shows intensive agricultural development.

The eruption of Popocatepetl was captured by the Sea-viewing Wide Field-of-View Sensor (SeaWiFS) on December 19, 2000, at a spatial resolution of 1 by 1 km (0.62 by 0.62 mi). This global-scale imager pictured the width of southern Mexico from the Pacific to the Caribbean in true color. Popocatepetl (P) is shown very near the center of the image, emitting a smoke plume moving south and east. A large cloud bank lies to the east and obscures much of the right-hand side of the image. High, thin cirrus clouds overlie part of the plume. Mexico City (M) is visible north and west of the volcano as a gray-brown patch flanked by north-south mountain ranges.

Popocatepetl imaged by Landsat-7 *(Image courtesy Ron Beck, EROS Data Center.)*

Eruption of Popocatepetl imaged by SeaWiFS *(Image courtesy SeaWiFS Project, NASA/Goddard Space Flight Center, and ORBIMAGE.)*

14.11 Anticlines and synclines *Structural diagrams for the Jura Mountains, France and Switzerland. A cross section shows the folds and ground surface (a). The landscape developed on the folds is shown in the block diagram (b). (After E. Raisz.)*

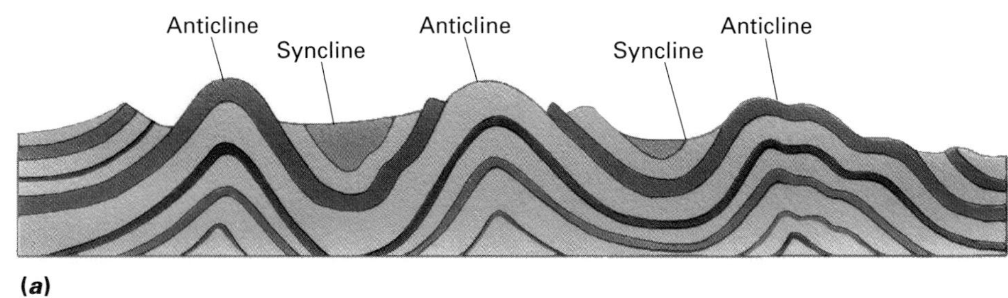

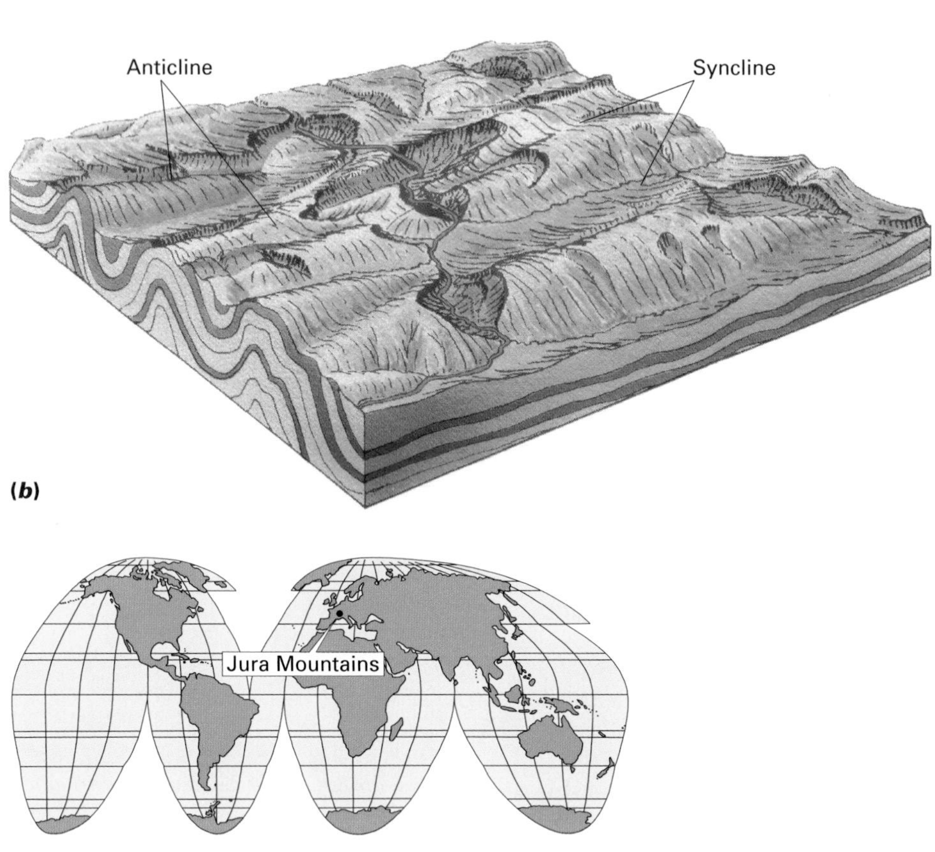

Normal faults are not usually isolated features. Commonly, they occur in multiple arrangements, often as a set of parallel faults. These arrangements give rise to a grain or pattern of rock structure and topography. A narrow block dropped down between two normal faults is a *graben* (Figure 14.14). A narrow block elevated between two normal faults is a *horst.* Grabens make conspicuous topographic trenches, with straight, parallel walls. Horsts make block-like plateaus or mountains, often with a flat top but steep, straight sides.

In rifted zones of the continents, regions where normal faulting occurs on a grand scale, mountain masses called *block mountains* are produced. The up-faulted mountain blocks can be described as either tilted or lifted (Figure 14.15). A tilted block has one steep face—the fault scarp—and one gently sloping side. A lifted block, which is a type of horst, is bounded by steep fault scarps on both sides.

TRANSCURRENT FAULTS Recall that lithospheric plates also slide past one another horizontally along major transform faults and that these features comprise one type of lithospheric plate boundary. Long before the principles of plate tectonics became known, geologists referred to such faults as **transcurrent faults** (Figure 14.12*c*), or sometimes as *strike-slip faults.* In a transcurrent fault, the movement is predominantly horizontal. Where the land surface is nearly flat, no scarp, or a very low one at most, results. Only a thin fault line is traceable across the surface. In some places a narrow trench, or rift, marks the fault.

The best known of the active transcurrent faults is the great San Andreas Fault, which can be followed for a distance of about 1000 km (about 600 mi) from the Gulf of California to Cape Mendocino, a location well north of the San Francisco area, where it heads out to sea. It is a transform fault that marks the active boundary between the Pacific plate and the North American plate (see Figure 13.4). The Pacific plate is moving toward the northwest, which means that a great portion of the state of California and all of Lower (Baja) California is moving bodily northwest with respect to the North American mainland.

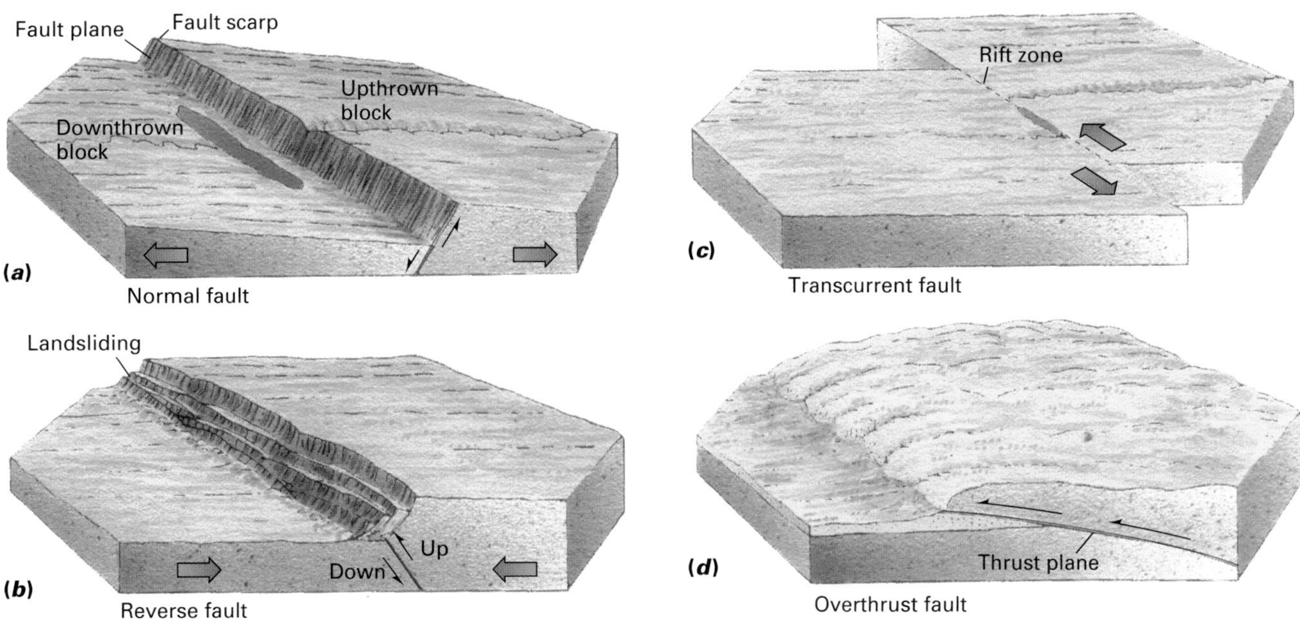

14.12 Four types of faults

Throughout many kilometers of its length, the San Andreas Fault appears as a straight, narrow scar. In some places this scar is a trench-like feature, and elsewhere it is a low scarp (Figure 14.16). Frequently, a stream valley takes an abrupt jog—for example, first right, then left— when crossing the fault line. This offset in the stream's course shows that many meters of movement have occurred in fairly recent time.

14.13 Fault scarp *This fault scarp was formed during the Hebgen Lake, Montana, earthquake of 1959. In a few moments, a displacement of 6 m (20 ft) took place on a normal fault.*

REVERSE AND OVERTHRUST FAULTS In a *reverse fault,* the inclination of the fault plane is such that one side rides up over the other and a crustal shortening occurs (Figure 14.12*b*). Reverse faults produce fault scarps similar to those of normal faults, but the possibility of landsliding is greater because an overhanging scarp tends to be formed. The San Fernando, California, earthquake of 1971 was generated by slippage on a reverse fault.

The *low-angle overthrust fault* (Figure 14.12*d*) involves predominantly horizontal movement. One slice of rock rides over the adjacent ground surface. A thrust slice may be up to 50 km (30 mi) wide. The evolution of low-angle thrust faults was explained in Chapter 13 and illustrated in Figure 13.11.

14.14 Initial landforms of normal faulting *A graben is a downdropped block, often forming a long, narrow valley. A horst is an upthrown block, forming a plateau, mesa, or mountain. (A. N. Strahler)*

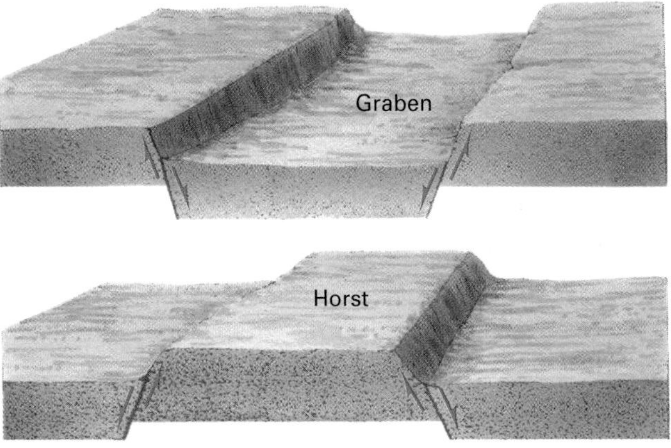

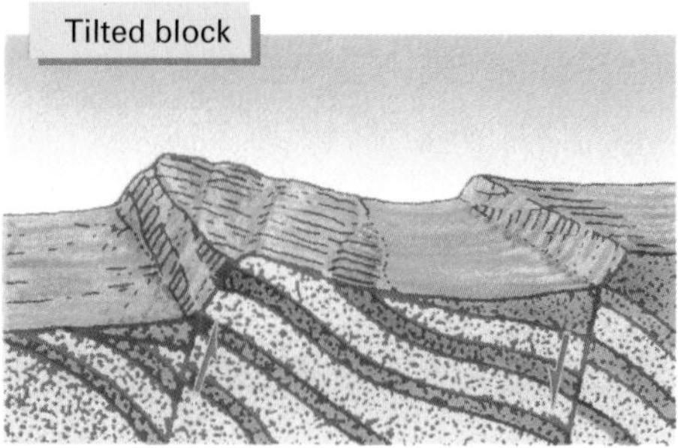

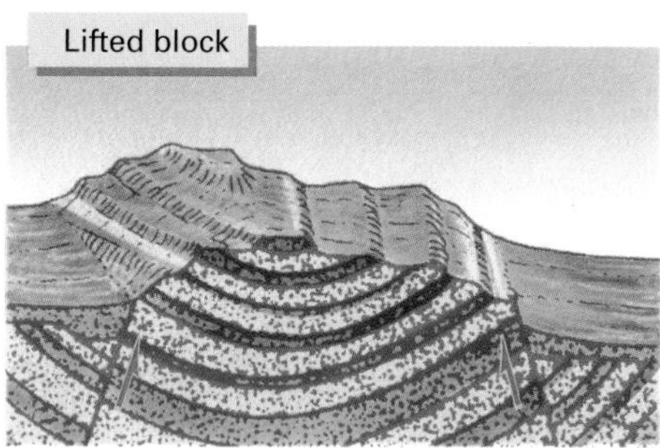

14.15 Fault block mountains *Fault block mountains may be of tilted type (left) or lifted type (right). (After W. M. Davis.)*

14.16 The San Andreas Fault in southern California *The fault is marked by a narrow trough. The fault is slightly offset in the middle distance.*

The Rift Valley System of East Africa

Rifting of continental lithosphere is the very first stage in the splitting apart of a continent to form a new ocean basin. The process is beautifully illustrated by the East African Rift Valley system. This region has attracted the attention of geologists since the early 1900s. They gave the name *rift valley* to what is basically a graben but with a more complex history that includes the building of volcanoes on the graben floor.

Figure 14.17 is a sketch map of the East African Rift Valley system. It is about 3000 km (1900 mi) long and extends from the Red Sea southward to the Zambezi River. Along this axis, the Earth's crust is being lifted and spread apart in a long, ridge-like swell. The Rift Valley system consists of a number of graben-like troughs. Each is a separate rift valley ranging in width from about 30 to 60 km (20 to 40 mi). As geologists noted in early field surveys of this system, the rift valleys are like keystone blocks of a masonry arch that have slipped down between neighboring blocks because the arch has spread apart somewhat. Thus, the floors of the rift valleys are above the elevation of most of the African continental surface. Major rivers and several long, deep lakes—Lake Nyasa and Lake Rudolph, for example—occupy some of the valley floors.

The sides of the rift valleys typically consist of multiple fault steps (Figure 14.18). Sediments, derived from the high plateaus that form the flanks of troughs, make thick fills in the floors of the valleys. Two great stratovolcanoes have been built close to the Rift Valley east of Lake Victoria. One is Mount Kilimanjaro, whose summit rises to over 6000 m (about 19,000 ft). The other, Mount Kenya, is only a little lower and lies right on the equator.

EARTHQUAKES

You have probably seen television news accounts of disastrous earthquakes and their destructive effects (Figure 14.19). Californians know about severe earthquakes from first-hand experience, but several other areas in North America have also experienced strong earthquakes, and a few of these have been very severe. An **earthquake** is a motion of the ground surface, ranging from a faint tremor to a wild motion capable of shaking buildings apart.

The earthquake is a form of energy of wave motion transmitted through the surface layer of the Earth. Waves move outward in widening circles from a point of sudden

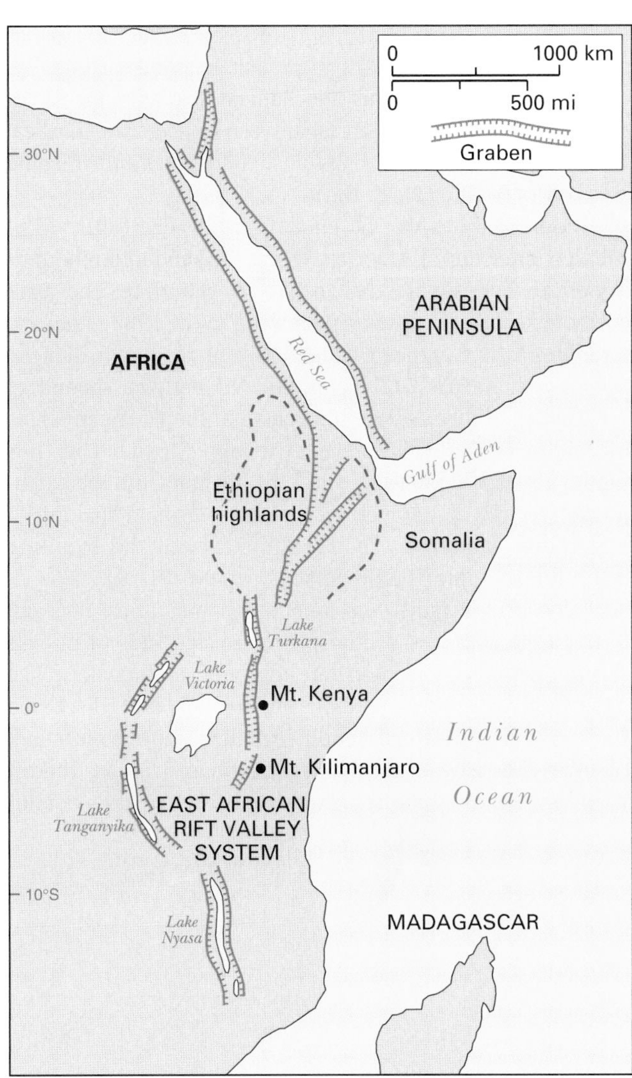

14.17 East African Rift Valley *This sketch map shows the East African Rift Valley system and the Red Sea to the north.*

energy release, called the *focus.* Like ripples produced when a pebble is thrown into a quiet pond, these seismic waves gradually lose energy as they travel outward in all directions. (The term *seismic* means "pertaining to earthquakes.")

Most earthquakes are produced by sudden slip movements along faults. They occur when rock on both sides of the fault is slowly bent over many years by tectonic forces. Energy accumulates in the bent rock, just as it does in a bent archer's bow. When a critical point is reached, the strain is relieved by slippage on the fault, and the rocks on opposite sides of the fault move in different directions. A large quantity of energy is instantaneously released in the form of seismic waves, which shake the ground. In the case of a transcurrent fault, on which movement is in a horizontal direction, slow bending of the rock that precedes the shock takes place over many decades. Sometimes a slow, steady displacement known as *fault creep* occurs, which tends to reduce the accumulation of stored energy.

The devastating San Francisco earthquake of 1906 resulted from slippage along the San Andreas Fault, which is dominantly a transcurrent fault. This fault also passes about 60 km (about 40 mi) inland of Los Angeles, placing the densely populated metropolitan Los Angeles region in great jeopardy. Associated with the San Andreas Fault are several important parallel and branching transcurrent faults, all of which are capable of generating severe earthquakes.

14.18 Eye on the Landscape The Rift Valley wall in Ethiopia *Multiple fault scarps give the landscape a stepped appearance.* **What else would the geographer see?...Answers at the end of the chapter.**

Earthquakes can also be produced by volcanic activity, as when magma rises or recedes within a volcanic chamber.

A scale of earthquake magnitudes was devised in 1935 by the distinguished seismologist Charles F. Richter. Now called the *Richter scale,* it describes the quantity of energy released by a single earthquake. Scale numbers range from 0 to 9, but there is really no upper limit other than nature's own energy release limit. For each whole unit of increase (say, from 5.0 to 6.0), the quantity of energy release increases by a factor of 32. A value of 9.5 is the largest observed to date—the Chilean earthquake of 1960. The great San Francisco earthquake of 1906 is now rated as magnitude 7.9. *Working It Out 14.4 • The Richter Scale* provides more information about the Richter scale and the energy released by some great earthquakes.

Earthquakes and Plate Tectonics

Seismic activity—the repeated occurrence of earthquakes—occurs primarily near lithospheric plate boundaries. Figure 14.20, which shows the location of all earthquake centers during a typical seven-year period, clearly reveals this pattern. The greatest intensity of seismic activity is found along converging plate boundaries where oceanic plates are undergoing subduction. Strong pressures build up at the downward-slanting contact of the two plates, and these are relieved by sudden fault slippages that generate earthquakes of large magnitude. This mechanism explains the great earthquakes experienced in Japan, Alaska, Central America, Chile, and other narrow zones close to trenches and volcanic arcs of the Pacific Ocean Basin.

Good examples can be cited from the Pacific coast of Mexico and Central America, where the subduction boundary of the Cocos plate lies close to the shoreline. The great earthquake that devastated Mexico City in 1986 was centered in the deep trench offshore. Two great shocks in close succession, the first of magnitude 8.1 and the second of 7.5, damaged cities along the coasts of the Mexican states of Mïchocoan and Guerrero. Although Mexico City lies inland about 300 km (about 185 mi) distant from the earthquake epicenters, it experienced intense ground shaking of underlying saturated clay formations, with the resulting death toll of some 10,000 persons (Figure 14.19).

Transcurrent faults on transform boundaries that cut through the continental lithosphere are also sites of intense seismic activity, with moderate to strong earthquakes. The most familiar example is the San Andreas Fault, discussed further in a following section. Another transcurrent fault often in the news is the North Anatolian Fault in Turkey, where the Persian subplate is moving westward at its bound-

14.19 Earthquake devastation *This building did not survive the Mexico City earthquake of September 1985.*

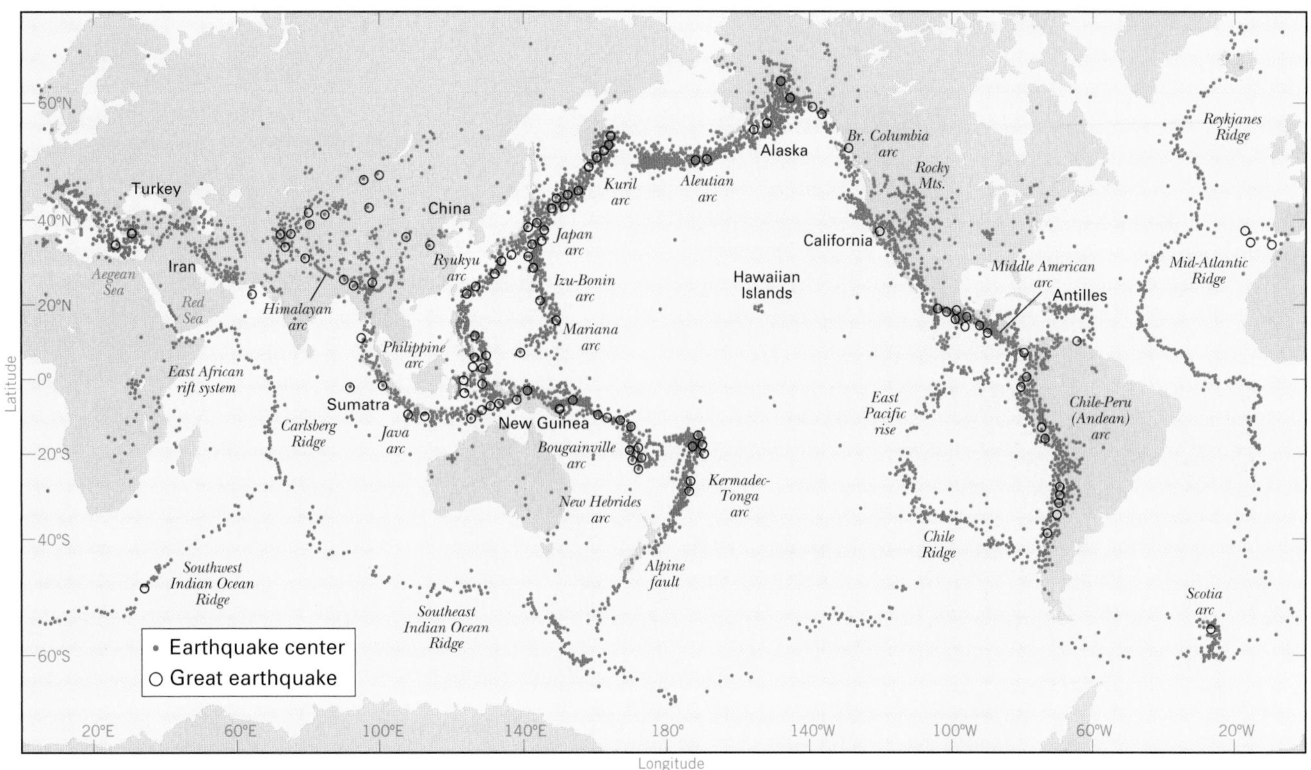

14.20 Earthquake locations *This world map plots earthquake center locations and centers of great earthquakes. Center locations of all earthquakes originating at depths of 0 to 100 km (62 mi) during a six-year period are shown by red dots. Each dot represents a single location or a cluster of centers. Black circles identify centers of earthquakes of Richter magnitude 8.0 or greater during an 80-year period. The map clearly shows the pattern of earthquakes occurring at subduction boundaries. (Compiled by A. N. Strahler from data of U.S. government. Copyright © A. N. Strahler.)*

ary with the European plate. The year 1999 saw a major earthquake there on August 17, centered near the city of Izmit, which measured 7.4 on the Richter scale and killed more than 15,000. A few months later, a quake of magnitude 7.2 occurred not far away on the same fault, killing hundreds more. Central and southeast Turkey shook again on related faults in 2002 and 2003, with death tolls in the hundreds.

Spreading boundaries are the location of a third class of narrow zones of seismic activity related to lithospheric plates. Most of these boundaries are identified with the midoceanic ridge and its branches. For the most part, earthquakes in this class are limited to moderate intensities.

Earthquakes also occur at scattered locations over the continental plates, far from active plate boundaries. In many cases, no active fault is visible, and the geologic cause of the earthquake is uncertain. For example, the great New Madrid earthquake of 1811 was centered in the Mississippi River floodplain in Missouri. It produced three great shocks in close succession, rated from 8.1 to 8.3 on the Richter scale. In southeastern Canada, significant earthquakes occur from time to time in southern Quebec and along the St. Lawrence River valley. The most recent of these is the Saguenay earthquake of 1988, which rated 5.9 on the Richter scale.

Seismic Sea Waves

An important environmental hazard often associated with a major earthquake centered on a subduction plate boundary is the *seismic sea wave,* or **tsunami,** as it is known to the Japanese. A train of these water waves is often generated in the ocean by a sudden movement of the sea floor at a point near the earthquake source. The waves travel over the ocean in ever-widening circles, but they are not perceptible at sea in deep water. (Seismic sea waves are sometimes referred to as "tidal waves," but since they have nothing to do with tides, the name is quite misleading.)

When a tsunami arrives at a distant coastline, it causes a rise in water level. Normal wind-driven waves, superimposed on the heightened water level, attack places inland that are normally above their reach. For example, in this century several destructive seismic sea waves in the Pacific Ocean attacked ground as high as 10 m (30 ft) above normal high-tide level, causing widespread destruction and many deaths by drowning in low-lying coastal areas. It is thought that the coastal flooding that occurred in Japan in 1703, with an estimated loss of life of 100,000 persons, may have been caused by seismic sea waves.

Working It Out | 14.4

The Richter Scale

The magnitude of earthquakes is most commonly assessed by the *Richter scale,* which was devised by the seismologist Charles F. Richter in 1935, then modified in 1956 by Richter and his colleague, Beno Gutenberg. This scale can be related to the energy released at the earthquake center, and thus it can be used as an estimate of the severity of a particular earthquake. The scale has neither a fixed maximum nor a minimum, but the highest-magnitude earthquakes thus far measured have been rated as 8.9 on the Richter scale. Earthquakes of magnitude 2.0 are the smallest normally detected by human senses, but instruments can detect quakes as small as –3.0 on the scale. The Richter scale is logarithmic and is based on the amplitude of the largest earthquake waves measured for a particular earthquake. For each unit of increase in the Richter scale, the amplitude of the earthquake wave increases by a factor of 10.

How is the Richter scale related to the amount of energy released by an earthquake? The Richter scale is proportional to the amplitude of the largest wave, but that is not a direct measure of energy released. However, geophysicists use the following relationship between the Richter scale number and energy release:

$$\log_{10}E = 4.8 + 1.5M$$

where E is the energy in joules and M is the Richter scale magnitude.

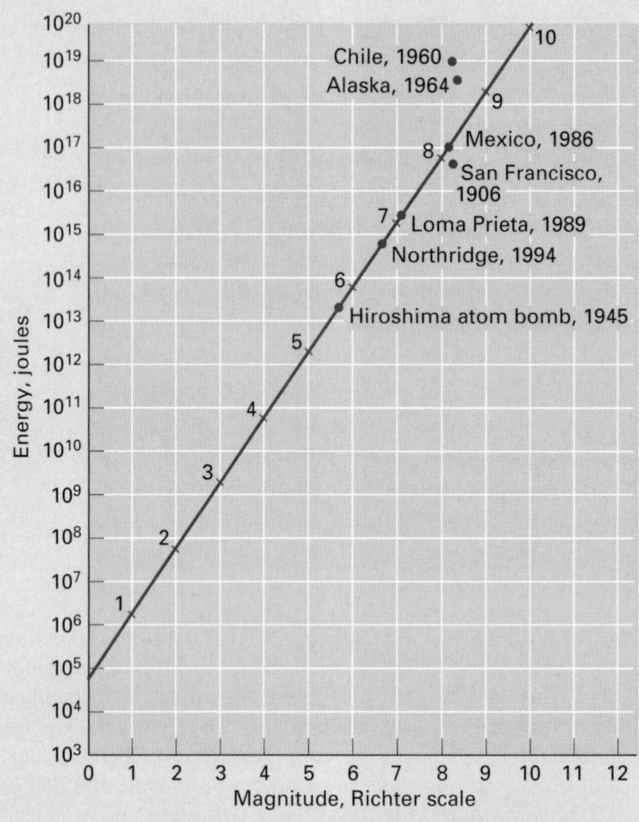

The Richter scale and energy released by earthquakes *(Copyright © by A. N. Strahler.)*

The figure above plots this relationship using a logarithmic scale for energy release. By taking the exponential of both sides, we can write this expression as

$$E = 10^{(4.8 + 1.5M)} = 10^{4.8} \times 10^{1.5M}$$

From this form, we can see that if the Richter scale number increases by 1 unit, then the energy release will increase by a factor of $10^{1.5 \times 1}$, or 31.6. This means that for each unit increase in the Richter scale, say from 4.0 to 5.0, about 32 times as much energy is released. For a two-unit increase, say from 4.0 to 6.0, the energy released will be $10^{1.5 \times 2} = 10^3 = 1000$ times as large!

Earthquakes along the San Andreas Fault

Almost a hundred years have passed since the great San Francisco earthquake of 1906 was generated by movement on the San Andreas Fault. The maximum horizontal displacement of the ground was about 6 m (20 ft). Since then, this sector of the fault has been locked—that is, the rocks on the two sides of the fault have been held together without sudden slippage (Figure 14.21). In the meantime, the two lithospheric plates that meet along the fault have been moving steadily with respect to one another. This means that a huge amount of unrelieved strain energy has already accumulated in the crustal rock on either side of the fault.

On October 17, 1989, the San Francisco Bay area was severely jolted by an earthquake with a Richter magnitude of 7.1. The earthquake's epicenter was located near Loma Prieta peak, about 80 km (50 mi) southeast of San Francisco, at a point only 12 km (7 mi) from the city of Santa Cruz, on

Richter Magnitude and Energy Release

Magnitude, Richter Scale*	Energy Release (joules)	Comment
2.0	6×10^7	Smallest quake normally detected by humans.
2.5–3.0	10^8–10^9	Quake can be felt if it is nearby. About 100,000 shallow quakes of this magnitude per year.
4.5	4×10^{11}	Can cause local damage.
5.7	2×10^{13}	Energy released by Hiroshima atom bomb.
6.0	6×10^{13}	Destructive in a limited area. About 100 shallow quakes per year of this magnitude.
6.7	7×10^{14}	Northridge earthquake of 1994.
7.0	2×10^{15}	Rated a major earthquake above this magnitude. Quake can be recorded over whole earth. About 14 per year this great or greater.
7.1	3×10^{15}	Loma Prieta earthquake of 1989.
8.25 (7.9)	4.5×10^{16}	San Francisco earthquake of 1906.
8.1	10^{17}	Mexican earthquake of 1986.
8.4 (9.2)	4×10^{18}	Alaskan earthquake of 1964.
8.3 (9.5)	10^{19}	Chilean earthquake of 1960, near the border of Equador.

* () indicates magnitude as adjusted by Kanamori.

News media accounts of earthquakes often get this wrong and state that an increase of one unit in the Richter scale produces a tenfold increase in energy released. These accounts confuse the amplitude of the earthquake wave, which does increase tenfold per scale unit, with the energy release, which increases by the factor of 32.

The relationship between energy release and Richter scale magnitude shown above works best for small and medium earthquakes. For large earthquakes, however, the energy release is underestimated. In 1977, the seismologist Hiroo Kanamori of California Institute of Technology developed a method to calculate earthquake magnitude that more accurately reflected the energy release. After examining the seismograms of great earthquakes of the twentieth century, he assigned larger magnitudes to many of them. Some of these are shown in the table. When they are plotted on the graph according to their Richter magnitude, they fall above the line, showing that they release more energy than expected by their Richter rating.

However they are rated, great earthquakes are among the most powerful natural phenomena known. The Alaskan earthquake of 1964 released a quantity of energy equal to that of about 200,000 atomic bombs of the size exploded at Hiroshima in 1945. As human habitation of large cities near active faults continues to increase, the potential for death and destruction from such powerful earthquakes increases as well.

Monterey Bay. The city of Santa Cruz suffered severe structural damage to older buildings. In the distant San Francisco Bay area, destructive ground shaking proved surprisingly severe. Buildings, bridges, and viaducts on landfills were particularly hard hit (Figure 14.22). Altogether, 62 lives were lost in this earthquake, and the damage was estimated to be about \$6 billion. In comparison, the 1906 earthquake took a toll of 700 lives and property damage equivalent to about 30 billion present-day dollars.

The displacement that caused the Loma Prieta earthquake occurred deep beneath the surface not far from the San Andreas Fault, which has not slipped since the great San Francisco earthquake of 1906. The slippage on the Loma Prieta fault amounted to about 1.8 m (6 ft) horizontally and 1.2 m (4 ft) vertically, but did not break the ground surface above it. Geologists state that the Loma Prieta slippage, though near the San Andreas Fault, probably has not relieved more than a small portion of the strain

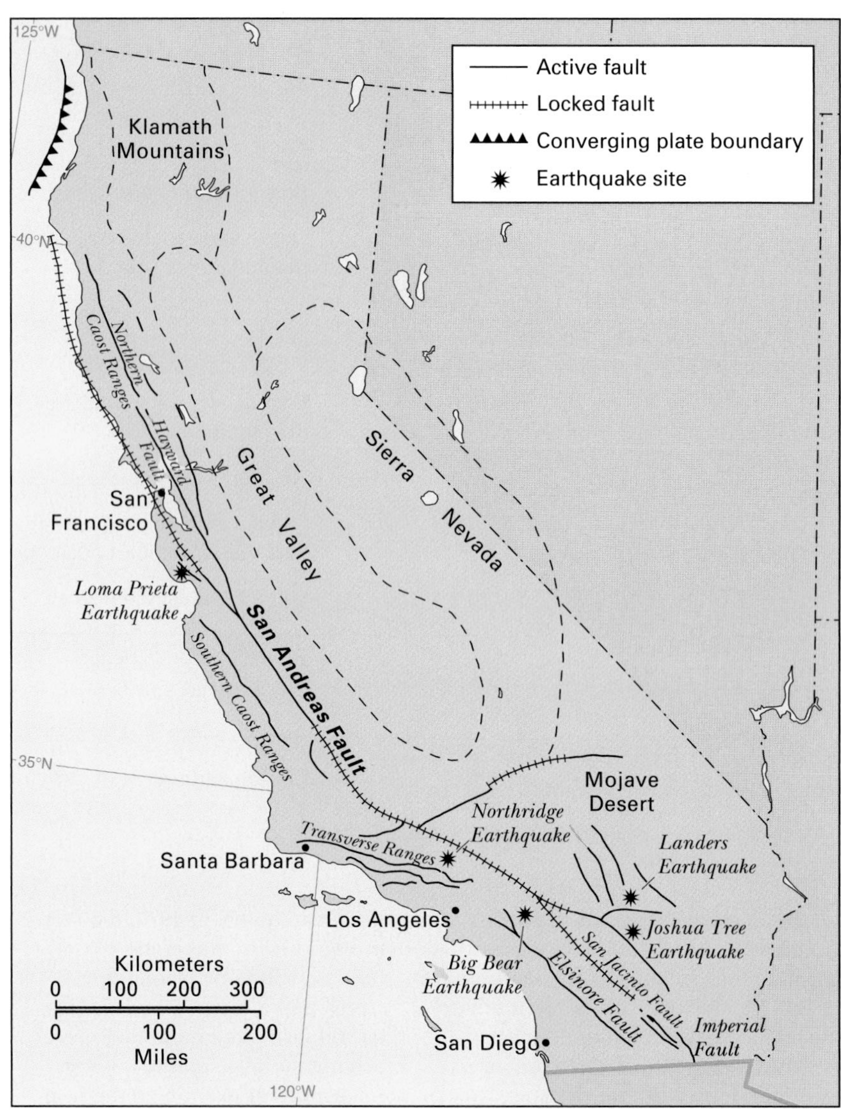

14.21 San Andreas fault system in California *A sketch map of the San Andreas Fault and Transverse Ranges showing the locked sections of the fault alternating with active sections. (Based on data of the U.S. Geological Survey.)*

on the San Andreas. While the occurrence of another major earthquake in the San Francisco region cannot be predicted with precision, it is inevitable. As each decade passes, the probability of that event becomes greater.

We may need only to look to Japan for a scenario of what the citizens of the San Francisco Bay area could experience when a branch of the San Andreas Fault lets go. The Hyogo-ken Nanbu earthquake that devastated the city of Kobe in January of 1995 (Figure 14.23) occurred on a short side branch of a major transcurrent fault quite similar tectonically to the San Andreas. A similar side branch of the main San Andreas Fault, called the Hayward Fault, runs through the East Bay region of San Francisco. Seismologists estimate that a slip on the Hayward Fault would generate a quake of magnitude 7.0—about the same as the Kobe quake—and has a 28-percent probability of occurring by the year 2018. Although the nation of Japan prides itself on its earthquake preparedness, the Kobe catastrophe left 5000 dead and 300,000 homeless, and did property damage estimated at 10 times that of the Northridge earthquake of 1994, which was the most costly earthquake in American history.

Along the southern California portion of the San Andreas Fault, a recent estimate placed at about 50 percent the likelihood that a very large earthquake will occur within the next 30 years. In 1992 three severe earthquakes occurred in close succession along active local faults a short distance north of the San Andreas Fault in the southern Mojave Desert (Figure 14.21). The second of these, the Landers earthquake, a powerful 7.5 on the Richter scale, occurred on a transcurrent fault trending north-northwest. It caused a 80-km (50-mi) rupture across the desert landscape. These three events have led to speculation that the likelihood of a major slip on the nearby San Andreas Fault in the near future has substantially increased.

For residents of the Los Angeles area, an additional serious threat lies in the large number of active faults close at hand. Movements on these local faults have produced more than 40 damaging earthquakes since 1800, including the Long Beach earthquakes of the 1930s and the San Fernando earthquake of 1971. The San Fernando earthquake measured 6.6 on the Richter scale and produced severe structural damage near the earthquake center. The brief but intense primary shock and aftershocks that followed damaged beyond repair many older structures built of unreinforced brick masonry. In 1987 an earthquake of magnitude 6.1 struck the vicinity of Pasadena and Whittier, located within about 20 km (12 mi) of downtown Los Angeles. Known as the Whittier Narrows earthquake, it was generated along a local fault system that had not previously shown significant seismic activity. The Northridge earthquake of 1994, described in *Eye on the Environment 14.5 • The Northridge Earthquake,* measured 6.7 on the Richter scale and produced severe structural damage near the earthquake center.

A slip along the San Andreas Fault, some 50 km (31 mi) to the north of the densely populated region of Los Angeles, will release an enormously larger quantity of energy than local earthquakes, such as the Northridge or San Fernando earthquakes. On the other hand, the destructive effects of a San Andreas earthquake in downtown Los Angeles will be somewhat moderated by the greater travel distance. Although the intensity of ground shaking might not be much different from that of the San Fernando earthquake, for example, it will last much longer and cover a much wider area of the Los Angeles region. The potential for damage and loss of life is enormous.

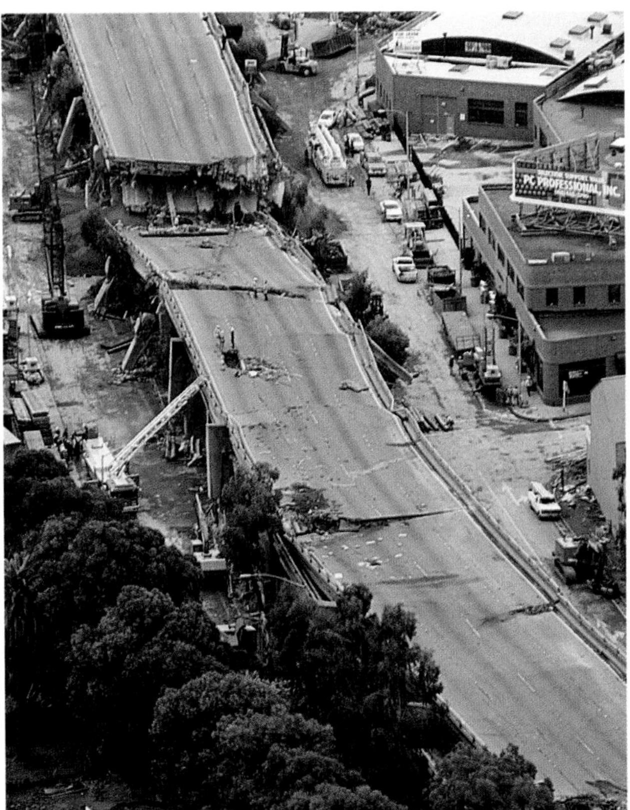

14.22 Earthquake damage in Oakland, California *This section of the double-decked Nimitz Freeway (Interstate 880) in Oakland, California, collapsed during the Loma Prieta earthquake, crushing at least 39 people in their cars.*

14.23 Kobe earthquake *This area within the city of Kobe, Japan, was reduced to rubble by the earthquake of January 1995 and the fires that followed in its aftermath.*

Eye on the Environment | 14.5

The Northridge Earthquake

The January 17, 1994, Northridge earthquake struck a heavily populated suburb of Los Angeles, California, at 4:30 A.M., PST. Assigned a Richter scale rating of 6.7, it produced the strongest ground motions ever recorded in an urban setting in North America and the greatest financial losses from a natural disaster in the United States since the great San Francisco earthquake of 1906.

Northridge lies in an east-west trending fault-rich valley close to the base of the lofty San Gabriel Mountains. This range is an uplifted fault-block that has produced earthquakes at many locations along the base of its steep southern flank. One of these was the equally severe San Fernando earthquake of 1971, centered not far to the north in the same valley. These active earthquake faults belong to a tectonic belt known as the Transverse Ranges, which runs more or less east-west through the Los Angeles Basin and heads out across the Pacific coastline beyond Santa Barbara (see Figure 14.21). Although the San Andreas Fault lies not far distant to the north, these two fault systems have different trends and operate quite independently.

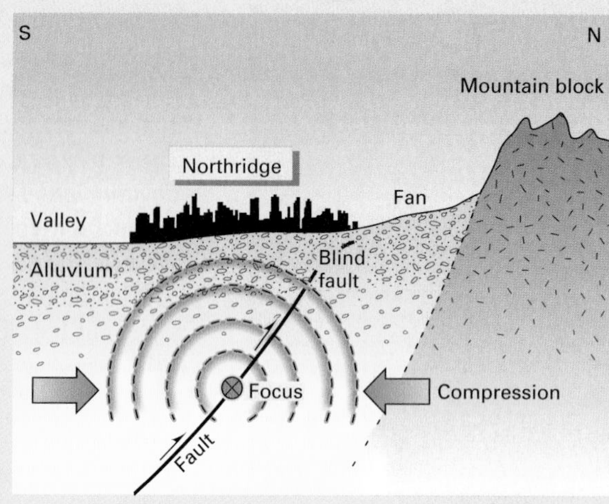

Geologic setting of the Northridge earthquake

The Northridge and San Fernando earthquake faults are of the reverse type (see Figure 14.12*b*). Our sketch shows how it works. Large arrows pointing toward each other indicate that along the Transverse Ranges the Earth's crust is under a strong, north-south, horizontal compressional force. The fault plane slants steeply upward. When the fault suddenly starts to slip at the focus, located about 20 km below the valley surface, the upper block, on the right, delivers a powerful upward punch that sharply lifts the ground surface. As with a boxer's "uppercut," what counts is the acceleration of that first upward slip motion. Seismic waves also spread out radially from the focus, generating strong ground shaking with horizontal motions. Notice that the displacement along the earthquake fault did not extend upward to the ground surface; that is, it is a "blind fault." Nevertheless, disturbances of the ground surface occurred over a wide area.

A LOOK AHEAD In the last three chapters, we have surveyed the composition, structure, geologic activity, and initial landforms of the Earth's crust. We began with a study of the rocks and minerals that make up the Earth's crust and core. We saw how the cycle of rock transformation involves a continuous cycling and recycling of rocks and minerals that has occurred over some 3 billion years or more of geologic time. In the second chapter, we developed plate tectonics as the mechanism powering the rock cycle. The global pattern of plate tectonics also explains the geographical distribution of mountain ranges, ocean basins, and continental shields occurring on the Earth's surface. In this chapter, we described the initial landforms that result directly from volcanic and tectonic activity, occurring primarily at the boundaries of spreading or colliding lithospheric plates.

With this survey of the Earth's crust and the geologic processes that shape it now completed, we can turn to other landform-creating processes in the following chapters. First will be the processes of weathering, which breaks rock into small particles, and mass wasting, which moves them downhill as large and small masses under the influence of gravity. Then we will turn to running water in three chapters that describe the behavior of rivers and streams, their work in shaping landforms, and how running water dissects rock layers to reveal their inner structures. In our two concluding chapters, we will examine landforms created by waves, wind, and glacial ice.

Northridge earthquake damage *This building suffered severe structural damage in the Northridge earthquake.*

Northridge had more severe structural damage than San Fernando, although both were Richter 6.7 energy releases. In the Northridge case, the focus lay directly beneath the city, whereas the focus of the San Fernando earthquake was located some distance back under the mountain block. Northridge casualties included 33 persons dead as a direct result of the earthquake, more than 7000 injuries treated at local hospitals, and over 20,000 people left homeless. Total financial losses were estimated as high as $20 billion. Sections of three freeways were closed, including the busiest highway in the country, I-5.

In the wake of the severe structural damages of the 1971 San Fernando earthquake, a program of strengthening (retrofitting) of buildings, freeways, and dams had been in progress for some two decades throughout the Los Angeles region by the time of the Northridge quake. While the retrofitting may have reduced some forms of damage, it was soon discovered that the long established practice of making welded connections of the steel frames of buildings and freeway bridges failed badly during the quake. The welded joints proved brittle and easy to fracture. In contrast, the few modern buildings designed to move sidewise easily on a fixed base performed well and without damage. The implications of these revelations are obviously of enormous consequence to the residents of the California "Southland." Is it now too late to correct a monstrous design error? There will be many more earthquakes like those of Northridge and San Fernando, occurring unexpectedly at widely separated points within the Transverse Ranges.

Chapter Summary

- **Landforms** are the surface features of the land, and **geomorphology** is the scientific study of landforms. *Initial landforms* are shaped by volcanic and tectonic activity, while *sequential landforms* are sculpted by agents of denudation, including running water, waves, wind, and glacial ice.
- **Volcanoes** are landforms marking the eruption of lava at the Earth's surface. **Stratovolcanoes**, formed by the emission of thick, gassy, felsic lavas, have steep slopes and tend toward explosive eruptions that can form *calderas*. Most active stratovolcanoes lie along the Pacific rim, where subduction of oceanic lithospheric plates is occurring.
- At *hotspots*, rising mantle material provides mafic magma that erupts as basaltic lavas. Because these lavas are more fluid and contain little gas, they form broadly rounded **shield volcanoes**. Hotspots occurring beneath continental crust can also provide vast areas of *flood basalts*. Some basaltic volcanoes occur along the midoceanic ridge.
- The two forms of tectonic activity are compression and extension. Compression occurs at lithospheric plate collisions. At first, the compression produces **folding**—*anti-*

clines (upfolds) and *synclines* (downfolds). If compression continues, folds may be overturned and eventually overthrust faulting can occur.

- Extension occurs where lithospheric plates are spreading apart, generating **normal faults**. These can produce upthrown and downdropped blocks that are sometimes as large as mountain ranges or *rift valleys*. **Transcurrent faults** occur where two rock masses move horizontally past each other.
- **Earthquakes** occur when rock layers, bent by tectonic activity, suddenly fracture and move. The sudden motion at the fault produces earthquake waves that shake and move the ground surface in the adjacent region. The energy released by an earthquake is measured by the *Richter scale*. Large earthquakes occurring near developed areas can cause great damage. Most severe earthquakes occur near plate collision boundaries.
- The San Andreas Fault is a major transcurrent fault located near two great urban areas—Los Angeles and San Francisco. The potential for a severe earthquake on this fault is high, and the probability of a major Earth movement increases every year.

Key Terms

landform	volcano	folding	transcurrent fault
geomorphology	stratovolcano	fault	earthquake
denudation	shield volcano	normal fault	tsunami

Review Questions

1. Distinguish between initial and sequential landforms. How do they represent the balance of power between internal Earth forces and external forces of denudation agents?
2. What is a stratovolcano? What is its characteristic shape, and why does that shape occur? Where do stratovolcanoes generally occur and why?
3. What is a shield volcano? How is it distinguished from a stratovolcano? Where are shield volcanoes found, and why? Give an example of a shield volcano. How are flood basalts related to shield volcanoes?
4. How can volcanic eruptions become natural disasters? Be specific about the types of volcanic events that can devastate habitations and extinguish nearby populations.
5. Briefly describe the Rift Valley system of East Africa as an example of normal faulting.
6. How does a transcurrent fault differ from a normal fault? What landforms are expected along a transcurrent fault? How are transcurrent faults related to plate tectonic movements?
7. What is an earthquake, and how does it arise? How are the locations of earthquakes related to plate tectonics?
8. Describe the tsunami, including its origin and effects.
9. Briefly summarize the geography and recent history of the San Andreas Fault system in California. What are the prospects for future earthquakes along the San Andreas Fault?

Focus on Systems 14.1 • The Life Cycle of a Hotspot Volcano

1. How does a life cycle contrast with a repeating time cycle?
2. Describe the stages in the life cycle of a basaltic shield volcano of the Hawaiian type.
3. What is a hotspot? What produces it? How is it related to the life cycle?

Eye on the Environment 14.2 • Geothermal Energy Sources

1. What is the ultimate source of geothermal power? Where would you go, and why, to find a geothermal power source?
2. How is geothermal energy extracted? What environmental concerns arise in this process?

Eye on the Environment 14.5 • The Northridge Earthquake

1. Explain the geologic setting of the Northridge Earthquake of 1994. What type of fault was involved? Where was it located? What type of motion occurred?
2. What discovery was made concerning construction practices as a result of the Northridge quake?

Visualizing Exercises

1. Sketch a cross section through a normal fault, labeling the fault plane, upthrown side, downthrown side, and fault scarp.
2. Sketch a cross section through a foreland fold belt showing rock layers in different colors or patterns. Label anticlines and synclines.

Essay Questions

1. Write a fictional news account of a volcanic eruption. Select a type of volcano—composite or shield—and a plausible location. Describe the eruption and its effects as it was witnessed by observers. Make up any details you need, but be sure they are scientifically correct.
2. How are mountains formed? Provide the plate tectonic setting for mountain formation, and then describe how specific types of mountain landforms arise.

Problems

Working It Out 14.4 • The Richter Scale

1. After measuring the amplitude and distance of an earthquake on a seismogram, a geophysicist determines that it rates 5.2 on the Richter scale. How much energy will the earthquake release?
2. Suppose that the energy of a magnitude 6.0 earthquake is released evenly over a 60-second period. Recalling that one watt (w) is equal to a joule per second (J/sec), how many 100-watt light bulbs would be powered by this rate of energy flow during that minute? The average U.S. electric power consumption rate is about 3×10^{12} W. What fraction of the U.S. power consumption could be sustained by this energy flow rate?
3. After recalculation, a geophysicist revises the magnitude of an earthquake upward from 7.1 to 7.3 on the Richter scale. By what proportion has the estimate of the energy released changed?

Eye on the Landscape

Chapter Opener Mutnovsky Volcano, Kamchatka Peninsula The structure of this volcanic cone is shown clearly at the edge of the crater **(A)**. The cone is built of layers of ash and lava, created by successive eruptions, which appear here as beds in the crater wall. Note also the volcanic bombs, or large rocks flung into the air, which are strewn across the slopes of the cone **(B)**. Volcanic activity is the subject of this chapter. From the liquid state of the crater lake **(C)** and patchy condition of the snow, this must be high summer. Kamchatka stretches from 50° to 60° N latitude and spans the moist continental (10) and boreal forest climates (11). These climates were covered in Chapter 11.

14.6 Flood basalts The Columbia River and its tributaries drain a large mountainous area reaching as far east as the Bitterroot Mountains of Idaho and as far north as Jasper National Park in Alberta. The river has been extensively developed for hydropower generation with the building of many large dams. Here we see the still waters **(A)** of a reservoir and a narrow lake shore **(B)** at the water's edge. Note also the sparse vegetation **(C)** characteristic of this location in the rain shadow of the Cascades. Lakes and dams are discussed in Chapters 16 and 17. Rain shadow effects on precipitation were covered in Chapter 6.

14.18 Ethiopian Rift Valley The local agriculture exploits the stepped landscape at the edge of the Rift Valley with cultivated fields **(A)** laid out on the flat tops of terraces. Given the mix of fallow and still-green fields, this photo is probably from the end of the growing season. Note also the flat faces of the slopes on the fault scarps between gullies **(B)**. These are likely to be remnants of the original planes of motion, given the recent geologic age of the Rift Valley. Chapter 24 shows the natural vegetation of the Ethiopean highlands as rainforest; precipitation is over 100 cm (40 in.) per year (Chapter 10). Fault planes and scarps are covered in this chapter.

Chapters in Part 4

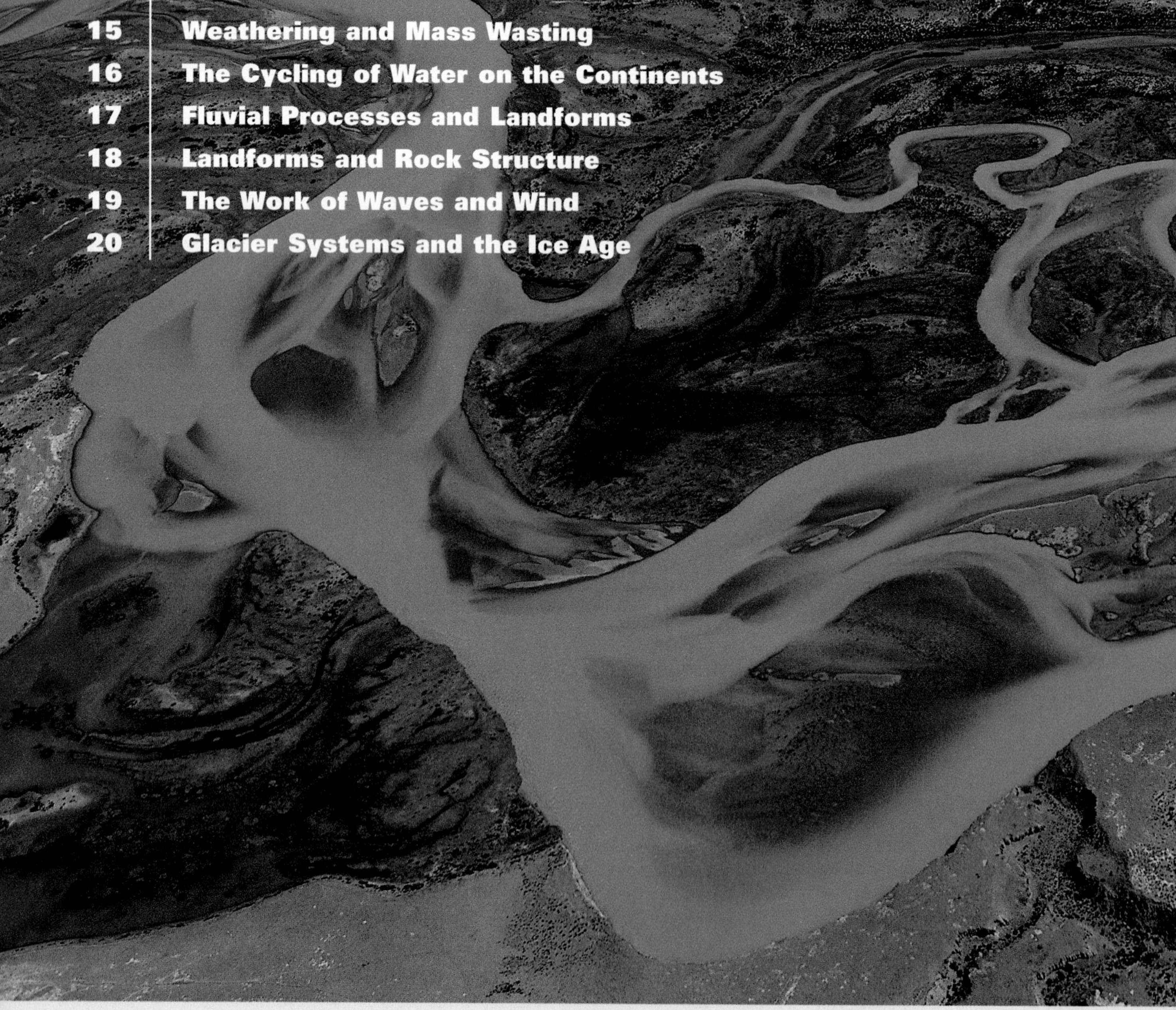

Part 4 | Systems of Landform Evolution

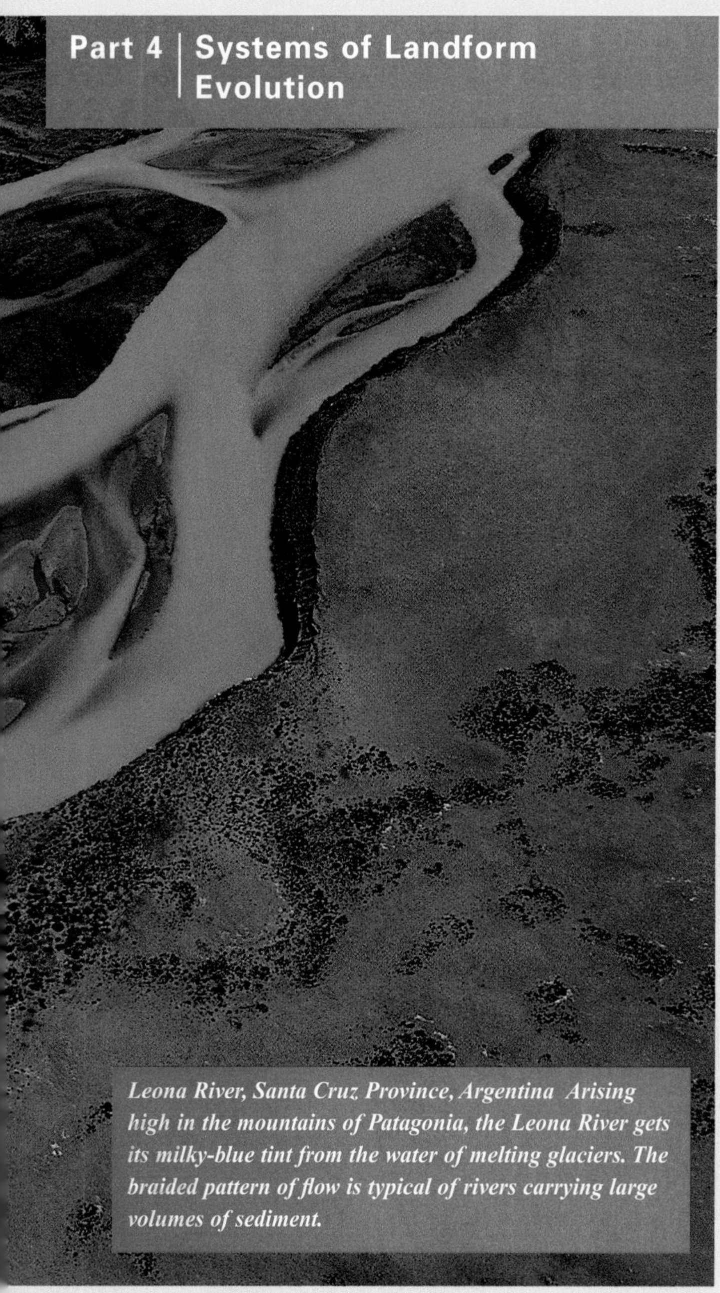

Leona River, Santa Cruz Province, Argentina Arising high in the mountains of Patagonia, the Leona River gets its milky-blue tint from the water of melting glaciers. The braided pattern of flow is typical of rivers carrying large volumes of sediment.

Physical geography is focused on the life layer—the zone of interactions among the atmosphere, hydrosphere, lithosphere, and biosphere in which we live, breathe, and carry out our daily lives. After examining the matter and energy flow systems of the solid Earth in Part 3, we now return our focus to the life layer to examine the matter and energy flow systems that shape the surface of the land. These systems operate largely on time scales intermediate between those of the atmosphere and solid Earth—on the order of hundreds to thousands to a few millions of years. Systems of landform evolution are mostly powered by gravity—the constant attraction of the Earth's mass for solids (rock particles and soil) and fluids (water and glacial ice) that moves them downhill, creating landforms in the process. But gravity only releases potential energy that has been stored in positioning these solids and fluids above a base level, such as the ocean. The potential energy is ultimately derived from two sources—solar energy, which places liquid and solid water atop the continents through precipitation, and the Earth's internal heat, which generates the Earth forces that uplift masses of rock and soil above base level. ■ Chapter 15 begins Part 4 by examining the processes of weathering that break up rock material in place, then turns to movements of masses of weathered material under gravity—landslides and earthflows, for example. Chapter 16 focuses on water at or near the land surface, in lakes, rivers, and as ground water. It is followed by Chapter 17, which moves from the study of water as surface water and ground water to consider its role as a landform-creating agent of erosion and deposition. In Chapter 18 we show how fluvial action interacts with rock structures—folds, faults and the like—to produce distinctive configurations of landforms related to those structures. Chapter 19 turns to two other landform-creating agents powered not by gravity, but by atmospheric motion—waves and wind. Chapter 20 concludes Part 4 by examining landforms created by glacial ice, both as high mountain glaciers and as vast continental ice sheets. In Part 5, we will complete our study of physical geography by focusing on the life layer from the viewpoint of the biosphere.

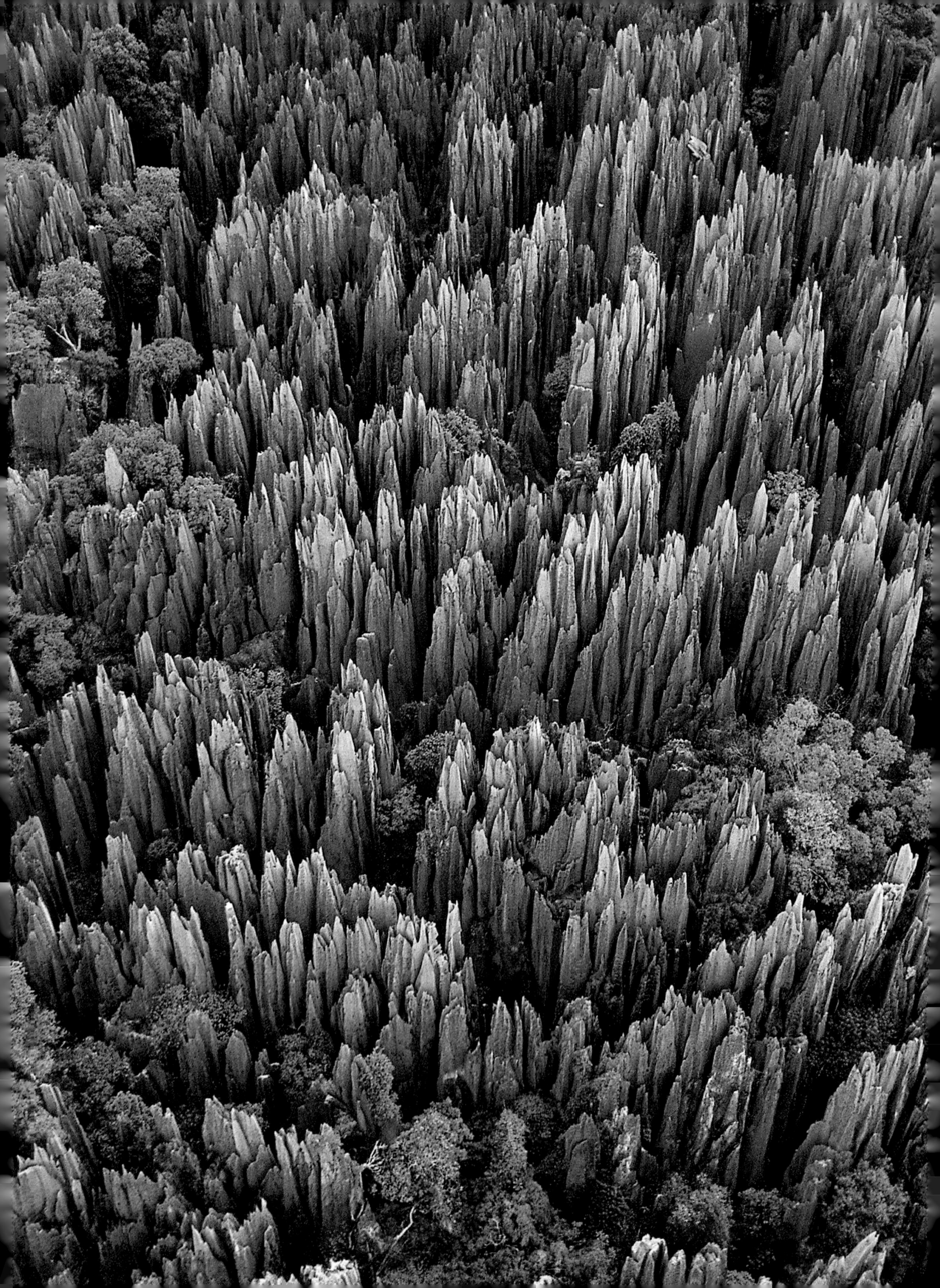

15 WEATHERING AND MASS WASTING

Tsingy of Bemaraha, Morondava region, Madagascar. These sharp pinnacles, some as tall as 30 m (95 ft), are formed by solution of pure limestone exposed to the natural carbonic acid in rainwater.

Now that we have completed our study of the Earth's crust—including its mineral composition, its moving lithospheric plates, and its tectonic and volcanic landforms—we can focus on the shallow life layer itself. At this sensitive interface, the external processes of denudation carve sequential landforms from the rocks uplifted by the Earth's internal processes. Our investigation of what happens to rock upon exposure at the surface began in Chapter 12 with a description of mineral alteration of rock and the production of sediment, which is an essential part of the cycle of rock change. In this chapter, we look further at the softening and breakup of rock, termed weathering, and how the resulting particles move downhill under the force of gravity, a process termed *mass wasting.*

Weathering is the general term applied to the combined action of all processes that cause rock to disintegrate physically and decompose chemically because of exposure near the Earth's surface. We can recognize two types of weathering. In *physical weathering,* rocks are fractured and broken apart, primarily by the growth of ice or salt crystals along rock planes and mineral contacts that are penetrated by watery solutions. In *chemical weathering,* rock minerals are transformed from types that were stable when the rocks were formed to types that are now stable at surface temperatures and pressures. As we saw in Chapter 12, these chemical processes include oxidation, hydrolysis, and acid solution. Weathering acts to produce **regolith**—a surface layer of weathered rock particles that lies above solid, unaltered rock. Weathering also leads to a number of distinctive landforms, as we will see in this chapter.

This chapter also examines how gravity acts on rock fragments, created by weathering, to produce landforms. In this process, gravity induces the spontaneous downhill movement of soil, regolith, and rock fragments, but without the action of moving water, air, or ice. The downhill movement is referred to as **mass wasting.** Movement of a mass of soil or weathered rock to lower levels takes place when the internal strength of the soil or weathered rock declines to a critical point below which the force of gravity cannot be resisted. This failure of strength under the ever-present force of gravity takes many forms and scales, and we will see that human activity causes or aggravates several forms of mass wasting.

We conclude our chapter with a look at a suite of special landforms and geomorphic processes that are found in the arctic lands. They are created primarily by the freezing and thawing of water, acting in concert with gravity.

15.1 Bedrock disintegration *Joint-block separation and granular disintegration are two common forms of bedrock disintegration. (Drawn by A. N. Strahler.)*

PHYSICAL WEATHERING

Physical weathering produces regolith from massive rock by the action of forces strong enough to fracture the rock. The physical weathering processes discussed in this chapter include frost action, salt-crystal growth, unloading, and wedging by plant roots. The term *mechanical weathering* is also used for physical weathering.

Frost Action

One of the most important physical weathering processes in cold climates is *frost action,* the repeated growth and melting of ice crystals in the pore spaces of soil and in rock fractures. In contrast to most other liquids, water expands when it freezes. Perhaps you've left a bottle of water chilling in the freezer overnight and observed this phenomenon first hand, removing a mass of ice surrounded by broken glass the next morning. The expansion of water in freezing can fragment even extremely hard rocks, given many cycles of freeze and thaw. Frost action and ice crystal growth produce a number of conspicuous effects and forms in all climates that have cold winters. Features caused by frost action and the buildup of ice below the surface are particularly visible in the tundra climate of arctic coasts and islands, and above the timberline in high mountains.

Almost everywhere, bedrock is cut through by systems of fractures called *joints.* These fractures are thought to occur as rocks exposed to heat and pressure cool and contract. Joints typically occur in parallel and intersecting planes, creating surfaces of weakness along which weathering can act to break rock into individual blocks. Since there is no relative movement of rock along joints, they cannot be considered to be faults. Joints are important to the physical weathering of rocks because they admit water to the rock. As we will see shortly, this allows ice and salt crystal growth to further fracture the rock, creating rock fragments and regolith.

In sedimentary rocks, the planes of stratification, or bedding planes, comprise another natural set of planes along which water can penetrate. Often these are cut at right angles by sets of joints. Comparatively weak stresses can separate joint blocks, while strong stresses are required to make fresh fractures through solid rock. The process of separating rock along joints and bedding planes can be called *block separation* (Figure 15.1).

As coarse-grained igneous rock becomes weakened by chemical decomposition, water is able to penetrate the contact surfaces between mineral grains. In this location, the water can freeze and exert forces strong enough to separate the grains. This form of breakup is termed *granular disintegration* (Figure 15.1). The end product is a fine gravel or coarse sand in which each grain consists of a single mineral particle separated from the others along its original crystal or grain boundaries.

The effects of frost action can be seen in all climates that have a winter season with many alternations of freeze and thaw. Where bedrock is exposed on knolls and mountain summits, joint blocks are pried apart by water that freezes in joint cracks (Figure 15.2a). Under the most favorable conditions, seen on high mountain summits and in the arctic tundra, large angular rock fragments accumulate in a layer that completely blankets the bedrock beneath. The German name *felsenmeer* ("rock sea") has been given to such expanses of broken rock (Figure 15.2b).

In high mountains, frost action on cliffs of bare rocks detaches rock fragments that fall to the cliff base. These loose fragments are known as *talus.* Where production of fragments is rapid, they accumulate to form *talus slopes.* Most cliffs are notched by narrow ravines that funnel the rock fragments into separate tracks. Each track, or chute, feeds a growing, cone-like talus body. Talus cones are arranged side by side along the base of the cliff (Figure 15.2c). Fresh talus slopes are unstable, so that the disturbance created by walking across the slope or dropping a large rock fragment from the cliff above will easily set off a sliding or rolling motion within the surface layer of fragments.

In fine-textured soils and sediments, composed largely of silt and clay, soil water freezes in horizontal layers or lens-shaped bodies. As these ice layers thicken, the overlying soil layer is heaved upward. Prolonged frost heaving can produce minor irregularities and small mounds on the soil surface. A rock fragment at the surface can sometimes conduct heat from the soil to the cold night air and sky in such a way as to cause perpendicular ice needles to grow beneath the fragment and lift it above the surface (Figure 15.2d). The same process acting on a rock fragment below the soil surface can eventually push the fragment to the surface.

Frost action is a dominant process in arctic and high-mountain tundra environments, where it is a factor in the formation of a wide variety of unique landforms. We investigate these landforms in the last section of this chapter.

Salt-Crystal Growth

Closely related to the growth of ice crystals is the weathering process of rock disintegration by growth of salt crystals in rock pores. This process, *salt-crystal growth,* operates extensively in dry climates and is responsible for many of the niches, shallow caves, rock arches, and pits seen in sandstone formations. During long drought periods, ground water is moved to the surface of the rock by *capillary action.* In this process, water is drawn into fine openings and passages in the rock by the same surface tension that gives a water droplet its rounded shape. As water evaporates from this porous outer zone of the sandstone, tiny crystals of salts such as halite (sodium chloride), calcite (calcium carbonate), or gypsum (calcium sulfate) are left behind. Over time, the growth force of these crystals produces grain-by-grain breakup of the sandstone, which crumbles into a sand and is swept away by wind and rain.

Zones of rock lying close to the base of a cliff are especially susceptible to breakup by salt-crystal growth because at that location the ground water seeps downward and outward to reach the rock surface (Figure 15.3). In the southwestern United States, many of the deep niches or cave-like recesses formed in this way were occupied by Native Americans. Their cliff dwellings gave them protection from the elements and safety from armed attack (Figure 15.4).

Salt crystallization also damages masonry buildings, as well as concrete sidewalks and streets. Brick and concrete in contact with moist soil are highly susceptible to grain-by-grain disintegration from salt crystallization. On damp basement floors and walls, the salt crystals can be seen as a soft, white, fibrous layer. The deicing salts spread on streets and highways can be quite destructive. Sodium chloride (rock salt), widely used for this purpose, is particularly damaging to concrete pavements and walks, curbstones, and other exposed masonry structures.

Salt-crystal growth occurs naturally in arid and semiarid regions. In humid climates, abundant rainfall dissolves salts and carries them downward to ground water. (Chapter 16 describes how rainwater infiltrates soils and moves to ground water below.)

Unloading

A widespread process of rock disruption related to physical weathering results from *unloading,* a process that relieves the confining pressure on underlying rock. *Exfoliation* is another term used for unloading. Unloading occurs as rock is brought near the surface by erosion of overlying layers. Rock formed at great depth beneath the Earth's surface (particularly igneous and metamorphic rock) is in a slightly compressed state because of the confining pressure of overlying rock. As the rock above is slowly worn away, the pressure is reduced, and the rock expands slightly in volume. This causes the rock to crack in layers that are more or less parallel to the surface, creating a type of jointing called *sheeting structure.* In massive rocks like granite or marble, thick curved layers or shells of rock break free in succession from the parent mass below, much as you might peel the layers of an onion.

Where sheeting structure has formed over the top of a single large knob or hill of massive rock, an *exfoliation dome* is produced (Figure 15.5). Domes are among the largest of the landforms shaped primarily by weathering. In Yosemite Valley, California, where domes are spectacularly displayed, the individual rock sheets may be as thick as 15 m (50 ft).

Other Physical Weathering Processes

Most rock-forming minerals expand when heated and contract when cooled. Where rock surfaces are exposed daily to the intense heating of the Sun alternating with nightly cooling, the resulting expansion and contraction exert powerful disruptive forces on the rock. Although first-hand evidence is lacking, it seems likely that daily temperature changes can cause the breakup of a surface layer of rock already weakened by other agents of weathering.

Another mechanism of rock breakup is the growth of plant roots, which can wedge joint blocks apart. You have probably observed concrete sidewalk blocks uplifted and fractured by the growth of tree roots. This process is also active when roots grow between rock layers or joint blocks. Even fine rootlets in joint fractures can cause the loosening of small rock fragments and grains.

Chemical Weathering and Its Landforms

We investigated **chemical weathering** processes in Chapter 12 under the heading of mineral alteration. Recall that the dominant processes of chemical change affecting silicate minerals are oxidation, hydrolysis, and carbonic acid action. Oxidation and hydrolysis change the chemical structure of minerals, turning them into new minerals that are typically softer and bulkier and therefore more susceptible to erosion and mass movement. Carbonic acid action dissolves minerals, washing them away in runoff. Note also that chemical reactions proceed more rapidly at warmer temperatures. Thus, chemical weathering is most effective in warm, moist climates.

15.2 Frost Action

a...Frost-shattered blocks *Ice crystal growth within the joint planes of rock can cause the rock to split apart. This split boulder, on the high country of the Sierra Nevada of California, is an example.*

b...Quartzite blocks *These frost-shattered blocks of quartzite are on the summit of the Snowy Range, Wyoming, at an elevation of about 3700 m (about 12,000 ft). (A. N. Strahler.)*

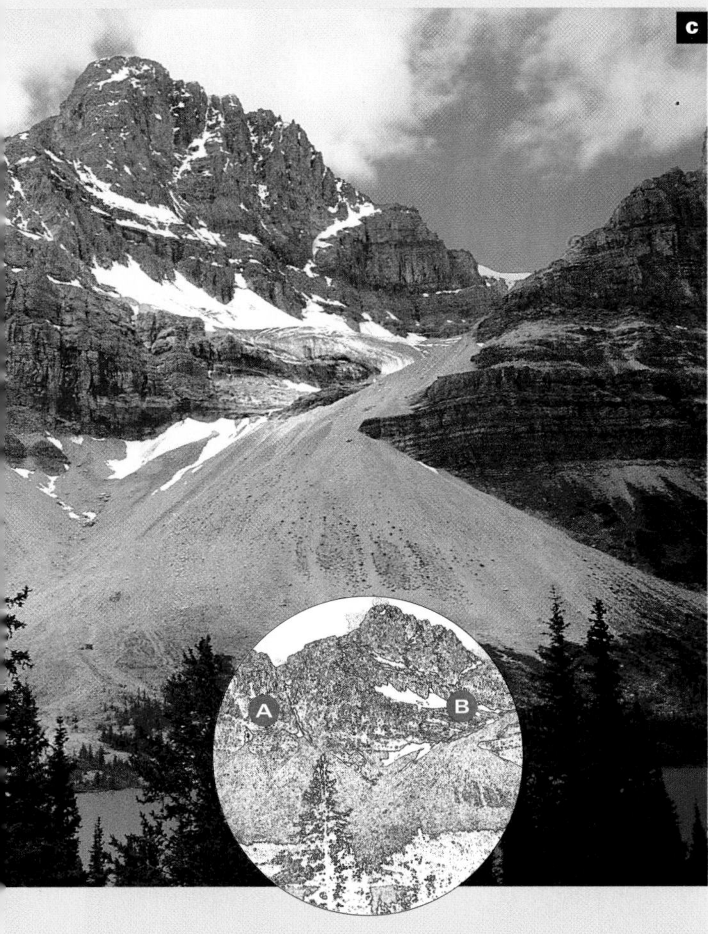

c...*Eye on the Landscape* ***Talus cones*** *Frost action has caused these cliffs to shed angular blocks of rock that accumulate in talus cones. In the foreground is the shore of Lake Louise in the Canadian Rockies.* **What else would the geographer see?...Answers at the end of the chapter**

d...***Needle ice growth*** *At night, water in the soil freezes at the surface, creating ice needles that can lift particles of soil or move larger stones.*

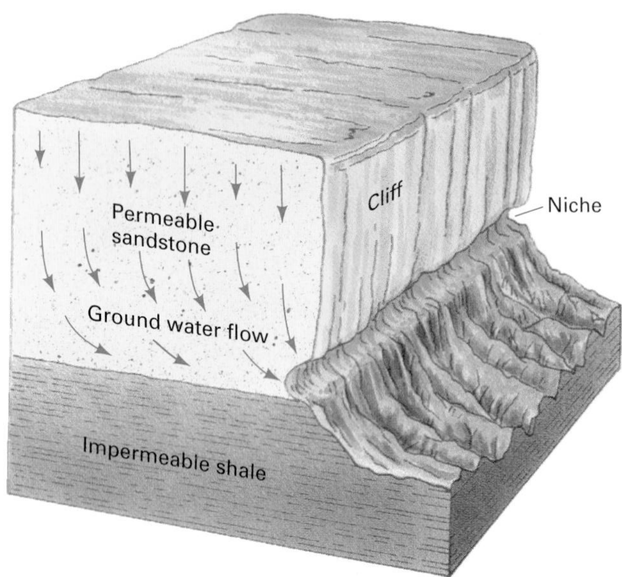

15.3 Niche formation *In dry climates there is a slow seepage of water from the cliff base. Salt crystal growth separates the grains of permeable sandstone, breaking them loose and creating a niche.*

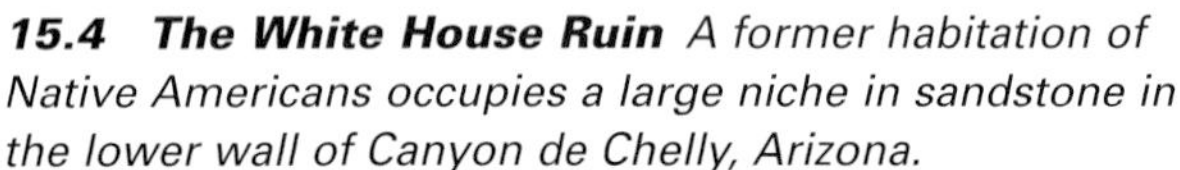

15.4 The White House Ruin *A former habitation of Native Americans occupies a large niche in sandstone in the lower wall of Canyon de Chelly, Arizona.*

15.5 Exfoliation domes *The sheeting structure of these exfoliation domes is visible in the lower part of the photo, where successive shells of rock have fallen away. North Dome and Royal Arches, Yosemite National Park, California.*

Hydrolysis and Oxidation

Decomposition by hydrolysis and oxidation changes the minerals of strong rock into weaker forms that are rich in clay minerals and oxides. In warm, humid climates of the equatorial, tropical, and subtropical zones, hydrolysis and oxidation over thousands of years have resulted in the decay of igneous and metamorphic rocks to depths as great as 100 m (about 300 ft). The decayed rock material is soft, clay-rich, and easily eroded. To the construction engineer, deeply weathered rock is of major concern in the building of highways, dams, or other heavy structures. Although the weathered rock is soft and easy to move, its high clay content reduces its strength, and foundations built on the weathered rock can fail under heavy loads.

In dry climates, exposed granite weathers by hydrolysis to produce many interesting boulder and pinnacle forms (Figure

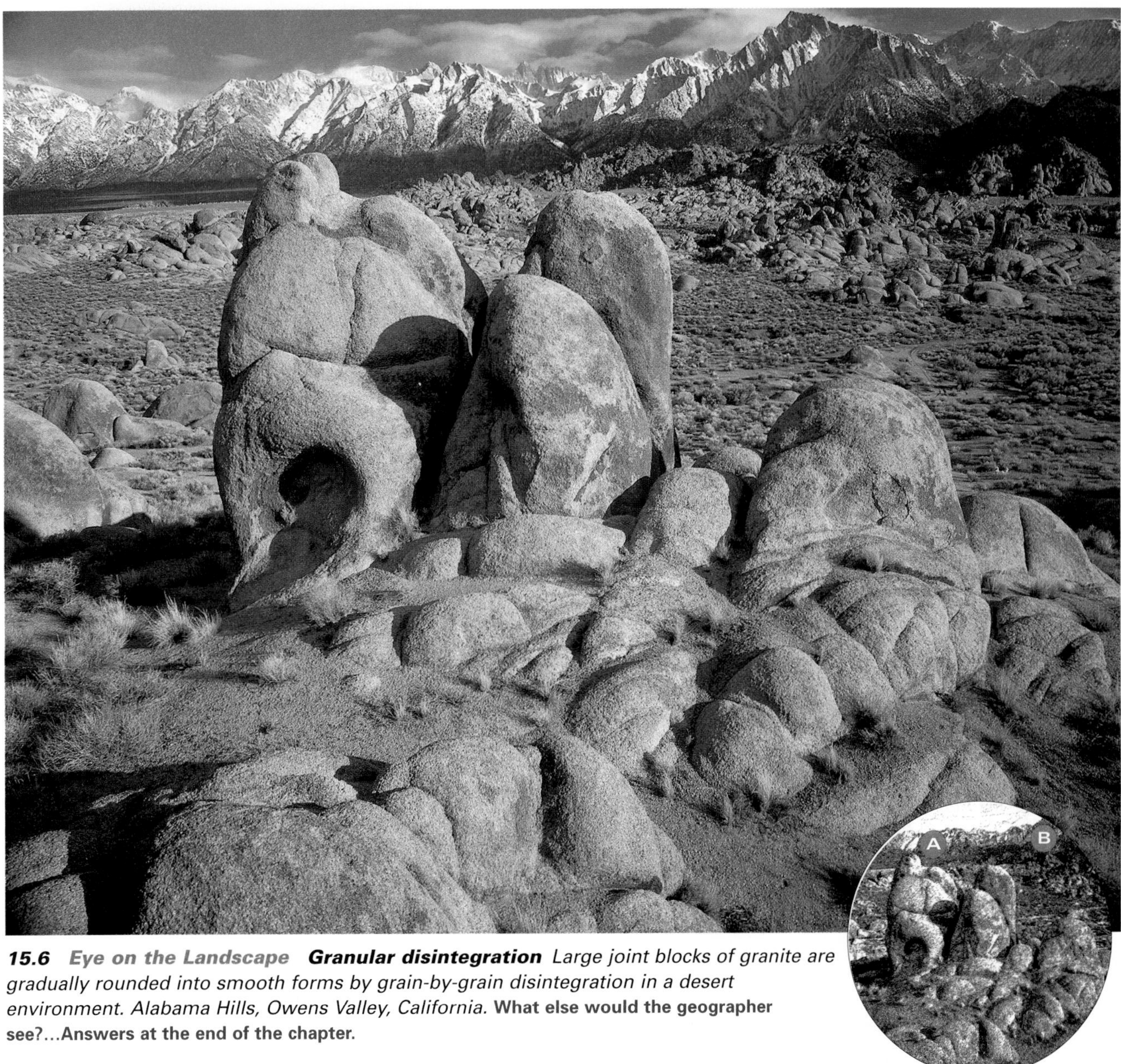

15.6 *Eye on the Landscape* ***Granular disintegration*** *Large joint blocks of granite are gradually rounded into smooth forms by grain-by-grain disintegration in a desert environment. Alabama Hills, Owens Valley, California.* **What else would the geographer see?...Answers at the end of the chapter.**

15.6). Although rainfall is infrequent, water penetrates the granite along planes between crystals of quartz and feldspar. Chemical weathering of these surfaces then breaks individual crystal grains away from the main mass of rock, leaving the rounded forms shown in the photo. The grain-by-grain breakup forms a fine desert gravel consisting largely of quartz and partially decomposed feldspar crystals.

Acid Action

Chemical weathering is also produced by *acid action,* largely that of *carbonic acid.* This weak acid is formed when carbon dioxide dissolves in water. Rainwater, soil water, and stream water all normally contain dissolved carbon dioxide. Carbonic acid slowly dissolves some types of minerals. Carbonate sedimentary rocks, such as limestone and marble, are particularly susceptible to the acid action. In this process, the mineral calcium carbonate is dissolved and carried away in solution in stream water.

Carbonic acid reaction with limestone produces many interesting surface forms, mostly of small dimensions. Outcrops of limestone typically show cupping (formation of rounded cavities), rilling (formation of surface valleys), grooving, and fluting in intricate designs (Figure 15.7). In a few places, the scale of deep grooves and high wall-like rock fins reaches proportions that keep people and animals from passing through. Carbonic acid in ground water can dissolve limestone to produce underground caverns as well as distinctive landscapes that form when underground caverns collapse. These landforms and landscapes will be described in Chapter 16.

In urban areas, air is commonly polluted by sulfur and nitrogen oxides. When these gases dissolve in rainwater, the result is acid precipitation (see Chapter 6). The acids

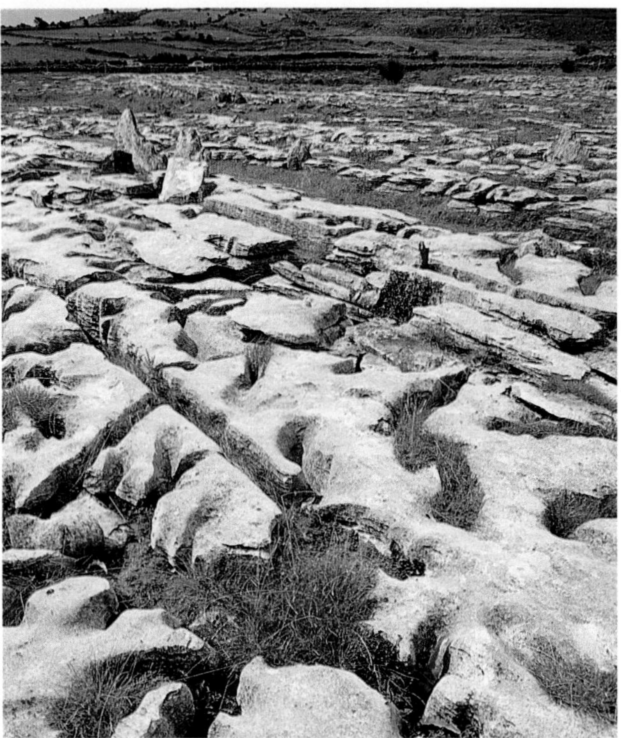

15.7 Solution features in limestone *This outcrop of pure limestone shows grooves and cavities formed by carbonic acid action. County Clare, Ireland.*

rapidly dissolve limestone and chemically weather other types of building stones. The result can be very damaging to stone sculptures, building decorations, and tombstones (Figure 15.8).

In the wet low-latitude climates, mafic rock, particularly basaltic lava, dissolves rapidly under attack by soil acids. The effects of solution removal of basaltic lava are displayed in spectacular grooves, fins, and spires on the walls of deep alcoves in part of the Hawaiian Islands (Figure 15.9). The landforms produced are quite similar to those formed by carbonic acid action on massive limestones in the moist climates of the midlatitudes.

Mass Wasting

With our discussion of weathering, we have described an array of processes that act to alter rock chemically and break up rock into fragments. These rock fragments are subjected to gravity, running water, waves, wind, and the flow of glacial ice in landform-making processes. In the remainder of this chapter, we consider the first of these landform agents—gravity. We will return to the others in the following chapters.

Everywhere on the Earth's surface, gravity pulls continuously downward on all materials. Bedrock is usually so strong and well supported that it remains fixed in place. However, when a mountain slope becomes too steep,

15.8 Chemical weathering of tombstones *These weathered tombstones are from a burying ground in Boston, Massachusetts. The marker on the left, carved in marble, has been strongly weathered, weakening the lettering. The marker on the right, made of slate, is much more resistant to erosion.*

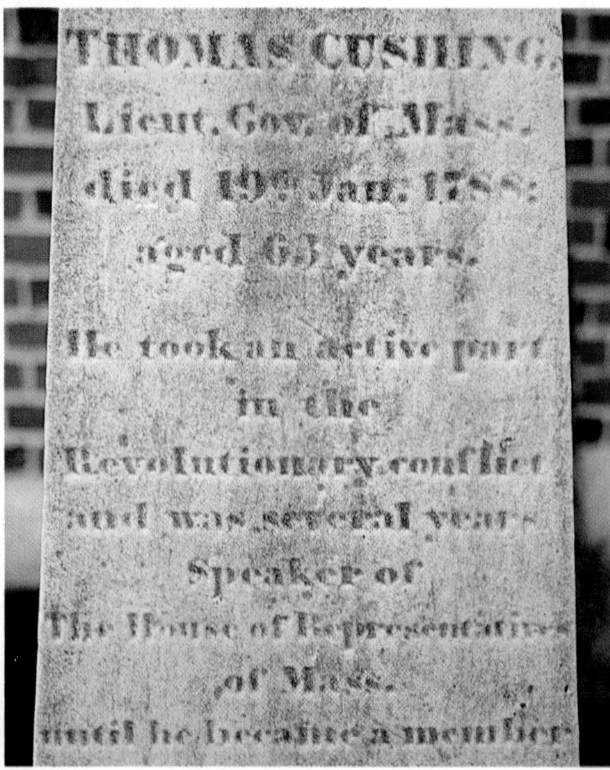

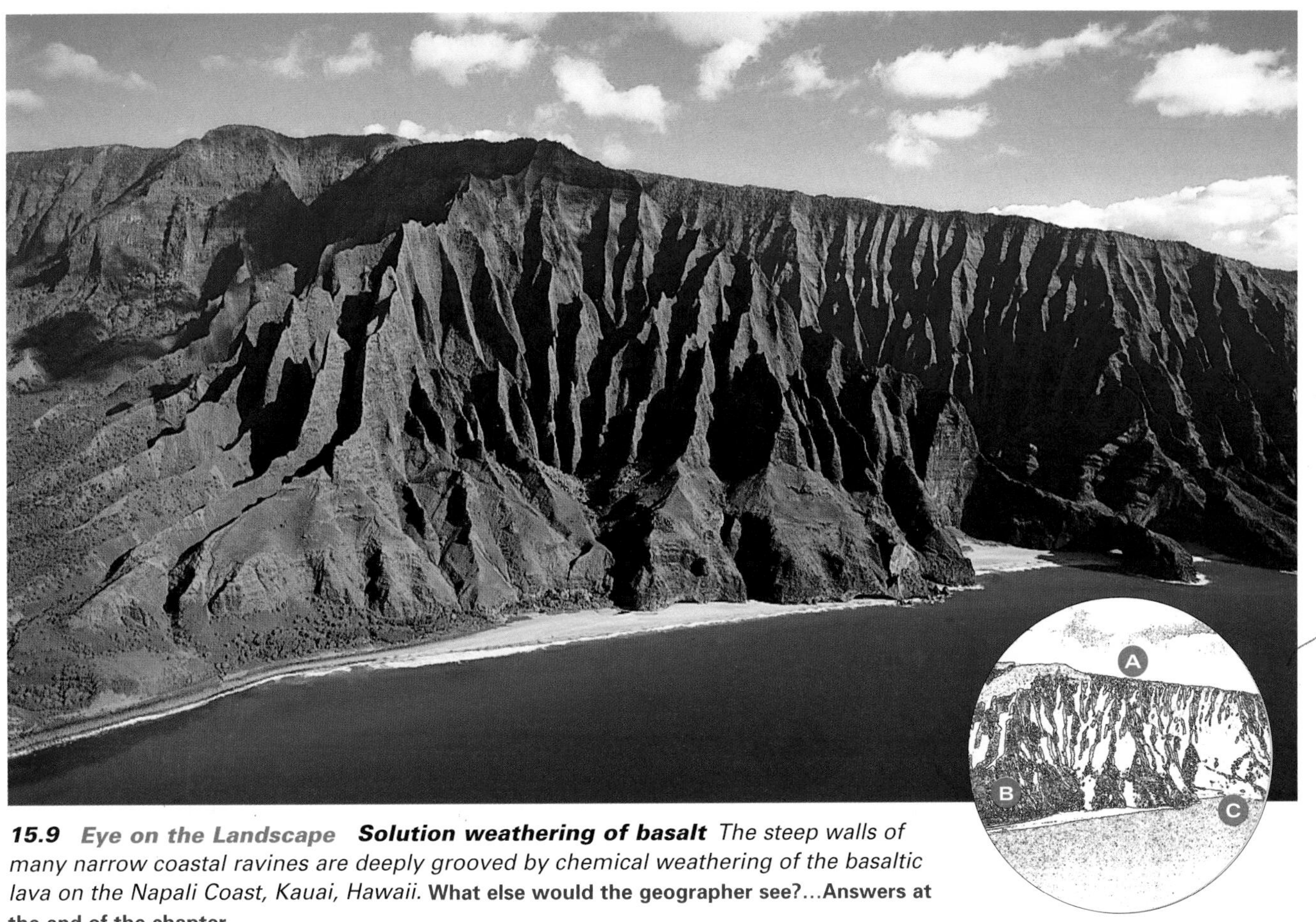

15.9 *Eye on the Landscape* ***Solution weathering of basalt*** *The steep walls of many narrow coastal ravines are deeply grooved by chemical weathering of the basaltic lava on the Napali Coast, Kauai, Hawaii.* **What else would the geographer see?...Answers at the end of the chapter.**

bedrock masses can break free and fall or slide to new positions of rest. In cases where huge masses of bedrock are involved, the result can be catastrophic to towns and villages in the path of the slide. Such slides are a major environmental hazard in mountainous regions. They are but one form of *mass wasting,* which we defined earlier as spontaneous downhill movement of soil, regolith, and rock under the influence of gravity.

Because soil, regolith, and many forms of sediment are held together poorly, they are much more susceptible to movement under the force of gravity than hard, massive bedrock. On most slopes, at least a small amount of downhill movement is going on constantly. Although much of this motion is imperceptible, the regolith sometimes slides or flows rapidly.

The processes of mass wasting and the landforms they produce are extremely varied and tend to grade one into another. We have selected only a few of the most important forms of mass wasting to present in this chapter—they are shown in Figure 15.10. At the top of the diagram, three categories of information are provided, including the Earth material involved, the properties of the material, and the kind of motion that occurs. Boxes in the body of the diagram contain the names of several forms of mass move-

15.10 Processes and forms of mass wasting

Kinds of earth materials:	Rock (dry)	Regolith, soil, alluvium, clays + water	Water + sediment
Physical properties:	Hard, brittle, solid	Plastic substance	Fluid
Kinds of motion:	Falling, rolling, sliding	Flowage within the mass	Fluid flow

Forms of mass movement: Very slow → Very fast

ROCK CREEP
TALUS CREEP
SOIL CREEP
SOLIFLUCTION
LANDSLIDES:
BEDROCK SLUMP
ROCKSLIDE
EARTHFLOW (slump or flowage)
MUDFLOW
STREAM FLOW
ROCKFALL
ALPINE DEBRIS AVALANCHE
DEBRIS FLOOD

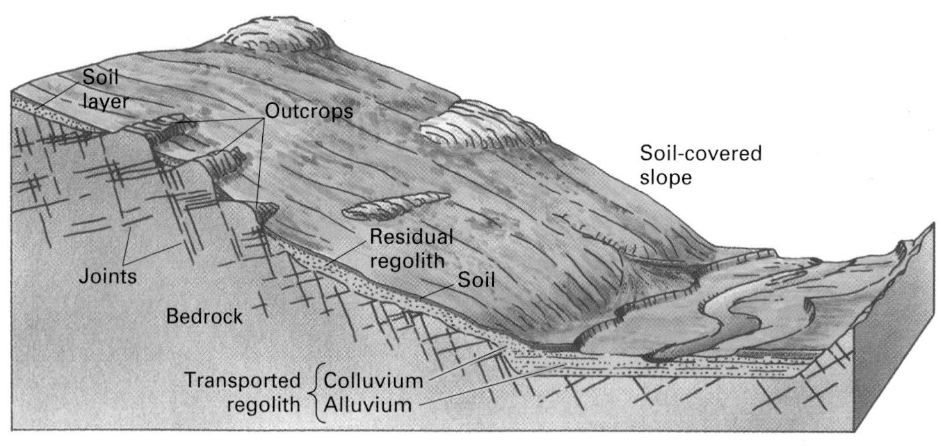

15.11 Soil, regolith, and outcrops on a hillslope *Colluvium accumulates at the foot of the slope, while alluvium, or transported regolith, lies in the floor of an adjacent stream valley. (Drawn by A. N. Strahler.)*

ments, arranged according to composition (left to right) and speed (top to bottom). As we describe each of these forms, this diagram will help you understand how it compares to the others.

Slopes

Mass wasting occurs on slopes. As used in physical geography, the term *slope* designates a small strip or patch of the land surface that is inclined from the horizontal. Thus, we speak of "mountain slopes," "hill slopes," or "valley-side slopes" to describe some of the inclined ground surfaces that we might encounter in traveling across a landscape. Slopes guide the downhill flow of surface water and fit together to form drainage systems within which surface-water flow converges into stream channels (Chapter 17). Nearly all natural surfaces slope to some degree. Very few are perfectly horizontal or vertical.

Most slopes are mantled with regolith, which grades downward into solid, unaltered rock, known simply as **bedrock.** Regolith provides the source for **sediment,** which consists of rock and mineral particles that are transported and deposited in a fluid medium. This fluid may be water, air, or even glacial ice. Both regolith and sediment comprise parent materials for the formation of *soil,* which is a surface layer of mineral and organic matter capable of supporting the growth of plants. (We'll return to soil in Chapter 21.)

Figure 15.11 shows a typical hill slope that forms one wall of the valley of a small stream. Soil and regolith blanket the bedrock, except in a few places where the bedrock is particularly hard and projects in the form of *outcrops. Residual regolith* is derived directly from the rock beneath and moves very slowly down the slope toward the stream. Accumulations of regolith at the foot of a slope are called *colluvium.* Beneath the valley bottom are layers of transported regolith, called **alluvium,** which is sediment transported and deposited by the stream. This sediment had its source in regolith prepared on hill slopes many kilometers or miles upstream. All accumulations of sediment on the land surface, whether deposited by streams, waves and currents, wind, or glacial ice, can be designated *transported regolith,* in contrast to residual regolith.

The thickness of soil and regolith is quite variable. Although the soil is rarely more than 1 or 2 m thick (about 3 to 6 ft), residual regolith on decayed and fragmented rock may extend down to depths of 5 to 100 m (about 15 to 300 ft), or more at some locations. On the other hand, soil or regolith, or both, may be missing. In some places, everything is stripped off down to the bedrock, which then appears at the surface as an outcrop. In other places, following cultivation or forest fires, the fertile soil is partly or entirely eroded away, and severe erosion exposes the regolith.

Soil Creep

On almost any soil-covered slope, there is extremely slow downhill movement of soil and regolith, a process called **soil creep.** Figure 15.12 shows some of the signs indicating soil creep. Joint blocks of distinctive rock types are found moved far downslope from the outcrop. In some layered rocks such as shales or slates, edges of the strata seem to "bend" in the downhill direction (Figure 15.13). This is not true plastic bending but is the result of downhill creep of many rock pieces on small joint cracks. Creep causes fence posts and

15.12 Indicators of soil creep *The slow, downhill creep of soil and regolith shows up in many ways on a hillside. (After C. F. S. Sharpe.)*

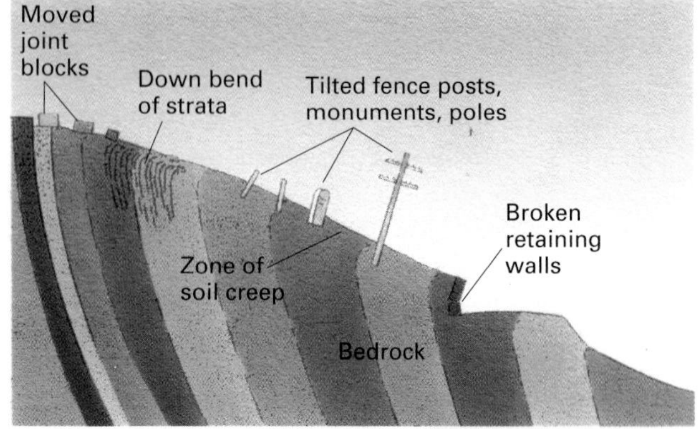

15.13 Soil creep *Slow, downhill creep of regolith on this mountainside near Downieville, California, has caused vertical rock layers to seem to bend rightward.*

utility poles to lean downslope and even shift measurably out of line. Roadside retaining walls can buckle and break under the pressure of soil creep.

Soil creep is caused by disturbance of the soil under the influence of gravity. Alternate drying and wetting of the soil, growth of ice needles and lenses, heating and cooling of the soil, trampling and burrowing by animals, and shaking by earthquakes all produce some disturbance of the soil and regolith. Because gravity exerts a downhill pull on every such rearrangement, the particles very gradually work their way downslope.

Earthflow

In regions of humid climate, a mass of water-saturated soil, regolith, or weak shale may move down a steep slope during a period of a few hours in the form of an **earthflow,** sketched in Figure 15.14. At the top, the material slumps away, leaving a curved, wall-like scarp. Often the soil farther from the scarp moves more rapidly, leaving a series of steps as the soil mass slips downward. At the bottom, the sodden soil flows in a sluggish mass that piles up in ridges or lobes to form a bulging toe.

Shallow earthflows, affecting only the soil and regolith, are common on sod-covered and forested slopes that have been saturated by heavy rains. An earthflow may affect a few square meters, or it may cover an area of a few hectares (several acres). If the bedrock of a mountainous region is rich in clay (derived from shale or deeply weathered volcanic rocks), earthflows sometimes involve millions of metric tons of bedrock moving by plastic flowage like a great mass of thick mud.

Earthflows are a common cause of blockage of highways and railroad lines, usually during periods of heavy rains. Generally, the rate of flowage is slow, so that the flows are not a threat to life. However, property damage to buildings, pavements, and utility lines is often severe where construction has taken place on unstable soil slopes. GEODISCOVERIES

Environmental Impact of Earthflows

One special form of earthflow has proved to be a major environmental hazard in parts of Norway and Sweden and along the St. Lawrence River and its tributaries in Quebec Province of Canada. This type of flowage involves horizontally layered clays, sands, and silts accumulated during the

15.14 Sketch of an earthflow *An earthflow with slump terraces well developed in the upper part. Flowage has produced a bulging toe. (After A. N. Strahler.)*

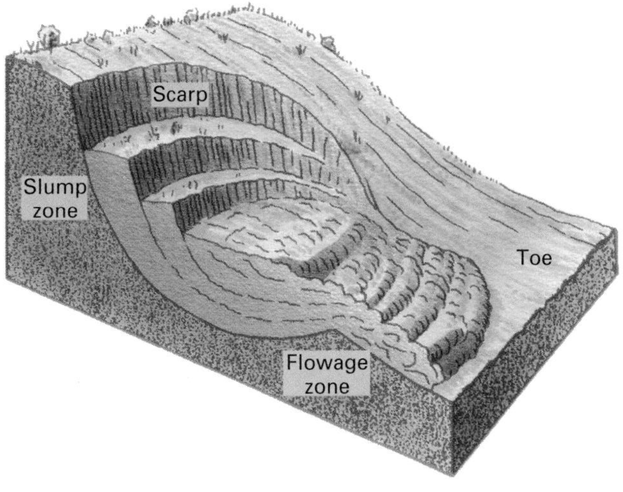

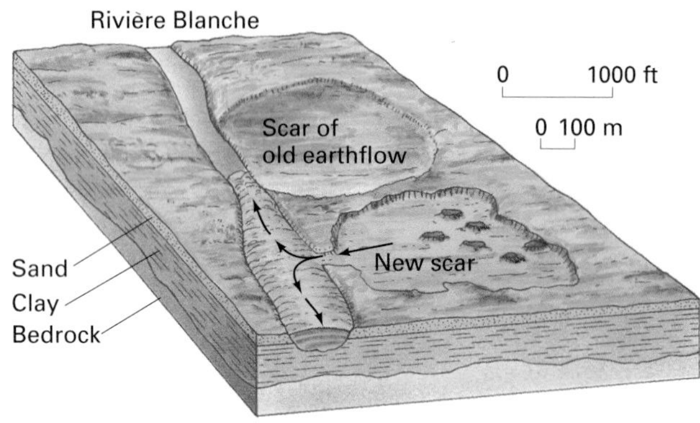

15.15 Earthflow at St. Thuribe *This block diagram depicts the earthflow of 1898 near St. Thuribe, Quebec. (After C. F. S. Sharpe, Landslides and Related Phenomena, Columbia Univ. Press, New York.)*

Ice Age that form low, flat-topped terraces adjacent to rivers and lakes. Over a large area, which may be 600 to 900 m (about 2000 to 3000 ft) across, a layer of silt and sand 6 to 12 m (about 20 to 40 ft) thick begins to move toward the river, sliding on a layer of soft clay that has spontaneously turned into a near-liquid state. The moving mass also settles downward and breaks into step-like masses. Carrying along houses or farms, the layer ultimately reaches the river, into which it pours as a great disordered mass of mud. Figure 15.15 provides a block diagram of a classic example that occurred in 1898 near St. Thuribe, Quebec.

This type of earthflow is caused by *quick clays*—clays that spontaneously change from a solid condition to a near-liquid condition when subjected to a shock or disturbance. Quick clays are thought to have formed in the shallow waters of saltwater bays near the end of the Ice Age. When deposited, the thin plates of clay in these layers have a "house of cards" structure with a large proportion of water-filled void space between plates. The salt water provides positively and negatively charged atoms that help to bind the structure and give it strength. But with the regional uplift that often occurs after ice sheets melt away, the quick clay is elevated above sea level, and fresh ground water replaces the salt water. A mechanical shock, such as an Earth tremor, then causes the house of cards structure to collapse. Because such a large proportion of water (from 45 to 80 percent by volume) is present, the clay–water mixture behaves as a liquid, with almost no strength remaining.

A particularly spectacular example of this type of earthflow occurred in Nicolet, Quebec, in 1955. A clay layer beneath the town liquefied, carrying much of the town into the Nicolet River (Figure 15.19). Fortunately, only three lives were lost, but the damage to buildings and a bridge ran into millions of dollars. A similar tragedy was repeated in St. Jean Vianney in 1971, with 31 deaths, when nearly 8 million cubic meters of clay liquefied and flowed down a stream valley and into the Saguenay River.

Mudflow and Debris Flood

One of the most spectacular forms of mass wasting and a potentially serious environmental hazard is the **mudflow.** This mud stream of fluid consistency pours swiftly down canyons in mountainous regions (Figure 15.17). In deserts, where vegetation does not protect the mountain soils, local thunderstorms produce rain much faster than it can be absorbed by the soil. As the water runs down the slopes, it forms a thin mud that flows down to the canyon floors and then follows the stream courses. As it flows, it picks up addi-

15.16 The Nicolet earthflow *Much of Nicolet, Quebec, flowed into the Nicolet River when an underlying layer of quick clay liquefied, causing an earthflow.*

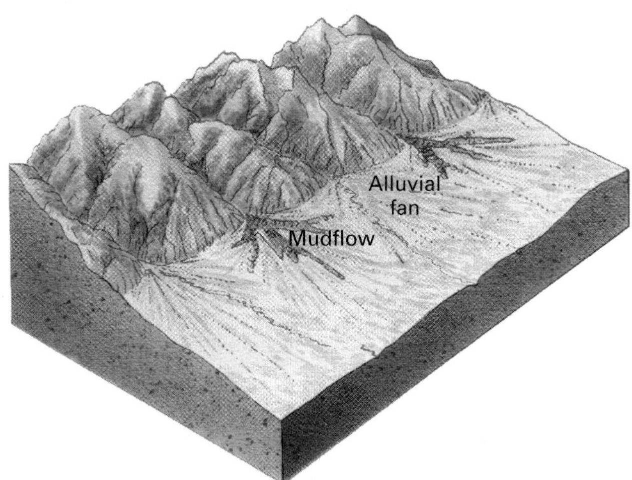

15.17 Mudflows in an arid environment *Thin, streamlike mudflows issue occasionally from canyon mouths in arid regions. The mud spreads out on the fan slopes below in long, narrow tongues. (Drawn by A. N. Strahler.)*

tional sediment, becoming thicker and thicker until it is too thick to flow further. Great boulders are carried along, buoyed up in the mud. Roads, bridges, and houses in the canyon floor are engulfed and destroyed. Where the mudflow emerges from a canyon and spreads across an alluvial fan, severe property damage and even loss of life may be the result (Figure 15.18).

Note that mudflows are rapid events in which water, sediment, and debris follow slopes and river valleys to lower elevations. In contrast, earthflows are slower and are confined to collapsing slopes. This difference is shown by the relative positions of these two types of mass wasting in Figure 15.10.

As explained in Chapter 14, mudflows that occur on the slopes of erupting volcanoes are called *lahars*. Freshly fallen volcanic ash and dust are turned into mud by heavy rains or melting snows and flow down the slopes of the volcano. Herculaneum, a city at the base of Mount Vesuvius, was destroyed by a mudflow during the eruption of A.D. 79. At the same time, the neighboring city of Pompeii was buried under volcanic ash.

Mudflows vary in consistency, from a mixture like concrete emerging from a mixing truck to consistencies similar to that of turbid river floodwaters. The watery type of mudflow is called a *debris flood* or *debris flow* in the western United States, and particularly in southern California, where it occurs commonly and with disastrous effects. The material carried in a debris flood ranges from fine particles to boulders to tree trunks and limbs. In mountainous regions on steep slopes, these flows are termed *alpine debris avalanches*. The intense rainfall of hurricanes striking the eastern United States often causes debris avalanches in the hollows and valleys of the Blue Ridge and Smoky Mountains.

Landslide

A **landslide** is the rapid sliding of large masses of bedrock or regolith. Wherever mountain slopes are steep, there is a possibility of large, disastrous landslides (Figure 15.19). In Switzerland, Norway, or the Canadian Rockies, for example, villages built on the floors of steep-sided valleys have been destroyed by the sliding of millions of cubic meters of rock set loose without warning.

15.18 Mudflow deposit *This mudflow, carrying numerous large boulders, issued from a steep mountain canyon in the Wasatch Mountains, Utah.*

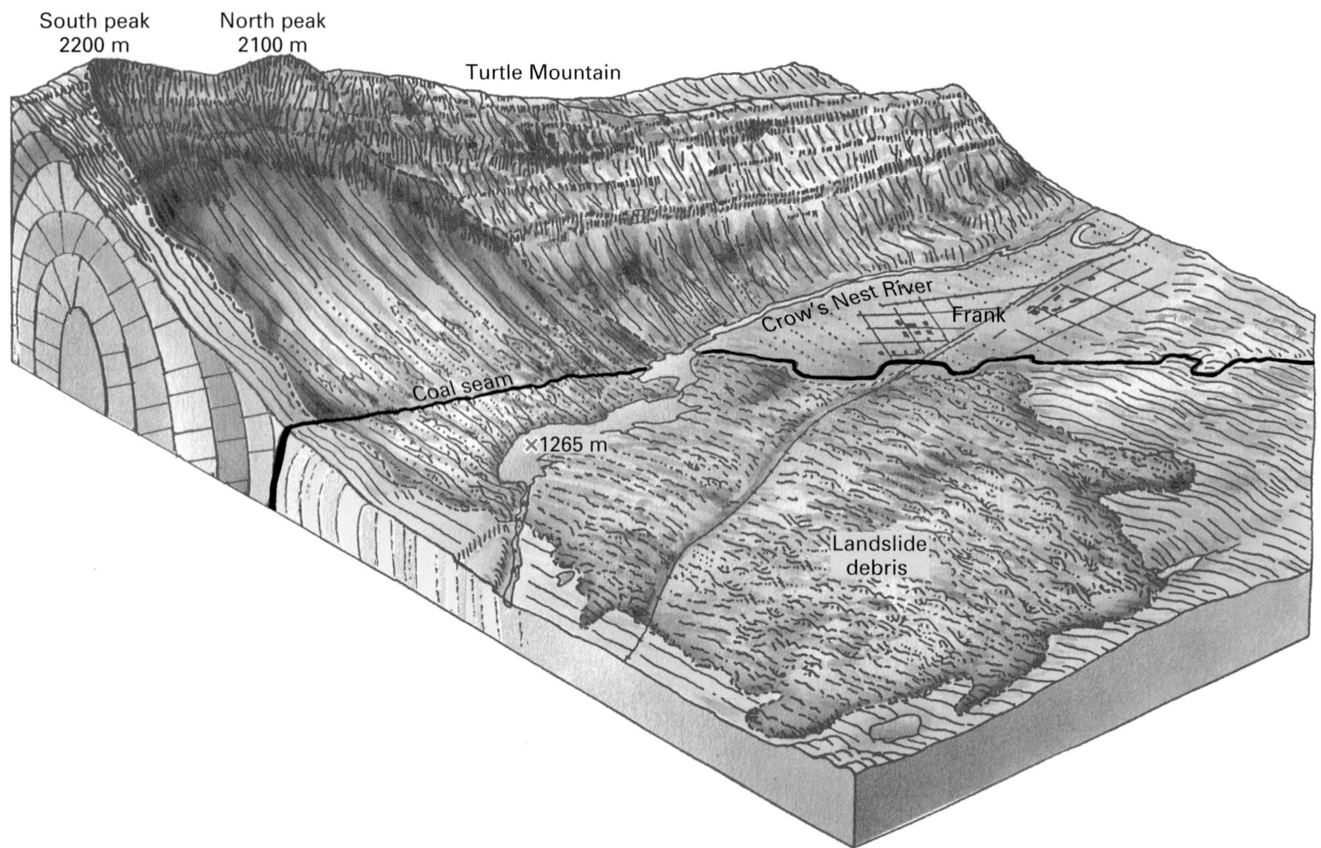

15.19 The Turtle Mountain slide *A classic example of an enormous, disastrous landslide is the Turtle Mountain slide, which took place near Frank, Alberta. A huge mass of limestone slid from the face of Turtle Mountain between South and North peaks, descended to the valley, then continued up the low slope of the opposite valley side until it came to rest as a great sheet of bouldery rock debris. (Data from Canadian Geological Survey, Department of Mines.) (Drawn by A. N. Strahler. Copyright © A. N. Strahler.)*

Landslides are triggered by earthquakes or sudden rock failures rather than by heavy rains or sheet floods. Thus, they are different from earthflows and mudflows, which are typically induced by heavy rains. Landslides can also result when the base of a slope is oversteepened by excavation or river erosion. As shown in Figure 15.13, landslides can range from *rockslides* of jumbled bedrock fragments to *bedrock slumps* in which most of the bedrock remains more or less intact as it moves. The amazing speed with which rockslide rubble travels down a mountainside is thought to be due to the presence of a layer of compressed air trapped between the slide and the ground surface. The air layer reduces frictional resistance and may keep the rubble from making contact with the surface beneath.

Severe earthquakes in mountainous regions are a major cause of landslides. An example is the landslide of Santa Tecla, El Salvador, that killed hundreds of people on January 13, 2001 (Figure 15.20). The earthquake was centered off the southern coast of El Salvador and measured 7.6 on the Richter scale. It was felt as far away as Mexico City. In the shaking, a steep slope above the neighborhood of Las Colinas collapsed, creating a wave of earth that swept across the ordered grid of houses and streets, burying hundreds of homes and their inhabitants. The same earthquake also triggered other landslides in the region, with additional loss of life and property. *Eye on the Environment 15.1 • The Great Hebgen Lake Disaster* presents the history of the Madison Slide, another earthquake-triggered landslide that dammed a river to create a new, natural lake. GEODISCOVERIES

Aside from occasional local catastrophes, landslides have rather limited environmental influence because they occur only sporadically and usually in thinly populated mountainous regions. Small slides can, however, repeatedly block or break important mountain highways or railway lines. GEODISCOVERIES

A large landslide releases a vast amount of gravitational energy. Some simple calculations indicate that the energy released by the Madison Slide within the minute or two of its occurrence was more than enough to equal American electrical power consumption for a day. *Working It Out 15.2 • The Power of Gravity* documents the enormous power of gravity and shows how this estimate was made.

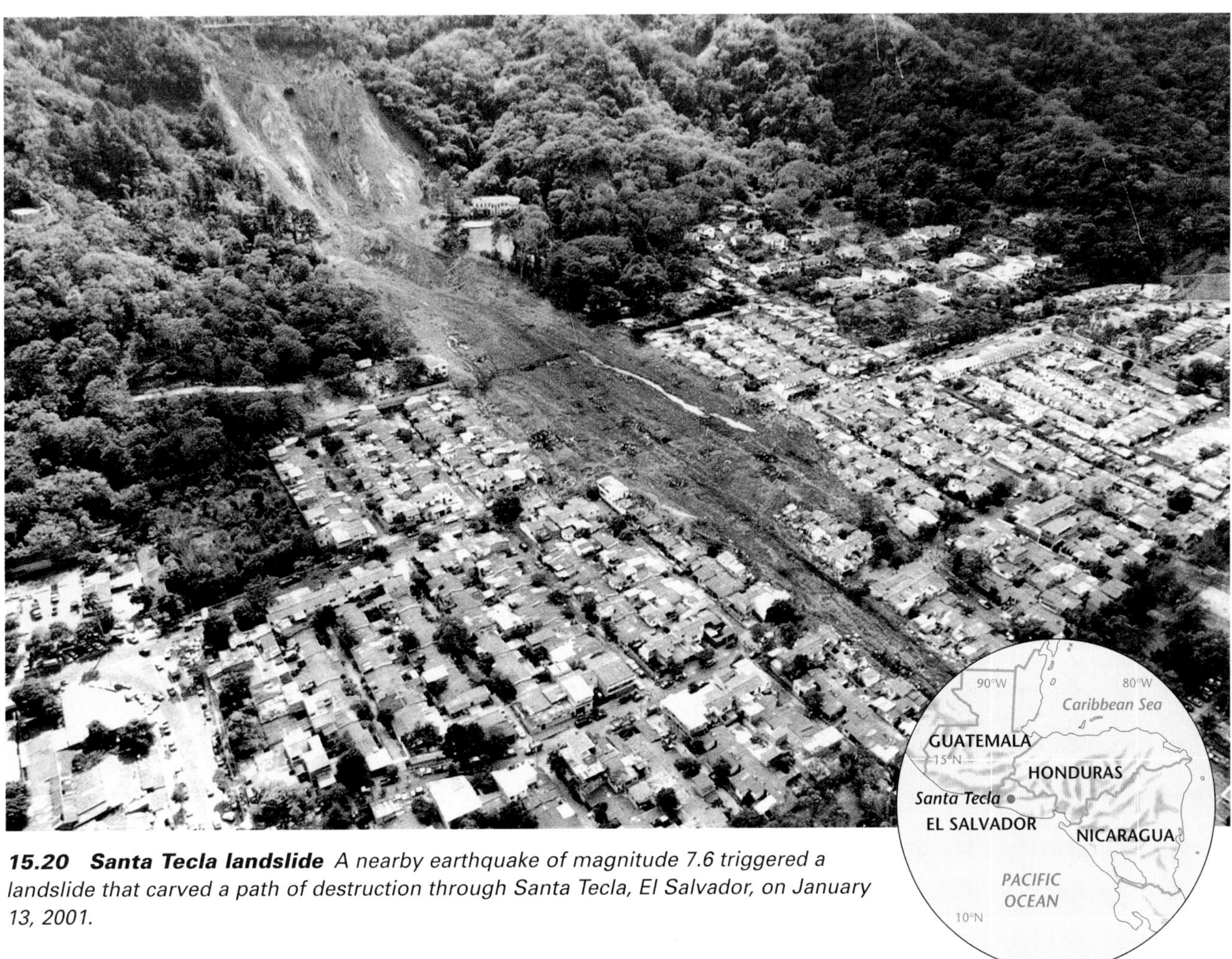

15.20 Santa Tecla landslide *A nearby earthquake of magnitude 7.6 triggered a landslide that carved a path of destruction through Santa Tecla, El Salvador, on January 13, 2001.*

Induced Mass Wasting

Human activities can induce mass wasting in forms ranging from mudflow and earthflow to landslide. These activities include (1) piling up of waste soil and rock into unstable accumulations that can move spontaneously, and (2) removing the underlying support of natural masses of soil, regolith, and bedrock. We can refer to mass movements produced by human activities as *induced mass wasting.*

In Los Angeles County, California, real estate development has been carried out on very steep hillsides and mountainsides. Roads and home sites have been bulldozed out of the deep regolith. The excavated regolith is piled up to form nearby embankments. When saturated by heavy winter rains, these embankments can give way. This produces earthflows, as well as mudflows and debris floods that travel far down the canyon floors and spread out on the alluvial fan surfaces below, burying streets and houses in mud and boulders.

Many debris floods of this area are also produced by heavy rains falling on mountain slopes denuded of vegetation by fire in the preceding dry season. Some of these fires are set carelessly or deliberately by humans.

Induced Earthflows

Examples of both large and small earthflows induced or aggravated by human activities are found in the Palos Verdes Hills of Los Angeles County, California (Figure 15.21). These movements occur in shales that tend to become plastic when water is added. The upper part of the earthflow subsides and slumps backward, as illustrated in Figure 15.14. The interior and lower parts of the mass move forward by slow flowage, producing a toe of flowage material.

The largest of the earthflows in the Palos Verdes Hill area was the Portuguese Bend "landslide," which affected an area of about 160 hectares (about 400 acres). It was caused by the slippage of sedimentary rock layers on an underlying layer of clay. The total movement over a three-year period was about 20 m (about 70 ft). Damage to residential and other structures totaled some $10 million. Geologists attributed the earthflow to infiltration of water from septic tanks and from irrigation water applied to lawns and gardens. A discharge of over 115,000 liters (about 30,000 gallons) of water per day from some 150 homes is believed to have sufficiently weakened the clay layer to start and sustain the flowage.

Eye on the Environment | 15.1

The Great Hebgen Lake Disaster

For some 200 vacationers camping near the Madison River in a deep canyon just downstream from Hebgen Lake, not far west of Yellowstone National Park, the night of August 17, 1959, began quietly, with almost everyone safely bedded down in their tents or camping trailers. Up to a certain point, it was everything a great vacation should be—that point in time was 11:37 P.M., Mountain Standard Time. At precisely that instant, not one but four terrifying forms of disaster were set loose on the sleeping vacationers—earthquake, landslide, hurricane-force wind, and raging flood. The earthquake, which measured 7.1 on the Richter scale, was the cause of it all. The first shocks, lasting several minutes, rocked the campers violently in their trailers and tents. Those who struggled to go outside could scarcely stand up, let alone run for safety.

Then came the landslide. A dentist and his wife watched through the window of their trailer as a mountain seemed to move across the canyon in front of them, trees flying from its surface like toothpicks in a gale. Then, as rocks began to bang against the sides and top of their trailer, they got out and raced for safer ground. Later, they found that the slide had stopped only about 25 m (80 ft) from the trailer. Pushed by the moving mountain was a vicious blast of wind. It swept upriver, tumbling trailers end over end.

Then came the flood. Two women schoolteachers, sleeping in their car only about 5 m (15 ft) from the river bank, awoke to the violent shaking of the earthquake. Like other campers, they first thought they had a marauding bear on their hands. Puzzled and frightened, they started the engine and headed the car for higher ground. As they did so, they were greeted by a great roar coming from the mountainside above and behind them. An instant later, the car was completely engulfed by a wall of water that surged up the river bank, then quickly drained back. With the screams of drowning campers in their ears, the two women managed to drive the car to safe ground, high above the river.

Madison slide *Seen from the air, the Madison slide forms a great dam of rubble across the Madison River Canyon. The slide was triggered by a nearby earthquake.*

After the first surge of water, generated as the landslide mass hit the river, the river began a rapid rise. This rise was aided by great surges of water topping the Hebgen Dam, located upstream, as earthquake aftershocks rocked the water of Hebgen Lake back and forth along its length of about 50 km (30 mi). In the darkness of night, the terrified victims of the flood had no idea what was happening. In the panic that ensued, a 71-year-old man performed an almost unbelievable act of heroism to save his wife and himself from drowning. As the water rose inside their house trailer, he forced open the door, pulled his wife with him to the trailer roof, then carried her up on the branches of a nearby pine tree. Here they were finally able to reach safety. The water had risen 10 m (about 30 ft) above ground level in just minutes.

The Madison Slide, as the huge earth movement was later named, had a bulk of 28 million m^3 (37 million yd^3) of rock (see photo). It consisted of a chunk of the south wall of the canyon, measuring over 600 m (about 2000 ft) in length and 300 m (about 1000 ft) in thickness. The mass descended over a half of a kilometer (about a third of a mile) to the Madison River, its speed estimated at 160 km (100 mi) per hour. Pulverized into bouldery debris, the slide crossed the canyon floor, its momentum carrying it over 120 m (about 400 ft) in vertical distance up the opposite canyon wall. At least 26 persons died beneath the slide, and their bodies have never been recovered. Acting as a huge dam, the slide caused the Madison River to back up, forming a new lake. In three weeks' time, the lake was nearly 100 m (330 ft) deep. Today it is a permanent feature, named Earthquake Lake.

Working It Out | 15.2

The Power of Gravity

Directly or indirectly, gravity powers many of the processes that shape the landscape, from the erosion of running water in carving landforms (Chapter 17) to the processes of mass wasting—earthflows, mudflows, solifluction, and landslides—that are the subject of this chapter. By acting to move mineral matter and water downhill, gravity does work over a span of time and is thus a source of power.

Let's briefly review gravitation, gravity, force, work, and power. *Gravitation* is the simultaneous attraction that occurs between two physical bodies. The strength of the attraction depends on the mass of each body and the distance between the two bodies. The attraction of the Earth for an object near the Earth's surface is called *gravity.* It is described as an acceleration that acts on the mass of the object to produce a force according to the simple relationship

$$F = m \times g,$$

where F is the force, m is the mass of the object, and g is the acceleration due to gravity (about 9.8 m/s^2). If the mass is measured in kg and the acceleration is measured in m/sec^2, then the force is in units of newtons *(N).*

Suppose now that gravity actually does some work—that is, it moves a mass through a distance. The work is measured as force times distance:

$$W = F \times d,$$

where W is the work and d is the distance. If the force is measured in newtons and the distance in meters, then the work done is measured in joules (J).

Power describes the time rate at which work is accomplished—that is, work per unit time, or

$$P = \frac{W}{T}$$

where P is the power, W is the work, and T is the time. If work is measured in joules (J) and time is measured in seconds (s), then power is in unit of watts (W). Recall from *Focus on Systems 4.1 • Forms of Energy* that work and energy have the same units and that energy is defined as the ability to do work. Thus, power is a measure of both a rate of energy flow and a rate at which work is being done.

As an example of the power of gravity, let's take the Madison Slide—a catastrophic landslide described in *Eye on the Environment 15.1 • The Great Hebgen Lake Disaster.* In this event, a rock body with a volume of about 28 million cubic meters (28×10^6 m^3) moved a vertical distance downward of about 500 m into the valley of the Madison River, attaining a velocity of about 150 km/hr. Propelled by its own inertia, the flow then moved uphill on the far side of the river valley by a vertical distance of about 120 m. In this slide, the potential energy of position of the mass above the valley bottom was converted to kinetic energy during the slide, and the kinetic energy was dissipated as heat produced by friction during the sliding motion.

Note that the slide moved material downhill through a vertical distance of 500 m, converting potential energy to kinetic energy and friction, then uphill a vertical distance of 120 m. On the uphill leg, some kinetic energy was converted back to potential energy as the mass moved upward against the acceleration of gravity. So the net release of potential energy was produced by a vertical distance change of 500 – 120 = 380 m.

How much energy was released? Or alternatively, how much work was done? From the formulas above, we can see that this requires knowing the mass, acceleration, and distance. Of these, only the mass is unknown. However, we know the volume of the landslide—28×10^6 m^3—so what is missing is the density—the mass of the rock per unit of volume. Let's assume that the rock has the density of granite, which is about 2700 kg/m^3 (2.7×10^3 kg/m^3) (Figure 12.5). Then we have

$$m = 28 \times 10^6 \text{ m}^3 \times \frac{2.7 \times 10^3 \text{ kg}}{\text{m}^3}$$
$$= 7.56 \times 10^{10} \text{ kg}$$

Gravity will act on this mass with a force

$$F = m \times g = 7.56 \times 10^{10} \text{ kg} \times \frac{9.8 \text{ m}}{\text{s}^2}$$
$$= 7.41 \times 10^{11} \text{ N}$$

This force is applied through a net distance of 380 m, so the work done is

$$W = F \times d = 7.41 \times 10^{11} \text{ N} \times 380 \text{ m}$$
$$= 2.82 \times 10^{14} \text{ J}$$

This work was converted to heat through the friction of the slide.

The amount of work done is rather similar to the amount of electrical energy consumed in a day in the United States, which is about 2.33×10^{14} J/day. Thus, the energy released by the Madison Slide, if converted completely to electric power, could satisfy U.S. electrical needs for slightly more than one day.

To calculate the power expended by gravity in the slide, we need to know how long the motion took. With some simple assumptions about the angles of the mountain slopes and the magnitude of friction in the earthflow, a reasonable guess for the time required is about 70 seconds. The power applied by gravity in the downward motion would then be

$$P = \frac{W}{T} = \frac{2.82 \times 10^{14} \text{ J}}{70 \text{ s}} = 4.02 \times 10^{12} \text{ W}$$

Expressed in watts, the average rate of U.S. electric power consumption is about 2.70×10^9 W. The rate of energy release in the Madison Slide is therefore about 1500 times greater than the rate at which the entire United States consumes electricity. This large number demonstrates the awesome power of gravity to shape and reshape the landscape.

15.21 Induced earthflow *This aerial view shows houses on Point Fermun, in Palos Verdes, California, disintegrating as they slide downward toward the sea.*

Scarification of the Land

Industrial societies now possess enormous machine power and explosives capable of moving great masses of regolith and bedrock from one place to another. This technology is used to extract mineral resources. Another use is to move earth in the construction of highway grades, airfields, building foundations, dams, canals, and various other large structures. Both activities involve removal of Earth materials, a process that destroys the preexisting ecosystems and habitats of plants and animals. When the materials are used to build up new land on adjacent surfaces, ecosystems and habitats are also destroyed—by burial. What distinguishes artificial forms of mass wasting from the natural forms is that machinery is used to raise Earth materials against the force of gravity. Explosives used in blasting can produce disruptive forces many times more powerful than the natural forces of physical weathering.

Scarification is a general term for excavations and other land disturbances produced to extract mineral resources. The rock waste, known as *spoil* or *tailings,* accumulates at the site. Scarification takes the form of open-pit mines, strip mines, quarries for structural materials, sand and gravel pits, clay pits, phosphate pits, scars from hydraulic mining, and stream gravel deposits reworked by dredging.

Strip mining is a particularly destructive scarification activity. Strip mining of coal is practiced where coal seams lie close to the surface or actually appear as outcroppings along hillsides (Figure 15.22). Earth-moving equipment removes the covering strata, or *overburden,* to bare the coal, which is lifted out by power shovels. The piled-up overburden remains as spoil. When saturated by heavy rains and melting snows, the spoil can generate earthflows and mudflows that descend on houses, roads, and forest. The spoil also supplies sediment that clogs stream channels far down the valleys.

As a result of these problems, strip mining is under strict control in most locations. Mine operators are not permitted to create hazardous spoil slopes, and they must restore spoil banks and ridges to a natural condition. In this process, termed *reclamation,* topsoil is removed and saved

15.22 Contour strip mining *This strip mine near Lynch, Kentucky, follows an outcrop of coal along a hillside contour. A highway makes use of the winding bench at the base of the high rock wall.*

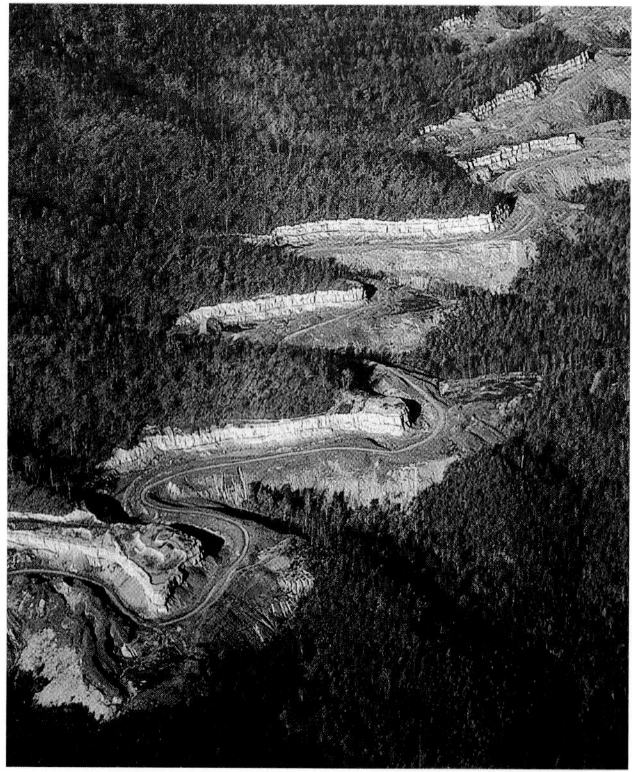

as the mine is opened and worked. When spoil is returned to the mine cavity after extraction of the coal, it is covered with the topsoil and suitable plants are introduced. With good management and proper planning, the resulting reclaimed land becomes usable again. But there are also many reclamation failures in which the land has been damaged beyond repair.

Scarification is on the increase. Driven by an ever increasing human population, the demand for coal and industrial minerals used in manufacturing and construction is on the rise. At the same time, as the richer and more readily available mineral deposits are consumed, industry turns to poorer and less easily accessible grades of ore. As a result, the rate of scarification is further increased, making the success of reclamation efforts increasingly important.

Processes and Landforms of Arctic and Alpine Tundra

The landscape of the treeless arctic and alpine tundra environment shows many distinctive effects of weathering and mass wasting under a regime of severe winter cold. In the tundra, soil water is frozen throughout a winter of many months duration. During the short summer season, however, surface thawing occurs, leaving the soil saturated and vulnerable to mass wasting and water erosion. With the return of cold temperatures, the freezing of soil water exerts a strong mechanical influence on the surface layer.

This intense frost action creates a distinctive set of landforms and landforming processes that we may term the *periglacial system.* Here, the adjective **periglacial** (*peri,* meaning "near") denotes an environment of intense frost action near the margins of active glaciers and ice sheets or in cold regions (which may actually be quite far from permanent ice and snow cover). The periglacial system is driven by a very large annual range in temperature, which creates a strong annual cycle in the flow of energy into and out of the surface layer of the ground. This cycle changes water from liquid water to solid ice, with an increase in volume, and back to liquid water again, with a decrease in volume. The expansion and contraction of water as it changes state, coupled with the pressure that ice crystal growth can exert, provide the mechanism for the movement of soil particles individually and collectively, thus creating the distinctive processes that operate in the periglacial environment. We preface our discussion of periglacial processes with a look at permafrost.

Arctic Permafrost

Ground and bedrock that are perennially (year-round) below the freezing point of fresh water (0°C; 32°F) are called **permafrost.** By "ground" we mean to include mineral matter falling within the full range of particle sizes described in Chapter 12—that is, clay, silt, sand, pebbles, and boulders. The term *permafrost* strictly refers only to ground temperature. Thus, solid bedrock perennially below the freezing point is also included in permafrost. Water is commonly present in pore spaces in the ground, and in the frozen state it is known as **ground ice.**

Permafrost prevails over most of the tundra climate region and in a wide area of boreal forest climate bordering the tundra. A distinctive feature of permafrost terrains is a shallow surface layer subject to seasonal thawing called the *active layer.* It ranges in thickness from about 15 cm (6 in.) to about 4 m (13 ft), depending on the latitude and nature of the ground. Below the active layer is the upper surface of the perennially frozen zone, called the *permafrost table.* Note that most lakes and rivers do not have permafrost beneath them, since the water at the bottom of these features is at or above 0°C throughout the year.

The distribution of permafrost in the northern hemisphere is shown in Figure 15.23. Four zones are shown on the map. *Continuous permafrost,* which extends under nearly all surface features, coincides largely with the tundra climate, but it also includes a large part of the boreal forest climate in Siberia. *Discontinuous permafrost,* which occurs in patches separated by unfrozen zones, occupies much of the boreal forest climate zone of North America and Eurasia. A third zone—*sub-sea permafrost*—lies beneath sea level in a shallow offshore zone. On our map

15.23 Permafrost map *Distribution of permafrost in the northern hemisphere (Adapted from Troy L. Péwé, Geotimes, vol. 29, no. 2, p. 11, copyright, 1984, by the American Geologic Institute. Used by permission.)*

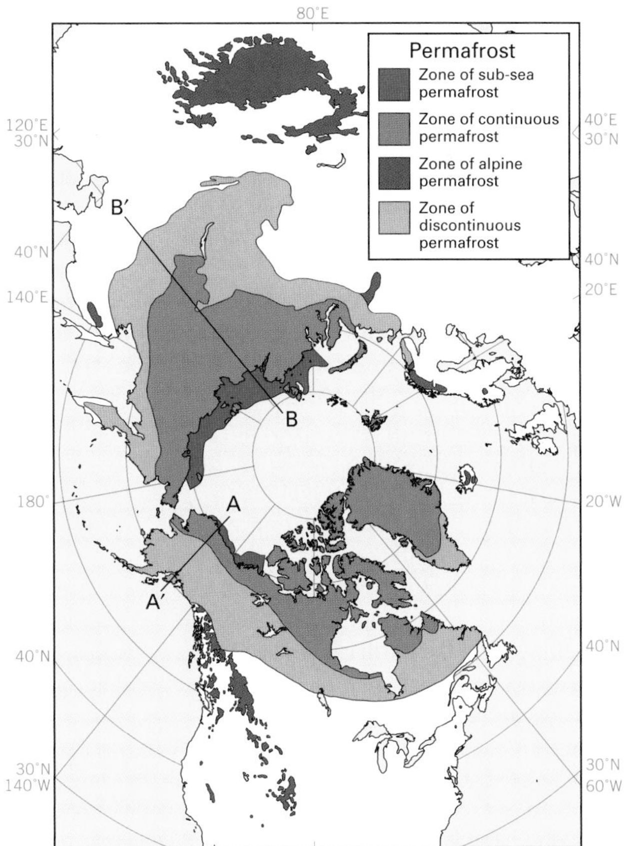

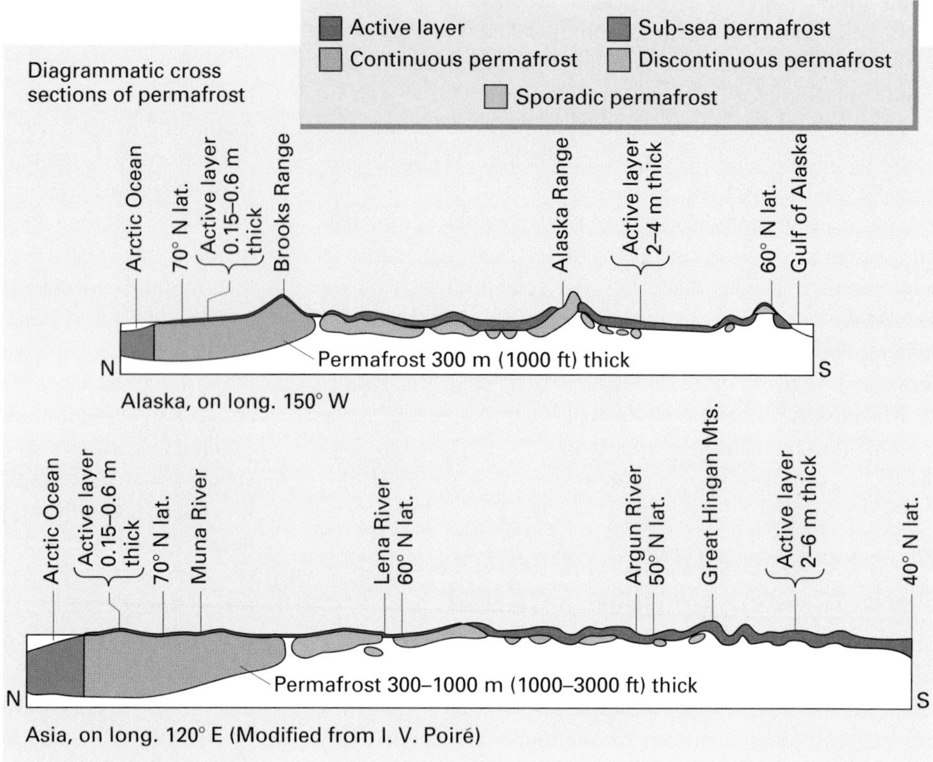

15.24 Permafrost map cross sections *These north-south cross sections of permafrost in Alaska and Asia are located on Figure 15.26. (From Robert F. Black, "Permafrost," Chapter 14 of P. D. Trask's Applied Sedimentation. Copyright © by John Wiley & Sons. Reprinted by permission of John Wiley & Sons, Inc.)*

you will find it beneath the Arctic sea offshore of the Asian coast and the coasts of Alaska, Yukon, and Northwest Territories in North America. Last is *alpine permafrost,* which is found at high elevations where temperatures are always below freezing.

Cross sections of the permafrost zone in Alaska and Asia are shown in Figure 15.24. In this diagram, the active layer includes the layer of annual freezing and thawing over all terrain, not just permafrost. Figure 15.24 also shows *sporadic permafrost,* a zone to the south of discontinuous permafrost in which occasional patches of permafrost occur.

Permafrost reaches to depths of roughly 1000 m (roughly 3000 ft) near the poleward end of its distribution and to 300 to 450 m (about 1000 to 1500 ft) in the continuous zone near lat. 70° N (Figure 15.24). Much of this permanently frozen zone is an inheritance from the more severe conditions of the last Ice Age, but some permafrost bodies may be growing under existing climate conditions.

Permafrost is an example of an energy flow system in which the depth of the active layer and the base of the permafrost are determined by the balance of heat energy flowing into the ground from the surface and the heat energy upwelling from the Earth's interior. *Focus on Systems 15.3 • Permafrost as an Energy Flow System* describes this energy flow system in more detail.

Even in the areas of deepest continuous permafrost, there exist isolated pockets that never fall below the freezing point. The Siberian word *talik* has been applied to these pockets. One kind of talik lies beneath a deep lake that freezes over in winter and thaws in summer. The depth to which such a talik extends depends on the size of the lake. For a large lake, the talik may extend to the bottom of the permafrost layer, making the talik a "hole" in the permafrost.

Forms of Ground Ice

The amount of ground ice present below the permafrost table varies greatly from place to place. Near the surface, it can take the form of a body of almost 100 percent ice. At increasing depth, the percentage of ice becomes lower and approaches zero at great depth.

A common form of ground ice is the **ice wedge.** In silty alluvium, such as that formed on river floodplains and delta plains in the arctic environment, ice accumulates in vertical wedge-forms in deep cracks in the sediment (Figure 15.25). Ice wedges originate in cracks that form when permafrost shrinks during extreme winter cold. As shown in Figure 15.26, surface meltwater enters the crack during the spring and freezes almost immediately, widening the crack. Repeated cracking and filling with new ice cause the ice wedge to thicken until it becomes as wide as 3 m (about 10 ft) and as deep as 30 m (about 50 ft). Ice wedges are typically interconnected into a system of polygons, called *ice-wedge polygons* (Figure 15.27).

Another remarkable ice-formed feature of the arctic tundra is a conspicuous conical mound, called a *pingo* (Figure 15.28). The pingo has a core of ice and grows in height as more ice accumulates, forcing up the overlying sediment. In extreme cases, pingos reach a height of 50 m

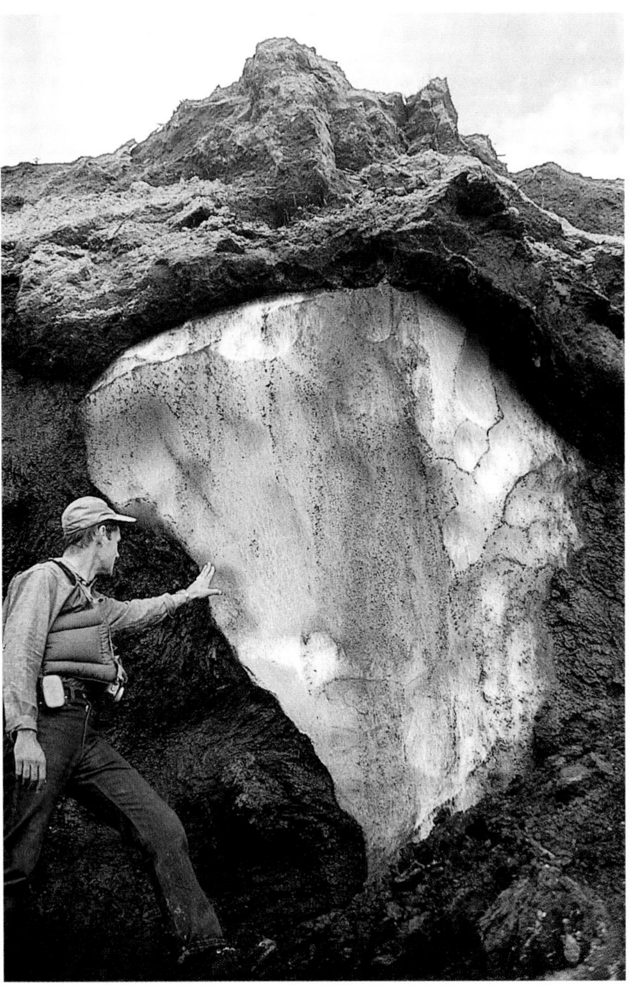

15.25 Ice wedge *Exposed by river erosion, this great wedge of solid ice fills a vertical crack in organic-rich floodplain silt along the Yukon River, near Galena in western Alaska.*

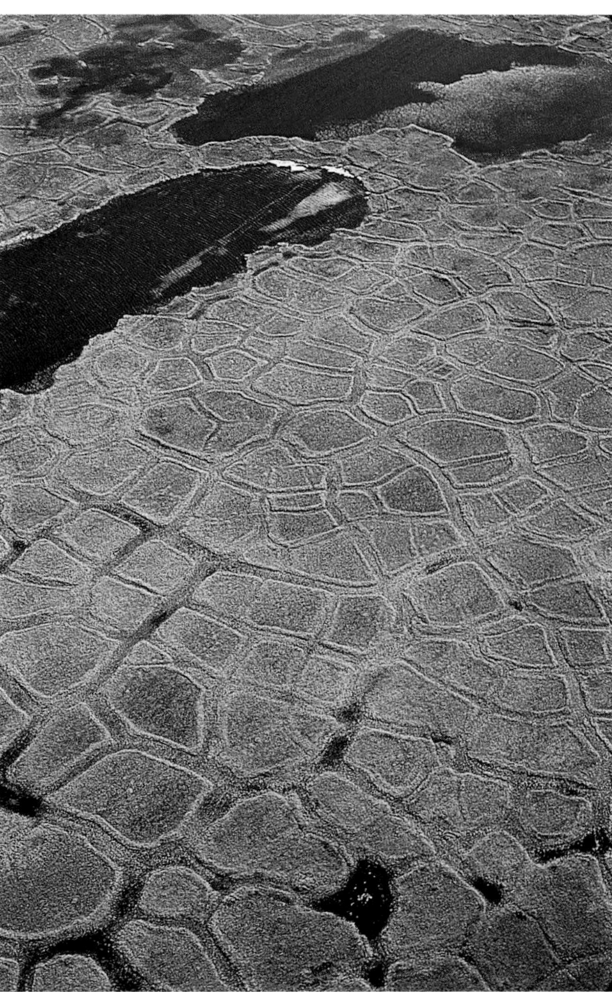

15.27 Ice-wedge polygons *These polygons are formed by the growth of ice wedges. Alaskan north slope, near the border of Alaska and Yukon Territory, Canada.*

15.26 Formation of an ice wedge *(a) In the beginning stage, an open crack appears during the winter and is filled with meltwater, which freezes into ice during the following autumn. (b) After several hundred winters, the ice wedge has grown, and continues to grow as the same seasonal sequence is repeated. (Adapted by permission from A. H. Lachenbruch in Rhodes W. Fairbridge, Ed., The Encyclopedia of Geomorphology, New York, Reinhold Publishing Corp.)*

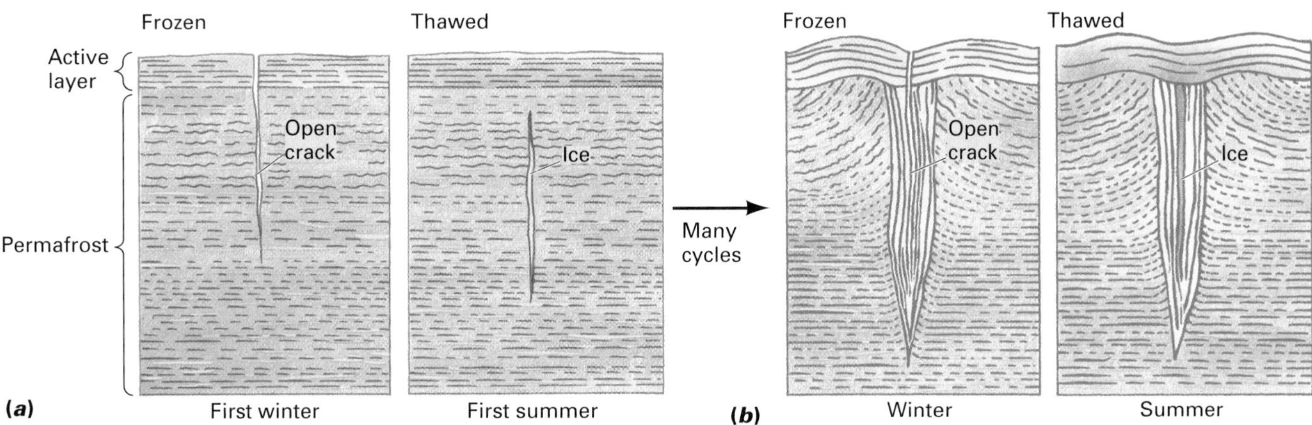

Focus on Systems 15.3

Permafrost as an Energy Flow System

The upper few hundred meters of the solid Earth receives flows of heat from two directions—downward from the atmosphere (or ocean) above and upward from the ground below. The flow of energy into and out of the ground at the surface depends on the surface energy balance. In winter, when the air is, on average, colder than the ground, heat energy will flow out of the ground to the atmosphere. In summer, when the air is, on average, warmer, heat energy will flow into the ground from the atmosphere. (*Focus on Systems 5.1 • The Surface Energy Balance Equation* described the surface energy balance in detail.)

The surface flow of heat energy into and out of the ground occurs mostly by conduction. Because the ground resists the penetration of heat, the ground temperature changes most rapidly close to the surface. At increasing depth in the ground, the temperature changes more slowly and to a lesser degree. Even so, the temperature of a layer 10 to 20 m (33 to 66 ft) deep in the ground can still show seasonal fluctuations.

Permafrost is defined as ground that is perennially at or below 0°C (32°F). Permafrost occurs in regions where the mean annual surface temperature is below freezing. Over very long periods of time (hundreds to thousands of years), the ground at a depth of 15 to 20 m tends to take on the average annual air temperature and can thus acquire a year-around temperature well below freezing.

The figure to the right shows this effect for a mean annual surface temperature of -10°C (14°F). In the winter (left), air temperatures drop and heat energy flows out of the ground, as shown by the color bar. In this example, we assume that the minimum annual surface temperature during the winter is –25°C (–13°F). Summer conditions are shown on the right side of the graph. Here, the maximum monthly surface temperature is 5°C (41°F) and heat energy flows into the ground.

The two upper curves on the graph show minimum and maximum annual temperatures reached by each layer according to its depth. At the surface, the minimum and maximum values correspond to the annual values of –25°C (–13°F) and 5°C (41°F) and the difference between the minimum and maximum values is large. With depth, the difference decreases as the minimum and maximum values converge. At about 14 m (46 ft) in this example, annual variation ceases and the temperature remains near, but slightly above, the mean annual surface temperature of –10°C (14°F).

Note that the uppermost layer of the ground is heated above freezing. This, of course, is the active layer of permafrost that freezes and thaws each year.

In contrast to the seasonally varying flow of heat near the surface is the steady flow of heat upward from the Earth's interior. As we saw in *Focus on Systems 12.4 • Powering the Rock Cycle,* crustal temperature increases with increasing depth due to the constant flow of heat from the interior. This increase is shown by a straight slanting line on the lower graph. The temperature gradient varies somewhat, but a typical value is about +3°C per 100 m (+1.6°F per 100 ft) of ground depth. Since the temperature increases, eventually it exceeds 0°C. This defines the base of the permafrost layer. It is reached, in this example, at a depth of about 333 m (1100 ft).

Note that the minimum and maximum curves are not actually plots of temperature with depth at one time. Because the ground resists heating, air temperature changes take time to penetrate the ground. The time lag increases with depth. In fact, the coldest temperature in the ground at 5 m (16 ft ft) commonly occurs about six months after the coldest surface conditions.

Suppose that the mean annual temperature is somewhat warmer—say –5°C instead of –10°C. In this case, we would expect the active layer to be thicker, as summer temperatures would be warmer and the soil would thaw more deeply. We would also expect the bottom of the permafrost to be encountered at a level that is not quite as deep. Thus, both the thickness of the active layer and the depth of per-

(164 ft) and a basal diameter as great as 600 m (about 2000 ft). GEODISCOVERIES

Patterned Ground and Solifluction

In areas of coarse-textured regolith consisting of rock particles in a wide range of sizes, the annual cycle of thawing and freezing causes the coarsest fragments—pebbles and cobbles—to move horizontally as well as vertically and thus become sorted out from the finer particles. This type of sorting produces ring-like arrangements of coarse fragments. The linking of adjacent rings produces a net-like pattern to form a system of *stone polygons* (also called "stone rings" or "stone nets") (Figure 15.29). On steep slopes, the process

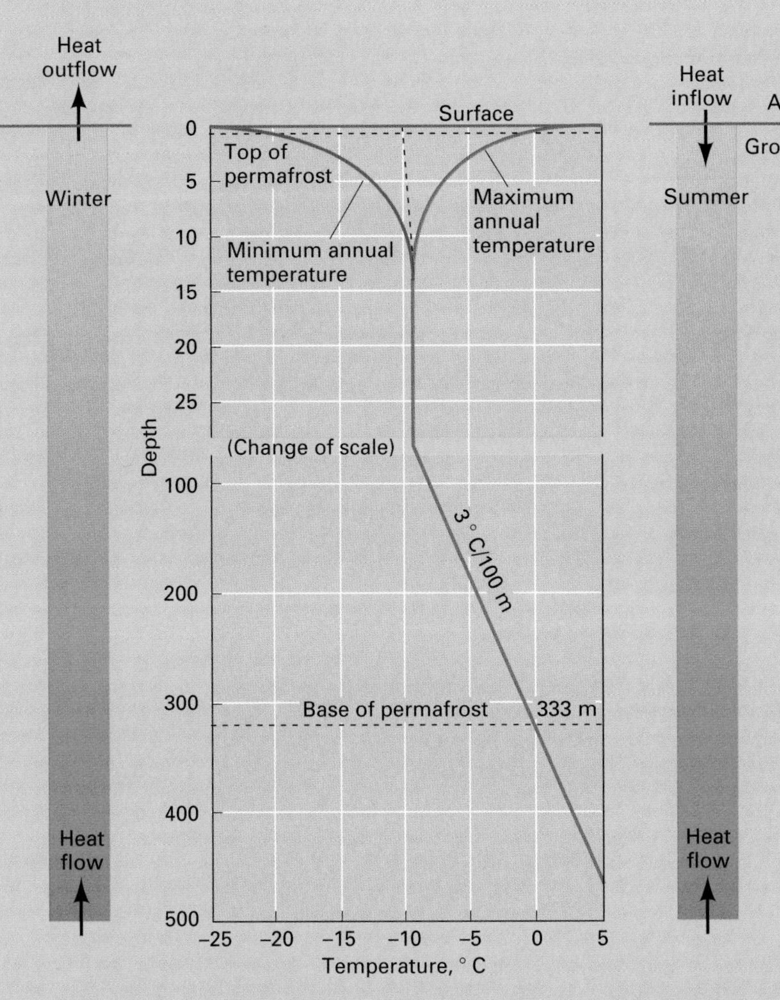

Maximum and minimum permafrost temperature profiles *Mean annual maximum and minimum ground temperatures plotted with depth for an example of permafrost. (Adapted by permission from A. H. Lachenbruch in Rhodes W. Fairbridge, Ed., The Encyclopedia of Geomorphology, New York, Reinhold Publishing Corp.)*

mafrost are related to the mean annual surface temperature—the colder the temperature, the thinner is the active layer and the deeper is the permafrost base.

Permafrost is another example of an energy flow system in which inflows and outflows of energy tend to reach a balance, or equilibrium, over time. However, because heat flows so slowly through the ground, it takes a long time for permafrost to reach an equilibrium when a climate change occurs—perhaps thousands of years in some cases. Researchers have speculated that permafrost bodies in regions where mean annual surface temperature is now well above 0°C may be relics of the Ice Age, slowly wasting away, but still in place after 12,000 years of postglacial climate.

of soil creep causes the polygons to be elongated in the downslope direction and to become transformed into parallel stone stripes (Figure 15.30). Both stone polygons and ice-wedge polygons are examples of a class of features called **patterned ground.**

A special variety of earthflow characteristic of arctic permafrost tundra regions is **solifluction** (from Latin words meaning "soil" and "to flow"). It is active in late summer, when thawing has penetrated to the bottom of the active layer. Moving almost imperceptibly, the saturated soil is deformed into *solifluction terraces* and *solifluction lobes* that give the tundra slope a stepped appearance (Figure 15.31).

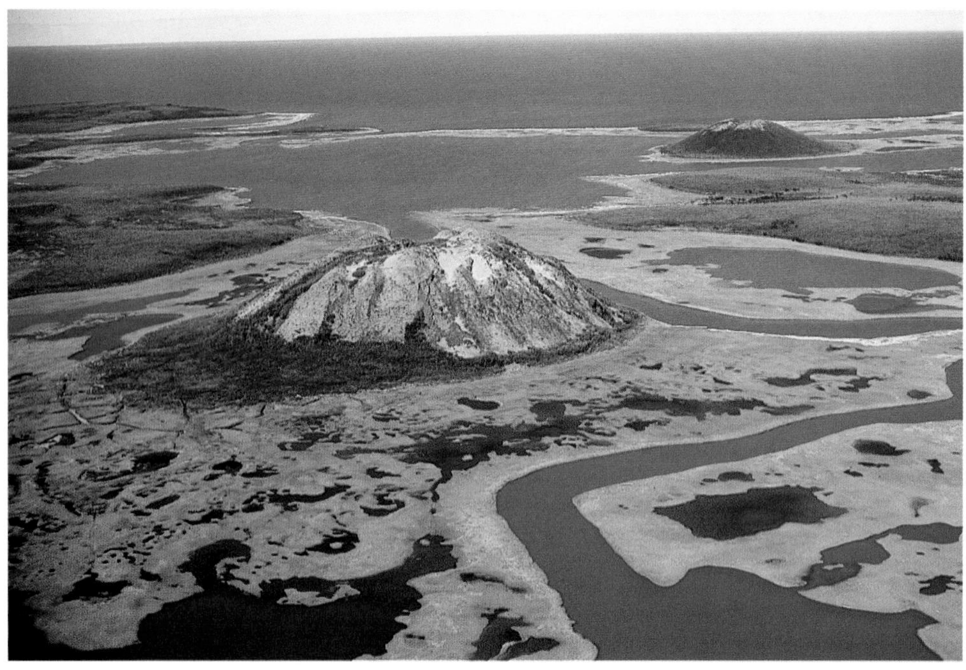

15.28 Ibyuk pingo
Located on the arctic coast near Tuktoyaktuk, Northwest Territories, this classic pingo began to form about 1200 years ago in a drained lake bed. Since that time, coastline retreat has flooded the former lake basin and it is now connected to Kugmallit Bay of the Beaufort sea.

15.29 Stone polygons *Sorted circles of gravel form a network of stone rings on this nearly flat land surface where water drainage is poor because of underlying permafrost. The circles in the foreground are 3 to 4 m (10 to 13 ft) across; the gravel ridges are 20 to 30 cm (8 to 12 in.) high. Broggerhalvoya, western Spitsbergen, latitude 78° N.*

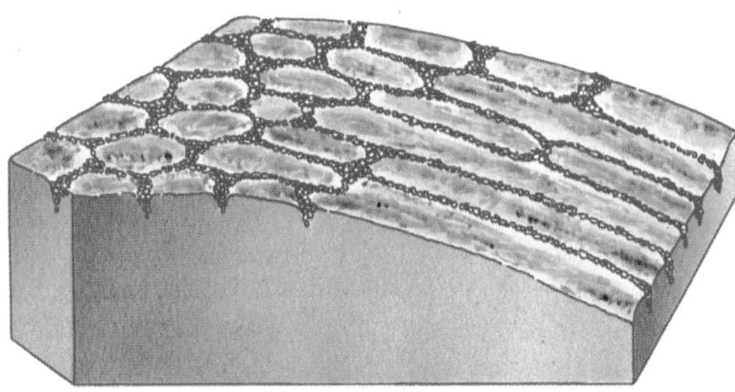

15.30 Stone stripes *This diagram shows how stone rings are often drawn out downslope into stone stripes. (Adapted from C. F. S. Sharpe, Landslides and Related Phenomena, p. 37, Figure 5. New York, Columbia Univ. Press. Used by permission of the publisher.)*

Alpine Tundra

Most of the periglacial processes and forms of the low arctic tundra are also found in the high alpine tundra throughout high mountains of the middle and high latitudes. Major regions of *alpine permafrost* in the northern hemisphere are shown on our map, Figure 15.23. In the alpine tundra, there is a dominance of steep mountainsides with large exposures of hard bedrock that has been strongly abraded by glacial ice. Postglacial talus slopes and cones formed of large angular blocks are conspicuously developed (Figure 15.2c). Patterned ground and solifluction terraces occupy relatively small valley floors where slopes are low and finer sediment tends to accumulate.

Alpine permafrost is restricted to elevations at which the mean annual temperature is below freezing. The general elevation at which mean annual temperatures fall to 0°C (32°F) will, of course, be related to latitude. In subtropical latitudes, elevations of 4000 m (about 13,000 ft) or greater are required, while at latitudes above 70° permafrost can be encountered at sea level. Figure 15.32 is a schematic graph showing the boundary between continuous and discontinuous permafrost as it varies with elevation and latitude. While the boundary at lat. 40° lies at about 3500 m (about 11,500 ft), it has fallen to 2200 m (about 6600 ft) at lat. 50°, and by lat. 68° it has descended to sea level.

Environmental Problems of Permafrost

Environmental degradation of permafrost regions arises from surface changes produced by human activity. One such activity is the destruction or removal of an insulating surface cover that typically consists of a layer of decaying organic

15.31 Solifluction *Solifluction has created this landscape of soil mounds in the Richardson Mountains, Northwest Territories, Canada. Bulging masses of water-saturated regolith have slowly moved downslope, overriding the permafrost below, while carrying their covers of plants and soil.*

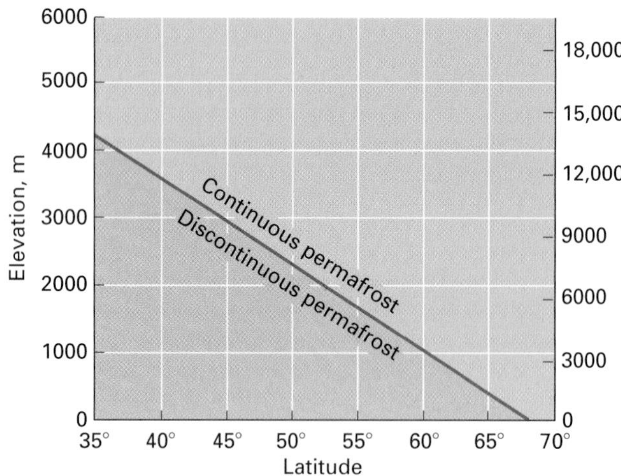

15.32 Boundary between continuous and discontinuous permafrost *This graph shows how the elevation of the boundary between continuous and discontinuous alpine permafrost changes with the latitude.*

matter in combination with living plants of the tundra or arctic forest. When this layer is scraped off or burned away, the summer thaw is extended to a greater depth, with the result that ice wedges and other ice bodies melt in the summer and waste downward. Meltwater mixes with silt and clay to form mud, which is then eroded and transported by water streams, leaving trench-like morasses. This total activity is called *thermal erosion.*

Over large expanses of nearly flat arctic tundra, removal of the natural surface cover is followed by widespread subsidence of the ground and the formation of depressions. Water erosion tends to enlarge these depressions, and thus numerous new lakes come into existence. The resulting terrain is described as *thermokarst.* (The word "karst," explained in Chapter 16, means a hummocky terrain developed by uneven solution removal of limestone bedrock.)

The consequences of disturbance of permafrost terrain became evident in World War II, when military bases, airfields, and highways were constructed hurriedly without regard for the maintenance of the natural protective surface insulation. In extreme cases, scraped areas turned into mud-filled depressions and even into small lakes that expanded in area with successive seasons of thaw, engulfing nearby buildings. Engineering practices now call for placing buildings on piles with an insulating air space below or for the deposition of a thick insulating pad of coarse gravel over the surface prior to construction. Steam and hot-water lines are placed above ground to prevent thaw of the permafrost layer.

Another serious engineering problem of arctic regions is the behavior of streams in winter. As the surfaces of streams and springs freeze over, the water beneath bursts out from place to place, freezing into huge accumulations of ice. If this phenomenon occurs at a highway bridge or culvert, the roadway may become impassible.

The lessons of superimposing our technology on a highly sensitive natural environment were learned the hard way—by encountering undesirable and costly effects that were not anticipated. Continued economic development of arctic regions is likely to teach additional hard lessons.

A LOOK AHEAD In this chapter, we have examined two related topics—weathering and mass wasting. In the weathering process, rock near the surface is broken up into smaller fragments and often altered in chemical composition. In the mass wasting process, weathered rock and soil move downhill in slow to sudden mass movements. The landforms of mass wasting are produced by gravity acting directly on soil and regolith. Gravity also powers another landform-producing agent—running water—which we take up in the next three chapters. The first deals with water in the hydrologic cycle, in soil, and in streams. The second deals specifically with how streams and rivers erode regolith and deposit sediment to create landforms. The third describes how stream erosion strips away rock layers of different resistance, providing large landforms that reveal underlying rock structures.

Chapter Summary

- **Weathering** is the action of processes that cause rock near the surface to disintegrate and decompose into **regolith**. **Mass wasting** is the spontaneous downhill motion of soil, regolith, or rock under gravity.
- **Physical weathering** produces regolith from solid rock by breaking bedrock into pieces. *Frost action* breaks rock apart by the repeated growth and melting of ice crystals in rock fractures and *joints*, as well as between individual mineral grains. In mountainous regions of vigorous frost, fields of angular blocks accumulate as *felsenmeers*. Slopes of rock fragments form *talus cones*. In soils and sediments, needle ice and ice lenses push rock and soil fragments upward. *Salt-crystal growth* in dry climates breaks individual grains of rock free, and can damage brick and concrete. *Unloading* of the weight of overlying rock layers can cause some types of rock to expand and break loose into thick shells, producing *exfoliation domes*. Daily temperature cycles in arid environments are thought to cause rock breakup. Wedging by plant roots also forces rock masses apart.
- **Chemical weathering** results from mineral alteration. Igneous and metamorphic rocks can decay to great depths through *hydrolysis* and *oxidation*, producing a regolith that is often rich in clay minerals. *Carbonic acid* action

dissolves limestone. In warm, humid environments, basaltic lavas can also show features of solution weathering produced by acid action.

- Mass wasting occurs on slopes that are mantled with regolith. **Soil creep** is a process of mass wasting in which regolith moves down slopes almost imperceptibly under the influence of gravity. In an **earthflow**, water-saturated soil or regolith slowly flows downhill. Quick clays, which are unstable and can liquefy when exposed to a shock, have produced earthflows in previously glaciated regions. A **mudflow** is much swifter than an earthflow. It follows stream courses, becoming thicker as it descends and picks up sediment. A watery mudflow with debris ranging from fine particles to boulders to tree trunks and limbs is called a *debris flow*. A **landslide** is a rapid sliding of large masses of bedrock, sometimes triggered by an earthquake.
- The **scarification** of land by human activities, such as mining of coal or ores, can heap up soil and regolith into unstable masses that produce earthflows or mudflows. Mass wasting can also be caused by removal of supporting layers, undermining the natural support of soil and regolith. These actions are termed *induced mass wasting*.
- The tundra environment is dominated by the **periglacial** *system* of distinctive landforms and processes related to freezing and thawing of water in the active layer of **permafrost**. *Continuous permafrost* and *sub-sea permafrost* occur at the highest latitudes, flanked by a band of *discontinuous permafrost* that is transitional to warmer regions.
- *Pingos* and **ice wedges** are forms of **ground ice** found in permafrost terrains. **Patterned ground** occurs when ice growth and melting in freeze–thaw cycles create *stone polygons* and *ice-wedge polygons*. During the brief summer thaw, saturated soils flow to form **solifluction** *terraces* and *lobes*. Alpine tundra and permafrost extend to lower latitudes in high-mountain environments. Removal of the surface layer for construction in permafrost terrains can cause semipermanent thawing of the upper permafrost, creating *thermal erosion* and *thermokarst*.

Key Terms

weathering	bedrock	mudflow	ground ice
regolith	sediment	landslide	ice wedge
mass wasting	alluvium	scarification	patterned ground
physical weathering	soil creep	periglacial	solifluction
chemical weathering	earthflow	permafrost	

Review Questions

1. What is meant by the term *weathering?* What types of weathering are recognized?
2. Define the terms *regolith, bedrock, sediment,* and *alluvium.*
3. How does frost action break up rock? Describe some landforms created by frost action and how they are formed.
4. How does salt-crystal growth break up rock? Give an example of a landform that arises from salt-crystal growth.
5. What is an exfoliation dome, and how does it arise? Provide an example.
6. Name three types of chemical weathering. Describe how limestone is often altered by a chemical weathering process.
7. Define mass wasting and identify the processes it includes.
8. What is soil creep, and how does it arise?
9. What is an earthflow? What features distinguish it as a landform?
10. Contrast earthflows and mudflows, providing an example of each.
11. Define the term *landslide.* How does a landslide differ from an earthflow?
12. Define and describe induced mass wasting. Provide some examples.
13. Explain the term *scarification.* Provide an example of an activity that produces scarification.
14. What is meant by the term *periglacial?*
15. Define and describe permafrost and some of its features, including ground ice, active layer, and permafrost table.
16. Identify and describe two forms of ground ice.
17. What is patterned ground? Identify two types of patterned ground.
18. What environmental problems are associated with human activity in regions of permafrost?

Eye on the Environment 15.1 • The Great Hebgen Lake Earthquake

1. Describe the formation of Earthquake Lake. What four forms of natural environmental disaster were involved?

Focus on Systems 15.3 • Permafrost as an Energy Flow System

1. Contrast the energy flows reaching the ground through the surface above and rising from below.
2. How does the profile of temperature with depth change from winter to summer in a permafrost region?
3. How is mean annual surface temperature related to the thickness of the active layer and the depth of permafrost?

Visualizing Exercises

1. Define the terms *regolith, bedrock, sediment,* and *alluvium.* Sketch a cross section through a part of the landscape showing these features and label them on the sketch.
2. Copy or trace Figure 15.10; then identify and plot on the diagram the mass movement associated with each of the following locations: Turtle Mountain, Palos Verdes Hills, Nicolet, Mount St. Helens, Madison River, and Herculaneum.

Essay Questions

1. A landscape includes a range of lofty mountains elevated above a dry desert plain. Describe the processes of weathering and mass wasting that might be found on this landscape and identify their location.
2. Imagine yourself as the newly appointed director of public safety and disaster planning for your state or province. One of your first jobs is to identify locations where human populations are threatened by potential disasters, including those of mass wasting. Where would you look for mass wasting hazards and why? In preparing your answer, you may want to consult maps of your state or province.

Problems

Working It Out 15.2 • The Power of Gravity

1. According to the account, the slide reached a velocity of at least 150 km/hr. What was the kinetic energy of the slide at that point? Recall from *Focus on Systems 4.1 • Forms of Energy* that the formula for kinetic energy is $E = {}^1/_2 mv^2$, where m is the mass (kg) and v is the velocity (m/s). How does this value compare to the total energy released in the slide?
2. Physics predicts the velocity, v (m/s), of a falling body in the absence of friction to be $v = \sqrt{2gd}$, where g (m/s^2) is the acceleration of gravity and d (m) is the distance of fall from rest. What is the velocity of a object falling without friction for a distance of 500 m? How does this compare with the velocity observed for the Madison Slide? Why are the two values different?

Eye on the Landscape

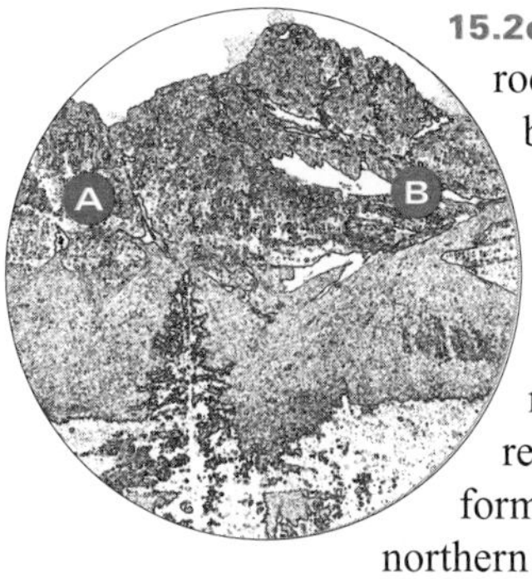

15.2c Talus cones Individual rock strata **(A)** are readily visible in this photo. They are ancient sedimentary rocks that were thrust over and above younger rocks of the plains during an arc-continent collision. Strong and resistant to erosion, they now form the magnificent peaks of the northern Rockies in Alberta. At **(B)** are the remains of a small glacier, coated with gray blocks of talus. Chapter 13, on plate tectonics, describes overthrust faults and arc-continent collisions. Alpine glaciers are covered in Chapter 20.

15.6 Granular disintegration The backdrop for this photo is the eastern side of the Sierra Nevada Range. The steep triangular facets of the mountain slopes at **(A)** mark a fault scarp along which the rocks have been uplifted by many successive motions. Faults and fault scarps are treated in Chapter 18. The peak at **(B)** shows recent glaciation, with knife-edge slopes and rounded basins. Compare with Figure 20.5 in Chapter 20, where the landforms of mountain glaciers are discussed.

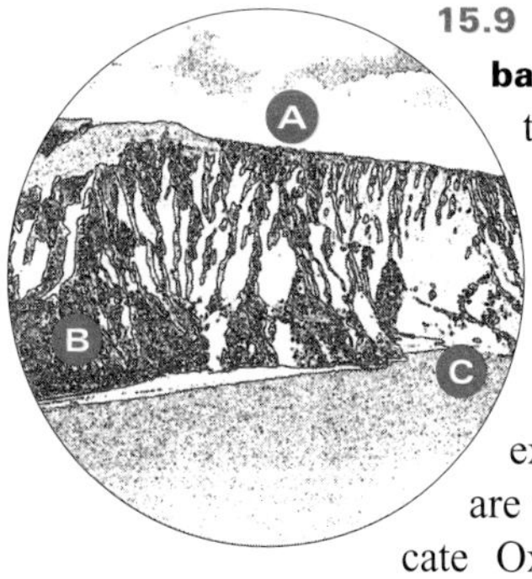

15.9 Solution weathering of basalt Kauai is the oldest of the Hawaiian Islands. Note the rounded dome shape of its outline **(A)**, which is characteristic of shield volcanoes (Chapter 14). The red-brown colors of soil, exposed on lower slopes **(B)**, are those of iron oxides and indicate Oxisols (Chapter 21). Pocket beaches and an arch at **(C)** are products of wave action (Chapter 19).

16 THE CYCLING OF WATER ON THE CONTINENTS

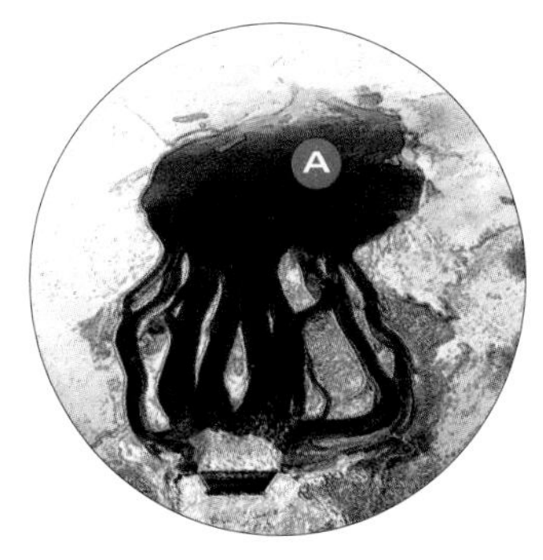

EYE ON THE LANDSCAPE

Waste water outflow from a desalination plant, Jahra region, Kuwait, into the Sea of Al-Doha. Human activities require fresh water, and where it is scarce, desalination of sea water can produce potable water. However, the cost is high, both in energy and in the environmental impact of salt brine released to the ocean. **What else would the geographer see?...Answers at the end of the chapter.**

Water is essential to life. Nearly all organisms require a constant flux of water or at least a water-rich environment for survival. Humans are no exception. Our activity depends on a constant supply of fresh water that is provided by precipitation over the lands. Some of this water infiltrates the land surface and is stored in soils, regolith, and pores in bedrock. An even smaller portion is in motion as flowing fresh water in streams and rivers. In this chapter, we focus on these two parts of the hydrologic cycle—water at the land surface and water that lies within the ground.

Recall from Chapter 6 that fresh water on the continents in surface and subsurface water is only a very small part—about 3 percent—of the hydrosphere's total water. Most of this fresh water is locked into ice sheets and mountain glaciers. Ground water accounts for a little more than half of 1 percent. Although this is a very small fraction of global water, it is still many times larger than the amount of fresh water in lakes, streams, and rivers, which account for only three-hundredths of 1 percent of the total water. Note also that ground water can be found at almost every location on land that receives rainfall. In contrast, fresh surface water varies widely in abundance. In many arid regions, streams and rivers are nonexistent for most or all of the year.

Let's review the hydrologic cycle, focusing on surface and ground water. In Chapter 6, we discussed atmospheric moisture and precipitation, describing the part of the hydrologic cycle in which water evaporates from ocean and land surfaces and then precipitates, falling to Earth as rain or snow. What happens then to this precipitation? Figure 16.1 provides the answer. As the diagram shows, a portion of the precipitation returns directly to the atmosphere through evaporation from the soil. Another portion travels downward,

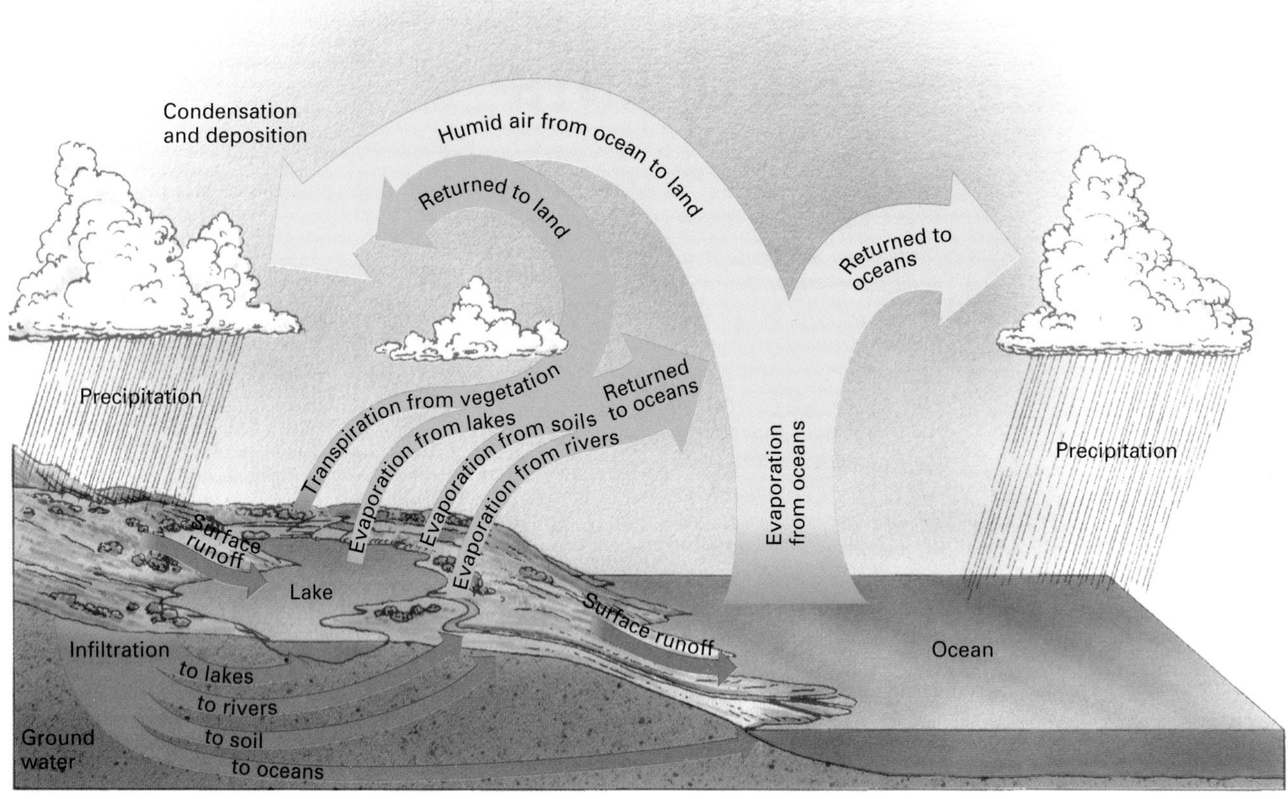

***16.1 The hydrologic cycle** The hydrologic cycle traces the various paths of water from oceans, through the atmosphere, to land, and its return to oceans.*

moving through the soil under the force of gravity to become part of the underlying ground water. Following underground flow paths, this subsurface water eventually emerges to become surface water, or it may emerge directly in the shore zone of the ocean. A third portion flows over the ground surface as runoff to lower levels. As it travels, the water flow becomes collected into streams, which eventually conduct the running water to the ocean.

In this chapter, we will trace the parts of the hydrologic cycle that include both the subsurface and surface pathways of water flow. The study of these flows is part of the science of *hydrology,* which is the study of water as a complex but unified system on the Earth.

Figure 16.2 shows what happens to water from precipitation as it first reaches the land surface. Most soil surfaces in their undisturbed, natural states are capable of absorbing the water from light or moderate rains by **infiltration.** In this process, water enters the small natural passageways between irregularly shaped soil particles, as well as the larger openings in the soil surface. These openings are formed by the borings of worms and animals, earth cracks produced by soil drying, cavities left from decay of plant roots, or spaces made by the growth and melting of frost crystals. A mat of decaying leaves and stems breaks the force of falling water drops and helps to keep these openings clear.

The precipitation that infiltrates the soil is temporarily held in the soil layer as soil water, occupying the *soil water belt* (Figure 16.2). Water within this belt can be returned to the surface and then to the atmosphere through a process that combines two components—direct evaporation from soil and transpiration by vegetation. As we explained in Chapter 5, these two forms of water vapor transport are combined in the term *evapotranspiration.*

When rain falls too rapidly to be passed downward through soil openings, **runoff** occurs and a surface water layer runs over the surface and down the direction of ground slope. This surface runoff is called **overland flow.** In periods of heavy, prolonged rain or rapid snowmelt, overland flow feeds directly to streams.

Overland flow also occurs when soil that is already saturated receives rainfall or snowmelt. Since soil openings and pores are already filled, water cannot infiltrate the soil and drain down to deeper layers. Under these conditions, nearly all of the precipitation or snowmelt will run off.

Runoff as overland flow moves surface particles from hills to valleys, and so it is an agent that shapes landforms.

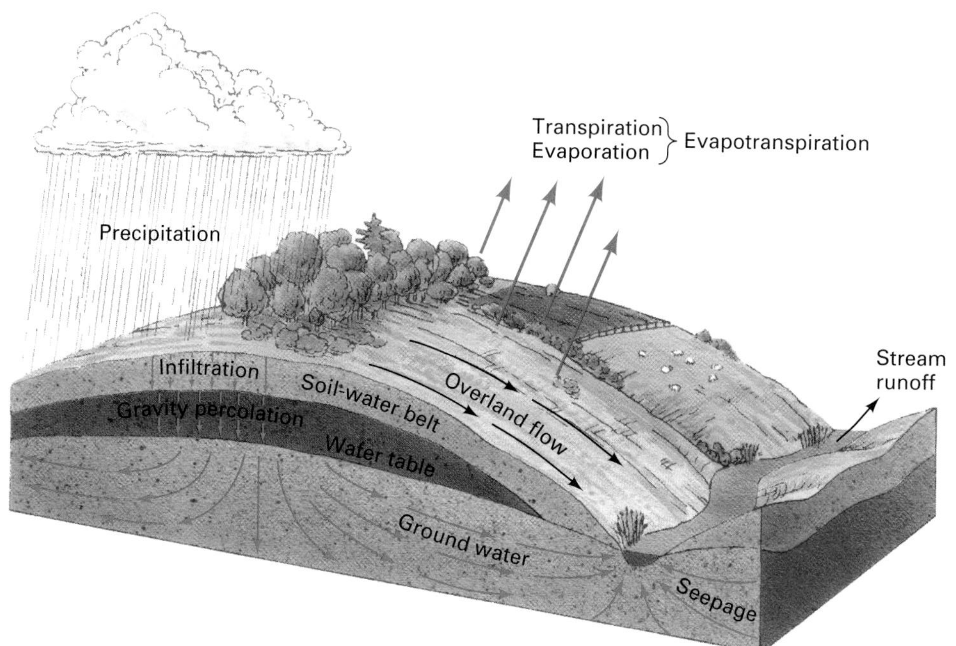

16.2 Paths of precipitation *Precipitation falling on the land follows three primary paths. Some precipitation is returned to the atmosphere through evapotranspiration, while some runs off the soil surface. The remainder sinks into the soil water belt, where it is accessible to plants. A portion of the infiltrating water passes through the soil water belt and percolates down to the ground water zone. (A. H. Strahler.)*

Because runoff supplies water to streams and rivers, it also allows rivers to cut canyons and gorges, and carry sediment to the ocean—but we are getting ahead of our story. We'll return to landforms carved by running water in Chapter 17.

Ground Water

As shown in Figure 16.2, water derived from precipitation can continue to flow downward beyond the soil water belt. This slow downward flow under the influence of gravity is termed *percolation*. Eventually, the percolating water reaches ground water. **Ground water** is the part of the subsurface water that fully saturates the pore spaces in bedrock, regolith, or soil, and so occupies the *saturated zone* (Figure 16.3). The **water table** marks the top of this zone. Above it is the *unsaturated zone* in which water does not fully saturate the pores. Here, water is held in thin films adhering to mineral surfaces. This zone also includes the soil water belt.

Ground water moves slowly along deep flow paths, eventually emerging by seepage into streams, ponds, lakes, and marshes. In these places the land surface dips below the water table. Streams that flow throughout the year—perennial streams—derive much of their water from ground water seepage. GEODISCOVERIES

The Water Table Surface

Where there are many wells in an area, the position of the water table can be mapped in detail (Figure 16.4). This is

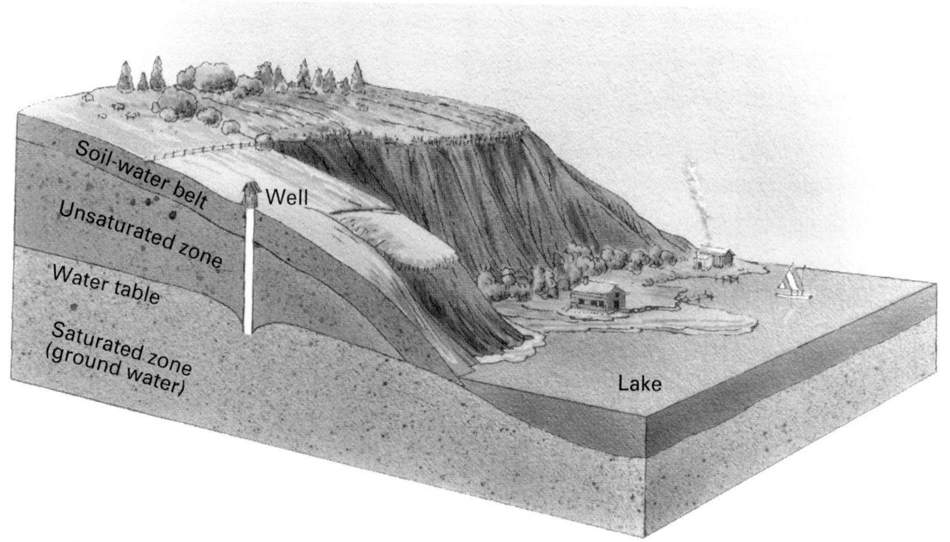

16.3 Zones of subsurface water *Water in the soil water belt is available to plants. Water in the unsaturated zone percolates downward to the saturated zone of ground water, where all pores and spaces are filled with water.*

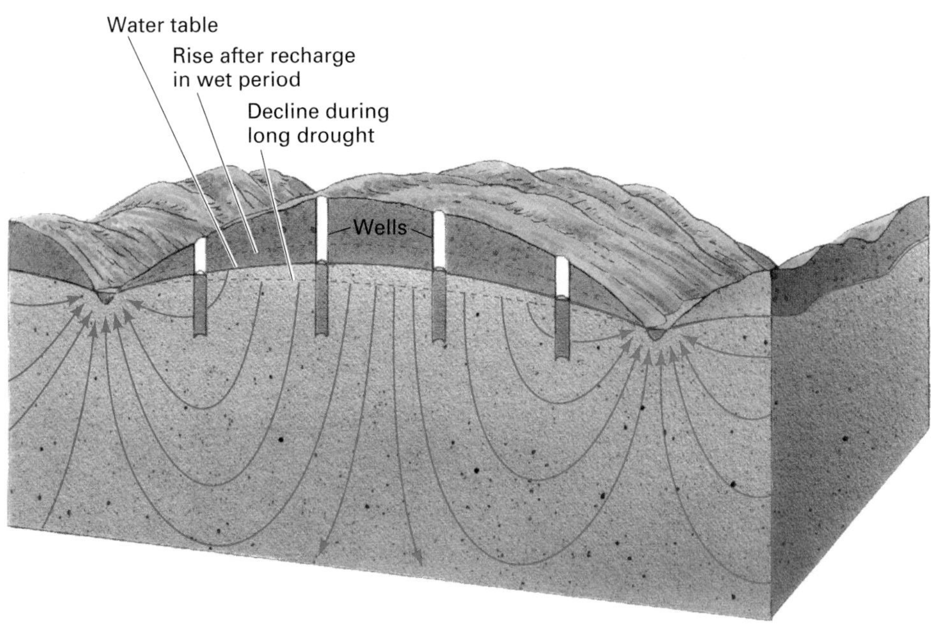

***16.4 Water table surface** The configuration of the water table surface conforms broadly with the land surface above it. It varies in response to prolonged wet and dry periods. Ground water flow paths circulate water to deep levels in a very slow motion and eventually feed streams by seepage.*

done by plotting the water heights in the wells and noting the trend of change in elevation from one well to the other. The water table is highest under the highest areas of land surface—hilltops and divides. The water table declines in elevation toward the valleys, where it appears at the surface close to streams, lakes, or marshes.

The reason for this water table configuration is that water percolating down through the unsaturated zone tends to raise the water table, while seepage into lakes, streams, and marshes tends to draw off ground water and to lower its level. These differences in water table level are built up and maintained because ground water only moves extremely slowly through the fine chinks and pores of bedrock and regolith. In periods of high precipitation, the water table rises under hilltops or divide areas. In periods of water deficit, or during a drought, the water table falls (Figure 16.4).

Figure 16.4 also shows paths of ground water flow. The direction of flow at any point depends on the direction of pressure at that point. Gravity always exerts a downward pressure, while the difference in the height of the water table between hilltop and streambed produces a sideward pressure force. In the presence of resistance to flow, the result of these downward and sideward pressures causes ground water to flow in curving paths. Water that enters the hillside midway between the hilltop and the stream flows rather directly toward the stream. Water reaching the water table midway between the two streams, however, flows almost straight down to great depths before recurving and rising upward again. Progress along these deep paths is incredibly slow, while flow near the surface is much faster. The most rapid flow is close to the stream, where the arrows converge. Over time, the level of the water table tends to remain stable, and the flow of water released to streams and lakes must balance the flow of water percolating down into the water table.

Aquifers

Sedimentary layers often exert a strong control over the storage and movement of ground water. For example, clean, well-sorted sand—such as that found in beaches, dunes, or stream alluvium—can hold an amount of ground water about equal to about one-third of its bulk volume. A bed of sand or sandstone is thus often a good *aquifer*—that is, a layer of rock or sediment that contains abundant, freely flowing ground water. In contrast, beds of clay and shale are relatively impermeable and hold little free water. They are known as *aquicludes*. Figure 16.5 shows a situation in which a shale bed is overlain by a bed of sandstone. Ground water moves freely through the sandstone aquifer and across the top of the shale aquiclude, emerging in springs along the canyon shown at the right. Also shown is a small lens of shale, which accumulates a shallow *perched water table* above it.

***16.5 Ground water in horizontal strata** Water flows freely within the sandstone aquifer but only very slowly in the shale aquiclude. A lens of shale creates a perched water table above the main water table. (Copyright © A. N. Strahler.)*

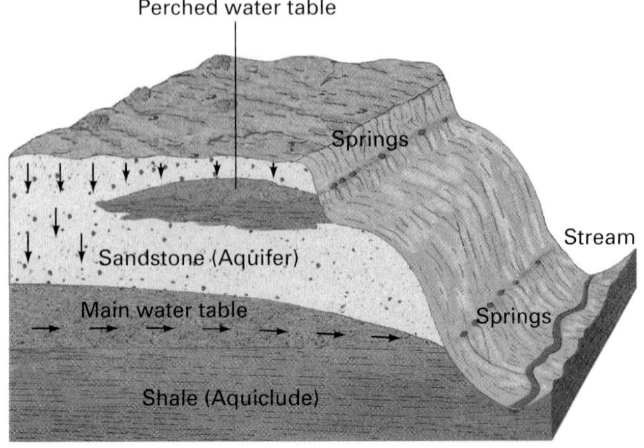

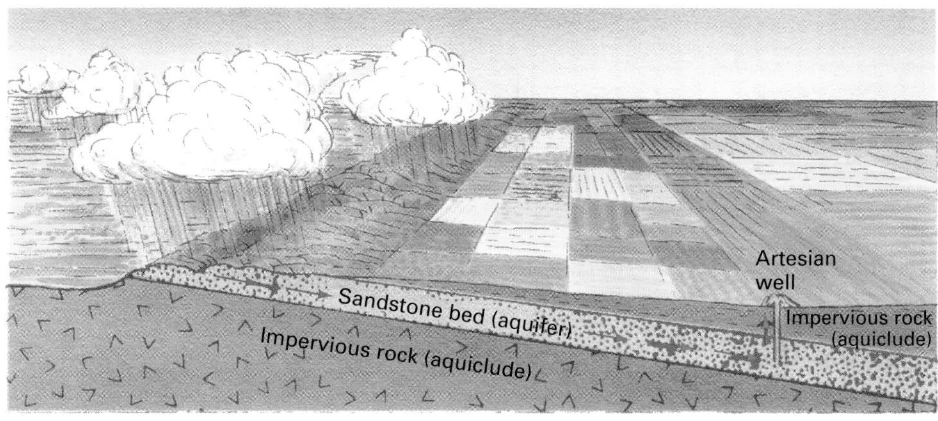

16.6 Artesian well *A dipping sandstone layer interleaved between impervious layers provides a source of water under pressure that flows naturally to the surface. (Drawn by Erwin Raisz. Copyright © A. N. Strahler.)*

When an aquifer is situated between two aquicludes, ground water in the aquifer may be under pressure and thus flow freely from a well. This type of self-flowing well is an **artesian well.** Figure 16.6 illustrates this situation, in which a porous sandstone bed (aquifer) is sandwiched between two impervious rock layers (aquicludes). Precipitation on the hills where the sandstone outcrops provides water that saturates the sandstone layer, filling it to that elevation. Since the elevation of the well that taps the aquifer is below that of the range of hills feeding the aquifer, hydrostatic pressure causes water to rise in the well.

Limestone Solution by Ground Water

We saw in Chapter 15 that in moist climates limestone at the surface is slowly dissolved by carbonic acid action, producing lowland areas. The slow flow of ground water in the saturated zone can also dissolve limestone below the surface, producing deep underground caverns. These features can then collapse, producing a sinking of the ground above and the development of a unique type of landscape.

Limestone Caverns

You are probably familiar with such famous caverns as Mammoth Cave or Carlsbad Caverns. *Limestone caverns* are interconnected subterranean cavities in bedrock formed by the corrosive action of circulating ground water on limestone. Figure 16.7 shows how caverns develop. As shown in the upper diagram, the action of carbonic acid is particularly concentrated in the saturated zone just below the water table. This removal process forms many kinds of underground "landforms," such as tortuous tubes and tunnels, great open chambers, and tall chimneys. Subterranean streams can be found flowing in the lowermost tunnels, and these carry the products of solution to emerge along the banks of surface streams and rivers.

In a later stage, shown in the right diagram, the stream has deepened its valley, and the water table has been correspondingly lowered to a new position. The cavern system previously formed is now in the unsaturated zone. Deposition of carbonate matter, known as *travertine,* takes place on exposed rock surfaces in the caverns. Encrustations of travertine take many beautiful forms—stalactites (hanging rods), stalagmites (upward pointing rods), columns, and drip curtains (Figure 16.8).

16.7 Cavern development *Limestone dissolves at the top of the ground water zone. When rapid erosion of streams lowers the water table, caverns result in the unsaturated zone. Water flow through the caverns results in travertine deposition. (Copyright © A. N. Strahler.)*

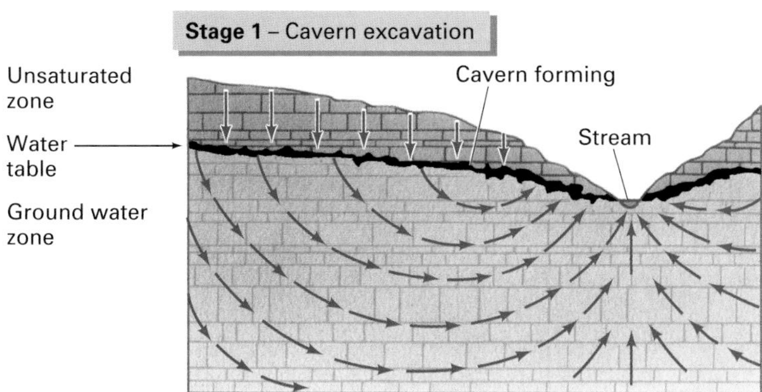

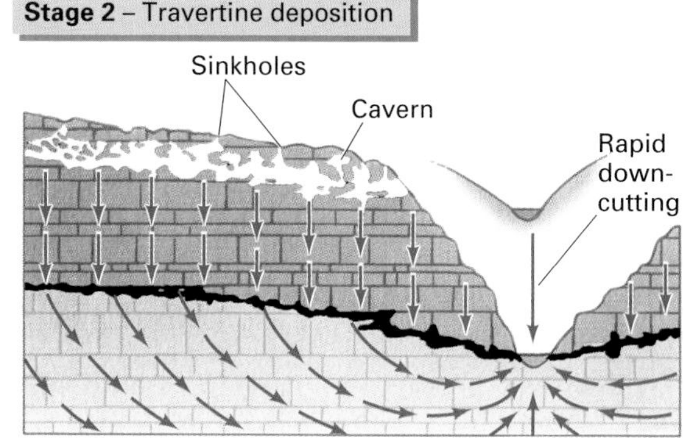

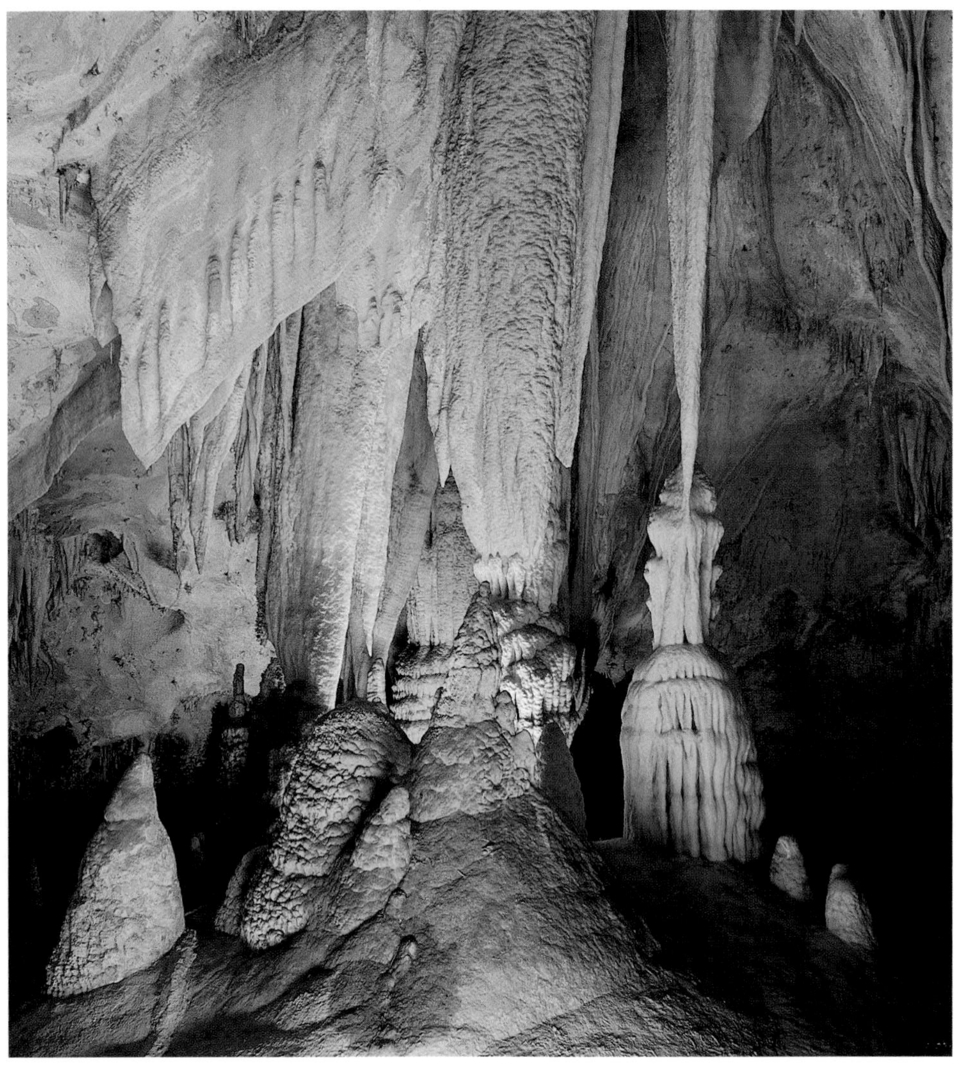

16.8 Inside a cavern *Travertine deposits in the Papoose Room of Carlsbad Caverns in New Mexico. Deposits include stalactites (slender rods hanging from the ceiling), stalagmites (upward pointing rods), and sturdy columns, all formed when dripping ground water evaporates, leaving travertine deposits of calcium carbonate behind.*

Karst Landscapes

Where limestone solution is very active, we find a landscape with many unique landforms. This is especially true along the Dalmatian coastal area of Croatia, where the landscape is called *karst.* Geographers apply that term to the topography of any limestone area where sinkholes are numerous and small surface streams are nonexistent. A *sinkhole* is a surface depression in a region of cavernous limestone (Figure 16.9). Some sinkholes are filled with soil washed from nearby hillsides, while others are steep-sided, deep holes. They develop where the limestone is more susceptible to solution weathering, or where an underground cavern near the surface has collapsed.

Development of a karst landscape is shown in Figure 16.10. In an early stage, funnel-like sinkholes are numerous. Later, the caverns collapse, leaving open, flat-floored valleys. Examples of some important regions of karst or karst-like topography are the Mammoth Cave region of Kentucky, the Yucatan Peninsula, and parts of Cuba and Puerto Rico. In regions such as southern China and west Malaysia, the karst landscape is dominated by steep-sided, conical limestone hills or towers, 100 to 500 m (about 300 to 1500 ft) high (Figure 16.11). The towers are sometimes capped by beds of more resistant rock or by impure limestones that dissolve more slowly. They are often riddled with caverns and passageways.

Problems of Ground Water Management

Rapid withdrawal of ground water has seriously impacted the environment in many places. Increased urban populations and industrial developments require larger water supplies—needs that cannot always be met by constructing new surface water reservoirs. To fill these needs, vast numbers of wells using powerful pumps draw huge volumes of ground water to the surface, greatly altering nature's balance of ground water discharge and recharge.

In dry climates, agriculture is often heavily dependent on irrigation water from pumped wells—especially since major river systems are likely to be already fully utilized for irrigation. Wells are also convenient water sources. They can be drilled within the limits of a given agricultural

16.9 ***Eye on the Landscape*** ***Sinkholes*** *Sinkholes in limestone are created by solution. These sinkholes are near Roswell, New Mexico.* **What else would the geographer see? ... Answers at the end of the chapter.**

16.10 Evolution of a karst landscape *(a) Rainfall enters the cavern system through sinkholes in the limestone. (b) Extensive collapse of caverns reveals surface streams flowing on shale beds beneath the limestone. Some parts of the flat-floored valleys can be cultivated. (Drawn by Erwin Raisz. Copyright © A. N. Strahler.)*

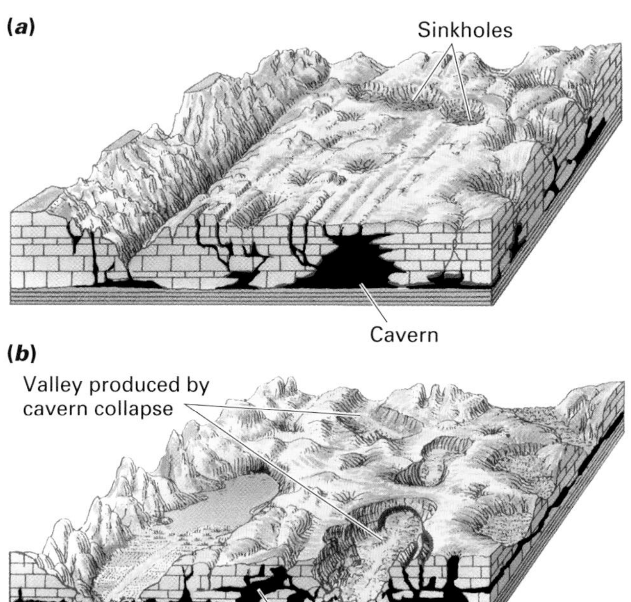

or industrial property and can provide immediate supplies of water without any need to construct expensive canals or aqueducts.

In earlier times, the small well that supplied the domestic and livestock needs of a home or farmstead was actually dug by hand and sometimes lined with masonry. By contrast, a modern well supplying irrigation and industrial water is drilled by powerful machinery that can bore a hole 40 cm (16 in.) or more in diameter to depths of 300 m (about 1000 ft) or more. Drilled wells are lined with metal casings that exclude impure near-surface water and prevent the walls from caving in and clogging the tube. Near the lower end of the hole, in the ground water zone, the casing is perforated to admit the water. The yield of a single drilled well ranges from as low as a few hundred liters or gallons per day in a domestic well to many millions of liters or gallons per day for a large industrial or irrigation well.

Water Table Depletion

As water is pumped from a well, the level of water in the well drops. At the same time, the surrounding water table is lowered in the shape of a downward-pointing cone, termed the *cone of depression* (Figure 16.12). The difference in height between the cone tip and the original water table is the *drawdown.* The cone of depression may extend out as far

***16.11 Eye on the Landscape* Tower karst** *White limestone can be seen exposed in the nearly vertical sides of these towers. Near Guilin (Kweilin), Guanxi Province, southern China at lat. 25° N.* **What else would the geographer see?...Answers at the end of the chapter.**

16.12 Drawdown and cone of depression in a pumped well *As the well draws water, the water table is depressed in a cone shape centered on the well. (A. H. Strahler.)*

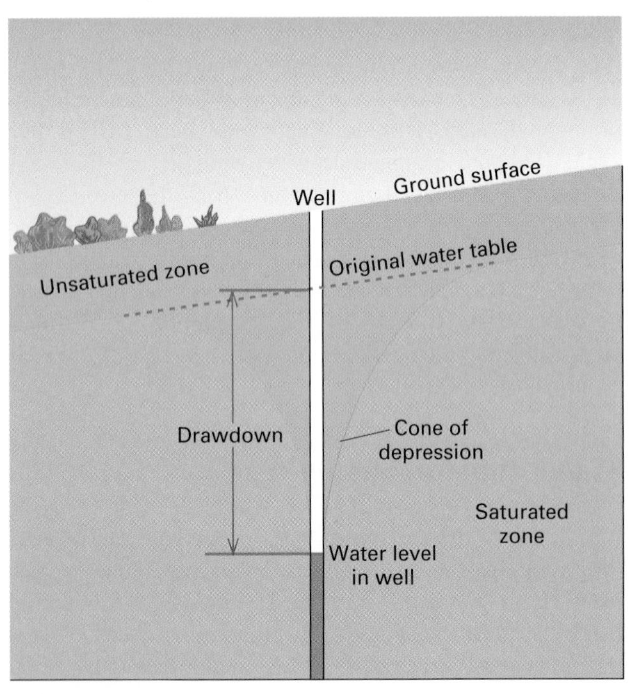

as 16 km (10 mi) or more from a well where heavy pumping is continued. Where many wells are in operation, their intersecting cones will produce a general lowering of the water table.

Water table depletion often greatly exceeds recharge—the rate at which infiltrating water moves downward to the saturated zone. In arid regions, much of the ground water for irrigation is drawn from wells driven into thick aquifers of sand and gravel. These deposits are often recharged by the seasonal flow of streams that originate high in adjacent mountains. Fanning out across the dry lowlands, the streams lose water, which sinks into the sands and gravels and eventually percolates to the water table below. The extraction of ground water by pumping can greatly exceed this recharge by stream flow, lowering the water table. Deeper wells and more powerful pumps are then required. The result is exhaustion of a natural resource that is not renewable except over very long periods of time.

Another byproduct of water table depletion is subsidence—a sinking of the land in response to the removal of water from underlying sediments. This problem has plagued a number of major cities that rely heavily on ground water wells for their water supplies. *Eye on Global Change 16.1* • *Sinking Cities* provides more information on this topic.

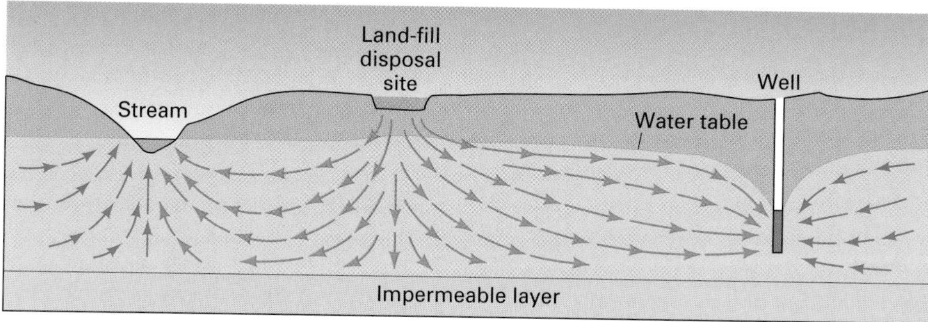

16.13 Movement of polluted ground water *Polluted water, leached from a waste disposal site, moves toward a supply well (right) and a stream (left). (Copyright © A. N. Strahler.)*

Contamination of Ground Water

Another major environmental problem related to ground water withdrawal is contamination of wells by pollutants that infiltrate the ground and reach the water table. Both solid and liquid wastes are responsible. Disposal of solid wastes poses a major environmental problem in developed countries because their advanced industrial economies provide an endless source of garbage and trash. Traditionally, these waste products were trucked to the town dump and burned there in continually smoldering fires that emitted foul smoke and gases. The partially consumed residual waste was then buried under earth.

In recent decades, a major effort has been made to improve solid-waste disposal methods. One method is high-temperature incineration, but it often leads to air pollution. Another is the sanitary landfill method in which waste is not allowed to burn. Instead, layers of waste are continually buried, usually by sand or clay available on the landfill site. The waste is thus situated in the unsaturated zone. Here it can react with rainwater that infiltrates the ground surface. This water picks up a wide variety of chemical compounds from the waste body and carries them down to the water table (Figure 16.13).

Once in the water table, the pollutants follow the flow paths of the ground water. As the arrows in the figure indicate, the polluted water may flow toward a supply well that draws in ground water from a large radius. Once the polluted water has reached the well, the water becomes unfit for human consumption. Polluted water may also move toward a nearby valley, causing pollution of the stream flowing there (left side of Figure 16.13).

Another source of contamination in coastal wells is *saltwater intrusion.* Since fresh water is less dense than saltwater, a coastal aquifer can be underlain by a layer of saltwater from the ocean. When the aquifer is depleted, the level of saltwater rises and eventually reaches the well from below, rendering the well unusable.

SURFACE WATER

So far, we have examined how water moves below the land surface. Now, we turn to tracing the flow paths of surplus water that runs off the land surface and ultimately reaches the sea. Here, we will be concerned primarily with rivers and streams. (In general usage, we speak of "rivers" as large watercourses and "streams" as smaller ones. However, the word "stream" is also used as a scientific term designating the channeled flow of surface water of any amount.) GEODISCOVERIES

Overland Flow and Stream Flow

As we saw earlier in this chapter, runoff that flows down the slopes of the land in broadly distributed sheets is overland flow. We can distinguish overland flow from *stream flow,* in which the water occupies a narrow channel confined by lateral banks. Overland flow can take several forms. Where the soil or rock surface is smooth, the flow may be a continuous thin film, called *sheet flow* (Figure 16.14). Where the ground

16.14 Overland flow *Overland flow, in the form of a thin sheet of water, covers the nearly flat plain in the middle distance. The water converges into stream flow in a narrow, steep-sided gully (left). This photograph was taken shortly after a summer thunderstorm had deluged the area. The locality, near Raton, New Mexico, shows steppe grassland vegetation.*

Eye on Global Change 16.1

Sinking Cities

Ground water is a resource on which human civilization now depends to meet much of its demand for fresh water. However, it may take thousands of years to accumulate large reservoirs of underground water. When this water is removed rapidly by pumping, the rate of removal far exceeds the rate of recharge. The volume of water in the reservoir is reduced, and the water table falls. What are the effects of this change?

One important environmental effect of excessive ground water withdrawal is subsidence of the ground surface. Venice, Italy, provides a dramatic example of this side effect. Venice was built in the eleventh century A.D. on low-lying islands in a coastal lagoon, sheltered from the ocean by a barrier beach. Underlying the area are some 1000 m (about 3300 ft) of layers of sand, gravel, clay, and silt, with some layers of peat. Compaction of these soft layers has been going on gradually for centuries under the heavy load of city buildings. However, ground water withdrawal, which has been greatly accelerated in recent decades, has aggravated the condition.

Many ancient buildings in Venice now rest at lower levels and have suffered severe damage as a result of flooding during winter storms on the adjacent Adriatic Sea. Coastal storms normally raise sea level, and when high tides occur at the same time, water rises even higher. The problem of flooding during storms is aggravated by the fact that many of the canals of Venice receive raw sewage, so that the floodwater is contaminated.

Most of the subsidence in recent decades has been attributed to withdrawals of large amounts of ground water from industrial wells at Porto Marghere, the modern port of Venice, located a few kilometers distant on the mainland shore. This pumping has now been greatly curtailed, reducing the rate of subsidence to a very small natural rate (about 1 mm per year). However, the threat of flooding and damage to churches and other buildings of great historical value remains. Flood control now depends on the construction of seawalls and floodgates on the barrier beach that lies between Venice and the open ocean.

By the early 1980s the port city of Bangkok, located on the Chao Phraya River a short distance from the Gulf of Thailand, had become the world's most rapidly sinking city as a result of massive ground water withdrawals from soft marine sediments beneath the city. During the 1980s the rate of subsidence reached 14 cm (6 in.) per year, and major floods were frequent. Some reduction in the rate of ground water withdrawal has produced a modest decrease in the rate of subsidence, but as in the case of Venice, the flood danger remains.

Ground water withdrawal has affected several regions in California, where ground water for irrigation has been pumped from basins filled with alluvial sediments. Water table levels in these basins have dropped over 30 m (98 ft), with a maximum drop of 120 to 150 m (394 to 492 ft) being recorded in one locality of California's San Joaquin Valley. In the Los Bañnos–Kettleman City area, ground subsidence of as much as 4.2 to 4.8 m (14 to 16 ft) was measured at some locations in a 35-year period.

Another important area of ground subsidence accompanying water withdrawal is beneath Hous-

is rough or pitted, flow may take the form of a series of tiny rivulets connecting one water-filled hollow with another. On a grass-covered slope, overland flow is subdivided into countless tiny threads of water, passing around the stems. Even in a heavy and prolonged rain, you might not notice overland flow in progress on a sloping lawn. On heavily forested slopes, overland flow may pass entirely concealed beneath a thick mat of decaying leaves.

Overland flow eventually contributes to a stream, which is a much deeper, more concentrated form of runoff. We can define a **stream** as a long, narrow body of flowing water occupying a trenchlike depression, or channel, and moving to lower levels under the force of gravity. The **channel** of a stream is a narrow trough. The forces of flowing water shape the trough to its most effective form for moving the quantities of water and sediment supplied to the stream (Figure 16.15). A channel may be so narrow that you can easily jump across it, or, in the case of the Mississippi River, as wide as 1.5 km (about 1 mi) or more.

As a stream flows under the influence of gravity, the water encounters resistance—a form of friction—with the channel walls. As a result, water close to the bed and banks moves slowly, while water in the central part of the flow moves faster. If the channel is straight and symmetrical, the single line of maximum velocity is located in midstream. If the stream curves, the maximum velocity shifts toward the bank on the outside of the curve.

Actually, the arrows in Figure 16.15 only show average velocity. In all but the most sluggish streams, the water is affected by *turbulence,* a system of countless eddies that are

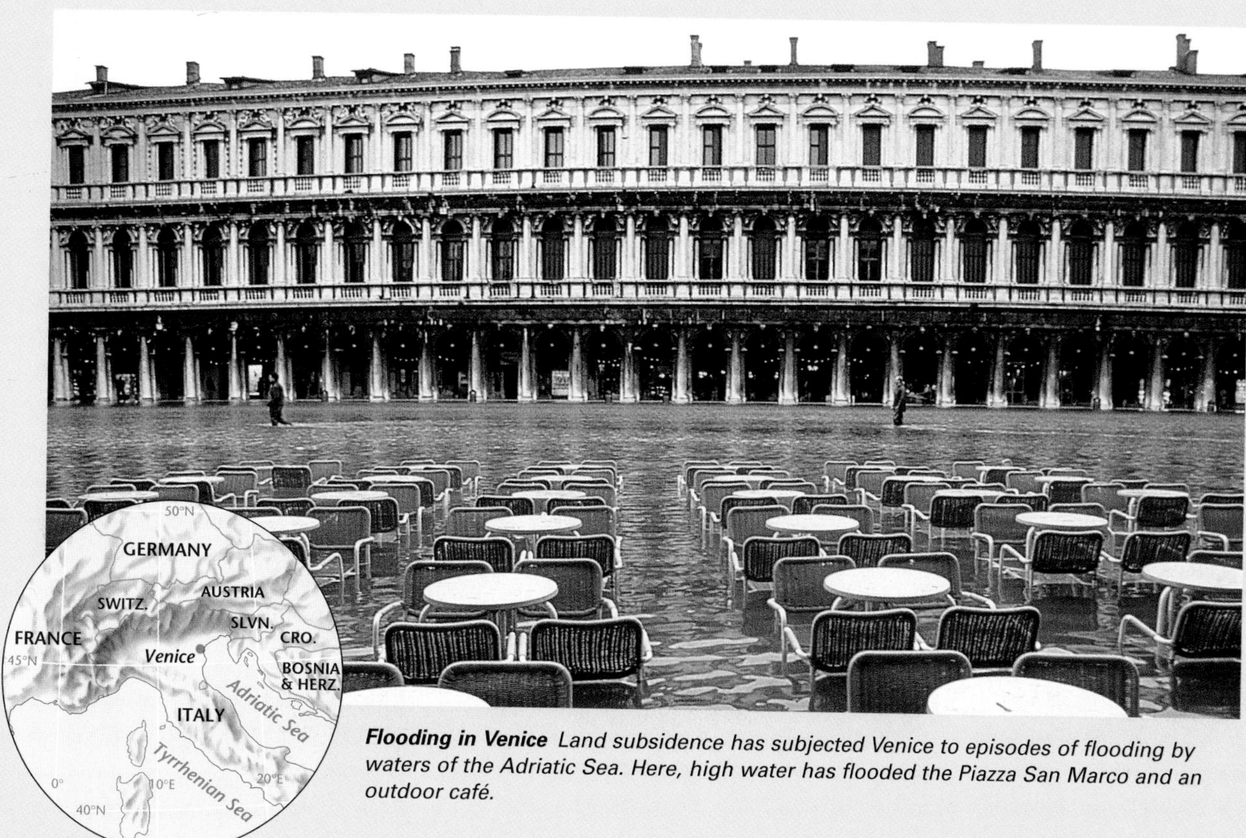

Flooding in Venice *Land subsidence has subjected Venice to episodes of flooding by waters of the Adriatic Sea. Here, high water has flooded the Piazza San Marco and an outdoor café.*

ton, Texas, where the ground surface has subsided from 0.3 to 1 m (1 to 3 ft) in a metropolitan area 50 km (31 mi) across. Damage to buildings, pavements, airport runways, and other structures has resulted.

Perhaps the most celebrated case of ground subsidence is that affecting Mexico City. Carefully measured ground subsidence has ranged from about 4 to 7 m (13 to 23 ft). The subsidence has resulted from the withdrawal of ground water from an aquifer system beneath the city and has caused many serious engineering problems. As water has been drained out, the volume of clay beds overlying the aquifer has contracted greatly. To combat the ground subsidence, recharge wells were drilled to inject water into the aquifer. In addition, new water supplies from sources outside the city area were developed to replace local ground water use.

16.15 Stream flow within a channel *Stream flow is most rapid near the center of the stream, where it encounters less friction with the bed and banks.*

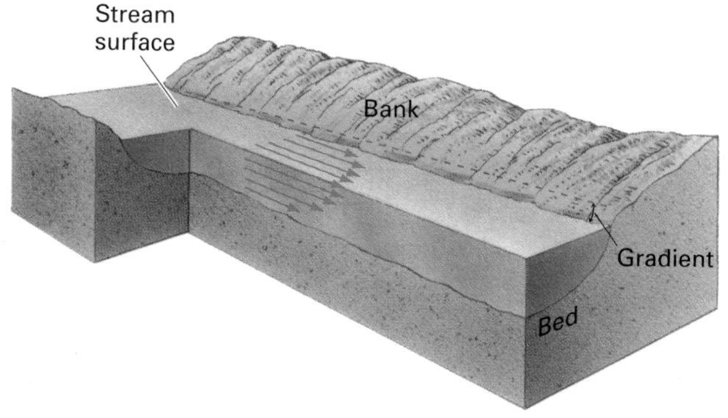

continually forming and dissolving. If we follow a particular water molecule, it will travel a highly irregular, corkscrew path as it is swept downstream. Motions include upward, downward, and sideward directions. Only if we measure the water velocity at a certain fixed point for a long period of time, say, several minutes, will the average motion at that point be downstream and in a line parallel with the surface and bed.

Stream Discharge

Stream flow at a given location is measured by its **discharge,** which is defined as the volume of water per unit time passing through a cross section of the stream at that location. It is measured in cubic meters (cubic feet) per

Geographers at Work | Energy and Water Flows in the High Arctic

by Kathy Young *York University, Toronto*

Global change is coming to the delicate high arctic environment. As we saw in our interchapter feature, *A Closer Look: Eye on Global Change 11.4 • Regional Impacts of Climate Change on North America*, most of the high arctic is predicted to become warmer, with more precipitation, yet more summer drought stress. How will global climate change affect the delicate ecosystems of arctic regions?

The hydrology of the arctic is delicately balanced on a local level. Vegetation and microclimate can change strikingly over only a few tens of meters, creating large local variations in soil water availability. Much of this variation is driven by perennial snowbanks, which release water during the summer, and by the depth and pattern of ground thaw. The surface energy balance, which determines how and where energy will be available to evaporate water, melt snow, and thaw frozen ground, also changes radically over short distances as substrates—snow, vegetation, soil—vary.

Kathy Young (shown here with an Arctic char) is an expert in the climate and hydrological processes of large perennial snowbeds in the Canadian high arctic.

To determine how the high arctic will be impacted by climate change, we must first understand the energy and water flows within these delicate environments. That's where Kathy Young, a geographer at York University, comes in.

Kathy has always been intrigued by the high arctic environment. She began her research career by developing surface energy balance models driven by high arctic weather data, showing that in many cases, slight changes in atmospheric conditions, such as varying cloud cover, air temperature, and atmospheric dust, can have strong effects on local hydrology. Through field study at sites on Ellesmere Island, located in the Canadian arctic near 80° N latitude, she confirmed the conclusions of those models—that atmospheric factors were more important than terrestial conditions of plant cover, surface reflectivity, and surface resistance to water flow in controlling the local hydrologic balance.

Kathy is presently studying the environmental factors that control vegetation growth near perennial snowbanks on Cornwallis Island, at about 75° N latitude. To find out more about her work and its implications for Arctic global change, visit our web site.

second. The cross-sectional area and average velocity of a stream can change within a short distance, even though the stream discharge does not change. These changes occur because of changes in the *gradient* of the stream channel. The gradient is the rate of fall in elevation of the stream surface in the downstream direction (Figure 16.15).

GEODISCOVERIES

When the gradient is steep, the force of gravity will act more strongly and flow velocity will be greater. When the stream channel has a gentle gradient, the velocity will be slower. However, in a short stretch of stream, the discharge will remain constant. This means that in stretches of rapids, where the stream flows swiftly, the stream channel will be shallow and narrow. In pools, where the stream flows more slowly, the stream channel will be wider and deeper to maintain the same discharge. Sequences of pools and rapids can be found along streams of all sizes.

Stream discharge at a location on a stream is determined by noting the height, or *stage,* of the surface of the stream above its bed or above a fixed level near the bed, such as a marker on a bridge abutment. Stage is measured by a *stream gauge,* which uses a float inside a stilling well to record the height of the water surface (Figure 16.16). Stage is then converted to discharge by consulting a table made from flow measurements acquired at that location at various stages in the past.

The discharges of streams and rivers change from day to day, and records of daily and flood discharges of major streams and rivers are important information. They are used in planning the development and distribution of surface waters, as well as in designing flood-protection structures and generating models of how floods progress down a particular river system. An important activity of the U.S. Geological Survey is the measurement, or gauging, of stream discharge in the United States. In cooperation with states and municipalities, this organization maintains over 6000 gauging stations on principal streams and their tributaries. In Canada, the Water Survey of Canada monitors water levels in rivers and lakes using a network of about 2500 gauges, many of which are operated in cooperation with individual provinces.

Figure 16.17 is a map showing the relative discharge of major rivers of the United States. The mighty Mississippi with

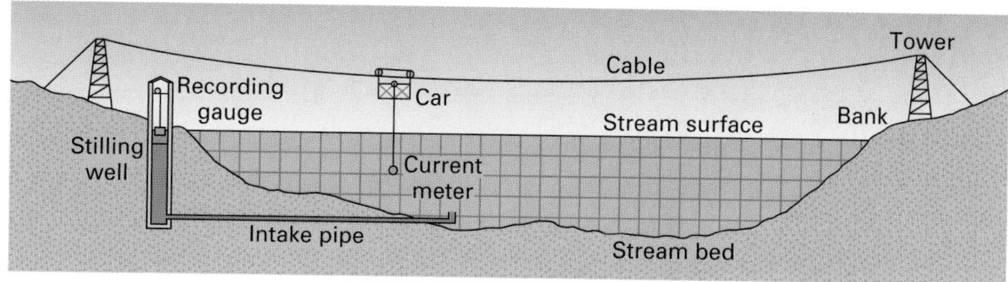

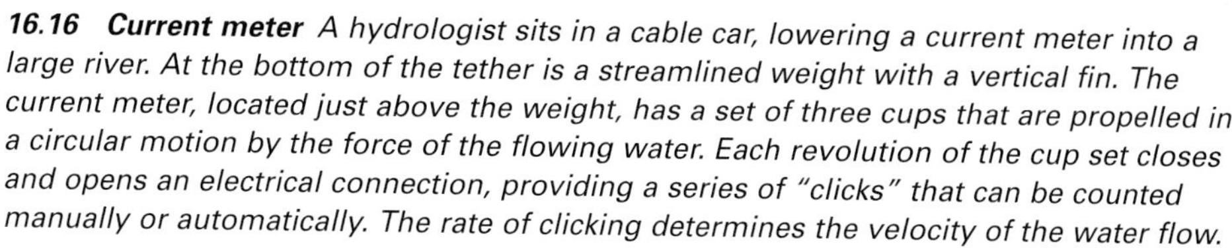
16.16 Current meter *A hydrologist sits in a cable car, lowering a current meter into a large river. At the bottom of the tether is a streamlined weight with a vertical fin. The current meter, located just above the weight, has a set of three cups that are propelled in a circular motion by the force of the flowing water. Each revolution of the cup set closes and opens an electrical connection, providing a series of "clicks" that can be counted manually or automatically. The rate of clicking determines the velocity of the water flow.*

its tributaries dwarfs all other North American rivers. Two other discharges are of major proportions—the Columbia River, draining a large segment of the Rocky Mountains in southwestern Canada and the northwestern United States, and the Great Lakes, discharging through the St. Lawrence River. The Colorado River, a much smaller stream, has its origin in the snowmelt of the Rocky Mountains and then crosses a vast semiarid and arid region that adds little tributary flow.

As the figure shows, the discharge of major rivers increases downstream as a natural consequence of the way streams and rivers combine to deliver runoff and sediment to the oceans. The gradient also changes in a downstream direction. The general rule is the larger the cross-sectional area of the stream, the lower the gradient. Great rivers, such as the Mississippi and Amazon, have gradients so low that they can be described as "flat." For example, the water surface of the lower Mississippi River falls in elevation about 3 cm for each kilometer of downstream distance (1.9 in. per mi).

Rivers with headwaters in high mountains have characteristics that are especially desirable for utilizing river flow for irrigation and for preventing floods. The higher ranges serve as snow storage areas, slowly releasing winter and

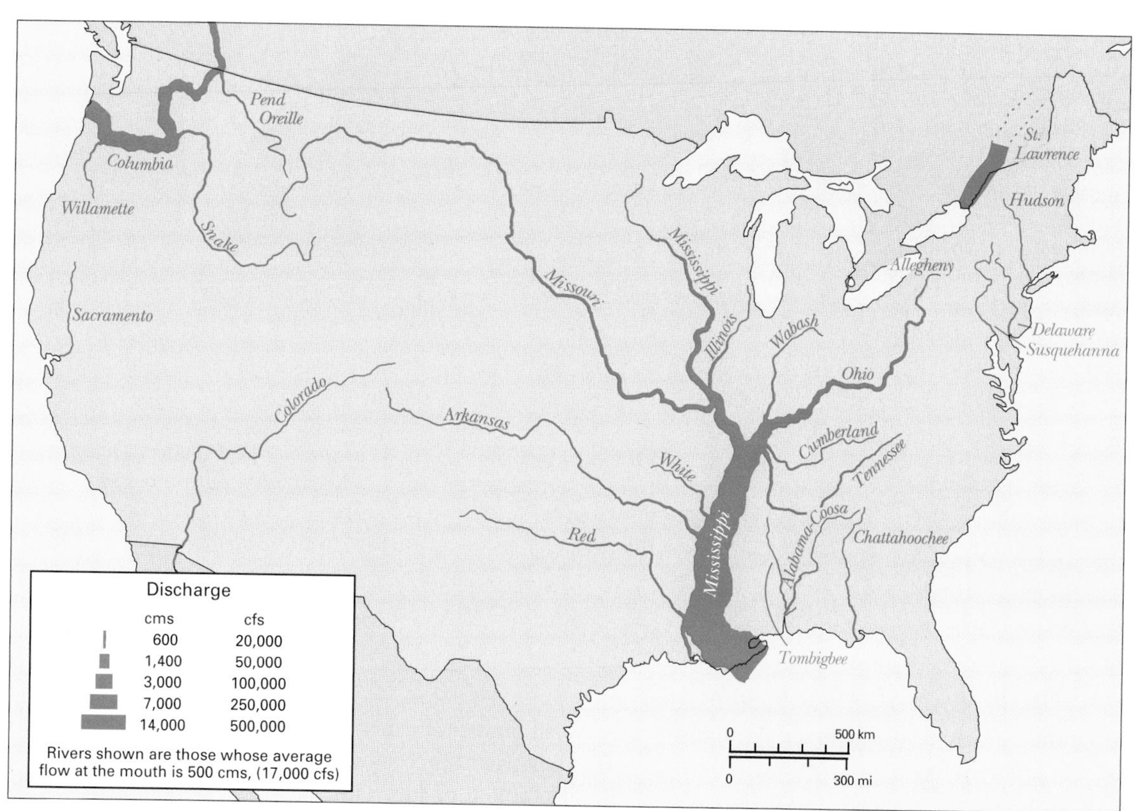

16.17 River discharge *This schematic map shows the relative magnitude of the discharge of U.S. rivers. Width of the river as drawn is proportional to mean annual discharge. (After U.S. Geological Survey.)*

Focus on Systems | 16.2

Energy in Stream Flow

Watching a flowing stream from a perch on its bank can be a fascinating experience. If it's a small stream, you may hear splashing and gurgling as the water flows around obstacles in its bed. If it's a large, deep stream—what we usually call a river—it may be flowing silently but with large eddies that roil the surface as they move by. In either case, energy is constantly being dissipated as the river flows. That is our theme here.

First, we need to examine briefly the geometric elements of a stream channel—what hydraulic engineers call the "hydraulic geometry." These are shown in the diagram at right. In *(a)*, notice the velocity arrows that show the flow pattern at the stream surface. Velocity is nearly zero at either bank, then increases as the center line of the channel is approached. Part *(b)* shows a vertical cross section along the center line of the channel. Velocity increases from the bottom of the channel to the water surface.

The discharge of a stream depends on the mean water velocity and the cross-sectional area of the stream. For example, if the mean velocity of a stream passing a certain point on the bank is 2 m/s and if the cross-sectional area of the stream at that point is 20 m^2, then in one second $2 \times 20 = 40$ m^3 will flow past the point. This simple relation is expressed by the equation

$$Q = A \times V$$

where Q is the discharge, A is the cross-sectional area, and V is the mean velocity.

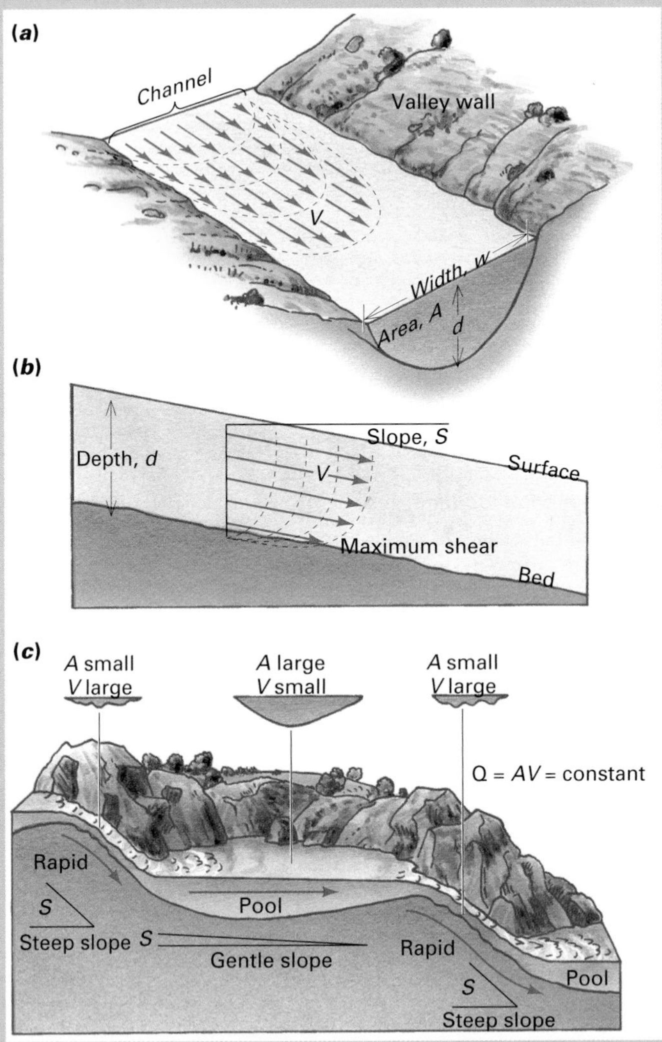

Characteristics of stream flow *The velocity of flow is greatest in the middle (a) and at the top (b) of the stream. (c) Mean velocity, cross-sectional area and slope change in the pools and rapids of a stream section of uniform discharge. (b, Copyright © A. N. Strahler.)*

Another important variable describing a stream at a particular point is the slope of the stream *(S)*. This value measures the inclination, or gradient, of the entire channel in the downstream direction. As you might expect, the slope is related to the velocity of the

spring precipitation through early or midsummer. As summer progresses, melting proceeds to successively higher levels. In this way, a continuous, sustained river flow is maintained. Among the snow-fed rivers of the western United States and Canada are the Fraser, Mackenzie, Columbia, Snake, Missouri, Arkansas, and Colorado.

What powers streams? The answer, of course, is gravity. This pervasive force moves water downhill and is opposed by the friction of the water with the bed and banks of the channel. *Focus on Systems 16.2 • Energy in Stream Flow* explains how gravitational power is related to the gradient of the stream and how that power is dissipated in stream flow.

stream. If the slope is steep, the water will flow more quickly and the velocity will be greater.

How are these quantities related from point to point along the stream? Imagine walking some distance along a stream. As you watch the stream, you will encounter pools, where the slope of the stream is low and the water moves slowly. Between these you will find stretches of rapids, where the slope is steeper and the water moves more swiftly. This situation is shown in part *(c)* of the figure.

Suppose that no new tributaries enter the stream in the distance that you traverse. In this case, Q will be constant through this stretch. If the slope steepens, velocity will increase, and we can see from the equation $Q = A \times V$ that if the velocity increases, then the cross-sectional area of the stream must decrease if Q is unchanged. In this way, we see that in pools, the slope will be low, velocity will be low, and cross-sectional area will be large. In the rapids, the situation is reversed—slope is steep, velocity is high, and cross-sectional area is small.

Let's turn now to the flow of energy accompanying stream flow. The driving force for stream flow is, of course, gravity. As water moves downslope in a stream channel, its potential energy is converted to kinetic energy of mass in motion, and the kinetic energy in turn is changed into heat by the friction of flow.

Every molecule of water in a stream tends to be drawn vertically down under the force of gravity, so that the water exerts a pressure—a force per unit area—upon the enclosing channel walls. This force increases with depth. A component of that force acts in the downstream direction, parallel with the stream bed, and tends to cause flow of one water layer over the next layer in a type of slipping motion known as *shear.* These layers can be thought of as thin sheets of molecules, each sheet slipping over the one below it, much as playing cards slip over one another when the deck is gently pushed across a table top. The layer immediately in contact with the solid stream bed does not slip, but each higher layer slips over the one below, so that the velocity *(V)* of forward motion increases upward from the bed. The same effect applies to the steep sides of the channel banks, where the layers in contact with the banks do not slip but successive layers toward the center do and so velocity increases. This type of flow is known as *laminar flow.* Moving away from the bed and banks, however, this smoothly layered form of shear gives way to *turbulent flow,* in which shear occurs in eddies of a wide range of sizes, intensities, and orientations.

Shear of both types meets with internal resistance, which requires the conversion of much of the potential energy into sensible heat, and this in turn raises the temperature of the water. The heat is eventually lost by conduction, radiation, or evaporation to the air above, and by conduction into the solid channel walls. As you might imagine, turbulent flow creates much more friction than laminar flow, so that much more energy is dissipated when water flow is turbulent.

How does the energy dissipated in stream flow differ in pools and rapids? In pools, slope is low, flow velocity is low, and less energy is dissipated as water moves a given distance downstream. In rapids, the slope is steeper, velocity is greater, and the energy dissipated will be much higher, for a given distance downstream.

You may be curious about just how much heat is released in the friction of flow of a stream. Since the potential energy released will be converted to heat, we only need to calculate the potential energy loss, which we can easily do using the same principles as in *Working It Out 15.2 • The Power of Gravity,* where we calculated the energy released by a landslide.

Suppose a cubic meter of stream water flows a horizontal distance of 1 km while descending 10 meters. This amounts to a slope of 10/1000 = 0.01 or 1 percent. The force of gravity acting on the cubic meter is $F = mg = 10^3$ kg $\times$ 9.8 m/s^2 = 9.8 $\times$ 10^3 N. The potential energy released in falling 10 meters is then $F \times d$ = 9.8 $\times$ 10^3 N $\times$ 10 m = 9.8 $\times$ 10^4 J. Let's further suppose that the mean velocity of the water flow is 1 m/s. Then the water will cover the distance of 1 km = 10^3 m in 10^3 s. The power produced is then energy per unit time, or 9.8 $\times$ 10^4 J $\div$ 10^3 s = 98 W. That is, stream flow will release a flow of energy equal to 98 watts per cubic meter, per degree of slope, per meter per second of velocity.

For example, Rock Creek near Red Lodge, Montana, discharges about 17 m^3/s at a velocity of about 0.85 m/s when it completely fills its channel, and its slope there is about 0.021 or 2.1 percent. Using the factor of 98 watts per cubic meter, per degree of slope, and per meter per second of velocity, this stream flow will release about 2980 W of power, which is enough to meet the demands of a few suburban households.

Drainage Systems

As runoff moves to lower and lower levels and eventually to the sea, it becomes organized into a **drainage system.** The system consists of a branched network of stream channels, as well as the adjacent sloping ground surfaces that contribute overland flow to those channels. Between the channels on the crests of ridges are *drainage divides,* which mark the boundary between slopes that contribute water to different streams or drainage systems. The entire system is bounded by an outer drainage divide that outlines a more-or-less pear-shaped **drainage basin** or **watershed** (Figure 16.18).

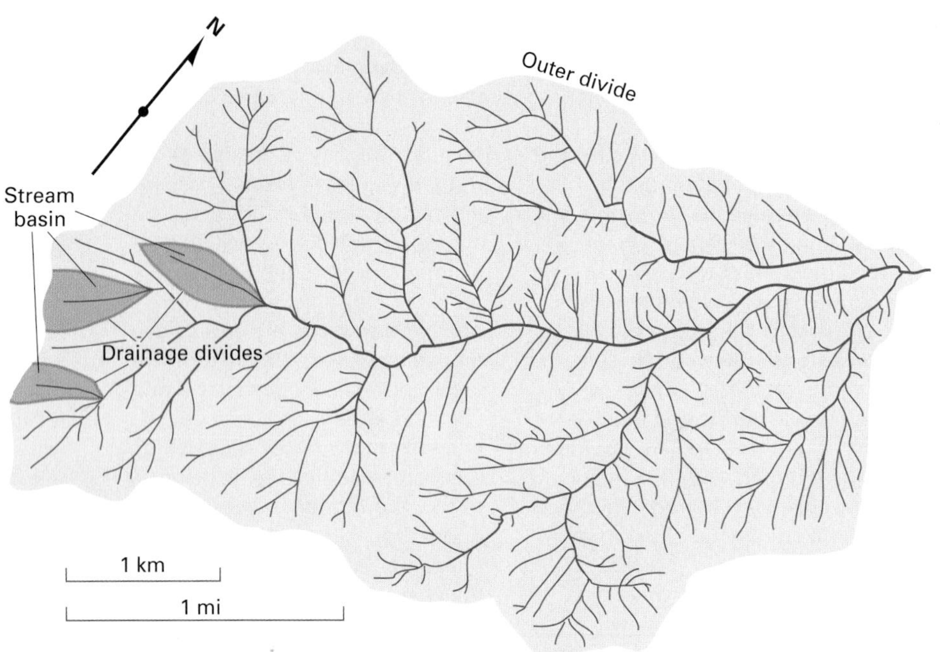

16.18 Channel network of a stream *Smaller and larger streams merge in a network that carries runoff downstream. Each small tributary has its own small drainage basin, bounded by drainage divides. An outer drainage divide delineates the stream's watershed at any point on the stream. (Data of U.S. Geological Survey and Mark A. Melton.)*

A typical stream network within a drainage basin is shown in Figure 16.18. Each fingertip tributary receives runoff from a small area of land surface surrounding the channel. This runoff is carried downstream and merged with runoff from other small tributaries as they join the main stream. The drainage system thus provides a converging mechanism that funnels overland flow into streams and smaller streams into larger ones.

Stream Flow

The discharge of a stream will increase in response to a period of heavy rainfall or snowmelt. However, the response is delayed, as the movement of water into stream channels takes time. The length of delay depends on a number of factors. The most important factor is the size of the drainage basin feeding the stream. Larger drainage basins show a longer delay.

The relationship between stream discharge and precipitation is best studied by means of a simple graph, called a *hydrograph,* which plots the discharge of a stream with time at a particular stream gauge. Figure 16.19 is a hydrograph for a drainage basin of about 800 km^2 (300 mi^2) in area located in Ohio within the moist continental climate ⑩. The graph shows the discharge of Sugar Creek (smooth line), the main stream of the drainage basin, during a four-day period that included a rainstorm. Rainfall for the 12-hour storm is shown by a bar graph giving the number of centimeters of precipitation in each two-hour period. The average total rainfall over the watershed of Sugar Creek was about 15 cm (6 in.). About half of this amount passed down the stream within three days' time. Some rainfall was held in the soil as soil water, some evaporated, and some infiltrated to the water table to be held in long-term storage as ground water.

Studying the rainfall and runoff graphs in Figure 16.19, we see that prior to the onset of the storm, Sugar Creek was car-

16.19 Sugar Creek hydrograph *Four days of precipitation and stream flow at Sugar Creek, Ohio, following a heavy rainstorm in August. (After Hoyt and Langbein, Floods, copyright © Princeton University Press. Used by permission.)*

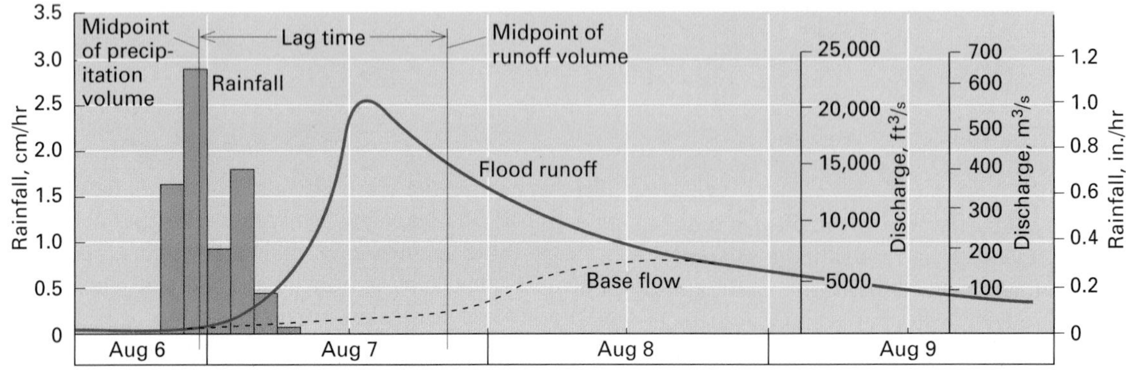

rying a small discharge. This flow, which is supplied by the seepage of ground water into the channel, is termed *base flow.* After the heavy rainfall began, several hours elapsed before the stream gauge began to show a rise in discharge. This interval, called the *lag time,* indicates that the branching system of channels was acting as a temporary reservoir. The channels were at first receiving inflow more rapidly than it could be passed down the channel system to the stream gauge.

Lag time is measured as the difference between the time at which half of the precipitation has occurred and that at which half the runoff has passed downstream. In the Sugar Creek example, the lag time was about 18 hours, with the peak flow reached almost 24 hours after the rainfall began. Note also that the stream's discharge rose much more abruptly than it fell. In general, the larger a watershed, the longer is the lag time between peak rainfall and peak discharge, and the more gradual is the rate of decline of discharge after the peak has passed. Another typical feature of a flood hydrograph is the slow but distinct rise in the amount of discharge contributed by base flow.

As the proportion of impervious surface increases, overland flow from the urbanized area generally increases. This change acts to increase the frequency and height of flood peaks during heavy storms for small watersheds lying largely within the urbanized area. Recharge to the ground water body beneath is also reduced, and this reduction, in turn, decreases the base flow contribution to channels in the same area. Thus, the full range of stream discharges, from low stages in dry periods to flood stages in wet periods, is made greater by urbanization.

A second change caused by urbanization is the introduction of storm sewers that quickly carry storm runoff from paved areas directly to stream channels for discharge. Thus, runoff travel time to channels is shortened, while the proportion of runoff is increased by the expansion in impervious surfaces. The two changes together act to reduce the lag time of urban streams and increase their peak discharge levels. Many rapidly expanding suburban communities are finding that low-lying, formerly flood-free, residential areas now experience periodic flooding as a result of urbanization.

How Urbanization Affects Stream Flow

The growth of cities and suburbs affects the flow of small streams in two ways. First, an increasing percentage of the surface becomes impervious to infiltration as it is covered by buildings, driveways, walks, pavements, and parking lots. In a closely built-up residential area with small lot sizes, the percentage of impervious surface may run as high as 80 percent.

The Annual Flow Cycle of a Large River

In regions of humid climates, where the water table is high and normally intersects the important stream channels, the hydrographs of larger streams will show clearly the effects of two sources of water—base flow and overland flow. Figure 16.20 is a hydrograph of the Chattahoochee River in Georgia, a fairly large river draining a watershed of 8700 km^2 (3350 mi^2), much of it in the southern Appalachian

16.20 Chattahoochee River hydrograph *This hydrograph shows the fluctuating discharge of the Chattahoochee River, Georgia, throughout a typical year. The high peaks are caused by runoff from streams that produced heavy overland flow. The inset graph expands the graph for the month of January. (Data of U.S. Geological Survey.)*

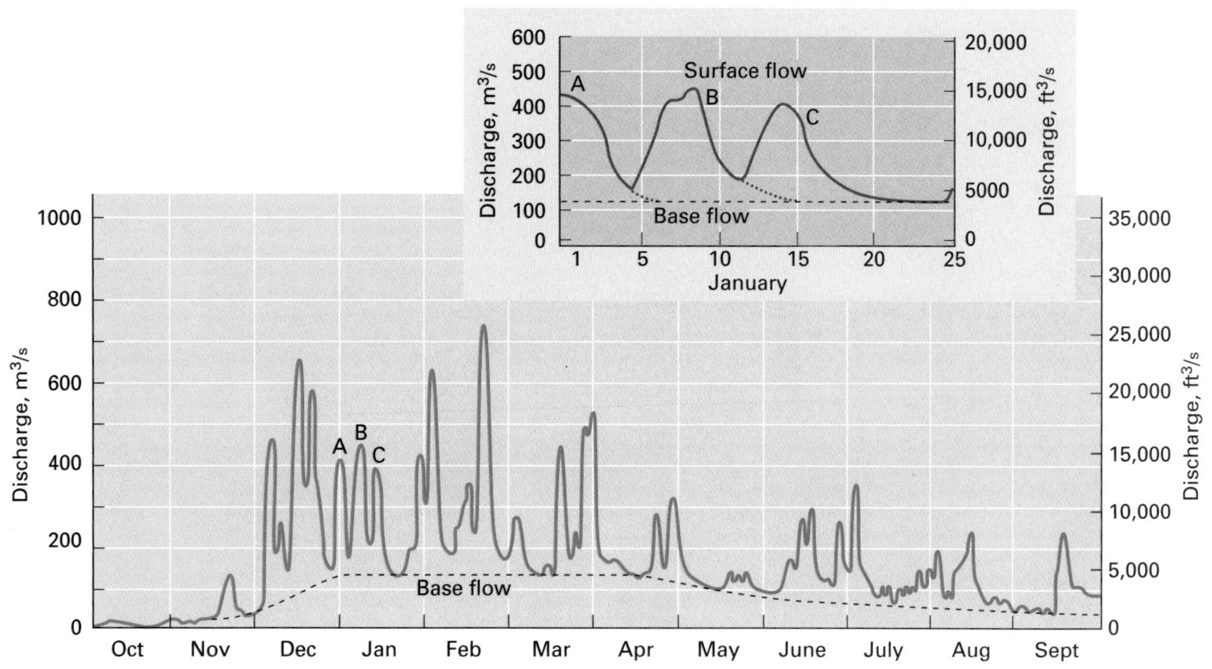

Mountains. The sharp, abrupt fluctuations in discharge are produced by overland flow following rain periods of one to three days' duration. These are each similar to the hydrograph of Figure 16.19, except that here they are shown much compressed by the time scale.

After each rain period the discharge falls off rapidly, but if another storm occurs within a few days, the discharge rises to another peak. The enlarged inset graph in Figure 16.20 shows details for the month of January. Here, three storms occur in rapid succession. Runoff drops between each storm but not completely to the level of base flow. When a long period intervenes between storms, the discharge falls to a low value, the base flow, at which it levels off.

Throughout the year the base flow, which represents ground water inflow into the stream, undergoes a marked annual cycle. During winter and early spring, water table levels are raised, and the rate of inflow into streams is increased. For the Chattahoochee River, the rate of base flow during January, February, March, and April holds uniform at about 100 m^3/s (about 3500 ft^3/s). The base flow begins to decline in spring, as heavy evapotranspiration losses reduce soil water and therefore cut off the recharge of ground water. The decline continues through the summer, reaching a low of about 30 m^3/s (about 1000 ft^3/s) by the end of October.

River Floods

You've probably seen enough media coverage of river floods to have a good idea of the appearance of floodwaters and the damage caused by flood erosion and deposition of silt and clay. We can define the term **flood** as the condition that exists when the discharge of a river cannot be accommodated within its normal channel. As a result, the water spreads over the adjoining ground, which is normally cropland or forest. Sometimes, however, the ground is occupied by houses, factories, or transportation corridors.

Most rivers of humid climates have a *floodplain,* a broad belt of low, flat ground bordering the channel on one or both sides that is flooded by stream waters about once a year (Figure 16.21). This flood usually occurs in the season when abundant surface runoff combines with the effects of a high water table to supply more runoff than can be carried in the channel. Annual inundation is considered a flood, even though its occurrence is expected and does not prevent the cultivation of crops after the flood has subsided. Annual flooding does not interfere with the growth

16.21 Floodplain *This aerial view shows the floodplain of the vast Amazon River in the foreground and far distance.*

of dense forests, which are widely distributed over low, marshy floodplains in all humid regions of the world. The National Weather Service, which provides a flood-warning service, designates a particular river stage at a given place as the *flood stage*. Above this critical level, inundation of the floodplain will occur. Still higher discharges of water, the rare and disastrous floods that may occur as seldom as once in 30 or 50 years, inundate ground lying well above the floodplain. Figure 16.22 shows rivers in flood.

Flash floods are characteristic of streams draining small watersheds with steep slopes. These streams have short lag times—perhaps only an hour or two—so when an intense rainfall event occurs, the stream rises very quickly to a high level. The flood arrives as a swiftly moving wall of turbulent water, sweeping away buildings and vehicles in its path. In arid western watersheds, great quantities of coarse rock debris are swept into the main channel and travel with the floodwater, producing debris floods (Chapter 15). In forested landscapes, tree limbs and trunks, soil, rocks, and boulders are swept downstream in the floodwaters. Because flash floods often occur too quickly to warn affected populations, they can cause significant loss of life.

Flood Prediction

The magnitude of a flood is usually measured by the peak discharge or highest stage of a river during the period of flooding. As the flood history of any river shows, large floods occur less frequently than smaller ones—that is, the greater the discharge or higher the stage, the less likely is the flood. *Working It Out 16.3 • Magnitude and Frequency of Flooding* presents a graph of flood frequency for a river and shows you how to determine the probability of various discharges given a record of peak annual discharges.

Another tool used to present the flood history of a river is the flood expectancy graph. Figure 16.23 shows expectancy graphs for two rivers. The meaning of the bar symbols is explained in the key. The Mississippi River at Vicksburg illustrates a great river responding largely to spring floods, with high waters in April and May and low flows in September and October. The Sacramento River shows the effects of the Mediterranean climate, with its winter wet season and long severe summer drought. Winter floods in January and February are caused by torrential rainstorms and snowmelt in the mountain watersheds of the Sierra Nevada and southern Cascades. By midsummer, river flow has shrunk to a very low stage.

The National Weather Service operates a River and Flood Forecasting Service through 85 offices located at strategic points along major river systems of the United States. When a flood threatens, forecasters analyze precipitation patterns and the progress of high waters moving downstream. Examining the flood history of the rivers and streams concerned, they develop specific flood forecasts. These are delivered to communities within the associated district, which usually covers one or more large watersheds. Flood warnings are publicized by every possible means, and various agencies cooperate closely to plan the evacuation of threatened areas and the removal or protection of property.

The Mississippi Flood of 1993

The Mississippi River Flood of 1993 provides an example of one of the most serious and damaging floods to strike the United States in recent years. Although the usual time for the flooding of the Mississippi is in spring (Figure 16.23), this extreme event occurred in midsummer. The 1993 flood was triggered by a succession of unprecedented rainfalls in the upper Mississippi Basin. In June of that year, rainfall totaled over 30 cm (12 in.) for large areas of the region. Southern Minnesota and western Wisconsin received even larger amounts. Monthly rainfall totals were the highest on many records dating back more than 100 years. The heavy rainfall continued into July, concentrated in Iowa, Illinois, and Missouri. Inundated by huge volumes of rain, the vast wet landscape of the upper Midwest came to resemble a sixth Great Lake when viewed from the air.

As the water drained from the landscape, so the Mississippi and its tributaries rose, creating a flood crest of swiftly moving turbid water that worked its way slowly downstream through July and early August. At nearly every community along the length of the Mississippi and Missouri rivers, citizens fought to raise and reinforce the levees that protected their lands, homes, and businesses from the ravages of the flood. Where levees failed to keep them in check, the rivers filled their broad floodplains from bluff to bluff, as shown in the two Landsat images of Figure 16.22b. By the time it was over, the waters had crested at 16.8 m (49.4 ft) at St. Louis, nearly 6 m (20 ft) above flood stage.

The Mississippi River Flood of 1993 was an environmental disaster of the first magnitude. An area twice the size of New Jersey was flooded, taking 50 lives and driving nearly 70,000 people from their homes. Property damage, including loss of agricultural crops, was estimated at $12 billion. Of the region's 1400 levees, at least 800 were breached or overtopped. Some flood engineers classed the event as a 500-year flood, meaning that only once in a 500-year interval is a flood of this magnitude likely to occur. However, this does not mean that a similar flood, or an even larger one, could not occur in the near future. For the people living on its banks and bottomlands, the mighty Mississippi is a sleeping giant that awakens with devastating consequences.

LAKES

A **lake** is a water body that has an upper surface exposed to the atmosphere and no appreciable gradient. The term *lake* includes a wide range of water bodies. Ponds (which are small, usually shallow water bodies), marshes, and swamps with standing water can all be included under the definition of a lake. Lakes receive water input from streams, overland flow, and ground water, and so are included as parts of drainage systems. Many lakes lose water at an outlet, where water drains over a dam (natural or constructed) to become

16.22 Rivers in Flood

a...***Red River in flood*** *This photo shows the Red River, near Fargo, North Dakota, in flood in April 1997. The normal channel is marked by the sinuous bands of trees running across the lower part of the photo. Dikes have kept the water from flooding the town in the center of the image.*

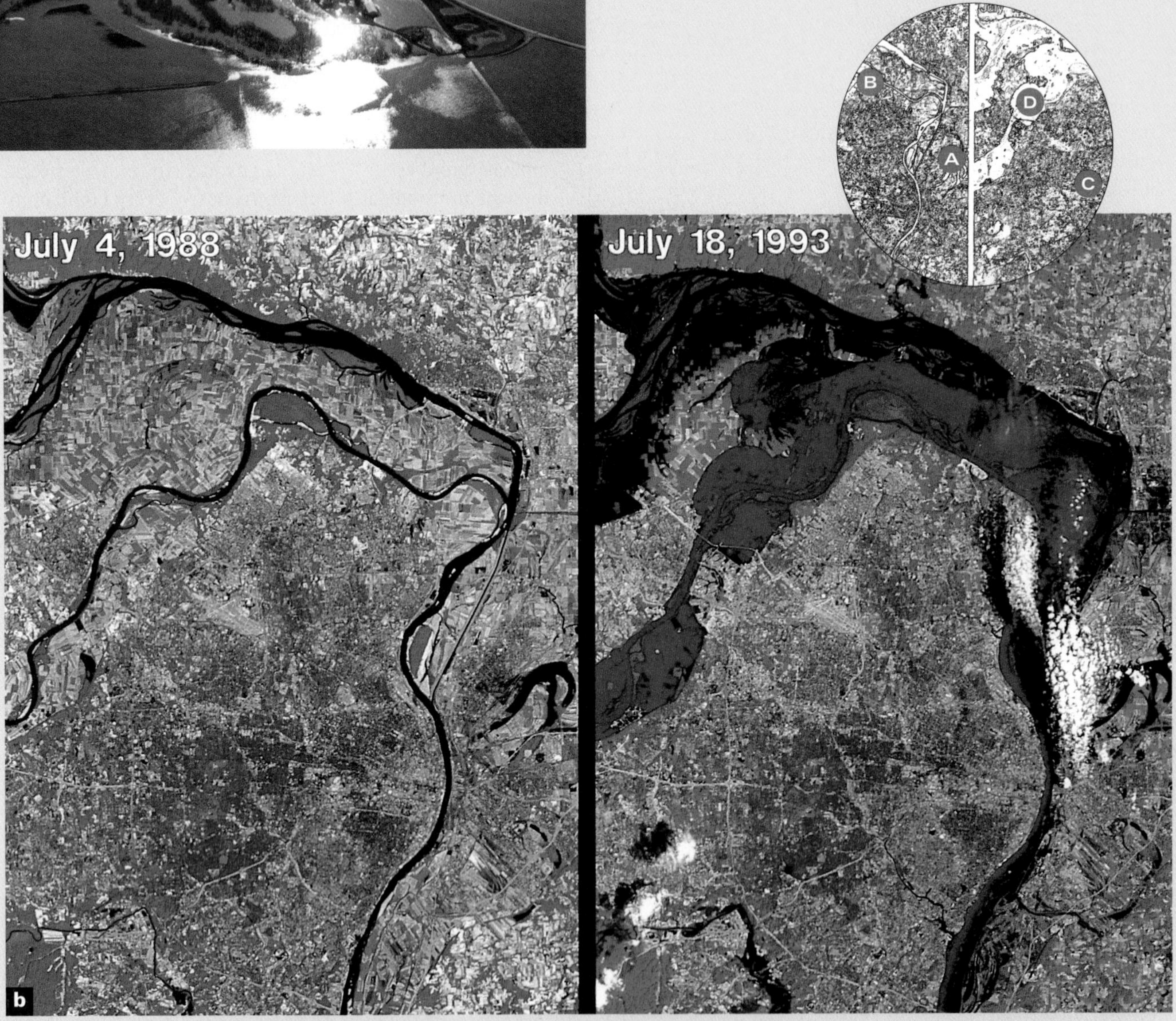

B D A C

July 4, 1988

July 18, 1993

b...***Eye on the Landscape*** ***Mississippi Flood of 1993*** *These two Landsat satellite images show St. Louis and vicinity during a normal year and during the flood of 1993. In these images, vegetation appears green, and urban areas appear in pink and purple tones. Some clouds appear in the 1993 image. The Mississippi River (topmost) joins the Missouri River to the north of the city. In 1993, the two rivers left their banks to spread into the bottomlands of their floodplains. The floodwaters crested at an even higher level at St. Louis on August 1, 1993.* **What else would the geographer see?...Answers at the end of the chapter.**

c...***River Power*** *A river in flood possesses enormous power to erode and move sediment. Here the Tuolumne River careens down a steep slope in Yosemite National Park. Over time, rivers like this can carve deep canyons, even in very resistant rock.*

d...***Local flooding*** *These houses in Lakeview, Ohio, are partially submerged following heavy rains in July 2003. Up to 38 cm (15 in.) of rain was reported in the region during this event.*

Working It Out | 16.3

Magnitude and Frequency of Flooding

It is a simple fact of nature that more extreme events happen less frequently. For example, hundreds or even thousands of small earthquakes may occur within a region during the period of a century, but only a few earthquakes are really large. In other words, the greater the magnitude of an event, the lesser the frequency at which it recurs.

The figure below shows this principle for the flood frequency of the Clearwater River at Kamaiah, Idaho. Each dot on the graph plots the maximum discharge of the river recorded within a particular year—that is, the peak flow, or largest flood, within the year. The numbers on the bottom horizontal scale indicate the probability that the given discharge will be equaled or exceeded in a given year. For example, a discharge associated with the value of 20 percent is interpreted to mean "a discharge of this magnitude (about 2000 m^3/s) can be expected to be equaled or exceeded in 20 out of 100 years." The numbers on the top horizontal scale show the recurrence interval (or return period). This value is simply the probability percentage divided into 100. For example, the return period for the probability of 20 percent is 100/20 = 5 years.

Note that the scale on the horizontal axis is not uniform. For this type of graph, the spacing is adjusted to follow a mathematical probability function such that the points will tend to plot on a straight line. The data for the Clearwater River follow the straight line fairly well.

As an example of using this graph, suppose that we would like to know how often a flow of 1500 m^3/s is likely to occur. Reading across from the 1500 m^3/s value on the vertical axis, we intersect the straight line at about 50 percent, meaning that in 50 years out of 100, this flow will be equaled or exceeded. Reading the scale at the top of the graph, we note that the return period for this event is two years. For a higher flow, say 2500 m^3/s, we can read about 6 percent probability for a return period of about 17 years.

How is the recurrence interval determined for a particular flood? The procedure is quite simple. First, assemble the maximum discharges for each year in a list. Next, reorder the values from greatest to least, assigning a rank of 1 to the largest value, 2 to the next largest, and so on. Then determine the recurrence interval for each flow from its rank, using the formula

$$I = \frac{N + 1}{R}$$

where I is the recurrence interval in years, N is the number of years in the record, and R is the rank of the value. Suppose that a flow is the largest in a set of 44 years of observations. Then its recurrence interval is $I = (N + 1)/R = (44 + 1)/1 = 45/1 = 45$ years. Suppose the flow is the fifth largest value. Its recurrence interval would then be $I = (44 + 1)/5 = 45/5 = 9$ years.

Keep in mind that the recurrence interval is only a way of expressing a probability. If a community experiences a "20-year flood" in one year, it does not mean that it will be 20 years until another flood of this magnitude comes along. There is nothing to prevent two 20-year floods from occurring in successive years. It's just not very likely.

Flood frequency data for the Clearwater River at Kamaiah, Idaho *Each dot is a measured maximum yearly discharge in a 53-year record. (Data of U.S. Geological Survey from R. K. Linsley, M. A. Kohler, and J. L. Paulhus, Hydrology for Engineers, second edition, McGraw-Hill, New York.)*

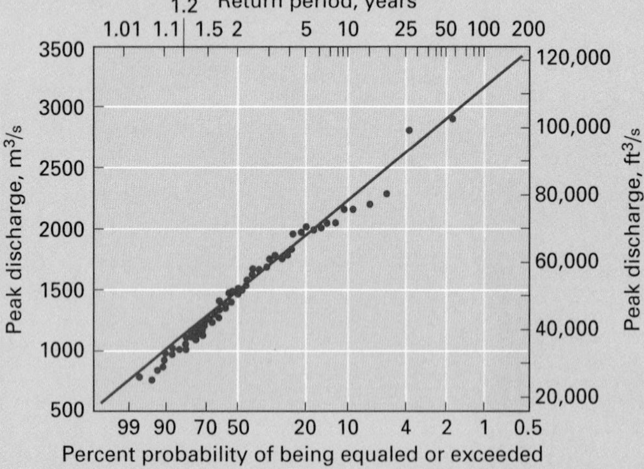

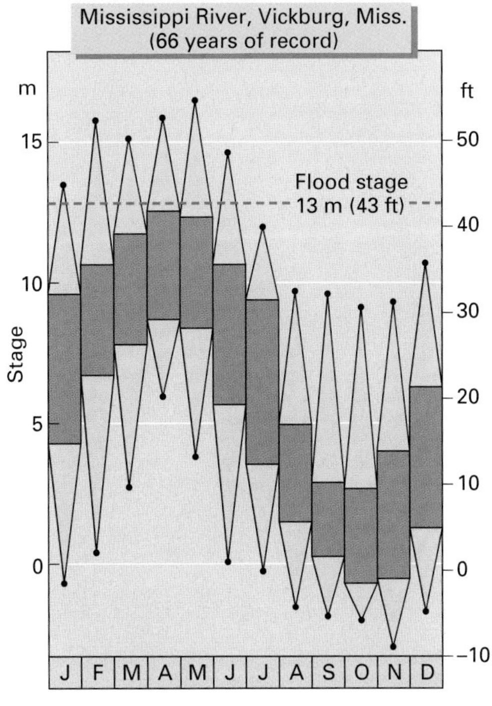

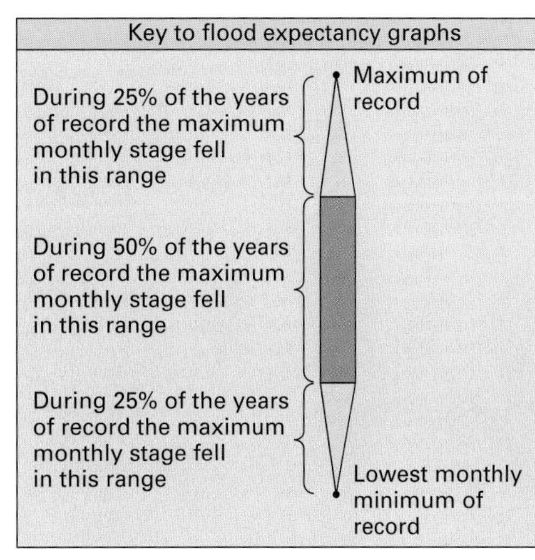

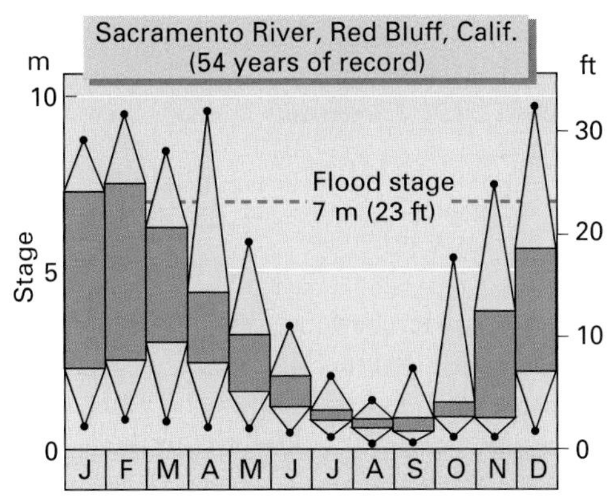

16.23 Flood expectancy graphs *Maximum and minimum monthly stages of the Mississippi River at Vicksburg, Mississippi, and the Sacramento River at Red Bluff, California.*

an outflowing stream. Lakes also lose water by evaporation. Lakes, like streams, are landscape features but are not usually considered to be landforms.

Lakes are quite important from the human viewpoint. They are frequently used as sources of fresh water and food, such as fish. Where dammed to a high level above the outlet stream, they can provide hydroelectric power as well. Lakes and ponds are also important recreation sites and sources of natural beauty.

Where lakes are not naturally present in the valley bottoms of drainage systems, we create lakes as needed by placing dams across the stream channels. Many regions that formerly had almost no natural lakes are now abundantly supplied. Some are small ponds built to serve ranches and farms, while others cover hundreds of square kilometers. In some areas, the number of artificial lakes is large enough to have significant effects on the region's hydrologic cycle.

Basins occupied by lakes show a wide range of origins as well as a vast range in dimensions. Lake basins, like stream channels, are true landforms. Basins are created by a number of geologic processes. For example, the tectonic process of crustal faulting creates many large, deep lakes. Lava flows often form a dam in a river valley, causing water to back up as a lake. Landslides suddenly create lakes, as we saw in the case of the Madison Slide. (See *Eye on the Environment 15.1 • The Great Hebgen Lake Disaster.*)

An important point about lakes in general is that they are short-lived features on the geologic time scale. Lakes disappear by one of two processes, or a combination of both. First, lakes that have stream outlets will be gradually drained as the outlets are eroded to lower levels. Where a strong bedrock threshold underlies the outlet, erosion will be slow but nevertheless certain. Second, lakes accumulate inorganic sediment carried by streams entering the lake and organic matter produced by plants and animals within the lake. Eventually, they fill up, forming a boggy wetland with little or no free water surface (Figure 16.24).

Lakes can also disappear when climate changes. If precipitation is reduced within a region, or temperatures and net radiation increase, evaporation can exceed input and the lake will dry up. Many former lakes of the southwest-

16.24 A freshwater pond in Wisconsin *Vegetation is slowly growing inward at the edges, and eventually the pond will become a bog supporting wet forest.*

ern United States flourished in moister periods of glacial advance during the Ice Age. Today, they have shrunk greatly or have disappeared entirely under the present arid regime.

In moist climates, the water level of lakes and ponds coincides closely with the water table in the surrounding area. Seepage of ground water into the lake, as well as direct runoff of precipitation, maintains these water surfaces permanently throughout the year. Examples of such freshwater ponds are found widely distributed in glaciated regions of North America and Europe. Here, plains of glacial sand and gravel contain natural pits and hollows left by the melting of stagnant ice masses that were buried in the sand and gravel deposits (see Chapter 20). Figure 16.25 is a block diagram showing small freshwater ponds on Cape Cod. The surface elevation of these ponds coincides closely with the level of the surrounding water table.

Many former freshwater water table ponds have become partially or entirely filled by organic matter from the growth and decay of water-loving plants (Figure 16.24). The ultimate result is a bog with its surface close to the water table. Freshwater marshes and swamps, in which water stands at or close to the ground surface over a broad area, also represent the appearance of the water table at the

16.25 Freshwater ponds *A sketch of freshwater ponds in sandy glacial deposits on Cape Cod, Massachusetts. The water level of the ponds corresponds closely to the water table. (Redrawn from A Geologist's View of Cape Cod, copyright © A. N. Strahler, 1966. Used by permission of Doubleday, a division of Bantam Doubleday Dell Publishing Group, Inc.)*

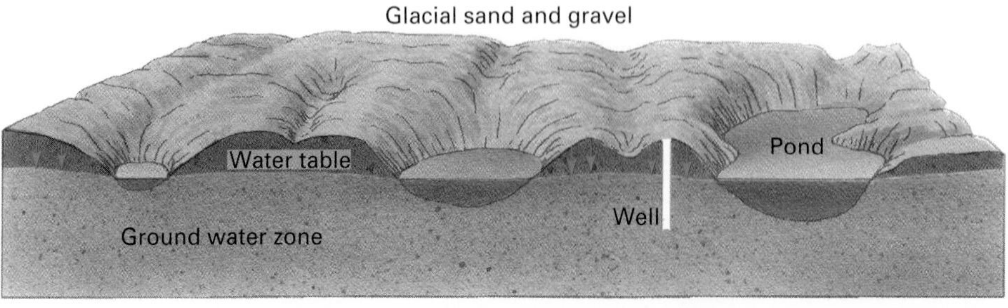

surface. Such areas of poor surface drainage are included under the name of *wetlands.*

Saline Lakes and Salt Flats

Lakes with no surface outlet are characteristic of arid regions. Here, the average rate of water loss by evaporation balances the average rate of stream inflow. If the rate of inflow increases, the lake level will rise. At the same time, the lake surface will increase in area, allowing a greater rate of evaporation. A new balance can then be achieved. Similarly, if the region becomes more arid, reducing input and increasing evaporation, the water level will fall to a lower level. (We noted this phenomenon in Chapter 2, where we discussed the inland lake as a matter flow system in equilibrium.)

Lakes without outlets often show salt buildup. Dissolved solids are brought into the lake by streams—usually streams that head in distant highlands where a water surplus exists. Since evaporation removes only pure water, the salts remain behind and the salinity of the water slowly increases. Salinity, or degree of "saltiness," refers to the abundance of certain common ions in the water. Eventually, salinity levels reach a point where salts are precipitated as solids (Figure 16.26).

Sometimes the surfaces of such lakes lie below sea level. An example is the Dead Sea, with a surface elevation of –396 m (–1299 ft). The largest of all lakes, the Caspian Sea, has a surface elevation of –25 m (82 ft). Both of these large lakes are saline.

In regions where climatic conditions consistently favor evaporation over input, the lake may be absent. Instead, a shallow basin covered with salt deposits (see Figure 17.29*b*), otherwise known as a *salt flat* or dry lake, will occur. In Chapter 17 we will describe dry lakebeds as landforms. On rare occasions, these flats are covered by a shallow layer of water, brought by flooding streams heading in adjacent highlands.

A number of desert salts are of economic value and have been profitably extracted from salt flats. In shallow coastal estuaries in the desert climate, sea salt for human consumption is commercially harvested by allowing it to evaporate in shallow basins. One well-known source of this salt is the Rann of Kutch, a coastal lowland in the tropical desert of westernmost India, close to Pakistan. Here the evaporation of shallow water of the Arabian Sea has long provided a major source of salt for inhabitants of the interior.

The Aral Sea is an inland lake that is rapidly becoming saline through human activity. The two major rivers that feed this vast central Asian water body have been largely diverted into irrigation of agricultural lands, and with its inflow greatly reduced, the lake has shrunk and become increasingly salty. Our interchapter feature, *A Closer Look: 16.4 • The Aral Sea—A Dying Saline Lake* tells the story of this environmental disaster.

Desert Irrigation

Human interaction with the tropical desert environment is as old as civilization itself. Two of the earliest sites of civilization—Egypt and Mesopotamia—lie in the tropical deserts. Successful occupation of these deserts requires irrigation with

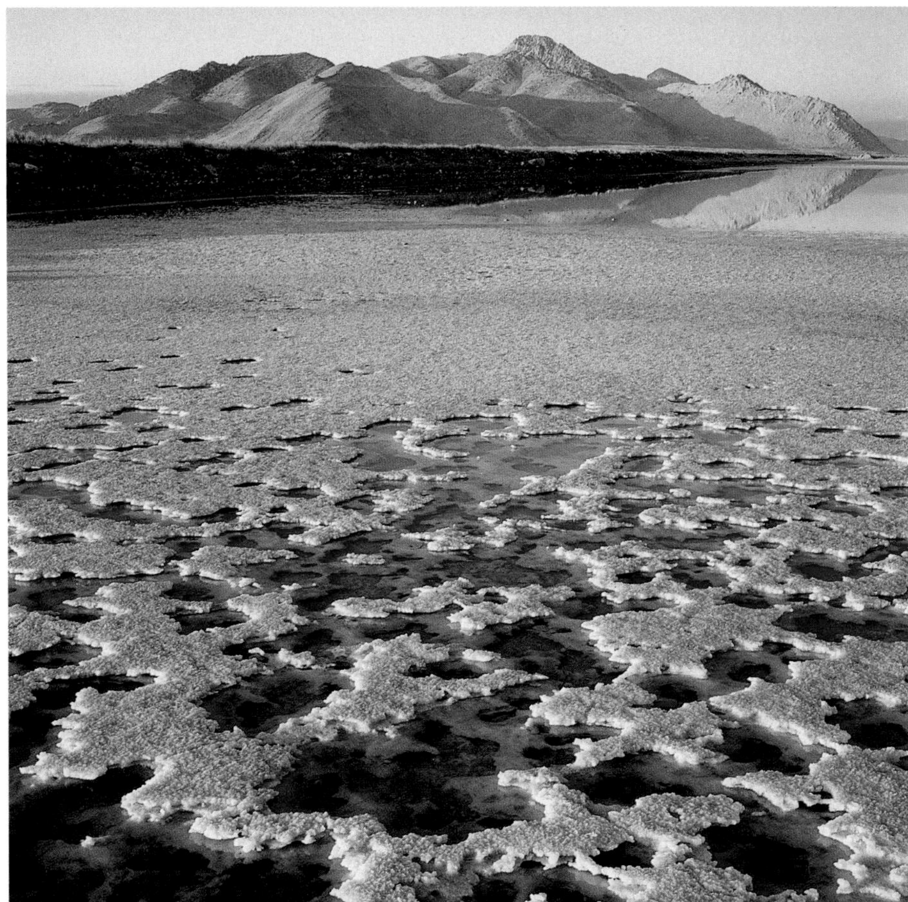

16.26 Salt encrustations *These salt encrustations at the edge of Great Salt Lake, Utah, were formed when the lake level dropped during a dry period.*

large supplies of water from nondesert sources. For Egypt and Mesopotamia, the water sources of ancient times were the rivers that cross the desert but derive their flow from regions that have a water surplus. These are referred to as *exotic rivers* because their flows are derived from an outside region.

Irrigation systems in arid lands divert the discharge of an exotic river such as the Nile, Indus, Jordan, or Colorado into a distribution system that allows the water to infiltrate the soil of areas under crop cultivation. Ultimately, such irrigation projects can suffer from two undesirable side effects: salinization and waterlogging of the soil.

Salinization occurs when salts build up in the soil to levels that inhibit plant growth. This happens because an irrigated area within a desert loses large amounts of soil water through evapotranspiration. Salts contained in the irrigation water remain in the soil and increase to high concentrations. Salinization may be prevented or cured by flushing the soil salts downward to lower levels by the use of more water. This remedy requires greater water use than for crop growth alone. In addition, new drainage systems must be installed to dispose of the excess saltwater.

Waterlogging occurs when irrigation with large volumes of water causes a rise in the water table, bringing the zone of saturation close to the surface. Most food crops cannot grow in perpetually saturated soils. When the water table rises to the point at which upward movement under capillary action can bring water to the surface, evaporation is increased and salinization is intensified.

Agricultural areas of major salinization include the Indus River Valley in Pakistan, the Euphrates Valley in Syria, the Nile delta of Egypt, and the wheat belt of western Australia. In the United States, extensive regions of heavily salinized agriculture are found in the San Joaquin and Imperial valleys of California. Other areas of salinization occur throughout the entire semiarid and arid regions of the western United States. In Chapter 21, *Eye on the Environment 21.2 • Death of a Civilization* documents the effects of salinization and waterlogging on the civilizations of ancient Mesopotamia.

Pollution of Surface Water

Streams, lakes, bogs, and marshes are specialized habitats of plants and animals. Their ecosystems are particularly sensitive to changes induced by human activity in the water balance and in water chemistry. Our industrial society not only makes radical physical changes in water flow by construction of engineering works (dams, irrigation systems, canals, dredged channels), but also pollutes and contaminates our surface waters with a large variety of wastes.

The sources of water pollutants are many and varied. Some industrial plants dispose of toxic metals and organic compounds by discharging them directly into streams and lakes. Many communities still discharge untreated or partly treated sewage wastes into surface waters. In urban and suburban areas, pollutant matter entering streams and lakes includes deicing salt and lawn conditioners (lime and fertilizers), which can also contaminate ground water. In agricultural regions, important sources of pollutants are fertilizers and livestock wastes. Mining and processing of mineral deposits are also major sources of water pollution. Even contamination by radioactive substances released from nuclear power and processing plants can occur.

Among the common chemical pollutants of both surface water and ground water are sulfate, chloride, sodium, nitrate, phosphate, and calcium ions. (Recall from simple chemistry that an ion is the charged form of a molecule or an atom and that many chemical compounds dissolve in water by forming ions. Chapter 21 provides more details.) Sulfate ions enter runoff both by fallout from polluted urban air and as sewage effluent. Chloride and sodium ions are contributed both by fallout from polluted air and by deicing salts used on highways. In some locations with snowy winters, community water supplies located close to highways have become polluted from deicing salts. Important sources of nitrate ions are fertilizers and sewage effluent. Excessive concentrations of nitrate in freshwater supplies are highly toxic, and, at the same time, their removal is difficult and expensive. Phosphate ions are contributed in part by fertilizers and by detergents in sewage effluent.

Phosphate and nitrate are plant nutrients and can lead to excessive growth of algae and other aquatic plants in streams and lakes. Applied to lakes, this process is known as *eutrophication,* which is often described as the "aging" of a lake. In eutrophication, the accumulation of nutrients stimulates plant growth, producing a large supply of dead organic matter in the lake. Microorganisms break down this organic matter but require oxygen in the process. However, oxygen dissolves only slightly in water, and so it is normally present only in low concentrations. The added burden of oxygen use by the decomposers reduces the oxygen level to the point where other organisms, such as desirable types of fish, cannot survive. After a few years of nutrient pollution, the lake can take on the characteristics of a shallow pond that results when a lake is slowly filled with sediment and organic matter over thousands of years by natural "aging" processes.

A particular form of chemical pollution of surface water goes under the name of *acid mine drainage.* It is an important form of environmental degradation in parts of Appalachia where abandoned coal mines and strip-mine workings are concentrated (Figure 16.27). Ground water emerges from abandoned mines and as soil water percolating through strip-mine waste banks. This water contains sulfuric acid and various salts of metals, particularly of iron. Acid of this origin in stream waters can have adverse effects on animal life. In sufficient concentrations, it is lethal to certain species of fish and has at times caused massive fish kills.

Toxic metals, among them mercury, along with pesticides and a host of other industrial chemicals, are introduced into streams and lakes in quantities that are locally damaging or lethal to plant and animal communities. In addition, sewage introduces live bacteria and viruses that are classed as biological pollutants. These pose a threat to the health of humans and animals alike.

Another form of pollution is *thermal pollution,* which refers to the discharge of heat into the environment from

16.27 Strip-mine water pollution *This small stream is badly polluted by acid waters that have percolated through strip-mine waste.*

combustion of fuels and from the conversion of nuclear energy into electric power. Thermal pollution of water takes the form of discharges of heated water into streams, estuaries, and lakes, which can have drastic effects on local aquatic life. The impact may be quite large in a small area.

Surface Water As a Natural Resource

Fresh surface water is a basic natural resource essential to human agricultural and industrial activities. Runoff held in reservoirs behind dams provides water supplies for great urban centers, such as New York City and Los Angeles. When diverted from large rivers, it provides irrigation water for highly productive lowlands in arid lands, such as the Sacramento and San Joaquin valleys of California and the Nile Valley of Egypt. To these uses of runoff are added hydroelectric power, where the gradient of a river is steep, or routes of inland navigation, where the gradient is gentle.

Our heavily industrialized society requires enormous supplies of fresh water for its sustained operation. Urban dwellers consume water in their homes at rates of 150 to 400 liters (50 to 100 gallons) per person per day (Figure 16.28). Large quantities of water are used for cooling in air conditioning units and power plants. Much of this water is obtained from surface water supplies, and the demand increases daily.

Unlike ground water, which represents a large water storage body, fresh surface water in the liquid state is stored only in small quantities. (An exception is the Great Lakes system.) Recall from Chapter 6 that the global quantity of available ground water is about 20 times as large as that stored in freshwater lakes and that the water held in streams is only about one one-hundredth of that in lakes. Because of small natural storage capacities, surface water can be drawn only at a rate comparable with its annual renewal through precipitation. Dams are built to develop useful storage capacity for runoff that would otherwise escape to the sea, but once the reservoir has been filled, water use must be scaled to match the natural supply rate averaged over the year. Development of surface water supplies brings on many environmental changes, both physical and biological, and

16.28 Domestic water use *This pie chart shows how water is used in an average home in Akron, Ohio. (U. S. Geological Survey.)*

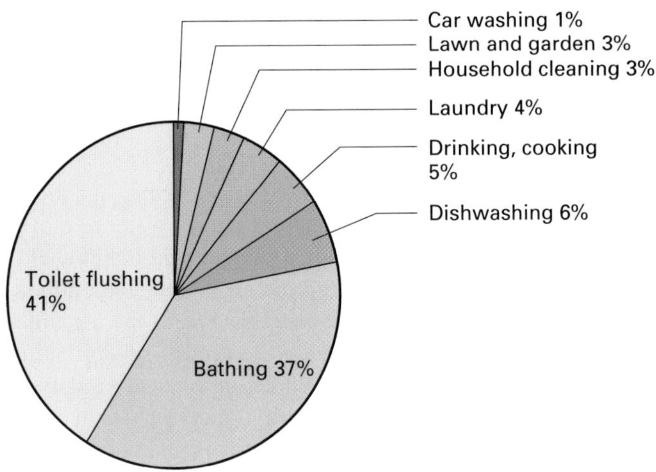

these must be taken into account in planning for future water developments.

A LOOK AHEAD This chapter has focused on water, including the precipitation that runs off the land and flows into the sea. Think for a moment about the drainage system of a stream that conducts this flow of runoff. The smallest streams catch runoff from slopes, carrying the runoff into larger streams. The larger streams, in turn, receive runoff from their side slopes and also pass their flow on to still larger streams. However, this cannot happen unless the gradients of slopes and streams are adjusted so that water keeps flowing downhill. This means that the landscape is shaped and organized into landforms that are an essential part of the drainage system. The shaping of landforms within the drainage system occurs as running water erodes the landscape, which is the subject of the next chapter.

Chapter Summary

- The fresh water of the lands accounts for only a small fraction of the Earth's water. Since it is produced by precipitation over land, it depends on the continued operation of the hydrologic cycle for its existence. The soil layer plays a key role in determining the fate of precipitation by diverting it in three ways: to the atmosphere as *evapotranspiration*, to ground water through *percolation*, and to streams and rivers as **runoff**.
- **Ground water** occupies the pore spaces in rock and regolith. The water table marks the upper surface of the *saturated zone* of ground water, where pores are completely full of water. Ground water moves in slow paths deep underground, recharging rivers, streams, ponds, and lakes by upward seepage and thus contributing to runoff. Solution of limestone by ground water can produce *limestone caverns* and generate *karst* landscapes.
- Wells draw down the water table and, in some regions, lower the water table more quickly than it can be recharged. Ground water contamination can occur when precipitation percolates through contaminated soils or waste materials. Landfills and dumps are common sources of ground water contaminants.
- Runoff includes *overland flow*, moving as a sheet across the land surface, and *stream flow* in **streams** and rivers, which is confined to a **channel**. Rivers and streams are organized into a **drainage system** that moves runoff from slopes into channels and from smaller channels into larger ones. The **discharge** of a stream measures the flow rate of water moving past a given location. Discharge increases downstream as tributary streams add more runoff.
- The *hydrograph* plots the discharge of a stream at a location through time. Since it takes time for water to move down slopes and into progressively larger stream channels, peak discharge differs from peak precipitation by a *lag time*. The larger the stream, the longer the lag time and the more gradual the rate of decline of discharge after the peak. Because urbanization typically involves covering ground surfaces with impervious materials, urban streams exhibit shorter lag times and higher peak discharges. Annual hydrographs of streams from humid regions show an annual cycle of *base flow* on which are superimposed discharge peaks related to individual rainfall episodes.
- **Floods** occur when river discharge increases and the flow can no longer be contained within the river's usual channel. Water spreads over the *floodplain*, inundating low fields and forests adjacent to the channel. When discharge is high, floodwaters can rise beyond normal levels to inundate nearby areas of development, causing damage and sometimes taking lives. *Flash floods* occur in small, steep watersheds and can be highly destructive. The Mississippi Flood of 1993, occurring at an unusual time—during the summer rather than the spring—was an extreme flood event that caused widespread destruction and damage.
- **Lakes** are especially important parts of the drainage system because they are sources of fresh water. They are also used for recreation, and, in many cases, they can supply hydroelectric power. Where lakes occur in inland basins, they are often saline. When climate changes, such lakes can dry up, creating *salt flats*.
- Irrigation is the diversion of fresh water from streams and rivers to supply the water needs of crops. In desert regions, where irrigation is most needed, problems of *salinization* and *waterlogging* can occur, reducing productivity and eventually creating unusable land.
- Water pollution arises from many sources, including industrial sites, sewage treatment plants, agricultural activities, mining, and processing of mineral deposits. Sulfate, nitrate, phosphate, chloride, sodium, and calcium ions are frequent contaminants. Toxic metals, pesticides, and industrial chemicals are also hazards.
- Ground water and surface water resources are essential for human activities. Human civilization is dependent on abundant supplies of fresh water for many uses. But because the fresh water of the continents is such a small part of the global water pool, utilization of water resources takes careful planning and management.

Key Terms

infiltration
runoff
overland flow
ground water
water table
artesian well
stream
channel
discharge
drainage system
drainage basin
watershed
flood
lake

Review Questions

1. What happens to precipitation falling on soil? What processes are involved?
2. How and under what conditions does precipitation reach ground water?
3. How do caverns come to be formed in limestone? Describe the key features of a karst landscape.
4. How do wells affect the water table? What happens when pumping exceeds recharge?
5. How is ground water contaminated? Describe how a well might become contaminated by a nearby landfill dump.
6. Define discharge (of a stream) and the two quantities that determine it. How does discharge vary in a downstream direction? How does gradient vary in a downstream direction?
7. What is a drainage system? How are slopes and streams arranged in a drainage basin?
8. Define the term *flood*. What is the floodplain? What factors are used in forecasting floods?
9. Why was the Mississippi Flood of 1993 so unusual?
10. How are lakes defined? What are some of their characteristics? What factors influence the size of lakes?
11. Describe some of the problems that can arise in long-continued irrigation of desert areas.
12. Identify common surface water pollutants and their sources.
13. How is surface water utilized as a natural resource?

Eye on Global Change 16.1 • Sinking Cities

1. Why has land subsidence occurred in Venice? What are the effects?
2. Identify four other locations that have suffered land subsidence caused by ground water withdrawal.

Focus on Systems 16.2 • Energy in Stream Flow

1. How are the cross-sectional area and velocity of a stream related to the flow?
2. How do cross-sectional area, velocity, and slope change in pools and rapids?
3. What two types of flow occur in a stream, and how do they differ?
4. How does the energy dissipated by stream flow differ in pools and rapids?

A Closer Look:
Eye on Global Change 16.4 • The Aral Sea—A Dying Saline Lake

1. What changes have occurred in the Aral Sea in the past 30 years?
2. Why have these changes occurred? What would be required to reverse them?
3. Describe the new project underway to rehabilitate the northern portion of the Aral Sea. What will the project change for the better? for the worse?

Visualizing Exercises

1. Sketch a cross section through the land surface showing the position of the water table and indicating flow directions of subsurface water motion with arrows. Include the flow paths of ground water. Be sure to provide a stream in your diagram. Label the saturated and unsaturated zones.
2. Why does water rise in an artesian well? Illustrate with a sketched cross-sectional diagram showing the aquifer, aquicludes, and the well.

Essay Questions

1. A thundershower causes heavy rain to fall in a small region near the headwaters of a major river system. Describe the flow paths of that water as it returns to the atmosphere and ocean. What human activities influence the flows? in what ways?
2. Imagine yourself a recently elected mayor of a small city located on the banks of a large river. What issues might you be concerned with that involve the river? In developing your answer, choose and specify some characteristics for this city—such as its population, its industries, its sewage systems, and the present uses of the river for water supply or recreation.

Problems

Working It Out 16.3 • Magnitude and Frequency of Flooding

1. The data to the right are peak flows for the West River, near Newfane, Vermont, for a 33-year period (1929–1961). (Following 1961, the river was controlled by a reservoir upstream). Rank the data from greatest to least and determine the recurrence interval for each flow using the formula. If two values are tied, use the average of the two ranks in determining the recurrence interval for the flow. What are the magnitudes of floods with recurrence intervals closest to 1, 2, 5, 10, and 35 years? (*Hint:* Many word processing programs and spreadsheets have the ability to sort data and thus can be used to order the list quickly and easily.)

Year	Flow, m^3/s	Year	Flow, m^3/s	Year	Flow, m^3/s
1929	234	1940	292	1951	312
1930	113	1941	119	1952	309
1931	217	1942	229	1953	337
1932	184	1943	234	1954	139
1933	245	1944	266	1955	279
1934	209	1945	210	1956	337
1935	225	1946	181	1957	164
1936	677	1947	357	1958	264
1937	216	1948	419	1959	172
1938	487	1949	711	1960	351
1939	193	1950	278	1961	194

Source: Data from U.S. Geological Survey.

2. Note that the percent probability that a flow will be equaled or exceeded is simply $100/I$, where I is the recurrence interval. What is the percent probability that a flood flow in any one year will exceed 250 m^3/s? 500 m^3/s? What flow will be equaled or exceeded in 25 percent of all years?

Eye on the Landscape

Chapter Opener Waste water outflow, desalination plant Notice how the brine plume remains distinct after entering the ocean. The brine is significantly saltier, and therefore more dense, than the ocean water, so it tends to flow under the lighter-blue water of the ocean for some distance before mixing. At **(A)**, you can see surface waves overtopping the deep blue of the waste brine. Relative density of ocean waters was mentioned in Chapter 7 in our discussion of thermohaline circulation.

16.9 Sinkholes Note the amount of bare rock visible in many large patches on this arid plain **(A)**. The lack of soil suggests that the limestone is so pure that its dissolution leaves little or no residual material behind. Also, look at the vegetation ringing the sinkholes **(B)**. The plants are probably drawing on ground water. From this observation, we might conclude that ground water is not very far below the vegetated region between the two roads **(C)**. Solution weathering of limestone was covered in Chapter 15. Phreatophytes, which are plants that survive in an arid environment by tapping the water table, are discussed in Chapter 23.

16.11 Tower karst As we saw in Chapter 15, solution weathering of certain types of bedrock in warm and wet environments can produce a landscape of steep, vertical slopes. Compare these towers **(A)**, formed by solution of limestone, with the fins and grooves of Kauai in Figure 15.9, formed by solution of basaltic lava. Note also the flooded fields **(B)**. They are probably rice paddies in the spring, just before planting with young rice stalks.

16.22b Mississippi Flood of 1993 Note the former river channels that have been cut off by migration of the river channel **(A)**. These are called ox-bow lakes and are described in Chapter 17. These images also demonstrate a type of false-color image in which the red color shows shortwave infrared reflectance, green shows near-infrared reflectance, and blue shows red reflectance. Recall from Chapter 3 that vegetation reflects very strongly in near-infrared wavelengths, and thus vegetation appears bright green **(B)**. Urban surfaces are bright in the shortwave infrared and thus appear red **(C)**. Sediment-laden river water is brown and so provides a strong red signal that translates as blue in the image **(D)**. *A Closer Look: Geographer's Tools 4.8 • Remote Sensing for Physical Geography* provides more detail.

A CLOSER LOOK

Eye on Global Change 16.4 The Aral Sea—A Dying Saline Lake

East of the Caspian Sea, astride the former Soviet republics of Kazakhstan and Uzbekistan, lies an immense saline lake—the Aral Sea. Fed by meltwaters of high glaciers and snowfields in the lofty Hindu Kush, Pamir, and Tien Shan Ranges, the lake endured through thousands of years as an oasis for terrestrial and aquatic wildlife deep in the heart of the central Asian desert.

But in the last 30 years, the Aral Sea, once larger than Lake Huron, has shrunk to a shadow of its former extent. The volume of its waters has decreased by 66 percent, and its salinity has increased from 1 percent to over 3 percent, making it more salty than sea water. Twenty of the 24 fish species native to the lake have disappeared. Its catch of commercial fish, which once supplied 10 percent of the total for the Soviet Union, has dwindled to zero. The deltas of the Amu Darya and Syr Darya rivers, which enter the south and east sides of the lake, were islands of great ecological diversity, teeming with fingerling fishes, birds, and their predators. Now only about half of the species of nesting birds remain. Many species of aquatic plants, shrubs, and grasses have vanished. Commercial hunting and trapping have almost ceased.

What caused this ecological catastrophe? The answer is simple—the lake's water supply was cut off. As an inland lake with no outlet, the Aral Sea receives water from the Amu Darya and Syr Darya, as well as a small amount from direct precipitation, but it loses water by evaporation. Its gains balanced its losses, and, although these gains and losses varied from year to year, the area, depth, and volume of the lake remained nearly constant until about 1960.

In the late 1950s the Soviet government embarked on the first phases of a vast irrigation program, using water from the Amu Darya and Syr Darya for cotton cropping on the region's desert plain. The diversion of water soon became significant as more and more land came under irrigation. As a result, the influx fell to nearly zero by the early 1980s. The surface level of the Aral was sharply lowered and its area reduced. The sea became divided into two separate parts.

As the lake's shoreline receded, the exposed lakebed became encrusted with salts. The once-flourishing fishing port of Muynak became a ghost town, 50 km (30 mi) from the new lake shoreline. Strong winds now blow salt particles and mineral dusts in great clouds southwestward over the irrigated cotton fields and westward over grazing pastures. These salts—particularly the sodium chloride and sodium sulfate components—are toxic to plants. The salt dust permanently poisons the soil and can only be flushed away with more irrigation water.

The dust also contains residues of pesticides and other agrochemical wastes. These airborne poisons have produced a high incidence of stillbirths, anemia, and eye and lung disease among Kazakhstan's 16 million people. In fact, some ecological journalists have called the lake a "liquid Chernobyl."

The future of the lake appears grim indeed. With the present fresh water inflow nearly zero, the salinity of the remaining lake has exceeded the level of the ocean. Without the sacrifice of agricultural production for

E16.4.1 The Aral Sea shrinks *This pair of satellite images shows the Aral Sea in 1976 and 1997. About two-thirds of the sea's volume has been lost in the last 30 years.*

16.4.2 Graveyard of ships *As the Aral Sea shrank, it left behind the hulks of abandoned fishing vessels that were made useless by the loss of fish populations.*

water to fill the lake, or the importation of more water from vast distances to the north, there is little that can be done to save the lake.

However, a plan is now underway to rehabilitate at least a portion of this vast ecological ruin. The plan makes use of the fact that the flow of the Syr Darya reaching the lake is still significant—in fact, it is enough to maintain the smaller northern section of the lake, called the Small Sea (see map), in a productive state. Simple in concept, the plan provides for a dike some 13 km (8 mi) long to separate the Small Sea from its larger brother, now termed the Big Sea. The dike will trap the inflowing waters of the Syr Darya and raise the level of the Small Sea by some 4 m (13 ft), returning about 600 km^2 (230 mi^2) of the surrounding salt plain to lake.

Salinity in the new Small Sea will drop from 35 parts per thou-

16.4.3 Blowing dust *Vast expanses of lake bottom, left high and dry by the shrinking of the Aral Sea, provide a source for wind-blown dust, here displayed along a road in the village of Kyzylkum.*

16.4.4 ***Salinization*** *Accumulation of salts in agricultural soils accompanies the shrinking of the Aral Sea. Salty windblown dust and saline ground water drawn to the surface concentrate salt in the top layer of soil.*

sand to a value between 4 and 17 parts per thousand. This will allow the 24 local species of fish to return and will revive the local fishing industry. In addition, the increased water surface area will increase local precipitation and humidity, expanding local pastures and cropping areas. The wetter soil will be less likely to blow in the chronic dust storms that have plagued the region. The expanded lake area should also reduce the problem of high salinity in local ground water.

The idea of a dike to separate the two arms of the Aral was first tested in the early 1990s. Desperate to improve the lives of the citizens of Aralsk, a seaport city of 35,000 now located 80 km (50 mi) from the lake shore, the mayor of Aralsk obtained a $2.5 million allocation from the central government to build a dike to enclose the flow of the Syr Darya within the Small Sea. The dike was built and at first worked well in containing the rising waters of the Small Sea. But since it was constructed of sand without a core of clay or a rock facing to protect it from erosion, it began to erode. Lacking a sluiceway to release water to the Big Sea, it was overtopped and breached on several occasions.

The success of the dike did not go unnoticed, however. In 2001, a World Bank loan to Kazakhstan for $64.5 million to build a permanent dike was approved, and by 2002, the Kazakhstan government had ratified the loan. Construction began in 2003, with completion due in mid-2004. Added along the way were funds for improvements in the irrigation infrastructure of the Syr Darya that would enhance runoff to the Small Sea. The new dike will have a much more gentle slope on the Small Sea side and will also have a proper sluiceway to convey high waters to the Big Sea. In four years, the Small Sea will rise by 4 m (13 ft).

But what of the Big Sea? Unfortunately, its prospects are dim. Presently at a salinity level of 85 parts per thousand (more than twice as salty as ocean water), the last of its native fish species are dying. Still, there is some hope for the Big Sea's fishermen. When the salinity reaches 110 parts per thousand, conditions will be favorable for brine shrimp. These tiny creatures are used as food for young fishes raised in fish farms worldwide and can provide a valuable cash crop. However, there will be no respite from the toxic dust storms of salt and pesticide residues that sweep over the region downwind from the Big Sea.

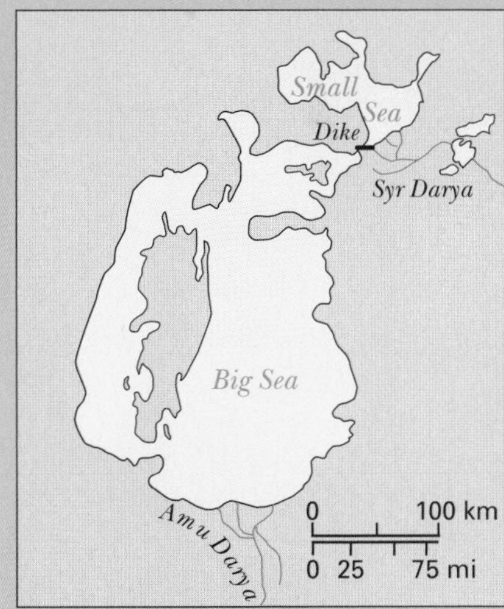

E16.4.5 *With the construciton of the new dike, the Aral Sea will be divided into a relatively healthy Small Sea and a ruined Big Sea.*

17 FLUVIAL PROCESSES AND LANDFORMS

Junction of Rio Uruguay and a tributary, Misiones, Argentina. Laden with red-brown sediment, the Rio Uruguay is joined here by a tributary. Flowing south, it will deposit its load about 900 km (620 mi) downstream at the Rio de la Plata estuary, near Buenos Aires.

Most of the world's land surface has been sculpted by running water. Flowing as a sheet across a land surface, running water picks up particles and moves them downslope into a stream channel. When rainfall is heavy, streams and rivers swell, lifting large volumes of sediment and carrying them downstream. In this way, running water erodes mountains and hills, carves valleys, and deposits sediment. This chapter describes the work of running water and the landforms it creates.

Running water is one of four flowing substances that erode, transport, and deposit mineral and organic matter. The other three are waves, glacial ice, and wind. These four fluid agents carry out the processes of *denudation,* which we discussed in Chapter 14. Recall that denudation is the total action of all processes by which the exposed rocks of the continents are worn away and the resulting sediments are transported to the sea or closed inland basins.

Denudation is an overall lowering of the land surface. If left unchecked to operate over geologic time, denudation will reduce a continent to a nearly featureless, sea-level surface. However, recall that plate tectonics has always kept continental crust elevated well above the ocean basins. The result is that running water, waves, glacial ice, and wind have always had plenty of raw material available to create the many landforms that we see around us.

Thanks to denudation processes and plate tectonics, the land environments of life are always in constant change, even as plants and animals undergone evolutionary development. Wind, water, waves, and ice produce and maintain a wide variety of landforms that are the habitats for evolving life forms. In turn, the life forms become adapted to those habitats and diversify to a degree that matches the diversity of the landforms themselves.

Fluvial Processes and Landforms

Landforms shaped by running water are described as **fluvial landforms.** They are shaped by the **fluvial processes** of overland flow and stream flow, which we described in Chapter 16. Weathering and the slower forms of mass wasting (Chapter 15), such as soil creep, operate hand in hand with overland flow, providing the rock and mineral fragments that are carried into stream systems.

Fluvial landforms and fluvial processes dominate the continental land surfaces the world over. Throughout geologic history, glacial ice has been present only in continental areas located in mid- and high latitudes and in high mountains. Landforms made by wind action are found only in very small parts of the continental surfaces. And landforms made by waves and currents are restricted to the narrow contact zone between oceans and continents. That is why, in terms of area, the fluvial landforms dominate the environment of terrestrial life.

Areas of fluvial landforms are also exploited for agriculture. Except for areas in the northern hemisphere that were formerly occupied by glacial ice, most land areas used in crop cultivation or for grazing have been shaped by fluvial processes. For this reason, fluvial landforms are very important to the support of the human species.

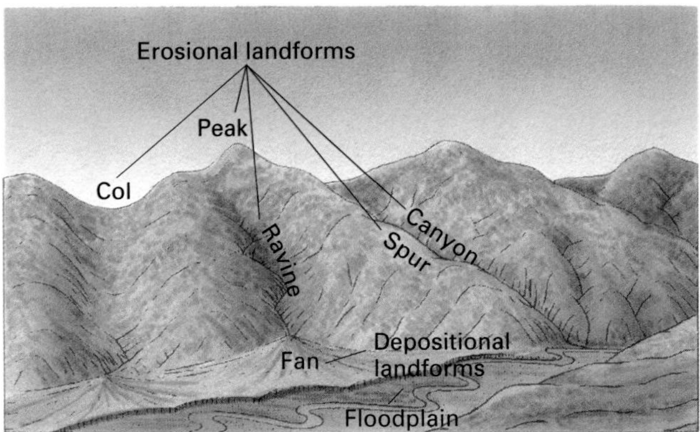

17.1 Erosional and depositional landforms *(A. N. Strahler.)*

Erosional and Depositional Landforms

All agents of denudation perform the geological activities of erosion, transportation, and deposition. Consequently, there are two major groups of landforms—erosional landforms and depositional landforms. When an initial landform, such as an uplifted crustal block, is created, it is attacked by the processes of denudation and especially by fluvial action. Valleys are formed where rock is eroded away by fluvial agents. Between the valleys are ridges, hills, or mountain summits, representing the remaining parts of the crustal block that are as yet uncarved by running water. These sequential landforms, shaped by progressive removal of the bedrock mass, are *erosional landforms.* Fragments of soil, regolith, and bedrock that are removed from the parent rock mass are transported and deposited elsewhere to make an entirely different set of surface features—the *depositional landforms.*

Figure 17.1 illustrates the two groups of landforms as produced by fluvial processes. The ravine, canyon, peak, spur, and col are erosional landforms. The fan, built of rock fragments below the mouth of the ravine, is a depositional landform. The floodplain, built of material transported by a stream, is also a depositional landform.

Slope Erosion

Fluvial action starts on the uplands as *soil erosion.* By exerting a dragging force over the soil surface, overland flow picks up particles of mineral matter ranging in size from fine colloidal clay to coarse sand or even gravel. The size grade selected depends on the speed of the current and the degree to which the particles are bound by plant rootlets or held down by a mat of leaves. Added to this solid matter is dissolved mineral matter in the form of ions produced by acid reactions or direct solution.

This ongoing removal of soil is part of the natural geological process of denudation. It occurs everywhere that precipitation falls on land. Under stable natural conditions in a humid climate, the erosion rate is slow enough that a soil with distinct horizons is formed and maintained. Each year a small amount of soil is washed away, while a small amount of solid rock material becomes altered to new regolith and soil. These conditions also enable plant communities to maintain themselves in a stable equilibrium. Soil scientists refer to this state of activity as the *geologic norm.*

Accelerated Erosion

In contrast, the rate of soil erosion may be enormously speeded up by human activities or by rare natural erosional events to produce a state of *accelerated erosion.* What happens then is that the soil is removed much faster than it can be formed, and the uppermost soil horizons are progressively exposed. Accelerated erosion arises most commonly when the plant cover and the physical state of the ground surface change. Destruction of vegetation by the clearing of land for cultivation or by forest fires sets the stage for a series of drastic changes. No foliage remains to intercept rain, and the protection of a ground cover of fallen leaves and stems is removed. Consequently, the raindrops fall directly on the mineral soil.

The direct force of falling drops on bare soil causes a geyserlike splashing in which soil particles are lifted and then dropped into new positions. This process is termed *splash erosion* (Figure 17.2). Soil scientists estimate that a torrential rainstorm has the ability to disturb as much as 225 metric tons of soil per hectare (about 100 U.S. tons per acre). On a sloping ground surface, splash erosion shifts the soil

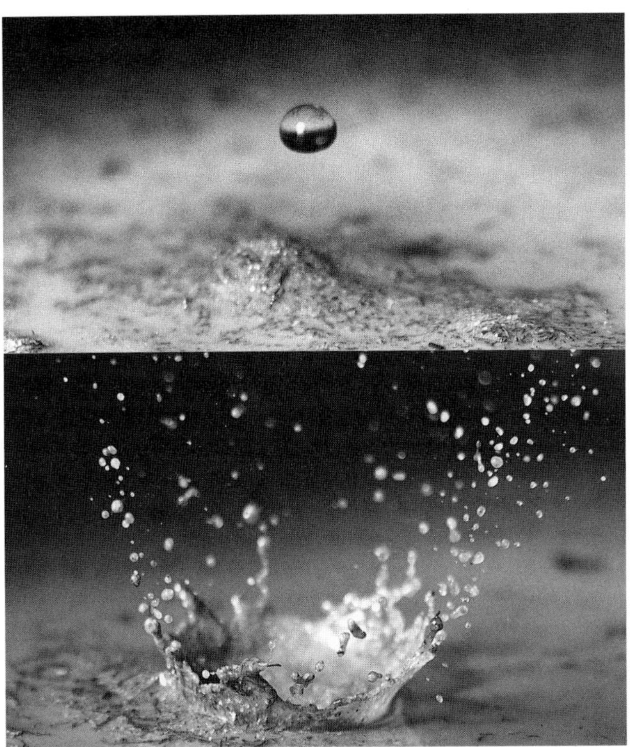

17.2 Soil erosion by rain splash *A large raindrop (above) lands on a wet soil surface, producing a miniature crater (below). Grains of clay and silt are thrown into the air, and the soil surface is disturbed.*

slowly downhill. An even more important effect is to cause the soil surface to become much less able to absorb water. This occurs because the natural soil openings become sealed by particles shifted by raindrop splash. Reduced infiltration, in turn, permits a much greater depth of overland flow to occur from a given amount of rain. So, the rate of soil erosion is intensified.

Another effect of the destruction of vegetation is to reduce greatly the resistance of the ground surface to the force of erosion under overland flow. On a slope covered by grass sod, even a deep layer of overland flow causes little soil erosion. This protective action is present because the energy of the moving water is dissipated in friction with the grass stems, which are tough and elastic. On a heavily forested slope, the surface layer of leaves, twigs, roots, and even fallen tree trunks take up the force of overland flow. Without such a cover, the eroding force is applied directly to the bare soil surface, easily dislodging the grains and sweeping them downslope.

We can get a good appreciation of the contrast between normal and accelerated erosion rates by comparing the quantity of sediment derived from cultivated surfaces with that derived from naturally forested or reforested surfaces. The comparison is made within a single region in which climate, soil, and topography are fairly uniform. *Sediment yield* is a technical term for the quantity of sediment removed by overland flow from a unit area of ground surface in a given unit of time. Yearly sediment yield is stated in metric tons per hectare, or tons per acre.

Figure 17.3 gives data of annual average sediment yield and runoff by overland flow from several types of upland surfaces in northern Mississippi. Notice that both surface runoff and sediment yield decrease greatly with the increased effectiveness of the protective vegetative cover. Sediment yield from cultivated land undergoing accelerated erosion is over ten times greater than that from pasture and about one thousand times greater than that from pine plantation land.

Sheet Erosion and Rilling

Accelerated soil erosion is a constant problem in cultivated regions with a substantial water surplus. When the natural cover of forest or prairie grasslands is first removed and the soil is plowed for cultivation, little erosion will occur until the action of rain splash has broken down the soil aggregates and sealed the larger openings. Then, however, overland flow begins to remove the soil in rather uniform thin layers, a process termed *sheet erosion.* Because of seasonal cultivation, the effects of sheet erosion are often little noticed until the upper horizons of the soil are removed or greatly thinned.

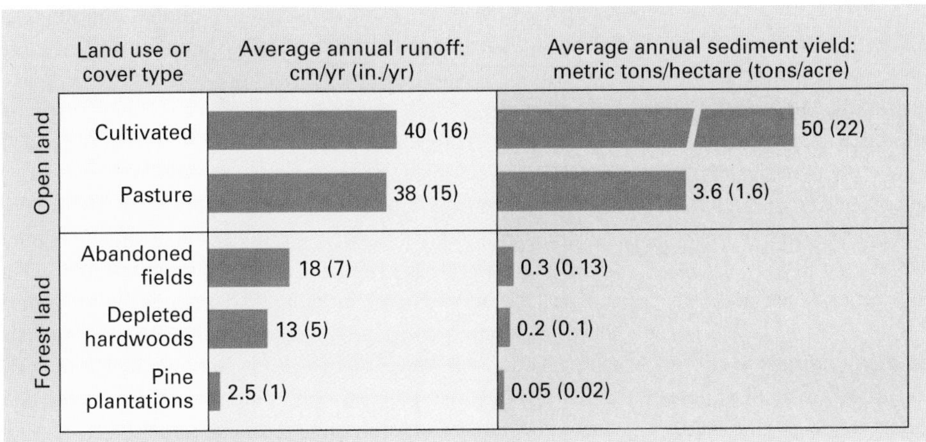

17.3 Runoff and sediment yield *This bar graph shows that both runoff and sediment yield are much greater for open land than for land covered by shrubs and forest. (Data of S. J. Ursic, U.S.D.A.)*

17.4 Gullies *Deep branching gullies have carved up an overgrazed pasture near Shawnee, Oklahoma. Contour terracing and check dams have halted the headward growth of the gullies.*

Where land slopes are steep, runoff from torrential rains produces a more destructive activity, *rill erosion,* in which many closely spaced channels are scored into the soil and regolith. If the rills are not destroyed by soil tillage, they may soon begin to join together into still larger channels. These deepen rapidly and soon become *gullies*—steep-walled, canyonlike trenches whose upper ends grow progressively upslope (Figure 17.4). Ultimately, a rugged, barren topography results from accelerated soil erosion that is allowed to proceed unchecked.

Colluvium and Alluvium

With an increased depth of overland flow and no vegetation cover to absorb the eroding force, soil particles are easily picked up and moved downslope. Eventually, they reach the base of the slope, where the surface slope becomes more gentle and meets the valley bottom. There the particles come to rest and accumulate in a thickening layer termed *colluvium* (see Figure 15.11). Because this deposit is built by overland flow, it has a sheetlike distribution and may be little noticed, except where it eventually buries fence posts or tree trunks.

If not deposited as colluvium, sediment carried by overland flow eventually reaches a stream in the adjacent valley floor. Once in the stream, it is carried farther downvalley and may accumulate as alluvium in layers on the valley floor. As we saw in Chapter 16, the term **alluvium** is used to describe any stream-laid sediment deposit. Deposition of alluvium can bury fertile floodplain soil under infertile, sandy layers. Coarse alluvium chokes the channels of small streams and can cause the water to flood broadly over the valley bottoms.

Slope Erosion in Semiarid and Arid Environments

Thus far, we have discussed slope erosion in moist climates with a natural vegetation of forest or a dense prairie grassland. Conditions are quite different in a midlatitude semiarid climate with summer drought. Here, the natural plant cover consists of short-grass prairie (steppe). Although it is sparse and provides a rather poor ground cover of plant litter, the grass cover is normally strong enough that a slow pace of erosion can be sustained.

Much the same conditions are also found in the tropical savanna grasslands. In these semiarid environments, however, the natural equilibrium is highly sensitive and easily upset. Depletion of the plant cover by fires or the grazing of herds of domesticated animals can easily set off rapid erosion. These sensitive, marginal environments require cautious use because they lack the potential to recover rapidly from accelerated erosion once it has begun.

17.5 Badlands *Clay beds form badlands at Zabriskie Point, Death Valley National Monument, California. (A. H. Strahler.)*

Erosion at a very high rate by overland flow is actually a natural process in certain favorable locations in semiarid and arid lands. Here, the erosion produces *badlands*. Badlands are underlain by clay formations, which are easily eroded by overland flow. Erosion rates are too fast to permit plants to take hold, and no soil can develop. A maze of small stream channels is developed, and ground slopes are very steep (Figure 17.5).

One well-known area of badlands in the semiarid shortgrass prairie is the Big Badlands of South Dakota, along the White River. Badlands such as these are self-sustaining and have been in existence on continents throughout much of geologic time. Badlands can also result from poor agricultural processes, especially when the vegetation cover of clay formations is disturbed by plowing or overgrazing.

The Work of Streams

The work of streams consists of three closely related activities—erosion, transportation, and deposition. **Stream erosion** is the progressive removal of mineral material from the floor and sides of the channel, whether bedrock or regolith. **Stream transportation** consists of movement of the eroded particles dragged over the stream bed, suspended in the body of the stream, or held in solution as ions. **Stream deposition** is the accumulation of transported particles on the stream bed and floodplain, or on the floor of a standing body of water into which the stream empties. Erosion cannot occur without some transportation taking place, and the transported particles must eventually come to rest. Thus, erosion, transportation, and deposition are simply three phases of a single activity.

Stream Erosion

Streams erode in various ways, depending on the nature of the channel materials and the tools with which the current is armed. The force of the flowing water not only sets up a dragging action on the bed and banks, but also causes particles to impact the bed and banks. Dragging and impact can easily erode alluvial materials, such as gravel, sand, silt, and clay. This form of erosion, called *hydraulic action,* can excavate enormous quantities in a short time. The undermining of the banks causes large masses of alluvium to slump into the river, where the particles are quickly separated and become part of the stream's load. This process of bank caving is an important source of sediment during high river stages and floods.

Where rock particles carried by the swift current strike against bedrock channel walls, chips of rock are detached. The large, strong fragments become rounded as they travel. The rolling of cobbles and boulders over the stream bed further crushes and grinds the smaller grains to produce a wide assortment of grain sizes. This process of mechanical wear is called *abrasion.* It is the principal means of erosion in bedrock too strong to be affected by simple hydraulic action. A striking example of abrasion is the erosion of a *pothole.* This occurs when a shallow depression in the bedrock of a

17.6 Potholes *These potholes in lava bedrock attest to the abrasion that takes place on the bed of a swift mountain stream. McCloud River, California. (A. H. Strahler.)*

stream bed acquires one or several grinding stones, which are spun around and around by the flowing water and carve a deep depression (Figure 17.6).

Finally, the chemical processes of rock weathering—acid reactions and solution—are effective in removing rock from the stream channel. The process is called *corrosion.* Effects of corrosion are conspicuous in limestone, which develops cupped and fluted surfaces.

Stream Transportation

The solid matter carried by a stream is the **stream load.** It is carried in three forms (Figure 17.7). *Dissolved matter* is transported invisibly in the form of chemical ions. All streams carry some dissolved ions resulting from mineral alteration. Sand, gravel, and larger particles move as *bed load* close to the channel floor by rolling or sliding. Clay and silt are carried in *suspension*—that is, they are held within the water by the upward elements of flow in turbulent eddies in the stream. This fraction of the transported matter is the *suspended load* (Figure 17.8). Of the three forms, suspended load is generally the largest.

A large river such as the Mississippi carries as much as 90 percent of its load in suspension. Most of the suspended load comes from its great western tributary, the Missouri River, which is fed from semiarid lands, including the Dakota Badlands. (Figure 16.17 shows the Mississippi River system and its major branches.)

The Yellow (Huang) River of China heads the world list in annual suspended sediment load, and its watershed sediment yield is one of the highest known for a large river basin. This is because much of its basin consists of cultivated upland surfaces of wind-deposited silt that is very easily eroded. (See Chapter 19 and Figure 19.28.) In addition, much of the river's upper watershed is in a semiarid climate with dry winters. Vegetation is sparse, and the runoff from heavy summer rains sweeps up a large amount of sediment.

Capacity of a Stream to Transport Load

The maximum solid load of debris that can be carried by a stream at a given discharge is a measure of the *stream capacity.* This load is usually measured in units of metric tons per day passing downstream at a given location. Total solid load includes both bed load and suspended load.

A stream's capacity to carry suspended load increases sharply with an increase in the stream's velocity because the

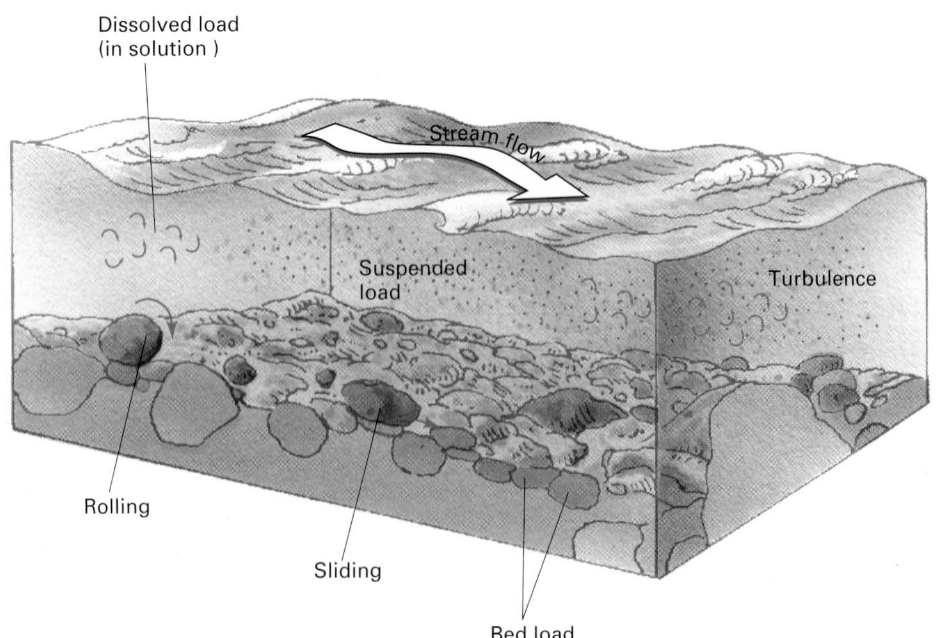

17.7 Sediment load *Streams carry their load as dissolved, suspended, and bed load. Suspended load is kept in suspension by turbulence. Bed load moves by sliding or rolling.*

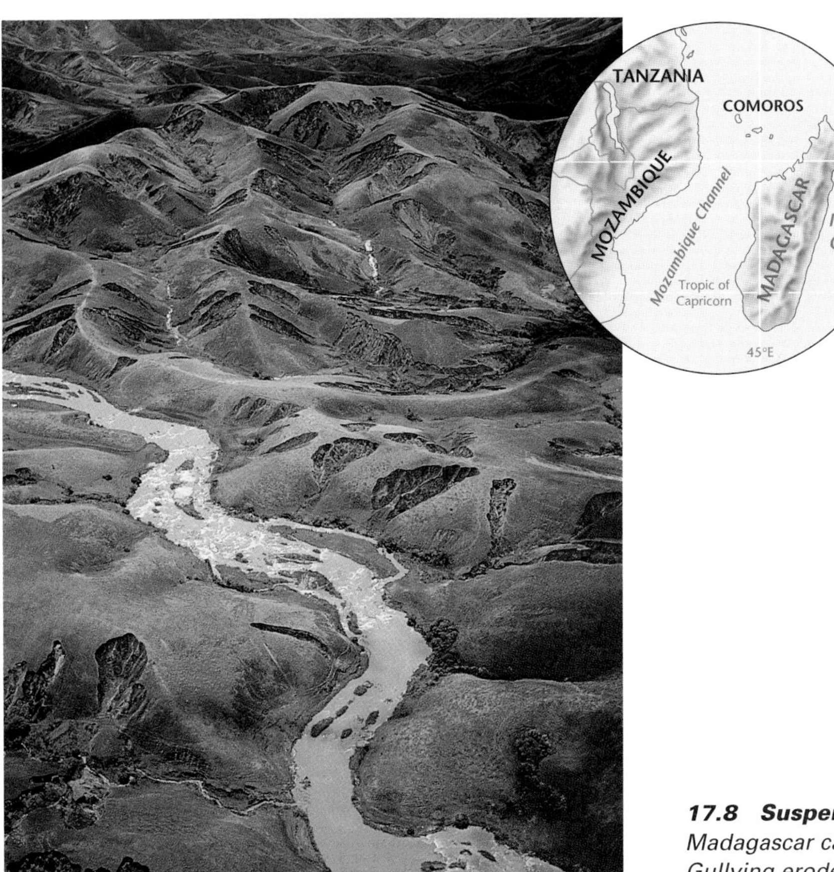

17.8 Suspended sediment *This turbulent stream in Madagascar carries a heavy load of suspended sediment. Gullying erodes the denuded hills in the foreground.*

swifter the current, the more intense the turbulence. The capacity to move bed load also increases with velocity—the faster water motion produces a stronger dragging force against the bed. In fact, the capacity to move bed load increases according to the third to fourth power of the velocity. In other words, when a stream's velocity is doubled in times of flood, its ability to transport bed load is increased from eight to sixteen times. Thus, most of the conspicuous changes in the channel of a stream occur in a flood stage. *Working It Out 17.1 • River Discharge and Suspended Sediment* shows how the suspended sediment load of a stream increases with discharge.

When water flow increases, a stream flowing in a channel that is cut into thick layers of silt, sand, and gravel will easily widen and deepen its channel. When the flow slackens, the stream will deposit material in the bed, filling the channel again. Where a stream flows in a channel of hard bedrock, the channel cannot be quickly deepened in response to rising waters and may not change much during a single flood. Such conditions exist in streams that occupy deep canyons and have steep gradients.

Stream Gradation

Most major stream systems have gone through thousands of years of runoff, erosion, and deposition. Over time, the gradients of stream segments tend become adjusted so that they just carry the average load of sediment that they receive from slopes and inflowing channels. How does this come about? GEODISCOVERIES

Visualize a small stream basin in which runoff and overland flow carry sediment to a stream channel. If more sediment accumulates each year in the stream channel than can be carried away, the surface of the channel will be built up and the slope of the stream will increase. But with increased slope comes increased stream velocity and increased ability to carry sediment. Eventually, the slope will reach a point at which the stream just carries away the sediment that it receives. If sediment flow to the stream is reduced, the stream will gradually erode its channel downward. This will reduce its slope and also reduce its ability to carry sediment until it can only carry the reduced amount it receives from the hill slopes. Since every stream channel experiences this process, eventually the whole stream will tend toward a state in which the slopes of all its segments form a coordinated network that just carries the sediment load contributed by the drainage basin. A stream in this equilibrium condition is referred to as a **graded stream.**

Figure 17.9 shows how a graded stream might develop on a landscape that is rapidly uplifted, perhaps by a series of fault steps or blocks, or possibly uncovered after erosion by continental glaciers. The side of the block shows a series of

Working It Out | 17.1

River Discharge and Suspended Sediment

As you might imagine, a river carries more sediment in flood than during periods of normal flow. This happens for a number of reasons.

- First, as the discharge increases, more water is available to carry sediment downstream.
- Second, as discharge increases, so does the velocity of the stream flow. And as the velocity of the water flow increases, so does the intensity of the turbulence that keeps suspended sediment in motion. Thus, more sediment can be carried in a cubic meter of swiftly moving water.
- Third, as the water rises and increases in velocity, the force of the stream flow on the bed increases. The stream then erodes sediment in its bed, scouring and deepening the channel, a process that also increases the sediment available for transportation.
- Fourth, the heavy precipitation that causes a flood generates overland flow that drags more sediment into upstream river channels. Thus, more sediment is available for transport.

The figure to the right is a graph plotting discharge against the suspended sediment load for the Powder River at Arvada, Wyoming. The discharge is measured in cubic meters per second (m^3/s) and the suspended sediment load in metric tons per day (t/d). The points are individual measurements collected over a period of time at a single location. The graph easily shows that as discharge increases, so does suspended sediment. For example, when the discharge goes from 1 to 10 m^3/s, the suspended sediment load increases from about 200 to 10,000 t/d. Thus, when discharge increases by a factor of 10, sediment load increases by a factor of about 50.

This type of increase follows a power function—that is, it fits a function of the form

$$y = ax^b$$

where $b > 1$. On normal (arithmetic) graph paper, this function plots as an ascending curve. On a log-log plot, however, it is a straight line. In the log-log plot, the scale positions a value based on the logarithm of the value rather than the value itself. (Note that the logarithm used here is the common logarithm, which uses the base 10.) The right and upper axes on the graph demonstrate this point. They show the logarithm of the values changing on a uniform scale from −2 to +2 for discharge and from 0 to 6 for suspended sediment load, while the log axes change by powers of 10.

This formulation works because, if we take the log of both sides of the power function $y = ax^b$, we have

$$\log y = \log a + b \log x$$

By substituting Y' for $\log y$, x' for $\log x$, and a' for $\log a$, we have

$$y' = a' + bx'$$

which is the equation for a straight line. Thus, by plotting the logs of x and y values, rather than the values themselves, we get a straight line. Using the log scale on the graph does this conversion automatically.

In the case of the relation between suspended sediment load and discharge for the Powder River at Arvada, Wyoming, the measurements come close to fitting a line with the equation

$$S = 178Q^{1.75}$$

where S is the suspended sediment load (t/d) and Q is the discharge (m^3/s). This can also be written in log-log form, as a straight-line function:

$$\log S = \log(178) + 1.75 \log Q$$
$$= 2.25 + 1.75 \log Q$$

The rapid increase in suspended load with increasing discharge is

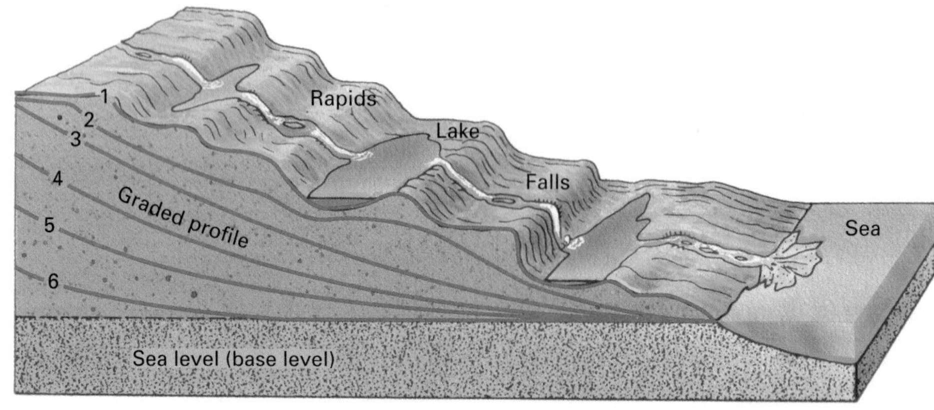

17.9 Stream gradation *Schematic diagram of gradation of a stream. Originally, the channel consists of a succession of lakes, falls, and rapids. (Copyright © A. N. Strahler.)*

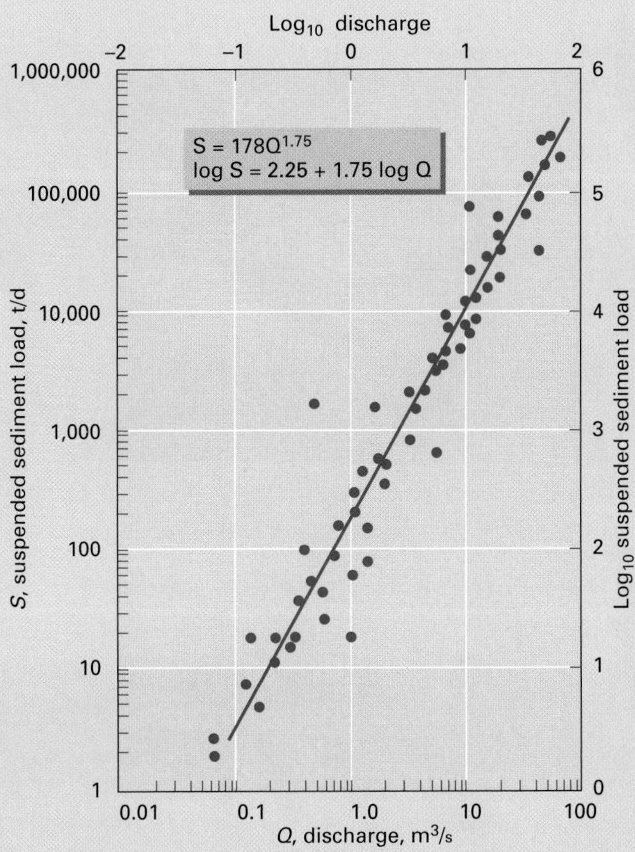

Sediment load and discharge *Increase of suspended sediment load with increase in discharge plotted on logarithmic scales. The dots are individual observations scattered about the line of best fit. Data are for the Powder River at Arvada, Wyoming. (Data of U.S. Geological Survey.)*

characteristic of nearly all rivers. In fact, the rate of increase for many rivers is actually larger than that for the Powder River at Arvada—values of 2–3 are typical for the power to which Q is raised.

The important point to carry home from this examination of the behavior of rivers is that transportation of suspended sediment increases very rapidly with discharge. Therefore, most of the transportation work of rivers is carried out in times of high flow. Many times more sediment may be carried downstream during a single small flood than over many months of normal river flow. And a large flood may move vast volumes of sediment that has been accumulating in the channel and floodplain in wait for a rare and extreme event.

stream profiles—plots of elevation of the stream with distance from the sea. At first, the stream is ungraded, with large fluctuations in the slope profile. Water accumulates in shallow depressions to form lakes, which overflow from higher to lower levels. Rapids and falls are abundant. As time passes, the landscape is slowly eroded by fluvial action. Each stream segment seeks its own equilibrium slope, and the stream profile is smoothed out into a uniform curve, shown as profile line 3. The profile has now been graded. From that time forward, this *graded profile* is steadily lowered in elevation as the landscape is further eroded (curves 4 through 6).

Landscape Evolution of a Graded Stream

Figure 17.10 illustrates the stream gradation process over a landscape in a series of block diagrams. In *(a),* we see *waterfalls* and *rapids,* which are simply portions of the channel with steep gradients. Flow velocity at these points is greatly increased, and abrasion of bedrock is therefore most intense. As a result, the falls are cut back and the rapids are trenched. At the same time, the ponded stretches of the stream are first filled by sediment and are later lowered in level as the lake outlets are cut down. In time, the lakes disappear and the falls are transformed into rapids.

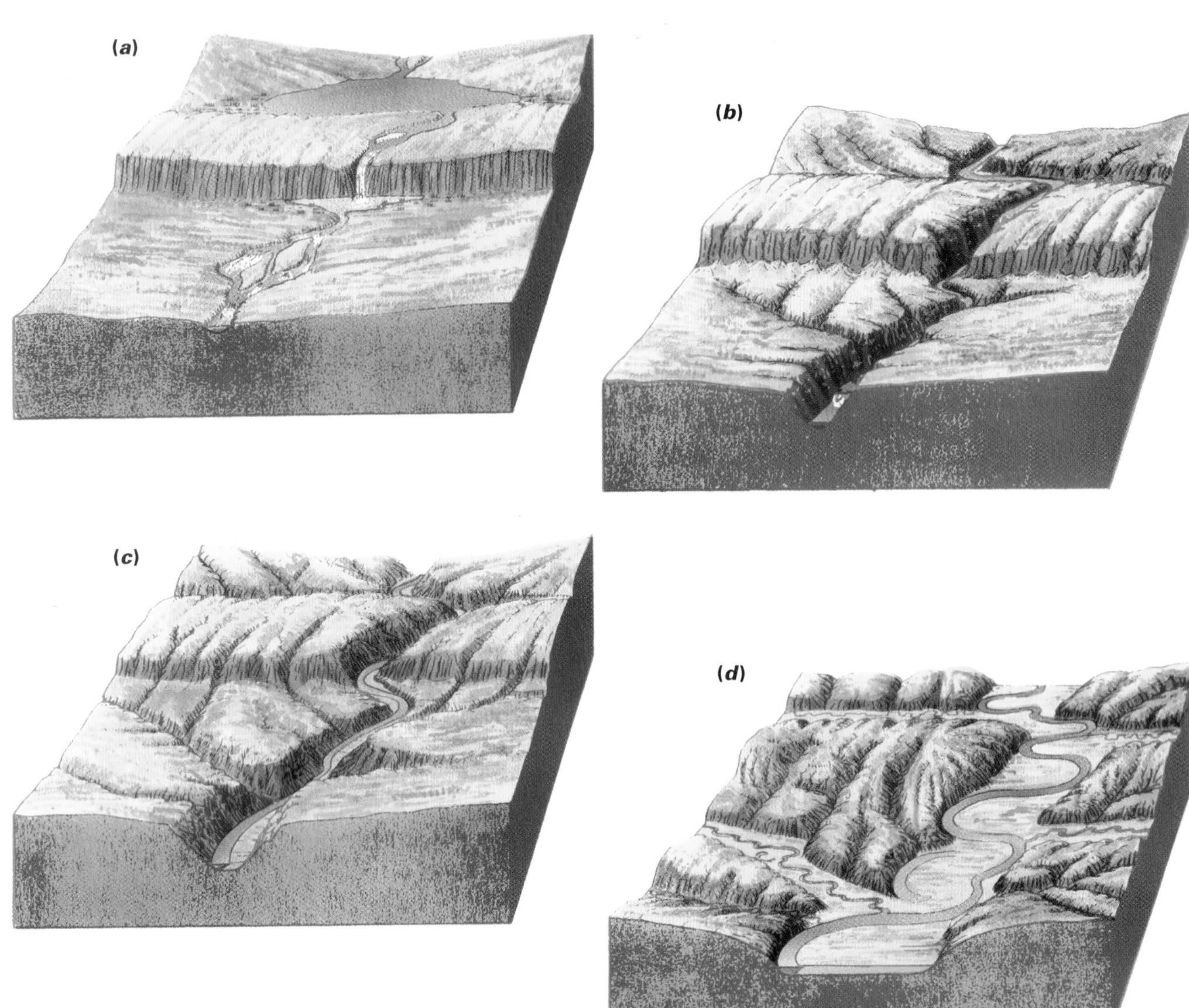

17.10 Evolution of a graded stream and its valley *(Drawn by E. Raisz. Copyright © A. N. Strahler.) (a) Stream established on a land surface dominated by landforms of recent tectonic activity. (b) Gradation in progress. The lakes and marshes drained. The gorge is deepening, and the tributary valleys are extending. (c) Graded profile attained. Floodplain development is beginning, and the widening of the valley is in progress. (d) Floodplain widened to accommodate meanders. Floodplains now extend up tributary valleys.*

In the early stages of gradation, the capacity of the stream exceeds the load supplied to it, so that little or no alluvium accumulates in the channels. Abrasion continues to deepen the major channels, with the result that they come to occupy steep-walled *gorges* or *canyons* (Figure 17.11). *Focus on Remote Sensing 17.2 • A Canyon Gallery* shows how some of the world's more spectacular canyons and gorges are imaged by sensors on orbiting spacecraft. Weathering and mass wasting of canyon walls contribute an increasing supply of rock debris to the channels. Also on the increase is debris shed from land surfaces that contribute overland flow to the newly developed branches.

The erosion of rapids reduces the gradient to a slope angle that more closely approximates the average gradient of that section of the stream (Figure 17.10*b*). At the same time, branches of the main stream are being extended into higher parts of the original land mass. These carve out many new small drainage basins. Thus, the original tectonic landscape is transformed into a complete fluvial landform system.

As landscape change continues, a gradual decrease in a stream's capacity to move bed load results from the gradual reduction in the channel gradient. Simultaneously, the load supplied to the stream from the entire upstream area is on the increase. So, the time comes when the supply of load exactly matches the stream's capacity to transport it. Figure 17.10*c* shows the landscape at this point in time. Now, all the major streams have achieved the graded con-

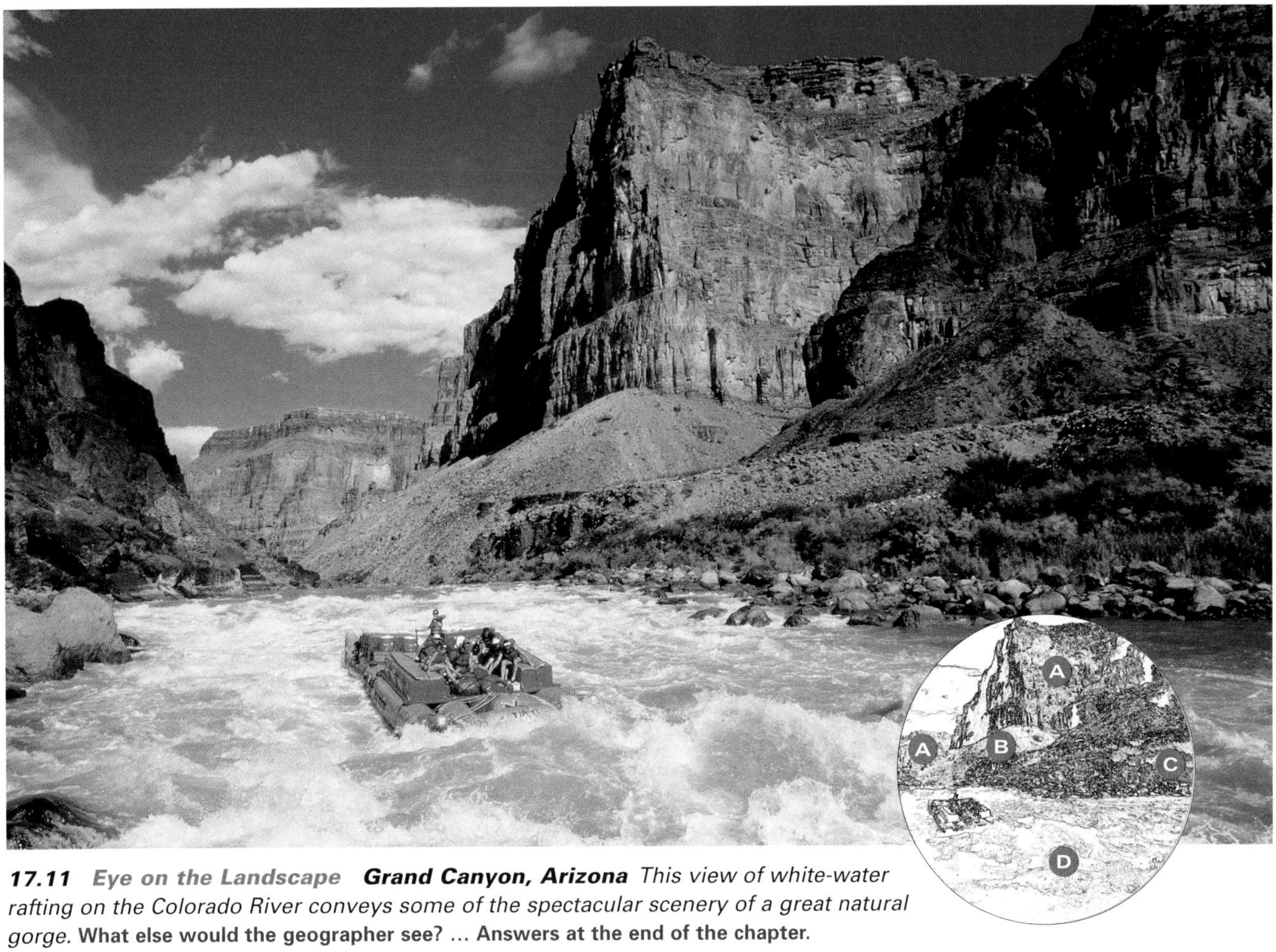

17.11 ***Eye on the Landscape*** ***Grand Canyon, Arizona*** *This view of white-water rafting on the Colorado River conveys some of the spectacular scenery of a great natural gorge.* **What else would the geographer see? ... Answers at the end of the chapter.**

dition and possess graded profiles that descend smoothly and uniformly.

The first indication that a stream has attained a graded condition is the beginning of floodplain development. For reasons that are not completely understood, the river begins to wander sidewards, cutting into the side slopes flanking its channel. A sinuous path of arcs and curving bends develops. On the inside of each bend, alluvium accumulates as a long, curving deposit of sediment—termed a point bar. Widening of the bar deposit produces a crescent-shaped area of low ground, which is the first stage in floodplain development. This stage is illustrated in Figure 17.10*c*. As lateral cutting continues, the floodplain strips are widened, and the channel develops sweeping bends (Figure 17.12). These winding river bends are called **alluvial meanders** *(d)*. In this way, the floodplain is widened into a continuous belt of flat land between steep valley walls. GEODISCOVERIES

Floodplain development reduces the frequency with which the river attacks and undermines the adjacent valley wall. Weathering, mass wasting, and overland flow can then act to reduce the steepness of the valley-side slopes (Figure (17.13). As a result, in a humid climate, the gorgelike aspect of the valley gradually disappears and eventually gives way to an open valley with soil-covered slopes protected by a dense plant cover. The time required for the stream to reach a graded condition and erode a broad valley will be a few tens of millions of years.

As we have noted, the gradient and channel form of each portion of the stream network is adjusted to carry the sediment load it receives. The gradient and channel form varies, however, from the small streams at the headwaters of the

17.12 **Meander bend** *Idealized map and cross-profile of a meander bend of a large alluvial river, such as the lower Mississippi. Arrows show the position of the swiftest current. (Copyright © A. N. Strahler.)*

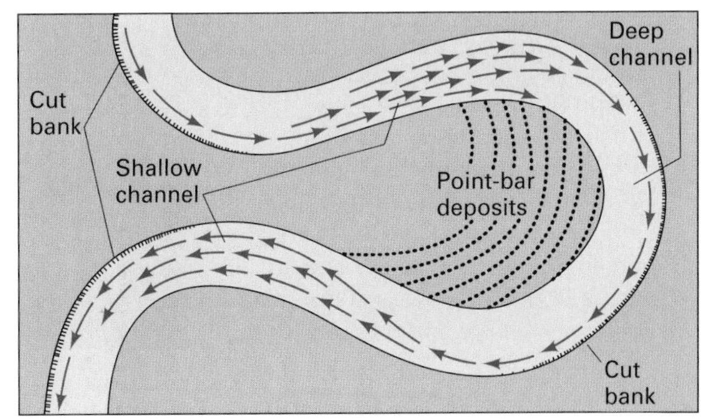

Focus on Remote Sensing | 17.2

A Canyon Gallery

Deep canyons, carved by powerful rivers crossing high terrain, are among the most dramatic features of the landscape. They provide a stunning testament to the power of fluvial systems to shape the Earth. Who can forget the awesome spectacle of a yawning canyon, seen from a precarious perch on a high rim? Here's a gallery of satellite images of canyons as acquired by the MISR and ASTER instruments.

The Grand Canyon of the Colorado River is among the most famous in the world. Spanning a length of about 450 km (about 280 mi) with vertical drops up to about 1500 m (about 5000 ft), it is indeed spectacular. Image (*a*), in true color, was acquired by MISR on December 31, 2000. Trace the path of the Colorado from Lake Powell, at the upper right, through narrow Marble Canyon, until the canyon broadens to the southeast of the snow-covered Kaibab Plateau. Here the Grand Canyon begins, revealing a dissected landscape between the two canyon rims as the river curves around the plateau. Note the sharp black shadow at the south rim, indicating that the rim is steep indeed. The canyon continues as the river flows westward to the edge of the frame. To the south of the canyon are cloud streaks that have formed from condensation trails of jet aircraft. Looking closely, you can see the shadow of each cloud as a dark streak to the north of the cloud.

Image (*b*), acquired by ASTER on May 12, 2000, is a computer-generated perspective view of the Grand Canyon looking north up Bright Angel Canyon from the South Rim. In this false color image, vegetation appears green and water appears blue, but rocks are not shown in their true colors. The blue and black patches on the green Kaibab Plateau are burned areas from a fire that is still smoldering. A pall of smoke stretches to the east. Compare this with Figure 18.5, which is a ground-level photo of Bright Angel Canyon.

Image (*c*), another MISR image, shows the coast of southern Peru in the Ariquipa region. Here the Pacific coast runs nearly east-west. The coastline is obscured by low clouds and fog at the bottom of the picture. Note the two vast canyons

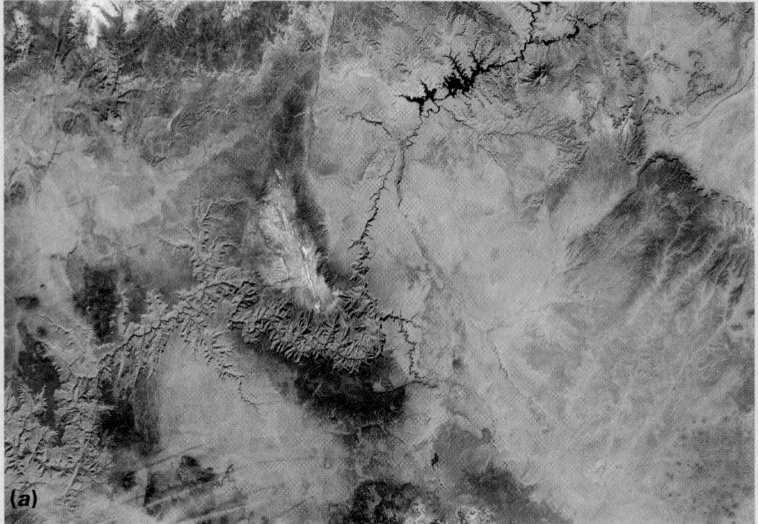

Grand Canyon, Arizona, imaged by MISR *(Courtesy NASA/GSFC/LARC/JPL, MISR team.)*

Grand Canyon perspective view from ASTER *(Courtesy NASA/GSFC/MITI/ERSDAC/JAROS and U.S. Japan ASTER Science Team.)*

(c)

Canyons of the Andes as seen by MISR *(Courtesy NASA/GSFC/LARC/JPL, MISR team.)*

reaching from the sea deep into the Andean Plateau. They are the canyons of the Rio Ocoña to the west and Rio Camaná to the east. Dwarfing the American Grand Canyon, they are far wider and deeper. In fact, the canyon of the middle branch of the Rio Ocoña (Rio Cotahuasi) reaches a depth of 3354 m (11,001 ft) below the plateau, which is more than twice as deep as the Grand Canyon. The white patch between the two canyon systems is the snow-covered Nevado Coropuna, at 6425 m (21,074 ft) elevation. Peaks to the west and east are Nevado Solimana (6117 m, 20,064 ft) and Nevado Ampato (5795 m, 18,909 ft).

Our last image (*d*) is another ASTER scene, a color infrared image of the Yangtze River canyon in the Three Gorges region of the provinces of Hubei and Sichuan, China. Although not a candidate for the deepest canyon, it is among the most scenic of the world's canyons, with steep limestone cliffs and forest-covered slopes separating clusters of quaint villages where tributaries meet the main channel. The inset image shows the construction site for the Three Gorges Dam, which will create a reservoir 175 m (574 ft) deep and about 600 km (about 375 mi) long. The dam will provide vast amounts of hydroelectric power and sharply reduce extreme flooding of the Yangtze. However, two of the three gorges will be flooded, and many local residents will be displaced. The Chinese government is making a major effort to forecast and mitigate environmental problems that will be created by the huge project.

Three Gorges region of the Yangtze River imaged by ASTER *(Courtesy NASA/GSFC/MITI/ERSDAC/JAROS and U.S. Japan ASTER Science Team.)*

(d)

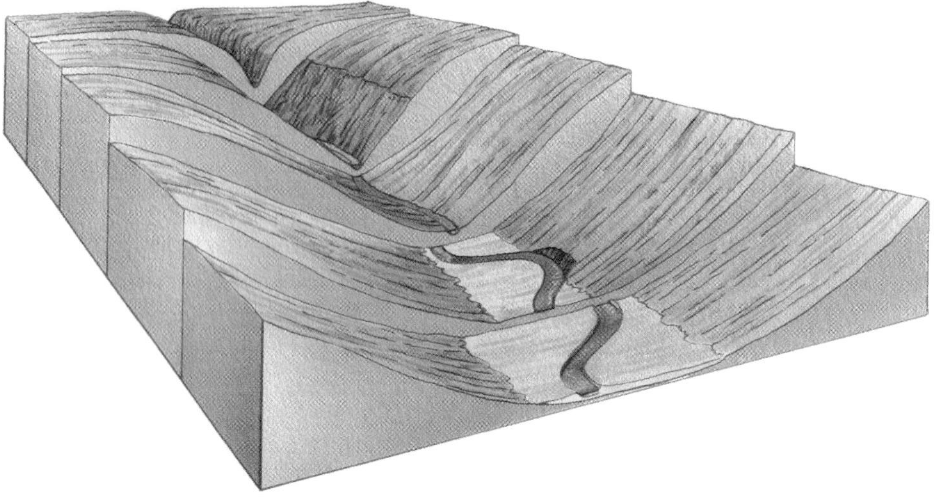

17.13 Evolution of side slopes *Following stream gradation, the valley walls become gentler in slope, and the bedrock is covered by soil and weathered rock. (After W. M. Davis. Copyright © A. N. Strahler.)*

network to the broad river channels found farther downstream. One way to study the properties of streams is to organize them by stream order, in which stream segments are ordered from those draining the smallest of slopes to those formed far downstream from the progressive joining of many upstream segments. This system is described further in *Focus on Systems 17.3 • Stream Networks as Trees.*

Great Waterfalls

As we see from Figure 17.9, the stream gradation process acts to smooth the profile of a stream by draining lakes and removing falls and rapids. Thus, large waterfalls on major rivers are comparatively rare the world over. Faulting and dislocation of large crustal blocks have caused spectacular waterfalls on several east African rivers (Victoria Falls, shown in Figure 17.16, is a prime example). As we explained in Chapter 14, this is the Rift Valley region, where block faulting has been taking place.

Another class of large waterfalls involves new river channels resulting from glacial activity in the Ice Age. Erosion and deposition by large moving ice sheets greatly disrupted drainage patterns in northern continental regions, creating lakes and causing river courses to be shifted to new locations. Niagara Falls is a good example (Figure 17.14). The outlet from Lake Erie into Lake Ontario, the Niagara River, is situated over a gently inclined layer of limestone, beneath which lies easily eroded shale. The river has gradually

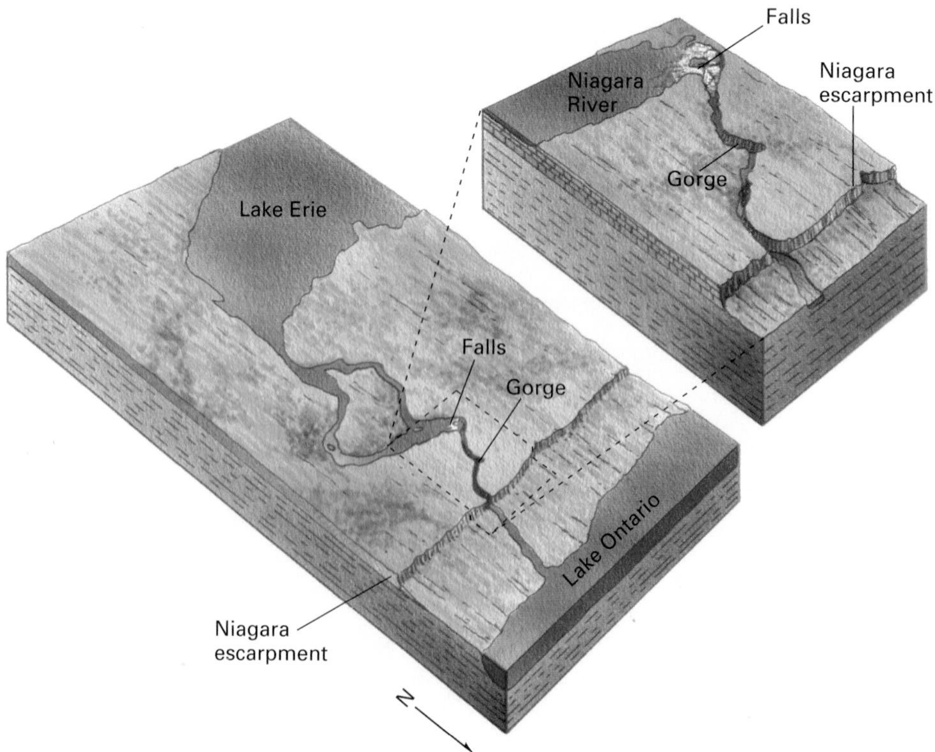

17.14 Niagara River topographic setting *A bird's-eye view of the Niagara River with its falls and gorge carved in strata of the Niagara Escarpment. View is toward the southwest from a point over Lake Ontario. (After a sketch by G. K. Gilbert. Redrawn from A. N. Strahler. Copyright © A. N. Strahler.)*

17.15 Niagara Falls *Niagara Falls is formed where the river passes over the eroded edge of a massive limestone layer.*

eroded the edge of the limestone layer, producing a steep gorge marked by Niagara Falls at its head (Figure 17.15). The height of the falls is now 52 m (171 ft), and its discharge is about 5700 m^3 per second (about 200,000 ft^3/s). The drop of Niagara Falls is utilized for the production of hydroelectric power by the Niagara Power Project. Water is withdrawn upstream from the falls and carried in tunnels to generating plants located about 6 km (4 mi) downstream from the falls.

GEODISCOVERIES

Dams and Resources

Because most large rivers of steep gradient do not have falls, dams are necessary to create the vertical drop required to spin the turbines of electric power generators. Hydroelectric power is cheap, nonpolluting, and renewable, and large

Geographers at Work | On the road to "Dam Nation"

by Molly Marie Pohl, ***San Diego State University***

For some, a high dam in a deep gorge is a natural wonder in its own right. Towering above a once-mighty river from canyon wall to canyon wall, its face bisected by a ribbon of cascading white water, the dam stands as a monument to both human engineering and control of the natural environment. Behind the dam, the azure waters of its lake stretch far upstream, flanked by hills and cliffs and lined with beaches. By generating hydroelectric power, providing irrigation and recreation, and protecting downstream communities from flooding, the dam turns the wild river into a humble human servant.

Yet today there is a movement underway to remove dams from rivers. Why? After living with dams for decades, their bad behavior has come to light. First, they stop sediment. This causes the river to erode its bed below the dam. Beaches and bars become scarce. Second, dams alter the flow characteristics of their rivers. By trapping and storing spring floodwaters for use in the dry summer and fall, the dam keeps the river from flooding in a natural way, which changes the habitat of freshwater plants and animals living in and near the river. The dam makes it much harder for fish, such as salmon, to migrate upstream to spawn. Another effect is to change the river's thermal regime, since water released downstream by a major dam is largely drawn from the cold depths of the lake.

Molly Marie Pohl researches dam construction, dam destruction, and the fate of American rivers.

Molly Marie Pohl is a "dam" geographer at San Diego State who specializes in the impacts of dams on rivers. She has studied how two dams affected the Elwha River, located on the Olympic Peninsula of Washington, using river gage records, repeat air photos, and historical documents and photographs coupled with detailed field work. But the best part is that the dams are slated for removal in the near future, and Molly will have the opportunity to see when and how the Elwha responds to its new freedom. She is also inventorying dam removals to study public attitudes about dams and river restoration. To read more about Molly and what she has learned about environmental "dam"age, visit our web site.

Focus on Systems | 17.3

Stream Networks as Trees

This is an essay about trees. Why do we discuss trees in a chapter on streams? Mathematicians give a special meaning to "tree": For them it is a branching system of lines and points, and it belongs in a branch of mathematics called *topology.* Look again at Figure 16.18, our map of a complete drainage system. It consists of two kinds of information. First is the "skeleton," a set of connected line segments that represent the stream channel system. A second set of connected lines consists of all the drainage divides—those topographic crests that direct the overland flow from precipitation into diverging slopes. When the divides are mapped, they delineate individual drainage basins. Three such basins are shown in Figure 16.18. We will need to treat this network of basins as a special and separate topological set.

Topology, though a geometrical subject, is unlike the Euclidean geometry you are probably familiar with. Euclid required that all lines in space be perfect in form—a perfect circle or ellipse, and perfectly straight lines, for examples. Topology feeds on curving and twisting lines, but it does use the Euclidean point. It also assumes that both lines and points have no breadth or thickness. (We only draw them that way.)

The ingredients of topology are shown in Figure 1. What we usually call a "point" is a *node;* a "line" is an *arc* (or *edge*). Nodes are positioned at both ends of an arc. Arcs can form a closed loop containing a *region.* An assemblage of arcs connected by nodes is referred to as a *network.* All these forms can lie in a space of any dimension, although we are concerned here only with topologies in real spaces of either two dimensions (planes) or three dimensions (volumes).

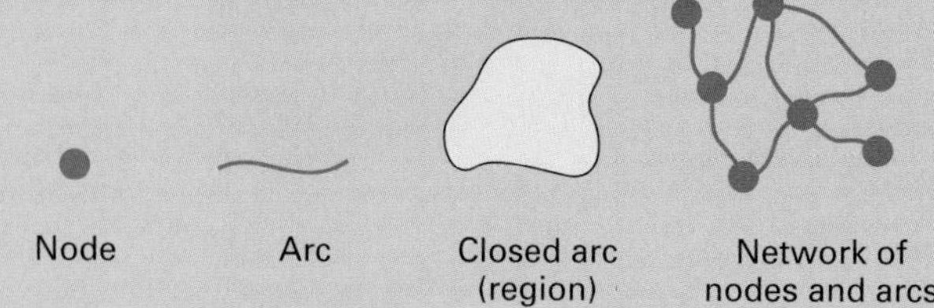

Figure 1 *Basic forms used in topology.*

Figure 2 is a formal topological diagram of a type of network called a *rooted tree.* It consists solely of nodes and arcs. Two types of nodes are recognized: *inner nodes* and *outer nodes.* Each inner node connects with three arcs, and each outer node connects only with one inner node. A third kind of node, called the *root,* serves as a starting point for the construction. This topological system serves well as a model not only of a stream system but also of several kinds of biological systems, among them the branching patterns of higher plants (trees, shrubs, veins of leaves), and even the respiratory pathways of the human lung. In these natural systems, a gaseous or fluid agent (air, water, sap) flows into or out of the system (or alternately in and out). Another example is a municipal storm sewer system, which is a tree of pipes designed to gather and dispose of rainwater.

Figure 2 *Topological diagram of a rooted tree.*

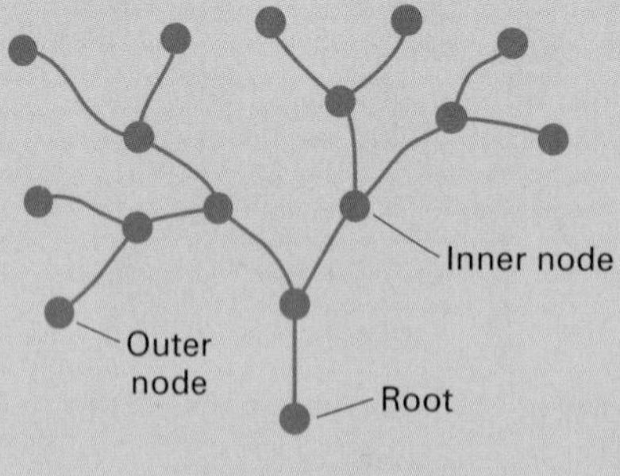

In Figure 3 we have adapted the rooted tree to a map of a natural stream system. (Note the resemblance to Figure 16.18.) Not only is the skeleton shown, but also the area of ground surface that encloses each stream arc. This surface slopes down from its perimeter—a drainage divide—to meet each stream channel and thus provides the channel with its water and sediment. In our topological analysis, it is a *region,* consisting of an area bounded by a closed loop of nodes and arcs. In this case, the loop bounding the region is composed of the arcs and nodes of the drainage divide network. All of these features—stream tree, drainage divide network, and regions that contribute flow to the channels—lie above a reference level and constitute a three-dimensional system. In the figure, they are projected upon a flat base, so our map is described as *planimetric.*

Geomorphologists make use of a convenient ordering system of the arcs and nodes within a tree. A single arc with nodes at each end is called a *stream segment.* The outermost (terminal) segments are designated in magnitude as *first-order segments* (or *first-order streams*). Where two first-order segments join, they generate a *second-order segment,* and so forth, until the single root segment is reached. That terminal node probably joins an existing stream channel of the same or higher order, but it may instead terminate at the shoreline of a lake or ocean. Our map also shows that for each stream segment there is an enclosing drainage basin of the same order. One example of a *first-order basin* is selected to show the overland trajectories (paths) of surface runoff.

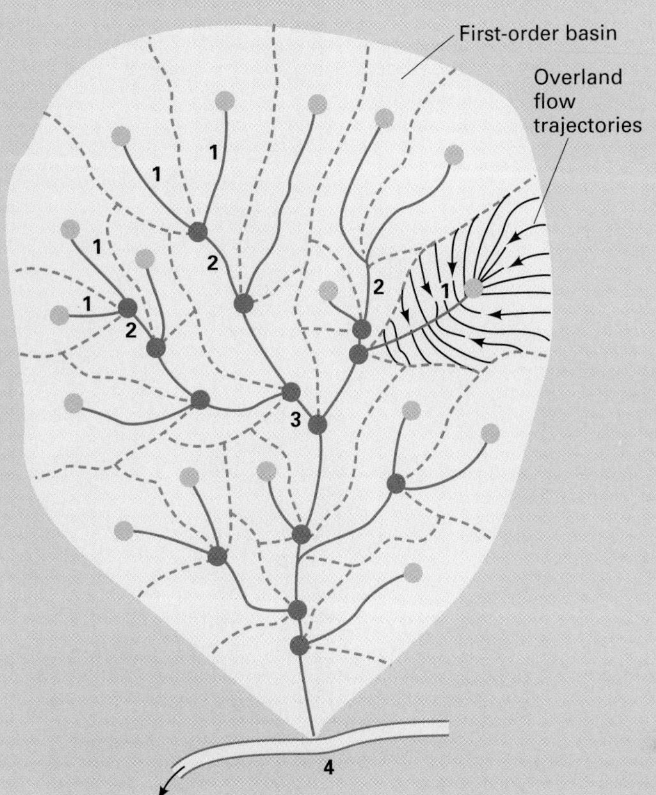

Figure 3 *Schematic map of a third-order drainage basin showing both the stream channel system and the drainage basin network. (Copyright © 1996 by A. N. Strahler.)*

We are now prepared to derive a major principle of stream system geomorphology. Consider a landscape that has reached a stable condition in which all of the upland surface has been occupied by watersheds. (See blocks *a, b,* and *c* of Figure 17.10.) Given time, the streams in the flow network will be graded, as we have described in the text. This means that each watershed is nicely adjusted in its form and slope so as to pass along toward the outlet all of the precipitation that falls on it, along with all of the sediment that is picked up from the ground surface and entrained in the channels. In this state of operation, each stream channel segment has the channel form and gradient (slope) necessary for denudation of its drainage area and for transportation to the next segment of the erosional debris that it receives.

Now, when two segments of the first order join to form one second-order segment, what adjustment must have been made? Below this junction, the discharge will (on average) have been doubled. At the next downstream junction, that of two second-order segments, discharge has again been doubled. There is a law of channel hydrology that the larger the channel cross section, the more efficient is the stream. Efficiency increases because the proportion of energy expended through friction with the bed and banks is reduced. In compensation, the stream has diminished its gradient, an adjustment that operates by a natural feedback process. This analysis helps to explain how the graded profile of a stream (illustrated in Figure 17.9) is developed and maintained.

The idea of connected networks distributed in real space is an important one in many subfields of geography. In the case of stream channel networks, the topological ordering of stream segments has provided a useful and productive way of studying and organizing information about streams and about the way that landscapes are eroded by fluvial action. If you go on to study geomorphology, the behavior and characteristics of stream networks will be a subject that you will return to in much greater detail.

17.16 Eye on the Landscape **Victoria Falls** *Located on the Zambezi River at the border of Zambabwe and Zambia in southern Africa, Victoria Falls is one of the world's scenic wonders.* **What else would the geographer see? ... Answers at the end of the chapter.**

dams provide fresh water for urban use and irrigation. However, lakes behind dams drown river valleys and can inundate the gorges, rapids, and waterfalls of major rivers. Scenic and recreational resources such as these can be lost. White-water boating and rafting, for example, are now popular sports on our wild rivers. The Grand Canyon of the Colorado River, probably more than any single product of fluvial processes, demonstrates the scenic and recreational value of a great river gorge.

Dam construction also destroys ecosystems adapted to the river environment. In addition, deposition of sediment behind the dam rapidly reduces the holding capacity of the lake and within a century or so may fill the lake basin. This reduces the ability of the lake to provide consistent supplies of water and hydroelectric power. It is small wonder, then, that new dam projects can meet with stiff opposition from concerned local citizens' groups and national environmental organizations.

Aggradation and Alluvial Terraces

A graded stream, delicately adjusted to its supply of water and rock waste from upstream sources, is highly sensitive to changes in those inputs. Changes in climate or vegetation cover bring changes in discharge and load at downstream points, and these changes in turn require channel readjustments. One kind of change is the buildup of alluvium in the valley floors.

17.17 Braided stream *The braided channel of the Chitina River, Wrangell Mountains, Alaska, shows many distinct channels separating and converging on a floodplain filled with glacial debris.*

Consider first what happens when bed load increases, exceeding the transporting capacity of the stream. Along a section of channel where the excess load is introduced, the coarse sediment accumulates on the stream bed in the form of bars of sand, gravel, and pebbles. These deposits raise the elevation of the stream bed, a process called *aggradation.* As more bed materials accumulate, the stream channel gradient is steepened and flow velocity increases. This increase enables bed materials to be dragged downstream and spread over the channel floor at more and more distant downstream sections. In this way, sediment introduced at the head of a stream will be gradually spread along the whole length of the stream.

Aggradation typically changes the channel cross section from a narrow and deep form to a wide and shallow one. Because bars are continually being formed, the flow is divided into multiple threads. These rejoin and subdivide repeatedly to give a typical *braided stream* (Figure 17.17). The coarse channel deposits spread across the former floodplain, burying fine-textured alluvium under the coarse material.

What processes cause alluvium to build up on valley floors and induce stream aggradation? One way is for alluvium to accumulate as a result of accelerated soil erosion. Other causes of aggradation are related to major changes in global climate, such as the onset of an ice age. Figure 17.17 illustrates one natural cause of aggradation—glacial activity—that has been of major importance in stream systems of North America and Eurasia during the recent Ice Age. In our photo example, a modern valley glacier has provided a large quantity of coarse rock debris at the head of the valley. Valley aggradation of a similar kind was widespread in a broad zone near the edges of the great ice sheets of the Ice Age. The accumulated alluvium filled most valleys to depths of several tens of meters. Figure 17.18*a* shows a valley filled in this manner by an aggrading stream. The case could represent any one of a large number of valleys in New England or the Middle West.

Suppose, next, that the source of bed load is cut off or greatly diminished. In the case illustrated in Figure 17.18, the ice sheets have disappeared from the distant headwater areas and, with them, the supplies of coarse rock debris. Reforestation of the landscape restores a protective cover to valley-side and hill slopes of the region, keeping coarse mineral particles from entering the stream through overland flow. Now the streams have abundant water discharges but little bed load. In other words, they are operating below their transporting capacity. The result is channel scour. The channel

17.18 Alluvial terrace formation *Alluvial terraces form when a graded stream slowly cuts away the alluvial fill in its valley. (Drawn by A. N. Strahler.)*

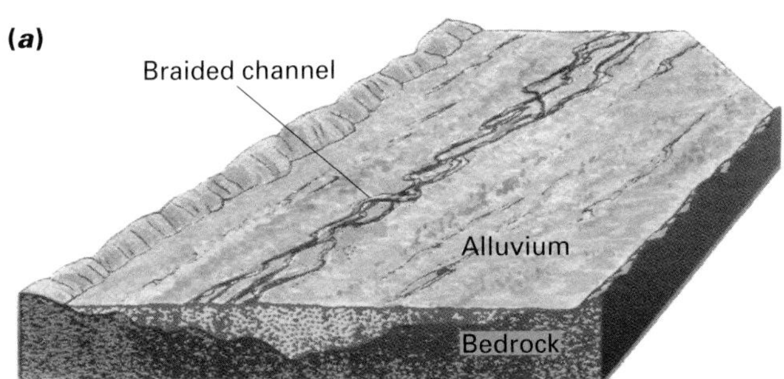

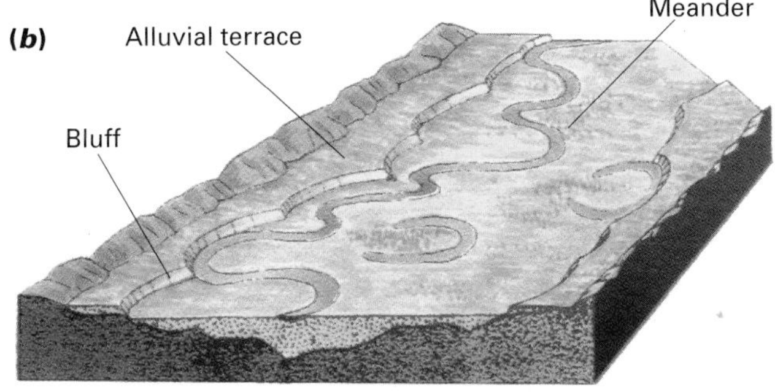

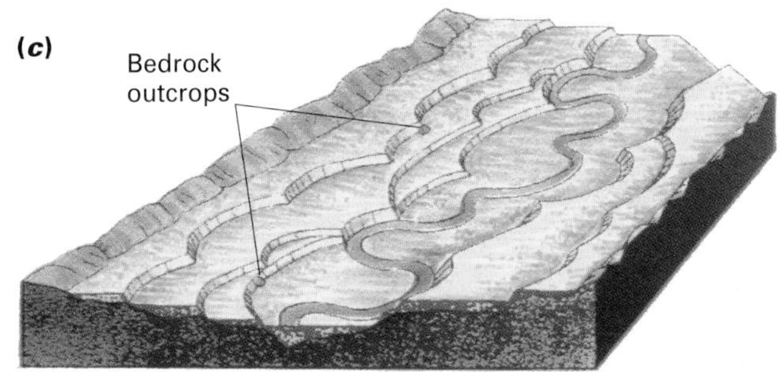

form becomes both deeper and narrower, and it also begins to develop meanders.

Gradually, the stream profile level is lowered, in a process called *degradation.* Because the stream is very close to being in the graded condition at all times, its dominant activity is lateral (sidewise) cutting by growth of meander bends, as shown in Figure 17.18*b.* By this process, the valley alluvium is gradually excavated and carried downstream, but not all of it can be removed because the channel encounters hard bedrock masses in many places. These obstructions prevent the cutting away of more alluvium. Consequently, as shown in *(c),* steplike alluvial surfaces remain on both sides of the valley. The treads of these steps are called *alluvial terraces.*

Alluvial terraces have always attracted human settlement because of the advantages over both the valley-bottom floodplain—which is subject to annual flooding—and the hill slopes beyond—which may be too steep and rocky to cultivate. Besides, terraces are easily tilled and make prime agricultural land (Figure 17.19). Towns are also easily laid out on the flat ground of a terrace. Roads and railroads can be constructed along the terrace surfaces with little difficulty.

Alluvial Rivers and Their Floodplains

We turn now to the graded river with its floodplain. As time passes, the floodplain is widened, so that broad areas of floodplain lie on both sides of the river channel. Civil engineers have given the name **alluvial river** to a large river of very low channel gradient. It flows on a thick floodplain accumulation of alluvium constructed by the river itself in earlier stages of its activity. Characteristically, an alluvial river experiences overbank floods each year or two. These floods occur during the season of large water surplus over the watershed. Overbank flooding of an alluvial river normally inundates part or all of a floodplain that is bounded on either side by rising steep slopes, called *bluffs.*

Typical landforms of an alluvial river and its floodplain are illustrated in Figure 17.20. Dominating the floodplain is the meandering river channel itself and abandoned stretches of former channels. Meanders develop narrow necks, which

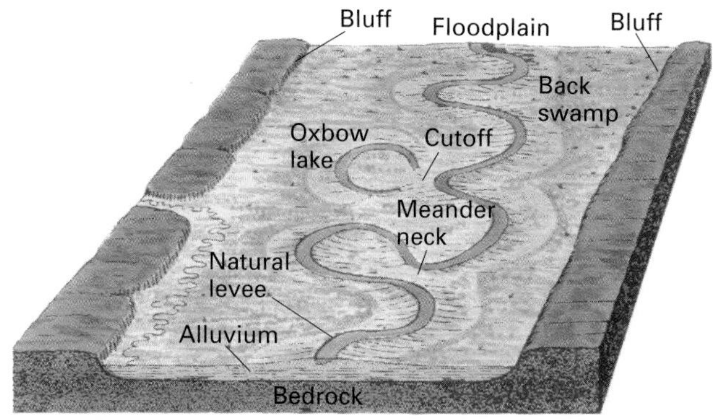

17.20 Floodplain landforms of an alluvial river *As meanders wander downriver, they create a variety of landforms, including ox-bow lakes, cutoffs, and natural levees. (Drawn by A. N. Strahler.)*

17.19 Alluvial terraces *These terraces line the Rakaia River gorge on the South Island of New Zealand. The flat terrace surface in the foreground is used as pasture for sheep. Two higher terrace levels can be seen at the left.*

17.21 Ox-bow lakes *This aerial photo of the Mississippi River floodplain shows two ox-bow lakes—former river bends that were cut off as the river shifted its course. At the bottom is a thin former meander channel still receiving river water. If flood deposits seal it off from the river, it will become an ox-bow lake as well. Vegetation appears bright red in this color-infrared photo.*

are cut through, thus shortening the river course and leaving a meander loop abandoned. This event is called a *cutoff.* It is quickly followed by deposition of silt and sand across the ends of the abandoned channel, producing an *ox-bow lake.* The ox-bow lake is gradually filled in with fine sediment brought in during high floods and with organic matter produced by aquatic plants. Eventually, the ox-bows are converted into swamps, but their identity is retained indefinitely (Figure 17.21).

During periods of overbank flooding, when the entire floodplain is inundated, water spreads from the main channel over adjacent floodplain deposits (Figure 17.22). As the current slackens, sand and silt are deposited in a zone adjacent to the channel. The result is an accumulation of higher land on either side of the channel known as a **natural levee.** Because deposition is heavier closest to the channel and decreases away from the channel, the levee surface slopes away from the channel (Figure 17.20). Between the levees and the bluffs is lower ground, called the *backswamp.* In Figure 17.22, the small settlements are located on the higher ground of the levee, next to the river, while the agricultural fields occupy the backswamp area.

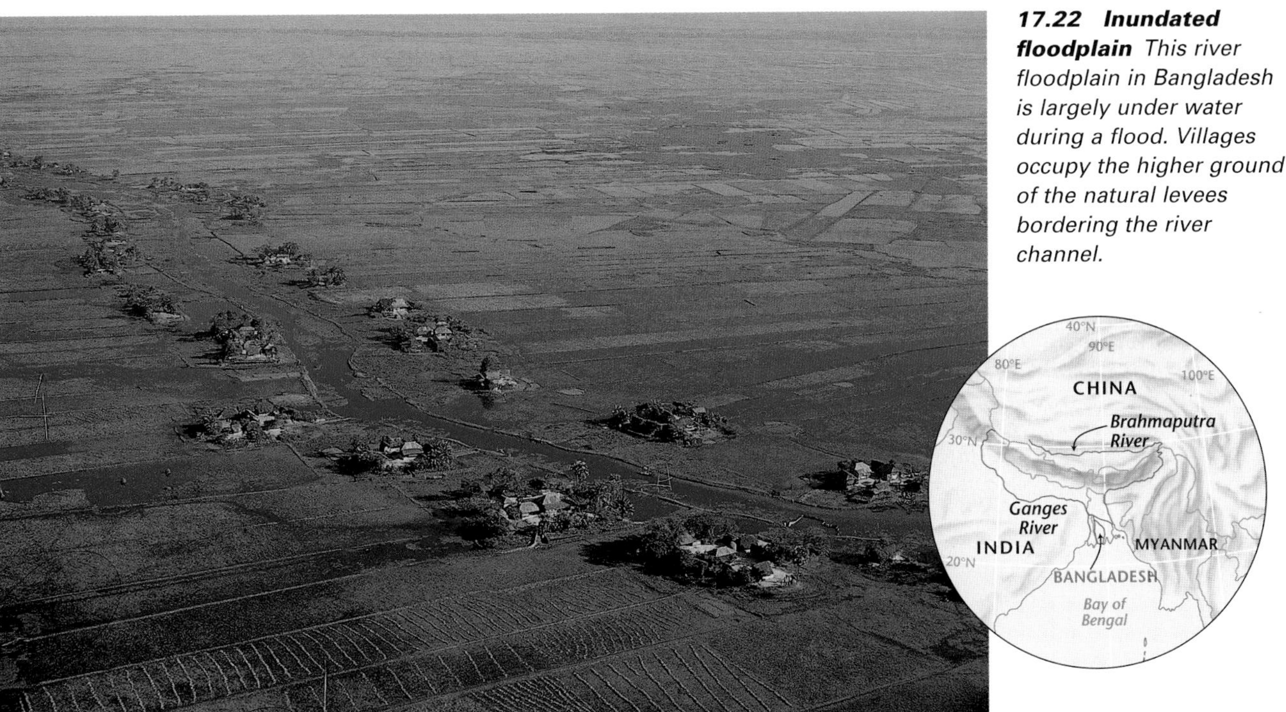

17.22 Inundated floodplain *This river floodplain in Bangladesh is largely under water during a flood. Villages occupy the higher ground of the natural levees bordering the river channel.*

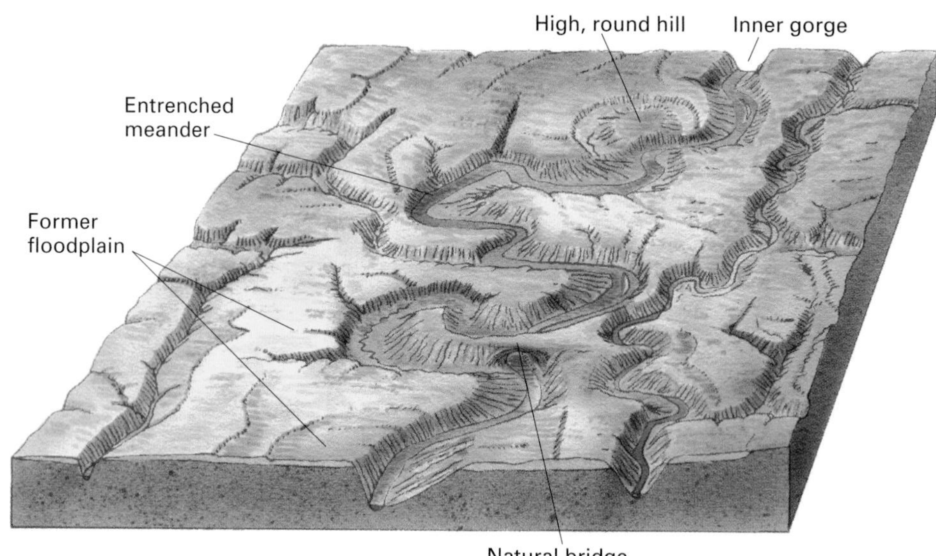

17.23 Block diagram of entrenched meanders *Uplift of a meandering stream has produced entrenched meanders. One meander neck has been cut through, forming a natural bridge. (Drawn by E. Raisz. Copyright © A. N. Strahler.)*

Overbank flooding not only results in the deposition of a thin layer of silt on the floodplain, but also brings an infusion of dissolved mineral substances that enter the soil. As a result of the resupply of nutrients, floodplain soils retain their remarkable fertility, even though they are located in regions of rainfall surplus from which these nutrients are normally leached away.

Entrenched Meanders

What happens when a broadly meandering river is uplifted by rapid tectonic activity? The uplift increases the river's gradient, so that it cuts downward into the bedrock below. This forms a steep-walled inner gorge. On either side lies the former floodplain, now a flat terrace high above river level. Any river deposits left on the terrace are rapidly stripped off by runoff because floods no longer reach the terraces to restore eroded sediment.

Uplift may cause the meanders to become impressed into the bedrock and give the inner gorge a meandering pattern (Figure 17.23). These sinuous bends are termed *entrenched meanders* to distinguish them from the floodplain meanders of an alluvial river (Figure 17.24). Although entrenched meanders are not free to shift about as floodplain meanders do, they can enlarge slowly so as to produce cutoffs. Cutoff of an entrenched meander leaves a high, round hill separated from the valley wall by the deep abandoned river channel and the shortened river course (Figure 17.23). As you might guess, such hills formed ideal natural fortifications. Many European fortresses of the Middle Ages were built on such cutoff meander spurs. Under unusual circumstances, where the bedrock includes a strong, massive sandstone formation, meander cutoff leaves a *natural bridge* formed by the narrow meander neck.

The Geographic Cycle

The Earth's fluvial landscapes are quite diverse. They range from mountain regions of steep slopes and rugged peaks to regions of gentle hills and valleys to nearly flat plains that stretch from horizon to horizon. One way to view these landscapes is to consider them as stages of evolution in a cycle that begins with rapid uplift and follows with long erosion by streams in a graded condition. This cycle, called the *geographic cycle,* was first described by William Morris Davis, a prominent geographer and geomorphologist of the late

17.24 Entrenched meanders *The Goosenecks of the San Juan River in Utah are deeply entrenched river meanders in horizontal sedimentary rock layers. The canyon, carved from sandstones and limestones, is about 370 m (1399 ft) deep.*

nineteenth and early twentieth centuries. Let's look at this idea in more detail.

Consider a landscape made up of many drainage basins and their branching stream networks that has been rapidly uplifted by tectonic forces. The region is rugged, with steep mountainsides and high, narrow crests (Figure 17.25*a*). We refer to this landscape as being in a *youthful stage.*

After initial uplift, the main streams draining the region establish a graded condition. Rock debris is transported out of each drainage basin at the same average rate as the debris is being contributed from the land surfaces within the basin. Eventually, the export of debris lowers the land surface generally, and the average altitude of the land surface steadily declines. This decline must be accompanied by a reduction in the average gradients of all streams *(b)*. As the sharp mountain peaks and gorges of the youthful stage give way to rounded hills and broad valleys, we reach the *mature stage* of the geographic cycle.

As time passes, the streams and valley-side slopes of the drainage basins undergo gradual change to lower gradients. In theory, the ultimate goal of the denudation process is to reduce the land mass to a featureless plain at sea level. In this process, a sea-level surface imagined to lie beneath the entire land mass represents the lower limiting level, or *base level,* of the fluvial denudation (labeled in Figures 17.9 and 17.25). But because the rate of denudation becomes progressively slower, the land surface approaches the base-level surface of zero elevation at a slower and slower pace. Under this scenario, the ultimate goal can never be reached. Instead, after the passage of some millions of years, the land surface is reduced to a gently rolling surface of low elevation, called a *peneplain (c)*. You can think of this strange term as meaning an "almost-plain." With the evolution of the peneplain, the landscape has reached *old age.*

Production of a peneplain requires a high degree of crustal and sea-level stability for a period of many millions of years. One region that has been cited as a possible example of a contemporary peneplain is the Amazon-Orinoco Basin of South America. This vast region is a stable continental shield of ancient rock with very low relief.

What happens if a peneplain is uplifted? Figure 17.25*d* shows the peneplain of *(c)* uplifted to an elevation of several hundred meters. The base level is now far below the land surface. Soon streams begin to trench the land mass and to carve deep, steep-walled valleys, shown in *(e)*. This process is called *rejuvenation.* With the passage of many millions of years, the landscape will be carved into the rugged stage shown in *(a)*, and the later stages of *(b)* and *(c)* will follow.

Equilibrium Approach to Landforms

While Davis's idealized geographic cycle is useful for understanding landscape evolution over very long periods of time, it does little to explain the diversity of the features observed in real landscapes. Most geomorphologists now approach landforms and landscapes from the viewpoint of *equilibrium.* This approach explains a fluvial landform as

(a)

(b)

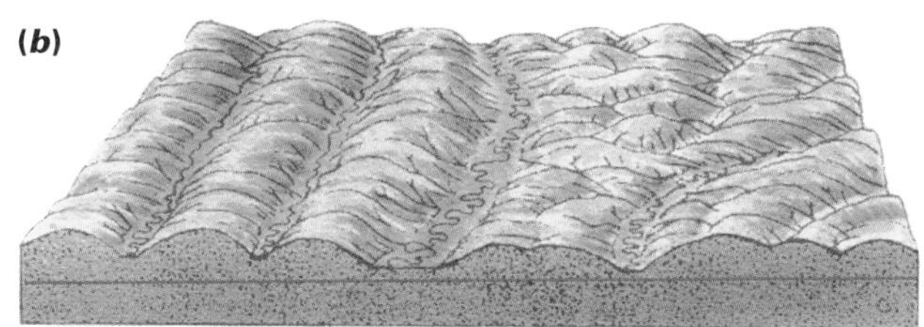

(c)

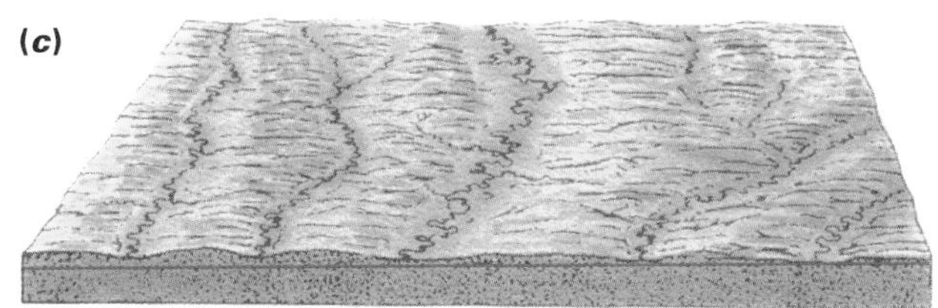

(d)

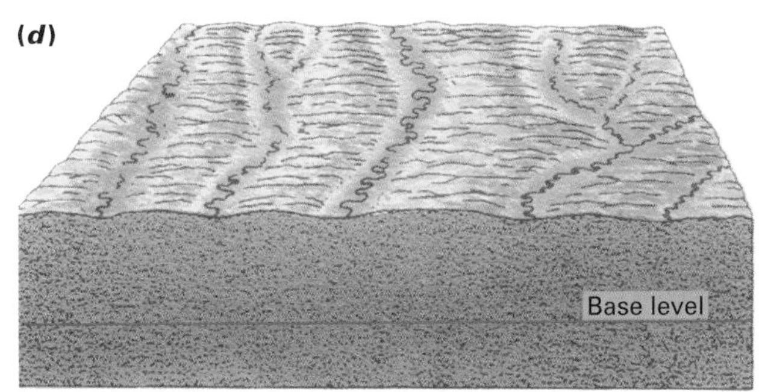

(e)

17.25 The geographic cycle *(a) In the youthful stage, relief is great, slopes are steep, and the rate of erosion is rapid. (b) In the mature stage, relief is greatly reduced, slopes are gentle, and rate of erosion is slow. Soils are thick over the broadly rounded hill summits. (c) In old age, after many millions of years of fluvial denudation, a peneplain is formed. Slopes are very gentle, and the landscape is an undulating plain. Floodplains are broad, and the stream gradients are extremely low. All of the land surface lies close to base level. (d) The peneplain is uplifted. (e) Streams trench a new system of deep valleys in the phase of land-mass rejuvenation. (Drawn by A. N. Strahler.)*

the product of forces acting upon it, including both forces of uplift and denudation, with the characteristics of the rock material playing an important role. Thus, we find steep slopes and high relief where the underlying rock is strong and highly resistant to erosion. Even a "youthful" landscape may be in a long-lived equilibrium state in which hill slopes and stream gradients remain steep in order to maintain a graded condition while eroding a strong rock like massive granite. In the next chapter, we will provide many examples of landforms that result from erosive processes acting on both strong and weak rock within the same region.

Another problem with Davis's geographic cycle is that it only applies where the land surface is stable over long periods of time. As we know from our study of plate tectonics in Chapter 13, crustal movements are frequent on the geologic time scale, and few regions of the land surface remain untouched by tectonic forces in the long run. Recall also that continental lithosphere floats on a soft asthenosphere. As layer upon layer of rock is stripped from a land mass by erosion, the land mass becomes lighter and is buoyed upward. The process of crustal rise in response to unloading is known as *isostatic compensation.* (*Focus on Systems 18.3 • A Model Denudation System* provides a more detailed look at this process.) The proper model, then, is one of uplift as an ongoing process to which erosional processes are constantly adjusting rather than as a sudden event followed by denudation.

FLUVIAL PROCESSES IN AN ARID CLIMATE

Desert regions look strikingly different from humid regions in both vegetation and landforms. Obviously, the lower precipitation makes the difference. Vegetation is sparse or absent, and land surfaces are mantled with mineral material—sand, gravel, rock fragments, or bedrock itself.

Although deserts have low precipitation, rain falls in dry climates as well as in moist, and most landforms of desert regions are formed by running water. A particular locality in a dry desert may experience heavy rain only once in several years. But when rain does fall, stream channels carry water and perform important work as agents of erosion, transportation, and deposition. Fluvial processes are especially effective in shaping desert landforms because of the sparse vegetation cover. The few small plants that survive offer little or no protection to soil or bedrock. Without a thick vegetative cover to protect the ground and hold back the swift downslope flow of water, large quantities of coarse rock debris are swept into the streams. A dry channel is transformed in a few minutes

17.26 Eye on the Landscape ***Flash flood*** *A flash flood has filled this desert channel in the Tucson Mountains of Arizona with raging, turbid waters. A distant thunderstorm produced the runoff.* **What else would the geographer see?...Answers at the end of the chapter.**

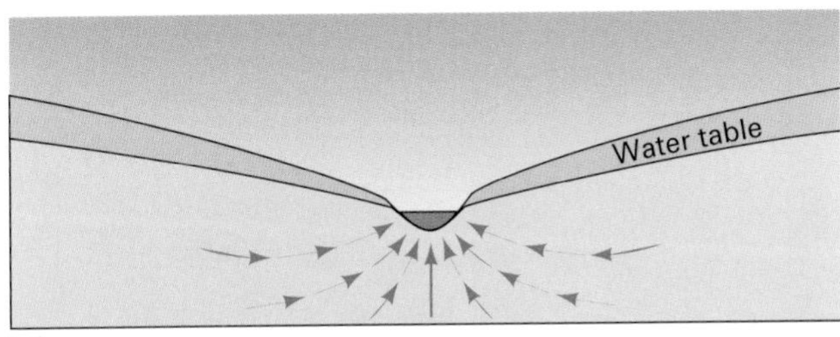

***17.27 Stream water flow to ground water** In humid regions (a), a stream channel receives ground water through seepage. In arid regions (b), stream water seeps out of the channel and into the water table below. (Copyright © A. N. Strahler.)*

into a raging flood of muddy water heavily charged with rock fragments (Figure 17.26).

An important contrast between regions of arid and humid climates lies in the way in which the water enters and leaves a stream channel (Figure 17.27). In a humid region *(a)* with a high water table sloping toward a stream channel, ground water moves steadily toward the channel and seeps into the stream bed, producing permanent (perennial) streams. In arid regions *(b)*, the water table normally lies far below the channel floor. Where a stream flows across a plain of gravel and sand, water is lost from the channel by seepage. Loss of discharge by seepage and evaporation strongly depletes the flow of streams in alluvium-filled valleys of arid regions. As a result, aggradation occurs and braided channels are common. Streams of desert regions are often short and end in alluvial deposits or on the floors of shallow, dry lakes.

Alluvial Fans

One very common landform built by braided, aggrading streams is the **alluvial fan,** a low cone of alluvial sands and gravels resembling in outline an open Japanese fan (Figure 17.28). The apex, or central point of the fan, lies at the mouth of a canyon or ravine. The fan is built out on an adjacent plain. Alluvial fans are of many sizes. In fact, some desert fans are many kilometers across (Figure 17.29*a*).

Fans are built by streams carrying heavy loads of coarse rock waste from a mountain or an upland region. The braided channel shifts constantly, but its position is firmly fixed at the canyon mouth. The lower part of the channel, below the apex, sweeps back and forth. This activity accounts for the semicircular fan form and the downward slope in all radial directions away from the apex.

Large, complex alluvial fans also include mudflows (Figure 17.28). Mud layers are interbedded with sand and gravel layers. Water infiltrates the fan at its head, making its way to lower levels along sand layers. The mudflow layers serve as barriers to ground water movement. The trapped ground water is under pressure from water higher in the fan apex. When a well is drilled into the lower slopes of the fan, water rises spontaneously as artesian flow. (See Chapter 16 and Figure 16.6.)

Alluvial fans are the primary sites of ground water reservoirs in the southwestern United States. In many fan areas, sustained heavy pumping of these reserves for irrigation has lowered the water table severely. The rate of recharge is extremely slow in comparison. At some locations, recharge is increased by building water-spreading structures and infiltrating basins on the fan surfaces. A serious side effect of excessive ground water withdrawal is subsidence (sinking) of the land surface.

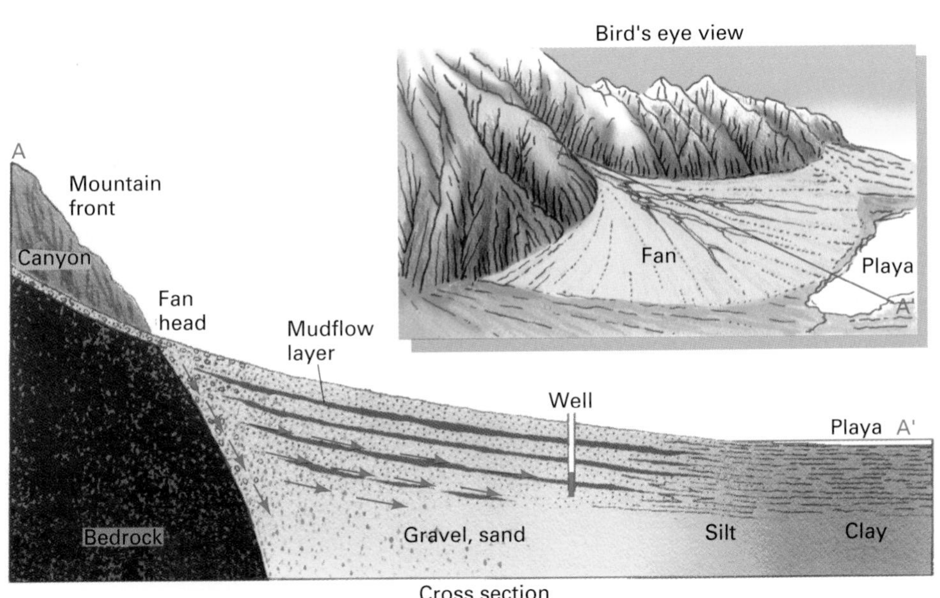

***17.28 Features of an alluvial fan** A cross section shows mudflow layers interbedded with sand layers, providing water (arrows) for a well in the fan. (Copyright © A. N. Strahler.)*

17.29 Landforms of Mountainous Deserts

a...***Alluvial fans*** *Great alluvial fans extend out upon the floor of Death Valley. The canyons from which they originate have carved deeply into a great uplifted fault block.*

c...***Pediment*** *A pediment formerly carved in granite at the foot of a desert mountain slope is now seen in profile as a straight, sloping line. Later erosion has exposed the bedrock, and weathering has formed the joint blocks into boulders (intermediate distance). Little Dragoon Mountains, near Adams Peak, about 20 km (about 12 mi) northeast of Benson, Arizona.*

b...***Desert playa*** *Racetrack Playa, a flat, white plain, is surrounded by alluvial fans and rugged mountains. This desert valley lies in the northern part of the Panamint Range, not far west of Death Valley, California. In the distance rises the steep eastern face of the Inyo Mountains, a great fault block.*

The Landscape of Mountainous Deserts

Where tectonic activity has recently produced block faulting in an area of continental desert, the assemblage of fluvial landforms is particularly diverse. The basin-and-range region of the western United States is such an area. It includes large parts of Nevada and Utah, southeastern California, southern Arizona and New Mexico, and adjacent parts of Mexico. The uplifted and tilted blocks, separated by downdropped tectonic basins, provide an environment for the development of spectacular erosional and depositional fluvial landforms.

Figure 17.30 demonstrates some landscape features of mountainous deserts. Figure 17.30*a* shows two uplifted fault blocks with a downdropped block between them. Although denudation acts on the uplifted blocks as they are being raised, we have shown them as very little modified at the time tectonic activity has ceased. At first, the faces of the fault block are extremely steep. They are scored with deep ravines, and talus blocks form cones at the bases of the blocks.

Figure 17.30*b* shows a later stage of denudation in which streams have carved up the mountain blocks into a rugged landscape of deep canyons and high divides. Rock waste furnished by these steep mountain slopes is carried from the mouths of canyons to form numerous large alluvial fans. The fan deposits form a continuous apron extending far out into the basins.

In the centers of desert basins lie the saline lakes and dry lake basins mentioned in Chapter 16. Accumulation of fine sediment and precipitated salts produces an extremely flat basin floor, called a **playa** in the southwestern United States and in Mexico. Salt flats are found where an evaporite layer forms the surface. In some playas, shallow water forms a salt lake.

Figure 17.29*b* is an air photograph of a mountainous desert landscape in the Death Valley region of eastern California. Three environmental zones can be seen in the photo. First is a region of rugged mountain masses dissected into canyons with steep rocky walls. At lower elevations, a zone of coalescing alluvial fans lines the mountain front. Last is the white playa occupying the central part of the basin.

In this type of mountainous desert, fluvial processes are limited to local transport of rock particles from a mountain range to the nearest adjacent basin, which receives all the sediment. The basin gradually fills as the mountains diminish in elevation. Because there is no outflow to the sea, the concept of a base level of denudation has no meaning. Each arid basin becomes a closed system as far as transport of sediment is involved. Only the hydrologic system is open, with water entering as precipitation and leaving as evaporation.

As the desert mountain masses are lowered in height and reduced in extent, a gently sloping rock floor, called a *pediment,* develops close to the receding mountain front (Figure 17.29*c*). As the remaining mountains shrink further in size, the pediment surfaces expand to form wide rock platforms thinly veneered with alluvium. This advanced stage is shown in Figure 17.30. The desert land surface produced in an advanced stage of fluvial activity is an undulating plain consisting largely of areas of pediment surrounded by areas of alluvial fan and playa surfaces.

17.30 Mountainous desert landforms *Idealized diagrams of landforms of the mountainous deserts of the southwestern United States. (a) Initial uplift of two blocks, with a downdropped valley between them. (b) Stage of rapid filling of tectonic basins with debris from high, rugged mountain blocks. (c) Advanced stage with small mountain remnants and broad playa and fan slopes. (Drawn by A. N. Strahler.)*

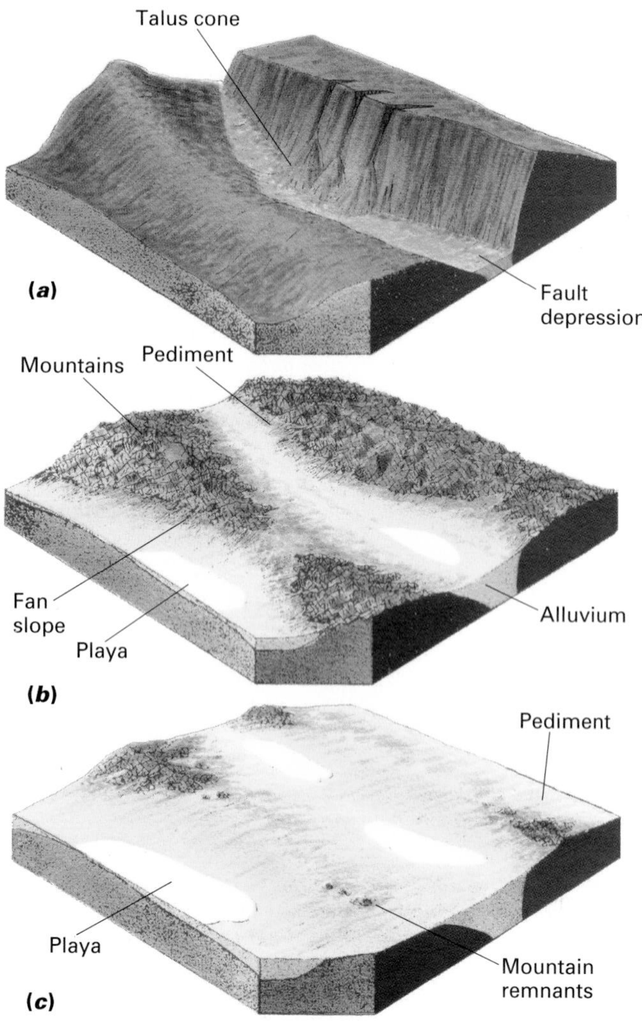

A LOOK AHEAD Running water, the primary agent of denudation, does not act equally on all types of rocks. Some rocks are more resistant to erosion, while others are less so. As a result, fluvial erosion can create unique and interesting landscapes that reveal rock structures of various kinds. This will be the subject of our next chapter.

Chapter Summary

- This chapter has covered the landforms and land-forming processes of running water, one of the four active agents of *denudation*. Like the other agents, running water erodes, transports, and deposits rock material, forming both *erosional* and *depositional* landforms.
- The work of running water begins on slopes, producing *colluvium* where overland flow moves soil particles downslope, and producing **alluvium** when the particles enter stream channels and are later deposited. In most natural landscapes, *soil erosion* and soil formation rates are more or less equal, a condition known as the *geologic norm*. *Badlands* are an exception in which natural erosion rates are very high.
- The work of streams includes **stream erosion** and **stream transportation**. Where stream channels are carved into soft materials, large amounts of sediment can be obtained by *hydraulic action*. Where stream channels flow on bedrock, channels are deepened only by the abrasion of bed and banks by mineral particles, large and small. Both the *suspended load* and *bed load* of rivers increase greatly as velocity increases. Velocity, in turn, depends on gradient.
- Over time, streams tend to a **graded** condition, in which their gradients are adjusted to move the average amount of water and sediment supplied to them by slopes. Lakes and *waterfalls*, created by tectonic, volcanic, or glacial activity, are short-lived events, geologically speaking, that give way to a smooth, *graded* stream *profile*. Grade is maintained as landscapes are eroded toward base level.
- When provided with a sudden inflow of rock material, as, for example, by glacial action, streams build up their beds by *aggradation*. When that inflow ceases, streams resume downcutting, leaving behind *alluvial terraces*.
- Large rivers with low gradients that move large quantities of sediment are termed **alluvial rivers**. The meandering of these rivers forms *cutoff* meanders, *ox-bow lakes*, and other typical landforms. Alluvial rivers are sites of intense human activity. Their fertile floodplains yield agricultural crops and provide easy transportation paths. When a region containing a meandering alluvial river is uplifted, *entrenched meanders* can result.
- The *geographic cycle* organizes fluvial landscapes according to their age in a cycle of uplift that forms mountains and subsequent erosion to nearly flat surfaces called peneplains. In the *equilibrium* approach, landforms are viewed as products of uplift and erosion as continuous processes acting on rocks of varying resistance to erosion.
- Although rainfall is scarce in deserts, running water is very effective there in producing **fluvial landforms**. Desert streams, subject to flash flooding, build **alluvial** fans at the mouths of canyons. Water sinks into the fan deposits, creating local ground water reservoirs. Eventually, desert mountains are worn down into gently sloping *pediments*. Fine sediments and salts, carried by streams, accumulate in **playas**, from which water evaporates, leaving sediment and salt behind.

Key Terms

fluvial landforms
fluvial processes
alluvium
stream erosion
stream transportation
stream deposition
stream load
graded stream
alluvial meander
alluvial river
natural levee
alluvial fan
playa

Review Questions

1. List and briefly identify the four flowing substances that serve as agents of denudation.
2. Describe the process of slope erosion. What is meant by the geologic norm?
3. Contrast the two terms *colluvium* and *alluvium*. Where on a landscape would you look to find each one?
4. What special conditions are required for badlands to form?
5. When and how does sheet erosion occur? How does it lead to rill erosion and gullying?
6. In what ways do streams erode their bed and banks?
7. What is stream load? Identify its three components. In what form do large rivers carry most of their load?
8. How is velocity related to the ability of a stream to move sediment downstream?
9. How does stream degradation produce alluvial terraces?
10. Define the term *alluvial river*. Identify some characteristic landforms of alluvial rivers. Why are alluvial rivers important to human civilization?
11. Describe the evolution of a fluvial landscape according to the geographic cycle. What is meant by rejuvenation?
12. What is the equilibrium approach to landforms? How does it differ from interpretation using the geographic cycle?

13. Why is fluvial action so effective in arid climates, considering that rainfall is scarce? How do streams in arid climates differ from streams of moist climates?
14. Describe the evolution of the landscape in a mountainous desert. Use the terms *alluvial fan, playa,* and *pediment* in your answer.

Focus on Systems 17.3 • Stream Networks as Trees

1. Identify the components of a rooted tree and use them to sketch an example.
2. How do geomorphologists order the arcs and nodes of a stream as a rooted tree?
3. How does the efficiency of a stream change with stream order? How does this explain the gradual reduction in slope of a graded stream in a downstream direction?

Visualizing Exercises

1. Compare erosional and depositional landforms. Sketch an example of each type.
2. What is a graded stream? Sketch the profile of a graded stream and compare it with the profile of a stream draining a recently uplifted set of landforms.
3. Sketch the floodplain of a graded, meandering river. Identify key landforms on the sketch. How do they form?

Essay Questions

1. A river originates high in the Rocky Mountains, crosses the high plains, flows through the agricultural regions of the Midwest, and finally reaches the sea. Describe the fluvial processes and landforms you might expect to find on a journey along the river from its headwaters to the ocean.
2. What would be the effects of climate change on a fluvial system? Choose either the effects of cooler temperatures and higher precipitation in a mountainous desert, or warmer temperatures and lower precipitation in a humid agricultural region.

Problems

Working It Out 17.1 • River Discharge and Suspended Sediment

1. What suspended load would you project for the Powder River at Arvada when the discharge is 50 m^3/s? 5 m^3/s? (*Hint:* Confirm your answers by examining the graph.)
2. If the discharge of the Powder River at Arvada doubles, what is the effect on its suspended sediment load?
3. A hydrologist measures discharge and sediment load for a river and plots the points as logarithms on ordinary graph paper. She fits a line to the points with the following equation:

$$\log S = 2.50 + 2.12 \log Q$$

Rewrite this equation as a power function of the form $S = aQ^b$.

Eye on the Landscape

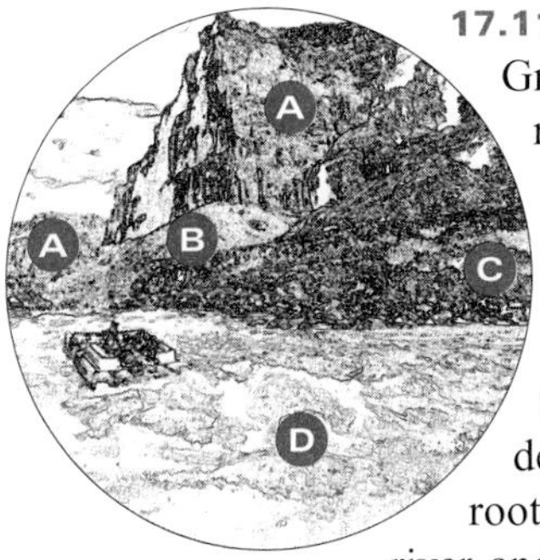

17.11 Grand Canyon, Arizona Grand Canyon is an entrenched river gorge cutting through layers of flat-lying sandstone and shale **(A)**. Note the large accumulations of talus at the base of the cliffs **(B)**. Most of the canyon shows only sparse desert vegetation, but deep-rooted plants appear near the river and on sedimentary deposits of side streams **(C)**. Surprisingly, the river water in the canyon **(D)** is quite cold. Nearly all of it is released from the cold, deep waters of Lake Powell behind the Glen Canyon Dam, which is upstream. We will see more of landforms of the Grand Canyon in Chapter 18. Plants and water are covered in Chapter 23, and dams were discussed in the preceding chapter (16).

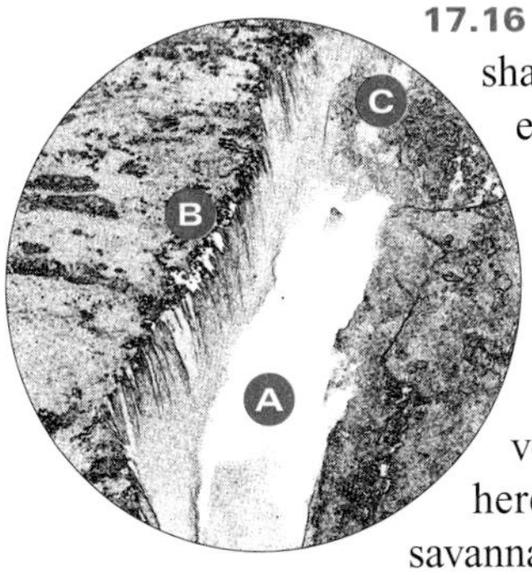

17.16 Victoria Falls A fault has shattered and broken rock layers, creating a zone of weakness that has been eroded by the Zambezi River to form the gorge of Victoria Falls **(A)**. Note the resistant rock layer that keeps the waterfall vertical **(B)**. Native vegetation here is thorntree-tall grass savanna, which is visible in the near distance **(C)**. Faults were covered in Chapter 14, while the vegetation of Africa is mapped and described in Chapter 24.

17.26 Flash flood One of the world's more vegetated deserts, the Sonoran Desert is known for its unique plants. At **(A)**, the distinctive spikes of the saguaro cactus are visible, projecting above the cover of streamside vegetation (probably ocotillo). Deserts were covered in Chapters 10 and 11; vegetation of the Sonoran Desert is mentioned in Chapter 24. Note the sediment at **(B)**. It is probably not deposited by the stream, because it is unsorted and the rock fragments are quite angular. It appears to be desert pavement (Chapter 19), flooded by runoff from the left that is flowing toward the channel.

18 LANDFORMS AND ROCK STRUCTURE

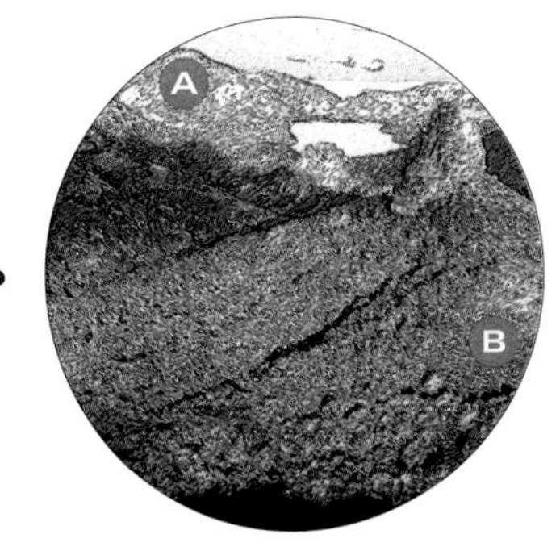

EYE ON THE LANDSCAPE
Mountains of Villa Nequén Province, Argentina. Autumn colors of the southern beech cloak an eroded volcanic landscape at the eastern edge of the Andes. The pinnacle in the foreground is the stock of an ancient volcano; the circular lake beyond may be a small caldera. **What else would the geographer see?...Answers at the end of the chapter.**

Denudation acts to wear down all rock masses exposed at the land surface. However, different types of rocks are worn down at different rates. Some are easily eroded, while others are highly resistant to erosion. Generally, weak rocks will underlie valleys, and strong rocks will underlie hills, ridges, and uplands. This means that the pattern of landforms on a landscape can reveal the sequence of rock layers underneath it. The sequence, in turn, is determined by the type of rock structure present—for example, a series of folds, or perhaps a dome, in a set of rock layers. Thus, there is often a direct relationship between landforms and rock structure.

Over the world's vast land area, many types of rock and rock structures appear. Repeated episodes of uplift, followed by periods of denudation, can bring to the surface rocks and rock structures formed deep within the crust. In this way, even ancient mountain roots, formed during continental collisions, may eventually appear at the surface and be subjected to denudation. Batholiths and other plutons—igneous rock structures produced by magmas that cool at depth—can similarly be exposed.

As we will see in this chapter, rock structure controls not only the locations of uplands and lowlands, but also the placement of streams and the shapes and heights of the intervening divides. A distinctive assemblage of landforms and stream patterns develops on each of the major types of crustal structures. Recall that in Chapter 14 we classified all landforms as either initial or sequential in origin. Initial landforms are produced directly by volcanic activity, folding, and faulting. Denudation soon converts these initial landforms into sequential landforms, which are controlled in shape, size, and arrangement by the underlying rocks and their structure. The subject of this chapter is therefore the sequential landforms that arise through erosion of rock structures.

Rock Structure as a Landform Control

As denudation takes place, landscape features develop according to patterns of bedrock composition and structure. Figure 18.1 shows an arrangement of five layers of sedimentary rock, together with a mass of much older igneous rock. The sediments were originally deposited in horizontal layers atop the igneous rock, but the block of igneous and sedimentary rock has been bent and tilted in an ancient orogeny. The diagram shows the usual landform shape of each rock type and whether it forms valleys or mountains. The cross section on the front of the block diagram shows conventional symbols used by geologists for each rock type—for example, fine dots for sandstone or "bricks" for limestone.

As indicated by the diagram, shale is a weak rock that is easily eroded and forms the low valley floors of the region. Limestone, subjected to solution by carbonic acid in rainwater and surface water, also forms valleys in humid climates. In arid climates, on the other hand, limestone is a resistant rock and usually stands high to form ridges and cliffs. Sandstone and conglomerate are typically resistant to denudation, and they form ridges or uplands. As a group, the igneous rocks are also resistant—they typically form uplands or mountains rising above adjacent areas of shale and limestone. Although metamorphic rocks in general are more resistant to denudation, individual metamorphic rock types—marble or schist, for example—vary in their resistance.

Figure 18.2 provides a preview of the important rock structures and their accompanying landforms that we will discuss in this chapter. In these sketches, resistant beds form ridges or belts of hills, while weaker rocks form lowlands or valleys. The upper four sketches show landforms and rock structures of sedimentary rock layers. The structures include horizontal layers, gently sloping layers, and layers that are folded or warped into domes. Other structures shown are fault blocks, plutons, metamorphic belts, and eroded volcanoes.

18.1 Landforms and rock types *Landforms evolve through the slow erosional removal of weaker rock, leaving the more resistant rock standing as ridges or mountains. (Drawn by A. N. Strahler.)*

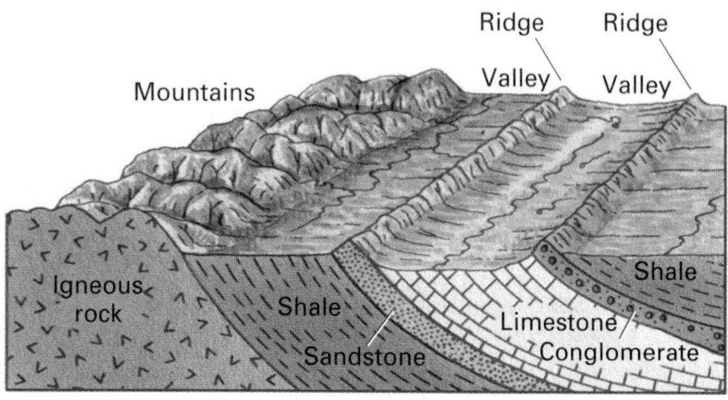

The broad features of landscapes controlled by the rock structures shown in Figure 18.2 are particularly visible from the perspective of orbiting spacecraft. *Focus on Remote Sensing 18.1 • Landsat Views Rock Structures* provides some satellite images in which rock structures are especially striking.

Strike and Dip

The surfaces of most rock layers are not flat but are tipped away from the horizontal. In addition, internal planes—such as joints—occur within rock layers and are likely to be tilted. We need a system of geometry to enable us to measure and describe the position in which these natural planes are held and to indicate them on maps. Examples of such planes include the bedding layers of sedimentary strata, the sides of a dike, and the joints in granite.

The tilt and orientation of a natural rock plane are measured with reference to a horizontal plane. Figure 18.3 shows a water surface resting against tilted sedimentary strata. Since it is horizontal, we can use the water surface as the horizontal plane. The acute angle formed between the rock plane and the horizontal plane is termed the *dip*. The amount of dip is stated in degrees, ranging from 0 for a horizontal rock plane to 90 for a vertical rock plane. Next, we examine the line of intersection between the inclined rock plane and the horizontal water plane. This horizontal line will have an orientation with respect to the compass. This orientation is termed the *strike*. In Figure 18.3, the strike is along a north-south line.

Landforms of Horizontal Strata and Coastal Plains

Extensive areas of the ancient continental shields are covered by thick sequences of horizontal sedimentary rock layers. At various times in the 600 million years following the end of Precambrian time, these strata were deposited in shallow inland seas. Following crustal uplift, with little disturbance other than minor crustal warping or faulting, these areas became continental surfaces undergoing denudation.

Arid Regions

In arid climates, where vegetation is sparse and the action of overland flow especially effective, sharply defined landforms develop on horizontal sedimentary strata (Figure 18.4). The normal sequence of landforms is a sheer rock wall, called a *cliff*, which develops at the edge of a resistant rock layer. At the base of the cliff is an inclined slope, which flattens out into a plain beyond.

In these arid regions, erosion strips away successive rock layers, leaving behind a broad platform capped by hard rock layers. This platform is usually called a **plateau.** Cliffs retreat as near-vertical surfaces because the weak clay or shale formations exposed at the cliff base are rapidly washed away by storm runoff and channel erosion. When undermined, the rock in the upper cliff face repeatedly breaks away along vertical fractures.

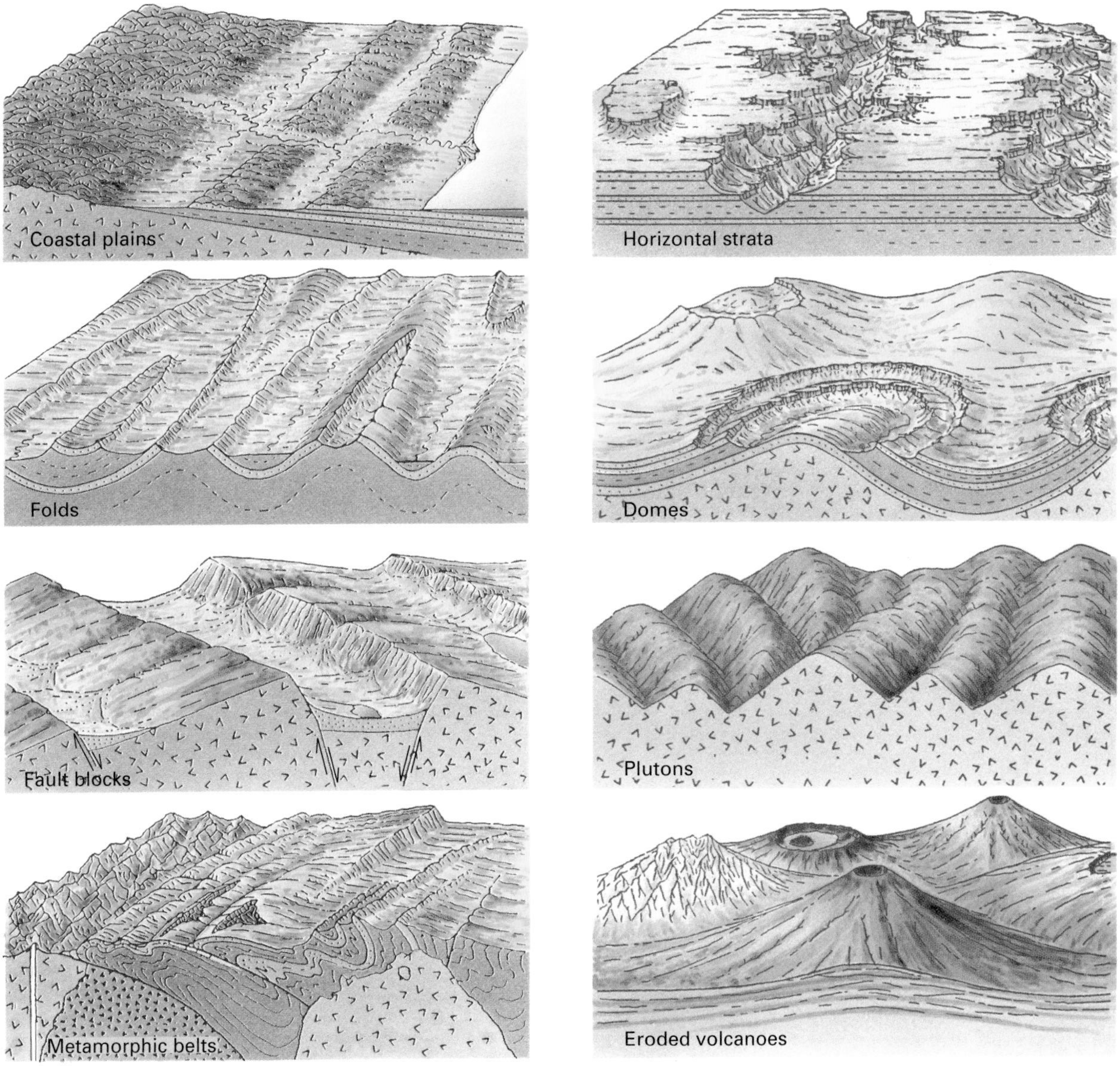

18.2 Landforms associated with various rock structures *(Drawn by A. N. Strahler.)*

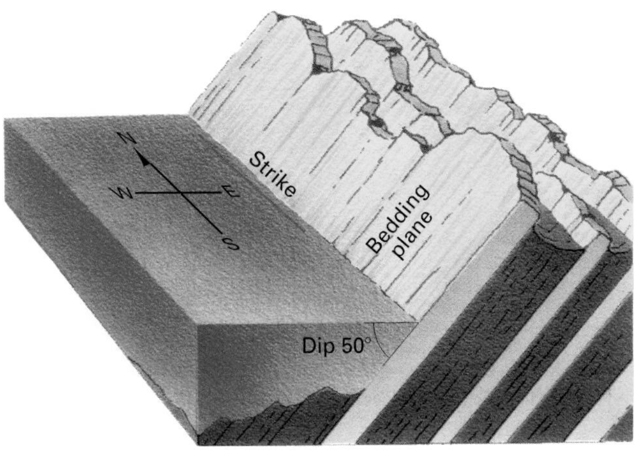

18.3 Strike and dip *(Drawn by A. N. Strahler.)*

Focus on Remote Sensing | 18.1

Landsat Views Rock Structures

With its 30 by 30 m (98 by 98 ft) spatial resolution and its 175 by 175 km (109 by 109 mi) image size, Landsat provides a perspective from space that captures a large region in a single view. This scale is ideal for revealing the relationship between landforms and rock structure that is the subject of this chapter.

Image (*a*) shows the ridge-and-valley country of south-central Pennsylvania in a color-infrared presentation. Like an old wooden plank deeply etched by drifting sand to reveal the grain, the land surface shows zigzag ridges formed by bands of hard quartzite. The strata were crumpled into folds during a continental collision that took place over 200 million years ago to produce the Appalachians. In this September image, the red color depicts vegetation cover, while the blue colors identify mainly agricultural fields. The cloud bands are formed by condensation from passing aircraft; note their shadows above and to the left.

Image (*b*) is a color-infrared view of Canyonlands National Park, Utah, and the surrounding area of the Colorado Plateau. The Colorado River makes its way across the scene from upper right to lower left, joined by the Green River near the center of the image. Notice the dendritic drainage pattern that is typical of flat-lying sedimentary rocks. As the river and its tributaries have eroded the landscape, individual rock layers have been stripped from large areas, producing regions dominated by different colors. The river system has cut valleys through beds of different color and thickness, creating a series of colored outlines around the drainage channels. Vegetation is nearly absent in this arid region, but deep red tones indicate a sparse tree cover on Abajo Peak, in the southeast corner of the image.

(*a*)

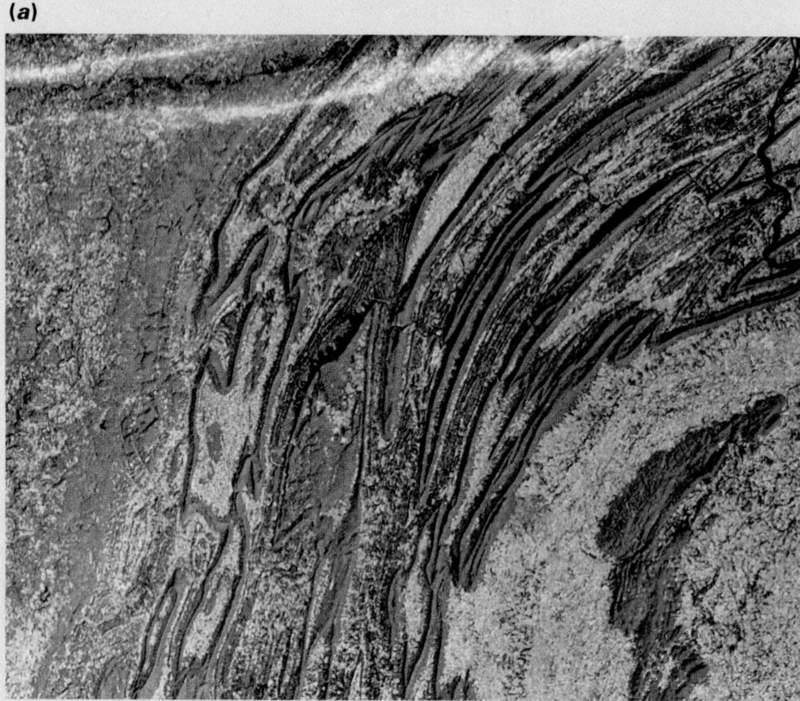

Ridge-and-valley landscape *(Courtesy Earth Satellite Corporation.)*

(*b*)

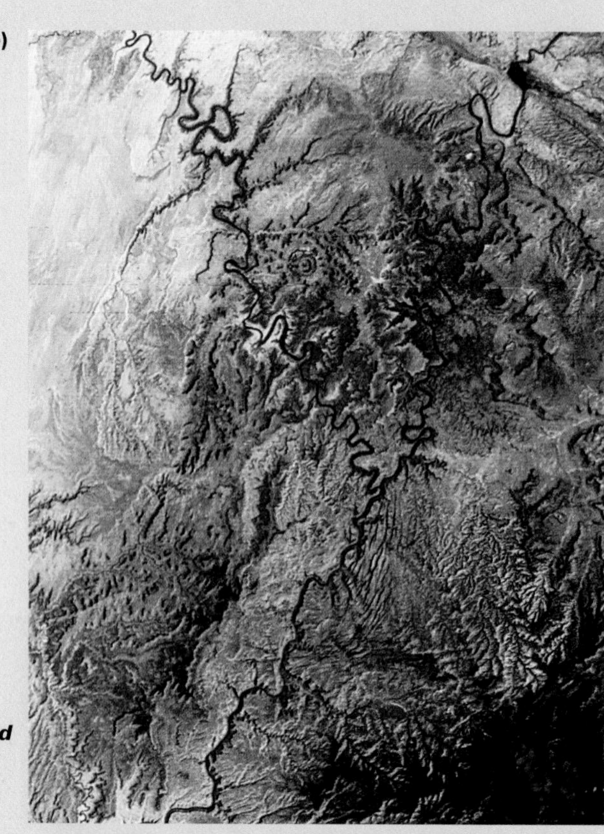

Canyonlands National Park and vicinity, near Moab, Utah *(NASA/JPL.)*

(c)

Southern Sierra Nevada, Central Valley, and Owens Valley of California *(NASA/JPL.)*

The southern portion of California's Sierra Nevada Mountains is the region imaged in (*c*). Bounded by faults on the east and west, the range is a huge uplifted block eroded to reveal a core of resistant granite. To the east is the down-dropped Owens Valley, with Owens Lake shrunken by diversion of surface and ground water to meet the needs of Los Angeles. To the west is the fertile Central Valley, with its lush irrigated croplands appearing bright red in this color infrared image. Note the valley of the Kern River, which has carved a straight, north-south gorge bisecting the southern end of the mountain mass. Near the head of the valley and to the west is snow-capped Mount Whitney. At 4418 M (14,491 ft), it is the highest peak in the conterminous United States.

Kauai (image *d*) is the oldest and most heavily dissected of the shield volcanoes that make up the island chain of Hawaii. The radial drainage pattern of streams and ridge crests leading away from the central summit is a primary feature of this image. The intense red colors of the northern and western slopes identify lush vegetation, watered by orographic rainfall from the prevailing northeast trade winds. Note the tiny popcorn cumulus clouds that have appeared on the lower slopes and ridge crests where the orographic effect is strong. On the far side of the island is an arid region of rain shadow, with bright green tones identifying the famous Waimea Canyon. The intense red color of the rocks here (see Figure 17.26), produced by weathered iron compounds from the basaltic lava, appears bright green in this color infrared image.

Kauai, Hawaii *(NASA/JPL)*

(*d*)

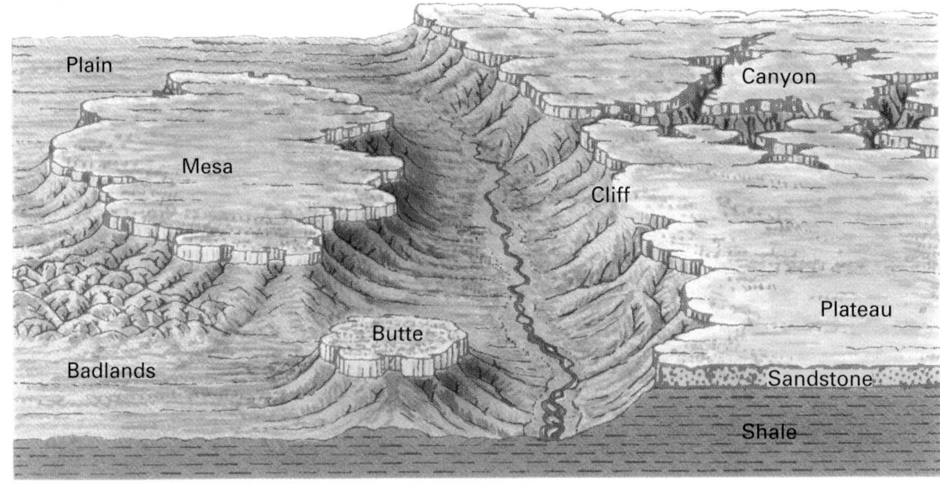

18.4 Arid climate landforms *In arid climates, distinctive erosional landforms develop in horizontal strata. (Drawn by A. N. Strahler.)*

Cliff retreat produces a **mesa,** which is a table-topped plateau bordered on all sides by cliffs. Mesas represent the remnants of a formerly extensive layer of resistant rock. As a mesa is reduced in area by retreat of the rimming cliffs, it maintains its flat top. Eventually, it becomes a small steep-sided hill known as a **butte.** Further erosion may produce a single, tall column before the landform is totally consumed.

In the walls of the great canyons of the Colorado Plateau region, these landforms are wonderfully displayed. Figure 18.5 shows a view of the Grand Canyon. The flat surface of the canyon rim marks the edge of the vast plateau surrounding the canyon. Within the canyon, resistant rock layers crop out at different levels, producing a series of steps. Cliffs of resistant rock form the risers, while weaker rocks form the sloping treads below. Benches or platforms develop atop resistant strata, extending the treads and producing a series of levels that break the descent into the canyon. Small mesas and buttes remain within the canyon, isolated by erosion of surrounding rock strata.

Drainage Patterns on Horizontal Strata

Imagine making a map of a region that showed only its stream channels. This map would show the stream pattern, or **drainage pattern,** of the region. The drainage pattern that develops on each land-mass type is often related to the underlying rock type and structure.

Regions of horizontal strata show broadly branching stream networks formed into a *dendritic drainage pattern.* Figure 18.2 is a map of the stream channels in a small area within the Appalachian Plateau region of Kentucky. The smaller streams in this pattern take a great variety of compass directions. Because the rock layers lie flat, no one direction is favored.

18.5 Stairstep landscape *Bright Angel Canyon, Grand Canyon National Park, Arizona. Note the "stairstep" landscape in which resistant sandstones and limestones form flat-topped benches and mesas, while weaker shales form slopes connecting them.*

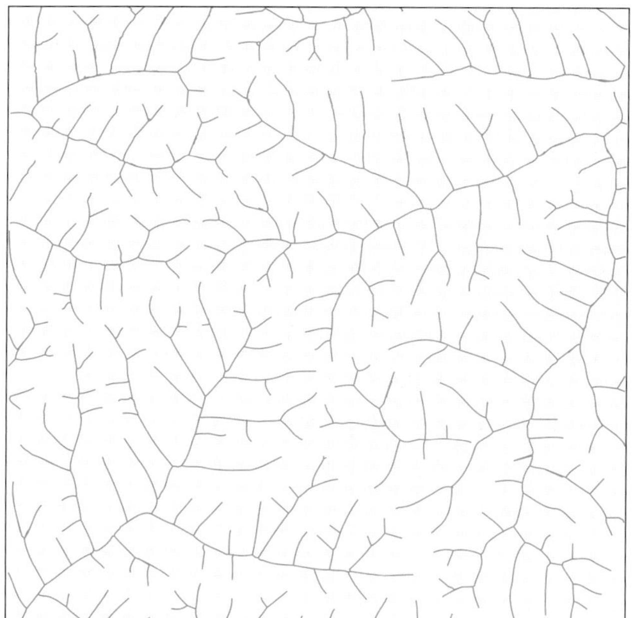

18.6 Dendritic drainage *This dendritic drainage pattern is from an area of horizontal strata in the Appalachian Plateau of northern Kentucky. (Derived from data of the U.S. Geological Survey.)*

Drainage patterns of stream networks, like the dendritic pattern in Figure 18.6, have a number of interesting properties that geomorphologists have studied extensively. *Working It Out 18.2 • Properties of Stream Networks* describes two of these properties—the number and mean length of streams of increasing order.

Coastal Plains

Coastal plains are found along passive continental margins that are largely free of tectonic activity. They are underlain by nearly horizontal strata that slope gently toward the ocean. Figure 18.7*a* shows a coastal zone that has recently emerged from beneath the sea. Formerly, this zone was a shallow continental shelf that accumulated successive layers of sediment brought from the land distributed by currents. On the newly formed land surface, streams flow directly seaward, down the gentle slope. A stream of this origin is a *consequent stream.* It is a stream whose course is controlled by the initial slope of a new land surface. Consequent streams form on many kinds of initial landforms, such as volcanoes, fault blocks, or beds of drained lakes.

In an advanced stage of coastal-plain development, a new series of streams and topographic features has developed (Figure 18.7*b*). Where more easily eroded strata (usually clay or shale) are exposed, denudation is rapid, making *lowlands.* Between them rise broad belts of hills called **cuestas.** Cuestas are commonly underlain by layers of sand, sandstone, limestone, or chalk that dip away from the cuesta at a low angle. The lowland lying between the area of older rock—the oldland—and the first cuesta is called the *inner lowland.*

Streams that develop along the trend of the lowlands, parallel with the shoreline, are of a class known as *subsequent streams.* They take their position along any belt or zone of weak rock and therefore follow closely the pattern of rock exposure. Subsequent streams occur in many regions; we will mention them again in the discussion of folds, domes, and faults.

The drainage lines on a fully dissected coastal plain combine to form a *trellis drainage pattern.* In this type of

18.7 Development of a broad coastal plain *(a) Early stage—plain recently emerged. (b) Advanced stage—cuestas and lowlands developed. (Drawn by A. N. Strahler.)*

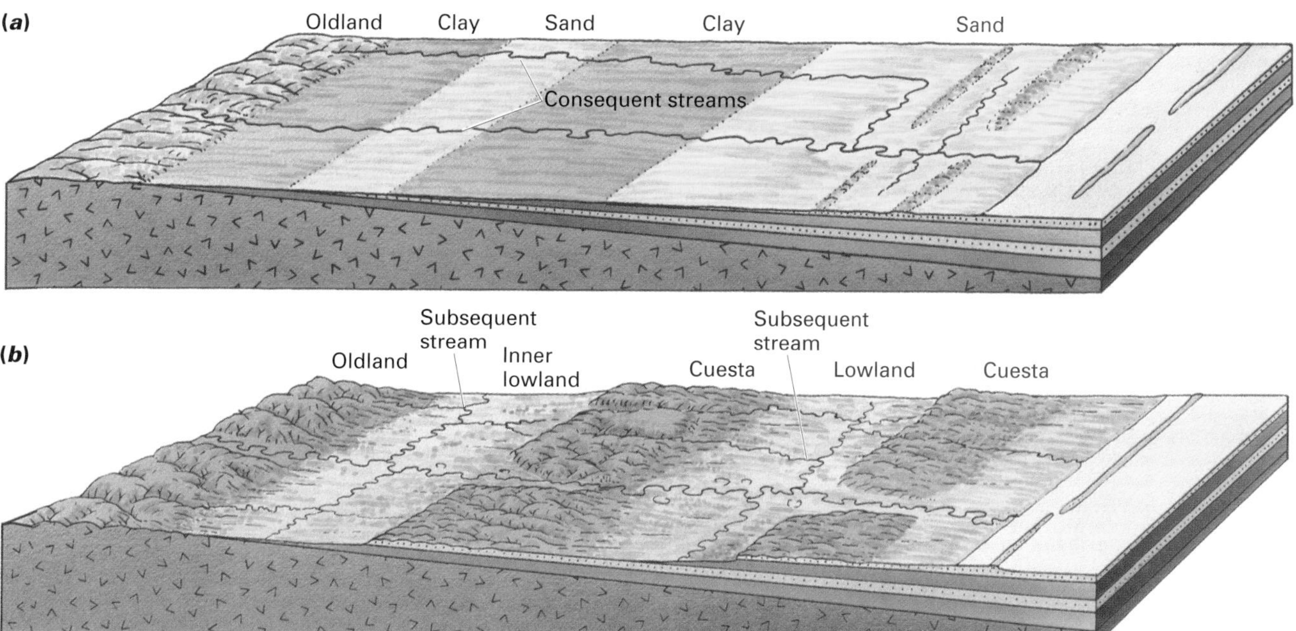

Working It Out 18.2

Properties of Stream Networks

Fluvial landforms are produced by the erosion and deposition of streams that are connected into networks. For this reason, the properties of stream networks have always been of interest to geomorphologists studying the landform-making process. Many of these properties are studied with respect to stream order. Recall from Chapter 17 that the finest stream segments are of first order. When two first-order streams join, the downstream segment is of second order. Third-order stream segments are formed by the junction of second-order segments, and so forth.

Two interesting properties of stream networks are the number of segments of each order and the average length of the segments of each order. As you might imagine, there are many more segments of first order than of second order; of second order than of third order; and so on. Thus, the number of segments decreases with increasing stream order. As to average length, we find that first-order stream segments are shortest, while second-order segments are longer, third-order are longer still, and so forth. This observation stands to reason, since there will be fewer streams of higher order. Thus, a higher-order stream segment will have to flow for a longer distance before being joined by another segment of the same order.

The table (right) provides some data from the Allegheny River, in McKean County, Pennsylvania. The Allegheny ultimately joins the Ohio River, but for this study, only that portion upstream from a particular point on a seventh-order segment of the Allegheny was examined. To provide the information shown in the table, detailed topographic maps were carefully studied over a large area. Each stream segment was identified, and its length was measured. The results were tabulated by stream order.

Looking first at the number of segments, note that the value drops very quickly as order increases. However, the ratio of number of segments of one order to the next seems roughly constant. This ratio is called the *bifurcation ratio* and is shown in the third column. For example, the ratio of first- to second-order segment number is 5966/1529 = 3.9, meaning that each second-order segment branches to form 3.9 first-order segments, on the average. Looking at the values in the column, we can see that the bifurcation ratio varies from 3.0 to 5.7; the average value is 4.37. In algebraic notation, we can write the bifurcation ratio, R_b as

$$R_b = \frac{N_u}{N_{u+1}}$$

where N_u is the number of segments of stream order u and N_{u+1} is the number of segments of stream order $u + 1$.

Part (*a*) of the figure plots the number of segments against stream order for the Allegheny. This graph is *semilogarithmic* (*semilog,* for short)—that is, the *x*-axis is linear, while the *y*-axis is logarithmic. Note that each point falls very close to a straight line. This line follows an equation of the form

$$N_u = R_b^{(k-u)} \qquad (1)$$

Here, k is the order of the largest stream in the network being considered—7 in the case of the Allegheny River drainage network in the table. We can then write the equation for the Allegheny as

$$N_u = 4.37^{(7-u)}$$

using the mean bifurcation ratio, 4.37, for R_b. With this formula, the number of segments expected for stream orders 1 to 7 will then be 4.37^6, 4.37^5, 4.37^4, 4.37^3, 4.37^2, 4.37^1, 4.37^0, or 6964, 1594, 365, 83, 19, 4, 1.

The figure also shows data for a fifth-order stream network in the Big Badlands region of South Dakota. For this stream network, the average bifurcation ratio is 3.47. The Big Badlands region differs considerably from the environment of the Allegheny River in that

Allegheny River Drainage Basin Characteristics

Stream Order u	Number of Segments N_u	Bifurcation Ratio R_b	Mean Segment Length (km) L_u	Cumulative Mean Segment Length (km) L_u^*	Length Ratio R_L
1	5966		0.15	0.15	
		3.9			3.2
2	1529		0.48	0.63	
		4.0			2.7
3	378		1.29	1.9	
		5.7			3.1
4	68		4.0	5.9	
		5.3			2.8
5	13		11.3	17.2	
		4.3			2.8
6	3		32.2	49.4	
		3.0			
7	1		13+		
		$\bar{R}_b = 4.37$			$\bar{R}_L = 2.92$

a given area of easily eroded badland terrain has many more stream channels (see Figure 17.5). However, the lines plotted for Big Badlands and the Allegheny River are close to parallel, indicating that they have similar average bifurcation ratios. In fact, bifurcation ratios for all natural stream networks tend to range between 3 and 5, no matter what type of substrate is being subjected to fluvial action. This remarkable observation was first made by the hydraulic engineer Robert E. Horton, and thus equation (1) above is sometimes referred to as Horton's law of stream numbers.

Turning now to average stream length, we anticipated that the average length of stream segments increases with stream order. The table confirms this fact, with mean lengths for the Allegheny ranging from 0.15 km for first-order segments to 32.2 km for sixth-order segments. As in the case of the bifurcation ratio for stream segment numbers, we can define a length ratio R_L for mean length of segments from two adjacent orders as

$$R_L = \frac{L_{u+1}}{L_u}$$

where L_u is the mean length of segments of order u. As the table shows, values of R_L are rather similar to values of R_b. For the Allegheny data, the average length ratio is $R_L = 2.92$.

Like the number of stream segments, the progression of lengths also closely follows a straight line on a semilog plot, provided that it is plotted as the cumulative mean length. This value, shown in the fifth column of the table, results when mean length values are added cumulatively. That is, the second value is the sum of the first two, the third value is the sum of the first three values, and so on. Stating this algebraically,

$$L_u^* = \sum_{i=1}^{u} L_u$$

where L_u^* is the cumulative mean length of segments of order u, and i is a whole number counting from 1 to u.

The equation for the straight line that connects values of L_u^* on a semilog graph is

$$L_u^* = L_1 \, R_L^{(u-1)} \qquad (2)$$

where L_1 is the mean length of first-order segments. For the case of the Allegheny, the table shows that $L_1 = 0.15$ km. Using the average length ratio $R_L = 2.92$, we have

$$L_u^* = 0.15 \times [2.92^{(u-1)}]$$

Part (*b*) of the figure plots the cumulative mean lengths for the first six orders of the Allegheny stream network. The points lie very close to a straight line. Also shown are data from a fourth-order stream network at Fern Canyon, California. The fact that both lines are nearly parallel shows that cumulative mean length tends to increase by the same factor in spite of differences in climate, substrate, and rock type. Equation (2) was also derived by Horton and is referred to as Horton's law of stream lengths.

Number and mean length of stream segments are just two interesting properties of stream networks studied by geomorphologists. Many other properties have been studied as well. If you continue the study of physical geography, you will learn more of the interesting behavior of streams.

Stream network properties *Numbers of segments (a) and mean segment length (b) plotted against stream order. (Data from M. E. Morisawa, K. G. Smith, and J. C. Maxwell.)*

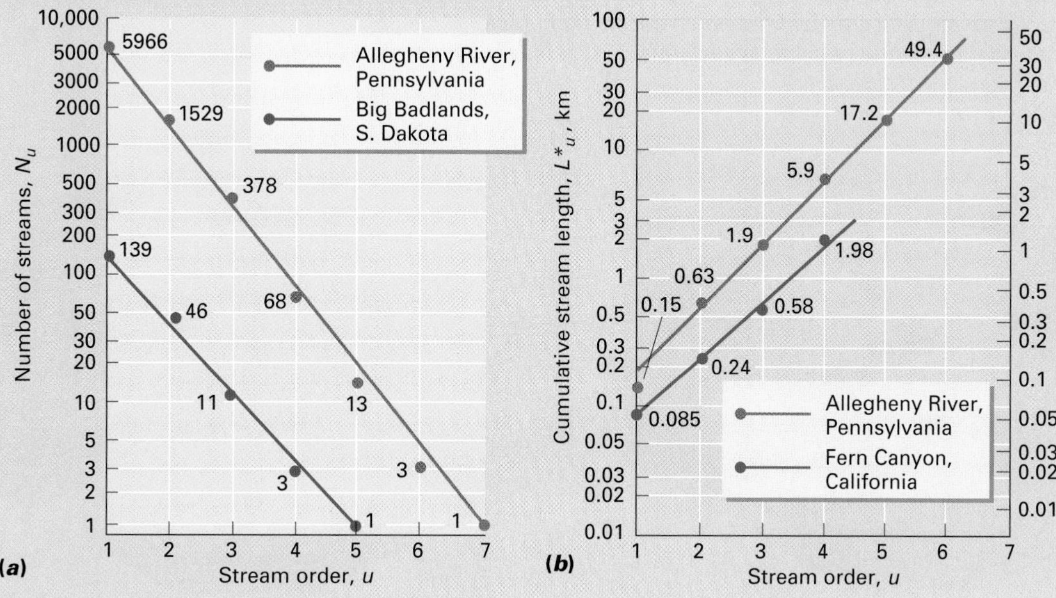

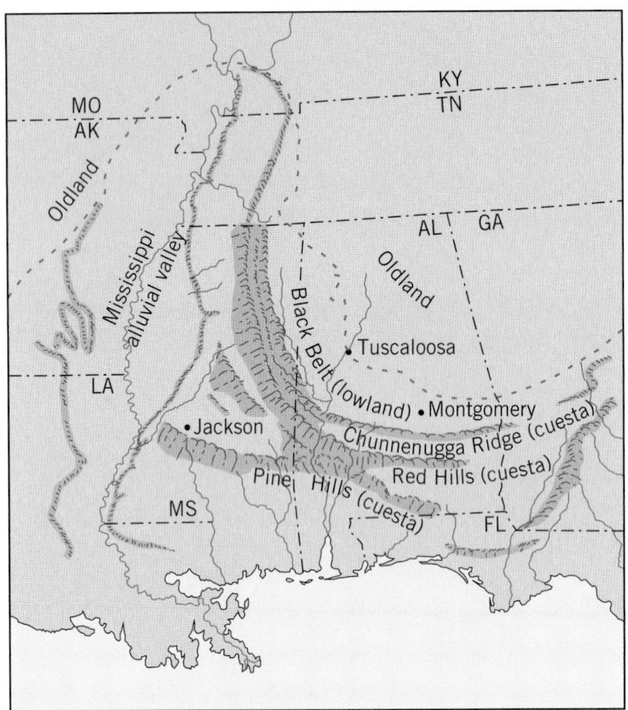

18.8 ***The Alabama–Mississippi coastal plain*** *This coastal plain is belted by a series of sandy cuestas and shale lowlands. (After A. K. Lobeck.)*

pattern, a main stream has tributaries that are arranged at right angles. You can see this pattern in the streams of Figure 18.7*b*. The subsequent streams trend at about right angles to the consequent streams.

The coastal plain of the United States is a major geographical region, ranging in width from 160 to 500 km (about 100 to 300 mi) and extending for 3000 km (about 2000 mi) along the Atlantic and Gulf coasts. In Alabama and Mississippi (Figure 18.8), the coastal plain shows many of the features of Figure 18.7. In this region, the coastal plain sweeps far inland, as the oldland boundary shows. Cuestas and lowlands run in belts curving to follow the ancient coast. The cuestas, named "hills" or "ridges," are underlain by sandy formations and support pine forests. Limestone forms lowlands, such as the Black Belt in Alabama. This belt is named for its dark, fertile soils.

Landforms of Warped Rock Layers

Although sedimentary rocks are horizontal or gently sloping in some regions, in other areas rock layers may be upwarped or downwarped. In some cases, these form domes and basins. In others, they form anticlines and synclines—wave-like forms of crests and troughs—as we noted in Chapter 14.

Sedimentary Domes

A distinctive land-mass type is the **sedimentary dome,** a circular or oval structure in which strata have been forced upward into a domed shape. Sedimentary domes occur in various places within the covered shield areas of the continents. Igneous intrusions at great depth may have been responsible for some of these uplifts. For others, upthrusting on deep faults may have been the cause.

Erosion features of a sedimentary dome are illustrated in the two block diagrams of Figure 18.9. Strata are first removed from the summit region of the dome, exposing older strata beneath. Eroded edges of steeply dipping strata form sharp-crested sawtooth ridges called *hogbacks* (Figure

18.9 ***Dome erosion*** *Erosion of sedimentary strata from the summit of a dome structure. (a) The strata are partially eroded, forming an encircling hogback ridge. (b) The strata are eroded from the center of dome, revealing a core of older igneous or metamorphic rock. (Drawn by A. N. Strahler.)*

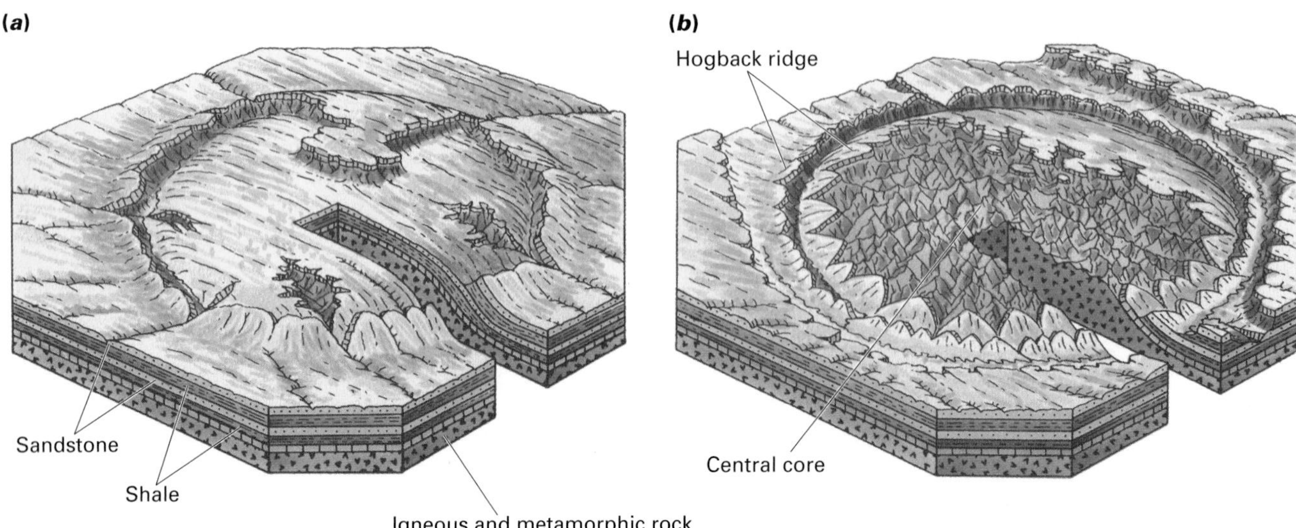

18.10 *Eye on the Landscape* ***Hogbacks*** *Sandstone hogbacks lie along the eastern base of the Colorado Front Range, which is a large domed structure. The view is southward toward the city of Boulder. The rugged forested terrain in the distance is developed on igneous and metamorphic rock exposed in the core of the uplift.* **What else would the geographer see? ... Answers at the end of the chapter.**

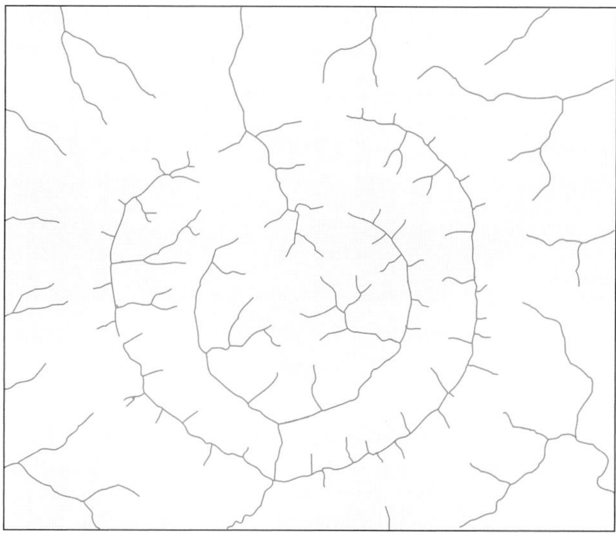

18.11 ***Dome drainage*** *The drainage pattern on an eroded dome combines annular and radial elements.*

18.12 ***The Black Hills dome*** *The Black Hills consist of a broad, flat-topped dome deeply eroded to expose a core of igneous and metamorphic rocks. (Drawn by A. N. Strahler.)*

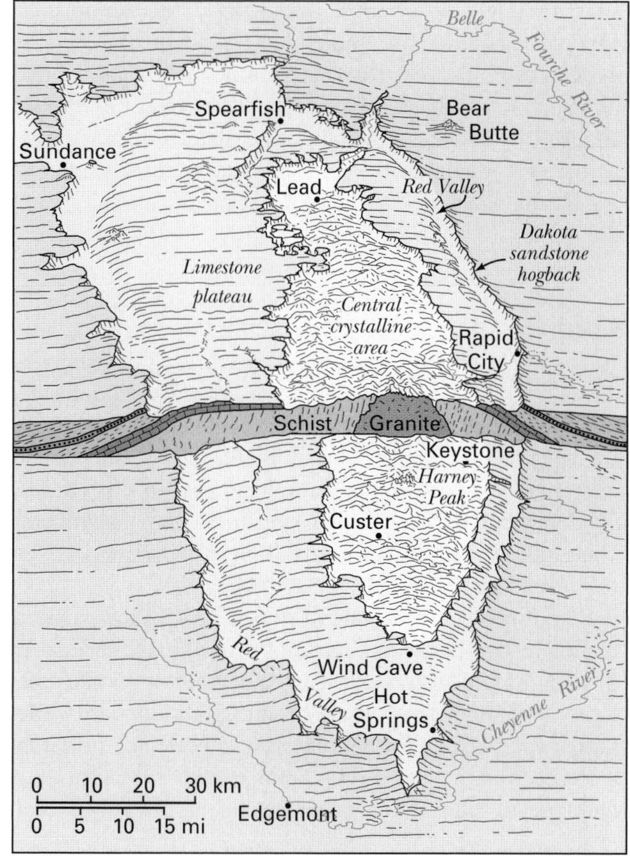

18.10). When the last of the strata have been removed, the ancient shield rock is exposed in the central core of the dome, which then develops a mountainous terrain.

The stream network on a deeply eroded dome shows dominant subsequent streams forming a circular system, called an annular drainage pattern (Figure 18.11). ("Annular" means "ring-shaped.") The shorter tributaries make a radial arrangement. The total pattern resembles a trellis pattern bent into a circular form.

THE BLACK HILLS DOME A classic example of a large and rather complex sedimentary dome is the Black Hills dome of western South Dakota and eastern Wyoming (Figure

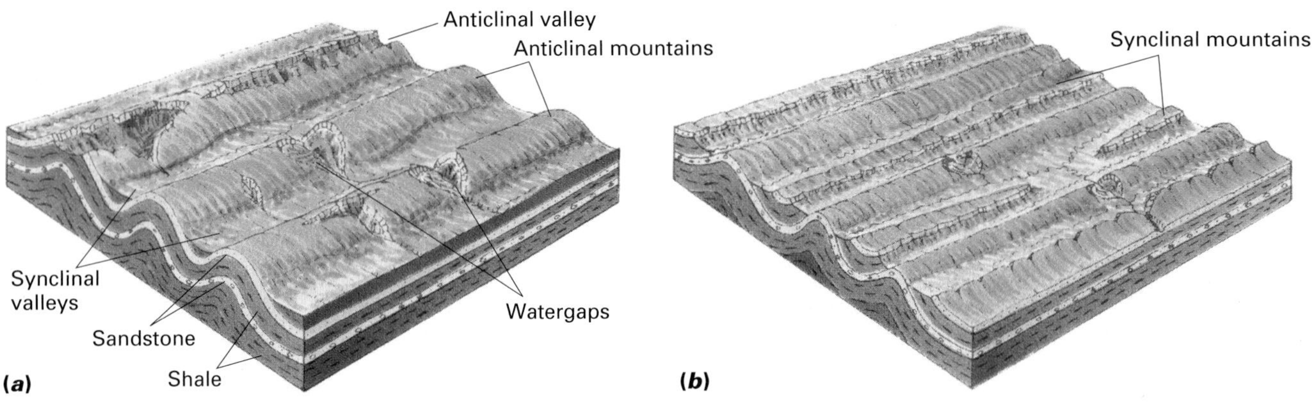

18.13 Stages in the erosional development of folded strata *(a) Erosion exposes a highly resistant layer of sandstone or quartzite, which controls much of the ridge-and-valley landscape. (b) Continued erosion partly removes the resistant formation but reveals another below it. (Drawn by A. N. Strahler.)*

18.12). The Red Valley is continuously developed around the entire dome and is underlain by a weak shale, which is easily washed away. On the outer side of the Red Valley is a high, sharp hogback of Dakota sandstone, known as Hogback Ridge, rising some 150 m (about 500 ft) above the level of the Red Valley. Farther out toward the margins of the dome, the strata are less steeply inclined and form a series of cuestas.

The eastern central part of the Black Hills consists of a mountainous core of intrusive and metamorphic rocks. These mountains are richly forested, while the surrounding valleys are beautiful open parks. Thus, the region is very attractive as a summer resort area. In the northern part of the central core, there are valuable ore deposits. At Lead is the fabulous Homestake Mine, one of the world's richest gold-producing mines. The west-central part of the Black Hills consists of a limestone plateau deeply carved by streams. The plateau represents one of the last remaining sedimentary rock layers to be stripped from the core of the dome.

Fold Belts

According to the model of mountain-making developed in Chapter 13, strata of the continental margins are deformed into folds along narrow belts during continental collision. As we explained in Chapter 14, a troughlike downbend of strata is a *syncline,* and the archlike upbend next to it is an *anticline.* In a belt of folded strata, synclines alternate with anticlines, like a succession of wave troughs and wave crests on the ocean surface. GEODISCOVERIES

The two block diagrams of Figure 18.13 show some of the distinctive landforms resulting from fluvial denudation of a belt of folded strata. Deep erosion of these simple, open folds in this land-mass type produces a **ridge-and-valley landscape.** Weaker formations such as shale and limestone are eroded away, leaving hard strata, such as sandstone or quartzite, to stand in bold relief as long, narrow ridges. Note that anticlines, or upwarps, are not always ridges. If a resistant rock type at the center of the anticline is eroded through to reveal softer rocks underneath, an *anticlinal valley* may form. An *anticlinal valley* is shown in Figure 18.13*a.* A *synclinal mountain* is also possible. It occurs when a resistant rock type is exposed at the center of a syncline, and the rock stands up as a ridge (Figure 18.13*b*).

On eroded folds, the stream network is distorted into a trellis drainage pattern (Figure 18.14). The principal elements of this pattern are long, parallel subsequent streams occupying the narrow valleys of weak rock. In a few places, a larger stream cuts across a ridge in a deep, steep-walled *water gap.*

The folds illustrated in Figure 18.13 are continuous and even-crested. They produce ridges that are approximately parallel in trend and continue for great distances. In some fold regions, however, the folds are not continuous and level-crested. Instead, the fold crests rise or descend from place to place. A descending fold crest is said to "plunge." Plunging folds give rise to a zigzag line of ridges (Figure 18.15). Excellent examples of plunging folds can be seen

18.14 Trellis drainage *A trellis drainage pattern on deeply eroded folds in the Central Appalachians.*

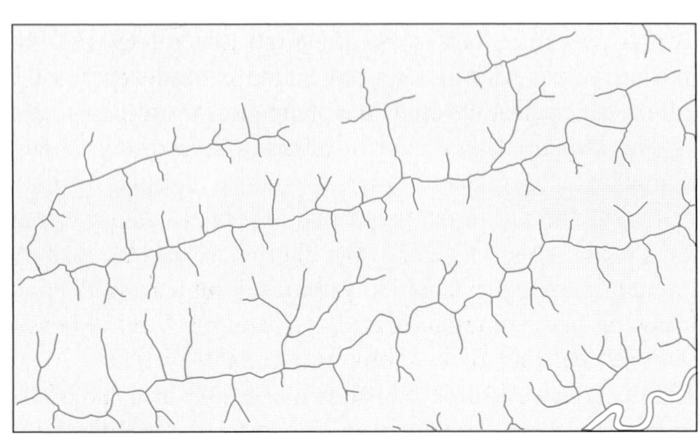

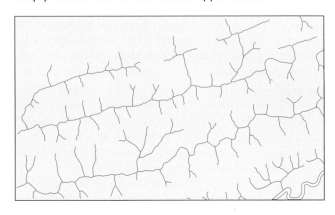

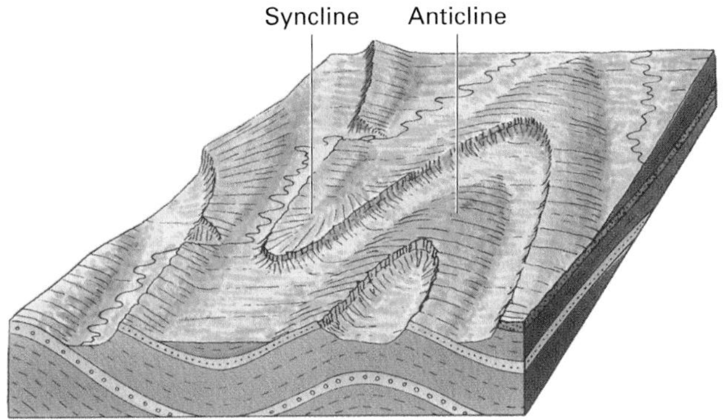

18.15 *Plunging folds* *Folds with crests that plunge downward give rise to zigzag ridges following erosion. (Drawn by Erwin Raisz. Copyright © by A. N. Strahler.)*

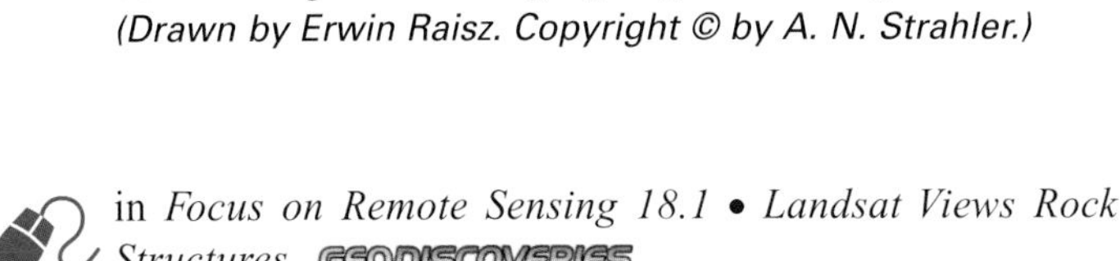

in *Focus on Remote Sensing 18.1 • Landsat Views Rock Structures.* GEODISCOVERIES

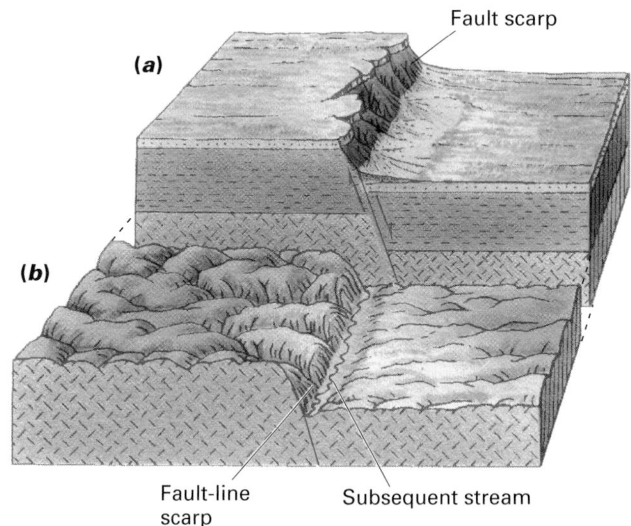

18.16 *Fault scarp evolution* *(a) A recently formed fault scarp. (b) Despite continental denudation over several million years, the fault plane causes a long narrow valley bounded by a scarp. (Drawn by A. N. Strahler. Copyright © A. N. Strahler.)*

Landforms Developed on Other Land-Mass Types

We now turn to the landforms developed on a few remaining land-mass types. First are erosional forms that develop on a land mass exposed to faulting. Then we discuss the landforms that develop on metamorphic rocks and on exposed igneous batholiths. Last are the landforms of eroded volcanoes.

Erosion Forms on Fault Structures

In Chapter 14, we found that active normal faulting produces a sharp surface break called a *fault scarp* (see Figure 14.13). Repeated faulting may produce a great rock cliff hundreds of meters high. Erosion quickly modifies a fault scarp, but because the fault plane extends hundreds of meters down into the bedrock, its effects on erosional landforms persist for long spans of geologic time. Figure 18.16 shows both the original fault scarp (above) and its later landform expression (below), known as a *fault-line scarp.* Even though the cover of sedimentary strata has been completely removed, exposing the ancient shield rock, the fault continues to produce a landform. Because the fault plane is a zone of weak rock that has been crushed during faulting, it is occupied by a subsequent stream. A scarp persists along the upthrown side. Such scarps on ancient fault lines are numerous in the exposed and covered continental shields.

Figure 18.17 shows some erosional features of a large, tilted fault block. The freshly uplifted block has a steep mountain face, but it is rapidly dissected into deep canyons. The upper mountain face becomes less steep as it is removed, and rock debris accumulates in the form of alluvial fans adjacent to the fault block. Vestiges of the fault plane are preserved in a line of triangular facets. Each facet represents the snubbed end of a ridge between two canyon mouths.

In addition to creating faults, tectonic activity is also capable of uplifting land masses well above base level. The land mass is often elevated much more rapidly than it can be eroded away, producing a plateau or mountain range that can persist for millions of years. *Focus on Systems 18.3 • A Model Denudation System* explores rates of uplift and erosion in more detail.

Metamorphic Belts

Where strata have been tightly folded and altered into metamorphic rocks during continental collision, denudation eventually develops a landscape with a strong grain of ridges and valleys. These features lack the sharpness and straightness of ridges and valleys in belts of open folds. Even so, the principle that governs the development of ridges and valleys in open folds applies to metamorphic landscapes as well—namely, that rocks resistant to denudation form highlands and ridges, while rocks that are more susceptible to denudation form lowlands and valleys.

Figure 18.18 shows typical erosional forms associated with parallel belts of metamorphic rocks, such as schist, slate, quartzite, and marble. Marble forms valleys, while slate and schist make hill belts. Quartzite stands out boldly and may produce conspicuous narrow hogbacks. Areas of gneiss form highlands.

Parts of New England, particularly the Taconic and Green Mountains of New Hampshire and Vermont, illustrate the landforms eroded on an ancient metamorphic belt. The larger valleys trend north and south and are underlain by

Focus on Systems 18.3

A Model Denudation System

As land masses are eroded by fluvial action, streams move mineral matter downhill from uplands to lowlands and ultimately to the ocean or a closed inland basin. As we saw in *Focus on Systems 16.2 • Energy in Stream Flow,* this motion is powered by gravity. When erosion strips upland surfaces, it also lightens the upland land mass. In Chapter 13, we established the principle that the Earth's outer shell of hard, brittle rock—the lithosphere—rests on the asthenosphere, a plastic rock layer below. Geologists have shown that the lithosphere literally floats on the plastic asthenosphere, much as an iceberg floats on denser sea water. In the case of the iceberg, if you were to remove the part showing above the water, the iceberg would rise and bring more ice above the water level. If you placed an added load of ice on the iceberg, it would sink lower in the water but would come to a new position of rest with its summit somewhat higher than before. This flotation principle, as applied to the lithosphere, is called *isostasy.*

When bedrock is eroded from a continent, the lithosphere rises beneath it according to isostasy. However, the rise is somewhat less than the thickness of material removed. When the eroded sediment accumulates in undersea layers on the continental margin, the added load causes the lithosphere to sink lower, but again, the amount of sinking is somewhat less than the thickness of sediment accumulating. A rising or sinking of the crust in response to denudation or sediment accumulation is referred to as *isostatic compensation.*

Isostatic compensation applies to the erosion of land masses lifted above a base level of erosion. What is responsible for the initial lifting? In many cases, it is the tectonic activity associated with lithospheric plate interactions. During arc-continent and continent-continent collisions, thickening of the continental lithosphere may occur rapidly, producing a high alpine chain—the Himalayas or Andes, for example.

The rate of uplift by tectonic and isostatic processes can be much greater than the rate of denudation by fluvial erosion. Based on direct geologic evidence, rates of recent tectonic uplift in various parts of the world are on the order of 4 to 12 m per 1000 years (about 15 to 40 ft/1000 yr). Rates of denudation are much lower. In rugged terrain, observed rates are on the order of 1 to 1.5 m/1000 yr (about 3 to 5 ft/1000 yr).

The rate of denudation will be strongly influenced by the average summit elevation of the land mass. The higher a mountain block rises, the steeper will be the gradients, on the average, of streams carving into that land mass. As we saw in Chapter 17, both the intensity of stream abrasion and the ability of a stream to transport load increase strongly as channel gradients increase. Valley wall slopes are also steeper, so mass wasting and erosion by overland flow will be more intense. As gradients diminish with time, however, the rate of denudation will also diminish.

Average denudation rates over thousands of years are determined by measuring sediment accumulation in lowland areas. If we had been able to monitor surface elevations over geologic time, we would have recorded the effect of isostatic compensation. For example, if erosion strips 1 m of sediment from an uplifted land mass, isostatic compensation might uplift the mass by 0.8 m in response, leaving a net reduction of elevation of only 0.2 m. Thus, the land mass as a whole will lower in elevation at a much slower rate than the rate at which it is stripped by erosion.

Figure 1 diagrams the erosion of an uplifted land mass. Here we assume that a large crustal mass is raised as a single block from a previous position close to sea level to a mean surface elevation of 5000 m (about 16,000 ft). The uplifted block has a very broadly domed summit and steep sides, and is bordered by low areas, at or below sea level, that serve as receiving sites for sediment carried out from the mountain block by streams. During the uplift, streams are active, and so the land mass shown in stage *a* has already been carved into a maze of steep-walled gorges that are organized into a fluvial system of steep-gradient streams. In stage *b,* after 15 m.y., the average elevation is reduced by half, to 2500 m (about 8000 ft). Slopes and stream gradients are still steep but have declined significantly. In stage *c,* at 30 m.y., the elevation is reduced by half again, to 1250 m (about 4000 ft). By now the slopes are much more gentle, as are stream gradients. Stage *d* marks another half-reduction, as does stage *e.* At this time (60 m.y. after initial uplift), the land mass has been eroded to a low-lying peneplain with a mean surface elevation of about 300 m (about 1000 ft).

This description follows a simple model for land-mass denudation—exponential decay. This is the same model that describes the decay of radioactive isotopes, which we presented in *Working It Out 12.1 • Radioactive Decay.* Figure 2 shows how the mean elevation decreases with time following the exponential decay model. For each 15 m.y. period, the mean land-mass elevation is reduced by half. The net denudation rate also drops by half with each 15 m.y. interval. Thus, the land mass is eroded at an ever-slower rate as it approaches a peneplain.

The initial rate of denudation, 231 m/m.y. (758 ft/m.y.), translates to 0.231 m/1000 yr (0.758 ft/1000 yr). Note that this rate is the net lowering rate, which includes the

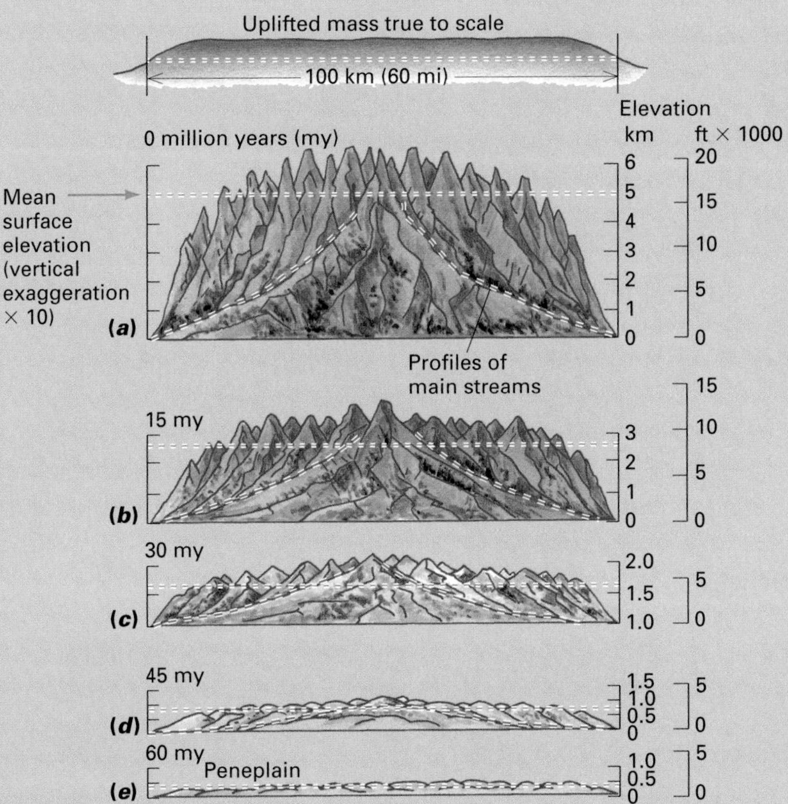

Figure 1 *Schematic diagram of land-mass denudation. In this model, the average surface elevation is reduced by one-half every 15 million years. (Copyright © A. N. Strahler.)*

effect of isostatic compensation. The true rate at which rock material is stripped from the land mass will be much greater. Assuming that isostatic compensation recovers 80 percent of the lost elevation, the erosion rate will be five times greater. Thus, the initial erosion rate in our example would be about 1.15 m/1000 yr (3.79 m/1000 yr), which agrees well with the rates of 1.0–1.5 m/1000 yr (about 3–5 ft/yr) observed for mountainous terrain. As noted, this rate will decrease as the land mass is eroded.

Note also that tectonic processes may interrupt the slow denudation of the land mass at any time. The land mass may be uplifted, or rejuvenated, as shown in Figure 17.25. Or, it may be downdropped and subjected to sedimentation from surrounding highland areas, or even lowered below sea level. Furthermore, climate may change, causing erosion rates to change. In fact, there are few places on Earth where the simple model presented above will actually apply. However, it establishes the principles of uplift, isostatic compensation, and denudation that are always in operation, even though a particular landscape may have a complex history.

What powers the denudation of the landscape? Tectonic processes, including isostatic rebound, provide gravitational potential energy as an input when the land mass is lifted above base level. Solar energy also powers the process by providing the precipitation that drives the fluvial processes of erosion.

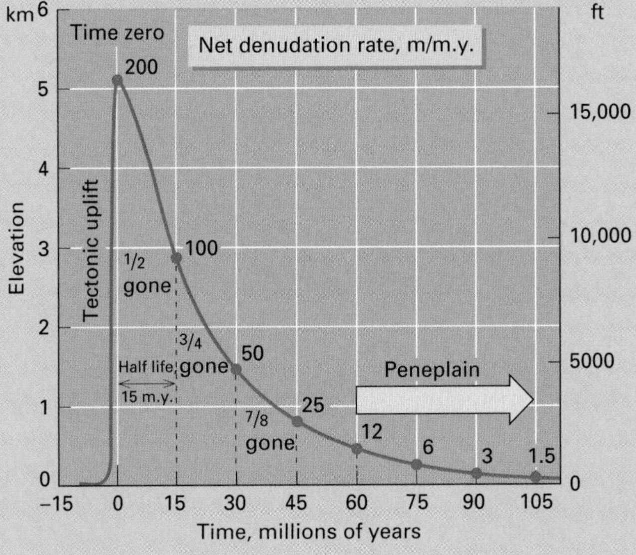

Figure 2 *Graph of decrease in average surface elevation with time, as shown in Figure 1. (Copyright © A. N. Strahler.)*

18.17 Eye on the Landscape Erosion of an uplifted fault block *The near-vertical faces of the foreground peaks in this photo mark a fault scarp along which the rocks have been uplifted by many successive motions. Talus slopes of rock debris line the base of the scarp. Eastern Sierra Nevada, California.* **What else would the geographer see?...Answers at the end of the chapter.**

marble. These are flanked by ridges of gneiss, schist, slate, or quartzite.

18.18 Metamorphic belts *Metamorphic rocks form long. narrow parallel belts of valleys and mountains. (Drawn by A. N. Strahler.)*

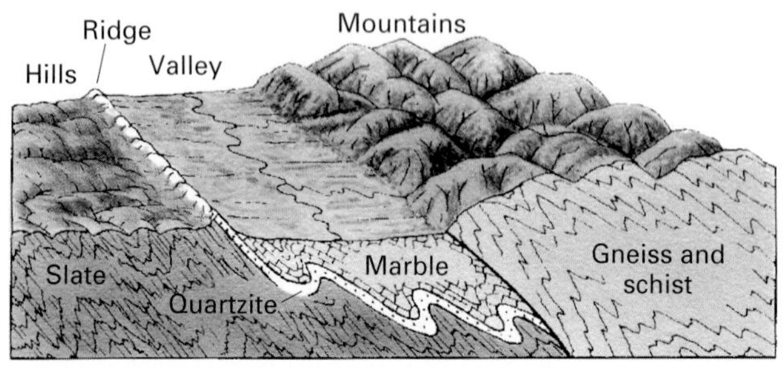

Exposed Batholiths and Monadnocks

Recall from Chapter 14 that batholiths—huge plutons of intrusive igneous rock—are formed deep below the Earth's surface. Some are eventually uncovered by erosion and appear at the surface. Because batholiths are typically composed of resistant rock, they are eroded into hilly or mountainous uplands.

Batholiths of granitic composition are a major ingredient in the mosaic of ancient rocks comprising the continental shields. A good example is the Idaho batholith, a granite mass exposed over an area of about 40,000 sq km (about 16,000 sq mi)—a region almost as large as New Hampshire and Vermont combined. Another example is the

18.19 Dendritic drainage *This dendritic drainage pattern is developed on the dissected Idaho batholith.*

Sierra Nevada batholith of California, which makes up most of the high central part of that great mountain block. The Canadian shield includes many batholiths, large and small.

A dendritic drainage pattern is often well developed on an eroded batholith (Figure 18.19). Note how similar this pattern is to that on horizontal strata (Figure 18.6).

Small bodies of granite, representing domelike projections of batholiths that lie below, are often found surrounded by ancient metamorphic rocks into which the granite was intruded (Figure 18.20). A good example is Sugar Loaf Mountain, which rises above the city of Rio de Janiero, Brazil. Sugar Loaf Mountain and other monolithic domes are unique landforms that are discussed further in *Eye on the Environment 18.4 • Marvelous, Majestic, Monolithic Domes.* The name **monadnock** has been given to an isolated mountain or hill that rises conspicuously above a peneplain. A monadnock develops because the rock within the monadnock is much more resistant to denudation processes than the bedrock of the surrounding region. The name is taken from Mount Monadnock in southern New Hampshire.

Deeply Eroded Volcanoes

Recall from Chapter 14 that volcanoes can be divided into two types—stratovolcanoes and shield volcanoes. Stratovolcanoes are typically constructed by explosive eruptions accumulating layers of thick, gummy andesite lava flows and airborne tephra, while shield volcanoes are built primarily of thin, runny flows of basalt. Thus, these two types of volcanoes produce different types of landforms as they erode away.

Figure 18.21 shows successive stages in the erosion of stratovolcanoes, lava flows, and a caldera. Shown in block (*a*) are active volcanoes in the process of building. These are initial landforms. Lava flows issuing from the volcanoes have spread down into a stream valley, following the downward grade of the valley and forming a lake behind the lava dam.

In block (*b*) some changes have taken place. The most conspicuous change is the destruction of the largest volcano to produce a caldera. A lake occupies the caldera, and a small cone has been built inside. One of the other volcanoes, formed earlier, has become extinct. It has been dissected by streams and has lost its smooth conical form. Smaller, neighboring volcanoes are still active, and the contrast in form is marked.

The system of streams on a dissected volcano cone is a *radial drainage pattern.* Because these streams take their position on a slope of an initial land surface, they are of the consequent variety. It is often possible to recognize volcanoes from a drainage map alone because of the perfection of the radial pattern (Figure 18.22). A good example of a partly dissected volcano is Mount Shasta in California (Figure 18.23*a*).

In Figure 18.21 all volcanoes are extinct and have been deeply eroded. The caldera lake has been drained, and the rim has been worn to a low, circular ridge. The lava flows have been able to resist erosion far better than the rock of the surrounding area and have come to stand as mesas high above the general level of the region.

Block (*d*) of Figure 18.21 shows an advanced stage of erosion of stratovolcanoes. All that remains now is a small, sharp peak, called a *volcanic neck*—which is the remains of lava that solidified in the pipe of the volcano. Radiating from it are wall-like dikes, formed of magma, which filled radial fractures around the base of the volcano. Perhaps the finest illustration of a volcanic neck with radial dikes is Ship Rock, New Mexico (Figure 18.23*b*).

18.20 Exposed batholiths *Batholiths appear at the land surface only after long-continued erosion has removed thousands of meters of overlying rocks. Small projections of the granite intrusion appear first and are surrounded by older rock. (Drawn by A. N. Strahler.)*

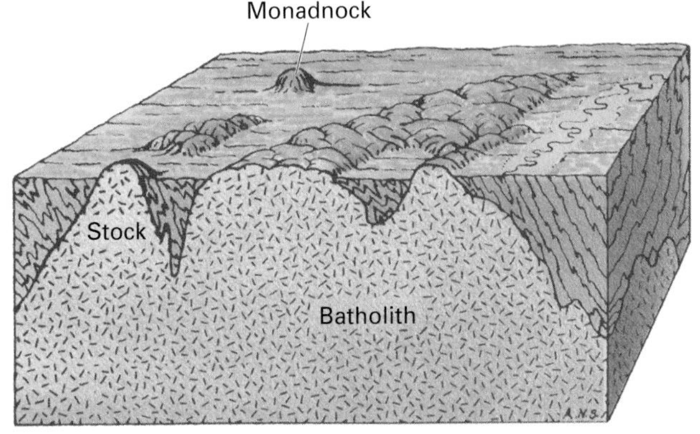

Eye on the Environment | 18.4

Marvelous, Majestic, Monolithic Domes

There is something awesome about a granite dome. Perhaps it's the way it projects above its surroundings, emerging abruptly from the ground and rising so steeply to its rounded summit. Perhaps it's the smooth, barren surface, hardly marred by a crack or crevice. Or maybe it's just the size—a huge, single, uniform object that dwarfs an observer like no other. In any event, these geological curiosities have aroused the interest of observers for centuries—perhaps even millennia.

Probably the most famous of all granite domes is Rio de Janeiro's Sugar Loaf, a monument that serves as a symbol for that cosmopolitan Brazilian city (*a*). Of course, Australia has its Ayres Rock, a gigantic rock mass in the desert of Northern Territory. But, alas, Ayres Rock is mere sandstone, not solid granite. Australia actually has some fine granite domes, located on the Eyre Peninsula of South Australia, but they have not achieved the international fame of Ayres Rock. Splendid granite domes also occur in the Nubian Desert of North Africa. Even Pasadena, California, has its own Little League contender in Eagle Rock, a small but prominent dome of massive conglomerate. These monolithic domes of uniform rock owe their rounded shapes to grain-by-grain disintegration of their rock surfaces. Because the domes lack the joints found in most other rocks, oxidation and hydrolysis act uniformly to weather the rock surface, producing a smooth, rounded form.

And what about the spectacular domes of Yosemite National Park in California's Sierra Nevada Range? Yosemite Valley has several splendid domes high up on either side of the deep canyon (*b*). As we mentioned in Chapter 14, these are exfoliation domes, bearing thick rock shells. As the dome weathers, the outermost shells break apart and slide off as new ones are generated within. Beneath their shells, however, the Yosemite domes seem to be genuine monoliths.

Another fine granite dome is Stone Mountain, near Atlanta (*c*). Like Sugar Loaf, it is a uniform mass of granite seemingly without joints or fissures. It is the surface remains of an ancient, knob-like granite pluton intruded into older metamorphic rocks of the region. For millions of years, this region has been undergoing denudation, and most of it has been reduced to a rolling upland. Numerous monadnocks of quartzite with elongate, ridge-like forms rise above the upland, but Stone Mountain is unique in terms of its compact outline and smoothly rounded form.

(*a*)

Sugar Loaf Mountain *The monolithic rock dome of Sugar Loaf graces the harbor of Rio de Janiero, Brazil.*

(b)

Yosemite Domes *These fine examples of monolithic granite domes are found in the Upper Yosemite Valley, Yosemite National Park, California.*

(c)

Stone Mountain *This dome is a striking erosional remnant about 2.4 km (1.5 mi) long that rises 193 m (650 ft) above the surrounding Piedmont plain.*

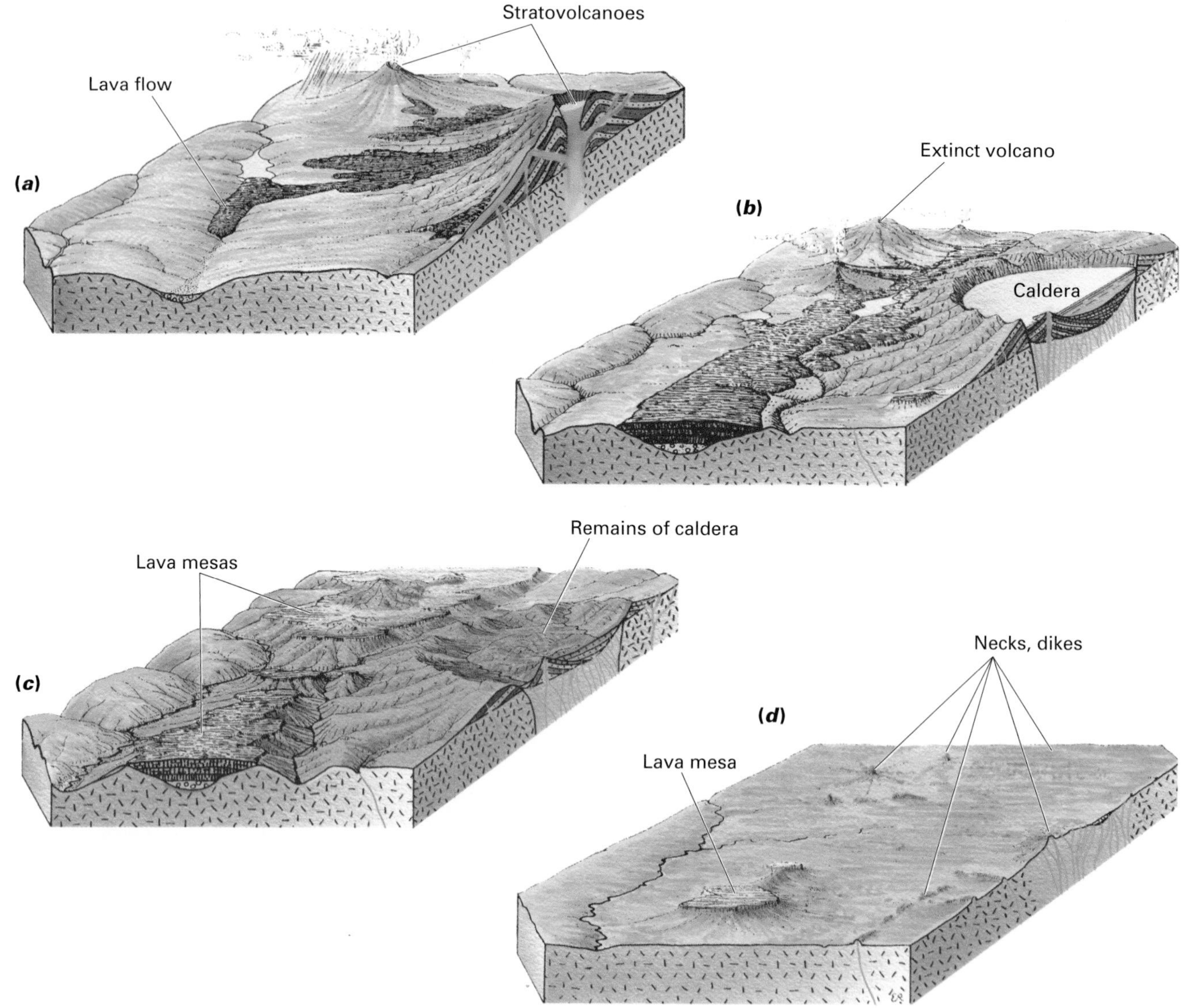

Stratovolcanoes

Lava flow

(*a*)

Extinct volcano

(*b*)

Caldera

Remains of caldera

Lava mesas

(*c*)

Necks, dikes

(*d*)

Lava mesa

18.21 Erosion in stratovolcanoes *Stages in the erosional development of stratovolcanoes and lava flows. (Drawn by Erwin Raisz. Copyright © by A. N. Strahler.)*

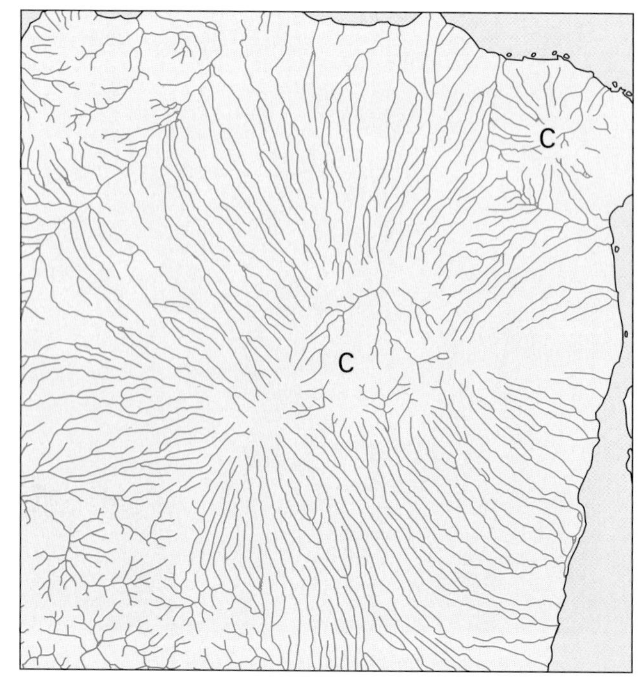

18.22 Radial drainage *Radial drainage patterns of volcanoes in the East Indies. The letter "C" shows the location of a crater.*

18.23 Dissected Volcanoes

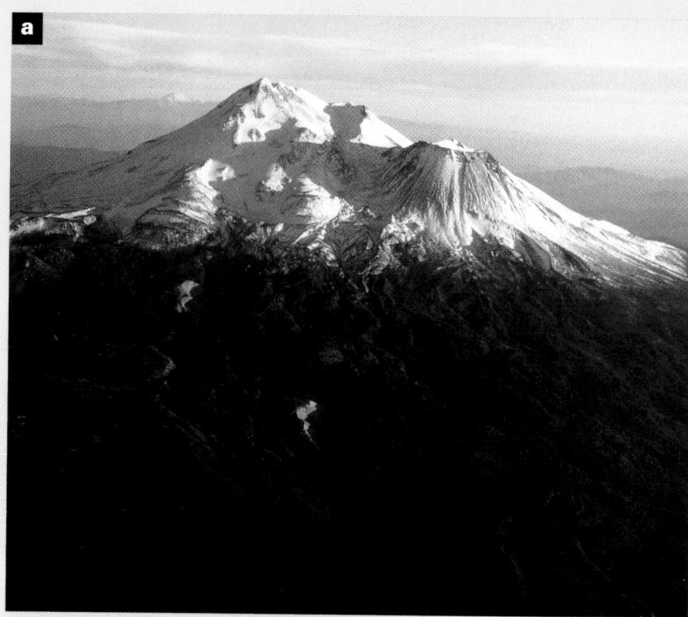

a...**Partly dissected volcano** *Mount Shasta in the Cascade Range, northern California, is a partly dissected composite volcano. It has been eroded by streams and small alpine glaciers. Partway up the right-hand side is a more recent subsidiary volcanic cone, called Shastina, with a sharp crater rim.*

c...**Advanced erosion of a shield volcano** *Waimaea Canyon, over 760 m (about 2500 ft) deep, has been eroded into the flank of an extinct shield volcano on Kauai, Hawaii. Gently sloping layers of basaltic lava flows are exposed in the canyon walls.*

b...**Volcanic neck** *Ship Rock, New Mexico, is a volcanic neck enclosed by a weak shale formation. The peak rises about 520 m (about 1700 ft) above the surrounding plain. In the foreground and to the left, wall-like dikes extend far out from the central peak.*

Shield volcanoes show erosion features that are quite different from those of stratovolcanoes. Figure 18.24*a* shows the first stage of erosion of Hawaiian shield volcanoes, beginning in the upper diagram with the active volcano and its central depression. These are initial landforms. Radial consequent streams cut deep canyons into the flanks of the extinct shield volcano (*b*), and these canyons are opened out into deep, steep-walled amphitheaters such as Waimea Canyon, on the island of Kauai (Figure 18.23*c*). In the last stages (*c*), the original surface of the shield volcano is entirely obliterated, leaving a rugged mountain mass made up of sharp-crested divides and deep canyons.

A Look Ahead With this chapter, we conclude our series of three chapters on running water as a landforming-making agent. Because the landscapes of most regions on the Earth's land surface are produced by fluvial processes acting on differing rock types, running water is by far the most important agent in creating the variety of landforms that we see around us. The three remaining agents of denudation are waves, wind, and glacial ice. These are the subjects of the final two chapters of Part 4. In the first, we describe how waves shape the landforms of beaches and coasts, and how wind shapes the dunes of coasts and deserts. In the second, we relate how glacial ice scrapes, pushes, and deposits rock materials to form the distinctive landforms of glaciated regions.

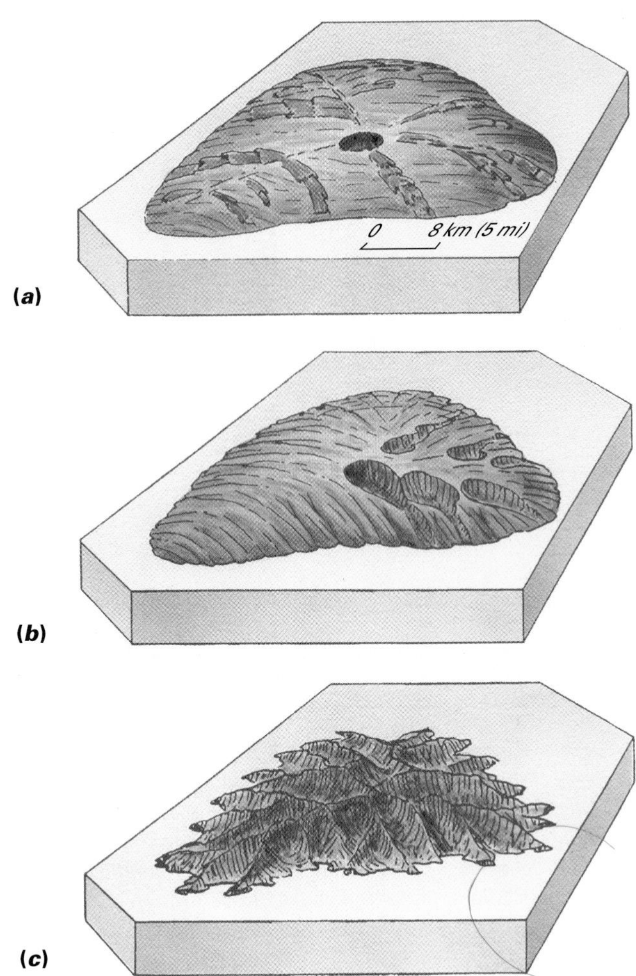

18.24 Erosion of shield volcanoes *Shield volcanoes in various stages of erosion make up the Hawaiian Islands. (a) Newly formed dome with central depression. (b) Early stage of erosion with deeply eroded valley heads. (c) Advanced erosion stage with steep slopes and mountainous relief. (Drawn by A. N. Strahler.)*

Chapter Summary

- This chapter has added a new dimension to the realm of landforms produced by fluvial denudation—the variety and complexity introduced by differences in rock composition and crustal structure. Because the various kinds of rocks offer different degrees of resistance to the forces of denudation, they exert a controlling influence on the shapes of landforms. We have seen, too, that streams carving up a land mass are controlled to a high degree by the structure of the rock on which they act. In this way, distinctive drainage patterns evolve.
- In arid regions of horizontal strata, resistant rock layers produce vertical *cliffs* separated by gentler slopes on less resistant rocks. The **drainage pattern** is *dendritic*. Where strata are gently dipping, as on a **coastal plain**, the more resistant rock layers stand out as **cuestas**, interspersed with lowland valleys on weaker rocks. *Consequent streams* cut across the cuestas toward the ocean, and *subsequent streams* follow the valleys. This forms a *trellis drainage pattern.*
- Where rock layers are arched upward into a **dome**, erosion produces a circular arrangement of rock layers outward from the center of the dome. Resistant strata form *hogbacks*, and weaker rocks form lowlands. Igneous rocks are often revealed in the center. The drainage pattern is *annular*.
- In fold belts, the sequence of *synclines* and *anticlines* brings a linear pattern of rock layers to the surface. Resistant strata form ridges, and weaker strata form valleys. Like the coastal plain, the drainage pattern will be trellised.
- Faulting provides an initial surface along which rock layers are moved—the *fault scarp*. This feature can persist as a fault-line scarp long after the initial scarp is gone. In metamorphic belts, weak layers of shale, slate, and marble underlie valleys, while gneiss and schist form uplands. Quartzite stands out as a ridge-former.
- Exposed batholiths are often composed of uniform, resistant igneous rock. They erode to form a dendritic drainage pattern. **Monadnocks** of intrusive igneous rock stand up above a plain of weaker rocks.
- Stratovolcanoes produce lava flows that initially follow valleys but are highly resistant to erosion. After erosion of the surrounding area, they can remain as highlands, lava ridges, or mesas. At the last stages of erosion, all that remains of stratovolcanoes are necks and dikes. Hawaiian shield volcanoes are eroded by streams that form deeply carved valleys with steeply sloping heads. Eventually, these merge to produce a landscape of steep slopes and knifelike ridges.

Key Terms

plateau
mesa
butte
drainage pattern
coastal plain
cuesta
sedimentary dome
ridge-and-valley landscape
monadnock

Review Questions

1. Why is there often a direct relationship between landforms and rock structure? How are geologic structures formed deep within the Earth exposed at the surface?
2. Which of the following types of rocks—shale, limestone, sandstone, conglomerate, igneous rocks—tends to form lowlands? uplands?
3. How are the tilt and orientation of a natural rock plane measured?
4. How are coastal plains formed? Identify and describe two landforms found on coastal plains. What drainage pattern is typical of coastal plains?
5. What types of landforms are associated with sedimentary domes? How are they formed? What type of drainage pattern would you expect for a dome and why? Provide an example of an eroded sedimentary dome and describe it briefly.
6. How does a ridge-and-valley landscape arise? Explain the formation of the ridges and valleys. What type of drainage pattern is found on this landscape? Where in the United States can you find a ridge-and-valley landscape?
7. What type of landform(s) might you expect to find in the presence of a fault? Why?
8. Which of the following types of rocks—schist, slate, quartzite, marble, gneiss—tend to form lowlands? uplands?
9. Do shield volcanoes erode differently from stratovolcanoes? How so?

Focus on Systems 18.3 • A Model Denudation System

1. What is isostasy? Explain isostatic compensation and why it occurs.
2. How do rates of uplift and denudation compare? How is the net rate of denudation affected by isostatic compensation?
3. What principle explains how the mean elevation of a land mass and the net rate of its denudation change with time?

Visualizing Exercises

1. Identify and sketch a typical landform of flat-lying rock layers found in an arid region.
2. What types of landforms and drainage patterns develop on batholiths? Provide a sketch of a batholith. What is the difference between a batholith and a monadnock?
3. Describe and sketch the stages through which a landscape of stratovolcanoes is eroded. What distinctive landforms remain as the last remnants?

Essay Questions

1. Imagine the following sequence of sedimentary strata—sandstone, shale, limestone, and shale. What landforms would you expect to develop in this structure if the sequence of beds is (*a*) flat-lying in an arid landscape; (*b*) slightly tilted as in a coastal plain; (*c*) folded into a syncline and an anticline in a fold belt; (*d*) fractured and displaced by a normal fault? Use sketches in your answer.
2. A region of ancient mountain roots, now exposed at the surface, includes a central core of plutonic rocks surrounded on either side by belts of metamorphic rocks. What landforms would you expect for this landscape and why?

Problems

Working It Out 18.2 • Properties of Stream Networks

1. A sixth-order stream network has an average bifurcation ratio of 3.5. How many stream segments would you expect for streams of order 1–6?
2. A fourth-order stream network has an average mean segment length ratio of 3.2, and the mean length of first-order stream segments is 0.12 km. What are the expected cumulative mean lengths for stream segments of order 2–4? Note that the (noncumulative) mean length for order u is simply the difference between the cumulative length at order u and the cumulative length at order u-1. That is, $L_u = L^*_u - L^*_{u-1}$. Using this relation, derive the expected (noncumulative) mean lengths for segments of orders 2–4.

Eye on the Landscape

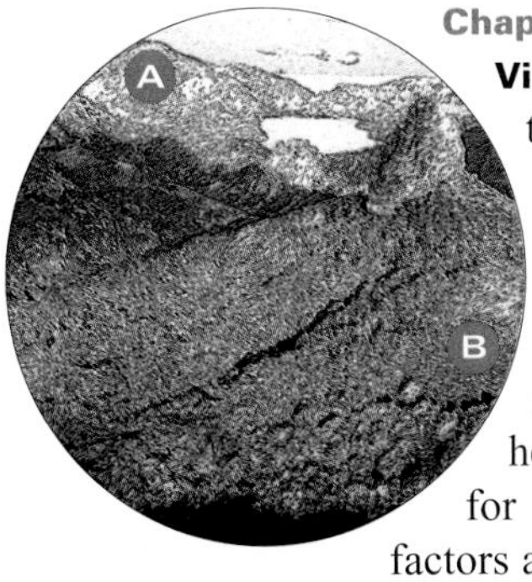

Chapter Opener Mountains of Villa Traful Note the assymetry of the vegetation on the hills and slopes of this landscape **(A)**. The beech seems to favor the southwest slopes, which are cooler and moister here in the southern hemisphere. (See Chapter 23 for more details on geomorphic factors affecting ecosystems.) At **(B)**, there is an interesting striped pattern to the shrubs on the slope. This is probably due to grazing, with the animals following trails along the contour. Grazing is an interaction among species that is covered in Chapter 23.

18.10 Hogbacks The city of Boulder **(A)** stands out from the surrounding region because of its green vegetation, which is uncharacteristic of the semiarid plain east of the Front Range. The city relies on water from Rocky Mountain runoff, some of which is actually transported by tunnel from the wetter west side of the range. Beyond Boulder are a series of alluvial fans where rivers exit from the mountain mass **(B)**. They, too, are better vegetated due to both irrigation and readily available ground water. Irrigation of arid lands and ground water are topics of Chapter 16.

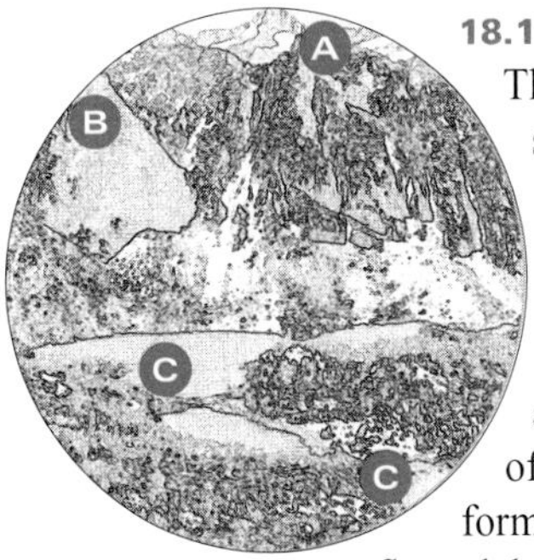

18.17 Uplifted fault block This photo of the dramatic fault scarp at the eastern edge of the Sierra Nevada shows a landscape with abundant evidence of glacial action. Note the knife-edge ridges and peaks that crown the top of the range **(A)**. These were formed as ice capping the range flowed down and out of the region. The entire basin in the foreground, from the near peaks **(B)** to the lakes **(C)**, was filled with ice that scoured loose rock from the basin, leaving an irregular topography of hills and depressions that are now filled with water. Chapter 20 describes landforms created by moving ice.

19 The Work of Waves and Wind

Sand dune crossing a road, Nile Valley, Egypt. This is a barchan dune, which is highly mobile, moving from right to left. Note how the power pole projecting from the dune has been extended upward. In the distance, small longitudinal dunes are visible.

In Chapter 7, we saw how the rotation of the Earth on its axis, combined with unequal solar heating of the Earth's surface, produces a system of global winds in which air moves as a fluid across the Earth's surface. This motion of air produces a frictional drag that can move surface materials. When winds blow over broad expanses of water, waves are generated. These waves then expend their energy at coastlines, eroding rock, moving sediment, and creating landforms. When winds blow over expanses of soil or alluvium that are not protected by vegetation, fine particles can be picked up and carried long distances before coming to rest. Coarser particles, such as sand, can also be moved to build landforms, such as sand dunes. Because wave action on beaches provides abundant sand, we often find landforms of both wind and waves near coastlines.

This chapter describes the processes of erosion and deposition, as well as the landforms that result when wind power moves surface materials either directly as wind or indirectly as waves. Unlike fluvial action or mass wasting, wind and breaking waves can move materials uphill against the force of gravity. For this reason, wind power is a unique land-forming agent. Another agent that acts in coastal regions is the tide—the rhythmic rise and fall in sea level that is generated by the gravitational attraction of the Sun and Moon. This subject is also covered in our chapter.

Throughout this chapter we will use the term **shoreline** to mean the shifting line of contact between water and land. The broader term **coastline,** or simply **coast,** refers to a zone in which coastal processes operate or have a strong influence. The coastline includes the shallow water zone in which waves perform their work, as well as beaches and cliffs shaped by waves, and coastal dunes. Also often present are **bays**—bodies of water that are sheltered by the configuration of the coast from strong wave action. Where a river empties into an ocean

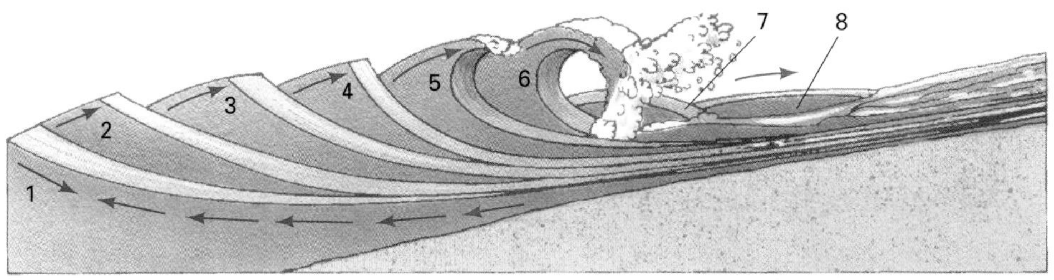

19.1 A breaking wave *As the wave approaches the beach (1–3), it steepens (4–5) and finally falls forward (6–7), rushing up the beach slope (8). (After W. M. Davis.)*

bay, the bay is termed an **estuary.** In an estuary, fresh and ocean water mix, creating a unique habitat for many plants and animals that is neither fresh water nor ocean.

As you read this chapter, keep in mind that the coastlines of large lakes exhibit many features and processes similar to those of oceanic coastlines. Given strong winds and a long stretch of open water, waves of oceanic intensity can build and attack the shoreline, much like storm waves on the ocean. Wave action, coupled with rivers flowing into lakes, can also provide abundant sediment—sand and silt—for windborne transportation and deposition in the nearshore environment. In North America, the Great Lakes provide many examples of coastal processes independent of the Atlantic and Pacific shores.

The Work of Waves

The most important agent shaping coastal landforms is wave action. The energy of waves is expended primarily in the constant churning of mineral particles and water as waves break at the shore. This churning erodes shoreline materials, moving the shoreline landward. But, as we will see shortly, the action of waves and currents can also move sediment along the shoreline for long distances. This activity can build beaches outward as well as form barrier islands just off shore.

Waves travel across the deep ocean with little loss of energy. When waves reach shallow water, the drag of the bottom slows and steepens the wave. However, the wave top maintains its forward velocity and eventually falls down onto the face of the wave, creating a *breaker* (Figure 19.1). Many tons of water surge forward, riding up the beach slope.

Where weak or soft materials—various kinds of regolith, such as alluvium—make up the coastline, the force of the forward-moving water alone easily cuts into the coastline. Here, erosion is rapid, and the shoreline may recede rapidly. Under these conditions, a steep bank—a *marine scarp*—is the typical coastal landform (Figure 19.2*a*). It retreats steadily under attack of storm waves. 

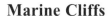

Marine Cliffs

Where a **marine cliff** lies within reach of the moving water, it is impacted with enormous force. Rock fragments of all sizes, from sand to cobbles, are carried by the surging water and thrust against the bedrock of the cliff. The impact breaks away new rock fragments, and the cliff is undercut at the base. In this way, the cliff erodes shoreward, maintaining its form as it retreats. The retreat of a marine cliff formed of hard bedrock is exceedingly slow, when judged in terms of a human life span.

Figure 19.3 illustrates some details of a typical marine cliff. A deep basal indentation, the *wave-cut notch,* marks the line of most intense wave erosion. The waves find points of weakness in the bedrock and penetrate deeply to form crevices and *sea caves.* Where a more resistant rock mass projects seaward, it may be cut through to form a picturesque *sea arch.* After an arch collapses, a rock column, known as a *stack,* remains (Figure 19.2*b*). Eventually, the stack is toppled by wave action and is leveled. As the sea cliff retreats landward, continued wave abrasion forms an *abrasion platform* (Figure 19.2*c*). This sloping rock floor continues to be eroded and widened by abrasion beneath the breakers. If a beach is present, it is little more than a thin layer of gravel and cobblestones atop the abrasion platform.

Beaches

Where sand is in abundant supply, it accumulates as a thick, wedge-shaped deposit, or **beach.** Beaches absorb the energy of breaking waves. During short periods of storm activity, the beach is cut back, and sand is carried offshore a short distance by the heavy wave action. However, the sand is slowly returned to the beach during long periods when waves are weak. In this way, a beach may retain a fairly stable but alternating configuration over many years' time.

Although beaches are most often formed of particles of fine to coarse quartz sand, some beaches are built from rounded pebbles or cobbles. Still others are formed from fragments of volcanic rock (Hawaii) or even shells (Florida).

Beaches are shaped by alternate landward and seaward currents of water generated by breaking waves. After a breaker has collapsed, a foamy, turbulent sheet of water rides up the beach slope. This *swash* is a powerful surge that causes a landward movement of sand and gravel on the beach. When the force of the swash has been spent against the slope of the beach, a return flow, or *backwash,* pours down the beach (Figure 19.4*a*). This "undercurrent" or "undertow," as it is popularly called, can be strong enough to sweep unwary bathers off their feet and carry them seaward beneath the next oncoming breaker.

Littoral Drift

The unceasing shifting of beach materials with swash and backwash of breaking waves also results in a sidewise move-

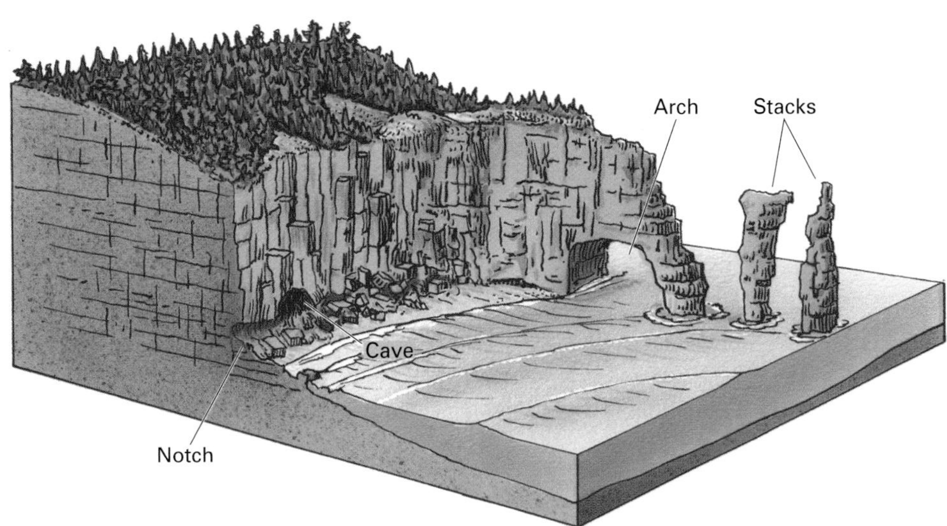

19.3 Landforms of sea cliffs *Wave action undercuts the marine cliff, maintaining its form. (Drawn by E. Raisz.)*

ment known as *beach drift* (Figure 19.4*a*). Wave fronts usually approach the shore at less than a right angle, so that the swash and its burden of sand ride obliquely up the beach. After the wave has spent its energy, the backwash flows down the slope of the beach in the most direct downhill direction. The particles are dragged directly seaward and come to rest at a position to one side of the starting place. This movement is repeated many times, and individual rock particles travel long distances along the shore. Multiplied many thousands of times to include the numberless particles of the beach, beach drift becomes a very significant form of sediment transport.

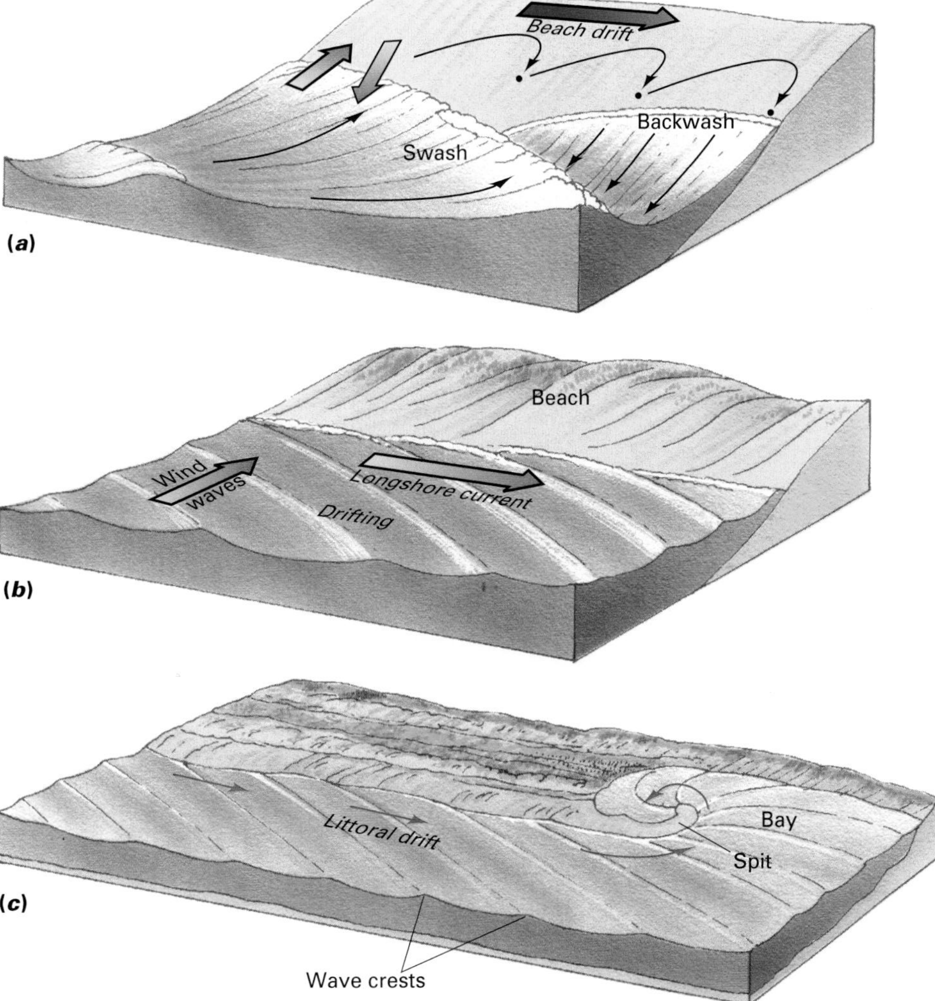

19.4 How waves move sediment by littoral drift *(a) Swash and backwash move particles along the beach in beach drift. (b) Waves set up a longshore current that move particles by longshore drift. (c) Littoral drift, produced by these two processes, creates a sandspit.*

19.2 Landforms of Coastal Erosion

a...*Eye on the Landscape*
Retreating shoreline *Mohegan Bluffs, Block Island, Rhode Island. This marine scarp of Ice Age sediments is being rapidly eroded, threatening historic Southeast Lighthouse. In 1993 the lighthouse was moved to a safer spot 73.5 m (245 ft) away.* **What else would the geographer see?...Answers at the end of the chapter.**

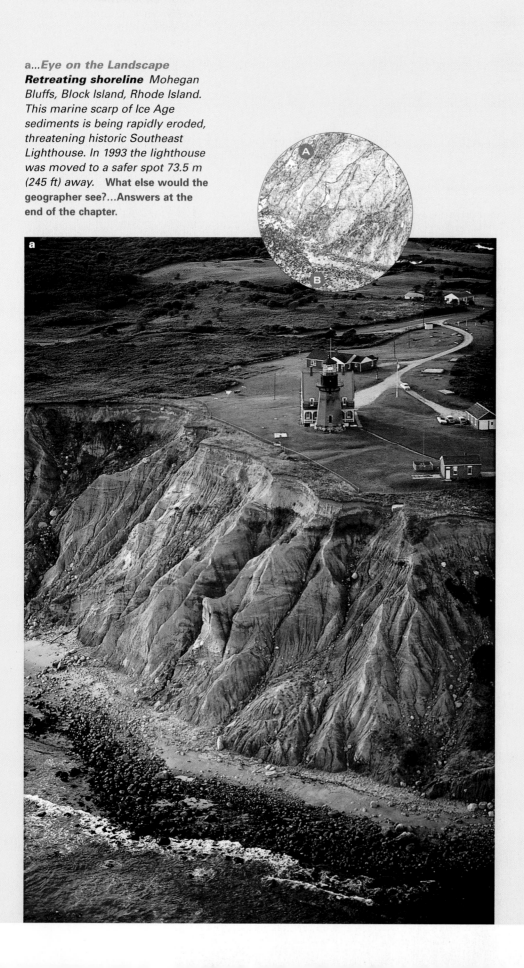

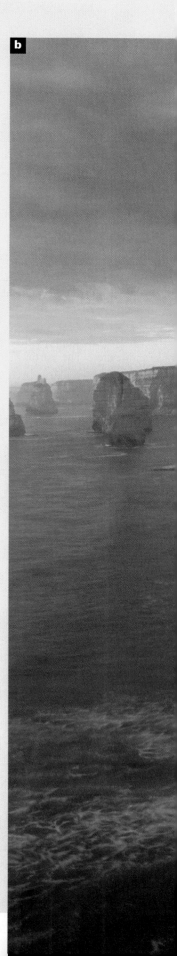

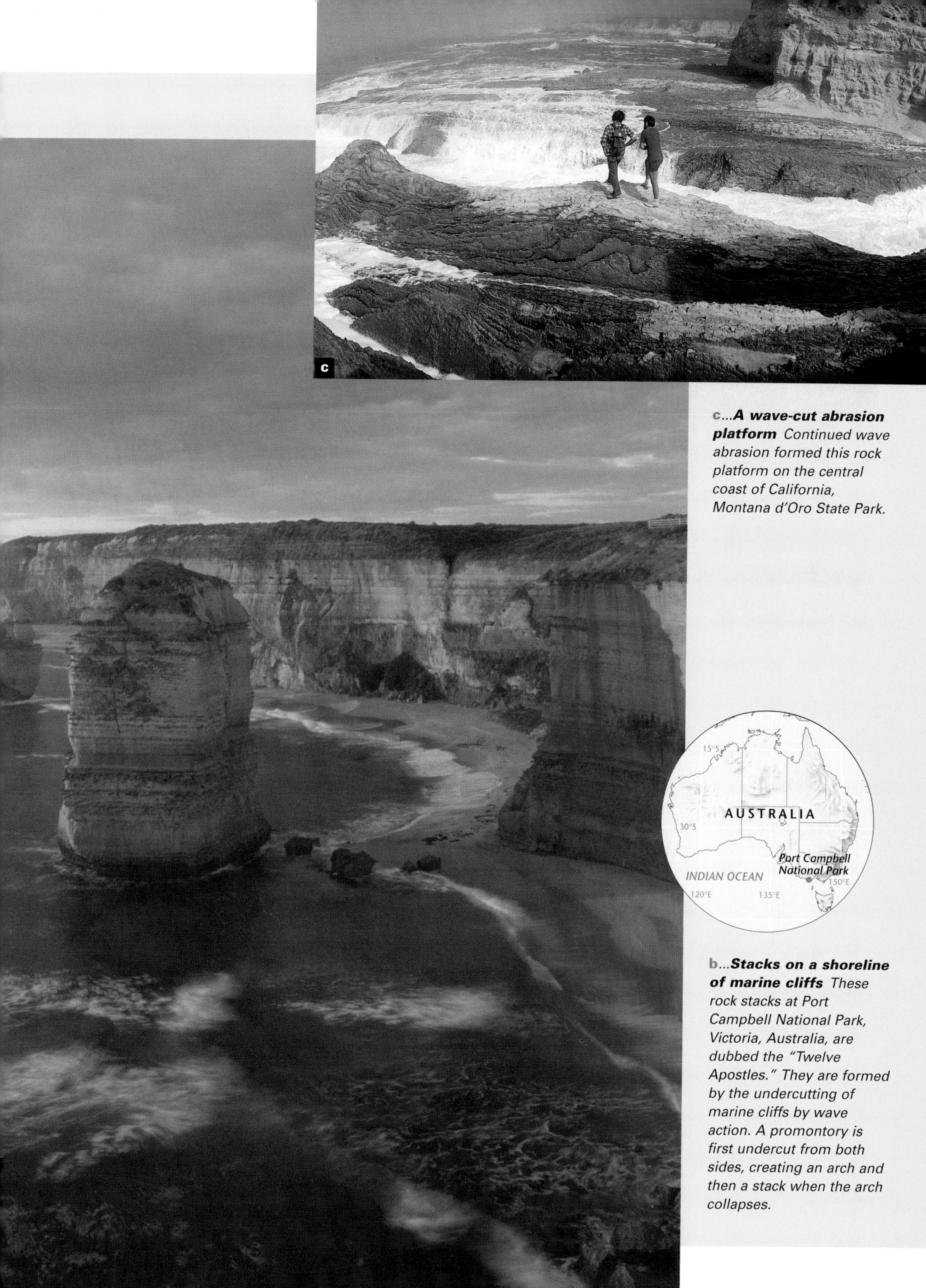

c...*A wave-cut abrasion platform* *Continued wave abrasion formed this rock platform on the central coast of California, Montana d'Oro State Park.*

b...*Stacks on a shoreline of marine cliffs* *These rock stacks at Port Campbell National Park, Victoria, Australia, are dubbed the "Twelve Apostles." They are formed by the undercutting of marine cliffs by wave action. A promontory is first undercut from both sides, creating an arch and then a stack when the arch collapses.*

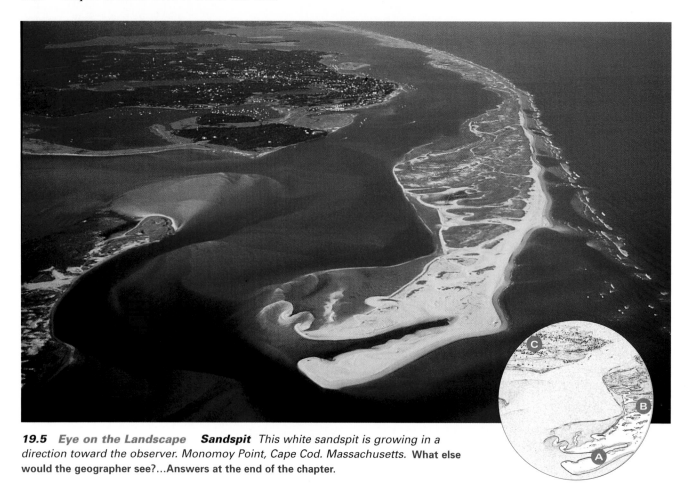

19.5* Eye on the Landscape** ***Sandspit *This white sandspit is growing in a direction toward the observer. Monomoy Point, Cape Cod. Massachusetts.* **What else would the geographer see?...Answers at the end of the chapter.**

Sediment in the shore zone is also moved along the beach in a related, but different, process. When waves approach a shoreline at an angle to the beach, a current is set up parallel to the shore in a direction away from the wind. This is known as a *longshore current* (Figure 19.4*b*). When wave and wind conditions are favorable, this current is capable of carrying sand along the sea bottom. The process is called *longshore drift.* Beach drift and longshore drift, acting together, move particles in the same direction for a given set of onshore winds. The total process is called **littoral drift** (Figure 19.4*c*. ("Littoral" means "pertaining to a coast or shore.")

Littoral drift operates to shape shorelines in two quite different situations. Where the shoreline is straight or broadly curved for many kilometers at a stretch, littoral drift moves the sand along the beach in one direction for a given set of prevailing winds. This situation is shown in Figure 19.4*c*. Where a bay exists, the sand is carried out into open water as a long finger, or *sandspit* (Figure 19.5). As the sandspit grows, it forms a barrier, called a *bar,* across the mouth of the bay.

A second situation is shown in Figure 19.6. Here the coastline consists of prominent headlands, projecting seaward, and deep bays. Approaching wave fronts slow when the water becomes shallow, and this slowing effect causes the wave front to wrap around the headland. High, wave-cut cliffs develop shoreward of an abrasion platform. Sediment from the eroding cliffs is carried by littoral drift along the sides of the bay, converging on the head of the bay. The result is a crescent-shaped beach, often called a *pocket beach.* GEODISCOVERIES

Beach erosion and deposition is one part of a larger system involving the movement of sediment provided by streams through the shallow water zone by littoral drift. In *Focus on Systems 19.1 • The Coastal Sediment Cell as a Matter Flow System,* we provide a systems view of this sediment transport.

19.6 Pocket beaches *On an embayed coast, sediment is carried from eroding headlands to the bayheads, where pocket beaches are formed. (Copyright © A. N. Strahler.)*

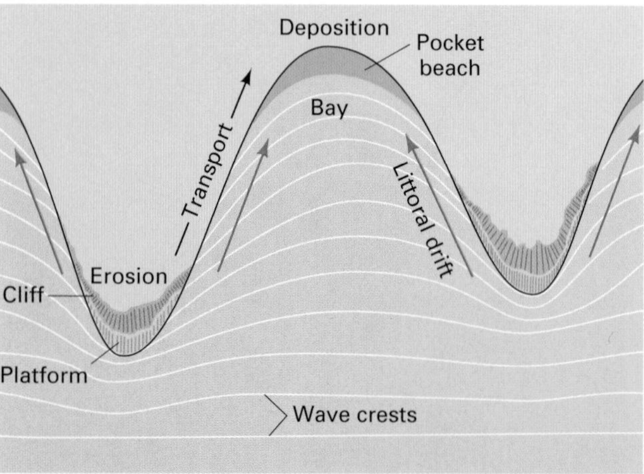

Focus on Systems | 19.1

The Coastal Sediment Cell as a Matter Flow System

Regular beachgoers are familiar with the changes in beaches that occur from day to day and season to season. During languid summer days of light winds and waves, the beach builds outward toward the ocean and establishes a steep face exposed to the rise and fall of the waves with the tide. Afternoon sea breezes move sand grains along the beach and landward into the dunes nearby.

With the storms and surf of winter, the waves work the beach intensively, scouring sand away from the beach and building offshore bars. Waves break farther out, and their swash traverses a long span of beach. At high tides, storm waves attack coastal cliffs, undercutting and releasing sediment or loose fragments of bedrock. Beyond the breakers, a strong littoral drift current, powered by winter's larger waves, moves large quantities of sediment parallel to the shore.

If you walk the beach for some distance, you will also notice changes. The beach will be very narrow at headlands and promontories, where wave action is intense and littoral drift quickly moves sediment away. In sheltered locations, littoral drift provides sediment that accumulates in pocket beaches. At the mouth of a stream or river, the beach broadens in response to an influx of sediment, and as you walk in the direction of littoral drift, the beach continues to be wider for some distance beyond the river mouth.

What you cannot see, however, is the presence of submarine canyons crossing the continental shelf and extending out to great depth. Sediment moving offshore in the littoral drift zone is carried into these canyons, where currents of turbid, sediment-laden water carry the sediment to undersea fans and cones deposited along the continental rise (see Figure 13.9). Many submarine canyons were cut by streams during the Ice Age, when sea level was 100 meters or more below present levels.

The movement of sediment along a typical stretch of the North American Pacific coast can be described as an open matter flow system, which we term here a coastal sediment cell. The figure below is a sketch of a coastal sediment cell. Wave approach at an angle to the beach generates a littoral drift that moves sediment parallel to the beach. Cliff erosion provides a source of sediment sufficient for narrow beaches to form in the direction of the drift current. Streams provide the major portion of the sediment, however, with very little derived from cliff erosion in most locations. Beach sediment forms a storage pool that is increased in summer and reduced in winter. Coupled to the beach is another pool of sediment in storage—sand in dunes that are built by wind and eroded by waves in intense storms. The system output is a flow of sediment through the submarine canyon to the continental rise.

The matter flow system of the coastal sediment cell underscores the importance of the flow of sediment from rivers in nourishing beaches. Many coastal communities have experienced beach recession because the rivers that feed their coastlines have been dammed, trapping sediment upstream and keeping it out of the flow system. With less sediment flowing through the coastal system as a whole, less is available for storage on the beach, and rock cliffs are more rapidly cut back. The California coast, with so many of its rivers dammed for water supply, provides many good examples of this accelerated wastage of cliffs and the loss of increasing numbers of oceanside homes each year.

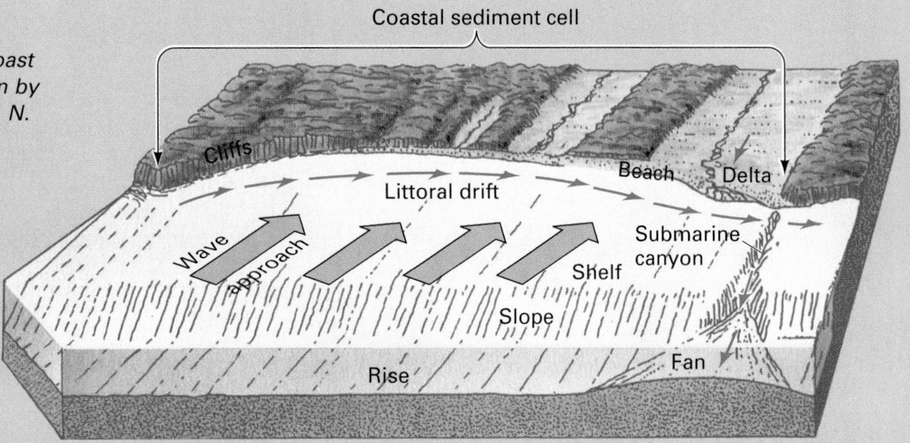

Costal sediment cell *A pictorial diagram of a coastal sediment cell typical of the coast of southern California. (Drawn by A. N. Strahler. Copyright © A. N. Strahler.)*

Littoral Drift and Shore Protection

When sand arrives at a particular section of the beach more rapidly than it is carried away, the beach is widened and built oceanward. This change is called **progradation** (building out). When sand leaves a section of beach more rapidly than it is brought in, the beach is narrowed and the shoreline moves landward. This change is called **retrogradation** (cutting back).

Along stretches of shoreline affected by retrogradation, the beach may be seriously depleted or even entirely disappear, destroying valuable shore property. In some circumstances, structures can be installed that will cause progradation, and so build a broad, protective beach. This is done by installing groins at close intervals along the beach. A *groin* is simply a wall or embankment built at right angles to the shoreline. It may be constructed of large rock masses, of concrete, or of wooden pilings. The groins act to trap sediment moving along the shore as littoral drift (Figure 19.7).

In some cases, the source of beach sand is sediment delivered to the coast by a river. Construction of dams far upstream on the river may drastically reduce the sediment load of the river, cutting off the source of sand for littoral drift. Retrogradation can then occur on a long stretch of shoreline.

Tidal Currents

Most marine coastlines are influenced by the *ocean tide,* a rhythmic rise and fall of sea level under the influence of changing attractive forces of Moon and Sun on the rotating Earth. Where tides are great, the effects of changing water level and the tidal currents thus set in motion are of major importance in shaping coastal landforms.

The tidal rise and fall of water level is graphically represented by the *tide curve.* Figure 19.8 is a tide curve for Boston Harbor covering a day's time. The water reached its maximum height, or high water, at 3.6 m (11.8 ft) on the tide staff, and then fell to its minimum height, or low water, at 0.8 m (2.6 ft), occurring about $6^{1}/_{4}$ hours later. A second high water occurred about $12^{1}/_{2}$ hours after the previous high water, completing a single tidal cycle. In this example, the tidal range, or difference between heights of successive high and low waters, is 2.8 m (9.2 ft).

In bays and estuaries, the changing tide sets in motion currents of water known as *tidal currents.* The relationships between tidal currents and the tide curve are shown in Figure 19.9. When the tide begins to fall, an *ebb current* sets in. This flow ceases about the time when the tide is at its lowest point. As the tide begins to rise, a landward current, the *flood current,* begins to flow. GEODISCOVERIES

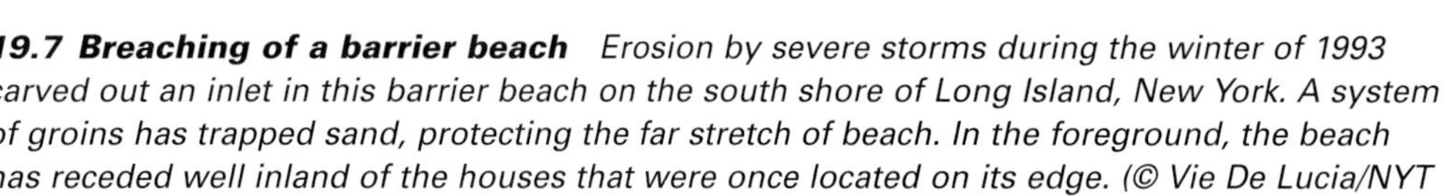

19.7 Breaching of a barrier beach *Erosion by severe storms during the winter of 1993 carved out an inlet in this barrier beach on the south shore of Long Island, New York. A system of groins has trapped sand, protecting the far stretch of beach. In the foreground, the beach has receded well inland of the houses that were once located on its edge. (© Vie De Lucia/NYT Pictures.)*

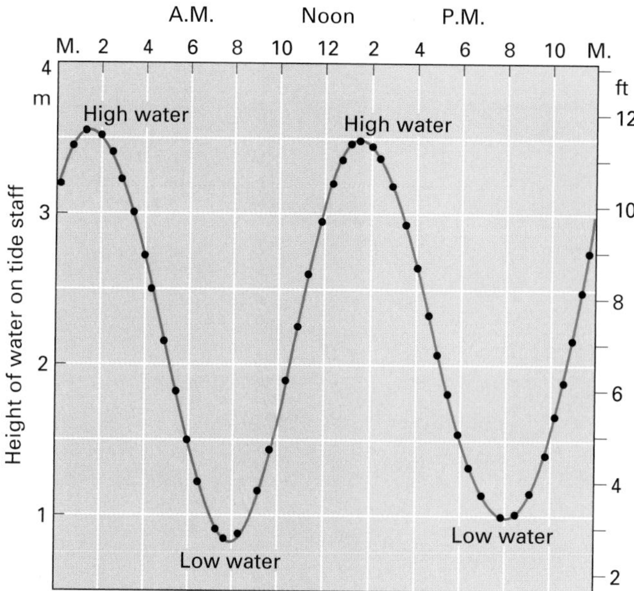

19.8 Tide curve *Height of water at Boston Harbor measured every half hour.*

Tidal Current Deposits

Ebb and flood currents generated by tides perform several important functions along a shoreline. First, the currents that flow in and out of bays through narrow inlets are very swift and can scour the inlet strongly. This keeps the inlet open, despite the tendency of shore-drifting processes to close the inlet with sand.

19.9 Tidal currents *The ebb current flows seaward as the tide level falls. The flood current flows landward as the tide level rises. Tidal currents are strongest when water level is changing most rapidly, in the middle of the cycle.*

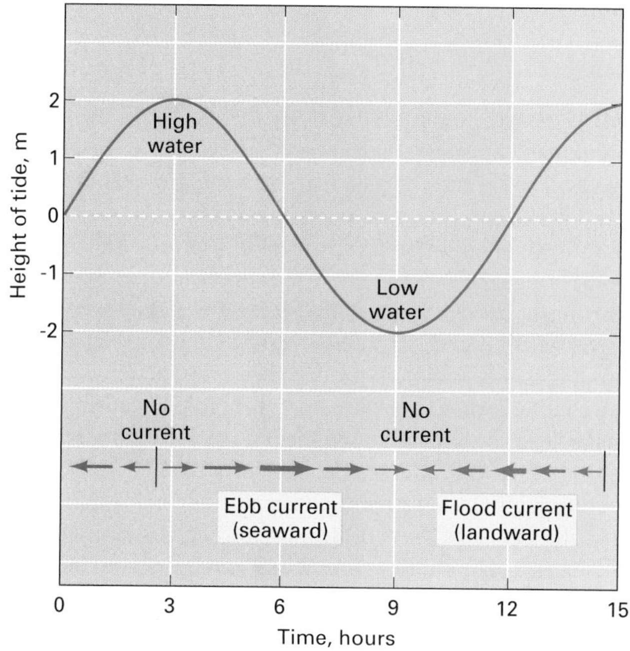

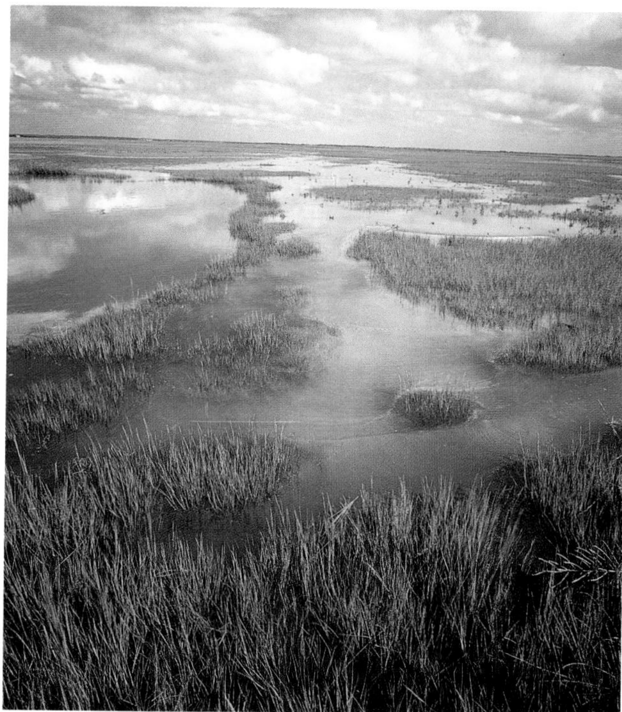

19.10 A tideland salt marsh *At high tide, most of this salt marsh is flooded. Chincoteague National Wildlife Refuge, Virginia.*

Second, tidal currents carry large amounts of fine silt and clay in suspension. This fine sediment is derived from streams that enter the bays or from bottom muds agitated by storm wave action. It settles to the floors of the bays and estuaries, where it accumulates in layers and gradually fills the bays. Much organic matter is present in this sediment.

In time, tidal sediments fill the bays and produce mud flats, which are barren expanses of silt and clay. They are exposed at low tide but covered at high tide. Next, a growth of salt-tolerant plants takes hold on the mud flat. The plant stems trap more sediment, and the flat is built up to approximately the level of high tide, becoming a *salt marsh* (Figure 19.10). A thick layer of peat is eventually formed at the surface. Tidal currents maintain their flow through the salt marsh by means of a highly complex network of winding tidal streams.

Types of Coastlines

The world's coastlines present a number of different coastline types. Each type is unique because of the distinctive land mass against which the ocean water has come to rest. One group of coastline types derives its qualities from **submergence,** the partial drowning of a coast by a rise of sea level or a sinking of the crust. Another group derives its qualities from **emergence,** the exposure of submarine landforms by a falling of sea level or a rising of the crust. Another group of coastline types results when new land is built out into the ocean by volcanoes and lava flows, by the growth of river deltas, or by the growth of coral reefs.

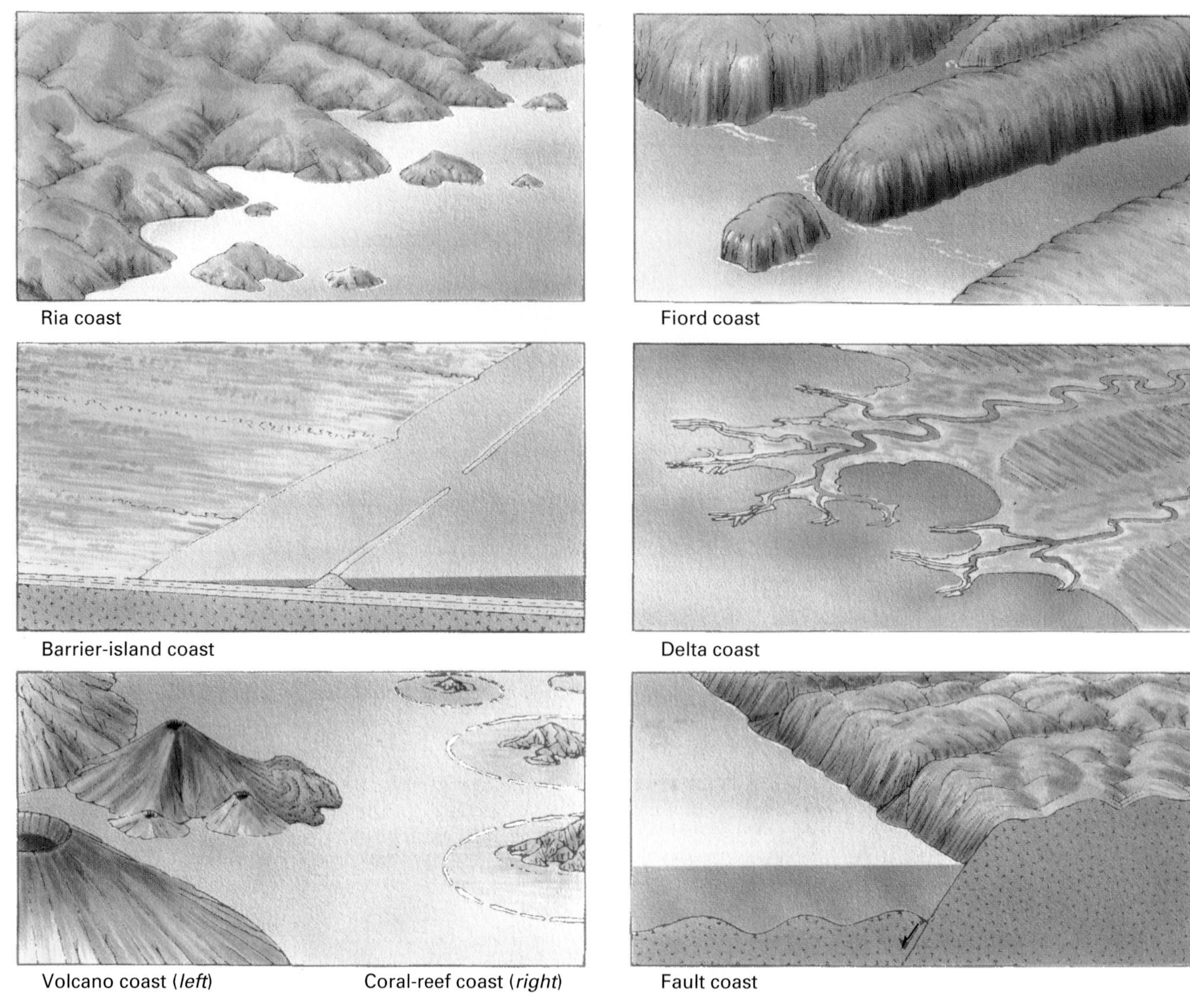

19.11 Seven common kinds of coastlines *These examples have been selected to illustrate a wide range in coastal features. (Drawn by A. N. Strahler.)*

Several important types of coastlines are illustrated in Figure 19.11. The first two are the result of submergence. The *ria coast* is a deeply embayed coast resulting from submergence of a land mass dissected by streams. The *fiord coast* is deeply indented by steep-walled fiords, which are submerged glacial troughs (discussed further in Chapter 20). The *barrier-island coast* is associated with a recently emerged coastal plain. Here, the offshore slope is very gentle, and a barrier island of sand, lying a short distance from the coast, is created by wave action. Large rivers build elaborate deltas, producing *delta coasts.* The *volcano coast* is formed by the eruption of volcanoes and lava flows, partly constructed below water level. Reef-building corals create new land and make a *coral-reef coast.* Down-faulting of the coastal margin of a continent can allow the shoreline to come to rest against a fault scarp, producing a *fault coast.*

Shorelines of Submergence

Shorelines of submergence include ria coasts and fiord coasts. The ria coast, which takes its name from the Spanish word for estuary, *ria,* has many offshore islands. A ria coast is formed when a rise of sea level or a crustal sinking (or both) brings the shoreline to rest against the sides of valleys previously carved by streams (Figure 19.12*a*). Wave attack forms cliffs on the exposed seaward sides of islands and headlands (*b*). Sediment produced by wave action accumulates in the form of beaches along the cliffed headlands and at the heads of bays. This sediment is carried by littoral drift and is often built into sandspits across bay mouths and as connecting links between islands and mainland (*c*). Eventually, the sandspits seal off the bays, forming estuaries (*d*). If sea level remains at the same height with respect to the land for a long time, the coast may evolve into a cliffed shoreline of narrow beaches (*e*).

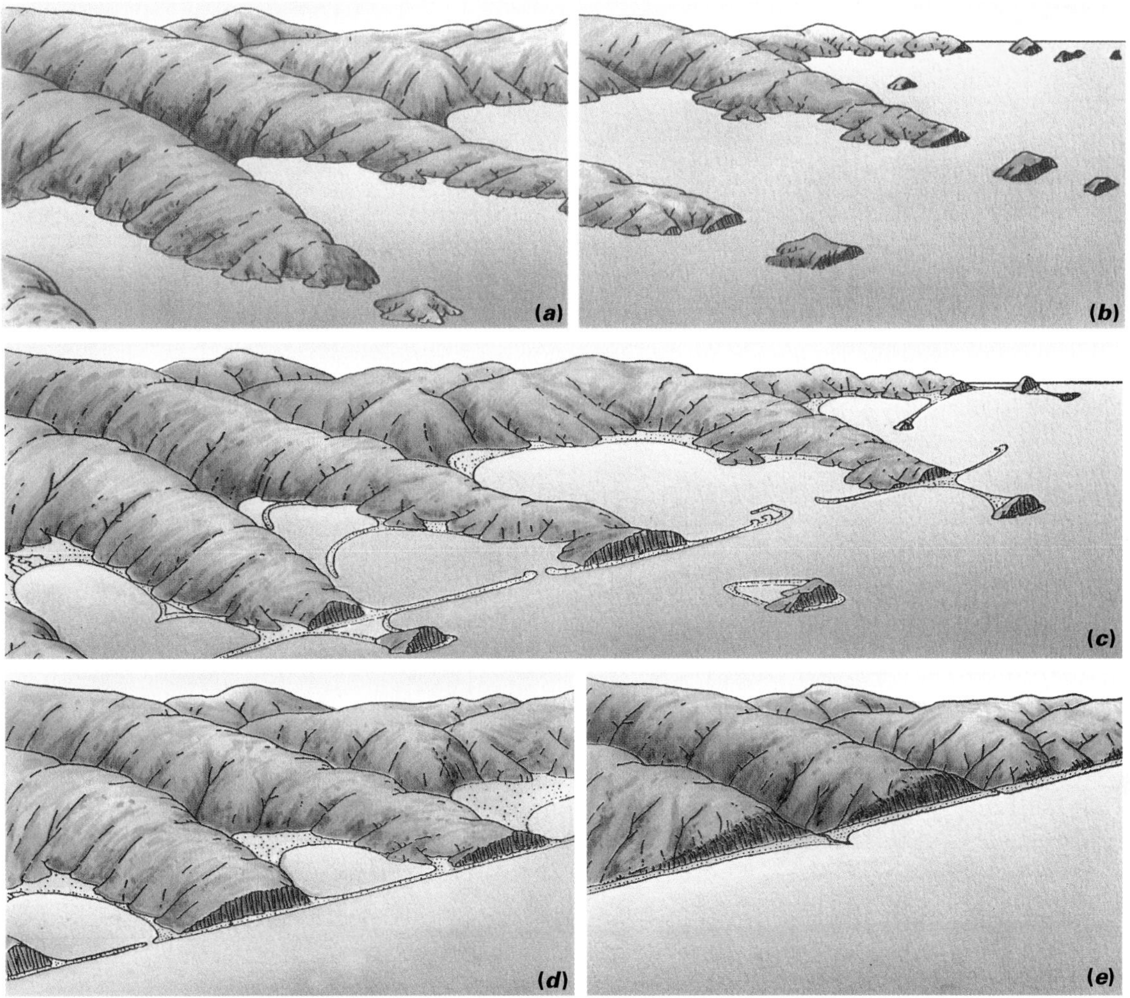

19.12 Stages in the evolution of a ria coastline *Wave action on a shoreline of submergence creates many interesting coastal landforms. (Drawn by A. N. Strahler.)*

The fiord coast is similar to the ria coast. However, the submerged valleys were carved by flowing glaciers instead of streams. As a result, the valleys are deep, with straight, steep sides. Because sediment rapidly sinks into the deep water, beaches are rare. Fiords are described further in Chapter 20 (see Figures 20.7 and 20.8).

Many glaciated coastlines, such as those of New England and the Canadian Maritime Provinces, are shorelines of submergence. During the Ice Age, the weight of ice sheets depressed the crust of these regions, and they are still slowly rising now that the ice sheets are gone. Meanwhile, ocean waves and currents are rapidly eroding their rocky shorelines, creating bays, bars, and estuaries, as shown in Figure 19.12.

Barrier-Island Coasts

In contrast to the bold relief and deeply embayed outlines of coastlines of submergence are low-lying coasts from which the land slopes gently beneath the sea. The coastal plain of the Atlantic and Gulf coasts of the United States presents a particularly fine example of such a gently sloping surface. As we explained in Chapter 18, this coastal plain is a belt of relatively young sedimentary strata, formerly accumulated beneath the sea as deposits on the continental shelf. During the latter part of the Cenozoic Era and into recent time, the coastal plain emerged from the ocean as a result of repeated crustal uplifts.

Along much of the Atlantic Gulf coast there exist *barrier islands,* low ridges of sand built by waves and further increased in height by the growth of sand dunes (Figure 19.13). Behind the barrier island lies a *lagoon.* It is a broad expanse of shallow water in places largely filled with tidal deposits.

A characteristic feature of barrier islands is the presence of gaps, known as *tidal inlets.* Strong currents flow alternately landward and seaward through these gaps as the tide rises and falls. In heavy storms, the barrier island may be breached by new inlets (Figure 19.7). After that occurs, tidal current scour will tend to keep a new inlet open. In some cases, the inlet is closed later by shore drifting of beach sand. Perhaps the finest example of a barrier-island coast is the Gulf coast of Texas (Figure 19.14).

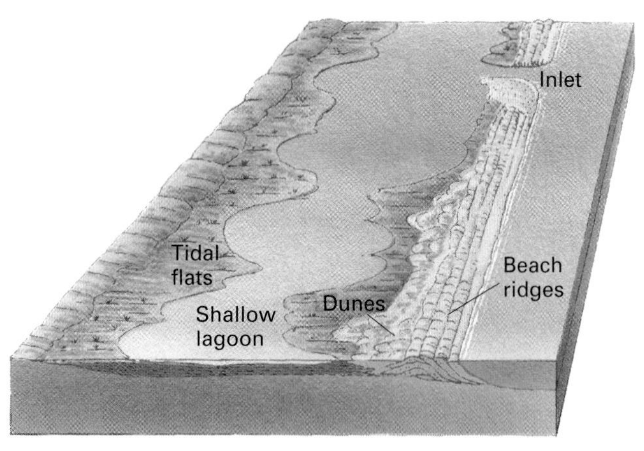

19.13 Barrier island *A barrier island is separated from the mainland by a wide lagoon. Sediments fill the lagoon, while dune ridges advance over the tidal flats. An inlet allows tidal flows to pass in and out of the lagoon. (Drawn by A. N. Strahler.)*

Delta and Volcano Coasts

The deposit of clay, silt, and sand made by a stream or river where it flows into a body of standing water is known as a **delta.** Deposition is caused by rapid reduction in velocity of the current as it pushes out into the standing water. Typically, the river channel divides and subdivides into lesser channels called *distributaries.* The coarser sand and silt particles settle out first, while the fine clays continue out farthest and eventually come to rest in fairly deep water. Contact of fresh with saltwater causes the finest clay particles to clot together and form larger particles that settle to the sea floor.

Deltas show a wide variety of outlines. The Nile delta has the basic triangular shape of the Greek letter delta. In outline, it resembles an alluvial fan. The Mississippi delta has a different shape. Long, branching fingers grow far out into the Gulf of Mexico at the ends of the distributaries, giving the impression of a bird's foot. A satellite image of the delta, Figure 19.15, shows the great quantity of suspended sediment—clay and fine silt—being discharged by the river into the Gulf. It amounts to about 1 million metric tons per day.

19.14 Barrier-island coast *(a) A Landsat image of the Texas barrier-island coast. Red colors on the Landsat image indicate vegetation. Bright white areas are dunes or beach sand. Lighter blue tones mark the presence of sediment in water inside of the barrier beach. (b) Index map of the location.*

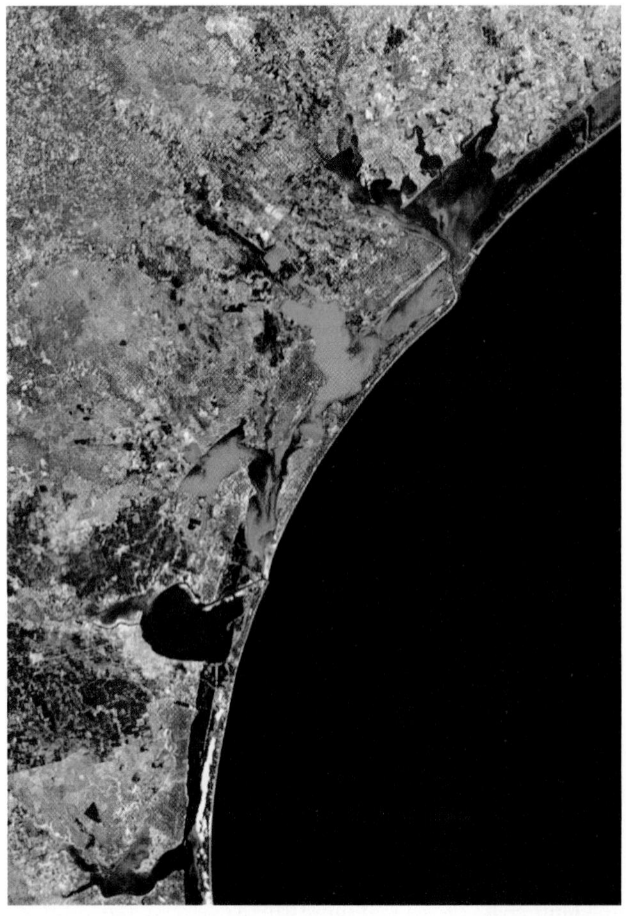

(a)

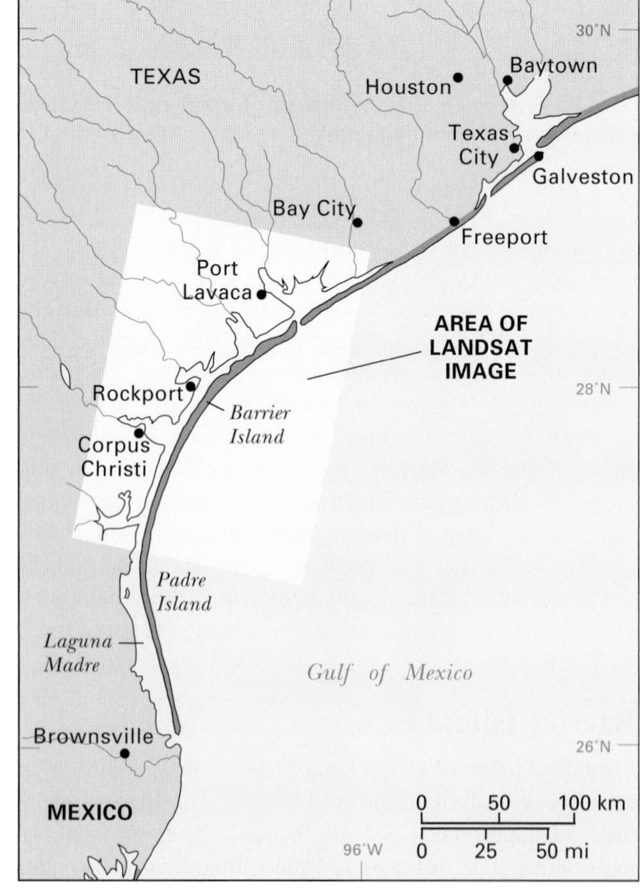

(b)

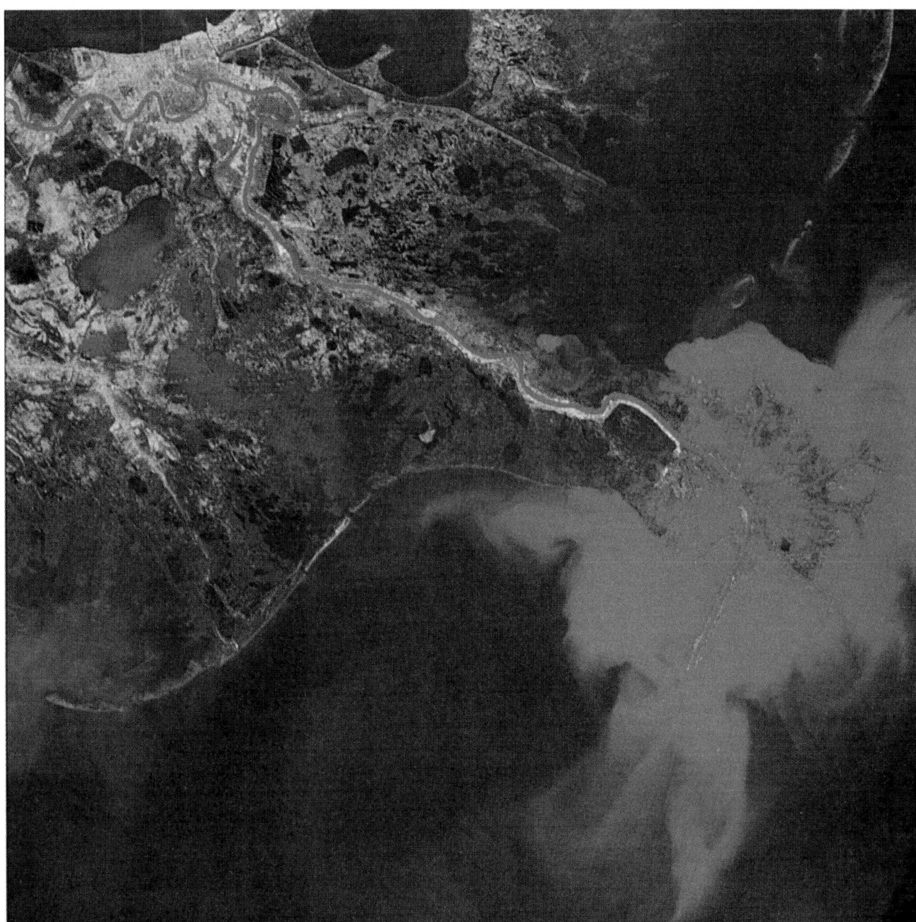

19.15 The Mississippi delta viewed by Landsat *The natural levees of the bird-foot delta appear as lace-like filaments in a great pool of turbid river water. New Orleans can be seen at the upper left, occupying the region between the natural levees of the Mississippi River and the southern shore of Lake Pontchartrain.*

Delta growth is often rapid, ranging from 3 m (about 10 ft) per year for the Nile to 60 m (about 200 ft) per year for the Mississippi delta. Some cities and towns that were at river mouths several hundred years ago are today several kilometers inland.

Volcano coasts arise where volcanic deposits—lava and ash—flow from active volcanoes into the ocean. Low cliffs occur when wave action erodes the fresh deposits. Beaches are typically narrow, steep, and composed of fine particles of the extrusive rock.

Coral-Reef Coasts

Coral-reef coasts are unique in that the addition of new land is made by organisms—corals and algae. Growing together, these organisms secrete rocklike deposits of mineral carbonate, called **coral reefs.** As coral colonies die, new ones are built on them, accumulating as limestone. Coral fragments are torn free by wave attack, and the pulverized fragments accumulate as sand beaches.

Coral-reef coasts occur in warm, tropical and equatorial waters between the limits of lat. 30° N and 25° S. Water temperatures above 20°C (68°F) are necessary for dense growth of coral reefs. Reef corals live near the water surface. The sea water must be free of suspended sediment and well aerated for vigorous coral growth. For this reason, corals thrive in positions exposed to wave attack from the open sea. Because muddy water prevents coral growth, reefs are missing opposite the mouths of muddy streams. Coral reefs are remarkably flat on top. They are exposed at low tide and covered at high tide.

There are three distinctive types of coral reefs—fringing reefs, barrier reefs, and atolls. *Fringing reefs* are built as platforms attached to shore (Figure 19.16). They are widest in front of headlands where wave attack is strongest. *Barrier reefs* lie out from shore and are separated from the mainland by a lagoon (Figure 19.16). Narrow gaps occur at intervals in barrier reefs. Through these openings, excess water from breaking waves is returned from the lagoon to the open sea.

Atolls are more or less circular coral reefs enclosing a lagoon but have no land inside. On large atolls, parts of the reef have been built up by wave action and wind to form low island chains connected by the reef. Most atolls are built on a foundation of volcanic rock that has subsided below sea level. (See *Focus on Systems 14.1 • The Life Cycle of a Hotspot Volcano.*)

Raised Shorelines and Marine Terraces

The active life of a shoreline is sometimes cut short by a sudden rise of the coast. When this tectonic event occurs, a *raised shoreline* is formed. If present, the marine cliff and abrasion platform are abruptly raised above the level of

19.16 Fringing reefs *Coral reefs fringe the Island of Moorea, Society Islands, South Pacific Ocean. The island is a deeply dissected volcano with a history of submergence. Tahiti lies in the background.*

wave action. The former abrasion platform has now become a *marine terrace* (Figure 19.17). Of course, fluvial denudation acts to erode the terrace as soon as it is formed. The terrace may also undergo partial burial under alluvial fan deposits.

Raised shorelines are common along the continental and island coasts of the Pacific Ocean because here tectonic processes are active along the mountain and island arcs. Repeated uplifts result in a series of raised shorelines in a steplike arrangement. Fine examples of these multiple marine terraces are seen on the western slope of San Clemente Island, off the California coast (Figure 19.18).

Rising Sea Level

As we noted in our interchapter feature, *A Closer Look: Eye on Global Change 5.5 • The IPCC Report of 2001,* global warming will result in a rise of sea level estimated at 9 to 88 cm (3.5 to 34.6 in.) between now and 2100. Some of the rise is due to the simple thermal expansion of the upper layers of the ocean, which will grow warmer. The remainder is contributed by the melting of glaciers and snowpacks as air temperatures rise. Sea-level rise will have effects ranging from displacement of estuaries to enhanced coastal erosion. Depending on the amount of rise, some low-lying islands will disappear along with their inhabitants. The causes and effects of sea level rise are documented more fully in our interchapter feature, *A Closer Look: Eye on Global Change 19.3 • Global Change and Coastal Environments.*

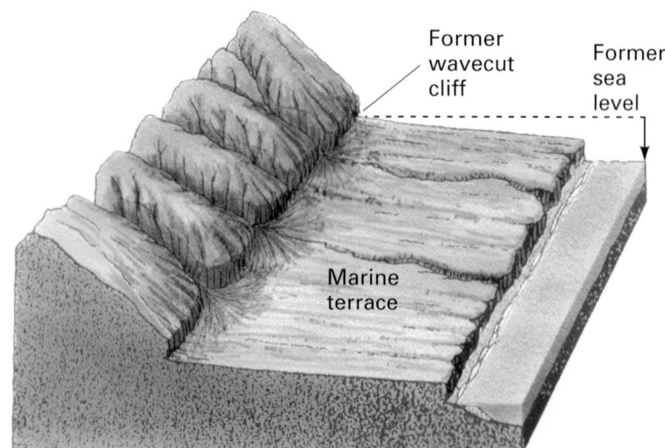

19.17 A marine terrace *A raised shoreline becomes a cliff parallel with the newer, lower shoreline. The former abrasion platform is now a marine terrace. (Drawn by A. N. Strahler.)*

19.18 A raised shoreline *Marine terraces on the western slope of San Clemente Island, off the southern California coast. More than 20 terraces have been identified in this series. The highest has an elevation of about 400 m (about 1300 ft).*

WIND ACTION

Transportation and deposition of sand by wind is an important process in shaping certain coastal landforms. We have already mentioned coastal sand dunes, which are derived from beach sand. In the remainder of this chapter, we investigate the transport of mineral particles by wind and the shaping of dune forms. Our discussion also provides information about dune forms far from the coast—in desert environments, where the lack of vegetation cover allows dunes to develop if a source of abundant sand particles is present.

Wind blowing over the land surface is one of the active agents of landform development. Ordinarily, wind is not strong enough to dislodge mineral matter from the surfaces of unweathered rock, or from moist, clay-rich soils, or from soils bound by a dense plant cover. Instead, the action of wind in eroding and transporting sediment is limited to land surfaces where small mineral and organic particles are in a loose, dry state. These areas are typically deserts and semiarid lands (steppes). An exception is the coastal environment, where beaches provide abundant supplies of loose sand. In this environment, wind action shapes coastal dunes, even where the climate is humid and the land surface inland from the coast is well protected by a plant cover.

Erosion by Wind

Wind performs two kinds of erosional work: abrasion and deflation. Loose particles lying on the ground surface may be lifted into the air or rolled along the ground by wind action. In the process of *wind abrasion,* wind drives sand and dust particles against an exposed rock or soil surface. This causes the surface to be worn away by the impact of the particles. Abrasion requires cutting tools—mineral particles—carried by the wind, while deflation is accomplished by air currents alone.

The sandblasting action of wind abrasion against exposed rock surfaces is limited to the basal meter or two of a rock mass that rises above a flat plain. This height is the limit to which sand grains can rise high into the air. Wind abrasion produces pits, grooves, and hollows in the rock. Wooden utility poles on windswept sandy plains are quickly cut through at the base unless a protective metal sheathing or heap of large stones is placed around the base.

The removal of loose particles from the ground is termed **deflation.** Deflation acts on loose soil or sediment. Dry river courses, beaches, and areas of recently formed glacial deposits are susceptible to deflation. In dry climates, much of the ground surface is subject to deflation because the soil or rock is largely bare of vegetation.

Wind is selective in its deflational action. The finest particles, those of clay and silt sizes, are lifted and raised into the air—sometimes to a height of a thousand meters (about 3300 ft) or more. Sand grains are moved only when winds are at least moderately strong and usually travel within a meter or two (about 3 to 6 ft) of the ground. Gravel fragments and rounded pebbles can be rolled or pushed over flat ground by strong winds, but they do not travel far. They

19.19 Desert pavement *A desert pavement is formed of closely fitted rock fragments. Lying on the surface are fine examples of wind-faceted rocks, which attain their unusual shapes by long-continued sandblasting.*

become easily lodged in hollows or between other large grains. Consequently, where a mixture of size of particles is present on the ground, the finer sized particles are removed and the coarser particles remain behind.

A landform produced by deflation is a shallow depression called a **blowout.** The size of the depression may range from a few meters (10 to 20 ft) to a kilometer (0.6 mi) or more in diameter, although it is usually only a few meters deep. Blowouts form in plains regions of dry climate. Any small depression in the surface of the plain, especially where the grass cover has been broken or disturbed, can form a blowout. Rains fill the depression and create a shallow pond or lake. As the water evaporates, the mud bottom dries out and cracks, leaving small scales or pellets of dried mud. These particles are lifted out by the wind.

Deflation is also active in semidesert and desert regions. In the southwestern United States, playas often occupy large areas on the flat floors of tectonic basins (see Figure 17.29*b*). Deflation has reduced many playas several meters in elevation.

Rainbeat, overland flow, and deflation may be active for a long period on the gently sloping surface of a desert alluvial fan or alluvial terrace. These processes remove fine particles, leaving coarser, heavier materials behind. As a result, rock fragments ranging in size from pebbles to small boulders become concentrated into a surface layer known as a **desert pavement** (Figure 19.19). The large fragments become closely fitted together, concealing the smaller particles—grains of sand, silt, and clay—that remain beneath. The pavement acts as an armor that effectively protects the finer

19.20 Eye on the Landscape Dust storm *A cloud of fine dust sweeps across this savanna plain in eastern Kenya.* **What else would the geographer see?...Answers at the end of the chapter.**

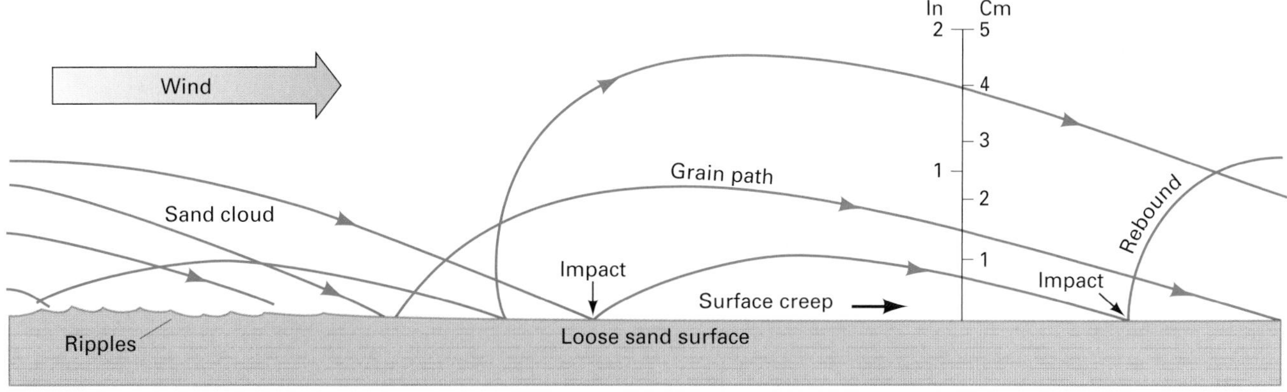

19.21 Saltation *Sand particles travel in a series of long leaps. (After R. A. Bagnold.)*

particles from rapid removal by deflation. However, the pavement is easily disturbed by the wheels of trucks and motorcycles, exposing the finer particles and allowing severe deflation and water erosion to follow.

Dust Storms

Strong, turbulent winds blowing over barren surfaces lift great quantities of fine dust into the air, forming a dense, high cloud called a **dust storm.** In semiarid grasslands, a dust storm is generated where ground surfaces have been stripped of protective vegetation cover by cultivation or grazing. Strong winds cause soil particles and coarse sand grains to hop along the ground. This motion breaks down the soil particles and disturbs more soil. With each impact, fine dust is released that can be carried upward by turbulent winds.

A dust storm approaches as a dark cloud extending from the ground surface to heights of several thousand meters (Figure 19.20). Typically, the advancing cloud wall represents a rapidly moving cold front. Within the dust cloud there is deep gloom or even total darkness. Visibility is cut to a few meters, and a fine choking dust penetrates everywhere.

SAND DUNES

A **sand dune** is any hill of loose sand shaped by the wind. Active dunes constantly change form under wind currents. Dunes form where there is a source of sand—for example, a sandstone formation that weathers easily to release individual grains, or perhaps a beach supplied with abundant sand from a nearby river mouth. Dunes must be free of a vegetation cover in order to form and move. They become inactive when stabilized by a vegetation cover or when patterns of wind or sand sources change.

Dune sand is most commonly composed of the mineral quartz, which is extremely hard and largely immune to chemical decay. The grains are beautifully rounded by abrasion (see Figure 12.12). Figure 19.21 shows how sand grains are moved by strong winds—in long, low leaps, bouncing after impact with other grains. Rebounding grains rarely rise more than half a centimeter above the dune surface. Grains struck by bouncing grains are pushed forward, and, in this way, the surface sand layer creeps downwind. This type of hopping, bouncing movement is termed *saltation.* GEODISCOVERIES

Types of Sand Dunes

One common type of sand dune is an isolated heap of free sand called a *barchan,* or *crescentic dune.* This type of dune has the outline of a crescent, and the points of the crescent are directed downwind (Figure 19.22). On the upwind side of the crest, the sand slope is gentle and smoothly rounded. On the downwind side of the dune, within the crescent, is a steep dune slope, the *slip face.* This face maintains a more or less constant angle from the horizontal (Figure 19.23*a*), which is known as the *angle of repose.* The slip face is oversteepened slightly by the accumulation of individual wind-carried sand grains until it becomes unstable and the outermost layer of sand on the face slips down the dune slope, restoring the angle of repose. For loose sand, this angle is about 35°. *Working It Out 19.2 • Angle of Repose of Dune Sands* describes this

19.22 Barchan dunes *The arrow indicates wind direction. (Drawn by A. N. Strahler.)*

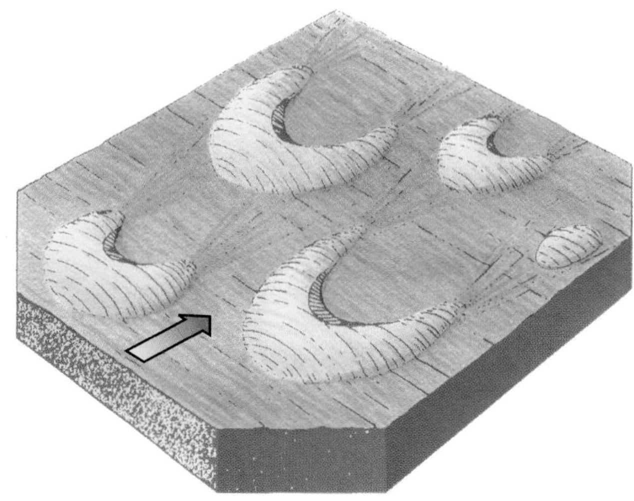

19.23 Dune Gallery

a...***Migrating barchan dunes*** *This aerial view shows a large barchan dune moving from right to left. At its apex is a smaller barchan dune that is overtaking it.*

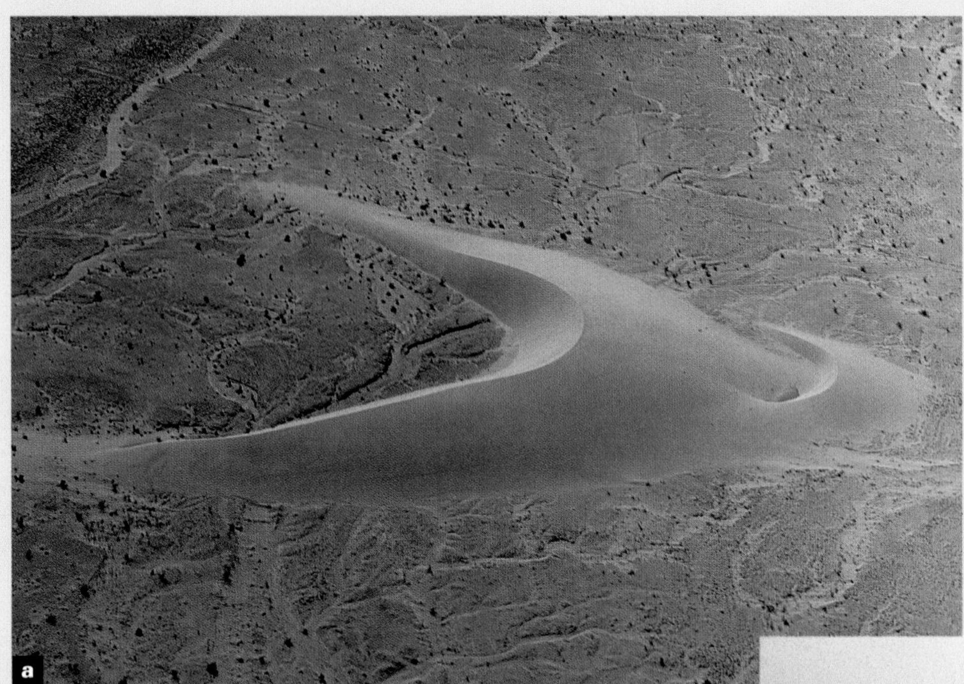

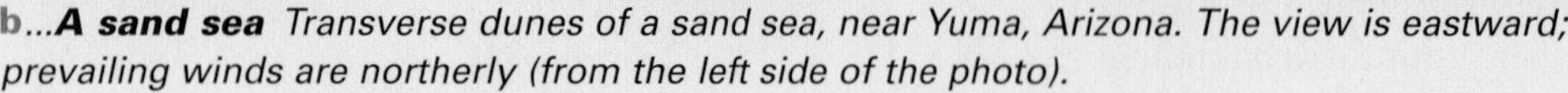

b...***A sand sea*** *Transverse dunes of a sand sea, near Yuma, Arizona. The view is eastward; prevailing winds are northerly (from the left side of the photo).*

c..***Coastal blowout dune*** *This coastal blowout dune is advancing over a coniferous forest, with the slip face gradually burying the tree trunks. Pacific coast, near Florence, Oregon.*

d...***Coastal foredunes*** *Beachgrass thriving on coastal foredunes has trapped drifting sand to produce a dune ridge. Queen's County, Prince Edward Island, Canada.*

Working It Out | 19.2

Angle of Repose of Dune Sands

In North Africa during World War II, German panzer tank forces under General Rommel were driving the British forces eastward. Battles were being fought on difficult desert terrain that included desert pavement as well as large areas of active sand dunes. Sudden, violent sandstorms that might rage for many hours severely tried both men and machines. A British Royal Army Engineer, Major R. A. Bagnold, was sent to Egypt to study this phenomenon. He had just completed in England a laboratory study of the motions of sand grains driven by strong winds. His treatise on dune dynamics, published in 1941 and reprinted in 1954, still remains a classic.

We reproduce here one of Bagnold's illustrations, showing how a new sand dune with a broadly rounded surface is transformed into a barchan dune (Figure 1). Leaping sand grains (Figure 19.21) transfer the sand downwind to form a high, sharp crest. From this crest, the grains leap out on to the steep slip face of the dune, where they come to rest. This activity tends to steepen the slope of the slip face, so that there comes a point when a thin layer of sand grains slides down to the base. In this way the slope is alternately diminished and steepened. Our question now is: What is the maximum slope angle that the grains can hold before the next slip occurs? Does this angle—called the *angle of repose*—vary according to the particle size or to mineral composition of the sand grains? or perhaps to some other factor?

To investigate this question, we perform a lab experiment that makes use of two sand samples. One, called "Ottawa sand," consists of beautifully rounded grains of pure quartz. In fact, they were deposited as dune sands back in Paleozoic time and are now being quarried for industrial uses. The sand has been screened so as to include only a very small range of diameters. Take a good look at these grains—they are shown in Chapter 12 as Figure 12.12. Our second sample consists of a common kind of beach sand scooped up at low tide. Both samples have been passed through a series of screens to eliminate grains larger than 2 mm and smaller than 1 mm, leaving coarse sand.

To carry out our experiment, we use a small box with vertical glass sides and a removable end piece (Figure 2). Sand is poured gently into the box until it is almost full. The end plate is then pulled out, allowing the sand to pour out into a container below. The angle of the sand surface can now be measured through the side glass. Each reading is observed to the nearest one-half degree. Ten such trials were made with the Ottawa sand and ten with the beach sand. The bar graph (Figure 3) shows how frequently each slope angle occurred for each type of sand. There is a clear difference between the two samples, for a wide gap separates the two samples.

Why does the beach sand show the higher mean? Because both samples were carefully screened, difference in size of particles is probably not a factor. But there are other differences. Because the beach sand has been worked by waves as well as wind, it may include grains that are more angular and thus able to resist movement better. The beach sand may also contain mineral particles other than quartz—heavy minerals, perhaps—that have higher density and would move farther downslope before stopping. Also, some of these other minerals might come in sharp-edged grains that serve to pack the mass more tightly together. Only a careful microscopic analysis would tell for sure.

Our two sets of data offer an opportunity to apply some simple mathematical statistics as an example of how physical geographers use statistics to ask and answer questions about data. Consider the bar graph for a moment.

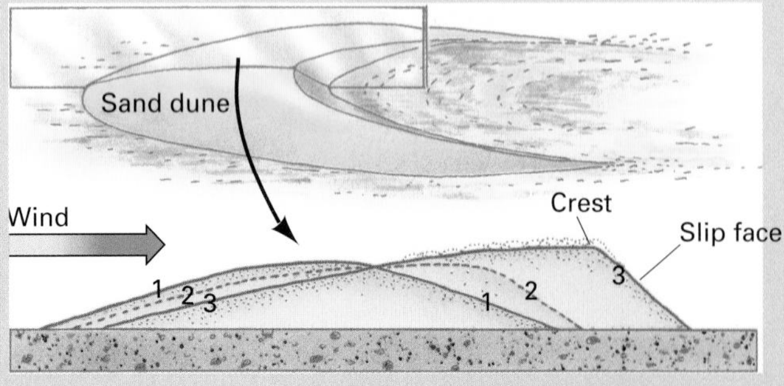

Figure 1 *Growth of a dune and development of a slip face. (After R. A. Bagnold, The Physics of Blown Sand and Desert Dunes, London, Methuen, 1941.)*

Do you suppose it's possible that both types of sand actually do have the same angle of repose but that by some accident of chance the data show they are separated? Statisticians have developed a test to see whether this might be the case—a *t*-test of two means. In this test, a sample statistic, *t*, is calculated. The greater the value of *t*, the more likely it is that the two means are different.

The formula for the *t*-statistic is

$$t = \frac{\left|\overline{X}_1 - \overline{X}_2\right|}{s_{\overline{X}_1 - \overline{X}_2}} \qquad (1)$$

The numerator of this expression is the absolute value (value without respect to sign) of the difference between the means of the two samples, here designated by $\overline{X}_1$ and $\overline{X}_2$. The greater is the absolute difference between the means, the larger will be the value of *t*. The denominator, $s_{\overline{X}_1 - \overline{X}_2}$, is an estimate of the standard deviation of this difference, based on the standard deviations of both samples. It is calculated from the individual sample standard deviations as shown in *Working It Out 11.1 • Standard Deviation and Coefficient of Variation.* The formula for this type of standard deviation is

$$s_{\overline{X}_1 - \overline{X}_2} = \sqrt{\frac{s_1^2 + s_2^2}{n}} \qquad (2)$$

where s_1 and s_2 are the sample standard deviations for the two groups. From the formula for *t* (1), we see that if the sample values have a small standard deviation, that will make the standard deviation of the difference small as well, and that in turn will make *t* larger.

For the comparison between Ottawa sand and beach sand, we have $\overline{X}_1 = 32.6$ and $s_1 = 0.32$, and $\overline{X}_2 = 35.1$ and $s_2 = 0.55$. Finding the standard deviation of the difference from (2),

$$s_{\overline{X}_1 - \overline{X}_2} = \sqrt{\frac{(0.32)^2 + (0.55)^2}{10}} = \sqrt{\frac{0.102 + 0.303}{10}} = \sqrt{0.041} = 0.20$$

Then, applying this to the calculation of *t* in (1), we have

$$t = \frac{\left|\overline{X}_1 - \overline{X}_2\right|}{s_{\overline{X}_1 - \overline{X}_2}} = \frac{|32.6 - 35.1|}{0.20} = \frac{2.5}{0.20} = 12.5$$

Consulting a statistical table, we note that if *t* is greater than 3.610 for the case of two samples of size 10, there is less than 1 chance in 1000 that the means are the same, but the difference has arisen by chance alone. Thus, we can conclude with confidence that the angles of repose of the two sand samples are different.

The *t*-test of two means is just one of many types of statistical tests that geographers use in examining data and drawing conclusions. If you continue your study of physical geography, you will undoubtedly learn more about such statistics and how they are used in geography and studies of the environment.

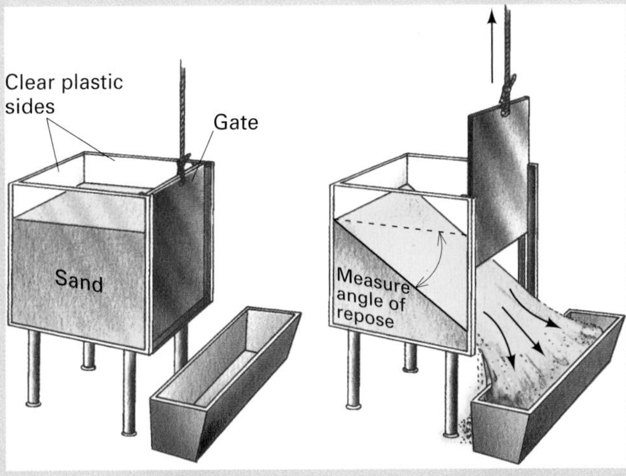

Figure 2 *A simple experiment to measure the angle of repose of sand.*

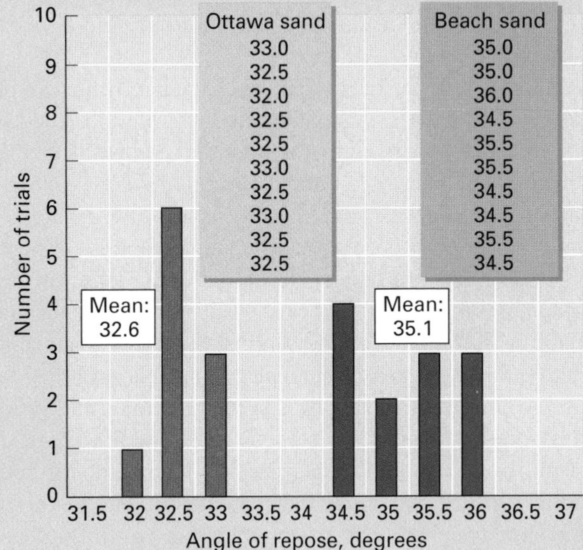

Figure 3 *Frequency graph counting the number of occurrences of repose angles observed for Ottawa and beach sand. (Copyright © A. N. Strahler.)*

process and develops a simple statistical test to determine whether two different samples of sand have different repose angles. GEODISCOVERIES

Barchan dunes usually rest on a flat, pebble-covered ground surface. The life of a barchan dune may begin as a sand drift in the lee of some obstacle, such as a small hill, rock, or clump of brush. Once a sufficient mass of sand has formed, it begins to move downwind, taking on the crescent form. For this reason, the dunes are usually arranged in chains extending downwind from the sand source.

Where sand is so abundant that it completely covers the solid ground, dunes take the form of wave-like ridges separated by trough-like furrows. These dunes are called *transverse dunes* because, like ocean waves, their crests trend at right angles to the direction of the dominant wind (Figure 19.23*a*). The entire area may be called a *sand sea* because it resembles a storm-tossed sea suddenly frozen to immobility. The sand ridges have sharp crests and are asymmetrical, the gentle slope being on the windward and the steep slip face on the lee side. Deep depressions lie between the dune ridges. Sand seas require enormous quantities of sand, supplied by material weathered from sandstone formations or from sands in nearby alluvial plains. Transverse dune belts also form adjacent to beaches that supply abundant sand and have strong onshore winds.

Wind is a major agent of landscape development in the Sahara Desert. Enormous quantities of reddish dune sand have been derived from weathering of sandstone formations. The sand is formed into a great sand sea, called an *erg*. Elsewhere, there are vast flat-surfaced sheets of sand that are armored by a layer of pebbles that forms a desert pavement. A surface of this kind in the Sahara is called a *reg*.

Some of the Saharan dunes are elaborate in shape. For example, the *star dune* (heaped dune), is a large hill of sand whose base resembles a many-pointed star in plan. The Arabian star dunes remain fixed in position and have served for centuries as reliable landmarks for desert travelers. The star dune also occurs in the deserts of the border region between the United States and Mexico.

Another group of dunes belongs to a family in which the curve of the dune crest is bowed outward in the downwind direction. (This curvature is the opposite of the barchan dune.) These are termed *parabolic dunes*. A common type of parabolic dune is the *coastal blowout dune*, formed adjacent to beaches. Here, large supplies of sand are available, and the sand is blown landward by prevailing winds (Figure 19.24*a*). A saucer-shaped depression is formed by deflation, and the sand is heaped in a curving ridge resembling a horseshoe in plan. On the landward side is a steep slip face that advances over the lower ground and buries forests, killing the trees (Figure 19.23*c*). Coastal blowout dunes are well displayed along the southern and eastern shore of Lake Michigan. Dunes of the southern shore have been protected for public use as the Indiana Dunes State Park.

On semiarid plains, where vegetation is sparse and winds are strong, groups of parabolic blowout dunes develop to the lee of shallow deflation hollows (Figure 19.24*b*). Sand is

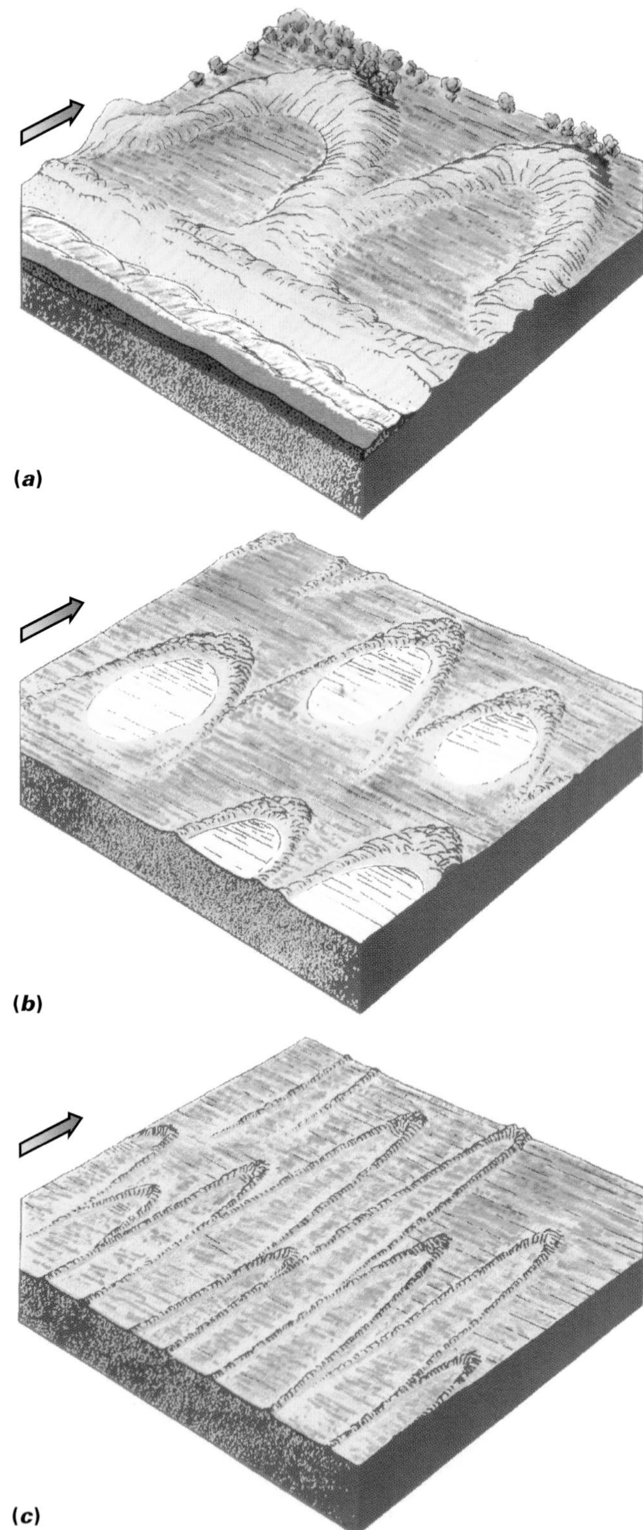

19.24 Three types of parabolic dunes *The prevailing wind direction is the same for all three types. (a) Coastal blowout dunes. (b) Parabolic dunes on a semiarid plain. (c) Parabolic dunes drawn out into hairpin forms. (Drawn by A. N. Strahler.)*

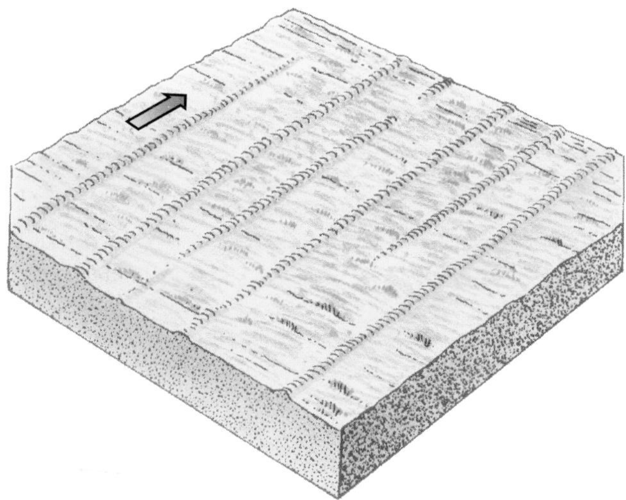

19.25 Longitudinal dunes *Longitudinal dunes run parallel to the direction of the wind. (Drawn by A. N. Strahler.)*

caught by low bushes and accumulates on a broad, low ridge. These dunes have no steep slip faces and may remain relatively immobile. In some cases, the dune ridge migrates downwind, drawing the dune into a long, narrow form with parallel sides resembling a hairpin in outline (Figure 19.24*c*).

Another class of dunes, described as *longitudinal dunes,* consists of long, narrow ridges oriented parallel with the direction of the prevailing wind (Figure 19.25). These dune ridges may be many kilometers long and cover vast areas of tropical and subtropical deserts in Africa and Australia.

Coastal Foredunes

Landward of sand beaches, we usually find a narrow belt of dunes in the form of irregularly shaped hills and depressions. These are the *foredunes.* They normally bear a cover of beachgrass and a few other species of plants capable of survival in the severe environment (Figure 19.23*d*). On coastal foredunes, the sparse cover of beachgrass and other small plants acts as a baffle to trap sand moving landward from the adjacent beach. As a result, the foredune ridge is built upward to become a barrier standing several meters above high-tide level.

Foredunes form a protective barrier for tidal lands that often lie on the landward side of a beach ridge or barrier island. In a severe storm, the swash of storm waves cuts away the upper part of the beach. Although the foredune barrier may then be attacked by wave action and partly cut away, it will not usually yield. Between storms, the beach is rebuilt, and, in due time, wind action restores the dune ridge, if plants are maintained.

If the plant cover of the dune ridge is reduced by vehicular and foot traffic, a blowout will rapidly develop. The new cavity can extend as a trench across the dune ridge. With the onset of a storm that brings high water levels and intense wave action, swash is funneled through the gap and spreads out on the tidal marsh or tidal lagoon behind the ridge. Sand swept through the gap is spread over the tidal deposits. If eroded, the gap can become a new tidal inlet for ocean water to reach the bay beyond the beach. For many coastal communities of the eastern United States seaboard, the breaching of a dune ridge with its accompanying overwash may bring unwanted change to the tidal marsh or estuary.

Loess

In several large midlatitude areas of the world, the surface is covered by deposits of wind-transported silt that has settled out from dust storms over many thousands of years. This material is known as **loess.** (The pronunciation of this German word is somewhere between "lerse" and "luss.") It generally has a uniform yellowish to buff color and lacks any visible layering. Loess tends to break away along vertical cliffs wherever it is exposed by the cutting of a stream or grading of a roadway (Figure 19.26). It is also very easily eroded by running water and is subject to

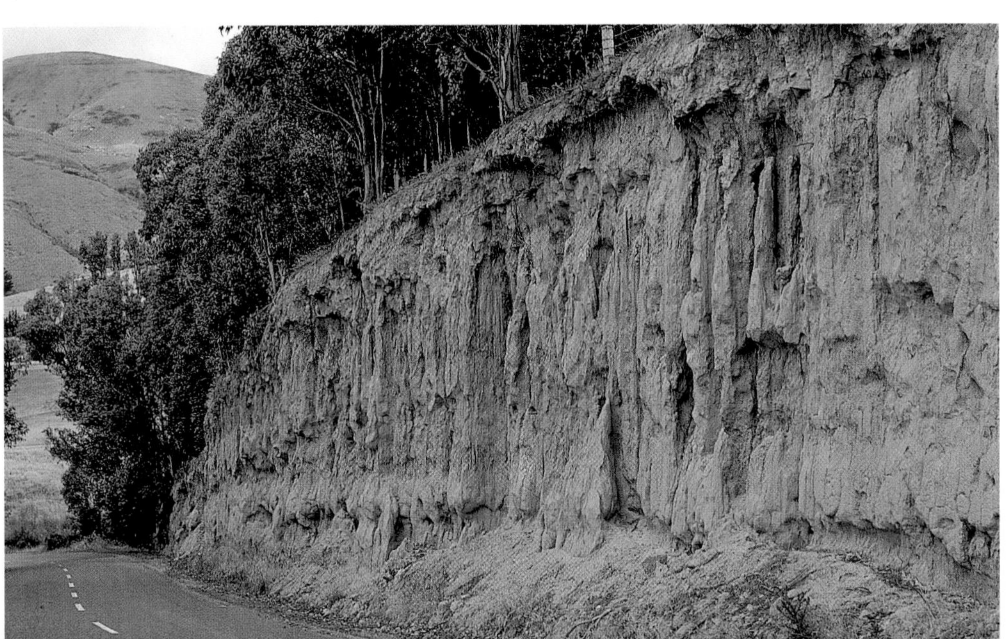

19.26 Wind-transported silt *This thick layer of loess in New Zealand was deposited during the Ice Age. Loess has excellent cohesion and often forms vertical faces as it wastes away.*

rapid gullying when the vegetation cover that protects it is broken. Because it is easily excavated, loess has been widely used for cave dwellings both in China and in Central Europe.

The thickest deposits of loess are in northern China, where a layer over 30 m (about 100 ft) thick is common and a maximum thickness of 100 m (about 300 ft) has been measured. This layer covers many hundreds of square kilometers and appears to have been brought as dust from the interior of Asia. Loess deposits are also of major importance in the United States, Central Europe, Central Asia, and Argentina.

In the United States, thick loess deposits lie in the Missouri-Mississippi Valley (Figure 19.27). Large areas of the prairie plains region of Indiana, Illinois, Iowa, Missouri, Nebraska, and Kansas are underlain by loess ranging in thickness from 1 to 30 m (about 3 to 100 ft). Extensive deposits also occur in Tennessee and Mississippi in areas bordering the lower Mississippi River floodplain. Still other loess deposits are in the Palouse region of northeast Washington and western Idaho.

The American and European loess deposits are directly related to the continental glaciers of the Pleistocene Epoch. At the time when the ice covered much of North America and Europe, a generally dry winter climate prevailed in the land bordering the ice sheets. Strong winds blew southward and eastward over the bare ground, picking up silt from the floodplains of braided streams that discharged the meltwater from the ice. This dust settled on the ground between streams, gradually building up a smooth, level ground surface. The loess is particularly thick along the eastern sides of the valleys because of prevailing westerly winds. It is well exposed along the bluffs of most streams flowing through these regions today.

Loess is of major importance in world agricultural resources. Loess forms the parent matter of rich black soils (Mollisols, Chapter 21) especially suited to cultivation of grains. The highly productive plains of southern Russia, the Argentine pampa, and the rich grain region of north China are underlain by loess. In the United States, corn is extensively cultivated on the loess plains in Kansas, Iowa, and Illinois, where rainfall is sufficient. Wheat is grown farther west on loess plains of Kansas and Nebraska and in the Palouse region of eastern Washington.

The thick loess deposit covering a large area of north-central China in the province of Shanxi and adjacent provinces poses a difficult problem of severe soil erosion. Although the loess is capable of standing in vertical walls, it also succumbs to deep gullying during the period of torrential summer rains. From the steep walls of these great scars, fine sediment is swept into streams and carried into tributaries of the Huang He (Yellow River). The Chinese government has implemented an intensive program of slope stabilization by using artificial contour terraces (seen in Figure 19.28) in combination with tree planting. Valley bottoms have been dammed so as to trap the silt to form flat patches of land suitable for cultivation.

19.27 Map of loess distribution in the central United States *(Data from Map of Pleistocene Eolian Deposits of the United States, Geological Society of America.)*

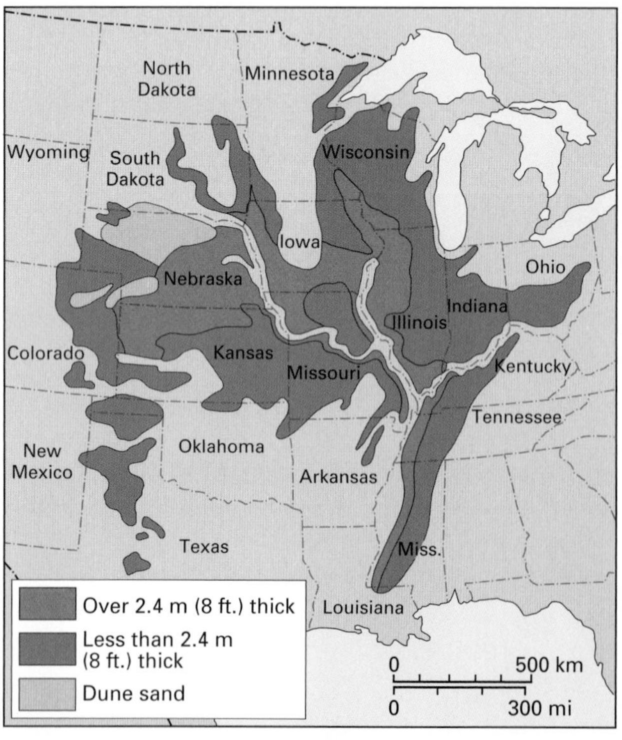

Induced Deflation

Induced deflation is a frequent occurrence when shortgrass prairie in a semiarid region is cultivated without irrigation. Plowing disturbs the natural soil surface and grass cover, and in drought years, when vegetation dies out, the unprotected soil is easily eroded by wind action. Much of the Great Plains region of North America has suffered such deflation, experiencing dust storms generated by turbulent winds. Strong cold fronts frequently sweep over this area and lift dust high into the troposphere at times when soil moisture is low. The "Dust Bowl" of the 1930s is an example (see Chapter 11).

Human activities in very dry, hot deserts contribute measurably to the raising of high dust clouds. In the desert of northwest India and Pakistan (the Thar Desert bordering the Indus River), the continued trampling of fine-textured soils by hooves of grazing animals and by human feet produces a dust cloud that hangs over the region for long periods. It extends to a height of 9 km (about 30,000 ft).

A LOOK AHEAD This chapter has described the processes and landforms associated with wind action, either directly or through the medium of wind-driven ocean waves. The power source for wind action is, of course, the Sun. By heating the Earth's surface in a nonuniform pattern, the flow of solar energy produces

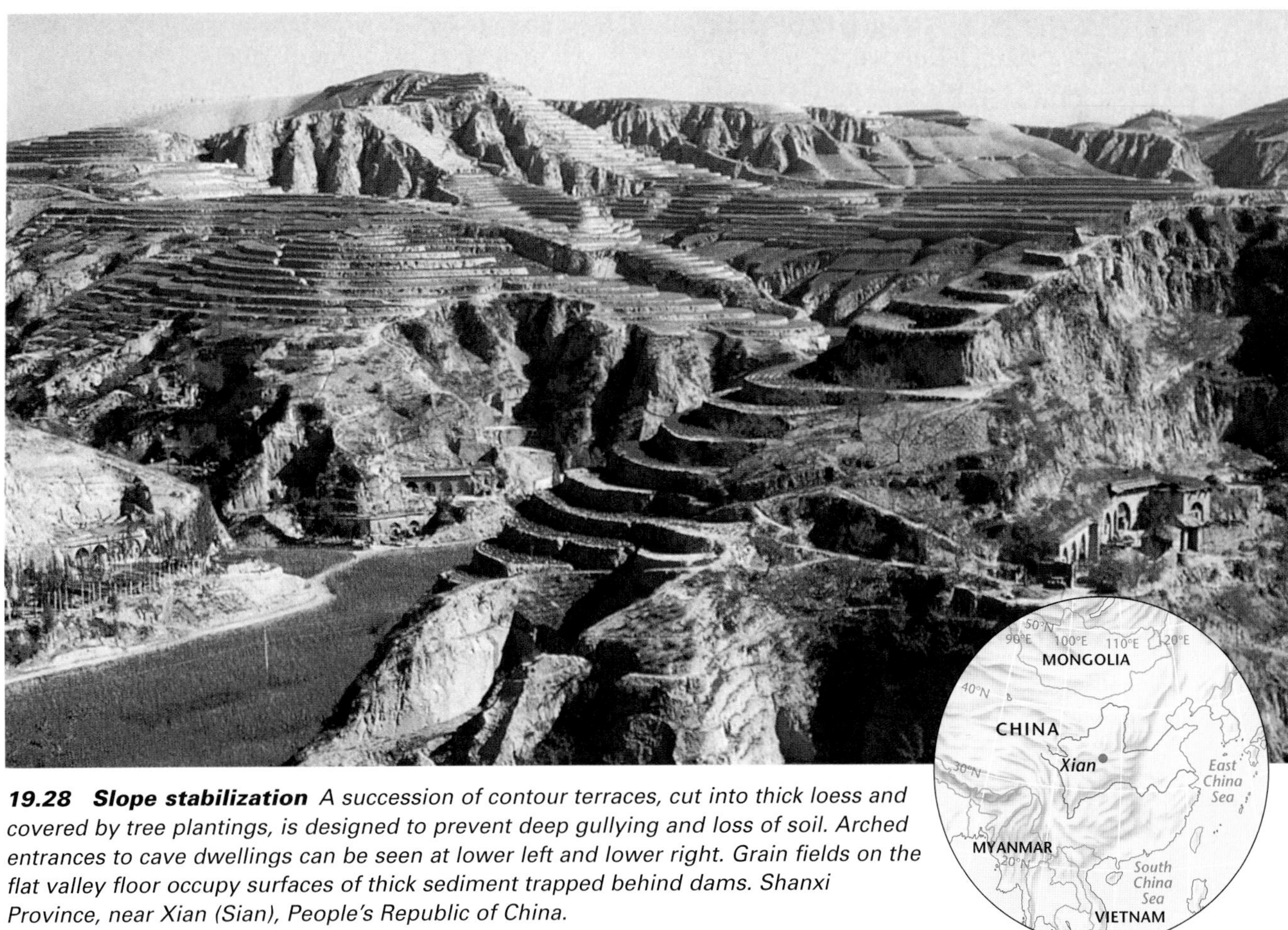

19.28 Slope stabilization *A succession of contour terraces, cut into thick loess and covered by tree plantings, is designed to prevent deep gullying and loss of soil. Arched entrances to cave dwellings can be seen at lower left and lower right. Grain fields on the flat valley floor occupy surfaces of thick sediment trapped behind dams. Shanxi Province, near Xian (Sian), People's Republic of China.*

pressure gradients that cause wind. The Sun also powers the last active landform-making agent of erosion, transportation, and deposition on our list—glacial ice. By evaporating water from the oceans and returning that water to the lands as snow, solar power creates the bodies of solid ice that we distinguish as mountain glaciers and ice sheets.

Compared to wind and water, glacial ice moves much more slowly but is far steadier in its motion. Like a vast conveyor belt, glacial ice moves sediment forward relentlessly, depositing the sediment at the ice margin, where the ice melts. By plowing its way over the landscape, glacial ice also shapes the local terrain—bulldozing loose rock from hillsides and plastering sediments underneath its vast bulk. This slow but steady action is very different from that of water, wind, and waves, and produces a set of landforms that is the subject of our next text chapter.

Chapter Summary

- This chapter has described the landforms of waves and wind, both of which are indirectly powered by the Earth's rotation and the unequal heating of its surface by the Sun. Waves act at the **shoreline**—the boundary between water and land. Waves expend their energy as *breakers*, which erode hard rock into **marine cliffs** and create *marine scarps* in softer materials.
- **Beaches**, usually formed of sand, are shaped by the swash and backwash of waves, which continually work and rework beach sediment. Wave action produces **littoral drift**, which moves sediment parallel to the beach. This sediment accumulates in bars and sandspits, which further extend the beach. Depending on the nature of *longshore currents* and the availability of sediment, shorelines can experience **progradation** or **retrogradation**.
- Tidal forces cause sea level to rise and fall rhythmically, and this change of level produces *tidal currents* in bays and estuaries. Tidal flows redistribute fine sediments within bays and estuaries, which can accumulate with the help of vegetation to form *salt marshes*.
- Coastlines of **submergence** result when coastal lands sink below sea level or sea level rises rapidly. Scenic *ria* and *fiord coasts* are examples. Coastlines of **emergence** include *barrier-island coasts* and *delta coasts*. **Coral-reef** coasts occur in regions of warm tropical and equatorial waters. Along some coasts, rapid uplift has occurred, creating *raised shorelines* and *marine terraces*.
- Global sea level is predicted to rise sharply in the twenty-first century due to both volume expansion of warmer sea water and increased melting of glaciers and snowpacks. Future rises may be very costly to human society as estuaries are displaced, islands are submerged, and coastal zones are subjected to frequent flooding.
- Wind is a landform-creating agent that acts by moving sediment. **Deflation** occurs when wind removes mineral particles—especially clay and silt, which can be carried long distances. Deflation creates **blowouts** in semidesert regions and lowers playa surfaces in deserts. In arid regions, deflation produces **dust storms**.
- **Sand dunes** form when a source, such as a sandstone outcrop or a beach, provides abundant sand that can be moved by wind action. *Barchan dunes* are arranged individually or in chains leading away from the sand source. *Transverse dunes* form a sand sea of frozen "wave" forms arranged perpendicular to the wind direction. *Parabolic dunes* are arc-shaped—*coastal blowout dunes* are an example. *Longitudinal dunes* parallel the wind direction and cover vast desert areas. Coastal *foredunes* are stabilized by dune grass and help protect the coast against storm wave action.
- **Loess** is a surface deposit of fine, wind-transported silt. It can be quite thick, and it typically forms vertical banks. Loess is very easily eroded by water and wind. In eastern Asia, the silt forming the loess was transported by winds from extensive interior deserts located to the north and west. In Europe and North America, the silt was derived from fresh glacial deposits during the Pleistocene Epoch.
- Human activities can hasten the action of deflation by breaking protective surface covers of vegetation and desert pavement.

Key Terms

shoreline
coastline
coast
bay
estuary
marine cliff
beach
littoral drift
progradation
retrogradation
submergence
emergence
delta
coral reef
atoll
deflation
blowout
desert pavement
dust storm
sand dune
loess

Review Questions

1. What is the energy source for wind and wave action?
2. What landforms can be found in areas where bedrock meets the sea?
3. What is littoral drift, and how is it produced by wave action?
4. Identify progradation and retrogradation. How can human activity influence retrogradation?
5. How are salt marshes formed? How can they be reclaimed for agricultural use?
6. What key features identify a coastline of submergence? Identify and compare the two types of coastlines of submergence.
7. Under what conditions do barrier-island coasts form? What are the typical features of this type of coastline? Provide and sketch an example of a barrier-island coast.
8. What conditions are necessary for the development of coral reefs? Identify three types of coral-reef coastlines.
9. How are marine terraces formed?
10. What is deflation, and what landforms does it produce? What role does the dust storm play in deflation?
11. How do sand dunes form? Describe and compare barchan dunes, transverse dunes, star dunes, coastal blowout dunes, parabolic dunes, and longitudinal dunes.
12. What is the role of coastal dunes in beach preservation? How are coastal dunes influenced by human activity? What problems can result?
13. Define the term *loess*. What is the source of loess, and how are loess deposits formed?

Focus on Systems 19.1 • The Coastal Sediment Cell as a Matter Flow System

1. What are the typical changes from summer to winter that you might expect for midlatitude beaches?
2. Identify and describe the pathways of flow in a coastal sediment cell, including inputs, outputs, and locations of sediment in storage.
3. How does the damming of coastal rivers affect the coastal sediment cell and the storage of sediment in beaches?

A Closer Look:
Eye on Global Change 19.3 • Global Change and Coastal Environments

1. Review the observed and predicted changes in global climate that will impact coastal environments.
2. Identify the global change factors that will affect coastal erosion and describe their impacts.
3. How have human activities induced land subsidence in wetlands? Why are wetlands of delta coasts particularly at risk?
4. What impact will global climate change have on coral reefs?
5. Why is global warming expected to cause coastal recession of arctic shorelines?

Visualizing Exercises

1. Describe the features of delta coasts and their formation. Sketch and compare the shapes of the Mississippi and Nile deltas.
2. Take a piece of paper and let it represent a map with winds coming from the north, at the top of the page. Then sketch the shapes of the following types of dunes: barchan, transverse, parabolic, and longitudinal.

Essay Questions

1. Consult an atlas to identify a good example of each of the following types of coastlines: ria coast, fiord coast, barrier-island coast, delta coast, coral-reef coast, and fault coast. For each example, provide a brief description of the key features you used to identify the coastline type.
2. Wind action moves sand close to the ground in a bouncing motion, whereas silt and clay are lifted and carried longer distances. Compare landforms and deposits that result from wind transportation of sand with those that result from wind transportation of silt and finer particles.

Problems

Working It Out 19.2 • Angle of Repose of Dune Sands

1. A geomorphologist is investigating the effect of grain size on the angle of repose. She obtains two samples of small faceted glass beads (rhinestones) with diameters of 3 mm and 4 mm from a jewelry manufacturer. Using these beads, she conducts 20 angle of repose trials for each size with the following results:

Angle of repose, degrees

Size	$\overline{X}$	s	32.0	32.5	33.0	33.5	34.0	34.5	35.0	35.5	36.0
3 mm	34.4	0.91			2	4	3	3	5	1	2
4 mm	33.3	0.72	1	4	7	2	4	2			

Plot the data on a graph following the example of the graph in *Working It Out 19.2.* Do the two samples appear to be as well separated as those of Ottawa and beach sand?

2. Calculate the t-statistic for the difference in the two means. Compare the result with the values in the table below, shown for two samples of size 20 each. Can you conclude that the size has an effect on the angle of repose?

t value	Chance
>2.42	1 in 100
>2.70	1 in 200
>3.31	1 in 1000

Eye on the Landscape

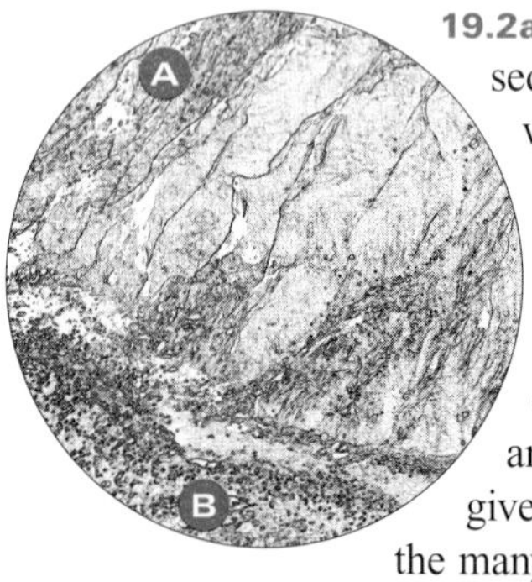

19.2a Retreating shoreline The sediments exposed in the bluff were laid down by streams fed by melting stagnant continental ice sheets at the end of the Ice Age. Because the ice melted irregularly, the deposits are not very uniform, and the exposed bedding **(A)** gives that impression. Note also the many large blocks of rock present within the deposits. They were let down from melting ice and were too large to be moved by the streams laying the sediments. A large lag deposit **(B)** of these stones remains at the water's edge, where it is being worked by waves. Deposits of continental ice sheets are covered in Chapter 20.

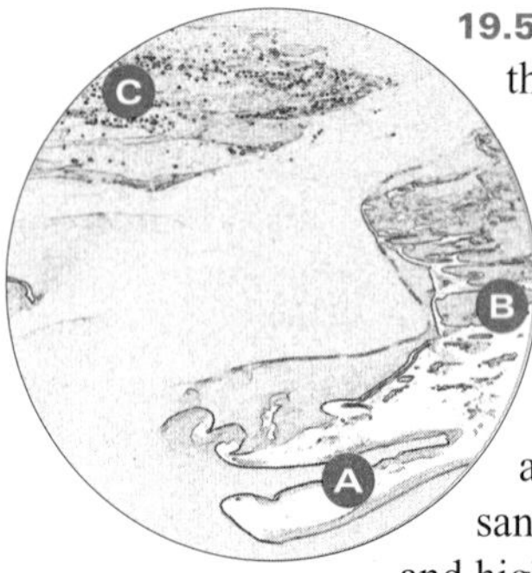

19.5 Sandspit The freshness of the sand here **(A)** indicates that Monomoy Point is growing very rapidly. The sand is supplied by the erosion of marine cliffs at Nauset Beach to the north. Note the overwash channels **(B)** that are cut through the vegetated sand dunes by the high waters and high surf of severe storms. To the upper left is the town of Chatham **(C)**, its low hills formed from rock debris shed by melting ice sheets during the Ice Age. The coastline has the "drowned" look of a coastline of submergence, with its many lakes and bays. See this chapter for more details.

19.20 Dust Storm Savanna vegetation is shown nicely in this photo **(A)**. It consists of scattered trees with an understory of shrubs and grasses (here in a brown and dormant stage during the dry season). Although the tree in **(B)** appears to be dead, it may be a small-leaved acacia in silhouette. Acacia is a plant genus that includes many common species of semiarid environments. Savanna vegetation is described in Chapters 10 and 24.

A CLOSER LOOK

Eye on Global Change 19.3 | Global Change and Coastal Environments

Global climate change over the remainder of the twenty-first century will have major impacts on coastal environments.[1] The changes include increases in sea-surface temperature and sea level, decreases in sea-ice cover, and changes in salinity, wave climate, and ocean circulation.

What changes have already occurred? According to the 2001 report of the Intergovernmental Panel on Climate Change, the global heat content of the ocean has been increasing since at least the late 1950s, and the increase in sea-surface temperature between 1950 and 1993 was about half that of the land-surface temperature. Sea level rose between 10 and 20 cm (4 and 8 in.). Sea-ice cover in the spring and summer of the northern hemisphere has decreased by 10 to 15 percent since the 1950s. After 1970, El Niño episodes, which affect ocean circulation as well as severe storm tracks and intensities, became more frequent, more intense, and more persistent than during the past 100 years.

What changes are in store? Between 1990 and 2100, global average surface temperature will increase 1.4 and 5.8°C (2.5 and 10.4°F), and if the present pattern persists, average sea-surface temperatures will account for one-third of that elevation. Sea level will rise from 9 to 88 cm (3.5 to 34.6°F), depending on the scenario. Snow and ice cover will continue to decrease, and mountain glaciers and icecaps will continue their retreat of the twentieth century. Tropical cyclone peak wind and peak precipitation intensities will increase, and El Niño extremes of flood and drought will be exaggerated.

Coastal Erosion

These changes are bad news for coastal environments. Let's begin with coastal erosion. Most coastal erosion occurs in severe storms, when high winds generate large waves and push water up onto the land in storm surges. Global warming will increase the frequency of high winds and heavy precipitation events, thus amplifying the effects of severe storms. More frequent and longer El Niños will increase the severity and frequency of Pacific storms on the North American coast, leading to increased sea-cliff erosion along southern California's south- and southwest-facing coastlines. During opposing La Niña events, Atlantic hurricane frequency and intensity will increase, with higher risk of damage to structures and coastal populations.

How will the rise in sea-level impact coastlines? Studies of sandy shorelines, which account for about 20 percent of the global coastline, have shown that over the past 100 years or so, about 70 percent of these shorelines have retreated and 10 percent have advanced, with the remainder stable. Sea-level rise enhances coastal erosion in storms, but in the long term, its effect is to push beaches, salt marshes, and estuaries landward. In an unaltered landscape, the migration of these features landward and seaward with the rise and fall of sea level over thousands of years is a natural process without significant ecological impact. However, it is a serious problem when the rise is rapid and

Coastal erosion *Wave action has undermined the bluff beneath these two buildings, depositing them on the beach below.*

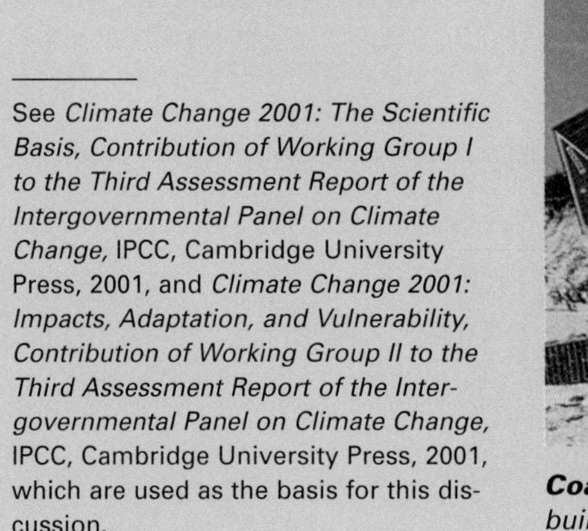

[1] See *Climate Change 2001: The Scientific Basis, Contribution of Working Group I to the Third Assessment Report of the Intergovernmental Panel on Climate Change,* IPCC, Cambridge University Press, 2001, and *Climate Change 2001: Impacts, Adaptation, and Vulnerability, Contribution of Working Group II to the Third Assessment Report of the Intergovernmental Panel on Climate Change,* IPCC, Cambridge University Press, 2001, which are used as the basis for this discussion.

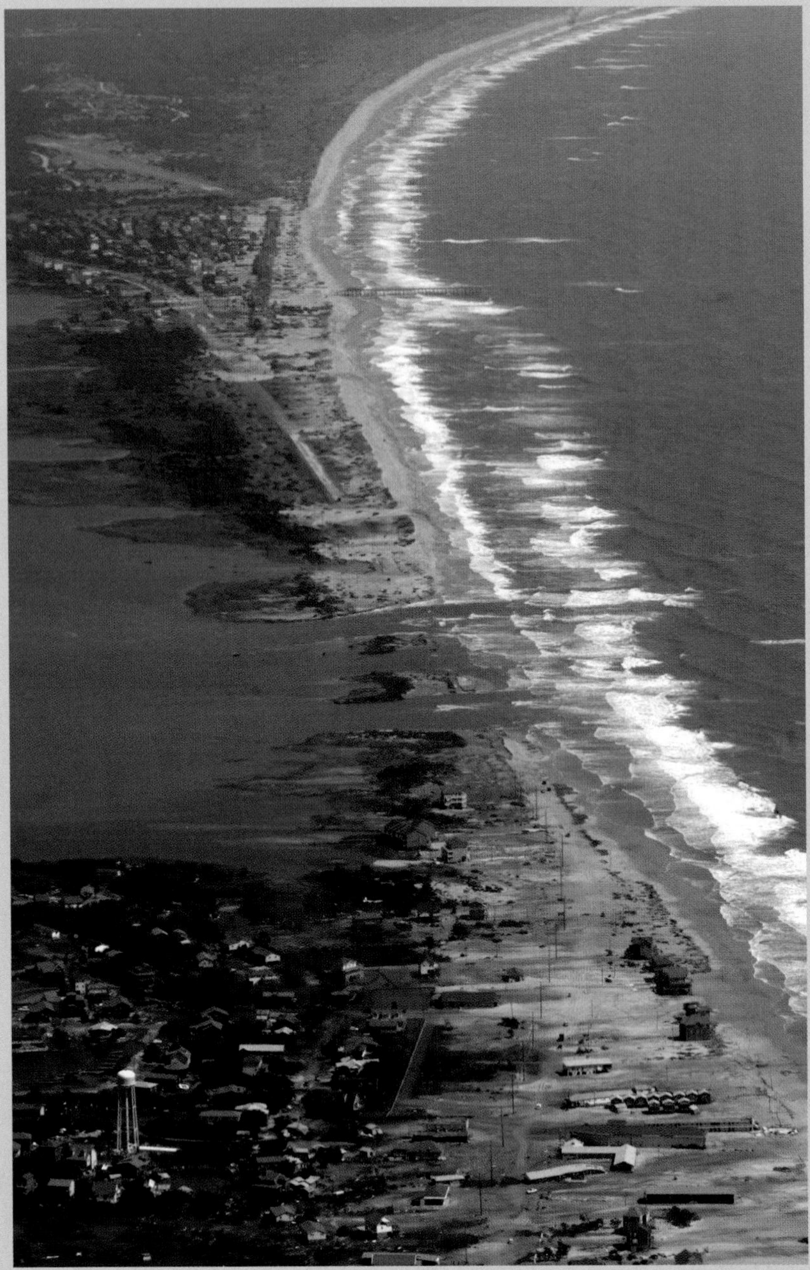

A new inlet *Storm waves from Hurricane Isabel breached the North Carolina barrier beach to create a new inlet, as shown in this aerial photo from September, 2003. Notice also the widespread destruction of the shoreline in the foreground, with streaks of sand carried far inland by wind and wave action. As sea level rises, ocean waves will attack barrier beaches with increasing frequency and severity.*

the coastline is developed. Beaches disappear and are replaced by sea walls. Salt marshes are drained to reduce inland flooding. Estuaries become shallower and more saline. In this way, the most productive areas of the coast are squeezed between a rising ocean and a water's edge that is increasingly defended.

It is also important to realize that sea-level rise will not be uniform. Modifying factors of waves, currents, tides, and offshore topography can act to magnify the rise, depending on the location. For example, some models predict doubled rates of sea-level rise for portions of the eastern United States, North American Pacific coast, and the western North American arctic shoreline.

Subsidence and Sea-Level Rise

Land subsidence is a contributing factor to the impact of sea-level rise. In an unaltered environment, rivers bring fine sediment to the coastline that settles in estuaries and is also carried by waves, currents, and tides into salt marshes and mangrove swamps. As the fine sediment accumulates, it slowly compacts, forming rich, dense layers of silt and clay mixed with fine organic particles. However, many coastlines are now fed by rivers that have been dammed at multiple points on their courses, which reduces that amount of fine sediment brought to the coast. Without a constant inflow of new sediment, coastal wetlands slowly sink as the older sediment that supports them compacts. This subsidence increases the effects of sea-level rise. Note also that estuaries are affected by changes in river flows as a result of changes in the frequency of severe precipitation events and summer droughts.

Delta coasts are especially sensitive to sediment starvation and subsidence. Here, rates of subsidence can reach 2 cm/yr (0.8 in./yr). The Mississippi has lost about half of its natural sediment load, and sediment transport by such rivers as the Nile and Indus has been reduced by 95 percent. Extracting ground water from deltas also increases subsidence. The Chao Phraya delta near Bangkok and the old deltas of the Huang and Changjing rivers in China are important examples.

According to recent estimates, sea-level rise and subsidence could cause the loss of as much as 22 percent of the world's coastal wetlands by the 2080s. Coupled with losses directly related to human activity, coastal wetlands could decrease by 30 percent. This level of reduction would have major effects on commercially important

Mississippi Delta marshland *With rising sea level and increasing subsidence, marshlands of many river deltas are endangered. Shown here is a marshland of the Mississippi delta in Louisiana, near the river's southwest pass.*

fish and shellfish populations, as well as on other organisms comprising the marine food chain.

Coral Reefs

Coral reefs, like coastal wetlands, perform important ecological functions, such as harboring marine fish and nursing their progeny. They are highly biodiverse, with some reefs containing more major plant and animal groups than rainforests. They also serve as protective barriers to coastlines that reduce the effects of storm waves and surges. However, more than half of the total area of living coral reefs is thought to be threatened by human activities ranging from water pollution to coral mining.

How will coral reefs be affected by global change? It appears that simple sea-level rise will not be a factor because healthy coral reefs are able to grow upward at a rate equal to or greater than projected sea-level rise. However, the increase in sea-surface temperature that will accompany global warming is of major concern. Many coral reefs appear to be at or near their upper temperature limits. When stressed, for example by a rise in temperature, many corals respond by "bleaching." In this process, they expel the algae that live symbiotically inside their structures, leaving the coral without color. However, the algae are necessary for the continued survival of the coral. The bleaching may be temporary if the stress subsides, but if permanent, the corals die. Major episodes of coral bleaching have been associated with the strong El Niños where water temperatures were increased by at least 1°C (1.8°F). During the very strong El Niño of 1997–1998, the Indian Ocean experienced a major coral bleaching event that was especially prominent on Australia's Great Barrier Reef.

Another concern is the effect of increased atmospheric CO_2 levels on corals. With higher concentrations of atmospheric CO_2, more CO_2 dissolves in sea water, causing the water to become slightly less alkaline. This in turn shifts the solubility of calcium carbonate, $CaCO_3$, making it less available to the corals to use in building their skeleton structure. Whether this will create yet another source of stress for coral reefs is still the subject of scientific study.

High-Latitude Coasts

Many pristine stretches of arctic shoreline are threatened by global warming. In these regions, the shoreline is sealed off for much of the year by sea ice. The shoreline is also buttressed by permanent ground ice that bonds unconsolidated sediment into a rocklike mass, much like the calcium carbonate or quartz cement of a sedimentary rock. This frozen ground is resistant to wave action and thus helps to protect the shoreline from erosion. In some areas, massive ice beds underlie major portions of the coastline.

With global warming, the shoreline is less protected by these

Bleached anemones *These anemones, observed in the shallow waters of the Maldive Islands, exhibit bleaching—a process in which coral animals expel symbiotic algae as a response to stress. The bleaching can be fatal.*

mechanisms. Sea ice melts earlier and returns later, increasing the season for wave action. Greater expanses of open sea allow larger waves to build. Ground ice thaws to a greater depth, releasing more surface sediment and allowing waves to scour the shore more deeply. Near-shore sediments, stored in cliffs or bluffs, thaw as well, releasing them to shoreline processes. And, recall from our interchapter feature, *A Closer Look: Eye on Global Change 11.4 • Regional Impacts of Climate Change on North America* that global warming will be especially severe at high altitudes. Rapid coastal recession under wave attack is already reported for many ice-rich coasts along the Beaufort Sea.

It is apparent that global climate change will have major impacts on coastal environments, with very broad implications for ecosystems and natural resources. It will take careful management of our coastlines to reduce those impacts on both human and natural systems.

Arctic shoreline *Pristine arctic shorelines, such as this beach on the coast of Spitsbergen Island, Svalbard archipelago, Norway, will be subjected to rapid change as global climate warms. Thawing of ground ice will release beach sediments and the early retreat of sheltering sea ice will expose the shoreline to enhanced summer wave attack. The yellow flowering plant in the foreground is the bog saxifrage.*

20 GLACIER SYSTEMS AND THE ICE AGE

Icebergs off the Adélie coast, Antarctica. This large tabular iceberg has recently detached from the Antarctic ice sheet. Note the layered structure visible on fresh ice faces.

In this chapter we turn to the last of the active agents that create landforms—glacial ice. Not long ago, during the Ice Age, much of northern North America and Eurasia was covered by massive sheets of glacial ice. As a result, glacial ice has played a dominant role in shaping landforms of large areas in midlatitude and subarctic zones. Glacial ice still exists today in two great accumulations of continental dimensions—the Greenland and Antarctic Ice Sheets—and in many smaller masses in high mountains.

The glacial ice sheets of Greenland and Antarctica strongly influence the radiation and heat balance of the globe. Because of their intense whiteness, they reflect much of the solar radiation they receive. Their intensely cold surface air temperatures contrast with temperatures at more equatorial latitudes. This temperature difference helps drive the system of meridional heat transport that we described in Chapters 4 and 6. In addition, these enormous ice accumulations represent water in storage in the solid state. They figure as a major component of the global water balance. When the volume of glacial ice increases, as during an ice age, sea levels must fall. When ice sheets melt away, sea level rises. Today's coastal environments evolved during the rising sea level that followed the melting of the last ice sheets of the Ice Age.

GLACIERS

Most of us know ice only as a brittle, crystalline solid because we are accustomed to seeing it in small quantities. Where a great thickness of ice exists, the pressure on the ice at the bottom makes the ice lose its rigidity. That is, the ice becomes plastic. This allows the ice mass to flow in response

20.1 Eye on the Landscape An alpine glacier *Aerial view of the Grand Plateau Glacier, Saint Elias Mountains, Glacier Bay National Park, Alaska. The dark stripes are medial moraines.* **What else would the geographer see?...Answers at the end of the chapter.**

to gravity, slowly spreading out over a larger area or moving downhill. On steep mountain slopes, the ice can also move by sliding. Movement is the key characteristic of a *glacier,* defined as any large natural accumulation of land ice affected by present or past motion.

Glacial ice accumulates when the average snowfall of the winter exceeds the amount of snow that is lost in summer by ablation. The term **ablation** means the loss of snow and ice by evaporation and melting. When winter snowfall exceeds summer ablation, a layer of snow is added each year to what has already accumulated. As the snow compacts by surface melting and refreezing, it turns into a granular ice and is then compressed by overlying layers into hard crystalline ice. When the ice mass is so thick that the lower layers become plastic, outward or downhill flow starts, and the ice mass is now an active glacier.

Glacial ice forms where temperatures are low and snowfall is high. These conditions can occur both at high elevations and at high latitudes. In mountains, glacial ice can form even in tropical and equatorial zones if the elevation is high enough to keep average annual temperatures below freezing. Orographic precipitation encourages the growth of glacial ice. In high mountains, glaciers flow from small high-elevation collecting grounds down to lower elevations, where temperatures are warmer. Here the ice disappears by ablation. Typically, mountain glaciers are long and narrow because they occupy former stream valleys. These **alpine glaciers** are a distinctive type (Figure 20.1).

In arctic and polar regions, prevailing temperatures are low enough that snow can accumulate over broad areas, eventually forming a vast layer of glacial ice. Accumulation starts on uplands that intercept heavy snowfall. The uplands become buried under enormous volumes of ice, which can reach a thickness of several thousand meters. The ice then spreads outward, over surrounding lowlands, and covers all landforms it encounters. This extensive type of ice mass is called an **ice sheet.** As already noted, ice sheets exist today in Greenland and Antarctica.

Glacial ice normally contains abundant rock fragments ranging from huge angular boulders to pulverized rock flour. Some of this material is eroded from the rock floor on which the ice moves. In alpine glaciers, rock debris is also derived from material that slides or falls from valley walls onto the ice.

Glaciers are capable of eroding and depositing great quantities of sediment. *Glacial abrasion* is a glacial erosion process caused by rock fragments that are held within the ice and scrape and grind against bedrock (Figure 20.2). Erosion also occurs by *plucking,* as moving ice lifts out blocks of bedrock that have been loosened by the freezing and expansion of water in joint fractures. Abrasion and plucking act to smooth the bed of a glacier as glacial flow continues through time. Rock debris brought into a glacier is eventually deposited at the lower end of a glacier, where the ice melts. Both erosion and deposition result in distinctive glacial landforms.

ALPINE GLACIERS.

Figure 20.3 illustrates a number of features of alpine glaciers. The illustration shows a simple glacier occupying a sloping valley between steep rock walls. Snow collects at the upper end in a bowl-shaped depression, the **cirque.** The upper end lies in a zone of accumulation. Layers of snow in the process of compaction and recrystallization are called *firn.*

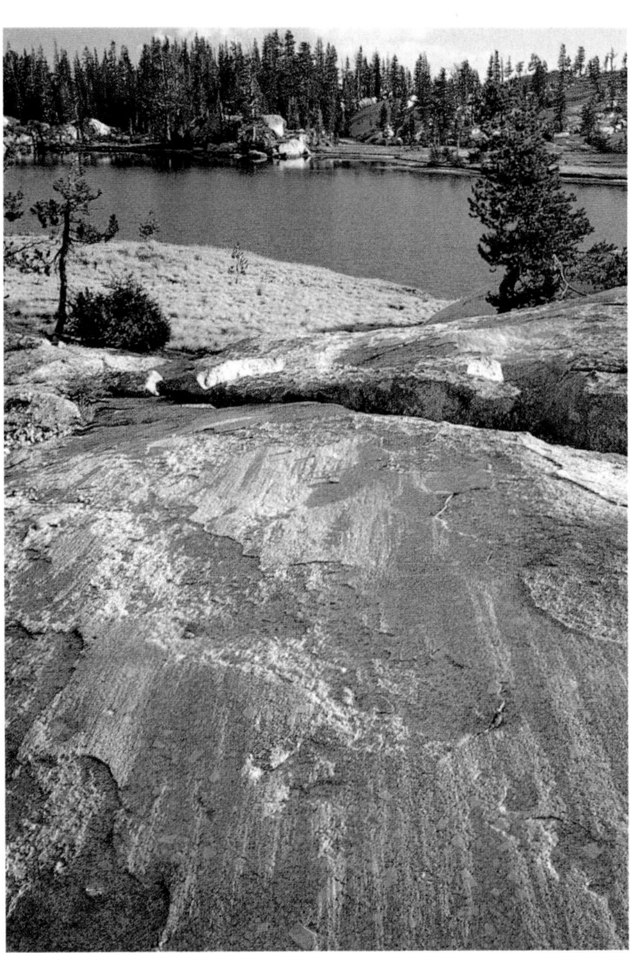

20.2 Glacial abrasion *This grooved and polished surface, now partly eroded, marks the former path of glacial ice. Cathedral Lakes, Yosemite National Park, California.*

The smooth firn field is slightly bowl-shaped in profile. Flowage in the glacial ice beneath the firn carries the ice downvalley out of the cirque. The rate of ice flow is accelerated at a steep rock step, where deep crevasses (gaping fractures) mark an ice fall. The lower part of the glacier lies in the zone of ablation. In this area, the rate of ice wastage is rapid, and old ice is exposed at the glacier surface. As the ice thins by ablation, it loses its plasticity and may develop deep crevasses. At its lower end, or terminus, the glacier carries abundant rock debris. As the downward-flowing ice melts, the debris accumulates.

Although the uppermost layer of a glacier is brittle, the ice beneath behaves as a plastic substance and moves by slow flowage (Figure 20.4). Like stream flow, glacier flow is most rapid far from the glacier's bed—near the midline and toward the top of the glacier's surface. Alpine glaciers also move by basal sliding. In this process, the ice slides downhill, lubricated by meltwater and mud at its base.

A glacier establishes a dynamic balance in which the rate of accumulation at the upper end balances the rate of ablation at the lower end. This balance is easily upset by changes in the average annual rates of accumulation or ablation, causing the glacier's terminus to move forward or melt.

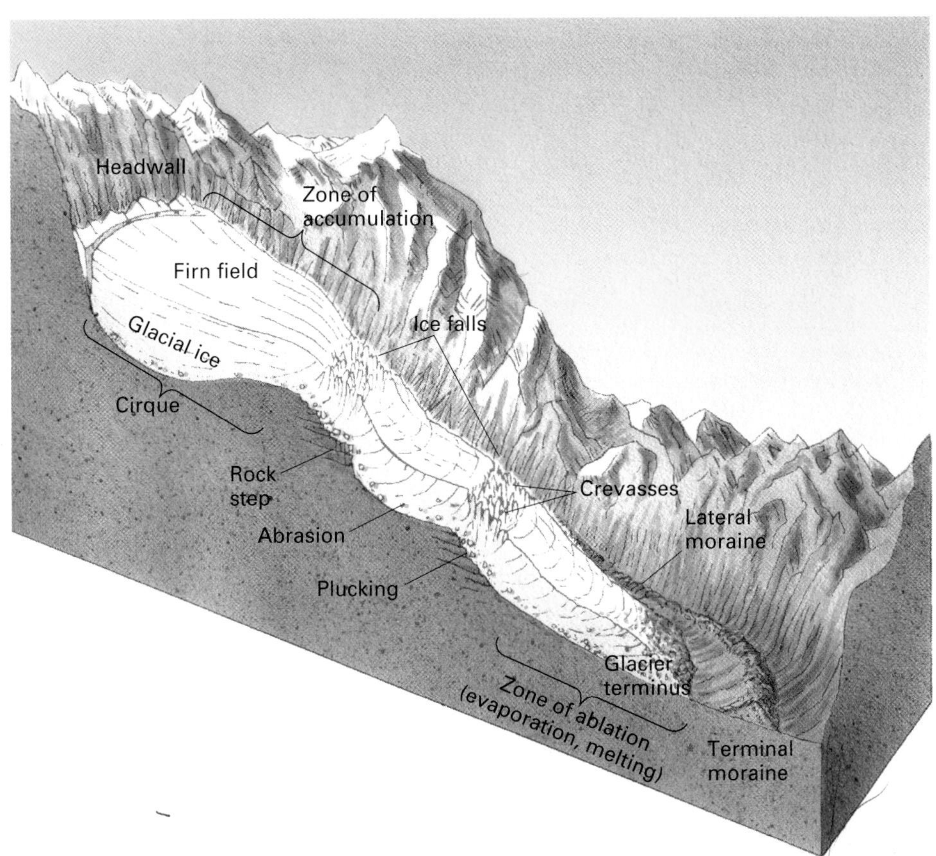

20.3 Cross section of an alpine glacier *Ice accumulates in the glacial cirque, then flows downhill, abrading and plucking the bedrock. Glacial debris accumulates at the glacier terminus. (After A. N. Strahler.)*

Focus on Systems | 20.1

A Glacier as a Flow System of Matter and Energy

A glacier provides a good example of a system of coupled matter and energy flow. For this analysis, we will consider an alpine glacier flowing from its head in high mountain cirques to its terminus in the valley below.

Part *a* of the figure below sketches the pathways of matter flow. In the zone of accumulation, snowfall provides input to the glacier as the snow becomes compacted and recrystallized to solid ice. During warm summer periods, some snow is lost to melting, evap-

Matter flow system

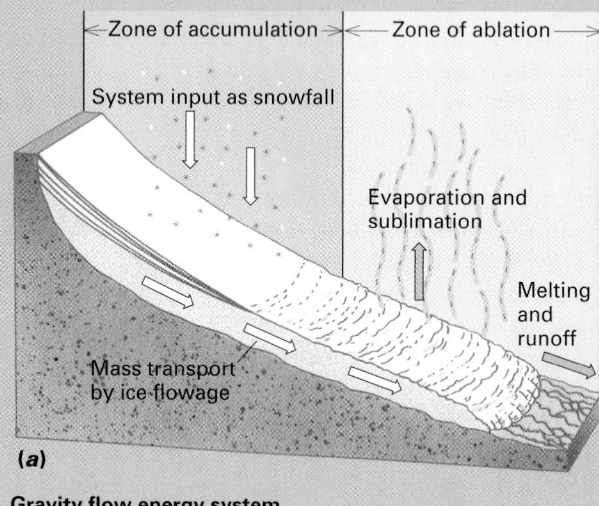

Matter and energy flow in an alpine glacier system *(Copyright © A. N. Strahler.)*

Gravity flow energy system

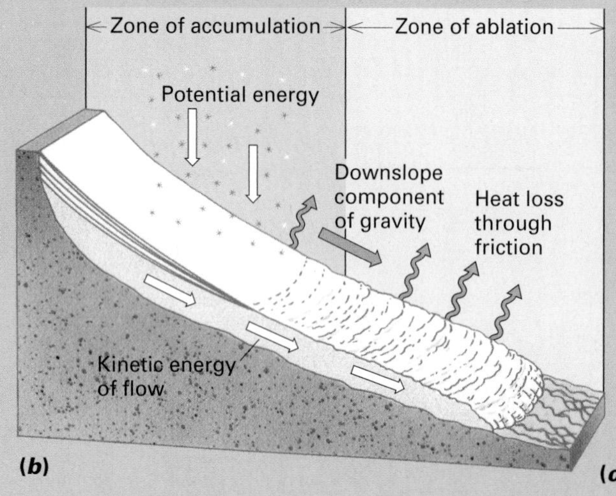

Thermal flow energy system

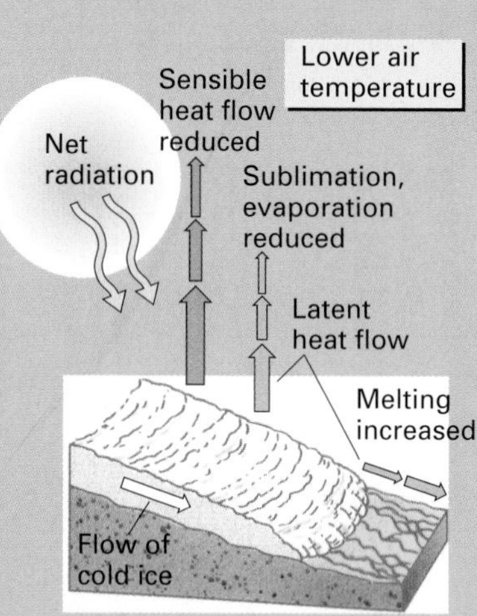

Glacial flow is usually very slow. It amounts to a few centimeters per day for large ice sheets and the more sluggish alpine glaciers, but as fast as several meters per day for an active alpine glacier. However, some alpine glaciers experience episodes of very rapid movement, termed *surges.* A surging glacier may travel downvalley at speeds of more than 60 m (about 200 ft) per day for several months. The reasons for surging are not well understood but probably involve mechanisms that increase the amount of meltwater beneath the ice, enhancing basal sliding. Most glaciers do not experience surging.

An alpine glacier is a good example of a flow system of matter and energy. Matter flows as ice and rock debris move downhill under the power of gravity. Potential energy is converted to kinetic energy in the motion of flow, then dissipated as heat through friction. More details are provided in *Focus on Systems 20.1 • A Glacier as a Flow System of Matter and Energy.* GEODISCOVERIES

oration, and sublimation, but the annual balance is on the side of accumulation.

As the glacier reaches lower elevations, loss of water from the glacier by evaporation, sublimation, and melting increases. The balance shifts to a net loss of ice over the year, and the glacier enters the zone of ablation. With further descent, the rate of loss increases with increasing mean annual temperature and the glacier's cross section shrinks rapidly. At the terminus, no more ice remains.

Over a period of years, the glacier flow system reaches a steady state in which the excess accumulation of water as ice in the zone of accumulation is balanced by the loss of water in the zone of ablation. Within the glacier, there is a continuous downhill flow of ice, and each year an increment equal to the amount of new ice formed moves across the boundary between the zone of accumulation and zone of ablation. The shape and appearance of the glacier remains unchanged, although there may be some small variation from year to year.

Consider what would happen if temperatures cooled and snowfall increased. In that event, more ice would form annually in the zone of accumulation. The boundary between the zone of accumulation and zone of ablation would move to a lower elevation. The glacier would take longer to melt at lower elevations, so the terminus would extend farther down the valley. Eventually, a new steady state would be reached in which a larger, thicker glacier terminates farther from its source. If temperatures warmed, the changes would reverse, producing a new steady state in which a thinner glacier terminates at a higher elevation.

As the glacier flows downhill, it also transports rock debris. Scouring its bed, it breaks off rock fragments and creates a valley with a distinctive U-shape. The glacier also receives material from sideslopes as rock fragments fall onto the surface of the glacier. This debris, ranging from large blocks to fine rock flour, is transported to the lowlands below and shaped into different types of depositional landforms. (These are discussed in the main portion of this chapter.) Note that the transport of debris by the glacier can be treated as a separate open flow system powered by gravity.

Let's turn now to the flows of energy coupled to the matter flow system. We can consider these as two subsystems—a gravity flow subsystem (part *b* of the figure) and a thermal flow subsystem (part *c*). The power sources of the gravity flow subsystem are solar energy and gravity. Solar energy flow provides the water at high elevations through precipitation, thus contributing potential energy as an input. Gravity powers the downhill motion of the ice. In this motion, potential energy is converted to kinetic energy, and the kinetic energy is dissipated as heat in the friction of glacial flow. Gravity also powers the landform-making processes of glacial erosion, transportation, and deposition.

The thermal flow subsystem, shown in part *c*, is easier to understand if we think of cold as a negative quantity, as the absence of heat. The flow of glacial ice, then, serves to export "cold" from higher elevations to lower ones. How does this occur? When ice descends to low elevations, its surface is subjected to shortwave heat flows from solar radiation. This input provides a positive net radiation balance during the day. The energy flow goes largely to melting the ice, which means that less energy is available to warm the overlying air through sensible heat transfer. Also, less energy remains to sublimate the snow and evaporate the meltwater, thus reducing the flow of latent heat to the atmosphere. The effect is that more incoming energy is converted to latent heat of melting, leaving less available to warm the air. This makes the local climate colder.

These three interlinked systems demonstrate the important point that real energy and matter flow systems can be complex when examined in detail. Analysis of energy and matter budgets of systems thus requires careful study if the analysis is to include all important factors.

Landforms Made by Alpine Glaciers

Landforms made by alpine glaciers are shown in a series of diagrams in Figure 20.5. Mountains are eroded and shaped by glaciers, and after the glaciers melt, the remaining landforms are exposed to view. Diagram *a* shows a region sculptured entirely by weathering, mass wasting, and streams. The mountains have a smooth, rounded appearance. Soil and regolith are thick.

Imagine now that a climatic change results in the accumulation of snow in the heads of the higher valleys. An early stage of glaciation is shown at the right side of diagram *b*, where snow is collecting and cirques are being carved by the grinding motion of the ice. Deepening of the cirques is aided by intensive frost shattering of the bedrock near the masses of compacted snow. At a later stage (center), glaciers have filled the valleys and are integrated into a system of tributaries that feed a trunk glacier. Tributary

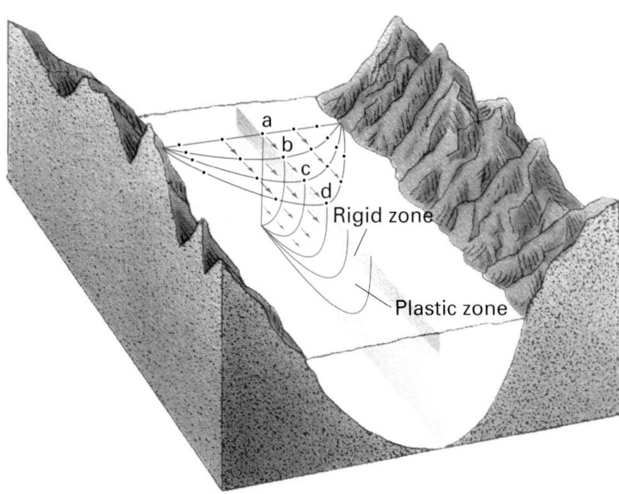

20.4 Motion of glacial ice *Ice moves most rapidly on the glacier's surface at its midline. Movement is slowest near the bed, where the ice contacts bedrock or sediment.*

glaciers join the main glacier smoothly. The cirques grow steadily larger. Their rough, steep walls soon replace the smooth, rounded slopes of the original mountain mass. Where two cirque walls intersect from opposite sides, a jagged, knifelike ridge, called an *arête,* is formed. Where three or more cirques grow together, a sharp-pointed peak is formed. Such peaks are called *horns.* On the left side of the diagram, a portion of the landscape is drawn as unglaciated.

A ridge or pile of rock debris left by glacial action that marks the edge of a glacier is termed a **moraine.** A *lateral moraine* is a debris ridge formed along the edge of the ice adjacent to the trough wall (Figure 20.3). Where two ice streams join, this marginal debris is dragged along to form a narrow band riding on the ice in midstream (Figures 20.1, 20.5*b*), called a *medial moraine.* At the terminus of a glacier, rock debris accumulates in a *terminal moraine,* an embankment curving across the valley floor and bending upvalley along each wall of the trough (Figure 20.6).

Glacial Troughs and Fiords

Glacier flow constantly deepens and widens its rock channel, so that after the ice has finally melted, a deep, steep-walled **glacial trough** remains (Figure 20.5*c*). The trough typically has a U-shape in cross-profile (Figure 20.7). Tributary glaciers also carve U-shaped troughs, but they are smaller in cross section and less deeply eroded by their smaller glaciers. Because the floors of these troughs lie high above the level of the main trough, they are called *hanging valleys.* Streams later occupy the abandoned valleys, providing scenic waterfalls that cascade over the lip of the hanging valley to the main trough below. High up in the smaller troughs, the bedrock is unevenly excavated, so that the floors of troughs and cirques contain rock basins and rock steps. The rock basins are occupied by small lakes, called *tarns* (Figure 20.5*c*). Major troughs sometimes hold large, elongated trough lakes. GEODISCOVERIES

Many large glacial troughs now are filled with alluvium and have flat floors. Aggrading streams that issued from the receding ice front were heavily laden with rock fragments so that the deposit of alluvium extended far downvalley. Figure 20.7 shows a comparison between a trough with little or no fill (*b*) and another with an alluvial-filled bottom (*c*).

When the floor of a trough open to the sea lies below sea level, the sea water enters as the ice front recedes. The result is a deep, narrow estuary known as a **fiord** (Figure 20.7*d*). Fiords are opening up today along the Alaskan coast, where some glaciers are melting back rapidly and ocean waters are filling their troughs. Fiords are found largely along mountainous coasts between lat. 50° and 70° N and S (Figure 20.8). On these coasts, glaciers were nourished by heavy orographic snowfall, associated with the marine west-coast climate ⑧.

As large, distinctive features in areas of rugged terrain, glaciers are readily viewed from the perspective of space. *Focus on Remote Sensing 20.2 • Remote Sensing of Glaciers* shows some spectacular images of glaciers acquired from Earth orbit. One of the images uses a special radar imaging technique to portray the motion of an ice stream. GEODISCOVERIES

Ice Sheets of the Present

In contrast to alpine glaciers are the enormous ice sheets of Antarctica and Greenland. These are huge plates of ice, thousands of meters thick in the central areas, resting on land masses of subcontinental size. The Greenland Ice Sheet has an area of 1.7 million sq km (about 670,000 sq mi) and occupies about seven-eighths of the entire island of Greenland (Figure 20.9). Only a narrow, mountainous coastal strip of land is exposed. The Antarctic Ice Sheet covers 13 million sq km (about 5 million sq mi) (Figure 20.10). Both ice sheets are developed on large, elevated land masses in high latitudes. No ice sheet exists near the north pole, which is positioned in the vast Arctic Ocean. Ice there occurs only as floating sea ice.

The surface of the Greenland Ice Sheet has the form of a very broad, smooth dome. Underneath the ice sheet, the rock floor lies near or slightly below sea level under the central region but is higher near the edges. The Antarctic Ice Sheet is thicker than the Greenland Ice Sheet—as much as 4000 m (about 13,000 ft) at maximum. At some locations, ice sheets extend long tongues, called outlet glaciers, to reach the sea at the heads of fiords. From the floating edge of the glacier, huge masses of ice break off and drift out to open sea with tidal currents to become icebergs. An important glacial feature of Antarctica is the presence of great plates of floating glacial ice, called *ice shelves* (Figure 20.10). Ice shelves are fed by the ice sheet, but they also accumulate new ice through the compaction of snow. GEODISCOVERIES

Sea Ice and Icebergs

Free-floating ice on the sea surface is of two types—sea ice and icebergs. *Sea ice* (Figure 20.11) is formed by direct

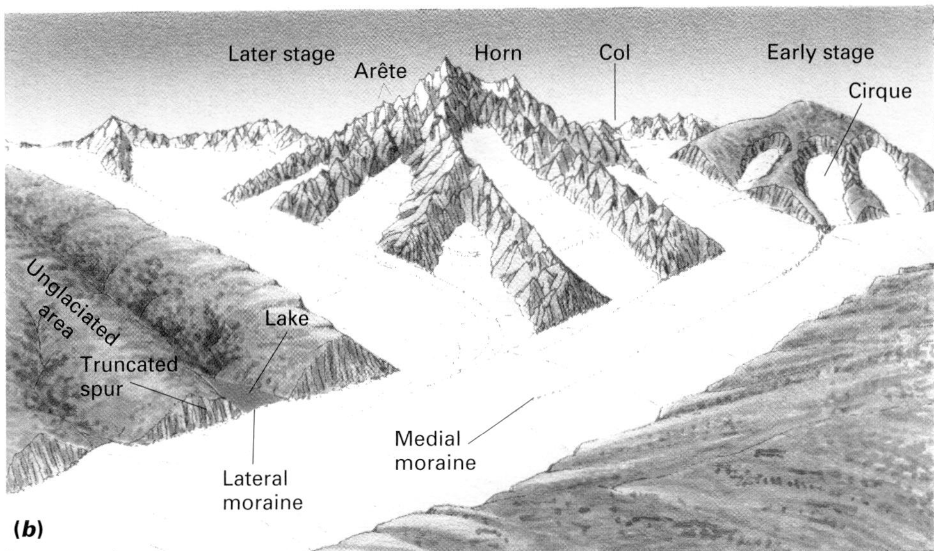

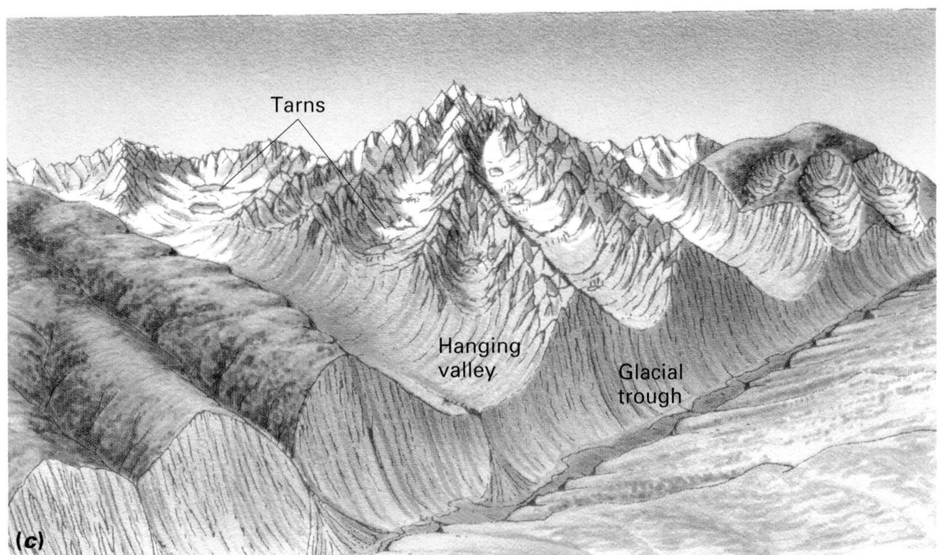

20.5 Landforms produced by alpine glaciers *(a) Before glaciation sets in, the region has smoothly rounded divides and narrow, V-shaped stream valleys. (b) After glaciation has been in progress for thousands of years, new erosional forms are developed. (c) With the disappearance of the ice, a system of glacial troughs is exposed. (Drawn by A. N. Strahler.)*

Focus on Remote Sensing | 20.2

Remote Sensing of Glaciers

Because glaciers are often found in inaccessible terrain or in extremely cold environments, they are difficult to survey and monitor. Satellite remote sensing provides an invaluable tool for studying both continental and alpine glaciers in spite of these difficulties.

Some of the world's more spectacular alpine glaciers are found in South America, along the crest of the Andes in Chile and Argentina, where mountain peaks reach as high as 3700 m (about 12,000 ft). Image *(a)* is a true-color image, acquired by astronauts aboard the International Space Station in December, 2000 of Cerro San Lorenzo (San Lorenzo Peak). The peak itself is a glacial horn. Leading away from the horn to the south is a long, sharp ridge, or arête. To the left of the peak is a cirque, now only partly filled with glacial ice. These features were carved during the Ice Age, when the alpine glaciers were larger and filled the now-empty glacial troughs behind the ridge to the right.

Northwest of Cerro San Lorenzo lies the San Quentin glacier, shown in ASTER image *(b)*, acquired on May 2, 2000. This color-infrared image was acquired at 15-m spatial resolution and shows the glacier in fine detail. The snout of the glacier ends in a shallow lake of sediment-laden water that drains by fine streams into the Golfo de Penas at lower left. Note the low, semicircular ridge a short distance from the lake—this is a terminal moraine, marking a stand of the ice tongue in the recent past. A high cloud partly obscures the southern end of the snout and nearby coastline. The intense red color indicates a thick vegetation cover.

The world's largest glacier is, of course, the ice cap that covers nearly all of Antarctica. At the edges of the ice cap are outlet glaciers, where glacial flow into the ocean is quite rapid. Image *(c)* shows the Lambert Glacier, one of the largest and longest of Antarctica's outlet glaciers. The image was acquired by Canada's Radarsat radar imager. Because this type of radar system can see through the clouds, it is ideal for mapping a large area in a short period of time. It also allows the measurement of the velocity of glacier flow by comparing paired images acquired at different times (in this case, 24 days apart) using a technique called radar interferometry. The image uses color to show the velocity. Brown tones indicate little or no motion, and show both exposed mountains and stationary ice. Green, blue, and red tones indicate increasing velocity. Velocity and direction are also shown by the arrows superimposed on the image. Glacial flow is most rapid at the left, where the flow is channeled through a narrow valley, and at the right, where the glacier spreads out and thins to feed the Amery Ice Shelf.

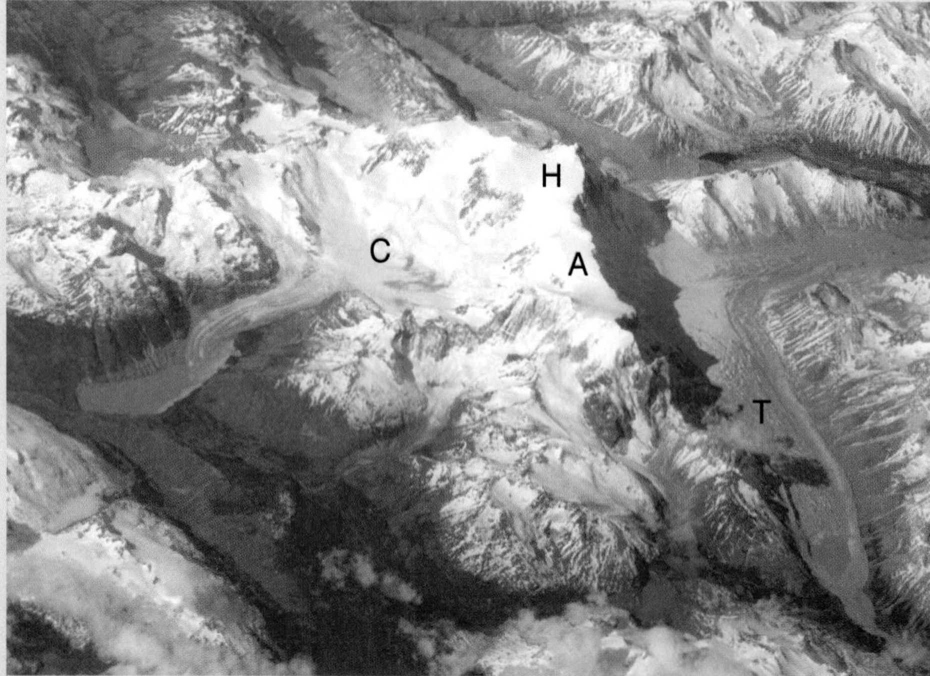

(a)

Andean alpine glacial features *H, horn (San Lorenzo Peak); A, arête; C, cirque; T, glacial trough. (Courtesy NASA.)*

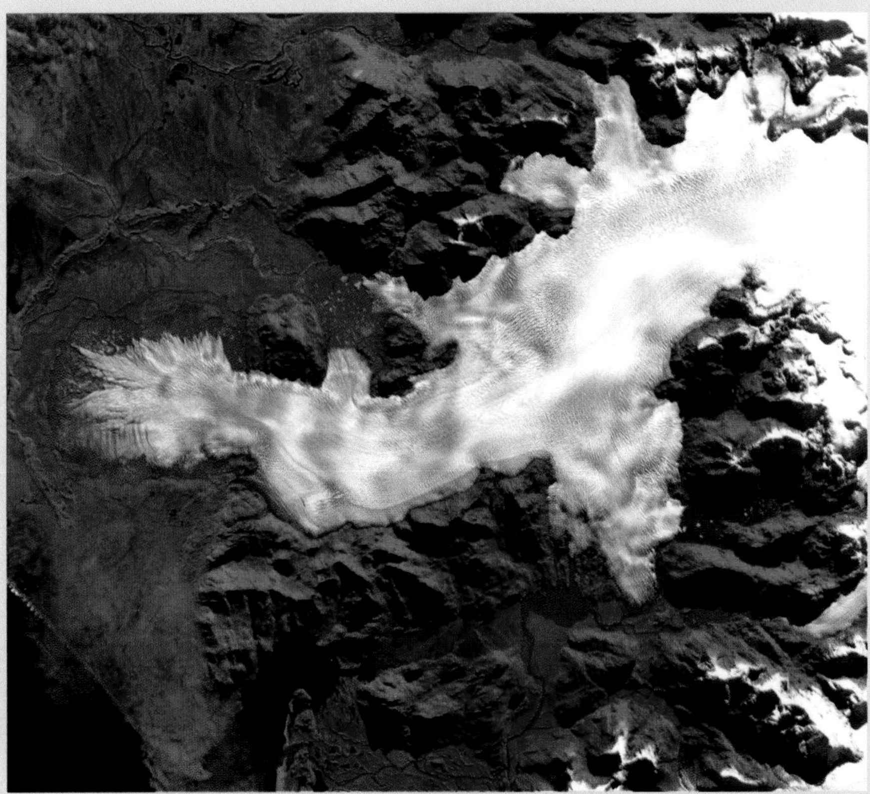

(*b*)

San Quentin Glacier, Chile *The San Quentin Glacier as imaged by ASTER. (Image courtesy NASA/GSFC/MITI/ERSDAC/JAROS and U.S./Japan ASTER Science Team.)*

Lambert Glacier, Antarctica *The Lambert Glacier as imaged by Radarsat during the 2000 Antarctic Mapping Mission. (Image courtesy Canadian Space Agency/NASA/Ohio State University, Jet Propulsion Laboratory, Alaska SAR Facility.)*

(*c*)

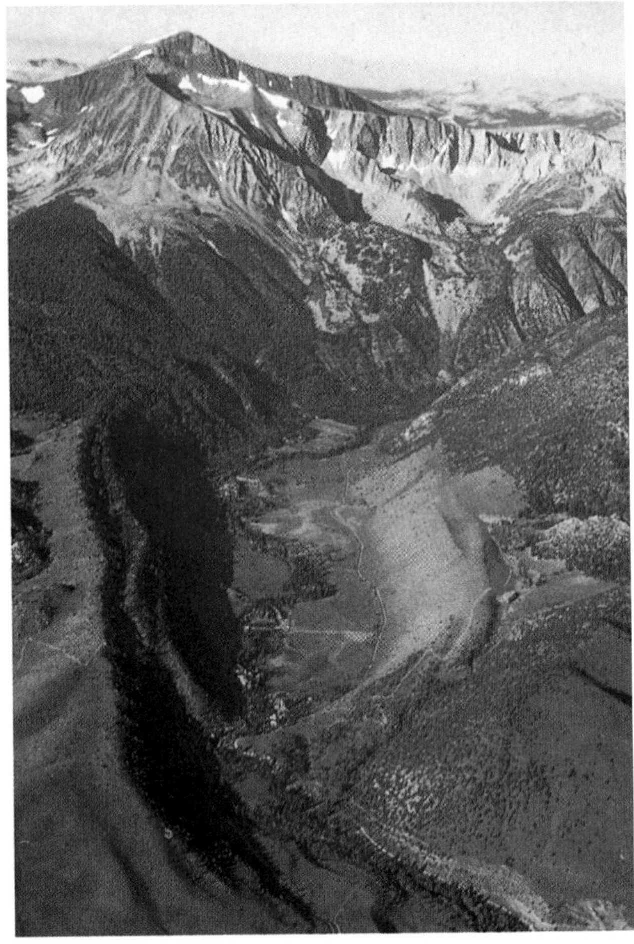

20.6 Alpine terminal moraine *A terminal moraine, shaped like the bow of a great canoe, lies at the mouth of a deep glacial trough on the east face of the Sierra Nevada. Cirques can be seen in the distance. Near Lee Vining, California.*

20.7 Development of a glacial trough *(a) During maximum glaciation, the U-shaped trough is filled by ice to the level of the small tributaries. (b) After glaciation, the trough floor may be occupied by a stream and lakes. (c) If the main stream is heavily loaded, it may fill the trough with alluvium. (d) Should the glacial trough have been deepened below sea level, it will be occupied by an arm of the sea, or fiord. (Drawn by E. Raisz.)*

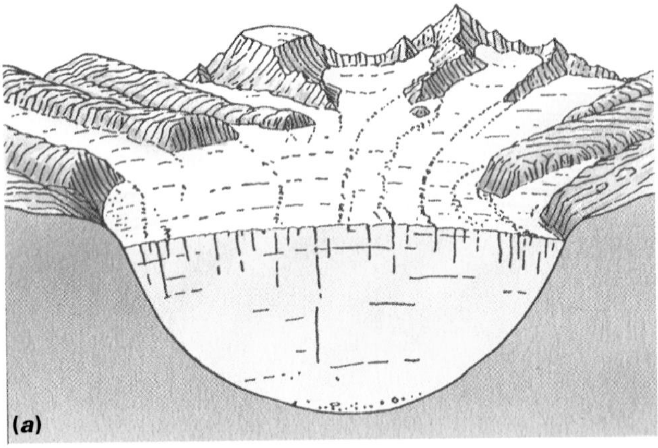

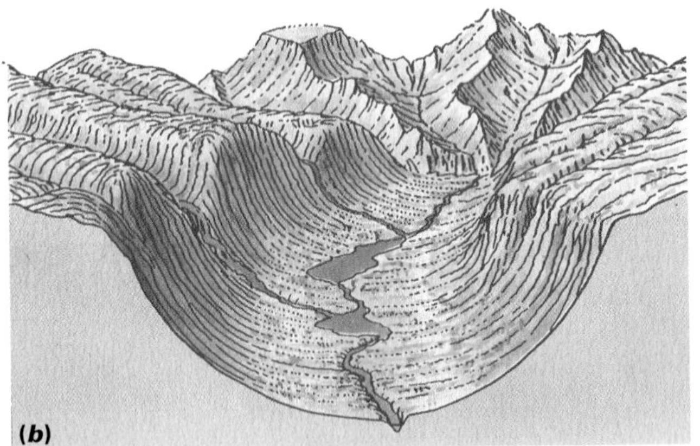

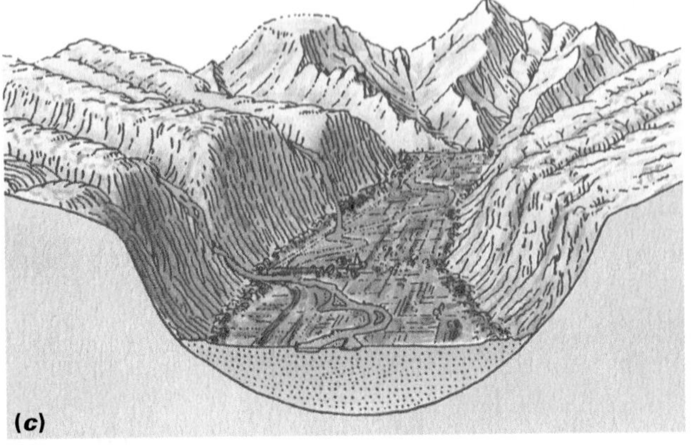

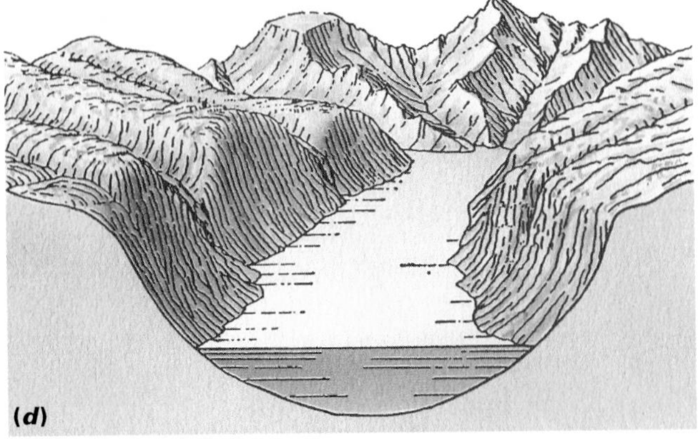

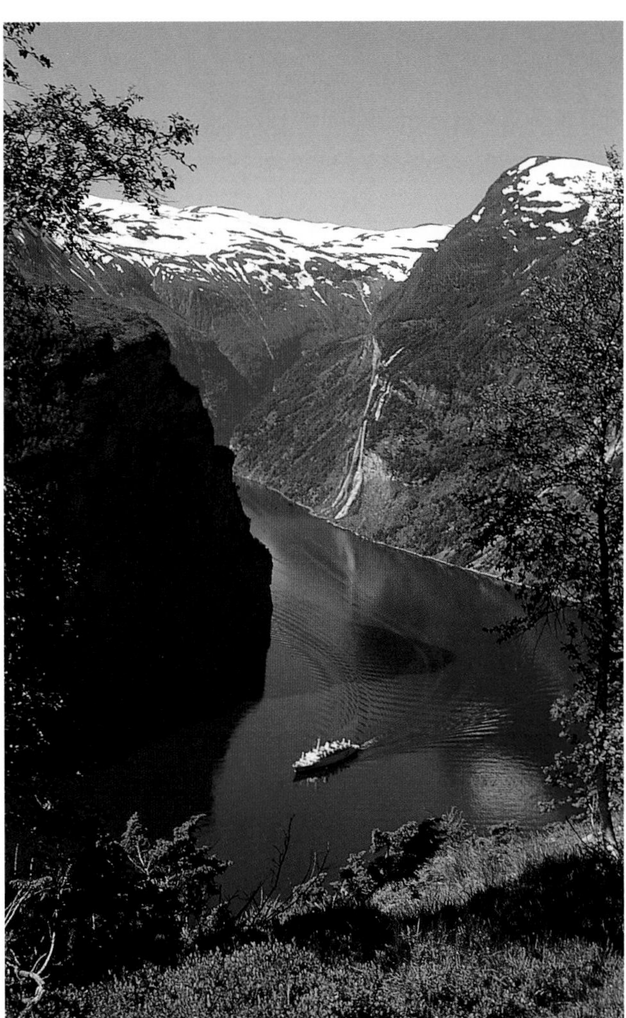

20.8 A glacial trough *Geirangerfjord, Norway, is a deeply carved glacial trough occupied by an arm of the sea.*

freezing of ocean water. In contrast, *icebergs* are bodies of land ice that have broken free from glaciers that terminate in the ocean. Aside from differences in origin, a major difference between sea ice and icebergs is thickness. Sea ice does not exceed 5 m (15 ft) in thickness, while icebergs may be hundreds of meters thick.

Pack ice is sea ice that completely covers the sea surface. Under the forces of wind and currents, pack ice breaks up into individual patches called ice floes. The narrow strips of open water between such floes are known as *leads.* Where ice floes are forcibly brought together by winds, the ice margins buckle and turn upward into pressure ridges that often resemble walls of ice. Travel on foot across the polar sea ice is extremely difficult because of such obstacles. The surface zone of sea ice is composed of fresh water, while the deeper ice is salty.

When a valley glacier or tongue of an ice sheet terminates in sea water, blocks of ice break off to form icebergs (Figure 20.12). Because they are only slightly less dense than sea water, icebergs float very low in the water. About five-sixths of the bulk of an iceberg is submerged. The ice is composed of fresh water since it is formed from compacted and recrystallized snow.

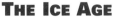

The Ice Age

The period during which continental ice sheets grow and spread outward over vast areas is known as a **glaciation.** Glaciation is associated with a general cooling of average air temperatures over the regions where the ice sheets originate. At the same time, ample snowfall must persist over the growth areas to allow the ice masses to build in volume.

20.9 The Greenland Ice Sheet *Contours show elevations of the ice sheet surface. (Based on data of R. F. Flint, Glacial and Pleistocene Geology, John Wiley & Sons, New York.)*

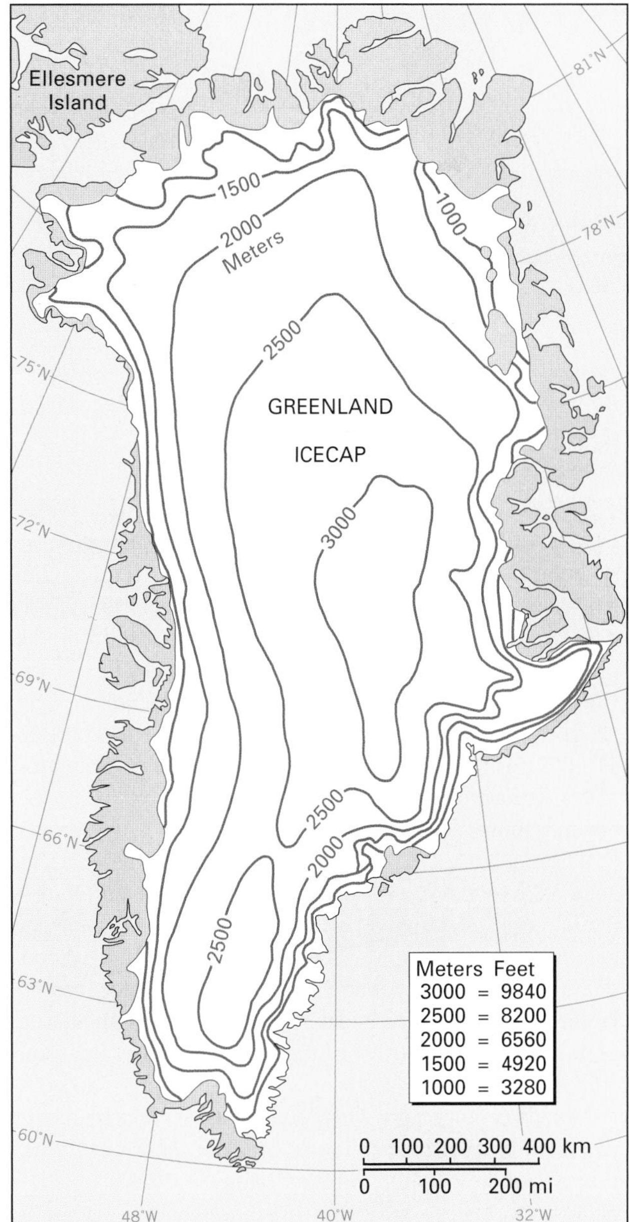

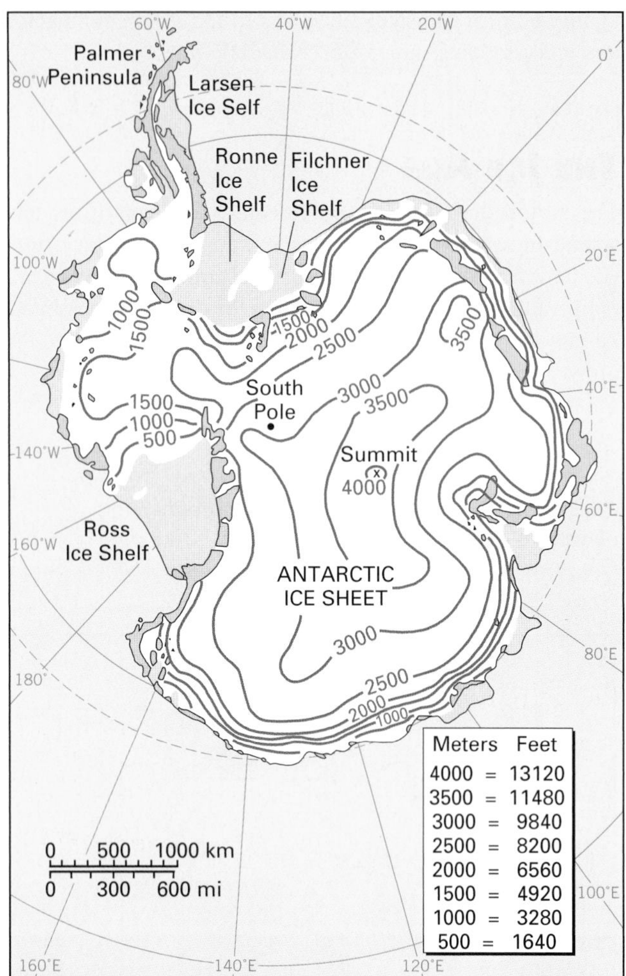

20.10 The Antarctic Ice Sheet and its ice shelves *Contours show elevations of the ice sheet surface. (Based on data of American Geophysical Union.)*

When the climate warms or snowfall decreases, ice sheets become thinner and cover less area. Eventually, the ice sheets may melt completely. This period is called a *deglaciation.* Following a deglaciation, but preceding the next glaciation, is a period in which a mild climate prevails—an **interglaciation.** The last interglaciation began about 140,000 years ago and ended between 120,000 and 110,000 year ago. A succession of alternating glaciations and interglaciations, spanning a period of 1 to 10 million years or more, constitutes an *ice age.*

Throughout the past 3 million years or so, the Earth has been experiencing the **Late-Cenozoic Ice Age** (or, simply, the **Ice Age**). As you may recall from Chapter 13, the Cenozoic Era has seven epochs (see Table 13.1). The Ice Age falls within the last three epochs: Pliocene, Pleistocene, and Holocene. These three epochs comprise only a small fraction—about one-twelfth—of the total duration of the Cenozoic Era.

A half-century ago, most geologists associated the Ice Age with the Pleistocene Epoch, which began about 1.6 million years before the present. However, new evidence obtained from deep-sea sediments shows that the glaciations of the Ice Age began in late Pliocene time, perhaps 2.5 to 3.0 million years ago.

At present, we are within an interglaciation of the Late-Cenozoic Ice Age, following a deglaciation that set in quite rapidly about 15,000 years ago. In the preceding glaciation, called the *Wisconsinan Glaciation*, ice sheets covered much of North America and Europe, as well as parts of northern Asia and southern South America. The maximum ice advance of the Wisconsinan Glaciation was reached about 18,000 years ago.

Glaciation During the Ice Age

Figures 20.13 and 20.14 show the maximum extent to which North America and Europe were covered during the last advance of the ice. Most of Canada was engulfed by the vast Laurentide Ice Sheet. It spread south into the United States, covering most of the land lying north of the Missouri and Ohio rivers, as well as northern Pennsylvania and all of New York and New England. Alpine glaciers of the western ranges coalesced into a single ice sheet that spread to the Pacific shores and met the Laurentide sheet on the east. Notice that an area in southwestern Wisconsin escaped inundation. Known as the Driftless Area, it was apparently bypassed by glacial lobes moving on either side.

20.11 Sea ice *A Landsat image of a portion of the Canadian arctic archipelago. An ice cap on a landmass is visible in the center of the photo, with exposed mountainous ridges on the uncovered portions of the landmass. A branching glacial trough, now a water-filled fiord, is seen in the lower part of the image. In the upper left are huge chunks of free-floating sea ice.*

20.12 Iceberg *Penguins enjoy an outing on an iceberg off the coast of Antarctica near Elephant Island.*

20.13 Maximum glaciation in North America *Continental glaciers of the Ice Age in North America at their maximum extent reached as far south as the present Ohio and Missouri rivers. Note that during glaciations sea level was much lower. The present coastline is shown for reference only. (Based on data of R. F. Flint, Glacial and Pleistocene Geology, John Wiley & Sons, New York.)*

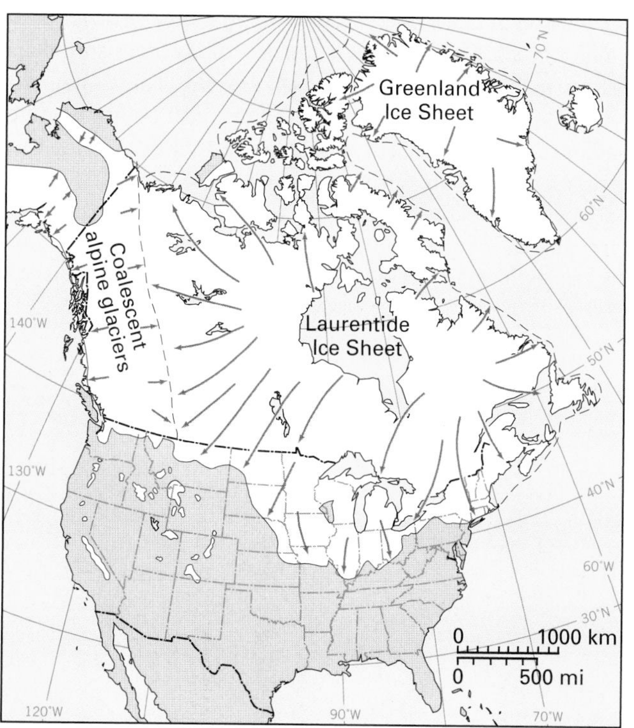

20.14 Maximum glaciation in Europe *The Scandinavian Ice Sheet dominated Northern Europe during the Ice Age glaciations. As noted in Figure 20.13, the present coastline is far inland from the coastline that prevailed during glaciations. (Based on data of R. F. Flint, Glacial and Pleistocene Geology, John Wiley & Sons, New York.)*

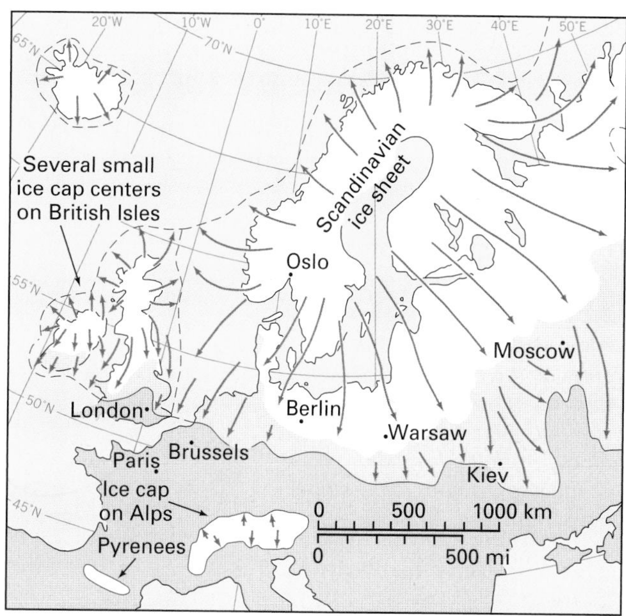

In Europe, the Scandinavian Ice Sheet centered on the Baltic Sea, covering the Scandinavian countries. It spread south into central Germany and far eastward to cover much of Russia. In north-central Siberia, large ice caps formed over the northern Ural Mountains and highland areas farther east. Ice from these centers grew into a large sheet covering much of central Siberia. The European Alps were capped by enlarged alpine glaciers. The British Isles were mostly covered by a small ice sheet that had several centers on highland areas and spread outward to coalesce with the Scandinavian Ice Sheet.

At the maximum spread of these ice sheets, sea level was as much as 125 m (410 ft) lower than today, exposing large areas of the continental shelf on both sides of the Atlantic Basin. The shelf supported a vegetated landscape populated with animal life, including Pleistocene elephants (mastodons and mammoths). The drawdown of sea level explains why the ice sheets shown on our maps extend far out into what is now the open ocean.

South America, too, had an ice sheet. It grew from ice caps on the southern Andes Range south of about latitude 40°S and spread westward to the Pacific shore, as well as eastward, to cover a broad belt of Patagonia. It covered all of Tierra del Fuego, the southern tip of the continent. The South Island of New Zealand, which today has a high spine of alpine mountains with small relict glaciers, developed a massive ice cap in late Pleistocene time. All high mountain areas of the world underwent greatly intensified alpine glaciation at the time of maximum ice sheet advance. Today, most remaining alpine glaciers are small ones. In less favorable locations, the Ice Age alpine glaciers are entirely gone.

The weights of the continental ice sheets, covering vast areas with ice masses several kilometers thick, exerted downward forces on the crust, causing depressions of the crust of hundreds of meters at some locations. *Working It Out 20.3 • Isostatic Rebound* provides a more detailed look at this phenomenon in two locations from Canada and one from Norway. GEODISCOVERIES

LANDFORMS MADE BY ICE SHEETS

Landforms made by the last ice advance and recession are very fresh in appearance and show little modification by erosion processes. It is to these landforms that we now turn our attention.

Erosion by Ice Sheets

Like alpine glaciers, ice sheets are highly effective eroding agents. The slowly moving ice scraped and ground away much solid bedrock, leaving behind smoothly rounded rock masses. These bear countless grooves and scratches trending in the general direction of ice movement (see Figure 20.2). Sometimes the ice polishes the rock to a smooth, shining surface. The evidence of ice abrasion is common throughout glaciated regions of North America and may be seen on almost any hard rock surface that is freshly exposed. Conspicuous knobs of solid bedrock shaped by the moving ice are also common features (Figure 20.15). The side from which the ice approached is usually smoothly rounded. The lee side, where the ice plucked out angular joint blocks, is irregular and blocky.

The ice sheets also excavated enormous amounts of rock at locations where the bedrock was weak and the flow of ice was channeled by the presence of a valley trending in the direction of ice flow. Under these conditions, the ice sheet behaved like a valley glacier, scooping out a deep, U-shaped trough. The Finger Lakes of western New York State are fine examples (Figure 20.16). Here, a set of former stream valleys lay largely parallel to the southward spread of the ice, and a set of long, deep basins was eroded. Blocked at their north ends by glacial debris, the basins now hold lakes. Many hundreds of lake basins were created by glacial erosion and deposition over the glaciated portion of North America.

Deposits Left by Ice Sheets

The term **glacial drift** includes all varieties of rock debris deposited in close association with glaciers. Drift is of two major types. *Stratified drift* consists of layers of sorted and stratified clays, silts, sands, or gravels. These materials were deposited by meltwater streams or in bodies of water adjacent to the ice. **Till** is an unstratified mixture of rock fragments, ranging in size from clay to boulders, that is deposited directly from the ice without water transport. As the glacial ice melts in a stagnant marginal zone, the rock particles it holds are lowered to the solid surface beneath,

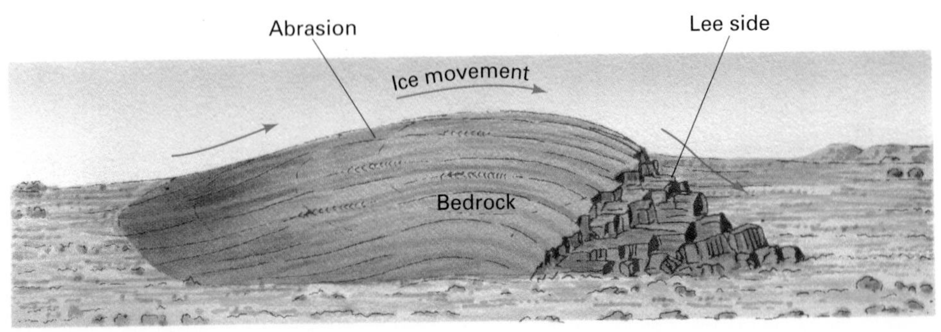

20.15 A glacially abraded rock knob *Glacial action abrades the rock into a smooth form as it rides over the rock summit, then plucks bedrock blocks from the lee side, producing a steep, rocky slope. (Copyright © A. N. Strahler.)*

20.16 Eye on the Landscape Ice sheet excavation *The Finger Lakes region of New York is shown in this photo taken by astronauts aboard the Space Shuttle. The lakes occupy valleys that were eroded and deepened by glacial ice.* **What else would the geographer see?...Answers at the end of the chapter.**

where they form a layer of debris (Figure 20.17). This *ablational till* shows no sorting and often consists of a mixture of sand and silt, with many angular pebbles and boulders. Beneath this residual layer there may be a basal layer of dense *lodgment till,* consisting of clay-rich debris previously dragged forward beneath the moving ice. Where till forms a thin, more or less even cover, it is referred to as *ground moraine.*

Over those parts of North America formerly covered by late-Cenozoic ice sheets, glacial drift thickness averages from 6 m (about 20 ft) over mountainous terrain, such as New England, to 15 m (about 50 ft) and more over the lowlands of the north-central United States. Over Iowa, drift thickness is from 45 to 60 m (about 150 to 200 ft), and over Illinois it averages more than 30 m (about 100 ft). In some places where deep stream valleys existed prior to glacial advance, as in parts of Ohio, drift is much thicker.

To understand the form and composition of deposits left by ice sheets, it will help to examine the conditions prevailing at the time of the ice sheet's existence. Figure 20.18*a* shows a region partly covered by an ice sheet with a stationary front edge. This condition occurs when the rate of ice ablation balances the amount of ice brought forward by spreading of the ice sheet. Although the ice fronts of the Ice Age advanced and receded in many minor and major fluctuations, there were long periods when the front was essentially stable and thick deposits of drift accumulated.

MORAINES The transportational work of an ice sheet resembles that of a huge conveyor belt. Anything carried on the belt is dumped off at the end and, if not constantly removed, will pile up in increasing quantity. Rock fragments brought within the ice are deposited at its outer edge

20.17 Glacial till *(a) As ice passes over the ground, sediment and coarse rock fragments are pressed into a layer of lodgement till. (b) When the overlying ice stagnates and melts, it leaves a residual deposit of ablation till. (Copyright © A. N. Strahler.)*

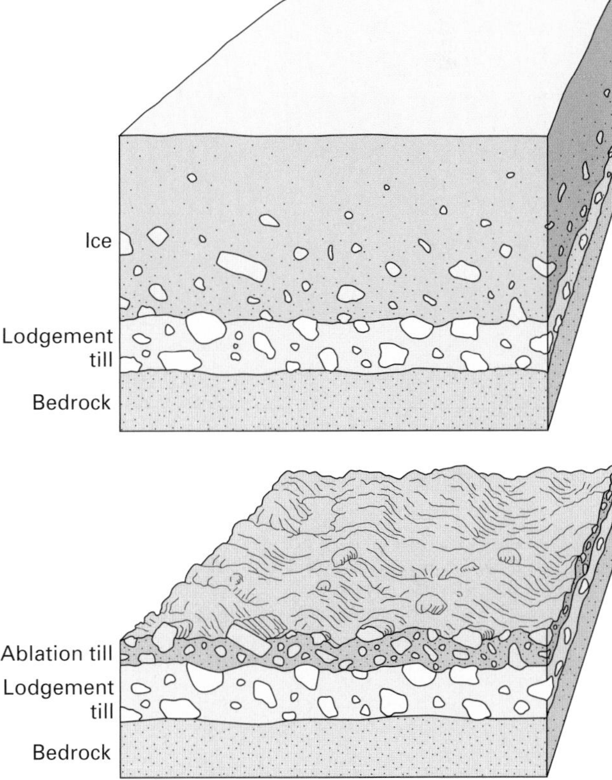

Working It Out 20.3

Isostatic Rebound

In *Focus on Systems 18.3 • A Model Denudation System,* we described the principle of *isostasy,* which states that the crust floats on a plastic asthenosphere. According to this principle, a crustal block rises as erosion strips away its rock mass through time, a phenomenon we termed *isostatic compensation.*

During the Ice Age, large continental ice sheets of several kilometers in thickness covered large areas of North America, Europe, and Asia. The weight of these ice sheets depressed the crust, causing sinking. Figure 1 diagrams this process. With the onset of the most recent glaciation at about 80,000 years before present, ice sheets grew rapidly, depressing the crust underneath. At the close of the last glacial period, about 12,000 years ago, deglaciation was rapid, requiring only a few thousand years. Relieved of the weight of the ice, the crust began to move upward, quickly at first, then more slowly. This crustal movement is known as *isostatic rebound.* Upward movement is still occurring today in some locations.

An important line of evidence for isostatic rebound is the position of ancient shorelines, now raised above sea level by the crustal motion. How did these shorelines form? As the ice sheets melted, sea level rose rapidly to near its present position. The crust, still depressed, was flooded with ocean waters, providing wave action at the shoreline that created beaches. As the crust rebounded upward, the beaches were elevated, leaving them hundreds of meters above present levels. By dating these ancient shorelines, geologists have established the amount and rate of uplift occurring in response to the unloading of ice. Many elevated shorelines can be found today in the coastal belts of Hudson Bay, located in north-central Canada, and in the Bay of Bothnia, which separates Sweden and Finland.

Figure 2 provides plots of shoreline elevations with age for three locations. The points are taken from published graphs. North Bay is in Ontario on Lake Huron, located at the south end of Georgian Bay. James Bay is the southernmost projection of Hudson Bay and lies near the center of crustal depression for the Laurentide Ice Sheet (see Figure 20.13). Oslofjord is the narrow glacial trench that provides ocean access to Oslo, Norway (see Figure 20.14). Note that both axes on the graphs are numerically reversed. For the *x*-axis, time is in years before present (B.P.), so time increases from left to right. For the *y*-axis, shoreline height is taken as the crustal depression at the time of formation of the shoreline. Upward crustal movement is shown by a higher position on this graph.

The upward motion of the crust in isostatic rebound is most rapid immediately following the melting of the ice. With time, rebound continues, but the rate of uplift decreases as the present level is approached. North Bay, located significantly south of the center of the ice sheet, shows the earliest rebound, followed by James Bay and then Oslofjord. Note the steeper rebound for James Bay. Located near the center of the ice sheet, this region was depressed most by the ice and so experienced the strongest rebound immediately after deglaciation.

The middle part of the figure shows the rebound plotted on a semilog graph. The points show a

Figure 1 *Crustal depression and rise in response to the growth and decay of continental ice sheets. (After A. N. Strahler.)*

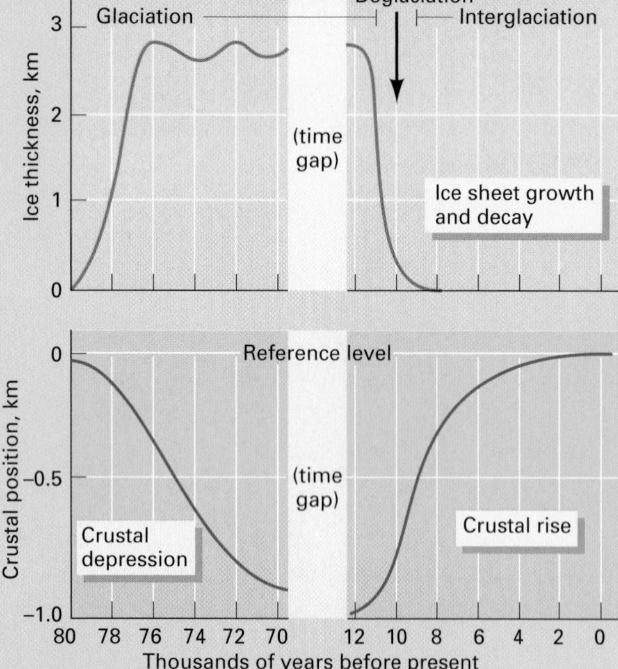

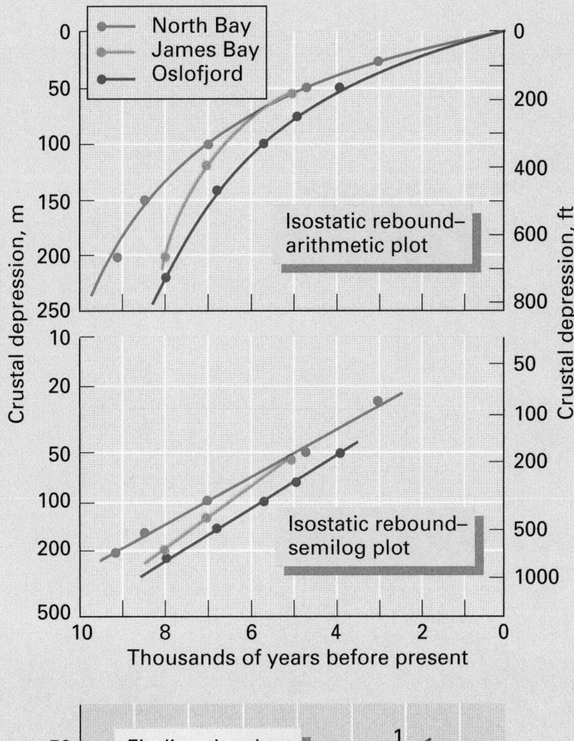

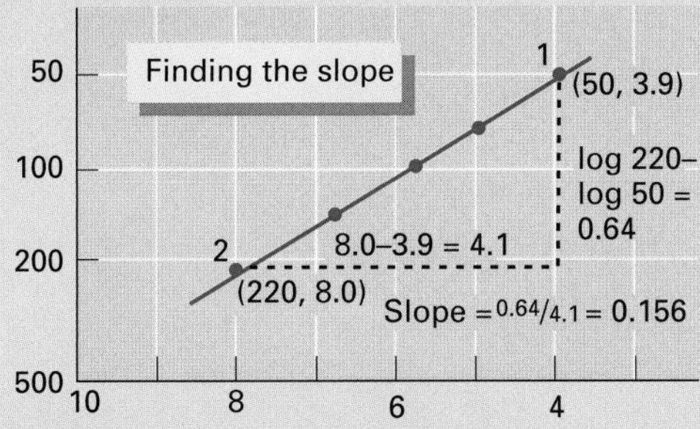

Figure 2 *Plots of isostatic rebound. Upper plots: arithmetic and semilogarithmic. The lowest plot shows how the slope of the semilog line is determined. (Data from R. F. Flint, W. Ferrand, and W. A. Heiskanen.)*

good fit to a straight line. For this to occur, the points follow a line with the equation

$$y = ba^x \qquad (1)$$

where *b* and *a* are constants. (Note that in evaluating this expression, *a* is raised to the power *x* before the result is multiplied by *b*.) By taking the log transform, we can see how this equation becomes a straight line:

$$\log y = \log b + x \log a$$

If we substitute $y' = \log y$, $b' = \log b$, and $a' = \log a$, the equation becomes

$$y' = b' + a'x \qquad (2)$$

which is a straight line. Since $y' = \log y$, we use a logarithmic scale on the *y*-axis. But since *x* is untransformed, we just use a normal arithmetic scale for the *x*-axis.

It is not difficult to determine the equation for a line that best fits points in a semilog graph. A simple way is to use two points that lie on or close to the line, and from their values determine the slope, which will give *a′*, and the intercept, which will give *b′*. The process is diagrammed in the bottom graph for the Oslofjord data.

Two points, 1 and 2, are compared: 50 m at 3900 years B.P., and 220 m at 8000 years B.P.

The slope of the line is simply the vertical change divided by the horizontal change for the two points. Normally, we would determine the change by subtracting the coordinates of point 2 from those of point 1, but in this case the axes are reversed, changing the sign of the values. So here we subtract the coordinates of point 1 from point 2. Note also that we will need to use log values for *y*, the shoreline height. In addition, it will be simpler to work in units of thousands of years, as shown on the *x*-axis. The slope is then

$$a' = \frac{\log 220 - \log 50}{8.0 - 3.9} = \frac{0.64}{4.1} = 0.156$$

For the intercept, *b′*, we can simply solve (2) for *b′*, giving $b' = y' - a'x$, and substitute the values of *y′* and *x* from one of the points, for example, point 1, given the value of *a′*:

$$b' = y' - a'x = \log 50 - 0.156 \times 3.9 = 1.09$$

In semilog form, the equation is then

$$\log y = 1.09 + 0.156x$$

To obtain the exponential form of the equation (1), we first find *a*. Since $a' = \log a$, $a = 10^{a'} = 10^{0.156} = 1.43$. Similarly, $b' = \log b$, so $b = 10^{b'} = 10^{1.09} = 12.3$. The equation is then

$$y = 12.3 \times (1.43)^x$$

Plotting data and fitting functions to the points is one of the important tools of science, including physical geography. Geophysicists have derived more complex formulas for crustal rebound using the theory of physics of elastic solids, but for many purposes, including exploring data and comparing trends, simple methods like the one above are often all that is needed.

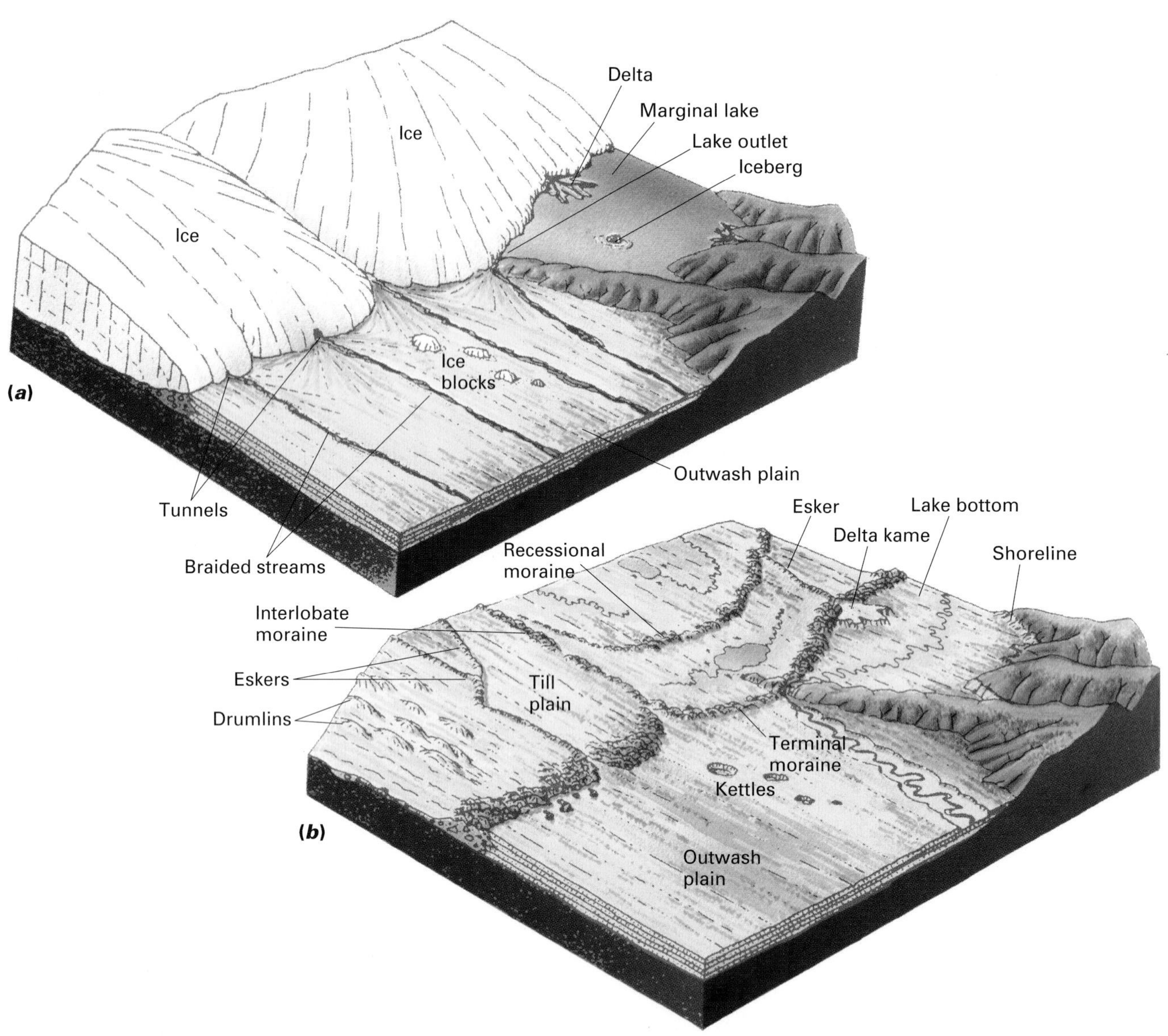

20.18 Marginal landforms of continental glaciers *(a) With the ice front stabilized and the ice in a wasting, stagnant condition, various depositional features are built by meltwater. (b) The ice has wasted completely away, exposing a variety of new landforms made under the ice. (Drawn by A. N. Strahler.)*

as the ice evaporates or melts. Glacial till that accumulates at the immediate ice edge forms an irregular, rubbly heap—the *terminal moraine.* After the ice has disappeared (Figure 20.18*b*), the moraine appears as a belt of knobby hills interspersed with basinlike hollows, or kettles, some of which hold small lakes (Figure 20.19*a*). The name *knob-and-kettle* is often applied to morainal belts.

Terminal moraines form great curving patterns. The outward curvature is southward and indicates that the ice advanced as a series of great *ice lobes,* each with a curved front (Figure 20.20). Where two lobes come together, the moraines curve back and fuse together into a single *interlobate moraine* pointed northward. In its general recession accompanying disappearance, the ice front paused for some time along a number of positions, causing morainal belts similar to the terminal moraine belt to be formed. These belts are known as recessional moraines (Figure 20.18*b*). They run roughly parallel with the terminal moraine but are often thin and discontinuous.

OUTWASH AND ESKERS Figure 20.18 shows a smooth, sloping plain lying in front of the ice margin. This is the *outwash plain,* formed of stratified drift left by braided streams issuing from the ice. The plain is built of layer upon layer of sands and gravels.

Large streams carrying meltwater issue from tunnels in the ice. These form when the ice front stops moving for many kilometers back from the front. After the ice has gone, the position of a former ice tunnel is marked by a long, sinuous ridge of sediment known as an esker (Figure 20.19*b*). The esker is the deposit of sand and gravel laid on the floor of the former ice tunnel. After the ice has melted away, only the streambed deposit remains, forming a ridge. Many eskers are several kilometers long.

DRUMLINS AND TILL PLAINS Another common glacial form is the *drumlin,* a smoothly rounded, oval hill resembling the bowl of an inverted teaspoon. It consists in most cases of glacial till (Figure 20.19*c*). Drumlins invariably lie in a zone behind the terminal moraine. They commonly occur in groups or swarms and may number in the hundreds. The long axis of each drumlin parallels the direction of ice movement. Drumlins are typically steeper at the broad end, which faces oncoming ice. The origin of drumlins is not well understood. They seem to have been formed under moving ice by a plastering action in which layer upon layer of bouldery clay was spread on the drumlin.

Between moraines, the surface overridden by the ice is covered by glacial till. This cover is often inconspicuous since it forms no prominent landscape feature. The till layer may be thick and may obscure, or entirely bury, the hills and valleys that existed before glaciation. Where thick and smoothly spread, the layer forms a level *till plain.* Plains of this origin are widespread throughout the central lowlands of the United States and southern Canada.

MARGINAL LAKES AND THEIR DEPOSITS When the ice advanced toward higher ground, valleys that may have opened out northward were blocked by ice. Under such conditions, marginal glacial lakes formed along the ice front (see Figure 20.18*a*). These lakes overflowed along the lowest available channel between the ice and the rising ground slope, or over some low pass along a divide. Streams of meltwater from the ice built *glacial deltas* into these marginal lakes.

When the ice withered away, the lakes drained, leaving a flat floor exposed. Here, layers of fine clay and silt had accumulated. Glacial lake plains often contain extensive areas of marshland. The deltas are now curiously isolated, flat-topped landforms known as *delta kames,* composed of well-washed and well-sorted sands and gravels (Figures 20.18*b,* 20.19*d*).

PLUVIAL LAKES During the Ice Age, some regions experienced a cooler, moister climate. In the western United States, closed basins filled with water, forming pluvial lakes. The largest of these, glacial Lake Bonneville, was about the size of Lake Michigan and occupied a vast area of western Utah. With the warmer and drier climate of the present interglacial period, these lakes shrank greatly in volume. Lake Bonneville became the present-day Great Salt Lake. Many other lakes dried up completely, forming desert playas. The history of these pluvial lakes is known from their ancient shorelines, some as high as 300 m (about 1000 ft) above present levels.

Environmental Aspects of Glacial Deposits

Because much of Europe and North America was glaciated by the Pleistocene ice sheets, landforms associated with the ice are of major environmental importance. Agricultural influences of glaciation are both favorable and unfavorable, depending on preglacial topography and the degree and nature of ice erosion and deposition.

In hilly or mountainous regions, such as New England, the glacial till is thinly distributed and extremely stony. Till cultivation is difficult because of countless boulders and cobbles in the clay soil. Till accumulations on steep mountain or roadside slopes are subject to mass movement as earthflows when clay in the till becomes weakened after absorbing water from melting snows and spring rains. Along moraine belts, the steep slopes, the irregularity of knob-and-kettle topography, and the abundance of boulders hinder crop cultivation but are suitable for use as pasture.

Flat till plains, outwash plains, and lake plains, on the other hand, can sometimes provide very productive agricultural land. Bordering the Great Lakes, fertile soils have formed on till plains and on exposed lakebeds. This fertility is enhanced by a blanket of wind-deposited silt (loess) that covers these plains (Chapter 19).

Stratified drift deposits are of great commercial value. The sands and gravels of outwash plains, deltas, and eskers provide necessary materials for both concrete manufacture and highway construction. Where it is thick, stratified drift forms an excellent aquifer and is a major source of ground water supplies (see Chapter 16).

Investigating the Ice Age

A great scientific breakthrough in the study of Ice Age glacial history came in the 1960s. First, scientists learned how to measure the absolute age of certain types of water-laid sediments by means of ancient magnetism. The Earth's magnetic field experienced many sudden reversals of polarity in Cenozoic time, and the absolute ages of these reversals have been firmly established. Second, techniques were developed to take long sample cores of undisturbed fine-textured sediments of the deep ocean floor. Within each core, scientists could determine the age of sediment layers at various control points by identifying magnetic polarity reversals. By further studying the composition and chemistry of the layers within the core, a record of ancient temperature cycles in the air and ocean could be established.

Deep-sea cores reveal a long history of alternating glaciations and interglaciations going back at least as far as 2 million years and possibly 3 million years before present. The cores show that in late-Cenozoic time more than 30 glaciations occurred, spaced at time intervals of about 90,000 years. How much longer this sequence will continue into the future is not known, but perhaps for 1 or 2 million years, or even longer.

20.19 Landforms of Continental Ice Sheets

a...***Recessional moraine*** *The lower left portion of this aerial scene from Langlade County, Wisconsin, shows a recessional moraine covered with forest vegetation. Note the bumpy, irregular topography of sediments piled up at the former ice edge. To the upper right is the now-cultivated outwash plain of sediments laid down by rivers and streams draining the melting ice front.*

b...***Esker*** *The curving ridge of sand and gravel in this photo is an esker, marking the bed of a river of meltwater flowing underneath a continental ice sheet near its margin. Kettle-Moraine State Park, Wisconsin.*

c...***Drumlin*** *This small drumlin, located south of Sodus, New York, shows a tapered form from upper right to lower left, indicating that the ice moved in that direction (north to south).*

d...***Kame*** *This tree-covered hill, rising above the surrounding plain, is a kame—a deposit of sand and gravel built out from the front of a retreating ice sheet, possibly as a delta accumulating in a short-lived lake. As the ice melted and the lake drained, the deposit lost its lateral support and slumped down under the force of gravity, forming a hill of roughly conical shape.*

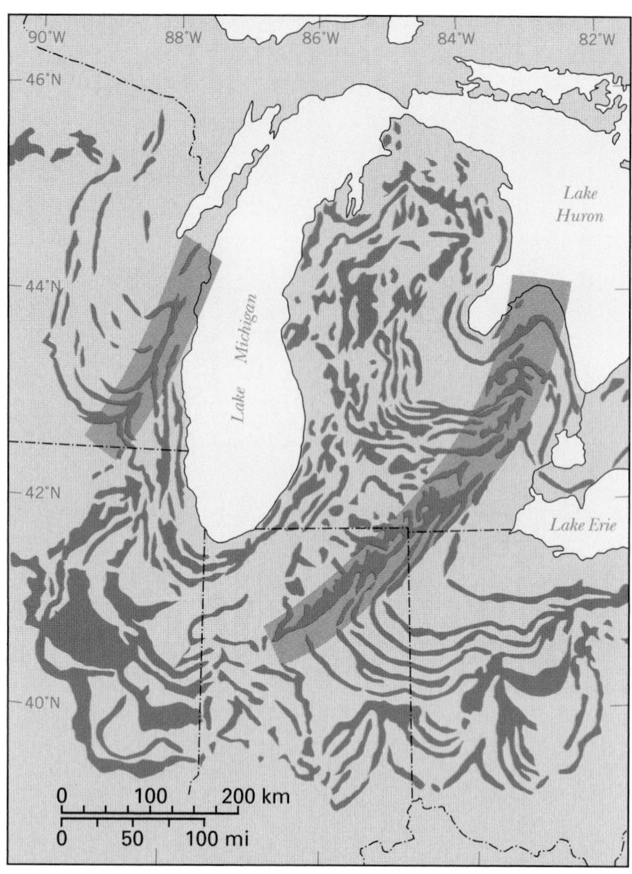

20.20 Map of Midwest moraines *Moraine belts of the north-central United States have a curving pattern left by ice lobes. Some regions of interlobate moraines are shown by the color overlay. (Based on data of R. F. Flint and others, Glacial Map of North America, Geological Society of America.)*

Possible Causes of the Late-Cenozoic Ice Age

What caused the Earth to enter into an Ice Age with its numerous cycles of glaciation and interglaciation? Three causes seem possible. First is a change in the placement of continents on the Earth's surface through plate tectonic activity. Second is an increase in the number and severity of volcanic eruptions. Third is a reduction in the Sun's energy output.

Perhaps the answer lies in plate tectonics, through the motions of lithospheric plates following the breakup of Pangea. Recall from Chapter 13 that in Permian time, only the northern tip of the Eurasian continent projected into the polar zone. But as the Atlantic Basin opened up, North America moved westward and poleward to a position opposite Eurasia, while Greenland took up a position between North America and Europe.

Geographers at Work

Ice Coring: Looking for Evidence of Climate Change

by Kurt M. Cuffey ***University of California, Berkeley***

Maybe the first time you saw a glacier was out of the window of an airplane—while approaching a landing at Seattle, you looked northward to Mount Ranier and squinted at the white mass that encased its lofty summit. Or maybe it was at Glacier National Park, where you walked up to the snout of Grinnell Glacier and kicked at its cheesy edge, wondering how this soggy pile of dirt-covered ice had lasted so long. Or maybe it was on that Alaska cruise, where the fog and low clouds lifted long enough for a few quick glimpses of white patches on the flanks of Glacier Bay.

Curious things, those glaciers. But one nice thing about them is that they're *old*. Thousands of years old. In fact, an ice sheet glacier, like that of Greenland or Antarctica, can hold layers of ice that fell as snow several hundred thousand years ago. By coring the ice sheet to a great depth, and then extracting and assaying the air trapped in successive layers with very sensitive instruments, it is possible to reconstruct the climates of long ago.

Kurt Cuffey, a geographer at UC Berkeley, couples the data from ice cores with mathematical models to study how ice sheets have responded to climate changes in the past. For example, his analysis has shown that the Greenland Ice Sheet melted back significantly during the last warm climate period, about 125,000 years ago. And, that during the height of the last glaciation, Greenland was as much as 20°C (36°F) colder than at present! To find out more about Kurt and his interests, visit our web site.

Kurt M. Cuffey's research focuses on the growth and shrinkage of the great ice sheets of Greenland and Antarctica. (P. N. Lombard)

The effect of these plate motions was to bring an enormous land-mass area to a high latitude and to surround a polar ocean with land. Because the flow of warm ocean currents into the polar ocean was greatly reduced, or at times totally cut off, this arrangement was favorable to the growth of ice sheets. The polar ocean was ice-covered much of the time, and average air temperatures in high latitudes were at times lowered enough to allow ice sheets to grow on the encircling continents. Furthermore, Antarctica moved southward during the breakup of Pangea and took up a position over the south pole. In that location, it was ideally situated to develop a large ice sheet. Some scientists have also proposed that the uplift of the Himalayan Plateau, a result of the collision of the Austral-Indian and Eurasian plates, could have modified weather patterns sufficiently to trigger the Ice Age.

The second geological mechanism suggested as a basic cause of the Ice Age is increased volcanic activity on a global scope in late-Cenozoic time. Volcanic eruptions produce dust veils that linger in the stratosphere and reduce the intensity of solar radiation reaching the ground (Chapter 4). Temporary cooling of near-surface air temperatures follows such eruptions. Although the geologic record shows periods of high levels of volcanic activity in the Miocene and Pliocene epochs, their role in initiating the Ice Age has not been convincingly demonstrated on the basis of present evidence.

Another possible cause of the Ice Age is a slow decrease in the Sun's energy output over the last several million years, perhaps as part of a cycle of slow increase and decrease over many million years' duration. As yet, data are insufficient to identify this mechanism as a possible basic cause. However, research on this topic is being stepped up as new knowledge of the Sun is acquired from satellites that probe the Sun's atmosphere and monitor its changing surface.

Possible Causes of Glaciation Cycles

What timing and triggering mechanisms are responsible for the many cycles of glaciation and interglaciation that the Earth is experiencing during the present Ice Age? Although many causes for glacial cycles have been proposed, we will limit our discussion here to one major contender called the **astronomical hypothesis.** It has been under consideration for about 40 years and is now widely accepted.

The astronomical hypothesis is based on well established motions of the Earth in its orbit around the Sun. Recall from Chapter 3 that the Earth's orbit around the Sun is an ellipse, not a circle, and that we refer to the point in the orbit nearest the Sun as perihelion, and the point farthest as aphelion. Presently, perihelion occurs around December 5 and aphelion around July 5. Astronomers have observed that the orbit slowly rotates on a 108,000 year cycle, thus shifting the absolute time of perihelion and aphelion by a very small amount each year. In addition, the orbit's shape varies on a cycle of 92,000 years, becoming more and less elliptical. This changes the Earth–Sun distance and therefore the amount of solar energy the Earth receives at each point of the annual cycle.

The Earth's axis of rotation also experiences cyclic motions. The tilt angle of the axis varies from about 22 to 24 degrees on a 40,000 year cycle. The axis also "wobbles" on a 26,000 year cycle, moving in a slow circular motion much like that of a spinning top or toy gyroscope.

As a result of these cycles in axial rotation and solar revolution, the annual insolation experienced at each latitude changes from year to year. Figure 20.21 shows a graph of summer insolation received at 65° N latitude for the last 500,000 years as calculated from these cycles. The graph is known as the Milankovitch curve, named for Milutin Milankovich, the astronomer who first calculated it in 1938. The dominant cycle of the curve has a period of about 40,000 years, but note that every second or third peak seems to be higher. Peaks at about 12,000, 130,000, 220,000, 285,000, and 380,000 years ago have been associated with rapid melting of ice sheets and the onset of deglaciations, as revealed by other dating methods involving ancient ice cores and deep lake sediment cores. Most

20.21 The Milankovitch curve *The vertical axis shows fluctuations in summer daily insolation at lat. 65° N for the last 500,000 years. These are calculated from mathematical models of the change in Earth–Sun distance and change in axial tilt with time. The zero value represents the present value. (Based on calculations by A. D. Vernekar, 1968. Copyright © A. N. Strahler.)*

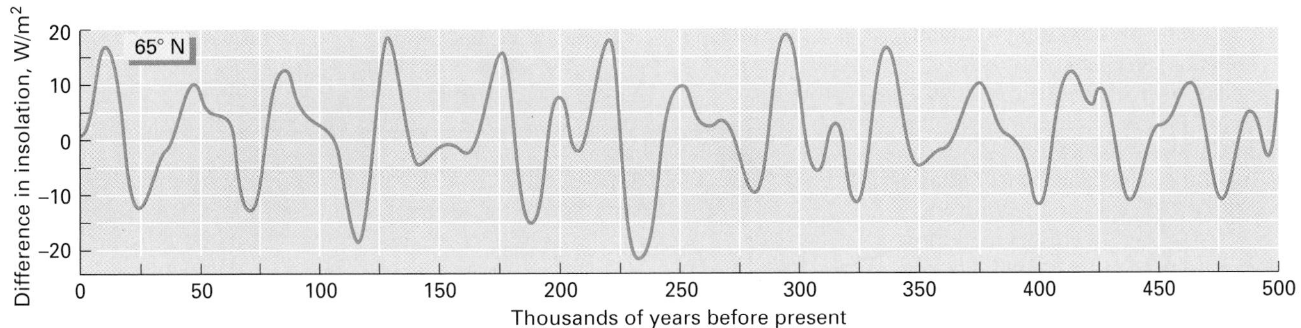

Eye on Global Change | 20.4

Ice Sheets and Global Warming

What effects will global warming have on the Earth's ice sheets? The Antarctic Ice Sheet holds 91 percent of the Earth's ice. If it were to melt entirely, the Earth's mean sea level would rise by about 40 m (about 200 ft). Additional water could be released by the melting of the Greenland Ice Sheet, which holds most of the remaining volume of land ice. Although global warming will accelerate glacial melting and thinning of the ice at the edges of ice sheets, increased precipitation, held by warmer air over the ice sheets, will likely produce a net growth in ice sheet thickness. Thus, it seems most unlikely that global warming will melt these ice sheets completely, unless the warming is much greater than anticipated. However, there may be other factors at work than snow buildup and marginal melting in the case of individual ice sheets.

In 1998 NASA researchers repeated measurements of the height of the southern half of the Greenland Ice Sheet that were originally made in 1993 using a laser altimeter mounted on an aircraft. They discovered that marginal thinning had reduced the volume of the ice sheet by about 42 km^3 (about 10 mi^3) in that five-year period. The thinning was primarily on the eastern side of the Ice Sheet. On the western side, melting was more in balance with new snow accumulation. The loss of this ice volume is not large enough to have much effect on global sea level. But a substantial flow of fresh meltwater into the ocean could influence the thermohaline circulation of the northern Atlantic (see Chapter 7), triggering an ocean cycle of several centuries' duration that would bring very cold winter weather to Europe.

At the other pole, recent attention has focused on the Western Antarctic Ice Sheet, shown in the map to the right. Much of this vast expanse of Antarctic ice is "grounded" on a bedrock base that is well below sea level—that is, a large portion of the ice sheet rests directly on deep bedrock without sea water underneath. Attached to the grounded ice sheet are the Ross, Ronne, and Filchner ice shelves, which are underlain by sea water. They are also grounded at some points where the shelves enclose islands and overlay higher parts of undersea topography. At present, the grounded ice shelves act to hold back the flow of the main part of the ice sheet.

Geophysicists regard this as an unstable situation. A rapid melting or deterioriation of the ice shelves, perhaps in response to global warming, would release the back pressure on the main part of the ice sheet, which would then move forward and thin rapidly. With a reduced thickness, reduced pressure at the bottom of the ice would allow sea water to enter, and soon most of the sheet would be floating in ocean water. The added bulk would then raise sea level by as much as 6 m (about 20 ft).

A key feature of the Western Antarctic Ice Sheet is the presence of a number of ice streams within the sheet. Ice in these streams flows at a much more rapid rate than in the surrounding ice mass, probably because geothermal activity under the ice streams provides enough heat to melt the ice at the base. This creates a liquid bottom layer that lubricates the ice motion. The ice streams blend gradually into the ice shelves, providing a main source for their supply of ice.

The flow of the ice streams is essential to maintaining the ice shelves. If these streams slowed, the ice shelf might retreat, which could release the unstable western ice sheet, producing catastrophic flooding. In 2002, Drs. Ian Joughin and Slawek Tulacyzk reported in *Science* magazine that the ice streams feeding the Ross Ice Shelf

West Antarctic Ice Sheet *A map of Antarctica showing the Western Antarctic Ice Sheet. (Adapted with permission from C.R. Bentley, Science Vol. 275, p. 1077. Copyright © American Association for the Advancement of Science.)*

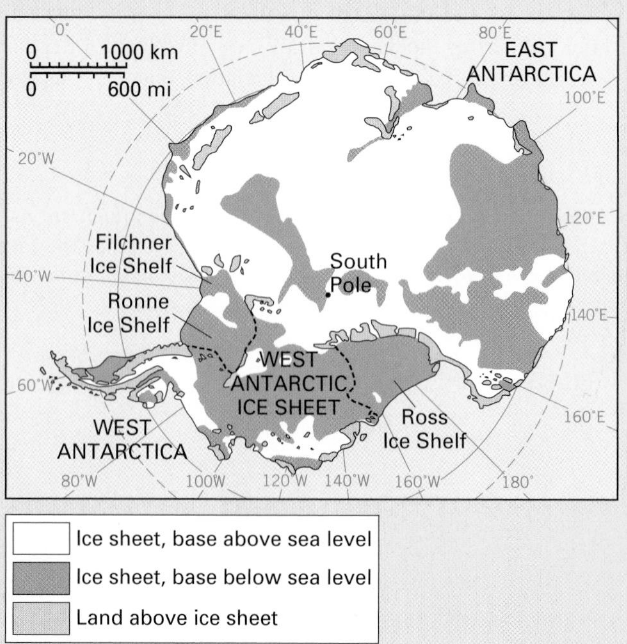

are getting thicker and flowing more slowly. They attribute the change to gradual thinning of the ice sheet since the end of the last glaciation, a situation that allows cold Antarctic temperatures to penetrate to the base of the ice and freeze the liquid bottom layer. If the slowing of the ice streams continues at the present rate for another 70 to 80 years, it would stop altogether, denying the West Antarctic Ice Sheet an influx of ice about equal to the average annual flow of the Missouri River. However, ice flow into other parts of the West Antarctic Ice Sheet seems to be speeding up, even as the ice streams feeding the Ross Ice Shelf are slowing. Whether this increasing flow will offset the effects of slowing Ross Shelf ice streams is not known.

Either way, new evidence suggests that there has been at least one collapse of the West Antarctic Ice Sheet during an interglacial period. Sediments extracted from below the ice sheet show the presence of marine organisms that indicate freely circulating ocean water at some time within the last two million years. Moreover, temperatures at the time of the collapse are inferred to be not much greater than those of today.

Meanwhile, the climatic warming of the past few decades seems to be causing the ice on the shelves to thin and fracture more easily, freeing huge icebergs to drift near the shelf front. The photo below shows six satellite images of Ross Island, at the western edge of the Ross Ice Shelf, acquired in 2000 and 2001 by the MISR instrument. The island is about 75 km (47 mi) long and includes Mount Erebus (leftmost peak) with an elevation of 3743 m (12, 277 ft). The first three show a vast iceberg, designated C-16, as it rotates a quarter-turn counterclockwise in a month's time and then stops. Slowly moving in from the left is another huge berg, designated B-15A. In the last image, sea ice from the previous winter has expanded to surround C-16, and B-15A presses snugly against it.

The motion of these huge chunks of floating glacial ice has affected local penguin rookeries. The extension of the sea ice coupled with the icebergs has trapped large numbers of penguins in their feeding grounds on the open sea, making it difficult for them to get to their breeding areas on land. One colony is in danger of extinction.

The possible effects of climate change on global ice sheets are far from certain. At best, we can hope that human-induced global warming will result in slow, progressive changes to which human civilization can readily adapt.

Icebergs and Ross Island seen by MISR *This sequence of images, acquired by NASA's MISR instrument, shows the motion of huge icebergs near Ross Island, Antarctica, at the edge of the Ross Ice Shelf.*

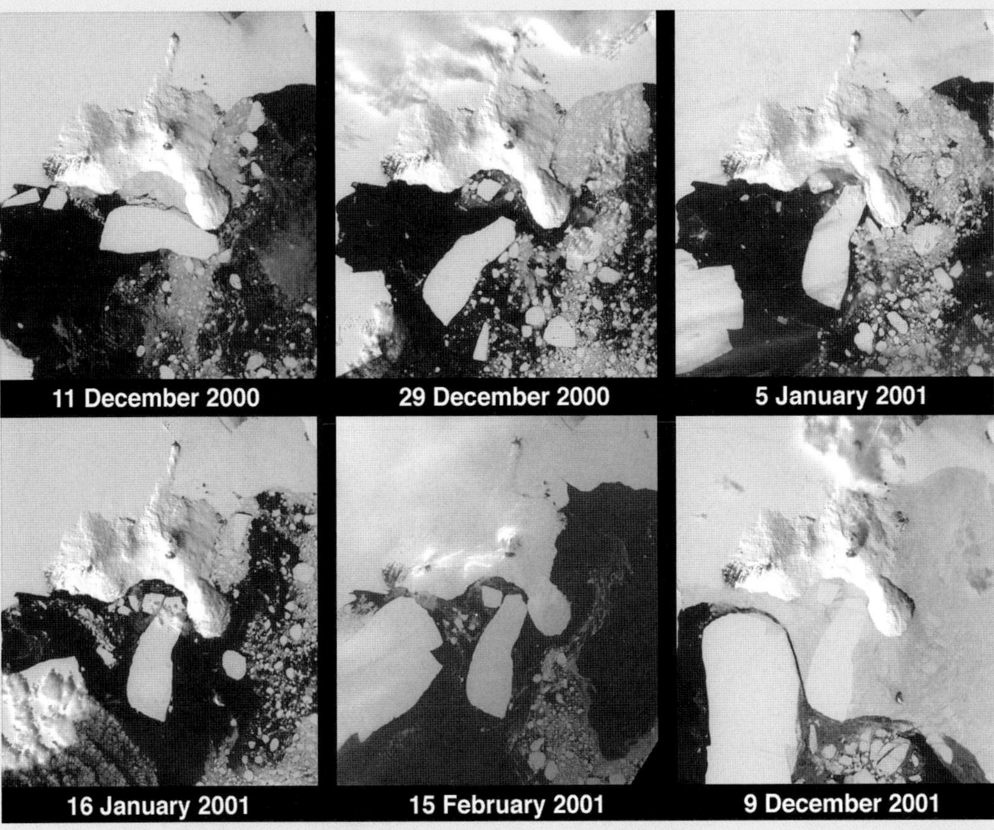

scientists studying climate change during the Ice Age now agree that cyclic insolation changes are the primary factor explaining the cycles of glaciation of the Ice Age.

The actual mechanisms by which insolation changes cause ice sheets to grow or to disappear are unclear. The entire subject is so complex that it is difficult for even a research scientist to grasp fully. Interactions between the atmosphere, the oceans, and the continental surfaces (including the ice sheets) are numerous and closely interrelated with many threads of cause and effect. Changes that occur in one Earth realm are fed back to the other realms in a most complex manner.

Holocene Environments

The elapsed time span of about 10,000 years since the Wisconsinan Glaciation ended is called the *Holocene Epoch.* It began with a rapid warming of ocean surface temperatures. Continental climate zones then quickly shifted poleward, and plants soon became reestablished in glaciated areas.

Three major climatic periods occurred during the Holocene Epoch leading up to the last 2000 years. These periods are inferred from studies of changes in vegetation cover types as observed in fossil pollen and spores preserved in glacial bogs. The earliest of the three is known as the Boreal stage and was characterized by boreal forest vegetation in midlatitude regions. There followed a general warming until the Atlantic stage, with temperatures somewhat warmer than today, was reached about 8000 years ago (–8000 years). Next came a period of temperatures that were below average, the Subboreal stage. This stage spanned the age range –5000 to –2000 years.

Through the availability of historical records and of more detailed evidence, the climate of the past 2000 years can be described on a finer scale. A secondary warm period occurred in the period A.D. 1000 to 1200 (–1000 to –800 years). This warm episode was followed by the Little Ice Age, A.D. 1450–1850 (–550 to –50 years). During the latter, valley glaciers made new advances and extended to lower elevations. More recent temperature cycles are discussed in Chapter 5.

Within the last century, global temperatures have slowly increased and are projected to increase more rapidly for at least the next 100 years. What will be the effect of global warming on the existing continental ice sheets of Greenland and Antarctica? *Eye on Global Change 20.4 • Ice Sheets and Global Warming* discusses the possibilities.

Cycles of glaciation and interglaciation, as well as the lesser climatic cycles of the Holocene Epoch, have proceeded without human influence for millions of years. They demonstrate the power of natural forces to make drastic swings from cold to warm climates, and they confound our efforts to understand human impact on climate.

A LOOK AHEAD The group of chapters we have now completed has reviewed landform-making processes that operate on the surface of the continents. Human influence on landforms is felt most strongly on surfaces of fluvial denudation because of the severity of surface changes caused by agriculture and urbanization. Landforms shaped by wind and by waves and currents are also highly sensitive to changes induced by human activity. Only glaciers maintain their integrity and are thus far largely undisturbed by human activity. Perhaps even this last realm of nature's superiority will eventually fall prey to human interference through climate changes induced by industrial activity.

Chapter Summary

- Glaciers form when snow accumulates to a great depth, creating a mass of ice that is plastic in lower layers and flows outward or downhill from a center in response to gravity. As they move, glaciers can deeply erode bedrock by abrasion and plucking. The eroded fragments, incorporated into the flowing ice, leave depositional landforms when the ice melts.
- **Alpine glaciers** develop in **cirques** in high mountain locations. Alpine glaciers flow downvalley on steep slopes, picking up rock debris and depositing it in *lateral* and *terminal* **moraines**. Through erosion, glaciers carve U-shaped **glacial troughs** that are distinctive features of glaciated mountain regions. They become **fiords** if later submerged by rising sea level.
- **Ice sheets** are huge plates of ice that cover vast areas. They are present today in Greenland and Antarctica. The Antarctic Ice Sheet includes *ice shelves*—great plates of floating glacial ice. Icebergs form when glacial ice flowing into an ocean breaks into great chunks and floats free. *Sea ice*, which is much thinner and more continuous, is formed by direct freezing of ocean water and accumulation of snow.
- An ice age includes alternating periods of **glaciation**, *deglaciation*, and **interglaciation**. During the past 2 to 3 million years, the Earth has experienced the **Late-Cenozoic Ice Age**. During this ice age, continental ice sheets have grown and melted as many as 30 times. The most recent glaciation is the *Wisconsinan Glaciation*, in which ice sheets covered much of North America and Europe, as well as parts of northern Asia and southern South America.
- Moving ice sheets create many types of landforms. Bedrock is grooved and scratched. Where rocks are weak, long valleys can be excavated to depths of hundreds of meters. The melting of glacial ice deposits **glacial drift**, which may be stratified by water flow or deposited directly as **till**. Moraines accumulate at ice edges. *Outwash plains* are built up by meltwater streams. Tunnels within the ice leave streambed deposits as *eskers*. Till may be spread smooth and thick under an ice sheet, leaving a *till plain*. This may be studded with elongated till mounds, termed *drumlins*. Meltwater streams build *glacial deltas* into lakes formed at the ice margin and line lake bottoms with clay and silt. When the lakes drain, these features remain.
- Several factors have been proposed to explain the cause of present ice age glaciations and interglaciations. These factors include ongoing change in the global position of continents, an increase in volcanism, and a reduction in the Sun's energy output. Individual cycles of glaciation seem strongly related to cyclic changes in Earth–Sun distance and axial tilt.

Key Terms

ablation
alpine glacier
ice sheet
cirque
moraine
glacial trough
fiord
glaciation
interglaciation
Late-Cenozoic Ice Age
glacial drift
till
astronomical hypothesis

Review Questions

1. How does a glacier form? What factors are important? Why does a glacier move?
2. Distinguish between alpine glaciers and ice sheets.
3. What is a glacial trough and how is it formed? What is its basic shape? In what ways can a glacial trough appear after glaciation is over?
4. Where are ice sheets present today? How thick are they?
5. Contrast sea ice and icebergs, including the processes by which they form.
6. Identify the Late-Cenozoic Ice Age. When did it begin? What was the last glaciation in this cycle? When did it end?
7. What areas were covered with ice sheets by the last glaciation? How was sea level affected?
8. What are moraines? How are they formed? What types of moraines are there?
9. Identify the landforms and deposits associated with stream action at or near the front of an ice sheet.
10. Identify the landforms and deposits associated with deposition underneath a moving ice sheet.
11. Identify the landforms and deposits associated with lakes that form at ice sheet margins.
12. What cycles are known to affect the amount of solar radiation received by polar regions of the Earth?
13. What is the Milankovitch curve? What does it show about warm and cold periods during the last 500,000 years?
14. How have environments changed during the Holocene Epoch? What periods are recognized, and what are their characteristics?

Focus on Systems 20.1 • A Glacier as a Flow System of Matter and Energy

1. Describe the matter flow system of an alpine glacier. How can a glacier achieve a steady state?
2. What two energy flow subsystems can be recognized in the flow of an alpine glacier? Describe each.

Eye on Global Change 20.4 • Ice Sheets and Global Warming

1. What process seems to be offsetting the rate of melting of ice sheets as climate warms?
2. Why is the Western Antarctic Ice Sheet regarded as unstable? What is the role of ice streams in maintaining the present size of the ice sheet?
3. What change has recently been observed in the flow of ice streams into the Ross Ice Shelf? What are the implications?

Visualizing Exercises

1. What are some typical features of an alpine glacier? Sketch a cross section along the length of an alpine glacier and label it.
2. Refer to Figure 20.10, which shows the Antarctic continent and its ice cap. Identify the Ross, Filchner, and Larsen ice shelves. Use the scale to measure the approximate area of each in square kilometers or miles. Consulting an atlas or almanac, identify the state or province of the United States or Canada that is nearest in area to each.

Essay Questions

1. Imagine that you are planning a car trip to the Canadian Rockies. What glacial landforms would you expect to find there? Where would you look for them?
2. At some time during the latter part of the Pliocene Epoch, the Earth entered an ice age. Describe the nature of this ice age and the cycles that occur within it. What explanations are proposed for causing an ice age and its cycles? What cycles have been observed since the last ice sheets retreated?

Problems

Working It Out 20.3 • Isostatic Rebound

1. The two extreme points for the James Bay data are 200 m at 8000 yr B.P. and 55 m at 5000 yr B.P. Find the equation fitting these points, and express it in both semilogarithmic and exponential form. How does the slope compare with that of Oslofjord? What does the comparison show?
2. For the Oslofjord equation, $b = 12.3$. What is the physical meaning of this value? (*Hint:* For the equation $y = ba^x$, when $x = 0$, $a^x = a^0 = 1$, so $y = b$. Thus, b is the value of y when $x = 0$, or in this case, the position on the uplift scale for the present time.)
3. A geomorphologist observes that isostatic rebound at a location fits the equation

$$y = 9.2 \times (1.7)^x$$

where y is the elevation of the shoreline in meters and x is the age, in thousands of years B.P. What is the slope of this function on a semilog plot? Is the rebound rate he observes greater or lesser than Oslofjord? Than James Bay?

Eye on the Landscape

20.1 Alpine glacier This alpine glacial landscape shows many of the glacial landforms and features described in this chapter. In the near distance are sharp peaks, or horns **(A)**, and knife-edge ridges, or arêtes **(B)**. In the middle foreground is a bowl-shaped cirque **(C)** that no longer bears glacial ice, although the glacier from the adjacent cirque spills into the empty cirque across a low pass (col). Glacial flow features are especially evident, including medial moraines **(D)**, lateral moraines **(E)**, and crevasses **(F)** that open as the ice falls across a rock step.

20.16 Finger Lakes The Finger Lakes of western New York State have been called "inland fiords" for their resemblance to the fiords of glaciated regions. (Fiords are discussed in this chapter.) As explained in our main text, the lakes were eroded from preexisting stream valleys by ice action during the multiple continental glaciations of the Ice Age. The two largest lakes in the center of the image **(A)** are Cayuga Lake (upper) and Seneca Lake (lower). The pattern of agricultural fields in the center and upper left part of the image **(B)** marks the intensive agricultural development of the Lake Erie lowlands and foothills of the Appalachian Plateau, which are mantled with productive soils of glacial origin (Udalfs; see Chapter 21). To the south, at the bottom of the image **(C)**, the terrain becomes more dissected, with more pronounced valleys and ridges, as it slopes upward to the higher elevations of the plateau. Stream dissection of the landscape was covered in Chapter 17.

Chapters in Part 5

Mangroves in Everglades National Park, Florida. The low islands separating these sinuous tidal channels where the fresh water of the Everglades meets the salt water of the Gulf of Mexico are covered by mangrove swamps—dwarf forests of tough, tall shrubs found along warm ocean coasts in brackish waters.

Part 5 | Systems and Cycles of Soils and the Biosphere

Part 5, the final part of our examination of physical geography, focuses directly on the life layer, where biological and physical processes interact. Soils are formed by the physical processes that weather and transport rock material as well as the biological processes that result in the decay of organic matter. These processes are strongly influenced by the abundance of water at the surface and by temperature, and thus they are functions of climate. Climate also describes the availability of sunlight, water, and heat, which condition the photosynthetic process and thus determine the rapidity and abundance of plant growth. Moreover, individual landforms present distinct habitats for the development of local soil types as well as local biotic communities. ■ From these observations, you can see why we have chosen to conclude our study of physical geography with soils systems, biospheric systems and cycles, and global biogeography. Because these topics refer to many of the systems, cycles, and processes that we have introduced in earlier chapters, we can finally do them justice. Chapter 21 provides an overview of the key characteristics of soils, the important processes of soil formation, the classification of soils, and the global distribution of soil types. Chapter 22 describes ecosystems, focusing on how ecosystems are powered and on the cycling of major nutrient elements by ecosystems and their environments. Chapter 23 documents the biogeographic processes that govern the distribution patterns of plants and animals at multiple space and time scales. Finally, Chapter 24 examines the nature of biomes and formation classes of vegetation and their global distribution patterns, especially as related to climate types.

21 SOIL SYSTEMS

Planted fields on the banks of the Rio Uruguay, Misiones Province, Argentina. The crops of this region, in various states of maturity, provide an abstract painting of striking yellows and greens on a canvas of red soil.

This chapter is devoted to soil systems, including the processes that form soils and give them their distinctive characteristics, soil classification, and global distribution of soil types. **Soil** is the uppermost layer of the land surface that plants use and depend on for nutrients, water, and physical support. Soils can vary greatly from continent to continent, region to region, and even from field to field. This is because they are influenced by factors and processes that can vary widely from place to place. For example, a field near a large river that floods regularly may acquire a layer of nutrient-rich silt, making its soil very productive. A nearby field at a higher elevation, without the benefit of silt enrichment, may be sandy or stony and require the addition of fertilizers to grow crops productively.

Vegetation is an important factor in determining soil qualities. For example, some of America's richest soils developed in the Middle West under a cover of thick grass sod. The deep roots of the grass, in a cycle of growth and decay, deposited nutrients and organic matter throughout the thick soil layer. In the Northeast, conifer forests provided a surface layer of decaying needles that kept the soil quite acid. This acidity allowed nutrients to be washed below root depth, out of the reach of plants. When farmed today, these soils need applications of lime to reduce their acidity and enhance their fertility.

Climate, measured by precipitation and temperature, is also an important determinant of soil properties. Precipitation controls the downward movement of nutrients and other chemical compounds in soils. If precipitation is abundant, water tends to wash soluble compounds, including nutrients, deeper into the soil and out of reach of plant roots.

Temperature acts to control the rate of decay of organic matter that falls to the soil from the plant cover or that is provided to the soil by the death of roots. When conditions are warm and moist, decay organisms work efficiently, consuming organic matter readily. Thus, organic matter and nutrients in soils of the tropical and equatorial zones are generally low. Where conditions are cooler, decay proceeds more slowly, and organic matter is more abundant in the soil. Of course, if the climate is very dry or desert-like, then vegetation growth is slow or absent. No matter what desert temperatures are like, organic matter will be low.

Time is also an important factor. The characteristics and properties of soils require time for development. For example, a fresh deposit of mineral matter, like the clean, sorted sand of a dune, may require hundreds to thousands of years to acquire the structure and properties of a sandy soil.

Geographers are keenly interested in the differences in soils from place to place over the globe. The ability of the soils and climate of a region to produce food largely determines the size of the population it will support. In spite of the growth of cities, most of the world's inhabitants still live close to the soil that furnishes their food. And many of those same inhabitants die prematurely when the soil does not furnish enough food for all.

21.1 A cross section through the land surface *In this cross section, vegetation and forest litter lie atop the soil. Below is regolith, produced by the breakup of the underlying bedrock.*

The Nature of the Soil

Soil, as the term is used in soil science, is a natural surface layer that contains living matter and can support plants. The soil consists of matter in all three states—solid, liquid, and gas. It includes both *mineral matter* and *organic matter.* Mineral matter is largely derived from rock material, whereas organic matter is of biological origin and may be living or dead. Living matter in the soil consists not only of plant roots, but also of many kinds of organisms, including microorganisms.

Soil scientists use the term *humus* to describe finely divided, partially decomposed organic matter in soils. Some humus rests on the soil surface, and some is mixed through the soil. Humus particles of the finest size are gradually carried downward to lower soil layers by rainfall that sinks in and moves through the soil. When abundant, humus particles can give the soil a brown or black coloration.

Both air and water are found in soil. Water may tend to contain high levels of dissolved substances, such as nutrients. Air in soils may have high levels of such gases as carbon dioxide or methane, and low levels of oxygen.

The solid, liquid, and gaseous matter of the soil are constantly changing and interacting through chemical and physical processes. This makes the soil a very dynamic layer. Because of these processes, soil science, often called *pedology,* is a highly complex body of knowledge.

Although we may think of soil as occurring everywhere, large expanses of continents possess a surface layer that cannot be called soil. For example, dunes of moving sand, bare rock surfaces of deserts and high mountains, and surfaces of fresh lava near active volcanoes do not have a soil layer.

The characteristics of soils are developed over a long period of time through a combination of many processes acting together. Physical processes act to break down rock fragments of regolith into smaller and smaller pieces. Chemical processes alter the mineral composition of the original rock, producing new minerals. Taken together, these physical and chemical processes are referred to as *weathering,* which we described in Chapter 15. Weathering occurs in soils and is part of the process by which soils develop their properties and characteristics.

In most soils, the inorganic material of the soil consists of fine particles of mineral matter. The term **parent material** describes all forms of mineral matter that are suitable for transformation into soil. Parent material may be derived from the underlying *bedrock,* which is solid rock below the soil layer (Figure 21.1). Over time, weathering processes soften, disintegrate, and break bedrock apart, forming a layer of *regolith,* or residual mineral matter. Regolith is one of the common forms of parent material. Other kinds of regolith consist of mineral particles transported to a place of rest by the action of streams, glaciers, waves and water currents, or winds. For example, dunes formed of sand transported by wind are a type of regolith on which soil may be formed.

Soil Color and Texture

The most obvious feature of a soil or soil layer is probably its color. Some color relationships are quite simple. For example, the soils of the Midwest prairies have a black or dark brown color because they contain abundant particles of humus. Red or yellow colors often mark the soils of the Southeast. These colors are created by the presence of iron-containing oxides.

In some areas, soil color may be inherited from the mineral parent material, but, more generally, soil color is generated by soil-forming processes. For example, a white surface layer in soils of dry climates often indicates the presence of mineral salts brought upward by evaporation. A pale, ash-gray layer near the top of soils of the boreal forest climate results when organic matter and various colored minerals are washed downward, leaving only pure, light-colored mineral matter behind. As we explain soil-forming processes and describe the various classes of soils, soil color will take on more meaning.

The mineral matter of the soil consists of individual mineral particles that vary widely in size. The term **soil texture** refers to the proportion of particles that fall into each of three size grades—*sand, silt,* and *clay.* The diameter range of each of these grades is shown in Figure 21.2. Millimeters are the standard units. Each unit on the scale represents a power of ten, so that clay particles of 0.000,001 millimeter diameter are one-millionth the size of sand grains 1 mm in diameter. The finest of all soil particles are termed *colloids.* In measuring soil texture, gravel and larger particles are eliminated, since these play no important role in soil processes.

Soil texture is described by a series of names that emphasize the dominant particle size, whether sand, silt, or clay (including colloids). Figure 21.3 gives examples of five soil textures with typical percentage compositions. A *loam* is a mixture containing a substantial proportion of each of the three grades. Loams are classified as sandy, silty, or clay-rich when one of these grades is dominant.

Why is soil texture important? Texture largely determines the ability of the soil to retain water. Coarse-textured (sandy) soils have many small passages between touching mineral grains that quickly conduct water through to deeper layers. If the soil consists of fine particles, passages and spaces are much smaller. Thus, water will penetrate more slowly and also tend to be retained. We will return to the important topic of the water-holding ability of soils in a later section.

21.2 Mineral particle sizes *Size grades, which are names like sand, silt, and clay, refer to mineral particles within a specific size range. They are defined using the metric system. English equivalents are also shown.*

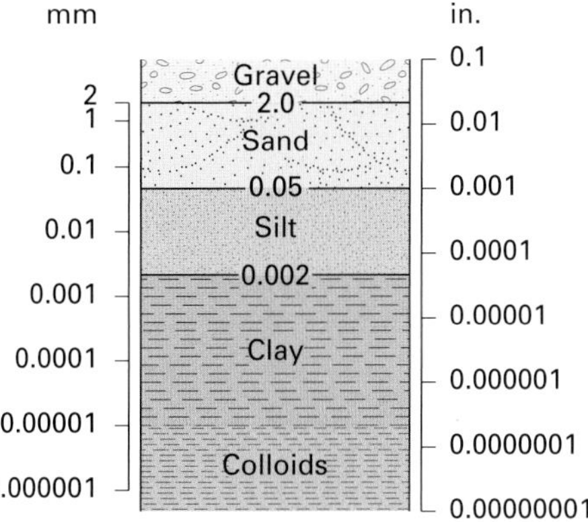

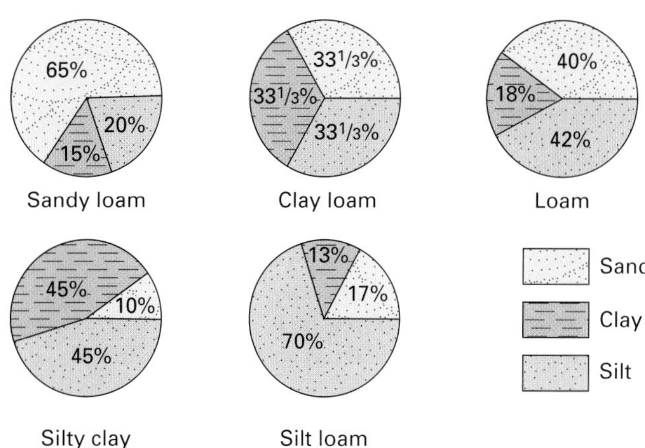

21.3 Soil textures *These diagrams show the proportion of sand, silt, and clay in five different soil texture classes. For description of soil texture, clay includes colloids.*

Soil Colloids

Soil colloids consist of particles smaller than one hundred-thousandth of a millimeter (0.000,01 mm, 0.000,000,4 in.). Like other soil particles, some colloids are mineral, while others are organic. Mineral colloids are usually very fine particles of clay minerals. If you examine mineral colloids under a microscope, you will find that they consist of thin, platelike bodies (Figure 21.4). When well mixed in water, particles this small remain suspended indefinitely, giving the water a murky appearance. Organic colloids are tiny bits of organic matter that are resistant to decay.

21.4 Mineral colloids *Seen here enlarged about 20,000 times are tiny flakes of clay minerals of colloidal dimensions. These particles have settled from suspension in San Francisco Bay.*

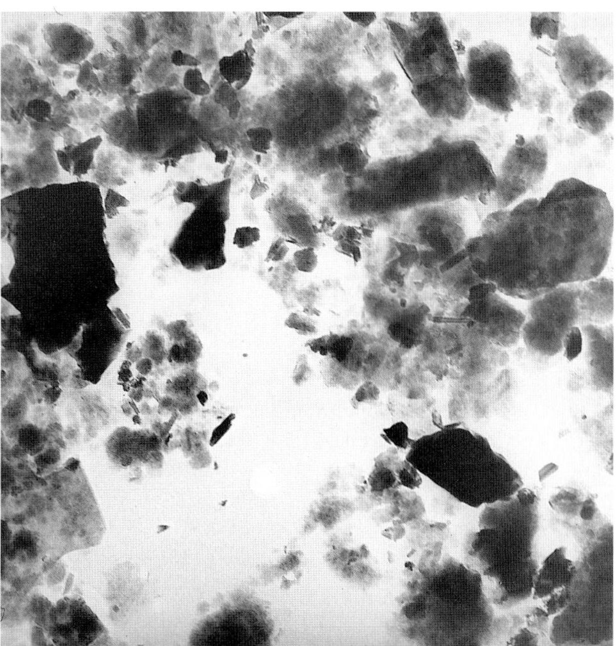

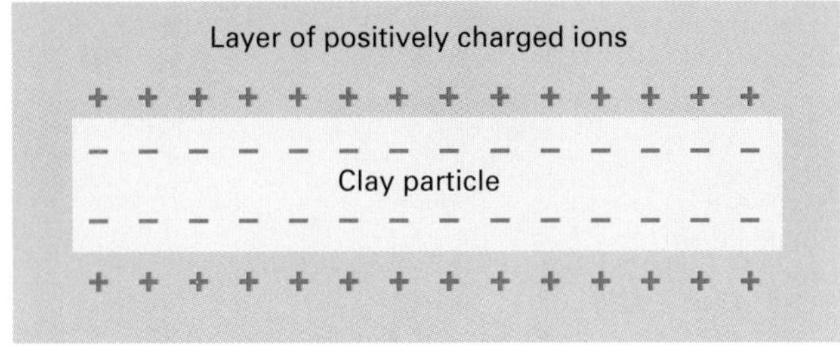

21.5 Schematic diagram of a colloid particle *The thin, flat colloidal particle has negative surface charges and a layer of positively charged ions held to the surface.*

Soil colloids are important because their surfaces attract soil nutrients, which are in the form of ions dissolved in soil water. Figure 21.5 diagrams a colloidal particle, showing this effect. Colloid surfaces tend to be negatively charged because of their molecular structure, and thus attract and hold positively charged ions. Among the many ions in soil water, one important group consists of **bases,** which are ions of four elements: calcium (Ca^{++}), magnesium (Mg^{++}), potassium (K^{+}), and sodium (Na^{+}). Because plants require these elements, they are among the *plant nutrients*—ions or chemical compounds that are needed for plant growth. Colloids hold these ions, but also give them up to plants when in close contact with root membranes. Without this ion-holding ability of soil colloids, most of the vital nutrients would be carried out of the soil by percolating water and would be taken out of the region in streams, eventually reaching the sea. This leaching process goes on continually in moist climates, but loss is greatly retarded by the ion-holding capacity of soil colloids.

Soil Acidity and Alkalinity

The soil solution also contains hydrogen (H^{+}) and aluminum (Al^{+++}) ions. But unlike the bases, they are not considered to be plant nutrients. The presence of these acid ions in the soil solution tends to make the solution acid in chemical balance.

An important principle of soil chemistry is that the acid ions have the power to replace the nutrient bases clinging to the surfaces of the soil colloids. As acid ions accumulate, the bases are released to the soil solution. They are gradually washed downward below rooting level, reducing soil fertility. When this happens, the soil acidity is increased.

The degree of acidity or alkalinity of a solution is designated by the pH value. The lower the pH value, the greater the degree of acidity. A pH value of 7 represents a neutral state—for example, pure water has a pH of 7. Lower values are in the acid range, while higher values are in the alkaline range.

Table 21.1 shows the natural range of acidity and alkalinity found in soils. High soil acidity is typical of cold, humid climates. In arid climates, soils are typically alkaline. Acidity can be corrected by the application of lime, a compound of calcium, carbon, and oxygen ($CaCO_3$), which removes acid ions and replaces them with the base calcium.

Soil Structure

Soil structure refers to the way in which soil grains are grouped together into larger masses, called *peds.* Peds range in size from small grains to large blocks. They are bound together by soil colloids. Small peds, roughly shaped like spheres, give the soil a granular structure or crumb structure (see Figure 21.6). Larger peds provide an angular, blocky structure. Peds form when colloid-rich clays shrink in volume as they dry out. Shrinkage results in formation of soil cracks, which define the surfaces of the peds.

Soils with a well-developed granular or blocky structure are easy to cultivate. This is an important agricultural factor in lands where primitive plows, drawn by animals, are still widely used. Soils with a high clay content can lack peds. These soils are sticky and heavy when wet and are difficult to cultivate. When dry, they become too hard to be worked.

Minerals of the Soil

Soil scientists recognize two classes of minerals abundant in soils: primary minerals and secondary minerals. The *pri-*

Table 21.1 | **Soil Acidity and Alkalinity**

<table>
<tr><th>pH</th><th>4.0</th><th>4.5</th><th>5.0</th><th>5.5</th><th>6.0 6.5</th><th>6.7 7.0</th><th></th><th>8.0</th><th>9.0</th><th>10.0 11.0</th></tr>
<tr><td>Acidity</td><td colspan="2">Very strongly acid</td><td>Strongly acid</td><td>Moderately acid</td><td>Slightly acid</td><td>Neutral</td><td>Weakly alkaline</td><td>Alkaline</td><td>Strongly alkaline</td><td>Excessively alkaline</td></tr>
<tr><td>Lime requirements</td><td colspan="2">Lime needed except for crops requiring acid soil</td><td colspan="2">Lime needed for all but acid-tolerant crops</td><td>Lime generally not required</td><td colspan="5">No lime needed</td></tr>
<tr><td>Occurrence</td><td>Rare</td><td>Frequent</td><td colspan="2">Very common in cultivated soils of humid climates</td><td colspan="4">Common in subhumid and arid climates</td><td colspan="2">Limited areas in deserts</td></tr>
</table>

Source: Based on data of C. E. Millar, L. M. Turk, and H. D. Foth, *Fundamentals of Soil Science,* John Wiley & Sons, New York.

21.6 Soil structure *This soil shows a granular structure. The grains are referred to as peds.*

mary minerals are compounds present in unaltered rock. These are mostly silicate minerals—compounds of silicon and oxygen, with varying proportions of aluminum, calcium, sodium, iron, and magnesium. (The silicate minerals were described more fully in Chapter 12.) Primary minerals form a large fraction of the solid matter of many kinds of soils, but they play no important role in sustaining plant or animal life.

When primary minerals are exposed to air and water at or near the Earth's surface, they are slowly altered in chemical composition. This process is part of *mineral alteration,* a chemical weathering process that was explained in more detail in Chapter 12. The primary minerals are altered into **secondary minerals,** which are essential to soil development and to soil fertility.

In terms of the properties of soils, the most important secondary minerals are the *clay minerals.* They form the majority of fine mineral particles in soils. From the viewpoint of soil fertility, the ability of a clay mineral to hold base ions is its most important property. This ability varies with the particular type of clay mineral; some hold bases tightly, and others loosely.

The nature of the clay minerals in a soil determines its *base status.* If the clay minerals can hold abundant base ions, the soil is of *high base status* and generally will be highly fertile. If the clay minerals hold a smaller supply of bases, the soil is of *low base status* and is generally less fertile. Humus colloids have a high capacity to hold bases so that the presence of humus is usually associated with potentially high soil fertility.

Mineral oxides are secondary minerals of importance in soils. They occur in many kinds of soils, particularly those that remain in place in areas of warm, moist climates over very long periods of time (hundreds of thousands of years). Under these conditions, minerals are ultimately broken down chemically into simple oxides, compounds in which a single element is combined with oxygen.

Oxides of aluminum and iron are the most important oxides in soils. Two atoms of aluminum are combined with three atoms of oxygen to form the *sesquioxide* of aluminum (Al_2O_3). (The prefix *sesqui-* means "one and a half" and refers to the chemical composition of one and one-half atoms of oxygen for every atom of aluminum.) In soils, aluminum oxide forms the mineral *bauxite,* which is a combination of aluminum sesquioxide and water molecules bound together. It occurs as hard, rocklike lumps and layers below the soil surface. Where bauxite layers are thick and uniform, they are sometimes strip-mined as aluminum ore.

Sesquioxide of iron (Fe_2O_3), again held in combination with water molecules, is *limonite,* a yellowish to reddish mineral that supplies the typical reddish to chocolate-brown colors of soils and rocks. Some shallow accumulations of limonite were formerly mined as a source of iron. Limonite and bauxite occur in close association in soils of warm, moist climates in low latitudes.

Soil Moisture

Besides providing nutrients for plant growth, the soil layer serves as a reservoir for the moisture that plants require. Soil moisture is a key factor in determining how the soils of a region support vegetation and crops.

The soil receives water from rain and from melting snow. Where does this water go? First, some of the water can run off the soil surface and not sink in. Instead, it flows into brooks, streams, and rivers, eventually reaching the sea. What of the water that sinks into the soil? Some of this water is returned to the atmosphere as water vapor. This happens when soil water evaporates and when transpiration by plants lifts soil water from roots to leaves, where it evaporates. Taken together, we can term these last two losses *evapotranspiration.* Some water can also flow completely through the soil layer to recharge supplies of ground water at depths below the reach of plant roots.

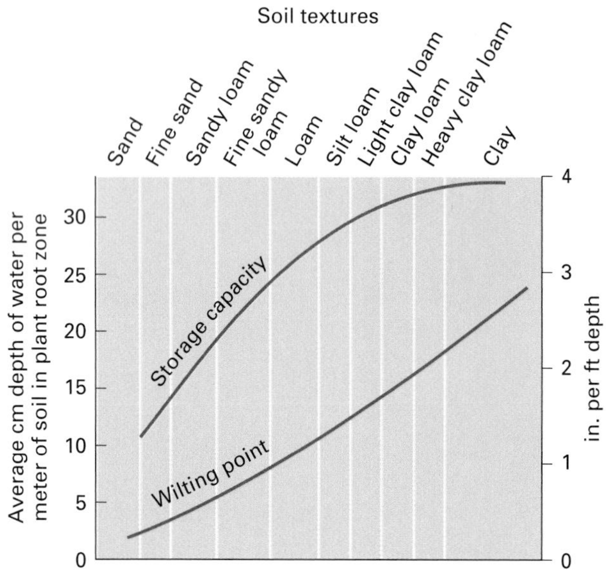

21.7 Storage capacity and wilting point according to soil texture *Finer textured soils hold more water. They also hold water more tightly, so that plants wilt more quickly.*

When precipitation infiltrates the soil, the water wets the soil layer. This process is called *soil water recharge.* Eventually, the soil layer holds the maximum possible quantity of water, even though the larger pores may remain filled with air. Water movement then continues downward.

Suppose now that no further water enters the soil for a time. Excess soil water continues to drain downward, but some water clings to the soil particles. This water resists the pull of gravity because of the force of *capillary tension.* To understand this force, think about a droplet of condensation that has formed on the cold surface of a glass of ice water. The water droplet seems to be enclosed in a "skin" of surface molecules, drawing the droplet together into a rounded shape. The "skin" is produced by capillary tension. This force keeps the drop clinging to the side of the glass indefinitely, defying the force of gravity. Similarly, tiny films of water adhere to soil grains, particularly at the points of grain contacts. They remain until they evaporate or are absorbed by plant rootlets.

When a soil has first been saturated by water and then allowed to drain under gravity until no more water moves downward, the soil is said to be holding its *storage capacity* of water. For most soils, drainage takes no more than two or three days. Most excess water is drained out within one day.

Storage capacity is measured in units of depth, usually centimeters or inches, as with precipitation. It depends largely on the texture of the soil, as shown in Figure 21.7. Finer textures hold more water than coarser textures. This occurs because fine particles have a much larger surface area in a unit of volume than coarse particles. Thus, a sandy soil has a small storage capacity, while a clay soil has a large storage capacity.

The figure also shows the *wilting point,* which is the water storage level below which plants will wilt. The wilting point depends on soil texture. Because fine particles hold water more tightly, it is more difficult for plants to extract moisture from fine soils. Thus, plants can wilt in fine-textured soils even though more soil water is present than in coarse-textured soils. The difference between the storage capacity of a soil and its wilting point is the available water capacity—that is, the maximum amount of water available to plants when the soil is at storage capacity. The available water capacity is greatest in loamy soils.

The Soil Water Balance

Water in the soil is a critical resource needed for plant growth. The amount of water available at any given time is determined by the *soil water balance,* which includes the gain, loss, and storage of soil water. Figure 21.8 is a pictorial flow diagram that illustrates the components of the balance. Water held in storage in the soil water zone is increased by recharge during precipitation but decreased by use through evapotranspiration. Surplus water is disposed of by downward percolation to the ground water zone or by overland flow.

To proceed, we must recognize two ways to define evapotranspiration. First is *actual evapotranspiration (Ea),* which is the true or real rate of water vapor return to the atmosphere from the ground and its plant cover. Second is *potential evapotranspiration (Ep),* representing the water vapor loss under an ideal set of conditions. One condition is that there is present a complete (or closed) cover of uniform vegetation consisting of fresh green leaves and no bare ground exposed through that cover. A second condition is that there is an adequate water supply, so that the storage

21.8 Schematic diagram of the soil water balance in a soil column *(Copyright © A. N. Strahler.)*

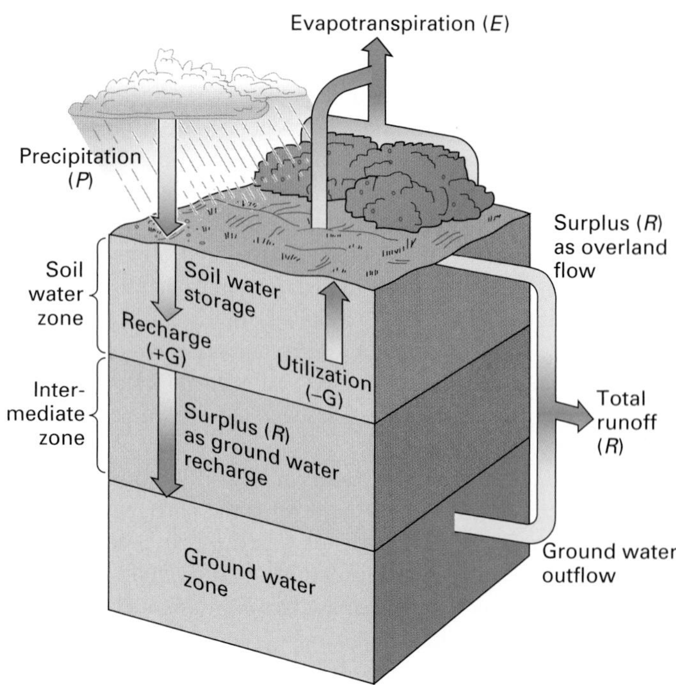

capacity of the soil is maintained at all times. This condition can be fulfilled naturally by abundant and frequent precipitation, or artificially by irrigation.

To simplify the ponderous terms we have just defined, they may be transformed as follows:

Actual evapotranspiration *(Ea)* is **water use.**
Potential evapotranspiration *(Ep)* is **water need.**

The word "need" signifies the quantity of soil water needed if plant growth is to be maximized for the given conditions of solar radiation and air temperature and the available supply of nutrients. The most important factor in determining water need is temperature. In warmer months, water need will be greater, while in cooler months, it will be less.

The difference between water use and water need is the *soil water shortage,* or *deficit.* This is the quantity of water that must be furnished by irrigation to achieve maximum crop growth within an agricultural system.

A Simple Soil Water Budget

We now turn to a simple accounting of the monthly and annual quantities of the components of the soil water balance. The numerical accounting is called a *soil water budget,* and it involves only simple addition and subtraction of monthly mean values for a given observing station. All terms of the soil water budget are stated in centimeters of water depth, the same as for precipitation.

A simplified soil water budget is shown in Figure 21.9. The seven terms we need for a complete budget are listed as follows, along with abbreviations used on the graph:

Precipitation, *P*
Water need, *Ep*
Water use, *Ea*
Storage withdrawal, *–G*
Storage recharge, *+G*
Soil water shortage, *D*
Water surplus, *R*

Points on the graph represent average monthly values of precipitation and water need. They are connected by smooth curves to enhance annual cycles of change. In our example, precipitation *(P)* is much the same in all months, with no strong annual cycle. In contrast, water need *(Ep)* shows a strong seasonal cycle, with low values in winter and a high summer peak. This example would fit a typical moist midlatitude climate with mild winters.

At the start of the year, precipitation greatly exceeds water use and a large water surplus *(R)* exists. This surplus is disposed of by runoff. By May, water use exceeds precipitation, and a water deficit occurs. In this month plants begin to withdraw soil water from storage.

Storage withdrawal (–G) is represented by the difference between the water-use curve and the precipitation curve. As storage withdrawal continues, however, plants draw soil water only with increasing difficulty. Thus, water use *(Ea)* is less than water need *(Ep)* during this period. Storage withdrawal continues throughout the summer. The deficit period lasts through September. The area labeled soil water shortage *(D)* is the difference between water need and water use. It represents the total quantity of water needed by irrigation to ensure maximum growth throughout the deficit period.

In October, precipitation *(P)* again begins to exceed water need *(Ep),* but the soil must first absorb an amount equal to the summer storage withdrawal. So there follows a period of *storage recharge* (+*G*) that lasts through November. In December the soil reaches its full storage capacity, arbitrarily fixed at 30 cm (11.8 in.). Now, a water surplus *(R)* again sets in, lasting through the winter.

The soil water budget was developed by C. Warren Thornthwaite, a distinguished climatologist and geographer who was concerned with practical problems of crop irrigation. He developed the calculation of the soil water budget in order to place irrigation on a precise, accurate basis. Only in the equatorial zone and in a few parts of the tropical and midlatitude zones is precipitation ample to fulfill the water need during the growing season. In a world beset by severe and prolonged food shortages, the Thornthwaite concepts and calculations are of great value in assessing the benefits to be gained by increased irrigation.

Calculating the annual soil water budget for a station is not difficult, given monthly values of precipitation, water use, and water need. *Working It Out 21.1 • Calculating a Simple Soil Water Budget* shows you how.

21.9 *A simplified soil water budget* *This soil water budget is typical of a moist climate in middle latitudes.*

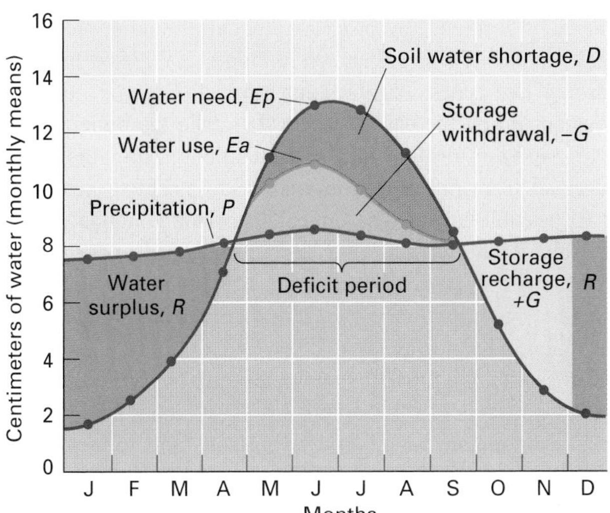

SOIL DEVELOPMENT

How do soils develop their distinctive characteristics? Let's turn to the processes that act to form soils and soil layers. We begin with soil horizons.

Working It Out | 21.1

Calculating a Simple Soil Water Budget

The soil water balance is actually a matter flow system in which an input of precipitation flows through soil pathways to be output as evapotranspiration or runoff (see Figure 21.8). The amount of water returned to the atmosphere through evapotranspiration *(Ea)* is determined largely by temperature and by the amount of vegetation cover. With warm temperatures and a thick layer of transpiring leaves, water use will be larger. The amount of runoff will depend in the long term on how water use *(Ea)* compares with precipitation *(P)*. If precipitation is greater, then runoff will occur, both as overland flow and as ground water outflow.

An important characteristic of the soil water matter flow system is the storage of water *(S)* in the soil layer. This storage provides a reserve of water for plants to draw upon, so that water use can exceed precipitation for some period of the year. Later, when precipitation exceeds water use, the reserve can be recharged, and only after the reserve is replenished can runoff occur again in the annual cycle.

The calculation of a monthly soil water budget is a simple exercise, once given the values of precipitation, water use, and water need for a station. Precipitation is easily observed by measurement with a rain gauge. Water need *(Ep)* is calculated from a formula that uses temperature and atmospheric moisture content. Water use is either measured directly or estimated from similar formulas calibrated by actual measurements.

The table and figure demonstrate the calculation of the monthly soil water budget for a station in the marine west-coast climate ⑧, which exhibits wet, cool winters and warm, drier sum-

A model soil water budget *Bars are correctly scaled to agree with values in the table.*

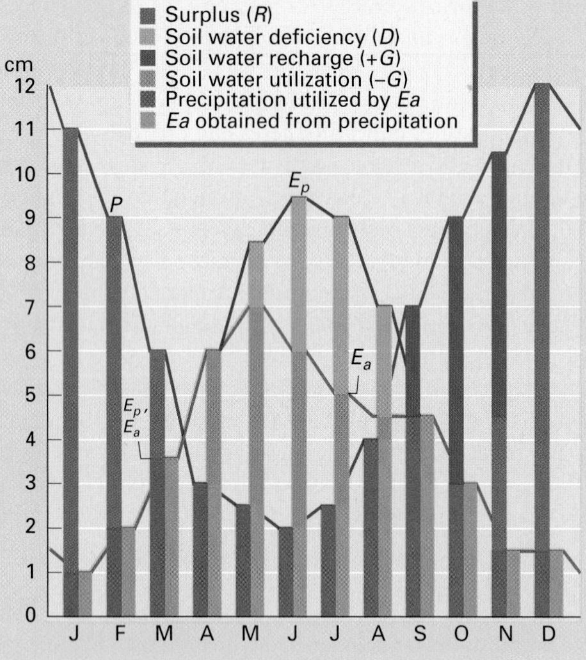

Soil Horizons

Most soils possess **soil horizons**—distinctive horizontal layers that differ in physical composition, chemical composition, or organic content or structure (Figure 21.10). Soil horizons are developed by the interactions through time of climate, living organisms, and the configuration of the land surface. Horizons usually develop by either selective removal or accumulation of certain ions, colloids, and chemical compounds. The removal or accumulation is normally produced by water seeping down through the soil profile from the surface to deeper layers. Horizons are often distinguished by their color. The display of horizons on a cross section through the soil is termed a **soil profile.** GEODISCOVERIES

Let's review briefly the main types of horizons and their characteristics. For now, this discussion will apply to the types of horizons and processes that are found in moist forest climates. These are shown in Figure 21.11. Soil horizons are of two types: organic and mineral. Organic horizons, designated by the capital letter *O,* overlie the mineral horizons and are formed from accumulations of organic matter derived from plants and animals. Soil scientists recognize two possible layers. The upper O_i horizon contains decomposing organic matter that is recognizable as leaves or twigs. The lower O_a horizon contains material that is broken down beyond recognition by eye. This material is humus, which we mentioned earlier.

Mineral horizons lie below the organic horizons. Four main horizons are important—*A, E, B,* and *C.* Plant roots readily penetrate *A, E,* and *B* horizons and influence soil development within them. Soil scientists limit the term *soil* to refer to the *A, E,* and *B* horizons. The *A horizon* is the mers. For each month, three items of basic data are first provided—precipitation *(P)*, water use *(Ea)*, and water need *(Ep)*. From these, the other necessary quantities can be found. In any month, precipitation can only follow three pathways—to the atmosphere as water use *(Ea)*, to streams and rivers as runoff *(R)*, or to storage (+*G*) within the soil layer. In January, precipitation is 11.0 cm, and water use is 1.0 cm, leaving 10.0 cm to go to storage or runoff. At this time of the year, storage is full, however, so this entire amount goes to runoff.

Simplified Example of a Soil Water Budget

Equation	P = Ea		+G	+R	Ep	(Ep – Ea) = D
January	11.0 = 1.0			+10.0	1.0	0.0
February	9.0 = 2.0			+7.0	2.0	0.0
March	6.0 = 3.5			+2.5	3.5	0.0
April	3.0 = 6.0	–3.0			6.0	0.0
May	2.5 = 7.0	–4.5			8.5	1.5
June	2.0 = 6.0	–4.0			9.5	3.5
July	2.5 = 5.0	–2.5			9.0	4.0
August	4.0 = 4.5	–0.5			7.0	2.5
September	7.0 = 4.5		+2.5		4.5	0.0
October	9.0 = 3.0		+6.0		3.0	0.0
November	10.5 = 1.5		+6.0	+ 3.0	1.5	0.0
December	12.0 = 1.5			+10.5	1.5	0.0
Totals	78.5 = 45.5	–14.5	+14.5	+33.0	57.0	11.5
	78.5 = 78.5					

The flow of runoff persists until April, when water use (6.0 cm) exceeds precipitation (3.0 cm) by 3.0 cm. The needed 3.0 cm is withdrawn from storage, giving the value of –3.0 for –*G*. (Separate columns are used for –*G* and +*G*.) Storage withdrawals continue through August, yielding a total withdrawal of 14.5 cm.

In September, precipitation (7.0 cm) again exceeds water use (4.5 cm), leaving a surplus (2.5 cm) for +*G*. This surplus begins to recharge the soil water storage. In October, soil water storage is also increased. The increase continues into November, when it is 9.0 cm. However, only 6.0 cm is required to balance the total summer withdrawal of 14.5 cm, leaving 9.0 – 6.0 = 3.0 cm for runoff *(R)*. In December, the surplus goes entirely to runoff.

The soil water shortage *(D)* is simply the difference between water need *(Ep)* and water use *(Ea)*. The two quantities are equal until May, when water use drops below water need as significant water storage withdrawals begin to occur. This condition persists until September, when abundant precipitation allows water use to equal water need.

These principles can be used to calculate the soil water budget for any station. It is important to note, however, that the depth of storage of water available to plants in the soil is limited. In calculating soil water budgets, a limit of 30 cm is normally used for field crops. Thus, total withdrawal and recharge may not exceed this amount.

The soil water balance not only influences human agricultural activities, but is also strongly linked to natural vegetation and is a key factor in climate classification. The soil water budget is an example of how systems principles of identifying flow pathways and budgeting flow rates can be used to describe and quantify phenomena of such great importance.

uppermost mineral horizon. It is rich in organic matter, consisting of numerous plant roots and downwashed humus from the organic horizons above. Next is the *E horizon.* Clay particles and oxides of aluminum and iron are removed from the *E* horizon by downward-seeping water, leaving behind pure grains of sand or coarse silt.

The *B horizon* receives the clay particles, aluminum, and iron oxides, as well as organic matter washed down from the *A* and *E* horizons. It is made dense and tough by the filling of natural spaces with clays and oxides. Transitional horizons are present at some locations.

Beneath the *B horizon* is the *C horizon*. It consists of the parent mineral matter of the soil and has been described earlier in this chapter as regolith (see Figure 21.1). Below the regolith lies bedrock or sediments of much older age than the soil.

Soil-Forming Processes

There are four classes of soil-forming processes. The first includes processes of *soil enrichment,* which add material to the soil body. For example, inorganic enrichment occurs when sediment is brought from higher to lower areas by overland flow. Stream flooding also deposits fine mineral particles on low-lying soil surfaces. Wind is another source of fine material that can accumulate on the soil surface. Organic enrichment occurs when humus accumulating in *O* horizons is carried downward to enrich the *A* horizon below.

The second class includes processes that remove material from the soil body. This *removal* occurs when surface erosion carries sediment away from the uppermost layer of soil. Another important process of loss is *leaching,* in which seeping water dissolves soil materials and moves them to deep levels or to ground water.

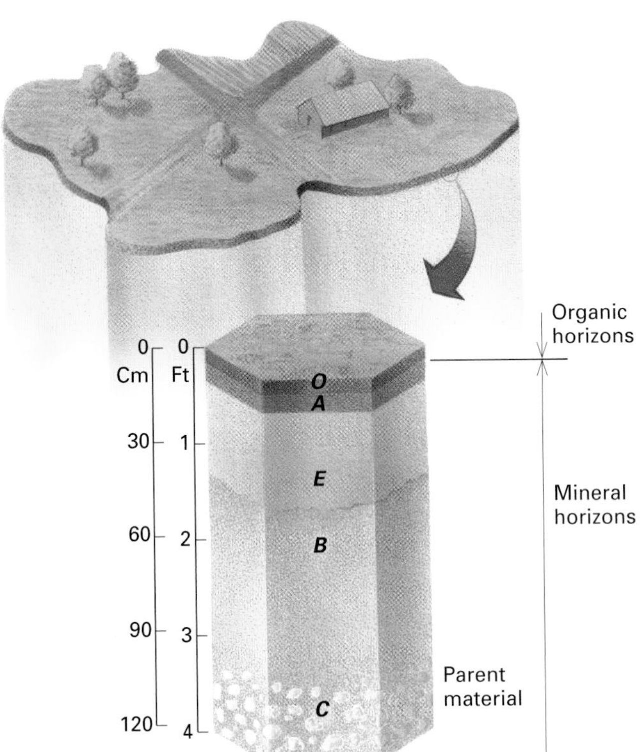

21.10 Soil horizons *A column of soil will normally show a series of horizons, which are horizontal layers with different properties.*

The third class of soil-forming processes involves *translocation,* in which materials are moved within the soil body, usually from one horizon to another. Two processes of translocation that operate simultaneously are eluviation and illuviation. **Eluviation** consists of the downward transport of fine particles, particularly the clays and colloids, from the uppermost part of the soil. Eluviation leaves behind grains of sand or coarse silt, forming the *E* horizon. **Illuviation** is the accumulation of materials that are brought down downward, normally from the *E* horizon to the *B* horizon. The materials that accumulate may be clay particles, humus, or sesquioxides of iron and aluminum.

Figure 21.12 is a soil profile developed under a cool, humid forest climate. It shows the effects of both soil enrichment and translocation processes. The topmost layer of the soil is a thin deposit of wind-blown silt and dune sand, which has enriched the soil profile. In translocation processes, eluviation has removed colloids and sesquioxides from the whitened *E* horizon. Illuviation has added them to the *B* horizon, which displays the orange-red colors of iron sesquioxide.

The translocation of calcium carbonate is another important process. In pure form, this secondary mineral is calcite ($CaCO_3$). In many areas, the parent material of the soil contains a substantial proportion of calcium carbonate derived from the disintegration of limestone, a common variety of bedrock. Carbonic acid, which forms when carbon dioxide gas dissolves in rainwater or soil water, readily reacts with calcium carbonate. The products of this reaction remain dissolved in solution as ions.

In moist climates, a large amount of surplus soil water moves downward to the ground water zone. This water movement leaches calcium carbonate from the entire soil in a process called *decalcification.* Soils that have lost most of their calcium are also usually acid in chemical balance and so are low in bases. Addition of lime or pulverized limestone not only corrects the acid condition, but also restores the calcium, which is used as a plant nutrient.

In dry climates, calcium carbonate is dissolved in the upper layers of the soil during periods of rain or snowmelt when soil water recharge is taking place. The dissolved carbonate matter is carried down to the *B* horizon, where water penetration reaches its limits. Here, the carbonate matter is precipitated (deposited in crystalline form) in the *B* horizon, a process called *calcification.* Calcium carbonate deposition takes the form of white or pale-colored grains, nodules, or plates in the *B* or *C* horizons.

21.11 Soil horizons of moist forest climates *A sequence of horizons that might appear in a forest soil developed under a cool, moist climate. (Natural Resources Conservation Service, U.S. Department of Agriculture.)*

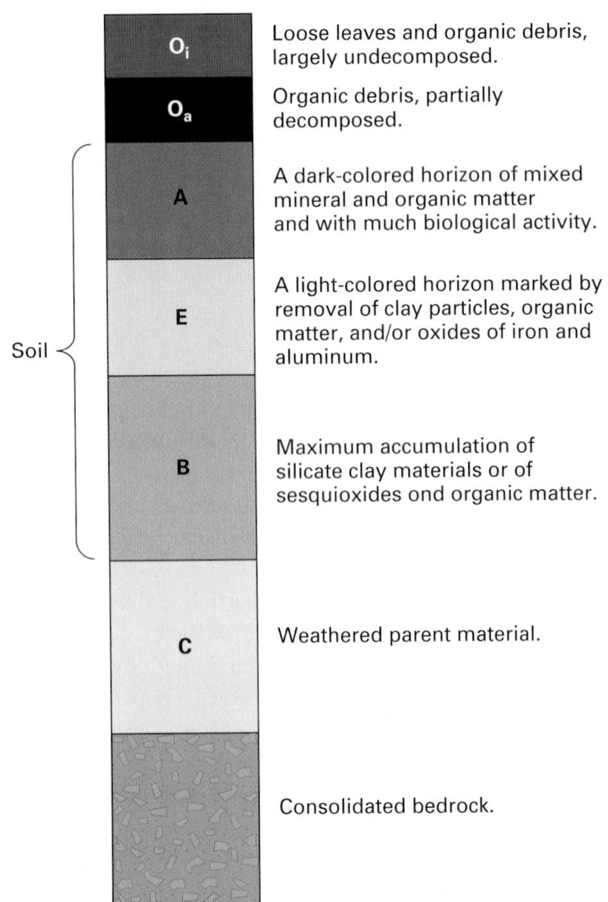

21.12 *A forest soil profile on outer Cape Cod* *The pale grayish E horizon overlies a reddish B horizon. A thin layer of wind-deposited silt and dune sand (pale brown layer) has been deposited on top.*

A last process of translocation occurs in desert climates. In some low areas, a layer of ground water lies close to the surface, producing a flat, poorly drained area. Evaporation of water at or near the soil surface draws up a continual flow of ground water by capillary tension, much like a cotton wick draws oil upward in an oil lamp. Moreover, the ground water is often rich in dissolved salts. When evaporation occurs, the salts precipitate and accumulate as a distinctive *salic horizon.* This process is called *salinization.* Most of the salts are compounds of sodium, of which ordinary table salt (sodium chloride, or halite, NaCl) is a familiar example. Sodium in large amounts is associated with highly alkaline conditions and is toxic to many kinds of plants. When salinization occurs in irrigated lands in a desert climate, the soil can be ruined for further agricultural use. *Eye on the Environment 21.2 • Death of a Civilization* shows how the ancient cultures of Babylon and Sumer declined as their irrigated desert soils deteriorated from salinization and waterlogging.

The last class of soil-forming processes involves the *transformation* of material within the soil body. An example is the conversion of minerals from primary to secondary types, which we have already described. Another example is decomposition of organic matter to produce humus, a process termed *humification.* In warm, moist climates, humification can decompose organic matter completely to yield carbon dioxide and water, leaving virtually no organic matter in the soil.

Soil Temperature and Other Factors

Soil temperature is another important factor in determining the chemical development of soils and the formation of horizons. Temperature acts as a control over biologic activity and also influences the intensity of chemical processes affecting soil minerals. Below 10°C (50°F), biological activity is slowed, and at or below the freezing point (0°C, 32°F), biological activity stops. Chemical processes affecting minerals are inactive. The root growth of most plants and germination of their seeds require soil temperatures above 5°C (41°F). For plants of the warm, wet low-latitude climates, germination of seeds may require a soil temperature of at least 24°C (75°F).

The temperature of the uppermost soil layer and the soil surface strongly affects the rate at which organic matter is decomposed by microorganisms. Thus, in cold climates, where decomposition is slow, organic matter in the form of fallen leaves and stems tends to accumulate to form a thick *O* horizon. As we have already described, this material becomes humus, which is carried downward to enrich the *A* horizon.

In warm, moist climates of low latitudes, the rate of decomposition of plant material is rapid, so that nearly all the fallen leaves and stems are disposed of by bacterial activity. Under these conditions, the *O* horizon may be missing and the entire soil profile will contain very little organic matter.

SOIL AND SURFACE CONFIGURATION The configuration, or shape, of the ground surface is an important factor in soil formation. Configuration includes the steepness of the ground surface, or *slope,* as well as its compass orientation, or *aspect.* Generally speaking, soil horizons are thick on gentle slopes but thin on steep slopes. This is because the soil is more rapidly removed by erosion processes on the steeper slopes.

Aspect acts to influence soil temperatures and the soil water regime. Slopes facing away from the sun are sheltered from direct insolation and tend to have cooler, moister soils. Slopes facing toward the sun are exposed to direct solar rays, raising soil temperatures and increasing evapotranspiration.

BIOLOGICAL PROCESSES IN SOIL FORMATION The presence and activities of living plants and animals, as well as their nonliving organic products, have an important influence on soil. We have already noted the role that organic matter as humus plays in soil fertility. The colloidal structure of humus holds bases, which are needed for plant growth. In this way, humus helps keep nutrients cycling through plants and soils, and ensures that nutrients will not be lost by leaching to groundwater. Humus also helps bind the soil into crumbs and clumps. This structure allows water and air to

Eye on the Environment | 21.2

Death of a Civilization

What causes a civilization to collapse after flourishing for thousands of years? Wars and conquests come to mind, but sometimes the cause is less obvious. A case in point is the decline of the civilization of Sumer, and its successor, Babylon. Perhaps you have already learned something of these ancient Middle Eastern cultures, which were nourished by the Tigris and Euphrates rivers thousands of years ago. These two famous rivers drain the Zagros and Taurus Mountains of Turkey and Iran, and flow out through a broad valley across the deserts of Iraq to the Persian Gulf.

The Sumerian civilization evolved in the lower part of the Tigris–Euphrates Valley, beginning with a village culture dating to about five millennia B.C. By 3000 B.C., the Sumerian civilization was well established. The agriculture that supported Sumer depended on a system of irrigation canals utilizing the waters of the Tigris and Euphrates. Supported by this highly productive agricultural base, urban culture progressed to a high level, including the development of a written language. Skilled craftsmen produced pottery, jewelry, and ornate weapons of copper, silver, and gold (see figure at right).

Trouble set in for the Sumerians in about 2400 B.C. It came in the form of a deterioration of their croplands caused by an accumulation of salt in the soil. In this process, today called salinization, a small increment of salt is added to the soil each year by the evaporation of irrigation water containing a low concentration of salts. Over time, the salt content of the soil increases until it reaches levels high enough to affect crop plants. Wheat, which was a staple crop of Sumer, is quite sensitive to salt, and its yield declines sharply as the salt concentration rises in the soil. Barley, another staple crop of that era, is somewhat less sensitive to salt.

Archaeologists and historians have inferred the onset of salinization in Sumerian agricultural lands by a shift in the proportions of wheat and barley in the national agricultural output. It seems that around 3500 B.C. these two grains were grown in about equal amounts. By 2500 B.C., however, wheat accounted for only about one-sixth of the total grain production. Evidently, the more salt-tolerant barley had replaced wheat over much of the area.

At this point in time, the Sumerian grain yield also began to decline seriously. Ancient records show that in 2400 B.C., around the Sumerian city of Girsu, fields were yielding an average of about 300 kg of grain per hectare (270 lb/acre)—a high rate, even by modern standards. By 2100 B.C. the grain yield had declined to 180 kg per hectare (160 lb/acre), and by 1700 B.C., to a mere 100 kg/ha (90 lb/acre). Cities so declined in population that they finally dwindled into villages. As Sumerian civilization withered in the south, Babylonia, in the northern part of the valley, became more important. Soon, political control of the region passed from Sumer to Babylon.

Historians chronicle an event that also seems to have played a part in the decline of the Sumerian civilization. A group of Sumerian cities along the Euphrates River needed more irrigation water than they had. After a long and fruitless series of conflicts with upstream cities to gain additional water, the problem was solved by running a canal across the valley from the Tigris River. Now there was plenty of water. However, it was used so abundantly that the ground water table began to rise. As the water table came closer to the ground

penetrate the soil freely, providing a healthy environment for plant roots.

Animals living in the soil include many species and come in many individual sizes—from bacteria to burrowing mammals. The total role of animals in soil formation is extremely important in soils that are warm and moist enough to support large animal populations. For example, earthworms continually rework the soil not only by burrowing, but also by passing soil through their intestinal tracts. They ingest large amounts of decaying leaf matter, carry it down from the surface, and incorporate it into the mineral soil horizons. Many forms of insect larvae perform a similar function. Tubelike openings are made by larger animals—moles, gophers, rabbits, badgers, prairie dogs, and other species.

Human activity also influences the physical and chemical nature of the soil. Large areas of agricultural soils have been cultivated for centuries. As a result, both the structure and composition of these agricultural soils have undergone great changes. These altered soils are often recognized as distinct soil classes that are just as important as natural soils.

An ancient enameled panel commemorating the victory in war of a king in the first dynasty of Ur, Mesopotamia. Note the rich agricultural spoils depicted in the second and third rows—cattle, sheep, fishes, sacks of grain.

surface, capillary tension drew the ground water to the surface, where it evaporated rapidly in the hot sun. The result was an even faster accumulation of salt. When the salty water table came up still further, it saturated the roots of the barley plants, devastating the crops. Unknowingly, the Sumerians destroyed their own land in their quest for water.

The rising civilization of Babylon was doomed to fall in turn. Along the Tigris River, east of the modern city of Baghdad, an extremely elaborate system of irrigation canals was built. This system was begun in a pre-Babylonian period, 3000 to 2400 B.C. After a long history of abandonments and reconstructions, it was superseded between A.D. 200 and 500 by a final irrigation system based on a central canal, the Nahrwan Canal.

A problem soon developed, however. The irrigation system featured long, branching channels that proved to be traps for silt, which settled out of muddy river water. Frequent cleaning of silt from the channels was required, and silt was piled up in great embankments and mounds beside the channels. Gradually, silt from the mounds was washed by rains into nearby fields, adding layer upon layer to the land surface. According to one estimate, about 1 m (3 ft) of silt accumulated over the fields in a 500-year period. This rise in level of the farmland by silting made its irrigation all the more difficult. Until the strong central government authority collapsed, things went along well enough. Afterward, however, the entire system gradually broke down, and by the twelfth century it was abandoned completely. Not long after, Mongol hordes invaded the valley. By then, little remained of the once prosperous civilization that had dominated the region for over 4000 years.

Salinization, waterlogging, and silt accumulation have affected nearly every major irrigation system developed in a desert climate. Today, Pakistan struggles with salinization and waterlogging in a vast irrigation system in the Indus Valley. The lower Colorado River lands, including areas in Mexico, have suffered from salinization. Israel faces the threat of salinization of vital agricultural lands that have only recently been placed in irrigation. Some new methods may help. For example, soaking the ground through tubes, while keeping the soil covered under a plastic sheet, can reduce the buildup of salt.

Small wonder, then, that new proposals for desert irrigation systems arouse little enthusiasm. Perhaps a record of nearly total failure, spread over the entire span of civilization, is beginning to get its message across.

The Global Scope of Soils

An important aspect of soil science for physical geography is the classification of soils into major types and subtypes that are recognized in terms of their distribution over the Earth's land surfaces. Geographers are particularly interested in the linkage of climate, parent material, time, biologic process, and landform with the distribution of types of soils. Geographers are also interested in the kinds of natural vegetation associated with each of the major soil classes. The geography of soils is thus essential in determining the quality of environments of the globe. It is important because soil fertility, along with availability of fresh water, is a basic measure of the ability of an environmental region to produce food for human consumption.

The soils of the world have been classified according to a system developed by scientists of the U.S. Soil Conservation Service, in cooperation with soil scientists of many other nations. Here we are concerned only with the two highest levels of this classification system. The top level contains 11 **soil orders** summarized in Table 21.2. The second level consists of *suborders,* of which we need to mention only a few.

Table 21.2 | Soil Orders

Group I	
Soils with well-developed horizons or with fully weathered minerals, resulting from long-continued adjustment to prevailing soil temperature and soil water conditions.	
Oxisols	Very old, highly weathered soils of low latitudes, with a subsurface horizon of accumulation of mineral oxides and very low base status.
Ultisols	Soils of equatorial, tropical, and subtropical latitude zones, with a subsurface horizon of clay accumulation and low base status.
Vertisols	Soils of subtropical and tropical zones with high clay content and high base status. Vertisols develop deep, wide cracks when dry, and the soil blocks formed by cracking move with respect to each other.
Alfisols	Soils of humid and subhumid climates with a subsurface horizon of clay accumulation and high base status. Alfisols range from equatorial to subarctic latitude zones.
Spodosols	Soils of cold, moist climates, with a well-developed *B* horizon of illuviation and low base status.
Mollisols	Soils of semiarid and subhumid midlatitude grasslands, with a dark, humus-rich epipedon and very high base status.
Aridisols	Soils of dry climates, low in organic matter, and often having subsurface horizons of accumulation of carbonate minerals or soluble salts.
Group II	
Soils with a large proportion of organic matter.	
Histosols	Soils with a thick upper layer very rich in organic matter.
Group III	
Soils with poorly developed horizons or no horizons, and capable of further mineral alteration.	
Entisols	Soils lacking horizons, usually because their parent material has accumulated only recently.
Inceptisols	Soils with weakly developed horizons, having minerals capable of further alteration by weathering processes.
Andisols	Soils with weakly developed horizons, having a high proportion of glassy volcanic parent material produced by erupting volcanoes.

Soil orders and suborders are often distinguished by the presence of a diagnostic horizon. Each diagnostic horizon has some unique combination of physical properties (color, structure, texture) or chemical properties (minerals present or absent). The two basic kinds of diagnostic horizons are (1) a horizon formed at the surface and called an *epipedon* (from the Greek, word *epi,* meaning "over" or "upon") and (2) a subsurface horizon formed by the removal or accumulation of matter. In our descriptions of soil orders, we will refer to a number of diagnostic horizons.

We can recognize three groups of soil orders. The largest group includes seven orders with well-developed horizons or fully weathered minerals. A second group includes a single soil order that is very rich in organic matter. The last group includes three soil orders with poorly developed horizons or no horizons.

The world soils map presented in Figure 21.13 shows the major areas of occurrence of the soil orders. The Alfisols (Table 21.2) have been subdivided into four important suborders that correspond well to four basic climate zones. The map is quite general, indicating those areas where a given soil order is likely to be found. The map does not show many important areas of Entisols, Inceptisols, Histosols, and Andisols (Table 21.2). These orders are largely of local occurrence, since they are found on recent deposits such as floodplains, glacial landforms, sand dunes, marshlands, bogs, or volcanic ash deposits. The map also shows areas of highlands. In these regions, the soil patterns are too complex to show at a global scale.

Soil Orders

Table 21.3 explains the names of the soil orders. The formative element is a syllable used in the names of suborders and lower groups. Although each order has several suborders, we will refer to only a few. GEODISCOVERIES 

Three soil orders dominate the vast land areas of low latitudes: Oxisols, Ultisols, and Vertisols. Soils of these orders have developed over long time spans in an environment of warm soil temperatures and soil water that is abundant in a wet season or lasts throughout the year. We will discuss these orders first.

OXISOLS **Oxisols** have developed in equatorial, tropical, and subtropical zones on land surfaces that have been stable over long periods of time. During soil development, the climate has been moist, with a large water surplus. Oxisols have developed over vast areas of South America and Africa in the wet equatorial climate ①. Here the native vegetation is rainforest. The wet-dry tropical climate ③ with its large seasonal water surplus is also associated with Oxisols in South America and Africa.

Oxisols usually lack distinct horizons, except for darkened surface layers. Soil minerals are weathered to an extreme degree and are dominated by stable sesquioxides of aluminum and iron. Red, yellow, and yellowish-brown colors are normal (Figures 21.14*a* and 21.15). The base status of the Oxisols is very low, since nearly all the bases required by plants have been removed from the soil profile. A small store of nutrient bases occurs very close to the soil surface. The soil is quite easily broken apart and allows easy penetration by rainwater and plant roots.

ULTISOLS **Ultisols** are quite closely related to the Oxisols in outward appearance and environment of origin. Ultisols are reddish to yellowish in color (Figures 21.14*b* and 21.16). They have a subsurface horizon of clay accumulation, called an *argillic horizon,* which is not found in the Oxisols. It is a *B* horizon and has developed through accumulation of clay in the process of illuviation. Although forest is the characterstic native vegetation, the base status of the Ultisols is low. As in the Oxisols, most of the bases are found in a shallow surface where they are released by the decay of plant matter. They are quickly taken up and recycled by the shallow roots of trees and shrubs.

In a few areas, the Ultisol profile contains a subsurface horizon of sesquioxides. This horizon is capable of hardening to a rock-like material if it becomes exposed at the surface and is subjected to repeated wetting and drying. This material is referred to as *plinthite* (from the Greek word *plinthos,* meaning "brick"). In the hardened state, plinthite is referred to as *laterite* (from the Latin *later,* or brick). In Southeast Asia, plinthite is quarried and cut into building blocks (Figure 21.17). These blocks harden into laterite blocks when they are exposed to the air.

Ultisols are widespread throughout Southeast Asia and the East Indies. Other important areas are in eastern Australia, Central America, South America, and the southeastern United States. Ultisols extend into the lower midlatitude zone in the United States, where they correspond quite closely in extent with the area of moist subtropical climate ⑥. In lower latitudes, Ultisols are identified with the wet-dry tropical climate ③ and the monsoon and trade-wind coastal climate ②. Note that all these climates have a dry season, even though it may be short.

Both Oxisols and Ultisols of low latitudes were used for centuries under shifting agriculture prior to the advent of modern agricultural technology. This primitive agricultural method, known as slash-and-burn, is still widely practiced. Without fertilizers, these soils can sustain crops on freshly cleared areas for only two or three years, at most, before the nutrient bases are exhausted and the garden plot must be abandoned. Substantial use of lime, fertilizers, and other industrial inputs is necessary for high, sustained crop yields. Furthermore, the exposed soil surface of the Ultisols is vulnerable to devastating soil erosion, particularly on steep hill slopes.

VERTISOLS **Vertisols** have a unique set of properties that stand in sharp contrast to the Oxisols and Ultisols. Vertisols are black in color and have a high clay content (Figures 21.14*c* and 21.18). Much of the clay consists of a particular mineral that shrinks and swells greatly with seasonal changes in soil water content. Wide, deep vertical cracks develop in the soil during the dry season. As the dry soil blocks are wetted and softened by rain, some fragments of surface soil drop into the cracks before they close, so that the soil "swallows itself" and is constantly being mixed.

Vertisols typically form under grass and savanna vegetation in subtropical and tropical climates with a pronounced dry season. These climates include the semiarid subtype of the dry tropical steppe climate ④ and the wet-dry tropical climate ③. Because Vertisols require a particular clay mineral as a parent material, the major areas of occurrence are scattered and show no distinctive pattern on the world map. An important region of Vertisols is the Deccan Plateau of western India, where basalt, a dark variety of igneous rock, supplies the silicate minerals that are altered into the necessary clay minerals.

Vertisols are high in base status and are particularly rich in such nutrient bases as calcium and magnesium. The soil solution is nearly neutral in pH, and a moderate content of organic matter is distributed through the soil. The soil retains large amounts of water because of its fine texture, but much of this water is held tightly by the clay particles and is not available to plants. Where soil cultivation depends on human or animal power, as it does in most of the developing nations where the soil occurs, agricultural yields are low. This is because the moist soil becomes highly plastic and is difficult to till with primitive tools. For this reason, many areas of Vertisols have been left in grass or shrub cover, providing grazing for cattle. Soil scientists think that the use of modern technology, including heavy farm machinery, could result in substantial production of food and fiber from Vertisols that are not now in production.

ALFISOLS The **Alfisols** are soils characterized by an argillic horizon, produced by illuviation. Unlike the argillic horizon of the Ultisols, this *B* horizon is enriched by silicate clay minerals that have an adequate capacity to hold bases such as calcium and magnesium. The base status of the Alfisols is therefore generally quite high.

Above the *B* horizon of clay accumulation is a horizon of pale color, the *E* horizon, that has lost some of the original bases, clay minerals, and sesquioxides by the process of elu-

21.13 Soils of the world

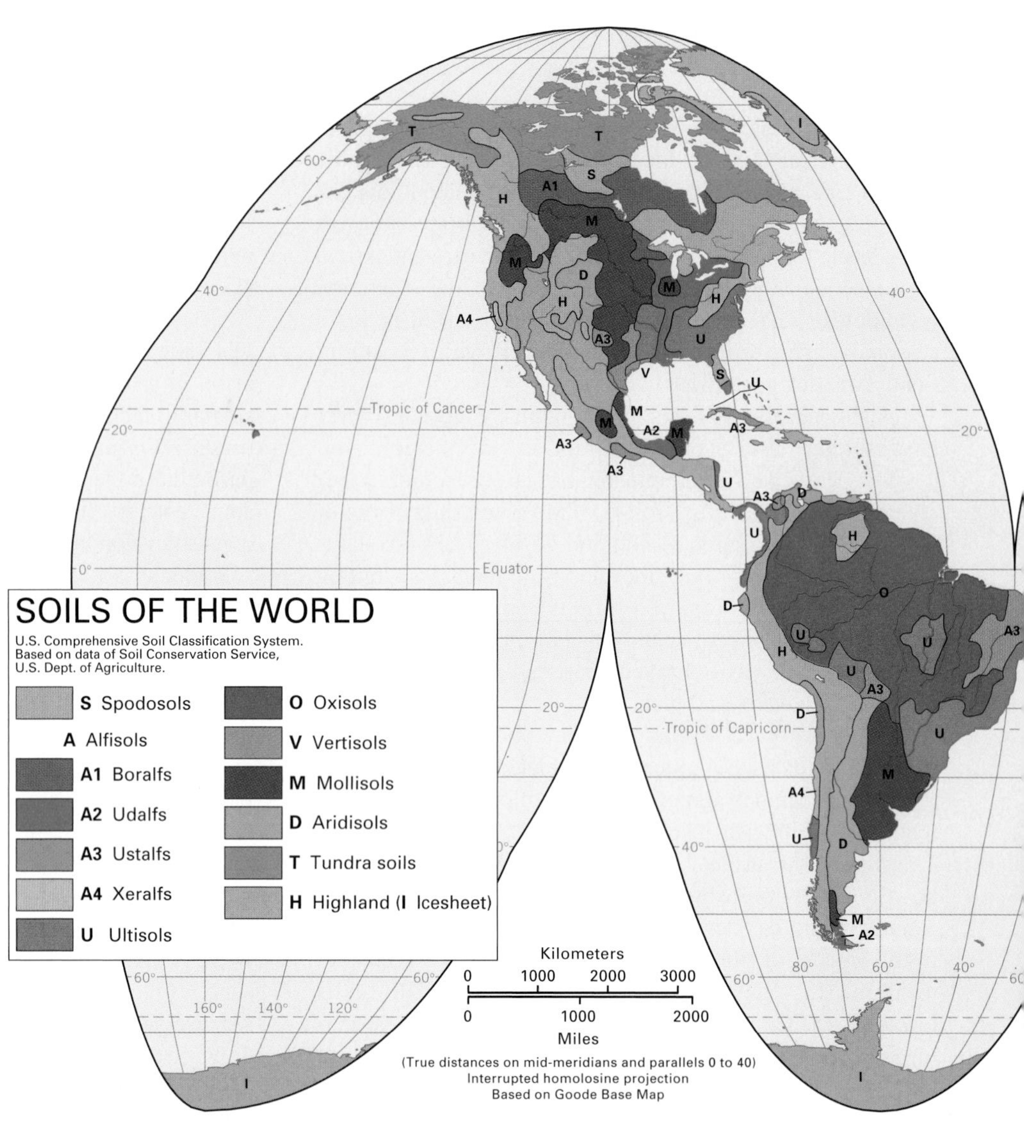

Table 21.3 | **Formative Elements in Names of Soil Orders**

Name of Order	Formative Element	Derivation of Formative Element	Pronunciation of Formative Element
Entisol	ent	Meaningless syllable	recent
Inceptisol	ept	L. *inceptum,* beginning	inept
Histosol	ist	Gr. *histos,* tissue	histology
Oxisol	ox	F. *oxide,* oxide	ox
Ultisol	ult	L. *ultimus,* last	ultimate
Vertisol	ert	L. *verto,* turn	invert
Alfisol	alf	Meaningless syllable	alfalfa
Spodosol	od	Gr. *spodos,* wood ash	odd
Mollisol	oll	L. *mollis,* soft	mollify
Aridisol	id	L. *aridus,* dry	arid
Andisol	and	Eng. *andesite,* a volcanic rock type	and

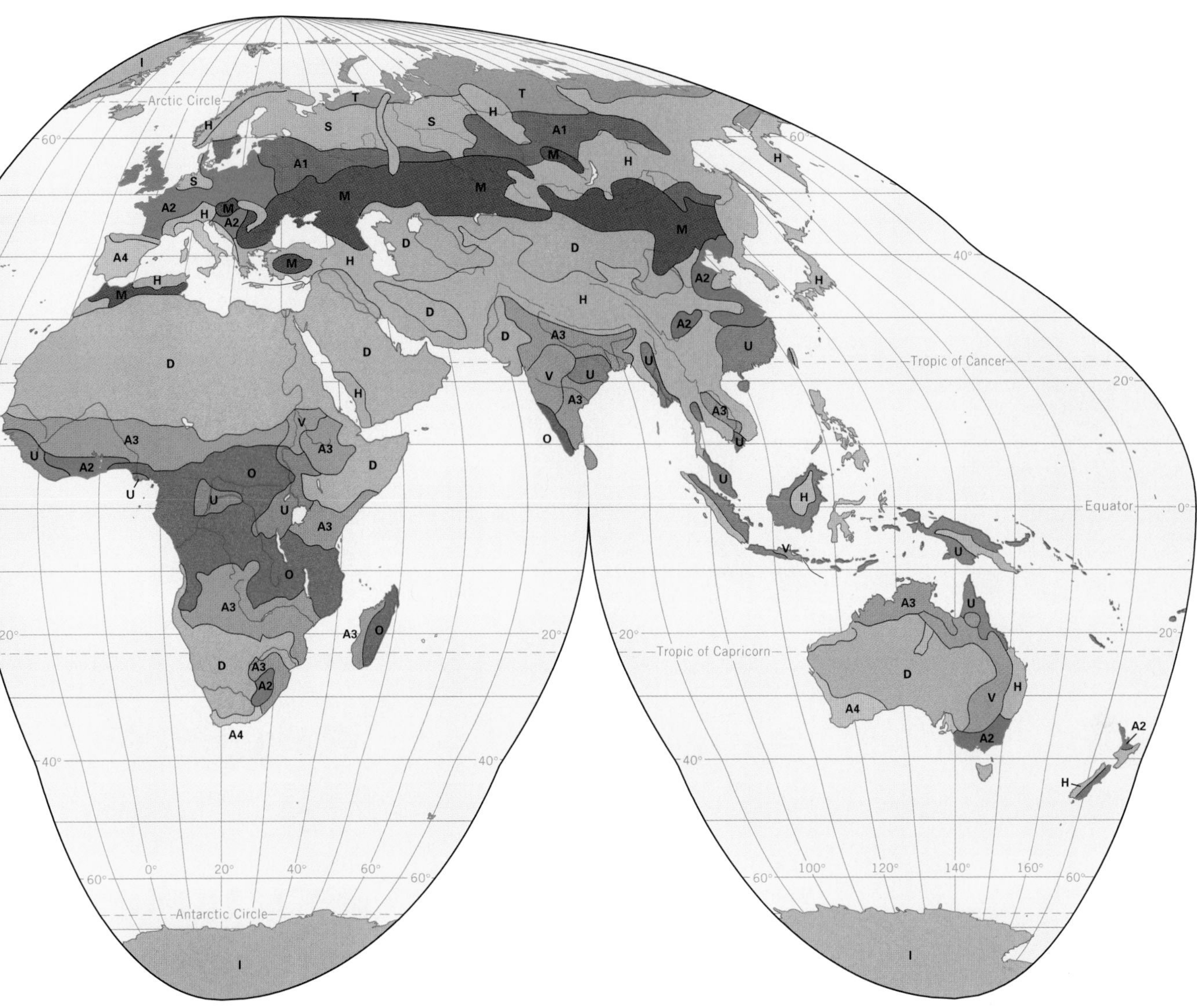

viation. These materials have become concentrated by illuviation in the *B* horizon. Alfisols also have a gray, brownish, or reddish surface horizon.

The world distribution of Alfisols is extremely wide in latitude (see Figure 21.13). Alfisols range from latitudes as high as 60° N in North America and Eurasia to the equatorial zone in South America and Africa. Obviously, the Alfisols span an enormous range in climate types. For this reason, we need to recognize four of the important suborders of Alfisols, each with its own climate affiliation.

Boralfs are Alfisols of cold (boreal) forest lands of North America and Eurasia. They have a gray surface horizon and a brownish subsoil. *Udalfs* are brownish Alfisols of the midlatitude zone. They are closely associated with the moist continental climate ⑩ in North America, Europe, and eastern Asia (Figure 21.14*d*).

Ustalfs are brownish to reddish Alfisols of the warmer climates (Figure 21.14*e*). They range from the subtropical zone to the equator and are associated with the wet-dry tropical climate ③ in Southeast Asia, Africa, Australia, and South America. *Xeralfs* are Alfisols of the Mediterranean climate ⑦, with its cool moist winter and dry summer. The Xeralfs are typically brownish or reddish in color.

SPODOSOLS Poleward of the Alfisols in North America and Eurasia lies a great belt of soils of the order **Spodosols,** formed in the cold boreal forest climate ⑪ beneath a needleleaf forest. Spodosols have a unique property—a *B* horizon of accumulation of reddish mineral matter with a low capacity to hold bases (Figure 21.14*f*). This horizon is called the spodic horizon (Figure 21.19). It is made up of a dense mixture of organic matter and compounds of aluminum and iron, all brought downward by eluviation from an overlying *E* horizon. Because of the intensive removal of matter from the *E* horizon, it has a bleached, pale gray to white appearance (Figure 21.12). This conspicuous feature led to the naming of the soil as *podzol* (ash-soil) by Russian peasants. In modern terminology, this pale layer is an *albic horizon.* The *O* horizon, a thin, very dark layer of organic matter, overlies the *A* horizon.

Spodosols are strongly acid and are low in plant nutrients such as calcium and magnesium. They are also low in

OXISOLS

A Torrox, Hawaii

ULTISOLS

B Udult, Virginia

VERTISOLS

C Ustert, India

ALFISOLS

D Udalf, Michigan

ALFISOLS

E Ustalf, Texas

SPODOSOLS

F Orthod, France

21.14 Soil profiles of several soil orders

MOLLISOLS

G Boroll, USSR

MOLLISOLS

H Udoll, Argentina

MOLLISOLS

I Ustoll, Colorado

MOLLISOLS

J Rendoll, Argentina

ARIDISOLS

K Argid, Colorado

HISTOSOLS

L Fibrist, Minnesota

21.15 An Oxisol in Hawaii *Sugarcane is being cultivated here.*

humus. Although the base status of the Spodosols is low, forests of pine and spruce are supported through the process of recycling of the bases.

Spodosols are closely associated with regions recently covered by the great ice sheets of the Pleistocene Epoch. These soils are therefore very young. Typically, the parent material is coarse sand consisting largely of the mineral quartz. This mineral cannot weather to form clay minerals.

Spodosols are naturally poor soils in terms of agricultural productivity. Because they are acid, application of lime is essential. Heavy applications of fertilizers are also required. With proper management and the input of the required industrial products, Spodosols can be highly productive, if the soil texture is favorable. High yields of potatoes from Spodosols in Maine and New Brunswick are examples. Another factor unfavorable to agriculture is the shortness of the growing season in the more northerly parts of the Spodosol belt.

HISTOSOLS Throughout the northern regions of Spodosols are countless patches of **Histosols.** This unique soil order has a very high content of organic matter in a thick, dark upper layer (Figure 21.14*l*). Most Histosols go by such common names as peats or mucks. They have formed in shallow lakes and ponds by accumulation of partially decayed plant matter. In time, the water is replaced by a layer of organic matter, or *peat,* and becomes a *bog* (Figure 21.20). Peat bogs are used extensively for cultivation of cranberries (cranberry bogs). Sphagnum peat from bogs is dried and baled for sale as a mulch for use on suburban lawns and shrubbery beds. For centuries, Europe has used dried peat from bogs of glacial origin as a low-grade fuel.

Some Histosols are *mucks*—organic soils composed of fine black materials of sticky consistency. These are agriculturally valuable in midlatitudes, where they occur as beds of former lakes in glaciated regions. After appropriate drainage and application of lime and fertilizers, these mucks are remarkably productive for garden vegetables (Figure 21.21). Histosols are also found in low latitudes, where conditions of poor drainage have favored thick accumulations of plant matter.

ENTISOLS Entisols have in common the combination of a mineral soil and the absence of distinct horizons. Entisols are soils in the sense that they support plants, but they may be found in any climate and under any vegetation. Entisols lack distinct horizons for two reasons. It may be the result of a parent material, such as quartz sand, in which horizons do not readily form. Or it may be the result of lack of time for

21.16 Ultisol profile in North Carolina *The thin, pale layer at the top is an E horizon, showing the effects of removal of materials by eluviation. Below the base of the thick, reddish B horizon is a blotchy-colored zone of plinthite.*

21.17 Laterite in India *Laterite, formed by hardening of plinthite, is being quarried for building stone in this scene from India.*

21.18 A Vertisol in Texas *The clay minerals that are abundant in Vertisols shrink when they dry out, producing deep cracks in the soil surface.*

horizons to form in recent deposits of alluvium or on actively eroding slopes.

Entisols occur from equatorial to arctic latitude zones. From the standpoint of agricultural productivity, Entisols of the subarctic zone and tropical deserts (along with arctic areas of Inceptisols) are the poorest of all soils. In contrast, Entisols and Inceptisols of floodplains and delta plains in warm and moist climates are among the most highly productive agricultural soils in the world because of their favorable texture, ample nutrient content, and large soil water storage.

INCEPTISOLS **Inceptisols** are soils with horizons that are weakly developed, usually because the soil is quite young. These areas occur within some of the regions shown on the world map as Ultisols and Oxisols. Especially important are the Inceptisols of river floodplains and delta plains in Southeast Asia that support dense populations of rice farmers.

In these regions, annual river floods cover low-lying plains and deposit layers of fine silt. This sediment is rich in primary minerals that yield bases as they weather chemically over time. The constant enrichment of the soil

21.19 Diagram of a Spodosol profile *The spodic horizon, a dense mixture of organic matter, aluminum, and iron oxides, and the gray to white albic horizon, are diagnostic features of a Spodosol profile.*

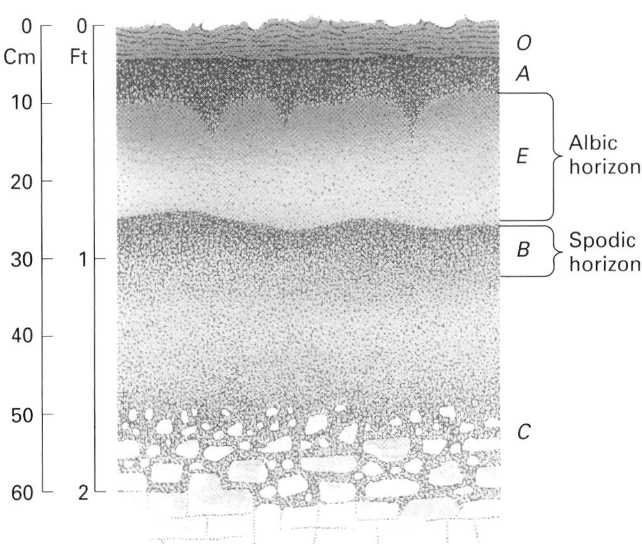

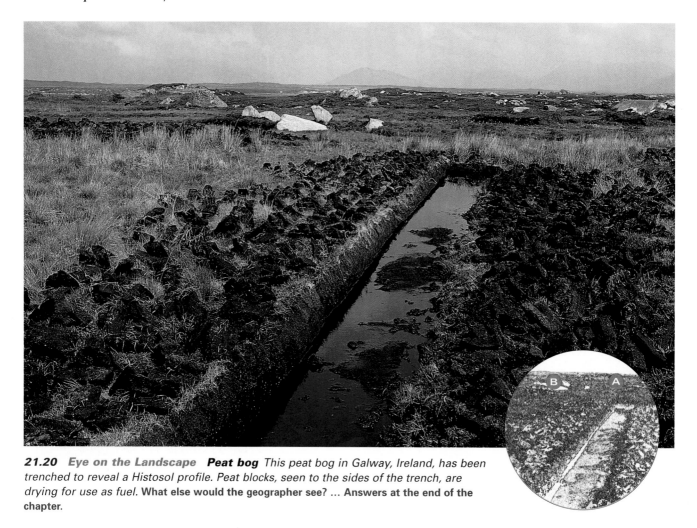

21.20 Eye on the Landscape Peat bog *This peat bog in Galway, Ireland, has been trenched to reveal a Histosol profile. Peat blocks, seen to the sides of the trench, are drying for use as fuel.* **What else would the geographer see? ... Answers at the end of the chapter.**

explains the high soil fertility in a region where uplands develop only Ultisols of low fertility. Inceptisols of these floodplain and delta lands are of a suborder called *Aquepts*—Inceptisols of wet places. Much closer to home is another prime example of Aquepts within the domain of the Ultisols—the lower Mississippi River floodplain and delta plain.

ANDISOLS **Andisols** are soils in which more than half of the parent mineral matter is volcanic ash, spewed high into the air from the craters of active volcanoes and coming to rest in layers over the surrounding landscape. The fine ash particles are glass-like shards. A high proportion of carbon, formed by the decay of plant matter, is also typical, so that the soil usually appears very dark in color. Andisols form

21.21 A Histosol in cultivation *Garden crops cultivated on a Histosol of a former glacial lake bed, Southern Ontario, Canada.*

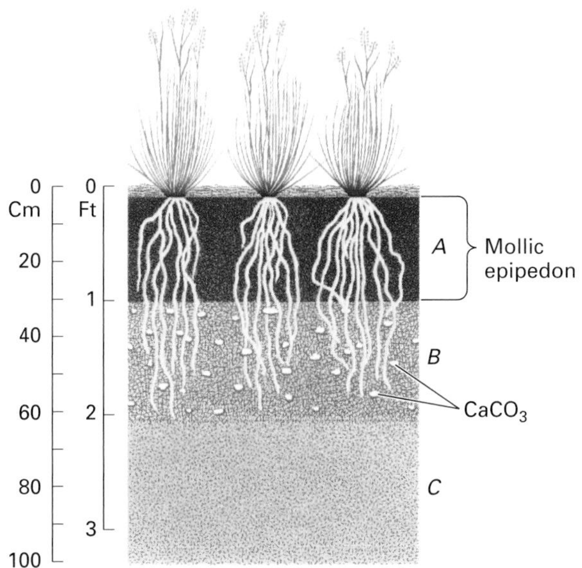

21.22 Schematic diagram of a Mollisol profile *The thick, dark brown to black mollic epipedon is a diagnostic feature of the Mollisol.*

over a wide range of latitudes and climates. They are for the most part fertile soils, and in moist climates they support a dense natural vegetation cover.

Andisols do not appear on our world map because they are found in small patches associated with individual volcanoes that are located mostly in the "Ring of Fire"—the chain of volcanic mountains and islands that surrounds the great Pacific Ocean. Andisols are also found on the island of Hawaii, where volcanoes are presently active.

MOLLISOLS **Mollisols** are soils of grasslands that occupy vast areas of semiarid and subhumid climates in midlatitudes. Mollisols are unique in having a very thick, dark brown to black surface horizon called the *mollic epipedon* (Figures 21.14*g* to *j* and 21.22). This layer lies within the *A* horizon and is always more than 25 cm (9.8 in.) thick. The soil has a loose, granular structure (Figure 21.6) or a soft consistency when dry. Other important qualities of the Mollisols are the dominance of calcium among the bases of the *A* and *B* horizons and the very high base status of the soil.

Most areas of Mollisols are closely associated with the semiarid subtype of the dry midlatitude climate ⑨ and the adjacent portion of the moist continental climate ⑩. In North America, Mollisols dominate the Great Plains region, the Columbia Plateau, and the northern Great Basin. In South America, a large area of Mollisols covers the Pampa region of Argentina and Uruguay. In Eurasia, a great belt of Mollisols stretches from Romania eastward across the steppes of Russia, Siberia, and Mongolia. Russians refer to the Mollisols as *chernozems,* a term that has gained widespread use throughout the Western world as well.

Because of their loose texture and very high base status, Mollisols are among the most naturally fertile soils in the world. They now produce most of the world's commercial grain crop. Most of these soils have been used for crop production only in the last century. Prior to that time, they were used mainly for grazing by nomadic herds. The Mollisols have favorable properties for growing cereals in large-scale mechanized farming and are relatively easy to manage. Production of grain varies considerably from one year to the next because seasonal rainfall is highly variable.

A brief mention of four suborders of the Mollisols as they occur in the United States and Canada will help you to understand important regional soil differences related to climate. *Borolls,* the cold-climate suborder of the Mollisols, are found in a large area extending on both sides of the U.S.-Canadian border east of the Rocky Mountains (Figure 21.14*g*). *Udolls* are Mollisols of a relatively moist climate, as compared with the other suborders. Formerly, the Udolls supported tall-grass prairie, but today they are closely identified with the corn belt in the American Midwest (Figure 21.14*h*).

Ustolls are Mollisols of the semiarid subtype of the dry midlatitude climate ⑨, with a substantial soil water shortage in the summer months (Figure 21.14*i*). The Ustolls underlie much of the short-grass prairie region east of the Rockies (Figure 21.23). *Xerolls* are Mollisols of the Mediterranean climate ⑦, with its tendency to cool, moist winters and rainless summers.

21.23 A Mollisol developed on loess *Dry prairie grasses are seen on the surface.*

21.24 ***Salic horizon*** *Appearing as a white layer, a salic horizon lies close to the surface in this Aridisol profile in the Nevada desert. The scale is marked in feet (1 ft = 30.5 cm).*

Desert and Tundra Soils

Desert and tundra soils are soils of extreme environments. Aridisols characterize the desert climate. As might be expected, they are low in organic matter and high in salts. Tundra soils are poorly developed because they are formed on very recent parent material, left in place by glacial activity during the Ice Age. Cold temperatures have also restricted soil development in tundra regions.

ARIDISOLS **Aridisols,** soils of the desert climate, are dry for long periods of time. Because the climate supports only a very sparse vegetation, humus is lacking and the soil color ranges from pale gray to pale red (Figure 21.14*k*). Soil horizons are weakly developed, but there may be important subsurface horizons of accumulated calcium carbonate (petrocalcic horizon) or soluble salts (salic horizon) (Figure 21.24). The salts, of which sodium is a dominant constituent, give the soil a very high degree of alkalinity. The Aridisols are closely correlated with the arid subtype of the dry tropical ④, dry subtropical ⑤, and dry midlatitude ⑨ climates.

Most Aridisols are used for nomadic grazing, as they have been through the ages. This use is dictated by the limited rainfall, which is inadequate for crops without irrigation. Locally, where water supplies from mountain streams or ground water permit, Aridisols can be highly productive for a wide variety of crops under irrigation (Figures 21.25, 21.26). Great irrigation systems, such as those of the Imperial Valley of the United States, the Nile Valley of Egypt, and the Indus Valley of Pakistan, have made Aridisols highly productive, but not without problems of salt buildup and waterlogging.

TUNDRA SOILS Soils of the arctic tundra fall largely into the order of Inceptisols, soils with weakly developed horizons that are usually associated with a moist climate. Inceptisols of the tundra climate belong to the suborder of Aquepts, which as we noted earlier are Inceptisols of wet places. More specifically, the tundra soils can be assigned to the *Cryaquepts,* a subdivision within the Aquepts. The prefix *cry* is derived from the Greek word *kryos,* meaning "icy cold." We may refer to these soils simply as *tundra soils.*

21.25 ***Desert Aridisol*** *This gray desert soil, an Aridisol, has proved highly productive when cultivated and irrigated. The locality is near Palm Springs, California, in the Coachella Valley.*

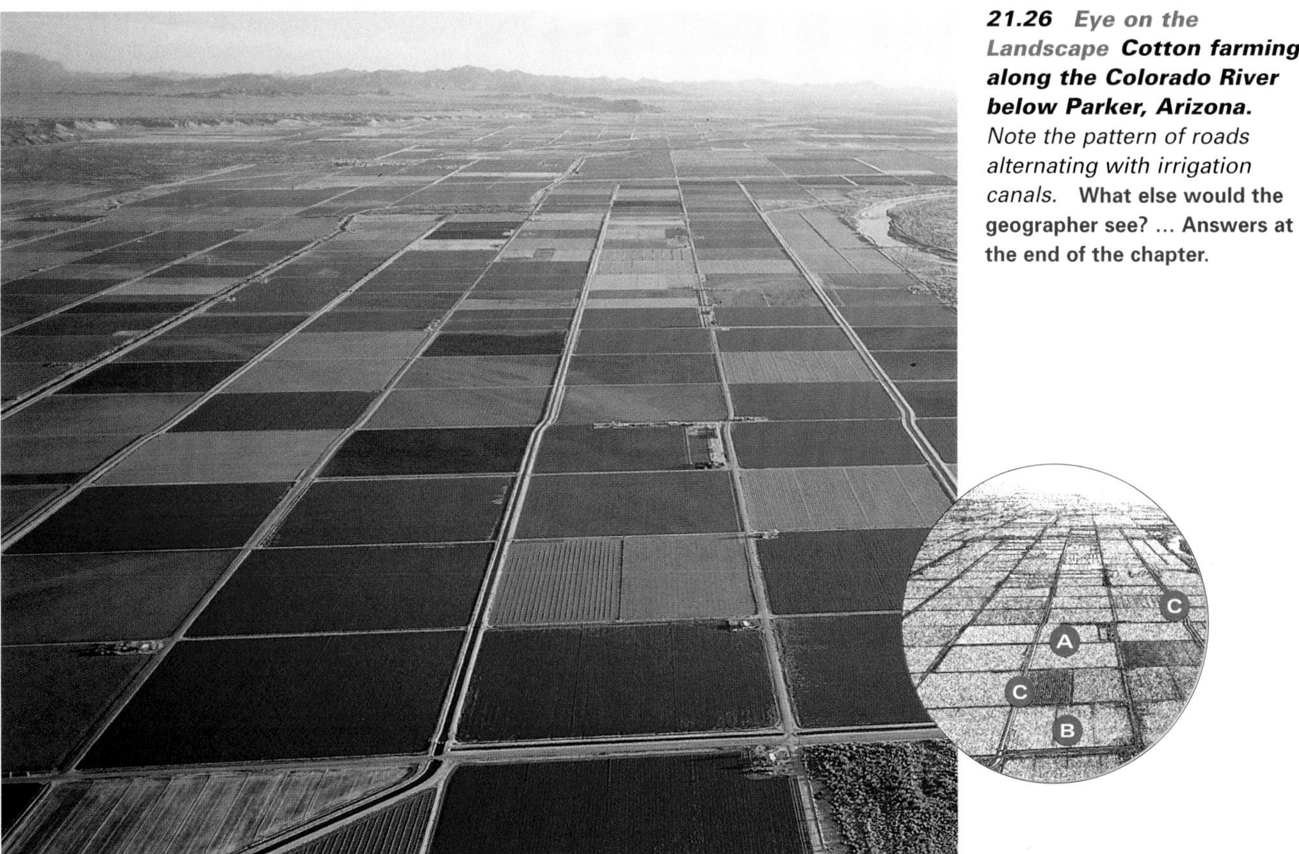

21.26 ***Eye on the Landscape*** ***Cotton farming along the Colorado River below Parker, Arizona.*** *Note the pattern of roads alternating with irrigation canals.* **What else would the geographer see? ... Answers at the end of the chapter.**

Tundra soils are formed largely of primary minerals ranging in size from silt to clay that are broken down by frost action and glacial grinding. Layers of peat are often present between mineral layers. Beneath the tundra soil lies perennially frozen ground (permafrost), described in Chapter 15. Because the annual summer thaw affects only a shallow surface layer, soil water cannot easily drain away. Thus, the soil is saturated with water over large areas. Repeated freezing and thawing of this shallow surface layer disrupts plant roots, so that only small, shallow-rooted plants can maintain a hold.

A Midcontinental Transect from Aridisols to Alfisols

Figure 21.27 provides a diagram showing generalized changes in soil profile and soil type as they might occur on a transect from a cool, dry desert to a cool moist climate, starting in the northern desert of the Great Basin and crossing the midwest. Starting with the Aridisols in the west, horizons are thin but show an accumulation of clay minerals in the *B* horizon. As precipitation increases, grasses dominate and Mollisols are formed. The *B* horizon thickens and

21.27 ***Soil profile transect*** *A schematic diagram of the changing soil profile from a cool dry desert on the west to a cool moist climate on the east. (After C. E. Millar, L. M. Turk, and H. D. Foth, Fundamentals of Soil Science, John Wiley and Sons, New York.)*

includes calcium accumulation. In the east, where alfisols occur, enhanced precipitation generates an *E* horizon that takes its place between the *A* and *B* horizons.

Global Climate Change and Agriculture

Soils are the foundation of agriculture, supporting and nourishing crops as well as native plants that yield food for humans and animals alike. However, our food supply is also dependent on the weather and climate that crops and plants experience during their growing seasons. As we saw earlier, global climate change is predicted to increase global temperatures and summer droughts, change rainfall patterns, and produce more extreme events. All of these changes will have impacts on agriculture.

In general, the impacts on crop yields are expected to be at first positive, but, as temperatures rise, then negative. An important factor is that CO_2 rises will act to fertilize plants, somewhat offsetting the effects of higher temperatures and drought stresses. According to crop growth models linked to climate models and economic models, food costs will rise as increases in agricultural outputs fall behind increases in demand. However, widespread adaptation by farmers and farm managers to new and changing climate conditions may be effective enough to compensate for the changes. Our interchapter feature, *A Closer Look: Eye on Global Change 21.3 • Global Climate Change and Agriculture*, at the end of this chapter, presents more details.

A Look Ahead The soil layer is a complex body subject to many influences, including the parent materials from which it is derived, the vegetation that it harbors, and the water regime of precipitation and evapotranspiration that it experiences. In many environments, soil-forming processes operate very slowly. Thus, some soils can be the products of complex histories involving climatic changes. By inducing soil erosion (Chapter 17), human activities can rapidly strip uppermost soil horizons, leaving less productive horizons at the surface. Soil horizons and properties can also vary strongly over short distances, thereby exhibiting important local variation. Soil science is an interesting and complex topic, and it is difficult to do justice to it in a single textbook chapter.

From the topic of soils, we turn next to ecosystems and biogeography in Chapters 22, 23, and 24. Like soil science, these fields are both broad and deep. But it is not difficult to highlight the ecological systems and biogeographic processes of importance to geographers. This is the objective of the following chapters.

Chapter Summary

- The **soil** layer is a complex mixture of solid, liquid, and gaseous components. It is derived from **parent material**, or *regolith*, that is produced from rock by *weathering*. The major factors influencing soil and soil development are parent material, climate, vegetation, and time.
- **Soil texture** refers to the proportions of *sand, silt*, and *clay* that are present. *Colloids* are the finest particles in soils and are important because they help retain nutrients, or bases, that are used by plants. Soils show a wide range of *pH values*, from acid to alkaline. Soils with granular or blocky structures are most easily cultivated.
- In soils, primary minerals are chemically altered to **secondary minerals**, which include oxides and *clay minerals*. The nature of the clay minerals determines the soil's base status. If *base status* is *high*, the soil retains nutrients. If *low*, the soil can lack fertility. When a soil is fully wetted by heavy rainfall or snowmelt and allowed to drain, it reaches its *storage capacity*. Evaporation from the surface and transpiration from plants draws down the soil water store until precipitation occurs again to recharge it.
- The **soil water balance** describes the gain, loss, and storage of soil water. It depends on **water need** *(potential evapotranspiration)*, **water use** *(actual evapotranspiration)*, and precipitation. Monitoring these values on a monthly basis provides a *soil water budget* for the year.
- Most soils possess distinctive horizontal layers called **horizons**. These layers are developed by processes of *enrichment, removal, translocation,* and *transformation*. In downward translocation, materials such as humus, clay particles, and mineral oxides are removed by **eluviation** from an upper horizon and accumulate by **illuviation** in a lower one. In *salinization*, salts are translocated upward by evaporating water to form a *salic horizon*. In *humification*, a transformation process, organic matter is broken down by bacterial decay. Where soil temperatures are warm, this process can be highly effective, leaving a soil low in organic content. Animals, such as earthworms, can be very important in soil formation where they are abundant.
- Global soils are classified into 11 **soil orders**, often by the presence of one of more diagnostic horizons. **Oxisols** are old, highly weathered soils of low latitudes. They have a horizon of mineral oxide accumulation and a low base status. **Ultisols** are also found in low latitudes. They have a horizon of clay accumulation and are also of low base status. **Vertisols** are rich in a type of clay mineral that expands and contracts with wetting and drying, and has a high base status. **Alfisols** have a horizon of clay accumulation like Ultisols, but they are of high base status. They are found in moist climates from equatorial to subarctic zones. **Spodosols**, found in cold, moist climates, exhibit a horizon of illuviation and low base status. **Histosols** have a thick upper layer formed almost entirely of organic matter.

- Three soil orders have poorly developed horizons or no horizons—**Entisols, Inceptisols**, and **Andisols**. Entisols are composed of fresh parent material and have no horizons. The horizons of Inceptisols are only weakly developed. Andisols are weakly developed soils occurring on young volcanic deposits.
- **Mollisols** have a thick upper layer rich in humus. They are soils of midlatitude grasslands. **Aridisols** are soils of arid regions, marked by horizons of accumulation of carbonate minerals or salts.
- **Tundra soils** are largely wet, cold Inceptisols.
- Global climate change will eventually reduce crop yields as temperatures and drought stress increase. Food prices will probably rise, but adjustments in farming practices may compensate.

Key Terms

soil	secondary minerals	eluviation	Vertisols	Inceptisols
parent material	water use	illuviation	Alfisols	Andisols
soil texture	water need	soil orders	Spodosols	Mollisols
soil colloids	soil horizons	Oxisols	Histosols	Aridisols
bases	soil profile	Ultisols	Entisols	

Review Questions

1. Which important factors condition the nature and development of the soil?
2. Soil color, soil texture, and soil structure are used to describe soils and soil horizons. Identify each of these three terms, showing how they are applied.
3. Explain the concepts of acidity and alkalinity as they apply to soils.
4. Identify two important classes of secondary minerals in soils and provide examples of each class.
5. How does the ability of soils to hold water vary, and how does this ability relate to soil texture?
6. Define water need (potential evapotranspiration) and water use (actual evapotranspiration). How are they used in the soil water balance?
7. Identify the following terms as used in the soil water budget: storage, withdrawal, storage recharge, soil water shortage, water surplus.
8. What is a soil horizon? How are soil horizons named? Provide two examples.
9. Identify four classes of soil-forming processes and describe each.
10. What are translocation processes? Identify and describe four translocation processes.
11. How many soil orders are there? Try to name them all.
12. Name three soil orders that are especially associated with low latitudes. For each order, provide at least one distinguishing characteristic and explain it.
13. Compare Alfisols and Spodosols. What features do they share? What features differentiate them? Where are they found?
14. Where are Mollisols found? How are the properties of Mollisols related to climate and vegetation cover? Name four suborders within the Mollisols.
15. Desert and tundra are extreme environments. Which soil order is characteristic of each environment? Briefly describe desert and tundra soils.

Eye on the Environment 21.2 • Death of a Civilization

1. Describe the processes of salinization, waterlogging, and siltation and their impact on agriculture.
2. A technological "fix' for siltation is to build a dam upstream of an irrigation canal complex, so that silt will be trapped behind the dam. What other effects might building such a dam produce, both desirable and undesirable?

A Closer Look:
Eye on Global Change 21.3 • Global Climate Change and Agriculture

1. How will increasing temperature affect agricultural crop yields? What will be the effect of increasing atmospheric CO_2 concentrations?
2. What actions can farmers and farm managers take to mitigate the effects of climate change?
3. How will climate change influence the quality of the agricultural environment?
4. What is the predicted impact of global climate change on food prices? Who will be most affected?

Visualizing Exercises

1. Sketch the profile of a Spodosol, labeling *O, A, E, B,* and *C* horizons. Diagram the movement of materials from the zone of eluviation to the zone of illuviation.
2. Examine the world soils map (Figure 21.13) and identify three soil types that are found near your location. Develop a short list of characteristics that would help you tell them apart.

Essay Questions

1. Document the important role of clay particles and clay mineral colloids in soils. What is meant by the term *clay?* What are colloids? What are their properties? How does the type of clay mineral influence soil fertility? How does the amount of clay influence the water-holding capacity of the soil? What is the role of clay minerals in horizon development?

2. Using the world maps of global soils and global climate, compare the pattern of soils on a transect along the 20° E longitude meridian with the patterns of climate encountered along the same meridian. What conclusions can you draw about the relationship between soils and climate? Be specific.

Problems

Working It Out 21.1 • Calculating a Simple Soil Water Budget

1. Below are monthly values and annual totals for precipitation, water need *(Ep),* and water use *(Ea)* at Urbana, Illinois, in the moist continental climate ⑩. Prepare and plot a soil water budget similar to the figure in *Working It Out • 21.1* for this station. This will involve determining soil water shortage *(D),* soil water utilization (–*G*), soil water recharge (+*G*), and water surplus *(R)* for each month following the method described in the text of the feature.

Soil Water Budget for Urbana, Illinois

Month	*P*	= *Ea*		+*G*	+*R*	*Ep*	*D*
Jan	5.7	0.0				0.0	
Feb	4.5	0.0				0.0	
Mar	8.2	1.4				1.4	
Apr	10.0	4.4				4.4	
May	9.9	8.8				8.8	
Jun	8.4	12.4				12.6	
Jul	8.0	13.4				14.9	
Aug	9.0	11.6				13.0	
Sep	8.3	7.8				7.8	
Oct	6.6	4.8				4.8	
Nov	5.7	1.4				1.4	
Dec	5.5	0.0				0.0	
Total	89.8	66.0	–12.0	+12.1	23.7	69.1	3.1

Eye on the Landscape

21.20 Peat bog Peat bogs are typical of glacial terrain. Moving ice sheets remove and take up loose soil and rock, then leave sediment behind as they melt. The low, hummocky terrain **(A)** at the edge of this bog may be part of a moraine, a landform that occurs at the melting edge of an ice sheet as wasting ice leaves piles of sediment behind in a chaotic and disorganized fashion. The isolated angular boulders **(B)** may be erratics—large rocks that are carried in the ice and moved long distances from their sources. Landforms of glacial ice sheets are described in Chapter 20.

21.26 Irrigated Aridisols The broad floodplain of the Colorado River (right) shows how productive Aridisols can be when properly managed and irrigated. A consistent climate with abundant sunlight, coupled with abundant river water, makes for high productivity and crop yields. Fallow fields **(A)** show typical Aridisol colors of gray to brown, while cotton fields are vibrant green **(B)**. Note the network of canals **(C)** serving the fields. Aridisols are covered in this chapter.

A CLOSER LOOK

Eye on Global Change 21.3 | Global Climate Change and Agriculture

For the remainder of the century, and probably well beyond, our global climate will change. The Earth will become warmer, especially in mid- and high latitudes. Most areas will have more precipitation, although higher temperatures will often bring more summer drought stress. Extreme events—heavy rainfalls and high winds— will be more frequent. How will global climate change impact agriculture?

Young corn field, Iowa *Climate change will impact how crops grow. In midlatitudes, earlier seasonal warming will allow a longer growing season, but hotter summers with higher drought potential may reduce later growth. However, increasing atmospheric CO_2 will tend to increase growth.*

Warming Temperatures and CO_2 Fertilization

Let's look first at the effects of warmer temperatures. In general, higher temperatures will increase the productivity of most mid- and high-latitude crops by shortening the growing season. But as you might imagine, once temperatures get too high, the effects will be negative. For crops on the edge of their high-temperature range, any warming will reduce productivity. Recall also that global warming is expected to reduce minimum temperatures. This will be beneficial for most mid- and high-latitude crops but detrimental to some tropical and equatorial crops.

On the other hand, a significant CO_2 fertilization effect is expected to occur as atmospheric CO_2 concentrations double. Recall that plants take in atmospheric CO_2 for photosynthesis and that because CO_2 is present in the atmosphere only in low concentrations, CO_2 limits the photosynthetic process in many cases. With higher concentrations, plants will be able to photosynthesize more rapidly and become more productive. This fertilization effect is well documented in many studies, both in greenhouses and by free-air release of CO_2 gas upwind from agricultural fields. The fertilization effect appears to be greater when plants are under greater stress, for example, from warmer temperatures and drought.

Crop Yields and Adaptation

What are the combined effects of changing temperatures and increasing CO_2 on crop yields? In general, if warming is less than about 2.5°C (4.5°F), yields increase, but if greater than that amount, yields decrease.[1] However, the predicted changes for different crops and regions vary significantly. For example, studies cited in the IPCC Global Change Report predict decreasing yields for corn, wheat, and rice in Egypt; a slight decrease for corn in Romania; increasing yields for wheat in Australia; and increasing yields for rice in temperate Asia.

An important factor affecting yields is *adaptation*. Adaptation describes actions taken by farmers to respond to climate change, such as changing planting and harvesting dates or selecting different strains of crops that are better adapted to changed conditions. Higher levels of adaptation include changing the crops that are planted, adding or modifying irrigation, using more or different fertilizers and pesticides, and other actions.

Studies show that adaptation can mitigate many of the negative effects of global change. For example, the U.S. Global Change Research Program Climate Change Report,[2] which includes adaptation

[1]See *Climate Change 2001: The Scientific Basis, Contribution of Working Group I to the Third Assessment Report of the Intergovernmental Panel on Climate Change,* IPCC Cambridge University Press, 2001, and *Climate Change 2001: Impacts, Adaptation, and Vulnerability, Contribution of Working Group II to the Third Assessment Report of the Intergovernmental Panel on Climate Change,* IPCC, Cambridge University Press, 2001, which are used as the basis for this discussion.

[2]*Climate Change Impacts on the United States: The Potential Consequences of Climate Variability and Change,* National Assessment Synthesis Team, U.S. Global Change Research Program, Washington, DC, 2000.

Cereal crop harvest, Novosibirsk region, Russia *With a changing climate, crop yields will change, but how they change will depend on the crop, the region, and the degree of adaptation employed by growers.*

as well as temperature change and CO_2 fertilization in its study, predicts that the productivity of American cotton, corn, soybeans, sorghum, barley, sugar beets, and citrus fruits will increase over the next century. However, yields for wheat, rice, oats, hay, sugar cane, potatoes, and tomatoes will increase under some conditions and decrease under others.

Other Factors

Although global climate change will affect future agricultural yields, other factors may be important as well. Soil degradation, which includes erosion of productive soil layers, chemical depletion of nutrients, waterlogging, and salinization, is a major challenge. A United Nations report estimated that 23 percent of the world's agricultural lands, pastures, forests, and woodlands have been degraded to a least some extent since World War II. Irrigated land, which provides about 40 percent of the world's food from 16 percent of the world's cropland, is increasing, but so is irrigated land that is degraded by salinization and waterlogging. Erosion can be expected to increase as well, as more frequent high-rainfall events induce rilling and gullying. Wind deflation of productive soils will also increase if global warming enhances wind speeds.

Insect pests will be affected by climate change, but the exact effects seem to depend on the pest, crop, and type of climate change. For example, the IPCC report cites a study showing that warmer temperatures increase the severity of rice leaf blast in cool subtropical regions, while reducing the severity in warm and humid subtropics. In the United States, the ranges of several important crop pests have been shown to expand in patterns that follow climate changes noted since the 1970s. Studies have also shown that weeds are stimulated by warm temperatures and higher levels of CO_2, and in some cases, they become more effective competitors, thus reducing yields.

Livestock are affected by climate change both directly and indirectly. Higher temperatures produce more stress in animals, which can impact growth, reproduction, and yield of milk or wool. Dairy cows will be less productive during the hotter summer months,

Soil erosion *Soil erosion, which strips off the most fertile part of the soil, presents an important challenge to increasing crop production and yield. More frequent and more intense rainfall events, which are predicted for midlatitude summers, will enhance soil loss by rilling and gullying of unprotected surfaces.*

Grazing in Burkina Faso *Livestock will be affected by climate change. Higher temperatures will produce more summer stress on grazing animals, but less winter stress. The quality and availability of forage will also be affected.*

and enhanced heat waves or winter storms will increase death rates in vulnerable animals. Indirectly, climate change will change the nature and availability of forage, grain, diseases, and parasites affecting livestock.

Food Prices and Hunger Risks

What are the likely effects of global climate change on food availability and food prices? Keep in mind that, so far, agricultural production has expanded to meet the expanding needs of the world's population. Thus, the question is not simply whether production will increase in the future, but whether it will keep pace with demand. Most studies seem to predict that under scenarios of global climate change greater that 2.5°C (4.5°F), demand will exceed supply and food prices will rise. Associated with this change will be small positive changes in income for developed regions, with small negative changes in income for developing regions. Vulnerable populations, such as marginal farmers and poor urban dwellers, will be placed at greater risk of hunger. However, if adaptation to climate change is more effective, these consequences may not occur.

Agriculture is a human activity that is essential to us all, and from the discussion above, it is obvious that the effects of global change on agriculture will be significant as our climate warms and CO_2 levels rise. Let's hope that our species is smart enough to adapt to the changes and provide an abundance of food for all.

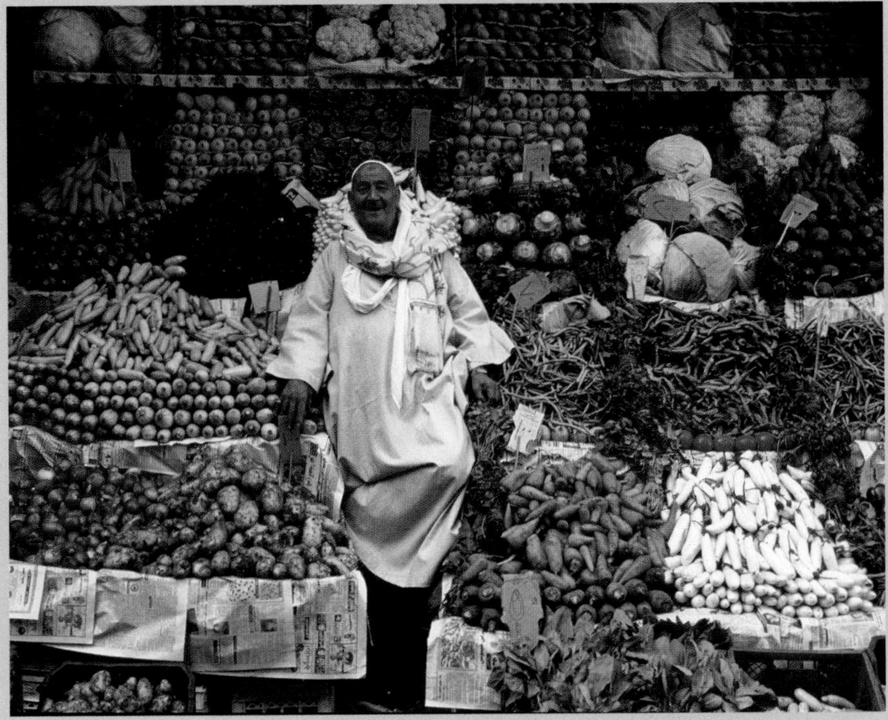

Vegetable vendor, Cairo, Egypt *Climate change will affect the price and availability of food from region to region. With global warming greater than 2.5°C (4.5°F), most studies predict that future supply will not meet future demand at present levels and prices will rise.*

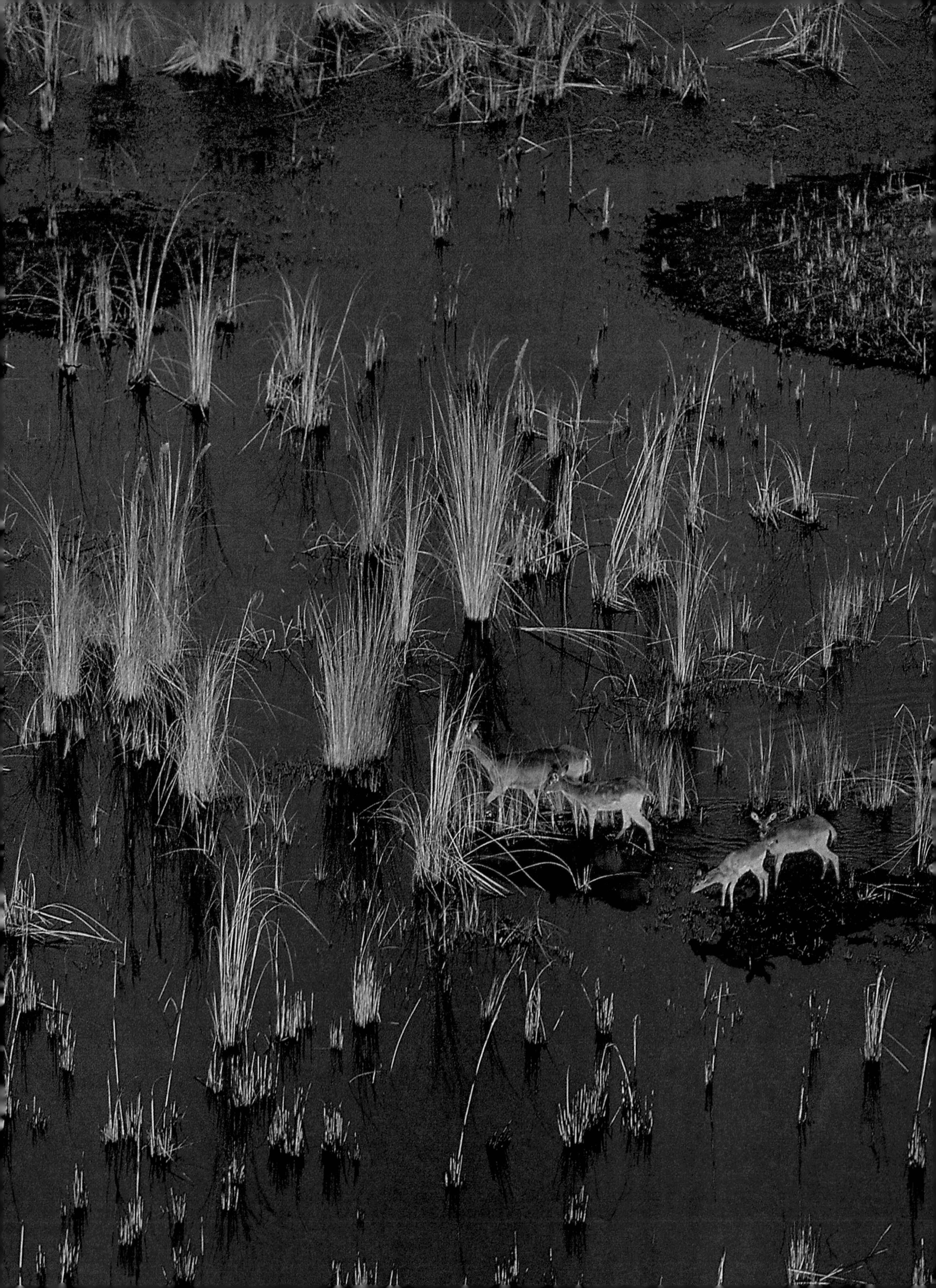

22 SYSTEMS AND CYCLES OF THE BIOSPHERE

***Lechwe in the Okavango delta, Botswana.** A vast labyrinth of swamps in landlocked Botswana marks the terminus of the Okavango River, which drains much of southern Angola. A watery oasis on the edge of the Kalahari Desert, the Okavango delta is home to hordes of teeming wildlife, including the lechwe, an antelope of swampy habitats.*

This chapter begins a series of three chapters that are concerned with biogeography. **Biogeography** is a branch of geography that focuses on the distribution of plants and animals over the Earth. It attempts to identify and describe the processes that influence plant and animal distribution patterns at varying scales of space and time. We can think of biogeography as encompassing two major views or themes. *Ecological biogeography* is concerned with how the distribution patterns of organisms are affected by the environment, including both the physical environment experienced by organisms and the biological environment created by interaction with other organisms. *Historical biogeography* focuses on how spatial distribution patterns of organisms arise over time and space. Evolution of species, migration of organisms and their methods of dispersal, and extinction of species are processes of interest in historical biogeography.

The essential concepts of ecological and historical biogeography are the subject of Chapter 23. In this chapter, we will examine some ideas from the domain of **ecology,** which is the study of the interactions between life-forms and their environment. These ideas focus on how organisms live and interact as ecosystems and how energy and matter are cycled by ecosystems.

We can define an **ecosystem** as a group of organisms and the environment with which they interact. Ecosystems have inputs of matter and energy that plants and animals use to grow, reproduce, and maintain life. Matter and energy are also exported from ecosystems.

ENERGY FLOW IN ECOSYSTEMS

We begin our examination of systems and cycles of the biosphere by focusing on the flows of energy that take place within ecosystems. These flows form the basis for studying the global productivity of ecosystems, a topic of concern to biogeographers and global ecologists alike.

The Food Web

A salt marsh provides a good example of an ecosystem (Figures 22.1, 22.2*b*). A variety of organisms are present—algae and aquatic plants, microorganisms, insects, snails, and crayfish, as well as such larger organisms as fishes, birds, shrews, mice, and rats. Inorganic components will be found as well—water, air, clay particles and organic sediment, inorganic nutrients, trace elements, and light energy. Energy transformations in the ecosystem occur by means of a series of steps or levels, referred to as a **food chain** or **food web.**

The plants and algae in the food web are the **primary producers.** They use light energy to convert carbon dioxide and water into carbohydrates (long chains of sugar molecules) and eventually into other biochemical molecules needed for the support of life. This process of energy conversion is called *photosynthesis,* and we will return to it in more detail shortly. Organisms engaged in photosynthesis form the base of the food web.

The primary producers support the **consumers**—organisms that ingest other organisms as their food source. At the lowest level of consumers are the *primary consumers* (the snails, insects, and fishes). At the next level are the *secondary consumers* (the mammals, birds, and larger fishes), which feed on the primary consumers. Still higher levels of feeding occur in the salt-marsh ecosystem as marsh hawks and owls consume the smaller animals below them in the food web. The **decomposers** feed on *detritus,* or decaying organic matter, derived from all levels. They are largely microscopic organisms (microorganisms) and bacteria.

The food web is really an energy flow system, tracing the path of solar energy through the ecosystem. Solar energy is absorbed by the primary producers and stored in the chemical products of photosynthesis. As these organisms are eaten and digested by consumers, chemical energy is released. This chemical energy is used to power new biochemical reactions, which again produce stored chemical energy in the bodies of the consumers.

At each level of energy flow in the food web, energy is lost to *respiration.* Respiration can be thought of as the burning of fuel to keep the organism operating. It will be discussed in more detail in the next section. Energy expended in respiration is ultimately lost as waste heat and cannot be stored for use by other organisms higher up in the food chain. This means that, generally, both the numbers of

22.1 Energy flow diagram of a salt-marsh ecosystem in winter *The arrows show how energy flows from the Sun to producers, consumers, and decomposers. (Food chain after R. L. Smith, Ecology and Field Biology, Harper and Row, New York.)*

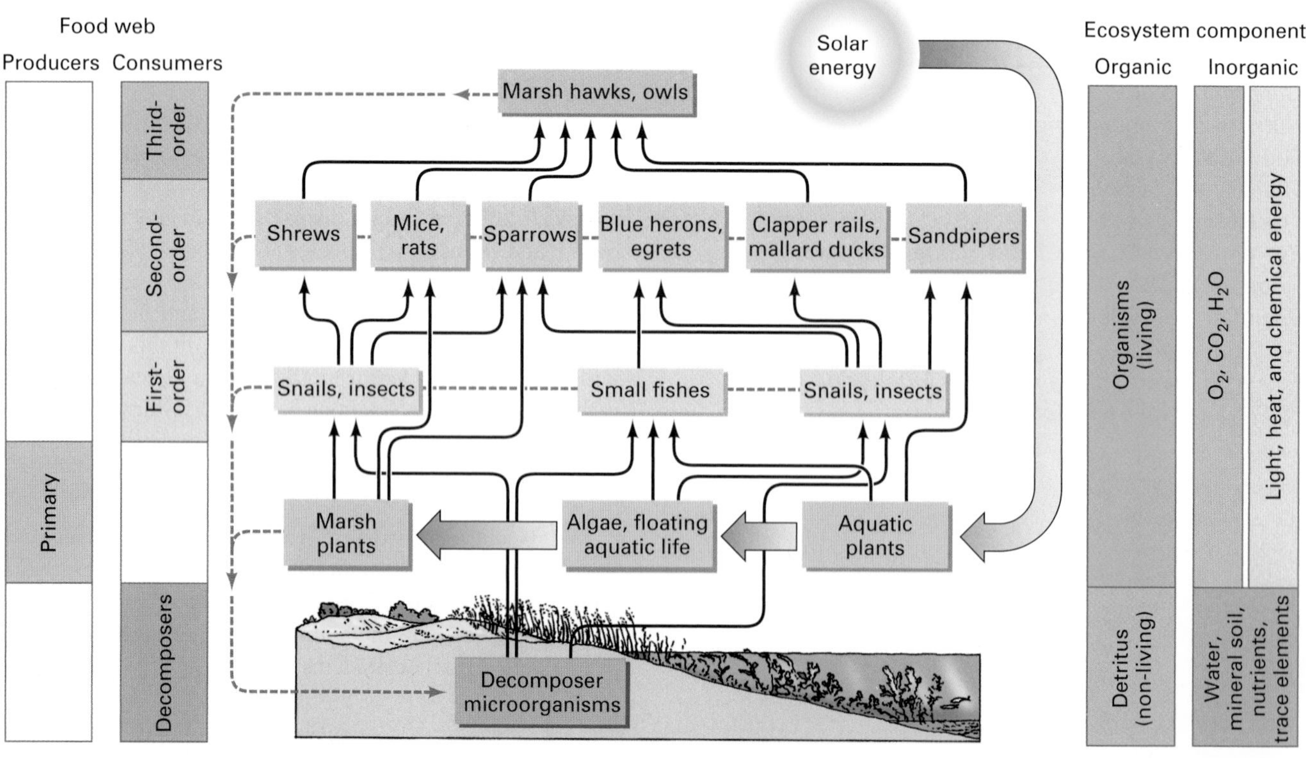

organisms and their total amount of living tissue must decrease greatly up the food chain. In general, only 10 to 50 percent of the energy stored in organic matter at one level can be passed up the chain to the next level. Normally, there are about four levels of consumers.

Figure 22.3 is a bar graph showing the percentage of energy passed up the chain when only 10 percent moves from one level to the next. The horizontal scale is in powers of 10. In ecosystems of the lands, the mass of organic matter and the number of individuals of the consuming animals decrease with each upward step. In the food chain shown in Figure 22.1, there are only a few marsh hawks and owls in the third level of consumers, while countless individuals are found in the primary level.

For individual species, the number of individuals of a species present in an ecosystem ultimately depends on the level of resources available to support the species population. If these resources provide a steady supply of energy, the population size will normally attain a steady level. In some cases, however, resources vary with time, for example, in an annual cycle. The population size of a species depending on these resources may then also fluctuate in a corresponding cycle. Population sizes and the change in size of species populations with time have been studied extensively by ecologists. *Working It Out 22.1 • Logistic Population Growth* provides more information on a type of population growth that approaches and slowly reaches a carrying capacity.

Photosynthesis and Respiration

Stated in the simplest possible terms, **photosynthesis** is the production of carbohydrate. *Carbohydrate* is a general term for a class of organic compounds consisting of the elements carbon, hydrogen, and oxygen. Carbohydrate molecules are composed of short chains of carbon bonded to one another. Also bonded to each carbon are hydrogen (H) atoms and hydroxyl (OH) molecules. We can symbolize a single carbon atom with its attached hydrogen atom and hydroxyl molecule as –CHOH–. The leading and trailing dashes indicate that the unit is just one portion of a longer chain of connected carbon atoms.

Photosynthesis of carbohydrate requires a series of complex biochemical reactions using water (H_2O) and carbon dioxide (CO_2) as well as light energy. A simplified chemical reaction for photosynthesis can be written as follows:

$$H_2O + CO_2 + \text{light energy} \rightarrow -CHOH- + O_2$$

Oxygen in the form of gas molecules (O_2) is a byproduct of photosynthesis. Photosynthesis is also referred to as *carbon fixation,* since in the process gaseous carbon as CO_2 is "fixed" to a solid form in carbohydrate.

Respiration is the process opposite to photosynthesis in which carbohydrate is broken down and combined with oxygen to yield carbon dioxide and water. The overall reaction is as follows:

$$-CHOH- + O_2 \rightarrow CO_2 + H_2O + \text{chemical energy}$$

As in the case of photosynthesis, the actual reactions are far from simple. The chemical energy released is stored in several types of energy-carrying molecules in living cells and used later to synthesize all the biological molecules necessary to sustain life.

At this point, it is helpful to link photosynthesis and respiration in a continuous cycle involving both the primary producer and the decomposer. (For now, we omit consumers from the cycle.) Figure 22.4 shows one closed loop for hydrogen (H), one for carbon (C), and two loops for oxygen (O). We are not taking into account that there are two atoms of hydrogen in each molecule of water and carbohydrate, or that there are two atoms of oxygen in each molecule of carbon dioxide and oxygen gas. Only the flow pattern counts in this representation.

A good place to start is the soil, from which water is drawn up into the body of a living plant. In the green leaves of the plant, photosynthesis takes place while light energy is absorbed by the leaf cells. Carbon dioxide is brought in from the atmosphere at this point. Oxygen is also liberated here and begins its atmospheric cycle. The plant tissue then dies and falls to the ground, where it is acted on by the decomposer. Through respiration, oxygen is taken out of the atmosphere or soil air and combined with the decomposing carbohydrate. Energy is now liberated. Here both carbon dioxide and water enter the atmosphere as gases.

An important concept emerges from this flow diagram. Energy passes through the system. It comes from the Sun and eventually returns to outer space. On the other hand, the material components—hydrogen, oxygen, and carbon—are recycled within the total system. Of course, many other material components are recycled in the same way. These are plant nutrients, essential in the growth of plants. Nutrients are constantly recycled. Because the Earth as a planet is a closed system, the material components never leave the total system. However, they can be stored in other ways and forms where they are unavailable for use by organisms for prolonged periods of geologic time. We will develop this concept more fully later in this chapter.

Net Photosynthesis

Because both photosynthesis and respiration occur simultaneously in a plant, the amount of new carbohydrate placed in storage is less than the total carbohydrate being synthesized. We must thus distinguish between gross photosynthesis and net photosynthesis. **Gross photosynthesis** is the total amount of carbohydrate produced by photosynthesis. **Net photosynthesis** is the amount of carbohydrate remaining after respiration has broken down sufficient carbohydrate to power the plant. Stated as an equation,

$$\text{Net photosynthesis} = \text{Gross photosynthesis} - \text{Respiration}$$

Because both photosynthesis and respiration occur in the same cell, gross photosynthesis cannot be measured readily. Instead, we will deal with net photosynthesis. In most cases, respiration will be held constant, so use of the net instead of the gross will show the same trends.

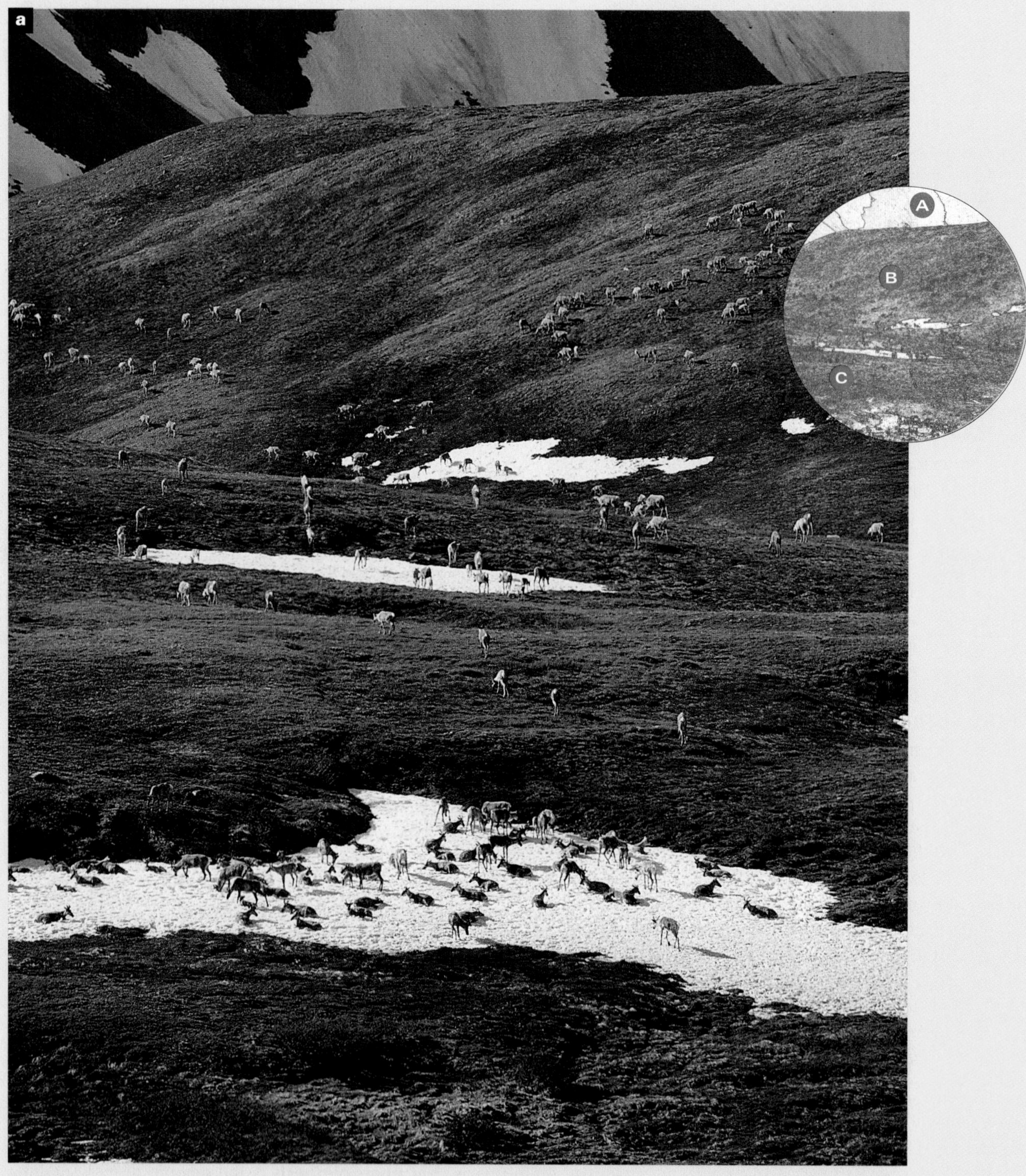

a...*Eye on the Landscape* ***Caribou in the foothills of the Brooks Range*** *The caribou, a large grazing mammal, is one of the important primary consumers of the tundra ecosystem.* **What else would the geographer see?...Answers at the end of the chapter.**

b...***Salt marsh*** *The salt marsh ecosystem supports a wide variety of life forms, both plant and animal. Here, the white ibis, great egret, and other wading bird species forage at the Merritt Island National Wildlife Refuge, Florida.*

c...***Savanna*** *The savanna ecosystem, with its abundance of grazing mammals, has a rich and complex food web. Here a top predator, a lioness, preys on a herd of zebra. Maasai Mara, Kenya.*

Working It Out | 22.1

Logistic Population Growth

The size of a population of plants or animals in an ecosystem is determined by the *growth rate* of the population. If the growth rate is positive, the population is increasing. If it is negative, the population is decreasing. If the growth rate is zero, the population size is stable, neither increasing nor decreasing. Growth rates are usually expressed as the percent or proportion of increase in the population in a given unit of time. As we saw in Chapter 5, this type of growth is *exponential growth,* which can be positive or negative. For example, a population of mosquitoes in the late spring might be increasing at a growth rate of 10 percent per week. Or, a population of alligators in the Everglades might be decreasing at 5 percent per year.

In most natural ecosystems, populations tend to be of a stable size and therefore exhibit a growth rate near zero, when taken over a long time period. Why does this stability occur? One important reason is that individuals of the same species are in competition with each other for the same resources. These resources could be light and nutrients in the case of plants, or individuals of a prey species for predators higher up on the food chain. Normally, the population will expand exponentially until the resources become increasingly scarce. As this happens, an increasing proportion of the population is unable to sustain itself, and so population growth slows and eventually stops. This type of growth, in which the population growth rate slows and eventually goes to zero is called *logistic growth.* At this point, the *carrying capacity* has been reached.

As an example of a population experiencing logistic growth, consider a microecosystem—a population of yeast bacteria growing in a culture medium. The table shows the results of an experiment in which yeast bacteria were grown over a period of 18 hours. Each hour a sample of yeast was removed, allowing calculation of the biomass of the whole population, shown in the second column. (Since it is too laborious to count the yeast bacteria in a sample to obtain the number of individuals in the culture medium, the biomass of the bacteria was used as a measure of the population size.) A scan down the column shows how the population increased rapidly at first and then stabilized at a value approaching about 665 grams. This value corresponds to the carrying capacity.

Column 3 shows the increase in biomass during the period. A scan of this column shows that biomass increments grew and then became smaller as the population biomass reached the carrying capacity. The largest increment to the population occurred in hour 8.

Column 4 shows the increase expressed as a percentage of the population at the start of each period. These values are quite high at first, showing that the population grew at rates of about 50 to 90 percent per hour during the first five hours. After that time the growth rate slowed, eventually decreasing to less than 1 percent per hour for hours 16–18.

The left-hand graph shows the yeast biomass plotted with time. The graph reveals an "S" shape that is characteristic of logistic growth. The population grew exponentially at first, then leveled off as it approached the carrying capacity. Plotted as a semilogarithmic function (right-hand graph), the growth follows a straight line at first, indicating initial exponential growth. However, the line curves smoothly toward the horizontal and levels out at the carrying capacity.

Logistic growth follows this equation:

$$P(t) = \frac{C}{1+Ae^{-kt}}$$

Growth of Yeast in a Culture Medium

(1) Time (hours)	(2) Biomass (grams)	(3) Increase (grams)	(4) Increase (percent)	(5) Logistic Model	(6) Difference
0	9.6			9.6	0.0
1	18.3	8.7	90.6	16.3	2.0
2	29.0	10.7	58.5	27.5	1.5
3	47.2	18.2	62.8	45.8	1.4
4	71.1	23.9	50.6	74.9	–3.8
5	119.1	48.0	67.5	119.0	0.1
6	174.6	55.5	46.6	181.0	–6.4
7	257.3	82.7	47.4	260.0	–2.7
8	350.7	93.4	36.3	348.5	2.2
9	441.0	90.3	25.7	434.9	6.1
10	513.3	72.3	16.4	508.3	5.0
11	559.7	46.4	9.0	563.7	–4.0
12	594.8	35.1	6.3	602.0	–7.2
13	629.4	34.6	5.8	626.8	2.6
14	640.8	11.4	1.8	642.2	–1.4
15	651.1	10.3	1.6	651.5	–0.4
16	655.9	4.8	0.7	657.1	–1.2
17	659.6	3.7	0.6	660.4	–0.8
18	661.8	2.2	0.3	662.3	–0.5

Source: Data of T. Carlson, *Biochemische Zeitschrift,* vol. 57.

$P(t)$ is the population size at time t, which depends on the value of t and three constants: C, A, and k. C is the carrying capacity, or the stable level of the population. For the yeast example, this would be about 665 grams.

The constant A is found by the relation

$$A = \frac{C - P_0}{P_0}$$

where P_0 is the initial population, which for the yeast example is 9.6 grams. This constant scales the factor e^{-kt} according to the range between the starting value, P_0, and the final value, C, the carrying capacity. For the yeast example,

$$A = \frac{C - P_0}{P_0} = \frac{665 - 9.6}{9.6} = 68.3$$

The constant k in the exponential expression of the denominator is the exponential growth rate of the population before the decrease sets in. It is related to the growth observed in a single time period by the equation

$$k = \ln(1 + R)$$

where R is the growth rate experienced at the beginning of growth, when the population size is far from the carrying capacity. Note that this formula uses the growth rate as a proportion rather than as a percent. This formula actually applies only to a very small time interval at the start of growth of a very small population, but it provides a useful approximation for other cases. Scanning column 4 of the table shows that the population increased at rates ranging from 51 to 91 percent in the first five hours. As it turns out, a value of 72 percent for R, or 0.72 expressed as a proportion, fits the data best. The value of k is then

$$k = \ln(1 + R) = \ln(1 + 0.72)$$
$$= \ln(1.72) = 0.542$$

With the values of C, A, and k known, we can now write the equation for the logistic growth of the yeast culture as

$$P(t) = \frac{C}{1 + Ae^{-kt}} = \frac{665}{1 + (68.3)e^{-0.542t}}$$

Column 5 of the table shows the values of biomass predicted by this logistic equation, and column 6 shows the difference between these predictions and the observations in column 2. The differences are very small, indicating that the model fits the data very well.

Logistic growth, in which a population increases exponentially and then slows its growth to reach a sustained carrying capacity, is only a very simple model of population growth. Other growth patterns characterize many natural populations. For example, a common pattern called *J-shaped growth* provides rapid growth followed by a destructive population crash to low levels. Populations then increase only to crash again and again. At present, the human population is increasing exponentially at an ever-increasing rate. Let's hope that human population growth will be logistic, not J-shaped!

Growth of yeast in a culture medium *(left) Growth plotted linearly, fitting the logistic model. (right) Growth plotted semilogarithmically. (Data of T. Carlson, Biochemische Zeitschrift, vol. 57.)*

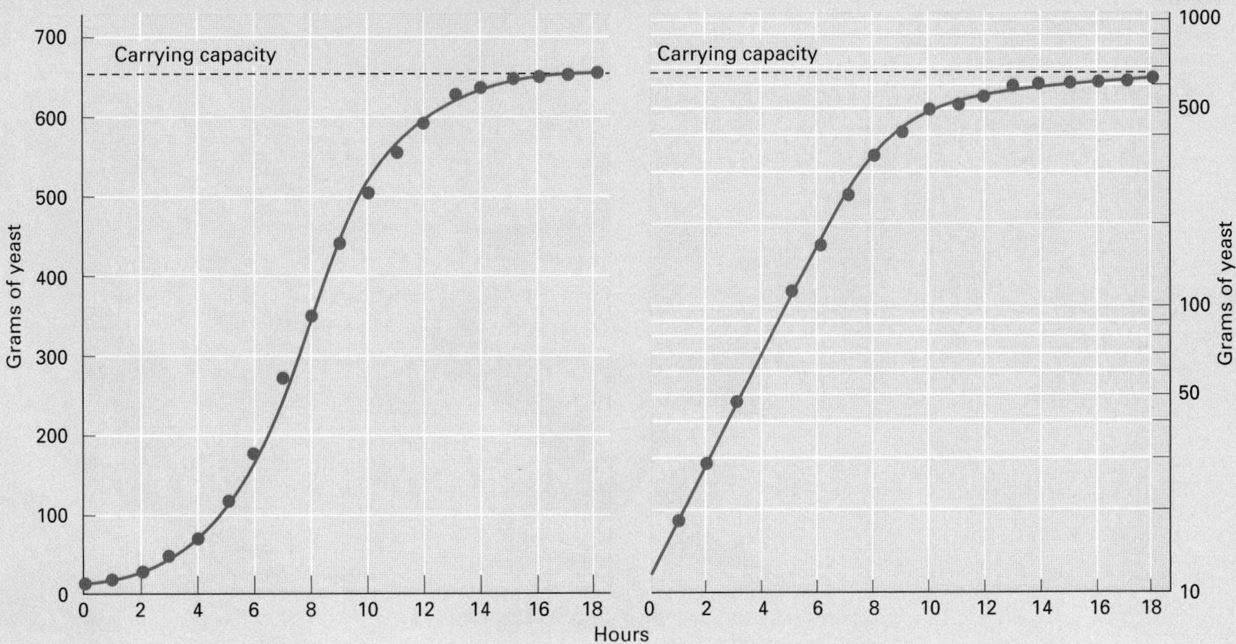

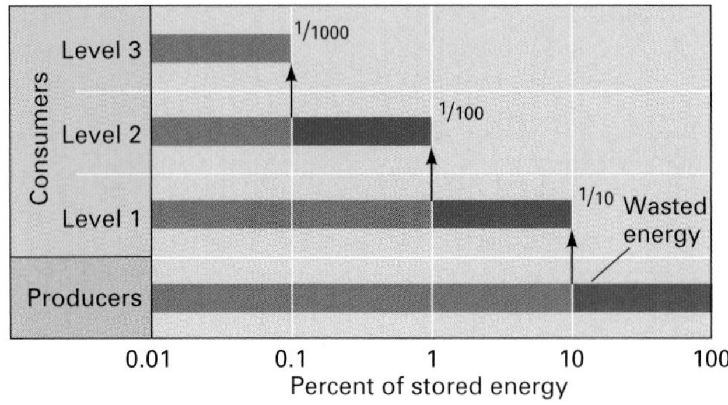

22.3 Energy loss *Percentage of energy passed up the steps of the food chain, assuming 90 percent is lost energy at each step.*

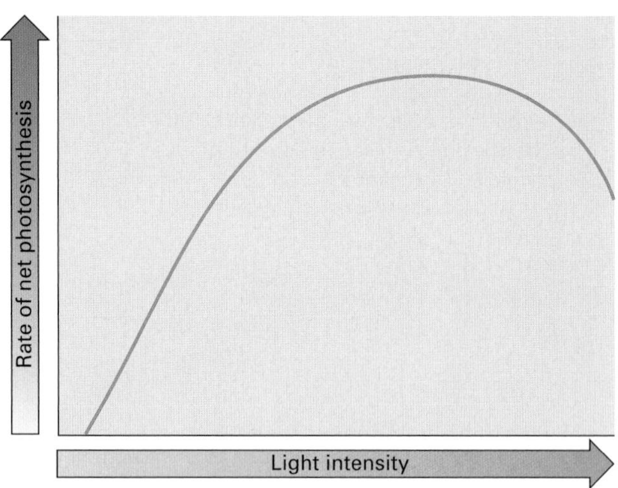

22.5 Net photosynthesis *The curve of net photosynthesis shows a steep initial rise, then levels off as light intensity rises.*

The rate of net photosynthesis is strongly dependent on the intensity of light energy available, up to a limit. Figure 22.5 shows this principle. The rate of net photosynthesis is indicated on the vertical axis by the rate at which a plant takes up carbon dioxide. On the horizontal axis, light intensity increases from left to right. At first, net photosynthesis rises rapidly as light intensity increases. The rate then slows and reaches a maximum value, shown by the plateau in the curve. Above this maximum, the rate falls off because the incoming light is also causing heating. This heating increases the rate of respiration, which offsets gross production by photosynthesis and decreases the net.

Light intensity sufficient to allow maximum net photosynthesis is only 10 to 30 percent of full summer sunlight for most green plants. Additional light energy is simply ineffective. Duration of daylight then becomes the important factor in the rate at which products of photosynthesis accumulate as plant tissues. On this subject, you can draw on your knowledge of the seasons and the changing angle of the Sun's rays with latitude. Figure 22.6 shows the duration of the daylight period with changing seasons for a wide range of latitudes in the northern hemisphere. At low latitudes, days are not far from the average 12-hour length throughout the year. At high latitudes, days are short in winter but long in summer. The seasonal contrast in day length increases with latitude. In subarctic latitudes, pho-

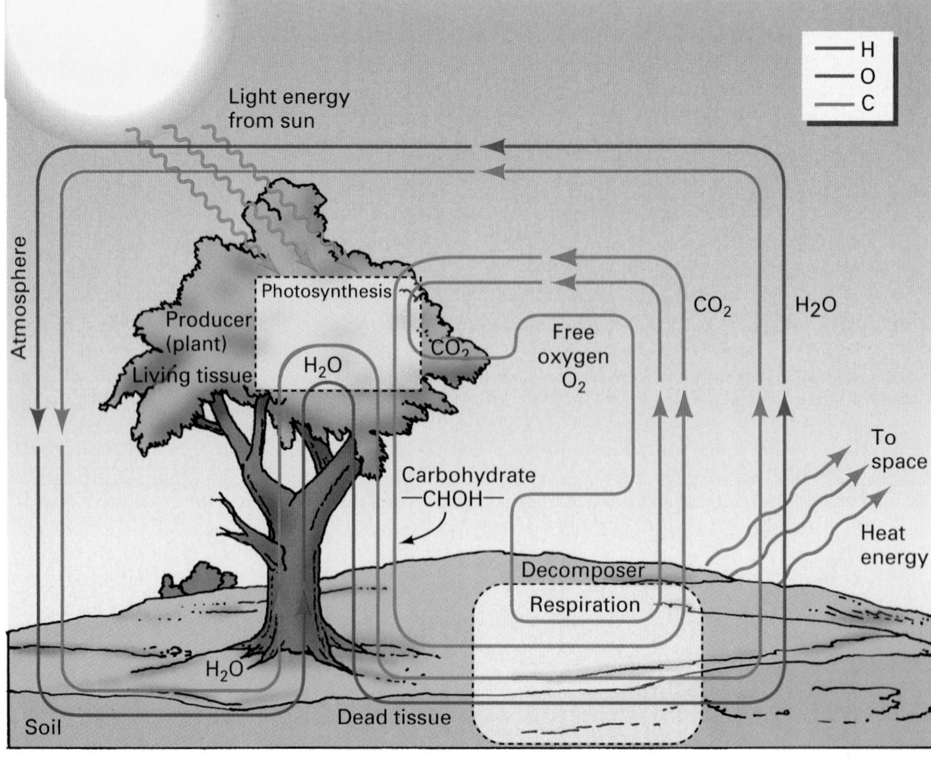

22.4 Photosynthesis and respiration cycles *A simplified flow diagram of the essential components of photosynthesis and respiration through the biosphere.*

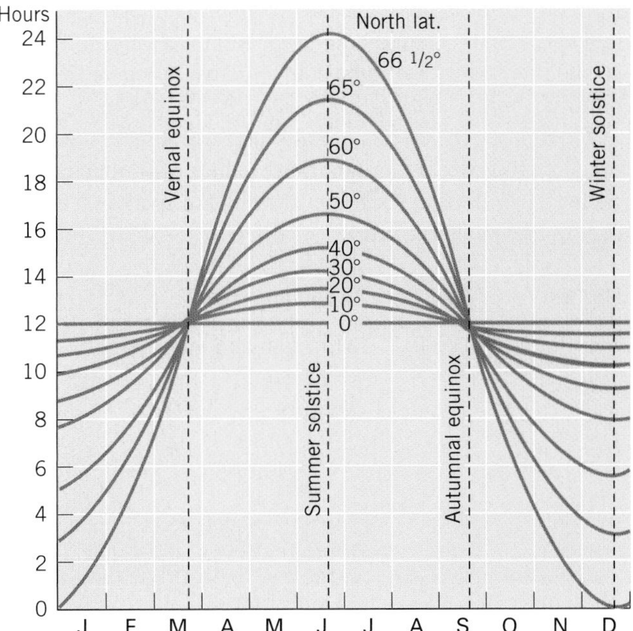

22.6 Day length variation *Duration of the daylight period at various latitudes throughout the year. The vertical scale gives the number of hours the sun is above the horizon.*

tosynthesis can go on in summer during most of the 24-hour day, a factor that can compensate significantly for the shortness of the growing season.

The rate of photosynthesis also increases as air temperature increases, up to a limit. Figure 22.7 shows the results of a laboratory experiment in which sphagnum moss was grown under constant illumination. Gross photosynthesis increased rapidly to a maximum at about 20°C (68°F), then leveled off. Respiration increased quite steadily to the limit of the experiment. Net photosynthesis, which is the difference between the values in the two curves, peaked at about 18°C (64°F), then fell off rapidly.

22.7 Temperature and energy flow *Respiration and gross and net photosynthesis vary with temperature. (Data of Stofelt, in A. C. Leopold, Plant Growth and Development, McGraw-Hill, New York.)*

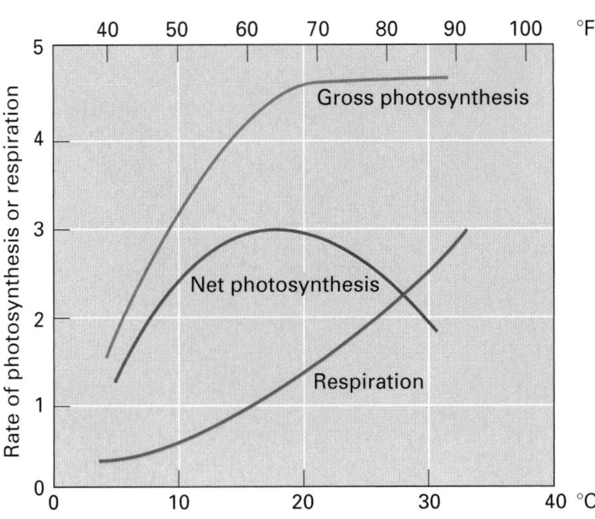

Net Primary Production

Plant ecologists measure the accumulated net production by photosynthesis in terms of the **biomass,** which is the dry weight of organic matter. This quantity could, of course, be stated for a single plant or animal, but a more useful measurement is the biomass per unit of surface area within the ecosystem—that is, kilograms of biomass per square meter or (metric) tons of biomass per hectare (1 hectare = 10^4 m^2). Of all ecosystems, forests have the greatest biomass because of the large amount of wood that the trees accumulate through time. The biomass of grasslands and croplands is much smaller in comparison. For fresh water bodies and the oceans, the biomass is even smaller—on the order of one-hundredth that of the grasslands and croplands.

Although the amount of biomass present per unit area is an important indicator of the amount of photosynthetic activity, it can be misleading. In some ecosystems, biomass is broken down very quickly by consumers and decomposers, so the amount maintained is less. From the viewpoint of ecosystem productivity, what is important is the annual yield of useful energy produced by the ecosystem, or the *net primary production.*

Table 22.1 provides the net primary production of various ecosystems in units of grams of dry organic matter produced annually from one square meter of surface. The figures are rough estimates, but they are nevertheless highly meaningful. Note that the highest values are in two quite

Table 22.1 Net Primary Production for Various Ecosystems

	Grams per Square Meter per Year	
	Average	**Typical Range**
Lands		
Rainforest of the equatorial zone	2000	1000–5000
Freshwater swamps and marshes	2500	800–4000
Midlatitude forest	1300	600–2500
Midlatitude grassland	500	150–1500
Agricultural land	650	100–4000
Lakes and streams	500	100–1500
Extreme desert	3	0–10
Oceans		
Algal beds and reefs	2000	1000–3000
Estuaries (tidal)	1800	500–4000
Continental shelf	360	300–600
Open ocean	125	1–400

unlike environments: forests and wetlands (estuaries). Agricultural land compares favorably with grassland, but the range is very large in agricultural land, reflecting many factors such as availability of soil water, soil fertility, and use of fertilizers and machinery.

Productivity of the oceans is generally low. The deep water oceanic zone, which comprises about 90 percent of the world ocean area, is the least productive of the marine ecosystems. Continental shelf areas are a good deal more productive and support much of the world's fishing industry (Figure 22.8).

Upwelling zones are also highly productive. Upwelling of cold water from ocean depths brings nutrients to the surface and greatly increases the growth of microscopic floating plants known as *phytoplankton*. These, in turn, serve as food sources for marine animals in the food chain. Consequently, zones of upwelling near habitable coastlines are highly productive fisheries. An example is the Peru Current off the west coast of South America. Here, countless individuals of a single species of small fish, the anchoveta, provide food for larger fish and for birds. The birds, in turn, excrete their wastes on the mainland coast of Peru. The accumulated deposit, called guano, is a rich source of nitrate fertilizer that is now severely depleted.

Net Production and Climate

What climatic factors control net primary productivity? We have already identified light intensity and duration, as well as temperature, as influencing net photosynthesis. Another important factor is the availability of water. A shortage or surplus of soil water might be the best climatic factor to examine, but data are not available. Ecologists have related net annual primary production to mean annual precipitation, as shown in Figure 22.9. The production values are for plant structures above the ground surface. Although the productivity increases rapidly with precipitation in the lower range from desert through semiarid to subhumid climates, it seems to level off in the humid range. Apparently, a large soil water surplus carries with it some counteractive influence, such as removal of plant nutrients by leaching.

Combining the effects of light intensity, temperature, and precipitation, we can assign rough values of productivity to each of the climates as follows (units are grams of carbon per square meter per year):

Highest (over 800)	Wet equatorial ①
Very high (600–800)	Monsoon and trade-wind coastal ② Wet-dry tropical ③
High (400–600)	Wet-dry tropical ③ (Southeast Asia) Moist subtropical ⑥ Marine west coast ⑧
Moderate (200–400)	Mediterranean ⑦ Moist continental ⑩
Low (100–200)	Dry tropical, semiarid ④s Dry midlatitude, semiarid ⑨s Boreal forest ⑩
Very low (0–100)	Dry tropical, desert ④d Dry midlatitude, desert ⑨d Boreal forest ⑩ Tundra ⑫

For natural ecosystems, productivity is largely dependent on climate and soils. For agricultural ecosystems, however, productivity is strongly influenced by the flow of energy, in the form of fertilizer, agricultural chemicals, and irrigation water, which is provided to the crop by human agents. Much if not all of this energy is derived from the burning of fossil fuels and so represents a conversion of fossil fuel energy to

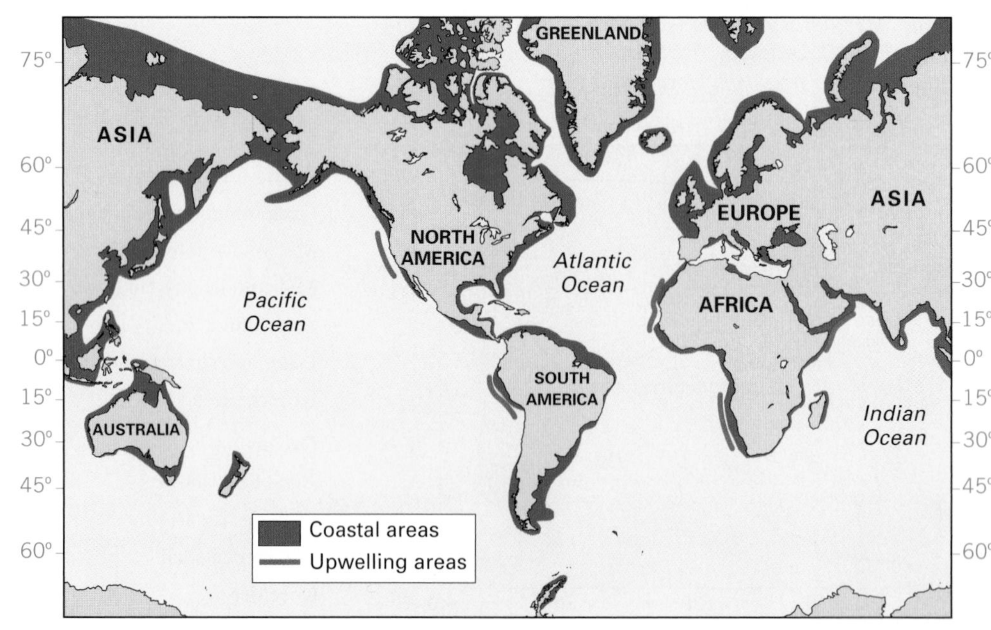

22.8 Distribution of world fisheries *Distribution of world fisheries. Coastal areas and upwelling areas together supply over 99 percent of world production. (Compiled by the National Science Board, National Science Foundation.)*

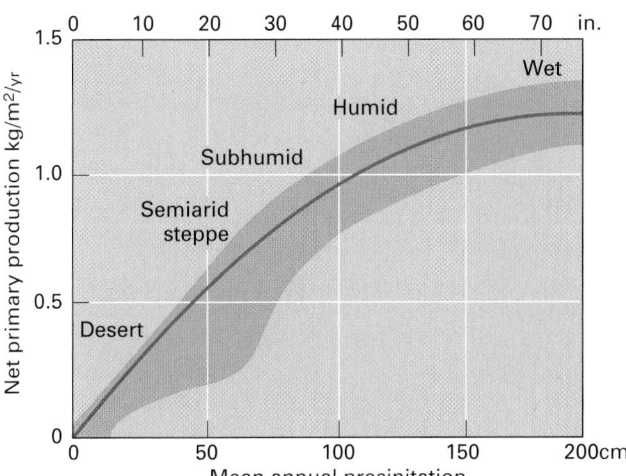

***22.9 Precipitation and net production** Net primary production increases rapidly with increasing precipitation, but levels off in the higher values. Observed values fall mostly within the shaded zone. (Data of Whittaker, 1970.)*

human foodstuffs that is not always very efficient. *Focus on Systems 22.2 • Agricultural Ecosystems* provides more information on this topic.

Within the past decade or so, remote sensing has come into use as a tool for mapping primary productivity on a global scale. Our interchapter feature, *A Closer Look: Focus on Remote Sensing 22.4 • Monitoring Global Productivity from Space* provides more information on how this is done. Using remote sensing, scientists have concluded that terrestial productivity has increased since about 1980, while ocean productivity has decreased. These changes are most likely linked to global climate changes, including global warming at higher latitudes, reductions in cloud cover in equatorial regions, and decreasing winds over oceans.

BIOMASS AS AN ENERGY SOURCE

Net primary production represents a source of renewable energy derived from the Sun that can be exploited to fill human energy needs. The use of biomass as an energy source involves releasing solar energy that has been fixed in plant tissues through photosynthesis. This process can take place in a number of ways—the simplest is direct burning of plant matter as fuel, as in a campfire or a wood-burning stove. Other approaches involve the generation of intermediate fuels from plant matter—methane gas, charcoal, and alcohol, for example. Biomass energy conversion is not highly energy efficient. Typical values of net annual primary production of plant communities range from 1 to 3 percent of available solar energy. However, the abundance of terrestrial biomass is so great that biomass utilization could provide the energy equivalent to 3 million barrels of oil per day for the United States with proper development.

One important use of biomass energy is the burning of firewood for cooking (and some space heating) in developing nations. The annual growth of wood in the forest of developing countries totals about half the world's energy production—plenty of firewood is thus available. However, fuelwood use exceeds production in many areas, creating local shortages and severe strains on some forest ecosystems. The forest-desert transition areas of thorntree, savanna, and desert scrub in central Africa south of the Sahara Desert are examples.

Even in closed stoves, wood burning is not very efficient, ranging from 10 to 15 percent for cooking. However, the conversion of wood to charcoal or gas can boost efficiencies to values as high as 70 to 80 percent with appropriate technology. In this process, termed *pyrolysis,* controlled partial burning in an oxygen-deficient environment reduces carbohydrate to free carbon (charcoal), and yields flammable gases such as carbon monoxide and hydrogen. Charcoal is more energy efficient than wood, burns more cleanly, and is easier to transport. As an added advantage, charcoal can be made from waste fibers and agricultural residues that would normally be discarded. Thus, charcoal is an efficient fuel that can help extend the firewood supply in areas where wood is in high demand.

A second method of extracting energy from biomass uses anaerobic digestion to produce *biogas.* In this process, animal and human wastes are fed into a closed digesting chamber, where anaerobic bacteria break down the waste to produce a gas that is a mixture of methane and carbon dioxide. The biogas can be easily burned for cooking or heating, or it may be used to generate electric power. The digested residue is a sweet-smelling fertilizer. China now maintains a vigorous program of construction of biogas digesters for the use of small family units. The benefits include better sanitation and reduced air pollution, as well as more efficient fuel usage.

Another use of biomass that is increasing in importance is the conversion of agricultural wastes to alcohol. In this process, yeast microorganisms are used to convert the carbohydrate to alcohol through fermentation. An advantage of alcohol is that it can serve as a substitute and extender for gasoline. Gasohol, a mixture of up to 10 percent alcohol in gasoline, can be burned in conventional engines without adjustment.

Brazil, a country without adequate petroleum production, has relied heavily on alcohol fuel derived from sugarcane. In a recent year, for example, alcohol provided 63 percent of Brazil's automotive fuel needs. Distillation of alcohol, however, requires heating, thus greatly reducing the net energy yield. Alcohol, charcoal, and firewood are all alternatives to fossil fuels that will become increasingly important as petroleum becomes scarcer and more costly in the coming decades.

Relying on biomass energy can also yield important benefits in reducing carbon dioxide emissions. However, burning biomass does not reduce the CO_2 flow to the atmosphere directly. The burning of biomass quickly releases CO_2 that would normally be released more slowly, as the biomass decays. But the energy obtained in the biomass burning will in all likelihood substitute for some fossil fuel burning. Because this fossil fuel is not burned, its CO_2 is not released to the atmosphere, thus reducing overall carbon dioxide emissions.

Focus on Systems | 22.2

Agricultural Ecosystems

The principles of energy use and flow in natural ecosystems also apply to agricultural ecosystems. Important differences exist, however, between natural ecosystems and those that are highly managed for agriculture. The first major difference is the reliance of agricultural ecosystems on inputs of energy that are ultimately derived from fossil fuels. The most obvious of these inputs is the fuel that runs the machinery used to plant, cultivate, and harvest crops. Another is the application of fertilizers and pesticides, which require large expenditures of fuel to extract, synthesize, and transport. Yet another is the use of electricity, powered by fossil fuel burning, for irrigation pumps to bring ground water or surface water to agricultural fields. Fossil fuels also indirectly power such activities as the breeding of plants that have higher yields and are resistant to disease, as well as the development of new chemicals to combat insect pests. Fuel is expended in transporting crops to distant sources of consumption, thus enabling large areas of similar climate and soils to be used for the same crop. In these and many other ways, the high yields obtained today are brought about only at the cost of a large energy input derived from fossil fuels.

A schematic flow diagram below shows how fuel-energy input enters into both the purchased inputs and the operations performed on the farm to raise and harvest crops. Solar energy of photosynthesis is, of course, a "free" input. But to be delivered to the humans or animals that consume it, the raw food or feed product of this flow system requires the expenditure of fossil fuel energy.

Agricultural ecosystems, unlike natural ecosystems, are simple in structure and function. They often consist of one genetic strain of one species. Such ecosystems are overly sensitive to attacks by one or two well-adapted insects that can multiply rapidly to take advantage of an abundant food source. Thus, pesticides are constantly needed to reduce insect populations. Weeds, too, are a problem, adapted as they are to rapid growth on disturbed soil in sunny environments. Weeds can divert much of the productivity to undesirable forms. Herbicides are often the immediate solution to this problem.

In natural ecosystems, nutrient elements are returned to the soil following the death of the plants that concentrate them. In agricultural ecosystems, this recycling is usually interrupted by harvesting the crop for consumption at a distant location. Continuous removal of nutrients in crop biomass intro-

Energy inputs in agriculture *A schematic diagram of the inputs of cultural energy into various stages in the agriculture production system. (After G. H. Heichel, American Scientist, Vol. 64.)*

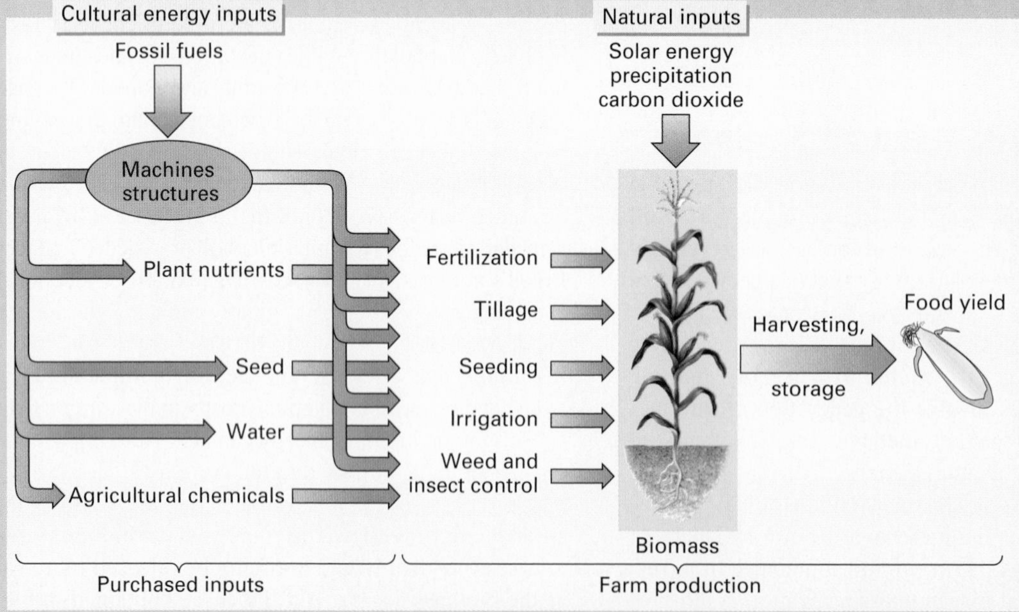

duces a new pathway in the nutrient cycle. To preserve soil fertility, still another pathway must be introduced—from the fertilizer plant to the farm. Keeping this new nutrient subcycle functioning requires considerable fossil fuel energy input.

Excluding the solar energy of photosynthesis, energy expended on the production of a raw food or feed crop can be referred to as *cultural energy.* Let us examine the cultural energy input needed to produce various feed and food crops. The upper graph below shows the relative efficiency of food production of 24 crops. The horizontal scale shows cultural energy expended in terms of thousands of joules per square meter per year (kJ/m²/yr). The vertical scale shows the ratio of food energy produced to input of cultural energy—the higher the ratio, the greater the efficiency (and the lower the cultural input required). Notice that field crops, such as sorghum and corn, which are used largely as animal feeds, have the highest levels of efficiency, whereas foods consumed directly by humans have relatively low efficiencies. Garden crops require a very high cultural energy input per unit area of land and also have low efficiencies.

The lower graph shows the protein derived from each unit of cultural energy on the vertical scale. Alfalfa and soybeans rate very high on this scale. Notice that oats, wheat, and corn have intermediate values of protein yield, but that rice ranks very low. Protein deficiency is a serious health problem for peoples subsisting largely on rice. A small area of the graph at the lower left is labeled "chicken, beef, pork." This insertion serves to show that protein obtained from meat has an energy efficiency only about one-tenth as great as that from soybeans. Indeed, the energy available from edible meat represents only about 10 percent of the energy expended in animal feed, a fact previously pointed out in our discussion of energy flow in the food chain.

The data thus indicate clearly that our food production system is not efficient in terms of cultural energy expended to furnish food to humans. The most highly coveted part of the Western diet, meat and garden vegetables, is extremely wasteful of cultural energy as compared with, say, diets based largely on grain foods (bread, cereals) and soybean products. In a future world where energy will be more costly and the human population will be larger, the human diet may well be forced to rely more heavily on these fuel-efficient foods.

Agricultural energy yields *(a) Cultural energy used to produce certain food crops in relation to yield of food energy. (b) Protein yield in relation to cultural energy for several kinds of crops. (After G. H. Heichel, American Scientist, Vol. 64.)*

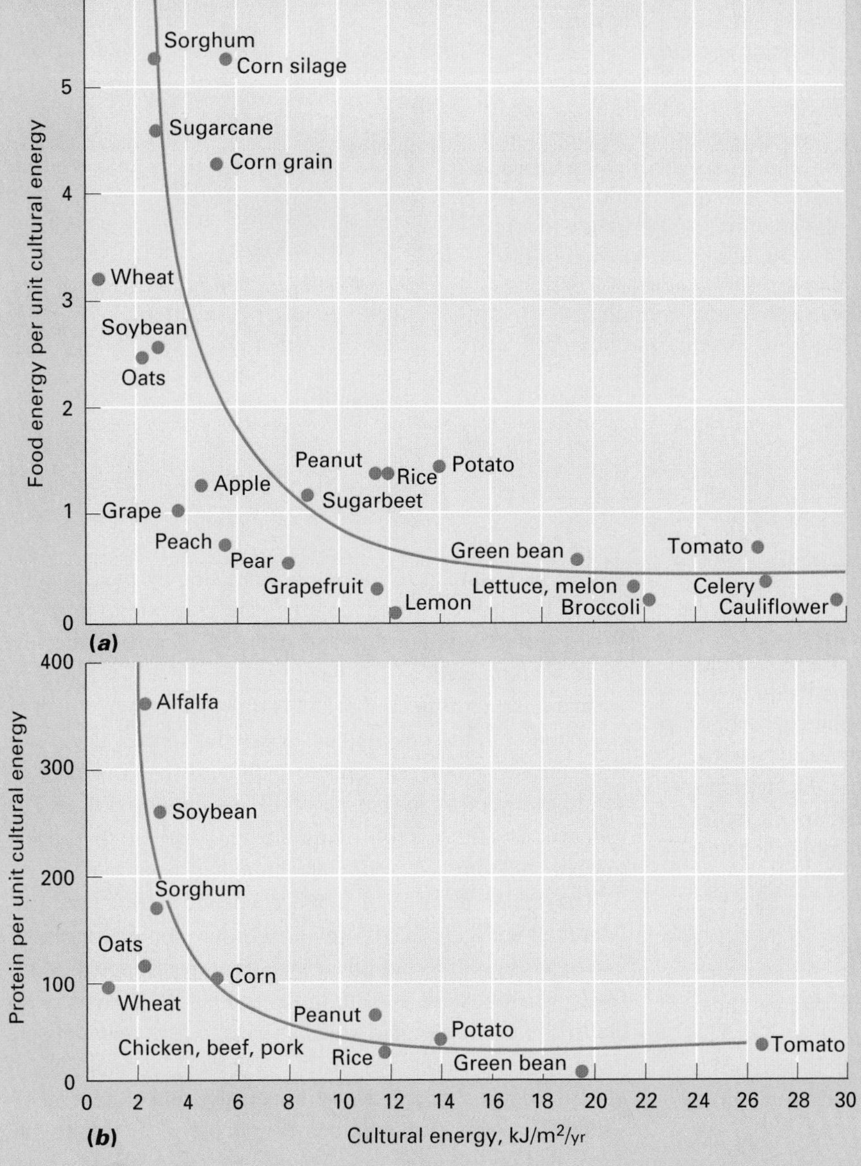

Geographers at Work

What Would Happen If...? Building Models of Landscapes and Ecological Processes

by Christina Tague *San Diego State University*

One of the key themes of this book has been systems—systems in which matter and energy flow from place to place and over time. The systems we have described in depth have been mostly explained in simple terms. However, real systems can be very complex, with many links and connections.

Take the water budget of a forested watershed, for example. Water enters as rainfall, which may not be uniform over the whole drainage basin. Some water runs downhill into streams, while some sinks into the soil and resupplies ground water. Trees pull soil water out of the ground and release it to the air as transpiration. Different trees at different locations transpire at different rates. Water also evaporates from soil directly. Both evaporation and transpiration depend on air temperature and relative humidity. And, because water is subject to gravity flow, it moves from one point on the landscape to the next.

Imagine having a computer model of a watershed that would describe the matter and energy flows of a watershed accurately. Then we could ask questions like "How will the water quality of the stream be affected by logging in the northeast corner of the watershed?" Or, "Which has a greater effect on the amount of stream flow—the removal of the trees or the building of logging roads?"

Christina Tague specializes in answering these kinds of questions. A geographer at San Diego State, her undergraduate training was as a systems design engineer, and she has assembled an impressive array of computer-based tools to model and visualize landscapes and their ecosystems. To find out how Christina designs computer models to answer these "What if's," visit our web site.

Christina Tague builds computer models to ask the "what if" questions about watershed management in forest and other ecosystems.

Biogeochemical Cycles in the Biosphere

We have seen how energy of solar origin flows through ecosystems, passing from one part of the food chain to the next, until it is ultimately lost from the biosphere as energy radiated to space. Matter also moves through ecosystems, but because gravity keeps surface material Earthbound, matter cannot be lost in the global ecosystem. As molecules are formed and reformed by chemical and biochemical reactions within an ecosystem, the atoms that compose them are not changed or lost. Thus matter is conserved within an ecosystem, and atoms and molecules can be used and reused, or cycled, within ecosystems.

Atoms and molecules move through ecosystems under the influence of both physical and biological processes. The pathways of a particular type of matter through the Earth's ecosystem comprise a **biogeochemical cycle** (sometimes referred to as a *material cycle,* or *nutrient cycle*).

We can recognize two types of biogeochemical cycles—gaseous and sedimentary. In the *sedimentary cycle,* the compound or element is released from rock by weathering, then follows the movement of running water either in solution or as sediment to the sea. Eventually, by precipitation and sedimentation, these materials are converted into rock. When the rock is uplifted and exposed to weathering, the cycle is completed.

In a *gaseous cycle,* a shortcut is provided—the element or compound can be converted into a gaseous form. The gas diffuses throughout the atmosphere and thus arrives over land or sea, to be reused by the biosphere, in a much shorter time. The primary constituents of living matter—carbon, hydrogen, oxygen, and nitrogen—all move through gaseous cycles.

The major features of a biogeochemical cycle are diagrammed in Figure 22.10. Any area or location of concentration of a material is a **pool.** There are two types of pools: *active pools,* where materials are in forms and places easily accessible to life processes, and *storage pools,* where materials are more or less inaccessible to life. A system of pathways of material flows connects the various active and storage pools within the cycle. Pathways between active pools are usually controlled by life processes, whereas pathways between storage pools are usually controlled by physical processes.

The magnitudes of the total storage and total active pools can be very different. In many cases, the active pools are

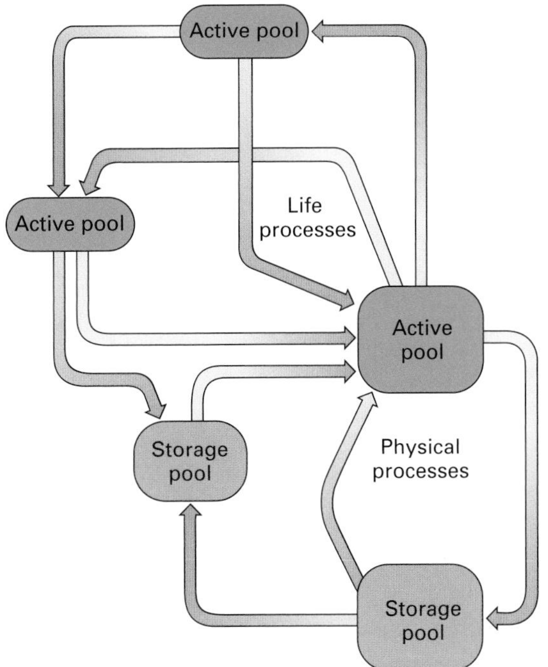

22.10 General features of a biogeochemical cycle

much smaller than storage pools, and materials move more rapidly between active pools than between storage pools or in and out of storage. Taking an example from the carbon cycle, photosynthesis and respiration will cycle all the carbon dioxide in the atmosphere (active pool) through plants in about 10 years. But it may be many millions of years before the carbonate sediments (storage pool) now forming as rock will be uplifted and decomposed to release carbon dioxide.

Nutrient Elements in the Biosphere

Since all elements are more or less available at the Earth's surface, we could actually identify cycles for them all. However, only a limited number of elements are important to life forms. Table 22.2 lists the 15 elements that are most abundant in global living matter and show their percentage representation. The three principal components of carbohydrate—hydrogen, carbon, and oxygen—account for 99.5 percent of all living matter and are called **macronutrients.** Macronutrients are elements required in substantial quantities for organic life to thrive. The remaining one-half percent is divided among 12 elements. Six of these are also macronutrients: nitrogen, calcium, potassium, magnesium, sulfur, and phosphorus. The first three macronutrients—hydrogen, carbon, oxygen—are materials whose pathways we have already followed in the photosynthesis-respiration circuits (Figure 22.4). We will undertake a more detailed analysis of the gaseous cycles of carbon and oxygen because they are influenced by human combustion of hydrocarbon compounds. We will also give special attention to nitrogen, the fourth most abundant element in the composition of living matter. Of the remaining macronutrients, three—calcium, potassium, and magnesium—are elements derived from silicate rocks through mineral weathering. Two other macronutrients derived from rock weathering are sulfur and phosphorus.

The Carbon Cycle

The movements of carbon through the life layer are of great importance because all life is composed of carbon compounds of one form or another. Of the total carbon available, most lies in storage pools as carbonate sediments below the Earth's surface. Only about two-tenths of 1 percent are readily available to organisms as CO_2 or as decaying biomass in active pools.

Some details of the *carbon cycle* are shown in a schematic diagram, Figure 22.11. In the gaseous portion of the cycle, carbon moves largely as carbon dioxide (CO_2), which is a free gas in the atmosphere and a dissolved gas in fresh and salt water. In the sedimentary portion of its cycle, carbon resides in carbohydrate molecules in organic matter, as hydrocarbon compounds in rock (petroleum, coal), and as mineral carbonate compounds such as calcium carbonate ($CaCO_3$). The world supply of atmospheric carbon dioxide is represented in Figure 22.11 by a box. It is a small portion of the carbon in active pools, constituting less than 2 percent. This atmospheric pool is supplied by plant and animal respiration in the oceans and on the lands. Under natural conditions, some new carbon enters the atmosphere each year from volcanoes by outgasing in the form of CO_2 and carbon monoxide (CO). Industry injects substantial amounts of carbon into the atmosphere through combustion of fossil fuels. This increment from fuel combustion and its effects on global air temperatures were discussed in Chapter 5.

Table 22.2 Elements Comprising Global Living Matter

Basic Carbohydrate	Percent*
Hydrogen (M)	49.74
Carbon (M)	24.90
Oxygen (M)	24.83
Subtotal	99.47
Other Nutrients	
Nitrogen (M)	0.272
Calcium (M)	0.072
Potassium (M)	0.044
Silicon	0.033
Magnesium (M)	0.031
Sulfur (M)	0.017
Aluminum	0.016
Phosphorus (M)	0.013
Chlorine	0.011
Sodium	0.006
Iron	0.005
Manganese	0.003

M—macronutrient

*Based on total of 15 most abundant elements.
Source: E. S. Deevey, Jr., *Scientific American,* Vol. 223.

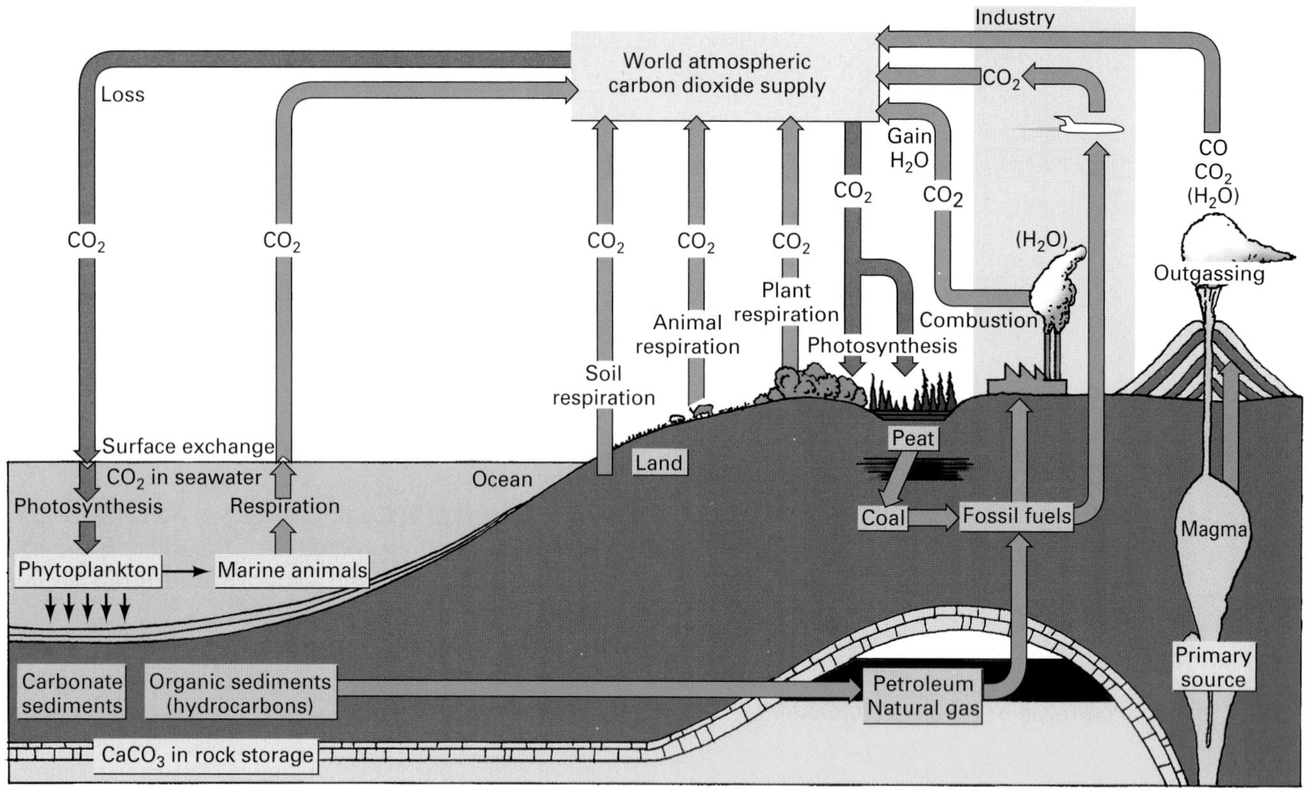

***22.11 The carbon cycle** (Copyright © A. N. Strahler.)*

Carbon dioxide leaves the atmospheric pool to enter the oceans, where it is used in photosynthesis by phytoplankton. These organisms are primary producers in the ocean ecosystem and are consumed by marine animals in the food chain. Phytoplankton also build skeletal structures of calcium carbonate. This mineral matter settles to the ocean floor to accumulate as sedimentary strata, an enormous storage pool not available to organisms until released later by rock weathering. Organic compounds synthesized by phytoplankton also settle to the ocean floor and eventually are transformed into the hydrocarbon compounds making up petroleum and natural gas. On the lands, plant matter accumulating over geologic time forms layers of peat that are ultimately transformed into coal. Petroleum, natural gas, and coal comprise the fossil fuels, and these represent huge storage pools of carbon. GEODISCOVERIES

Human activity is presently affecting the carbon cycle very significantly. Through the burning of fossil fuels, CO_2 is being released to the atmosphere at a rate far beyond that of any natural process. *Eye on Global Change 22.3 • Human Impact on the Carbon Cycle* documents how human activity has influenced the major flows within the carbon cycle. GEODISCOVERIES

The Oxygen Cycle

Details of the *oxygen cycle* are shown in schematic form in Figure 22.12. The complete picture of the cycling of oxygen also includes its movements and storages when combined with carbon as carbon dioxide and as organic and inorganic compounds. These we have covered in the carbon cycle.

The world supply of atmospheric free oxygen is shown in Figure 22.12 by a box at the top of the diagram. Oxygen enters this active pool through release in photosynthesis, both in the oceans and on the lands. Each year a small amount of new oxygen comes from volcanoes through outgasing, principally as CO_2 and H_2O (shown in Figure 22.11). Balancing the input to the atmospheric pool is loss through organic respiration and mineral oxidation. Adding to the withdrawal from the atmospheric oxygen pool is industrial activity through the combustion (oxidation) of wood and fossil fuels. Forest fires and grass fires (not shown) are another means of oxygen consumption. The oceans also contain a small, active pool of dissolved gaseous oxygen. Some oxygen is continuously placed in storage in mineral carbonate form in ocean-floor sediments.

Human activity reduces the amount of oxygen in the air by (1) burning fossil fuels; (2) clearing and draining land, which speeds the oxidation of soils and soil organic matter; and (3) reducing photosynthesis by clearing forests for agriculture and by paving and covering previously productive surfaces. The importance of urbanization can be appreciated by the fact that every six months a land area about the size of Rhode Island is covered by new construction in the United States alone. Fortunately, the oxygen

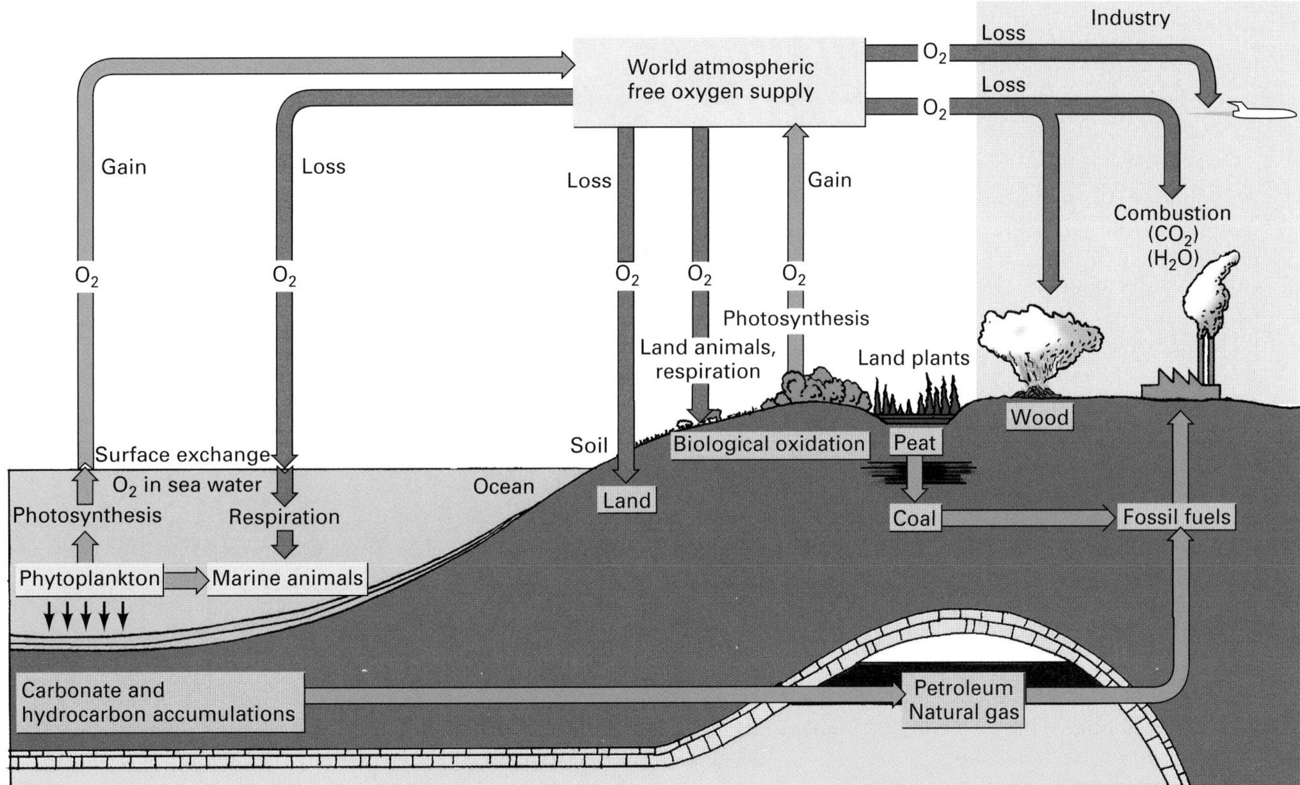

22.12 The oxygen cycle *(Copyright © A. N. Strahler.)*

pool is so large that the human impact or potential impact is very small, at least at this time.

The Nitrogen Cycle

Nitrogen moves through the biosphere in a gaseous *nitrogen cycle* in which the atmosphere, containing 78 percent nitrogen as N_2 by volume, is a vast storage pool (Figure 22.13). Nitrogen in the atmosphere in the form of N_2 cannot be assimilated directly by plants or animals. Only certain microorganisms possess the ability to utilize N_2 directly, a process termed *nitrogen fixation.* One class of such microorganisms consists of certain species of free-living soil bacteria. Some blue-green algae can also fix nitrogen.

Another class consists of the symbiotic nitrogen fixers. In a symbiotic relationship, two species of organisms live in close physical contact, each contributing to the life processes or structures of the other. Symbiotic nitrogen fixers are bacteria of the genus *Rhizobium.* These bacteria are associated with some 190 species of trees and shrubs as well as almost all members of the legume family. Legumes important as agricultural crops are clover, alfalfa, soybeans, peas, beans, and peanuts. *Rhizobium* bacteria infect the root cells of these plants in root nodules produced jointly by action of the plant and the bacteria. The bacteria supply the nitrogen to the plant through nitrogen fixation, while the plant supplies nutrients and organic compounds needed by the bacteria. Crops of legumes are often planted in seasonal rotation with other food crops to ensure an adequate nitrogen supply in the soil. Both the action of nitrogen-fixing crops and that of soil bacteria are shown in the nitrogen cycle diagram (Figure 22.13).

Nitrogen is lost to the biosphere by *denitrification,* a process in which certain soil bacteria convert nitrogen from usable forms back to N_2. This process is also shown in the diagram. Denitrification completes the organic portion of the nitrogen cycle, as nitrogen returns to the atmosphere.

At the present time, nitrogen fixation is far exceeding denitrification, and usable nitrogen is accumulating in the life layer. This excess of fixation is produced almost entirely by human activities. Human activity fixes nitrogen in the manufacture of nitrogen fertilizers and by oxidizing nitrogen in the combustion of fossil fuels. Widespread cultivation of legumes has also greatly increased worldwide nitrogen fixation. At present rates, nitrogen fixation attributable to human activity nearly equals all natural biological fixation.

Much of the nitrogen fixed by human activities is carried from the soil into rivers and lakes and ultimately reaches the ocean. Major water pollution problems can arise when nitrogen stimulates the growth of algae and phytoplankton. The respiration of these organisms can then reduce quantities of dissolved oxygen to levels that are detrimental to desirable forms of aquatic life. These problems will be accentuated in years to come because industrial fixation of nitrogen in fertilizer manufacture is doubling about every six years at present. The global impact of such large amounts of nitrogen

Eye on Global Change | 22.3

Human Impact on the Carbon Cycle

Carbon is an element that is abundant at the Earth's surface and is also essential for life. As noted earlier in this chapter, carbon cycles continuously among the land surface, atmosphere, and ocean in many complex pathways. However, these flows are now strongly influenced by human activity. The most important human impact on the carbon cycle is the burning of fossil fuels. Another important human impact lies in changing the Earth's land covers— for example, in clearing forests or abandoning agricultural areas. Let's look at these impacts in more detail.

The figure below shows a simple diagram of the major flows within the carbon cycle for the period 1989–1998. The magnitudes of the annual flows are shown in gigatons (Gt) of carbon per year (1 gigaton = 10^9 metric tons = 10^{12} kg = 1.1×10^9 English tons = 1.1 English gigatons). These flows are estimates, and a second value after each value indicates its uncertainty. For example,fossil fuel burning liberates 6.3 ± 0.6 Gt/yr, which we can interpret as a flow that is most likely to be in the range $6.3 - 0.6 = 5.7$ to $6.3 + 0.6 = 6.9$ Gt/yr.

By comparing the flows, we see that about half of the output of carbon by fossil fuel burning is taken up by the atmosphere (3.3 ± 0.2 Gt/yr). Of the remaining amount, about 2/3 (2.3 ± 0.8 Gt/yr) is absorbed by the oceans. This leaves unaccounted an amount of about $6.3 - 3.3 - 2.3 = 0.7$ Gt/yr. Since there are no other significant pathways, this carbon must be flowing into the biosphere. In other words, ecosystems are a *sink* for CO_2, accepting about 0.7 Gt/yr of carbon.

Ecosystems cycle carbon in photosynthesis, respiration, decomposition, and combustion. Photosynthesis and respiration are basic physiological processes that fix and release CO_2. Decomposition is the process in which bacteria and fungi digest dead organic matter, and is actually a form of respiration. Combustion refers to uncontrolled combustion, as when an ecosystem burns. It is very hard to account for these processes globally in such a way as to know their net effect, but by applying the logic of budgeting, we know that they must sum to the value of 0.7 Gt/yr of carbon build-up in land ecosystems mentioned above.

If the value of 0.7 Gt/yr in carbon uptake is correct, the amount

The global carbon cycle *Values are in gigatons of carbon per year. (Data from UN/IPCC.)*

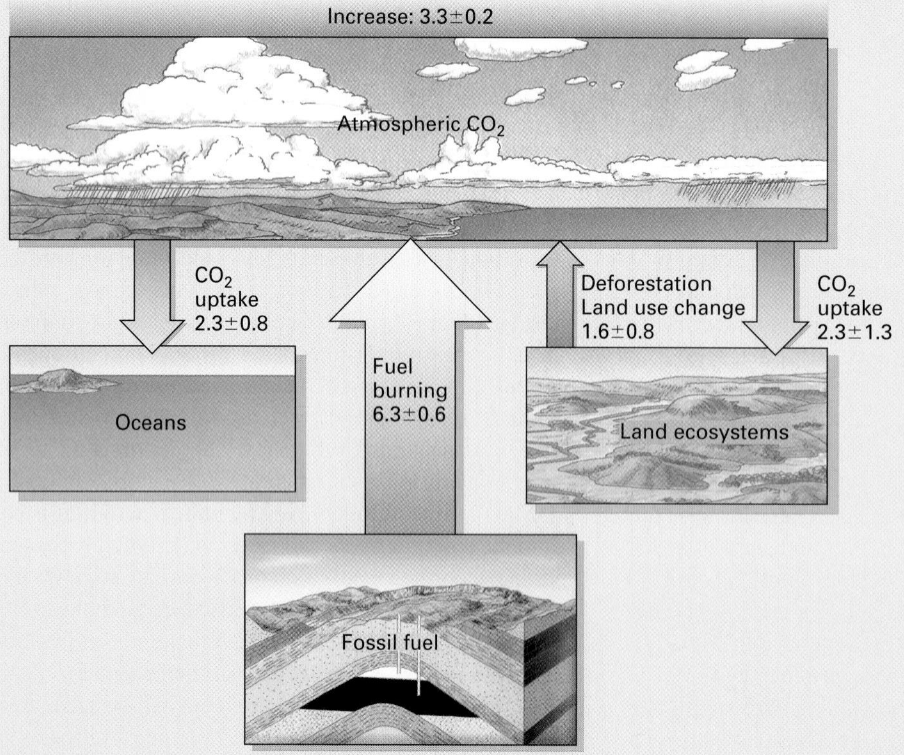

of terrestrial biomass must be increasing at that rate. However, forests are presently diminishing in area as they are logged or converted to farmland or grazing land. This conversion is primarily occurring in tropical and equatorial regions, and it is estimated to release about 1.6 ± 0.8 Gt/yr of carbon to the atmosphere. Since this release is included in the net land ecosystem uptake of 0.7 Gt/yr, the remainder of the world's forests must be taking up the 1.6 Gt/yr loss from deforestation as well as an additional 0.7 Gt/yr. So we can estimate that mid- and high-latitude forests are increasing in area or biomass to fix carbon at a rate of 1.6 + 0.7 = 2.3 Gt/yr.

Independent evidence seems to confirm this conclusion. In Europe, for example, forest statistics show an increase of growing stock—the volume of living trees—of about 25 percent from 1970 to 1990. This increase has been sustained in spite of damage to forests by air pollution, especially in eastern Europe. In North America, forest areas are increasing in many regions as agricultural production has abandoned marginal areas to natural forest growth. New England is a good example of this trend. A century ago, only a small portion of New England was forested. Now only a small portion is cleared.

Some of the increase in global biomass may also be the effect of enhancement of photosynthesis by warmer temperatures and increased CO_2 concentrations (discussed in more detail shortly). Another factor proposed to account for increased ecosystem productivity is nitrogen fertilization of soils by washout of nitrogen pollutant gases in the atmosphere.

Some foresters have observed that harvesting mature forests and replacing them with young, fast-growing timber should increase the rate of withdrawal Of CO_2 from the atmosphere. Since the lumber of the mature forests goes into semipermanent storage in dwellings and structures where it is protected from decay and oxidation to CO_2, it represents a withdrawal of CO_2 from the atmosphere. The young forests that replace the mature ones grow quickly, fixing carbon at a much faster rate than the older, mature forest, in which annual growth has slowed.

A report of research scientists at the College of Forestry of Oregon State University showed, however, that the conversion of old-growth forests to young, fast-growing forests will not significantly decrease atmospheric CO_2. They calculated that while 42 percent of the harvested timber goes into comparatively long term storage (greater than 5 years) in building structures, much of the remainder is directly discarded on the logging site where it is burned or rapidly decomposes. In addition, some biomass becomes waste in factory processing of the lumber, where sawdust and scrap are burned as fuel. Similarly, the manufacture of paper also results in short-term conversion of a large proportion of the harvested trees to CO_2. In sum, harvesting of old-growth forests as now practiced actually contributes substantially to atmospheric CO_2.

Some environmentalists have advocated increased tree planting as a way of enhancing CO_2 fixation. To take up the quantity of carbon now being released by fossil fuel burning would require some 7 million square kilometers of new closed-crown broadleaved deciduous forest—an area about the size of Australia. To absorb the net increase in atmospheric carbon would require about half that area. This would be a daunting task at best.

Another factor is that increased CO_2 concentration in the atmosphere might enhance photosynthesis and thus increase the rate of carbon fixation. The enhancement of photosynthesis by increased CO_2 concentrations has been observed for many plants, and demonstrated as a way of increasing yields of some crops. However, CO_2 is only one factor in photosynthesis—light, temperature, nutrients, and water are also needed, and restrictions in any of these will reduce photosynthesis. In one research study, forest photosynthesis under enriched CO_2 conditions was stimulated, but no net increase in carbon storage was observed. On the other hand, the CO_2 enrichment seemed to enhance root growth and nitrogen fixation by root nodules, thus making the trees more able to withstand stress.

While the dynamics of forests are important in the global carbon cycle, soils may be even more important. Recent inventories estimate that about four times as much carbon resides in soils than in above-ground biomass. The largest reservoir of soil carbon is in the boreal forest. In fact, there is about as much carbon in boreal forest soils as in all above-ground vegetation. This soil carbon has accumulated over thousands of years under cold conditions that have retarded its decay. However, there is now great concern that global warming, which is acting more strongly at high latitudes, will increase the rate of decay of this vast carbon pool and that boreal forests, which are presently a sink for CO_2, will become a source.

Reducing the rate of carbon dioxide buildup in the atmosphere is a matter of great international concern. As we noted in Chapter 5, an international treaty limiting emissions Of CO_2 and other greenhouse gases was signed at the Rio de Janeiro Earth Summit in 1992 by nearly 150 nations. Since that time, the world's nations have been struggling with the implementation of a plan to control these emissions. While much good progress has been made, more work is needed. An effective global commitment to reduction of CO_2 releases and control of global warming still awaits us.

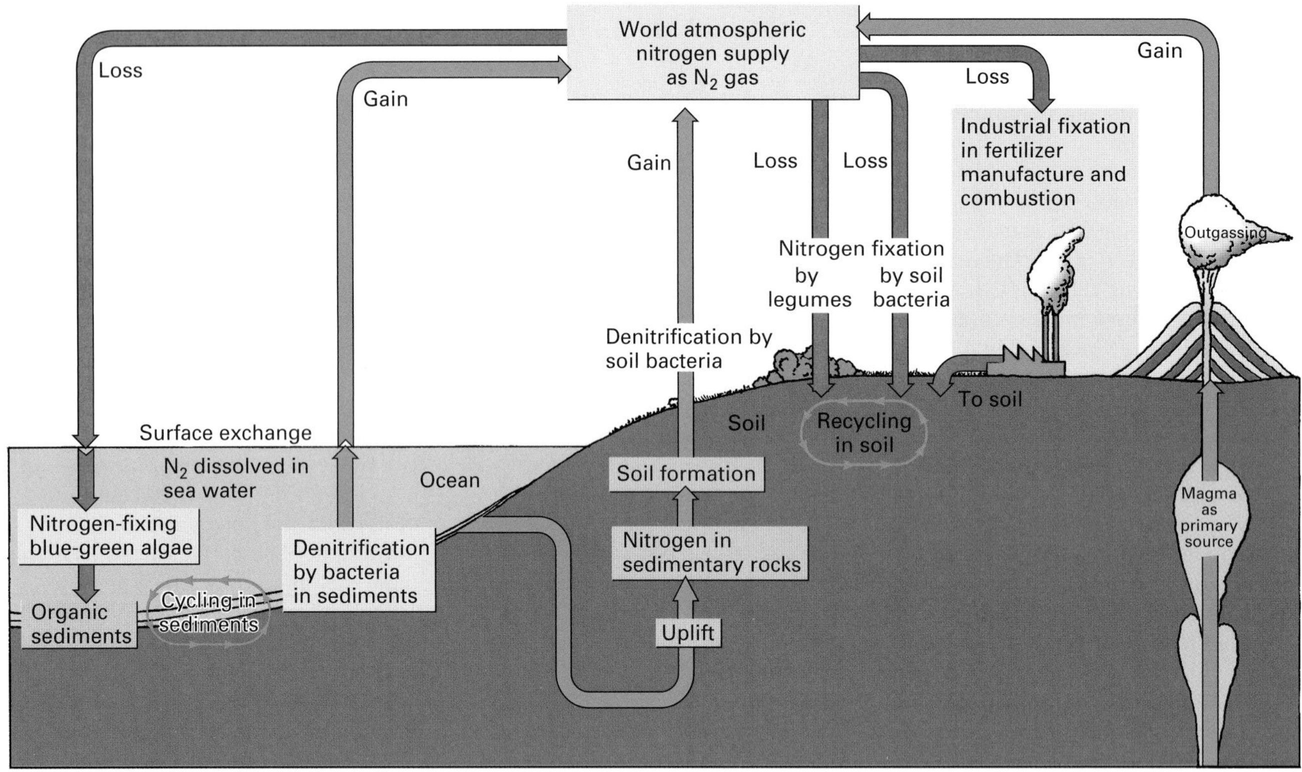

22.13 The nitrogen cycle

reaching rivers, lakes, and oceans on the Earth's global ecosystem remains uncertain.

Sedimentary Cycles

The oxygen, carbon, and nitrogen cycles are all referred to as gaseous cycles because they possess a gaseous phase in which the element involved is present in significant quantities in the atmosphere. Many other elements move in sedimentary cycles, that is, from the land to ocean in running water, returning after millions of years as uplifted terrestrial rock. These elements are not present in the atmosphere except in small quantities as blowing dust or condensation nuclei in precipitation.

Figure 22.14 shows how some important macronutrients move in sedimentary cycles. Within the large box representing the lithosphere are smaller compartments representing the parent matter of the soil and the soil itself. In the soil, nutrients are held as ions on the surfaces of soil colloids and are readily available to plants. (This was explained in Chapter 21.)

The nutrient elements are also held in enormous storage pools, where they are unavailable to organisms. These storage pools include sea water (unavailable to land organisms), sediments on the sea floor, and enormous accumulations of sedimentary rock beneath both lands and oceans. Eventually, elements held in the geologic storage pools are released into the soil by weathering. Soil particles are lifted into the atmosphere by winds and fall back to Earth or are washed down by precipitation. Chlorine and sulfur are shown as passing from the ocean into the atmosphere and entering the soil by the same mechanisms of fallout and washout.

The organic realm, or biosphere, is shown in three compartments: producers, consumers, and decomposers. Considerable element recycling occurs between organisms of these three classes and the soil. The elements used in the biosphere, however, are continually escaping to the sea as ions dissolved in stream runoff and ground water flow.

A LOOK AHEAD The organizing principles of energy flow and matter cycling can greatly aid our understanding of the processes of the biosphere and their implications for human activity. Just as solar energy is the driving force for the circulation of global atmosphere and ocean fluids, so it also provides the power source for photosynthesis, on which all the world's organisms ultimately depend for sustenance. Humans are no different from other organisms. Without this input of solar energy and its conversion to consumable biomass, we would soon perish.

Of course, we are also dependent on the smooth functioning of material cycles. Without the tiny fraction of the Earth's atmosphere that is carbon dioxide, no terrestrial photosynthesis would occur. If that component is enhanced by human activity through CO_2 release, it is sure to impact the productivity of the biosphere directly, as well as indirectly through climate change. Similarly, we influence the nitrogen and oxygen cycles without knowing or understanding the consequences.

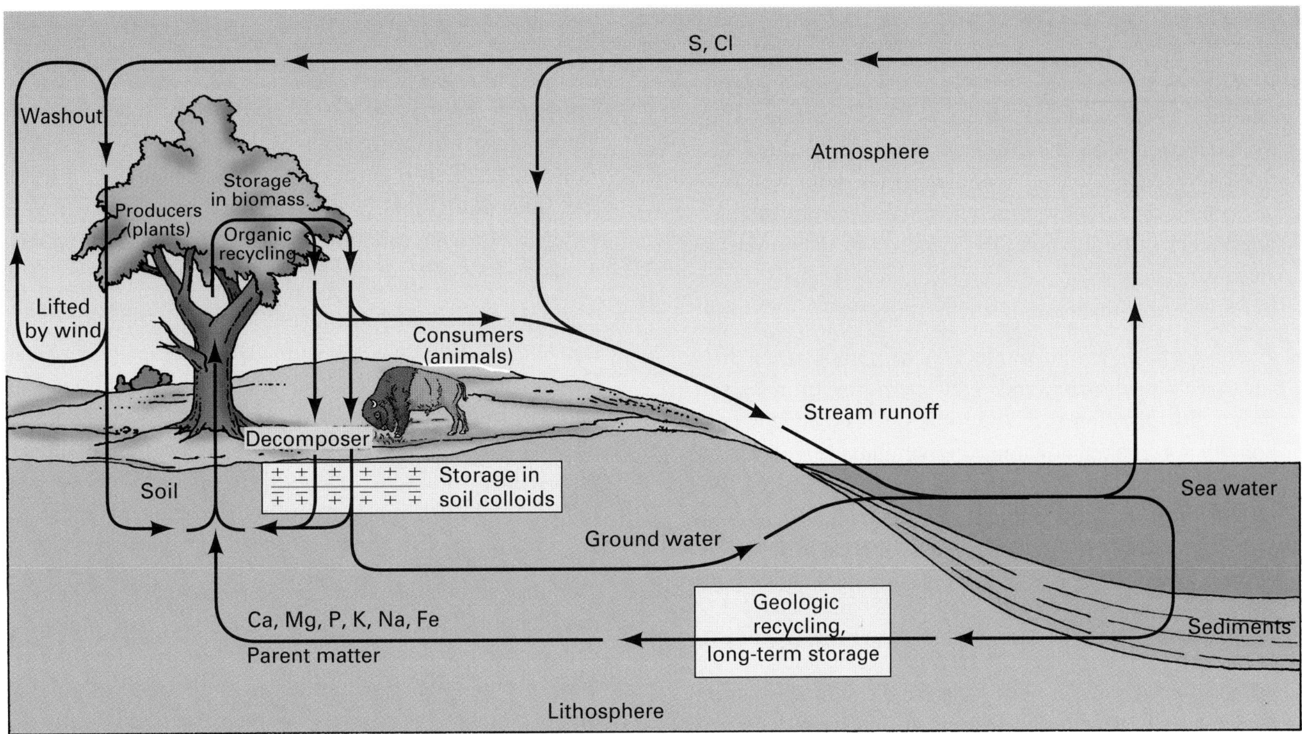

***22.14 Sedimentary cycles** A flow diagram of the sedimentary cycle of materials in and out of the biosphere and within the inorganic realm of the lithosphere, hydrosphere, and atmosphere.*

In our final two chapters, we examine the biosphere from a different perspective. Rather than focus on the cycling of energy and matter, we look at the spatial patterns of plants and animals at scales ranging from local to global. In Chapter 23, we examine the processes of biogeography that determine the distributions of individuals and species. Organism–environment relationships are important at the local and regional scale, but at the global scale historical processes such as evolution, dispersal, migration, and extinction of species are the primary determiners of spatial patterns of plant and animal species. In our closing chapter, we survey the world's major biome types, their human utilization, and how they are impacted by human activities.

Chapter Summary

- **Ecology** is the science of interactions among organisms and their environment. Its focus is the **ecosystem**, which by interaction among components provides pathways for flows of energy and cycles of matter. The **food web** of an ecosystem details how food energy flows from **primary producers** through **consumers** and on to **decomposers**. Because energy is lost at each level, only a relatively few top-level consumers are normally present.
- **Photosynthesis** is the production of carbohydrate from water, carbon dioxide, and light energy by primary producers. **Respiration** is the opposite process, in which carbohydrate is broken down into carbon dioxide and water to yield chemical energy and thus power organisms. **Net photosynthesis** is the amount of carbohydrate remaining after respiration has reduced **gross photosynthesis**. Net photosynthesis increases with increasing light and temperature, up to a point.
- Forests and estuaries are ecosystems with high rates of net primary production, while grasslands and agricultural lands are generally lower. Oceans are most productive in coastal and upwelling zones near continents. Among climate types, those with abundant rainfall and warm temperatures are most productive.
- *Biomass* is an attractive form of solar-powered energy. Charcoal, biogas, and alcohol are biomass products that can be used as fuels.
- **Biogeochemical cycles** are of two types—*gaseous*, in which the element has an important gaseous phase and moves within the atmosphere, and *sedimentary*, when no important gaseous phase is involved. Biogeochemical cycles consist of *active pools* and *storage pools* linked by flow paths. Of most concern are biogeochemical cycles of the **macronutrients**, which include carbon, hydrogen, oxygen, nitrogen, calcium, potassium, magnesium, potassium, and magnesium.
- The *carbon cycle* includes an active pool of biospheric carbon and atmospheric CO_2, with a large storage pool of carbonate in sediments. Human activities have provided a pathway from storage to active pools by the burning of fossil fuel. The *oxygen cycle* features an active pool of atmospheric O_2, which is increased by photosynthesis and reduced by respiration, combustion, and mineral oxidation.
- The *nitrogen cycle* also has an important gas phase, but the nitrogen is largely held in the form of N_2, which cannot be used directly by most organisms. *Nitrogen fixation* occurs when N_2 is converted to more useful forms by bacteria or blue-green algae, often in symbiosis with higher plants. Human activity has doubled the rate of nitrogen fixation, largely through fertilizer manufacture.
- Sedimentary cycles involve macronutrients that do not have an important gas phase. These elements are held in active pools in living and decaying organisms and in soils. Storage pools include sea water, sediments, and sedimentary rocks.

Key Terms

ecology
ecosystem
food chain
food web
primary producers
consumers
decomposers
gross photosynthesis
net photosynthesis
respiration
biomass
biogeochemical cycle
pool
macronutrients

Review Questions

1. Define the terms *biogeography, ecology,* and *ecosystem.*
2. What is a food web or food chain? What are its essential components? How does energy flow through the food web of an ecosystem?
3. Compare and contrast the processes of photosynthesis and respiration. What classes of organisms are associated with each?

4. How is net primary production related to biomass? Identify some types of terrestrial ecosystems that have a high rate of net primary production and some with a low rate.
5. Which areas of oceans and land are associated with high net primary productivity? How is net primary production on land related to climate?
6. Why is biomass energy a desirable energy source? Identify and describe the most useful forms of biomass energy as fuel.
7. What is a *biogeochemical cycle?* What are its essential features? Identify and compare two types of biogeochemical cycles.
8. List nine macronutrients and identify those associated with gaseous and with sedimentary cycles.
9. What are the essential features and flow pathways of the carbon cycle? How have human activities impacted the carbon cycle?
10. What are the essential features and flow pathways of the oxygen cycle? What are the effects of human activity on the oxygen cycle?
11. What are the essential features and flow pathways of the nitrogen cycle? What role do bacteria play? How has human activity modified the nitrogen cycle?
12. What are the essential features and flow pathways of macronutrients in sedimentary cycles?

Focus on Systems 22.2 • Agricultural Ecosystems

1. In what ways is agricultural production dependent on fossil fuels?
2. How do agricultural ecosystems differ from natural ecosystems?
3. Use the concept of *cultural energy* to discuss the efficiency of production of different food crops and the implications for human diets.

Eye on the Environment 22.3 • Human Impact on the Carbon Cycle

1. As a visualizing exercise, diagram the global carbon cycle, indicating the magnitudes of flows between compartments.
2. How is land use change affecting the global carbon balance? Will replacing old-growth forests with younger faster-growing forests help remove CO_2 from the atmosphere?
3. Will increasing levels of atmospheric CO_2 have an effect on the rate of carbon fixation by ecosystems? How will increases in temperature affect the release of boreal soil carbon?

A Closer Look:
Focus on Remote Sensing 22.4 • Monitoring Global Productivity from Space

1. What factors control primary production at any point on the globe? Identify each factor and relate it to the process of photosynthesis.
2. What is a leaf-area index? How and why can it be sensed remotely? How does it relate to the fraction of photosynthetically active radiation absorbed by the plants on the surface?
3. What factor is most important in determining ocean primary productivity? How is it mapped using remote sensing?
4. What are the primary patterns of ocean primary productivity over the oceans? over land?
5. What changes have occurred over the past two decades in net primary productivity over land? How are they related to the controlling factors identified in Question 1?
6. What changes have occurred over the past two decades in net primary productivity over oceans? What are the possible causes of these changes?
7. What are the implications of productivity changes for the carbon cycle? How will remote sensing be useful in assessing the impact of human activity on the global carbon cycle?

Visualizing Exercises

1. Sketch a graph showing the relationship among gross photosynthesis, net photosynthesis, respiration, and temperature. How is net photosynthesis obtained from gross photosynthesis and respiration for each temperature?
2. Diagram the general features of a biogeochemical cycle, in which storage pools and active pools are linked by life processes and physical processes.

Essay Questions

1. Select one of the cycles described in the text (carbon, oxygen, nitrogen, sedimentary). Identify and describe the power sources for each of the major pathways in the cycle.

2. Suppose atmospheric carbon dioxide concentration doubles. What will be the effect on the carbon cycle? How will flows change? Which pools will increase? decrease?

Problems

Working It Out 22.1 • Logistic Population Growth

1. An ecologist repeats the experiment of yeast growing in a culture medium, but this time alters the medium to see what effect the change may have on population growth. She plots the data, shown in columns 1 and 2 below, and observes that the carrying capacity *(C)* is about 440 grams. She then calculates the increase in each time period in columns 3 and 4, and estimates an initial growth rate *(R)* of about 60 percent. She uses these values to model the growth using the logistic model, calculating the values in columns 5 and 6. The following table omits the results of her calculations for times of 3, 7, 9, 13, and 17 hours. Find the missing values. For increases, calculate these by comparing the observed biomass with the observed biomass of the previous period. For the logistic model, find the constants A and k, then evaluate the logistic equation to provide the modeled result. Subtract the modeled biomass from the observed biomass to find the difference.

Yeast Growth, Alternative Medium

(1) Time (hours)	(2) Biomass (grams)	(3) Increase (grams)	(4) Increase (percent)	(5) Logistic Model	(6) Difference
0	10.0			10.0	0.0
1	16.2	6.2	62.0	15.8	0.4
2	26.2	10.0	61.7	24.7	1.5
3	37.4				
4	61.0	23.6	63.1	58.2	2.8
5	85.7	24.7	40.5	86.3	–0.6
6	127.2	41.5	48.4	123.5	3.7
7	165.8				
8	222.0	56.2	33.9	219.9	2.1
9	271.7				
10	316.8	45.1	16.6	316.3	0.5
11	350.4	33.6	10.6	353.6	–3.2
12	383.8	33.4	9.5	381.7	2.1
13	399.9				
14	419.1	19.2	4.8	415.2	3.9
15	428.0	8.9	2.1	424.2	3.8
16	430.5	2.5	0.6	430.0	0.5
17	433.6				
18	443.5	9.9	2.3	436.0	7.5

2. (*a*) Compare the pattern of increase in grams with the pattern of increase in percent. (*b*) Do the data seem to fit the model well? (*c*) Compare the outcome of this experiment with that in the *Working It Out* box. Do the two different culture media seem to be different? What other explanations might there be for this result?

Eye on the Landscape

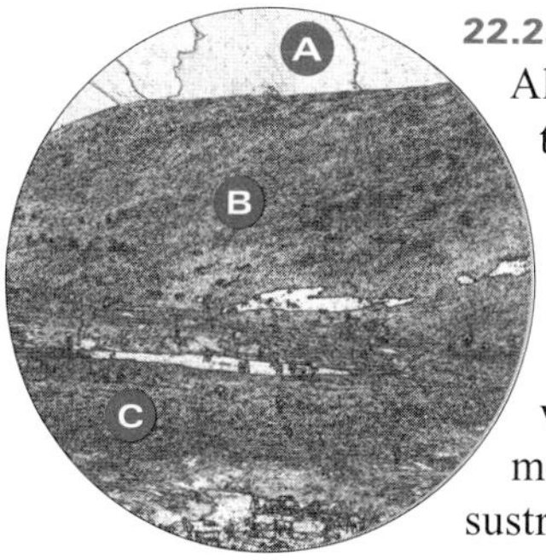

22.2a Caribou, Brooks Range Although not clearly shown in this photo, the steep slopes of the Brooks Range are not far, judging from the snow-covered talus cones at **(A)**. At **(B)** we see solifluction lobes, where thawed soil has slowly moved downhill on a frozen sustrate. At **(C)**, we see the grasses, sedges, and other low plants of the arctic tundra that are the forage for the caribou. Talus slopes and solifluction lobes were treated in Chapter 15; Chapter 24 provides more detail on the tundra.

A CLOSER LOOK

Focus on Remote Sensing 22.4 | Monitoring Global Productivity from Space

As human activity drives carbon dioxide concentrations in the atmosphere to ever higher levels, it becomes ever more important to understand and model the photosynthetic activity of the biosphere. As we saw in *Eye on Global Change 22.3 • Human Impact on the Carbon Cycle,* analysis of the global carbon budget shows that the biosphere must be a *carbon sink*—that is, for carbon flows to balance, global photosynthesis must exceed global respiration by a significant amount. This means that fixed carbon is accumulating in the biosphere, reducing the amount of CO_2 buildup in the atmosphere. However, this conclusion is uncertain without some sort of direct measurement. Until recently, there was no way to measure the photosynthetic activity of the globe, but new techniques using remote sensing, meteorological observations, and models of biological productivity now make it possible.

What factors control primary production at any point on the globe? They include:

- *The amount of photosynthetic material present.* For terrestrial plants, this is measured by the surface area of leaves above a square meter of ground. For oceanic phytoplankton, it is the chlorophyll concentration within a cubic meter of ocean surface. Chlorophyll is a plant pigment that converts light energy into the chemical energy needed to fix carbon.
- *Light.* Photosynthesis requires light. What is important is the amount of light absorbed by the photosynthetic material, which depends on two things: (1) the amount of illumination from the sun and sky, which is determined by such factors as season, latitude, and cloud cover, and (2) the amount of photosynthetic material present to absorb the available light energy.
- *Temperature.* The biochemical process of fixing carbon is sensitive to temperature, as we noted in the main chapter text. Warmer temperatures favor photosynthesis, but if temperatures are too high, photosynthesis shuts down.
- *Water.* Recall that terrestrial plants transport water from their roots to their leaves in the process of transpiration. As the water evaporates from leaf pores, it cools the leaves and allows them to maintain high levels of photosynthesis even under intense sunlight. Of course, water is freely available to phytoplankton at all times.
- *Nutrients.* Nutrients can play an indirect role by reducing the health of organisms or limiting their development. In many cases, the productivity of phytoplankton is limited by a scarcity of nutrients, especially iron.

Land Productivity

Given these factors, how can we use remote sensing to monitor global productivity? First, we use color—that is, the "greenness" of the surface—to measure the amount of photosynthetic material. Recall from our interchapter feature *A Closer Look: Geographer's Tools 4.8 • Remote Sensing for Physical Geography* that land plants strongly absorb red light and strongly reflect near-infrared light. Thus, locations that appear darker in a red band image and brighter in a near-infrared band image will have more leaf area.

Figure 1*a* shows an image of leaf-area index for an area of North America showing the conterminous United States and a portion of southeastern Canada. Leaf-area index (LAI) is simply the ratio of the area of (one-sided) leaf surface to the area of ground below. Thus, a leaf-area index of 2 indicates that each square meter of ground surface is covered by 2 square meters of leaf area. Typical leaf-area indexes range from 0 to about 6 or 7. The map of leaf-area index is derived from MODIS images acquired during May 1–10, 2003. At this time of year, the southeast, midwest, and west coast regions are abundantly green.

Figure 1*b* converts the leaf-area measurement into another measurement—the fraction of photosynthetically active radiation (FPAR) absorbed by the leaf area. As shown on the scale, this fraction ranges from 0 to 100 percent. It expresses the proportion of useful solar radiation (visible light) that is actually absorbed by the leaf canopy. With this proportion in hand, and an estimate of the solar energy falling on each point during each day, it is possible to estimate the energy available for photosynthesis.

Given that amount of energy, how much photosynthesis actually occurs? That depends on both temperature and water availability, as noted above. Another factor is the efficiency of the vegetation-cover type—given the same environmental conditions and the same leaf area, some vegetation types are more productive than others. If we combine temperature and rainfall from meteorological data sources and efficiency based on a map of vegetation cover type, we can finally calculate gross primary productivity (Figure 1*c*). The map shows the same general patterns as the others, with maximum gross primary productivity in the southeast and west coast regions, but the fine detail shows how rainfall, cloudiness, temperature, and vegetation-cover type affect each small area.

The last step is to estimate respiration—the rate at which carbon is released to the atmosphere in plant metabolism. This will depend on many of the same factors as photosynthesis, including leaf area,

(a)

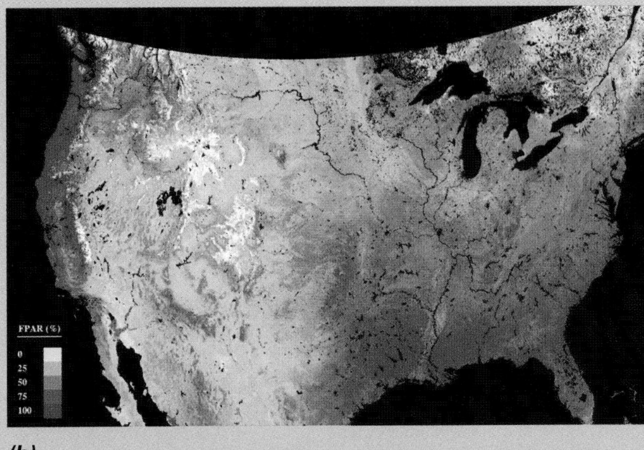

(b)

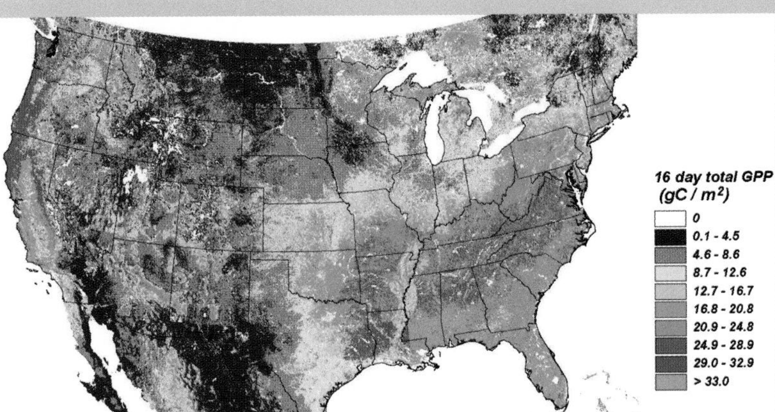

(c)

Figure 1 Mapping primary productivity in North America *(a) Leaf-area index from MODIS during the period May 1–10, 2003. (b) Fraction of photosynthetically active radiation (FPAR) absorbed by vegetation, also derived from MODIS, for the same period. (c) Gross primary production, from MODIS and ancillary data, for the same period. (R. Nemani and S. W. Running, University of Montana NTSG/NASA.)*

temperature, and moisture. The result is net primary productivity, which is shown globally in later figures.

Ocean Productivity

For oceans, the amount of photosynthetic material present at a location is measured in terms of chlorophyll concentration in ocean water. This quantity is expressed in units of milligrams of chlorophyll per cubic meter of sea water. As with terrestrial plants, color determines this quantity, with satellite instruments using special narrow spectral bands to detect the presence and amount of chlorophyll.

Figure 2*a* shows a true-color image of the Puget Sound area acquired by the SeaWiFS instrument, which is designed specifically to image the color of ocean water. Note the band of greenish water along the coast—the greenish color indicates the presence of phytoplankton, just as green inland images indicates the presence of leafy plants. Note also that the Puget Sound appears dark and seems to lack phytoplankton.

Figure 2*b*, however, tells a different story. By carefully analyzing the spectral information acquired by SeaWiFS, it is possible to estimate chlorophyll concentration across the image. It turns out that the concentration of chlorophyll is as high, or higher, in the areas of dark water as in the areas of blue-green water. Apparently, there are two types of phytoplankton communities here—one that absorbs all the blue and green light from the Sun, leaving little to be reflected back to the imager (black water), and another that reflects some blue and green light back to space (blue-green water).

Chlorophyll concentration, like leaf-area index, is not sufficient to model photosynthesis accurately. Just because chlorophyll is present, it doesn't mean that active photosynthesis is proceeding at a maximum rate. For example, lack of nutrients such as iron may be slowing the photosynthetic process. How can we analyze the efficiency of photosynthesis by phytoplankton from afar? That's where MODIS comes in. This satellite instrument has special spectral bands that measure chlorophyll fluorescence. In the phenomenon of fluorescence, chlorophyll "glows," emitting light at particular wavelengths. Chlorophyll fluorescence occurs primarily when absorbed sunlight is not being used to fix carbon. In other words, when chlorophyll fluorescence is strong, photosynthesis is weak. Therefore, we can use the strength of the fluorescence leaving ocean waters to measure the efficiency of the chlorophyll that is present. With this missing link in the chain of information, mapping primary productivity of the ocean becomes possible.

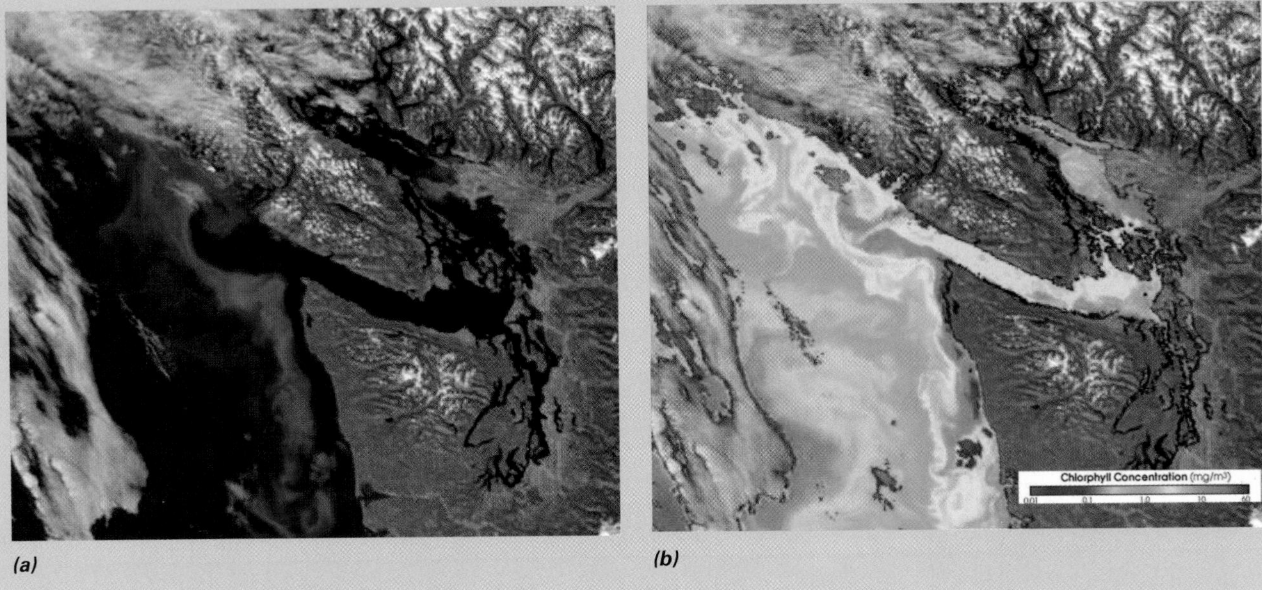

Figure 2 ***SeaWiFS views Puget Sound*** *(a) A true-color image of Puget Sound, Washington, and British Columbia, acquired on July 9, 2003. (b) Chlorophyll concentration on the same day. (SeaWiFS Project, NASA/GSFC/ORBIMAGE.)*

Global Productivity

Figure 3 shows a global map of net primary production of both land and oceans for 2002. The first thing to note is that net primary production is generally higher on land than in oceans. However, there is a lot more ocean surface than land surface. When this fact is taken into account, terrestrial and oceanic net primary production are about equal.

Looking at the ocean pattern, we see that large central areas of ocean are unproductive (dark purple). This is largely due to lack of essential nutrients in surface waters. Near land, and along

Figure 3 ***Global net primary productivity from MODIS for 2002*** *This image maps global productivity, including both land and oceans, as viewed by the MODIS instruments on NASA's Terra and Aqua satellite platforms. (MODIS Science Team/NASA.)*

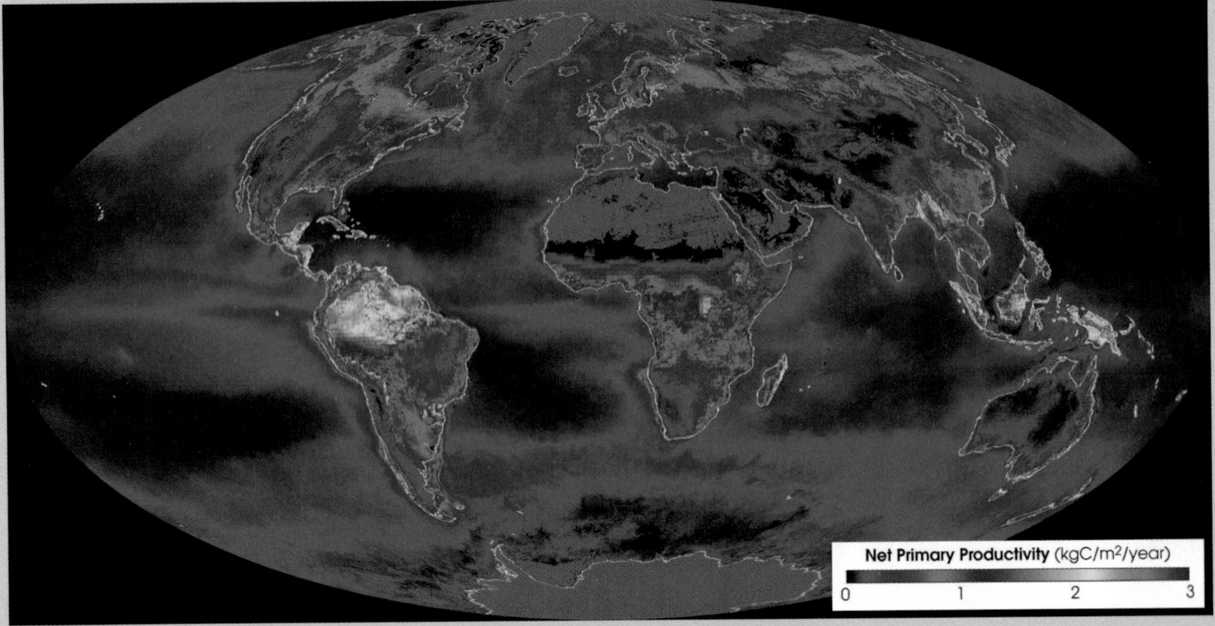

coasts where nutrient-rich deep water upwells from below, nutrients are in greater supply and oceans are more productive. As shown in Figure 22.8, upwelling is particularly important along west coasts—for example, the west coast of South America from Peru southward shows the dark and light blue tones that are more characteristic of land than ocean. Higher levels are also seen at high latitudes. In these regions, the thermocline (Chapter 7) that inhibits mixing of surface and deep waters is absent for much of the year, so nutrients are more abundant and net primary production is enhanced.

On land, the highest values of net primary production are shown by the yellow and red tones of the Amazon Basin and equatorial Asia. Lesser, but still high, values are encountered in eastern North America, central Africa, eastern Asia, and Scandinavia. Low values characterize arid and semiarid regions, such as the Kalahari Desert of southern Africa, the interior deserts of Australia, semiarid western North America, the Sahara, and the interior steppelands of Asia. (Gray areas show regions for which data are not available.)

Figure 4 shows net primary productivity for June 2002 *(a)* and December 2002 *(b)*. Comparing the two images, we see that ocean productivity patterns are much the same at low and midlatitudes, but quite different at high latitudes. Here, waters are significantly more productive in summer in each hemisphere. Over land, the June pattern shows the highest values of net primary productivity in the boreal and eastern continental zones of the northern hemisphere. These values are produced by the long days and warm temperatures of June. Productivity remains high in the Amazon and equatorial regions. In December, net primary production is high in South America and sub-Saharan Africa. Australia shows a significant green-up as equatorial Asia maintains high values. Meanwhile, most of northern North America and Eurasia are dormant.

Figure 4 Global net primary productivity with the seasons *(a) Net primary productivity for June 2002 as derived from MODIS. (b) Net primary productivity for December 2002. (MODIS Science Team/NASA.)*

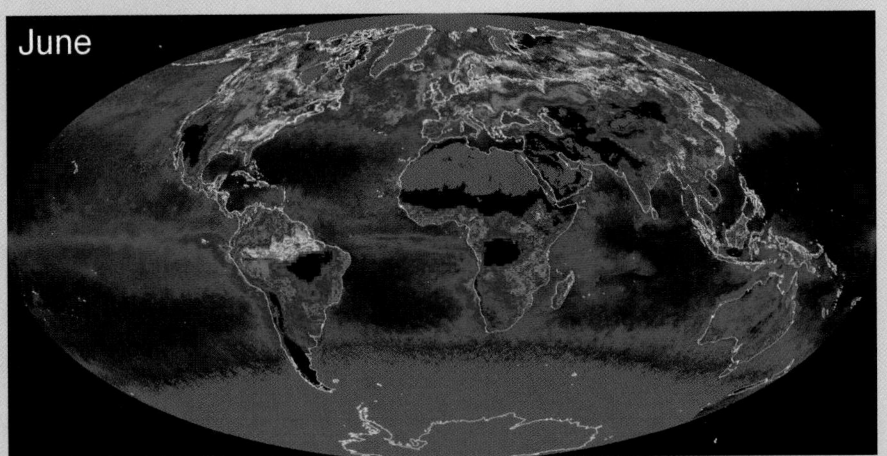

(a)

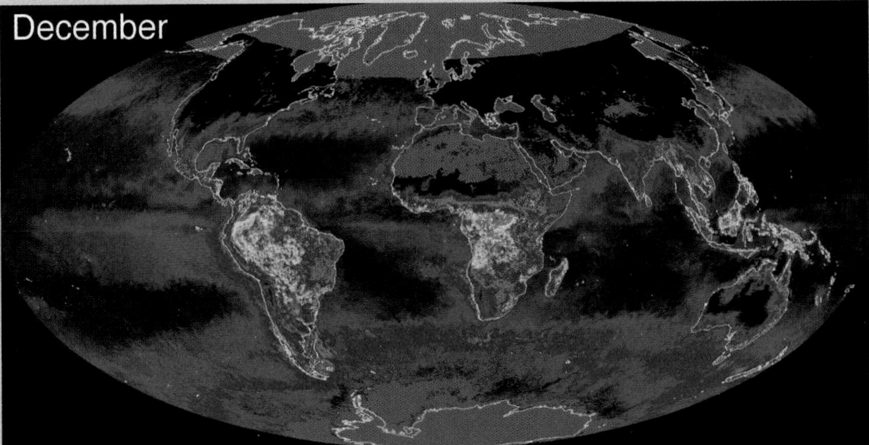

(b)

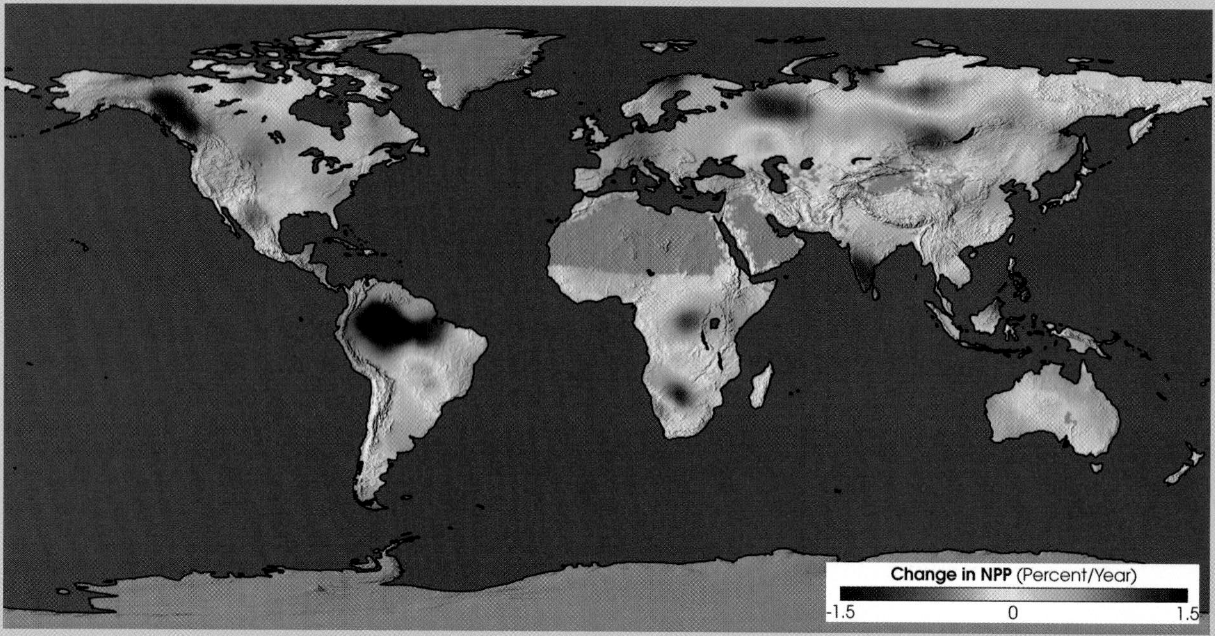

Figure 5 **Change in net primary productivity for land, 1982–1999** *This image, constructed from images of NOAA's AVHRR instruments acquired from 1982 to 1999, shows the change in net primary productivity in units of percent per year. (R. Nemani and S. W. Running, University of Montana NTSG/NASA.)*

Recent Changes in Global Productivity

In a recent study analyzing satellite images acquired by NOAA's Advanced Very High Resolution Radiometer (AVHRR) between 1982 and 1999, a team of scientists were able to map changes in land net primary productivity over the past two decades (Figure 5).[1] Overall, they showed that land productivity increased by about 6 percent during the two decades. As shown in the figure, the greatest increases occurred in equatorial and subtropical South America, Africa, and India. Large increases were also noted in northwestern North America and northern Russia, west of the Ural Mountains. By comparing these results with meteorological data, the scientists also showed that the increase at low latitudes was due to reduced cloud cover, which allowed more light for photosynthesis. At high latitudes, the causes were increased temperature, and, to some extent, increased water availability.

Over about the same time period, large changes in ocean phytoplankton concentration took place as well. Remote sensing scientists recently compared average chlorophyll concentrations during the months of July through September for the years 1979–1986 with years 1997–2000.[2] Although the data from the two periods were acquired by different satellite instruments, the researchers blended both sets of data with observations from ocean buoys and research vessels to make them comparable. Figure 6 shows an image of changes in chlorophyll concentration that they observed. Note that the color scale is logarithmic, with deeper tones of blue and red showing much larger changes than their lighter tones. The most striking feature the scientists observed was the decline in phytoplankton in northern oceans (blue), which amounted to about 30 percent in the North Pacific and 14 percent in the North Atlantic. In the equatorial zones, the researchers observed increases of up to 50 percent (red tones) at some locations, but the increases were not large enough to account for the high-latitude decreases. The scientists concluded that global concentrations of phytoplankton had decreased overall.

What could be the cause of such a decline? One possibility is that increasing sea-surface temperature

[1]Nemani, R. R., Keeling, C. D., Hashimoto, H., Jolly, W. M., Piper, S. C., Tucker, C. J., Myneni, R. B. and Running, S. W., 2003, Climate-driven increases in global terrestrial net primary production from 1982 to 1999. *Science* (June 6, 2003).

[2]Gregg, W. W. and Conkright, M. E., 2002, Decadal changes in global ocean chlorophyll, *Geophys. Res. Lett.*, 29 (15), doi:10.1029/2002GL014689, 2002; Gregg, W. W., Conkright, M. E., Ginoux, P., O'Reilly, J. E., and Casey, N. W., 2003, Ocean primary production and climate: Global decadal changes, *Geophys. Res. Lett.*, 30 (15), 1809, doi:10.1029/2003GL016889, 2003.

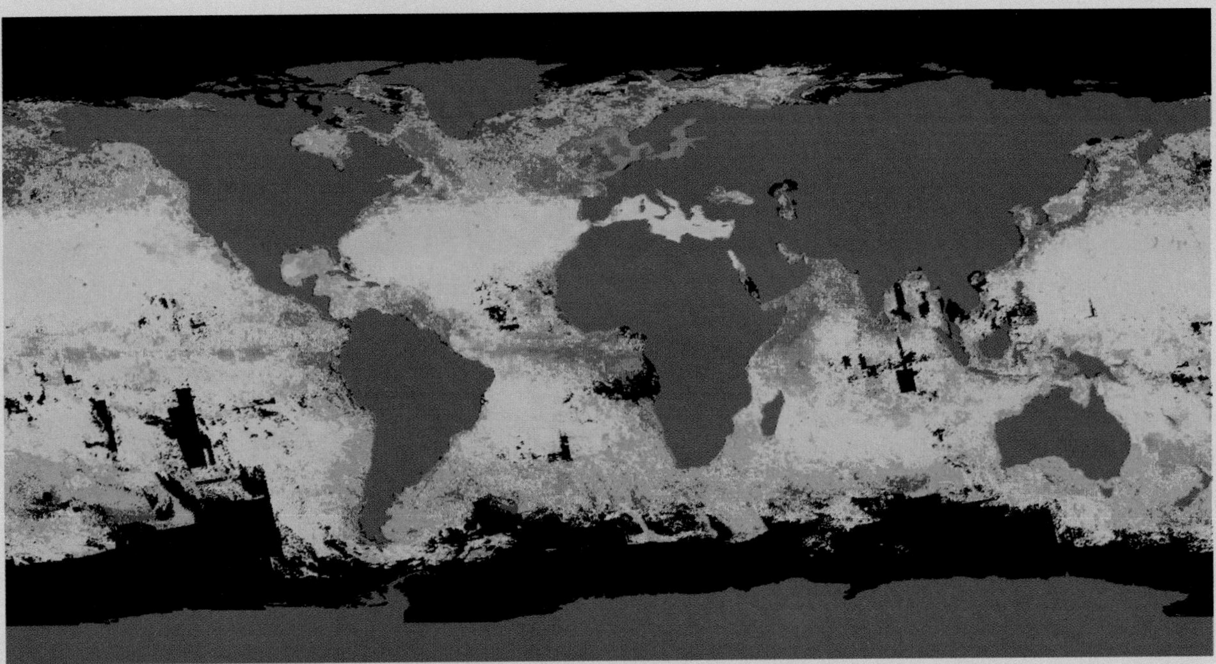

Figure 6 Change in ocean chlorophyll concentration from 1979–1986 to 1997–2000 *This image shows how summer ocean chlorophyll concentrations have changed over the past two decades. Data are for July–September as acquired by NOAA's Coastal Zone Color Scanner (1979–1986) and SeaWiFs instrument (1997–2000). (NASA/NOAA.)*

is increasing the duration and strength of the thermocline at high latitudes, which inhibits the mixing of nutrient-rich deep water with nutrient-poor surface water and thus keeps the phytoplankton population in a nutrient-limited condition. Another possibility is that wind speeds are decreasing, which will also reduce mixing. Actually, both of these changes have already been observed. Summer sea-surface temperatures in northern regions increased by 0.4°C (0.7°F) from the early 1980s to 2000, and average spring wind stresses on the sea surface have decreased by about 8 percent. However, it is not certain that these changes are the result of global warming rather than a multiyear ocean cycle yet to be discovered.

Implications for the Global Carbon Budget

What are the implications of these changes in global productivity for the global carbon budget? First, it would seem that the balance in global productivity between land and ocean is shifting toward land. Although this may increase the amount of terrestrial biomass and decrease the rate of CO_2 buildup, it also means that ocean productivity is probably declining at the cost of damage to oceanic ecosystems.

Second, it is important to keep in mind that net primary productivity isn't the whole story, at least on land. It doesn't include soil respiration, which could be very important. By soil respiration, we mean the decay of organic matter in soils, which releases CO_2 to the atmosphere. Soil respiration is quite temperature-dependent, and small increases in temperature can stimulate large increases in respiration. Considering that boreal forest soils are very rich in organic matter and that global temperatures are increasing most rapidly at high latitudes, increased release of CO_2 by soil respiration could well overwhelm increasing fixation due to higher temperatures.

Third, we don't know if the increase in land productivity will continue. For example, with more sunlight and higher temperatures, moisture may become limiting in equatorial and tropical forests, causing net primary productivity to plateau or even decrease. Moreover, the large changes in temperature forecast for high-latitude regions will ultimately lead to reduced productivity as boreal forests come under increasing stress and trees at their southern limits die.

All these uncertainties emphasize the importance of being able to map and monitor global productivity using remote sensing. While we may not yet be able to predict exactly how the carbon cycle will behave in the future, at least we can track the changes that are occurring. With time, that will allow us to refine carbon cycle models so that we will indeed come to understand the full impact of human activity on the global carbon cycle.

23 BIOGEOGRAPHIC PROCESSES

***Ashes of a tree in Bouna, Côte d'Ivoire.** A creeping brushfire in this semiarid region of west Africa has slowly consumed a fallen tree, leaving a striking pattern of white ash. In this region, traditional agricultural practice is to burn fields and brush during the dry season.*

In the preceding chapter, we viewed ecosystems from the perspective of energy and showed how the biosphere fixes sunlight to power global cycles of such elements as carbon, oxygen, and nitrogen. But ecosystems are composed of individual organisms that utilize and interact with their environment in different ways. From fungi digesting organic matter on a forest floor to ospreys fishing in a coastal estuary, each organism has a range of environmental conditions that limits its survival as well as a set of characteristic adaptations that it exploits to obtain the energy it needs to live.

In this chapter, we first explore ecological biogeography, which examines how relationships between organisms and environment help determine when and where organisms are found. This branch of biogeography explains the spatial pattern of organisms on local and regional scales and on the time scale of a few generations of each organism.

But the distribution patterns of organisms on these time and spatial scales are only half the story. To understand continental- and global-scale patterns over longer time periods, we will turn to the ideas of historical biogeography. This branch of biogeography describes such processes as evolution, dispersal, and extinction of species through time. When complete, our story will help you understand and appreciate the diversity of our biosphere, as well as how that diversity is threatened by human activity.

ECOLOGICAL BIOGEOGRAPHY

We begin our discussion with the relationship between organisms and their physical environment. As we travel through a hilly, wooded area, it is easy to see that the ecosystems we encounter are strongly influenced by landform and

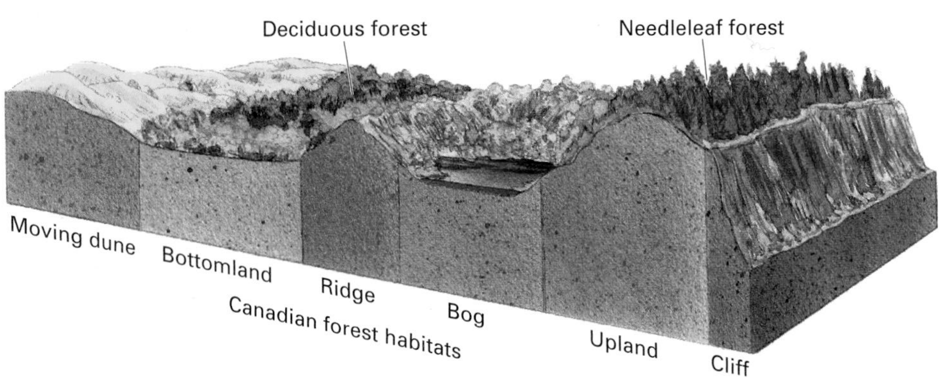

23.1 Habitats within the Canadian boreal forest *Habitats of the Canadian boreal forest are quite varied and include moving dune, bottomland, ridge, bog, and upland. (After P. Dansereau.)*

soil. For example, the upland forest of oak and hickory on the Blue Ridge Mountains of Virginia gives way to hemlock, birch, and maple in small valleys and low places. Upland soils are thick, stony, and well-drained, while the soils of the valleys and swales are finer, richer in organic matter, and wetter more of the time. Forest communities are also strikingly different in form on rocky ridges and on steep cliffs, where pines and scrubby oaks abound. Here water drains away rapidly, and soil is thin or largely absent. These variations illustrate the concept of the **habitat**—a subdivision of the environment according to the needs and preferences of organisms or groups of organisms. Figure 23.1 presents an example taken from a Canadian boreal forest. Here, there are six distinctive habitats: upland, bog, bottomland, ridge, cliff, and active sand dune. Each habitat supports a different type of ecosystem.

A concept related to the habitat, the *ecological niche,* includes the functional role played by an organism as well as the physical space it inhabits. If the habitat is the individual's "address," then the niche is its "profession," including how and where it obtains its energy and how it influences other species and the environment around it. Included in the ecological niche are the organism's tolerances and responses to changes in moisture, temperature, soil chemistry, illumination, and other factors. Although many different species may occupy the same habitat, only a few of these species will ever share the same ecological niche, for, as we'll see shortly, evolution will tend to separate them.

As we move from habitat to habitat, we find that each is the home of a group of organisms, each occupying different but interrelated ecological niches. We can define a *community* as an assemblage of organisms that live in a particular habitat and interact with one another. Although every organism must adjust to variations in the environment on its own, we find that similar habitats often contain similar assemblages of organisms. Biogeographers and ecologists recognize specific types of communities, called *associations,* in which typical organisms are likely to be found together. Sometimes these associations are defined by species, as in the beech-birch-maple forest that is found in the Great Lakes region and that stretches to New England. Other times they are defined more generally by the life-form of the vegetation cover, as in the boreal forest biome (sometimes facetiously termed the "spruce-moose" biome), which consists of the broad circumpolar band of coniferous forest found in the northern hemisphere and includes many similar and related species of plants, animals, and microbes. (We'll say more about biomes in Chapter 24.)

What physical environmental factors are most important in determining where organisms, as individuals and species, are found? In general, moisture and temperature are most important. Although organisms are sometimes present under conditions of extreme temperatures or dryness as spores or cysts, nearly all organisms have limits that are exceeded at least somewhere on Earth at some time. At the global scale, temperature and moisture patterns translate into climate. For this reason, as we will see in Chapter 24, there is a very strong relationship between climate and vegetation.

Other environmental factors that influence plant and animal distribution patterns and life cycles are light and wind. Light varies within communities, as, for example, between the top and bottom of a forest canopy, as well as with time, as in the changes in daylight length that occur with the seasons. Wind exposes plants to drying and can cause plants to be stunted on the windward side. Let's look at environmental factors in more detail.

Water Need

The availability of water to terrestrial organisms at a particular point in time or space is determined by the balances among precipitation, evaporation, runoff, and infiltration (Chapter 21). This balance is, in turn, affected by organisms—mainly the plant cover. Through transpiration, plants return much of the soil water to the atmosphere. By obstructing overland flow and increasing soil porosity, plants reduce runoff and increase infiltration. Burrowing animals enhance infiltration as well. Although these local flows of water are important within individual habitats, their effects are small compared to those of the physical processes that control the major features of the water cycle. Thus, the major pattern of variation in water from place to place is still determined by the overall dynamics of the atmosphere and oceans in the form of global climate.

Both plants and animals show a variety of adaptations that enable them to cope with the abundance or scarcity of water. In plants, many of these adaptations affect the transpiration mechanism. Evaporation at the leaf is controlled by specialized leaf pores, which provide openings in the outer layer of cells. When soil water is depleted, the pores close and evaporation is greatly reduced. Plants that are adapted to drought conditions are termed **xerophytes.** The word "xerophyte" comes from the Greek roots *xero-*, meaning "dry," and *phyton,* meaning "plant." Some xerophytes are adapted to habitats that dry quickly following rapid drainage of precipitation—for example, sand dunes, beaches, and bare rock surfaces. Others are adapted to habitats in which rainfall is simply scarce, such as deserts.

In some xerophytes, water loss is reduced by a thick layer of wax or wax-like material on leaves and stems. The wax helps to seal water vapor inside the leaf or stem. Still other xerophytes adapt to a desert environment by greatly reducing their leaf area or by bearing no leaves at all. Needlelike leaves, or spines in place of leaves, are also adaptations of plants to conserve water. In cactus plants, the foliage leaf is not present, and transpiration is limited to thickened, water-filled stems that store water for use during long, dry periods (Figure 23.2).

Adaptations of plants to water-scarce environments also include improved abilities to obtain and store water. Roots may extend deeply to reach soil moisture far from the surface. In cases where the roots reach to the ground water zone, a steady supply of water is assured. Plants drawing from ground water are termed *phreatophytes* and may be found along dry stream channels and valley floors in desert regions. In these environments, ground water is usually near the surface. Other desert plants produce a widespread, but shallow, root system. This system enables them to absorb water from short desert downpours that saturate only the uppermost soil layer. Leaves and stems of desert plants are commonly greatly thickened by a spongy tissue in which much water can be stored. Plants employing this adaptation are called *succulents* (see also Chapter 10).

Another adaptation to extreme aridity is a very short life cycle. Many small desert plants will germinate from seed, then leaf out, bear flowers, and produce seed in the few weeks immediately following a heavy rain shower. In this way, they complete their life cycle when soil moisture is available, and they survive the dry period as seeds that require no moisture.

Certain climates, such as the wet-dry tropical climate ③ and the moist continental climate ⑩, have a yearly cycle with one season in which water is unavailable to plants because of lack of precipitation or because the soil water is frozen. This season alternates with one in which there is abundant water. Plants adapted to such regimes are called *tropophytes,* from the Greek word *trophos,* meaning "change" or "turn." Tropohytes respond to this pattern by dropping their leaves at the close of the moist season and becoming dormant during the dry season. When water is again available, they leaf out and grow at a rapid rate. Trees and shrubs that shed their leaves seasonally are termed *deciduous,* while *evergreen* plants retain most of their leaves in a green state through one or more years.

23.2 Prickly pear cactus (Opuntia) *The prickly pear cactus is widely distributed in the western hemisphere. This clump of cactus is in the Sonoran desert, Arizona.*

The Mediterranean climate ⑦ also has a strong seasonal wet-dry alternation, with dry summers and wet winters. Plants in this climate are often xerophytic and characteristically have hard, thick, leathery leaves. An example is the live oak, which holds most of its leaves through the dry season (Figure 23.3). As we saw in Chapter 11, such hard-leaved evergreen trees and woody shrubs are called *sclerophylls.* (The prefix *scler-* is from the Greek root for "hard" and is combined with the Greek word for leaf, *phyllon.*) Plants that hold their leaves through a dry or cold season have the advantage of being able to resume photosynthesis immediately when growing conditions become favorable, whereas the deciduous plants must grow a new set of leaves.

To cope with water shortages, **xeric animals** have evolved methods that are somewhat similar to those used by the plants. Many of the invertebrates exhibit the same pattern as ephemeral annuals—evading the dry period in dormant states. When rain falls, they emerge to take advantage of the new and short-lived vegetation that often results. For example, many species of birds regulate their behavior to nest only when the rains occur, the time of most abundant food for their offspring. The tiny brine shrimp of the Great Basin may wait many years in dormancy until normally dry lake beds fill with water, an event that occurs perhaps three or four times a century. The shrimp then emerge and complete their life cycles before the lake evaporates.

Mammals are by nature poorly adapted to desert environments, but many survive through a variety of mechanisms that enable them to avoid water loss. Just as plants reduce transpiration to conserve water, so many desert mammals do not sweat through skin glands. Instead they rely on other methods of cooling, such as avoiding the Sun and becoming active only at night. In this respect, they are joined by most of the rest of the desert fauna, spending their days in cool burrows in the soil and their nights foraging for food (Figure 23.4).

Temperature

The temperature of the air and soil, the second of the important climatic factors in ecology, acts directly on organisms

23.3 Live oak *This California live oak is an example of a sclerophyll—a plant with thick, leathery leaves that is adapted to an environment with a very dry season.*

through its influence on the rates at which physiological processes take place in plant and animal tissues. In general, each plant species has an optimum temperature associated with each of its functions, such as photosynthesis (see Chapter 22), flowering, fruiting, or seed germination. There are also limiting lower and upper temperatures for these individual functions and for the total survival of the plant itself.

Temperature can also act indirectly on plants and animals. For example, higher air temperatures reduce the relative humidity of the air (Chapter 6), thus enhancing transpiration from plant leaves as well as increasing direct evaporation of soil water.

In general, the colder the climate, the fewer the species that are capable of surviving. A large number of tropical plant species cannot survive below-freezing temperatures for more than a few hours. In the severely cold arctic and alpine environments of high latitudes and high altitudes, only a few plant and animal species are found. This principle explains why an equatorial rainforest contains such a diverse array of plants and animals, whereas a forest of the subarctic zone will be dominated by only a few. Tolerance to cold in plants is closely tied to the physical disruption that accompanies the growth of ice crystals inside cells. Cold-tolerant plant species can expel excess water from cells to spaces between cells, where freezing does no damage.

23.4 Desert reptile *The noctural activity of the night snake,* Hypsiglena torquata, *is captured in this flash photo from Baja, Mexico.*

The effects of temperature variations on animals are moderated by their physiology and by their ability to seek sheltered environments. Most animals lack a physiological mechanism for internal temperature regulation. These animals, including reptiles, invertebrates, fish, and amphibians, are *cold-blooded animals*—their body temperature passively follows the environment. With a few exceptions (notably fish and some social insects), these animals are active only during the warmer parts of the year. They survive the cold weather of the midlatitude zone winter by becoming dormant. Some vertebrates enter a dormant state termed *hibernation* in which metabolic processes virtually stop and body temperatures closely parallel those of the surroundings. Most hibernators seek out burrows, nests, or other environments where winter temperatures do not reach extremes or fluctuate rapidly. Because the annual range of soil temperatures is greatly reduced below the uppermost layers, soil burrows are particularly suited to hibernation.

Other animals maintain their tissues at a constant temperature by internal metabolism. This group includes the birds and mammals. These *warm-blooded animals* possess a variety of adaptations to maintain a constant internal temperature. Fur, hair, and feathers act as insulation by trapping dead air spaces next to the skin surface, reducing heat loss to the surrounding air or water. A thick layer of fat will also provide excellent insulation (Figure 23.5). Other adaptations are for cooling—for example, sweating or panting uses the high latent heat of vaporization of water to remove heat. Heat loss is also facilitated by exposing blood-circulating tissues to the cooler surroundings. The seal's flippers and bird's feet serve this function.

***23.5 Alaskan brown bear** This brown bear, snacking on a salmon, is well-insulated against winter cold by a heavy coat and a layer of fat.*

Adaptations to temperature extremes can allow a species to exploit a potentially harsh environment. The caribou (Figure 23.6) is but one of several species of large grazing mammals that can endure very cold temperatures. During the mild season, caribou graze the far northern tundra, then migrate southward to pass the winter in a less hostile environment.

Other Climatic Factors

The factor of light is also important in determining local plant distribution patterns. Some plants are adapted to bright sunlight, while others require shade (Figure 23.7). The amount of light available to a plant will depend in large part on the plant's position. Tree crowns in the upper layer of a forest receive maximum light but correspondingly reduce the amount available to lower layers. In extreme cases, forest trees so effectively cut off light that the forest floor is almost free of shrubs and smaller plants. In certain deciduous forests of midlatitudes, the period of early spring, before the trees are in leaf, is one of high light intensity at ground level, permitting the smaller plants to go through a rapid growth cycle. In summer these plants will largely disappear as the leaf canopy is completed. Other low plants in the same habitat require shade and do not appear until the leaf canopy is well developed.

Treated on a global basis, the factor of light available for plant growth varies by latitude. Duration of daylight in summer increases rapidly with higher latitude and reaches its maximum poleward of the arctic and antarctic circles, where the Sun may be above the horizon for 24 hours (see Figure 22.6). Although the growing season for plants is greatly shortened at high latitudes by frost, the rate of plant growth in the short frost-free summer is greatly accelerated by the prolonged daylight.

***23.6** Eye on the Landscape **Grazing caribou** A herd of caribou grazes on tundra in the foothills of the Brooks Range, Alaska.* **What else would the geographer see? ... Answers at the end of the chapter.**

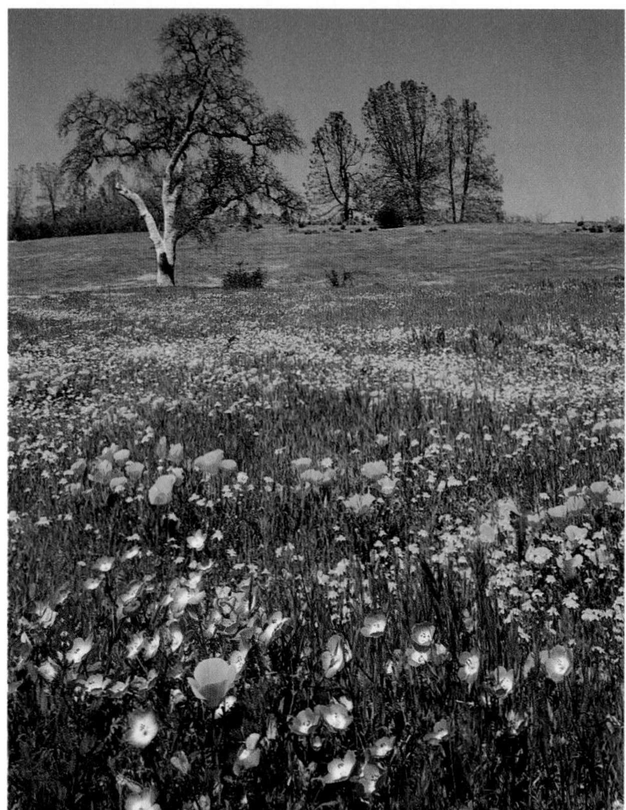

23.7 Sun-loving and shade-loving plants *Sun-loving spring flowers, including baby blue-eyes and California poppies, bloom in this open field near San Luis Opisbo, California (left). In contrast, the forest floor in this scene in the Great Smoky Mountains National Park is carpeted with shade-loving trillium flowers (right).*

In midlatitudes, where vegetation is of a deciduous type, the annual rhythm of increasing and decreasing periods of daylight determines the timing of budding, flowering, fruiting, leaf shedding, and other vegetation activities. As to the importance of light intensity itself, even on overcast days there is sufficient light to permit most plants to carry out photosynthesis at their maximum rates.

Light also influences animal behavior. The day–night cycle controls the activity patterns of many animals. Birds, for example, are generally active during the day, whereas small foraging mammals, such as weasels, skunks, and chipmunks, are more active at night. Light also controls seasonal activity through *photoperiod,* or daylight length, in midlatitudes. As autumn days grow shorter and shorter, squirrels and other rodents hoard food for the coming winter season. Later, increasing photoperiod will trigger such activities as mating and reproduction in the spring.

Wind is also an important environmental factor in the structure of vegetation in highly exposed positions. Close to timberline in high mountains and along the northern limits of tree growth in the arctic zone, trees are deformed by wind in such a way that the branches project from the lee side of the trunk only (flag shape) (see Figure 24.24). Some trees may show trunks and branches bent to near-horizontal attitude, facing away from the prevailing wind direction. The effect of wind is to cause excessive drying, damaging the exposed side of the plant. The tree limit on mountainsides thus varies in elevation with degree of exposure of the slope to strong prevailing winds and will extend higher on lee slopes and in sheltered pockets.

Bioclimatic Frontiers

Taken separately or together, climatic factors of moisture, temperature, light, and wind can act to limit the distribution of plant and animal species. Biogeographers recognize that there is a critical level of climatic stress beyond which a species cannot survive and that there will be a geographic boundary marking the limits of the potential distribution of a species. Such a boundary is sometimes referred to as a **bioclimatic frontier.** Although the frontier is usually marked by a complex of climatic elements, it is sometimes possible to single out one climatic element related to soil water or temperature that coincides with it.

The distribution of ponderosa pine *(Pinus ponderosa)* in western North America provides an example (Figure 23.8). In this mountainous region, annual rainfall varies sharply with elevation. The 50 cm (20 in.) isohyet (rainfall contour) of annual total precipitation encloses most of the upland areas having the yellow pine. It is the parallelism of the isohyet with forest boundary, rather than actual degree of coincidence, that is significant. The sugar maple *(Acer*

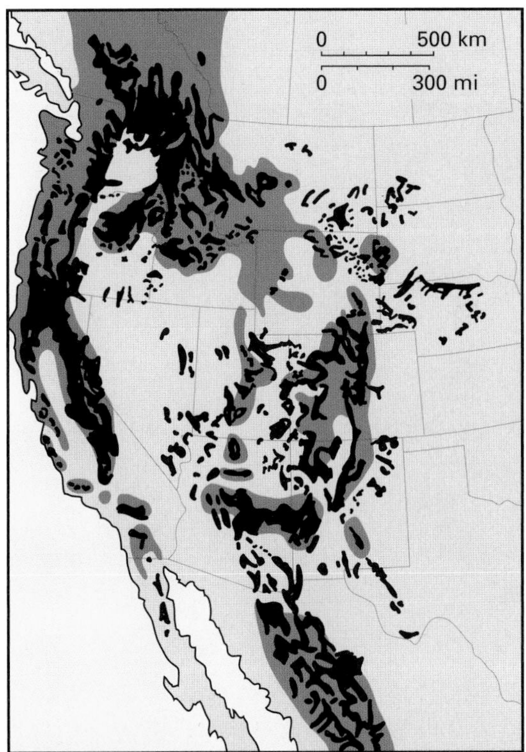

23.8 Distribution of ponderosa pine in western North America *Areas of ponderosa pine (Pinus ponderosa) are shown in black. The edge of the shaded area marks the rainfall contour (isohyet) of 50 cm (20 in.).*

saccharum) is a somewhat more complex case (Figure 23.9). Here the boundaries on the north, west, and south coincide roughly with selected values of annual precipitation, mean annual minimum temperature, and mean annual snowfall.

Although bioclimatic limits must exist for all species, no plant or animal need necessarily be found at its frontier. Many other factors may act to keep the spread of a species in check. A species may be limited by diseases or predators found in adjacent regions. In another example, a species (especially a plant species) may migrate slowly and may still be radiating outward from the location in which it evolved. (More details will be given later.) Or a species may be dependent on another species and therefore be limited by the latter's distribution.

Geomorphic Factors

Geomorphic factors (landform factors) influencing ecosystems include such elements as slope steepness (the angle that the ground surface makes with respect to the horizontal), slope aspect (the orientation of a sloping ground surface with respect to geographic north), and relief (the difference in elevation of divides and adjacent valley bottoms). In a much broader sense, geomorphic factors include the entire sculpturing of the landforms of a region by processes of erosion, transportation, and deposition via streams, waves, wind, and ice, and by forces of volcanism and mountain building. These are topics covered in detail in Chapters 14 through 20.

Slope steepness acts indirectly by influencing the rate at which precipitation is drained from the surface. On steep slopes, surface runoff is rapid, and soil water recharge by infiltration is reduced. On gentle slopes, much of the precipitation can penetrate the soil and be retained. More rapid erosion on steep slopes may result in thin soil, whereas that on gentler slopes is thicker. Slope aspect has a direct influence on plants by increasing or decreasing the exposure to sunlight and prevailing winds. Slopes facing the Sun have a warmer, drier environment than slopes that face away from the Sun and therefore lie in shade for much longer periods of the day. In midlatitudes, these slope-aspect contrasts may be so strong as to produce quite different biotic communities on north-facing and south-facing slopes (Figure 23.10).

Geomorphic factors are in part responsible for the dryness or wetness of the habitat within a region that has essentially the same overall climate. Each community has its own microclimate. On divides, peaks, and ridge crests, the soil tends to dryness because of rapid drainage and because the surfaces are more exposed to sunlight and drying winds. By contrast, the valley floors are wetter because surface runoff over the ground and into streams causes water to converge there. In humid climates, the ground water table in the val-

23.9 Bioclimatic limits of sugar maple *The shaded area shows the distribution of sugar maple (Acer saccharum) in eastern North America. Line 1, 75 cm (30 in.) annual precipitation. Line 2, –40°C (–40°F) mean annual minimum temperature. Line 3, eastern limit of yearly boundary between arid and humid climates. Line 4, 25 cm (10 in.) mean annual snowfall. Line 5, –10°C (16°F) mean annual minimum temperature.*

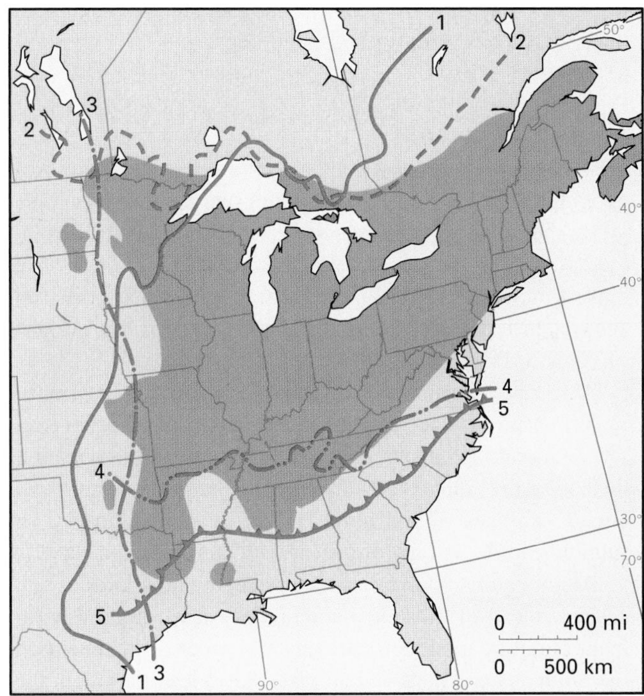

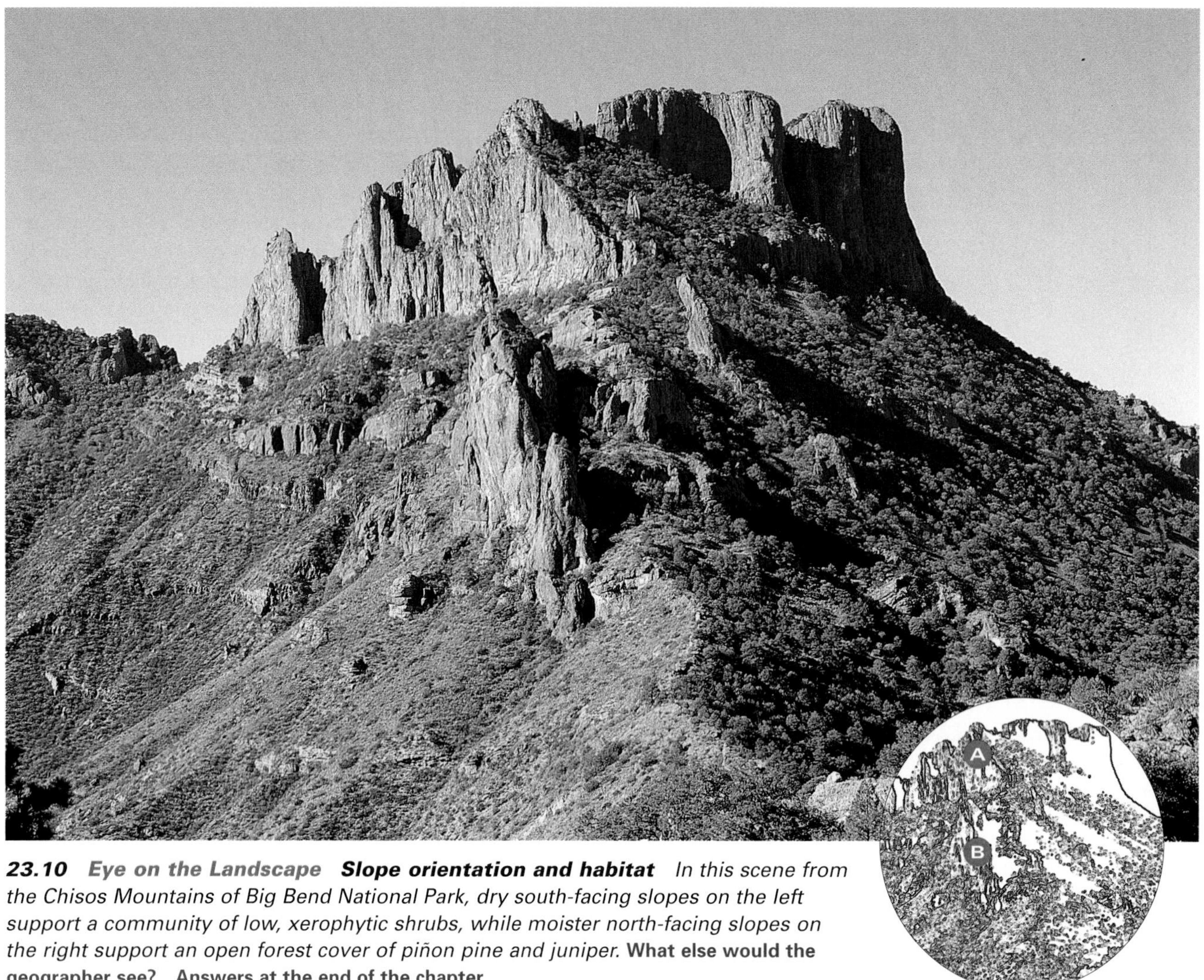

23.10 Eye on the Landscape ***Slope orientation and habitat*** *In this scene from the Chisos Mountains of Big Bend National Park, dry south-facing slopes on the left support a community of low, xerophytic shrubs, while moister north-facing slopes on the right support an open forest cover of piñon pine and juniper.* **What else would the geographer see?...Answers at the end of the chapter.**

ley floors may lie close to or at the ground surface to produce marshes, swamps, ponds, and bogs.

Edaphic Factors

Edaphic factors are those related to the soil. In terms of biogeography, we can look at soils from two perspectives. One views the broad patterns of soil distribution as controlled by the climatic regimes. Patterns of soils and climates are closely correlated with the global patterns of terrestrial ecosystem types. These broad relationships are the subject of our next chapter.

A second viewpoint is in terms of habitats—the small-scale mosaic of place-to-place variations of the Earth's surface. Among the edaphic factors important in differentiating the habitat are soil texture and structure, humus content, presence or absence of soil horizons, soil alkalinity, acidity, or salinity, and the activity of bacteria and animals in the soil.

Although this book treats the systematic principles of soil science ahead of those of natural ecosystems, a good argument might be made for reversing this order of treatment on the grounds that plants and animals play a leading role in the development of soil characteristics. Given a barren habitat recently formed by some geologic event, such as the outpouring of lava or the emergence of a coastal zone from beneath the sea, the gradual evolution of a soil profile goes hand in hand with the occupance of the habitat by a succession of biotic communities. The plants profoundly alter the soil by such processes as contributing organic matter or producing acids that act on the mineral matter. Animal life, feeding on the plant life, also makes its contribution to the physical and chemical processes of soil evolution.

Disturbance

Another environmental factor affecting ecosystems is *disturbance,* which includes fire, flood, volcanic eruption, storm waves, high winds, and other infrequent catastrophic events that damage or destroy ecosystems and modify habitats. Although disturbance can greatly alter the nature of an ecosystem, it is often part of a natural cycle of regeneration that provides opportunities for short-lived or specialized species to grow and reproduce. In this way, disturbance is often a natural process to which many ecosystems are adapted.

For example, *fire* is a phenomenon that strikes most forests sooner or later. In many cases, the fire is beneficial. It cleans out the understory and consumes dead and decaying organic matter while leaving most of the overstory trees untouched. On the forest floor, mineral soil is exposed and fertilized with new ash, providing a productive environment for dormant seeds. Sunlight is also abundant, with shrubs and forbs no longer shading the soil. Among tree species, pines are typically well adapted to germinating under such conditions. In fact, the jack pine of eastern North America and the lodgepole pine of the intermountain West have cones that remain tightly closed until the heat of a fire opens them, allowing the seeds to be released. These species are thus directly dependent on fire to maintain their geographic range and importance in the ecosystem. In the Rockies and boreal forests, there are many patches, large and small, of jack pine and lodgepole pine of different ages that document a long history of burns. These stands also serve as specialized habitats for particular insects, birds, and mammals.

Fires are also important to the preservation of grasslands. Grasses have extensive root systems below ground and germinal buds located at or just below the surface, making them quite fire-resistant. However, woody plants are not so resistant and are usually killed by grassfires. Most grasslands are dependent to some degree on fire for their maintenance. In Mediterranean climates, chaparral vegetation is also adapted to regular burning.

In many regions, active fire suppression has reduced the frequency of burning to well below natural levels. In forests, this causes dead wood to build up on the forest floor. When a fire does start, it burns hotter and more rapidly and consumes the crowns of many overstory trees. Once beneficial, fire is now destructive, leaving only a charred landscape of blackened soil at the mercy of erosion. In most cases, the forest comes back to health, but it may take many years. *Eye on the Environment 23.1 • The Great Yellowstone Fire* describes the fires that swept the forests of Yellowstone National Park in 1988 and explains the role of fire in the dynamics of the western forest ecosystem.

Fires, both natural and human-induced, have effects well beyond the area burned. Smoke released from fires is a major source of air pollution in areas where there is a dry season of frequent fires. Gases released by combustion, including sulfur and nitrogen oxides, enhance the greenhouse effect and thus add to global climate warming. *Focus on Remote Sensing 23.2 • Remote Sensing of Fires* shows how remote sensing is used to monitor the global occurrence of fires.

Although fire is an obvious and frequent type of disturbance, other types are also important. Flooding, in addition to displacing animal communities, deprives plant roots of oxygen. Where flooding brings a swift current, mechanical damage rips limbs from trees and scours out roots. Frequent flooding, for example, along the banks of major rivers, often limits the vegetation cover to species that are resistant to such effects.

High winds can blow down large areas of forest, uprooting trees and setting up conditions for large fires after a few years of decay of the fallen trees. High winds also bring destructive waves to coral reefs, as well as sand and scouring to bays and coastal marshes. At a fine spatial scale, disturbance includes the fall of individual trees or large limbs within forests, creating light openings for shaded species in the understory to fill. Even animal wallows can provide unique habitats of disturbance.

Interactions Among Species

Species interactions can also be important factors in determining the distribution patterns of plants and animals. Two species that are part of the same ecosystem can interact with one another in three ways: interaction may be negative to one or both species; or the two species may be neutral, not affecting each other; or interaction may be positive, benefiting at least one of the species.

Competition between species, a negative interaction, occurs whenever two species require a common resource that is in short supply. Because neither species has full use of the resource, both populations suffer, showing growth rates lower than those when only one of the species is present. Sometimes one species will win the competition and crowd out its competitor. At other times, the two species may remain in competition indefinitely.

Competition between species is an unstable situation. If a genetic strain within one of the populations emerges that uses a substitute resource for which there is no competition, its survival rate will be higher than that of the remaining strain, which still competes. The original strain may become extinct. In this way, evolutionary mechanisms tend to reduce competition among species.

Predation and parasitism are negative interactions between species. *Predation* occurs when one species feeds on another (Figure 23.11). The benefits are obviously positive to the predator species, which obtains energy for survival, and negative to the prey species. *Parasitism* occurs when one species gains nutrition from another, typically when the parasite organism invades or attaches to the body of the host in some way.

Although we tend to think of predation and parasitism as negative processes that benefit one species at the expense of the other, it may well be that these interactions are really beneficial in the long term to the host or prey populations. A classic example is the growth of the deer herd on the Kaibab Plateau north of the Grand Canyon in Arizona (Figure 23.12). Initially at a population of about 4000, in an area of 283,000 hectares (700,000 acres), the herd grew to nearly 100,000 in the short span of 1907–1924 in direct proportion to a government predator control and game protection program. Wolves became extinct in the area, and populations of coyotes and mountain lions were greatly reduced. The huge deer population, however, proved too much for the land, and overgrazing led to a population crash. In one year, half the animals starved to death; by the late 1930s, the population had declined to a stable level near 10,000. Thus, predation maintained the deer population at levels that were in harmony with the supportive ability of the environment. In addition to maintenance of equilibrium population levels, predation and parasitism differen-

Eye on the Environment | 23.1

The Great Yellowstone Fire

Yellowstone is the largest and oldest American national park. Thanks in part to the magnificent photographs of its wondrous geysers and travertine terraces, taken by the pioneer photographer, William Henry Jackson, and brought back to Congress in Washington, it was declared a park in 1872. Almost a square in outline, its 900,000 hectares (2.2 million acres) occupy the northwest corner of Wyoming within the Rocky Mountain region. Volcanic rocks dominate the geology, and lava flows occupy much of the area, forming a high plateau. Because an ancient caldera underwent collapse here, geothermal activity continues to this day in many localities in the park. (See Chapter 14 and Figures 14.8 and 14.9.) Its rich, well-watered forests and parks support populations of bison, elk, and grizzly bear, along with many other mammal and bird species—a powerful attraction for millions of tourists. All in all, we have here a priceless forest ecosystem, little disturbed by natural catastrophe or human interference for at least the past two centuries.

Through August and September of 1988, many forest fires burned out of control in Yellowstone Park, casting a smoke pall over large areas of surrounding states. Of the 45 fires, most were started by lightning strikes, and this number was not unusual for the area. It had, however, been the driest summer for more than a century of record, and temperatures were persistently high. Once started, the fires spread rapidly and uncontrollably under the driving force of strong winds that in one fire reached 160 km (100 mi) per hour. Firefighters were powerless to check the spread of the flames. Dense forests of lodgepole pine that had not burned for 250 years were killed by flames that raced through the forest canopy. Fires at ground level spread rapidly through areas of older forest with ample fuel supplies accumulated as fallen trees and branches. The fires also consumed patches of young forest that rarely ignite and have typically served well as natural firebreaks.

Contrary to greatly exaggerated stories in the news media, a later study of air photographs showed that only 20 percent of the park land had been burned over in some degree. Fears of massive extermination of wildlife were also unfounded. Most of the large animals survived: only 9 bison of the herd of 2500 and 350 elk of the herd of 30,000 perished in the fires. Mammals large and small, along with insects and birds, returned quickly to the burned areas, with the herbivores taking advantage of grasses that grew vigorously among the ashes of the forest floor.

For a full century since the establishment of Yellowstone Park in 1872, the National Park Service had practiced its "no-burn" policy of not allowing any fire to spread unchecked. This policy was based on the concept that national parks must be maintained in a pristine condition, their plants being permitted to grow and die unaffected by any human activity or by fires. In 1972, this practice was reversed to a "let-burn" policy under which naturally occurring fires were to be allowed to spread unchecked. As expected, the Park Service immediately came under strong criticism for having practiced their let-burn policy by standing aside and allowing the July blazes to spread to mammoth proportions. By late July, park officials decided to suppress all fires, but a massive effort with national support was to no avail. In retrospect, however, it is question-

23.11 Predation *This giant anteater enjoys a lunch of Brazilian termites.*

Yellowstone fire and recovery *The fires that ravaged Yellowstone National Park in 1988 completely destroyed many stands of trees (left). After 10 years, however, natural regeneration had started new forests to take their place (right).*

able whether immediate action to put out the first fires could have made much difference over the vast area of the park. Another carefully drawn conclusion following study of the fire by forest ecologists was that the additional fuel accumulated in the 100-year "no-burn" period made only a minor difference in the extent and intensity of the 1988 fire.

Ecologists view the occurrence of a season of great fires in Yellowstone Park about every 250 to 300 years as part of a natural cycle of fire-driven ecological succession. Following such a fire, saplings of lodgepole pine occupy the burned area, growing tall and closely set to form a dense crown canopy, enduring between 150 and 250 years after the initial fire. During this phase, large canopy burns can occur, restarting the succession. If no fire occurs, the succession continues with a dying out of the pines and the growth of a new generation of other tree species, such as the spruce and fir. By this time—more than 300 years from the start—the forest floor becomes highly flammable and can be readily swept by fire to restart the succession.

Recent studies have established that the natural succession described above affects isolated patches of the entire forested region, resulting in a mosaic of patches in different stages of succession. In this mosaic, the younger patches serve as natural firebreaks to limit the spread of a fire occurring in adjacent older patches. Wildlife is thus able to escape the fires and then reinhabits the burned area, where a new ground cover of grasses and shrubs provides food for the grazing herbivores. This natural process of fire and renewal is important to the maintenance of other types of ecosystems as well, including the tall-grass prairies of Kansas, the pine forests of the southeastern United States, and much of the Canadian boreal forest.

23.12 Rise and fall of the Kaibab deer herd *This graph plots the population size of the deer herd in the Kaibab National Forest, Arizona, as it rose in response to the abolition of predators and then crashed when the population ran out of food. (After D. I. Rasmussen, Ecological Monographs, vol. 11, 1941, p. 237.)*

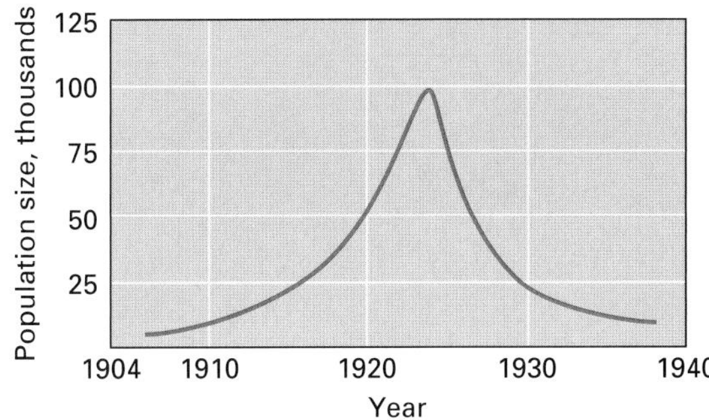

tially remove the weaker individuals and can improve the genetic composition of the species.

A third type of predation is *herbivory,* in which grazing of plants by animals reduces the viability of the plant species population. Some species are well-adapted to grazing and can maintain themselves well in the face of grazing pressure. Others may be quite sensitive to this process. When overgrazing occurs, these differing sensitivities can produce significant changes in the structure and composition of plant communities.

A fourth type of negative interaction is *allelopathy,* a phenomenon of the plant kingdom in which chemical toxins produced by one species serve to inhibit the growth of others. As an example, sage, a common shrub species in the California chaparral, produces leaves rich in volatile toxins (cineole and camphor). As the leaves fall and accumulate in the soil, the allelopathic toxins build to a level sufficient to

Focus on Remote Sensing | 23.2

Remote Sensing of Fires

Wildfires occur frequently on the Earth's land surface, and biomass burning has important effects on both local and global ecosystems. Biomass burns inefficiently, releasing not only carbon dioxide and water, but also a number of other greenhouse gases that absorb outgoing longwave radiation and enhance the greenhouse effect. Aerosols are another by-product of inefficient combustion that can effect atmospheric processes. Burning mobilizes such nutrients as nitrogen, phosphorus, and sulfur in ash so that they become available for a new generation of plants, but it also carries them upward in smoke and as gases. Fire affects ecosystems by changing species composition and creating a patchy structure of diversity on the landscape. It can also stimulate runoff and soil erosion where a significant layer of vegetation is lost to fire. For these reasons and more, monitoring of fires by remote sensing is a topic of intense interest to global change researchers.

Fires can be remotely sensed in several ways. Thermal imagers detect active fires as bright spots because they emit more heat energy than normal surfaces. However, fires may be obscured under clouds. Smoke plumes can also identify the location of fires, but are hard to distinguish from clouds in some images. Fire scars can also be detected, especially using mid-infrared bands.

Image (*a*) shows the Earth as imaged by MODIS in true color on August 23, 2000. Superimposed on the true color browse image are red squares indicating high temperatures as sensed by MODIS thermal bands. The inset images highlight fires in the Amazon basin, southern Africa, and Australia. Image (*b*) is a MODIS image of bush fires in northern Australia on October 2, 2000. Inset into the Gulf of Carpentaria is a close-up of one of the larger ones. The red outlines identify pixels where fire has been detected. The dark burn scar and smoke plumes are readily visible.

Images (*c*) and (*d*) show Los Alamos, New Mexico, from Landsat-7 before and after the Cerro Grande Fire of May, 2000. Over 20,000 people were forced to flee their homes as this disastrous fire progressed, and more than 200 homes were consumed by the flames. The images are composites of mid-infrared, near-infrared, and green Landsat bands assigned colors of red, green, and blue, respec-

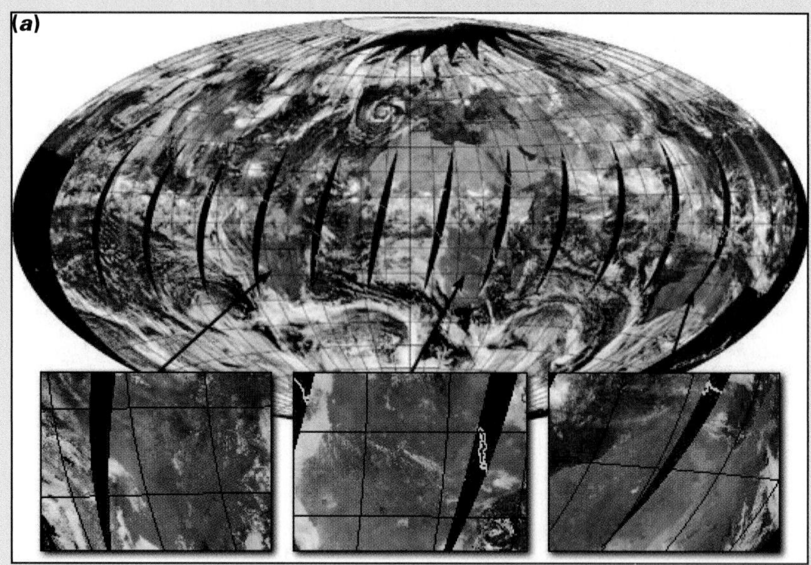

MODIS fire browse image *Global MODIS coverage for August 23, 2000, displayed in true color. (MODIS Land Team/C. Justice/L. Giglio/J. Descloitres/NASA.)*

Australian bush fires from MODIS *Fires in north-central Australia, shown in true color. Red outlines identify pixels flagged as burning by thermal bands. (MODIS Land Team/C. Justice/L. Giglio/J. Descloitres/NASA.)*

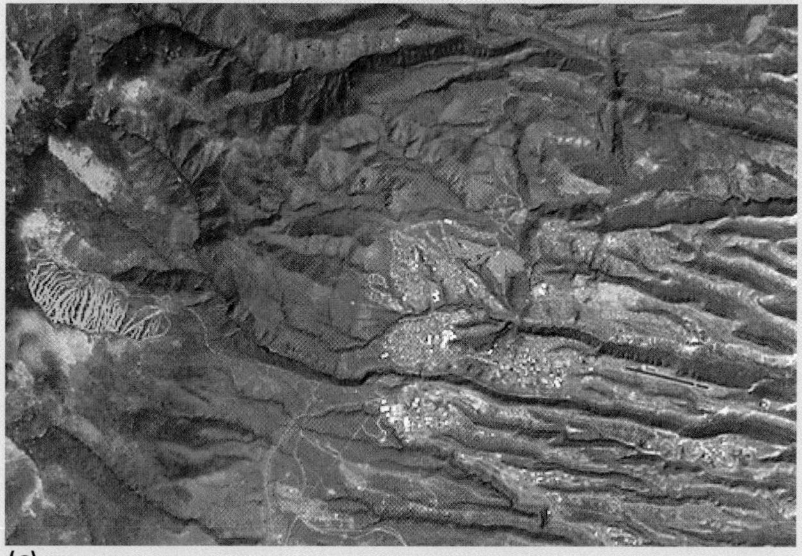
(*c*)

Los Alamos on April 14, 2000 *(Robert Simmon, NASA GSFC.)*

tively. Green colors show vegetation, with the brightest green identifying the dense grass of golf courses. Blue and purplish tones are developed areas. The strong red color in the June image identifies the burned area, which is very dark in green and near-infrared bands, but bright in the mid-infrared. The blue stripes on the left edge of the April image (*c*) are ski runs covered with snow. They appear bright green due to their grass cover in June.

Like MODIS, the SeaWiFS instrument produces a wide view of the Earth's surface that can detect the smoke from fires. Image (*e*) is a true color image of the southeastern United States acquired by the instrument from a track over west Texas on February 5, 2001. The view is to the east, along the Gulf coast from Texas to Florida. Dozens of smoke plumes are clearly visible. The Mississippi River valley crosses the center of the image from left to right as an area of light-colored fallow soils. Louisiana's Mississippi Delta is an obvious coastal feature, with a plume of fresh sediment extending out into the Gulf of Mexico.

(*d*)

Los Alamos on June 17, 2000, after the fire *(Robert Simmon, NASA GSFC.)*

(*e*)

Southeastern U.S. from SeaWiFS *This oblique view of the Gulf Coast from the SeaWiFS imager aboard the OrbView-2 satellite shows numerous smoke plumes. (Image courtesy SeaWiFS Project, NASA/Goddard Space Flight Center, and ORBIMAGE.)*

inhibit the growth of herbaceous plants, such as grasses, which are thus only found in adjacent areas. Still other chaparral shrubs produce water-soluble antibiotics that also inhibit the growth of nearby grasses. These chemical defenses, however, are broken down by periodic fires, which are events essential to maintenance of the chaparral ecosystem. The fires destroy the toxins and also trigger the germination of seeds of many species of annual herbaceous plants by breaking their seed coats. The annuals then dominate the area until the shrubs grow and force them out by allelopathy, beginning the cycle anew.

The term **symbiosis** includes three types of positive interactions between species: commensalism, protocooperation, and mutualism. In commensalism, one of the species is benefited and the other is unaffected. Examples of commensals include the epiphytic plants—such as orchids or Spanish moss—which live on the branches of larger plants (Figure 24.4*c*). These epiphytes depend on their hosts for physical support only. In the animal kingdom, small commensal crabs or fishes seek shelter in the burrows of sea worms; or the commensal remora fish attaches itself to a shark, feeding on bits of leftover food as its host dines.

When the relationship benefits both parties but is not essential to their existence, it is termed protocooperation. The attachment of a stinging coelenterate to a crab is an example of protocooperation. The crab gains camouflage and an additional measure of defense, while the coelenterate eats bits of stray food which the crab misses.

Where protocooperation has progressed to the point that one or both species cannot survive alone, the result is mutualism. A classic example is the association of the nitrogen-fixing bacterium *Rhizobium* with the root tissue of certain types of plants (legumes), in which the action of the bacteria converts nitrogen gas to a form directly usable by the plant. The association is mutualistic because *Rhizobium* cannot survive alone.

Ecological Succession

The phenomenon of change in plant and animal communities through time is a familiar one. A drive in the country reveals patches of vegetation in many stages of development—from open, cultivated fields through grassy shrublands to forests. Clear lakes, gradually filled with sediment from the rivers that drain into them, become bogs. These kinds of changes, in which biotic communities succeed one another, are referred to as **ecological succession.** GEODISCOVERIES

In general, succession leads to formation of the most complex community of organisms possible in an area, given its physical controlling factors of climate, soil, and water. The series of communities that follow one another on the way to the stable stage is called a *sere.* Each of the temporary communities is referred to as a *seral stage.* The stable community, which is the end point of succession, is the **climax.** If succession begins on a newly constructed deposit of mineral sediment, it is termed *primary succession.* If succession occurs on a previously vegetated area that has been recently disturbed by such agents as fire, flood, windstorm, or humans, it is referred to as *secondary succession.*

A new site on which primary succession occurs may have one of several origins—for example, a sand dune, a sand beach, the surface of a new lava flow or freshly fallen layer of volcanic ash, or the deposits of silt on the inside of a river bend that is gradually shifting. Such a site will not have a true soil with horizons. Rather, it may perhaps be little more than a deposit of coarse mineral fragments (Entisol). In other cases, such as that of floodplain silt deposits, the surface layer may represent redeposited soil endowed with substantial amounts of organic matter and nutrients.

The first stage of a succession is a *pioneer stage.* It includes a few plant and animal species unusually well adapted to adverse conditions of rapid water drainage and drying of soil and to excessive exposure to sunlight, wind, and extreme ground and lower-air temperatures. As pioneer plants grow, their roots penetrate the soil, and their subsequent decay adds humus to the soil. Fallen leaves and stems add an organic layer to the ground surface. Bacteria and animals begin to live in the soil in large numbers. Grazing mammals feed on the small plants. Birds forage the newly vegetated area for seeds and grubs.

Soon conditions are favorable for other species that invade the area and displace the pioneers. The new arrivals may be larger plant forms providing more extensive cover of foliage over the ground. In this case the climate near the ground is considerably altered toward one of less extreme air and soil temperatures, high humidities, and less intense insolation. Still other species now invade and thrive in the modified environment. When the succession has finally run its course, a climax community of plant and animal species in a more or less stable composition will exist.

The colonization of a sand dune provides an example of primary succession. Growing foredunes bordering the ocean or lake shore present a sterile habitat. The dune sand—usually largely quartz, feldspar, and other common rock-forming minerals—lacks such important nutrients as nitrogen, calcium, and phosphorus, and its water-holding ability is very low. Under the intense solar radiation of the day, the dune surface is a hot, drying environment. At night, radiation cooling in the absence of moisture produces low surface temperatures.

One of the first pioneers of this extreme environment is beachgrass (Figure 23.13, left). This plant reproduces by sending out rhizomes (creeping underground stems), and the plant thus slowly spreads over the dune. Beachgrass is well adapted to the eolian environment; it does not die when buried by moving sand, but instead puts up shoots to reach the new surface.

After colonization, the shoots of beachgrass act to form a baffle that suppresses movement of sand, and thus the dune becomes more stable. With increasing stabilization, plants that are more adapted to the dry, extreme environment but cannot withstand much burial begin to colonize the dune. Typically, these are low, matlike woody shrubs, such as beach wormwood or false heather.

On older beach and dune ridges of the central Atlantic coastal plain, the species that follow matlike shrubs are typi-

23.13 Dune succession at Sandy Hook, New Jersey *(left) Beachgrass is a pioneer on beach dunes and helps stabilize the dune against wind erosion. (right) The climax forest on dunes. Holly (Ilex opaca), seen on the left, is an important constituent of the climax forest. Note the leaves and decaying organic matter on the forest floor. (Alan Strahler.)*

cally larger woody plants and such trees as beach plum, bayberry, poison ivy, and choke cherry. These species all have one thing in common—their fruits are berries that are eaten by birds. The seeds from the berries are excreted as the birds forage among the low dune shrubs, thereby sowing the next stage of succession. As the scrubby bushes and small trees spread, they shade out the mat-like shrubs and any remaining beachgrass. Pines may also enter at this stage.

At this point, the soil begins to accumulate a significant amount of organic matter. No longer dry and sterile, it now possesses organic compounds and nutrients, and it has accumulated enough colloids to hold water for longer intervals. These soil conditions encourage the growth of such broadleaf species as red maple, hackberry, holly, and oaks, which shade out the existing shrubs and small trees (Figure 23.13, right). Once the forest is established, it tends to reproduce itself; the species of which it is composed are tolerant to shade, and their seeds can germinate on the organic forest floor. Thus, the climax is reached. The stages through which the ecosystem has developed constitute the sere, progressing from beachgrass to low shrubs to higher shrubs and small trees to forest.

Although this example has stressed the changes in plant cover, animal species are also changing as succession proceeds. Table 23.1 shows how some typical invertebrates appear and disappear through succession on the Lake Michigan dunes. Note that the seral stages shown in the table for these inland dunes are somewhat different from those described for the coastal environment.

Where disturbance alters an existing community, secondary succession can occur. *Old-field succession,* taking place on abandoned farmland, is a good example of secondary succession (Figure 23.14). In the eastern United States, the first stages of the sere often depend on the last use of the land before abandonment. If row crops were last cultivated, one set of pioneers, usually annuals and biennials, will appear. If small grain crops were cultivated, the pioneers are often perennial herbs and grasses. If pasture is abandoned, those pioneers that were not grazed will have a head start. Where mineral soil was freshly exposed by plowing, pines are often important following the first stages of succession because pine seeds favor disturbed soil and strong sunlight for germination. Although slower-growing than other pioneers, the pines will eventually shade the others out and become dominant. Their dominance is only temporary, however, because their seeds cannot germinate in shade and litter on the forest floor. Seeds of hardwoods such as maples and oaks can germinate under these conditions, and as the pines die, hardwood seedlings grow quickly to fill the holes produced in the canopy. The climax, then, is the hardwood forest. Figure 23.15 is a schematic diagram showing one example of old-field succession.

Succession, Change, and Equilibrium

The successional changes we have described result from the actions of the plants and animals themselves. One set of inhabitants paves the way for the next. As long as nearby populations of species provide colonizers, the changes lead automatically from bare soil or fallow field to climax forest. This type of succession is often termed *autogenic* (self-producing) *succession.*

In many cases, however, autogenic succession does not run its full course. Environmental disturbances, such as wind, fire, flood, or renewed clearing for agriculture, may occur, diverting the course of succession temporarily or even permanently. For example, autogenic succession on seaside dunes may proceed for some years, but sooner or later a severe storm will generate waves and wind that reduce the developing community to a sand bar once again. Or a mature forest may be destroyed by fire, inviting the succession of plants and animals that are specifically adapted to burned environments. (See *Eye on the Environment 23.1 • The Great Yellowstone Fire.*) In addition, habitat conditions such as site exposure, unusual bedrock, or impeded drainage can

23.14 Old-Field Succession

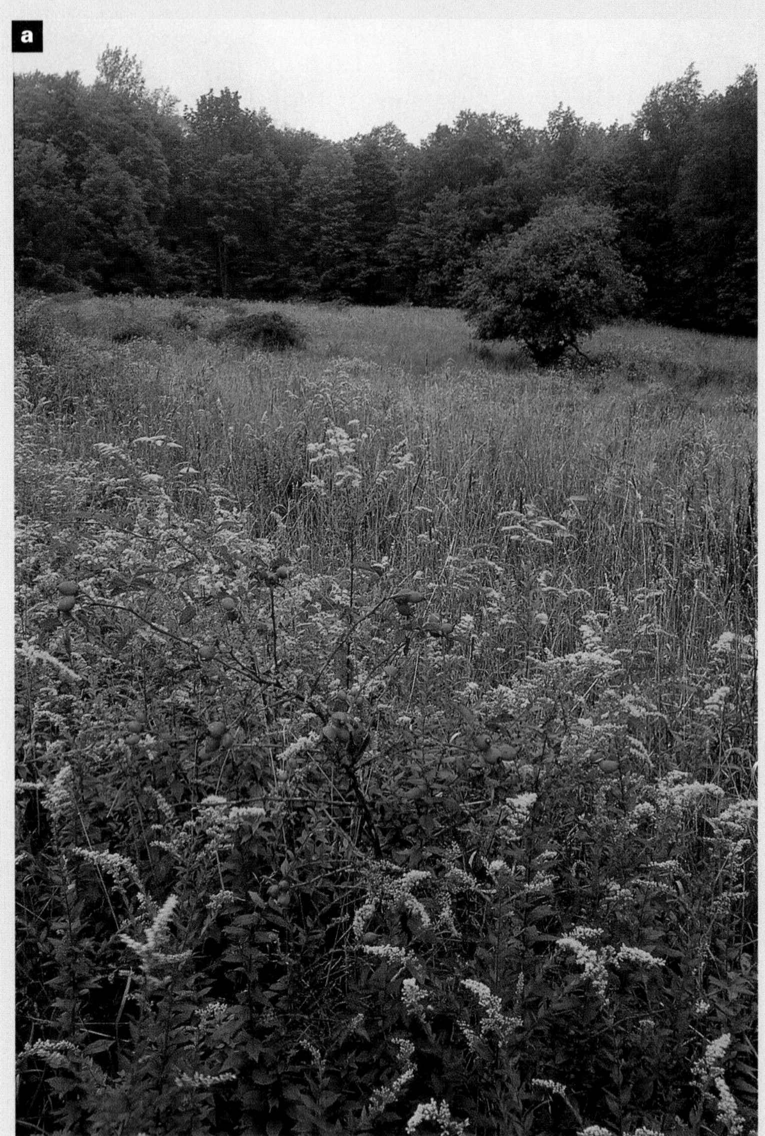

a...*Early stage* *When cultivation ceases, grasses and forbs colonize the bare soil. Invading tree and shrub species begin acquiring a foothold. This scene is about 10 years or more after abandonment, Pocono Mountains, Pennsylvania.*

b...*Woody pioneers* *In this scene from northeastern Pennsylvania, red cedar, locust, and pine are now established. About 20 years or more after abandonment.*

c...***Advance of the forest*** *A decade or two later, the cedars are larger and broadleaf deciduous species (wild cherry, red maple) are becoming more prominent. Delaware Water Gap National Recreation Area, Pennsylvania.*

d...***Mature second growth forest*** *After several more decades, the deciduous hardwoods shade out the conifers. Maple and birch, which are well adapted to a dense forest environment, increase, and the forest canopy becomes nearly closed. The old stone fence marks the boundary between former fields. Pennsylvania.*

Table 23.1 | Invertebrate Succession on the Lake Michigan Dunes

	Successional Stages				
Invertebrate	**Beachgrass–Cottonwood**	**Jack Pine Forest**	**Black Oak Dry Forest**	**Oak and Oak–Hickory Moist Forest**	**Beech–Maple Forest Climax**
White tiger beetle	x				
Sand spider	x				
Long-horn grasshopper	x	x			
Burrowing spider	x	x			
Bronze tiger beetle		x			
Migratory locust		x			
Ant lion			x		
Flatbug			x		
Wireworms			x	x	x
Snail			x	x	x
Green tiger beetle				x	x
Camel cricket				x	x
Sowbugs				x	x
Earthworms				x	x
Woodroaches				x	x
Grouse locust					x

Source: V. E. Shelford, as presented in E. P. Odum, *Fundamentals of Ecology* (Philadelphia: W. B Saunders Co., p. 259).

hold back or divert the course of succession so successfully that the climax is never reached.

Introduction of a new species can also greatly alter existing ecosystems and successional pathways. The parasitic chestnut blight, introduced from Europe to New York in about 1910, decimated populations of the American chestnut tree within a period of about 40 years. This tree species, which may have accounted for as many as one-fourth of the mature trees in eastern forests, is now found only as small blighted stems sprouting from old root systems.

These examples show that while succession is a reasonable model to explain many of the changes that we see in ecosystems with time, it may be more realistic to view the pattern of ecosystems on the landscape as reflecting a spatial dynamic equilibrium between autogenic forces of self-induced change and external forces of disturbance that reverse or rechannel autogenic change temporarily or permanently. From this viewpoint, the biotic landscape is thus a mosaic of patches of distinctive biotic communities with different biological potentials and different histories.

This view of the landscape assumes that all successional species are always available to colonize new space or invade seral communities. The nature of the biotic communities that we see will then be determined by varying environmental and ecological factors as they act within new space created by physical and human processes. But what about the continental and global scale, taken over longer spans of time? Here, not all species are available to colonize new

23.15 Old-field succession *A typical old-field succession sequence for the southeastern United States, following abandonment of cultivated fields. This is a pictorial graph of continuously changing plant composition spanning about 150 years.*

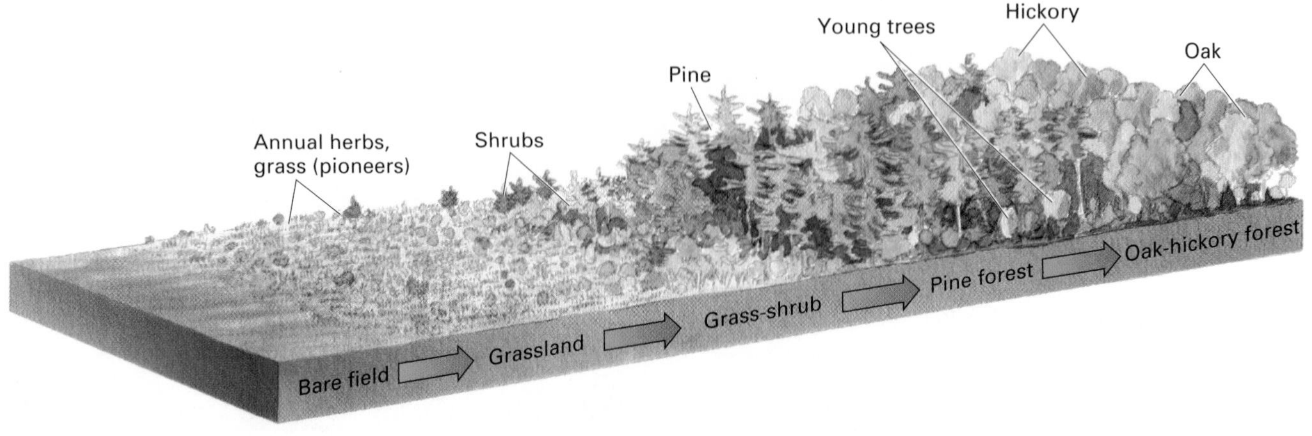

space. On this scale, the processes of migration, dispersal, evolution, and extinction are more important in determining the spatial patterns we observe. These processes are within the realm of historical biogeography, to which we turn next.

HISTORICAL BIOGEOGRAPHY

Historical biogeography examines four key processes that influence the distribution of species: evolution, speciation, extinction, and dispersal. Let's look at each of these in turn, beginning with evolution.

Evolution

An astonishing number of organisms exist on Earth, each adapted to the ecosystem in which it carries out its life cycle. About 40,000 species of microorganisms, 350,000 species of plants, and 2.2 million species of animals, including some 800,000 insect species, have been described and identified. However, many organisms remain unclassified. Estimates suggest that species of plants will ultimately number about 540,000. Of an estimated 2 million insect species, only slightly more than one-third have been classified.

How has life gained this astonishing diversity? Through the process of **evolution,** the environment itself has acted on organisms to create this diversity of life-forms, even as organic processes have been a prime factor in shaping the environment. You are probably familiar with the name of Sir Charles Darwin, whose monumental work, *The Origin of Species by Means of Natural Selection,* was published in 1859. Through exhaustive studies, Darwin showed that all life possessed *variation,* the differences that arise between parent and offspring. The environment acts on variation in organisms, Darwin observed, in much the same way that a plant or animal breeder does, selecting for propagation only those individuals with the best qualities, those best suited to their environment. Darwin termed this survival and reproduction of the fittest **natural selection.** He saw that variation could, when acted upon by natural selection through time, bring about the formation of new species whose individuals differed greatly from their ancestors. Thus, Darwin viewed the formation of new species as a product of variation acted upon by natural selection.

The weakness in Darwin's theory lay not in the origin of species as products of natural selection, but in the process of variation. He was at a loss to explain why variation occurred, and he simply accepted it as a natural and an automatic property of life. We know now that variation results from the interaction of two sources: *mutation,* which alters the genetic material and consequently the physical characteristics of organisms, and *recombination,* the pairing of parental genetic material in offspring in unique assortments.

Mutation results when the genetic material (DNA, or deoxyribonucleic acid) of a reproductive cell is changed. This requires breaking and reassembling chemical bonds in the DNA of chromosomes, which can come about when the cell is exposed to heat, ionizing radiation, or to certain types of chemical agents. Most mutations either have no effect or are deleterious, but a small proportion may have a positive effect on the genetic makeup of the individual. If that positive effect makes the individual organism more likely to survive and reproduce, then the altered gene is likely to survive as well and be passed on to offspring.

Recombination describes the process by which an offspring receives two slightly different copies, or *alleles,* of each gene from its parents. One allele may be dominant and suppress the other, or the two alleles may act simultaneously. Because each individual receives two alleles of each gene, and there are typically tens of thousands of genes in an organism, the possible number of genetic combinations is very large. Thus, recombination provides a constant source of variation that acts to make every offspring slightly different from the next.

Mutations act to change the nature of species through time. But just what is a species? For our purposes here, we can define a **species** (plural, *species*) as all individuals capable of interbreeding to produce fertile offspring. A *genus* (plural, *genera*) is a collection of closely related species that share a similar genetic evolutionary history. Note that each species has a scientific name, composed of a generic name and a specific name in combination. Thus, red oak, a common deciduous tree of eastern North America, is *Quercus rubra.* The related white oak is *Quercus alba* (Figure 23.16).

Although the true test of the species is the ability of all of its individuals to reproduce successfully with one another, this criterion is not always easily applied. Instead, the species is usually defined by *morphology*—the outward form and appearance of its individuals. The *phenotype* of an individual is the morphological expression of its gene set, or *genotype,* and includes all the physical aspects of its structure that are readily perceivable. Species, then, are usually defined by a characteristic phenotype or range of phenotypes.

Speciation

Biogeographers use the term **speciation** to refer to the process by which species are differentiated and maintained. Actually, speciation is not the result of a single process but arises from a number of component processes acting together through time. One of these component processes is mutation, which we have already described. The rate of mutation depends on the organism's exposure to mutagenic agents as well as the organism's sensitivity to those agents. Another process is natural selection, which as we have seen acts to favor individuals with genotypes that produce successful traits for a particular environment.

A third speciation process is *genetic drift.* In this process, chance mutations without any particular beneficial effect change the genetic composition of a breeding population until it diverges from other populations. Genetic drift is a weak factor in large populations, but in small populations, such as a colony of a few pioneers in a new habitat, random mutations are more likely to be preserved. *Gene flow* is an opposite process in which evolving

23.16 Red and white oaks *Although similar in general appearance, these two species of oak are easily separated. Red oak acorns have a flat cap and stubby nut, with pointed, bristle tips on the leaf lobes (left). White oak acorns have a deeper cap and a pointed nut, with rounded leaf-lobe tips (right).*

populations exchange alleles as individuals move among populations. This process helps maintain the uniformity of the gene pool of the species by moving new alleles into all populations.

Speciation often occurs when populations become isolated from one another. This **geographic isolation** can happen in several ways. For example, plate tectonics may uplift a mountain range that separates a population into two different subpopulations by a climatic barrier. Or a chance long-distance dispersal may establish a new population far from the main one. These are examples of *allopatric speciation,* in which populations are geographically isolated and gene flow between the populations does not take place. As genetic drift and natural selection proceed, the populations gradually diverge and eventually lose the ability to interbreed successfully.

The evolution of finch species on the Galapagos Islands is an example of allopatric speciation created by isolation and natural selection (Figure 23.17). This cluster of five major volcanic islands and nine lesser ones, located about 800 km (500 mi) from the coast of Ecuador, was visited by Charles Darwin and provided much raw material for his ideas about evolution. As the story has been reconstructed, the islands were first colonized by a single original finch species, the blue-black grassquit. Over time, individual populations became adapted to conditions on particular islands through natural selection, and, enhanced by their isolation on different islands, evolved into different species. Later, some of these species successfully reinvaded other islands, continuing the speciation and evolution process. Today 5 genera and 13 species of finches are found on the Galapagos, specializing in diets of seeds, buds, and insects.

In contrast to allopatric speciation is *sympatric speciation,* in which speciation occurs within a larger population. Imagine a species that has two different primary food sources. Eventually, mutations will arise that favor each food source at the expense of the other—for example, two different lengths or shapes of a bird's beak that facilitate feeding on two different types of fruit or seeds. As these mutations are exposed to natural selection, they will produce two different populations, each adapted to its own food source. Eventually, the populations may become separate species. For example, hundreds of species of cichlid fishes have evolved from a few founding stocks in the lakes of the African Rift Valley. Lake Victoria alone, which was dry as recently as about 12,000 years before present, supports more than 300 of these species. Although these populations have all been in contact in the same shallow lake, they have specialized and diverged through natural selection to the point of becoming individual species. This type of evolution, in which a new environment provides the opportunity for the formation of many new species adjusted to different habitats is termed *adaptive radiation.*

Another mechanism of sympatric speciation that is quite important in plants is **polyploidy.** Normal organisms have two sets of genes and chromosomes—that is, they are *diploid.* Through accidents in the reproduction process, two closely related species can cross in such a way that the offspring has both sets of genes from both parents. These *tetraploids* are fertile but cannot reproduce with the populations from which they arose, and so they are instantly isolated as new species. By some estimates, 70 to 80 percent of higher plant species have arisen in this fashion.

Extinction

Over geologic time, the fate of all species is **extinction.** When conditions change more quickly than populations can evolve new adaptations, population size falls. When that occurs, the population becomes increasingly more vulnerable to chance occurrences, such as a fire, a rare climatic event, or an outbreak of disease. Ultimately, the population succumbs, and the species becomes extinct.

Some extinctions are very rapid, particularly those induced by human activity. A classic example is that of the passenger pigeon, which was a dominant bird of eastern North America in the late nineteenth century. Flying in huge

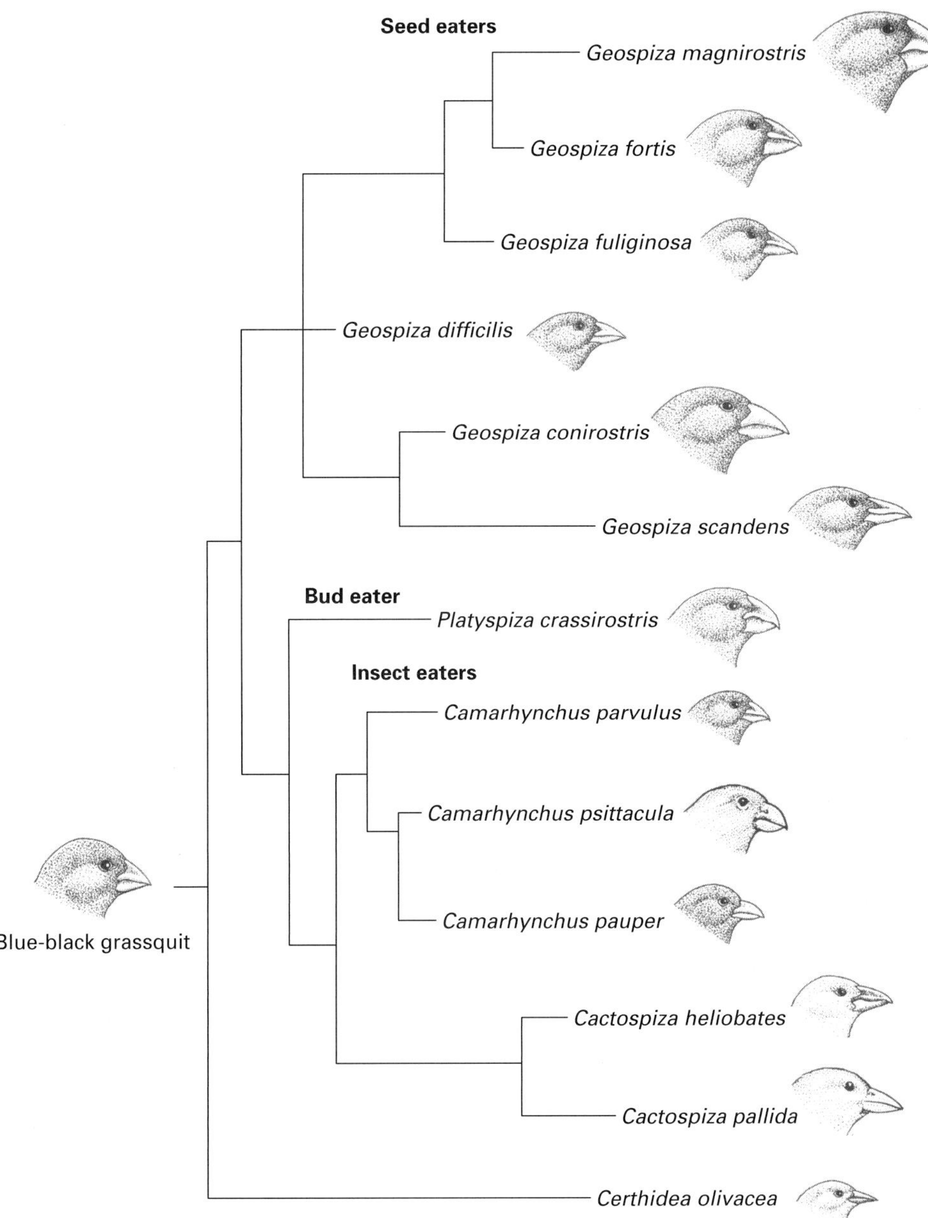

23.17 Adaptive radiation of finch species of the Galapagos Islands *Five genera and 13 species of finch evolved from a single ancestral population of the blue-black grassquit. Beak shapes are adapted to the primary food source: seeds, buds, and insects. (From J. H. Brown and M. V. Lomolino, Biogeography, second ed., 1998, Sinauer, Sunderland, Massachusetts, used by permission.)*

flocks and feeding on seeds and fruits such as beechnuts and acorns, these birds were easily captured in nets and shipped to markets for food. By 1890, they were virtually gone, and the last known passenger pigeon died in the Cincinnati Zoo in 1914.

Rare but extreme events can also cause extinctions. Many lines of evidence have converged to document that the Earth was struck by an asteroid about 65 million years ago. The impact, which occurred on the continental shelf near the Yucatan Peninsula, raised a global dust cloud that blocked sunlight from the surface, cooling the Earth's climate intensely for a period of perhaps several years. Dinosaurs and many other groups of terrestrial and marine organisms were wiped out. Less affected were organisms that were less sensitive to a brief, but intense, period of cold. These included birds and mammals, which have internal metabolic temperature regulation, as well as seed plants and insects that pass part of their life cycle in a dormant state.

Dispersal

Nearly all types of organisms have some mechanism of **dispersal**—that is, a capacity to move from a location of birth or origin to new sites. Often dispersal is confined to one life stage, as in the dispersal of higher plants as seeds. Even in animals that are inherently mobile, there is often a developmental stage when movement from one site to the next is more likely to occur.

Normally, dispersal does not change the geographic range of a species. Seeds fall near their sources, and animals seek out nearby habitats to which they are adjusted. Dispersal is thus largely a method for gene flow that helps to encourage the cross-breeding of organisms throughout a population. When land is cleared or new land is formed, dispersal moves colonists into the new environment. We documented this role of dispersal earlier in this chapter as part of succession. Species also disperse by *diffusion,* the slow extension of range from year to year.

23.18 Cattle egret *This small, white heron migrated from Africa to South America in the late 1880s. It follows grazing animals, such as this Texas steer, and feeds on insects and small invertebrates flushed by the grazing action.*

A rare, long-distance dispersal event can be very significant in establishing biogeographic patterns. We have already noted how a single ancestral colony of finches invaded the Galapagos Islands and underwent adaptive radiation to populate all of the islands successfully.

Some species have modes of propagation that are especially well-adapted to long-distance dispersal. Mangrove species, which line coastal estuaries in equatorial and tropical regions, have seeds that are carried thousands of miles on winds and ocean currents to populate far distant shores. Another example of a plant well adapted to oceanic dispersal is the coconut palm (Figure 10.7). Its large seed, housed in a floating husk, has made it a universal occupant of island beaches. Among the animals, birds, bats, and insects are frequent long-distance travelers. Generally, nonflying mammals, freshwater fishes, and amphibians are less likely to make long leaps, with rats and tortoises the exceptions.

The case of the cattle egret demonstrates both long-distance dispersal and diffusion (Figure 23.18). This small heron crossed the Atlantic, arriving in northeastern South America from Africa in the late 1880s. A hundred years later, it had become one of the most abundant herons of the Americas. Figure 23.19 maps its range extension over much of this period. Another example of diffusion is the northward colonization of the British Isles by oaks following the retreat of continental glaciers after the close of the Ice Age. As shown in Figure 23.20, oaks required about 3500 years, from about 9500 years before present to 6000 years before present, to reach their northern limit.

Dispersal often means surmounting *barriers*—that is, regions a species is unable to colonize or perhaps even occupy for a short period of time. Long-distance dispersal can mean bridging such obvious barriers as an ocean of salt water or an ice sheet by an unlikely accident. But other barriers are not so obvious. For example, the basin and range country of Utah, Nevada, and California presents a sea of desert with islands of forest to some diffusing species. While birds and bats may have no difficulty moving from one island to the next, a small mammal may never cross the desert sea, at least under its own power. In this case, the barrier is one in which the physiological limits of the species are exceeded. But there may be ecological barriers as well—for example, a zone of intense predation or a region occupied by strongly and successfully competing species.

Just as there are barriers to dispersal, so there are corridors that facilitate dispersal. Central America forms a present-day land bridge connecting North and South America, for example. It has been in place for about 3.5 million years. Other cor-

23.19 Diffusion of the cattle egret *After long-distance dispersal to northeastern South America, the cattle egret spread to Central and North America as well as to coastal regions of western South America. (From J. H. Brown and M. V. Lomolino, Biogeography, second ed., 1998, Sinauer, Sunderland, Massachusetts, used by permission.)*

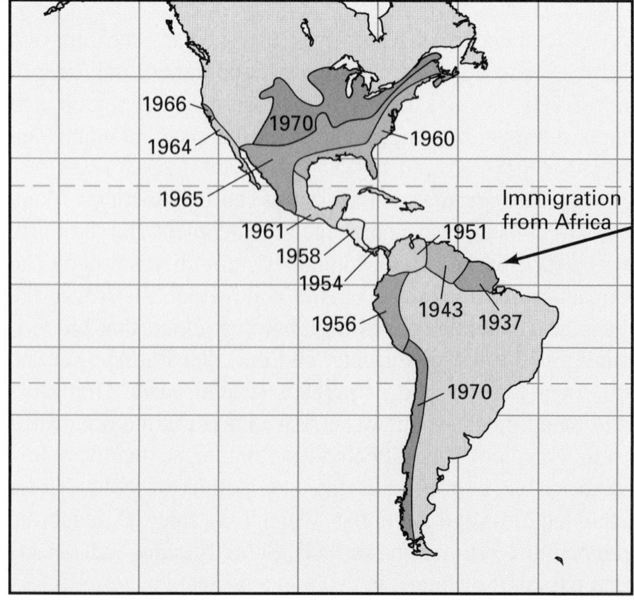

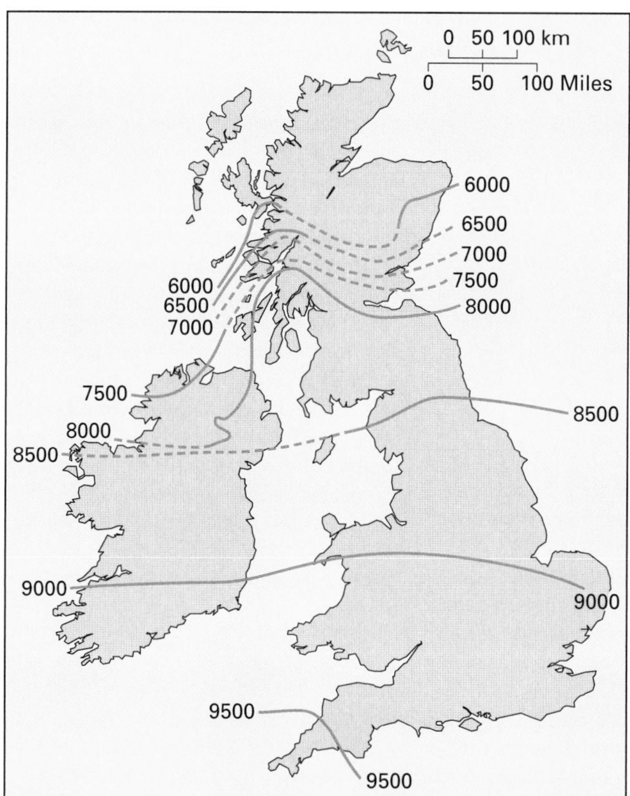

23.20 Diffusion of oaks *Following the retreat and melting of continental glaciers at the close of the Ice Age, oak species diffused northward across the British Isles. Contours indicate northern border at times in years before present. Dashed lines are less certain. (From H. J. B. Birks, J. Biogeography, vol. 16, pp. 503–540. Used by permission.)*

ridors of great importance to present-day species distribution patterns have existed in the recent past. For example, the Bering Strait region between Alaska and easternmost Siberia was dry land during the early Cenozoic Era and during the Ice Age, when sea level dropped by more than 100 m (325 ft). Many plant and animal species of Asia are known to have crossed this bridge and then spread southward into the Americas. One notable migrant species of the last continental glaciation was the aboriginal human, and there is substantial evidence to support the hypothesis that the skilled hunters who crossed the Bering land bridge were responsible for the extinction of many of the large animals, including wooly mammoths and ground sloths, that disappeared from the Americas about 10,000 years ago (Figure 23.21).

The dispersal process, coupled with extinction, has been studied intensively by ecologists and biogeographers, and the result has been the development of mathematical models explaining the number of species that might be expected within a region of a given size. *Working It Out 23.3 • Island Biogeography: The Species-Area Curve* shows how this works.

Distribution Patterns

The processes of evolution, speciation, extinction, and dispersal have over time produced many spatial patterns of species distribution on today's Earth. One of the simplest patterns is that of the **endemic** species, which is found in one region or location and nowhere else. An endemic distribution can arise in one of two ways—as the result of a contraction of a broader range or as the location of origin of a species that has not spread widely beyond the endemic region. Some endemic species are ancient relics of biological strains that have otherwise gone extinct. An example is the gingko tree (Figure 23.22), which was widespread throughout the Mesozoic Era but until recently was restricted to a small region in eastern China. It is now widely planted as an urban street tree, known for its hardiness.

In contrast to endemics are *cosmopolitan* species, which are found very widely. Among these is the human species, *Homo sapiens.* Bats and peregrine falcons are also very widely dispersed, cosmopolitan species. Very small organisms, or organisms with very small propagating forms, are often of cosmopolitan distribution, ditributed widely by atmospheric and oceanic circulations.

Another interesting pattern is **disjunction,** in which one or more closely related species are found in widely separated regions. An example is the distribution of the tinamous and flightless ratite birds, which includes such oddities as the ostrich, emu, cassowary, and kiwi (Figure 23.23). This disjunct pattern is thought to result from an ancestral species that was widespread across the ancient

23.21 Wooly mammoth *A reconstruction of the wooly mammoth, a huge tusked mammal that inhabited North America throughout the Ice Age and became extinct about 10,000 years ago, most likely from hunting by prehistoric humans.*

Working It Out | 23.3

Island Biogeography: The Species-Area Curve

"An island is certainly an intrinsically appealing study object. It is simpler than a continent or an ocean, a visibly discrete object that can be labeled with a name and its resident populations identified thereby. In the science of biogeography, the island is the first unit that the mind can pick out and begin to comprehend.[1]" These words introduce the first chapter of *The Theory of Island Biogeography,* a classic of biogeography written by R. H. MacArthur and E. O. Wilson in 1967. In this book, MacArthur and Wilson brought together elements of population biology, mathematical modeling, statistics, and geography to produce a new approach to understanding the processes that distribute the world's flora and fauna across the globe.

MacArthur and Wilson begin their book with an examination of a biogeographic principle that became clear as scientists first began to make lists of the species of plants and animals found on individual islands. The principle is simply that the larger the island, the more species it contains. That is, the number of species increases with island area in a systematic way. This finding is true whether studying lists of all plant and animal species or even just those species within a particular genus, family, or other group of related organisms.

Let's explore the relationship between species and area with two examples drawn from their book. The table above shows the areas of some islands of the Lesser and Greater Antilles as well as counts of species of amphibians and reptiles found on each.

Island	Area, km^2	Species*
Redonda	1.3	3
Saba	13	5
Montserrat	102	9
Puerto Rico	9104	40
Jamaica	10,991	39
Hispaniola	76,480	84
Cuba	110,860	76

**Data of P. J. Darlington, 1957.*

These data are plotted in the figure below. In graph *a,* we can see that as the area of the island increases, the species count increases steeply at first. But as we consider larger islands, the number of species increases more gradually. The effect is to generate a curve with a slope that is initially steep, but then becomes flatter. In fact, this *species-area curve* follows a power function—that is, the number of species is related to the area raised to a power. Expressed as an equation,

$$S = cA^z$$

where S is the number of species, A is the area of the island, and c is a constant. The variable z is a power to which A is raised, which must lie between zero and one to give a curve of this shape. Note that we've seen a power function before. In *Working It Out 17.1 • River Discharge and Suspended Sediment* we saw a similar power function showing that the suspended sediment load of a river is proportional to the river's discharge raised to a power.

Recall also that when a power function is plotted on logarithmic axes, it follows a straight line. In part *b* of the figure, species number is plotted as a function of area on logarithmic axes. The values follow a straight line closely, showing that the power law holds quite well for this example. By applying a statistical procedure to the data in the table, we can find the values of c and z that best fit this relationship. The resulting best-fit equation for the power function is

$$S = 2.47A^{0.30}$$

The line seen on both graphs follows this equation.

Our second example, shown in the graph at right, plots area against counts of land and fresh-water bird species on 23 islands of the Sunda group, the Republic of the Philippines, and New Guinea. These islands are located in a broad arc off the southeast coast of Asia. As in the Antilles, there is a strong straight-line relationship between species and area in the log-log plot. For these data, the best-fit values of c and z are

$$S = 4.61A^{0.34}$$

Note that the values of z are somewhat similar in these two examples. A number of studies of species-area curves for different islands and different

[1]MacArthur, R. H. and E. O. Wilson, 1967, *The Theory of Island Biogeography,* Princeton University Press, Princeton, NJ, p. 3.

Species counts of reptiles and amphibians on selected islands of the Lesser and Greater Antilles *(a) Data plotted on linear axes. (b) Data plotted on logarithmic axes.*

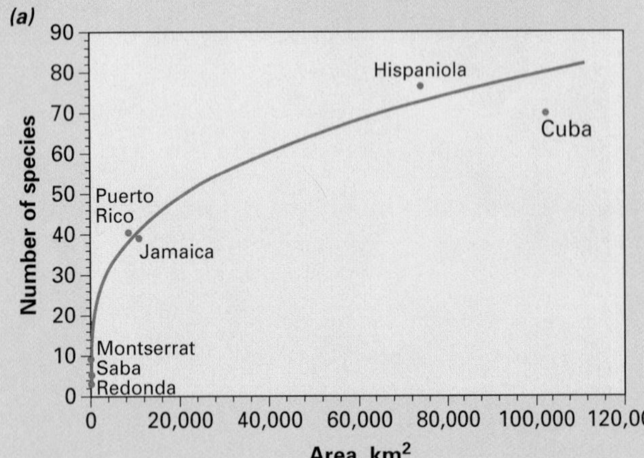

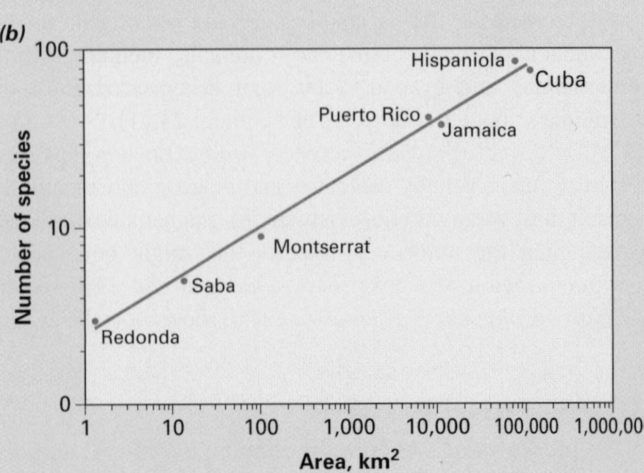

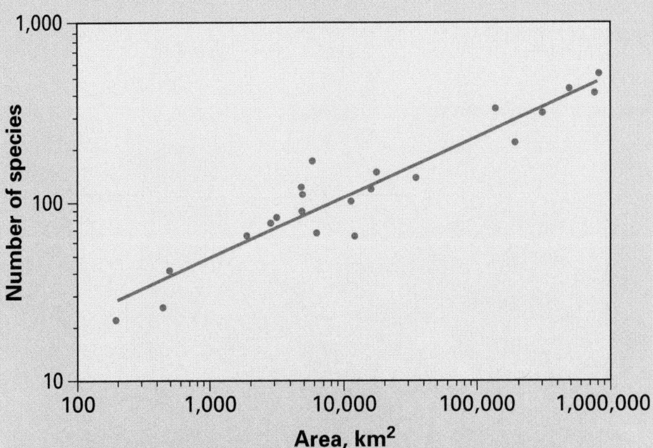

Species-area plot for land and fresh-water bird species *Data are shown for 23 islands of the Sunda group, Republic of the Philippines, and New Guinea. Christmas Island (CI) is omitted as an outlier in finding the best-fit line. Data are from sources as cited in R. H. MacArthur and E. O. Wilson, 1963, Evolution, vol. 17, p. 374.*

species groups shows that values of z are generally in the range of 0.20 to 0.35. Actually, mathematical ecologists have shown that in theory z is expected to be about 0.26 if species follow a log-normal rule of abundance. This rule is a way of stating that, in any given region, only a few species are relatively common, while most species are moderately rare or very rare. The rule has been observed to fit a broad range of data and holds for many situations.

The values of c, however, are less similar in these examples—the value of c for land and fresh-water birds of the Asian islands, at 4.61, is nearly twice that of amphibians and reptiles of the Antilles, at 2.47. These values depend on the number of species of a particular type that the island environment can support. For example, if we evaluate the two equations for a fixed area, say 1000 km^2, we get 19.6 species of amphibians and reptiles expected and 48.3 species of land and fresh-water birds expected. This shows that, given islands of comparable size, the Sunda Islands-Philippines-New Guinea group supports more bird species than the Antilles islands support amphibian and reptile species.

Let's think a bit more about how species come to be found on islands. First, species can immigrate to islands from the mainland or other islands. Over hundreds or thousands of years, individuals find their way to islands in accidents of chance—a typhoon, for example, may carry a bird to an island hundreds of kilometers from its normal habitat. Or a flood may sweep an animal out to sea on floating debris that comes to rest on a far island shore. In this way, an island receives a constant flow of new immigrants.

The rate of immigration will depend on the area of the island—a large island is more likely to receive more immigrants, simply because it has more area to intercept accidental travelers. Also, a large island is more likely to have a greater diversity of habitats, so that more immigrants can successfully establish themselves once they arrive. So if we define the rate of immigration I as the number of new species per year that an unpopulated island could support, that rate will be low for small islands and high for large islands.

At the same time, species on islands suffer extinction, which reduces the total number of species present. The extinction rate also depends on the area of the island. On a small island, the population of a species will also tend to be small, and a catastrophic event, like a typhoon or volcanic eruption, is more likely to wipe out all the individuals of species. On a larger island, there is a better chance for survival of enough individuals to keep the species existing there. Defining the extinction rate E as the number of species per year that are lost given that all possible species are present, that rate will be high for small islands and low for large islands. Thus, we have the situation shown in the first two rows of the table below:

	Small island	Large island
Immigration rate	low	high
Extinction rate	high	low
$I/(I + E)$	small	large

Let's now add the rate of immigration and rate of extinction together to get a *turnover rate* $(I + E)$. This rate expresses the number of species that either immigrate or become extinct in a given year. The expression $I/(I + E)$ is then the proportion of the turnover rate that arises from immigration. As shown in the last row of the table above, it will be small for small islands and large for large islands.

As you might deduce, this proportion determines the average size of the species pool present on the island. Thus, we can write

$$S = P\left(\frac{I}{I+E}\right) \tag{3}$$

where S is the expected number of species and P is the number of species in the pool available for colonization of the island (*i.e.*, on the mainland)[2].

The importance of this expression is that it gives us a reason for the species-area curve. It shows that the number of species depends on the balance between immigration and extinction—a balance that in turn depends on the size of the island. Small islands, then, have low immigration rates and high extinction rates, so have few species. Larger islands have higher immigration rates and lower extinction rates, so have more species.

Some questions remain. Exactly how is area related to immigration and extinction rates? What is the effect of distance between the island and the source of its species? What is the effect of habitat diversity on immigration, extinction, and species number? How is the immigration rate reduced as the island's ecological niches are filled? Does the species-area relationship hold for regions within continents as well as islands? If so, is it any different? If these questions interest you, the realm of biogeographic theory, which has expanded greatly in the decades since MacArthur and Wilson's *Theory of Island Biogeography*, awaits your study.

[2]Ibid., p. 26.

23.22 Gingko tree *The gingko tree is an eastern Chinese endemic that has survived millions of years with little evolutionary change. Thanks to human activity, it is now much more widely distributed. Shown here in an urban habitat (Brooklyn, New York), the gingko has a distinctive shape with long branches of closely-held leaves (right). The fan-shaped leaves show an uncommon arrangement of parallel veins radiating from the leaf stem (left).*

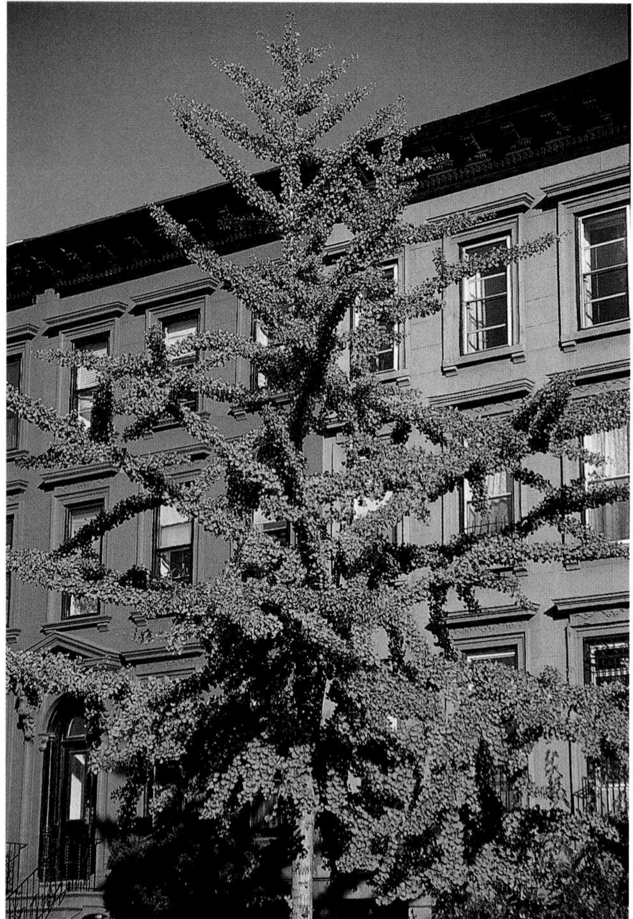

continent of Gondwana (see Chapter 13). As plate tectonics split Gondwana apart into North and South America, Africa, Australia, and New Zealand, isolation and evolution differentiated the ancestral lineage into the diverse array of related species that now inhabit these continents. Another disjunct pattern that is less well understood is the occurrence of closely related species of such common desert plants of the North American southwest as mesquite, creosote bush, and paloverde in the desert environments of Chile and Argentina.

23.24 Biogeographic regions *(a) Biogeographic regions of land plants. (After Goode, 1974.) (b) Biogeographic regions of land animals. (After Wallace, 1876.)*

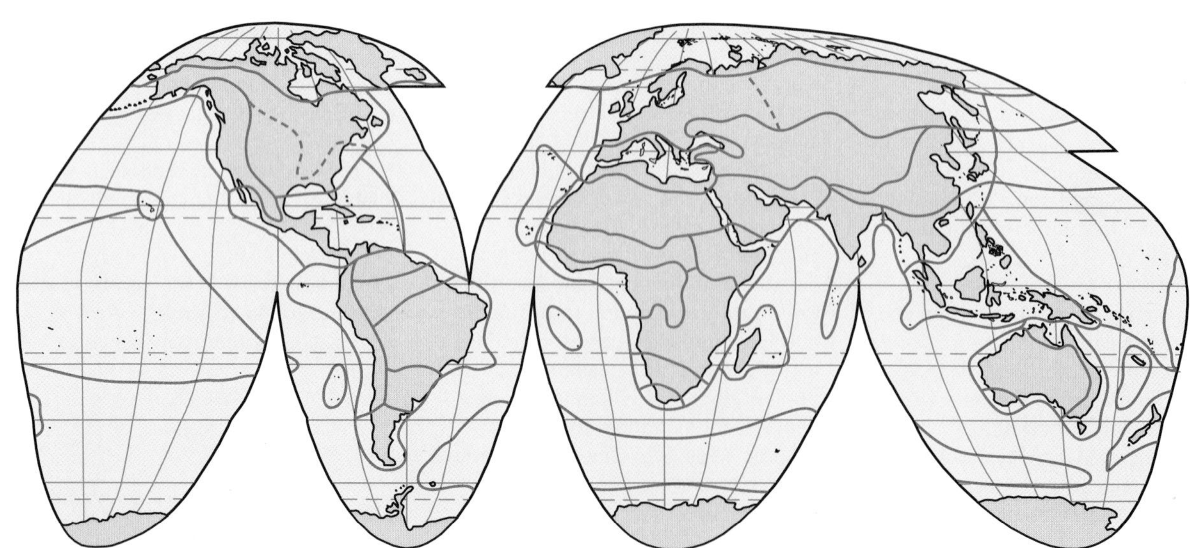

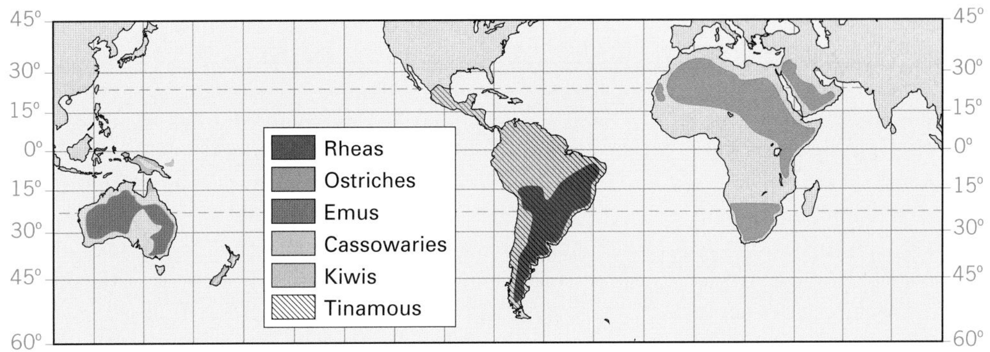

***23.23 Disjunct distribution** The distribution pattern of the ratite birds and tinamous shows disjunctions resulting from isolation by continental motions. (From J. H. Brown and M. V. Lomolino, Biogeography, second ed., 1998, Sinauer, Sunderland, Massachusetts, used by permission.)*

Biogeographic Regions

As spatial distributions of species are examined on a global scale, certain common patterns emerge. Although closely related species tend to be nearby or occupy similar regions, larger groups of organisms, such as families and orders, tend to show similar patterns that are often disjunct. For example, the South America–Africa–Australia–New Zealand pattern described for the ratite birds fits the distribution of many other ancient families of plants and animals and is a reflection of common ancestry on Gondwana. This example emphasizes the importance of history in determining these global patterns. Another important determinant of spatial patterns is global climate. Often, members of the same lineage have similar adaptations to environment and so are found in similar climatic regions.

As a result of these factors, biogeographers have been able to define **biogeographic regions** in which the same or closely related plants and animals tend to be found together. That is, when we cross the boundary between two biogeographic regions, we pass from one group of distinctive plants and animals to another. Figure 23.24 shows the major bigeographic regions recognized for plants and animals. Note that many of the boundaries on the two maps are very close, indicating that at the global scale, plants and animals have similar and related histories of evolution and environmental affinity.

BIODIVERSITY

From the principles of ecological and historical biogeography presented in this chapter, we can see that **biodiversity**—the variety of biological life on Earth—depends both on the variety of the Earth's environments and the processes of evolution, dispersal, and extinction as they have acted through geologic time. At the present time, the Earth's biodiversity is rapidly decreasing as a result of human activity. Our species, *Homo sapiens,* has come to utilize some 20 to 40 percent of global primary productivity as well as exploit 70 percent of its marine fisheries to provide its food. In this process, it has doubled the natural rate of nitrogen fixation, used more than half of the Earth's supply of surface water, and transformed more than 40 percent of the land surface.

Recent human activity has ushered in a wave of extinctions unlike any that has been seen for millions of years. In the last 40 years, biologists have documented the disap-

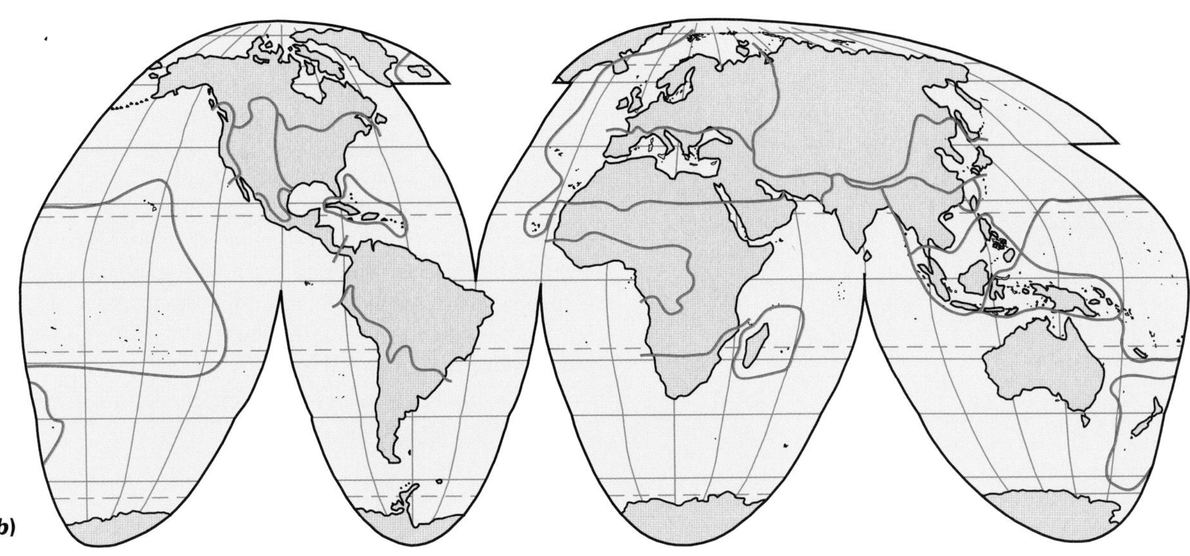

pearance of several hundred land animal species, including 58 mammals, 115 birds, 100 reptiles, and 64 amphibians. Aquatic species have also been severely impacted, with 40 species or subspecies of freshwater fish lost in North America alone in the last few decades. In the plant kingdom, botanists estimate that over 600 species have become extinct in the past four centuries. Figure 23.25 shows the status of some important groups of plant and animal species of the United States. Many species are extinct or imperiled. Only about a third of the species of crayfishes and freshwater mussels are considered secure, as are about half of the species of amphibians and freshwater fishes.

These documented extinctions may be only the tip of the iceberg. Many species remain as yet undiscovered and so may become extinct without ever being described. Many of these species will likely be important to a better understanding of evolutionary history, if recent experience is any guide, because they will include representatives of new biological divisions of the highest level. In the last decade alone, three new families of flowering plants and two new phyla of animals were described. (The *phylum,* plural *phyla,* is the highest division of higher animal life.) Among animal groups, insects, spiders, and other invertebrates are the most poorly known. Fungi and microbes are other groups in which a large proportion of species have yet to be determined.

23.25 Degree of endangerment of plant and animal groups in the United States *These bar graphs show the proportion of species within each group as secure, vulnerable, imperiled, or extinct. (Data from Nature Conservancy.)*

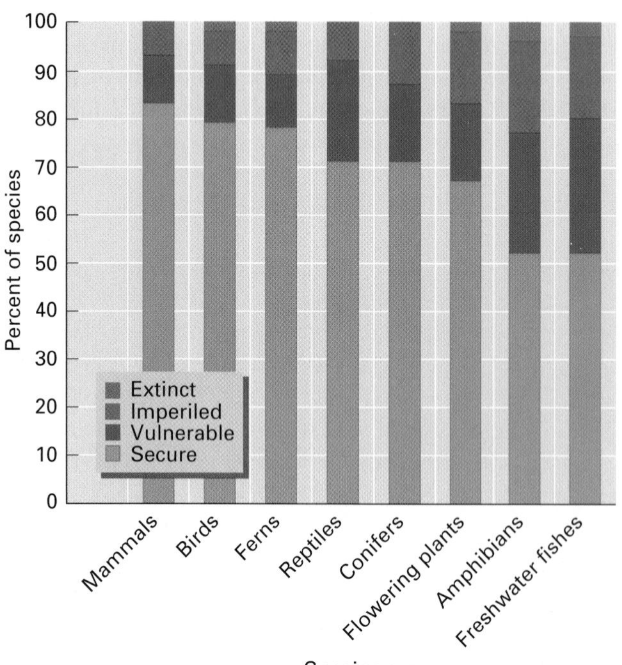

How has human activity created extinctions? One way has been to disperse new organisms that outcompete or predate existing organisms. Island populations have especially suffered from this process. Developing in isolation, island species have often not evolved defense mechanisms to protect themselves from predators, including humans. Many islands were subjected to waves of invading species, ranging from rats to weeds, brought first by prehistoric humans and later by explorers and conquerors. Hunting by prehistoric humans alone was sufficient to exterminate many species, not only from islands but from large continental regions as well. Another mechanism is the use of fire. As humans learned to use fire in hunting and to clear and maintain open land, large areas became subject to periodic burning. Habitat alteration and fragmentation is yet another cause of extinctions. By isolating plant and animal populations and altering their environment, human activities cause populations to shrink, making extinction more likely.

Biodiversity is not uniform over the Earth's surface. In general, tropical and equatorial regions have more species and more variation in species composition between different habitats. In areas of isolation, such as islands or mountaintops, species diversity tends to increase with the size of the isolated area and decrease with the degree of isolation from surrounding sources of colonists. Much of this island diversity is contributed by endemic species. Geographic areas in which biodiversity is especially high are referred to as *hotspots.* An important strategy for preservation of global biodiversity is to identify hotspots and take conservation measures to protect them. In this way, conservation efforts can be most effective.

Why is biodiversity important? Nature, operating over millennia, has provided an incredibly rich array of organisms that interact with one another in a seamless web of organic life. Humans, for all their cleverness, are still part of that web. When we cause the extinction of a species, we break a link in the web. Ultimately, the web will thin, with important consequences for both the human species and continued life on Earth. We have no way of knowing which organisms future humans will rely on and how. It seems prudent to keep as many of them around for as long as possible.

A Look Ahead

In this chapter we've focused on the processes that determine the spatial patterns of biota on Earth. We have seen how organisms adjust to their environments and how natural selection works in response to environmental pressures. We have also seen how evolution, dispersal, and extinction generate patterns of species distribution and determine biodiversity. Our last chapter takes a more functional view of the life layer by inventorying the global biomes—major divisions of ecosystems that are based largely on the dominant life-form of their vegetation covers.

Chapter Summary

- *Ecological biogeography* examines how relationships between organisms and environment help determine when and where organisms are found. *Historical biogeography* examines how, where, and when species have evolved and how they are distributed over longer times and broader scales.
- A *community* of organisms occupies a particular environment, or **habitat**. *Associations* are often defined by characteristic species or vegetation life-forms. Environmental actors influencing the distribution patterns of organisms include moisture, temperature, light, and wind.
- Organisms require water to live, and so they are limited by the availability of water. Organisms influence water availability when plants transpire, and soil organisms facilitate water absorption by the soil. **Xerophytes** are adapted to arid habitats. They reduce water loss by having waxy leaves, spines instead of leaves, or no leaves at all. *Phreatophytes* have deep roots; succulents store water in spongy tissues. *Tropophytes* are deciduous during the dry or cold season. *Sclerophylls* have thick, leathery leaves that resist drying during the Mediterranean summer climate. **Xeric animals** include vertebrates that are nocturnal and have various adaptations to conserve water. Invertebrates such as brine shrimp can adjust their life cycle to prolonged drought.
- Temperature acts on plants to trigger and control stages of plant growth as well as to limit growth at temperature extremes. Survival below freezing requires special adaptations, and so only a small proportion of plants are frost tolerant. *Cold-blooded animals* have body temperatures that follow the environment, but they can moderate these temperatures by seeking warm or cool places. Mammals and birds are warm-blooded animals that maintain constant internal temperatures through a variety of adaptation mechanisms.
- The light available to a plant depends on its position in the structure of the community. Duration and intensity of light vary with latitude and season and serve as a cue to initiate growth stages in many plants. The day–night cycle regulates much of animal behavior, as does *photoperiod* (daylight length). Wind deforms plant growth by dessicating buds and young growth on the windward side of the plant.
- A **bioclimatic frontier** marks the potential distribution boundary of a species. Other factors may also limit the distribution of a species.
- *Geomorphic factors* of slope steepness and orientation affect both the moisture and temperature environment of the habitat and serve to differentiate the microclimate of each community.
- Soil, or *edaphic, factors* such as soil texture, structure, acidity, alkalinity, and salinity can also limit the distribution patterns of organisms or affect community composition.
- *Disturbance* includes catastrophic events that damage or destroy ecosystems. Fire is a very common type of disturbance that influences forests, grasslands, and s Floods, high winds, and storm waves are others. Many ecosystems include specialized species that are well-adapted to disturbance.
- Species interact in a number of ways, including **competition**, *predation* and *parasitism*, and *herbivory*. The explosion and crash of the Kaibab Plateau deer herd provide an example of the beneficial nature of some predator–prey relationships. In *allelopathy*, plant species literally poison the soil environment against competing species. Positive (beneficial) interactions between species are termed *symbiosis*.
- **Ecological succession** comes about as ecosystems change in predictable ways through time. A series of stable communities follows a *sere* to a **climax**. *Primary succession* occurs on new soil substrate, while *secondary succession* occurs on disturbed habitats. Succession on coastal sand dunes follows a series of stages from dune grass to an oak and holly forest. *Old-field succession*, which occurs on abandoned farmland, also leads to a deciduous forest climax. Although succession is a natural tendency for ecosystems to change with time, it is opposed by natural disturbances and limited by local environmental conditions.
- *Historical biogeography* focuses on evolution, speciation, extinction, and dispersal as they influence the distribution patterns of species.
- Life has attained its astonishing diversity through evolution. In this process, **natural selection** acts on variation to produce populations that are progressively better adjusted to their environment. Variation arises from *mutation* and *recombination*.
- A **species** is best defined as a population of organisms that are capable of interbreeding successfully, but instead it is usually defined by a typical *morphology* or *phenotype*.
- **Speciation** is the process by which species are differentiated and maintained. It includes mutation, natural selection, *genetic drift*, and *gene flow*. **Geographic isolation** acts to isolate subpopulations of a species, allowing genetic divergence and speciation to occur. The finches of the Galapagos Islands provide an example of *allopatric speciation* by geographic isolation. In sympatric speciation, adaptive pressures force a breeding population to separate into different subpopulations that may become species. Sympatric speciation of plants has included **polyploidy**, which is an important mechanism of evolution for higher plants.
- **Extinction** occurs when populations become very small and thus are vulnerable to chance occurrences of fire, disease, or climate anomaly. Rare but very extreme events can cause mass extinctions. An example is the asteroid impact that the Earth suffered about 65 million years ago. The global dust cloud that lingered for several years caused extreme cold temperatures, which wiped out many important lineages of plants and animals.

- Species expand their ranges by **dispersal**. Plants are generally dispersed by seeds, while animals often disperse under their own power. Since most dispersal happens within the range of a species, it acts primarily to encourage gene flow between subpopulations. Long-distance dispersal, though very rare, may still be very important in establishing biogeographic patterns. *Barriers*, often climatic or topographic, inhibit dispersal and induce geographic isolation. Corridors provide pathways that facilitate dispersal.
- **Endemic** species are found in one region or location and nowhere else. They arise by either a contraction of the range of a species or by a recent speciation event. *Cosmopolitan* species are widely dispersed and nearly universal. **Disjunction** occurs when one or more closely related species appear in widely separated regions.
- **Biogeographic regions** capture patterns of occurrence in which the same or closely related plants and animals tend to be found together. They result because their species have common histories and similar environmental preferences.
- **Biodiversity** is rapidly decreasing as human activity progressively affects the Earth. Extinction rates for many groups of plants and animals are as high or higher today than they have been at any time in the past. Humans act to disperse predators, parasites, and competitors widely, disrupting long-established evolutionary adjustments of species with their environments. Hunting and burning have exterminated many species. Habitat alteration and fragmentation also produce extinctions. Preservation of global biodiversity includes a strategy of protecting *hotspots* where diversity is greatest.

Key Terms

habitat
xerophyte
xeric animals
bioclimatic frontier
competition
symbiosis
ecological succession
climax
evolution
natural selection
species
speciation
geographic isolation
polyploidy
extinction
dispersal
endemic
disjunction
biogeographic region
biodiversity

Review Questions

1. What is a *habitat?* What are some of the characteristics that differentiate habitats? Compare habitat with *niche.*
2. Contrast the terms *ecosystem, community,* and *association.*
3. Although water is a necessity for terrestrial life, many organisms have adapted to arid environments. Describe some of the adaptations that plants and animals have evolved to cope with the desert.
4. Terrestrial temperatures vary widely. How does the annual variation in temperature influence plant growth, development, and distribution? How do animals cope with variation in temperature?
5. How does the ecological factor of light affect plants and animals? How does wind affect plants?
6. What is meant by the term *bioclimatic frontier?* Provide an example.
7. How do geomorphic and edaphic factors influence the habitat of a community?
8. Identify several types of disturbance experienced by ecosystems. How does fire affect forests and grasslands?
9. Contrast the terms used to describe interactions among species. Provide an example of positive predation.
10. Describe ecological succession using the terms *sere, seral stage, pioneer,* and *climax.* Use dune succession as an example.
11. How do primary succession and secondary succession differ? Describe old-field succession as an example of secondary succession.
12. How does the pattern of ecosystems on the landscape reflect an equilibrium?
13. Explain Darwin's theory of evolution by means of natural selection. What key point was Darwin unable to explain?
14. What two sources of variation act to differentiate offspring from parents?
15. What is *speciation?* Identify and describe four component processes of speciation.
16. What is the effect of geographic isolation on speciation? Provide an example of allopatric speciation.
17. How does sympatric speciation differ from allopatric speciation? Provide an example of sympatric speciation.
18. What is *extinction?* Provide some examples of extinctions of species.
19. Describe the process of dispersal. Provide a few examples of plants and animals suited to long-distance dispersal.
20. Contrast barriers and corridors in the dispersal process.
21. How does an endemic distribution pattern differ from a cosmopolitan pattern? What is a *disjunction?*
22. How are biogeographic regions differentiated?
23. What is biodiversity? How has human activity impacted biodiversity?

Eye on the Environment 23.1 • The Great Yellowstone Fire

1. Describe the great fire that struck Yellowstone National Park in the summer of 1988. How much of the park actually burned? What types of habitats were affected?
2. What is the natural fire succession process for the park?
3. Would a "no-burn" policy have stopped the fires? Did 100 years of no-burn policy contribute significantly to the fires of 1988?

Visualizing Exercise

1. Carefully compare parts (*a*) and (*b*) of Figure 23.24. Which boundaries are similar, and which are different? Speculate on possible reasons for the similarities and differences.

Essay Questions

1. Select three distinctive habitats for plants and animals that occur nearby and with which you are familiar. Organize a field trip (real or imaginary) to visit the habitats. Compare their physical environments and describe the basic characteristics of the ecosystems found there.
2. Imagine yourself as a biogeographer discovering a new group of islands. Select a global location for your island group, including climate, geologic substrate, and proximity to nearby continents or land masses. What types of organisms would you expect to find within your island group and why?

Problems

Working It Out 23.3 • Island Biogeography: The Species-Area Curve

1. A zoogeographer assembles the following data for area and species counts of termites and butterflies on a set of (fictitious) islands several hundred kilometers off the coast of Brazil.

Island	Area, km²	Termites	Butterflies
Alhambra	156	11	23
Bonarote	845	16	41
Charleston	1746	26	47
Danica	3550	31	64
Edmund	14,323	42	101
Fonseca	71,420	68	151

Her analysis shows that the species-area curve for termites best fits the expression $S = 2.07A^{0.31}$, while the curve for butterflies best fits the expression $S = 4.56A^{0.32}$. Plot the data in two ways—on normal (linear) axes and on logarithmic axes. Show data for both termites and butterflies on each graph. Sketch smooth curves to fit the two datasets on normal axes and straight lines on the logarithmic axes.

2. Use the expressions for the two species-area curves to find the predicted number of species of termites and butterflies for Charleston Island. Compare these with the observed values in the table. Are the observed values higher or lower than predicted? Identify some possible reasons why this difference might occur.

Eye on the Landscape

23.6 Grazing caribou Notice the "blobs" of soil on this slope **(A)**, distinguished by color and shape. These are solifluction lobes, produced in the short arctic summer when the snow melts and the surface layer of soil thaws. The saturated soil creeps downhill in lobes until it dries out sufficiently and its motion stops. Chapter 15 describes this process as well as other surface processes characteristic of the arctic environment. Also note the snow patches **(B)**. In some years, these will persist through the summer.

23.10 Slope orientation and habitat The vertical faces of these cliffs **(A)** and pinnacles **(B)** are most likely joint planes—planes of fractures in the rock resulting from cooling or from release of the pressure of overlying rock removed by erosion. Joints are discussed in Chapter 14.

24 GLOBAL ECOSYSTEMS

Fur seals on a rock, Western Cape Province, Republic of South Africa. *Although fur seals are aquatic mammals that are more at home swimming in coastal waters, they gather annually on rocks or protected beaches to mate and reproduce in colonies of hundreds of indiviuals.*

In the last two chapters we focused on ecology and biogeography, examining the principles and processes that determine the distribution of plants and animals on the Earth's land surface. In the concluding chapter of our text, we survey the broad global distribution patterns of ecosystems, emphasizing the characteristics of the vegetation cover. We emphasize the vegetation because it is a visible and obvious part of the landscape, and because humans depend on plants for many life essentials, including food, medicines, fuel, clothing, and shelter. Also, the largest division of ecosystems—the biome—is defined primarily on the characteristics of the vegetation cover. Note that vegetation is often strongly related to climate, so we will find the occurrence of major biomes coincides in many cases with broad climate types. Thus, your knowledge of climate from Chapters 9–11 will come in handy as you study the world's vegetation. GEODISCOVERIES

NATURAL VEGETATION

Over the last few thousand years, human societies have come to dominate much of the land area of our planet. In many regions, humans have changed the natural vegetation—sometimes drastically, other times subtly. What do we mean by natural vegetation? **Natural vegetation** is a plant cover that develops with little or no human interference. It is subject to natural forces of modification and destruction, such as storms or fires. Natural vegetation can still be seen over vast areas of the wet equatorial climate, although the rainforests there are being rapidly cleared. Much of the arctic tundra and the boreal forest of the subarctic zones is in a natural state.

In contrast to natural vegetation is *human-influenced vegetation,* which is modified by human activities. Much of the land surface in midlatitudes is totally under human control, through intensive agriculture, grazing, or urbanization. Some areas of natural vegetation appear to be untouched but are actually dominated by human activity in a subtle manner. For example, most national parks and national forests have been protected from fire for many decades. When lightning starts a forest fire, the firefighters put out the flames as fast as possible. However, periodic burning is part of the natural cycle in many regions. One vital function of fire is to release nutrients that are stored in plant tissues. When the vegetation burns, the ashes containing the nutrients remain. These enrich the soil for the next cycle of vegetation cover. In recent years, managers of some parks and forests have stopped suppressing wildfires, allowing the return of more natural periodic burning. (See *Eye on the Environment 23.1 • The Great Yellowstone Fire.*)

Our species has influenced vegetation in yet another way—by moving plant species from their original habitats to foreign lands and foreign environments. The eucalyptus tree is a striking example. From Australia, various species of eucalyptus have been transplanted to such far-off lands as California, North Africa, and India. Sometimes exported plants thrive like weeds, forcing out natural species and becoming a major nuisance. Few of the grasses that clothe the coastal ranges of California are native species, yet a casual observer might think that these represent native vegetation.

Even so, all plants have limited tolerance to the environmental conditions of soil water, heat and cold, and soil nutrients. Consequently, the structure and outward appearance of the plant cover conforms to basic environmental controls, and each vegetation type is associated with a characteristic geographical region—whether forest, grassland, or desert.

STRUCTURE AND LIFE-FORM OF PLANTS

Plants come in many types, shapes, and sizes. Botanists recognize and classify plants by species. However, the plant geographer is less concerned with individual species and more concerned with the plant cover as a whole. In describing the plant cover, plant geographers refer to the **life-form** of the plant—its physical structure, size, and shape. Although the life-forms go by common names and are well understood by almost everyone, we will review them to establish a uniform set of meanings.

Both trees and shrubs are erect, woody plants (Figure 24.1). They are *perennial,* meaning that their woody tissues endure from year to year. Most have life spans of many years. *Trees* are large, woody perennial plants having a single upright main trunk, often with few branches in the lower part but branching in the upper part to form a crown. *Shrubs* are woody perennial plants that have several stems branching from a base near the soil surface, so as to place the mass of foliage close to ground level.

Lianas are also woody plants, but they take the form of vines supported on trees and shrubs. Lianas include not only the tall, heavy vines of the wet equatorial and tropical rainforests, but also some woody vines of midlatitude forests. Poison ivy and the tree-climbing form of poison oak are familiar North American examples of lianas.

Herbs comprise a major class of plant life-forms. They lack woody stems and so are usually small, tender plants. They occur in a wide range of shapes and leaf types. Some are *annuals,* living only for a single season—while others are *perennials*—living for multiple seasons. Some herbs are broad-leaved, and others are narrow-leaved, such as grasses. Herbs as a class share few characteristics in common except that they usually form a low layer as compared with shrubs and trees.

Forest is a vegetation structure in which trees grow close together. Crowns are in contact, so that the foliage largely shades the ground. Many forests in moist climates show at least three layers of life-forms (Figure 24.1). Tree crowns form the uppermost layer, shrubs an intermediate layer, and herbs a lower layer. There is sometimes a fourth, lowermost layer that consists of mosses and related very small plants. In **woodland,** crowns of trees are mostly separated by open areas that usually have a low herb or shrub layer.

24.1 Layers of a beech-maple-hemlock forest *In this schematic diagram, the vertical dimensions of the lower layers are greatly exaggerated. (After P. Dansereau.)*

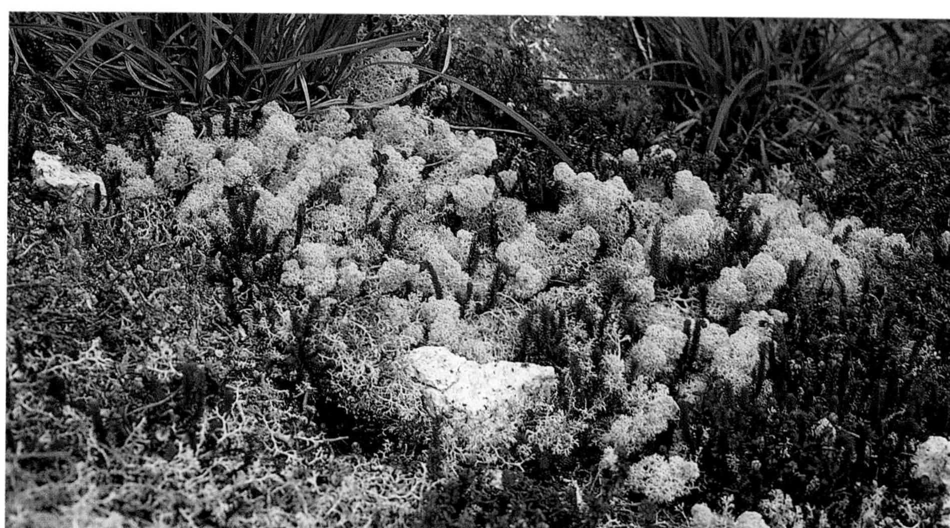

***24.2 Lichen** Reindeer moss, a white variety of lichen, seen here on rocky tundra of Alaska.*

Lichens are another life-form seen in a layer close to the ground. They are plant forms in which algae and fungi live together to form a single plant structure. In some alpine and arctic environments, lichens grow in profusion and dominate the vegetation (Figure 24.2).

Terrestrial Ecosystems—The Biomes

From the viewpoint of human use, ecosystems are natural resource systems. Food, fiber, fuel, and structural material are products of ecosystems and are manufactured by organisms using energy derived from the Sun. Humans harvest that energy by using ecosystem products. The products and productivity of ecosystems depend to a large degree on climate. Where temperature and rainfall cycles permit, ecosystems provide a rich bounty for human use. Where temperature or rainfall cycles restrict ecosystems, human activities can also be limited. Of course, humans, too, are part of the ecosystem. A persistent theme of human geography is the study of how human societies function within ecosystems, utilizing ecosystem resources and modifying ecosystems for human benefit.

Ecosystems fall into two major groups—aquatic and terrestrial. *Aquatic ecosystems* include marine environments and the fresh water environments of the lands. Marine ecosystems include the open ocean, coastal estuaries, and coral reefs. Fresh water ecosystems include lakes, ponds, streams, marshes, and bogs. Our survey of physical geography will not include these aquatic environments. Instead, we will focus on the **terrestrial ecosystems,** which are dominated by land plants spread widely over the upland surfaces of the continents. The terrestrial ecosystems are directly influenced by climate and interact with the soil. In this way, they are closely woven into the fabric of physical geography.

Within terrestrial ecosystems, the largest recognizable subdivision is the **biome.** Although the biome includes the total assemblage of plant and animal life interacting within the life layer, the green plants dominate the biome physically because of their enormous biomass, as compared with that of other organisms. Plant geographers concentrate on the characteristic life-form of the green plants within the biome. These life-forms are principally trees, shrubs, lianas, and herbs, but other life-forms are important in certain biomes.

There are five principal biomes. The **forest biome** is dominated by trees, which form a closed or nearly closed canopy. Forest requires an abundance of soil water, so forests are found in moist climates. Temperatures must also be suitable, requiring at least a warm season, if not warm temperatures the year round. The **savanna biome** is transitional between forest and grassland. It exhibits an open cover of trees with grasses and herbs underneath. The **grassland biome** develops in regions with moderate shortages of soil water. The semiarid regions of the dry tropical, dry subtropical, and dry midlatitude climates are the home of the grassland biome. Temperatures must also provide adequate warmth during the growing season. The **desert biome** includes organisms that can survive a moderate to severe water shortage for most, if not all, of the year. Temperatures can range from very hot to cool. Plants are often xerophytes, showing adaptations to the dry environment. The **tundra biome** is limited by cold temperatures. Only small plants that can grow quickly when temperatures warm above freezing in the warmest month or two can survive.

Biogeographers break the biomes down further into smaller vegetation units, called *formation classes,* using the life-form of the plants. For example, at least four and perhaps as many as six kinds of forests are easily distinguished within the forest biome. At least three kinds of grasslands are easily recognizable. Deserts, too, span a wide range in terms of the abundance and life-form of plants. The formation classes described in the remaining portion of this chapter are major, widespread types that are clearly associated with specific climate types. Figure 24.3 is a generalized world map of the formation classes. It simplifies the very

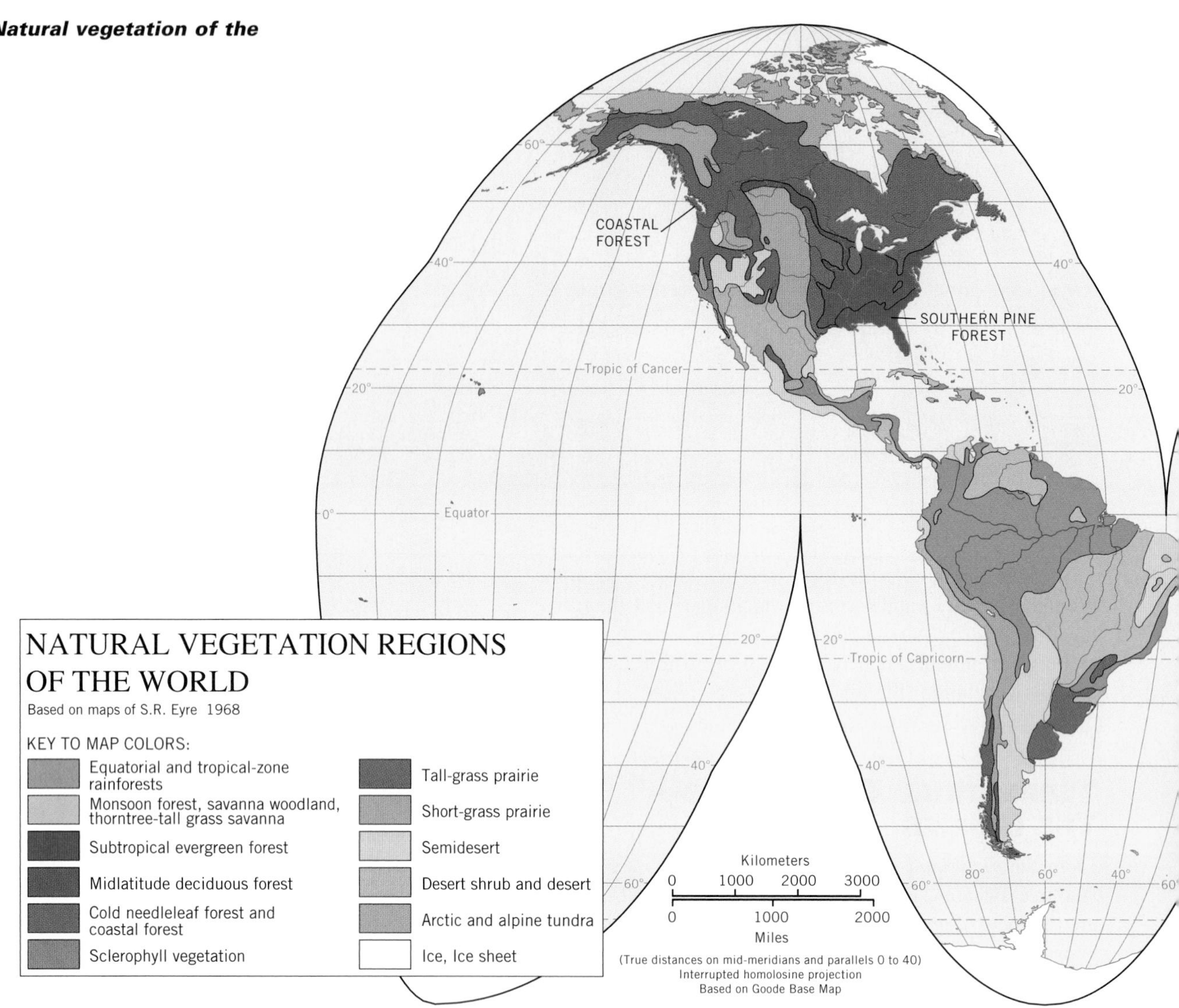

24.3 Natural vegetation of the world.

complex patterns of natural vegetation to show large uniform regions in which a given formation class might be expected to occur. With the advent of remote sensing, it is possible to map global land cover more accurately. *Focus on Remote Sensing 24.1 • Mapping Global Land Cover from Satellite* describes how data from NASA's MODIS instrument is used for this purpose. GEODISCOVERIES

Forest Biome

Within the forest biome, we can recognize six major formations: low-latitude rainforest, monsoon forest, subtropical evergreen forest, midlatitude deciduous forest, needleleaf forest, and sclerophyll forest.

LOW-LATITUDE RAINFOREST *Low-latitude rainforest,* found in the equatorial and tropical latitude zones, consists of tall, closely set trees. Crowns form a continuous canopy of foliage and provide dense shade for the ground and lower layers (Figure 24.4*a*). The trees are characteristically smooth-barked and unbranched in the lower two-thirds. Tree leaves are large and evergreen—thus, the equatorial rainforest is often described as *broadleaf evergreen forest.*

Crowns of the trees of the low-latitude rainforest tend to form two or three layers (Figure 24.5). The highest layer consists of scattered "emergent" crowns that protrude from the closed canopy below, often rising to 40 m (130 ft). Some emergent species develop wide buttress roots, which aid in their physical support (Figure 24.4*b*). Below the layer of emergents is a second, continuous layer, which is 15 to 30 m (about 50 to 100 ft) high. A third, lower layer consists of small, slender trees 5 to 15 m (about 15 to 50 ft) high with narrow crowns.

Typical of the low-latitude rainforest are thick, woody lianas supported by the trunks and branches of trees. Some are slender, like ropes, while others reach thicknesses of 20 cm (8 in.). They climb high into the trees to the upper canopy, where light is available, and develop numerous branches of their own. *Epiphytes* ("air plants") are also common in low-latitude rainforest. These plants attach themselves to the trunk, branches, or foliage of trees and lianas.

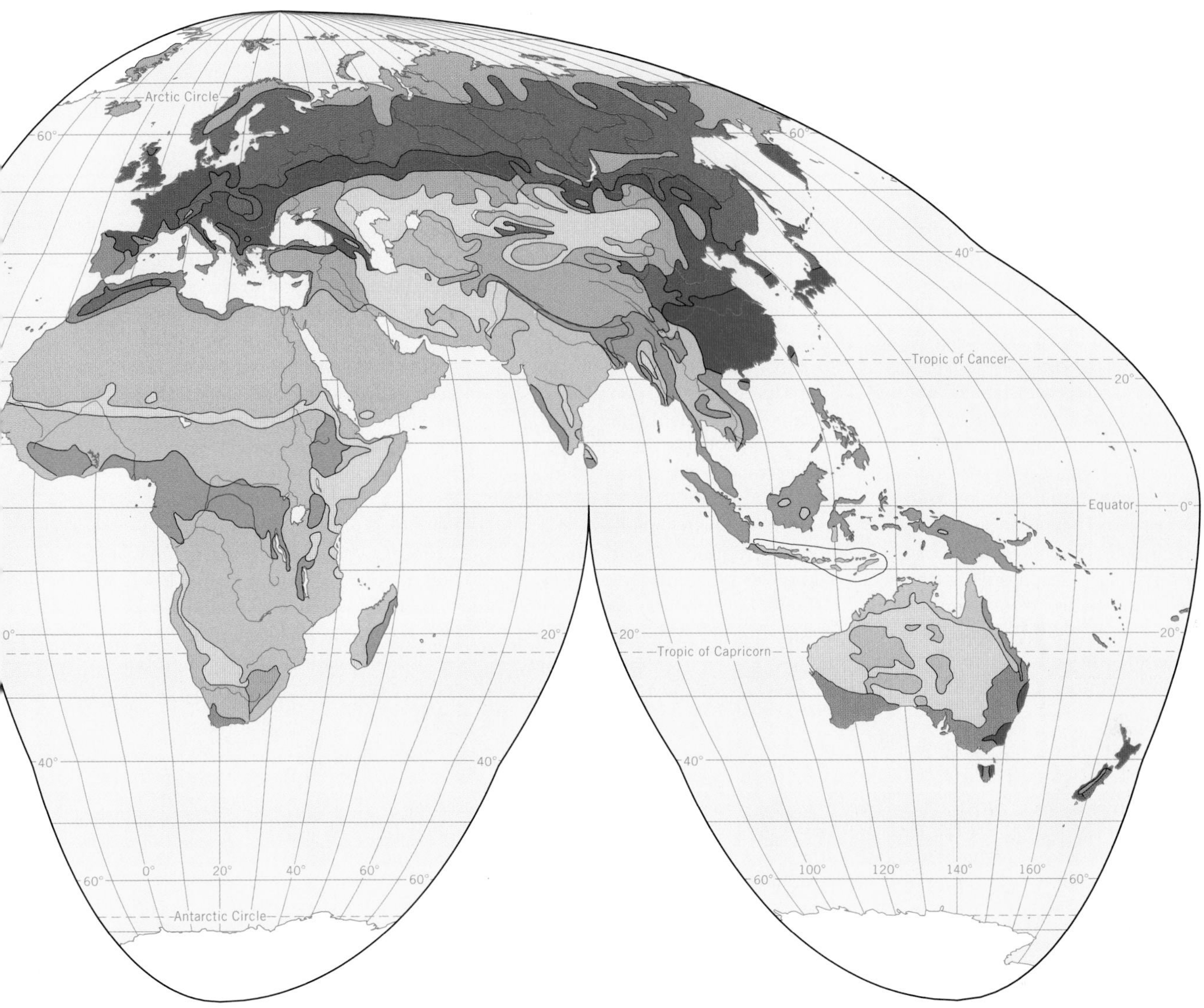

Their host is used solely as a means of physical support. Epiphytes include plants of many different types—ferns, orchids, mosses, and lichens (Figure 24.4*c*).

A particularly important characteristic of the low-latitude rainforest is the large number of species of trees that coexist. In equatorial regions of rainforest, as many as 3000 species may be found in a few square kilometers. Individuals of a given species are often widely separated. Many species of plants and animals in this very diverse ecosystem still have not been identified or named by biologists.

Equatorial and tropical rainforests are not jungles of impenetrable plant thickets. Rather, the floor of the low-latitude rainforest is usually so densely shaded that plant foliage is sparse close to the ground. This gives the forest an open aspect, making it easy to travel within its interior. The ground surface is covered only by a thin litter of leaves. Dead plant matter rapidly decomposes because the warm temperatures and abundant moisture promote its breakdown by bacteria. Nutrients released by decay are quickly absorbed by roots. As a result, the soil is low in organic matter.

Large herbivores are uncommon in the low-latitude rainforest. They include the African okapi and the tapir of South America and Asia. Most herbivores are climbers, and they include many primates—monkeys and apes. Toucans, parrots, and tinamous join fruit-eating bats as flying grazers of the forest. Tree sloths spend their lifetimes hanging upside down as they browse the forest canopy. There are few large predators. Notable are the leopards of African and Asian forests and the jaguars and ocelots of the South American forests.

Low-latitude rainforest develops in a climate that is continuously warm, frost-free, and has abundant precipitation in all months of the year (or, at most, has only one or two dry months). These conditions occur in the wet equatorial climate ① and the monsoon and trade-wind coastal climate ②. In the absence of a cold or dry season, plant growth goes on continuously throughout the year. In this uniform environment, some plant species grow new leaves continuously, shedding old ones continuously as well. Still other plant species shed their leaves according to their own seasons, responding to the slight changes in daylight length that occur with the seasons.

Focus on Remote Sensing | 24.1

Mapping Global Land Cover by Satellite

Imagine yourself as an astronaut living on an orbiting space station, watching the Earth turn underneath you. One of the first things that would strike you about the land surface is its color and how it changes from place to place and time to time. Deserts are in shades of brown, dotted with white salty playas and the black spots and streaks of recent volcanic activity. Equatorial forests are green and lush, dissected by branching lines of dark rivers. Shrublands are marked by earth colors, but with a greenish tinge. Some regions show substantial change throughout the year. In the midlatitude zone, forests and agricultural lands go from intense green in the summer to brown as leaves drop and crops are harvested. Snow expands equatorward in the fall and winter and retreats poleward in the spring and summer. In the tropical zones, grasslands and savannas go from brown to green to brown again as the rainy season comes and goes. Some features, such as lakes, remain nearly unchanged throughout the year. Thus, the color of the land surface and the change in color through time, as viewed from an orbiting Earth satellite, can provide information about the type of land cover present at a location.

Ever since the first satellite images of the Earth were received, scientists have used color—that is, the spectral reflectance of the surface—as an indicator of land cover type. For example, there is a thirty-year history of producing land cover maps for local areas using individual Landsat images from cloud-free dates. But global mapping of land cover requires instruments that can observe the surface on a daily or near-daily basis, thus maximizing the opportunity to image the land surface in parts of the globe that are frequently cloud-covered. These instruments have a coarser ground resolution than Landsat, with pixels in the range of 500 to 1100 meters on a side, but that is not a deficiency when working at a global scale.

Shown below is a map of global land cover produced from MODIS

Global land cover from MODIS *This map of global land cover types was constructed from MODIS data acquired largely during 2001. The map has a spatial resolution of 1 km^2—that is, each square kilometer of the Earth's land surface is independently assigned a land cover type label. Goode's homolosine projection. (A. H. Strahler, Boston University/NASA.)*

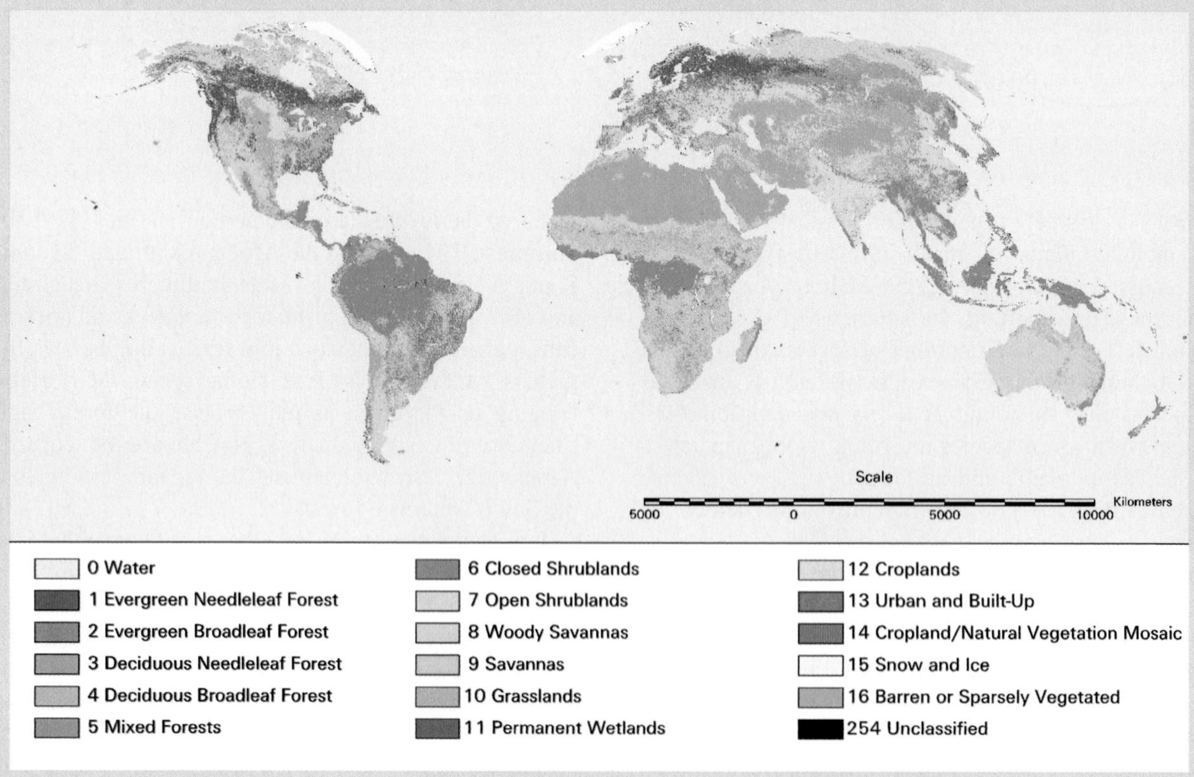

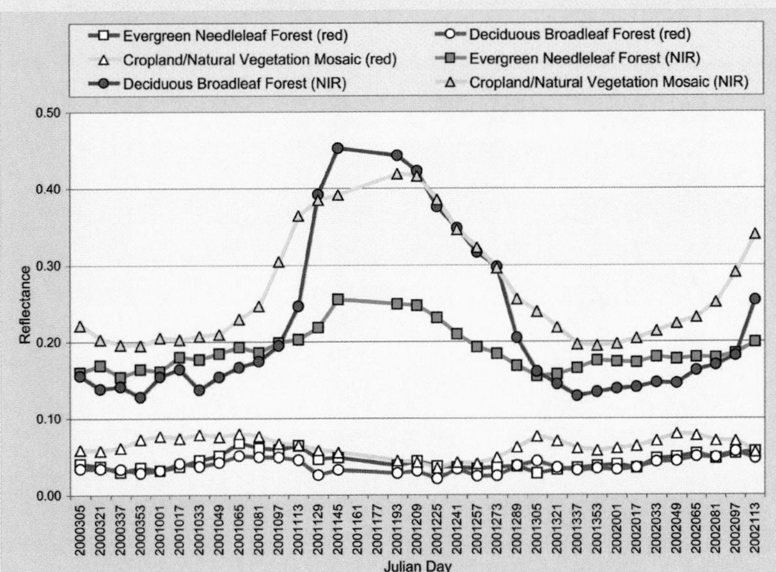

Spectral and temporal reflectance patterns *The plot shows how the spectral reflectance of three land cover types in the southeastern United States varies during the period 31 October 2000 to 23 April 2002. Dates are shown in Julian date format: the first four digits indicate the year, and the remaining three indicate the day of the year with 1 January taken as 001. (A. H. Strahler, Boston University/NASA.)*

images acquired largely in 2001. The legend recognizes 17 types of land covers, including forests, shrublands, savannas, grasslands, and wetlands.

The global pattern of land cover types is rather similar to that shown in Figure 24.3. Evergreen broadleaf forest dominates the equatorial belt, stretching from South America, through Central Africa, to south Asia. Adjoining the equatorial forest belt are regions of savannas and grasslands, which have strong wet-dry climates. The vast desert region running from the Sahara to the Gobi is barren or sparsely vegetated. It is flanked by grasslands on the west, north, and east. Broadleaf deciduous forests are prominent in eastern North America, western Europe, and eastern Asia. Evergreen needleleaf forests span the boreal zone from Alaska and northwest Canada to Siberia. Croplands are found throughout most regions of human habitation, except for dry desert regions and cold boreal zones.

The map was constructed using both spectral and temporal information. The graph above shows how these information sources are used. It depicts reflectance values in red and near-infrared (NIR) spectral bands for three land cover types as observed in the southeastern United States—evergreen needleleaf forest, cropland/natural vegetation mosaic, and deciduous broadleaf forest. The reflectances are shown as they change over the course of about a year and a half, from 31 October 2000 (2000305) to 23 April 2002 (2002113).

The top three curves, which are near-infrared values, show the patterns of the three types most clearly. During the winter, values are generally low, with deciduous broadleaf forest, evergreen needleleaf forest, and cropland in increasing order of reflectance. As spring begins, cropland reflectance rises before deciduous broadleaf forest but deciduous broadleaf forest reaches a higher peak. Evergreen needleleaf forest also shows a spring green-up, but it is later than the others and peaks at a lower value. Reflectance gradually drops during the fall, and the three types reach about the same reflectance levels as in the prior year. The three lower curves, which display the red band, also have distinctive features, but they are less obvious on this plot.

Given spectral and temporal information, how is the global map made? The process by which each global pixel is given a label is referred to as *classification*. In short, a computer program is presented with many examples of each land cover type. It then "learns" the examples and uses them to classify pixels depending on their spectral and temporal pattern. The MODIS global land cover map shown was prepared with more than 1500 examples of the 17 land cover types. It is estimated to be about 75-80 percent accurate.

Land cover mapping is a common application of remote sensing. Given the ability of spaceborne instruments to image the Earth consistently and repeatedly, classification of remotely sensed data is a natural way of extending our knowledge from the specific to the general to provide valuable new geographic information.

24.4 Low-latitude Rainforest Gallery

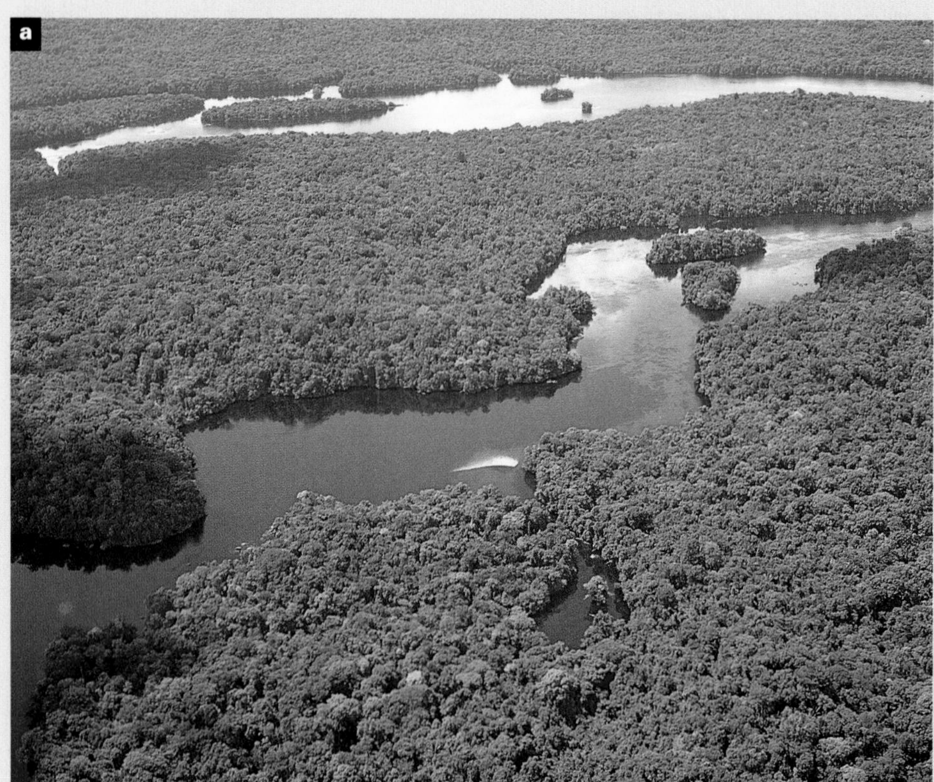

a...***Aerial view of low-latitude rainforest*** *Middle Mazaruni River near Bartica, Guyana.*

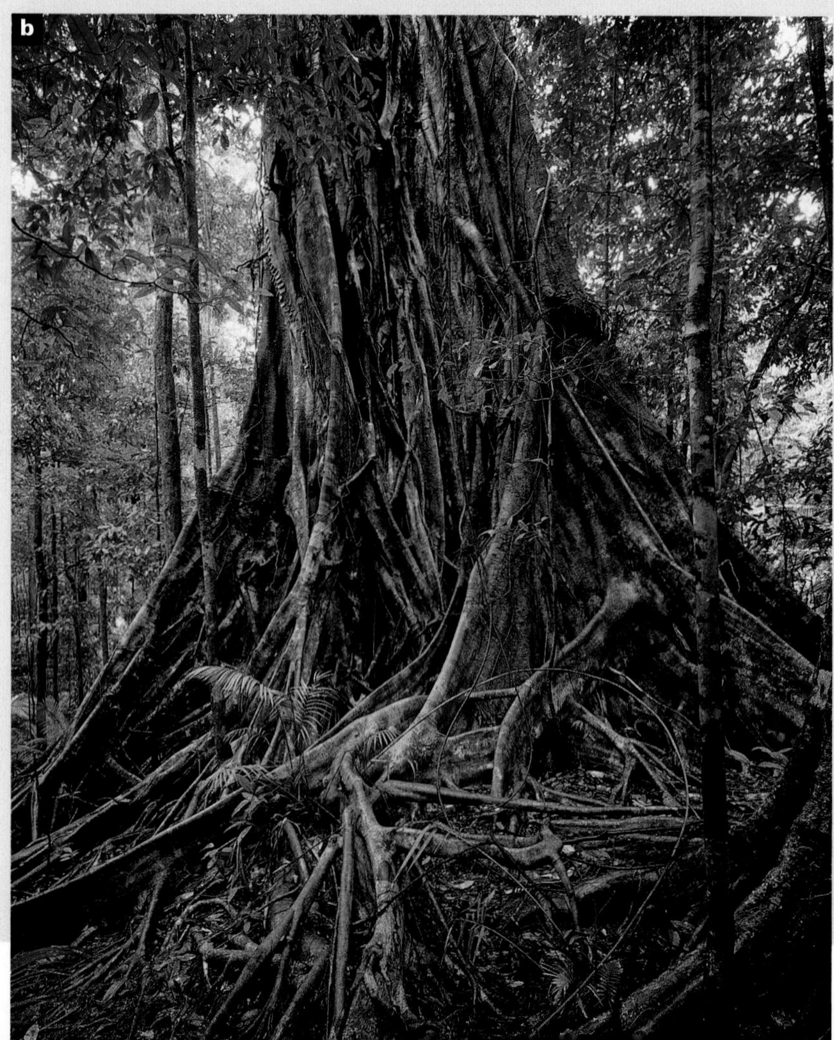

b...***Buttress roots at the base of a large rainforest tree*** *From Daintree National Park, Queensland, Australia.*

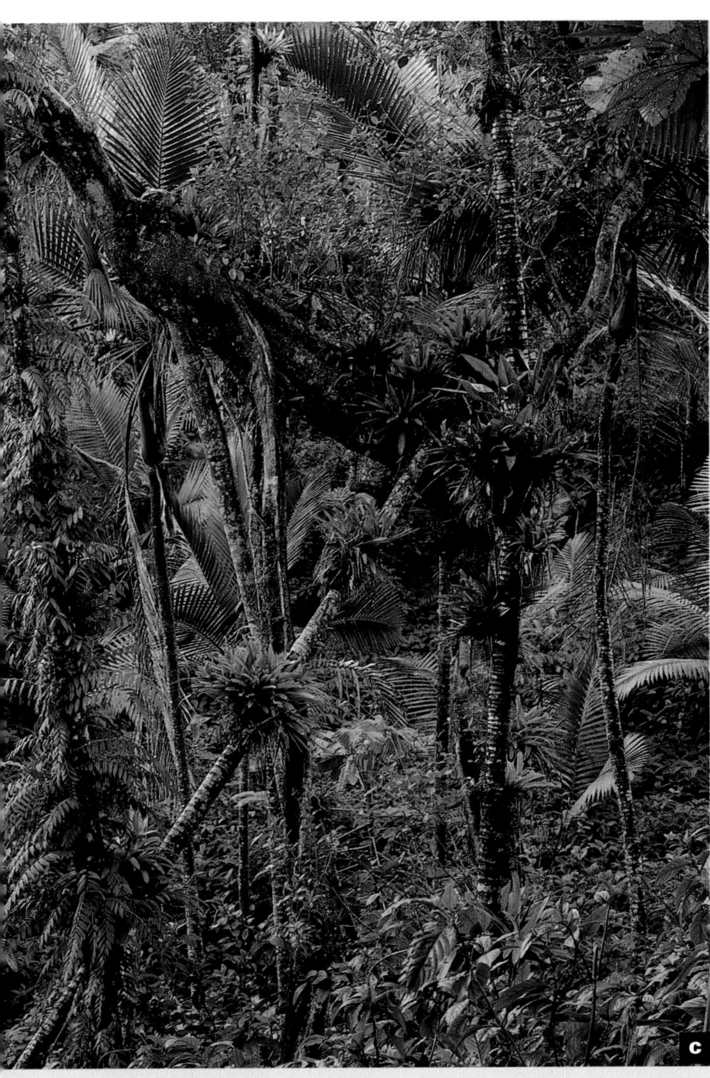

c...***Epiphytes*** *In this photo from the El Yunque rainforest, Caribbean National Forest, Puerto Rico, red-flowering epiphytes adorn the trunks of sierra palms.*

d...***Three-toed tree sloth*** *This slow-moving, amiable denizen of the low-latitude rainforest, here photographed in French Guyana, lives in the trees, foraging on the leaves and fruits of the many rainforest tree species. Once more widespread, the tree sloth is now endangered.*

24.5 Rainforest layers *This diagram shows the typical structure of equatorial rainforest. (After J. S. Beard, The Natural Vegetation of Trinidad, Clarendon Press, Oxford.)*

World distribution of the low-latitude rainforest is shown in Figure 24.6. A large area of rainforest lies astride the equator. In South America, this equatorial rainforest includes the Amazon lowland. In Africa, the *equatorial rainforest* is found in the Congo lowland and in a coastal zone extending westward from Nigeria to Guinea. In Indonesia, the island of Sumatra, on the west, bounds a region of equatorial rainforest that stretches eastward to the islands of the western Pacific.

The low-latitude rainforest extends poleward through the tropical zone (lat. 10° to 25° N and S) along monsoon and trade-wind coasts. The monsoon and trade-wind coastal climate in which this *tropical-zone rainforest* thrives has a short dry season. However, the dry season is not intense enough to deplete soil water.

In the northern hemisphere, trade-wind coasts that have rainforests are found in the Philippine Islands and also along the eastern coasts of Central America and the West Indies. These highlands receive abundant orographic rainfall in the belt of the trade winds. A good example of tropical-zone rainforest is the rainforest of the eastern mountains of Puerto Rico. In Southeast Asia, tropical-zone

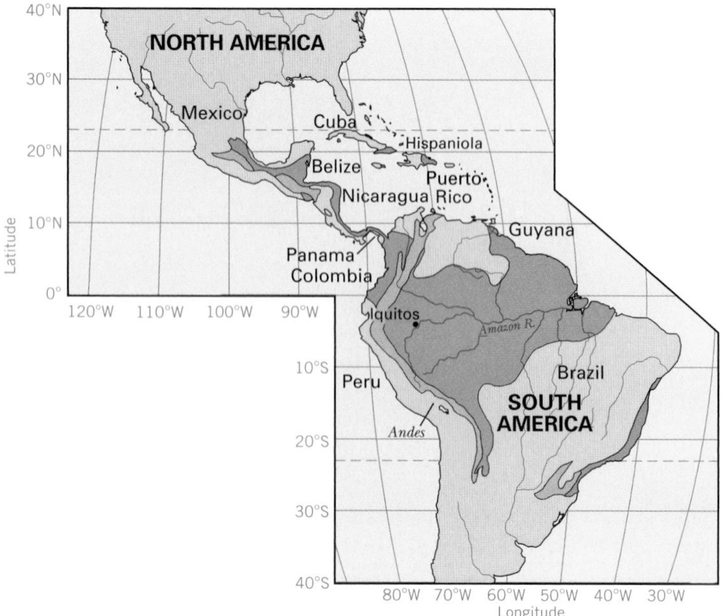

24.6 Low-latitude rainforest *World map of low-latitude rainforest, showing equatorial and tropical rainforest types. (Data source same as Figure 24.3. Based on Goode Base Map.)*

rainforest is extensive in monsoon coastal zones and highlands that have heavy rainfall and a very short dry season. These occur in Vietnam and Laos, southeastern China, and on the western coasts of India and Myanmar. In the southern hemisphere, belts of tropical-zone rainforest extend down the eastern Brazilian coast, the Madagascar coast, and the coast of northeastern Australia.

Within the regions of low-latitude rainforest are many island-like highland regions where climate is cooler and rainfall is increased by the orographic effect. Here, rainforest extends upward on the rising mountain slopes. Between 1000 and 2000 m (about 3300 and 6500 ft), the rainforest gradually changes in structure and becomes *montane forest* (Figure 24.6). The canopy of montane forest is more open, and tree heights are lower, than in the rainforest. With increasing elevation, the forest canopy height becomes even lower. Tree ferns and bamboos are numerous, and epiphytes are particularly abundant. As elevation increases, mist and fog become persistent, giving high-elevation montane forest the name *cloud forest.*

Low-latitude rainforest is under increasing human pressure. Slowly but surely, the rainforest is being conquered by logging, clearcutting, and conversion to grazing and farming. *Eye on the Environment 24.2 • Exploitation of the Low-Latitude Rainforest Ecosystem* describes this process in more detail.

MONSOON FOREST *Monsoon forest* of the tropical latitude zone differs from tropical rainforest because it is deciduous, with most of the trees of the monsoon forest shedding their leaves during the dry season. Shedding of leaves results from the stress of a long dry season that occurs at the time of low sun and cooler temperatures. In the dry season, the forest resembles the deciduous forests of the midlatitudes during their leafless winter season.

Monsoon forest is typically open. It grades into woodland, with open areas occupied by shrubs and grasses (Figure 24.7). Because of its open nature, light easily reaches the lower layers of the monsoon forest. As a result, these lower layers are better developed than in the rainforest. Tree heights are also lower. Typically, many tree species are present—as many as 30 to 40 species in a small tract—although the rainforest has many more. Tree trunks are massive, often with thick, rough bark. Branching starts at a comparatively low level and produces large, round crowns.

Figure 24.8 is a world map of the monsoon forest and closely related formation classes. Monsoon forest develops in the wet-dry tropical climate ③ in which a long rainy season alternates with a dry, rather cool season. These conditions, though most strongly developed in the Asiatic monsoon climate, are not limited to that area. The typical regions of monsoon forest are in Myanmar, Thailand, and Cambodia. In the monsoon forest of southern Asia, the teakwood tree was once abundant and was widely exported to the Western world to make furniture, paneling, and decking. Now this great tree is logged out, and the Indian elephant, once trained to carry out this logging work, is unemployed. Large areas of monsoon forest also occur in south central Africa and in Central and South America, bordering the equatorial and tropical rainforests.

SUBTROPICAL EVERGREEN FOREST *Subtropical evergreen forest* is generally found in regions of moist subtropical climate ⑥, where winters are mild and there is ample rainfall throughout the year. This forest occurs in two forms: broadleaf and needleleaf. The *subtropical broadleaf evergreen forest* differs from the low-latitude rainforests, which are also broadleaf evergreen types, in having relatively few species of trees. Trees are not as tall as in the low-latitude

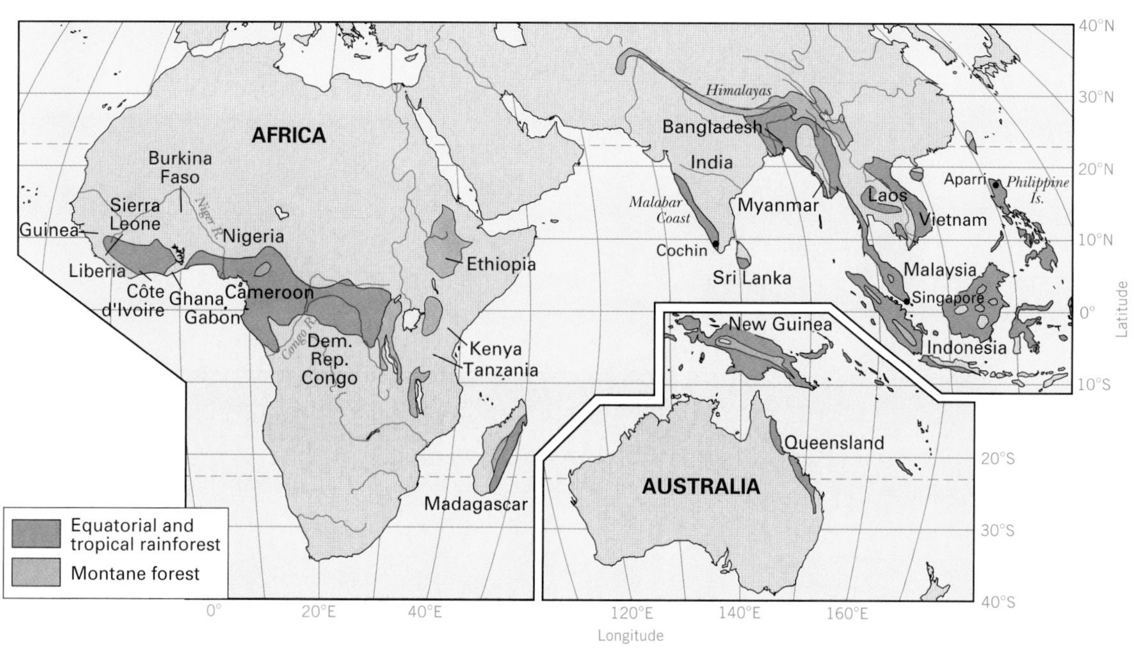

Eye on the Environment 24.2

Exploitation of the Low-Latitude Rainforest Ecosystem

Many of the world's equatorial and tropical regions are home to the rainforest ecosystem. This ecosystem is perhaps the most diverse on Earth. That is, it possesses more species of plants and animals than any other. Very large tracts of rainforest still exist in South America, south Asia, and some parts of Africa. Ecologists regard this ecosystem as a genetic reservoir of many species of plants and animals. But as human populations expand and the quest for agricultural land continues, low-latitude rainforests are being threatened with clearing, logging, cultivation of cash crops, and animal grazing.

In the past, low-latitude rainforests were farmed by native peoples using the *slash-and-burn* method—cutting down all the vegetation in a small area, then burning it (see photo below). In a rainforest ecosystem, most of the nutrients are held within living plants rather than in the soil. Burning the vegetation on the site releases the trapped nutrients, returning a portion of them to the soil. Here, the nutrients are available to growing crops. The supply of nutrients derived from the original vegetation cover is small, however, and the harvesting of crops rapidly depletes the nutrients. After a few seasons of cultivation, a new field is cleared, and the old field is abandoned. Rainforest plants reestablish their hold on the abandoned area. Eventually, the rainforest returns to its original state. This cycle shows that primitive, slash-and-burn agriculture is fully compatible with maintenance of the rainforest ecosystem.

On the other hand, modern intensive agriculture uses large areas of land and is not compatible with the rainforest ecosystem. When large areas are abandoned, seed sources are so far away that the original forest species cannot take hold. Instead, secondary species dominate, often accompanied by species from other vegetation types. These species are good invaders, and once they enter an area, they tend to stay. The dominance of these secondary species is permanent, at least on the human time scale. Thus, we can regard the rainforest ecosystem as a resource that, once cleared, will never return. The loss of low-latitude rainforest will result in the disappearance of thousands of species of organisms from the rainforest environment—a loss of millions of years of evolution, together with the destruction of the most complex ecosystem on Earth.

In Amazonia, transformation of large areas of rainforest into agricultural land uses heavy machinery to carve out major highways, such as the Trans-Amazon Highway in Brazil, and innumerable secondary roads and trails. Large fields for cattle pasture or commercial crops are created by cutting, bulldozing, clearing, and burning the vegetation. In some regions, the great broadleaved rainforest trees are removed for commercial lumber.

What are the effects of large-scale clearing? A recent study using a computer model by scientists of the University of Maryland and the Brazilian Space Research Institute indicated that when the Amazon rainforest is entirely removed and replaced with pasture, surface and soil temperatures will be increased 1 to 3°C (2 to 5°F).

24.7 Monsoon woodland *This woodland is in the Bandipur Wild Animal Sanctuary in the Nilgiri Hills of southern India. The scene is taken in the rainy season, with trees in full leaf.*

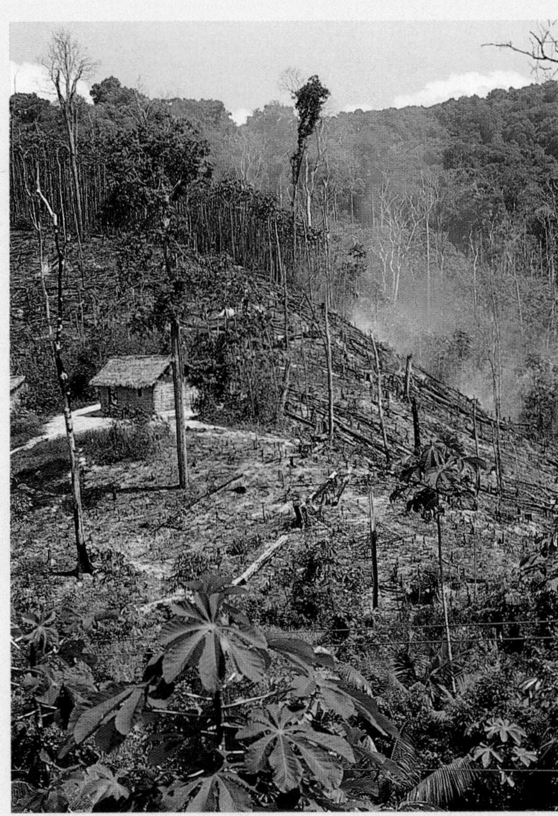

Slash-and-burn clearing *This rainforest in Maranhão, Brazil, has been felled and burned in preparation for cultivation.*

Precipitation in the region will decline by 26 percent and evaporation by 30 percent. The deforestation will change weather and wind patterns so that less water vapor enters the Amazon Basin from outside sources, making the basin even drier. In areas where a marked dry season occurs, that season will be lengthened. Although such models contain simplifications and are subject to error, the results confirm the pessimistic conclusion that once large-scale deforestation has occurred, artificial restoration of a rainforest comparable to the original one may be impossible to achieve.

According to a report by the United Nations Food and Agriculture Organization, about 0.6 percent of the world's rainforest is lost annually by conversion to other uses. More rainforest land, 2.2 million hectares (about 8500 mi^2), is lost annually in Asia than in Latin America and the Caribbean, where 1.9 million hectares (about 7300 mi^2) are converted every year. Africa's loss of rainforest is estimated at about 470,000 hectares per year (1800 mi^2). Among individual countries, Brazil and Indonesia are the loss leaders, accounting for nearly half of the rainforest area converted to other uses. Note that these values do not include even larger losses of moist deciduous forests in these regions. Deforestation in low-latitude dry deciduous forests and hill and montane forests is also very serious.

Although deforestation rates are very rapid in some regions, many nations are now working to reduce the rate of loss of rainforest environment. However, because the rainforest can provide agricultural land, minerals, and timber, the pressure to allow deforestation continues.

rainforests. Their leaves tend to be smaller and more leathery, and the leaf canopy less dense. The subtropical broadleaf evergreen forest often has a well-developed lower layer of vegetation. Depending on the location, this layer may include tree ferns, small palms, bamboos, shrubs, and herbaceous plants. Lianas and epiphytes are abundant.

Figure 24.9 is a map of the subtropical evergreen forests of the northern hemisphere. Here subtropical evergreen forest consists of broad-leaved trees such as evergreen oaks and trees of the laurel and magnolia families. The name "laurel forest" is applied to these forests, which are associated with the moist subtropical climate ⑥ in the southeastern United States, southern China, and southern Japan. However, these regions are under intense crop cultivation because of their favorable climate. In these areas, the land has been cleared of natural vegetation for centuries, and little natural laurel forest remains.

The *subtropical needleleaf evergreen forest* occurs only in the southeastern United States (Figure 24.9). Here it is referred to as the southern pine forest, since it is dominated by species of pine. It is found on the wide belt of sandy soils that fringes the Atlantic and Gulf coasts. Because the soils are sandy, water drains away quickly, leaving the soils quite dry. In infrequent drought years, these forests may burn. Since pines are well adapted to droughts and fires, they form a stable vegetation cover for large areas of the region. Timber companies have taken advantage of this natural preference for pines, creating many plantations yielding valuable lumber and pulp (Figure 24.10).

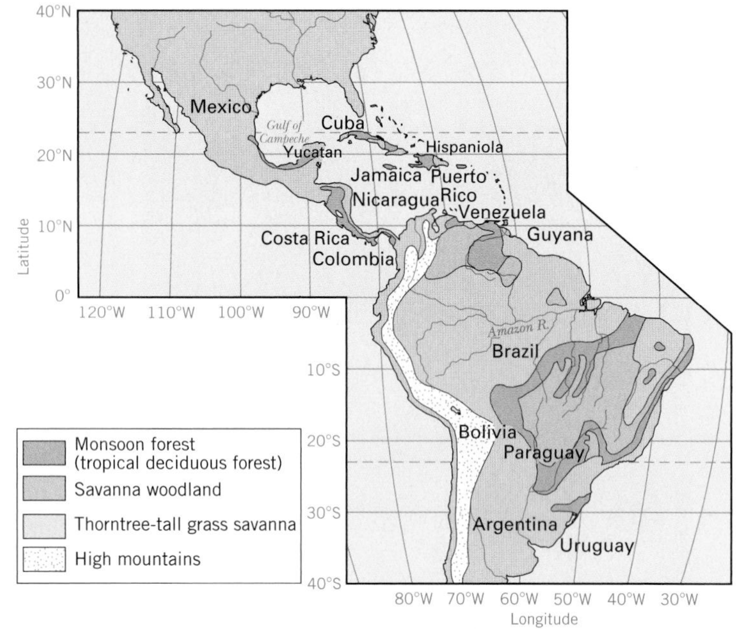

24.8 Monsoon forest *World map of monsoon forest and related types—savanna woodland and thorntree-tall grass savanna. (Data source same as Figure 24.3. Based on Goode Base Map.)*

MIDLATITUDE DECIDUOUS FOREST *Midlatitude deciduous forest* is the native forest type of eastern North America and Western Europe. It is dominated by tall, broadleaf trees that provide a continuous and dense canopy in summer but shed their leaves completely in the winter. Lower layers of small trees and shrubs are weakly developed. In the spring, a lush layer of lowermost herbs quickly develops but soon fades after the trees have reached full foliage and shaded the ground.

Figure 24.11 is a map of midlatitude deciduous forests, which are found almost entirely in the northern hemisphere. Throughout much of its range, this forest type is associated with the moist continental climate ⑩. Recall from Chapter 9 that this climate receives adequate precipitation in all months, normally with a summer maximum. There is a strong annual temperature cycle with a cold winter season and a warm summer.

Common trees of the deciduous forest of eastern North America, southeastern Europe, and eastern Asia are oak, beech, birch, hickory, walnut, maple, elm, and ash. A few needleleaf trees are often present as well—hemlock, for example (Figure 24.12). Where the deciduous forests have been cleared by lumbering, pines readily develop as second-growth forest.

In Western Europe, the midlatitude deciduous forest is associated with the marine west-coast climate ⑧. Here the

24.9 Subtropical evergreen forest *Northern hemisphere map of subtropical evergreen forests, including the southern pine forest. (Data source same as Figure 24.3. Based on Goode Base Map.)*

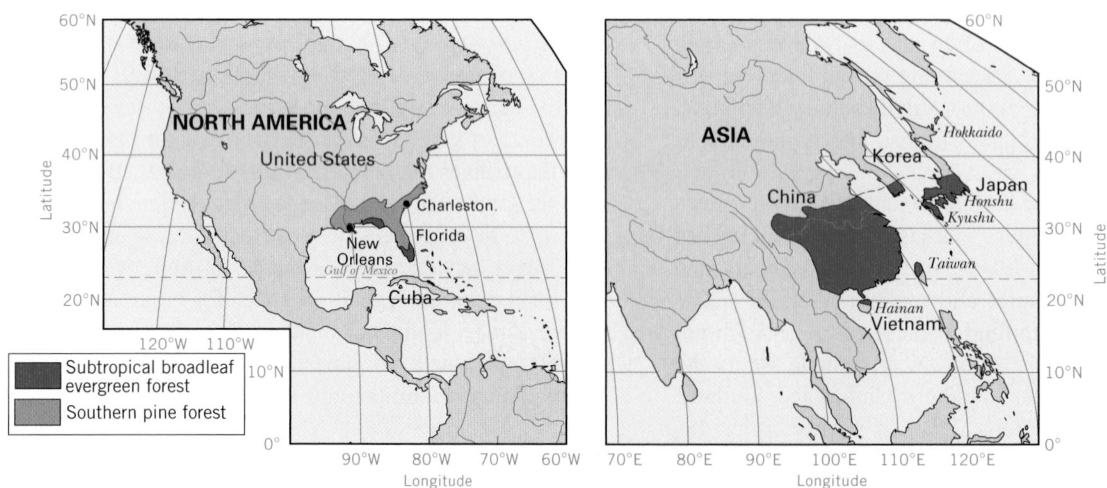

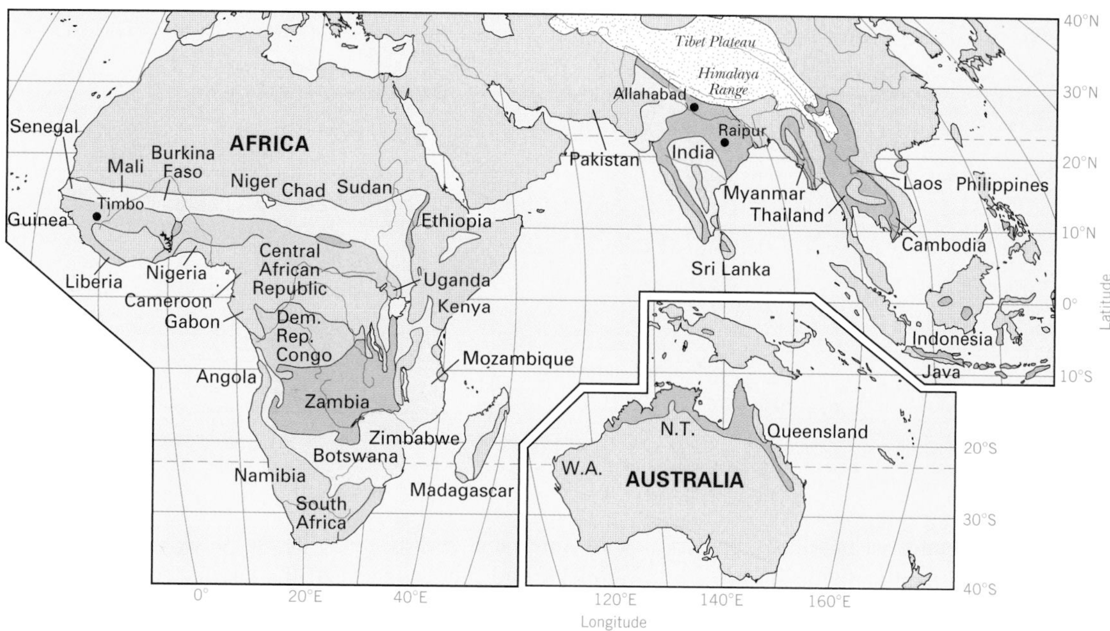

dominant trees are mostly oak and ash, with beech in cooler and moister areas. In Asia, the midlatitude deciduous forest occurs as a belt between the boreal forest to the north and steppelands to the south. A small area of deciduous forest is found in Patagonia, near the southern tip of South America.

The deciduous forest includes a great variety of animal life, much of it stratified according to canopy layers. As many as five layers can be distinguished: the upper canopy, lower canopy, understory, shrub layer, and ground layer. Because the ground layer presents a more uniform environment in terms of humidity and temperature, it contains

24.10 Pine plantation *This plantation of longleaf pine grows on the sandy soil of the southeastern coastal plain. Near Waycross, Georgia.*

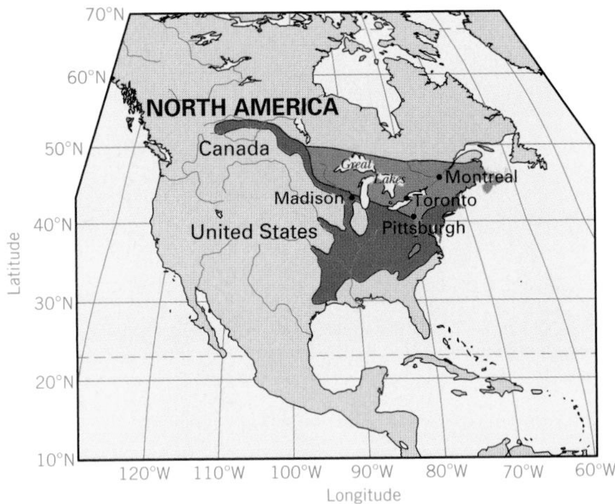

24.11 Midlatitude deciduous forest *Northern hemisphere map of midlatitude deciduous forests. (Data source same as Figure 24.3. Based on Goode Base Map.)*

the largest concentration of organisms and the greatest diversity of species. Many small mammals burrow in the soil for shelter or food in the form of soil invertebrates. Among this burrowing group are ground squirrels, mice, and shrews, as well as some larger animals—foxes, woodchucks, and rabbits. Most of the larger mammals feed on ground and shrub layer vegetation, except for some, such as the brown bear, which are omnivorous and prey upon the small animals as well.

24.12 Deciduous forest *In this deciduous forest of the Catskill Mountains, New York, some needleleaf trees—pine and hemlock—are also present.*

Even though birds possess the ability to move through the layers at will, most actually restrict themselves to one or more layers. For example, the wood peewee is found in the lower canopy, and the red-eyed vireo is found in the understory. Above them are the scarlet tanagers and blackburnian warblers, which are upper canopy dwellers. Below are the ground dwellers, such as grouse, warblers, and ovenbirds. Flying insects often show similar stratification patterns.

Among the large herbivores that graze in the deciduous forest are the red deer and roe deer of Eurasia and the white-tailed deer of North America. Smaller herbivores include voles, mice, and squirrels. Predators include bears, wildcats, lynx, wolves, foxes, weasels, and, among birds, owl species.

NEEDLELEAF FOREST *Needleleaf forest* refers to a forest composed largely of straight-trunked, cone-shaped trees with relatively short branches and small, narrow, needle-like leaves. These trees are conifers. Most are evergreen, retaining their needles for several years before shedding them. When the needleleaf forest is dense, it provides continuous and deep shade to the ground. Lower layers of vegetation are sparse or absent, except for a thick carpet of mosses that may occur. Species are few—in fact, large tracts of needleleaf forest consist almost entirely of only one or two species.

Boreal forest is the cold-climate needleleaf forest of high latitudes. It occurs in two great continental belts, one in North America and one in Eurasia (Figure 24.13). These belts span their land masses from west to east in latitudes 45° N to 75° N, and they closely correspond to the region of boreal forest climate ⑪. The boreal forest of North America, Europe, and western Siberia is composed of such evergreen conifers as spruce and fir (Figure 24.14). The boreal forest of north-central and eastern Siberia is dominated by larch.

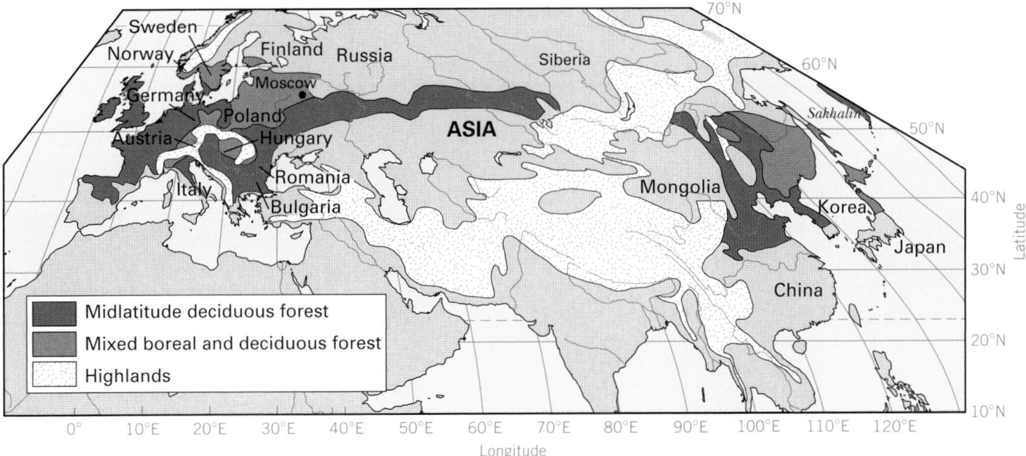

The larch tree sheds its needles in winter and is thus a deciduous needleleaf tree. Broadleaf deciduous trees, such as aspen, balsam poplar, willow, and birch, tend to take over rapidly in areas of needleleaf forest that have been burned over. These species can also be found bordering streams and in open places. Between the boreal forest and the midlatitude deciduous forest lies a broad transition zone of mixed boreal and deciduous forest.

Needleleaf evergreen forest extends into lower latitudes wherever mountain ranges and high plateaus exist. For example, in western North America this formation class extends southward into the United States on the Sierra Nevada and Rocky Mountain Ranges and over parts of the higher plateaus of the southwestern states (Figure 24.15, *left*). In Europe, needleleaf evergreen forests flourish on all the higher mountain ranges.

Geographers at Work

Mapping Forest Ecosystems and Their Biophysical Characteristics with Remote Sensing

by Paul Treitz ***Queen's University, Kingston, Ontario, Canada***

Did you ever hug a tree? If you've done any field work in forestry, you've probably hugged a lot of them. To measure the size of a tree trunk, you have to put your arms around the tree in order to pass a special measuring tape from one hand to the other. Pulling the tape tight, you read off the diameter indicated where the tape's end crosses the remainder of the tape.

How about the height of a tree? Well, you could climb the tree, carrying the end of a measuring tape to the top. Or you could cut the tree down and lay the tape along its length. Of course, there is an easier way—go a fixed distance away from the tree and then sight the angle to its top. From trigonometry, it's easy to find the height.

The height and trunk diameter of a tree are basic characteristics that are related both to the tree's biomass and the amount of photosynthetic material it bears in its leaves. Thus, these measurements are very important in assessing the health and productivity of forest ecosystems. But measuring height, diameter, and related quantities, such as tree spacing or crown diameter, is pretty tedious. Wouldn't it be useful if we could measure these characteristics indirectly using remote sensing?

That's where Paul Treitz's research comes in. He's developing techniques for using Landsat images, air photos, and even airborne lasers to survey large areas of Canadian forests. His goal is to provide the information forest managers need for sustainable development of forest resources while conserving the biodiversity and long-term health of forest ecosystems. You can find out more about his research and see photos of his students making tree measurements at our web site.

Paul Treitz's research focuses on the spectral, spatial, and temporal analysis of remote sensing data of forest ecosystems.

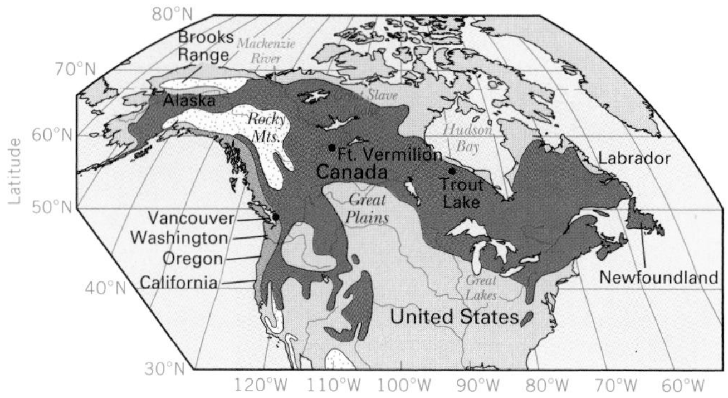

***24.13 Needleleaf forest** Northern hemisphere map of cold-climate needleleaf forests, including coastal forest. (Data source same as Figure 24.3. Based on Goode Base Map.)*

In its northernmost range, the boreal needleleaf forest grades into cold woodland. This form of vegetation is limited to the northern portions of the boreal forest climate ⑪ and the southern portions of the tundra climate ⑫. Trees are low in height and spaced well apart. A shrub layer may be well developed. The ground cover of lichens and mosses is distinctive. Cold woodland is often referred to as *taiga.* It is transitional into the treeless tundra at its northern fringe.

Mammals of the boreal needleleaf forest in North America include deer, moose, elk, black bear, marten, mink, wolf, wolverine, and fisher. Common birds include jays, ravens, chickadees, nuthatches, and a number of warblers. The caribou, lemming, and snowshoe rabbit inhabit both needleleaf forest and the adjacent tundra biome. The boreal forest often experiences large fluctuations in animal species populations, a result of the low diversity and highly variable environment.

Coastal forest is a distinctive needleleaf evergreen forest of the Pacific Northwest coastal belt, ranging in latitude from northern California to southern Alaska (Figure 24.13). Here, in a band of heavy orographic precipitation, mild temperatures, and high humidity, are perhaps the densest of all conifer forests, with magnificent specimens of cedar, spruce, and Douglas fir. At the extreme southern end, coastal forest includes the world's largest trees—redwoods (Figure 24.15, *right*). Individual redwood trees attain heights of over 100 m (about 330 ft) and girths of over 20 m (about 65 ft).

SCLEROPHYLL FOREST The native vegetation of the Mediterranean climate ⑦ is adapted to survival through the long summer drought. Shrubs and trees that can survive such drought are characteristically equipped with small, hard, or thick

***24.14 Boreal forest** A view of the boreal forest, Denali National Park, Alaska, pictured here just after the first snowfall of the season. At this location near the northern limits of the boreal forest, the tree cover is sparse. The golden leaves of aspen mark the presence of this deciduous species.*

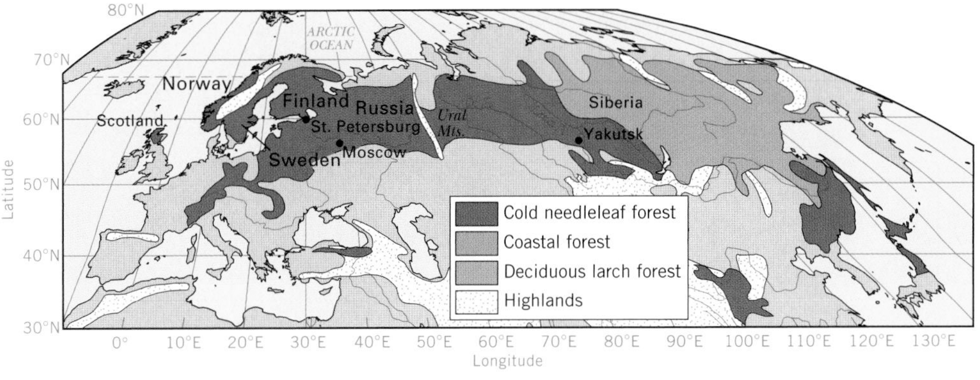

24.15 Two kinds of needleleaf forest of the western United States *(left) Open forest of western yellow pine (ponderosa pine), in the Kaibab National Forest, Arizona. (right) A grove of great redwood trees in Humboldt State Park, California.*

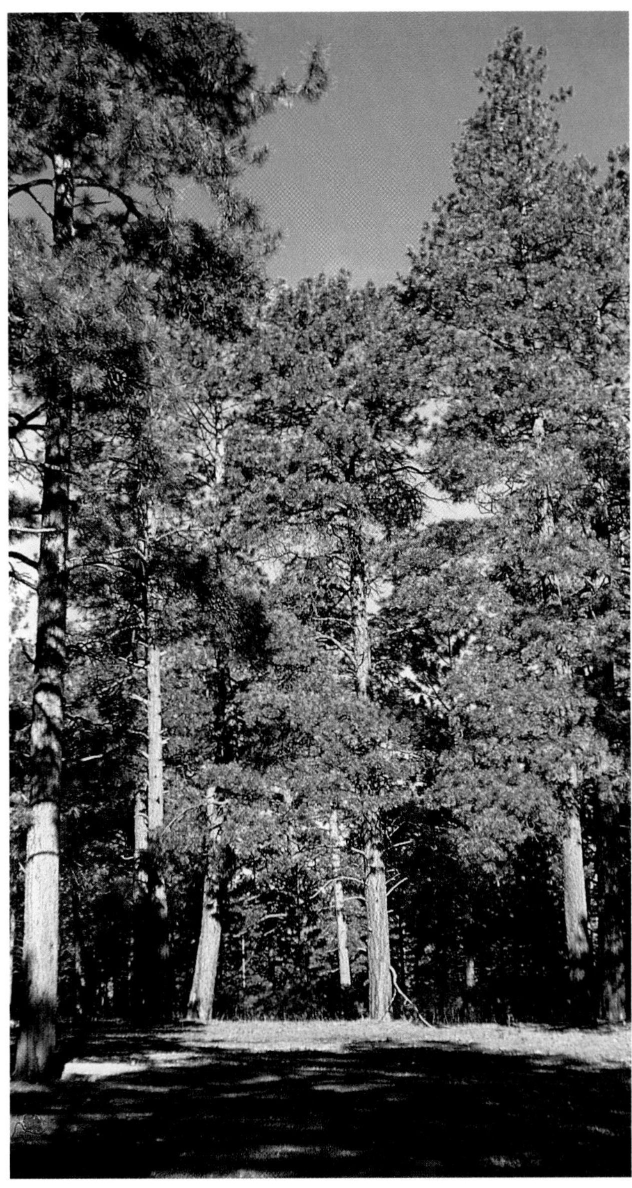

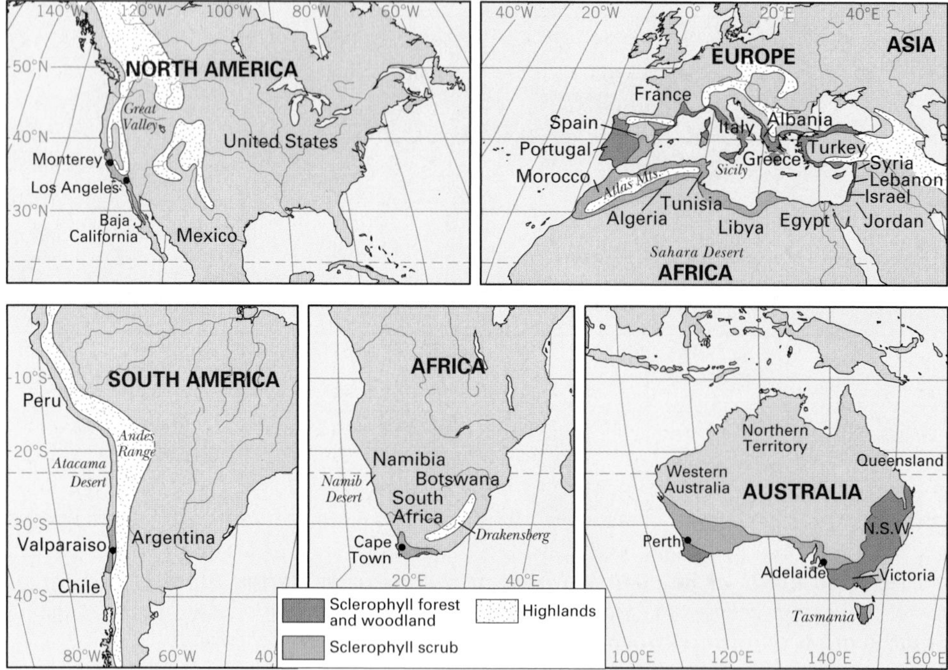

24.16 World map of sclerophyll vegetation *(Data source same as Figure 24.3. Based on Goode Base Map.)*

leaves that resist water loss through transpiration. As we noted earlier in the chapter, these plants are called sclerophylls.

Sclerophyll forest consists of trees with small, hard, leathery leaves. The trees are often low-branched and gnarled, with thick bark. The formation class includes *sclerophyll woodland,* an open forest in which only 25 to 60 percent of the ground is covered by trees. Also included are extensive areas of *scrub,* a plant formation type consisting of shrubs covering somewhat less than half of the ground area. The trees and shrubs are evergreen, retaining their thickened leaves despite a severe annual drought.

Our map of sclerophyll vegetation, Figure 24.16, includes forest, woodland, and scrub types. Sclerophyll forest is closely associated with the Mediterranean climate ⑦ and is narrowly limited to west coasts between 30° and 40° or 45° N and S latitude. In the Mediterranean lands, the sclerophyll forest forms a narrow, coastal belt ringing the Mediterranean Sea. Here, the Mediterranean forest consists of such trees as cork oak, live oak, Aleppo pine, stone pine, and olive. Over the centuries, human activity has reduced the sclerophyll forest to woodland or destroyed it entirely. Today, large areas of this former forest consist of dense scrub.

The other northern hemisphere region of sclerophyll vegetation is the California coast ranges. Here, the sclerophyll forest or woodland is typically dominated by live oak and white oak. Grassland occupies the open ground between the scattered oaks (see Figure 11.9*b*). Much of the remaining vegetation is sclerophyll scrub or "dwarf forest," known as *chaparral.* It varies in composition with elevation and exposure. Chaparral may contain wild lilac, manzanita, mountain mahogany, poison oak, and live oak.

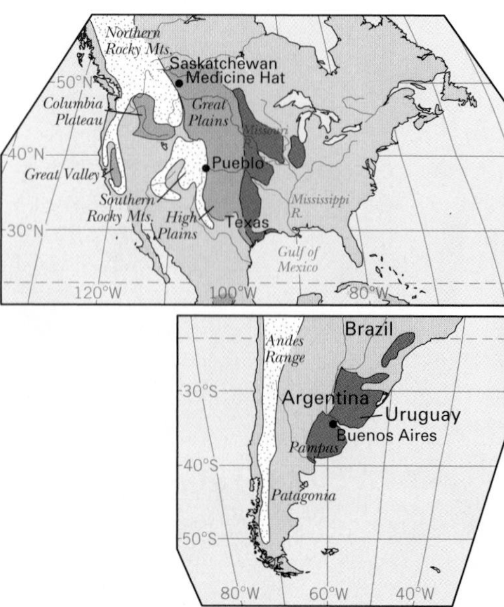

24.17 Grassland *World map of the grassland biome in subtropical and midlatitude zones. (Data source same as Figure 24.3. Based on Goode Base Map.)*

In central Chile and in the Cape region of South Africa, sclerophyll vegetation has a similar appearance, but the dominant species are quite different. Important areas of sclerophyll forest, woodland, and scrub are also found in southeast, south central, and southwest Australia, including many species of eucalyptus and acacia.

Savanna Biome

The savanna biome is usually associated with the tropical wet-dry climate ③ of Africa and South America. It includes vegetation formation classes ranging from woodland to grassland. In *savanna woodland,* the trees are spaced rather widely apart because soil moisture during the dry season is not sufficient to support a full tree cover. The open spacing permits development of a dense lower layer, which usually consists of grasses. The woodland has an open, park-like appearance. Savanna woodland usually lies in a broad belt adjacent to equatorial rainforest.

In the tropical savanna woodland of Africa, the trees are of medium height. Tree crowns are flattened or umbrella-shaped, and the trunks have thick, rough bark (see Figure 10.10). Some species of trees are xerophytic forms with small leaves and thorns. Others are broad-leaved deciduous species that shed their leaves in the dry season. In this respect, savanna woodland resembles monsoon forest.

Fire is a frequent occurrence in the savanna woodland during the dry season, but the tree species of the savanna are particularly resistant to fire. Many geographers hold the view that periodic burning of the savanna grasses maintains the grassland against the invasion of forest. Fire does not kill the underground parts of grass plants, but it limits tree growth to individuals of fire-resistant species. Many rainforest tree species that might otherwise grow in the wet-dry climate regime are prevented by fires from invading. The browsing of animals, which kills many young trees, is also a factor in maintaining grassland at the expense of forest.

The regions of savanna woodland are shown in Figure 24.8, along with monsoon forest (discussed earlier). In Africa, the savanna woodland grades into a belt of *thorntree-tall-grass savanna,* a formation class transitional to the desert biome. The trees are largely of thorny species. Trees are more widely scattered, and the open grassland is more extensive than in the savanna woodland. One characteristic tree is the flat-topped acacia, seen in Figure 10.10. Elephant grass is a common species. It can grow to a height of 5 m (16 ft) to form an impenetrable thicket.

The thorntree-tall-grass savanna is closely identified with the semiarid subtype of the dry tropical and subtropical climates (④s, ⑤s). In the semiarid climate, soil water storage is adequate for the needs of plants only during the brief rainy season. The onset of the rains is quickly followed by the greening of the trees and grasses. For this reason, vegetation of the savanna biome is described as rain-green, an adjective that also applies to the monsoon forest.

The African savanna is widely known for the diversity of its large grazing mammals, which include more than a dozen antelopes. Careful studies have shown that each species has a particular preference for different parts of the grasses it consumes—blade, sheath, and stem. Grazing stimulates the grasses to continue to grow, and so the ecosystem is more productive when grazed than when left alone. With these grazers comes a large variety of predators—lions, leopards, cheetahs, hyenas, and jackals. Elephants are the largest animals of the savanna and adjacent woodland regions.

Grassland Biome

The grassland biome includes two major formation classes that we will discuss here—tall-grass prairie and steppe (Figure 24.17). *Tall-grass prairie* consists largely of tall grasses. *Forbs,* which are broad-leaved herbs, are also present. Trees and shrubs are absent from the prairie but may occur in the same region as narrow patches of forest in stream valleys. The grasses are deeply rooted and form a thick and continuous turf (Figure 24.18).

Prairie grasslands are best developed in regions of the midlatitude and subtropical zones with well-developed winter and summer seasons. The grasses flower in spring and

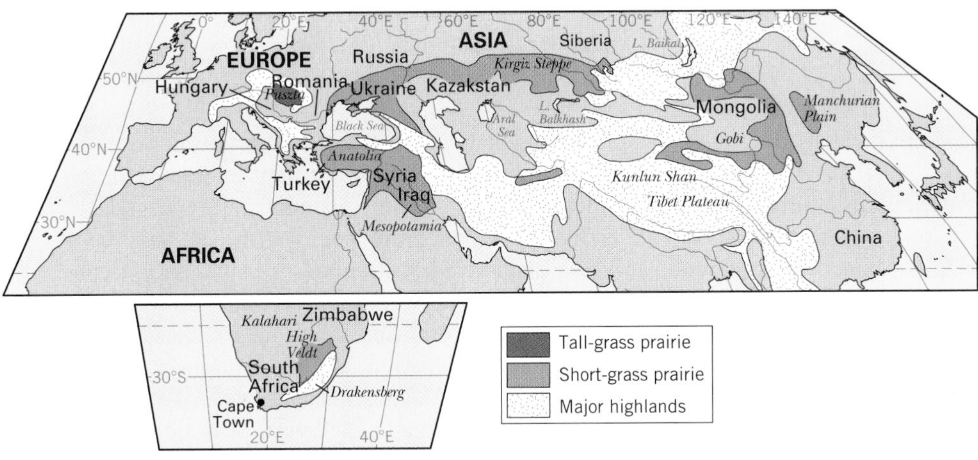

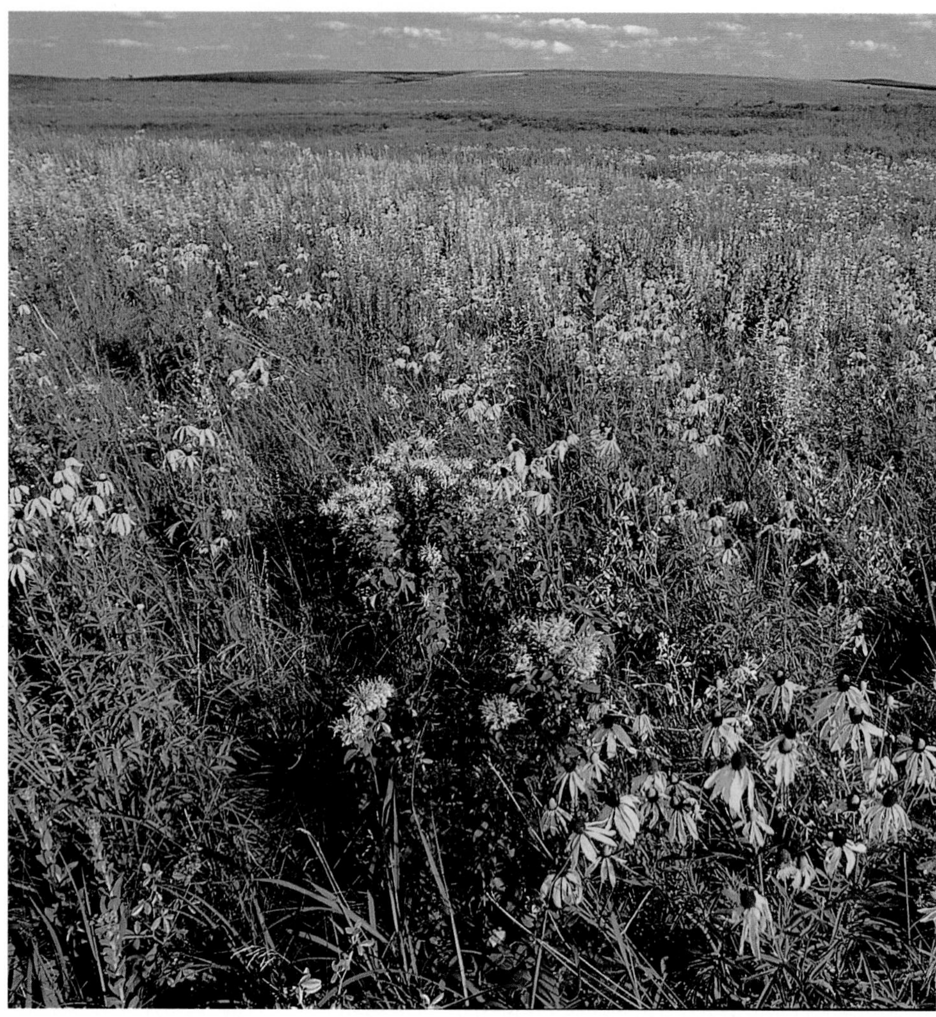

24.18 Tall-grass prairie, Iowa *In addition to grasses, tall-grass prairie vegetation includes many forbs, such as the flowering species shown in this photo.*

early summer, and the forbs flower in late summer. Tall-grass prairies are closely associated with the drier areas of moist continental climate ⑩. Here, soil water is in short supply during the summer months.

When European settlers first arrived in North America, the tall-grass prairies were found in a belt extending from the Texas Gulf coast northward to southern Saskatchewan (Figure 24.17). A broad peninsula of tall-grass prairie extended eastward into Illinois, where conditions are somewhat more moist. Since the time of settlement, these prairies have been converted almost entirely to agricultural land. Another major area of tall-grass prairie is the Pampa region of South America, which occupies parts of Uruguay and eastern Argentina. The Pampa region falls into the moist subtropical climate ⑥, with mild winters and abundant precipitation.

Recall from Chapter 9 that **steppe,** also called *short-grass prairie,* is a vegetation type consisting of short grasses occurring in sparse clumps or bunches. Scattered shrubs and low trees may also be found in the steppe. The plant cover is poor, and much bare soil is exposed. Many species of grasses and forbs occur. A typical grass of the American steppe is buffalo grass. Other typical plants are the sunflower and loco weed. Steppe grades into semi-desert in dry environments and into prairie where rainfall is higher.

Our map of the grassland biome (Figure 24.17) shows that steppe grassland is concentrated largely in the midlatitude areas of North America and Eurasia. The only southern hemisphere occurrence shown on the map is the "veldt" region of South Africa, a highland steppe surface in Orange Free State and Transvaal.

Steppe grasslands correspond well with the semiarid subtype of the dry continental climate ⑨. Spring rains nourish the grasses, which grow rapidly until early summer. By midsummer, the grasses are usually dormant. Occasional summer rainstorms cause periods of revived growth.

The animals of the grassland are distinctive. Before the exploitation of the North American grassland for cattle ranching, large grazing mammals were abundant. These included the pronghorn antelope, the elk, and the buffalo, which ranged widely from tall-grass prairie to steppe. Now nearly extinct, the buffalo *(Bison)* once numbered 60 million and roamed the grasslands from the Rockies to the Shenandoah Valley of Virginia. By 1889, however, this herd had been reduced to 800 individuals. Most of the present buffalo population is confined to Yellowstone National Park. Today, rodents, and rabbits join cattle as the major grazers in the grasslands ecosystem.

The grassland ecosystem supports some rather unique adaptations to life. A common adaptive mechanism is jumping or leaping locomotion, assuring an unimpeded view of the surroundings. Jackrabbits and jumping mice are examples of jumping rodents. The pronghorn combines the leap with great speed, which allows it to avoid predators and fire. Burrowing is also another common life habit, for the soil provides the only shelter in the exposed grasslands. Examples are burrowing rodents, including prairie dogs, gophers, and field mice. Rabbits exploit old burrows, using them for nesting or shelter. Invertebrates also seek shelter in the soil, and many are adapted to living with the burrows of rodents, where extremes of moisture and temperature are substantially moderated.

Desert Biome

The desert biome includes several formation classes that are transitional from grassland and savanna biomes into vegetation of the arid desert. Here we recognize two basic formation classes: semidesert and dry desert. Figure 24.19 is a world map of the desert biome.

Semidesert is a transitional formation class found in a wide latitude range—from the tropical zone to the midlatitude zone. It is identified primarily with the arid subtypes of all three dry climates. Semidesert consists of sparse xerophytic shrubs. One example is the sage-brush vegetation of the middle and southern Rocky Mountain region and Colorado Plateau (Figure 24.20). Recently, as a result of overgrazing and trampling by livestock, semidesert shrub vegetation seems to have expanded widely into areas of the western United States that were formerly steppe grasslands.

Thorntree semidesert of the tropical zone consists of xerophytic trees and shrubs that are adapted to a climate with a very long, hot dry season and only a very brief, but intense, rainy season. These conditions are found in the semiarid and arid subtypes of the dry tropical ④ and dry subtropical ⑤ climates. The thorny trees and shrubs are known locally as thorn forest, thornbush, or thornwoods (see Figure 10.16*a*). Many of these are deciduous plants that shed their leaves in the dry season. The shrubs may be closely intergrown to form dense thickets. Cactus plants are present in some localities.

Dry desert is a formation class of xerophytic plants that are widely dispersed over only a very small proportion of the ground. The visible vegetation of dry desert consists of small, hard-leaved, or spiny shrubs, succulent plants (such as cactus), or hard grasses. Many species of small annual plants may be present but appear only after a rare, but heavy, desert downpour. Much of the world map area assigned to desert vegetation has no plant cover at all because the surface consists of shifting dune sands or sterile salt flats.

Desert plants differ greatly in appearance from one part of the world to another. In the Mojave and Sonoran deserts of the southwestern United States, plants are often large, giving the appearance of a woodland (Figure 24.21). Examples are the tree-like saguaro cactus, the prickly pear cactus, the ocotillo, creosote bush, and smoke tree.

Desert animals, like the plants, are typically adapted to the dry conditions of the desert. We have already mentioned some adaptations of xeric animals to the water scarcity of the desert. Important herbivores in American deserts include kangaroo rats, jackrabbits, and grasshopper mice. Insects are abundant, as are insect-eating bats and birds such as the cactus wren. Reptiles, especially lizards, are also common.

Tundra Biome

Arctic tundra is a formation class of the tundra climate ⑫. (See Figure 11.23 for a polar map of the arctic tundra.) In this climate, plants grow during the brief summer of long days and short (or absent) nights. At this time, air temperatures rise above freezing, and a shallow surface layer of ground ice thaws. The permafrost beneath, however, remains frozen, keeping the meltwater at the surface. These conditions create a marshy environment for at least a short time over wide areas. Because plant remains decay very slowly within the cold meltwater, layers of organic matter can build up in the marshy ground. Frost action in the soil fractures and breaks large roots, keeping tundra plants small. In winter, wind-driven snow and extreme cold also injure plant parts that project above the snow.

Plants of the arctic tundra are mostly low herbs, although dwarf willow, a small woody plant, occurs in places. Sedges, grasses, mosses, and lichens dominate the tundra in a low layer (Figure 11.23). Typical plant species are ridge sedge, arctic meadow grass, cotton grasses, and snow lichen. There are also many species of forbs that flower brightly in the summer. Tundra composition varies greatly as soils range from wet to well drained. One form of tundra consists of sturdy hummocks of plants with low, water-covered ground between. Some areas of arctic scrub vegetation composed of willows and birches are also found in tundra.

Tundra vegetation is also found at high elevations. This *alpine tundra* develops above the limit of tree growth and below the vegetation-free zone of bare rock and perpetual snow (Figure 24.22). Alpine tundra resembles arctic tundra in many physical respects.

As is most often true in particularly dynamic environments, species diversity in the tundra is low, but the abundance of individuals is high. Among the animals, vast herds of caribou in North America and reindeer (their Eurasian relatives) roam the tundra, lightly grazing the lichens and plants and moving constantly. A smaller number of musk-oxen are also primary consumers of the tundra vegetation. Wolves and wolverines, as well as arctic foxes and polar bears, are predators. Among the smaller mammals, snowshoe rabbits and lemmings are important herbivores. Invertebrates are scarce in the tundra, except for a small number of insect species. Black flies, deerflies, mosquitoes, and "no-see-ums" (tiny biting midges) are all abundant and can make July on the tundra most uncomfortable. Reptiles and

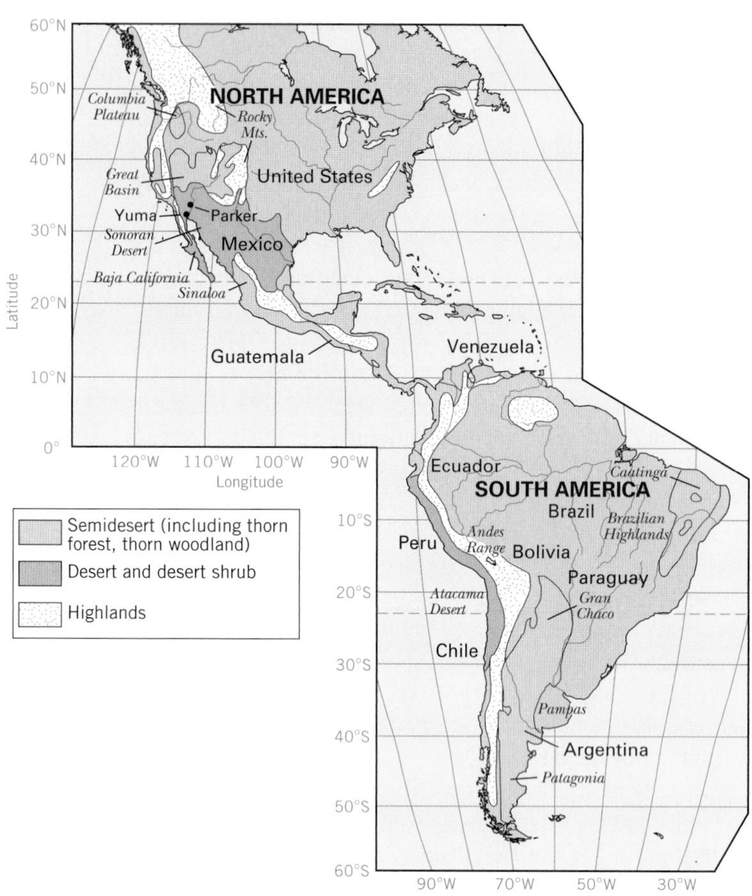

***24.19* Desert** *World map of the desert biome, including desert and semidesert formation classes. (Data source same as Figure 24.3. Based on Goode Base Map.)*

***24.20* Eye on the Landscape Sagebrush semidesert** *Sagebrush dominates the landscape near Monument Valley, Utah—a region of mesas, buttes, and pinnacles on the Colorado Plateau.* **What else would the geographer see? ... Answers at the end of the chapter.**

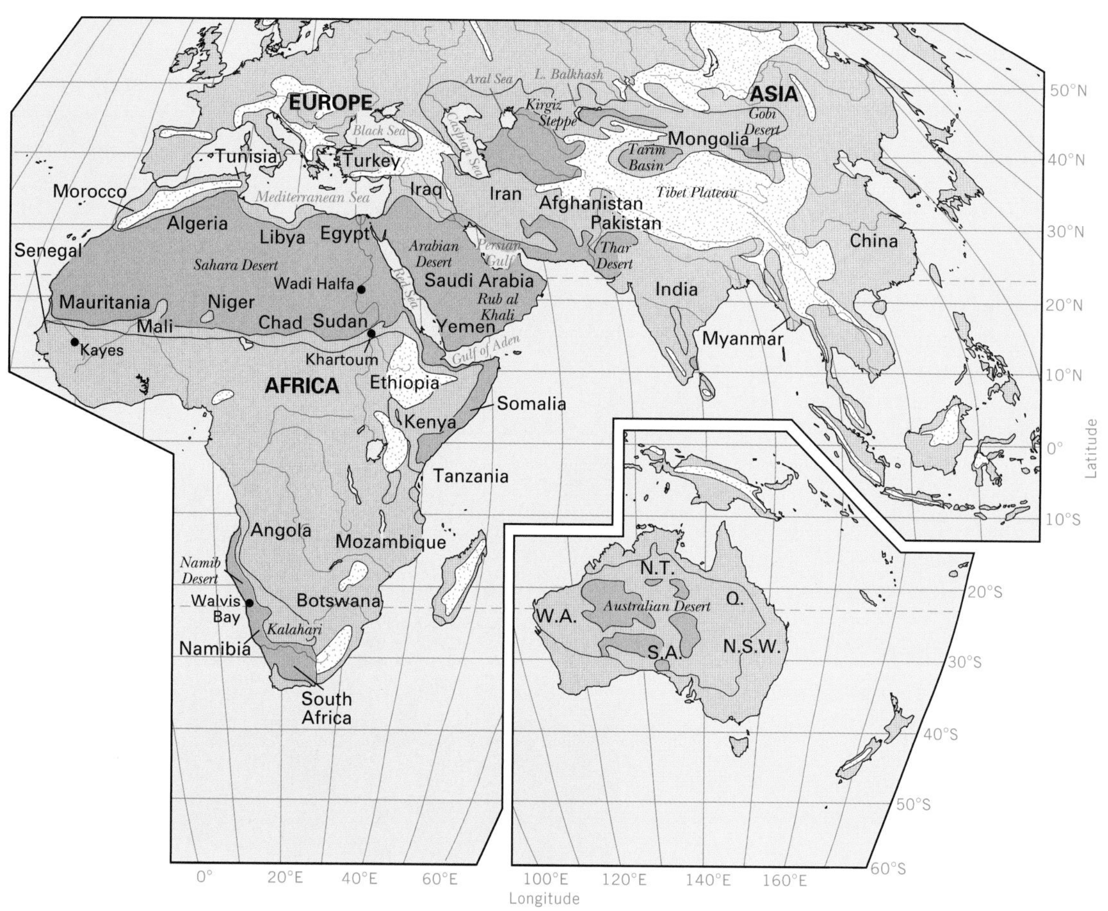

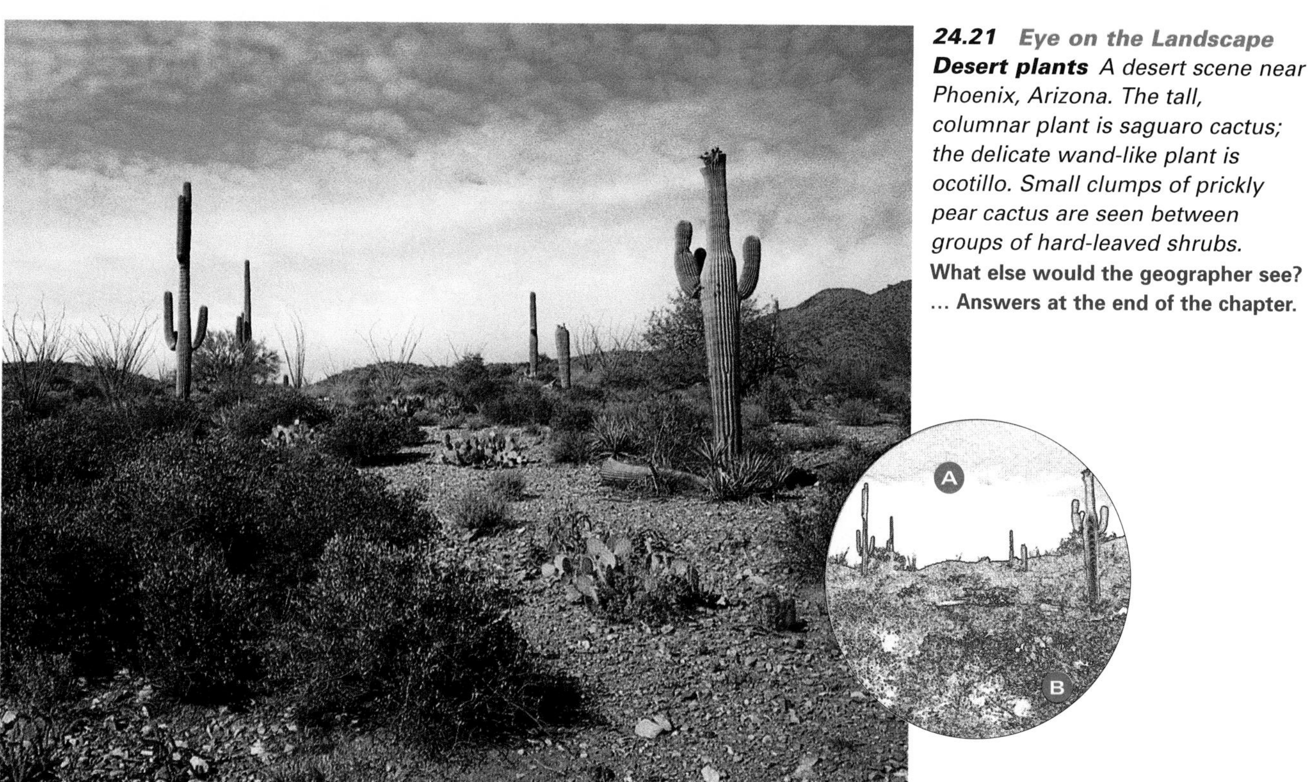

24.21 Eye on the Landscape
Desert plants *A desert scene near Phoenix, Arizona. The tall, columnar plant is saguaro cactus; the delicate wand-like plant is ocotillo. Small clumps of prickly pear cactus are seen between groups of hard-leaved shrubs.* **What else would the geographer see? ... Answers at the end of the chapter.**

24.22 Alpine tundra *Flag-shaped spruce trees (on left side of photo), shaped by prevailing winds, mark the upper limit of tree growth. Near the summit of the Snowy Range, Wyoming.*

amphibians are also rare. The boggy tundra, however, presents an ideal summer environment for many migratory birds such as waterfowl, sandpipers, and plovers.

The food web of the tundra ecosystem is simple and direct. The important producer is reindeer moss, the lichen *Cladonia rangifera.* In addition to the caribou and reindeer, lemmings, ptarmigan (arctic grouse), and snowshoe rabbits are important lichen grazers. The important predators are the fox, wolf, and lynx, although all these animals may feed directly on plants as well. During the summer, the abundant insects help support the migratory waterfowl populations.

Altitude Zones of Vegetation

In earlier chapters, we described the effects of increasing elevation on climatic factors, particularly air temperature and precipitation. We noted that, with elevation, temperatures decrease and precipitation generally increases. These changes produce systematic changes in the vegetation cover as well, yielding a sequence of vegetation zones related to altitude.

The vegetation zones of the Colorado Plateau region in northern Arizona and adjacent states provide a striking example of this altitude zonation. Figure 24.23 is a diagram showing a cross section of the land surface in this region. Elevations range from about 700 m (about 2300 ft) at the bottom of the Grand Canyon to 3844 m (12,608 ft) at the top of San Francisco Peak. The vegetation cover and rainfall

24.23 Altitude zone of vegetation in the arid southwestern United States *The profile shows the Grand Canyon-San Francisco Mountain district of northern Arizona. (Based on data of G. A. Pearson, C. H. Merriam, and A. N. Strahler.)*

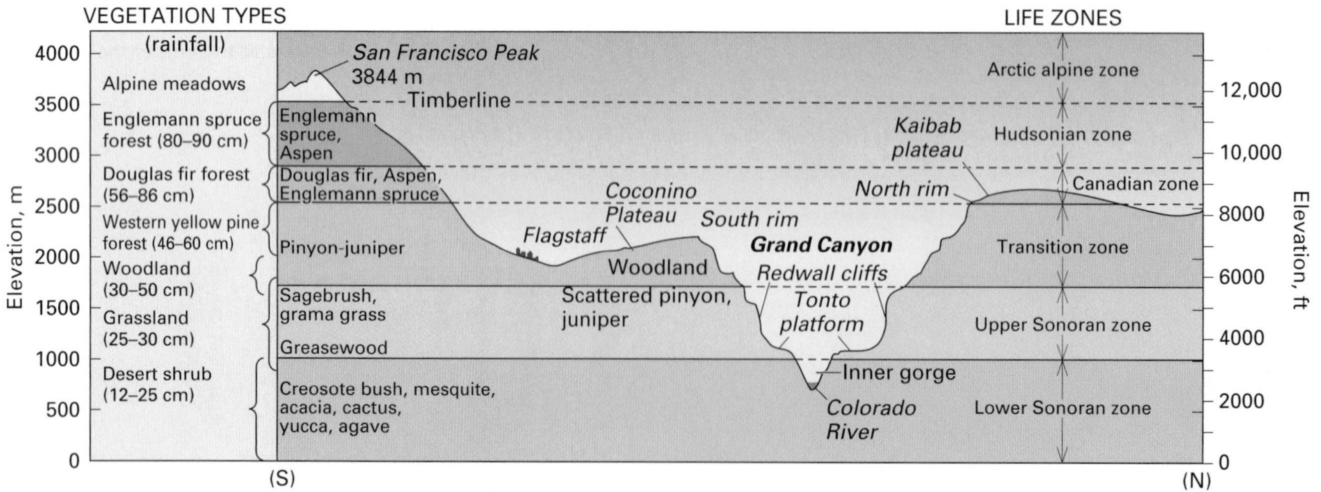

range are shown on the left. The vegetation zonation includes desert shrub, grassland, woodland, pine forest, Douglas fir forest, Engelmann spruce forest, and alpine meadow. Annual rainfall ranges from 12 to 25 cm (about 5 to 10 in.) in the desert shrub vegetation type to 80 to 90 cm (about 30 to 35 in.) in the Engelmann spruce forest.

Ecologists have used the vegetation zonation to set up a series of *life zones,* which are also shown on the figure. These range from the lower Sonoran life zone, which includes the desert shrubs typical of the Sonoran desert, to the arctic-alpine life zone, which includes the alpine meadows at the top of San Francisco Peak. Other life zones take their names from typical regions of vegetation cover. For example, the Hudsonian zone, 2900 to 3500 m (about 9500 to 11,500 ft), bears a needleleaf forest quite similar to needleleaf boreal forest of the subarctic zone near Hudson's Bay, Canada.

Climatic Gradients and Vegetation Types

In discussing the major formation classes of vegetation, we have emphasized the importance of climate. As climate changes with latitude or longitude, vegetation will also change. Figure 24.24 shows three transects across portions of continents that illustrate this principle. (For these transects, we will ignore the effects of mountains or highland regions on climate and vegetation.)

The upper transect stretches from the equator to the tropic of cancer in Africa. Across this region, climate ranges through all four low-latitude climates: wet equatorial ①, monsoon and trade-wind coastal ②, wet-dry tropical ③, and dry tropical ④. Vegetation grades from equatorial rainforest, savanna woodland, and savanna grassland to tropical scrub and tropical desert.

The middle transect is a composite from the tropic of cancer to the arctic circle in Africa and Eurasia. Climates include many of the mid- and high-latitude types: dry subtropical ⑤, Mediterranean ⑦, moist continental ⑩, boreal forest ⑪, and tundra ⑫. The vegetation cover grades from tropical desert through subtropical steppe to sclerophyll forest in the Mediterranean. Further north is the midlatitude deciduous forest in the region of moist continental climate ⑩, which grades into boreal needleleaf forest, subarctic woodland, and finally tundra.

The lower transect ranges across the United States, from Nevada to Ohio. On this transect, the climate begins as dry midlatitude ⑨. Precipitation gradually increases eastward, reaching moist continental ⑩ near the Mississippi River. The vegetation changes from midlatitude desert and steppe to short-grass prairie, tall-grass prairie, and midlatitude deciduous forest.

The changes on these transects are largely gradational rather than abrupt. Yet, the global maps of both vegetation and climate show distinct boundaries from one region to the next. Which is correct? The true situation is gradational rather than abrupt. Maps must necessarily have boundaries to communicate information. But climate and vegetation know no specific boundaries. Instead, they are classified into specific types for convenience in studying their spatial patterns. When studying any map of natural features, keep in mind that boundaries are always approximate and gradational.

A LOOK BACK This chapter concludes our overview of physical geography as the science of global environments. Our study began with an examination of weather and climate systems, powered largely by solar energy. We then turned to systems of the solid Earth, powered by Earth forces. Next we presented the systems of landforms and landscape evolution, driven both by Earth forces and solar power. Lastly, we covered systems of soils and ecosystems, grouped together by their dependence on climate and substrate, and powered largely by solar energy. Throughout, we have maintained a focus on environmental problems, especially those associated with global change, and stressed the advantages of viewing many of the phenomena we have described as the products of systems of matter and energy flow.

Physical geography is, of course, concerned with the Earth—our home planet, at least for the foreseeable future. As a discipline, physical geography stresses the interrelationships among its many component sciences and focuses on our environment in an integrated way. Yet it emphasizes processes that have, thus far, not felt the major effects of human activity. As the human population continues to grow, our influence on our home planet will grow more and more profound. This book has given you the tools to understand and assess the essential features of human impact on natural systems. We, as authors, hope that what you have learned will serve you, its readers, well as you face the challenges of world citizenship in the twenty-first century.

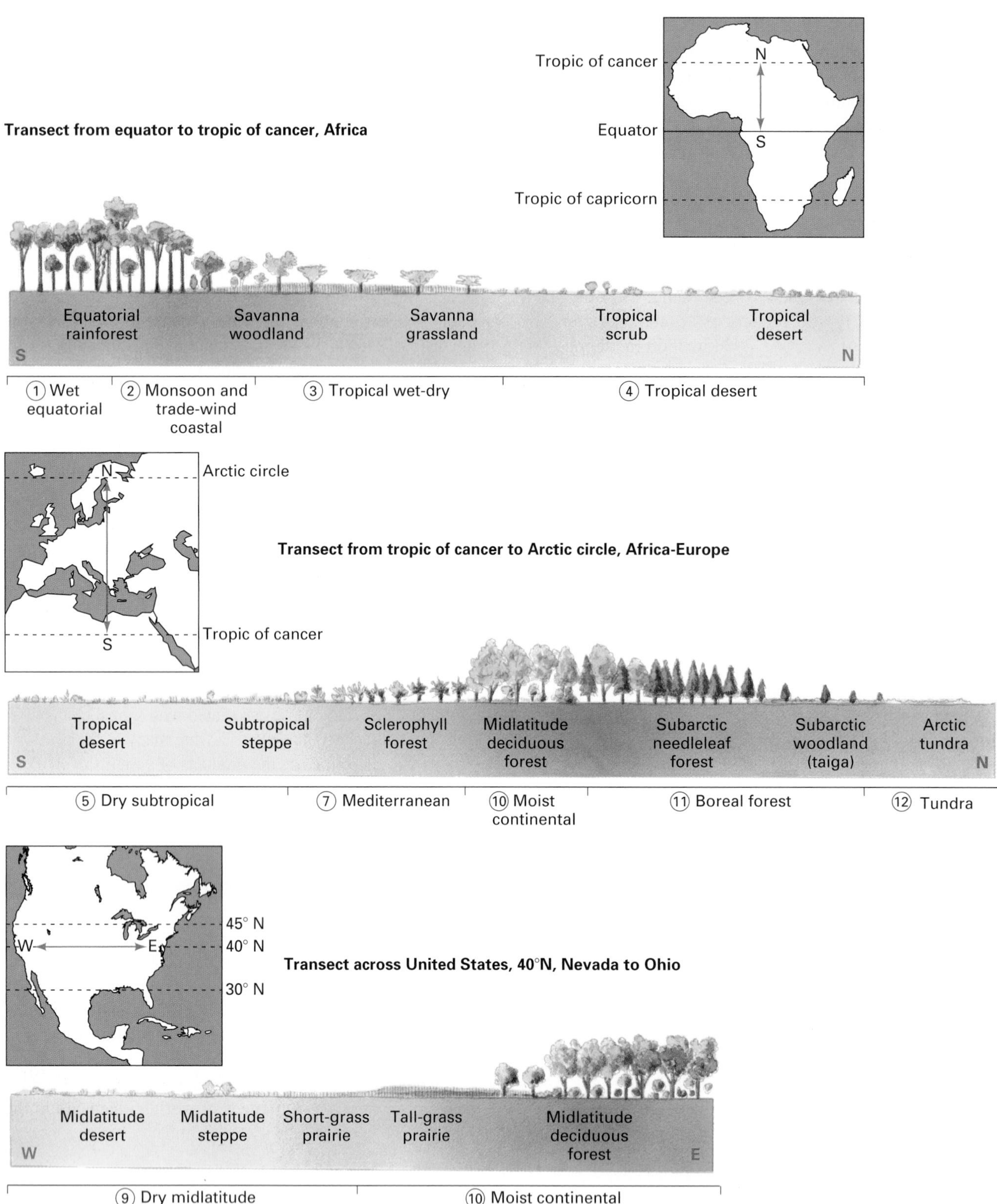

24.24 Vegetation transects *Three continental transects showing the succession of plant formation classes across climatic gradients.*

Chapter Summary

- **Natural vegetation** is a plant cover that develops with little or no human interference. Although much vegetation appears to be in a natural state, humans influence the vegetation cover by fire suppression and introduction of new species.
- The **life-form** of a plant refers to its physical structure, size, and shape. Life-forms include *trees, shrubs, lianas, herbs,* and *lichens*.
- The largest unit of terrestrial ecosystems is the **biome**: forest, grassland, savanna, desert, and tundra. The **forest biome** includes a number of important forest formation classes. The *low-latitude rainforest* exhibits a dense canopy and open floor with a very large number of species. *Subtropical evergreen forest* occurs in *broadleaf* and *needleleaf* forms in the moist subtropical climate ⑥. Monsoon forest is largely deciduous, with most species shedding their leaves after the wet season.
- *Midlatitude deciduous forest* is associated with the moist continental climate ⑩. Its species shed their leaves before the cold season. *Needleleaf forest* consists largely of evergreen conifers. It includes the *coastal forest* of the Pacific Northwest, the *boreal forest* of high latitudes, and needleleaved mountain forests. *Sclerophyll forest* is co[illegible] of trees with small, hard, leathery leaves and is found in the Mediterranean climate ⑦ region.
- The **savanna biome** consists of widely spaced trees with an understory, often of grasses. Dry-season fire is frequent in the savanna biome, limiting the number of trees and encouraging the growth of grasses.
- The **grassland biome** of midlatitude regions includes *tall-grass prairie*, in moister environments, and *short-grass prairie*, or **steppe**, in semiarid areas.
- Vegetation of the **desert biome** ranges from thorny shrubs and small trees to dry desert vegetation comprised of drought-adapted species.
- **Tundra biome** vegetation is limited largely to low herbs that are adapted to the severe drying cold experienced on the fringes of the Arctic Ocean.
- Since climate changes with altitude, vegetation typically occurs in altitudinal zones. Climate also changes gradually with latitude, and so biome changes are typically gradual, without abrupt boundaries.

Key Terms

terrestrial ecosystem	forest biome	grassland biome	tundra biome
biome	savanna biome	desert biome	steppe

Review Questions

1. What is natural vegetation? How do humans influence vegetation?
2. Plant geographers describe vegetation by its overall structure and by the life-forms of individual plants. Define and differentiate the following terms: forest, woodland, tree, shrub, herb, liana, perennial, deciduous, evergreen, broadleaf, and needleleaf.
3. What are the five main biome types that ecologists and biogeographers recognize? Describe each briefly.
4. Low-latitude rainforests occupy a large region of the Earth's land surface. What are the characteristics of these forests? Include forest structure, types of plants, diversity, and climate in your answer.
5. Monsoon forest and midlatitude deciduous forest are both deciduous but for different reasons. Compare the characteristics of these two formation classes and their climates.
6. Subtropical broadleaf evergreen forest and tall-grass prairie are two vegetation formation classes that have been greatly altered by human activities. How was this done and why?
7. Distinguish among the types of needleleaf forest. What characteristics do they share? How are they different? How do their climates compare?
8. Which type of forest, with related woodland and scrub types, is associated with the Mediterranean climate? What are the features of these vegetation types? How are they adapted to the Mediterranean climate?
9. How do traditional agricultural practices in the low-latitude rainforest compare to present-day practices? What are the implications for the rainforest environment?
10. What are the effects of large-scale clearing on the rainforest environment?

11. Describe the formation classes of the savanna biome. Where is this biome found and in what climate types? What role does fire play in the savanna biome?
12. Compare the two formation classes of the grassland biome. How do their climates differ?
13. Describe the vegetation types of the desert biome.
14. What are the features of arctic and alpine tundra? How does the cold tundra climate influence the vegetation cover?
15. How does elevation influence vegetation? Provide an example of how vegetation zonation is related to elevation.

Visualizing Exercises

1. Forests often contain plants of many different life-forms. Sketch a cross section of a forest including typical life-forms, and identify them with labels.
2. How does elevation influence vegetation? Sketch a hypothetical mountain peak in the southwestern U.S. desert that rises from a plain at about 500 m (about 1600 ft) elevation to a summit at about 4000 m (about 13,000 ft) and label the vegetation zones you might expect to find on its flanks.

Essay Questions

1. Figure 24.24 presents a vegetation transect from Nevada to Ohio. Expand the transect on the west so that it begins in Los Angeles. On the east, extend it northeast from Ohio through Pennsylvania, New York, western Massachusetts, and New Hampshire, to end in Maine. Sketch the vegetation types in your additions and label them, as in the diagram. Below your vegetation transect, draw a long bar subdivided to show the climate types.
2. Construct a similar transect of climate and vegetation from Miami to St. Louis, Minneapolis, and Winnipeg.

Eye on the Landscape

24.20 Sagebrush semidesert These tall columnar landforms are buttes **(A)**. Mesas are larger, isolated rock platforms **(B)**. Here in Monument Valley they are formed by a thick and uniform rock layer that weathers into rectangular blocks along joint planes. The blocks fall away, leaving a vertical cliff face behind. Buttes and mesas are discussed in Chapter 18, and joints are covered in Chapter 15.

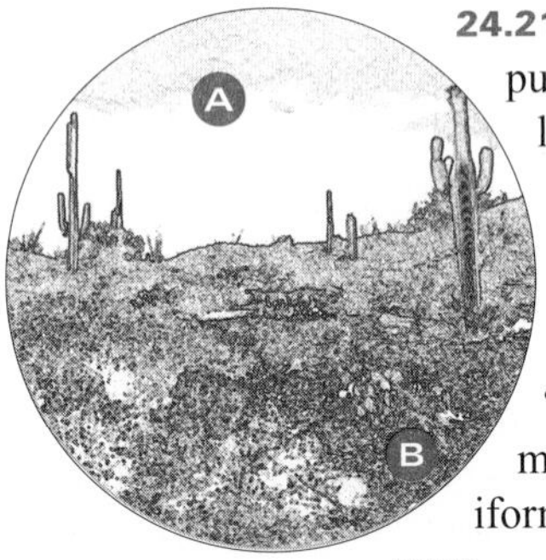

24.21 Desert plants This sky of puffy altocumulus clouds and layered altostratus **(A)** indicates moisture at higher levels in the troposphere. The photo, taken in July, reflects conditions of the local "monsoon" season, in which moist air from the Gulf of California moves into southern Arizona, often generating intense thunderstorms and flash floods. The gravel-covered ground surface **(B)** is characteristic of deserts, where wind and storm runoff remove fine particles and leave coarser rock fragments behind. Clouds are covered in Chapter 6, and the Arizona monsoon is mentioned in Chapter 7. Erosion by runoff and wind are treated in Chapters 17 and 19.

Appendix 1 THE CANADIAN SYSTEM OF SOIL CLASSIFICATION*

The formation of the National Soil Survey Committee of Canada in 1940 was a milestone in the development of soil classification and of pedology generally in Canada. Prior to that time, Canada used the 1938 USDA system. Although the Canadian experience showed that the concept of zonal soils was useful in the western plains, it proved less applicable in eastern Canada where parent materials and relief factors had a dominant influence on soil properties and development in many areas.

Canadian pedologists observed closely the evolution of the U.S. Comprehensive Soil Classification System (CSCS) during the 1950s and 1960s and ultimately adopted several important features of that system. Nevertheless, the special needs of a workable Canadian national system required that a completely independent classification system be established. Because Canada lies entirely in a latitude zone poleward of the 40th parallel, there was no need to incorporate those soil orders found only in lower latitudes. Furthermore, the vast expanse of Canadian territory lying within the boreal forest and tundra climates necessitated the recognition at the highest taxonomic level (the order) of soils of cold regions that appear only as suborders and even as great groups in the CSCS—for example, the Cryaquents and Cryorthents that occupy much of the soils map of northern Canada above the 50th parallel (Figure 21.13).

The overall philosophy of the Canadian system is pragmatic: the aim is to organize knowledge of soils in a reasonable and usable way. The system is a natural or taxonomic one in which the classes (taxa) are based on properties of the soils themselves and not on interpretations of the soils for various uses. Thus, the taxa are concepts based on generalization of properties of real, not idealized, bodies of soils. Although taxa in the Canadian system are defined on the basis of actual soil properties that can be observed and measured, the system has a genetic bias in that properties or combinations of properties that reflect genesis are favored in distinguishing among the higher taxa. Thus, the soils brought together under a single soil order are seen as the product of a similar set of dominant soil-forming processes resulting from broadly similar climatic conditions.

The Canadian system recognizes the *pedon* as the basic unit of soils; it is defined as in the CSCS. Major mineral horizons of the soil (A, B, C) are defined in much the same way as in the U.S. system. Thus, the Canadian system of soil taxonomy is more closely related to the U.S. system than to any other. Both are hierarchical, and the taxa are defined on the basis of measurable soil properties. However, they differ in several respects. The Canadian system is designed to classify only the soils that occur in Canada and is not a comprehensive system. The U.S. system includes the suborder, a taxon not recognized in the Canadian system. Because 90 percent of the area of Canada is not likely to be cultivated, the Canadian system does not recognize as diagnostic those horizons strongly affected by plowing and application of soil conditioners and fertilizers.

* Thoughout this section numerous sentences and phrases are taken verbatim or paraphrased from the following work: Canada Soil Survey Committee, *The Canadian System of Soil Classification,* Research Branch, Canada Department of Agriculture, Publication 1646, 1978. Table A1.1 and Figure A1.2 are also compiled from this source.

Table A1.1 | **Subhorizons and Organic Horizons of the Canadian System of Soil Classification**

Subhorizons; Lowercase Suffixes

b	Buried soil horizon.
c	Cemented (irreversible) pedogenic horizon.
ca	Horizon of secondary carbonate enrichment in which the concentration of lime exceeds that in the unenriched parent material.
e	Horizon characterized by the eluviation of clay, Fe, Al, or organic matter alone or in combination.
f	Horizon enriched with amorphous material, principally Al and Fe combined with organic matter; reddish near upper boundary, becoming yellower at depth.
g	Horizon characterized by gray colors, or prominent mottling, or both, indicative of permanent or intense reduction.
h	Horizon enriched with organic matter.
j	Used as a modifier of suffixes e, f, g, n, and t to denote an expression of, but failure to meet, the specified limits of a suffix it modifies.
k	Denotes presence of carbonate as indicated by visible effervescence when dilute HCl is added.
m	Horizon slightly altered by hydrolysis, oxidation, or solution, or all three to give a change in color or structure, or both.
n	Horizon in which the ratio of exchangeable Ca to exchangeable Na is 10 or less. It must also have the following distinctive morphological characteristics: prismatic or columnar structure, dark coatings on ped surfaces, and hard to very hard consistence when dry.
p	Horizon disturbed by human activities such as cultivation, logging, and habitation.
s	Horizon of salts, including gypsum, which may be detected as crystals or veins or as surface crusts of salt crystals.
sa	Horizon with secondary enrichment of salts more soluble than Ca and Mg carbonates; the concentration of salts exceeds that in the unenriched parent material.
t	Illuvial horizon enriched with silicate clay.
u	Horizon that is markedly disrupted by physical or faunal processes other than cryoturbation.
x	Horizon of fragipan character. (A fragipan is a loamy subsurface horizon of high bulk density and very low organic matter content. When dry it is hard and seems to be cemented.)
y	Horizon affected by cryoturbation as manifested by disrupted or broken horizons, incorporation of materials from other horizons, and mechanical sorting.
z	A frozen layer.

Organic Horizons

O	Organic horizon developed mainly from mosses, rushes, and woody materials.
L	Organic horizon characterized by an accumulation of organic matter derived mainly from leaves, twigs, and woody materials in which the organic structures are easily discernible.
F	Same as L, above, except that original structures are difficult to recognize.
H	Organic horizon characterized by decomposed organic matter in which the original structures are indiscernible.

Soil Horizons and Other Layers

The definitions of classes in the Canadian system are based mainly on kinds, degrees of development, and sequence of soil horizons and other layers in pedons. The major mineral horizons are A, B, and C. The major organic horizons are L, F, and H, which are mainly forest litter at various stages of decomposition, and O, which is derived mainly from bog, marsh, or swamp vegetation. Subdivisions of horizons are labeled by adding lowercase suffixes to the major horizon symbols—for example, Ah or Ae.

Besides the horizons, nonsoil layers are recognized. Two such layers are R, rock, and W, water. Lower mineral layers not affected by pedogenic processes are also identified. In organic soils, layers are described as tiers.

The principal mineral horizons, A, B, and C, are defined as follows:

A Mineral horizon found at or near the surface in the zone of leaching or eluviation of materials in solution or suspension, or of maximum *in situ* accumulation of organic matter, or both.

B Mineral horizon characterized by enrichment in organic matter, sesquioxides, or clay; or by the development of soil structure; or by change of color denoting hydrolysis, reduction, or oxidation.

C Mineral horizon comparatively unaffected by the pedogenic processes operative in A and B horizons. The processes of gleying and the accumulation of calcium and magnesium and more soluble salts can occur in this horizon.

Lowercase suffixes, used to designate subdivisions of horizons, are shown in Table A1.1.

Soil Orders of the Canadian System

Nine soil orders make up the highest taxon of the Canadian System of Soil Classification. Listed in alphabetical order, they are as follows.

Brunisolic	Gleysolic	Podzolic
Chernozemic	Luvisolic	Regosolic
Cryosolic	Organic	Solonetzic

Table A1.2 lists the great groups within each order.

Table A1.2 Great Groups of the Canadian Soil Classification System

Order	Great Group
Brunisolic	Melanic Brunisol Eutric Brunisol Sombric Brunisol Dystric Brunisol
Chernozemic	Brown Dark Brown Black Dark Gray
Cryosolic	Turbic Cryosol Static Cryosol Organic Cryosol
Gleysolic	Humic Gleysol Gleysol Luvic Gleysol
Luvisolic	Gray Brown Luvisol Gray Luvisol
Organic	Fibrisol Mesisol Humisol Folisol
Podzolic	Humic Podzol Ferro-Humic Podzol Humo-Ferric Podzol
Regosolic	Regosol Humic Regosol
Solonetzic	Solentz Solodized Solonetz Solod

Brunisolic Order

The central concept of the *Brunisolic order* is that of soils under forest having brownish-colored Bm horizons. Most Brunisolic soils are well to imperfectly drained. They occur in a wide range of climatic and vegetative environments, including boreal forest; mixed forest, shrubs, and grass; and heath and tundra. As compared with the Chernozemic soils, the Brunisolic soils show a weak B horizon of accumulation attributable to their moister environment. Brunisolic soils lack the diagnostic podzolic B horizon of the Podzolic soils, in which accumulation in the B horizon is strongly developed. The Melanic Brunisol shown in profile Figure A1.1 can be found in the St. Lawrence Lowlands, surrounded by Podzolic soils (Figure A1.2).

Chernozemic Order

The general concept of the *Chernozemic order* is that of well to imperfectly drained soils having surface horizons darkened by the accumulation of organic matter from the decomposition of xerophytic or mesophytic grasses and forms representative of grassland communities or of grassland-forest communities with associated shrubs and forbs. The major area of Chernozemic soils is the cool, subarid to subhumid interior plains of western Canada. Most Chernozemic soils are frozen during some period each winter, and the soil is dry at some period each summer. The mean annual temperature is higher than 0°C and usually less than 5.5°C. The associated climate is typically the semiarid (steppe) variety of the dry midlatitude climate (9) (Figure A1.2).

Essential to the definition of soils of the Chernozemic order is that they must have an A horizon (typically, Ah) in which organic matter has accumulated, and they must meet several other requirements. The A horizon is at least 10 cm thick; its color is dark brown to black. It usually has sufficiently good structure that it is neither massive and hard nor single-grained when dry. The profile shown in Figure A1.1 is that of the Orthic subgroup of the Brown great group; it shows a Bm horizon that is typically of prismatic structure. The C horizon is one of lime accumulation (Cca). Clearly, the Chernozemic soils can be closely correlated with the Mollisols of the CSCS.

Cryosolic Order

Soils of the *Cryosolic order* occupy much of the northern third of Canada where permafrost remains close to the surface of both mineral and organic deposits. Cryosolic soils predominate north of the tree line, are common in the subarctic forest area in fine-textured soils, and extend into the boreal forest in some organic materials and into some alpine areas of mountainous regions. Cryoturbation (intense disturbance by freeze–thaw activity) of these soils is common, and it may be indicated by patterned ground features such as sorted and nonsorted nets, circles, polygons, stripes, and Earth hummocks.

Cryosolic soils are found in either mineral or organic materials that have permafrost either within 1 m of the surface or within 2 m if more than one-third of the pedon has been strongly cryoturbated, as indicated by disrupted, mixed, or broken horizons. The profile shown in Figure A1.1 is that of the Orthic subgroup of the Static Cryosol great group. Note the presence of organic L, H, and O surface horizons and the thin Ah horizon. The Cryosolic soils are closely correlated with the Cryaquepts of the CSCS.

Gleysolic Order

Soils of the *Gleysolic order* have features indicating periodic or prolonged saturation with water and reducing conditions. They commonly occur in patchy association with other soils in the landscape. Gleysolic soils are usually associated with either a high groundwater table at some period of the year or temporary saturation above a relatively impermeable layer. Some Gleysolic soils may be submerged under shallow water throughout the year. The profile shown in Figure A1.1 is that of the Gleysol great group. It has a thick Ah horizon. The underlying Bg horizon is grayish and shows mottling typical of reducing conditions.

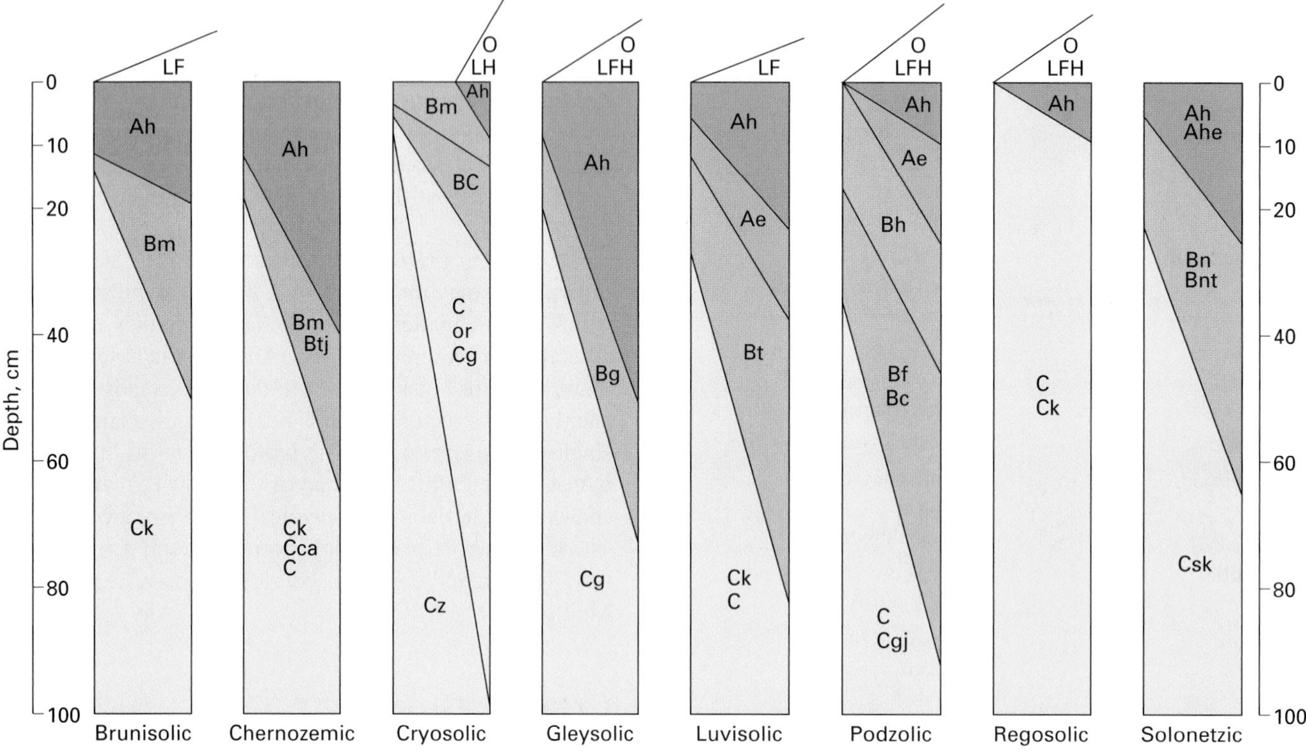

A1.1 Soil profiles *Representative schematic profiles of eight of the nine orders of the Canadian system of soil classification. Slanting lines show the range in depth and thickness of each horizon. (The horizon planes are actually approximately horizontal within the pedon.) See Table A1.1 for an explanation of symbols. (From Canada Soil Survey Committee, Research Branch, Canada Department of Agriculture, 1978.)*

Luvisolic Order

Soils of the *Luvisolic order* generally have light-colored, eluvial horizons (Ae), and they have illuvial B horizons (Bt) in which silicate clay has accumulated. These soils develop characteristically in well to imperfectly drained sites, in sandy loam to clay base-saturated parent materials under frost vegetation in subhumid to humid, mild to very cold climates. The genesis of Luvisolic soils is thought to involve the suspension of clay in the soil solution near the soil surface, downward movement of the suspended clay with the soil solution, and deposition of the translocated clay at a depth where downward motion of the soil solution ceases or becomes very slow. The representative profile shown in Figure A1.1 is that of the Orthic subgroup of the Gray Brown Luvisol great group.

Luvisolic soils occur from the southern extremity of Ontario to the zone of permafrost and from New Foundland to British Columbia. The largest area of these soils are Gray Luvisols occurring in the central to northern interior plains under deciduous, mixed, and coniferous forest. In this location they appear to correlate with the Boralfs of the Alfisol order in the CSCS. Gray-Brown Luvisolic soils of southern Ontario would correlate with the suborder of Udalfs.

Organic Order

Soils of the *Organic order* are composed largely of organic materials. They include most of the soils commonly known as peat, muck, or bog soils. Organic soils contain 17 percent or more organic carbon (30 percent organic matter) by weight. Most Organic soils are saturated with water for prolonged periods. They occur widely in poorly and very poorly drained depressions and level areas in regions of subhumid to humid climate and are derived from vegetation that grows in such sites. However, one group of Organic soils consists of leaf litter overlying rock or fragmental material; soils of this group may occur on steep slopes and may rarely be saturated with water. (No profile of the Organic soils is shown in Figure A1.1.) Organic soils can be correlated with the Histosols of the CSCS.

Podzolic Order

Soils of the *Podzolic order* have B horizons in which the dominant accumulation product is amorphous material composed mainly of humified organic matter in varying degrees with Al and Fe. Typically, Podzolic soils occur in coarse- to medium-textured, acid parent materials, under forest and heath vegetation in cool to very cold humid to

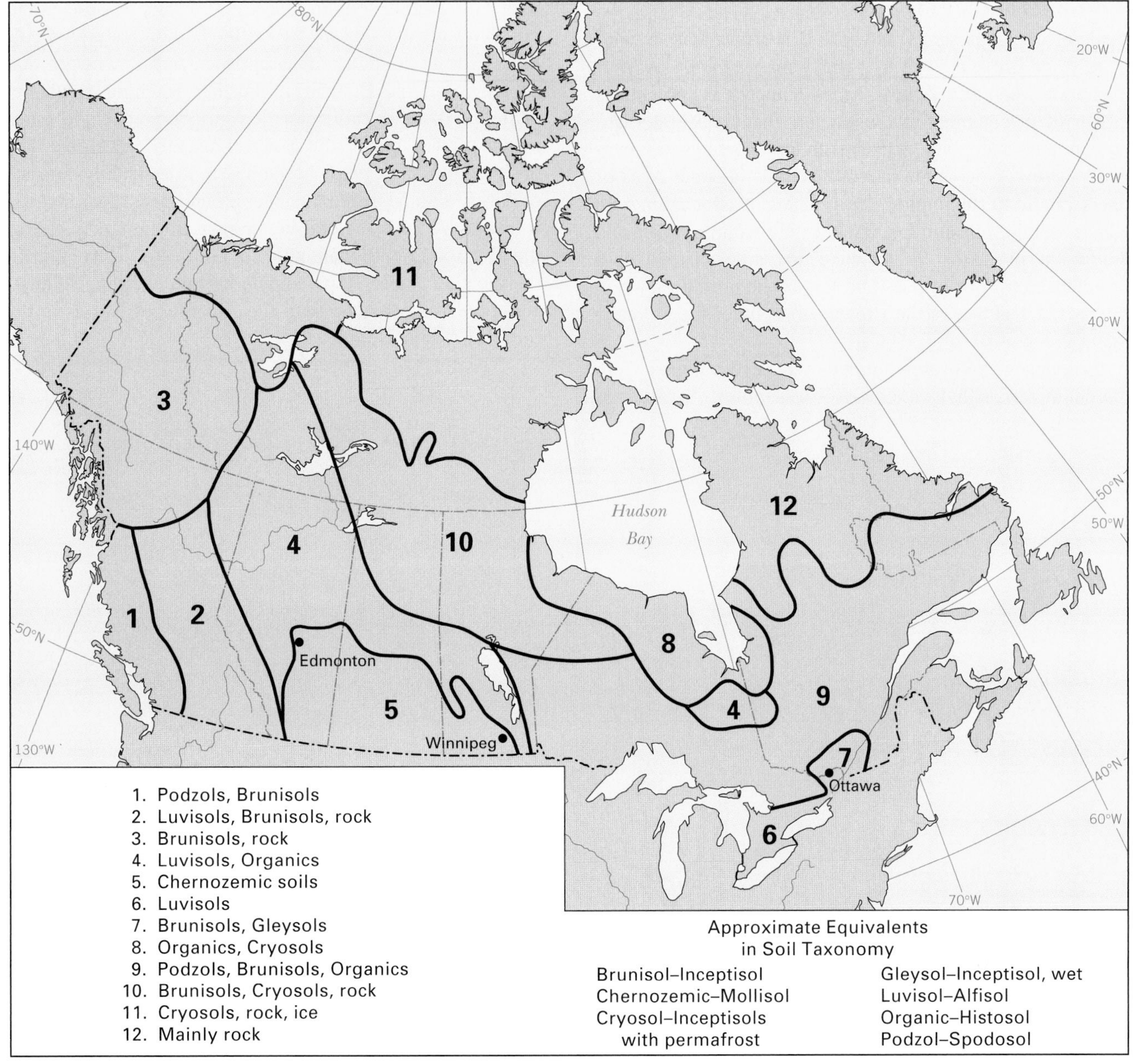

A1.2 Soil regions *Generalized map of soil regions of Canada. (Courtesy of Land Resources Research Institute, Agriculture Canada.) (Illustration is taken from Fundamentals of Soil Science, 7th ed., by Henry D. Foth, John Wiley & Sons.)*

very humid climates. Podzolic soils can usually be readily recognized in the field. Generally, they have organic surface horizons that are commonly L, F, and H. Most Podzolic soils have a reddish brown to black B horizon (Bh) with an abrupt upper boundary. The profile shown in Figure A1.1 is that of the Orthic subgroup of the Humic Podzol great group.

The Podzolic soils correspond closely to the Spodosols (Orthods) of the CSCS.

Regosolic Order

Regosolic soils have weakly developed horizons. The lack of development of genetic horizons may be due to any number of factors: youthfulness of the parent material, for example, recent alluvium; instability of the material, for example, colluvium on slopes subject to mass wasting; nature of the material, for example, nearly pure quartz sand; climate, for example, dry cold conditions. Regosolic soils are generally rapidly to imperfectly drained. They occur in a wide range of vegetation and climates. The profile shown in Figure A1.1 is that of the Orthic subgroup of the Regosol great group. It has only a thin humic A horizon (Ah) and a surface horizon of organic materials.

Regosolic soils correspond with the Entisols of the CSCS.

Solonetzic Order

Soils of the *Solonetzic order* have B horizons that are very hard when dry and swell to a sticky mass of very low permeability when wet. Typically, the Solonetzic B horizon has prismatic or columnar macrostructure that breaks into hard to extremely hard, blocky peds with dark coatings. Solonetzic soils occur on saline parent materials in some areas of the semiarid to subhumid interior plains in association with Chernozemic soils and to a lesser extent with Luvisolic and Gleysolic soils. Most Solonetzic soils are associated with a vegetative cover of grasses and forbs. The profile shown in Figure A1.1 is that of the Brown subgroup of the Solonetz great group.

Solonetzic soils are thought to have developed from parent materials that were more or less uniformly salinized with salts high in sodium. Leaching of salts by descending rainwater presumably mobilizes the sodium-saturated colloids. The colloids are apparently carried downward and deposited in the B horizon. Further leaching results in depletion of alkali cations in the A horizon, which becomes acidic, and a platy Ahe horizon usually develops. The underlying Solonetzic B horizon (Bn, Bnt) usually consists of darkly stained, fused, intact columnar peds. This stage is followed by the structural breakdown of the upper part of the B horizon and eventually its complete destruction in the most advanced stage, known as solodization. Solonetzic soils are correlated with the suborder of Argids in the order of Aridisols under the CSCS.

Appendix

2 CLIMATE DEFINITIONS AND BOUNDARIES

The following table summarizes the definitions and boundaries of climates and climate subtypes based on the soil-water balance, as described in Chapter 16 and shown on the world climate map, Figure 9.9. All definitions and boundaries are provisional.

Ep Water need (potential evapotranspiration)
D Soil water shortage (deficit)
R Water surplus (runoff)
S Storage (limited to 30 cm)

Group I: Low-Latitude Climates

1. Wet equatorial climate ①
Ep ≥ 10 cm in every month, and
S ≥ 20 cm in 10 or more months.

2. Monsoon and trade-wind coastal climate ②
Ep ≥ 4 cm in every month, or
Ep > 130 cm annual total, or both, and
S ≥ 20 cm in 6, 7, 8, or 9 consecutive months, or, if
S > 20 cm in 10 or more months, then Ep ≤ 10 cm in 5 or more consecutive months.

3. Wet-dry tropical climate ③
D ≥ 20cm, and
R ≥ 10 cm, and
Ep ≥ 130 cm annual total, or Ep ≥ 4 cm in every month, or both, and
S ≥ 20 cm in 5 months or fewer, or minimum monthly S < 3 cm.

4. Dry tropical climate ④
D ≥15 cm, and
R = 0, and
Ep ≥ 130 cm annual total, or Ep ≥ 4 cm in every month, or both.

Subtypes of dry climates (④, ⑤, ⑦, and ⑨)
s Semiarid subtype (Steppe subtype) At least 1 month with S > 2 cm.
a Desert subtype
No month with S > 2 cm.

Group II: Midlatitude Climates

5. Dry subtropical climate ⑤
D ≥ 15 cm, and
R = 0, and Ep < 130 cm annual total, and
Ep ≥ 0.8 cm in every month, and
Ep < 4 cm in 1 month.
(Subtypes ⑤a and ⑤s as defined under ④.)

6. Moist subtropical climate ⑥
D < 15 cm when R = 0, and
Ep < 4 cm in at least 1 month, and
Ep ≥ 0.8 cm in every month.

7. Mediterranean climate ⑦
D ≥ 15 cm, and
R ≥ 0, and
Ep ≥ 0.8 cm in every month, and storage index > 75%, or P/Ea × 100 < 40%. (Subtypes ⑦a and ⑦s as defined under ④.)

8. Marine west-coast climate ⑧
D < 15 cm, and
Ep < 80 cm annual total, and
Ep ≥ 0.8 cm in every month.

9. Dry midlatitude climate ⑨
D ≥ 15 cm, and
R = 0, and
Ep ≤ 0.7 cm in at least 1 month, and

Ep > 52.5 cm annual total.
(Subtypes ⑨a and ⑨s as defined under ④.)

10. Moist continental climate ⑩
D < 15 cm when R = 0, and
Ep ≤ 0.7 cm in at least 1 month, and
Ep > 52.5 cm annual total.

Group III: High-Latitude Climates

11. Boreal forest climate ⑪
52.5 cm > Ep > 35 cm annual total, and
Ep = 0 in fewer than 8 consecutive months.

12. Tundra climate ⑫
Ep < 35 cm annual total, and
Ep = 0 in 8 or more consecutive months.

13. Ice-sheet climate ⑬
Ep = 0 in all months.

Appendix

3 CONVERSION FACTORS

Metric to English

Metric Measure	Multiply by*	English Measure
LENGTH		
Millimeters (mm)	0.0394	Inches (in.)
Centimeters (cm)	0.394	Inches (in.)
Meters (m)	3.28	Feet (ft)
Kilometers (km)	0.621	Miles (mi)
AREA		
Square centimeters (cm^2)	0.155	Square inches (in^2)
Square meters (m^2)	10.8	Square feet (ft^2)
Square meters (m^2)	1.12	Square yards (yd^2)
Square kilometers (km^2)	0.386	Square miles (mi^2)
Hectares (ha)	2.47	Acres
VOLUME		
Cubic centimeters (cm^3)	0.0610	Cubic inches (in^3)
Cubic meters (m^3)	35.3	Cubic feet (ft^3)
Cubic meters (m^3)	1.31	Cubic yards (yd^3)
Milliliters (ml)	0.0338	Fluid ounces (fl oz)
Liters (l)	1.06	Quarts (qt)
Liters (l)	0.264	Gallons (gal)
MASS		
Grams (g)	0.0353	Ounces (oz)
Kilograms (kg)	2.20	Pounds (lb)
Kilograms (kg)	0.00110	Tons (2000 lb)
Tonnes (t)	1.10	Tons (2000 lb)

English to Metric

English Measure	Multiply by*	Metric Measure
LENGTH		
Inches (in.)	2.54	Centimeters (cm)
Feet (ft)	0.305	Meters (m)
Yards (yd)	0.914	Meters (m)
Miles (mi)	1.61	Kilometers (km)
AREA		
Square inches (in^2)	6.45	Square centimeters (cm^2)
Square feet (ft^2)	0.0929	Square meters (m^2)
Square yards (yd^2)	0.836	Square meters (m^2)
Square miles (mi^2)	2.59	Square kilometers (km^2)
Acres	0.405	Hectares (ha)
VOLUME		
Cubic inches (in^3)	16.4	Cubic centimeters (cm^3)
Cubic feet (ft^3)	0.0283	Cubic meters (m^3)
Cubic yards (yd^3)	0.765	Cubic meters (m^3)
Fluid ounces (fl oz)	29.6	Milliliters (ml)
Pints (pt)	0.473	Liters (l)
Quarts (qt)	0.946	Liters (l)
Gallons (gal)	3.79	Liters (l)
MASS		
Ounces (oz)	28.4	Grams (g)
Pounds (lb)	0.454	Kilograms (kg)
Tons (2000 lb)	907	Kilograms (kg)
Tons (2000 lb)	0.907	Tonnes (t)

* Conversion factors shown to 3 decimal-digit precision.

Answers to Problems

Working It Out 3.1 • Distances from Latitude and Longitude

1. Since the two cities are separated by 11° of latitude, and each latitude degree is 111 km or 69 mi, we have

$$11° \text{ lat} \times \frac{111 \text{ km}}{1° \text{ lat}} = 1221 \text{ km}$$

$$11° \text{ lat} \times \frac{69 \text{ mi}}{1° \text{ lat}} = 759 \text{ mi}$$

2. The two cities are nearly on the 45° parallel, and a degree of longitude there is equivalent to cos (45°) × 111 km = 0.707 × 111 = 78.5 km (0.707 × 69 mi = 48.8 mi). The longitude difference is 124° – 76° = 48°, so

$$48° \text{ long} \times \frac{78.5 \text{ km}}{1° \text{ long}} = 3768 \text{ km}$$

$$48° \text{ long} \times \frac{48.8 \text{ mi}}{1° \text{ long}} = 2342 \text{ mi}$$

3. Close to the equator, a degree of latitude is about equal to a degree of longitude, so both are about 111 km (or 69 mi). A square area 111 km (69 mi) on a side then has the following area:

$$111 \text{ km} \times 111 \text{ km} = 12{,}321 \text{ km}^2, \text{ or}$$

$$69 \text{ mi} \times 69 \text{ mi} = 4761 \text{ mi}^2.$$

At Winnipeg, a degree of longitude has length equal to cos (50°) × 111 km = 0.643 × 111 = 71.3 km (0.643 × 69 = 44.3 mi). Thus, at Winnipeg, a region of 1° longitude by 1° latitude will be of area

$$71.3 \text{ km} \times 111 \text{ km} = 7914 \text{ km}^2, \text{ or}$$

$$44.3 \text{ mi} \times 69 \text{ mi} = 3057 \text{ mi}^2.$$

Working It Out 3.4 • Global Timekeeping

1. Applying the formula, $D = Z_{HOME} - Z_{AWAY} = +6 - (-8) = +14$, so add 14 hours to Chicago time to get Beijing time. When you depart, at 3:30 P.M. on Saturday, June 14 in Chicago, it will be 5:30 A.M. on Sunday in Beijing. Add 16 hours and 30 minutes for the flight, and you get a scheduled arrival time of 10:00 P.M. Sunday, June 15, in Beijing.

2. From Figure 3.13, Anchorage is in zone +9, and Washington is in zone +5. Applying the formula, $D = Z_{HOME} - Z_{AWAY} = +9 - (+5) = +4$, so when you leave Anchorage at 1:59 A.M., the time in Washington will be 5:59 A.M.. If the plane arrives at 3:51 P.M., that will be 9 hours, 52 minutes later. Subtracting the 1 hour, 25 minute layover, the air time is then 8 hours, 27 minutes.

Working It Out 4.2 • Radiation Laws

1. The Stefan-Boltzmann Law describes the flow rate of energy leaving a surface, M, depending on its temperature. Since the temperature of the Sun's surface is 5950°K, the energy flow rate will be

$$\begin{aligned} M = \sigma T^4 &= (5.67 \times 10^{-8} \text{ W/m}^2\text{K}^4) \times (5950 \text{ K})^4 \\ &= 5.67 \times 10^{-8} \text{ W/m}^2\text{K}^4 \times 1.25 \times 10^{15} \text{ K}^4 \\ &= 7.11 \times 10^7 \text{ W/m}^2. \end{aligned}$$

The wavelength of greatest radiance, λ_{max}, follows Wein's Law. Thus,

$$\lambda_{\max} = \frac{b}{T} = \frac{2898 \text{ µm K}}{5950 \text{ K}} = 0.487 \text{ µm}.$$

2. This problem is the same as Problem 1, except that the surface temperature is different—15.4°C. In absolute temperature, this value is 15.4 + 273 = 288.4 K. So,

$$\begin{aligned} M = \sigma T^4 &= (5.67 \times 10^{-8} \text{ W/m}^2\text{K}^4) \times (288.4 \text{ K})^4 \\ &= 5.67 \times 10^{-8} \text{ W/m}^2\text{K}^4 \times 6.92 \times 10^9 \text{ K}^4 \\ &= 392 \text{ W/m}^2 \end{aligned}$$

and

$$\lambda_{max} = \frac{b}{T} = \frac{2898 \text{ µm K}}{288.4 \text{ K}} = 10.1 \text{ µm}.$$

3. From Problem 1, we have the Sun's surface energy emission rate as 7.11×10^7 W/m², and from Problem 2 we have the Earth's surface energy emission rate as 392 W/m². The ratio is therefore

$$\frac{7.11 \times 10^7 \text{ W/m}^2}{392 \text{ W/m}^2} = 1.81 \times 10^5 = 181{,}000.$$

Thus, the flow of energy from a square meter of the Sun's surface is about 181,000 times as large that from a square meter of the Earth's surface.

Working It Out 4.3 • Calculating the Global Radiation Balance

1. Since the radiation flow to Venus is 1.92 times greater than the radiation flow to the Earth, the solar constant for Venus will be 1.92 × 1.37 kW/m^2 = 2.63 kW/m^2. If the radius of Venus is 6050 km, the area the planet presents to the Sun will be that of a disk with area πr^2 = 3.14 × (6050 km)^2 = 3.14 × 3.65 × 10^7 km^2 = 1.15 × 10^8 km^2 = 1.15 × 10^{14} m^2. The total energy flow intercepted is equal to Venus's solar constant times the area it presents to the Sun, or = 2.63 kW/m^2 × 1.15 × 10^{14} m^2 = 3.03 × 10^{14} kW. For outflows, 65 percent of the incoming solar radiation is directly reflected back, providing a shortwave radiation flow rate to space of 0.65 × 3.03 × 10^{14} kW = 1.97 × 10^{14} kW. The remaining portion, 35 percent, or 0.35 × 3.03 × 10^{14} kW = 1.06 × 10^{14} kW, flows outward to space as longwave radiation emitted by the planet and its atmosphere.

Working It Out 5.2 • Temperature Conversion

1. Using the formula $F = \frac{9}{5}C + 32$, we have $F = \frac{9}{5}(38) + 32 = 68.4 + 32 = 100.4$°F for Toronto. For Buffalo, $C = \frac{5}{9}(F - 32) = \frac{5}{9}(38 - 32) = \frac{5}{9}(6) = 3.3$°C.

2. For 46°F, $C = \frac{5}{9}(46 - 32) = \frac{5}{9}(14) = 7.8$°C, and for 28°F, $C = \frac{5}{9}(28 - 32) = \frac{5}{9}(-4) = -2.2$. Thus, the range is 7.8 – (– 2.2) = 10°C. An easier way is to note that when comparing differences, only the fraction is needed for conversion. Thus, Δ18°F = $\Delta\frac{5}{9}(18)$°C = Δ10°C, where the symbol Δ denotes difference.

3. Let X be the unknown temperature in °F and °C. Since the temperature is the same, we substitute X for F and C and equate the two formulas:

$$\frac{9}{5}X + 32 = \frac{5}{9}(X - 32)$$

$$\frac{9}{5}X + 32 = \frac{5}{9}X - \frac{5}{9}(32)$$

$$\frac{9}{5}X - \frac{5}{9}X = -\frac{5}{9}(32) - 32$$

$$\frac{9}{5}X - \frac{5}{9}X = -\frac{5}{9}(32) - \frac{9}{9}(32)$$

$$\frac{81}{45}X - \frac{25}{45}X = -\frac{14}{9}(32)$$

$$\frac{56}{45}X = -\frac{448}{9}$$

$$X = -\frac{448 \times 45}{9 \times 56}$$

$$= -40$$

Thus, the thermometers will have an identical reading at –40°.

Working It Out 5.3 • Exponential Growth

1. For the 2 percent rate, the multiplier will be

$$M = e^{(R \times T)} = e^{(0.02 \times 50)} = 2.718^{1.00} = 2.72,$$

so 2.72 × 360 = 979 ppm. For the 3 percent rate,

$$M = e^{(0.03 \times 50)} = 2.718^{1.50} = 4.48,$$

and 4.48 × 360 = 1613 ppm.

2. Doubling time for Singapore is 70 ÷ 1.3 = 53.8 yrs, and for Republic of Congo is 70 ÷ 3.0 = 23.3 yrs. For Singapore,

$$M = e^{(0.013 \times 25)} = 2.718^{0.325} = 1.38,$$

and so the population will be 2.8 × 1.38 = 3.88 million. For Congo,

$$M = e^{(0.03 \times 25)} = 2.718^{0.75} = 2.12,$$

and the population will be 2.12 × 2.4 = 5.08 million.

Working It Out 6.1 • Energy and Latent Heat

1. Following the example, we can easily find that 4.19 × (100 – 15) + 2260 = 4.19 × 85 + 2260 = 2616 kJ/kg are required. Thus, we have

$$490 \text{ km}^3 \times \left[\frac{10^3 \text{ m}}{1 \text{ km}}\right]^3 \times \frac{10^3 \text{ kg}}{1 \text{ m}^3} \times \frac{2616 \text{ kJ}}{1 \text{ kg}} = 1.28 \times 10^{18} \text{ kJ}.$$

2. The ratio of annual energy in evaporation to annual energy consumption is then

$$\frac{1.28 \times 10^{18} \text{ kJ}}{8.5 \times 10^{13} \text{ kJ}} = 1.51 \times 10^4 = 15{,}100.$$

Thus, the annual evaporation is about 15 thousand times larger than U.S. annual energy consumption. Put another way,

$$\frac{8.5 \times 10^{13} \text{ kJ}}{1.28 \times 10^{18} \text{ kJ}} \times 100 = 0.00663\%,$$

or U.S. annual energy consumption is about 7 one-thousandths of one percent of annual evaporation.

Working It Out 6.3 • The Lifting Condensation Level

1.

$$H = 1000 \times \frac{25 - 18}{8.2} = 1000 \times \frac{7}{8.2} = 854 \text{ m}$$

$$T = 25 - 854 \times \frac{10}{1000} = 25 - 8.5 = 16.5°\text{C}$$

2.

$$H = 1000 \times \frac{30 - 18}{8.2} = 1000 \times \frac{12}{8.2} = 1463 \text{ m}$$

$$T = 25 - 1463 \times \frac{10}{1000} = 25 - 14.6 = 15.5°\text{C}$$

3. Equation (1) states

$$T_0 - H \times R_{DRY} = T_{DEW} - H \times R_{DEW}$$

Placing terms with T on the left and terms with H on the right,

$$T_0 - T_{DEW} = H \times R_{DRY} - H \times R_{DEW}$$

Factoring,

$$T_0 - T_{DEW} = H\,(R_{DRY} - R_{DEW})$$

Solving for H, we have

$$H = \frac{T_0 - T_{DEW}}{R_{DRY} - R_{DEW}} = \frac{T_0 - T_{DEW}}{\frac{10}{1000} - \frac{1.8}{1000}} = 1000\,\frac{T_0 - T_{DEW}}{8.2}$$

which is Equation (2).

Working It Out 7.1 • Pressure and Density in the Oceans and Atmosphere

1. Since P (in t/m^2) = D (in m) for the ocean, the pressure at the bottom of the diving pool will be 5 t/m^2. Adding atmospheric pressure (1 t/m^2) gives 6 t/m^2. The fraction due to water is then 5/6, while the fraction due to the atmosphere is 1/6. For the deep-sea diver, the pressure will be 100

+ 1 = 101, with fractions 100/101 for ocean water and 1/101 for the atmosphere.

2. For Mt. Washington, $Z = 1917$ m = 1.917 km, so

$$P_Z = 1014 \times [1 - (0.0226 \times 1.917)]^{5.26}$$
$$= 1014 \times (1 - 0.0433)^{5.26}$$
$$= 1014 \times (0.9567)^{5.26}$$
$$= 1014 \times 0.7925 = 804 \text{ mb.}$$

For Mt. Whitney, $Z = 4418$ m = 4.418 km, and

$$P_Z = 1014 \times [1 - (0.0226 \times 4.418)]^{5.26}$$
$$= 583 \text{ mb.}$$

The percentages are then 804/1014 × 100 = 79.3% for Mt. Washington and 583/1014 × 100 = 57.5% for Mt. Whitney. Reading from the graph should give about the same result.

3. At 350 m = 0.35 km, the barometer would read

$$P_Z = 1014 \times [1 - (0.0226 \times 0.35)]^{5.26}$$
$$= 973 \text{ mb.}$$

for a change of 1014 – 973 = 41 mb, so the answer is yes.

Working It Out 9.2 • Averaging in Time Cycles

1.

J	F	M	A	M	J	J	A	S	O	N	D
4.9	5.4	7.1	12.8	10.5	7.1	9.8	8.0	8.7	8.8	11.5	7.8

2. This sequence of years seems somewhat drier than the average. With the exception of March and July, all 5-year monthly means are lower than the long-term monthly means.

Working It Out 10.1 • Cycles of Rainfall in the Low Latitudes

1. The mean is 16.4 cm, and the mean deviation is 4.3 cm.

2. The relative variability is 0.26, which makes San Juan about the same as Padang, but less variable than Abbassia and Bombay.

Working It Out 11.1 • Standard Deviation and Coefficient of Variation

1.

Statistic	June	September
$\overline{P}$	18.0	11.1
s_P	8.46	4.63
CV_P	0.47	0.40

Working It Out 12.1 • Radioactive Decay

1. The half-life is 1.28 b.y., $k = 0.693/1.28 = 0.542$; so

$$P(t = 1) = e^{-0.542 \times 1} = e^{-0.542} = 0.582 = 58.2\%$$
$$P(t = 3) = e^{-0.542 \times 3} = e^{-1.626} = 0.197 = 19.7\%$$

2. Here, the half-life is 14.1 b.y., $k = 0.693/14.1 = 0.0491$, and

$$P(t = 5) = e^{-0.0491 \times 5} = e^{-0.245} = 0.782 = 78.2\%$$
$$P(t = 10) = e^{-0.0491 \times 10} = e^{-0.491} = 0.612 = 61.2\%$$
$$P(t = 15) = e^{-0.0491 \times 15} = e^{-0.737} = 0.479 = 47.9\%$$

3. For ^{14}C, $k = 0.693/5730 = 1.21 \times 10^{-4}$. Then

$$P(t) = e^{-1.21\times10^{-4}t}.$$

In this case, $P(t)$ is known and equal to 0.1, while t is unknown. Solving for t, we start with

$$P(t) = e^{-kt}$$

and taking the log to the base e of both sides, we obtain

$$\ln(P(t)) = -kt$$

Then we can simply solve for t:

$$t = -\frac{\ln(P(t))}{k}$$

Substituting values for $P(t)$ and k in this formula yields

$$t = -\frac{\ln(P(t))}{k} = -\frac{\ln(0.1)}{1.21 \times 10^{-4}} = -\frac{-2.30}{1.21 \times 10^{-4}} = 1.90 \times 10^4$$
$$= 19{,}000 \text{ yrs}$$

Working It Out 13.1 • Radiometric Dating

1. $$t = \frac{1}{k}\ln\left[\frac{D}{M} + 1\right] = \frac{1}{0.155}\ln[0.448 + 1] = \frac{0.370}{0.155} = 2.39 \text{ b.y.}$$

2. For this decay sequence, $H = 7.04 \times 10^8$ yr = 0.704 b.y., and $k = 0.693/H = 0.693/0.704 = 0.985$ for time in b.y. Thus,

$$t = \frac{1}{0.985}\ln[9.56 + 1] = \frac{2.36}{0.985} = 2.40 \text{ b.y.}$$

The results of the two analyses are therefore quite consistent.

Working It Out 14.4 • The Richter Scale

1. Using the formula, we have

$$\log_{10}E = 4.8 + 1.5M = 4.8 + 1.5 \times 5.2 = 12.6,$$

so $E = 10^{12.6} = 3.98 \times 10^{12}$ joules.

2. Again applying the formula,

$$\log_{10}E = 4.8 + 1.5 \times 6.0 = 13.8,$$

so $E = 10^{13.8} = 6.31 \times 10^{13}$ joules. If that energy is released in a 60 second period, then the flow rate is $6.31 \times 10^{13} \div 60 = 1.05 \times 10^{12}$ J/sec = 1.05×10^{12} W. Since each bulb consumes an energy flow of 100 W, $1.05 \times 10^{12} \div 10^2 = 1.05 \times 10^{10}$, or about 10 billion bulbs would be powered. Compared to the average U.S. electric power consumption rate, the flow rate of 1.05×10^{12} is about 1/3.

3. The ratio will be

$$\frac{E(7.3)}{E(7.1)} = \frac{10^{4.8+1.5\times7.3}}{10^{4.8+1.5\times7.1}} = \frac{10^{15.75}}{10^{15.45}} = 10^{15.75-15.45}$$
$$= 10^{0.3} = 2.00$$

That is, the energy release estimate has doubled.

Working It Out 15.1 • The Power of Gravity

1. To apply the formula, we first need to convert the velocity of the rock mass from km/hr to m/sec:

$$\frac{150 \text{ km}}{\text{hr}} \times \frac{1 \text{ hr}}{60 \text{ min}} \times \frac{1 \text{ min}}{60 \text{ s}} \times \frac{10^3 \text{ m}}{\text{km}} = 41.7 \text{ m/s}$$

Then we have

$$E = \frac{1}{2}mv^2 = \frac{1}{2} \times 7.56 \times 10^{10} \text{ kg} \times \left(\frac{41.7 \text{ m}}{\text{s}}\right)^2$$
$$= 6.57 \times 10^{13} \text{ J}$$

The total energy released is 2.82×10^{14} J, so the ratio of kinetic to total energy is

$$\frac{6.57 \times 10^{13} \text{ J}}{2.82 \times 10^{14} \text{ J}} = 0.223 = 22.3\%$$

2. The free-fall velocity will be

$$v = \sqrt{2gd} = \sqrt{2 \times \frac{9.8 \text{ m}}{\text{s}^2} \times 500 \text{ m}} = 99.0 \text{ m/s}$$

The ratio of the velocity of the Madison Slide to the free-fall velocity is then

$$\frac{41.7 \text{ m/s}}{99.0 \text{ m/s}} = .421 = 42.1\%$$

Thus, the slide moves less than half as fast as a free-falling body. The slide moves more slowly because it encounters friction in moving.

Working It Out 16.3 • Magnitude and Frequency of Flooding

1. The ranked flows with recurrence intervals are shown below. The magnitudes of floods with recurrence intervals closest to 1, 2, 5, 10, and 35 years are 113, 234, 337, 487, and 711 m³/s, respectively.

Rank	Flow, m³/s	Recurence, yrs	Rank	Flow, m³/s	Recurence, yrs
1	711	34.0	18*	234	2.0
2	677	17.0	19	229	1.8
3	487	11.3	20	225	1.7
4	419	8.5	21	217	1.6
5	357	6.8	22	216	1.5
6	351	5.7	23	210	1.5
7*	337	4.5	24	209	1.4
8*	337	4.5	25	194	1.4
9	312	3.8	26	193	1.3
10	309	3.4	27	184	1.3
11	292	3.1	28	181	1.2
12	279	2.8	29	172	1.2
13	278	2.6	30	164	1.1
14	266	2.4	31	139	1.1
15	264	2.3	32	119	1.1
16	245	2.1	33	113	1.0
17*	234	2.0			

* Indicates tied ranking.

2. For 250 m³/s, the table shows the recurrence interval for 245 m³/s to be 2.1 years, and 264 m³/s to be 2.3 years. The percent probabilities for these flows are then 100/2.1 = 47.6 percent and 100/2.3 = 43.5 percent. So the percent probability will be less than 47.6 percent but not as little as 43.5 percent. A good guess would be about 46.5 percent. In the case of 500 m³/s, the values for 487 and 677 m³/s are 8.85 and 5.88 percent, so a good guess would be about 8.5 percent. If a flow is to be equalled or exceeded in 25 percent of all years, its recurrence interval must be 100/25 = 4 yrs. From the table, we see that flows of 312 m³/s and 337 m³/s are associated with recurrence intervals of 3.8 and 4.5 years, respectively. So this flow will be more than 312 m³/s, but less than 337 m³/s. A reasonable guess would be about 320 m³/s.

Working It Out 17.1 • River Discharge and Suspended Sediment

1. Applying the formula for this river, we have for 50 m³/s

$$S = 178Q^{1.75} = 178(50)^{1.75} = 178 \times 940 = 167{,}000 \text{ t/d}$$

and for 5 m³/s,

$$S = 178(5)^{1.75} = 178 \times 16.7 = 2980 \text{ t/d}$$

2. Let's assume that the discharge is 1 m³/s and doubles to 2 m³/s. The ratio of suspended sediment transported at 2 m³/s to that at 1 m³/s is then

$$\frac{S(2)}{S(1)} = \frac{178(2)^{1.75}}{178(1)^{1.75}} = \frac{2^{1.75}}{1^{1.75}} = \frac{2^{1.75}}{1} = 3.36.$$

Thus, we would expect that the sediment load would more than triple with a doubling of discharge.

3. To convert the equation from logarithmic form to exponential form, raise 10 to the quantity on each side of the equation:

$$\log S = 2.50 + 2.12 \log Q$$
$$10^{\log S} = 10^{(2.50+2.12 \log Q)}$$
$$S = 10^{2.50} \times 10^{(2.12 \log Q)}$$
$$= 316 \times Q^{2.12}$$

Thus, $S = 316Q^{2.12}$.

Working It Out 18.1 • Properties of Stream Networks

1. From the problem statement, we have $R_b = 3.5$ and $k = 6$. Thus, the formula is

$$N_u = R_b^{(k-u)} = 3.5^{(6-u)},$$

so we have

$$N_1 = 3.5^{(6-1)} = 3.5^5 = 525$$
$$N_2 = 3.5^{(6-2)} = 3.5^4 = 150$$
$$N_3 = 3.5^{(6-3)} = 3.5^3 = 42.9$$
$$N_4 = 3.5^{(6-4)} = 3.5^2 = 12.3$$
$$N_5 = 3.5^{(6-5)} = 3.5^1 = 3.5$$
$$N_6 = 3.5^{(6-6)} = 3.5^0 = 1$$

2. Substituting $L_1 = 0.12$ and $R_L = 3.2$ into the formula for expected cumulative mean length from the text gives

$$L_u^* = L_1 R_L^{(u-1)} = 0.12 \times 3.2^{(u-1)},$$

so we have

$$L_1^* = 0.12 \times 3.2^{(1-1)} = 0.12 \times 3.2^0 = 0.12 \text{ km}$$
$$L_2^* = 0.12 \times 3.2^{(2-1)} = 0.12 \times 3.2^1 = 0.384 \text{ km}$$
$$L_3^* = 0.12 \times 3.2^{(3-1)} = 0.12 \times 3.2^2 = 1.23 \text{ km}$$
$$L_4^* = 0.12 \times 3.2^{(4-1)} = 0.12 \times 3.2^3 = 3.93 \text{ km}$$

Using the formula $L_u = L_u^* - L_{u-1}^*$ to obtain the (noncumulative) mean lengths, we have

$$L_1 = L_1^* = 0.12$$
$$L_2 = L_2^* - L_1^* = 0.384 - 0.12 = 0.264 \text{ km}$$
$$L_3 = L_3^* - L_2^* = 1.23 - 0.384 = 0.846 \text{ km}$$
$$L_4 = L_4^* - L_3^* = 3.93 - 1.23 = 2.70 \text{ km}$$

Working It Out 19.2 • Angle of Repose of Dune Sands

1. Your graph should look something like the one below. The samples are not as well separated, since there is significant overlap between them.

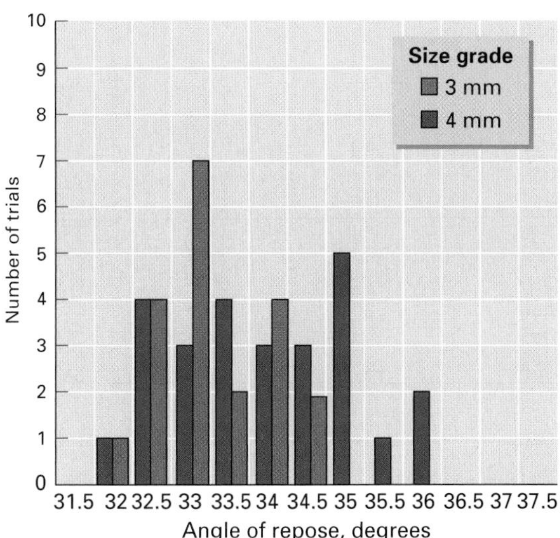

Frequency graph

2. First, find the standard deviation of the difference in sample means using the formula:

$$s_{\bar{X}_1-\bar{X}_2} = \sqrt{\frac{s_1^2+s_2^2}{n}} = \sqrt{\frac{(0.91)^2+(0.72)^2}{20}}$$

$$= \sqrt{\frac{0.828+0.518}{20}} = \sqrt{\frac{1.346}{20}} = \sqrt{0.0673} = 0.26$$

Then apply the formula for t:

$$t = \frac{|\bar{X}_1-\bar{X}_2|}{s_{\bar{X}_1-\bar{X}_2}} = \frac{|34.4-33.3|}{0.26} = \frac{1.1}{0.26} = 4.23$$

Comparing this value with those given in the problem, we see that there is less than 1 chance in 1000 that the two means are actually the same and the difference arose by chance. So we could conclude that grain size is definitely related to angle of repose, with the larger grain size having the smaller angle of repose.

Working It Out 20.3 • Isostatic Rebound

1. For the slope, we have

$$a' = \frac{\log 200 - \log 55}{8.0-5.0} = \frac{0.56}{3.0} = 0.187.$$

For the intercept,

$$b' = y' - a'x = \log 55 - 0.187 \times 5.0 = 0.805.$$

Thus, $\log y = 0.805 + 0.187\ x$. For the exponential form, $a = 10^{a'} = 10^{0.187} = 1.54$, $b = 10^{b'} = 10^{0.805} = 6.38$, and so $y = 6.38(1.54)^x$. Comparing the two slopes, we see that the slope for James Bay, 0.187, is steeper than that of Oslofjord, at 0.156. This agrees with the slopes as plotted in the figure. The comparison shows that the rate of uplift due to isostatic rebound was greater for James Bay than for Oslofjord.

2. If b is the position on the uplift scale for the present time, it means that 12.3 m of uplift remains to be accomplished.

3. To find the slope, $a' = \log a = \log 1.7 = 0.230$. This rebound rate is greater than those of both Oslofjord (0.156) and James Bay (0.187).

Working It Out 21.1 • Calculating a Simple Soil Water Budget

1.

Soil-water Budget for Urbana, Illinois

Month	P	Ea	$-G$	$+G$	R	Ep	D
Jan	5.7	0.0			+5.7	0.0	0.0
Feb	4.5	0.0			+4.5	0.0	0.0
Mar	8.2	1.4			+6.8	1.4	0.0
Apr	10.0	4.4			+5.6	4.4	0.0
May	9.9	8.8			+1.1	8.8	0.0
Jun	8.4	12.4	–4.0			12.6	0.2
Jul	8.0	13.4	–5.4			14.9	1.5
Aug	9.0	11.6	–2.6			13.0	1.4
Sep	8.3	7.8		+0.5		7.8	0.0
Oct	6.6	4.8		+1.8		4.8	0.0
Nov	5.7	1.4		+4.3		1.4	0.0
Dec	5.5	0.0		+5.5		0.0	0.0
Total	89.8	66.0	–12.0	+12.1	23.7	69.1	3.1

Working It Out 22.1 • Logistic Population Growth

1. Taking C at 440 and reading P_0 from the value in the table at $t = 0$, we have for A,

$$A = \frac{C - P_0}{P_0} = \frac{440-10.0}{10.0} = 43.0$$

For k, we have

$$k = \ln(1+R) = \ln(1+0.60) = \ln(1.60) = 0.470$$

The logistic equation for these data is then

$$P(t) = \frac{C}{1+Ae^{-kt}} = \frac{440}{1+(43)e^{-0.47t}}$$

Missing values for the table:

(1)	(3)	(4)	(5)	(6)
Time, hours	Increase, grams	Increase, percent	Logistic model	Difference
3	11.2	42.7	38.3	–0.9
7	38.6	30.3	169.1	–3.3
9	49.7	22.4	270.6	1.1
13	16.1	4.2	401.7	–1.8
17	3.1	0.7	433.7	–0.1

2. *(a)* For the increase in grams, values increase fairly steadily until a peak is reached at 8 hours, then decline to small values. In contrast, the increase in percent is high at first, then decreases to very small values. *(b)* Judging by the differences in column 6, the model fits the data quite well. Most differences are only a few grams. *(c)* Compared to the data in the box, the yeast growth pattern here shows a lower carrying capacity and lower growth rate. This could indicate that the medium is less nutritive. Other factors that might explain this result include the use of a different strain of yeast, or incubation of the yeast at a different temperature.

Working It Out 23.3 • Island Biogeography: The Species-Area Curve

1.

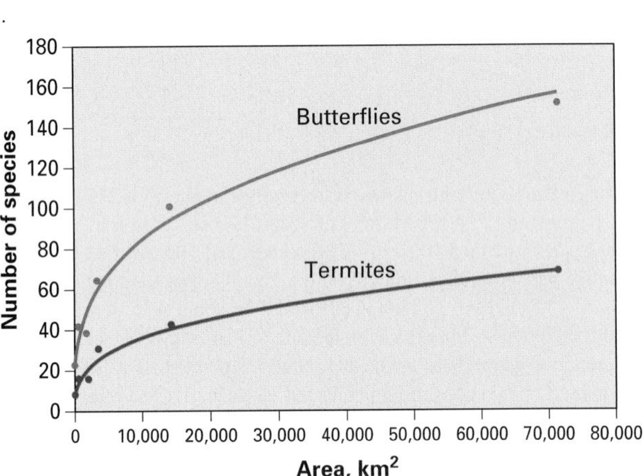

2.

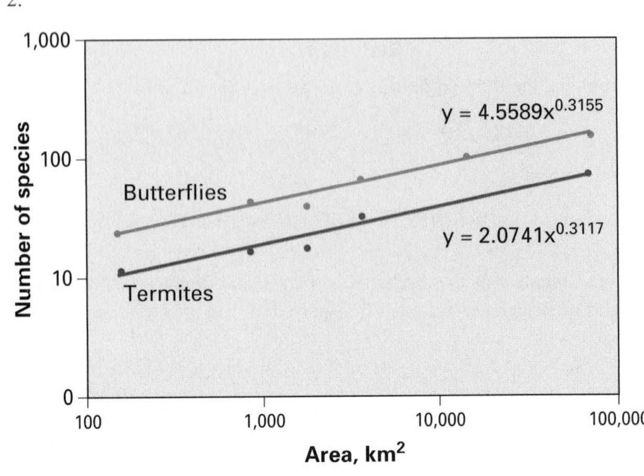

Termites:

$$\begin{aligned} S &= 2.07A^{0.31} \\ &= 2.07(1746)^{0.31} \\ &= 2.07(10.12) \\ &= 20.9 \end{aligned}$$

Butterflies:

$$\begin{aligned} S &= 4.56A^{0.32} \\ &= 4.56(1746)^{0.32} \\ &= 4.56(10.90) \\ &= 49.7 \end{aligned}$$

The two predicted values are somewhat higher than those actually observed. Perhaps Charleston Island has a less diverse environment than the other islands, given their sizes, and so has fewer termite and butterfly species. Or it may be that the island has not been fully explored and all species present might not have been found. Or perhaps a recent catastrophic event has wiped out some species and they have not yet returned by migration.

Glossary

This glossary contains definitions of terms shown in the text in italics or boldface. Terms that are *italicized* within the definitions will be found as individual entries elsewhere in the glossary.

A horizon *mineral* horizon of the *soil,* overlying the *E* and *B horizons.*

ablation a wastage of glacial ice by both *melting* and *evaporation.*

abrasion erosion of *bedrock* of a *stream channel* by impact of particles carried in a stream and by rolling of larger *rock* fragments over the stream bed; abrasion is also an activity of glacial ice, waves, and wind.

abrasion platform sloping, nearly flat *bedrock* surface extending out from the foot of a *marine cliff* under the shallow water of breaker zone.

absorption of radiation transfer of *electromagnetic energy* into heat *energy* within a *gas* or *liquid* through which the radiation is passing or at the surface of a *solid* struck by the radiation.

abyssal plain large expanse of very smooth, flat ocean floor found at depths of 4600 to 5500 m (15,000 to 18,000 ft).

abyssal hills small hills rising to heights from a few tens of meters to a few hundred meters above the deep ocean floor.

accelerated erosion *soil erosion* occurring at a rate much faster than *soil horizons* can be formed from the parent *regolith.*

accretion of lithosphere production of new *oceanic lithosphere* at an active *spreading plate boundary* by the rise and solidification of *magma* of basaltic composition.

accretionary prism mass of deformed trench *sediments* and ocean floor *sediments* accumulated in wedgelike slices on the underside of the overlying plate above a plate undergoing *subduction.*

acid deposition the *deposition* of acid raindrops and/or dry acidic dust particles on vegetation and ground surfaces.

acid mine drainage sulfuric acid effluent from *coal* mines, mine tailings, or spoil ridges made by *strip mining.*

acid action solution of minerals by acids occurring in soil and ground water.

acid rain rainwater having an abnormally low *pH,* between 2 and 5, as a result of air pollution by sulfur oxides and nitrogen oxides.

active continental margins continental margins that coincide with tectonically active plate boundaries. (See also *continental margins, passive continental margins.*)

active layer shallow surface layer subject to seasonal thawing in *permafrost* regions.

active pool type of pool in the *biogeochemical cycle* in which the materials are in forms and places easily accessible to life processes. (See also *storage pool.*)

active systems *remote sensing* systems that emit a beam of wave *energy* at a source and measure the intensity of that *energy* reflected back to the source.

actual evapotranspiration (water use) Actual rate of *evapotranspiration* at a given time and place.

adiabatic lapse rate (See *dry adiabatic lapse rate, wet adiabatic lapse rate.*)

adiabatic process change of temperature within a *gas* because of compression or expansion, without gain or loss of heat from the outside.

advection fog *fog* produced by *condensation* within a moist basal air layer moving over a cold land or water surface.

aerosols tiny particles present in the *atmosphere,* so small and light that the slightest movements of air keep them aloft.

aggradation raising of *stream channel* altitude by continued *deposition* of *bed load.*

air a mixture of gases that surrounds the Earth.

air mass extensive body of air within which upward gradients of temperature and moisture are fairly uniform over a large area.

air pollutant an unwanted substance injected into the *atmosphere* from the Earth's surface by either natural or human activities; includes *aerosols, gases,* and *particulates.*

air temperature temperature of air, normally observed by a *thermometer* under standard conditions of shelter and height above the ground.

albedo percentage of downwelling solar *radiation* reflected upward from a surface.

albic horizon pale, often sandy *soil horizon* from which *clay* and free iron oxides have been removed. Found in the profile of the *Spodosols.*

Alfisols *soil order* consisting of *soils* of humid and subhumid climates, with high *base status* and an *argillic horizon.*

allele specific version of a particular gene.

allelopathy interaction among *species* in which a plant secretes substances into the soil that are toxic to other organisms.

allopatric speciation type of *speciation* in which populations are geographically isolated and gene flow between the populations does not take place.

alluvial fan gently sloping, conical accumulation of coarse *alluvium* deposited by a *braided stream* undergoing *aggradation* below the point of emergence of the channel from a narrow *gorge* or *canyon.*

alluvial meanders sinuous bends of a *graded stream* flowing in the alluvial deposit of a *floodplain.*

alluvial river *stream* of low *gradient* flowing upon thick deposits of *alluvium* and experiencing approximately annual overbank flooding of the adjacent *floodplain.*

alluvial terrace benchlike landform carved in *alluvium* by a *stream* during *degradation.*

alluvium any stream-laid *sediment* deposit found in a *stream channel* and in low parts of a stream valley subject to flooding.

alpine chains high mountain ranges that are narrow belts of *tectonic activity* severely deformed by *folding* and thrusting in comparatively recent geologic time.

alpine debris avalanche *debris flood* of steep mountain *slopes,* often laden with tree trunks, limbs, and large boulders.

alpine glacier long, narrow, mountain *glacier* on a steep downgrade, occupying the floor of a troughlike valley.

alpine permafrost *permafrost* occurring at high altitudes equatorward of the normal limit of *permafrost.*

alpine tundra a plant *formation class* within the *tundra biome,* found at high altitudes above the limit of *tree* growth.

amphibole group *silicate minerals* rich in calcium, magnesium, and iron, dark in color, high in *density,* and classed as *mafic minerals.*

amplitude for a smooth wavelike curve, the difference in height between a crest and the adjacent trough.

andesite *extrusive igneous rock* of diorite composition, dominated by *plagioclase feldspar;* the extrusive equivalent of *diorite.*

Andisols a *soil order* that includes *soils* formed on volcanic ash; often enriched by organic matter, yielding a dark soil color.

anemometer weather instrument used to indicate *wind* speed.

aneroid barometer *barometer* using a mechanism consisting of a partially evacuated air chamber and a flexible diaphragm.

angle of repose natural surface inclination *(dip)* of a *slope* consisting of loose, coarse, well-sorted *rock* or *mineral fragments;* for example, the *slip face* of a *sand dune,* a *talus slope,* or the sides of a *cinder cone.*

annuals plants that live only a single growing season, passing the unfavorable season as a seed or spore.

annular drainage pattern a stream network dominated by concentric (ringlike) major *subsequent streams.*

antarctic circle *parallel of latitude* at 66 1/2°S.

antarctic front zone frontal zone of interaction between antarctic *air masses* and polar air masses.

antarctic zone *latitude* zone in the latitude range 60° to 75°S (more or less), centered on the *antarctic circle,* and lying between the *subantarctic zone* and the *polar zone.*

anticlinal valley valley eroded in weak *strata* along the central line or axis of an eroded *anticline.*

anticline upfold of *strata* or other layered *rock* in an archlike structure; a class of *folds.* (See also *syncline.*)

anticyclone center of high *atmospheric pressure.*

aphelion point on the Earth's elliptical orbit at which the Earth is farthest from the Sun.

aquatic ecosystem *ecosystem* of a *lake, bog,* pond, river, *estuary,* or other body of water.

Aquepts *suborder* of the *soil order Inceptisols;* includes Inceptisols of wet places, seasonally saturated with water.

aquiclude *rock* mass or layer that impedes or prevents the movement of *groundwater.*

aquifer *rock* mass or layer that readily transmits and holds *groundwater.*

arc curved line that forms a portion of a circle.

arc-continent collision collision of a volcanic arc with *continental lithosphere* along a *subduction* boundary.

arctic circle *parallel of latitude* at 66 1/2°N.

arctic front zone frontal zone of interaction between arctic *air masses* and polar air masses.

arctic tundra a plant *formation class* within the *tundra biome,* consisting of low, mostly herbaceous plants, but with some very small stunted *trees,* associated with the *tundra climate* ⑫.

arctic zone *latitude* zone in the latitude range 60° to 75°N (more or less), centered about on the *arctic circle,* and lying between the *subarctic zone* and the *polar zone.*

arête sharp, knifelike divide or crest formed between two *cirques* by alpine glaciation.

argillic horizon *soil horizon,* usually the B *horizon,* in which *clay minerals* have accumulated by *illuviation.*

arid (dry climate subtype) subtype of the dry climates that is extremely dry and supports little or no vegetation cover.

Aridisols *soil order* consisting of soils of dry climates, with or without *argillic horizons,* and with accumulations of *carbonates* or soluble salts.

artesian well drilled well in which water rises under hydraulic pressure above the level of the surrounding *water table* and may reach the surface.

aspect compass orientation of a *slope* as an inclined element of the ground surface.

association plant-animal community type identified by the typical organisms that are likely to be found together.

asthenosphere soft layer of the upper *mantle,* beneath the rigid *lithosphere.*

astronomical hypothesis explanation for glaciations and interglaciations making use of cyclic variations in the form of solar *energy* received at the Earth's surface.

atmosphere envelope of gases surrounding the Earth, held by *gravity.*

atmospheric pressure pressure exerted by the atmosphere because of the force of *gravity* acting upon the overlying column of air.

atoll circular or closed-loop *coral reef* enclosing an open *lagoon* with no island inside.

atomic mass number total number of protons and *neutrons* within the nucleus of an atom.

atomic number number of protons within the nucleus of an atom; determines element name and chemical properties of the atom.

autogenic succession form of *ecological succession* that is self-producing—that is, results from the actions of plants and animals themselves.

autumnal equinox *equinox* occurring on September 22 or 23.

average deviation difference between a single value and the mean of all values, taken without respect to sign.

axial rift narrow, trenchlike depression situated along the center line of the *mid-*

oceanic ridge and identified with active seafloor spreading.

axis of rotation center line around which a body revolves, as the Earth's axis of rotation.

B horizon mineral *soil horizon* located beneath the A *horizon*, and usually characterized by a gain of *mineral matter* (such as *clay minerals* and oxides of aluminum and iron) and organic matter *(humus)*.

backswamp area of low, swampy ground on the *floodplain* of an *alluvial river* between the *natural levee* and the *bluffs*.

backwash return flow of *swash* water under influence of gravity.

badlands rugged land surface of steep *slopes*, resembling miniature mountains, developed on weak *clay* formations or clay-rich *regolith* by fluvial erosion too rapid to permit plant growth and soil formation.

bar low ridge of *sand* built above water level across the mouth of a *bay* or in shallow water paralleling the shoreline. May also refer to embankment of sand or gravel on floor of a *stream channel*.

bar (pressure) unit of pressure equal to 10^5 Pa *(pascals)*; approximately equal to the pressure of the Earth's *atmosphere* at sea level.

barchan dune *sand dune* of crescentic base outline with a sharp crest and a steep lee *slip face*, with crescent points (horns) pointing downwind.

barometer instrument for measurement of *atmospheric pressure*.

barrier (to dispersal) a zone or region that a *species* is unable to colonize or perhaps even occupy for a short time, thus halting *diffusion*.

barrier island long narrow island, built largely of beach *sand* and dune sand, parallel with the mainland and separated from it by a *lagoon*.

barrier reef *coral reef* separated from mainland *shoreline* by a *lagoon*.

barrier-island coast *coastline* with broad zone of shallow water offshore (a *lagoon*) shut off from the ocean by a *barrier island*.

basalt *extrusive igneous rock* of *gabbro* composition; occurs as *lava*.

base flow that portion of the *discharge* of a *stream* contributed by *ground water* seepage.

base level lower limiting surface or level that can ultimately be attained by a *stream* under conditions of stability of the Earth's crust and sea level; an imaginary surface equivalent to sea level projected inland.

base status of soils quality of a *soil* as measured by the presence or absence of *clay minerals* capable of holding large numbers of *bases*. Soils of high *base status* are rich in base-holding *clay minerals*; soils of low *base status* are deficient in such minerals.

bases certain positively-charged *ions* in the *soil* that are also plant nutrients; the most important are calcium, magnesium, potassium, and sodium.

batholith large, deep-seated body of *intrusive igneous rock*, usually with an area of surface exposure greater than 100 km^2 (40 mi^2).

bauxite mixture of several *clay minerals*, consisting largely of aluminum oxide and water with impurities; a principal ore of aluminum.

bay a body of water sheltered from strong wave action by the configuration of the *coast*.

beach thick, wedge-shaped accumulation of *sand, gravel*, or cobbles in the zone of breaking waves.

beach drift transport of *sand* on a beach parallel with a *shoreline* by a succession of landward and seaward water movements at times when *swash* approaches obliquely.

bearing direction angle between a line of interest and a reference line, which is usually a line pointing north.

bed load that portion of the *stream load* moving close to the stream bed by rolling and sliding.

bedrock solid *rock* in place with respect to the surrounding and underlying *rock* and relatively unchanged by *weathering* processes.

bedrock slump *landslide* of *bedrock* in which most of the *bedrock* remains more or less intact as it moves.

bioclimatic frontier geographic boundary corresponding with a critical limiting level of climate stress beyond which a *species* cannot survive.

biodiversity the variety of biological life on Earth or within a region.

biogas mixture of methane and *carbon dioxide* generated by action of anaerobic bacteria in animal and human wastes enclosed in a digesting chamber.

biogeochemical cycle total system of *pathways* by which a particular type of *matter* (a given element, compound, or ion, for example) moves through the Earth's *ecosystem* or *biosphere*; also called a *material cycle* or *nutrient cycle*.

biogeographic region region in which the same or closely related plants and animals tend to be found together.

biogeography the study of the distributions of organisms at varying spatial and temporal *scales*, as well as the processes that produce these distribution patterns.

biomass dry weight of living organic matter in an *ecosystem* within a designated surface area; units are kilograms of organic matter per square meter.

biome largest recognizable subdivision of *terrestrial ecosystems*, including the total assemblage of plant and animal life interacting within the *life layer*.

biosphere all living organisms of the Earth and the environments with which they interact.

bitumen combustible mixture of hydrocarbons that is highly viscous and will flow only when heated; considered a form of petroleum.

bituminous sand (See *bitumen*.)

blackbody ideal object or surface that is a perfect radiator and absorber of *energy*; absorbs all radiation it intercepts and emits radiation perfectly according to physical theory.

block mountains class of mountains produced by block faulting and usually bounded by *normal faults*.

block separation separation of individual joint blocks during the process of *physical weathering*.

blowout shallow depression produced by continued *deflation*.

bluffs steeply rising ground slopes marking the outer limits of a *floodplain*.

bog a shallow depression filled with organic matter, for example a glacial lake or pond basin filled with *peat*.

Boralfs *suborder* of the *soil order Alfisols*; includes Alfisols of *boreal forests* or high mountains.

boreal forest variety of *needleleaf forest* found in the *boreal forest climate* ⑪ regions of North America and Eurasia.

boreal forest climate ⑪ cold climate of the *subarctic zone* in the northern *hemisphere* with long, extremely severe winters and several consecutive months of zero *potential evapotranspiration (water need)*.

Borolls *suborder* of the *soil order Mollisols*; includes Mollisols of cold-winter semiarid plants *(steppes)* or high mountains.

braided stream *stream* with shallow channel in coarse *alluvium* carrying multiple threads of fast flow that subdivide and rejoin repeatedly and continually shift in position.

breaker sudden collapse of a steepened water wave as it approaches the shoreline.

broadleaf deciduous forest *forest* type consisting of broadleaf *deciduous trees* and found in the *moist subtropical climate* ⑥ in parts of the *marine west-*

coast climate ⑧. (See also *midlatitude deciduous forest.*)

broadleaf evergreen forest *forest* type consisting of broadleaf *evergreen trees* and found in the wet equatorial and tropical climates. (See also *low-latitude rainforest.*)

Brunisolic order a class of *forest soils* in the Canadian soil classification system with brownish *B horizon.*

budget in flow systems, an accounting of *energy* and *matter* flows that enter, move within, and leave a system.

bush-fallow farming agricultural system practiced in the African *savanna woodland* in which *trees* are cut and burned to provide cultivation plots.

butte prominent, steep-sided hill or peak, often representing the final remnant of a resistant layer in a region of flat-lying *strata.*

C horizon *soil horizon* lying beneath the *B horizon,* consisting of *sediment* or *regolith* that is the *parent material* of the soil.

calcification accumulation of *calcium carbonate* in a soil, usually occurring in the *B* or *C horizons.*

calcite mineral having the composition *calcium carbonate.*

calcium carbonate compound consisting of calcium (Ca) and carbonate (CO_3) *ions,* formula $CaCo_3$, occurring naturally as the mineral *calcite.*

caldera large, steep-sided circular depression resulting from the explosion and subsidence of a *stratovolcano.*

canyon (See *gorge.*)

capillary action process by which *capillary tension* draws water into a small opening, such as a *soil* pore or a *rock joint.*

capillary tension a cohesive force among surface molecules of a *liquid* that gives a droplet its rounded shape.

carbohydrate class of organic compounds consisting of the elements carbon, hydrogen and oxygen.

carbon cycle *biogeochemical cycle* in which carbon moves through the *biosphere;* includes both *gaseous cycles* and *sedimentary cycles.*

carbon dioxide the chemical compound CO_2, formed by the union of two atoms of oxygen and one atom of carbon; normally a gas present in low concentration in the *atmosphere.*

carbon fixation (See *photosynthesis.*)

carbonates (carbonate minerals, carbonate rocks) *minerals* that are carbonate compounds of calcium or magnesium or both, i.e., *calcium carbonate* or magnesium carbonate. (See also *calcite.*)

carbonic acid a weak acid created when CO_2 gas dissolves in water.

carbonic acid action chemical reaction of *carbonic acid* in rainwater, *soil water,* and *ground water* with *minerals;* most strongly affects carbonate minerals and *rocks,* such as limestone and marble; an activity of *chemical weathering.*

cartography the science and art of making maps.

Celsius scale temperature scale in which the *freezing* point of water is 0°, and the boiling point is 100°.

Cenozoic Era last (youngest) of the *eras* of geologic time.

channel (See *stream channel.*)

chaparral sclerophyll scrub and dwarf *forest* plant *formation class* found throughout the coastal mountain ranges and hills of central and southern California.

chemical energy *energy* stored within an organic molecule and capable of being transformed into *heat* during metabolism.

chemical weathering chemical change in *rock-forming* minerals through exposure to atmospheric conditions in the presence of water; mainly involving *oxidation, hydrolysis, carbonic acid action,* or direct solution.

chemically precipitated sediment *sediment* consisting of *mineral matter* precipitated from a water solution in which the matter has been transported in the dissolved state as *ions.*

chernozem type of *soil order* closely equivalent to *Mollisol;* an order of the Canadian Soil Classification System.

Chernozemic order a class of grassland *soils* in the Canadian soil classification system with a thick *A horizon* rich in organic matter.

chert *sedimentary rock* composed largely of silicon dioxide and various impurities, in the form of nodules and layers, often occurring with *limestone* layers.

chinook wind a *local wind* occurring at certain times to the lee of the Rocky Mountains; a very dry wind with a high capacity to evaporate *snow.*

chlorofluorocarbons (CFCs) synthetic chemical compounds containing chlorine, fluorine, and carbon atoms that are widely used as coolant fluids in refrigeration systems.

cinder cone conical hill built of coarse *tephra* ejected from a narrow volcanic vent; a type of *volcano.*

circle of illumination great circle that divides the globe at all times into a sunlit *hemisphere* and a shadowed hemisphere.

circum-Pacific belt chains of andesite *volcanoes* making up mountain belts and *island arcs* surrounding the Pacific Ocean basin.

cirque bowl-shaped depression carved in *rock* by glacial processes and holding the *firn* of the upper end of an *alpine glacier.*

clast rock or mineral fragment broken from a parent *rock* source.

clastic sediment *sediment* consisting of particles broken away physically from a parent *rock* source.

clay *sediment* particles smaller than 0.004 mm in diameter.

clay minerals class of *minerals* produced by alteration of *silicate minerals,* having plastic properties when moist.

claystone *sedimentary rock* formed by lithification of *clay* and lacking *fissile* structure.

cliff sheer, near-vertical *rock* wall formed from flat-lying resistant layered *rocks,* usually *sandstone, limestone,* or *lava* flows; may refer to any near-vertical rock wall. (See also *marine cliff.*)

climate generalized statement of the prevailing weather conditions at a given place, based on statistics of a long period of record and including mean values, departures from those means, and the probabilities associated with those departures.

climatic frontier a geographical boundary that marks the limit of survival of a plant *species* subjected to climatic stress.

climatology the science that describes and explains the variability in space and time of the heat and moisture states of the Earth's surface, especially its land surfaces.

climax stable community of plants and animals reached at the end point of *ecological succession.*

climograph a graph on which two or more climatic variables, such as monthly mean temperature and monthly mean precipitation, are plotted for each month of the year.

closed flow system flow system that is completely self-contained within a boundary through which no *matter* or *energy* is exchanged with the external environment. (See also *open flow system.*)

cloud forest a type of low evergreen rainforest that occurs high on mountain slopes, where *clouds* and *fog* are frequent.

clouds dense concentrations of suspended water or ice particles in the diameter range 20 to 50 µm. (See *cumuliform clouds, stratiform clouds.*)

coal *rock* consisting of hydrocarbon compounds, formed of compacted, lithified, and altered accumulations of plant remains *(peat).*

coarse textured (rock) having *mineral* crystals sufficiently large that they are at least visible to the naked eye or with low magnification.

coast (See *coastline.*)

coastal and marine geography the study of the geomorphic processes that shape shores and coastlines and their application to coastal development and marine resource utilization.

coastal blowout dune high *sand dune* of the *parabolic dunes* class formed adjacent to a beach, usually with a deep *deflation* hollow *(blowout)* enclosed within the dune ridge.

coastal forest subtype of *needleleaf evergreen forest* found in the humid coastal zone of the northwestern United States and western Canada.

coastal plain coastal belt, emerged from beneath the sea as a former *continental shelf,* underlain by *strata* with gentle *dip* seaward.

coastline (coast) zone in which coastal processes operate or have a strong influence.

coefficient of variation in statistics, the ratio of the standard deviation to the mean.

cold front moving weather *front* along which a cold *air mass* moves underneath a warm air mass, causing the latter to be lifted.

cold-blooded animal animal whose body temperature passively follows the temperature of the environment.

cold-core ring circular eddy of cold water, surrounded by warm water and lying adjacent to a warm, poleward-moving *ocean current,* such as the Gulf Stream. (See also *warm-core ring.*)

colloids particles of extremely small size, capable of remaining indefinitely in suspension in water. May be mineral or organic in nature.

colluvium deposit of *sediment* or *rock* particles accumulating from overland flow at the base of a *slope* and originating from higher slopes where *sheet erosion* is in progress. (See also *alluvium.*)

community an assemblage of organisms that live in a particular *habitat* and interact with one another.

competition form of interaction among plant or animal *species* in which both draw resources from the same *pool.*

component in flow systems, a part of the system, such as a *pathway,* connection, or flow of *matter* or *energy.*

composite volcano volcano composed of layers of ash and lava. See *stratovolcano.*

compression (tectonic) squeezing together, as horizontal compression of crustal layers by *tectonic* processes.

condensation process of change of *matter* in the gaseous state *(water vapor)* to the liquid state (liquid water) or solid state (ice).

condensation nucleus a tiny bit of solid *matter (aerosol)* in the *atmosphere* on which *water vapor* condenses to form a tiny water droplet.

conduction of heat transmission of *sensible heat* through *matter* by transfer of *energy* from one atom or molecule to the next in the direction of decreasing temperature.

cone of depression conical configuration of the lowered *water table* around a well from which water is being rapidly withdrawn.

conformal projection *map* projection that preserves without shearing the true shape or outline of any small surface feature of the Earth.

conglomerate a *sedimentary rock* composed of pebbles in a matrix of finer *rock* particles.

conic projections a group of *map projections* in which the *geographic grid* is transformed to lie on the surface of a developed cone.

consequent stream *stream* that takes its course down the slope of an *initial landform,* such as a newly emerged *coastal plain* or a *volcano.*

consumers animals in the *food chain* that live on organic matter formed by *primary producers* or by other *consumers.* (See also *primary consumers, secondary consumers.*)

consumption (of a lithospheric plate) destruction or disappearance of a subducting *lithospheric plate* in the *asthenosphere,* in part by *melting* of the upper surface, but largely by softening because of heating to the temperature of the surrounding *mantle rock.*

continental collision event in *plate tectonics* in which subduction brings two segments of the *continental lithosphere* into contact, leading to formation of a *continental suture.*

continental crust crust of the continents, of felsic composition in the upper part; thicker and less dense than *oceanic crust.*

continental drift hypothesis, introduced by Alfred Wegener and others early in the 1900s, of the breakup of a parent continent, *Pangea,* starting near the close of the *Mesozoic Era,* and resulting in the present arrangement of *continental shields* and intervening *ocean-basin floors.*

continental lithosphere *lithosphere* bearing *continental crust* of *felsic igneous rock.*

continental margins (1) Topographic: one of three major divisions of the ocean basins, being the zones directly adjacent to the continent and including the *continental shelf, continental slope,* and *continental rise.* (2) Tectonic: marginal belt of continental crust and lithosphere that is in contact with *oceanic crust* and *lithosphere,* with or without an active plate boundary being present at the contact. (See also *active continental margins, passive continental margins.*)

continental rise gently sloping seafloor lying at the foot of the *continental slope* and leading gradually into the *abyssal plain.*

continental rupture crustal spreading apart affecting the *continental lithosphere,* so as to cause a *rift valley* to appear and to widen, eventually creating a new belt of *oceanic lithosphere.*

continental scale scale of observation at which we recognize continents and other large Earth surface features, such as ocean currents.

continental shelf shallow, gently sloping belt of seafloor adjacent to the continental shoreline and terminating at its outer edge in the *continental slope.*

continental shields ancient crustal *rock* masses of the continents, largely *igneous rock* and *metamorphic rock,* and mostly of *Precambrian age.*

continental slope steeply descending belt of seafloor between the *continental shelf* and the *continental rise.*

continental suture long, narrow zone of crustal deformation, including underthrusting and intense *folding,* produced by a *continental collision.* Examples: Himalayan Range, European Alps.

continuous permafrost *permafrost* that underlies more than 90 percent of the surface area of a region.

convection (atmospheric) air motion consisting of strong updrafts taking place within a *convection cell.*

convection cell individual column of strong updrafts produced by atmospheric *convection.*

convection loop circuit of moving *fluid,* such as *air* or water, created by unequal heating of the *fluid.*

convectional precipitation a form of *precipitation* induced when warm, moist air is heated at the ground surface, rises, cools, and condenses to form water droplets, raindrops, and eventually, rainfall.

converging boundary boundary between two crustal plates along which *subduction* is occurring and *lithosphere* is being consumed.

coral reef rocklike accumulation of *carbonates* secreted by corals and algae in shallow water along a marine shoreline.

coral-reef coast *coast* built out by accumulations of *limestone* in *coral reefs.*

core of Earth spherical central mass of the Earth composed largely of iron and consisting of an outer liquid zone and an interior solid zone.

Coriolis effect effect of the Earth's rotation tending to turn the direction of motion of any object or *fluid* toward the right in the northern *hemisphere* and to the left in the southern hemisphere.

corrosion erosion of *bedrock* of a *stream channel* (or other *rock* surface) by chemical reactions between solutions in stream water and *mineral* surfaces.

cosmopolitan species *species* that are found very widely.

counterradiation *longwave radiation* of atmosphere directed downward to the Earth's surface.

covered shields areas of *continental shields* in which the ancient *rocks* are covered beneath a thin layer of sedimentary *strata.*

crater central summit depression associated with the principal vent of a *volcano.*

crescentic dune (See *barchan dunes.*)

crevasse gaping crack in the brittle surface ice of a *glacier.*

crude oil liquid fraction of *petroleum.*

crust of Earth outermost solid shell or layer of the Earth, composed largely of *silicate minerals.*

Cryaquepts great group within the soil *suborder* of *Aquepts;* includes Aquepts of cold climate regions and particularly the *tundra climate* ⑫.

Cryosolic order a class of *soils* in the Canadian soil classification system associated with strong frost action and underlying *permafrost.*

cryoturbation movement of mineral particles of any size by freezing and thawing of ice.

cuesta *erosional landform* developed on resistant *strata* having low to moderate *dip* and taking the form of an asymmetrical low ridge or hill belt with one side a steep slope and the other a gentle slope; usually associated with a *coastal plain.*

cultural energy *energy* in forms exclusive of solar *energy* of *photosynthesis* that is expended on the production of raw food or feed crops in agricultural *ecosystems.*

cumuliform clouds *clouds* of globular shape, often with extended vertical development.

cumulonimbus cloud large, dense *cumuliform cloud* yielding *precipitation.*

cumulus cloud type consisting of low-lying, white cloud masses of globular shape well separated from one another.

cutoff cutting-through of a narrow neck of land, so as to bypass the stream flow in an *alluvial meander* and cause it to be abandoned.

cycle in flow systems, a closed flow system of *matter.* Example: *biogeochemical cycle.* (See *closed flow system.*)

cycle of rock change total cycle of changes in which *rock* of any one of the three major *rock* classes—*igneous rock, sedimentary rock, metamorphic rock*—is transformed into *rock* of one of the other classes.

cyclone center of low *atmospheric pressure.* (See *tropical cyclone, wave cyclone.*)

cyclonic precipitation a form of *precipitation* that occurs as warm moist air is lifted by air motion occurring in a *cyclone.*

cyclonic storm intense weather disturbance within a moving *cyclone* generating strong winds, cloudiness, and *precipitation.*

cylindric projections group of *map projections* in which the *geographic grid* is transformed to lie on the surface of a developed cylinder.

data acquisition component component of a *geographic information system* in which data are gathered together for input to the system.

data management component component of a *geographic information system* that creates, stores, retrieves, and modifies data layers and *spatial objects*

daughter product new *isotope* created by decay of an *unstable isotope.*

daylight saving time time system under which time is advanced by one hour with respect to the *standard time* of the prevailing *standard meridian.*

debris flood (debris flow) streamlike flow of muddy water heavily charged with *sediment* of a wide range of size grades, including boulders, generated by sporadic torrential rains upon steep mountain watersheds.

decalcification removal of *calcium carbonate* from a *soil horizon* as *carbonic acid* reacts with *carbonate mineral matter.*

December solstice (See *winter solstice.*)

deciduous plant *tree* or *shrub* that sheds its leaves seasonally.

declination of Sun latitude at which the Sun is directly overhead; varies from $-23^1/_2°$ ($23^1/_2°$ S lat.) to $+23^1/_2°$ N lat.)

décollement detachment and extensive sliding of a *rock* layer, usually *sedimentary,* over a near-horizontal basal *rock* surface; a special form of low-angle thrust *faulting.*

decomposers organisms that feed on dead organisms from all levels of the *food chain;* most are microorganisms and bacteria that feed on decaying organic matter.

deep sea cone a fan-shaped accumulation of undersea *sediment* on the *continental rise* produced by sediment-rich currents flowing down the *continental slope.*

deficit (soil-water shortage) in the soil-water budget, the difference between *water use* and *water need;* the quantity of irrigation water required to achieve maximum growth of agricultural crops.

deflation lifting and transport in *turbulent suspension* by wind of loose particles of *soil* or *regolith* from dry ground surfaces.

deglaciation widespread recession of *ice sheets* during a period of warming global climate, leading to an interglaciation. (See also *glaciation, interglaciation.*)

degradation lowering or downcutting of a *stream channel* by *stream erosion* in *alluvium* or *bedrock.*

degree of arc measurement of the angle associated with an *arc,* in degrees.

delta *sediment* deposit built by a stream entering a body of standing water and formed of the *stream load.*

delta coast *coast* bordered by a *delta.*

delta kame flat-topped hill of *stratified drift* representing a glacial *delta* constructed adjacent to an *ice sheet* in a marginal glacial lake.

dendritic drainage pattern *drainage pattern* of treelike branched form, in which the smaller streams take a wide variety of directions and show no parallelism or dominant trend.

denitrification biochemical process in which nitrogen in forms usable to plants is converted into molecular nitrogen in the gaseous form and returned to the atmosphere—a process that is part of the *nitrogen cycle.*

density of matter quantity of mass per unit of volume, stated in kg m^{-2}.

denudation total action of all processes whereby the exposed *rocks* of the continents are worn down and the resulting *sediments* are transported to the sea by the *fluid agents;* includes also *weathering* and *mass wasting.*

deposition (atmosphere) the change of state of a substance from a *gas (water vapor)* to a *solid* (ice); in the science of *meteorology,* the term sublimation is used to describe both this process and the change of state from solid to vapor. (See *sublimation.*)

deposition (of sediment) (See *stream deposition.*)

depositional landform *landform* made by *deposition* of *sediment.*

desert biome *biome* of the dry climates consisting of thinly dispersed plants that may be *shrubs,* grasses, or perennial *herbs,* but lacking in *trees.*

desert pavement surface layer of closely fitted pebbles or coarse *sand* from which finer particles have been removed.

desertification (See *land degradation.*)

detritus decaying organic matter on which *decomposers* feed.

dew point lapse rate rate at which the dew point of an air mass decreases with elevation; typical value is 1.8°C/1000 m (1.0°F/1000 ft)

dew point temperature temperature of an *air mass* at which the air holds its full capacity of water vapor.

diagnostic horizons *soil horizons,* rigorously defined, that are used as diagnostic criteria in classifying *soils.*

diffuse radiation solar radiation that has been *scattered* (deflected or reflected) by minute dust particles or cloud particles in the *atmosphere.*

diffuse reflection solar *radiation* scattered back to space by the Earth's atmosphere.

diffusion the slow extension of the range of a *species* by normal processes of dispersal.

digital image numeric representation of a picture consisting of a collection of numeric brightness values (pixels) arrayed in a fine grid pattern.

dike thin layer of *intrusive igneous rock,* often near-vertical or with steep *dip,* occupying a widened fracture in the surrounding *rock* and typically cutting across older *rock* planes.

dimictic in *limnology*, mixing twice each year, as a dimictic *lake.*

diorite *intrusive igneous rock* consisting dominantly of plagioclase feldspar and pyroxene; a *felsic igneous rock.*

dip acute angle between an inclined natural *rock* plane or surface and an imaginary horizontal plane of reference; always measured perpendicular to the *strike.* Also a verb, meaning to incline toward.

diploid having two sets of chromosomes, one from each parent organism.

discharge volume of flow moving through a given cross section of a stream in a given unit of time; commonly given in cubic meters (feet) per second.

discontinuous permafrost *permafrost* that underlies from 10 to 90 percent of the surface of a region.

disjunction geographic distribution pattern of *species* in which one or more closely related species are found in widely separated regions.

dispersal the capacity of a *species* to move from a location of birth or origin to new sites.

distributary branching *stream channel* that crosses a *delta* to discharge into open water.

diurnal adjective meaning "daily."

doldrums belt of calms and variable winds occurring at times along the *equatorial trough.*

dolomite carbonate mineral or *sedimentary rock* having the composition calcium magnesium carbonate.

dome (See *sedimentary dome.*)

drainage basin total land surface occupied by a *drainage system,* bounded by a *drainage divide* or watershed.

drainage divide imaginary line following a crest of high land such that overland flow on opposite sides of the line enters different *streams.*

drainage pattern the plan of a network of interconnected *stream channels.*

drainage system a branched network of *stream channels* and adjacent land *slopes,* bounded by a *drainage divide* and converging to a single channel at the outlet.

drainage winds *winds,* usually cold, that flow from higher to lower regions under the direct influence of *gravity.*

drawdown (of a well) difference in height between base of cone of depression and original water table surface.

drought occurrence of substantially lower-than-average *precipitation* in a season that normally has ample precipitation for the support of food-producing plants.

drumlin hill of glacial *till,* oval or elliptical in basal outline and with smoothly rounded summit, formed by plastering of till beneath moving, debris-laden glacial ice.

dry adiabatic lapse rate rate at which rising air is cooled by expansion when no *condensation* is occurring; 10°C per 1000 m (5.5°F per 1000 ft).

dry desert plant *formation class* in the *desert biome* consisting of widely dispersed xerophytic plants that may be small, hard-leaved or spiny *shrubs,* succulent plants (cacti), or hard grasses.

dry lake shallow basin covered with salt deposits formed when stream input to the basin is subjected to severe *evaporation*; may also form by evaporation of a saline lake when climate changes; see also *salt flat.*

dry midlatitude climate ⑨ dry climate of the *midlatitude zone* with a strong annual cycle of *potential evapotranspiration (water need)* and cold winters.

dry subtropical climate ⑤ dry climate of the *subtropical zone,* transitional between the *dry tropical climate* ④ and the *dry midlatitude climate* ⑨.

dry tropical climate ④ climate of the *tropical zone* with large total annual *potential evapotranspiration (water need).*

dune (See *sand dune.*)

dust bowl western Great Plains of the United States, which suffered severe wind deflation and soil drifting during the drought years of the middle 1930s.

dust storm heavy concentration of dust in a turbulent *air mass,* often associated with a *cold front.*

E horizon soil mineral horizon lying below the *A horizon* and characterized by the loss of *clay minerals* and oxides of iron and aluminum; it may show a concentration of *quartz* grains and is often pale in color.

Earth's crust (See *crust of Earth.*)

earth hummock low mound of vegetation-covered earth found in *permafrost* terrain, formed by cycles of *ground ice* growth and melting (see also *mud hummock*).

earthflow moderately rapid downhill flowage of masses of water-saturated *soil, regolith,* or weak *shale,* typically forming a steplike terrace at the top and a bulging toe at the base.

earthquake a trembling or shaking of the ground produced by the passage of *seismic waves.*

earthquake focus point within the Earth at which the *energy* of an *earthquake* is first released by rupture and from which *seismic waves* emanate.

easterly wave weak, slowly moving trough of low pressure within the belt of *tropical easterlies;* causes a weather disturbance with rain showers.

ebb current oceanward flow of *tidal current* in a *bay* or tidal stream.

ecological succession time-succession (sequence) of distinctive plant and animal communities occurring within a given area of newly formed land or land cleared of plant cover by burning, clear cutting, or other agents.

ecology science of interactions between life forms and their environment; the science of *ecosystems.*

ecosystem group of organisms and the environment with which the organisms interact.

edaphic factors factors relating to soil that influence a terrestrial ecosystem.

El Niño episodic cessation of the typical *upwelling* of cold deep water off the coast of Peru; literally, "The Christ Child," for its occurrence in the Christmas season once every few years.

electromagnetic radiation (electromagnetic energy) wavelike form of *energy* radiated by any substance possessing heat; it travels through space at the speed of light.

electromagnetic spectrum the total *wavelength* range of *electromagnetic energy.*

eluviation soil-forming process consisting of the downward transport of fine particles, particularly the *soil colloids* (both mineral and organic), carrying them out of an upper *soil horizon.*

emergence exposure of submarine landforms by a lowering of sea level or a rise of the crust, or both.

endemic species a *species* found only in one region or location.

energy the capacity to do work, that is, to bring about a change in the state or motion of *matter.*

energy balance (global) balance between *shortwave* solar *radiation* received by the Earth-atmosphere system and *radiation* lost to space by *shortwave* reflection and *longwave radiation* from the Earth-atmosphere system.

energy balance (of a surface) balance between the flows of *energy* reaching a surface and the flows of *energy* leaving it.

energy flow system *open system* that receives an input of *energy,* undergoes internal *energy* flow, *energy* transformation, and *energy* storage, and has an *energy* output.

Entisols *soil order* consisting of mineral soils lacking *soil horizons* that would persist after normal plowing.

entrenched meanders winding, sinuous valley produced by *degradation* of a *stream* with trenching into the *bedrock* by downcutting.

environmental temperature lapse rate rate of temperature decrease upward through the *troposphere;* standard value is 6.4 C°/km (3 F°/1000 ft).

epilimnion warm, less dense, upper layer of a *lake* that forms by solar heating and wind-induced circulation.

epipedon *soil horizon* that forms at the surface.

epiphytes plants that live above ground level out of contact with the soil, usually growing on the limbs of *trees* or *shrubs;* also called "air plants."

epoch a subdivision of geologic time.

equal-area projections class of *map projections* on which any given area of the Earth's surface is shown to correct relative areal extent, regardless of position on the globe.

equator *parallel of latitude* occupying a position midway between the Earth's poles of *rotation;* the largest of the parallels, designated as *latitude* 0°.

equatorial current westward-flowing *ocean current* in the belt of the *trade winds.*

equatorial easterlies upper-level easterly air flow over the *equatorial zone.*

equatorial rainforest plant *formation class* within the *forest biome,* consisting of tall, closely set broadleaf *trees* of evergreen or semideciduous habit.

equatorial trough atmospheric low-pressure trough centered more or less over the *equator* and situated between the two belts of *trade winds.*

equatorial zone *latitude* zone lying between lat. 10° S and 10° N (more or less) and centered upon the *equator.*

equilibrium in flow systems, a state of balance in which flow rates remain unchanged.

equinox instant in time when the *subsolar point* falls on the Earth's equator and the *circle of illumination* passes through both poles. *Vernal equinox* occurs on March 20 or 21; *autumnal equinox* on September 22 or 23.

era major subdivision of geologic time consisting of a number of geologic periods. The three *eras* following *Precambrian time* are *Paleozoic, Mesozoic,* and *Cenozoic.*

erg large expanse of active *sand dunes* in the Sahara Desert of North Africa.

erosional landforms class of the *sequential landforms* shaped by the removal of *regolith* or *bedrock* by agents of erosion. Examples: *gorge,* glacial *cirque, marine cliff.*

esker narrow, often sinuous embankment of coarse gravel and boulders deposited in the bed of a meltwater *stream* enclosed in a tunnel within stagnant ice of an *ice sheet.*

estuary *bay* that receives fresh water from a river mouth and salt water from the ocean.

Eurasian-Indonesian belt mountain arc system extending from southern Europe across southern Asia and Indonesia.

eustatic referring to a true change in sea level, as opposed to a local change created by upward or downward *tectonic* motion of land.

eutrophication excessive growth of algae and other related organisms in a *stream* or *lake* as a result of the input of large amounts of nutrient *ions,* especially phosphate and nitrate.

evaporation process in which water in liquid state or solid state passes into the vapor state.

evaporites class of *chemically precipitated sediment* and *sedimentary rock* composed of soluble salts deposited from saltwater bodies.

evapotranspiration combined water loss to the atmosphere by *evaporation* from the *soil* and *transpiration* from plants.

evergreen plant *tree* or *shrub* that holds most of its green leaves throughout the year.

evolution the creation of the diversity of life forms through the process of natural selection.

exfoliation see *unloading.*

exfoliation dome smoothly rounded *rock* knob or hilltop bearing *rock* sheets or shells produced by spontaneous expansion accompanying *unloading.*

exotic river *stream* that flows across a region of dry climate and derives its *discharge* from adjacent uplands where a *water surplus* exists.

exponential growth increase in number or value over time in which the increase is a constant proportion or percentage within each time unit.

exposed shields areas of *continental shields* in which the ancient basement *rock,* usually of *Precambrian* age, is exposed to the surface.

extension (tectonic) drawing apart of crustal layers by *tectonic activity* resulting in *faulting.*

extinction the event that the number of organisms of a *species* shrinks to zero so that the species no longer exists.

extrusion release of molten *rock magma* at the surface, as in a flow of *lava* or shower of volcanic ash.

extrusive igneous rock *rock* produced by the solidification of *lava* or ejected fragments of *igneous rock (tephra).*

Fahrenheit scale temperature scale in which the *freezing* point of water is 32°, and the boiling point 212°.

fair weather system a *traveling anticyclone,* in which the descent of air suppresses clouds and precipitation and weather is typically fair.

fallout *gravity* fall of atmospheric particles of *particulates* reaching the ground.

fault sharp break in *rock* with a displacement (slippage) of the block on one side with respect to an adjacent block. (See *normal fault, overthrust fault, strike-slip fault, transform fault.*)

fault coast *coast* formed when a *shoreline* comes to rest against a *fault scarp.*

fault creep more or less continuous slippage on a *fault plane,* relieving some of the accumulated strain.

fault plane surface of slippage between two Earth blocks moving relative to each other during faulting.

fault scarp clifflike surface feature produced by faulting and exposing the *fault plane;* commonly associated with a *normal fault.*

fault-line scarp erosion scarp developed upon an inactive *fault* line.

feedback in flow systems, a linkage between flow paths such that the flow in one *pathway* acts either to reduce or increase the flow in another *pathway.*

feldspar group of *silicate minerals* consisting of silicate of aluminum and one or more of the metals potassium, sodium, or calcium. (See *plagioclase feldspar, potash feldspar.*)

felsenmeer expanse of large blocks of *rock* produced by *joint* block separation and shattering by *frost action* at high altitudes or in high latitudes; from the German for "rock sea."

felsic igneous rock *igneous rock* dominantly composed of *felsic minerals.*

felsic minerals (felsic mineral group) *quartz* and *feldspars* treated as a mineral group of light color and relatively low *density.* (See also *mafic minerals.*)

fetch distance over water that wind blows, creating wind waves.

fine textured (rock) having *mineral* crystals too small to be seen by eye or with low magnification.

fiord narrow, deep ocean embayment partially filling a *glacial trough.*

fiord coast deeply embayed, rugged coast formed by partial *submergence* of *glacial troughs.*

firn granular old *snow* forming a surface layer in the zone of accumulation of a *glacier.*

fissile adjective describing a *rock,* usually *shale,* that readily splits up into small flakes or scales.

flash flood flood in which heavy rainfall causes a stream or river to rise very rapidly.

flood stream flow at a stream *stage* so high that it cannot be accommodated within the *stream channel* and must spread over the banks to inundate the adjacent *floodplain.*

flood basalts large-scale outpourings of basalt *lava* to produce thick accumulations of *basalt* over large areas.

flood current landward flow of a *tidal current.*

flood stage designated stream-surface level for a particular point on a *stream,* higher than which overbank flooding may be expected.

floodplain belt of low, flat ground, present on one or both sides of a *stream channel,* subject to inundation by a *flood* about once annually and underlain by alluvium.

flow system a physical *system* in which *matter, energy,* or both move through time from one location to another.

fluid substance that flows readily when subjected to unbalanced stresses; may exist as a *gas* or a *liquid.*

fluid agents *fluids* that erode, transport, and deposit *mineral matter* and organic matter; they are running water, waves and currents, glacial ice, and *wind.*

fluvial landforms *landforms* shaped by running water.

fluvial processes geomorphic processes in which running water is the dominant *fluid* agent, acting as *overland flow* and *stream flow.*

focus (See *earthquake focus.*)

fog cloud layer in contact with land or sea surface, or very close to that surface. (See *advection fog, radiation fog.*)

folding process by which *folds* are produced; a form of *tectonic activity.*

folds wavelike corrugations of *strata* (or other layered *rock* masses) as a result of crustal *compression.*

food chain (food web) organization of an *ecosystem* into steps or levels through which *energy* flows as the organisms at each level consume *energy* stored in the bodies of organisms of the next lower level.

forb broad-leaved *herb,* as distinguished from the grasses.

forearc trough in plate tectonics, a shallow trough between a *tectonic arc* and a continent; accumulates *sediment* in a basinlike structure.

foredunes ridge of irregular *sand dunes* typically found adjacent to *beaches* on low-lying *coasts* and bearing a partial cover of plants.

foreland folds *folds* produced by *continental collision* in *strata* of a *passive continental margin.*

forest assemblage of *trees* growing close together, their crowns forming a layer of foliage that largely shades the ground.

forest biome *biome* that includes all regions of *forest* over the lands of the Earth.

formation classes subdivisions within a *biome* based on the size, shape, and structure of the plants that dominate the vegetation.

fossil fuels naturally occurring hydrocarbon compounds that represent the altered remains of organic materials enclosed in *rock;* examples are, *coal, petroleum (crude oil),* and *natural* gas.

fractional scale (See *scale fraction.*)

freezing change from liquid state to solid state accompanied by release of *latent heat,* becoming *sensible heat.*

freezing front location at which freezing is occurring in the *active layer* of *permafrost* during the annual freeze-over; fronts may move downward from the top or upward from the bottom.

fringing reef *coral reef* directly attached to land with no intervening *lagoon* of open water.

front surface of contact between two unlike *air masses.* (See *cold front, occluded front, polar front, warm front.*)

frost action *rock* breakup by forces accompanying the *freezing* of water.

gabbro *intrusive igneous rock* consisting largely of pyroxene and *plagioclase feldspar,* with variable amounts of *olivine;* a *mafic igneous rock.*

gas (gaseous state) *fluid* of very low density (as compared with a liquid of the same chemical composition) that expands to fill uniformly any small container and is readily compressed.

gaseous cycle type of *biogeochemical cycle* in which an element or compound is converted into *gaseous* form, diffuses through the *atmosphere,* and passes rapidly over land or sea where it is reused in the *biosphere.*

gene flow speciation process in which evolving populations exchange alleles as individuals move among populations.

genetic drift *speciation* process in which chance mutations change the genetic composition of a breeding population until it diverges from other populations.

genotype the gene set of an individual organism or *species.*

genus a collection of closely related *species* that share a similar genetic evolutionary history.

geographic grid complete network of parallels and meridians on the surface of the globe, used to fix the locations of surface points.

geographic information system (GIS) a system for acquiring, processing, storing, querying, creating, and displaying *spatial data;* normally computer-based.

geographic isolation *speciation* process in which a breeding population is split into parts by an emerging geographic barrier, such as an uplifting mountain range or a changing climate.

geography the study of the evolving character and organization of the Earth's surface.

geography of soils the study of the distribution of soil types and properties and the processes of soil formation.

geologic norm stable natural condition in a moist climate in which slow *soil erosion* is paced by maintenance of *soil horizons* bearing a plant community in an equilibrium state.

geology science of the solid Earth, including the Earth's origin and history, materials comprising the Earth, and the processes acting within the Earth and upon its surface.

geomorphology science of Earth surface processes and *landforms,* including their history and processes of origin.

geostrophic wind *wind* at high levels above the Earth's surface blowing parallel with a system of straight, parallel *isobars.*

geyser periodic jetlike emission of hot water and steam from a narrow vent at a geothermal locality.

glacial abrasion *abrasion* by a moving *glacier* of the *bedrock* floor beneath it.

glacial delta *delta* built by meltwater streams of a *glacier* into standing water of a marginal glacial lake.

glacial drift general term for all varieties and forms of *rock* debris deposited in close association with *ice sheets* of the *Pleistocene Epoch.*

glacial plucking removal of masses of *bedrock* from beneath an *alpine glacier* or *ice sheets* as ice moves forward suddenly.

glacial trough deep, steep-sided *rock* trench of U-shaped cross section formed by *alpine glacier* erosion.

glaciation (1) general term for the total process of glacier growth and *landform* modification by *glaciers.* (2) single episode or time period in which *ice sheets* formed, spread, and disappeared.

glacier large natural accumulation of land ice affected by present or past flowage. (See *alpine glacier.*)

Gleysolic order a class of *soils* in the Canadian soil classification system characterized by indicators of periodic or prolonged water saturation.

global radiation balance the energy flow process by which the Earth absorbs shortwave solar radiation and emits longwave radiation. In the long run, the two flows must balance.

global scale scale at which we are concerned with the Earth as a whole, for example in considering Earth-Sun relationships.

gneiss variety of *metamorphic rock* showing banding and commonly rich in *quartz* and *feldspar.*

Gondwana a *supercontinent* of the Permian *Period* including much of the regions that are now South America, Africa, Antarctica, Australia, New Zealand, Madagascar, and peninsular India.

Goode projection an equal-area *map projection,* often used to display areal thematic information, such as *climate* or *soil* type.

gorge (canyon) steep-sided *bedrock* valley with a narrow floor limited to the width of a *stream channel.*

graben trenchlike depression representing the surface of a crustal block dropped down between two opposed, infacing *normal faults.* (See *rift valley.*)

graded profile smoothly descending profile displayed by a *graded stream.*

graded stream *stream* (or *stream channel*) with *stream gradient* so adjusted as to achieve a balanced state in which average *bed load* transport is matched to average bed load input; an average condition over periods of many years' duration.

gradient degree of *slope,* as the gradient of a river or a flowing glacier.

granite *intrusive igneous rock* consisting largely of *quartz, potash feldspar,* and *plagioclase feldspar,* with minor amounts of biotite and hornblende; a *felsic igneous rock.*

granitic rock general term for *rock* of the upper layer of the *continental crust,* composed largely of *felsic igneous* and *metamorphic rock; rock* of composition similar to that of *granite.*

granular disintegration grain-by-grain breakup of the outer surface of coarse-grained *rock,* yielding *sand* and gravel and leaving behind rounded boulders.

graphic scale *map* scale as shown by a line divided into equal parts.

grassland biome *biome* consisting largely or entirely of *herbs,* which may include grasses, grasslike plants, and *forbs.*

gravitation mutual attraction between any two masses.

gravity gravitational attraction of the Earth upon any small mass near the Earth's surface. (See *gravitation.*)

gravity gliding the sliding of a *thrust sheet* away from the center of an *orogen* under the force of *gravity.*

great circle circle formed by passing a plane through the exact center of a perfect sphere; the largest circle that can be drawn on the surface of a sphere.

greenhouse effect accumulation of heat in the lower *atmosphere* through the absorption of *longwave radiation* from the Earth's surface.

greenhouse gases atmospheric gases such as CO_2 and *chlorofluorocarbons (CFCs)* that absorb outgoing *longwave radiation,* contributing to the *greenhouse effect.*

groin wall or embankment built out into the water at right angles to the *shoreline.*

gross photosynthesis total amount of *carbohydrate* produced by *photosynthesis* by a given organism or group of organisms in a given unit of time.

ground ice frozen water within the pores of *soils* and *regolith* or as free bodies or lenses of solid ice.

ground moraine moraine formed of till distributed beneath a large expanse of land surface covered at one time by an ice sheet.

ground water *subsurface water* occupying the *saturated zone* and moving under the force of *gravity.*

growth rate (of a population) rate at which a population growths or shrinks with time; usually expressed as a percent or proportion of increase or decrease in a given unit of time.

gullies deep, V-shaped trenches carved by newly formed *streams* in rapid headward growth during advanced stages of *accelerated soil erosion.*

guyot sunken remnant of a volcanic island.

gyres large circular *ocean current* systems centered upon the oceanic subtropical *high-pressure cells.*

habitat subdivision of the environment according to the needs and preferences of organisms or groups of organisms.

Hadley cell atmospheric circulation cell in low latitudes involving rising air over the *equatorial trough* and sinking air over the *subtropical high-pressure belts.*

hail form of *precipitation* consisting of pellets or spheres of ice with a concentric layered structure.

half-life time required for an initial quantity at time-zero to be reduced by one-half in an exponential decay system.

hanging valley stream valley that has been truncated by marine erosion so as to appear in cross section in a *marine cliff,* or truncated by glacial erosion so as to appear in cross section in the upper wall of a *glacial trough.*

haze minor concentration of *pollutants* or natural forms of *aerosols* in the atmosphere causing a reduction in visibility.

hazards assessment a field of study blending *physical* and *human geography* to focus on the perception of risk of natural hazards and on developing public policy to mitigate that risk.

heat (See *sensible heat, latent heat.*)

heat island persistent region of higher air temperatures centered over a city.

hemisphere half of a sphere; that portion of the Earth's surface found between the *equator* and a pole.

herb tender plant, lacking woody stems, usually small or low; may be annual or perennial.

herbivory form of interaction among *species* in which an animal (herbivore) grazes on herbaceous plants.

heterosphere region of the *atmosphere* above about 100 km in which *gas* molecules tend to become increasingly sorted into layers by molecular weight and electric charge.

hibernation dormant state of some vertebrate animals during the winter season.

high base status (See *base status of soils.*)

high-latitude climates group of climates in the *subarctic zone, arctic zone,* and *polar zone,* dominated by arctic *air masses* and polar air masses.

high-level temperature inversion condition in which a high-level layer of warm air overlies a layer of cooler air, reversing the normal trend of cooling with altitude.

high-pressure cell center of high barometric pressure; an *anticyclone.*

Histosols *soil order* consisting of *soils* with a thick upper layer of organic matter.

hogbacks sharp-crested, often sawtooth ridges formed of the upturned edge of a resistant *rock* layer of *sandstone, limestone,* or *lava.*

Holocene Epoch last *epoch* of geologic time, commencing about 10,000 years ago; it followed the *Pleistocene Epoch* and includes the present.

homosphere the lower portion of the *atmosphere,* below about 100 km altitude, in which atmospheric *gases* are uniformly mixed.

horse latitudes *subtropical high-pressure belt* of the North Atlantic Ocean, coincident with the central region of the Azores high; a belt of weak, variable winds and frequent calms.

horst crustal block uplifted between two *normal faults.*

hot springs springs discharging heated *groundwater* at a temperature close to the boiling point; found in geothermal areas and thought to be related to a *magma* body at depth.

hotspot (biogeography) geographic region of high biodiversity.

hotspot (plate tectonics) center of intrusive *igneous* and *volcanic* activity thought to be located over a rising *mantle plume.*

human habitat the lands of the Earth that support human life.

human geography the part of *systematic geography* that deals with social, economic and behavioral processes that differentiate *places.*

human-influenced vegetation vegetation that has been influenced in some way by human activity, for example through cultivation, grazing, timber cutting, or urbanization.

humidity general term for the amount of *water vapor* present in the air. (See *relative humidity, specific humidity.*)

humification *pedogenic process* of transformation of plant tissues into *humus.*

humus dark brown to black organic matter on or in the *soil,* consisting of fragmented plant tissues partly digested by organisms.

hurricane *tropical cyclone* of the western North Atlantic and Caribbean Sea.

hydraulic action *stream erosion* by impact force of the flowing water upon the bed and banks of the *stream channel.*

hydrograph graphic presentation of the variation in *stream discharge* with elapsed time, based on data of stream gauging at a given station on a stream.

hydrologic cycle total plan of movement, exchange, and storage of the Earth's free water in gaseous state, liquid state, and solid state.

hydrology science of the Earth's water and its motions through the *hydrologic cycle.*

hydrolysis chemical union of water molecules with *minerals* to form different, more stable mineral compounds.

hydrosphere total water realm of the Earth's surface zone, including the oceans, surface waters of the lands, *groundwater,* and water held in the *atmosphere.*

hygrometer instrument that measures the *water vapor* content of the *atmosphere;* some types measure *relative humidity* directly.

hypolimnion cold, dense lower layer of a *lake.*

ice age span of geologic time, usually on the order of one to three million years, or longer, in which glaciations alternate with interglaciations repeatedly in rhythm with cyclic global climate changes. (See also *interglaciation, glaciation.*)

Ice Age (Late-Cenozoic Ice Age) the present ice age, which began in late Pliocene time, perhaps 2.5 to 3 million years ago.

ice lens more-or-less horizontal layer of *segregated ice* formed by capillary movement of soil water toward a freezing front.

ice lobes (glacial lobes) broad tonguelike extensions of an *ice sheet* resulting from more rapid ice motion where terrain was more favorable.

ice sheet large thick plate of glacial ice moving outward in all directions from a central region of accumulation.

ice shelf thick plate of floating glacial ice attached to an *ice sheet* and fed by the ice sheet and by *snow* accumulation.

ice storm occurrence of heavy glaze of ice on solid surfaces.

ice wedge vertical, wall-like body of ground ice, often tapering downward, occupying a shrinkage crack in *silt* of *permafrost* areas.

ice-sheet climate ⑬ severely cold climate, found on the Greenland and Antarctic *ice sheets,* with *potential evapotranspiration (water need)* effectively zero throughout the year.

ice-wedge polygons polygonal networks of *ice wedges.*

iceberg mass of glacial ice floating in the ocean, derived from a *glacier* that extends into tidal water.

iceberg mass of glacial ice floating in the ocean, derived from a glacier that extends into tidal water.

igneous rock *rock* solidified from a high-temperature molten state; *rock* formed by cooling of *magma.* (See *extrusive igneous rock, felsic igneous rock, intrusive igneous rock, mafic igneous rock, ultramafic igenous rock.*)

illuviation accumulation in a lower *soil horizon* (typically, the *B horizon*) of materials brought down from a higher horizon; a soil-forming process.

image processing mathematical manipulation of digital images, for example, to enhance contrast or edges.

Inceptisols *soil order* consisting of soils having weakly developed *soil horizons* and containing weatherable *minerals.*

induced deflation loss of *soil* by wind erosion that is triggered by human activity such as cultivation or overgrazing.

induced mass wasting *mass wasting* that is induced by human activity, such as creation of waste *soil* and *rock* piles or undercutting of *slopes* in construction.

infiltration absorption and downward movement of *precipitation* into the *soil* and *regolith.*

infrared imagery images formed by *infrared radiation* emanating from the ground surface as recorded by a remote sensor.

infrared radiation *electromagnetic energy* in the *wavelength* range of 0.7 to about 200 μm.

initial landforms *landforms* produced directly by internal Earth processes of *volcanism* and *tectonic activity.* Examples: *volcano, fault scarp.*

inner lowland on a *coastal plain,* a shallow valley lying between the first *cuesta* and the area of older *rock* (oldland).

input flow of *matter* or *energy* into a system.

insolation interception of solar *energy (shortwave radiation)* by an exposed surface.

inspiral horizontal inward spiral or motion, such as that found in a *cyclone.*

interglaciation within an *ice age,* a time interval of mild global climate in which continental *ice sheets* were largely absent or were limited to the Greenland and Antarctic ice sheets; the interval between two glaciations. (See also *deglaciation, glaciation.*)

interlobate moraine *moraine* formed between two adjacent lobes of an *ice sheet.*

International Date Line the 180° *meridian of longitude,* together with deviations east and west of that meridian, forming the time boundary between adjacent *standard time zones* that are 12 hours fast and 12 hours slow with respect to Greenwich standard time.

interrupted projection projection subdivided into a number of sectors (gores), each of which is centered on a different central meridian.

intertropical convergence zone (ITCZ) zone of convergence of *air masses* of *tropical easterlies (trade winds)* along the axis of the *equatorial trough.*

intrusion body of *igneous rock* injected as *magma* into preexisting crustal *rock;* example: *dike* or *sill.*

intrusive igneous rock *igneous rock* body produced by solidification of *magma* beneath the surface, surrounded by preexisting *rock.*

inversion (See *temperature inversion.*)

ion atom or group of atoms bearing an electrical charge as the result of a gain or loss of one or more electrons.

island arcs curved lines of volcanic islands associated with active *subduction* zones along the boundaries of *lithospheric plates.*

isobars lines on *map* passing through all points having the same *atmospheric pressure.*

isohyet line on a *map* drawn through all points having the same numerical value of *precipitation.*

isopleth line on a *map* or globe drawn through all points having the same value of a selected property or entity.

isostasy principle describing the flotation of the *lithosphere,* which is less dense, on the plastic *asthenosphere,* which is more dense.

isostatic compensation crustal rise or sinking in response to unloading by *denudation* or loading by sediment deposition, following the principle of *isostasy.*

isostatic rebound local crustal rise after the melting of ice sheets, following the principle of *isostasy.*

isotherm line on a *map* drawn through all points having the same air temperature.

isotope form of an element with a unique *atomic mass number.*

jet stream high-speed air flow in narrow bands within the *upper-air westerlies* and along certain other global *latitude* zones at high levels.

joints fractures within *bedrock,* usually occurring in parallel and intersecting sets of planes.

Joule unit of work or energy in the metric system; symbol, J.

June solstice (See *summer solstice.*)

karst landscape or topography dominated by surface features of *limestone* solution and underlain by a *limestone cavern* system.

Kelvin scale (K) temperature scale on which the starting point is absolute zero, equivalent to –273°C.

kinetic energy form of *energy* represented by *matter* (mass) in motion.

knob and kettle terrain of numerous small knobs of *glacial drift* and deep depressions usually situated along the *moraine* belt of a former *ice sheet.*

lag time interval of time between occurrence of precipitation and peak discharge of a *stream.*

lagoon shallow body of open water lying between a *barrier island* or a *barrier reef* and the mainland.

lahar rapid downslope or downvalley movement of a tonguelike mass of water-saturated *tephra* (volcanic ash) originating high up on a steep-sided volcanic cone; a variety of mudflow.

lake body of standing water that is enclosed on all sides by land.

laminar flow smooth, even flow of a *fluid* shearing in thin layers without *turbulence.*

land breeze local wind blowing from land to water during the night.

land degradation *degradation* of the quality of plant cover and *soil* as a result of overuse by humans and their domesticated animals, especially during periods of *drought.*

landforms configurations of the land surface taking distinctive forms and produced by natural processes. Examples: hill, valley, plateau. (See *depositional landforms, erosional landforms, initial landforms, sequential landforms.*)

landmass large area of *continental crust* lying above sea level (base level) and thus available for removal by *denudation.*

landmass rejuvenation episode of rapid fluvial *denudation* set off by a rapid crustal rise, increasing the available *landmass.*

landslide rapid sliding of large masses of *bedrock* on steep mountain slopes or from high *cliffs.*

lapse rate rate at which temperature decreases with increasing altitude (See *environmental temperature lapse rate, dry adiabatic lapse rate, wet adiabatic lapse rate.*)

large-scale map *map* with *fractional scale* greater than 1:100,000; usually shows a small area.

Late-Cenozoic Ice Age the series of *glaciations, deglaciations* and *interglaciations* experienced during the late *Cenozoic Era.*

latent heat heat absorbed and held in storage in a *gas* or *liquid* during the processes of *evaporation,* or *melting,* or *sublimation;* distinguished from *sensible heat.*

latent heat transfer flow of *latent heat* that results when water absorbs heat to change from a *liquid* or *solid* to a *gas* and then later releases that heat to new surroundings by *condensation* or *deposition.*

lateral moraine *moraine* forming an embankment between the ice of an *alpine glacier* and the adjacent valley wall.

laterite rocklike layer rich in *sequioxides* and iron, including the minerals *bauxite* and *limonite,* found in low latitudes in association with *Ultisols* and *Oxisols.*

latitude *arc* of a *meridian* between the *equator* and a given point on the globe.

Laurasia a *supercontinent* of the Permian *Period* including much of the regions that are now North America and western Eurasia.

lava *magma* emerging on the Earth's solid surface, exposed to air or water.

leaching *pedogenic process* in which material is lost from the *soil* by downward washing out and removal by percolating surplus soil water.

leads narrow strips of open ocean water between ice floes.

level of condensation elevation at which an upward-moving parcel of moist air cools to the *dew point* and *condensation* begins to occur.

liana woody vine supported on the trunk or branches of a *tree.*

lichens plant forms in which algae and fungi live together (in a symbiotic relationship) to create a single structure; they typically form tough, leathery coatings or crusts attached to *rocks* and tree trunks.

life cycle continuous progression of stages in a growth or development process, such as that of a living organism.

life layer shallow surface zone containing the *biosphere;* a zone of interaction between *atmosphere* and land surface, and between atmosphere and ocean surface.

life zones series of vegetation zones describing vegetation types that are encountered with increasing elevation, especially in the southwestern U.S.

life-form characteristic physical structure, size, and shape of a plant or of an assemblage of plants.

limestone nonclastic *sedimentary rock* in which *calcite* is the predominant *mineral,* and with varying minor amounts of other minerals and *clay.*

limestone caverns interconnected subterranean cavities formed in *limestone* by *carbonic acid action* occurring in slowly moving *groundwater.*

limnology study of the physical, chemical, and biological processes of *lakes.*

limonite mineral or group of *minerals* consisting largely of iron oxide and water, produced by *chemical weathering* of other iron-bearing minerals.

line type of *spatial object* in a *geographic information system* that has starting and ending *nodes;* may be directional.

liquid *fluid* that maintains a free upper surface and is only very slightly compressible, as compared with a *gas.*

lithosphere strong, brittle outermost *rock* layer of the Earth, lying above the *asthenosphere.*

lithospheric plate segment of *lithosphere* moving as a unit, in contact with adjacent lithospheric plates along plate boundaries.

littoral drift transport of *sediment* parallel with the *shoreline* by the combined action of *beach drift* and *longshore current* transport.

loam soil-texture class in which no one of the three size grades *(sand, silt, clay)* dominates over the other two.

local scale scale of observation of the Earth in which local processes and phenomena are observed.

local winds general term for *winds* generated as direct or immediate effects of the local terrain.

loess accumulation of yellowish to buff-colored, fine-grained *sediment,* largely of *silt* grade, upon upland surfaces after transport in the air in *turbulent suspension* (i.e., carried in a *dust storm*).

logistic growth growth according to a mathematical model in which the *growth rate* eventually decreases to near zero.

longitude *arc* of a parallel between the *prime meridian* and a given point on the globe.

longitudinal dunes class of *sand dunes* in which the dune ridges are oriented parallel with the prevailing wind.

longshore current current in the breaker zone, running parallel with the *shoreline* and set up by the oblique approach of waves.

longshore drift *littoral drift* caused by action of a *longshore current.*

longwave radiation *electromagnetic energy* emitted by the Earth, largely in the range from 3 to 50 μm.

low base status (See *base status of soils.*)

low-angle overthrust fault *overthrust fault* in which the *fault plane* or fault surface has a low angle of *dip* or may be horizontal.

low-latitude climates group of climates of the *equatorial zone* and *tropical zone* dominated by the subtropical high-pressure belt and the *equatorial trough.*

low-latitude rainforest evergreen broadleaf forest of the wet equatorial and tropical climate zones.

low-latitude rainforest environment low-latitude environment of warm temperatures and abundant *precipitation* that characterizes rainforest in the *wet equatorial* ① and *monsoon and trade-wind coastal* ② climates.

low-level temperature inversion atmospheric condition in which temperature near the ground increases, rather than decreases, with elevation.

low-pressure trough zone of low pressure between two *anticyclones.*

lowlands broad, open valleys between two *cuestas* of a *coastal plain.* (The term may refer to any low areas of land surface.)

Luvisolic order a class of *forest soils* in the Canadian soil classification system in which the *B horizon* accumulates *clay.*

mafic igneous rock *igneous rock* dominantly composed of *mafic minerals.*

mafic minerals (mafic mineral group) *minerals,* largely *silicate minerals,* rich in magnesium and iron, dark in color, and of relatively great density.

magma mobile, high-temperature molten state of *rock,* usually of *silicate mineral* composition and with dissolved *gases.*

manipulation and analysis component component of a *geographic information system* that responds to spatial queries and creates new data layers.

mantle *rock* layer or shell of the Earth beneath the *crust* and surrounding the *core,* composed of *ultramafic igneous rock* of *silicate mineral* composition.

mantle plume a columnlike rising of heated *mantle rock,* thought to be the cause of a *hot spot* in the overlying *lithospheric plate.*

map a paper representation of space showing point, line, or area data.

map projection any orderly system of parallels and meridians drawn on a flat surface to represent the Earth's curved surface.

marble variety of *metamorphic rock* derived from *limestone* or dolomite by recrystallization under pressure.

marine cliff *rock* cliff shaped and maintained by the undermining action of breaking waves.

marine scarp steep seaward *slope* in poorly consolidated *alluvium, glacial drift,* or other forms of *regolith,* produced along a coastline by the undermining action of waves.

marine terrace former *abrasion platform* elevated to become a steplike coastal *landform.*

marine west-coast climate ⑧ cool moist climate of west coasts in the *midlatitude zone,* usually with a substantial annual *water surplus* and a distinct winter *precipitation* maximum.

marl soft, white, carbonate-rich mud produced when calcium carbonate precipitates from a *lake* and accumulates in beds and banks on and near the lake shore.

mass number (See *atomic mass number.*)

mass wasting spontaneous downhill movement of *soil, regolith,* and *bedrock* under the influence of *gravity,* rather than by the action of *fluid* agents.

massive icy beds layers of ice-rich sediment, found in *permafrost* regions, formed by upwelling *ground water* that flows to a freezing front.

material cycle a closed matter flow system, in which matter flows endlessly, powered by energy inputs (See *biogeochemical cycle.*)

mathematical modeling using variables and equations to represent real processes and systems.

matter physical substance that has mass and density.

matter flow system total system of *pathways* by which a particular type of *matter* (a given element, compound, or ion, for example) moves through the Earth's *ecosystem* or *biosphere.*

mean annual temperature mean of daily air temperature means for a given year or succession of years.

mean daily temperature sum of daily maximum and minimum air temperature readings divided by two.

mean monthly temperature mean of daily air temperature means for a given calendar month.

mean velocity mean, or average, speed of flow of water through an entire stream cross section.

meanders (See *alluvial meanders.*)

mechanical energy *energy* of motion or position; includes *kinetic energy* and *potential energy.*

mechanical weathering (See *physical weathering.*)

medial moraine long, narrow deposit of fragments on the surface of a *glacier;* created by the merging of *lateral moraines* when two glaciers join into a single stream of ice flow.

Mediterranean climate ⑦ climate type of the *subtropical zone,* characterized by the alternation of a very dry summer and a mild, rainy winter.

melting change from solid state to liquid state, accompanied by absorption of *sensible heat* to become *latent heat.*

Mercator projection conformal *map projection* with horizontal parallels and vertical meridians and with *map* scale rapidly increasing with increase in *latitude.*

mercury barometer *barometer* using the Torricelli principle, in which *atmospheric pressure* counterbalances a column of mercury in a tube.

meridian of longitude north–south line on the surface of the global *oblate ellipsoid,* connecting the *north pole* and *south pole.*

meridional transport flow of *energy* (heat) or *matter* (water) across the *parallels of latitude,* either poleward or equatorward.

mesa table-topped *plateau* of comparatively small extent bounded by *cliffs* and occurring in a region of flat-lying *strata.*

mesopause Upper limit of the *mesosphere.*

mesosphere atmospheric layer of upwardly diminishing temperature, situated above the stratopause and below the mesopause.

Mesozoic Era second of three geologic *eras* following *Precambrian time.*

metalimnion layer of a *lake*, between the *epilimnion* and *hypolimnion*, in which temperature decreases with depth; contains the thermocline.

metamorphic rock *rock* altered in physical structure and/or chemical *(mineral)* composition by action of heat, pressure, *shearing* stress, or infusion of elements, all taking place at substantial depth beneath the surface.

meteorology science of the *atmosphere;* particularly the physics of the lower or inner atmosphere.

mica group aluminum-silicate *mineral* group of complex chemical formula having perfect cleavage into thin sheets.

microburst brief onset of intense *winds* close to the ground beneath the downdraft zone of a *thunderstorm* cell.

microcontinent fragment of *continental crust* and its *lithosphere* of subcontinental dimensions that is embedded in an expanse of *oceanic lithosphere.*

micrometer metric unit of length equal to one-millionth of a meter (0. 000001 m); abbreviated μm.

microwaves waves of the *electromagnetic radiation* spectrum in the *wavelength* band from about 0.03 cm to about 1 cm.

mid-oceanic ridge one of three major divisions of the ocean basins, being the central belt of submarine mountain topography with a characteristic *axial rift.*

midlatitude climates group of climates of the *midlatitude zone* and *subtropical zone,* located in the *polar front zone* and dominated by both tropical *air masses* and polar air masses.

midlatitude deciduous forest plant *formation class* within the *forest biome* dominated by tall, broadleaf deciduous *trees,* found mostly in the *moist continental climate* ⑩ and *marine west-coast climate* ⑧.

midlatitude zones latitude zones occupying the *latitude* range 35° to 55° N and S (more or less) and lying between the *subtropical zones* and the *subarctic (subantarctic) zones.*

millibar unit of *atmospheric pressure;* one-thousandth of a bar. *Bar* is a force of one million dynes per square centimeter.

mineral naturally occurring inorganic substance, usually having a definite chemical composition and a characteristic atomic structure. (See *felsic minerals, mafic minerals, silicate minerals.*)

mineral alteration chemical change of *minerals* to more stable compounds upon exposure to atmospheric conditions; same as *chemical weathering.*

mineral matter (soils) component of *soil* consisting of weathered or unweathered mineral grains.

mineral oxides (soils) secondary *minerals* found in *soils* in which original minerals have been altered by chemical combination with oxygen.

minute (of arc) 1/60 of a degree.

mistral local drainage wind of cold air affecting the Rhone Valley of southern France.

Moho contact surface between the Earth's *crust* and *mantle;* a contraction of Mohorovic, the name of the seismologist who discovered this feature.

moist continental climate ⑩ moist climate of the *midlatitude zone* with strongly defined winter and summer seasons, adequate *precipitation* throughout the year, and a substantial annual *water surplus.*

moist subtropical climate ⑥ moist climate of the *subtropical zone,* characterized by a moderate to large annual *water surplus* and a strongly seasonal cycle of *potential evapotranspiration (water need).*

mollic epipedon relatively thick, dark-colored surface *soil horizon,* containing substantial amounts of organic matter *(humus)* and usually rich in *bases.*

Mollisols *soil order* consisting of *soils* with a *mollic horizon* and high *base status.*

monadnock prominent, isolated mountain or large hill rising conspicuously above a surrounding *peneplain* and composed of a *rock* more resistant than that underlying the peneplain; a *landform* of *denudation* in moist climates.

monomictic in *limnology*, mixing once each year, as a monomictic *lake*.

monsoon and trade-wind coastal climate ② moist climate of low latitudes showing a strong rainfall peak in the season of high sun and a short period of reduced rainfall.

monsoon forest *formation class* within the *forest biome* consisting in part of deciduous *trees* adapted to a long dry season in the *wet–dry tropical climate* ③.

monsoon system system of low-level *winds* blowing into a continent in summer and out of it in winter, controlled by *atmospheric pressure* systems developed seasonally over the continent.

montane forest plant *formation class* of the *forest biome* found in cool upland environments of the *tropical zone* and *equatorial zone.*

moraine accumulation of *rock* debris carried by an *alpine glacier* or an *ice sheet* and deposited by the ice to become a *depositional landform.* (See *lateral moraine, terminal moraine.*)

morphology the outward form and appearance of individual organisms or *species.*

mountain arc curving section of an *alpine chain* occurring on a *converging boundary* between two crustal plates.

mountain roots erosional remnants of deep portions of ancient *continental sutures* that were once *alpine chains.*

mountain winds daytime movements of air up the *gradient* of valleys and mountain slopes; alternating with nocturnal *valley winds.*

mucks organic *soils* largely composed of fine, black, sticky organic matter.

mud *sediment* consisting of a mixture of *clay* and *silt* with water, often with minor amounts of *sand* and sometimes with organic matter.

mud hummock low mound of earth found in *permafrost* terrain, formed by cycles of *ground ice* growth and melting, with center of bare ground; vegetation may occur at edges (see also *earth hummock*).

mudflow a form of *mass wasting* consisting of the downslope flowage of a mixture of water and *mineral* fragments (*soil, regolith,* disintegrated *bedrock*), usually following a natural drainage line or *stream channel.*

mudstone *sedimentary rock* formed by the lithification of *mud.*

multipurpose map *map* containing several different types of information.

multispectral image image consisting of two or more images, each of which is taken from a different portion of the spectrum (e.g., blue, green, red, infrared).

multispectral scanner *remote sensing* instrument, flown on an aircraft or spacecraft, that simultaneously collects multiple *digital images (multispectral images)* of the ground. Typically, images are collected in four or more spectral bands.

mutation change in genetic material of a reproductive cell.

nappe overturned recumbent *fold* of *strata,* usually associated with *thrust sheets* in a collision *orogen.*

natural bridge natural *rock* arch spanning a *stream channel,* formed by cutoff of an *entrenched meander* bend.

natural flow systems flow systems of *energy* or naturally-occurring substances that are powered largely or completely by natural power sources.

natural gas naturally occurring mixture of hydrocarbon compounds (principally methane) in the gaseous state held within certain porous *rocks.*

natural levee belt of higher ground paralleling a meandering *alluvial river* on both sides of the *stream channel* and built up by *deposition* of fine *sediment* during periods of overbank flooding.

natural selection selection of organisms by environment in a process similar to selection of plants or animals for breeding by agriculturalists.

natural vegetation stable, mature plant cover characteristic of a given area of land surface largely free from the influences and impacts of human activities.

needleleaf evergreen forest *needleleaf forest* composed of evergreen tree species, such as spruce, fir, and pine.

needleleaf forest plant *formation class* within the *forest biome,* consisting largely of needleleaf *trees.* (See also *boreal forest.*)

needleleaf tree tree with long, thin or flat leaves, such as pine, fir, larch, or spruce.

negative exponential mathematical form of a curve that smoothly decreases to approach a steady value, usually zero.

negative feedback in flow systems, a linkage between flow paths such that the flow in one *pathway* acts to reduce the flow in another *pathway.* (See also *feedback, positive feedback.*)

net photosynthesis *carbohydrate* production remaining in an organism after *respiration* has broken down sufficient *carbohydrate* to power the metabolism of the organism.

net primary production rate at which *carbohydrate* is accumulated in the tissues of plants within a given *ecosystem;* units are kilograms of dry organic matter per year per square meter of surface area.

net radiation difference in intensity between all incoming *energy* (positive quantity) and all outgoing *energy* (negative quantity) carried by both *shortwave radiation* and *longwave radiation.*

neutron atomic particle contained within the nucleus of an atom; similar in mass to a proton, but without a magnetic charge.

nitrogen cycle *biogeochemical cycle* in which nitrogen moves through the *biosphere* by the processes of *nitrogen fixation* and *denitrification.*

nitrogen fixation chemical process of conversion of *gaseous* molecular nitrogen of the *atmosphere* into compounds or ions that can be directly utilized by plants; a process carried out within the *nitrogen cycle* by certain microorganisms.

node point marking the end of a *line* or the intersection of *lines* as *spatial objects* in a *geographic information system*

noon (See *solar noon.*)

noon angle (of the Sun) angle of the Sun above the horizon at its highest point during the day.

normal fault variety of *fault* in which the *fault plane* inclines *(dips)* toward the downthrown block and a major component of the motion is vertical.

north pole point at which the northern end of the Earth's *axis of rotation* intersects the Earth's surface.

northeast trade winds surface *winds* of low latitudes that blow steadily from the northeast. (See also *trade winds.*)

nuclei (atmospheric) minute particles of solid *matter* suspended in the *atmosphere* and serving as cores for *condensation* of water or ice.

nutrient cycle (See *biogeochemical cycle.*)

O_1 horizon surface *soil horizon* containing decaying organic matter that is recognizable as leaves, twigs, or other organic structures.

O_a horizon *soil horizon* below the O_1 horizon containing decaying organic matter that is too decomposed to recognize as specific plant parts, such as leaves or twigs.

oasis desert area where *groundwater* is tapped for crop irrigation and human needs.

oblate ellipsoid geometric solid resembling a flattened sphere, with polar axis shorter than the equatorial diameter.

occluded front weather *front* along which a moving *cold front* has overtaken a *warm front,* forcing the warm *air mass* aloft.

ocean basin floors one of the major divisions of the ocean basins, comprising the deep portions consisting of *abyssal plains* and low hills.

ocean current persistent, dominantly horizontal flow of ocean water.

ocean tide periodic rise and fall of the ocean level induced by gravitational attraction between the Earth and Moon in combination with Earth *rotation.*

oceanic crust crust of basaltic composition beneath the ocean floors, capping *oceanic lithosphere.* (See also *continental crust.*)

oceanic lithosphere *lithosphere* bearing *oceanic crust.*

oceanic trench narrow, deep depression in the seafloor representing the line of *subduction* of an oceanic *lithospheric plate* beneath the margin of a continental lithospheric plate; often associated with an *island arc.*

oil sand (See *bituminous sand.*)

old-field succession form of *secondary succession* typical of an abandoned field, such as might be found in eastern or central North America.

olivine *silicate mineral* with magnesium and iron but no aluminum, usually olive-green or grayish-green; a *mafic mineral.*

open flow system system of interconnected flow paths of *energy* and/or *matter* with a boundary through which that *energy* and/or *matter* can enter and leave the system.

organic matter (soils) material in *soil* that was originally produced by plants or animals and has been subjected to decay.

Organic order a class of *soils* in the Canadian soil classification system that is composed largely of organic materials.

organic sediment *sediment* consisting of the organic remains of plants or animals.

orogen the mass of tectonically deformed *rocks* and related *igneous rocks* produced during an *orogeny.*

orogeny major episode of *tectonic activity* resulting in *strata* being deformed by folding and faulting.

orographic pertaining to mountains.

orographic precipitation *precipitation* induced by the forced rise of moist air over a mountain barrier.

outcrop surface exposure of *bedrock.*

output the flow of *matter* or *energy* out of a system.

outspiral horizontal outward spiral or motion, such as that found in an *anticyclone.*

outwash glacial deposit of stratified drift left by *braided streams* issuing from the front of a *glacier.*

outwash plain flat, gently sloping plain built up of *sand* and gravel by the *aggradation* of meltwater *streams* in front of the margin of an *ice sheet.*

overburden *strata* overlying a layer or *stratum* of interest, as overburden above a *coal* seam.

overland flow motion of a surface layer of water over a sloping ground surface at times when the *infiltration* rate is exceeded by the *precipitation* rate; a form of *runoff.*

overthrust fault *fault* characterized by the overriding of one crustal block (or *thrust sheet*) over another along a gently inclined *fault plane;* associated with crustal *compression.*

overturn in *limnology*, mixing of water from the surface to the bottom a *lake*; occurs when a lake has a uniform temperature profile.

oxbow lake crescent-shaped lake representing the abandoned channel left by the *cutoff* of an *alluvial meander.*

oxidation chemical union of free oxygen with metallic elements in *minerals.*

oxide chemical compound containing oxygen; in *soils,* iron oxides and aluminum oxides are examples.

Oxisols *soil order* consisting of very old, highly weathered *soils* of low latitudes, with an oxic horizon and low *base status.*

oxygen cycle *biogeochemical cycle* in which oxygen moves through the *biosphere* in both *gaseous* and sedimentary forms.

ozone a form of oxygen with a molecule consisting of three atoms of oxygen, O_3.

ozone layer layer in the *stratosphere,* mostly in the altitude range 20 to 35 km (12 to 31 mi), in which a concentration of *ozone* is produced by the action of solar *ultraviolet radiation.*

pack ice floating *sea ice* that completely covers the sea surface.

Paleozoic Era first of three geologic *eras* comprising all geologic time younger than *Precambrian time.*

Pangea hypothetical parent continent, enduring until near the close of the *Mesozoic Era,* consisting of the *continental shields* of *Laurasia* and *Gondwana* joined into a single unit.

parabolic dunes isolated low *sand dunes* of parabolic outline, with points directed into the prevailing *wind.*

parallel of latitude east-west circle on the Earth's surface, lying in a plane parallel with the *equator* and at right angles to the *axis of rotation.*

parasitism form of negative interaction between *species* in which a small species (parasite) feeds on a larger one (host) without necessarily killing it.

parent material inorganic, *mineral* base from which the *soil* is formed; usually consists of *regolith.*

particulates *solid* and *liquid* particles capable of being suspended for long periods in the *atmosphere.*

pascal metric unit of pressure, defined as a force of one newton per square meter (1 N/m^2); symbol, Pa; 100 Pa = 1 mb, 10^5 Pa = 1 bar.

passive continental margins continental margins lacking active plate boundaries at the contact of *continental crust* with *oceanic crust.* A passive margin thus lies within a single *lithospheric plate.* Example: Atlantic continental margin of North America. (See also *continental margins, active continental margins.*)

passive systems electromagnetic remote sensing systems that measure radiant *energy* reflected or emitted by an object or surface.

pathway in an *energy flow system,* a mechanism by which *matter* or *energy* flows from one part of the system to another.

patterned ground general term for a ground surface that bears polygonal or ring-like features, including stone circles, nets, polygons, steps, and stripes; includes *ice wedge polygons*; typically produced by *frost action* in cold climates.

peat partially decomposed, compacted accumulation of plant remains occurring in a *bog* environment.

ped individual natural *soil* aggregate.

pediment gently sloping, rock-floored land surface found at the base of a mountain mass or *cliff* in an arid region.

pedogenic processes group of recognized basic soil-forming processes, mostly involving the gain, loss, *translocation,* or transformation of materials within the *soil* body.

pedology science of the *soil* as a natural surface layer capable of supporting living plants; synonymous with *soil science.*

peneplain land surface of low elevation and slight relief produced in the late stages of *denudation* of a *landmass.*

perched water table surface of a lens of *ground water* held above the main body of ground water by a discontinuous impervious layer.

percolation slow, downward flow of water by *gravity* through *soil* and subsurface layers toward the *water table.*

perennials plants that live for more than one growing season.

peridotite *igneous rock* consisting largely of olivine and pyroxene; an *ultramafic igneous rock* occurring as a pluton, also thought to compose much of the upper *mantle.*

periglacial in an environment of intense *frost action,* located in cold climate regions or near the margins of *alpine glaciers* or large *ice sheets.*

periglacial system a distinctive set of landforms and land-forming processes that are created by intense frost action.

perihelion point on the Earth's elliptical orbit at which the Earth is nearest to the Sun.

period in *limnology*, the time for a full wave of a water surface to occur at a point, as in the period of a *seiche.*

period of geologic time time subdivision of the *era,* each ranging in duration between about 35 and 70 million years.

permafrost *soil, regolith*, and *bedrock* at a temperature below 0°C (32°F), found in cold climates of arctic, subarctic, and alpine regions.

permafrost table in *permafrost,* the upper surface of perennially frozen ground; lower surface of the *active layer.*

petroleum (crude oil) natural liquid mixture of many complex hydrocarbon compounds of organic origin, found in accumulations (oil pools) within certain *sedimentary rocks.*

pH measure of the concentration of hydrogen ions in a solution. (The number represents the logarithm to the base 10 of the reciprocal of the weight in grams of hydrogen ions per liter of water.) Acid solutions have pH values less than 6, and basic solutions have pH values greater than 6.

phenotype the morphological expression of the *genotype* of an individual. It includes all the physical aspects of its structure that are readily perceivable.

photoperiod duration of daylight on a given day of the year at a given latitude.

photosynthesis production of carbohydrate by the union of water with *carbon dioxide* while absorbing light *energy.*

phreatophytes plants that draw water from the *ground water table* beneath *alluvium* of dry stream channels and valley floors in desert regions.

phylum highest division of higher plant and animal life.

physical geography the part of *systematic geography* that deals with the natural processes occurring at the Earth's surface that provide the physical setting for human activities; includes the broad fields of *climatology, geomorphology, coastal and marine geography, geography of soils,* and *biogeography.*

physical weathering breakup of massive *rock (bedrock)* into small particles through the action of physical forces acting at or near the Earth's surface. (See *weathering.*)

phytoplankton microscopic plants found largely in the uppermost layer of ocean or lake water.

pingo conspicuous conical mound or circular hill, having a core of ice, found on plains of the arctic tundra where *permafrost* is present.

pioneer stage first stage of an *ecological succession.*

pioneer plants plants that first invade an environment of new land or a *soil* that has been cleared of vegetation cover; often these are annual *herbs.*

place in geography, a location on the Earth's surface, typically a settlement or small region with unique characteristics.

plagioclase feldspar aluminum-silicate *mineral* with sodium or calcium or both.

plane of the ecliptic imaginary plane in which the Earth's orbit lies.

plant ecology the study of the relationships between plants and their environment.

plant nutrients *ions* or chemical compounds that are needed for plant growth.

plate tectonics theory of *tectonic activity* dealing with *lithospheric plates* and their activity.

plateau upland surface, more or less flat and horizontal, upheld by resistant beds of *sedimentary rock* or *lava* flows and bounded by a steep *cliff.*

playa flat land surface underlain by fine *sediment* or evaporite minerals deposited from shallow lake waters in a dry climate in the floor of a closed topographic depression.

Pleistocene Epoch *epoch* of the *Cenozoic Era,* often identified as the Ice Age; it preceded the *Holocene Epoch.*

plinthite iron-rich concentrations present in some kinds of *soils* in deeper *soil horizons* and capable of hardening into rock-like material with repeated wetting and drying.

plucking (See *glacial plucking.*)

pluton any body of *intrusive igneous rock* that has solidified below the surface, enclosed in preexisting *rock.*

pocket beach *beach* of crescentic outline located at a *bay* head.

podzol type of *soil order* closely equivalent to *Spodosol;* an order of the Canadian Soil Classification System.

Podzolic order a class of *forest* and heath *soils* in the Canadian soil classification system in which an amorphous material of humified organic matter with Al and Fe accumulates.

point *spatial object* in a *geographic information system* with no area.

point bar deposit of coarse bed-load *alluvium* accumulated on the inside of a growing *alluvial meander.*

polar easterlies system of easterly surface winds at high latitude, best developed in the southern *hemisphere,* over Antarctica.

polar front *front* lying between cold polar *air masses* and warm tropical air masses, often situated along a *jet stream* within the *upper-air westerlies.*

polar front jet stream *jet stream* found along the *polar front,* where cold polar air and warm tropical air are in contact.

polar front zone broad zone in midlatitudes and higher latitudes, occupied by the shifting *polar front.*

polar high persistent low-level center of high *atmospheric pressure* located over the *polar zone* of Antarctica.

polar outbreak tongue of cold polar air, preceded by a *cold front,* penetrating far into the *tropical zone* and often reaching the *equatorial zone;* it brings rain squalls and unusual cold.

polar projection *map projection* centered on Earth's *north pole* or *south pole.*

polar zones *latitude* zones lying between 75° and 90° N and S.

poleward heat transport movement of heat from equatorial and tropical regions toward the poles, occurring as *latent* and *sensible heat transfer.*

pollutants in air pollution studies, foreign matter injected into the lower *atmosphere* as *particulates* or as chemical pollutant *gases.*

pollution dome broad, low dome-shaped layer of polluted air, formed over an urban area at times when winds are weak or calm prevails.

pollution plume (1) The trace or path of pollutant substances, moving along the flow paths of *groundwater.* (2) Trail of polluted air carried downwind from a pollution source by strong winds.

polygon type of *spatial object* in a *geographic information system* with a closed chain of connected *lines* surrounding an area.

polymictic in *limnology*, mixing throughout the year, as a polymictic *lake*.

polyploidy mechanism of *speciation* in which entire chromosome sets of organisms are doubled, tripled, quadrupled, etc.

polypedon smallest distinctive geographic unit of the *soil* of a given area.

pool in flow systems, an area or location of concentration of *matter.* (See also *active pool, storage pool.*)

positive feedback in flow systems, a linkage between flow paths such that the flow in one *pathway* acts to increase the flow in another *pathway.* (See also *feedback, negative feedback.*)

potash feldspar aluminum-silicate *mineral* with potassium the dominant metal.

potential energy *energy* of position; produced by *gravitational* attraction of the Earth's mass for a smaller mass on or near the Earth's surface.

potential evapotranspiration (water need) ideal or hypothetical rate of *evapotranspiration* estimated to occur from a complete canopy of green foliage of growing plants continuously supplied with all the *soil water* they can use; a real condition reached in those situations where *precipitation* is sufficiently great or irrigation water is supplied in sufficient amounts.

pothole cylindrical cavity in hard *bedrock* of a *stream channel* produced by *abrasion* of a rounded *rock* fragment rotating within the cavity.

power source flow of *energy* into a *flow system* that causes *matter* to move.

prairie plant *formation class* of the *grassland biome,* consisting of dominant tall grasses and subdominant *forbs,* widespread in subhumid continental climate regions of the *subtropical zone* and *midlatitude zone.* (See *short-grass prairie, tall-grass prairie.*)

Precambrian time all of geologic time older than the beginning of the Cambrian Period, i.e., older than 600 million years.

precipitation particles of *liquid* water or ice that fall from the atmosphere and may reach the ground. (See *orographic precipitation, convectional precipitation, cyclonic precipitation.*)

predation form of negative interaction among animal *species* in which one species (predator) kills and consumes the other (prey).

preprocessing component component of a *geographic information system* that prepares data for entry to the system.

pressure gradient change of *atmospheric pressure* measured along a line at right angles to the *isobars.*

pressure gradient force force acting horizontally, tending to move air in the direction of lower *atmospheric pressure.*

prevailing westerly winds (westerlies) surface winds blowing from a generally westerly direction in the *midlatitude zone,* but varying greatly in direction and intensity.

primary consumers organisms at the lowest level of the *food chain* that ingest *primary producers* or *decomposers* as their *energy* source.

primary minerals in *pedology (soil science),* the original, unaltered *silicate minerals* of *igneous rocks* and *metamorphic rocks.*

primary producers organisms that use light *energy* to convert *carbon dioxide* and water to *carbohydrates* through the process of *photosynthesis.*

primary succession *ecological succession* that begins on a newly constructed substrate.

prime meridian reference meridian of zero *longitude;* universally accepted as the Greenwich meridian.

proton positively charged particle within the nucleus of a atom.

product generation component component of a *geographic information system* that provides output products such as maps, images, or tabular reports.

progradation shoreward building of a *beach, bar,* or *sandspit* by addition of coarse *sediment* carried by *littoral drift* or brought from deeper water offshore.

pyroxene group complex aluminum-silicate *minerals* rich in calcium, magnesium, and iron, dark in color, high in density, classed as *mafic minerals.*

quartz mineral of silicon dioxide composition.

quartzite *metamorphic rock* consisting largely of the mineral *quartz.*

quick clays *clay* layers that spontaneously change from a solid condition to a near-liquid condition when disturbed.

radar an active *remote sensing* system in which a pulse of radiation is emitted by an instrument, and the strength of the echo of the pulse is recorded.

radial drainage pattern stream pattern consisting of *streams* radiating outward from a central peak or highland, such as a *sedimentary dome* or a *volcano.*

radiant energy transfer net flow of radiant *energy* between an object and its surroundings.

radiation (See *electromagnetic radiation.*)

radiation balance condition of balance between incoming *energy* of solar *shortwave radiation* and outgoing *longwave radiation* emitted by the Earth into space.

radiation fog *fog* produced by radiation cooling of the basal air layer.

radioactive decay spontaneous change in the nucleus of an atom that leads to the emission of *matter* and *energy.*

radiogenic heat heat from the Earth's interior that is slowly released by the *radioactive decay* of *unstable isotopes.*

radiometric dating a method of determining the geologic age of a *rock* or *mineral* by measuring the proportions of certain of its elements in their different isotopic forms.

rain form of *precipitation* consisting of falling water drops, usually 0.5 mm or larger in diameter.

rain gauge instrument used to measure the amount of *rain* that has fallen.

rain-green vegetation vegetation that puts out green foliage in the wet season, but becomes largely dormant in the dry season; found in the *tropical zone,* it includes the *savanna biome* and *monsoon forest.*

rainshadow belt of arid climate to lee of a mountain barrier, produced as a result of adiabatic warming of descending air.

raised shoreline former *shoreline* lifted above the limit of wave action; also called an elevated *shoreline.*

rapids steep-*gradient* reaches of a *stream channel* in which *stream* velocity is high.

reclamation in *strip mining,* the process of restoring *spoil* banks and ridges to a natural condition.

recombination source of variation in organisms arising from the free interchange of *alleles* of genes during the reproduction process.

recumbent overturned, as a folded sequence of *rock* layers in which the folds are doubled back upon themselves.

reflection outward scattering of *radiation* toward space by the *atmosphere* and/or Earth's surface.

reg desert surface armored with a pebble layer, resulting from long-continued *deflation;* found in the Sahara Desert of North Africa.

regional geography that branch of *geography* concerned with how the Earth's surface is differentiated into unique *places.*

regional scale the scale of observation at which subcontinental regions are discernable.

regolith layer of *mineral* particles overlying the *bedrock;* may be derived by *weathering* of underlying bedrock or be transported from other locations by *fluid* agents. (See *residual regolith, transported regolith.*)

Regosolic order a class of *soils* in the Canadian soil classification system that exhibits weakly developed *horizons.*

relative humidity ratio of *water vapor* present in the air to the maximum quantity possible for *saturated air* at the same temperature.

relative variability ratio of a variability measure, such as the average deviation, to the mean of all observations.

remote sensing measurement of some property of an object or surface by means other than direct contact; usually refers to the gathering of scientific information about the Earth's surface from great heights and over broad areas, using instruments mounted on aircraft or orbiting space vehicles.

remote sensor instrument or device measuring *electromagnetic radiation* reflected or emitted from a target body.

removal in soil science, the set of processes that result in the removal of material from a *soil horizon,* such as surface erosion or *leaching.*

representative fraction (R.F.) (See *scale fraction.*)

residual regolith *regolith* formed in place by alteration of the *bedrock* directly beneath it.

resolution on a *map,* power to resolve small objects present on the ground.

respiration the oxidation of organic compounds by organisms that powers bodily functions.

retrogradation cutting back (retreat) of a *shoreline, beach, marine cliff,* or *marine scarp* by wave action.

retrogressive thaw slump slump and flowage of overlying sediment occurring where erosion exposes ice-rich *permafrost* or massive *ground ice* to thawing.

reverse fault type of *fault* in which one fault block rides up over the other on a steep *fault plane.*

revolution motion of a planet in its orbit around the Sun, or of a planetary satellite around a planet.

rhyolite *extrusive igneous rock* of *granite* composition; it occurs as *lava* or *tephra.*

ria coastal embayment or *estuary.*

ria coast deeply embayed *coast* formed by partial *submergence* of a *landmass* previously shaped by fluvial *denudation.*

Richter scale scale of magnitude numbers describing the quantity of *energy* released by an *earthquake.*

ridge-and-valley landscape assemblage of *landforms* developed by *denudation* of a system of open *folds* of *strata* and consisting of long, narrow ridges and valleys arranged in parallel or zigzag patterns.

rift valley trenchlike valley with steep, parallel sides; essentially a *graben* between two *normal faults;* associated with crustal spreading.

rill erosion form of *accelerated erosion* in which numerous, closely spaced miniature channels (rills) are scored into the surface of exposed *soil* or *regolith.*

rock natural aggregate of *minerals* in the solid state; usually hard and consisting of one, two, or more mineral varieties.

rockslide *landslide* of jumbled *bedrock* fragments.

rock terrace terrace carved in *bedrock* during the *degradation* of a *stream channel* induced by the crustal rise or a fall of the sea level. (See also *alluvial terrace, marine terrace.*)

Rodinia early *supercontinent,* predating *Pangea,* that was fully formed about 700 million years ago.

Rossby waves horizontal undulations in the flow path of the *upper-air westerlies;* also known as upper-air waves.

rotation spinning of an object around an axis.

runoff flow of water from continents to oceans by way of *stream flow* and *groundwater* flow; a term in the water balance of the *hydrologic cycle.* In a more restricted sense, runoff refers to surface flow by *overland flow* and channel flow.

Sahel (Sahelian zones) belt of *wet–dry tropical* ③ and *semiarid dry tropical* ④ climate in Africa in which *precipitation* is highly variable from year to year.

salic horizon *soil horizon* enriched by soluble salts.

salinity degree of "saltiness" of water; refers to the abundance of such ions as sodium, calcium, potassium, chloride, fluoride, sulfate, and carbonate.

salinization precipitation of soluble salts within the *soil.*

salt flat shallow basin covered with salt deposits formed when stream input to the basin is evaporated to dryness from the basin of a lake; may also form by evaporation of a saline lake when climate changes; see also *dry lake.*

salt marsh *peat*-covered expanse of *sediment* built up to the level of high tide over a previously formed tidal mud flat.

salt water intrusion occurs in a coastal well when an upper layer of fresh water is pumped out, leaving a salt water layer below to feed the well.

salt-crystal growth a form of *weathering* in which *rock* is disintegrated by the expansive pressure of growing salt crystals during dry weather periods when *evaporation* is rapid.

saltation leaping, impacting, and rebounding of sand grains transported over a *sand* or pebble surface by *wind.*

sample standard deviation in statistics, the square root of the average squared deviation from the mean for a sample.

sample statistics numerical values that give basic information about a sample and its variability.

sand *sediment* particles between 0.06 and 2 mm in diameter.

sand dune hill or ridge of loose, well-sorted *sand* shaped by *wind* and usually capable of downwind motion.

sand sea field of *transverse dunes.*

sandspit narrow, fingerlike embankment of *sand* constructed by *littoral drift* into the open water of a *bay.*

sandstone variety of *sedimentary rock* consisting largely of mineral particles of sand grade size.

Santa Ana easterly *wind,* often hot and dry, that blows from the interior desert region of southern California and passes over the coastal mountain ranges to reach the Pacific Ocean.

saturated air air holding the maximum possible quantity of *water vapor at* a given temperature and pressure.

saturated zone zone beneath the land surface in which all pores of the *bedrock* or *regolith* are filled with *groundwater.*

savanna a vegetation cover of widely-spaced *trees* with a grassland beneath.

savanna biome *biome* that consists of a combination of *trees* and grassland in various proportions.

savanna woodland plant *formation class* of the *savanna biome* consisting of a *woodland* of widely spaced *trees* and a grass layer, found throughout the *wet–dry tropical climate* ③ regions in a belt adjacent to the *monsoon forest* and *low-latitude rainforest.*

scale the magnitude of a phenomenon or system, as for example, global scale or local scale.

scale fraction ratio that relates distance on the Earth's surface to distance on a *map* or surface of a globe.

scale of globe ratio of size of a globe to size of the Earth, where size is expressed by a measure of length or distance.

scale of map ratio of distance between two points on a *map* and the same two points on the ground.

scanning systems *remote sensing* systems that make use of a scanning beam to generate images over the frame of surveillance.

scarification general term for artificial excavations and other land disturbances produced for purposes of extracting or processing mineral resources.

scattering turning aside of radiation by an atmospheric molecule or particle so that the direction of the scattered ray is changed.

schist foliated *metamorphic rock* in which mica flakes are typically found oriented parallel with foliation surfaces.

sclerophyll forest plant *formation class* of the *forest biome,* consisting of low sclerophyll *trees,* and often including sclerophyll woodland or *scrub,* associated with regions of *Mediterranean climate* ⑦.

sclerophyll woodland plant *formation class* of the *forest biome* composed of widely-spaced sclerophyll *trees* and *shrubs.*

sclerophylls hard-leaved evergreen *trees* and *shrubs* capable of enduring a long, dry summer.

scoria *lava* or *tephra* containing numerous cavities produced by expanding gases during cooling.

scrub plant *formation class* or subclass consisting of *shrubs* and having a canopy coverage of about 50 percent.

sea arch arch-like *landform* of a rocky, cliffed coast created when waves erode through a narrow headland from both sides.

sea breeze local wind blowing from sea to land during the day.

sea cave cave near the base of a *marine cliff,* eroded by breaking waves.

sea fog *fog* layer formed at sea when warm moist air passes over a cool ocean current and is chilled to the *condensation* point.

sea ice floating ice of the oceans formed by direct *freezing* of ocean water.

second of arc 1/60 of a minute, or 1/3600 of a degree.

secondary consumers animals that feed on *primary consumers.*

secondary minerals in *soil science, minerals* that are stable in the surface environment, derived by *mineral alteration* of the *primary minerals.*

secondary succession *ecological succession* beginning on a previously vegetated area that has been recently disturbed by such agents as fire, flood, windstorm, or humans.

sediment finely divided *mineral matter* and organic matter derived directly or indirectly from preexisting *rock* and from life processes. (See *chemically precipitated sediment, organic sediment.*)

sediment yield quantity of sediment removed by *overland flow* from a land surface of given unit area in a given unit of time.

sedimentary cycle type of *biogeochemical cycle* in which the compound or element is released from *rock* by *weathering,* follows the movement of running water either in solution or as *sediment* to reach the sea, and is eventually converted into *rock.*

sedimentary dome up-arched *strata* forming a circular structure with domed summit and flanks with moderate to steep outward *dip.*

sedimentary rock *rock* formed from accumulation of *sediment.*

segregated ice lenses of ice occurring as free masses in soil or regolith of *permafrost* terrain.

seiche standing wave on a lake causing oscillation in the water level during a *period* of minutes to hours; initiated by wind forcing water across the surface.

seismic sea wave (tsunami) train of sea waves set off by an *earthquake* (or other seafloor disturbance) traveling over the ocean surface.

seismic waves waves sent out during an *earthquake* by faulting or other crustal disturbance from an *earthquake focus* and propagated through the solid Earth.

semiarid (steppe) dry climate subtype subtype of the dry climates exhibiting a short wet season supporting the growth of grasses and *annual* plants.

semidesert plant *formation class* of the *desert biome,* consisting of xerophytic *shrub* vegetation with a poorly developed herbaceous lower layer; subtypes are semidesert scrub and *woodland.*

sensible heat heat measurable by a *thermometer;* an indication of the intensity of *kinetic energy* of molecular motion within a substance.

sensible heat transfer flow of heat from one substance to another by direct contact.

sequential landforms *landforms* produced by external Earth processes in the total activity of *denudation.* Examples: *gorge, alluvial fan, floodplain.*

seral stage stage in a *sere.*

sere in an *ecological succession,* the series of biotic communities that follow one another on the way to the stable stage, or *climax.*

sesquioxides oxides of aluminum or iron with a ratio of two atoms of aluminum or iron to three atoms of oxygen.

set-down temporary lowering of *lake* level when wind blows water toward the opposite side of the lake.

set-up temporary raising of *lake* level when wind blows water toward the lake shore.

shale fissile, *sedimentary rock* of *mud* or *clay* composition, showing lamination.

shearing (of rock) slipping motion between very thin *rock* layers, like a deck of cards fanned with the sweep of a palm.

sheet erosion type of *accelerated soil erosion* in which thin layers of *soil* are removed without formation of rills or *gullies.*

sheet flow overland flow taking the form of a continuous thin film of water over a smooth surface of *soil, regolith,* or *rock.*

sheeting structure thick, subparallel layers of massive *bedrock* formed by spontaneous expansion accompanying *unloading.*

shield volcano low, often large, domelike accumulation of basalt lava flows emerging from long radial fissures on flanks.

shoreline shifting line of contact between water and land.

short-grass praire plant *formation class* in the *grassland biome* consisting of short grasses sparsely distributed in clumps and bunches and some *shrubs,* widespread in areas of semiarid climate in continental interiors of North America and Eurasia; also called *steppe.*

shortwave infrared *infrared radiation* with wavelengths shorter than 3 μm.

shortwave radiation *electromagnetic energy* in the range from 0.2 to 3 μm, including most of the *energy* spectrum of solar radiation.

shrubs woody perennial plants, usually small or low, with several low-branching stems and a foliage mass close to the ground.

silica silicon dioxide in any of several mineral forms.

silicate minerals (silicates) *minerals* containing silicon and oxygen atoms, linked in the crystal space lattice in units of four oxygen atoms to each silicon atom.

sill *intrusive igneous rock* in the form of a plate where *magma* was forced into a natural parting in the *bedrock,* such as a bedding surface in a sequence of *sedimentary rocks.*

silt *sediment* particles between 0.004 and 0.06 mm in diameter.

sinkhole surface depression in *limestone,* leading down into *limestone caverns.*

slash-and-burn agricultural system, practiced in the *low-latitude rainforest,* in which small areas are cleared and the *trees* burned, forming plots that can be cultivated for brief periods.

slate compact, fine-grained variety of *metamorphic rock,* derived from *shale,* showing well-developed cleavage.

sleet form of *precipitation* consisting of ice pellets, which may be frozen raindrops.

sling psychrometer form of *hygrometer* consisting of a wet-bulb thermometer and a dry-bulb thermometer.

slip face steep face of an active *sand dune,* receiving sand by *saltation* over the dune crest and repeatedly sliding because of oversteepening.

slope (1) Degree of inclination from the horizontal of an element of ground surface, analogous to *dip* in the geologic sense. (2) Any portion or element of the Earth's solid surface. (3) Verb meaning "to incline."

small circle circle formed by passing a plane through a sphere without passing through the exact center.

small-scale map *map* with *fractional scale* of less than 1 : 100,000; usually shows a large area.

smog mixture of *aerosols* and chemical *pollutants* in the lower atmosphere, usually found over urban areas.

snow form of *precipitation* consisting of ice particles.

soil natural terrestrial surface layer containing living matter and supporting or capable of supporting plants.

soil colloids mineral particles of extremely small size, capable of remaining suspended indefinitely in water; typically they have the form of thin plates or scales.

soil creep extremely slow downhill movement of *soil* and *regolith* as a result of continued agitation and disturbance of the particles by such activities as *frost action,* temperature changes, or wetting and drying of the soil.

soil enrichment additions of materials to the *soil* body; one of the *pedogenic processes.*

soil erosion erosional removal of material from the *soil* surface.

soil horizon distinctive layer of the *soil,* more or less horizontal, set apart from other soil zones or layers by differences in physical and chemical composition, organic content, structure, or a combination of those properties, produced by soil-forming processes.

soil orders those eleven *soil* classes forming the highest category in the classification of soils.

soil profile display of *soil horizons* on the face of a freshly cut vertical exposure through the *soil.*

soil science (See *pedology.*)

soil solum that part of the *soil* made up of the A, E, and B *soil horizons;* the soil zone in which living plant roots can influence the development of soil horizons.

soil structure presence, size, and form of aggregations (lumps or clusters) of *soil* particles.

soil texture descriptive property of the *mineral* portion of the *soil* based on varying proportions of *sand, silt,* and *clay.*

soil water water held in the *soil* and available to plants through their root systems; a form of *subsurface water.*

soil water balance balance among the component terms of the *soil-water budget;* namely, *precipitation, evapotranspiration,* change in soil water storage, and water surplus.

soil water belt *soil* layer from which plants draw *soil water.*

soil water budget accounting system evaluating the daily, monthly, or yearly amounts of *precipitation, evapotranspiration,* soil-water storage, water deficit, and water surplus.

soil water recharge restoring of depleted *soil water* by *infiltration* of *precipitation.*

soil water shortage (See *deficit.*)

soil water storage actual quantity of water held in the *soil water belt* at any given instant; usually applied to a soil layer of given depth, such as 300 cm (about 12 in.).

solar constant intensity of solar radiation falling upon a unit area of surface held at right angles to the Sun's rays at a point outside the Earth's *atmosphere;* equal to an *energy* flow of about 1400 W/m^2.

solar day average time required for the Earth to complete one *rotation* with respect to the Sun; time elapsed between one solar noon and the next, averaged over the period of one year.

solar noon instant at which the *subsolar point crosses* the *meridian of longitude* of a given point on the Earth; instant at which the Sun's shadow points exactly due north or due south at a given location.

solids substances in the solid state; they resist changes in shape and volume, are usually capable of withstanding large unbalanced forces without yielding, but will ultimately yield by sudden breakage.

solifluction tundra (arctic) variety of *earthflow* in which sediments of the *active layer* move in a mass slowly downhill over a water-rich plastic layer occurring at the top of *permafrost*; produces *solifluction terraces* and *solifluction lobes.*

solifluction lobe bulging mass of saturated *regolith* with steep curved front moved downhill by *solifluction.*

solifluction terrace mass of saturated *regolith* formed by *solifluction* into a flat-topped terrace.

Solonetzic order a class of *soils* in the Canadian soil classification system with a *B horizon* of sticky *clay* that dries to a vary hard condition.

sorting separation of one grade size of *sediment* particles from another by the action of currents of air or water.

source region extensive land or ocean surface over which an *air mass* derives its temperature and moisture characteristics.

south pole point at which the southern end of the Earth's *axis of rotation* intersects the Earth's surface.

southeast trade winds surface *winds* of low latitudes that blow steadily from the southeast. (See also *trade winds.*)

southern pine forest pine forest that is typically found on sandy soils of the Atlantic and Gulf Coast coastal plains.

Southern Oscillation episodic reversal of prevailing barometric pressure differences between two regions, one centered on Darwin, Australia, in the eastern Indian Ocean, and the other on Tahiti in the western Pacific Ocean; a precursor to the occurrence of an El Niño event. (See also *El Niño.*)

southern pine forest subtype of *needleleaf forest* dominated by pines and occurring in the *moist subtropical climate* ⑥.

spatial data information associated with a specific location or area of the Earth's surface.

spatial object a geographic area, *line* or *point* to which information is attached.

speciation the process by which *species* are differentiated and maintained.

species a collection of individual organisms that are capable of interbreeding to produce fertile offspring.

specific heat physical constant of a material that describes the amount of heat energy in joules required to raise the temperature of one gram of the material by one Celsius degree.

specific humidity mass of *water vapor* contained in a unit mass of air.

spit (See *sandspit.*)

splash erosion *soil erosion* caused by direct impact of falling raindrops on a wet surface of *soil* or *regolith.*

spodic horizon *soil horizon* containing precipitated amorphous materials composed of organic matter and *sesquioxides* of aluminum, with or without iron.

Spodosols *soil order* consisting of *soils* with a *spodic horizon,* an *albic horizon,* with low *base status,* and lacking in *carbonate* materials.

spoil *rock* waste removed in a mining operation.

spreading plate boundary *lithospheric plate* boundary along which two plates of *oceanic lithosphere* are undergoing separation, while at the same time new lithosphere is being formed by *accretion.* (See also *transform plate boundary.*)

stable air mass *air mass* in which the *environmental temperature lapse rate* is less than the *dry adiabatic lapse rate,* inhibiting *convectional* uplift and mixing.

stack (marine) isolated columnar mass of *bedrock* left standing in front of a retreating *marine cliff.*

stage height of the surface of a river above its bed or a fixed level near the bed.

standard meridians *standard time* meridians separated by 15° of *longitude* and having values that are multiples of 15°. (In some cases meridians are used that are multiples of $7^1/_2$°.)

standard time system time system based on the local time of a *standard meridian* and applied to belts of *longitude* extending $7^1/_2$° (more or less) on either side of that meridian.

standard time zone zone of the Earth in which all inhabitants keep the same time, which is that of a *standard meridian* within the zone.

star dune large, isolated *sand dune* with radial ridges culminating in a peaked summit; found in the deserts of North Africa and the Arabian Peninsula.

statistics a branch of mathematical sciences that deals with the analysis of numerical data.

steppe semiarid grassland occurring largely in dry continental interiors. (See *short-grass prairie.*)

steppe climate (See *semiarid (steppe) dry climate subtype.*)

stone polygons linked ringlike ridges of cobbles or boulders lying at the surface of the ground in arctic and alpine tundra regions.

storage capacity maximum capacity of *soil* to hold water against the pull of *gravity.*

storage pool type of pool in a *biogeochemical cycle* in which materials are largely inaccessible to life. (See also *active pool.*)

storage recharge restoration of stored soil water during periods when *precipitation* exceeds *potential evapotranspiration (water need).*

storage withdrawal depletion of stored *soil water* during periods when *evapotranspiration* exceeds *precipitation,* calculated as the difference between *actual evapotranspiration (water use)* and *precipitation.*

storm surge rapid rise of coastal water level accompanying the onshore arrival of a *tropical cyclone.*

strata layers of *sediment* or *sedimentary rock* in which individual beds are separated from one another along bedding planes.

stratified drift *glacial drift* made up of sorted and layered *clay, silt, sand,* or gravel deposited from meltwater in *stream channels,* or in marginal lakes close to the ice front.

stratiform clouds clouds of layered, blanketlike form.

stratopause upper limit of the stratosphere.

stratosphere layer of *atmosphere* lying directly above the *troposphere.*

stratovolcano volcano constructed of multiple layers of *lava* and *tephra* (volcanic ash).

stratus cloud type of the low-height family formed into a dense, dark gray layer.

stream long, narrow body of flowing water occupying a *stream channel* and moving to lower levels under the force of *gravity.* (See *consequent stream, graded stream, subsequent stream.*)

stream capacity maximum *stream load* of solid matter that can be carried by a *stream* for a given *discharge.*

stream channel long, narrow, troughlike depression occupied and shaped by a *stream* moving to progressively lower levels.

stream deposition accumulation of transported particles on a *stream* bed, upon the adjacent *floodplain,* or in a body of standing water.

stream erosion progressive removal of mineral particles from the floor or sides of a *stream channel* by drag force of the moving water, or by *abrasion,* or by *corrosion.*

stream flow water flow in a *stream channel;* same as channel flow.

stream gradient rate of descent to lower elevations along the length of a *stream channel,* stated in m/km, ft/mi, degrees, or percent.

stream load solid matter carried by a *stream* in dissolved form (as *ions*), in *turbulent suspension,* and as *bed load.*

stream profile a graph of the elevation of a *stream* plotted against its distance downstream.

stream transportation downvalley movement of eroded particles in a *stream channel* in solution, in *turbulent suspension,* or as *bed load.*

strike compass direction of the line of intersection of an inclined *rock* plane and a horizontal plane of reference. (See *dip.*)

strike-slip fault variety of *fault* on which the motion is dominantly horizontal along a near-vertical *fault plane.*

strip mining mining method in which overburden is first removed from a seam of *coal,* or a sedimentary ore, allowing the coal or ore to be extracted.

structure (of a system) the pattern of the pathways and their interconnections within a flow system.

subantarctic low-pressure belt persistent belt of low *atmospheric pressure* centered about at lat. 65°S over the Southern Ocean.

subantarctic zone *latitude* zone lying between lat. 55° and 60°S (more or less) and occupying a region between the *midlatitude zone* and the *antarctic zone.*

subarctic zone *latitude* zone between lat. 55° and 60° N (more or less), occupying a region between the *midlatitude zone* and the *arctic zone.*

subduction descent of the downbent edge of a *lithospheric plate* into the *asthenosphere* so as to pass beneath the edge of the adjoining plate.

sublimation process of change of ice (solid state) to *water vapor* (gaseous state); in *meteorology,* sublimation also refers to the change of state from water vapor (liquid) to ice (solid), which is referred to as *deposition* in this text.

submergence inundation or partial drowning of a former land surface by a rise of sea level or a sinking of the *crust* or both.

suborder a unit of *soil* classification representing a subdivision of the *soil order.*

subsea permafrost *permafrost* lying below sea level, found in a shallow offshore zone fringing the arctic seacoast.

subsequent stream *stream* that develops its course by *stream erosion* along a band or belt of weaker *rock.*

subsolar point point on the Earth's surface at which solar rays are perpendicular to the surface.

subsurface water water of the lands held in *soil, regolith,* or *bedrock* below the surface.

subtropical broadleaf evergreen forest a formation class of the forest biome composed of broadleaf evergreen *trees;* occurs primarily in the regions of the *moist subtropical climate* ⑥.

subtropical evergreen forest a subdivision of the *forest biome* composed of both broadleaf and needleleaf evergreen *trees.*

subtropical high-pressure belts belts of persistent high *atmospheric pressure* trending east–west and centered about on lat. 30° N and S.

subtropical jet stream *jet stream* of westerly winds forming at the *tropopause,* just above the *Hadley cell.*

subtropical needleleaf evergreen forest a *formation class* of the *forest biome* composed of needleleaf evergreen *trees* occurring the *moist subtropical climate* ⑥ of the southeastern U.S. ; also referred to as the southern pine forest.

subtropical zones *latitude* zones occupying the region of lat. 25° to 35° N and S (more or less) and lying between the *tropical zones* and the *midlatitude zones.*

succulents plants adapted to resist water losses by means of thickened spongy tissue in which water is stored.

summer monsoon inflow of maritime air at low levels from the Indian Ocean toward the Asiatic low pressure center in the season of high Sun; associated with the rainy season of the *wet–dry tropical climate* ③ and the Asiatic monsoon climate.

summer solstice solstice occurring on June 21 or 22, when the *subsolar point* is located at 23 1/2°N.

Sun-synchronous orbit satellite orbit in which the orbital plane remains fixed in position with respect to the Sun.

supercontinent single world continent, formed when *plate tectonic* motions move continents together into a single, large land mass. (See also *Pangea.*)

supercooled water water existing in the liquid state at a temperature lower than the normal *freezing* point.

surface the very thin layer of a substance that received and radiates energy and conducts heat to and away from the substance.

surface energy balance equation equation expressing the balance among *heat* flows to and from a surface.

surface water water of the lands flowing freely (as *streams*) or impounded (as ponds, *lakes,* marshes).

surges episodes of very rapid downvalley movement within an *alpine glacier.*

suspended load that part of the *stream load* carried in *turbulent suspension.*

suspension (See *turbulent suspension.*)

suture (See *continental suture.*)

swash surge of water up the *beach* slope (landward) following collapse of a *breaker.*

symbiosis form of positive interaction between *species* that is beneficial to one of the species and does not harm the other.

sympatric speciation type of *speciation* in which speciation occurs within a larger population.

synclinal mountain steep-sided ridge or elongate mountain developed by erosion of a syncline.

synclinal valley valley eroded on weak *strata* along the central trough or axis of a *syncline.*

syncline downfold of *strata* (or other layered *rock*) in a troughlike structure; a class of *folds.* (See also *anticline.*)

system (1) a collection of things that are somehow related or organized; (2) a scheme for naming, as in a classification system; (3) a flow system of *matter* and *energy.*

systematic geography the study of the physical, economic, and social processes that differentiate the Earth's surface into *places.*

systems approach the study of the interconnections among natural processes by focusing on how, where, and when *matter* and *energy* flow in natural systems.

systems theory body of knowledge explaining how systems work.

taiga plant *formation class* consisting of *woodland* with low, widely spaced *trees* and a ground cover of lichens and mosses, found along the northern fringes of the region of *boreal forest climate* ⑪*;* also called cold woodland.

tailings (See *spoil.*)

talik pocket or region within permafrost that is unfrozen; ranges from small inclusions to large "holes" in permafrost under lakes.

tall-grass prairie a *formation class* of the *grassland biome* that consists of tall grasses with broad-leaved *herbs.*

talus accumulation of loose *rock* fragments derived by fall of *rock* from a *cliff.*

talus slope slope formed of *talus.*

tar sand (See *bitumin.*)

tarn small *lake* occupying a *rock* basin in a *cirque* of *glacial trough.*

tectonic activity process of bending (folding) and breaking (faulting) of crustal mountains, concentrated on or near active *lithospheric plate* boundaries.

tectonic arc long, narrow chain of islands or mountains or a narrow submarine ridge adjacent to a *subduction* boundary and its trench, formed by *tectonic processes,* such as the construction and rise of an *accretionary prism.*

tectonic crest ridgelike summit line of a *tectonic arc* associated with an *accretionary prism.*

tectonics branch of *geology* relating to tectonic activity and the features it produces. (See also *plate tectonics, tectonic activity.*)

temperature gradient rate of temperature change along a selected line or direction.

temperature inversion upward reversal of the normal *environmental temperature lapse rate,* so that the air temperature increases upward. (See *low-level temperature inversion, high-level temperature inversion.*)

temperature regime distinctive type of annual temperature cycle.

tephra collective term for all size grades of solid *igneous rock* particles blown out under gas pressure from a volcanic vent.

terminal moraine *moraine* deposited as an embankment at the terminus of an *alpine glacier* or at the leading edge of an *ice sheet.*

terrane continental crustal *rock* unit having a distinctive set of lithologic properties, reflecting its geologic history, that distinguish it from adjacent or surrounding *continental crust.*

terrestrial ecosystems *ecosystems* of land plants and animals found on upland surfaces of the continents.

tetraploid having four sets of chromosomes instead of a normal two sets.

thematic map *map* showing a single type of information.

theme category or class of information displayed on a *map.*

thermal erosion in regions of permafrost, the physical disruption of the land surface by melting of *ground ice,* brought about by removal of a protective organic layer.

thermal infrared a portion of the *infrared radiation wavelength* band, from approximately from 3 to 20 μm, in which objects at temperatures encountered on the Earth's surface (including fires) emit *electromagnetic radiation.*

thermal pollution form of water pollution in which heated water is discharged into a *stream* or *lake* from the cooling system of a power plant or other industrial heat source.

thermistor electronic device that measures (air) temperature.

thermocline water layer of a lake or the ocean in which temperature changes rapidly in the vertical direction.

thermokarst in arctic environments, an uneven terrain produced by thawing of the upper layer of *permafrost,* with settling of sediment and related water erosion; often occurs when the natural surface cover is disturbed by fire or human activity.

thermokarst lake shallow lake formed by the thawing and settling of permafrost, usually in response to disturbance of the natural surface cover by fire or human activity.

thermometer instrument measuring temperature.

thermometer shelter louvered wooden cabinet of standard construction used to hold *thermometers* and other weather-monitoring equipment.

thermosphere atmospheric layer of upwardly increasing temperature, lying above the *mesopause.*

thorntree semidesert *formation class* within the *desert biome,* transitional from *grassland biome* and *savanna biome* and consisting of xerophytic *trees* and *shrubs.*

thorntree-tall-grass savanna plant *formation class,* transitional between the *savanna biome* and the *grassland biome,* consisting of widely scattered *trees* in an open grassland.

thrust sheet sheetlike mass of *rock* moving forward over a *low-angle overthrust fault.*

thunderstorm intense, local convectional storm associated with a *cumulonimbus cloud* and yielding heavy *precipitation,* also with lightning and thunder, and sometimes the fall of *hail.*

tidal current current set in motion by the *ocean tide.*

tidal inlet narrow opening in a *barrier island* or baymouth *bar* through which *tidal currents* flow.

tide (See *ocean tide.*)

tide curve graphical presentation of the rhythmic rise and fall of ocean water because of *ocean tides.*

till heterogeneous mixture of *rock* fragments ranging in size from *clay* to boulders, deposited beneath moving glacial ice or directly from the *melting* in place of stagnant glacial ice.

till plain undulating, plainlike land surface underlain by glacial *till.*

time cycle in flow systems, a regular alternation of flow rates with time.

time zones zones or belts of given east–west *(longitudinal)* extent within which *standard time* is applied according to a uniform system.

topographic contour *isopleth* of uniform elevation appearing on a *map.*

tornado small, very intense wind vortex with extremely low air pressure in center, formed beneath a dense *cumulonimbus cloud* in proximity to a *cold front.*

trade winds (trades) surface winds in low latitudes, representing the low-level airflow within the *tropical easterlies.*

transcurrent fault *fault* on which the relative motion is dominantly horizontal, in the direction of the *strike* of the fault; also called a *strike-slip fault.*

transform fault special case of a *strike-slip fault* making up the boundary of two moving *lithospheric plates;* usually found along an offset of the *mid-oceanic ridge* where seafloor spreading is in progress.

transform plate boundary *lithospheric plate* boundary along which two plates are in contact on a *transform fault;* the relative motion is that of a *strike-slip fault.*

transform scar linear topographic feature of the ocean floor taking the form of an irregular scarp or ridge and originating at the offset *axial rift* of the *mid-oceanic ridge;* it represents a former *transform fault* but is no longer a plate boundary.

transformation (soils) a class of soil-forming processes that transform materials within the soil body; examples include *mineral alteration* and *humification.*

translocation a soil-forming process in which materials are moved within the soil body, usually from one horizon to another.

transpiration evaporative loss of water to the *atmosphere* from leaf pores of plants.

transportation (See *stream transportation.*)

transported regolith *regolith* formed of *mineral matter* carried by *fluid* agents from a distant source and deposited upon the *bedrock* or upon older regolith. Examples: floodplain silt, lake clay, beach sand.

transverse dunes field of wavelike *sand dunes* with crests running at right angles to the direction of the prevailing *wind.*

traveling anticyclone center of high pressure and *outspiraling* winds that travels over the Earth's surface; often associated with clear, dry weather.

traveling cyclone center of low pressure and *inspiraling* winds that travels over the Earth's surface; includes *wave cyclones, tropical cyclones,* and *tornadoes.*

travertine *carbonate mineral matter,* usually *calcite,* accumulating upon *limestone cavern* surfaces situated in the *unsaturated zone.*

tree large erect woody perennial plant typically having a single main trunk, few branches in the lower part, and a branching crown.

trellis drainage pattern *drainage pattern* characterized by a dominant parallel set of major *subsequent streams,* joined at right angles by numerous short tributaries; typical of *coastal plains* and belts of eroded *folds.*

tropic of cancer *parallel of latitude* at 23 1/2°N.

tropic of capricorn *parallel of latitude* at 23 1/2°S.

tropical cyclone intense *traveling cyclone* of tropical and subtropical latitudes, accompanied by high *winds* and heavy rainfall.

tropical easterlies low-latitude wind system of persistent air flow from east to west between the two *subtropical high-pressure belts.*

tropical easterly jet stream upper-air *jet stream* of seasonal occurrence, running east to west at very high altitudes over Southeast Asia.

tropical high-pressure belt a high-pressure belt occurring in tropical latitudes at a high level in the *troposphere;* extends downward and poleward to form the *subtropical high-pressure* belt, located at the surface.

tropical zones *latitude* zones centered on the *tropic of cancer* and the *tropic of capricorn,* within the latitude ranges 10° to 25° N and 10° to 25°S, respectively.

tropical-zone rainforest plant *formation class* within the *forest biome* similar to *equatorial rainforest,* but occurring farther poleward in tropical regions.

tropopause boundary between *troposphere* and *stratosphere.*

tropophyte plant that sheds its leaves and enters a dormant state during a dry or cold season when little soil water is available.

troposphere lowermost layer of the *atmosphere* in which air temperature falls steadily with increasing altitude.

tsunami (See *seismic sea wave.*)

tundra biome *biome* of the cold regions of *arctic tundra* and *alpine tundra,* consisting of grasses, grasslike plants, flowering *herbs,* dwarf *shrubs,* mosses, and *lichens.*

tundra climate ⑫ cold climate of the *arctic zone* with eight or more consecutive months of zero *potential evapotranspiration (water need).*

tundra soils soils of the arctic *tundra climate* ⑫ regions.

turbulence in *fluid* flow, the motion of individual water particles in complex eddies, superimposed on the average downstream flow path.

turbulent flow mode of *fluid* flow in which individual *fluid* particles (molecules) move in complex eddies, superimposed on the average downstream flow path.

turbulent suspension *stream transportation* in which particles of *sediment* are held in the body of the *stream* by turbulent eddies. (Also applies to wind transportation.)

typhoon *tropical cyclone* of the western North Pacific and coastal waters of Southeast Asia.

Udalfs suborder of the *soil order Alfisols;* includes Alfisols of moist regions, usually in the *midlatitude zone,* with deciduous forest as the natural vegetation.

Udolls suborder of the *soil order Mollisols;* includes Mollisols of the moist soil-water regime in the *midlatitude zone* and with no horizon of *calcium carbonate* accumulation.

Ultisols *soil order* consisting of *soils* of warm soil temperatures with an *argillic horizon* and low *base status.*

ultramafic igneous rock *igneous rock* composed almost entirely of *mafic minerals,* usually olivine or *pyroxene group.*

ultraviolet radiation *electromagnetic energy* in the *wavelength* range of 0.2 to 0.4 μm.

unloading process of removal of overlying *rock* load from *bedrock* by processes of *denudation,* accompanied by expansion and often leading to the development of *sheeting structure.*

unsaturated zone *subsurface water* zone in which pores are not fully saturated, except at times when *infiltration* is very rapid; lies above the *saturated zone.*

unstable air air with substantial content of *water vapor,* capable of breaking into spontaneous convectional activity leading to the development of heavy showers and *thunderstorms.*

unstable isotope elemental isotope that spontaneously decays to produce one or more new isotopes. (See also *daughter product.*)

upper-air westerlies system of westerly winds in the upper *atmosphere* over middle and high latitudes.

upwelling upward motion of cold, nutrient-rich ocean waters, often associated with cool equatorward currents occurring along *continental margins.*

Ustalfs suborder of the *soil order Alfisols;* includes Alfisols of semiarid and seasonally dry climates in which the *soil* is dry for a long period in most years.

Ustolls suborder of the *soil order Mollisols;* includes Mollisols of the semiarid climate in the *midlatitude zone,* with a horizon of *calcium carbonate* accumulation.

valley winds air movement at night down the *gradient* of valleys and the enclosing mountainsides; alternating with daytime *mountain winds.*

variability measure of the variation in an series of observations that center around a mean.

variation in the study of evolution, natural differences arising between parents and offspring as a result of *mutation* and *recombination.*

varve annual layer of *sediment* on the bottom of a *lake* or the ocean marked by a change in color or texture of the sediment.

veins small, irregular, branching network of *intrusive rock* within a preexisting *rock* mass.

vernal equinox *equinox* occurring on March 20 or 21, when the *subsolar point* is at the *equator.*

Vertisols *soil order* consisting of *soils* of the *subtropical zone* and the *tropical zone* with high *clay* content, developing deep, wide cracks when dry, and showing evidence of movement between aggregates.

visible light *electromagnetic energy* in the *wavelength* range of 0.4 to 0.7 μm.

void empty region of pore space in sediment; often occupied by water or water films.

volcanic bombs boulder-sized, semi-solid masses of *lava* that are ejected from an erupting *volcano.*

volcanic neck isolated, narrow steep-sided peak formed by erosion of *igneous rock* previously solidified in the feeder pipe of an extinct *volcano.*

volcanism general term for *volcano* building and related forms of extrusive igneous activity.

volcano conical, circular structure built by accumulation of *lava* flows and *tephra.* (See *stratovolcano, shield volcano.*)

volcano coast *coast* formed by *volcanoes* and *lava* flows built partly below and partly above sea level.

warm front moving weather *front* along which a warm *air mass* is sliding up over a cold air mass, leading to production of *stratiform clouds* and *precipitation.*

warm-blooded animal animal that possesses one or more adaptations to maintain a constant internal temperature despite fluctuations in the environmental temperature.

warm-core ring circular eddy of warm water, surrounded by cold water and lying adjacent to a warm, poleward moving ocean current, such as the Gulf Stream. (See also *cold-core ring.*)

washout downsweeping of atmospheric *particulates* by *precipitation.*

water gap narrow transverse *gorge* cut across a narrow ridge by a *stream,* usually in a region of eroded *folds.*

water need (See *potential evapotranspiration.*)

water resources a field of study that couples basic study of the location, distribution, and movement of water with the utilization and quality of water for human use.

water surplus water disposed of by *runoff* or percolation to the groundwater zone after the *storage capacity* of the *soil* is full.

water table upper boundary surface of the *saturated zone;* the upper limit of the *groundwater* body.

water use (See *actual evapotranspiration.*)

water vapor the gaseous state of water.

waterfall abrupt descent of a *stream* over a *bedrock* step in the *stream channel.*

waterlogging rise of a *water table* in *alluvium* to bring the zone of saturation into the root zone of plants.

watt unit of power equal to the quantity of work done at the rate of one joule per second; symbol, W.

wave cyclone *traveling cyclone* of the midlatitudes involving interaction of cold and warm *air masses* along sharply defined *fronts.*

wave-cut notch *rock* recess at the base of a *marine cliff* where wave impact is concentrated.

wavelength distance separating one wave crest from the next in any uniform succession of traveling waves.

weak equatorial low weak, slowly moving low-pressure center *(cyclone)* accompanied by numerous convectional showers and *thunderstorms;* it forms close to the *intertropical convergence zone* in the rainy season, or *summer monsoon.*

weather physical state of the *atmosphere* at a given time and place.

weather system recurring pattern of atmospheric circulation associated with characteristic weather, such as a *cyclone* or *anticyclone.*

weathering total of all processes acting at or near the Earth's surface to cause physical disruption and chemical decomposition of *rock.* (See *chemical weathering, physical weathering.*)

west-wind drift ocean drift current moving eastward in zone of *prevailing westerlies.*

westerlies (See *prevailing westerly winds, upper-air westerlies.*)

wet adiabatic lapse rate reduced *adiabatic lapse rate* when *condensation* is taking place in rising air; value ranges between 4 and 9°C per 1000 m (2.2 and 4.9°F per 1000 ft).

wet equatorial climate ① moist climate of the *equatorial zone* with a large annual *water surplus,* and with uniformly warm temperatures throughout the year.

wetlands land areas of poor surface drainage, such as marshes and swamps.

wet-dry tropical climate ③ climate of the *tropical zone* characterized by a very wet season alternating with a very dry season.

whiting precipitation of calcium carbonate in lake water, causing the water to have a milky appearance.

Wilson Cycle *plate tectonic* cycle in which continents rupture and pull apart, forming oceans and *oceanic crust,* then converge and collide with accompanying subduction of *oceanic crust.*

wilting point quantity of stored *soil water,* less than which the foliage of plants not adapted to *drought* will wilt.

wind air motion, dominantly horizontal relative to the Earth's surface.

wind abrasion mechanical wearing action of wind-driven *mineral* particles striking exposed *rock* surfaces.

wind vane weather instrument used to indicate *wind* direction.

winter monsoon outflow of continental air at low levels from the Siberian high, passing over Southeast Asia as a dry, cool northerly *wind.*

winter solstice solstice occurring on December 21 or 22, when the *subsolar point* is at 23 1/2°S.

Wisconsinan Glaciation last glaciation of the *Pleistocene Epoch.*

woodland plant *formation class,* transitional between *forest biome* and *savanna biome,* consisting of widely spaced *trees* with canopy coverage between 25 and 60 percent.

Xeralfs suborder of the *soil order Alfisols;* includes Alfisols of the *Mediterranean climate* ⑦.

xeric animals animals adapted to dry conditions typical of a *desert* climate.

Xerolls suborder of the *soil order Mollisols;* includes Mollisols of the *Mediterranean climate* ⑦.

xerophytes plants adapted to a dry environment.

zooplankton microscopic animals found largely in the uppermost layer of ocean or lake water.

PHOTO CREDITS

Chapter 1 Pages 2 & 4: Yann Arthus-Bertrand/Altitude. Page 6: Liaison Agency, Inc. Page 8 (top): David Olsen/Weatherstock. Page 8 (bottom): T.A. Wiewandt/DRK Photo. Pages 8 & 9 (center): Larry Ulrich/DRK Photo. Page 9 (top): Fletcher & Baylis/Photo Researchers. Page 9 (right): Michael Fogden/DRK Photo. Page 12 (top): USGS. Page 12 (bottom): Provided by the SeaWiFS Project, NASA/Goddard Space Flight Center & ORBIMAGE. Pages 12 & 13 (center): Selected graphic image supplied courtesy of Environmental Systems Research Institute, Inc. © 2000, Environmental Systems Research Institute, Inc. Page 14 (top left): Science Photo Library/Photo Researchers. Page 14 (top right): Michael Collier/DRK Photo. Page 14 (center): Michael Fogden/DRK Photo. Page 14 (bottom left): Mark Newman/Bruce Coleman, Inc. Page 14 (bottom right): ©AP Wide World Photos. Page 15: Dar Roberts.

Chapter 2 Page 18: Yann Arthus-Bertrand/Altitude. Page 22 (top): Tom Van Sant/Geosphere Project, Santa Monica/Science Photo Library/Photo Researchers. Page 22 (center): Photo Library International/ESA/Photo Researchers. Page 22 (bottom): Worldsat International Inc./Photo Researchers. Page 23 (top): Landsat image courtesy EROS Data Center/USGS. Page 23 (bottom): Gary Braasch/Corbis Images. Page 24: ©John Blaustein/Woodfin Camp & Associates. Page 25: Earth Satellite Corporation/Science Photo Library/Photo Researchers. Page 26: ©Ed Hille/Corbis Stock Market. Page 29: ©Douglas Kirkland/Corbis Sygma. Page 32: Yann Arthus-Bertrand/Altitude.

Chapter 3 Page 34: Yann Arthus-Bertrand/Altitude. Page 36 (top left): PLI/Science Photo Library/Photo Researchers. Page 36 (top right): NASA. Page 36 (bottom right): Richard L. Carlton/Photo Researchers. Page 39: ©National Maritime Museum, Greenwich, London. Page 50: Fred Hirschmann Wilderness Photography.

Chapter 4 Page 62: Yann Arthus-Bertrand/Altitude. Page 67: Courtesy Daedalus Enterprises, Inc. & National Geographic Magazine. Page 79: NASA/Goddard Space Flight Center. Page 81: Alan H. Strahler. Page 85 (top left): Simon Fraser/Science Photo Library/Photo Researchers. Page 85 (top right): Stephen J. Krasemann/DRK Photo. Page 85 (center left): Galaxy Contact/Explorer/Photo Researchers. Page 85 (bottom right): Stephen J. Krasemann/DRK Photo. Page 86: Joe Flores, Southern California Edison/NREL. Page 93: Courtesy NASA, compiled and annotated by John E. Estes and Leslie W. Senger. Page 94: Courtesy Geonex/Chicago Aerial Survey, Inc. Page 95: SAR image courtesy of Intera Technologies Corporation, Calgary, Alberta, Canada. Pages 96 & 97: Alan H. Strahler. Page 98 (top right): Photo courtesy of Eastman Kodak Company. Page 98 (bottom): Alan H. Strahler.

Chapter 5 Page 100: Yann Arthus-Bertrand. Page 105 (left): Environmental Analytical Systems. Page 105 (right): ©Steve McCutcheon. Page 108 (left): Tom Bean/DRK Photo. Page 108 (right): Jeremy Woodhouse/DRK Photo. Page 109: M. Roth/National University of Singapore. Page 110: Courtesy NASA/EPA. Provided by Dr. Dale Quattrochi, Marshall Space Flight Center. Page 112: ©George Wuerthner. Page 114: Christi Carter/Grant Heilman Photography. Page 126: Shawn Henry/SABA.

Chapter 6 Page 134: Yann Arthus-Bertrand/Altitude. Page 141: Arthur N. Strahler. Page 145 (top left): ©Darryl Torckler/Stone/Getty Images. Page 145 (top right): ©Kees van den Berg/Photo Researchers. Page 145 (bottom left): ©Trevor Mein/Stone/Getty Images. Page 145 (bottom right): ©Larry Ulrich/Stone/Getty Images. Page 147 (top): University of Wisconsin Space Science and Engineering Center. Page 147 (center): Courtesy NASA. Image produced by M. Jentoft-Nilsen, F. Hasler, D. Chesters, and T. Neilsen. Page 147 (bottom): ©Ron Sanford/Stone/Getty Images. Page 148 (top left): ©Larry West/Bruce Coleman, Inc. Page 148 (bottom right): Arthur N. Strahler. Page 153: Warren Faidley/Weatherstock. Page 154 (top): ©Nuridsany et Perennou/Photo Researchers. Page 154 (bottom): ©John H. Hoffman/Bruce Coleman, Inc. Page 156: Andrew Comrie. Page 162 (left): Data received by the Canada Centre for Remote Sensing. Image provided courtesy of RADARSAT International. Page 162 (right): David R. Frazier/Photo Researchers. Page 164: © John S. Shelton. Page 166: National Atmospheric Deposition Program (NRSP-3)/National Trends Network, Illinois State Water Survey. Page 167: ©Jim Strawser/Grant Heilman Photography.

Chapter 7 Page 168: Yann Arthus-Bertrand/Altitude. Page 173: Courtesy Taylor Instrument Company and Wards Natural Science Establishment, Rochester, New York. Page 176 (top): Katherine Klink. Page 176 (bottom): P.F. Bentley. Page 178: Arthur N. Strahler. Page 179: James Stevenson/Science Photo Library/Photo Researchers. Page 192: Courtesy NASA. Page 195 (top): Courtesy Otis B. Brown, Robert Evans, and M. Carle, University of Miami, Rosenstiel School of Marine and Atmospheric Science, Florida, and NOAA/Satellite Data Services Division. Page 195 (bottom left): Courtesy Otis B. Brown, University of Miami, and Gene Carl Feldman, NASA/Goddard Space Flight Center. Page 195 (bottom right): Courtesy Otis B. Brown, University of Miami, and Gene Carl Feldman, NASA/Goddard Space Flight Center.

Chapter 8 Page 204: Yann Arthus-Bertrand/Altitude. Page 208: Steve Pace/Envision. Page 210: The Image Bank/Getty Images. Page 214: ©Keith Brewster/Weatherstock. Page 215: © Christopher Smith /Getty Images News & Sport Services. Pages 216 & 217: Courtesy NASA. Page 220 (top): Courtesy NASA Goddard Space Fight Center/ A. F. Hasler, H. Pierce, K. Palaniappan, and M. Manyin. Page 220 (bottom): Courtesy NASA. Images and rendering by Marit Jentoft-Nilsen. Page 221 (top): Courtesy NOAA. Page 221 (bottom): NASA, Science Photo Library/Photo Researchers. Page 222: ©Robert Holland/The Image Bank. Page 223: ©Ted Levin/Animals Animals.

Chapter 9 Page 232: Yann Arthus-Bertrand/Altitude. Page 240 (top): ©Tom Bean. Page 240 (bottom): ©David Pollack/Corbis Stock Market.

Chapter 10 Page 256: Yann Arthus-Bertrand/Altitude. Page 262: Lisa Naughton. Page 263: ©Will & Deni McIntyre/Photo Researchers. Page 264: Arthur N. Strahler. Page 268 (top): ©Boyd Norton. Page 268 (bottom): ©M. Philip Kahl/Photo Researchers. Page 269: ©George Holton/Photo Researchers. Page 270: Alain Nogues/Corbis Sygma. Page 272: ©Marvin E. Newman/Woodfin Camp & Associates. Page 274 (top): Frans Lanting/Minden Pictures, Inc. Page 274 (center): Michael Fogden/DRK Photo. Page 274 (bottom): Walter Schmidt/Peter Arnold, Inc. Page 275 (top): Joel Bennett/Peter Arnold, Inc. Page 275 (bottom): Art Wolfe/Photo Researchers. Page 277: Galen Rowell/ Mountain Light Photography, Inc. Page 278: Miles Ertman/Masterfile. Page 279: ©Martin Rogers/Stone/Getty Images.

Chapter 11 Page 282: Yann Arthus-Bertrand/Altitude. Page 288: ©Fred Hirschmann. Page 291: Arthur N. Strahler. Page 294 (top): Porterfield-Chickering/Photo Researchers. Page 294 (bottom): ©D. Cavagnaro/DRK Photo. Page 295 (top): Tom McHugh/Photo Researchers. Page 295 (bottom): Arthur N. Strahler. Page 297: David Muench Photography. Page 300 (top): ©Tony Craddock/Stone/Getty

Images. Page 300 (bottom): ©Janet Foster/Masterfile. Page 301: ©John M. Roberts/Corbis Stock Market. Page 302: Library of Congress. Page 304: ©Porterfield/Chickering/Photo Researchers. Page 308: R. N. Drummond. Page 309: Warren Garst/Tom Stack & Associates. Page 310: Kirsty Duncan. Page 324 (top): Gregory G. Dimijian/Photo Researchers. Page 324 (center): AFP/Corbis Images. Page 324 (bottom): ©AP/Wide World Photos. Page 326: Yann Arthus-Bertrand/Altitude.

Chapter 12 Page 328: Yann Arthus-Bertrand/Altitude. Page 330: Roberto de Gugliemo/Photo Researchers. Page 334 (left): Ward's Natural Science Establishment. Page 334 (right): Ward's Natural Science Establishment. Page 334 (bottom): Travelpix/Taxi/Getty Images. Page 335: Ward's Natural Science Establishment. Page 336: Courtesy of Richard S. Williams, Jr., U. S. Geological Survey. Page 338: Arthur N. Strahler. Page 339: ©Carr Clifton/Minden Pictures, Inc. Page 340 (top): Andrew McIntyre/Columbia University . Page 340 (bottom) & 341: Arthur N. Strahler. Page 342 (top): Victor Englebert/Photo Researchers. Page 342 (bottom): ©Charles R. Belinky/Photo Researchers. Page 343: Arthur N. Strahler. Page 344: ©Freeman Patterson/Masterfile. Page 347: NASA/GSFC/MITI/ERSDA C/JAROS and U.S./Japan ASTER Science Team.

Chapter 13 Page 354: Yann Arthus-Bertrand/Altitude. Page 364: David T. Sandwell. Page 377: NASA. Page 380: Joe Meert.

Chapter 14 Page 386: Yann Arthus-Bertrand/Altitude. Page 390 (top left): James Mason/Black Star. Page 390 (top right): K. Hamdorf/Auscape International Ltd. Page 390 (bottom): Greg Vaughn/Tom Stack & Associates. Page 392 (top): Kevin West/Liaison Agency, Inc./Getty Images. Page 392 (bottom): Werner Stoy/Camera Hawaii, Inc. Page 393: Arthur N. Strahler. Page 396 (top): Arthur N. Strahler. Page 396 (bottom): ©Larry Ulrich. Page 397: Courtesy Pacific Gas & Electric Company. Page 398: Steve Vidler/Leo de Wys, Inc. Page 400 (top): Image courtesy NASA/GSFC/MITI/ERSDAC/JAROS and U.S./Japan ASTER Science Team. Page 400 (bottom): Image courtesy NASA/JPL/NIMA. Page 401 (top): Image courtesy Ron Beck, EROS Data Center. Page 401 (bottom): Image courtesy SeaWiFS Project, NASA/Goddard Space Flight Center, and ORBIMAGE. Page 403: Arthur N. Strahler. Page 404: James Balog/Black Star. Page 405: Georg Gerster/Comstock Images. Page 406: Exelsior/Sipa Press. Page 411 (top): G. Hall/Woodfin Camp & Associates. Page 411 (bottom): ©AP/Wide World Photos. Page 413: Rick Rickman/Matrix International, Inc. Page 416: Yann Arthus-Bertrand/Altitude.

Chapter 15 Page 418: Yann Arthus-Bertrand/Altitude. Page 422 (top): Arthur N. Strahler. Page 422 (bottom): ©Susan Rayfield/Photo Researchers. Page 423 (top): Arthur N. Strahler. Page 423 (bottom): ©Steve McCutcheon. Page 424 (left): ©Kunio Owaki/Corbis Stock Market. Page 424 (right): ©George Wuerthner. Page 425: ©David Muench Photography. Page 426 (top): ©Tom Bean. Page 426 (bottom): Alan H. Strahler. Page 427: ©Douglas Peebles Photography. Page 429: Mark A. Melton. Page 430: Courtesy Raymond Drouin. Page 431: Orlo E. Childs. Page 433: ©AP/Wide World Photos. Page 434: Arthur N. Strahler. Page 436 (top): Courtesy Los Angeles County Department of Public Works. Page 436 (bottom): ©Bill Davis/Black Star. Page 439 (left): Troy L. Pewe. Page 439 (right): Stephen J. Krasemann/DRK Photo. Page 442 (top): Steve McCutcheon. Page 442 (bottom): Bernard Hallet, Periglacial Laboratory, Quaternary Research Center. Page 443: ©Steve McCutcheon/Alaska Pictorial Service.

Chapter 16 Page 448: Yann Arthus-Bertrand/Altitude. Page 454: ©Laurence Parent. Page 455: © John S. Shelton. Page 456: ©Bruno Barbey/Magnum Photos, Inc. Page 457: Photo by Mark A. Melton and editing by Terri Breed. Page 459: ©M. Smith/Sipa Press. Page 460: Kathy Young. Page 461: David R. Frazier/Photo Researchers. Page 466: Jacques Jangoux/Peter Arnold, Inc. Page 468 (top): ©Dan Koeck/Liaison Agency, Inc./Getty Images. Page 469 (top): Galen Rowell/Corbis Images. Page 469 (bottom): Michael Williams/Getty Images News and Sport Services. Page 472: ©Tom Bean/DRK Photo. Page 473: ©Tom Till Photography. Page 475: National Audubon Society/Photo Researchers. Page 479: Courtesy of Worldsat International. Page 480 (top): Shepard Sherbell/SABA. Page 480 (bottom): Panos Pictures. Page 481: Dieter Telemans/Panos Pictures.

Chapter 17 Page 482: Yann Arthus-Bertrand/Altitude. Page 485: Offical U.S. Navy Photographs. Page 486: Mark A. Melton. Page 487: ©Joe Englander/Viesti Associates, Inc. Page 488: Alan H. Strahler. Page 489: ©Frans Lanting/Minden Pictures, Inc. Page 493: ©John Beatty/Stone/Getty Images. Page 494: Courtesy NASA/GSFC/LARC/JPL. Page 495 (top): Courtesy NASA/GSFC/LARC/JPL, MISR Team. Page 495 (bottom): Courtesy NASA/GSFC/MITI/ERSDACJAROS and U.S. Japan ASTER Science Team. Page 497 (top): Jan Kopec/Stone/Getty Images. Page 497 (bottom): Molly Pahl. Page 500: L.D. Gordon/The Image Bank. Page 501: ©Tom Bean. Page 502: ©F. Kenneth Hare. Page 503 (top): NASA/Science Source/Photo Researchers. Page 503 (bottom): ©Albert Moldvay/Eriako Associates. Page 504: Breck Kent/Animals Animals/Earth Scenes. Page 506: T. A. Wiewandt/DRK Photo. Page 508 (top): Mark A. Melton. Page 508 (bottom): © John S. Shelton. Page 509 (top): Mark A. Melton.

Chapter 18 Page 514: Yann Arthus-Bertrand/Altitude. Page 518 (top): Earth Satellite Corporation. Page 518 (bottom) & 519: Alan H. Strahler. Page 520: Larry Ulrich/DRK Photo. Page 525: © John S. Shelton. Page 530: J.A. Kraulis/Masterfile. Page 532: Reni Burri/Magnum Photos, Inc. Page 533 (top): ©Carr Clifton. Page 533 (bottom): Landis Aerial Photo. Page 536 (top left): © John S. Shelton. Page 536 (center): ©Larry Ulrich. Page 536 (bottom): Alex McLean/Landslides.

Chapter 19 Page 540: Yann Arthus-Bertrand/Altitude. Page 544 (top): Chad Ehlers/Stone/Getty Images. Page 544 (bottom): Alex McLean/Landslides. Page 545: Arthur N. Strahler. Page 546: Steve Dunnell/The Image Bank/Getty Images. Page 548: ©Cliff de Bear. Page 549: David Muench Photography. Pages 552 & 553: Courtesy NASA. Page 554: David Hiser. Page 555: © John S. Shelton. Page 556 (top): ©Tom Bean. Page 556 (bottom): M.J. Coe/Animals Animals/Earth Scenes. Page 558: John S. Shelton. Page 559 (top): ©William E. Ferguson. Page 559 (bottom): J. A. Kraulis/Masterfile. Page 563: ©G. R. Roberts. Page 565: Alan H. Strahler. Page 569: ©Stephen Rose/Liaison Agency, Inc./Getty Images. Page 570: Stephen Crowley/New York Times Pictures. Page 571 (top): C.C. Lockwood/DRK Photo. Page 572 (top): BIOS (P.Kobeh)/Peter Arnold, Inc. Page 572 (bottom): Tui De Roy/Minden Pictures, Inc.

Chapter 20 Page 574: Yann Arthus-Bertrand/Altitude. Page 576: Fred Hirschmann Wilderness Photography. Page 577: Carr Clifton. Page 582: Courtesy NASA. Page 583 (top): Image courtesy NASA/GSFC/MITI/ERSDAC/JAROS and U.S./Japan ASTER Science Team. Page 583 (bottom): Image courtesy Canadian Space Agency/NASA/Ohio State University/Jet Propulsion Laboratory, Alaska SAR Facility. Page 584: John S. Shelton. Page 585: Floyd L. Norgaard/Ric Ergenbright. Page 586: Courtesy NASA. Page 587: ©Wolfgang Kaehler. Page 589: NASA Media Services. Page 594 (top): Tom & Susan Bean, Inc. Page 594 (bottom): C. Wolinsky/Stock, Boston. Page 595 (top): Arthur N. Strahler. Page 595 (bottom): Tom & Susan Bean, Inc. Page 596: Kurt Cuffey. Page 599: Courtesy Alan H. Strahler. Page 604: Yann Arthus-Bertrand/Altitude.

Chapter 21 Page 606: Yann Arthus-Bertrand/Altitude. Page 609: Courtesy R. B. Krone, San Francisco District Corps. of Engineers, U. S. Army. Page 611: R. Schaetzl. Page 617: Arthur N. Strahler. Page 619: ©Erich Lessing/Art Resource. Pages 624 & 625: Henry D. Foth. Page 626 (top): Alan H. Strahler. Page 626 (bottom): Soil Conservation Service. Page 627 (left): Henry D. Foth. Page 627 (right): William E. Ferguson. Page 628 (top): Ric Ergenbright/Stone/Getty Images. Page 628 (bottom): Robin White/Fotolex Associates. Page 629: R. Schaetzl. Page 630 (top): Courtesy Soil Conservation Service. Page 630 (bottom): Ned L. Reglein. Page 631: M. Collier/DRK Photo. Page 635: D. Cavagnaro/DRK Photo. Page 636 (top): Sovfoto/eastfoto. Page 636 (bottom): Leonard Lee Rue III/Bruce Coleman, Inc. Page 637 (top): Mark Edwards/Peter Arnold, Inc. Page 637 (bottom): Charlie Waite/Stone/Getty Images.

Chapter 22 Page 638: Yann Arthus-Bertrand/Altitude. Page 642: Michio Hoshino/Minden Pictures, Inc. Page 643 (top): Marc Epstein/DRK Photo. Page 643 (bottom): Manoj Shah/Stone/Getty Images. Page 652: Christina Tague. Page 665 (top left): S.W. Running, University of Montana NTSG/ NASA. Page 665 (top right): S.W. Running, University of Montana NTSG/ NASA. Page 665 (bottom): S.W. Running, University of Montana NTSG/Courtesy NASA. Page 666 (top left): SeaWiFS Project/ORBIMAGES. Page 666 (top right): SeaWiFS Project/ORBIMAGES. Page 666 (bottom): MODIS Science Team/ NASA. Page 667 (top): MODIS Science Team/ NASA. Page 667 (bottom): MODIS Science Team/NASA. Page 668: R. Nemani, University of Montana NTSG/ NASA. Page 669: NASA/NOAA.

Chapter 23 Page 670: Yann Arthus-Bertrand/Altitude. Page 673: Michael P. Gadomski/Photo Researchers. Page 674 (top): ©Josef Muench. Page 674 (bottom): Dennis Sheridan. Page 675 (top): ©Kevin Schafer. Page 675 (bottom): Michio Hoshino/Minden Pictures, Inc. Page 676 (left): Tom Till/DRK Photo. Page 676 (right): Gregory G. Dimijian/Photo Researchers. Page 678: Gregory G. Dimijian/Photo Researchers. Page 680: Lewis Kemper/DRK Photo. Page 681 (left): M.P. Kahl/Photo Researchers. Page 681 (right): MODIS Land Team/C. Justice/L. Giglio/J. Descloitres/NASA. Page 682 (top): Alan H. Strahler. Page 682 (bottom): Robert Simmon, NASA, GSFC. Page 683 (top): Robert Simmon, NASA, GSFC. Page 683 (center): Image courtesy SeaWiFS Project, NASA/Goddard Space Flight Center, and ORBIMAGE. Page 683 (bottom): Tom Brakefield/DRK Photo. Page 685: Alan H. Strahler. Page 686 (top): Michael P. Gadomski/Photo Researchers. Page 686 (bottom): Michael P. Gadomski/Photo Researchers. Page 687: Michael P. Gadomski/Photo Researchers. Page 690 (left): Richard Parker/Photo Researchers. Page 690 (right): Wendy Neefus/Animals Animals/Earth Scenes. Page 692: Art Morris/Birds as Art/Visuals Unlimited. Page 693: Kenneth W. Fink/Photo Researchers. Page 696 (left): S.W. Carter/Photo Researchers. Page 696 (right): Runk/Schoenberger/Grant Heilman Photography.

Chapter 24 Page 702: Yann Arthus-Bertrand/Altitude. Page 705: ©Steve McCutcheon. Page 710 (top): ©M. Freeman/Bruce Coleman, Inc. Page 710 (bottom): ©Ferrero/Labat/Auscape International Pty. Ltd. Page 711 (top): ©Tom Bean. Page 711 (bottom): Michel Gunther/Peter Arnold, Inc. Page 714 (bottom): © John S. Shelton. Page 715: ©Jacques Jangoux/Peter Arnold, Inc. Page 717: ©Kenneth Murray/Photo Researchers. Page 718: ©Jake Rajs. Page 719: Paul Trietz. Page 720: ©Michael Townsend/Stone/Getty Images. Page 721 (left): Arthur N. Strahler. Page 721 (right): Alan H. Strahler. Page 724: ©Annie Griffiths Belt/DRK Photo. Page 726 (bottom): ©Brian A. Vikander. Pages 727 & 728: Arthur N. Strahler.

INDEX

Note: Page references followed by "f" indicate illustrations; while "t" indicates tables.

Topographic Map Symbols

BATHYMETRIC FEATURES

Area exposed at mean low tide; sounding datum line***

Channel***

Sunken rock***

BOUNDARIES

National

State or territorial

County or equivalent

Civil township or equivalent

Incorporated city or equivalent

Federally administered park, reservation, or monument (external)

Federally administered park, reservation, or monument (internal)

State forest, park, reservation, or monument and large county park

Forest Service administrative area*

Forest Service ranger district*

National Forest System land status, Forest Service lands*

National Forest System land status, non-Forest Service lands*

Small park (county or city)

COASTAL FEATURES

Foreshore flat — Mud

Coral or rock reef — Reef

Rock, bare or awash; dangerous to navigation

Group of rocks, bare or awash

Exposed wreck

Depth curve; sounding — 18 23

Breakwater, pier, jetty, or wharf

Seawall

Oil or gas well; platform

CONTOURS

Topographic

Index — 6000

Approximate or indefinite

Intermediate

Approximate or indefinite

Supplementary

Depression

Cut

Fill

SURFACE FEATURES

Levee — Levee

Sand or mud — Sand

Disturbed surface

Gravel beach or glacial moraine — Gravel

Tailings pond — Tailings Pond

BUILDINGS AND RELATED FEATURES

Building

School; house of worship

Athletic field

Built-up area

Forest headquarters*

Ranger district office*

Guard station or work center*

Racetrack or raceway

Airport, paved landing strip, runway, taxiway, or apron

Unpaved landing strip

Well (other than water), windmill or wind generator

Tanks

Covered reservoir

Gaging station

Located or landmark object (feature as labeled)

Boat ramp or boat access*

Roadside park or rest area

Picnic area

Campground

Winter recreation area*

Cemetery — Cem

MARINE SHORELINES

Shoreline

Apparent (edge of vegetation)***

Indefinite or unsurveyed

MINES AND CAVES

Quarry or open pit mine

Gravel, sand, clay, or borrow pit

Mine tunnel or cave entrance

Mine shaft

Prospect

Tailings — Tailings

Mine dump

Former disposal site or mine

RAILROADS AND RELATED FEATURES

Standard guage railroad, single track

Standard guage railroad, multiple track

Narrow guage railroad, single track

Narrow guage railroad, multiple track

Railroad siding

Railroad in highway

Railroad in road

Railroad in light duty road*

Railroad underpass; overpass

Railroad bridge; drawbridge

Railroad tunnel

Railroad yard

SUBMERGED AREAS AND BOGS

Marsh or swamp

Submerged marsh or swamp

Wooded marsh or swamp

Submerged wooded marsh or swamp

ROADS AND RELATED FEATURES

Please note: Roads on Provisional-edition maps are not classified as primary, secondary, or light duty. These roads are all classified as improved roads and are symbolized the same as light duty roads.

Primary highway

Secondary highway

Light duty road

Light duty road, paved*

Light duty road, gravel*

Light duty road, dirt*

Light duty road, unspecified*

Unimproved road

Unimproved road*

4WD road

4WD road*

Trail

Highway or road with median strip

Highway or road under construction — Under Const

Highway or road underpass; overpass

Highway or road bridge; drawbridge

Highway or road tunnel

Road block, berm, or barrier*

Gate on road*

Trailhead*

RIVERS, LAKES, AND CANALS

Perennial stream

Perennial river

Intermittent stream

Intermittent river

Disappearing stream

Falls, small

Falls, large

Rapids, small

Rapids, large

Perennial lake/pond

Intermittent lake/pond

Dry lake/pond — Dry Lake

Narrow wash

Wide wash — Wash

Canal, flume, or aqueduct with lock

Elevated aqueduct, flume, or conduit

Aqueduct tunnel

Water well, geyser, fumarole, or mud pot

Spring or seep

TRANSMISSION LINES AND PIPELINES

Power transmission line; pole; tower

Telephone line — Telephone

Aboveground pipeline

Underground pipeline — Pipeline

VEGETATION

Woodland

Shrubland

Orchard

Vineyard

GLACIERS AND PERMANENT SNOWFIELDS

Contours and limits

Formlines

Glacial advance

Glacial retreat